HANDBOOK OF STATISTICS
VOLUME 21

Handbook of Statistics

VOLUME 21

General Editor
C. R. Rao

ELSEVIER
AMSTERDAM • BOSTON • LONDON • NEW YORK • OXFORD
PARIS • SAN DIEGO • SAN FRANCISCO • SINGAPORE • SYDNEY • TOKYO

Stochastic Processes: Modelling and Simulation

Edited by

D. N. Shanbhag
Department of Probability and Statistics
The University of Sheffield, UK

C. R. Rao
Center for Multivariate Analysis
Department of Statistics, The Pennsylvania State University
University Park, PA, USA

2003

ELSEVIER
AMSTERDAM • BOSTON • LONDON • NEW YORK • OXFORD
PARIS • SAN DIEGO • SAN FRANCISCO • SINGAPORE • SYDNEY • TOKYO

ELSEVIER SCIENCE B.V.
Sara Burgerhartstraat 25
P.O. Box 211, 1000 AE Amsterdam, The Netherlands

© 2003 Elsevier Science B.V. All rights reserved

This work is protected under copyright by Elsevier Science, and the following terms and conditions apply to its use:

Photocopying
Single photocopies of single chapters may be made for personal use as allowed by national copyright laws. Permission of the Publisher and payment of a fee is required for all other photocopying, including multiple or systematic copying, copying for advertising or promotional purposes, resale, and all forms of document delivery. Special rates are available for educational institutions that wish to make photocopies for non-profit educational classroom use.

Permissions may be sought directly from Elsevier Science via their homepage (http://www.elsevier.com) by selecting 'Customer support' and then 'Permissions'. Alternatively you can send an e-mail to: permissions@elsevier.com, or fax to: (+44) 1865 853333.

In the USA, users may clear permissions and make payments through the Copyright Clearance Center, Inc., 222 Rosewood Drive, Danvers, MA 01923, USA; phone: (+1) (978) 7508400, fax: (+1) (978) 7504744, and in the UK through the Copyright Licensing Agency Rapid Clearance Service (CLARCS), 90 Tottenham Court Road, London W1P 0LP, UK; phone: (+44) 207 631 5555; fax: (+44) 207 631 5500. Other countries may have a local reprographic rights agency for payments.

Derivative Works
Tables of contents may be reproduced for internal circulation, but permission of Elsevier Science is required for external resale or distribution of such material.

Permission of the Publisher is required for all other derivative works, including compilations and translations.

Electronic Storage or Usage
Permission of the Publisher is required to store or use electronically any material contained in this work, including any chapter or part of a chapter.

Except as outlined above, no part of this work may be reproduced, stored in a retrieval system or transmitted in any form or by any means, electronic, mechanical, photocopying, recording or otherwise, without prior written permission of the Publisher.

Address permissions requests to: Elsevier Science Global Rights Department, at the fax and e-mail addresses noted above.

Notice
No responsibility is assumed by the Publisher for any injury and/or damage to persons or property as a matter of products liability, negligence or otherwise, or from any use or operation of any methods, products, instructions or ideas contained in the material herein. Because of rapid advances in the medical sciences, in particular, independent verification of diagnoses and drug dosages should be made.

First edition 2003

Library of Congress Cataloging in Publication Data
A catalog record from the Library of Congress has been applied for.

Bitish Library Cataloguing-in-Publication Data
Stochastic processes: modelling and simulation. –
    (Handbook of statistics; v. 21)
    1. Stochastic processes
    I. Shanbhag, D. N. II. Rao, C. Radhakrishna (Calyampudi Radhakrishna), 1920-
    519. 2'3
    ISBN: 0444500138

ISBN: 0-444-50013-8
ISSN: 0169-7161

⊗ The paper used in this publication meets the requirements of ANSI/NISO Z39.48-1992 (Permanence of Paper).
Printed in The Netherlands.

Preface

This is a sequel to Volume 19 of Handbook of Statistics on Stochastic Processes: Theory and Methods. The present volume is concerned mainly with the theme of reviewing the different lines of research and developments in stochasic processes of applied flavour.

This latter volume consists of 23 chapters which are devoted respectively to the following topics: Control Theoretic Approaches to Manufacturing Systems (Boukas and Liu), Models of Random Graphs and Applications (Cannings and Penman), Locally Self-similar Processes with Wavelet Analysis (Cavanaugh et al.), Models for DNA Replication (Cowan), Empirical Processes with Applications to Testing Certain Models (Ferreira), Patterns of Sequences of Random Events (Gani), Models in Telecommunications and Relevant Problems (Gautam), Epidemic Modeling and Simulation (Greenhalgh), Inference and Simulation of Random Fields (Greenwood and Wefelmeyer), Aspects of Self-similarity involving Fractals (Hambly), Numerical Methods for Queues (Heyman), Markov Chain Applications to studies of Runs and Patterns (Koutras), Applications of Markov Random Fields to Image Analysis (Li), Semi-Markov Processes in Reliability (Limnios and Oprişan), Departures and Related Characteristics for Queues (Manoharan et al.), Discrete Variate Time Series (McKenzie), Extreme Value Theory with aspects of modeling and simulation (Nadarajah), Branching Processes with Biological Applications (Pakes), Markov Chain Approaches to Damage Models (Rao, C. R. et al.), Point Processes in Astronomy (Scargle and Babu), Non-linear Non-Gaussian State-space Modeling with Monte Carlo Techniques (Tanizaki), Bilinear Random Processes (Subba Rao and Terdic), Markov Modeling in Studies of Ion Channels (Yeo et al.).

An attempt is made to cover in this volume, as in the case of its predecessor, as many topics as possible. As implied earlier, this volume concentrates mostly on dealing with items of applied nature. However, in doing so, it also includes one or two articles of theoretical nature, assuming that there are indications of these providing openings for future research to specialists working in applied areas.

We are grateful to all the contributors and the referees for their substantial help in completing this project successfully. Also, we would like to thank Drs. G. Wanrooy, N. van Dijk and E. Bomers, as well as Ms. A. Deelen, of Elsevier for their patience and encouragement. They have made every effort to make our job of editing this volume easier. Finally, we would like to take the opportunity of thanking the Department of Statistics at the Pennsylvania State University, USA, and the Department of Probability and Statistics at the University of Sheffield, UK, for providing us with facilities to edit this volume. This project is supported by the US Army Research Grant DAA H 04-96-1-0082.

D. N. Shanbhag
C. R. Rao

Table of contents

Preface v

Contributors xv

Ch. 1. Modelling and Numerical Methods in Manufacturing System Using Control Theory 1
E. K. Boukas and Z. K. Liu

1. Introduction 1
2. Controlled piecewise deterministic processes 3
3. Continuous flow model for production control 15
4. Preventive maintenance and production control model 19
5. Maintenance model without considering the machine aging 32
6. Robust controller for a class of production and maintenance 36
7. Conclusion 46
   References 47

Ch. 2. Models of Random Graphs and their Applications 51
C. Cannings and D. B. Penman

1. Introduction 51
2. Overview of the Erdős–Rényi model 57
3. Applications of the Erdős–Rényi model 63
4. Geometric random graphs 65
5. Random cluster models 69
6. Random randomly coloured graphs 72
7. Other models of random graphs 78
   References 87

Ch. 3. Locally Self-Similar Processes and their Wavelet Analysis 93
J. E. Cavanaugh, Y. Wang and J. W. Davis

1. Introduction 93
2. Locally self-similar processes 95
3. Generalized fractional Brownian motion 97
4. Estimating the scaling function 102
5. Implementation of the estimation procedure 104
6. Simulations 106
7. Applications 117
8. Conclusion 124
   Appendix 124
   References 133

Ch. 4. Stochastic Models for DNA Replication 137
R. Cowan

1. Variation in biology 137
2. Stochastic chemistry 138
3. Exponentially distributed waiting times 141
4. The biological cell 142
5. A glib mathematical abstraction 143
6. The spatial pattern of replication origins 144
7. The time to separation of a long DNA molecule 146
8. The proportion of origins initiated 150
9. Mean eye lengths and eye-to-eye distances 151
10. What is happening inside the eyes? 153
11. The Cowan–Chiu model of fragment formation 155
12. The expectations of N_t and P_t: renewal equations 155
13. The expectation of D_t: the quasi-renewal equation 157
14. Relationship between fragment length and primer-site spacing 159
15. Estimated spacing between primer sites 160
16. The competing theory 160
17. Notations for the competing theory 161
18. Analysis of the lagging strand for Model B 162
19. Analysis of the leading strand for Model B 163
20. Concluding remarks 164
    References 165

Ch. 5. An Empirical Process with Applications to Testing the Exponential and Geometric Models 167
J. A. Ferreira

1. Introduction 167
2. The empirical integrated lack-of-memory process 170
3. Connection with certain test statistics and empirical processes 172
4. Asymptotic behaviour of the process 181
5. Statement of results; examples and comparisons 190
6. Integral statistics. Applications to testing 196

7. Asymptotic efficiency in the continuous case 207
   Acknowledgements 223
   References 223

Ch. 6. Patterns in Sequences of Random Events 227
        J. Gani

1. Early encounters: random numbers and the theory of runs 227
2. Sequences of events with repetitions 230
3. Strings and string overlaps 232
4. Further classical and martingale methods 235
5. Markov chain techniques 238
   References 240

Ch. 7. Stochastic Models in Telecommunications for Optimal Design, Control and Performance Evaluation 243
        N. Gautam

1. Introduction 243
2. Traffic models 244
3. Network performance using traffic models 250
4. LAN (multiaccess communication) models 268
5. Other topics and models 272
   Acknowledgements 280
   References 281

Ch. 8. Stochastic Processes in Epidemic Modelling and Simulation 285
        D. Greenhalgh

1. Introduction 285
2. Chain binomial models 285
3. The simple and general stochastic epidemic models 288
4. Spatial models 301
5. Stochastic models for control of epidemics 308
6. Specific applications 312
7. Stochastic processes in parameter estimation and hypothesis testing 321
8. Summary and conclusions 328
   Acknowledgement 330
   References 330

Ch. 9. Empirical Estimators Based on MCMC Data 337
P. E. Greenwood and W. Wefelmeyer

1. Introduction 337
2. The asymptotic variance of empirical estimators for Markov chains 341
3. Efficient estimation for Markov chain models 345
4. Improving empirical estimators by conditioning 348
5. Asymptotic variance of empirical estimators for Gibbs samplers 351
6. Asymptotic variance bounds for Gibbs samplers 353
7. Improving empirical estimators for random fields with local interactions 359
8. Exploiting symmetries of random fields 363
   Acknowledgement 366
   References 366

Ch. 10. Fractals and the Modelling of Self-Similarity 371
B. M. Hambly

1. Introduction 371
2. Fractal geometry 373
3. Fractals and stochastic processes 388
4. Dynamic fractal models 391
5. Further applications and conclusion 403
   References 404

Ch. 11. Numerical Methods in Queueing Theory 407
D. Heyman

1. Introduction 407
2. Numerical inversion of Laplace transforms 410
3. The ubiquity of Markov chains 413
4. Finite Markov chains 414
5. Infinite Markov chains 418
6. The BMAP/G/1 queue 423
7. The quasi birth-and-death process 425
   References 428

Ch. 12. Applications of Markov Chains to the Distribution Theory of Runs and Patterns 431
M. V. Koutras

1. Introduction 431
2. The Markov Chain imbedding technique 433
3. Success runs and pattern distributions 435
4. Markov Chain imbeddable variables of binomial type 443
5. Waiting time distributions associated with *MVB*'s 446
6. The number of runs and patterns as members of the *MVB* family 448
7. Multivariate *MVB* distributions 454

8. Multivariate success runs distributions 457
9. Alternative methods for exact distribution evaluation 466
   References 470

Ch. 13. Modelling Image Analysis Problems Using Markov Random Fields 473
S. Z. Li

1. Introduction 473
2. Image labeling 475
3. Markov random fields and Gibbs distributions 480
4. Useful MRF models 489
5. The MAP–MRF framework 499
   Acknowledgement 507
   References 507

Ch. 14. An Introduction to Semi-Markov Processes with Application to Reliability 515
N. Limnios and G. Oprişan

1. Introduction 515
2. Semi-Markov kernel 516
3. Markov renewal processes (MRP) 518
4. Semi-Markov processes with an arbitrary state space 525
5. Markov renewal equation 528
6. The countable case 532
7. Classification of states 537
8. Asymptotic behavior 540
9. Some recent approaches to semi-Markov processes 542
10. Reliability modeling and estimation 551
    References 554

Ch. 15. Departures and Related Characteristics in Queueing Models 557
M. Manoharan, M. H. Alamatsaz and D. N. Shanbhag

1. Introduction 557
2. Characterization/identifiability via output processes 558
3. Characterization/identifiability via infinite divisibility property 564
4. Strong unimodality and other relevant properties 568
   References 570

Ch. 16. Discrete Variate Time Series 573
E. McKenzie

1. Introduction 573
2. Markov chains 575
3. The DARMA models 576
4. Models based on thinning 578
5. Regression models 594
6. State space and Bayesian models 597
7. The future 602
   References 602

Ch. 17. Extreme Value Theory, Models and Simulation 607
S. Nadarajah

1. Introduction 607
2. Limit laws in univariate extremes and characterizations 608
3. Rates of convergence 613
4. Generalized extreme value (GEV) distribution 618
5. Generalized Pareto (GP) distribution 620
6. Joint distribution of the r-largest order statistics 623
7. A point process characterization 624
8. Extremes of stochastic processes 625
9. Limit laws for multivariate extremes 635
10. Characterizations of the domain of attraction 637
11. Characterizations of multivariate extreme value distributions 649
12. Rates of convergence 652
13. Parametric families for bivariate extreme value distributions 654
14. Parametric families for multivariate extreme value distributions 663
15. Extremes of multivariate stochastic processes 677
    Acknowledgements 679
    References 680

Ch. 18. Biological Applications of Branching Processes 693
A. G. Pakes

0. Introduction 693
1. History, surnames, and sex 695
2. Genetics and evolution 703
3. Epidemic modelling 728
4. Ecology and conservation modelling 738
   References 762

Ch. 19. Markov Chain Approaches to Damage Models 775
C. R. Rao, M. Albassam, M. B. Rao and D. N. Shanbhag

1. Introduction 775
2. Modified versions of some basic results on damage models 776
3. Characterizations based on modified Rao–Rubin conditions 782
4. Characterization via conditional expectations 788
   References 793

Ch. 20. Point Processes in Astronomy: Exciting Events in the Universe 795
J. D. Scargle and G. J. Babu

1. Introduction: what's the point? 795
2. Unique features of astronomical point processes 797
3. Naive point process theory 798
4. The mystery of Gamma Ray bursts 801
5. Other examples of astronomical point processes 815
6. Conclusion 822
   Acknowledgement 823
   References 823

Ch. 21. On the Theory of Discrete and Continuous Bilinear Time Series Models 827
T. Subba Rao and Gy. Terdik

1. Introduction 827
2. Linear time series models and cumulant spectra 833
3. Volterra expansion and bilinear models 834
4. Higher-order moments and identification 837
5. Estimation of higher-order cumulants and the bilinear models 838
6. Multivariate nonlinear time series and higher-order cumulants of random vectors 843
7. Spurious regression and cointegration, nonlinearity 847
8. Time dependent nonlinear models 850
9. Long range dependence 852
10. Stationary bilinear process in continuous time 861
    Acknowledgement 866
    References 867

Ch. 22. Nonlinear and Non-Gaussian State-Space Modeling with Monte Carlo Techniques: A Survey and Comparative Study 871
H. Tanizaki

1. Introduction 871
2. State-space model 874
3. Nonlinear and non-Gaussian state-space modeling 887
4. Monte Carlo studies 906
5. Summary and concluding remarks 917
   Appendix A. Linear and normal system 919

Appendix B. Sampling methods 920
Appendix C. Recursive versus non-recursive algorithms 923
Acknowledgements 926
References 926

Ch. 23. Markov Modelling of Burst Behaviour in Ion Channels 931
G. F. Yeo, R. K. Milne, B. W. Madsen, Y. Li and R. O. Edeson

1. Introduction 931
2. Aggregated Markov chains 934
3. Theoretical bursts 939
4. Empirical bursts 945
5. A five-state ligand-activated ion channel model 949
6. A linear sequential model with drug blockade 951
7. A model showing biphasic drug effects 955
8. A model for a supergated double-barrelled chloride channel 958
9. Some comments on statistical inference 963
   Note added in proof 964
   Acknowledgements 965
   References 965

Subject Index 969

Contents of Previous Volumes 979

Contributors

M. H. Alamatsaz, *Department of Statistics, Isfahan University, Isfahan-81745, Iran, e-mail: mh_alamatsaz@yahoo.com* (Ch. 15)

M. Albassam, *Department of Statistics, King Abdulaziz University, Jeddah, P.O. Box 80203, Saudi Arabia, e-mail: malbassam@hotmail.com* (Ch. 19)

G. J. Babu, *Department of Statistics, Pennsylvania State University, University Park, PA 16802-2111, USA, e-mail: babu@stat.psu.edu* (Ch. 20)

E. K. Boukas, *Mechanical Engineering Department, École Polytechnique de Montréal, P.O. Box 6079, station "centre-ville", Montréal, Québec, H3C 3A7 Canada, e-mail: el-kebir.boukas@polymtl.ca* (Ch. 1)

C. Cannings, *Division of Genomic Medicine, University of Sheffield, Sheffield, S10 2JF, UK, e-mail: c.cannings@sheffield.ac.uk* (Ch. 2)

J. E. Cavanaugh, *Department of Statistics, University of Missouri, Columbia, MO 65211, USA, e-mail: cavanaug@stat.missouri.edu* (Ch. 3)

R. Cowan, *School of Mathematics and Statistics, University of Sydney, NSW 2006, Australia, e-mail: rcowan@mail.usyd.edu.au* (Ch. 4)

J. W. Davis, *Department of Statistics, University of Missouri, Columbia, MO 65211, USA, e-mail: wdavis@stat.missouri.edu* (Ch. 3)

R. O. Edeson, *Department of Anaesthesia, Sir Charles Gairdner Hospital, Nedlands 6009, Australia, e-mail: redeson@cyllene.uwa.edu.au* (Ch. 23)

J. A. Ferreira, *CWI, Centre for Mathematics and Computer Science, Amsterdam, The Netherlands, e-mail: jose.ferreira@cwi.nl* (Ch. 5)

J. Gani, *School of Mathematical Sciences, Australian National University, Canberra ACT 0200, Australia, e-mail: gani@wintermute.anu.edu.au* (Ch. 6)

N. Gautam, *Department of Industrial and Manufacturing Engineering, The Pennsylvania State University, 310 Leonhard Building, University Park, PA 16802, USA, e-mail: ngautam@psu.edu* (Ch. 7)

D. Greenhalgh, *Department of Statistics and Modelling Science, University of Strathclyde, Livingstone Tower, 26 Richmond Street, Glasgow G1 1XH, UK, e-mail: david@stams.strath.ac.uk* (Ch. 8)

P. Greenwood, *Department of Mathematics and Statistics, College of Liberal Art and Sciences, Arizona State University, Tempe, AZ 85287-1804, USA, e-mail: pgreenw@graph.la.asu.edu* (Ch. 9)

B. M. Hambly, *Mathematical Institute, University of Oxford, 24-29 St Giles, Oxford OX1 3LB, UK, e-mail: hambly@maths.ox.ac.uk* (Ch. 10)

D. Heyman, *AT&T Labs, Room D5-3C35, 200 Laurel Avenue, Middletown, NJ 07748, USA, e-mail: dheyman@att.com* (Ch. 11)

M. V. Koutras, *Department of Statistics and Insurance Science, University of Piraeus, Greece, e-mail: mkoutras@unipi.gr* (Ch. 12)

S. Z. Li, *Microsoft Research China,1 Beijing Sigma Center, Beijing 100080, China, e-mail: szli@microsoft.com* (Ch. 13)

Y. Li, *Environmetrics Group, CSIRO Mathematical and Information Sciences, Private Bag No. 5, Wembley 6913, Australia, e-mail: Yun.Li@csiro.au* (Ch. 23)

N. Limnios, *Equipe de Mathématiques Appliquées, Université de Technologie de Compiègne, France, e-mail: nikolaos.limnios@utc.fr* (Ch. 14)

Z. K. Liu, *Mechanical Engineering Department, École Polytechnique de Montréal, P.O. Box 6079, station "centre-ville", Montréal, Québec, H3C 3A7 Canada, e-mail: liuzikuan@yahoo.com* (Ch. 1)

B. W. Madsen, *School of Mathematics and Statistics, The University of Western Australia, Crawley 6009, Australia, e-mail: bwmadsen@bigpond.com* (Ch. 23)

M. Mancharan, *Department of Statistics, University of Calicut, Kerala-673635, India, e-mail: mano30@rediffmail.com* (Ch. 15)

E. McKenzie, *Department of Statistics & Modelling Science, University of Strathclyde, UK, e-mail: ed@stams.strath.ac.uk* (Ch. 16)

R. K. Milne, *School of Mathematics and Statistics, The University of Western Australia, Crawley 6009, Australia, e-mail: milne@maths.uwa.edu.au* (Ch. 23)

S. Nadarajah, *Department of Mathematics, University of South Florida, Tampa, Florida 33620, USA, e-mail: snadaraj@chuma1.cas.usf.edu* (Ch. 17)

G. Oprişan, *Department of Mathematics, 'Politehnica' University of Bucharest, Romania, e-mail: oprisan@euler.math.pub.ro* (Ch. 14)

A. G. Pakes, *Department of Mathematics, University of Western Australia, 35 Stirling Highway, Crawley, WA 6009, Australia, e-mail: pakes@maths.uwa.edu.au* (Ch. 18)

D. B. Penman, *Department of Mathematics, University of Essex, Wivenhoe Park, Colchester C04 3SQ, UK, e-mail: dbpenman@essex.ac.uk* (Ch. 2)

C. R. Rao, *Department of Statistics, The Pennsylvania State University, University Park, PA 16802, USA, e-mail: crr1@email.psu.edu* (Ch. 19)

M. B. Rao, *Department of Statistics, North Dakota State University, Fargo, ND 58102, USA, e-mail: MB_Rao@ndsu.nodak.edu* (Ch. 19)

J. D. Scargle, *Space Science Division, National Aeronautics and Space Administration, Ames Research Center, MS 245-3, Moffett Field, CA 94035-1000, USA, e-mail: jeffrey@sunshine.arc.nasa.gov* (Ch. 20)

D. N. Shanbhag, *Department of Probability and Statistics, University of Sheffield, Sheffield S3 7RH, UK, e-mail: d.shanbhag@btopenworld.com* (Chs. 15, 19)

T. Subba Rao, *Institute of Science and Technology, University of Manchester, Manchester, UK and University of Debrecen, Hungary, e-mail: tata.subbarao@umist.ac.uk* (Ch. 21)

H. Tanizaki, *Faculty of Economics, Kobe University, Rokkodai, Nadaku, Kobe 657-8501, Japan, e-mail: tanizaki@kobe-u.ac.jp* (Ch. 22)

Gy. Terdik, *Institute of Science and Technology, University of Manchester, Manchester, UK and University of Debrecen, Hungary, e-mail: Terdik@delfin.unideb.hu* (Ch. 21)

Y. Wang, *Department of Statistics, University of Connecticut, Storrs, CT 06269, USA, e-mail: yzwang@uconn.stat.edu* (Ch. 3)

W. Wefelmeyer, *Fachbereich 6 Mathematik, Universität Siegen, Walter-Flex-Str. 3, 57068 Siegen, Germany, e-mail: wefelm@mathematik.uni-siegen.de* (Ch. 9)

G. F. Yeo, *Mathematics and Statistics, DSE, Murdoch University, Murdoch 6150, Australia, e-mail: yeo@maths.uwa.edu.au* (Ch. 23)

Modeling and Numerical Methods in Manufacturing System Using Control Theory

E. K. Boukas and Z. K. Liu

1. Introduction

Theoretically, manufacturing is defined as the transformation of material in a factory using some resources into something useful and portable (see Gershwin, 1993). This definition excludes the construction of buildings and bridges, since they are built where they are to be used. Generally, manufacturing can be divided into two classes: discrete manufacturing and continuous manufacturing. The first one produces distinct items, such as car, TV set and pen. The second one produces quantity of material such as oil. However, sometimes the continuous manufacturing models are used to approximately characterize the discrete ones. A manufacturing system is a set of machines, transportation elements, computers, storage buffers and other items that are used together for manufacturing. People are also part of the system. In a manufacturing system changing a setup is referred to changing a machine from producing one part type to producing another one. If such changes can be made at no cost (or time), this system is called a flexible manufacturing system (FMS).

To model this type of systems, many approaches have been developed. Among them we quote the ones based on queuing theory, control systems, Petri nets, discrete event systems, etc. For more details about these techniques we refer the readers to Sethi and Zhang (1994), Gershwin (1993), Maimoun et al. (1998), Buzacott and Shanthikumar (1993), Viswanadham and Narahari (1992) and references therein.

During the past decades, the continuous flow model for manufacturing system has received considerable attention and thus vast achievements have been reported (see Sethi and Zhang, 1994; Gershwin, 1993; Maimoun et al., 1998 and references therein). In this model, the material is treated as a continuous flow. The system is modeled as a hybrid one, i.e., the state of the system is comprised of two components: a discrete part, specifying the production capacity of the machine and a continuous part denoting the inventory level/shortage. The dynamics of the machine is governed by a continuous-time Markov process with finite state space. When the state of the machine is fixed, the continuous part state evolves as a deterministic system. Olsder and Suri (1980) were the first to cast the production planning of FMS into Rishel's formalism of system with jump Markov disturbance (Rishel, 1975). Based on Rishel's results on system with jump Markov disturbance and Davis' (1993) results on piecewise deterministic process, the production planning theory of continuous flow model has been developed rigorously (see Gershwin, 1993; Sethi and Zhang, 1994; Maimoun et al., 1998 and references therein).

Usually, a manufacturing system is a large scale and complex one. There may be hundreds of machines and thousands of part types may be produced each year. Moreover, there exist a lot of discrete events which occur randomly or deterministically. For instance, the failures and repairs of the machines, the departure of workers, the variation of demand, changes of the design, etc. These factors make the modeling and control of manufacturing system a challenging problem and thus attract a lot of researchers from operation research and automatic control communities (see Buzacott and Shanthikumar, 1993; Gershwin, 1993; Sethi and Zhang, 1994; Maimoun et al., 1998 and references therein). Since the occurrence frequency of the discrete events in a manufacturing system may vary greatly, for example, the failure of the machine occurs much less often than the operation, it is natural to divide the control or management into a hierarchy consisting of a number of different levels. Each level is characterized by its discrete events. Gershwin (1989) investigated the hierarchy scheduling planning in a dynamic manufacturing system with machines failures, setups, demand changes, etc. In modeling the decisions at each level, quantities that vary slowly (variables corresponding to higher level) are treated as constants and the ones that vary much faster (variables at lower levels of the hierarchy) are modeled in such a way that ignores their variations and are treated as continuous flow, e.g., when considering the failures and repairs of machine, the long run decision is regarded as constant and the operation is regarded as a continuous flow. Once the production rate of this level is determined, at the lower level, the dispatching time of every part can be obtained by some scheduling policies, such as stair-cases policy (see Gershwin, 1993). There have been several hierarchical scheduling algorithms, some of them are quite practical and successful (see Gershwin, 1993).

Another setting of hierarchical control in manufacturing system is to model the uncertain production capacity and demand by a singular Markov process (see Sethi and Zhang, 1994 or Yin and Zhang, 1998). The objective is to seek short term decisions to meet the demand in the long run at the minimal discounted cost including the production cost, inventory cost, and the cost of other decisions such as investment, etc. The key point of this approach is to first establish a limiting control problem by replacing the stochastic production capacity by the average total capacity of the machines and appropriately changing the cost function, and then design a suboptimal controller from the limiting control problem, which is much easier to solve than the original one. The controller constructed from the limiting control problem can be proven to be asymptotically optimal to the original one.

Setup is of great importance to manufacturing system with small batches and multiple part types, which may cost time or money or both. A lot of achievements on setups scheduling planning have been reported. Connolly et al. (1992), Gershwin et al. (1988), Hu et al. (1994), and Sharifnia et al. (1991) have developed various possible heuristic policies and have carried out numerical computations and simulations. Sethi and Zhang (1994) considered the asymptotically optimal policy by using the hierarchical control. Yan and Zhang (1997) established an approximate optimality condition and developed a computation algorithm for solving the optimal production control and setup scheduling problem.

Since the machine is unreliable, the maintenance problem is of great importance. Costa and Davis (1989) modeled the maintenance activity by impulse control, in which the machine is assumed to break down with age-dependent failure rate and take a fixed time to repair it. It must be taken out of service when its aging reaches a fixed value but may be taken out service for maintenance at any time before that. When to do the maintenance is a control variable. For more detail on this model, the reader is referred to (Costa and Davis, 1989; Costa, 1991; Davis, 1993).

In the context of continuous flow model, some variations of the continuous flow model for production control involving maintenance control have been developed by Boukas and his co-workers. Boukas and Haurie (1990) modeled the preventive maintenance as a special state of the machine, the preventive rate is defined as the jump rate from the working state to the preventive maintenance state, which is dependent on the age of the machine. Instead of using the age of the machine, in Boukas and Liu (1999b) the machine is assumed to have four modes, three working modes: good, average and bad, and a failure state. The jump rates from average and bad mode to working mode are called the preventive maintenance rate and the one from failure mode to the working mode is the corrective maintenance rate. Boukas, Zhang, and Zhu (1994), Boukas and Yang (1996) consider a class of maintenance activities which are executed without stopping the machine and can reduce the machine failure rates. The maintenance control problem with setups included is considered by Boukas and Kenne (1997).

Other types of models have been reported in the literature (see Buzacott and Shanthikumar, 1993; Viswanadaham and Narahari, 1992; Desrochers, 1990 and the references therein). In these references, we can find other studies that talk about modeling and numerical methods in manufacturing systems that mostly based on queuing theory, simulation and different optimization techniques (e.g., linear programming, dynamic programming, stochastic programming, multi-objective optimization, etc.). There are also other techniques that are mainly based on Petri net theory. For more information about these studies, we refer the reader to (Viswanadham and Narahari, 1992; Desrochers, 1990) and the references therein.

This chapter will be devoted to introducing the maintenance and control models for flexible manufacturing systems. The rest of this chapter is organized as follows: In Section 2, the definition and optimal control problem of piecewise-deterministic process are addressed and the numerical computation techniques to solve the optimization problem are developed. Section 3 introduces the continuous flow model of production control. Section 4 addresses the preventive maintenance models of Boukas and Haurie (1990). Section 5 is devoted to the model developed by Boukas and Liu (1999b) that contains preventive and corrective maintenances. Section 6 addresses the model developed by Boukas, Zhang and Yin (1996).

2. Controlled piecewise deterministic processes

Continuous-time stochastic models are usually used in practice to model physical phenomena. Due to the fact that dynamic systems are often vulnerable to component failures and repairs, sudden environment disturbances, changing subsystem interconnection and abrupt variation of the operating point of a nonlinear plant, etc., there are more and more examples that show the importance of dynamic system subject to abrupt variations in their structure. Piecewise deterministic processes (PDPs) are a general class of non-diffusion stochastic models, which can be used to model such systems. Such models have a hybrid state vector with two components. The first component specifies the mode of the system and the second one is referred to as the continuous state, which denotes the physical state of the system. The first part of the state visits in a stochastic manner a finite set of modes. When the mode is fixed, the continuous state evolves as a deterministic dynamic system. When the system enters a new mode, some components of the second part of the state vector may have various discontinuities. A PDP can be described as follows (for more information of PDP, the reader is referred to Davis (1993)).

2.1. Definition and optimal control problem of PDP

Let $\mathcal{E}$ be a countable set and ℓ be a function mapping $\mathcal{E}$ into $\mathbb{N}$, i.e., $\ell : \mathcal{E} \to \mathbb{N}$. For each $\alpha \in \mathcal{E}$, E_α^0 denotes a Borel set of $\mathbb{R}^{\ell(\alpha)}$, i.e., $E_\alpha^0 \subset \mathbb{R}^{\ell(\alpha)}$. Define

$$E^0 = \bigcup_{\alpha \in S} E_\alpha^0 = \{(\alpha, z) : \alpha \in \mathcal{E}, \ z \in E_\alpha^0\},$$

which is a disjoint union of E_α^0's. For each $\alpha \in \mathcal{E}$, we assume the vector field

$$f^\alpha : E_\alpha^0 \to E_\alpha^0$$

is a locally Lipschitz continuous function, determining a flow $\phi_\alpha(x)$. For each $x = (\alpha, z) \in E^0$, define

$$t_*(x) = \begin{cases} \inf\{t > 0 : \phi_\alpha(t, z) \in \partial E_\alpha^0\}, \\ \infty, \quad \text{if no such time exists,} \end{cases}$$

where ∂E_α^0 is the boundary of E_α^0. Thus $t_*(x)$ is the boundary hitting time for the starting point x. If $t_\infty(x)$ denotes the explosion time of the trajectory $\phi_\alpha(\cdot, z)$, then we assume that $t_\infty(x) = \infty$ when $t_*(x) = \infty$, thus effectively ruling out explosions. Now define

$$\partial^\pm E_\alpha^0 = \{z \in \partial E_\alpha^0 : z = \phi_\alpha(\pm t, \xi) \text{ for some } \xi \in E_\alpha^0, \ t > 0\},$$

$$\partial_1 E_\alpha^0 = \partial^- E_\alpha^0 \setminus \partial^+ E_\alpha^0,$$

$$E_\alpha = E_\alpha^0 \cup \partial_1 E_\alpha^0.$$

With these definitions, the state space and boundary of PDP can be respectively defined as follows:

$$E = \bigcup_{\alpha \in \mathcal{E}} E_\alpha, \quad \text{state space,} \tag{1}$$

$$\Gamma^* = \bigcup_{\alpha \in \mathcal{E}} \partial^+ E_\alpha^0, \quad \text{boundary.} \tag{2}$$

Thus the boundary of the state space consists of all those points which can be hit by the state trajectory. The points on some ∂E_α^0 which cannot be hit by the state of the trajectory are also included in the state space. The boundary of E consists of all the active boundary points, i.e., points in ∂E_α^0 that can be hit by the state trajectory.

The evolution of a piecewise deterministic Markov process (PDP) taking values in E is characterized by its three local characteristics:

1. a Lipschitz continuous vector field $f^\alpha : E \to \mathbb{R}^n$, which determines a flow $\phi_\alpha(t, z)$ in E such that, for $t > 0$,

$$\frac{\mathrm{d}}{\mathrm{d}t}\phi_\alpha(t, z) = f^\alpha(t, z), \qquad \phi_\alpha(0, z) = z, \quad \forall x = (\alpha, z) \in E;$$

2. a jump rate $q : E \to \mathbb{R}_+$, which satisfies that for each $x \in E$, there is a $\varepsilon > 0$ such that

$$\int_0^\varepsilon q(\alpha, \phi_\alpha(t, z))\, \mathrm{d}t < \infty;$$

3. a transition measure $Q : E \to \mathcal{P}(E)$, where $\mathcal{P}(E)$ denotes the set of probability measures on E.

By using these characteristics, a right-continuous sample path $\{x_t : t > 0\}$ starting at $x = (\alpha, z) \in E$ can be constructed as follows. Define

$$x_t \triangleq (\alpha, \phi_\alpha(t, z)), \quad \text{if } 0 \leqslant t < \tau_1,$$

where τ_1 is the realization of the first jump time T_1 with the following generalized negative exponential distribution

$$P(T_1 > t) = \exp\left(-\int_0^t q(\alpha, \phi_\alpha(s, z))\, \mathrm{d}s\right).$$

Having realized $T_1 = \tau_1$, we have $x_{T_1^-} \triangleq (\alpha, \phi_\alpha(\tau_1, z))$ and the post-jump state x_{τ_1} which has the distribution given by

$$P\big((\alpha', z_{\tau_1}) \in A | T_1 = \tau_1\big) = Q(A, x_{\tau_1^-})$$

on a Borel set A in E.

Restarting the process at x_{τ_1} and proceeding recursively according to the same recipe, one obtains a sequence of jump-time realizations $\tau_1, \tau_2, \ldots$. Between each two consecutive jumps, $\alpha(t)$ remains constant and $z(t)$ follows the integral curves of f^α. Considering this construction as generic yields the stochastic process $\{x_t : t \geqslant 0, x_0 = x\}$

and the sequence of its jump times $T_1, T_2, \ldots$. It can be shown that x_t is a strong Markov process with right continuous, left-limited sample paths (see Davis, 1993).

Piecewise-deterministic processes include a variety of stochastic processes arising from engineering, operation research, management science, economics and inventory system, etc. Examples are queuing systems, insurance analysis (see Dassios and Embrechts, 1989), capacity expansion (see Davis et al., 1987), permanent health insurance model (Davis, 1993), inventory control model (see Sethi and Zhang, 1994), production and maintenance model (see Boukas and Haurie, 1990). Due to its extensive applications, the optimal control problem has received considerable attention. Gatarek (1992), Costa and Davis (1989), and Davis (1993) have studied the impulse control of PDPs. In the context of nonsmooth analysis, Dempster (1991) developed the condition for the uniqueness of the solution to the associated HJB equation of PDPs optimal control involving Clarke generalized gradient. The existence of relaxed controls for PDPs was proved by Davis (1993). Soner (1986), Lenhart and Liao (1988) used the viscosity solution to formulate the optimal control of PDPs. For more information on the optimal control of PDPs, the reader is referred to Davis (1993) and Boukas (1993).

In this chapter, the models for the production and maintenance control in manufacturing system will be presented as a special class of piecewise deterministic Markov processes without active boundary points in the state space and the state jump can be represented by a function g. The model can be described as follows:

$$\dot{z}(t) = f^{\alpha(t)}(z(t), u(t)), \quad \forall t \in [T_n, T_{n+1}), \tag{3}$$

$$z(T_n) = g^{\alpha(T_n)}(z(T_n^-)), \quad n = 0, 1, 2, \ldots, \tag{4}$$

where $z = [z_1, \ldots, z_p]^T \in \mathbb{R}^p$, $u = [u_1, \ldots, u_q]^T \in \mathbb{R}^q$ are respectively, the state and control vectors, $f^\beta = [f_1^\beta, \ldots, f_p^\beta]^T$ and $g^\beta = [g_1^\beta, \ldots, g_p^\beta]^T$ represent real valued vectors, and x^T denotes the transpose of x. The initial conditions for the state and for the jump disturbance, i.e., the mode, are $z(0) = z^0 \in \mathbb{R}^p$ and $\alpha(0) = \beta_0 \in \mathcal{E}$, respectively. The set $\mathcal{E}$ is referred as the index set.

$\alpha = \{\alpha(t) : t \geq 0\}$ represents a controlled Markov process with right continuous trajectories and taking values on the finite state space $\mathcal{E}$. When the stochastic process $\alpha(t)$ jumps from mode β to mode β', the derivatives in (3) change from $f^\beta(z, u)$ to $f^{\beta'}(z, u)$. Between consecutive jump times the state of the process $\alpha(t)$ remains constant. The evolution of this process is completely defined by the jump rates $q(\beta, z, u)$ and the transition probabilities $\pi(\beta'|\beta, z, u)$. The set $\mathcal{E}$ is assumed to be finite. T_n (random variable) is the time of the occurrence of the nth jump of the process α. For each $\beta \in \mathcal{E}$, let $q(\beta, z, u)$ be a bounded and continuously differentiable function. At the jump time T_n, the state z is reset at a value $z(T_n)$ defined by Eq. (4) where $g^\beta(\cdot) : \mathbb{R}^p \to \mathbb{R}^p$ is, for any value $\beta \in \mathcal{E}$, a given function.

REMARK 2.1. This description of the system dynamics generalizes the control framework studied in depth by Rishel (1975), Wonham (1971) and Sworder and Robinson (1974), etc. The generalization lies in the fact that the jump Markov disturbances are controlled, and also from the discontinuities in the z-trajectory generated by Eqs. (3)–(4).

For each $\beta \in \mathcal{E}$, let $f^\beta(\cdot,\cdot): \mathbb{R}^p \times \mathbb{R}^q \mapsto \mathbb{R}^p$ be a bounded and continuously differentiable function with bounded partial derivatives in z. Let $U(\beta)$, $\beta \in \mathcal{E}$ (a closed subset of $\mathbb{R}^q$) denote the control constraints. Any measurable function with values in $U(\beta)$, for each $\beta \in \mathcal{E}$, is called an admissible control. Let $\mathcal{U}$ be a class of stationary control functions $u_\beta(z)$, with values in $U(\beta)$ defined on $\mathcal{E} \times \mathbb{R}^p$, called the class of admissible policies. The continuous differentiability assumption is a severe restriction on the considered class of optimization problems, but it is the assumption which allows the simpler exposition that was given in Boukas and Haurie (1988). Later, in the practical models, the restriction will be removed by introducing the notion of viscosity solution of Hamilton–Jacobi–Bellman equation.

The optimal control problem may now be stated as follows: given the dynamical system described by Eqs. (3)–(4), find a control policy $u_\beta(z) \in \mathcal{U}$ such that the expected value of the cost functional

$$J(\beta, z, u) = \mathbf{E}_\mathbf{u} \left\{ \int_0^\infty e^{-\rho t} c(\alpha(t), z(t), u(t)) \, dt \,\middle|\, \alpha(0) = \beta, \; z(0) = z \right\} \qquad (5)$$

is minimized over $\mathcal{U}$.

In Eq. (5), ρ ($\rho > 0$) represents the continuous discount rate, and $c(\beta, \cdot, \cdot): \mathbb{R}^p \times \mathbb{R}^q \mapsto \mathbb{R}^+$, $\beta \in \mathcal{E}$, is the family of cost rate functions, satisfying the same assumptions as $f^\beta(\cdot, \cdot)$.

We now proceed to give more precise definition of the controlled stochastic process. Let $(\Omega, \mathcal{F})$ be a measure space. We consider a function $X(t, \omega)$ defined as:

$$X: \mathcal{D} \times \Omega \mapsto \mathcal{E} \times \mathbb{R}^p, \quad \mathcal{D} \subset \mathbb{R}^+,$$
$$X(t, \omega) = (\alpha(t, \omega), z(t, \omega))$$

which is measurable with respect to $\mathcal{B}_\mathcal{D} \times \mathcal{F}$ ($\mathcal{B}_\mathcal{D}$ is a σ-field).

Let $\mathcal{F}_t = \sigma\{X(s, \cdot): s \leqslant t\}$ be the σ-field generated by the past observations of X up to time t. We now assume the following:

ASSUMPTION 2.1. The behavior of the dynamical system (3)–(4) under an admissible control policy $u_\beta(\cdot) \in \mathcal{U}$ is completely described by a probability measure P_u on $(\Omega, \mathcal{F}_\infty)$. Thus the process $X_u = (X(t, \cdot), \mathcal{F}_t, P_u)$, $t \in \mathcal{D}$, is well defined. For a given $\omega \in \Omega$ with $z(0, \omega) = z^0$ and $\alpha(0, \omega) = \beta_0$, we define:

$$T_1(\omega) = \inf\{t > 0: \alpha(t, \omega) \neq \beta_0\},$$
$$\beta_1(\omega) = \alpha(T_1(\omega), \omega),$$
$$\vdots$$
$$T_{n+1}(\omega) = \inf\{t > T_n(\omega): \alpha(t, \omega) \neq \alpha(T_n, \omega)\},$$
$$\beta_{n+1}(\omega) = \alpha(T_{n+1}(\omega), \omega),$$
$$\vdots$$

ASSUMPTION 2.2. For any admissible control policy $u_\beta(\cdot) \in \mathcal{U}$, and almost any $\omega \in \Omega$, there exist a finite number of jump times $T_n(\omega)$ on any bounded interval $[0, T]$, $T > 0$. Thus the function $X_u(t, \omega) = (\alpha_u(t, \omega), z_u(t, \omega))$ satisfies:

$$\alpha_u(0, \omega) = \beta_0,$$

$$z_u(t, \omega) = z^0 + \int_0^t f^{\beta_0}(z_u(s, \omega), u_{\beta_0}(z(s, \omega))) \, ds, \quad \forall t \in [0, T_1(\omega)),$$

$$\vdots$$

$$\alpha_u(t, \omega) = \beta_n(\omega),$$

$$z_u(t, \omega) = g^{\beta_n(\omega)}(z_u(T_n^-(\omega), \omega)) + \int_{T_n(\omega)}^t f^{\beta_n(\omega)}(z_u(s, \omega), u_{\beta_n}(z(s, \omega))) \, ds,$$

$$\forall t \in [T_n(\omega), T_{n+1}(\omega)),$$

$$\vdots$$

ASSUMPTION 2.3. For any admissible control policy $u_\beta(\cdot) \in \mathcal{U}$, we have:

$$P_u(T_{n+1} \in [t, t + dt] \mid T_{n+1} \geq T_n, \; \alpha(t) = \beta_n, \; z(t) = z)$$
$$= q(\beta_n, z, u_{\beta_n}(z)) \, dt + o(dt),$$
$$P_u(\alpha(t) = \beta_{n+1} \mid T_{n+1} = t, \; \alpha(t^-) = \beta_n, \; z(t^-) = z) = \pi(\beta_{n+1} \mid \beta_n, x, u).$$

Given these assumptions and an initial state (β_0, z^0), the question which will be addressed in the rest of this section is to find a policy $u_\beta(\cdot) \in \mathcal{U}$ that minimizes the cost functional defined by (5) subject to the dynamical system (3)–(4).

REMARK 2.2. From the theory of the stochastic differential equations and the previous assumptions on the functions f^β and g^β for each β, we recall that the system (3)–(4) admits a unique solution corresponding to each policy $u_\beta(z) \in \mathcal{U}$. Let $z^\beta(s; t, z)$ denote the value of this solution at time s.

The class of control policies $\mathcal{U}$ is such that for each β, the mapping $u_\beta(\cdot) : z \to U(\beta)$ is sufficiently smooth. Thus for each control law $u(\cdot) \in \mathcal{U}$, there exists a probability measure P_u on $(\Omega, \mathcal{F})$ such that the process (α, z) is well defined and the cost (5) is finite. Let the value function $V(\beta, z)$ be defined by the following equation:

$$V(\beta, z) = \inf_{u \in \mathcal{U}} \mathbf{E}_u \left\{ \int_0^\infty e^{-\rho \tau} c(\alpha(\tau), z(\tau), u(\tau)) \, d\tau \mid \alpha(0) = \beta, \; z(0) = z \right\}.$$

Under the appropriate assumptions, the optimality conditions of the infinite horizon problem are given by the following theorem:

THEOREM 2.1. *A necessary and sufficient condition for a control policy $u_\beta(\cdot) \in \mathcal{U}$ to be optimal is that, for each $\beta \in \mathcal{E}$, its performance function $V(\beta, z)$ satisfies the nonlinear partial differential equation*:

$$\rho V(\beta, z)$$
$$= \min_{u(\cdot) \in U(\beta)} \left\{ c(\beta, z, u) + \sum_{i=1}^{p} \frac{\partial}{\partial z_i} V(\beta, z) f_i^\beta(z(t), u_\beta(z)) - q(\beta, z, u_\beta(z)) V(\beta, z) \right.$$
$$\left. + \sum_{\beta' \in \mathcal{E} - \{\beta\}} q(\beta, z, u) V(\beta', g^{\beta'}(z(t))) \pi(\beta' | \beta, z, u) \right\}, \quad \forall \beta \in \mathcal{E}, \tag{6}$$

where $\frac{\partial}{\partial z_i} V(\beta, z)$ stands for the partial derivative of the value function $V(\beta, z)$ with respect to the component z_i of the state vector z.

PROOF. The reader is referred to Boukas and Haurie (1988) for the proof of this theorem. □

As we can see the system given by (6) is not easy to solve since it combines a set of nonlinear partial derivatives equations and optimization problem. To overcome this difficulty, we can approximate the solution by using numerical methods. In the next section, we will develop two numerical methods to solve these optimality conditions and which we believe that they can be extended to other class of optimization problems especially the nonstationary case.

2.2. *Numerical approximations techniques*

To approximate the solution of the Hamilton–Jacobi–Bellman (HJB) equation corresponding to the deterministic or the stochastic optimal control problem, many approaches have been proposed. For this purpose, we refer the reader to Boukas (1995) and Kushner and Dupuis (1992).

In this section we will give an extension of some numerical approximation techniques which were used respectively by Kushner (1977), Kushner and Dupuis (1992) and by Gonzales and Roffman (1985) to approximate the solution of the optimality conditions corresponding to other class of optimization problems. Kushner has used his approach to solve an elliptic and parabolic partial differential system associated with a stochastic control problem with diffusion disturbances. Gonzales and Roffman have used their approach to solve a deterministic control problem. Our aim is to use these approaches to solve a set of coupled partial differential equations representing the optimality conditions of the optimization problem presented in last subsection. The idea behind these approaches consists, within a finite grid G_z^h with unit cell of lengths $(h_1, \ldots, h_p)$ for the state vector and a finite grid G_u^h with unit cell of lengths $(y_1, \ldots, y_q)$ for the control vector, of using an approximation scheme for the partial derivatives of the value function $V(\beta, z)$ which will transform the initial optimization problem to an auxiliary discounted Markov decision problem. This will

2.2.1. Discounted Markov decision process optimization

Before presenting the numerical methods, let us define the discounted Markov decision process (DMDP) optimization problem. Consider a Markov process X_t which is observed at time points $t = 0, 1, 2, \ldots$ to be in one of the possible states of some finite state space $S = \{1, 2, \ldots, N\}$. After observing the state of the process, an action must be chosen from a finite space action denoted by A.

If the process X_t is in state s at time t and action a is chosen, then two things occur: (i) we incur a cost $c(s, a)$ which is bounded and (ii) the next state of the system is chosen according to the transition probabilities $P_{ss'}(a)$.

The optimization problem assumes a discounted factor $\delta \in (0, 1)$, and attempts to minimize the expected discounted cost. The use of δ is necessary to make the costs incurred at future dates less important than the cost incurred today. A mapping $\gamma : S \to A$ is called a policy. Let $\mathcal{A}$ be set of all the policies. For a policy γ, let

$$V_\gamma(s) = \mathbf{E}_\gamma \left[\sum_{t=0}^{\infty} \delta^t c(X_t, a_t) | X_0 = s \right],$$

where $\mathbf{E}_\gamma$ stands for the conditional expectation given that the policy γ is used.

Let the optimal cost function be defined as:

$$V_\alpha(s) = \inf_\gamma V(s).$$

In the following, we will recall some known results on this class of optimization problems. The reader is referred to Haurie and L'Ecuyer (1986) for more information on the topic and for the proofs of these results.

LEMMA 2.1. *The expected cost satisfies the following equation*:

$$V_\alpha(s) = \min_{a \in A} \left\{ c(s, a) + \delta \sum_{s'=1}^{N} P_{ss'}(a) V_\alpha(s') \right\}, \quad \forall s \in S.$$

Let $B(I)$ denote the set of all bounded real-valued functions defined on the state space S. Let the mapping T_α be defined by:

$$T_\alpha : B(I) \to B(I),$$

$$(T_\alpha w)(s) = \min_{a \in A} \left\{ c(s, a) + \delta \sum_{s'=1}^{N} P_{ss'}(a) w(s') \right\}, \quad \forall s \in S. \tag{7}$$

Let T_α^k be the composition of the map T_α with itself k times.

LEMMA 2.2. *The mapping T_α defined by (7) is contractive.*

LEMMA 2.3. *The expected cost $V_\alpha(\cdot)$ is the unique solution of the following equation:*

$$V_\alpha(s) = \min_{a \in A} \left\{ c(s,a) + \delta \sum_{s'=1}^{N} P_{ss'}(a) V_\alpha(s') \right\}, \quad \forall s \in \mathcal{S}.$$

Furthermore, for any $w \in B(I)$ the mapping $T_\alpha^n w$ converges to V_α as n goes to infinity.

Let us now see how we can put our optimization problem in this formalism. Since our problem has a continuous state vector z and a continuous control vector u, we need first to choose an appropriate discretization of the state space and the control space. Let G_z^h and G_u^h denote respectively the corresponding discrete state space and discrete control space and assume that they have finite elements with respectively n_z points for G_z^h and n_u points for G_u^h.

For the mode of the piecewise deterministic system, we do not need any discretization. Let $\mathcal{S}$ denote the global state space, $\mathcal{S} = \mathcal{E} \times G_z^h$ and N its number of elements. As we will see later, the constructed approximating Markov process X_t will jump between these states ($s = (\alpha, z) \in \mathcal{S}$), with the transition probabilities $P_{ss'}(a)$, when the control action a is chosen from G_u^h. These transition probabilities are defined as:

$$P_{ss'}(a) = \begin{cases} p_h^\beta(z, z+h; a), & \text{if } z \text{ jumps,} \\ \tilde{p}_h^\beta(\beta, z; \beta', a), & \text{if } \alpha \text{ jumps,} \end{cases}$$

where $p_h^\beta(z, z+h; a)$ and $\tilde{p}_h^\beta(\beta, z; \beta', a)$ are the probability transition between state s when the action a is used. The corresponding instantaneous cost function $c(s,a)$ and the discount factor δ of the approximating DMDP depend on the used discretization approach. Their explicit expressions will be defined later.

2.2.2. First approach

Let h_i denote the finite difference interval, in the coordinate i, and e_i the unit vector in the ith coordinate direction. The approximation that we use for $\frac{\partial}{\partial z_i} V(\beta, z)$ for each $\beta \in \mathcal{E}$, will depend on the sign of $f_i^\beta(z, u)$. Let G_z^h denote the finite difference grid which is a subset of $\mathbb{R}^p$.

This approach was used by Kushner to solve some optimization problems and it consists of approximating the value function $V(\beta, z)$ by a function $V_h(\beta, z)$, and to replace the first derivative partial derivative of the value function, $\frac{\partial}{\partial z_i} V(\beta, z)$, by the following expressions:

$$\frac{\partial}{\partial z_i} V(\beta, z) = \begin{cases} \frac{1}{h_i}[V_h(\beta, z+e_i h_i) - V_h(\beta, z)], & \text{if } \dot{z}(t) \geqslant 0, \\ \frac{1}{h_i}[V_h(\beta, z) - V_h(\beta, z-e_i h_i)], & \text{otherwise.} \end{cases} \quad (8)$$

For each β, define the functions $p_h^\beta(\cdot\,;\,\cdot,\cdot)$, $\tilde{p}_h^\beta(\cdot,\cdot\,;\,\cdot,\cdot)$ and $Q_h^\beta(\cdot,\cdot)$ respectively as follows:

$$Q_h^\beta(z,u) = q(\beta,z,u) + \sum_{i=1}^{p}\left[|\dot{z}_i(t)|/h_i\right],$$

$$p_h^\beta(z;\,z\pm e_ih,u) = f_i^\pm(z,u)/\left[h_iQ_h^\beta(z,u)\right],$$

$$\tilde{p}_h^\beta(\beta,z;\,\beta',u) = q(\beta,z,u)\pi(\beta'|\beta,z,u)/Q_h^\beta(z,u),$$

$$f_i^+(z,u) = \max\left(0, f_i^\beta(z,u)\right),$$

$$f_i^-(z,u) = \max\left(0, -f_i^\beta(z,u)\right).$$

Let $p_h^\beta(z;\,z\pm h,u) = 0$ for all points z not in the grid.

Putting the finite difference approximation of the partial derivatives as defined in (8) into (6), and collecting coefficients of the terms $V_h(\beta,z)$, $V_h(\beta, z\pm e_ih_i)$, yields, for a finite difference interval h applying to z,

$$V_h(\beta,z) = \left\{\frac{c(\beta,z,u)}{Q_h^\beta(z,u)\left[1+\frac{\rho}{Q_h^\beta(z,u)}\right]}\right.$$

$$+ \frac{1}{\left[1+\frac{\rho}{Q_h^\beta(z,u)}\right]}\left[\sum_{z'\in G_h} p_h^\beta(z;z',u)V_h(\beta,z')\right.$$

$$\left.\left.+ \sum_{\beta'\in\mathcal{E}-\{\beta\}} \tilde{p}_h^\beta(\beta,z;\beta',u)V_h(\beta',g^{\beta'}(z))\right]\right\}. \tag{9}$$

Let us define $c(s,u)$ and δ as follows:

$$c(s,u) = \frac{c^\beta(z,u)}{Q_h^\beta(z,u)\left[1+\frac{\rho}{Q_h^\beta(z,u)}\right]},$$

$$\delta = \frac{1}{1+\frac{\rho}{Q_h^\beta(z,u)}}.$$

A careful examination of Eq. (9) reveals that the coefficient of $V_h(\cdot,\cdot)$ are similar to transition probabilities between points of the finite set $\mathcal{S}$ since they are nonnegative and sum to, at most, unity. $c(s,u)$ is also nonnegative and bounded. δ, as defined, represents really a discount factor with values in $(0,1)$. Then, Eq. (9) has the basic form of the cost equation of the discounted Markov decision process optimization for a given control action. The approximating optimization problem built on the finite state space $\mathcal{S}$ has then the following cost equation:

$$V_h(\beta, z) = \min_{u \in G_u^h} \left\{ \frac{c(\beta, z, u)}{Q_h^\beta(z, u)\left[1 + \frac{\rho}{Q_h^\beta(z,u)}\right]} \right.$$

$$+ \frac{1}{\left[1 + \frac{\rho}{Q_h^\beta(z,u)}\right]} \left[\sum_{z' \in G_h} p_h^\beta(z; z', u) V_h(\beta, z') \right.$$

$$\left. \left. + \sum_{\beta' \in \mathcal{E} - \{\beta\}} \tilde{p}_h^\beta(z, \beta; \beta', u) V_h(\beta', g^{\beta'}(z)) \right] \right\}. \quad (10)$$

Based on the results presented in Subsection 2.2.1, we claim the uniqueness and the existence of the solution of the approximating optimization problem. It is plausible that the algorithms used in the discounted Markov process optimization would be helpful in computing this solution.

2.2.3. Second approach

In this approach, we will use the directional derivative in our approximation of the partial differential of the value function with respect to the variable z_i. This approach is related to Gonzales and Roffman's (1985) work, and it consists of approximating the value function $V(\beta, z)$ by a function $V_h(\beta, z)$, and to replace the first partial derivative of the value function, $\frac{\partial}{\partial z_i} V(\beta, z)$, in the direction of $f^\beta(\cdot, \cdot)$ using the directional derivative. Such approximation is given by the following expressions:

$$\frac{\partial V_h(\beta, z)}{\partial z_f} \| f^\beta(z^k) \|$$

$$= \begin{cases} [V_h(\beta, r_k(u)) - V_h(\beta, z^k)] \frac{\|f^\beta(z^k, u)\|}{\|r_k(u) - z^k\|}, & \text{if } f^\beta(z^k, u) \neq 0, \\ 0, & \text{otherwise,} \end{cases} \quad (11)$$

where z^k is a point of the finite grid G_z^h and

$$r_k(u) = \sum_{j=1}^{n_h} m_j^\beta(z^k, u) z^j, \qquad m_j^\beta(z^k, u) \geq 0, \qquad \sum_{j=1}^{n_h} m_j^\beta(z^k, u) = 1.$$

Putting the directional derivative approximation of the partial derivatives of the value function as defined by Eq. (11) into Eqs. (6), and collecting coefficients of the terms $V_h(\beta, z)$, $V_h(\beta, z \pm e_i h_i)$, yields a similar cost equation as the one obtained in Eq. (9), which in turn can be rewritten as the one of Eq. (10) with

$$Q_h^\beta(z^k, u) = \frac{\|f^\beta(z^k, u)\|}{\|\sum_{j=1}^{n_h} m_j(z^k, u) z^j - z^k\|} + q(\beta, z^k, u),$$

$$p_h^\beta(z^k; z^k \pm h, u) = \frac{\|f^\beta(z^k, u)\| m_j(z^k, u)}{\|\sum_{j=1}^{n_h} m_j(z^k, u) z^j - z^k\| Q_h^\beta(z^k, u)},$$

$$\tilde{p}_h^\beta(\beta, z^k; \beta', u) = \frac{q(\beta, z^k, u) \pi(\beta'|\beta, z^k, u)}{Q_h^\beta(z^k, u)}.$$

The two approaches produce a similar expression, which can be interpreted as a discrete Markovian decision control problem with finite state space and finite action space. This approximating optimization problem satisfies all the assumptions of the existence and the uniqueness of the optimal solution. To obtain an approximation of this solution, we can use the successive approximation technique or policy iteration technique. In the following section, we present the policy iteration technique.

2.2.4. Policy iteration

For a given policy $u_\beta(\cdot)$, let us introduce the mappings T_u and T^* acting on $V_h = (V_h(\beta))_{\beta \in \mathcal{E}}$, and defined by

$$T_u(V_h(\beta, z)) = \frac{c(\beta, z, u)}{Q_h^\beta(z, u)\left[1 + \frac{\rho}{Q_h^\beta(z,u)}\right]}$$

$$+ \frac{1}{\left[1 + \frac{\rho}{Q_h^\beta(z,u)}\right]} \left[\sum_{z' \in G_h} p_h^\beta(z; z', u) V_h(\beta, z') \right.$$

$$\left. + \sum_{\beta' \in \mathcal{E} - \{\beta\}} \tilde{p}_h^\beta(\beta, z; \beta', u) V_h(\beta', g^{\beta'}(z)) \right], \tag{12}$$

$$T^*(V_h(\beta, z)) = \min_{u \in G_u^h} \{T_u(V_h(\beta, z))\}. \tag{13}$$

The policy iteration algorithm operates as follows. For a given finite difference interval h:

Step 1. (Initialization) Choose $\varepsilon \in \mathbb{R}_+$, set: $k := 1$, $V_h^k(\beta, z) := 0$, $\forall \beta \in \mathcal{E}$, $\forall z \in G_z^h$, and guess an initial stationary policy $u^k \in G_u^h$.

Step 2. (Policy Evaluation) Given the stationary policy $u^k \in G_u^h$, compute the corresponding cost function from Eq. (12).

Step 3. (Policy Improvement) Obtain a new stationary policy $u^{k+1} \in G_u^h$ by using Eq. (13).

Step 4. Test:

$$\bar{c} := \min_{\substack{\beta \in \mathcal{E} \\ z \in G_z^h}} \{V_h^k(\beta, z) - V_h^{k-1}(\beta, z)\},$$

$$\underline{c} := \max_{\substack{\beta \in \mathcal{E} \\ z \in G_z^h}} \{V_h^k(\beta, z) - V_h^{k-1}(\beta, z)\}$$

with $c_{\min} \triangleq \frac{\rho}{1-\rho}\bar{c}$, $c_{\max} \triangleq \frac{\rho}{1-\rho}\underline{c}$.

If $|c_{\max} - c_{\min}| \leqslant \varepsilon$, then stop $u^* = \gamma^k$; else let $k = k+1$, and go the Step 2.

The use of the policy iteration will give a solution of the approximating optimization problem. The results established by Kushner and Dupuis (1992) concerning the convergence of the approximating solution to the real solution when h goes to zero can be adapted to our case.

THEOREM 2.2. *If there exist a constant C and a constant k_1 such that*

$$0 \leqslant V_h(\beta, z) \leqslant C(1 + |z|^{k_1}), \quad \forall \beta \in \mathcal{E}.$$

Then we have:

$$\lim_{h \to 0} V_h(\beta, z) = V(\beta, z), \quad \forall \beta \in \mathcal{E}.$$

PROOF. The proof of this theorem can be adapted from the one presented in Boukas et al. (1996). □

REMARK 2.3. The numerical method provided in this section have been successfully used for small systems (see Boukas, 1995, 1987, 1990). However, for large scale systems, the numerical computation problem of the optimal control easily runs into the "curse of dimensionality" problem and therefore there has been on satisfactory approach at present.

3. Continuous flow model for production control

During the past decades, great progresses in the study of the continuous flow model of FMS have been achieved. In these models, the inventory/backlog of part is assumed to be a continuous flow and the stochastic changes in the capacity of the system is modeled as a continuous-time Markov process (see Akella and Kumar, 1986; Olsder and Suri, 1980). Rishel's formalism of system with Markov disturbances (see Rishel, 1975) and the piecewise-deterministic process (PDP) developed by Davis (1984) have provided the theoretical framework. One of the most important achievement in this line of research has been to identify the structure of the optimal feedback control policy.

Kimemia and Gershwin (1983) showed that the optimal control for such systems has a special structure, which is called the *hedging point policy*. The key point of this policy is that the optimal controlled system should have a non-negative production surplus (the hedging level) of part to hedge against the future possible shortages resulting from the machine failure. When the inventory deviates from this point, the controller should push it back as soon as possible. For the one machine one part-type system with constant demand rate, the optimal hedging point was first solved by Akella and Kumar (1986) for the discounted cost model and the setting of average cost was first solved by Bielecki and Kumar (1988). Their results are outlined as follows.

Let $x(t)$, $t \geq 0$, be the inventory or backlog of the system and $\alpha(t) \in \mathcal{S} = \{1, 2\}$ be a Markov process, representing the state of the machine, i.e., $\alpha(t) = 1$ denotes the machine is under working condition, and $\alpha(t) = 2$ means the machine is under repair. Let

$$Q = \begin{pmatrix} -q_1 & q_1 \\ q_2 & -q_2 \end{pmatrix} \tag{14}$$

be the generator of $\{\alpha(t), t \geq 0\}$. Let $u(t)$ be the production rate at time t, which is a control variable and satisfies the following constraints:

$$0 \leq u(t) \leq \bar{u} I_{\{\alpha(t)=1\}},$$

where $\bar{u}$, a constant, is the production capacity of the machine, and $I_{\{\cdot\}}$ is the indicator function.

Consider the following discounted cost function

$$\mathbf{E}\left[\int_0^\infty e^{-\rho t}\left[c^+ x^+(t) + c^- x^-(t)\right] dt\right],$$

where c^+, c^- are positive constants, $x^+ = \max(x, 0)$, $x^- = \max(-x, 0)$, $\rho > 0$ is the discounted rate.

For any admissible control law $u(\cdot)$ (see Akella and Kumar, 1986 for the discussion of admissible control), process $\{(\alpha(t), x(t)), t \geq 0\}$ is a PDP with local characteristics

(1) vector field: $f^\alpha(t, x) = u(t) I_{\{\alpha(t)=1\}} - d$;
(2) jump rate: $q_{\alpha(t)}$;
(3) transition measure: $Q((\alpha, x), \{\alpha'\} \times A) = \sigma_{\{\alpha' \times \{x\}\}}(\beta \times A)$, $\forall A \in \mathbb{R}$, $\alpha' \neq \alpha$ where $\sigma_{\{\alpha' \times \{x\}\}}(\cdot)$ is the Dirac measure on $\mathcal{S} \times \mathbb{R}$ with mass at $\alpha' \times \{x\}$, i.e.,

$$\sigma_{\{\alpha' \times \{x\}\}}(\beta \times A) = \begin{cases} 1, & \text{if } \beta = \alpha' \text{ and } x \in A, \\ 0, & \text{otherwise.} \end{cases}$$

Let $V_\alpha(x)$ be the value function corresponding to initial condition $\alpha(0) = \alpha$, $x(0) = x$, then it can be shown that $V_\alpha(x)$, $\alpha = 1, 2$, is the unique solution of the following Hamilton–Jaccobi–Bellman (HJB) equations (see Akella and Kumar, 1986)

$$\begin{pmatrix} \min_{u \in [0,\bar{u}]} (u-d) \frac{dV_1(x)}{dx} \\ -d \frac{dV_2(x)}{dx} \end{pmatrix}$$

$$= \begin{pmatrix} \rho + q_1 & -q_1 \\ -q_2 & \rho + q_2 \end{pmatrix} \begin{pmatrix} V_1(x) \\ V_2(x) \end{pmatrix} - \begin{pmatrix} 1 \\ 1 \end{pmatrix} c^+ x^+ + c^- x^-. \quad (15)$$

From (15), it follows that the optimal production rate is

$$u(\alpha(t), x(t)) = \begin{cases} \bar{u}, & \text{if } \frac{dV_1(x)}{dx} < 0, \\ d, & \text{if } \frac{dV_1(x)}{dx} = 0, \\ 0, & \text{otherwise,} \end{cases} \quad (16)$$

where the purpose of letting $u(t) = d$ when $dV_1(x)/dx = 0$ is to prevent the chattering phenomenon. Akella and Kumar (1986) proved that $V_\alpha(x)$ is convex and continuously differentiable, so the optimal control is of the following structure:

$$u(\alpha(t), x(t)) = \begin{cases} \bar{u}, & \text{if } x(t) < x^* \text{ and } \alpha(t) = 1, \\ d, & \text{if } x(t) = x^* \text{ and } \alpha(t) = 1, \\ 0, & \text{otherwise,} \end{cases} \quad (17)$$

where x^*, called the hedging point, is the solution of $dV_1(x)/dx = 0$. Akella and Kumar (1986) have solved (15) analytically and thus got the following expression for x^*:

$$x^* = \max \left\{ 0, \frac{1}{\lambda_-} \log \left[\frac{c^+}{c^+ + c^-} \cdot \left(1 + \frac{\rho d}{q_1 d - (\rho + q_2 + d\lambda_-)(\bar{u} - d)} \right) \right] \right\}, \quad (18)$$

where λ_- is the unique negative eigenvalue of the matrix

$$A_1 \triangleq \begin{pmatrix} \frac{\rho + q_1}{\bar{u} - d} & \frac{-q_1}{\bar{u} - d} \\ \frac{q_2}{d} & -\frac{q_2 + \rho}{d} \end{pmatrix}.$$

Bielecki and Kumar (1988) showed that for a manufacturing system with unreliable machine, the zero-inventory policy can still be optimal in some cases. Their optimization problem is to seek a feedback control policy to minimize the following long run average expected cost incurred per unit time

$$\lim_{T \to \infty} \frac{1}{T} \mathbf{E} \left[\int_0^T [c^+ x_t^+ + c^- x_t^-] dt \right].$$

They proved that in this case policy (18) is still optimal with optimal inventory level given by

$$x^* = \begin{cases} 0, & \text{if } \frac{\bar{u}q_1(c^++c^-)}{c^+(\bar{u}-d)(q_1+q_2)} \leqslant 1 \text{ and } \frac{\bar{u}-d}{q_1} > \frac{d}{q_2}, \\ \infty, & \text{if } \frac{\bar{u}-d}{q_1} < \frac{d}{q_2}, \\ \frac{1}{q_2/d-q_1/(\bar{u}-d)} \log\left[\frac{\bar{u}q_1(c^++c^-)}{c^+(\bar{u}-d)(q_1+q_2)}\right], & \text{otherwise.} \end{cases}$$

For the average cost problem, by using the discounted vanishing approach, Sethi et al. (1997) provided the existence condition for the optimal control and established the verification theorem in the case of the machine having multiple states. Veatch and Caramanis (1999) established the differentiability and convexity of the differential cost function and proved that under an optimal control policy the differential cost is continuously differentiable on attractive control switching boundaries.

In above models, the jump rates of the Markov process governing the mode switching of the system are assumed to be constant. However, a common knowledge is that the faster the machine processes the parts, the more prone to fail it is. Due to this observation, Hu et al. (1994) studied the production control problem with production rate dependent jump rate, i.e., in (14) jump rate $q_1(\cdot)$ is an increasing function of production u. Their results show that the linearity of the jump rate function is both necessary and sufficient for the optimality of the hedging point policy.

Since the introduction of hedging point policy, although in the general cases its optimality can not be proved, it has received considerable attention due to its simplicity. Some structural properties of the hedging points for single part type and multiple machine state system were established by Hu and Xiang (1994), Yu and Song (1999). Sharifnia (1988) showed how the optimal hedging point in the case of one part-type and multiple machine-states can be calculated. Liberopoulos and Caramanis (1994) showed that the technique developed by Sharifnia can be applied to compute the hedging point policy even when the transition rates of the machines state depend on the production rate.

Since solving the optimal production rate control problem often involves solving either systems of partial differential equations or dynamic programming equations, it is difficult to find an analytical solution for the multiple mode or multiple part-type systems. Evidently, from the fact that the hedging policy is completely determined by parameters it follows that restricted to the subset of hedging point policies, the optimal control problem is, in fact, a parameter optimization problem. Therefore several approximation procedures have been developed to get sub-optimal controllers. Among them we quote the works of Bremaud et al. (1997), Caramanis and Liberopoulos (1992) and Tu, Song and Lou (1993). They apply the Perturbation Analysis put forward by Ho and his co-workers to develop optimization techniques based on simulation to obtain the optimal hedging point policy.

Since the machine is unreliable, the maintenance problem is of great importance. The following sections will be devoted to introducing some continuous flow production models with maintenance considered.

4. Preventive maintenance and production control model

In this section, we address a preventive maintenance model (see Boukas and Haurie, 1990), in which the maintenance is modeled as a state of the machine and the preventive maintenance rate is the jump rate from the working mode to this mode.

4.1. Model description

Consider a manufacturing system consisting of m machines and producing p part types. The machines are assumed to have three modes: operational denoted by 1, break down denoted by 2 and under maintenance denoted by 3. When the machine is in the operational mode, it can produce any part type and when it is under repair or maintenance, nothing is produced. Let $\alpha_i(t) \in \mathcal{M} = \{1, 2, 3\}$ be the state and $a_i(t) \in [0, \infty)$ be the aging of the machine M_i at time t. Then the process $\alpha(t) = (\alpha_1(t), \ldots, \alpha_p(t))^\top \in \mathcal{S} = \mathcal{M}^m$ denotes the state of the working station. The demand rates of the parts are assumed to be constants, denoted by $d = (d_1, \ldots, d_p)^\top$. We will use $x_i(t)$, $i = 1, \ldots, p$, to denote the inventory level when positive or shortage when negative of part type i at time t. Let $u_{ij}(t)$ be the production rate of machine i producing part type j. Then the state equation of the surplus is given by

$$\frac{dx(t)}{dt} = u(t) - d, \qquad x(0) = x_0, \qquad (19)$$

where x_0 is the initial surplus value, $u(t) = (u_1(t), \ldots, u_p(t))^\top$ is the production rates of the system at time t, with $u_i(t) = \sum_{j=1}^{m} u_{ji}(t)$ being the production rate of part type i. For $i = 1, \ldots, m$, $j = 1, \ldots, p$, define function

$$\delta_{ij}(t) = \begin{cases} 1, & \alpha_i(t) = j, \\ 0, & \text{otherwise.} \end{cases}$$

Variable $u(t)$ will be a control variable, satisfying the constraints

$$0 \leq \tilde{u}_i(t) = \sum_{j=1}^{p} u_{ij} \leq \bar{u}_i \delta_{i1}(t),$$

where $\bar{u}_i$ is a positive constant, representing the maximal production capacity of machine i.

Suppose the aging of the machine at time t is an increasing function of its production rate. Thus the age of machine i is the solution of the following differential equation:

$$\frac{da_i(t)}{dt} = f_i\big(\tilde{u}_i(t)\big), \quad i = 1, \ldots, m, \ t > T_i, \qquad (20)$$

$$a_i(T_i) = 0, \quad i = 1, \ldots, m, \qquad (21)$$

where $\tilde{u}_i(t) = \sum_{j=1}^p u_{ij}(t)$ is the production rate of machine i, T_i is the last restart time of machine i, $f_i(\cdot)$ is an increasing function with $f_i(0) = 0$. For simplicity, we make the following assumption.

ASSUMPTION 4.1. For any i, $f_i(\tilde{u}_i) = \sum_{l=1}^p f_{il} u_{il}$, with $f_{il} \geqslant 0$ being constants.

Let $\lambda_{kl}^i(a_i(t))$ be the jump rates from state k to state l of machine i which are defined as follows:

$$\lambda_{12}^i(a(t)) = \lim_{\Delta \to 0} \frac{P(\alpha_i(t+\Delta) = 2 | \alpha_i(t) = 1)}{\Delta}, \tag{22}$$

$$\lambda_{21}^i(a(t)) = \lim_{\Delta \to 0} \frac{P(\alpha_i(t+\Delta) = 1 | \alpha_i(t) = 2)}{\Delta}, \tag{23}$$

$$\lambda_{13}^i(a(t)) \triangleq v_i(t) = \lim_{\Delta \to 0} \frac{P(\alpha_i(t+\Delta) = 3 | \alpha_i(t) = 1)}{\Delta}, \tag{24}$$

$$\lambda_{31}^i(a_i(t)) = \lim_{\Delta \to 0} \frac{P(\alpha_i(t+\Delta) = 1 | \alpha_i(t) = 3)}{\Delta}, \tag{25}$$

all other transition rates λ_{kl}^i, $l \neq k$, equal zero. Define

$$\lambda_{kk}^i(a_i) = -\sum_{l \neq k} \lambda_{kl}^i(a_i).$$

In Eq. (24), the jump rate $v_i(t) = \lambda_{13}^i(a)$ from working mode to the preventive maintenance mode is a control variable, called the preventive maintenance rate of machine i, which is assumed to satisfy the following constraints

$$0 \leqslant v_i(t) \leqslant \bar{v}_i \delta_{i1}(t),$$

where $\bar{v}_i$ is a positive constant. The inverse of this control variable can be interpreted as the expected delay between a call for the technician and his arrival.

Eq. (22) implies that failures occur as a Poisson process, the failure rate of the machine i depends on its age $a_i(t)$.

In Eqs. (23) and (25), $\lambda_{21}^i(a_i(t))$ and $\lambda_{31}^i(a_i(t))$ denote respectively the jump rates from repair mode and preventive maintenance mode to operational mode. Noting that $f(0) = 0$, thus when the machine is in the repair or preventive maintenance modes, its aging remains the same as that the last jump occurs. That is, during the repair or preventive maintenance, the aging of machine is $a_i(t) = a_i(T_i)$ with T_i being the last jump time. Therefore, the repair time and the preventive times are exponentially distributed with parameters $\lambda_{21}^i(a_i(T_i))$ and $\lambda_{31}^i(a_i(T_i))$ respectively.

Assume the independence between the machines. From the jump rates defined by (22)–(25), one can easily deduce the transition rates $q_{\beta\beta'}(a, v)$ as follows: for any

$\beta, \beta' \in \mathcal{S}$,

$$q_{\beta\beta'}(a, v) = \begin{cases} \lambda_{\beta_i \beta'_i}(a_i, v_i), & \text{if } \beta_i \neq \beta'_i, \beta_j = \beta'_j, j \neq i, \\ 0, & \text{otherwise.} \end{cases}$$

Define

$$q_{\beta\beta}(a, v) = -\sum_{\beta' \in \mathcal{S}(\beta)} q_{\beta\beta'}(a, v) = \sum_{i=1}^{m} \lambda_{\beta_i \beta_i}(a_i, v_i), \quad q_\beta(a, v) = -q_{\beta\beta}(a, v),$$

where $\mathcal{S}(\beta) = \{\beta'$: there exists only one $i \in \{1, \ldots, m\}$ s.t. $\beta'_i \neq \beta_i\}$. Put

$$Q(a, v) = (\pi_{\beta\beta'}(a, v)), \quad \beta, \beta' \in \mathcal{S}, \tag{26}$$

with

$$\pi_{\beta\beta'}(a, v) = \begin{cases} \dfrac{q_{\beta\beta'}(a,v)}{q_{\beta\beta}(a,v)}, & \beta' \in \mathcal{S}(\beta), \\ 0, & \text{otherwise.} \end{cases}$$

ASSUMPTION 4.2. Suppose $\lambda^i_{kl}(a)$, $k, l \in \mathcal{M}$, are bounded and Lipschitz in a, then $q_{\beta\beta'}(a, v)$ is bounded and local Lipschitz in a, in the following we assume that there exist a constant c_0, such that $q_\beta(a, v) \geq c_0$, $\forall \beta \in \mathcal{S}$.

The condition (21) implies that a repair or a preventive maintenance job restores the cumulative age to zero value. Therefore, at a jump time τ of the process $\alpha(t)$, we define a reset function

$$\phi(a, \alpha) = a' = (a'_1, \ldots, a'_m), \tag{27}$$

where

$$a'_i = \begin{cases} 0, & \text{if } \alpha_i(\tau^+) = 1 \text{ and } \alpha_i(\tau^-) \neq 1, \\ a_i(\tau^-), & \text{otherwise.} \end{cases}$$

This function describes the age discontinuity which may occur at a jump time of the operational state of a machine.

The variables α, x, a are the state variables of the system. We call $z \triangleq (x, a) \in \mathbb{R}^p \times \mathbb{R}^m_+$ the continuous part of the state. Let $\gamma(t) = (u(t), v(t))$ denote control variables.

Consider the following cost function

$$J(\alpha, x, a, \gamma) = \mathbf{E}\left[\int_0^\infty e^{-\rho t} \psi(\alpha(t), x(t), a(t), \gamma(t)) \, dt \,|\, \alpha(0) = \alpha, \right.$$

$$\left. x(0) = x, \ a(0) = a \right], \tag{28}$$

where $\rho > 0$ is the discounted rate, $\psi(\alpha(t), z(t), \gamma(t))$ is the cost rate function, for which the following assumption holds.

ASSUMPTION 4.3. There exist constants K_g and k_g such that

(1) $\forall \alpha, \gamma$,

$$0 \leqslant \psi(\alpha, x, a, \gamma) \leqslant K_g\left(1 + \|x\|^{k_g} + \|a\|^{k_g}\right).$$

(2) $\forall x_1, x_2 \in \mathbb{R}^p$, $a_1, a_2 \in \mathbb{R}^m$, $\alpha \in \mathcal{S}$,

$$\|\psi(\alpha, x_1, a_1) - \psi(\alpha, x_2, a_2)\|$$
$$\leqslant K_g\left(1 + \|x_1\|^{k_g} + \|x_2\|^{k_g} + \|a_1\|^{k_g} + \|a_2\|^{k_g}\right)$$
$$\times \left(\|x_1 - x_2\| + \|a_1 - a_2\|\right).$$

In the following we will formulate an optimization problem that allows us to seek the optimal control policy for the preventive maintenance control and production planning. Let us define a set of control law

$$\Gamma(\alpha(t))$$
$$= \left\{ (u(t), v(t)) \colon u(t) \in \mathbb{R}^p,\ u_i(t) = \sum_{j=1}^{m} u_{ji}(t),\ 0 \leqslant \sum_{j=1}^{p} u_{ij}(t) \leqslant \bar{u}_i \delta_{i1}(t), \right.$$
$$\left. v(t) = (v_1(t), \ldots, v_m(t))^\top,\ 0 \leqslant v_i(t) \leqslant \bar{v}_i \delta_{i1}(t) \right\}.$$

DEFINITION 4.1. A control $\gamma(\cdot) = \{(u(t), v(t)), t \geqslant 0\}$ is said to be admissible if

(i) $(u(\cdot), v(\cdot)) = \{(u(t), v(t)), t \geqslant 0\}$ is adapted to the σ-algebra generated by $\{(\alpha_t, x(t)), t \geqslant 0\}$, denoted by $\sigma\{(\alpha_s, x(s)), 0 \leqslant s \leqslant t\}$.
(ii) $(u(t), v(t)) \in \Gamma(\alpha(t))$.
(iii) $u(t), v(t)$ are right-continuous and have left-hand limits.

REMARK 4.1. In the above definition, the assumption that $u(t)$, $v(t)$ are right continuous and have left-hand limits is only used to prove that the value function is the viscosity solution of the Hamiltonian–Jacobi–Bellman equation.

Let $\mathcal{A}$ be the collection of all admissible control. It can be shown that with a given control law $(u(\cdot), v(\cdot))$, the process $\{(\alpha(t), z(t)), t \geqslant 0\}$ is a Piecewise deterministic process (see Davis, 1993). Its local characteristics are as follows:

(1) vector field: $g = ((u(t) - d)^\top, f^\top(u(t)))^\top$;
(2) jump rate: $q(\alpha, x, a) = q_\alpha(a, v)$;
(3) transition measure: $Q((\alpha, x, a), \{\alpha'\} \times A) = \sum_{\alpha' \neq \alpha} q_{\alpha\alpha'}(a, v) \sigma_{\{(x, \phi(a, \alpha'))\}}(A)$, for any Borel set A in $\mathbb{R}^p \times \mathbb{R}^m_+$, where $\sigma_z(\cdot)$ is a Dirac measure.

REMARK 4.2. By Assumption 4.2, one can verify that if $h(\omega) \in C(\mathcal{S} \times \mathbb{R}^p \times \mathbb{R}_+^m)$ has polynomial growth and is locally Lipschitz in (x, a), then

$$\int_{\mathcal{S} \times \mathbb{R}^p \times \mathbb{R}_+^m} h(\omega) Q(\alpha, x, a; d\omega) = \sum_{\alpha' \in \mathcal{S}(\alpha)} h(\alpha', x, \phi(a, \alpha')) q_{\alpha\alpha'}(a, v)$$

as a function of (α, x, a) is local Lipschitz in (x, a), i.e., there exist positive constants C_g, k_0 such that for any (x_1, a_1), (x_2, a_2)

$$\left| \int_{\mathcal{S} \times \mathbb{R}^p \times \mathbb{R}_+^m} h(\omega) \big[Q(\alpha, x_1, a_1; d\omega) - Q(\alpha, x_2, a_2; d\omega) \big] \right|$$
$$\leqslant C_g \big(1 + \|x_1\|^{k_0} + \|x_2\|^{k_0} + \|a_1\|^{k_0} + \|a_2\|^{k_0} \big) \big(\|x_1 - x_2\| + \|a_1 - a_2\| \big).$$

REMARK 4.3. Checking the definition of PDP provided in Section 2.1, one can conclude that for any $\alpha \in \mathcal{S}$, $E_\alpha^0 = \mathbb{R}^p \times \mathbb{R}_+^m$. Therefore any (α, x, a), with some $a_i = 0$, is on ∂E_α^0. However it is obviously such point $(\alpha, x, a) \in \partial^- E_\alpha^0$, which is not a boundary point from the definitions of boundary and the state space of PDP. Therefore there is no boundary point in the state space of $\{(\alpha(t), x(t), a(t)), t \geqslant 0\}$.

DEFINITION 4.2. A measurable function $(u(\alpha(t), z(t)), v(\alpha(t), z(t))) \in \mathcal{A}$, is called an admissible feedback control, or simply a feedback control, if

(i) for any $x \in \mathbb{R}^p$ and $\alpha \in \mathcal{S}$, the following equations have a unique solution

$$\dot{x}(t) = u\big(\alpha(t), x(t)\big) - d, \qquad x(0) = x,$$
$$\dot{a}(t) = f\big(u(t)\big), \qquad a(0) = a,$$

where $(\alpha(t), x(t), a(t))$ is a Piecewise deterministic Markov process with local characteristics g, q, Q as defined above.

(ii) $(u(\cdot), v(\cdot)) \in \mathcal{A}$.

Let the function $V(\alpha, z)$ be defined as follows:

$$V(\alpha, z) = \min_{(u(\cdot), v(\cdot)) \in \mathcal{A}} J\big(\alpha, z, u(\cdot), v(\cdot)\big).$$

Let H be a real-value function defined by

$$H\big(\alpha, x, a, w(\cdot, x, a), r_1, r_2, \gamma\big)$$
$$= (u - d)^\top \cdot r_1 + f^\top(u) r_2$$
$$+ \psi(\alpha, x, a, \gamma) + \sum_{\beta \neq \alpha} q_{\alpha\beta}(a, v) \big[w\big(\beta, x, \phi(a, \beta)\big) - w(\alpha, x, a) \big]$$

with $w \in C(\mathcal{S} \times \mathbb{R}^p \times \mathbb{R}_+^m)$, $r_1 \in \mathbb{R}^p$, $r_2 \in \mathbb{R}^m$, $\gamma = (u, v)$. Then the HJB equation associated with the problem can be formally written as follows (see Boukas, 1993; Sethi and Zhang, 1994)

$$\rho V(\alpha, x, a) = \min_{\gamma \in \Gamma(\alpha)} \{H(\alpha, x, a, V(\cdot, x, a), V_x(\alpha, z), V_a(\alpha, z), \gamma)\}, \qquad (29)$$

where $V_z(\alpha, z)$, $V_a(\alpha, z)$ denote the partial derivatives of V with respect to x and a respectively.

Note that the HJB equation (29) involves the first-order partial derivatives of the value function under no guarantee that those partial derivatives would exist. In fact, one can always construct an example in which the value function does not have partial derivatives at certain points. In order to deal with such possible nondifferentiability of the value function, the concept of viscosity solution is usually used. For more information and discussion on viscosity solution, the reader is referred to the books by Fleming and Soner (1992) or Sethi and Zhang (1994). In what follows, we briefly describe the concept. For $w \in C(\mathcal{S} \times \mathbb{R}^p \times \mathbb{R}_+^m)$, let

$$\Phi_w(\alpha, x, a, r_1, r_2, h_1, h_2)$$
$$\triangleq \frac{w(\alpha, x + h_1, a + h_2) - w(\alpha, x, a) - h_1 r_1 - h_2 r_2}{\|h_1\| + \|h_2\|}$$

for $r_1, h_1 \in \mathbb{R}^p$ and $r_2, h_2 \in \mathbb{R}^m$.

Define the superdifferential $D^+ w(x, a, \alpha)$ and subdifferential $D^- w(\alpha, x, a)$ as follows:

$$D^+ w(\alpha, x, a)$$
$$= \left\{(r_1, r_2) \in \mathbb{R}^p \times \mathbb{R}_+^m : \limsup_{(h_1, h_2) \to 0} \Phi_w(\alpha, x, a, r_1, r_2, h_1, h_2) \leq 0\right\},$$

$$D^- w(\alpha, x, a)$$
$$= \left\{(r_1, r_2) \in \mathbb{R}^p \times \mathbb{R}_+^m : \liminf_{(h_1, h_2) \to 0} \Phi_w(\alpha, x, a, r_1, r_2, h_1, h_2) \geq 0\right\}.$$

DEFINITION 4.3. Suppose $w(\alpha, x, a)$ is continuous in (x, a) and has at most polynomial growth, i.e., there exist positive constants C, k_0, such that

$$|w(\alpha, x, a)| \leq C(1 + \|x\|^{k_0} + \|a\|^{k_0}).$$

Then,

(a) $v(\alpha, x, a)$ is said to be a viscosity subsolution to (29), if for all $(r_1, r_2) \in D^+ w(\alpha, x, a)$,

$$\rho w(\alpha, x, a) \leq \min_{(u, v) \in \Gamma(\alpha)} \{H(\alpha, x, a, w(\cdot, x, a), r_1, r_2, u, v)\}, \qquad \alpha \in \mathcal{S}.$$

(b) $w(\alpha, x, a)$ is said to be a viscosity supersolution to (29), if for all $(r_1, r_2) \in D^-w(\alpha, x, a)$,

$$\rho w(\alpha, x, a) \geqslant \min_{(u,v) \in \Gamma(\alpha)} \{H(\alpha, x, a, w(\cdot, x, a), r_1, r_2, u, v)\}, \quad \alpha \in \mathcal{S}.$$

(c) $w(\alpha, x, a)$ is said to be a viscosity solution to (29), if (a) and (b) hold.

It is easy to see that if $w(\alpha, z)$ is differentiable at z_0, then both $D^+w(\alpha, z)$ and $D^-w(\alpha, z)$ are singleton and equal to $\{w_z(\alpha, z_0)\}$. Hence, the equality in the HJB equation (29) holds at points whenever the value function $w(\alpha, z)$ is differentiable.

4.2. Properties of the value function

For the optimization problem under consideration, we will establish a certain number of properties of the value function. These results are gathered in theorems.

THEOREM 4.1. *Suppose the previous assumptions hold. The following statements hold for the value function $V(\alpha, z)$:*

(i) *There exist constants C and k_0, such that*

$$\|V(\alpha, x, a)\| \leqslant C_g (1 + \|x\|^{k_0} + \|a\|^{k_0}).$$

(ii) $V(\alpha, x, a)$ *is locally Lipschitz in $z = (x, a)$, i.e., there exists constant C, k_0 such that*

$$\|V(\alpha, z_1) - V(\alpha, z_2)\| \leqslant C_g (1 + \|z_1\|^{k_0} + \|z_2\|^{k_0}) \|z_1 - z_2\|, \quad \forall \alpha \in \mathcal{S}.$$

(iii) $V(\alpha, z)$ *is the unique viscosity solution of the HJB equation* (29).

PROOF. Here we only give a sketch of the proof, the details can be adapted from Theorem 1 of Boukas et al. (1996).

In view of (19) and (20), we have

$$x_i(0) - d_i t \leqslant x_i(t) \leqslant x_i(0) + \left(\sum_{i=1}^{m} \bar{u}_i - d_i\right) t,$$

$$0 \leqslant a_i(t) \leqslant a_i(0) + \left(\sum_{l=1}^{p} f_{il} \bar{u}_i\right) t.$$

This implies that $x(t)$ and $a(t)$ have at most linear growth, which combined with Assumption 4.3 yields

$$J(\alpha, x, a, \gamma) = \mathbf{E}\left[\int_0^\infty e^{-\rho t} \psi(\alpha(t), x(t), a(t), \gamma(t)) \, dt \, | \alpha(0) = \alpha, \right.$$
$$\left. x(0) = x, \ a(0) = a\right]$$

$$\leqslant K_g \mathbf{E}\bigg[\int_0^\infty e^{-\rho t}\big(1+\|x(t)\|^{k_g}+\|a(t)\|^{k_g}\big)\,dt\,|\alpha(0)=\alpha,$$

$$x(0)=x,\ a(0)=a\bigg]$$

$$\leqslant C_g\big(1+\|x(t)\|^{k_g}+\|a(t)\|^{k_g}\big),$$

from which the proof of (i) follows.

Now we come to prove (ii). To this end, we first define

$$V^0(\alpha,x,a) = \inf_{\gamma\in\mathcal{A}} J^0(\alpha,x,a,\gamma),$$

where $J^0(\alpha,x,a,\gamma) = J(\alpha,x,a,\gamma)$ with no jump in $\alpha(t)$, i.e., $\alpha(t)\equiv\alpha$, $t\geqslant 0$. By using the locally Lipschitz property of ψ, it is easy to verify that $J^0(\alpha,x,a,\gamma)$ is locally Lipschitz in (x,a), from which it follows that $V^0(\alpha,x,a)$ is locally Lipschitz in (x,a).

Then for $k\geqslant 1$, functions $V^k(\cdot)$, $J^k(\cdot)$ are defined inductively as follows: for any $\gamma\in\mathcal{A}$, let τ be the first jump of $\alpha(t)$, which is dependent on the control law γ, define

$$J^k(\alpha,x,a,\gamma) = \mathbf{E}\bigg[\int_0^\tau e^{-\rho t}\psi\big(\alpha(t),x(t),a(t),\gamma(t)\big)\,dt$$

$$+ e^{-\rho\tau}V^{k-1}\big(\alpha(\tau),x(\tau),\phi(a(\tau),\alpha(\tau))\big)\bigg] \quad (30)$$

and

$$V^k(\alpha,x,a) = \inf_{\gamma\in\mathcal{A}} J^k(\alpha,x,a,\gamma).$$

The rest of the proof can be divided into three steps:

(a) $V^k(\alpha,x,a,)$ is locally Lipschitz in (x,a).
(b) $V(\alpha,x,a) = \lim_{k\to\infty} V^k(\alpha,x,a)$.
(c) $V(\alpha,x,a)$ is locally Lipschitz in (x,a).

Using Remark 4.2 and the fact that $\psi(\alpha,x,a,\gamma)$ is locally Lipschitz in (x,a), it is easy to verify that if $V^{k-1}(\alpha,x,a)$ is locally Lipschitz, then $V^N(\alpha,x,a)$ is locally Lipschitz also.

Since τ has failure rate function $q_\alpha(a,v)$, $J^k(\alpha,x,a,\gamma)$ can be rewritten as

$$J^k(\alpha,x,a,\gamma)$$

$$= \int_0^\infty e^{-\rho t}\big(e^{-\int_0^t q_\alpha(a(s),v(s))\,ds}\big)\bigg[\psi\big(\alpha(t),x(t),a(t),\gamma(t)\big)$$

$$+ \sum_{\beta\neq\alpha(t)} q_{\alpha\beta}\big(a(t),v(t)\big)V^{k-1}\big(\beta,x(t),\phi(a(t),\beta)\big)\bigg]dt. \quad (31)$$

On the function space $C(\mathcal{S} \times \mathbb{R}^p \times \mathbb{R}^m)$, define operator $\mathcal{T}$ as follows: for every $w(\alpha, x, a) \in C(\mathcal{S} \times \mathbb{R}^p \times \mathbb{R}^m)$

$$(\mathcal{T}w)(\alpha, x, a) = \inf_{\gamma \in \mathcal{A}} \int_0^\infty e^{-\rho t} \left(e^{-\int_0^t q_\alpha(a(s), \gamma(s)) ds} \right) \bigg[\psi(\alpha(t), x(t), a(t), \gamma(t)) \\ + \sum_{\beta \neq \alpha(t)} q_{\alpha\beta}(a(t), v(t)) w(\beta, x(t), \phi(a(t), \beta)) \bigg] dt.$$

It is easy to check that $\mathcal{T}$ is contractive.

Combining (31) with the definition of $\mathcal{T}$ implies $\mathcal{T}(V^{k-1}) = V^k$, from which it follows that $V(\alpha, x, a) = \lim_{N \to \infty} V^N(\alpha, x, a)$ and the Lipschitz property of $V(\alpha, x, a)$ follows from the Lipschitz property of V^N and (b). This completes the proof of (ii).

The proof of (iii) can be adapted from Theorem G.1 in Sethi and Zhang (1994). □

THEOREM 4.2 (Verification Theorem). *Suppose* $w(\alpha, z) \in C^1(\mathbb{R}^p \times \mathbb{R}^m)$, $\alpha \in \mathcal{S}$, *satisfy*

$$\|w(\alpha, z)\| \leq C(1 + \|z\|^{k_0}), \quad \forall \alpha \in \mathcal{S},$$

and satisfy the HJB equation (29). *Then the following assertions hold*:

(i) $w(\alpha, z) = V(\alpha, z) \leq J(\alpha, z, \gamma)$ *for any control* $\gamma \in \mathcal{A}$.
(ii) *Suppose there exists a control* $\gamma^* \in \mathcal{A}$, *such that* γ^*, $\alpha^*(t)$, $x^*(t)$, $a^*(t)$ *satisfy*

$$\dot{x}^*(t) = u^*(t) - d, \quad x_0^* = x,$$
$$\dot{a}^*(t) = f(u^*(t)), \quad a^*(0) = a,$$

and

$$\min_{\gamma(t) \in \Gamma(\alpha^*(t))} \{H(\alpha^*(t), x^*(t), a^*(t), w(\cdot, x^*(t), a^*(t)), r_1^*(t), r_2^*(t), \gamma(t))\}$$
$$= H(\alpha^*(t), x^*(t), a^*(t), w(\cdot, x^*(t), a^*(t)), r_1^*(t), r_2^*(t), \gamma^*(t))\} \quad (32)$$

a.e. in t *with probability one, where* $(\alpha^*(t), x^*(t), a^*(t))$ *is the state trajectory with control* γ^*, $r_1^*(t) = w_x(\alpha^*(t), x^*(t), a^*(t))$, $r_2^*(t) = w_a(\alpha^*(t), x^*(t), a^*(t))$. *Then* γ^* *is optimal, i.e.,*

$$w(\alpha, z) = V(\alpha, z) = J(\alpha, z, \gamma^*).$$

PROOF. A differentiable function solving the HJB equation (29) is in particular the viscosity solution to the HJB equation. So $w(x, i) = V(x, i)$ follows from the uniqueness of the viscosity solution of the HJB equation. By the definition of the value function and Theorem 4.1, we get

$$V(\alpha, x, a) = w(\alpha, x, a) \leq J(\alpha, x, a, \gamma(\cdot)), \quad \forall \gamma(\cdot) \in \mathcal{A}.$$

This proves (i).

By Dynkin's formula (see, e.g., Davis, 1993), we have $\forall T > 0$,

$$V(\alpha, x, a) = \mathbf{E}\Bigg[\int_0^T e^{-\rho t}\bigg(\rho V(\alpha^*(t), x^*(t), a^*(t))$$

$$- (u^*(t) - d)^\top \frac{\partial V(\alpha^*(t), x^*(t), a^*(t))}{\partial x}$$

$$- f^\top(u^*(t)) \frac{\partial V(\alpha^*(t), x^*(t), a^*(t))}{\partial a}$$

$$- \sum_{\beta \neq \alpha^*(t)} q_{\alpha^*(t)\beta}(a^*(t), v^*(t))[V(\beta, x^*(t), \phi(a^*(t), \beta))$$

$$- V(\alpha^*(t), x^*(t), a^*(t))]\bigg) dt$$

$$+ e^{-\rho T} V(\alpha^*(T), x^*(T), a^*(T))\Bigg]. \tag{33}$$

From (32) and (33), it follows that

$$V(\alpha, x, a) = \mathbf{E}\Bigg[\int_0^T e^{-\rho t} \psi(\alpha^*(t), x^*(t), a^*(t), \gamma^*(t)) dt$$

$$+ e^{-\rho T} V(\alpha^*(T), x^*(T), a^*(T))\Bigg]. \tag{34}$$

Noting the polynomial growth rate of $V(\alpha, x, a)$ and letting $T \to \infty$ in (34) produce (ii). This completes the proof of Theorem 4.2. □

Theorem 4.2 shows that to find the optimal production rate and the optimal maintenance rate, one needs only to solve the HJB equations (29). As is well known in the literature, the optimization problem representing the production planning we are dealing with is a large scale problem and thus the HJB equation often lacks closed solution. The only way to obtain its solution is to recourse to numerical methods (see Sethi and Zhang, 1994).

If the cost rate function $\psi(\cdot)$ does not contain the control variables, from the HJB equation (29), we can conclude that the production rate $u(t)$ will be chosen to minimize the trajectory derivative $\frac{d}{dt} V(\alpha, z(t))$. The maintenance rate $v(t)$ is determined independently from $u(t)$. When the failure rates are not age dependent, the production rates can be determined according to the hedging point policy. The hedging point is, in fact, the steady state of the inventory process. In the model considered herein, when a machine is in the operational mode, its aging is always increasing and there cannot be a steady state. This means that the concept of hedging point is not directly relevant any more. However, we can give a related notion called *hedging surfaces* which are defined by the mappings $x = \tilde{x}^\alpha(a)$, where

$$\min_x V(\alpha, x, a) = V\left(\alpha, \tilde{x}^\alpha(a), a\right).$$

The production rates will determine the age and the inventory trajectory. They will be chosen so as to reach as rapidly as possible the hedging surface corresponding to the current operational mode, and once on the hedging surface, the trajectory will be maintained on it as long as the mode remains the same. The notion of hedging surface can be regarded as a generalization of hedging point, which will be a convenient way to represent the optimal production rate when the aging of the machine is considered.

4.3. Illustrative example

To illustrate the usefulness of the numerical technique provided in Section 2.2 and show the importance of the model of this section, let us give a numerical example of the above model. This example is adapted from Boukas and Haurie (1990). Consider a manufacturing system that consists of two machines and produces one part type. Let

$$\psi(\alpha, x, a, \gamma) = c^+ x^+ + c^- x^- + c^\alpha(t), \tag{35}$$

where c^+, c^- are positive constants and $c^\alpha(t)$ is defined by

$$c^\alpha(t) = \sum_{i=1}^{2} c_{1i}\delta_{i2}(t) + c_{2i}\delta_{i3}(t)$$

with c_{1i}: cost rate (positive constant) applying to machine i's repair activity; c_{2i}: cost rate (positive constant) applying to machine i's preventive maintenance activity.

A. Data

1. Control constrains: $\bar{u}_1 = 0.18$, $\bar{u}_2 = 0.19$, $\bar{v}_1 = 0.5$, $\bar{v}_2 = 0.5$.
2. Demand rate: $d = 0.25$.
3. Jump rates: $\lambda^i_{12}(a, v) = 0.02(1 - e^{-2 \times 10^{-2} a_i})$; $\lambda^i_{21}(a, v) = 0.02$; $\lambda^i_{31}(a, v) = 0.2$.
4. Discounted rate: $\rho = 0.03$.
5. The coefficients of the cost function ψ given by (35) are provided in Table 1.

Table 1

c^+	c^-	$c^{(1,1)}$	$c^{(1,2)}$	$c^{(1,3)}$	$c^{(2,1)}$	$c^{(2,2)}$	$c^{(2,3)}$	$c^{(3,1)}$	$c^{(3,2)}$	$c^{(3,3)}$
1	10	0	10	1	10	20	11	1	11	2

B. The grid definition

The grid G associated with our example is a subset of $\mathbb{R}^3$. For a given finite difference interval h, this grid with $n_1 \times n_2 \times n_3$ points is defined by

$$G = \{z^i \in \mathbb{R}^3, i = (i_1, i_2, i_3) \mid$$
$$z_1^i = z_1^0 + (i_1 - 1)h_1 \colon z_1^0 = 0,\ z_1^{n_1} = 100,\ i_1 \in \{1, \ldots, n_1\}$$
$$z_2^i = z_2^0 + (i_2 - 1)h_2 \colon z_2^0 = 0,\ z_2^{n_2} = 100,\ i_2 \in \{1, \ldots, n_2\}$$
$$z_3^i = z_3^0 + (i_3 - 1)h_3 \colon z_3^0 = 0,\ z_3^{n_3} = 100,\ i_3 \in \{1, \ldots, n_3\}\}.$$

For each $\alpha \in \mathcal{S}$, when a component of z reaches its boundary, a control is imposed. This control forces the system to jump with probability one to a specified state, as described in Boukas (1987).

REMARK 4.4. To implement the approximation technique developed in Section 2.2, one has to impose some boundary conditions to describe the boundary behavior of the system at the border of G. These boundary conditions correspond to an additional approximation of the original problem. However, when G is large enough, it is felt that these boundary conditions are realistic and that the influence of this approximation will be negligible.

C. Results

The computation results are plotted as Figures 1–6. Based on the computation results, we can make the following comments.

(1) The value functions shown in Figures 1 and 2 are clearly almost convex. This confirms the appropriateness of a quadratic approximation as in Kimemia and Gershwin (1983) and followers.
(2) The hedging surface corresponding to the operational mode is shown in Figure 3. One can see that as the machine ages a_1, a_2 increase the value of the surplus variable x on the surface also increases. Therefore, the hedging value for the surplus is increased when the probability of failure of one of the machines becomes higher.
(3) The preventive maintenance actions are triggered according to an age limit policy described in Figures 4–6, which reveals that as the surplus level increases, the age limit decreases, i.e., tends to send the two machines to preventive maintenance more often.

REMARK 4.5. The models for maintenance control introduced in the this section is logic and intuitive. However, the introduction of the ages of the machines increases the dimension of the system greatly. On the other hand, it is well known that Kushner's approximation technique is very sensitive to the dimension of the system and easily runs into the "curse of dimensionality". Therefore the models introduced in this section are only suitable for the system with small size.

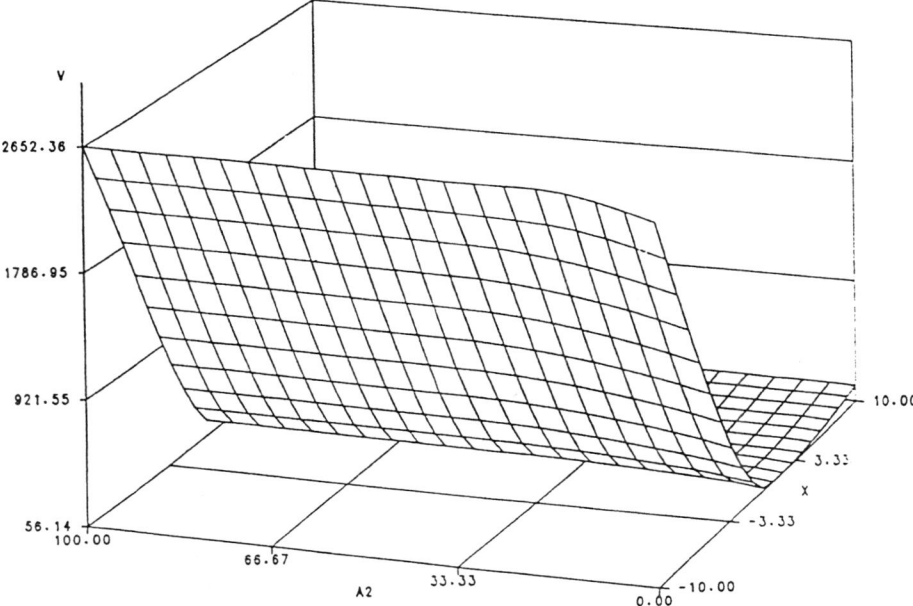

Fig. 1. The value function versus the stock level and age of machine # 2 in mode (1,1) when the age of machine # 1 is equal to 40.

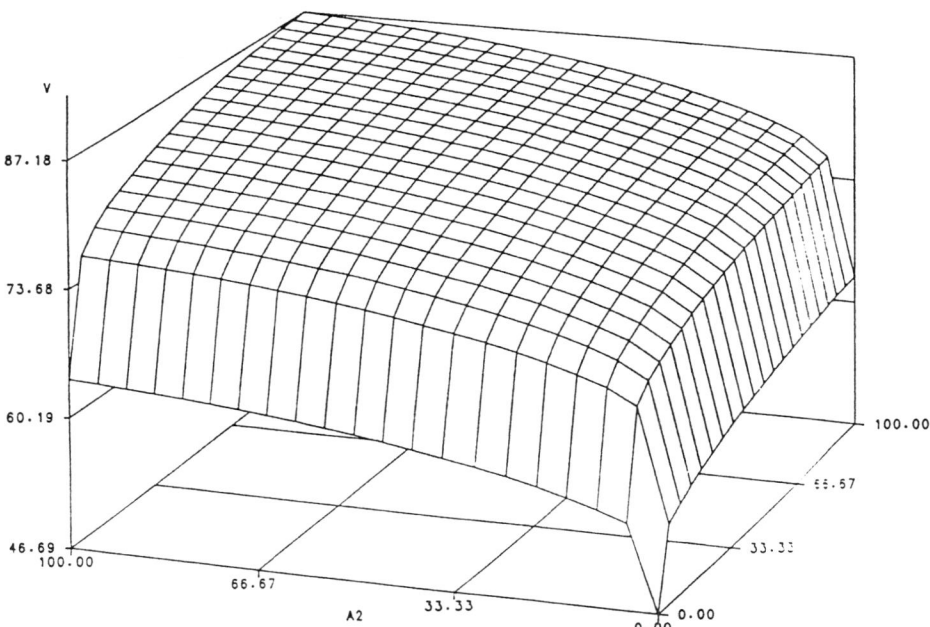

Fig. 2. Value function versus the ages of the two machines when the stock level is 3.

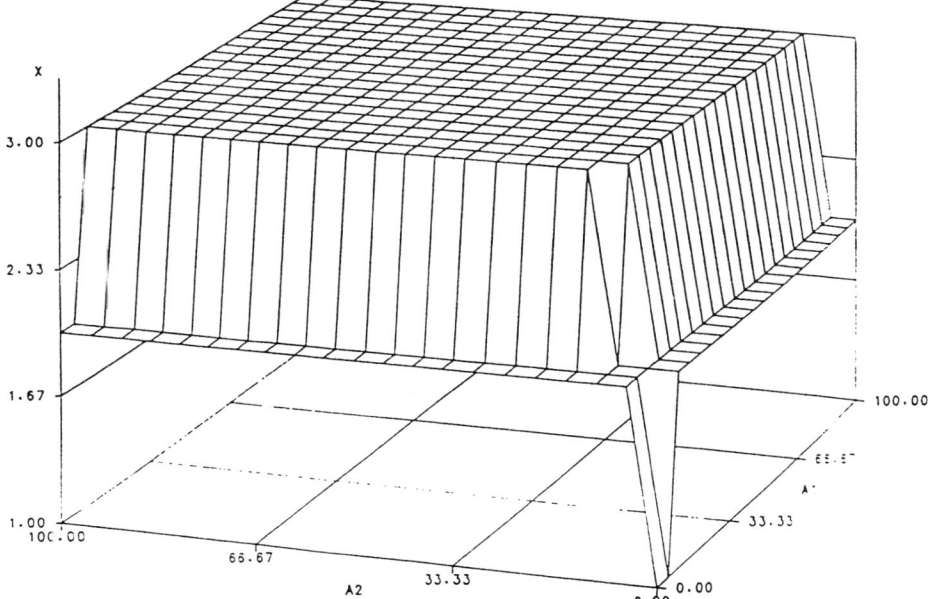

Fig. 3. Hedging surface versus the ages of the machines when the corrective maintenance is considered.

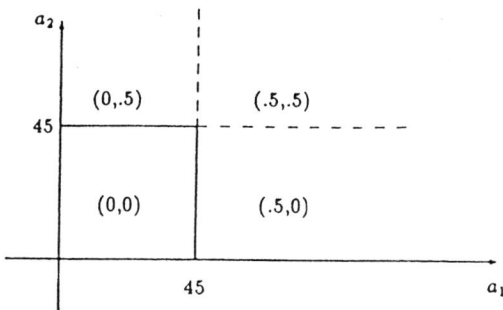

Fig. 4. Preventive maintenance policy when the stock level is equal to -3.

5. Maintenance model without considering the machine aging

To overcome the disadvantage of the dimension problem resulting from ages of the machines, this section will introduce a model developed by Boukas and Liu (1999b), in which the age of the machine is not considered. For simplicity, let us consider a manufacturing system consisting of one machine that producing one part type. Assume that the machine has multiple modes denoted by $S = \{1, 2, 3, 4\}$. Mode 4 means that the machine is under repair. Mode 1 means that the machine is in a good mode, mode 2 means that the machine is in an average state and mode 3 indicates that the machine is

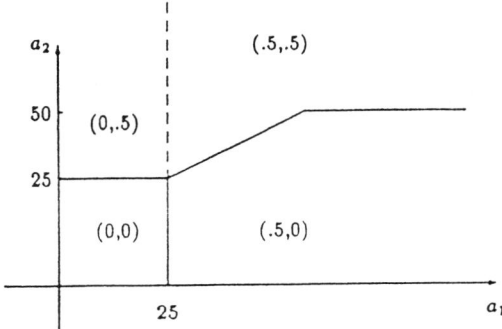

Fig. 5. Preventive maintenance policy when the stock level is zero.

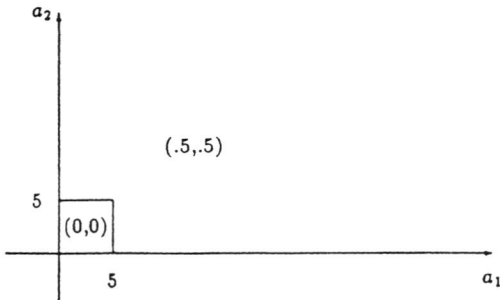

Fig. 6. Preventive maintenance policy when the stock level is equal to 3.

in bad state. In mode 4 the machine does not produce any parts. In modes 1, 2, and 3, the machine produces parts, but the rate of the rejected parts will depend on the state of the machine. The transition between the four states of the machine is governed by a continuous-time Markov process $\{\alpha(t), t \geq 0\}$ taking value in S with generator

$$Q = \begin{pmatrix} q_{11} & q_{12} & 0 & q_{14} \\ q_{21} & q_{22} & q_{23} & q_{24} \\ q_{31} & 0 & q_{33} & q_{34} \\ q_{41} & 0 & 0 & q_{44} \end{pmatrix},$$

where q_{i1}, $i = 2, 3, 4$, are control variables: q_{i1}, $i = 2, 3$, are preventive maintenance variables denoted by $v_p^1, v_p^2 \in [0, \bar{v}_p]$, and q_{41} is the corrective maintenance variable denoted by $v_r \in [0, \bar{v}_r]$, where $\bar{v}_p$, $\bar{v}_r$ are positive constants. The generator of the controlled Markov process can be rewritten as:

$$Q(v) = \begin{pmatrix} q_{11} & q_{12} & 0 & q_{14} \\ v_p^1 & q_{22}(v_p^1) & q_{23} & q_{24} \\ v_p^2 & 0 & q_{33}(v_p^2) & q_{34} \\ v_0 + v_r & 0 & 0 & q_{44}(v^r) \end{pmatrix}, \qquad (36)$$

where $q_{22}(v_p^1) = -(v_p^1 + q_{23} + q_{24})$, $q_{33}(v_p^2) = -(v_p^2 + q_{34})$, $q_{44}(v_r) = -(v_0 + v^r)$ and write $v = (v_p^1, v_p^2, v^r)$.

REMARK 5.1. In $Q(v)$, v_0 is a positive constant, which is the smallest jump rate from repair state to good state. This transition rate is assumed to be free of charge. If a greater transition rate is required, i.e., shorter repair time is required, then a cost has to be paid.

Let $x(t) \in \mathbb{R}$ denote the inventory when positive and backlogs when negative. The parts will be assumed to deteriorate with constant rate μ when they remain in the stock. The demand rate is assumed to be constant and denoted by d. When the inventory is negative and the machine is under repair, then the demand will be canceled with a constant rate $1 - \theta$. With these assumptions, the inventory dynamic equation of this manufacturing system can be written as follows:

$$\dot{x}(t) = f(\alpha(t), x(t), u(t)), \qquad x(0) = x_0 \text{ given},$$

with

$$f(i, x(t), u(t)) = -\mu I_{\{x(t) \geq 0\}} x(t) + \rho(i) u(t) - \theta(i, x) d \quad \text{when } \alpha(t) = i,$$

where I_A is the indicator function of A,

$$\theta(i, x) = \begin{cases} \theta, & \text{if } i = 4, x < 0, \\ 1, & \text{otherwise}. \end{cases}$$

Here $\rho(i)$ is a positive constant which can be interpreted in such a way that $(1 - \rho(i))$ is the rate of rejected parts when the mode in mode i.

The optimization problem is to seek a control law which minimizes the following cost function:

$$J(i_0, x_0, u(\cdot), v(\cdot))$$
$$= \mathbf{E}\left[\int_0^\infty e^{-\rho t} \psi(\alpha(t), x(t), u(t), v(t)) | \alpha(0) = i_0, \; x(0) = x_0 \right].$$

Here

$$\psi(i, x, u, v) = c_u(1 - \rho_i)u + c_p^1 v_p^1 I_{\{i=2\}} + c_p^2 v_p^2 I_{\{i=3\}} + c_r v^r I_{\{i=4\}}$$
$$+ c^+ x^+ + c^- x^-,$$

where c_u, c_p^1, c_p^2, c_r, c^+, c^- are positive constants which can be interpreted as follows: c^+ is the inventory holding cost rate; c^- is the shortage cost rate; c_p^1 and c_p^2 are the preventive maintenance cost rates; c_r is the corrective maintenance cost rate; $c_u(1 - \rho(i))u(t)$ denotes the cost resulting from the rejected parts. For more detail of this model, the reader is referred to Boukas and Liu (1999b).

EXAMPLE 5.1. To illustrate the usefulness of this model, we will give a simple example, in which only corrective maintenance control is considered. Consider a manufacturing system consisting of one machine and producing one part type. The demand rate is a constant denoted by d. The machine is assumed to have two models: operation denoted by 1, and under repair denoted by 2. The mode transition of the machine is assumed to be governed by a Markov process $\alpha = \{\alpha(t), t \geq 0\}$ with state space $\mathcal{S} = \{1, 2\}$ and generator

$$Q = \begin{pmatrix} -q_1 & q_1 \\ v_0 + v & -(v_0 + v) \end{pmatrix},$$

where $v \in [0, \bar{v}]$ is the corrective maintenance rate. $u(t) \in [0, \bar{u}]$ is the production rate. Consider cost function $\psi(x, v) = c^+ x^+ + c^- x^- + c_r v I_{\{\alpha(t)=2\}}$.

Let $V_i(x)$, $i \in \mathcal{S}$ be the value function with the initial condition $\alpha(0) = i$, $x(0) = x$. Then, it can be shown that $V_i(x)$, $i \in \mathcal{S}$, is the unique viscosity solution of the following HJB equations (see Boukas and Liu, 1999c for the proof).

$$\rho V_1(x) = \min_{u \in [0, \bar{u}]} \left\{ c^+ x^+ + c^- x^- + (u - d) \frac{dV_1(x)}{dx} \right. $$
$$\left. + q_1 (V_2(x) - V_1(x)) \right\}, \tag{37}$$

$$\rho V_2(x) = \min_{v \in [0, \bar{v}]} \left\{ c^+ x^+ + c^- x^- + c_r v - d \frac{dV_2(x)}{dx} \right.$$
$$\left. + (v_0 + v)(V_1(x) - V_2(x)) \right\}, \tag{38}$$

from which we get the optimal control as follows:

$$u^*(t) = \begin{cases} \bar{u}, & \text{if } \frac{dV_1(x)}{dx} < 0, \\ d, & \text{if } \frac{dV_1(x)}{dx} = 0, \\ 0, & \text{otherwise}, \end{cases} \tag{39}$$

$$v^*(t) = \begin{cases} \bar{v}, & \text{if } c_r + V_1(x) - V_2(x) \leq 0, \\ 0, & \text{otherwise}. \end{cases} \tag{40}$$

In (39), the purpose of letting $u(t) = d$ when $dV_1(x)/dx = 0$ is to keep the optimal trajectory at the level x^* with $dV_1(x)/dx|_{x=x^*} = 0$ as long as possible.

Table 2
Simulation data

c^+	c^-	c_r	q_1	ρ	d	$\bar{u}$	$\bar{v}$	v_0
1	10	10	0.05	0.01	0.2	0.27	0.1	0.15

In the case of the value function $V_1(x)$ being convex in x, from (39) it follows that the optimal production rate is still hedging point policy.

Evidently, the optimal corrective maintenance rate is a bang-bang control. From (40), it follows that if $V_1(x) - V_2(x)$ is increasing, then the optimal maintenance control is a threshold policy, i.e.,

$$v^*(t) = \begin{cases} \bar{v}, & \text{if } x < x_*, \\ 0, & \text{otherwise,} \end{cases}$$

where x_* is the solution of $c_r + V_1(x) - V_2(x) = 0$.

Generally, it is not easy to get the closed solution of HJB equations (37) and (38). Here, a numerical solution is given. The data of the system are provided as in Table 2.

A Matlab program has been developed to find the numerical solution of the HJB equation of this problem. The results are plotted in Figures 7–10. Figures 7 and 8 represent respectively the evolution of the value function in function of the stock level in mode 1 and mode 2. Figures 9 and 10 represent respectively the evolution of the production rate and the corrective maintenance rate in function of the stock level.

The numerical results show that the optimal production rate is a hedging point policy with hedging level $x^* = 4.200$. The optimal corrective maintenance rate is a threshold policy with threshold value $x_* = 3.700$.

When $v(t) \equiv \mu$, a constant, then this model becomes the one introduced in Section 3 and thus the hedging point is optimal. In the case of no corrective maintenance control, i.e., $v(t) \equiv 0$, by using (18) direct computation gives the optimal hedging point is 5.4988, which is greater than the one with corrective maintenance control.

The above results confirm the following conclusions:

(1) The optimal corrective maintenance control is a bang-bang control.
(2) The optimal production rate is still hedging point policy. With corrective maintenance control, the optimal hedging level is lowered and the performance is improved.

6. Robust controller for a class of production and maintenance

Boukas, Zhang, and Zhu (1994), Boukas and Yang (1996) investigated a class of maintenance activities such as lubrication, routine adjustments, etc. These maintenance actions are executed as the machine is working, which can reduce the machine failure rates and improves the productivity of the system.

6.1. Model description and the optimal control

Let us consider a manufacturing system that consists of M machines and produces p different part types. Assume that the machines are subject to random failures and repairs. The state of each machine can be classified as operational denoted by 1, and under repair denoted by 2.

When a machine is operational, it can produce any part type, and when it is under repair, nothing is produced. Let $\alpha_i(t)$ denote the state of machine i and $\mathcal{M} = \{1, 2\}$ the state space of the process $\alpha_i(t)$. Then, the vector process $\alpha(t) = (\alpha_1(t), \ldots, \alpha_M(t)) \in \mathcal{S} = \{1, 2\} \times \{1, 2\} \times \cdots \times \{1, 2\}$ describes the state process of the global manufacturing system. The set $\mathcal{S}$ has $r = 2^M$ states and is denoted by $\mathcal{S} = \{\alpha^1, \ldots, \alpha^r\}$.

Let $u(t) \in \mathbb{R}^p$ and $v(t) \in \mathbb{R}^M$ denote the vectors of the production rate and the maintenance rate of the system at time t, respectively. We use $a_i(t)$ to represent the age of the ith machine at time t and $a(t) = (a_1(t), \ldots, a_M(t))$ the vector age of the system. We use $x_j(t)$ to represent the inventory/shortage level of part type j at time t and $x(t) = (x_1(t), \ldots, x_p(t))$ the vector of inventory/shortage level of the system. Let $z(t) \in \mathbb{R}^p$ denote the vector representing the demand rate. The inventory/shortage levels and machine ages of the system are described by the following differential equations:

$$\dot{x}(t) = u(t) - z(t), \quad x(0) = x_0,$$
$$\dot{a}(t) = f(\alpha(t), u(t), v(t)), \quad a(0) = a_0, \quad (41)$$

where $u(t) = \sum_{i=1}^{M} u_i(t)$, and $u_i(t) = (u_{i1}, \ldots, u_{ip})^\top$ is the instantaneous vector production rate of the ith machine (u_{ij} represents the production rate of part type j of machine i). The function $f = (f_1, \ldots, f_M)^\top$ in (41) represents the aging rate of the machines as a function of $\alpha(t)$, $u(t)$, and $v(t)$. We assume that f is linear in (u, v) and $f_i(\alpha, u, v) = 0$ when $\alpha_i = 2$, i.e., the ith machine is under repair. More specifically, we consider

$$f(\alpha, u, v) = K_1(\alpha)u - K_2(\alpha)v \quad (42)$$

with matrices $K_1(\alpha)$ and $K_2(\alpha)$ of appropriate dimensions. The ith rows of $K_1(\alpha)$ and $K_2(\alpha)$ are 0 whenever $\alpha_i = 2$. Notice that the linear assumption on $f(\cdot)$ is used for simplifying the model and discussions. The results in this section can be extended to nonlinear cases.

Suppose the demand rate $z(\cdot)$ to be an unknown disturbance process fluctuating in a compact set of an Euclidean space. Let $\mathbb{Z}$ denote the set of all such demand rates.

The control variables under consideration are the production rate u and the maintenance rate v. The maintenance interrupts the production process and uses a fraction of machine up times. As discussed in Boukas et al. (1994, 1996), if the machine state is $\alpha = (\alpha_1, \ldots, \alpha_M)$ then the control constraints can be formulated as:

$$b_{i1}u_{i1} + \cdots + b_{ip}u_{ip} + h_i v_i \leq I_{\{\alpha_i = 1\}}, \quad i = 1, \ldots, M,$$

for nonnegative constants b_{ij} and h_i, $i = 1, 2, \ldots, M$, $j = 1, 2, \ldots, p$. Moreover, $u_{ij}(t) \geq 0$ and $v_i(t) \geq 0$. Namely, no negative production or maintenance will be allowed.

REMARK 6.1. In this section, we allow simultaneous production and maintenance. Part of the machines can be used to produce parts while the rest are undergoing preventive maintenance. The results in this section can be extended to models in which a fixed cost is required for switching between production and maintenance; see Sethi and Zhang (1994). Nevertheless, the model in this section represents an approximation to those models when the fixed switching costs are small.

Noticing that the aging rate is always nonnegative, in addition to the above constraints, we also assume that for each $\alpha \in \mathcal{S}$,

$$K_1(\alpha)u - K_2(\alpha)v \geq 0.$$

Let $\alpha(t) \in \mathcal{S}$ be a finite state Markov process. Let $\Gamma(\alpha(t))$ denote the following control set:

$$\Gamma(\alpha) = \Gamma_0(\alpha) \cap \{K_1(\alpha)u - K_2(\alpha)v \geq 0\},$$

where

$$\Gamma_0(\alpha) = \left\{(u, v): u = \sum_{i=1}^{M} u_i,\ u_i = (u_{i1}, \ldots, u_{ip})^\top,\ v = (v_1, \ldots, v_M)^\top,\right.$$
$$\left.\sum_{j=1}^{p} b_{ij} u_{ij} + h_i v_i \leq I_{\{\alpha_i = 1\}},\ u_{ij} \geq 0,\ v_i \geq 0,\ i = 1, 2, \ldots, M\right\}.$$

To proceed, we first define admissible controls and admissible feedback controls, respectively.

DEFINITION 6.1. A control $(u(\cdot), v(\cdot)) = \{(u(t), v(t)): t \geq 0\}$ is *admissible* if: (i) $(u(\cdot), v(\cdot))$ is adapted to the σ-algebra generated by $\{(\alpha(\cdot), x(t)), t \geq 0\}$, denoted as $\sigma\{(\alpha(s), x(t)): 0 \leq s \leq t\}$, and (ii) $(u(t), v(t)) \in \Gamma(\alpha(t))$ for all $t \geq 0$.

Let $\mathcal{A}$ denote the set of all admissible controls.

DEFINITION 6.2. A function $(u(\alpha, x, a), v(\alpha, x, a))$ is an admissible *feedback* control, if (i) for any given initial (x, a), the following equations have a unique solution $(x(\cdot), a(\cdot))$:

$$\dot{x}(t) = u(\alpha(t), x(t), a(t)) - z(t), \qquad x(0) = x,$$
$$\dot{a}(t) = f(\alpha(t), u(t), v(t)), \qquad a(0) = a,$$

and (ii) $(u(\cdot), v(\cdot)) = (u(\alpha(\cdot), x(\cdot), a(\cdot)), v(\alpha(\cdot), x(\cdot), a(\cdot))) \in \mathcal{A}$.

For any given $(u(\cdot), v(\cdot)) \in \mathcal{A}$ and $z(\cdot) \in \mathbb{Z}$, let

$$J(\alpha, x, a, u(\cdot), v(\cdot), z(\cdot)) = \mathbf{E}\left[\int_0^\infty e^{-\rho t} \psi(\alpha(t), x(t), a(t), u(t), v(t))\, dt\right].$$

Note that the dependence of $z(\cdot)$ in $J(\cdot)$ is through the dynamics of $x(t)$. So $x(t)$ above really should have been written as $x(t, z(t))$. Nevertheless for notational simplicity, we suppress the $z(\cdot)$ dependence.

The objective of the problem is to find a control policy $(u(\cdot), v(\cdot)) \in \mathcal{A}$ that minimizes the following cost functional:

$$\hat{J}(\alpha, x, a, u(\cdot), v(\cdot)) := \sup_{z(\cdot) \in \mathbb{Z}} J(\alpha, x, a, u(\cdot), v(\cdot), z(\cdot)).$$

Define the value function as the minimum of the cost over $(u(\cdot), v(\cdot)) \in \mathcal{A}$, i.e.,

$$V(\alpha, x, a) = \inf_{(u(\cdot), v(\cdot)) \in \mathcal{A}} \hat{J}(\alpha, x, a, u(\cdot), v(\cdot)). \tag{43}$$

The following assumptions are needed to formulate the optimization problem of this section.

(A1) $\psi(\alpha, x, a, u, v)$ satisfies Assumption 4.3.
(A2) $\alpha(t) \in \mathcal{M}$ is a Markov chain generated by $Q(a, u, v) = (q_{\alpha\beta}(a, u, v))$ such that $q_{\alpha\beta}(a, u, v) \geq 0$ for $\alpha \neq \beta$ and $q_{\alpha\alpha}(a, u, v) = -\sum_{\beta \neq \alpha} q_{\alpha\beta}(a, u, v)$. Moreover, $q_{\alpha\beta}(a, u, v)$ is bounded, Lipschitz in a, and $|q_{\alpha\alpha}(a, u, v)| \geq c_0 > 0$, for some constant $c_0 > 0$.
(A3) The demand $z(\cdot)$ is $\sigma\{\alpha(s): 0 \leq s \leq t\}$ adapted. For all $t \geq 0$, $z(t) \geq 0$ and $z(t) \in \Gamma_z$, a compact subset of $\mathbb{R}^P$.

Let H be a real-valued functional defined by:

$$H(\alpha, x, a, w(\cdot, x, a), r_1, r_2, u, v, z)$$
$$= (u - z)r_1 + f(\alpha, u, v)r_2 + \psi(\alpha, x, a, u, v)$$
$$+ \sum_{\beta \neq \alpha} q_{\alpha\beta}(a, u, v)\bigl[w(\beta, x, a) - w(\alpha, x, a)\bigr].$$

Then, the Hamilton–Jacobi–Isac (HJI) equation associated with the problem can be formally written as follows:

$$\rho w(\alpha, x, a)$$
$$= \min_{(u,v) \in \Gamma(\alpha)} \max_{z \in \Gamma_z} H\bigl(\alpha, x, a, w(\cdot, x, a), w_x(\alpha, x, a), w_a(\alpha, x, a), u, v, z\bigr). \tag{44}$$

With these assumptions, we have the following theorems. For the proofs the reader is referred to (Boukas et al., 1996).

THEOREM 6.1 (HJI equation). *There exist constants C and k_g, such that*

(i) $0 \leqslant V(\alpha, x, a) \leqslant C(1 + \|x\|^{k_g} + \|a\|^{k_g})$, $\forall \alpha \in \mathcal{S}$.
(ii) $\|V(\alpha, x, a) - V(\alpha, x', a')\| \leqslant C(1 + \|x\|^{k_g} + \|x'\|^{k_g} + \|a\|^{k_g} + \|a'\|^{k_g})(\|x - x'\| + \|a - a'\|)$.
(iii) $V(\alpha, x, a)$ *is the only viscosity solution to the HJI equation* (44).

THEOREM 6.2 (Verification Theorem). *Let $w(\alpha, x, a)$ denote a differentiable solution of the HJI equation* (44) *such that*

$$0 \leqslant w(\alpha, x, a) \leqslant C(1 + \|x\|^{k_g} + \|a\|^{k_g}).$$

Then:

(i) $w(\alpha, x, a) \leqslant \hat{J}(\alpha, x, a, u(\cdot), v(\cdot))$ *for all* $(u(\cdot), v(\cdot)) \in \mathcal{A}$.
(ii) *If* $(u^*(\alpha, x, a), v^*(\alpha, x, a))$ *is an admissible feedback control such that*

$$\min_{(u,v) \in \Gamma(\alpha)} \max_{z \in \Gamma_z} H\big(\alpha, x, a, V(\cdot, x, a), V_x(\alpha, x, a), V_a(\alpha, x, a), u, v, z\big)$$

$$= \max_{z \in \Gamma_z} H\big(\alpha, x, a, V(\cdot, x, a), V_x(\alpha, x, a), V_a(\alpha, x, a), u^*(\alpha, x, a),$$

$$v^*(\alpha, x, a), z\big), \tag{45}$$

then

$$\hat{J}\big(\alpha, x, a, u^*(\cdot), v^*(\cdot)\big) = V(\alpha, x, a) = w(\alpha, x, a),$$

where

$$\big(u^*(\cdot), v^*(\cdot)\big) = \big(u^*(\alpha(\cdot), x(\cdot), a(\cdot)), v^*(\alpha(\cdot), x(\cdot), a(\cdot))\big).$$

This means that $(u^*(\cdot), v^*(\cdot))$ *is optimal*.

To get the optimal production and maintenance rates, one has to solve the HJI equation (44), which lacks closed solution in the general case. Boukas, Zhang and Yin (1996) have developed a numerical technique for this problem. However, Boukas and Yang (1996) solved the optimal production and maintenance control problem in the following special case.

Suppose the manufacturing system consists one machine and produces one part-type with constant demand rate, denoted by d. The machine is assumed to have two modes: operational denoted by 1, and down denoted by 2. The effect of the preventive maintenance is to reduce the machine failure rate and to improve the productivity of the system. This is accomplished by spending a fraction of time for maintenance during the cycle of production. By choosing $u(\cdot)$ and $v(\cdot)$ properly, one can track the demand in a way that minimizes the total cost under consideration.

Let $\bar{u}$ and $\bar{v}$ denote respectively the maximum production rate and the maximum maintenance rate.

The state equation of the surplus and the aging are given by:

$$\dot{x}(t) = u(t) - d, \qquad x(0) = x^0,$$
$$\dot{a}(t) = f(u(t), v(t)), \qquad a(T) = 0,$$

where x^0 is a given initial surplus value, the real-valued function $f \geqslant 0$, $a \in \mathbb{R}^+$, $u(t) \in [0, \bar{u}]$, $v(t) \in [0, \bar{v}]$, and the random variable T ($T \geqslant 0$) is the last restart time of the machine.

We suppose that the solution $v^* = v^*(u)$ of $f(u, v) = 0$ satisfies $\bar{v} \geqslant v^*$. In the following we will suppose that $f(u, v) = k_1 u - k_2 v \geqslant 0$ (where $k_1 > 0, k_2 > 0$), this implies that $\bar{v} \leqslant \frac{k_1}{k_2}\bar{u}$, so this assumption means that $\bar{v} = \frac{k_1}{k_2}\bar{u}$. This assumption is realistic since in practice the velocity of the aging of the machine increases with the production rate u and decreases with the preventive maintenance rate v.

Let $\{\alpha(t), t \geqslant 0\}$ specify the mode switching of the machine. Suppose $\{\alpha(t), t \geqslant 0\}$ is a Markov process with state space $\mathcal{S} = \{1, 2\}$ and generator

$$Q = \begin{bmatrix} -q_1(a) & q_1(a) \\ q_2 & -q_2 \end{bmatrix},$$

where the jump rates $q_1(a)$ is assumed to be a continuous bounded and increasing function of a and q_2 is constant.

Let $\Gamma(\alpha(t))$ denote the following control set:

$$\Gamma(\alpha(t)) = \{(u(t), v(t)) \mid 0 \leqslant u(t) \leqslant \bar{u} I_{\{\alpha(t)=1\}}, \ 0 \leqslant v(t) \leqslant \bar{v} I_{\{\alpha(t)=1\}}$$
$$\text{and } k_1 u(t) - k_2 v(t) \geqslant 0\}.$$

DEFINITION 6.3. A control $\gamma(t) = (u(t), v(t))$ is said to be admissible if:

(i) $\gamma(t)$ is adapted with respect to the σ-algebra generated by the random process $(\alpha(t), x(t))$, $t \geqslant 0$, denoted as $\sigma\{(\alpha(s), x(s)): 0 \leqslant s \leqslant t\}$, and
(ii) $\gamma(t) \in \Gamma(\alpha(t))$ for all $t \geqslant 0$.

Let $\mathcal{A}$ denote the set of all admissible controls. Let $\gamma(\alpha, x, a) = (u, v)$ be a feedback control policy. Suppose that the corresponding cost is given by:

$$J^\gamma(\alpha, x, a) = E_\gamma \left\{ \int_0^\infty e^{-\rho t} \left[c^+ x^+ + c^- x^- + l(\alpha(t)) v \right] dt \,|\, \alpha(0) = \alpha, \right.$$
$$\left. x(0) = x, \ a(0) = a \right\},$$

where

$$l(\alpha) = \begin{cases} k, & \text{if } \alpha = 1, \\ 0, & \text{if } \alpha = 0. \end{cases}$$

We first consider a special case where the cost does not include the maintenance activity, i.e., $k = 0$. Let $\gamma(\alpha, x, a) = (u, v)$ be a feedback control policy, and suppose that the corresponding cost is given by:

$$J^\gamma(\alpha, x, a)$$
$$= \mathbf{E}_\gamma \left\{ \int_0^\infty e^{-\rho t} \left(c^+ x^+ + c^- x^- \right) dt \mid \alpha(0) = \alpha, \ x(0) = x, \ a(0) = a \right\}.$$

Since, in general, we know that the cost is an increasing function of the machine age a, it follows that $\frac{\partial J^\gamma}{\partial a}(\alpha, x, a) \geq 0$.

When a machine has a breakdown, it goes through a repair process. As before, we assume that a repaired machine is considered renewed, i.e., the age of the machine is resumed to 0.

REMARK 6.2. Note that in this model, the jump rate q_2 is a constant and the instantaneous cost function is independent of the age of the machine, so we can regard the age of the machine is reset to zero at the beginning of the repair. That is $a(t) = 0$ if $\alpha(t) = 2$. Consequently, when $\alpha(t) = 2$, $f(u, v) = 0$. This treatment will not influence the cost function.

In this special case, the following theorem gives the optimal policy for v:

THEOREM 6.3. *The optimal policy for the maintenance* $v^* = v^*(u)$ *is the solution of* $f(u, v) = 0$.

PROOF. Note that the Hamilton–Jacobi–Bellman (HJB) equation, in this special case, is similar to (15). Let

$$V_\alpha(x, a) = \inf_{\gamma \in \mathcal{A}} J^\gamma(\alpha, x, a), \quad \alpha \in \mathcal{S},$$

be the optimal cost function. Then, $V_i(x, a)$ is a solution to the HJB equations if it is continuously differentiable. Since $\frac{\partial}{\partial a} V_1(x, a) \geq 0$, we can choose v^* such that $f(u, v) = 0$ which minimizes the following expression:

$$\min_u \left[(u - d) \frac{d}{dx} V_1(x, a) + \frac{d}{da} V_1(x, a) f(u, v) \right]. \tag{46}$$

Hence the optimal policy for v is the solution of $f(u, v) = 0$. □

If $f(u, v) = k_1 u - k_2 v = 0$, then $v = \frac{k_1}{k_2} u$, which means that the maintenance rate v should be proportional to the production rate.

If we ignore the maintenance cost in this case (i.e., we suppose maintenance does not cost money), the best policy should be always to keep the machine as good as new through maintenance (i.e., $f(u, v) = 0$ implies $\dot{a}(t) = 0$, which means that the machine age should be kept as a constant). If we can keep the machine age as a constant through maintenance, the problem becomes similar to the one considered by Akella and Kumar (1986), so the solution of the optimal policy for the production rate u is similar to the one obtained in Akella and Kumar (1986). However, maintenance always costs money in practice, and we cannot ignore this fact.

To get the optimal control law, let us consider the HJB equations.

THEOREM 6.4. *The HJB equation in this case can be given by*:

$$\begin{pmatrix} \min_{u,v} \left[(u-d) \frac{\partial}{\partial x} V_1(x,a) + (k_1 u - k_2 v) \frac{\partial}{\partial a} V_1(x,a) + kv \right] \\ -d \frac{\partial}{\partial x} V_2(x,a) \end{pmatrix}$$

$$= \begin{pmatrix} \rho + q_1(a) & -q_1(a) \\ -q_2 & \rho + q_2 \end{pmatrix} \begin{pmatrix} V_1(x,a) \\ V_2(x,a) \end{pmatrix} - \begin{pmatrix} 1 \\ 1 \end{pmatrix} (c^+ x^+ + c^- x^-).$$

First, we try to find the optimal policy for v, and then we determine the corresponding optimal production rate. The following theorem gives the optimal control policy v^*.

THEOREM 6.5. *The optimal policy for the maintenance rate v is given by*:

$$v^* = \begin{cases} \frac{k_1}{k_2} u, & \text{if } \frac{\partial}{\partial a} V_1(x,a) \geqslant \frac{k}{k_2}, \\ 0, & \text{if } \frac{\partial}{\partial a} V_1(x,a) < \frac{k}{k_2}. \end{cases}$$

The result of Theorem 6.5 means that if the change of the machine aging increases the cost significantly, then we should do maintenance such that the machine is kept as good as new (since from $v = \frac{k_1}{k_2} u$, we have, $\dot{a}(t) = 0$, i.e., machine aging should be kept as a constant). If the change of the machine aging does not increase the cost too much, we do not need to do maintenance. In the case that the machine aging remains constant, Akella and Kumar (1986) proved that the optimal control is the hedging point policy. The corresponding hedging point z is a constant. This has no sense when the machine age changes. Evidently, a new machine has the capacity to respond to the demand with very high probability, but for an old machine the situation will be different. Based on this remark, the hedging level should be a function of the machine age. The following theorem gives the age dependent hedging point policy.

THEOREM 6.6. *Let*

$$A_1(a) = \begin{pmatrix} \frac{\rho + q_1(a)}{\bar{u} - d} & \frac{-q_1(a)}{\bar{u} - d} \\ \frac{q_2}{d} & -\frac{q_2 + \rho}{d} \end{pmatrix},$$

and let $\lambda_-(a)$ and $\lambda_+(a)$ denote the negative and positive eigenvalues of $A_1(a)$, respectively. In the case of $\frac{\partial}{\partial a} V_1(x,a) \geqslant k/k_2$, the optimal policy for the production rate is given by:

$$u^{z^*} = \begin{cases} \bar{u}, & \text{if } x(t) < Z^*(a), \\ d, & \text{if } x(t) = Z^*(a), \\ 0, & \text{if } x(t) > Z^*(a), \end{cases}$$

where the optimal inventory level $Z^*(a)$ is nonnegative and is given by:

$$Z^*(a) = \max\left\{0, \frac{1}{\lambda_-(a)} \log\left[\frac{c^+}{c^+ + c^-}\right.\right.$$
$$+ \frac{\rho}{c^+ + c^-} \left[\left(c^+ + \frac{kk_1}{k_2}(\rho + q_1(a))\right)\right) \frac{d}{q_1(a)d - (\bar{u} - d)(\lambda_-(a)d + \rho + q_2)}\right]$$
$$- \frac{\rho}{c^+ + c^-} \left[\frac{kk_1}{k_2} \frac{(\bar{u} - d)(\lambda_-(a)d + \rho + q_2)}{q_1(a)d - (\bar{u} - d)(\lambda_-(a)d + \rho + q_2)}\right.$$
$$\left.\left.\left.- \frac{kk_1}{k_2} \frac{q_1(a)\bar{u}}{q_1(a)d - (\bar{u} - d)(\lambda_-(a)d + \rho + q_2)}\right]\right]\right\}.$$

In the case $\frac{\partial}{\partial a} V_1(x,a) < k/k_2$, the optimal policy for the production rate u is given by:

$$u^* = \begin{cases} \bar{u}, & \text{if } \frac{\partial}{\partial x} V_1(x,a) + \frac{\partial}{\partial a} V_1(x,a) k_1 < 0, \\ d + a\frac{\partial x}{\partial a}, & \text{if } \frac{\partial}{\partial x} V_1(x,a) + \frac{\partial}{\partial a} V_1(x,a) k_1 = 0, \\ 0, & \text{if } \frac{\partial}{\partial x} V_1(x,a) + \frac{\partial}{\partial a} V_1(x,a) k_1 > 0, \end{cases}$$

where $\frac{\partial x}{\partial a} = \frac{\partial V_1(x,a)}{\partial a} / \frac{\partial V_1(x,a)}{\partial x}$ is calculated through the solution of

$$\frac{\partial}{\partial x} V_1(x,a) + \frac{\partial}{\partial a} V_1(x,a) k_1 = 0. \tag{47}$$

PROOF. For the proof, see Boukas and Yang (1996). □

Obviously, in the case of $\frac{\partial}{\partial a} V_1(x,a) < k/k_2$, the optimal control is completely characterized by the solution of Eq. (47), which is called the critical surface. Now we proceed to specify the critical surface in this case.

Let $A_2(a)$, b_1 and b_2 be defined as follows:

$$A_2(a) = \begin{pmatrix} -\frac{\rho + q_1(a)}{d} & \frac{q_1(a)}{d} \\ \frac{q_2}{d} & -\frac{q_2 + \rho}{d} \end{pmatrix},$$

$$b_1 = \begin{pmatrix} -\frac{1}{\bar{u} - d} \\ \frac{1}{d} \end{pmatrix}, \quad b_2 = \begin{pmatrix} \frac{1}{d} \\ \frac{1}{d} \end{pmatrix}.$$

The general solution of (47) is given by

$$V_1(x, a) = \eta(k_1 x - a),$$

where η is a differentiable function.

When the age is fixed, i.e., $a = a_0$, the situation becomes similar to the one discussed by Akella and Kumar (1986). From Akella and Kumar (1986), $Z^*(a_0)$ is given by

$$Z^*(a_0) = \max\left\{0, \frac{1}{\lambda_-(a_0)} \log\left[\frac{c^+}{c^+ + c^-}\right. \right.$$
$$\left.\left. \times \left[1 + \frac{\rho d}{q_1(a_0)d - (\rho + q_2 + d\lambda_-(a_0))(\bar{u} - d)}\right]\right]\right\}.$$

Moreover, we know that in the case $Z^*(a_0) = 0$, $G(x) = V_1(x, a_0)$ is given by:

$$G(x) = \begin{cases} [1,0][\frac{c^-}{\rho\lambda_+(a_0)} e^{\lambda_+(a_0)x} w^+ \\ \quad + A_1^{-1} b_1 c^- x + A_1^{-2} b_1 c^-], & \text{if } x \leq 0, \\ [1,0]\{e^{A_2 x}[\frac{c^-}{\rho\lambda_+(a_0)} w^+ + A_2^{-2} b_1 c^- + A_2^{-2} b_2 c^+] \\ \quad - [A_2^{-1} b_2 c^+ x + A_2^{-2} b_2 c^+]\}, & \text{if } x \geq 0, \end{cases} \quad (48)$$

where $w^+ = \binom{1}{w_2^+}$ is the eigenvector of $A_1(a)$ corresponding to $\lambda_+(a)$. In the case of $Z^*(a_0) > 0$, $G(x) = V_1(x, a_0)$ is given by:

$$G(x) = \begin{cases} [1,0]\{e^{A_1(x-z^*)}[A_1^{-2} b_1 c^+ - A_1^{-1}\binom{0}{1}\frac{c^+}{q_1(a_0)}] \\ \quad - e^{A_1 x} A_1^{-2} b_1(c^+ + c^-) \\ \quad + A_1^{-1} b_1 c^- x + A_1^{-2} b_1 c^-\}, & \text{if } x \leq 0, \\ [1,0]\{e^{A_1(x-z^*)}[A_1^{-2} b_1 c^+ - A_1^{-1} \\ \quad - A_1^{-1}\binom{0}{1}\frac{c^+}{q_1(a_0)}] - A_1^{-1} b_1 c^+ x \\ \quad - A_1^{-2} b_1 c^+\}, & \text{if } 0 \leq x \leq Z^*(a_0), \\ [1,0]\{e^{A_2(x-z^*)}[A_2^{-1} b_2 c^+ z^* - A_1^{-1}\binom{0}{1}\frac{c^+}{q_1(a_0)} \\ \quad - A_1^{-1} b_1 c^+ z^* + A_2^{-2} b_2 c^+] \\ \quad - A_2^{-1} b_2 c^+ x - A_2^{-2} b_2 c^+\}, & \text{if } x \geq Z^*(a_0). \end{cases} \quad (49)$$

Hence (47) satisfies the following initial condition:

$$V_1(x, a)|_{a=a_0} = G(x), \quad (50)$$

where the expression of $G(x)$ is given by Eq. (48) or (49).

Therefore, from above argument, we have the following result:

THEOREM 6.7. *The solution of* (47) *with initial condition* (50) *is given by:*

$$V_1(x, a) = G\left(x - \frac{a}{k_1} + \frac{a_0}{k_1}\right),$$

where the expression of $G(x)$ *is given by Eq.* (48) *or* (49).

7. Conclusion

This chapter addresses the continuous flow model for production and maintenance control in flexible manufacturing system and some computation methods to find the optimal control. The maintenance activities considered in this chapter can be divided into two classes. The first class treats the maintenance as a special state of the machine and the jump rate from the operation mode to maintenance mode is modeled as maintenance rate which is a control variable. The second one includes the activities as adjustment, lubrication, etc., which are executed when the machine is working and can reduce the aging rate of the machine. In the first class of maintenance, two models are introduced. In the first one provided in Section 4, the jump rates of the machine (including the maintenance rate) are dependent on the age of the machine. The use of the age of the machine is very intuitive but increases the dimension of the system and thus increases the computation burden. To overcome this disadvantage, Section 5 introduces another model in which instead of using the age of the machine, the machine is assumed to have four modes, three working modes: 'good', 'average' and 'bad', and a failure mode. The jump rates from 'average' and 'bad' to 'good' are defined as the preventive maintenance rates and the jump rate from failure to 'good' is treated as corrective maintenance rate. The optimal control is characterized and some illustrative examples are provided. For the second class of maintenance activities, Section 6 introduces a model in which the demand rate is assumed to be uncertain, which is varying in a compact set. The optimal controllers are characterized and in a special case, the optimal control problem is solved.

The generic procedure to tackle the optimal control problem of production and maintenance is as follows. First model the inventory and machine mode by a controlled piecewise deterministic process and develop some properties of the optimal cost function, such as polynomial growth, locally Lipschitz which are necessary to character the optimal control. Then prove that the value is the unique viscosity solution of the associate HJB equation. Finally, by using Kushner's approximation technique, develop a numerical algorithm to get an approximate solution of the optimal control.

Some future research directions are still available.

1. For the setting of diffusion process, Menaldi (1989) has proved that Kushner's approximation algorithm converges with rate $\sqrt{h}$, it is of interesting to consider the convergence rate of the algorithm provided in this chapter.
2. When the machine has multiple modes, the state of the machine may not be available for feedback. In this case, how to handle the optimal production and maintenance control problem is still an open question.

3. For a complex manufacturing system, the hierarchical control for the preventive and corrective maintenance is also an open question.
4. For a manufacturing system, it is critical to have a good market for its production. It is well known that marketing actions, such as advertisement, pricing, etc., can increase the demand rate. To model the demand rate as multiple levels, consider the jump rate of from one demand level to another as control variables and seek a short term decision to maximize the profit of the system in the long run. Especially, in some cases, the demand rate level is not perfect available, how to handle these cases are still open.

References

Akella, R., and P. R. Kumar (1986). Optimal control of production rate in a failure prone manufacturing system. *IEEE Trans. Automat. Control* **AC-31**(2), 116–126.

Algoet, P. H. (1988). Flow balance equations for the steady-state distribution of a flexible manufacturing system. *IEEE Trans. Automat. Control* **34**(8), 917–921.

Bertsekas, D. P. (1987). *Dynamic Programming: Deterministic and Stochastic Models*. Prentice-Hall, Inc.

Bielecki, T. and P. R. Kumar (1988). Optimality of zero-inventory policies for unreliable manufacturing systems. *Oper. Res.* **36**(4), 532–541.

Boukas, E. K. (1987). Commande optimale stochastique appliquée aux systèmes de production. Ph.D. thesis, Ecole Polytechnique de Montréal, Université de Montréal.

Boukas, E. K. (1993). Control of systems with controlled jump Markov disturbances. *Control Theory and Advanced Technology* **9**(2), 577–595.

Boukas, E. K. (1995). Numerical methods for HJB equations of optimization problems for piecewise deterministic systems. *Optim. Control Appl. Methods* **16**, 41–58.

Boukas, E. K. (1998). Hedging point policy improvement. *J. Optim. Theory Appl.* **97**(1).

Boukas, E. K. and A. Haurie (1988). Optimality conditions for continuous time systems with controlled jump Markov disturbances: application to production and maintenance scheduling. In *Analysis and Optimization of Systems*. Proceeding INRIA 8th International Conference. Springer-Verlag, Antibes.

Boukas, E. K. and A. Haurie (1990). Manufacturing flow control and preventive maintenance: a stochastic control approach. *IEEE Trans. Automat. Control* **35**(9), 1024–1031.

Boukas, E. K. and J. P. Kenne (1997). Maintenance and production control of manufacturing system with setups. *Lectures in Appl. Math.* **33**, 55–70.

Boukas, E. K. and Z. K. Liu (1999a). Jump linear quadratic regulator with controlled jump rates. *IEEE Trans. Automat. Control,* to appear.

Boukas, E. K. and Z. K. Liu (1999b). Production and maintenance control for manufacturing system. In *Proceeding of 38th IEEE CDC*.

Boukas, E. K. and Z.K. Liu (1999c). Production and corrective maintenance control for flexible manufacturing systems. Manuscript.

Boukas E. K., and H. Yang (1996). Manufacturing flow control and preventive maintenance: a stochastic control approach. *IEEE Trans. Automat. Control* **41**(6), 881–885.

Boukas, E. K., Q. Zhang and G. Yin (1996). On robust design for a class of failure prone manufacturing system. In *Recent Advances in Control and Optimization of Manufacturing Systems* (Eds. Yin and Zhang). Lecture Notes in Control and Inform. Sci., Vol. 214. Springer-Verlag, London.

Boukas, E. K., Q. Zhu and Q. Zhang (1994). Piecewise deterministic Markov process model for flexible manufacturing systems with preventive maintenance. *J. Optim. Theory Appl.* **81**(2), 258–275.

Bremaud, P., R. P. Malhame and L. Massoulie (1997). A manufacturing system with general failure process: stability and IPA of hedging control policies. *IEEE Trans. Automat. Control* **42**(2), 155–170.

Buzacott, J. A. and J. G. Shanthikumar (1993). *Stochastic Models of Manufacturing Systems*. Prentice-Hall, Englewood Cliffs, NJ.

Connolly, S., Y. Dallery and S. B. Gershwin (1992). A real-time policy for performing setup changes in a manufacturing system. In *Proceeding of the 31st IEEE Conference on Decision and Control*, Dec., Tucson, AZ.

Caramanis, M. and G. Liberopoulos (1992). Perturbation analysis for the design of flexible manufacturing system flow controllers. *Oper. Res.* **40**, 1107–1125.

Costa, O. L. V. (1989). Average impulse control of piecewise deterministic processes. *IMA J. Math. Control Inform.* **6**(4), 379–397.

Costa, O. L. V. (1991). Impulse control of piecewise-deterministic processes via linear programming. *IEEE Trans. Automat. Control* **36**(3), 371–375.

Costa, O. L. V. and M. H. A. Davis (1989). Impulse control of piecewise-deterministic processes. *Math. Control Signals Systems* **2**(3), 187–206.

Dassios, A. and P. Embrechts (1989). Martingales and insurance risk. *Comm. Statist. Stochastic Models* **5**(2), 181–217.

Davis, M. H. A. (1984). Piecewise deterministic Markov processes: a general class off non-diffusion stochastic models. *J. Roy. Statist. Soc.* **46**(3), 353–388.

Davis, M. H. A. (1986). Control of piecewise-deterministic processes via discrete-time dynamic programming. In *Stochastic Differential Systems,* pp. 140–150 (Eds. K. Helmes and C. Kohlman). Springer-Verlag.

Davis, M. H. A. (1993). *Markov Modeling and Optimization*. Chapman and Hall.

Davis, M. H. A., M. A. H. Dempster, S. P. Sethi and D. Vermes (1987). Optimal capacity expansion under uncertainty. *Adv. Appl. Probab.* **19**, 156–176.

Dempster, M. A. H. (1991). Optimal control of piecewise deterministic Markov processes. In *Applied Stochastic Analysis*, pp. 303–325. Gordon and Breach, New York.

Desrochers, A. A. (1990). *Modeling and Control Of Automated Manufacturing Systems*. IEEE Computer Society Press, Washington.

Duncan, T. E., B. Pasik-Duncan and Q. Zhang (1999). Adaptive control of stochastic manufacturing system with hidden Markovian demands and small noise. *IEEE Trans. Automat. Control* **44**(2), 427–431.

Fleming, W. H., and H. M. Soner (1992). *Controlled Markov Processes and Viscosity Solutions*. Springer-Verlag, New York.

Gatarek, D. (1992). Optimality conditions for impulsive control of piecewise-deterministic processes. *Math. Control Signals Systems* **5**(2), 217–232.

Gershwin, S. B. (1989). Hierarchical flow control: a framework for scheduling and planning discrete events in manufacturing systems. In *Proceeding of the IEEE, Special Issue on Discrete Event Dynamic Systems*, Vol. 77, No. 1, 195–209.

Gershwin, S. B. (1993) *Manufacturing Systems Engineering*. Prentice-Hall, Englewood Cliffs.

Gershwin, S. B., M. Caramanis and P. Murray (1988). Simulation experience with a hierarchical scheduling policy for a simple manufacturing system. In *Proceeding of the 27th IEEE Conference on Decision and Control,* pp. 1941–1849. Dec., Austtin, TX.

Gonzales, R. and E. Roffman (1985). On the deterministic control problems: an approximation procedure for the optimal cost I. The stationary problem. *SIAM Control Optim.* **23**(2), 242–266.

Haurie, A. and P. L'Ecuyer (1986). Approximation and bounds in discrete event dynamic programming. *IEEE Trans. Automat. Control* **31**(3), 227–235.

Hu, J. Q., and M. Caramanis (1994). Dynamic set-up scheduling of flexible manufacturing systems: design and stability of near optimal general round robin policies. In *Discrete Event Systems, IMA Volumes in Mathematics and Applications Series* (Eds. P. R. Kumar and P. Varaiya). Springer-Verlag.

Hu, J. Q., P. Vakili and G. X. Yu (1994). Optimality of hedging point policies in the production control of failure prone manufacturing systems. *IEEE Trans. Automat. Control* **39**(9), 1875–1880.

Hu, J. Q. and D. Xiang (1994). Structural properties of optimal flow control for failure prone production system. *IEEE Trans. Automat. Control* **39**(3), 640–642.

Kimemia, J. and S. B. Gershwin (1983). An algorithm for computer control of a flexible manufacturing system. *IIE Transactions* **15**(4), 353–362.

Kushner, H. J. (1977). *Probability Methods for Approximation in Stochastic Control and for Elliptic Equations*. Academic Press, New York.

Kushner, H. J. and P. G. Dupuis (1992). *Numerical Methods for Stochastic Control Problems in Continuous Time*. Springer-Verlag, New York.

Lenhart, M. S. and Y. C. Liao (1988). Switching control of piecewise-deterministic processes. *J. Optim. Theory Appl.* **59**(1), 99–115.

Liberopoulos, G. and M. Caramanis (1994). Production control of manufacturing system with production rate-dependent failure rates. *IEEE Trans. Automat. Control* **39**(4), 889–895.

Maimoun, O., E. Khenelnitsky, and K. Kogan (1998). *Optimal Flow Control in Manufacturing System: Production Planning and Scheduling.* Kluwer Academic Publishers, Dordrecht, The Netherlands.

Menaldi, J. L. (1989). Some estimates for finite difference approximations. *SIAM J. Control Optim.* **27**, 579–607.

Olsder, G. J. and R. Suri (1980). Time optimal of parts-routing in a manufacturing system with failure prone machines. In *Proc. 19th IEEE Conference on Decision and Control*, pp. 722–727. Albuquerque, New Mexico.

Rishel, R. (1975). Control of systems with jump Markov disturbances, *IEEE Trans. Automat. Control* **20**, 241–244.

Ross, S. M. (1970). *Applied Probability Methods with Optimization Applications*. Holden-Day, San Francisco.

Sethi, S. P., M. I. Taksar and Q. Zhang (1997). Optimal production planning in a stochastic manufacturing system with long-run run average cost. *J. Optim. Theory Appl.* **92**(1), 161–188.

Sethi, S. P., and Q. Zhang (1994). *Hierarchical Decision Making in Stochastic Manufacturing Systems.* Birkhäuser, Boston.

Sharifnia, A. (1988). Production control of a manufacturing system with multiple machine states. *IEEE Trans. Automat. Control* **33**(7), 620–625.

Sharifnia, A., M. Caramanis and S. B. Gershwin (1991). Dynamic setup scheduling and flow control in manufacturing systems. *Discrete Event Dynamic Systems: Theory and Applications* **1**, 149–175.

Soner, H. M. (1986). Optimal control with state-space constraint II. *SIAM J. Control Optim.* **24**(6), 1110–1122.

Tu, F. S., D. P. Song and S. X. C. Lou (1993). Preventive hedging point control policy and its realization. *Preprint of the XII World IFAC Congress,* July 18–23, Sydney, Australia, Vol. 5, pp. 13–16.

Sworder, D. D. and V. G. Robinson (1974). Feedback regulators for jump parameter systems with state and control dependent transition rates. *IEEE Trans. Automat. Control* **18**, 355–359.

Veatch, M. H. and M. C. Caramanis (1999). Optimal average cost manufacturing flow controllers: convexity and differentiability. *IEEE Trans. Automat. Control* **44**(4), 779–783.

Viswanadham, N. and Y. Narahari (1992). *Performance modeling of automated manufacturing systems*. Prentice-Hall, Englewood Cliffs.

Wonham, W. M. (1971). Random differential equations in control theory. In *Probabilistic Methods in Applied Mathematics,* Vol. 2 (Ed. A. T. Bharucha-reid). Academic Press, New York.

Yan, H. and Q. Zhang (1997). A numerical method in optimal production and setup scheduling of stochastic manufacturing system. *IEEE Trans. Automat. Control* **42**(10), 1452–1555.

Yin, G. and Q. Zhang (1998). *Continuous-Time Markov Chains and Applications: A Singular Perturbation Approach.* Springer-Verlag.

Yin, G., and Q. Zhang (1996). *Recent Advances in Control and Optimization of Manufacturing System.* Lecture Notes in Control and Inform. Sci., Vol. 214. Springer-Verlag, London.

Yu, X. Z. and W. Z. Song (1999). Further properties of optimal hedging points in a class of manufacturing system. *IEEE Trans. Automat. Control* **44**(2), 379–382.

Models of Random Graphs and their Applications

C. Cannings and D. B. Penman

1. Introduction

1.1. The need for random graphs

Networks are ubiquitous. They arise naturally as models of communication networks, networks of friends, in the communication of infection, rumours or information, as models of atoms and bonds between them in chemistry, as autocatalytic nets (Kauffman, 1993) and elsewhere. Mathematically the notion is captured in a graph: a finite set of vertices V and a set E of edges between some of the distinct vertices. For most of this paper, we discuss graphs which have finite vertex set, do not have multiple edges between two vertices or loops from a vertex to itself, and whose edges are undirected. We mostly study **labelled** graphs on n vertices $\{1, 2, \ldots, n\}$ (n will be reserved for the number of vertices throughout): so, for example, we regard the graph on $\{1, 2, 3\}$ whose only edge is $(1, 2)$ and the one whose only edge is $(1, 3)$ as distinct, despite the fact that they are isomorphic. Two meaty introductions to graph theory are (Bollobás, 1998) and (Diestel, 1997). A typical undergraduate course in graph theory is required to read this paper, but knowledge of random graphs is not assumed.

Let $\wp$ be a property a graph may or may not possess, such as being connected, having a given diameter, having a 5-cycle as a subgraph. (Frequently, a property of graphs on n vertices is identified with the set of graphs on n vertices having that property.) It is of interest to investigate whether most, few, or some intermediate proportion of graphs have $\wp$, and whether the various quantities vary little or greatly over examples. For small n, exact enumeration is possible, but as there are $2^{n(n-1)/2}$ labelled graphs on n vertices, for large n we can only talk about typical behaviour, and are thus lead to random graphs.

A model of random graphs is a stochastic mechanism for determining which of the $n(n-1)/2$ potential edges actually arise: for labelled graphs, this is a rule assigning probabilities to each of the $2^{n(n-1)/2}$ labelled graphs. There are obviously uncountably many ways to do this, so very little can be said in such generality; we must choose an amenable distribution to proceed. Since calculating probabilities is easier if events of interest are independent, and the basic question about a graph is whether or not each possible edge arises, it is natural to begin by studying models where each edge arises independently of all the others. In the easiest such model, each edge has the same probability $p(n)$ of arising. (Note the possible dependence of p on n: indeed the case of constant p is often dull.) This model, denoted $G(n, p(n))$ (or $G_{p(n)}$ if n is clear in context: and indeed the dependence of p on n is sometimes supressed) is called the Erdős–Rényi model, as these two mathematicians initiated the serious study of random graphs. It has been studied a great deal: the bible of the subject is (Bollobás, 1985), and the recent (Janson et al., 2000) concentrates on areas where there has been significant progress since Bollobás. Both these books give sharp and detailed forms of results: novices may find it easier to start with the deliberately introductory (Palmer, 1985) or the introductory chapters on random graphs in (Bollobás, 1998) or (Diestel, 1997). (Kolchin, 1999) is another recent book on random graphs (albeit with a rather different flavour), and a recent survey is (Karoński, 1995). For convenience, we often refer to such books rather than original papers: thus certain authors appear in our bibliography disproportionately to their contribution to the subject.

Closely related to $G(n, p)$ is $G(n, M)$ where all graphs on n vertices and M edges are equiprobable. Of course $G(n, M)$ is the same as $G(n, p)$ conditional on having M edges. The theory of $G(n, M)$ for M near $p(n)n(n-1)/2$ is, unsurprisingly, closely related to that of $G(n, p(n))$: see Theorem 2.1 of (Karoński, 1995) for Łuczak's generalisation of Bollobás' result to this effect. More generally, we can consider a **random graph process**, where at each time $n = 1, 2, \ldots, n(n-1)/2$ we add one edge, chosen uniformly and at random from those not yet in the graph, to it. We thus generate a random sequence (G_t) of graphs on n labelled vertices, G_t having t edges. An interesting random variable is then the **hitting time** of property $\wp$, the first t ($1 \leqslant t \leqslant n(n-1)/2$) for which G_t has $\wp$. Later we shall study restricted random graph processes, where additional conditions (e.g., that the graph be triangle-free) are imposed on the evolution.

A natural generalisation of Erdős–Rényi random graphs is when the edge between vertices v_1 and v_2 arises with probability $p_{v_1 v_2}$, independently of all other edges. The Erdős–Rényi model is recovered if all the p_{ij} are equal. Models of this sort are studied in papers such as (Kovalenko, 1975; Ivchenko, 1975; Juhász, 1991). For example, an easily stated result from (Shepp, 1989) is: if the infinite vertex set is $\{1, 2, \ldots\}$, and i and j are adjacent with probability $\lambda / \max\{i, j\}$, then $\lim_{n \to \infty} P\{\text{the graph is connected}\} = 1$ if and only if $\lambda > 1/4$.

Although the Erdős–Rényi model is mathematically tractable, there is mathematical interest in comparing it with alternative models. Additionally, in many real networks, edges will not in fact arise independently and equiprobably, so $G(n, p(n))$ may be an inadequate representation. There has thus been recent interest in other models of random graphs. We will discuss these below after introducing the key notion of a threshold.

1.2. Thresholds

Perhaps the single most important feature of the Erdős–Rényi model is the fact that, for many properties of a graph, the probability that the graph has that property changes rather suddenly from being close to 0 to being close to 1. More precisely, there is often a function $p^*(n)$ such that

$$\lim_{n \to \infty} \frac{p(n)}{p^*(n)} = \infty \Rightarrow \lim_{n \to \infty} P\{G_p \in \wp\} = 1,$$

but

$$\lim_{n \to \infty} \frac{p(n)}{p^*(n)} = 0 \Rightarrow \lim_{n \to \infty} P\{G_p \in \wp\} = 0.$$

($p^*(n)$ is not unique – but this is not usually a problem in practice.)

Not all interesting properties have a threshold (see, e.g., Thomason, 1988), but many do. One result supporting this is that every **increasing** property has a threshold (Bollobás and Thomason, 1987). ($\wp$ is monotone increasing if and only if, whenever a spanning subgraph of a graph H has $\wp$, then H has it also; if $\wp$ is monotone then $P\{G(n, p) \in \wp\}$ is a nondecreasing function of p for fixed n.) Recent work (Bourgain and Kalai, 1999) uses ideas from Fourier analysis to show that for monotone properties, the transition happens quite quickly: if, given $\varepsilon > 0$, $P\{G(n, p_0) \in \wp\} = \varepsilon$ and $P\{G(n, p_1) \in \wp\} = 1 - \varepsilon$, then $p_1 - p_0$ has order of magnitude $\log(n)^{\varepsilon-2}$. For information about how sharp thresholds for various kinds of properties are, see (Friedgut, 1999).

A second supporting result is that if the property $\wp$ can be expressed as a sentence of first-order logic (a **first-order** property) then (for p constant) $\lim_{n \to \infty} P\{G(n, p) \text{ has } \wp\}$ is 0 or 1. (An introduction to first-order logic in the context of graphs is (Cameron, 1999): novices should be aware that many interesting graph properties are not first-order.)

Motivated by this, we say that $G(n, p)$ has property $\wp$ **whp** (for 'with high probability') if

$$\lim_{n \to \infty} P\{G(n, p) \in \wp\} = 1.$$

Sometimes, e.g., in (Bollobás, 1985), the terminology '$\wp$ holds for a.e. graph' is used: other synonyms include '$\wp$ holds a.a.s.' (for 'asymptotically almost surely') or '$\wp$ holds a.s.' All the terms (including **whp**) are potentially misleading: we are really talking about convergence in probability.

The classic example is the probability that a graph is connected: for this property the threshold is $p^*(n) = \log(n)/n$. More precisely,

$$\lim_{n \to \infty} P\left\{G\left(n, \frac{\log(n) + c + o(1)}{n}\right) \text{ is connected}\right\} = e^{-e^{-c}}.$$

If we know the threshold for a property $\wp$, we thus understand the limiting probability that $G(n, p)$ has $\wp$. Of course in practice we are looking at graphs with some fixed large number of vertices. Thus one wants to ensure that the limiting approximation is accurate for not very large n. In (Godehardt, 1990a), a particular form of the threshold for connectedness, for which convergence to the limiting distribution is fast, is exhibited. It would be interesting to have similar results for other thresholds. The other situation in practice is graphs which do not have a particularly large number of vertices. Chapter XVI of (Bollobás, 1985) contains various calculations of the probability of various properties for small values of n: these are exact to the number of decimal places given. There are extensive tables of graphs on small numbers of vertices with given properties: see, for example, (Royle, 2001) or (McKay, 2001) and the links suggested there.

1.3. Other models of random graphs

We now consider some natural ways of generating random graphs where, unlike in the Erdős–Rényi model, there is some dependence between whether or not edges are present. As these models are just beginning to be developed, there is little uniformity of notation, and results are often harder to obtain, or less precise, than in the Erdős–Rényi model.

A first alternative approach starts from the fact that in many practical problems, the vertices are in fact randomly positioned in some geometric space (usually Euclidean), and two vertices are adjacent if and only if the distance between them (in some specified norm) is less than a certain quantity. In this kind of model, correlation between the presence of edges arises from the triangle inequality. Usually the points are taken to be uniformly distributed in (say) $[0, 1]^n$, but some literature also exists on points uniformly distributed on a circle.

A second approach is to start from the Erdős–Rényi model but, by biasing the formula for the probability of a set of edges in some way, to favour certain kinds of graphs arising. The best-known example of this is the **random-cluster model**, defined as follows. Given some graph $G = G(V, E)$ and some set $A \subseteq E$ of edges, let $c(V, A)$ denote the number of components of the graph whose vertex set is V and whose edge set is A. Then the probability that the edges which arise are exactly those in A is

$$\frac{p^{|A|}(1-p)^{|E|-|A|} q^{c(V,A)}}{\sum_{F \subseteq E} p^{|F|}(1-p)^{|E|-|F|} q^{c(V,F)}}.$$

Observe that when $G = K_n$ (the so-called mean field case) and $q = 1$, we recover the Erdős–Rényi model. If $q > 1$ graphs with many components are favoured and if $q < 1$ connected graphs are favoured. The study of this model is closely linked to percolation theory and statistical physics.

A third approach is to make the presence or absence of edges between two vertices dependent on the types of the vertices. The first idea along these lines is discussed in (Karoński et al., 1999) and (Fill et al., 2000). Here, each vertex v is assigned a random binomial subset S_v of $A = \{1, 2, \ldots, m\}$ (that is, each element of A is in S_v with probability p, independent of all other elements of A). The S_v are independent as v varies. Then two vertices v and w are adjacent if and only if $S_v \cap S_w \neq \emptyset$. Such a graph is called a **random intersection graph** or RIC graph, and denoted by $G(n, m, p)$.

Observe that the overall probability that any particular edge arises is $1 - (1 - p^2)^m$. To obtain the results on the subgraph problem discussed later, one usually takes $m = \lfloor n^\beta \rfloor$ (often in the literature, just $m = n^\beta$ is written: what we call β is called α in (Karoński et al., 1999) and (Fill et al., 2000), but we will use α elsewhere). We can consider the evolution of the model as p goes from $1/(n\sqrt{m})$ (which can easily be checked to be the threshold for an edge to appear) to $\sqrt{2\log(n)/m}$, the threshold for all edges to be present.

A more general model with vertex-type dependence was introduced in (Penman, 1998). Here one assigns to each vertex independently one of k types (so far the case where k is fixed has been studied most, but k growing suitably with n is likely to be interesting also). The types are represented by a colour: of course, this colouring has nothing to do with **proper colouring** as in chromatic numbers of graphs. Each vertex receives colour i with probability s_i (so that the numbers of vertices of the various colours are a multinomial random variable). Then an edge between vertices of colours i and j arises with probability p_{ij}. Thus the model is described by the vector $\mathbf{s} = (s_1, \ldots, s_k)$ and the symmetric $k \times k$ matrix P. We call such graphs **random randomly coloured graphs**, abbreviated to RRC graphs: the notation is $\Gamma(n, k, \mathbf{s}, P)$. One often compares the resulting graph with the Erdős–Rényi model $G(n, \alpha)$ where $\alpha = \sum_{i,j} s_i p_{ij} s_j$ so that the overall probability that any edge arises is the same in the random randomly coloured graph and in the Erdős–Rényi graph. Then the differences between the two models reflect the correlation structure in the RRC graph, which arises from the fact that the colour is hidden. Note that conditional on the colours, we recover a model with independent but not usually equiprobable edges. The models are designed to have enough independence to allow calculations, but enough correlation structure to provide novel behaviour. Certainly the behaviour does differ from that of $G(n, \alpha)$. For example, let $G_{p,q}$ be the case with two equiprobable colours, vertices the same colour being adjacent with probability p and those of different colours adjacent with probability q. Then $\alpha = (p + q)/2$. Then by simple calculation the probability of a triangle is $\alpha^3 + ((p - q)/2)^3$ so such a model is never the same as the Erdős–Rényi model.

It is easy to see that RIC graphs are a special case of RRC graphs: there are 2^m colours, one for each possible subset of A, and P is a matrix of zeroes and ones. This special case is in various ways more tractable than general RRC graphs: for example, there is an obvious notion of evolution for $G(n, m, p)$ but not an obvious one for general $\Gamma(n, k, \mathbf{s}, P)$.

There is no reason in principle why one should not study graphs which are (say) both geometric and randomly coloured, so that each of n vertices in some geometric space is independently assigned one of k colours, and two vertices of colours i and j are adjacent if and only if the distance between them is less than a_{ij}. Other ways to mix models are also possible. We are not aware of work in this direction.

1.4. The contents of this survey

Previous surveys have concentrated almost exclusively on the Erdős–Rényi model, in a largely theoretical way. Here we discuss and compare several models of random graphs, and say slightly more about applications. At the same time, we aim to give the main underlying theoretical ideas. An inevitable consequence of extra breadth is less depth. Our hopes are that someone who has found a problem for which some model of random graphs might be appropriate may, by reading this article, be directed to the literature relevant to that problem, and that those who are familiar with some of the models may find the references to recent literature and other models helpful. No real originality is claimed.

In Section 2, we discuss some basic theoretical results about the Erdős–Rényi model and other models with independent edges, which will inform the subsequent discussion of alternative models. Section 3 discusses some applications of the Erdős–Rényi model. Section 4 deals with geometric random graphs and their applications. Section 5 discusses the random cluster model and its relatives. Section 6 discusses random randomly coloured graphs (including RIC graphs) and their applications. In Section 7 we discuss several other models of random graphs.

The discussion in this chapter is almost exclusively concerned with the well-developed theory of asymptotic properties as the number of vertices $n \to \infty$. However readers should be aware that in addition random graphs can often be used to demonstrate the existence of graphs with given properties: the classic example is the proof by Erdős of the existence of graphs with arbitrarily large girth and chromatic number (this is noteworthy as, crudely speaking, locally such a graph looks like a tree and so looks 2-colourable). In recent years, there has been interest, partly motivated by computer science applications, in when such arguments can be **derandomized** so as to provide constructive existence proofs of the required objects. For more information on the ideas in this paragraph, we refer in the first instance to (Alon and Spencer, 1992).

We take a highly informal attitude to problems of computational complexity. From now on, we will only use the phrase 'easy to compute' to describe problems for which a solution can be **found** (not checked) in time polynomial in the number of vertices. Problems which are described as 'hard to compute' will usually be in the class NP (for decision problems) or $\#P$ (for counting problems): these are two classes of problems which are widely believed to be strictly harder than those soluble in polynomial time. Readers wishing to learn about computational complexity should consult, e.g., (Garey and Johnson, 1979).

1.5. Brief remarks on simulation

If we cannot prove analytic facts about a model of random graphs, the next best thing is to simulate from them. It should be emphasised this is a poor substitute: it is impossible to guess the threshold for some property from a simulation of graphs on even a large fixed number of vertices.

Often this will mean using MCMC to sample from the stationary distribution of some Markov chain. As usual in this context, the question of whether or not the chain is rapidly mixing (crudely, whether the distribution is within a reasonable distance of the stationary distribution in a reasonably short time) is crucial (and unfortunately often difficult to answer).

In (Tinhofer, 1993) methods are given to sample from various classes of random graphs on labelled vertices, including trees (both rooted and unrooted), connected graphs with given numbers of vertices and edges, Eulerian graphs, bipartite graphs, and graphs with given degree sequence. He also gives algorithms for many models of unlabelled graphs. For each algorithm, a statistical analysis is given. We will see how to simulate some other models later.

2. Overview of the Erdős–Rényi model

2.1. Toolbox

We introduce here, in the context of the Erdős–Rényi model, certain tools which are frequently used in the study of random graphs: as we shall see, they are useful in other models too.

The simplest technique is linearity of expectation, especially as applied to a random variable X (e.g., the number of subgraphs of our random graph isomorphic to G) which is the sum of indicator variables. It is often used in conjunction with the following triviality, the **first moment method**. Throughout this chapter, $\mathbf{E}$ denotes expectation.

LEMMA 2.1. *If X is a $\mathbf{N}$-valued random variable, $P\{X > 0\} \leqslant \mathbf{E}(X)$.*

PROOF. $P\{X > 0\} = \sum_{r=1}^{\infty} P\{X = r\} \leqslant \sum_{r=1}^{\infty} r P\{X = r\} = \mathbf{E}(X)$. □

In particular, if $\lim_{n \to \infty} \mathbf{E}(X) = 0$, then $\lim_{n \to \infty} P\{X > 0\} = 0$. This might suggest that the value for $p(n)$ for which $\mathbf{E}(X)$ tends to a finite non-zero constant is the threshold for when $X > 0$. This will turn out to be true for counting subgraphs isomorphic to a given graph H in the Erdős–Rényi model: however, it will not be true in some other models.

Another elementary fact is the **second moment method**:

LEMMA 2.2. *Let the $\mathbf{N}$-valued random variable X have mean μ and variance σ^2. Then*

$$P\{X = 0\} \leqslant \frac{\sigma^2}{\mu^2}.$$

PROOF. We have, using Chebychev's inequality,

$$P\{X = 0\} \leqslant P\{|X - \mu| \geqslant \mu\} \leqslant \frac{\sigma^2}{\mu^2}.$$ □

In practice, this is used when the variance is small, to show that X is **whp** non-zero. For more discussion, see (Janson et al., 2000), p. 34.

An area of great importance is the use of various concentration inequalities to show that a random variable is tightly concentrated around its mean. The commonest of these is the Chernoff upper bound (Bernstein's inequality). The version here (proved in McDiarmid, 1989) does not require the success probabilities to be equal, but is uniform in the success probabilities.

LEMMA 2.3. *Let $X_1, X_2, \ldots$ be Bernoulli random variables, with $P\{X_i = 1\} = p_i$ and $P\{X_i = 0\} = 1 - p_i$. Let $\bar{p} = \sum_{i=1}^{n} p_i / n$. Then*

$$P\left\{\sum_{i=1}^{n} X_i \geqslant n(\bar{p} + t)\right\} \leqslant \exp(-2nt^2)$$

and

$$P\left\{\sum_{i=1}^{n} X_i \leqslant n(\bar{p} - t)\right\} \leqslant \exp(-2nt^2).$$

Ideally one would like to have a large deviations principle for the random variables involved: roughly, this means a formula of the form

$$P\left\{\sum_{i=1}^{n} X_i \geqslant an\right\} = e^{-nf(a) + o(n)}$$

for a suitable function $f(a)$. In the i.i.d. case, one can extend the argument for the Chernoff bound to obtain this. The situation is more complex if the X_i are dependent (but see Biggins and Penman, 2003a).

Another technique for obtaining concentration results is the use of martingales. Recall that a sequence $(X_i)_0^n$ is a martingale sequence if and only if $\mathbf{E}(|X_n|) < \infty$ for all n and $\mathbf{E}(X_{n+1}|X_n, X_{n-1}, \ldots, X_0) = X_n$. The next result is standard; see, e.g., (McDiarmid, 1989).

LEMMA 2.4. *Let $X_0, X_1, \ldots$ be a martingale sequence with $|X_i - X_{i-1}| \leqslant c_i$ for all $1 \leqslant i \leqslant n$. Then*

$$P\{X_n - X_0 \geqslant t\} \leqslant 2 \exp\left(\frac{-t^2}{2 \sum_{k=1}^{n} c_k^2}\right).$$

A more recent technique is Talagrand's method (see (McDiarmid, 1998) for discussion). For example, McDiarmid uses this to obtain an upper bound on the probability that the weight of the minimum spanning tree in a complete graph on n vertices, when the weight of each edge is an independent uniform random variable on $[0, 1]$, differs substantially from $\zeta(3) \simeq 1.202$ (the value it takes **whp**).

An idea which in some sense extends concentration from individual graphs to graph processes is the **differential equation method**. Very loosely, suppose we have a restricted random graph process, consider some numerical invariant of the graphs, say Y_t on G_t, and let H_t be the history of the process up to time t. If we have, uniformly in t,

$$\mathbf{E}(Y_{t+1} - Y_t | H_t) = f(t/n, Y_t/n) + o(1),$$

where f satisfies conditions similar to those in the existence and uniqueness theorem for differential equations, then **whp** $Y_t = nz(t/n) + \text{o}(n)$ uniformly in t, where $z(t)$ is the unique solution of

$$\frac{\mathrm{d}z}{\mathrm{d}t} = f(t, z)$$

satisfying the initial conditions. We refer to (Wormald, 1995) for the precise statement of the result and some applications.

Ideally, of course, one would prefer to understand the full asymptotic distribution of the graph invariant X. The two distributions to which convergence is most common are (unsurprisingly) normal and Poisson, as we often deal with sums of weakly dependent indicators. For information on how to prove normality results, see (Ruciński, 1992): for the Poisson case, see (Barbour et al., 1990). We observe that Barbour's modification of the Stein–Chen method, which also gives error bounds on the convergence to the Poisson, is frequently useful.

Often we cannot obtain a very accurate estimate of a probability, but can at least obtain an inequality between it and some probability that is much easier to estimate. A typical result here is the **FKG inequality**: see, e.g., (Bollobás, 1986) for a proof.

THEOREM 2.5. *Suppose that μ is a probability measure on the subsets of some finite set S, satisfying the inequality*

$$\mu\{x \cup y\}\mu\{x \cap y\} \geqslant \mu\{x\}\mu\{y\} \quad \forall x, y \in L$$

*(such a μ is said to be **log-supermodular**). Then if f and g are non-negative non-decreasing functions on S we have*

$$\mathbf{E}(f)\mathbf{E}(g) = \sum_{x \in L} \mu(x)f(x) \sum_{x \in L} \mu(x)g(x) \leqslant \sum_{x \in L} \mu(x)f(x)g(x) = \mathbf{E}(fg).$$

In particular, if μ is a probability measure for which the FKG inequality holds and A and B are non-decreasing events,

$$P\{A \cap B\} \geqslant P\{A\}P\{B\}.$$

It is straightforward to see that if A is a random independent subset of $\{1, 2, \ldots, k\}$ (by which we mean elements of $\{1, 2, \ldots, k\}$ are in A independently, with $P\{i \in A\} = s_i$), then μ is log-supermodular.

Our final tool is the use of correlation inequalities to show that some collection of events which are weakly dependent are asymptotically independent. The best known example is the so-called Janson inequality:

THEOREM 2.6. *Suppose $\{A_i\}$ for $i \in \{1, 2, \ldots, n\}$ are events in a probability space such that, for $S \subset \{1, 2, \ldots, n\}$, we have*

(1) $\forall i$ and S with $i \notin S$, $P\{A_i | \bigcap_{j \in S} A_j^c\} \leqslant P\{A_i\}$.
(2) $\forall i \neq j$ and S with $i, j \notin S$, $P\{A_i \cap A_j | \bigcap_{k \in S} A_k^c\} \leqslant P\{A_i \cap A_j\}$.

Let $M = \prod_{i=1}^n P\{A_i^c\}$; then, if $P\{A_i\} \leqslant \varepsilon$ $\forall i$ and $\Delta = \sum P\{A_i \cap A_j\}$ where the sum is taken over pairs of events which are dependent but not identical, we have

$$M \leqslant P\left\{\bigcap_{i=1}^n A_i^c\right\} \leqslant M e^{\frac{\Delta}{2(1-\varepsilon)}}.$$

In particular, if $\varepsilon = o(1)$ and $\Delta = o(1)$ we get the asymptotic formula

$$P\left\{\bigcap A_i^c\right\} \sim M.$$

A proof due to Boppana and Spencer, somewhat more elementary than Janson's original one, is given in (Alon and Spencer, 1992). For further discussion, including the related Suen inequality, see (Janson et al., 2000).

2.2. The evolution of a random graph

In many models of random graph theory, it is possible to consider the evolution of the random graph as some parameter increases. Here we discuss the evolution of $G(n, p)$ as p rises from 0 to 1. Chapter 3 of (Karoński, 1995) is a good place to start filling in the details of the very simplistic account here.

When $p(n) = o(1/n)$, the graph is **whp** a forest (calculate the expected number of cycles, show it tends to zero and use the first moment method). More interesting behaviour starts when $p = c/n$.

THEOREM 2.7. *Consider $G(n, c/n)$.*

(1) *If $c < 1$, **whp** the graph consists of components of size $O(\log(n))$ which have at most one cycle.*

(2) *If $c = 1$, the largest component is of order $O(n^{2/3})$.*

(3) *If $c > 1$, **whp** $G(n, p)$ consists of a unique **giant component** which contains $(1 - x(c)/c + o(1))n$ vertices (here $x(c)$ is the unique root in $(0, 1)$ of $xe^{-x} = ce^{-c}$) and small components with at most one cycle.*

The technical details of this fundamental and influential result of Erdős and Rényi (which is often called the **phase transition**) are given in (Bollobás, 1985), Chapters V and VI. There is further discussion in (Janson et al., 2000). Extremely precise results are obtained using generating function techniques in (Janson et al., 1994). We mention here one more recent result about the giant component, namely a large deviations principle for its order (see O'Connell, 1998).

(Łuczak, 1994) discusses other contexts where a phase transition occurs: we discuss some of these later.

Some insight can be had from the following heuristic, which appears in expanded form in (Alon and Spencer, 1992). The mean number of neighbours of a vertex v is about $n(c/n) = c$. Consider them as offspring of v, the vertices at distance 2 from v as the grandchildren of v, and so on. Treating this as (roughly) a branching process, theory suggests it should die out if $c < 1$ and have a positive probability of survival if $c > 1$. What happens is essentially that all the processes (for each vertex), which do not die out early, merge to form the giant component.

We next discuss when the first subgraph isomorphic to a particular graph (on a fixed number of vertices) appears. The key quantity turns out to be the density of edges. If H is a subgraph of G, $d(H) = |E(H)|/|H|$ is its edge density: for a graph G, let $d^*(G) = \max_{H \subseteq G} d(H)$ be the maximum subgraph density. The next result is a composition of results by various authors, culminating with Ruciński and Vince. The second moment method plays a critical role.

THEOREM 2.8. *Let $X_n(G)$ be the number of copies of G in $G(n, p(n))$. Then, using $\to$ to denote convergence in distribution,*

(I) *$X_n(G) \to 0$ if $p(n) = o(n^{-1/d^*(G)})$.*
(II) *If $d(H) < d(G)$ for all proper subgraphs of G (that is, G is **strictly balanced**), then for $p = c/n^{1/d^*(G)}$, letting $\mathrm{Aut}(G)$ be the automorphism group of G,*

$$X_n(G) \to \mathrm{Po}(\lambda), \quad \text{where } \lambda = \frac{c^{|E(G)|}}{|\mathrm{Aut}(G)|}.$$

(III) *If $1/n^{1/d^*(G)} = o(p(n))$ and $n^{-2} = o(1 - p(n))$, then $X_n(G)$ has asymptotically normal distribution.*

Since subgraphs with a high density of edges will tend to occur in clusters, it is natural to ask whether a compound Poisson approximation might be better: (Stark, 2001) gives a result on this for certain balanced graphs.

Another key milestone is when $p = \log(n)/n$.

THEOREM 2.9. *Suppose $p(n) = (\log(n) + c + o(1))/n$. Then*

$$\lim_{n \to \infty} P\{G(n, p(n)) \text{ is connected}\} = e^{-e^{-c}}.$$

This is also the threshold for not having any isolated vertices. Indeed, the giant component has by now absorbed all small components except possibly some isolated vertices, so the graph is connected essentially if and only if it has no isolated vertex. (It also turns out to be the threshold for possessing a perfect matching if n is even.)

Only slightly more edges are required for a Hamiltonian cycle, a result of Bollobás and independently of Komlós and Szemerédi. See (Bollobás, 1985), Chapter VIII.

THEOREM 2.10. *Let $p(n) = (\log(n) + \log\log(n) + c)/n$. Then*

$$P\{G(n, p) \text{ is Hamiltonian}\} = e^{-e^{-c}}.$$

This says that having minimum degree δ at least 2 (obviously necessary for a Hamiltonian cycle) is also **whp** sufficient. The generalisation that in a graph process when the minimum degree is k the graph **whp** has $\lfloor k/2 \rfloor$ edge-disjoint Hamiltonian cycles was proved in (Bollobás and Frieze, 1985).

We now move on to the situation when the graph is connected, where there are at least four measures of connectivity; the minimum degree δ, the edge-connectivity λ, the vertex connectivity κ and Fiedler's algebraic connectivity μ (the second smallest eigenvalue of the Laplacian of the graph). The inequalities $\delta \geqslant \lambda \geqslant \kappa$ are easily proved, and any such triple is attained by some graph, but typically the three measures are equal.

THEOREM 2.11. *Let $p = p(n)$, $0 < p < 1$. Then* **whp** *$G(n, p(n))$ has $\kappa = \lambda = \delta$.*

In fact, one can prove much more precise results in terms of hitting times: see (Bollobás, 1985), Chapter VII. When p is constant, one can show that all these invariants are **whp** equal to $pn + o(n)$: then μ (which is always $\leqslant \kappa$) is also close to pn (see Juhász, 1991).

We next consider largest independent sets: their orders, by results of Matula and Frieze, turn out to be very tightly concentrated. Let $b = 1/(1-p)$.

THEOREM 2.12. *For $0 < p < 1$ constant,* **whp** *the independence number α of $G(n, p)$ satisfies*

$$\left| \alpha - 2\log_b(n) + 2\log_b \log_b(n) - 2\log_b\left(\frac{e}{2}\right) - 1 \right| < \frac{3}{2}.$$

For $np = c$ and $\varepsilon > 0$, if c is $o(n)$ but $\geqslant c_\varepsilon$ for some sufficiently large constant c_ε, then **whp**

$$\left| \alpha - \frac{2n}{c}\left(\log(c) - \log\log(c) - \log(2) + 1\right) \right| \leqslant \frac{\varepsilon n}{c}.$$

The proof is in Chapter IX of (Bollobás, 1985). Of course, these imply results for the clique number $\omega(G)$ by considering complements.

An obvious lower bound for the chromatic number of G is n/α (since the colour classes are independent sets). It turns out that in $G(n, p)$ this is asymptotically the right answer.

THEOREM 2.13. *Let $0 < p < 1$ be fixed. Then* **whp**

$$\chi(G(n, p)) \sim \frac{n}{2\log_b(n)}.$$

We refer to (Karoński, 1995) for an overview of Bollobás' proof, which makes heavy use of martingale arguments. There is an alternative proof using the Janson inequalities (see Alon and Spencer, 1992). (McDiarmid, 1990) refines the result to

$$\chi(G_{n,p}) = \frac{n}{2\log_b(n) - 2\log_b\log_b(n) + O(1)} \text{ whp}.$$

For a generalisation, see (Bollobás and Thomason, 2000).

We refer to (Bollobás, 1985) and (Janson et al., 2000) for discussion of several interesting topics not discussed in detail here.

3. Applications of the Erdős–Rényi model

Random graphs have for long been used as models of typical behaviour of actual networks. Here we deliberately restrict ourselves to discussing two less well known problems. Section 7.14 is also relevant here.

3.1. Epidemics

In (Barbour and Mollison, 1987), it is shown that $G(n, p)$ is essentially equivalent to a standard elementary model of an epidemic called the **Reed–Frost** model, a Markov chain with states $\{(i, r): i, r \geq 0, i + r \leq n\}$ where, for $0 \leq j \leq n - i - r$,

$$P\{(i, r) \to (j, i + r)\} = \binom{n - i - r}{j}\left(1 - (1 - p)^i\right)^j (1 - p)^{i(n - i - r - j)},$$

where i is the number of infected individuals, r the number removed. The construction is as follows; we take one or more vertices as the initial infected individuals, numbering $i(0)$; their neighbours in the graph are the $i(1)$ infectives at time 1, and then the infectives at time 0 are removed. In general, $i(t + 1)$ is the number of neighbours of the $i(t)$ infectives at time t who have not previously been infected. Note that this set-up emphasises the role of individuals, and the lists L_a of those infected by a particular individual a. This process appears to yield a directed graph; however, in such an epidemic, at most one of the two events a infects b and b infects a can occur, so we can essentially forget about orientation, and arrive at the (undirected) random graph $G(n, p)$. For more details of the construction see (Barbour and Mollison, 1987).

Because the graph is a model of the Reed–Frost process, it is now immediate from the results on existence of giant components that there is a genuine epidemic (as opposed to small clusters of disease) if and only if $p(n) \geq c/n$ with $c > 1$, and that if there is a genuine epidemic, the number not affected by the illness (i.e., the number of isolated vertices) is asymptotically Poisson. In fact Barbour and Mollison (1987) suggests that the interaction between the subjects works both ways: the conjecture that the distance from a random point in the giant component to one of the vertices furthest from it is $k \log(n) + O(1)$, with variability confined to the $O(1)$ term, is suggested more immediately from the epidemic viewpoint than the classical random graphs viewpoint.

Another way to use random graphs to model the spread of disease is to say that the social network of contacts between individuals is a graph G, so that now the edges of the graph represent relationships between individuals rather than possible channels for disease transmission. We then run some epidemic process on G, and ask what the typical behaviour is when G is a random graph from some model. For example, one could choose $m = n\mu$ initial infectives at random from the population of size n, say that each infectious individual remains so for a time period with exponential distribution with parameter γ, during which he makes 'close contacts' with each of his neighbours at the points (in time) of a Poisson process with parameter λ. If an individual thus contacted is still susceptible, he immediately becomes infectious. After the infectious period, the individual recovers and becomes immune to further infection. Note that in this process, the per capita infection rate is large, but only a few individuals may be contacted, whereas in the Reed–Frost model described above the infection rate is low but many individuals may be contacted. If this process takes place on a $G(n, \beta/n)$, then (for large n) τ, the proportion of the population who are eventually infected (including those initially infected), satisfies the equation

$$1 + \mu - \tau = \exp\left(-\frac{\beta\lambda\tau}{\lambda+\gamma}\right).$$

See (Andersson and Britton, 2000) for more details, and some other graph-theoretic models.

3.2. Evolutionary conflicts

See (Cannings, 1997) for a more detailed account of this subject. Let

$$\Delta_k = \left\{(s_1, \ldots, s_k) \text{ such that } s_i \geq 0 \; \forall 1 \leq i \leq k \text{ and } \sum_{i=1}^{k} s_i = 1\right\}.$$

If A is a real k by k matrix, an **evolutionarily stable strategy (ESS)** of A is a $\mathbf{p} \in \Delta_k$ such that

(1) $\mathbf{p}^T A \mathbf{p} \geq \mathbf{q}^T A \mathbf{p} \; \forall \mathbf{q} \in \Delta_k$.
(2) If $\mathbf{q} \in \Delta_k$, $\mathbf{q} \neq \mathbf{p}$ and $\mathbf{p}^T A \mathbf{p} = \mathbf{q}^T A \mathbf{p}$ then $\mathbf{p}^T A \mathbf{q} > \mathbf{q}^T A \mathbf{q}$.

The context is that of a biological population in which individuals compete for limited resources, with a finite number of different **pure strategies** available to each individual, and an individual playing strategy i against one playing strategy j receives a payoff a_{ij}. An ESS $\mathbf{p}$ is a specification of a mean strategy played by the population which does not allow any alternative strategy to invade. Define the **support** $R(\mathbf{s})$ of $\mathbf{s} \in \Delta_k$ to be $\{i : s_i \neq 0\}$, $1 \leq i \leq k$. Then the **pattern of ESSs** for A is the set of supports of the ESSs of A; it is a set of subsets of $\{1, 2, \ldots, k\}$ subject to certain combinatorial restrictions: for example, it is a Sperner family. We would like to classify what Sperner families can be the pattern of ESSs of some matrix A; here is one way to guarantee the existence of such an A.

THEOREM 3.1. *Suppose A is an n by n real symmetric matrix with $a_{ij} = 0$ for $i = j$, and all other $a_{ij} = \pm 1$. Define a labelled graph $G(A)$ on $\{1, 2, \ldots, n\}$ with an edge between i and j if and only if $a_{ij} = 1$. Then the set of supports of ESSs of A is the same as the set of cliques of $G(A)$.*

PROOF. See (Vickers and Cannings, 1988). □

Of course the cliques of a graph are in general hard to compute (though less difficult than finding general ESSs). However, if we can comment on the distribution of the number of cliques in some model of random graphs, this will lead to a result about probable patterns of ESSs for the corresponding matrices. Here, the results for $G(n, p)$ imply that for most such A the ESSs have support of size at most $2\log_{1/p}(n)$. The technique could also be useful for small values of n in that, if we can rule several patterns out by theoretical arguments, and then generate the others using random graphs in this way, we have then completed the classification of what patterns are attainable for that value of n.

4. Geometric random graphs

4.1. Basics

In this section we discuss geometric random graphs and their uses. A fuller and more general account of this topic will appear in the forthcoming book (Penrose, 2003).

Recall the basic idea. $X_1, X_2, \ldots$ are independent random variables uniformly distributed in the unit cube $[0, 1]^d \subseteq \mathbf{R}^d$. We assume $d \geqslant 2$ until further notice. Let (ρ_n) be a sequence of non-negative numbers. The **random geometric graph** $RGG(n, \rho_n)$ has vertices $X_1 \ldots X_n$ and an edge between X_i and X_j if and only if $d(X_i, X_j) \leqslant \rho_n$. Here d is some chosen metric: we will consider only l_p norms, that is

$$\|\mathbf{z}\|_p = \left(\sum_{i=1}^{d} z_i^p\right)^{1/p}$$

(including the l_∞ norm $\|\mathbf{z}\|_\infty = \max\{z_1 \ldots z_d\}$). We can consider the evolution of the graph as ρ_n increases.

Appel and Russo (1997a, 1997b) consider the minimum and maximum degrees of these graphs in the l_∞ norm. Note first that if

$$\rho_n = \left(\frac{b\log\log(n)}{n^2}\right)^{1/d}$$

then for $b > 2$, **whp** the graph has at least one edge: on the other hand, if $\rho_n = 1 - r\log\log(n)/n$, then if $r > 1/d$ G is incomplete **whp**, but if $r \leqslant 1/d$, G is complete **whp**. (Related criteria in terms of limits of series are also given.)

Now let

$$H(t) = \frac{\log(t)}{t} + \frac{1}{t} - 1, \quad 0 < t < \infty.$$

THEOREM 4.1. *Suppose that*

$$\lim_{n \to \infty} \left(\rho_n^d \frac{n}{\log n} \right) = c \in (0, \infty].$$

Then if $c < 1/(2d)$, **whp** $\delta = 0$. *If* $c = 1/(2d|H(2/a)|)$, *then* **whp**

$$\liminf_n \frac{\delta}{n\rho_n^d} \geqslant a.$$

A result of Appel and Russo shows that the threshold for having no isolated vertex is the same as the threshold c_n for connectedness, in the sense that $\lim_{n \to \infty} c_n/d_n = 1$. This does use $d \geqslant 2$: for $d = 1$, the two thresholds can be different. In (Penrose, 1999) this is generalised by showing that for any $d \geqslant 2$, $1 < p \leqslant \infty$ and $k \geqslant 0$, in an RGG with l_p norm the hitting times for $\kappa = k$ and $\delta = k$ are equal. This of course generalises a result for the Erdős–Rényi model: we see later similar results for random regular graphs and random subgraphs of the n-cube. The phenomenon is not universal: for example, in $G_{p,0}$ clearly δ is approximately $pn/2$ but $\kappa = 0$ **whp**, and it also fails for random graphs on the circle. It would be interesting to understand more generally when $\delta = \kappa$, perhaps in terms of a suitable measure of whether or not the graph tends to evolve in dense clumps which only gradually merge.

We now return to the l_∞ results of (Appel and Russo, 1997b), and examine the maximum degree Δ of the graph.

THEOREM 4.2. *Suppose that*

$$\lim_{n \to \infty} \frac{n\rho_n^d}{\log(n)} = c \in (0, \infty].$$

Let $b = \limsup_n \rho_n$ *and* $a(c)$ *be the unique root in* $[1, \infty)$ *of*

$$a\log(a) - a + 1 = \frac{1}{2^d c}$$

(with $a(\infty) = 1$*). Then,* **whp**

$$\min\{2^d, b^{-d}\} \leqslant \liminf_n \frac{\Delta}{n\rho_n^d} \leqslant \limsup_n \frac{\Delta}{n\rho_n^d} \leqslant a(c)2^d.$$

In particular, if $c = \infty$ *and* $b \leqslant 1/2$, **whp**

$$\lim_{n \to \infty} \frac{\Delta}{n\rho_n^d} = 2^d.$$

The theorem, together with the easy facts that for any graph G

$$\omega(G) \leqslant \chi(G) \leqslant \Delta(G) + 1$$

and the fact that

$$\Delta\big(RGG(n, l_\infty, \rho_n)\big) \leqslant \omega\big(RGG(n, l_\infty, 2\rho_n)\big) - 1$$

allows one to deduce that, for ρ_n such that

$$\lim_{n\to\infty} \frac{n\rho_n^d}{\log(n)} = c \in (0, \infty)$$

the order (in n) of each of Δ, ω and χ is $\log(n)$ (with different constants). We refer to (Appel and Russo, 1997b) for more details.

4.2. Spanning trees

A variant on the above is to say that all edges between the n points exist, and ask about the minimum length spanning tree of the graph. Let L_n denote the length of the longest edge in this. In (Penrose, 1997) it is shown that

$$\lim_{n\to\infty} P\{nL_n^2 - \log(n) \leqslant c\} = e^{-e^{-c}}.$$

If instead the points are from a standard d-dimensional normal (that is, the coordinates are independent $N(0, 1)$ random variables) then

$$\lim_{n\to\infty} P\{(2\log(n))^{1/2} L_n - b_n \leqslant c\} = e^{-e^{-c}},$$

where the b_n are explicit constants asymptotic to $\log\log(n)$.

Results on the structure of the minimum spanning tree are likely to be of interest in statistics, since given two samples $\{x_1, \ldots, x_m\}$ from a random variable with density f and $\{y_1, \ldots, y_n\}$ from a random variable with density g, there are grounds for believing that statistics which reject H_0: $f = g$ for small values of $R_{m,n}$, the number of edges in the minimum spanning tree whose endpoints are in different samples, are likely to be powerful against general alternative hypotheses. See (Penrose and Henze, 1999), where it is proven that $R_{m,n}$ has several statistically desirable properties.

4.3. Applications to layout problems on graphs

By a **layout problem** we mean a problem of ordering the vertices of a graph so that adjacent vertices are close in the ordering. For a given layout ϕ let $\sigma(e, \phi)$ be the modulus of the difference between the two integers associated with the two endpoints. Some examples are:

- The minimum bandwidth problem (MBW): find $\min_\phi \max_{e\in E} \sigma(e, \phi)$.

- Minimum linear arrangement problem (MLA): find $\min_\phi \sum_{e \in E} \sigma(e, \phi)$.
- Minimum bisection problem (MBIS): partition the vertices into two equally sized sets so as to minimize the number of edges between them.

In general such problems are hard: thus we want methods which will hopefully give a good approximation to the answer quickly. One way to test such a method is on random graphs. It turns out in this context that, in the Erdős–Rényi model, all orderings have roughly the same cost **whp**, so we instead turn to geometric graphs: we denote the MBW problem for a $RGG(n, \rho_n)$ by $\mathrm{MBW}(RGG(n, \rho_n))$, and similarly for the other problems discussed above. The idea is to take (ρ_n) such that either

$$\lim_{n \to \infty} n\rho_n^d = \lambda \in \mathbf{R}$$

or this limit is infinity. The result is as follows.

THEOREM 4.3. *There exists a constant K such that, with failure probability exponentially small in $n\rho_n$,*

$$\mathrm{MBW}(RGG(n, \rho_n)) \leqslant Kn\rho_n$$

and, with failure probability exponentially small in $\rho_n^{(1-d)/2} \log(\rho_n)^{-2}$,

$$\mathrm{MLA}(RGG(n, \rho_n)) \leqslant Kn^3 \rho_n^{d+1},$$
$$\mathrm{MBIS}(RGG(n, \rho_n)) \leqslant Kn^2 \rho_n^{d+1}.$$

We refer to (Penrose, 2000) and (Diaz et al., 2000) for more details of where such results are applied, and similar results for other layout problems.

4.4. Geometric graphs and cluster analysis

Suppose we are given a random sample of size n from some (now possibly 1-dimensional) distribution. We generate a graph by saying that they are the vertices, and that two are adjacent if and only if they are within distance $d(n)$ of each other. The objective is to produce the components of the graph. There is an obvious application to statistical cluster analysis. For general information on this, see (Godehardt, 1990b).

In most of the original papers, the graph thus generated is called a random interval graph, but we prefer to use that term for a slightly different notion discussed later.

Godehardt and Harris (1997) and Godehardt and Jaworski (1996) discuss a number of results on the case where the random sample is from the uniform distribution on $(0, 1)$, giving asymptotic results on the number of edges, number of complete graphs of order m, number of components of various kinds (including isolated vertices) and the degrees of vertices. McColm (2003) discusses existence of thresholds for increasing properties in this model, imitating Bollobás and Thomason's result for $G(n, p)$, though the situation here is more complex.

In practice, we do not want to use just one measure of similarity, but several, and are thus led to consider random multigraphs rather than merely random graphs. Godehardt (1990b) discusses such extensions.

4.5. Geometric random graphs on the circle

Other models with a geometric flavour can be considered. For example, we could have n points uniformly distributed on a circle of circumference 1, two being adjacent if and only if the distance between them (along the circumference rather than the chord) is less than $a(n)$. Most of the results which follow are from (Maehara, 1990) (see also Chapter 7 of Barbour et al., 1990) or from (Godehardt and Jaworski, 1996). The probability that two vertices are adjacent is then easily seen to be $2a(n)$. Let W be the number of spacings between the vertices, considered in consecutive order round the circle, which are greater than $a(n)$ in length. Then the number of components of the graph is W, except that when there are no such spacings the graph is connected. One can obtain a Poisson approximation for W: this gives that the threshold for connectedness is $a(n) = (\log(n) - \log(\lambda) + o(1))/n$. At the threshold, note that **whp** the graph is connected if and only if it has a Hamiltonian path, and the proportion of the connected graphs which have a Hamiltonian cycle is at least $\lambda/(1+\lambda)$ rather than the fraction $\lambda/(1+\lambda)$ given at present. Note that, in this model, the threshold for no isolated vertex is half the threshold for connectedness (which is why the hitting times of $\delta = k$ and $\kappa = k$ differ in this model). The point in the evolution at which any fixed graph first appears as a subgraph is also understood. These ideas also show that the graphs are perfect provided $a(n) = o(\log(n)/n)$, but for large $a(n)$ the chromatic number can substantially exceed the clique number. The first-order logic properties of this model are studied in (McColm, 1999).

5. Random cluster models

5.1. Basics

Recall that here, that given a graph $G = G(V, E)$ with vertex set $V = \{1, 2, \ldots, n\}$, the probability that some set $A \subseteq E$ of edges is exactly the set of edges of the random graph is

$$\mu(A) = \frac{p^{|A|}(1-p)^{|E|-|A|}q^{c(V,A)}}{\sum_{A \subseteq E} p^{|A|}(1-p)^{|E|-|A|}q^{c(V,A)}}.$$

We call the measure μ the random cluster measure. Such measures arise naturally in consideration of polymers, where there is a tendency for molecules to coagulate into large complex molecules, giving few components in the resulting graph.

Note that if $q = 1/k$, k a positive integer, we can think of it as the probability that, if we give each component of $G(n, p)$ one of k colours, we get a particular pattern of coloured components.

One of the principal problems with calculating μ is the fact that we need to understand the denominator term (which we refer to as Z). If $q = 1$, of course $Z = 1$, but in general it is more complex: one can easily show that the value of Z is closely related to the so-called Tutte polynomial of the graph G: see (Welsh, 1994), Proposition 2.1. It is known that the Tutte polynomial is usually hard to compute, so the same holds for Z.

The random cluster measure μ can be obtained as the stationary distribution of a reversible Markov chain. In this chain, at each stage, one selects $e \in E$ at random and (irrespective of whether or not the edge has actually arisen) lets it be present with probability p if its end vertices are connected through the rest of the graph, and with probability $p/(p + q(1 - p))$ otherwise.

In addition to μ, we are also interested in

$$\lambda(A) = \sum_{X: E \supseteq X \supseteq A} \mu(X)$$

which is the probability that the set A of edges is open (possibly along with other edges). The following theorem is a consequence of the FKG inequality.

THEOREM 5.1. *For fixed p, the function λ is a monotone non-increasing function of q for $q \geqslant 1$.*

See (Welsh, 1994) for the proof and the conjecture that the result extends to $q < 1$. A partial result in this direction is in (Grimmett, 1995).

5.2. The phase transition in random cluster models

The following results are for the mean field case, i.e., when $G = K_n$. They are from (Bollobás et al., 1996a), where the model is called $G(n, p, q)$, but we will call it $RCG(n, p\ q)$ to avoid confusion with notation introduced later. We then have the following result, analogous to the phase transition in the Erdős–Rényi model.

THEOREM 5.2. *Consider $RCG(n, \lambda/n, q)$ where $q > 0$ is constant (the case $q = 0$ is degenerate). Define*

$$\lambda_c(q) = \begin{cases} q, & \text{if } 0 < q \leqslant 2, \\ 2(q-1)/(q-2) \cdot \log(q-1), & \text{if } q > 2. \end{cases}$$

Define also $\theta(\lambda)$ to be 0 for $\lambda < \lambda_c(q)$, and to be the largest root of

$$e^{-\lambda \theta} = \frac{1 - \theta}{1 + (q-1)\theta}$$

otherwise. Then, using $X_n = O_p(f(n))$ to mean that, for any function $\omega(n)$ tending to infinity with n, we have

$$\lim_{n \to \infty} P\{|X_n| \leqslant f(n)\omega(n)\} = 1$$

and defining $X_n = o_p(f(n))$ similarly, we have the following.

(1) If $\lambda < \lambda_c(q)$, **whp** the graph consists of components with at most one cycle, the largest component having order

$$\frac{1}{-\log(\lambda/q) - 1 + \lambda/q} \log(n) + \mathrm{O}_p\big(\log\log(n)\big).$$

The number of edges is $\lambda n/(2q) + \mathrm{o}_p(n)$.

(2) If $\lambda = \lambda_c(q)$ and $1 \leqslant q \leqslant 2$, **whp** the graph consists of components with at most one cycle, and a bounded number of components with more than one cycle. The largest component has order $\mathrm{o}_p(n)$ and the largest tree $\mathrm{O}_p(n^{2/3})$. If however $q > 2$, then **whp** the graph is either like the graphs in (1) above or like those in (3) below. (It is not known if both cases are possible, or what happens if $q < 1$.)

(3) If $\lambda > \lambda_c(q)$, **whp** $RCG(n, p, q)$ has a giant component of order $\theta(\lambda)n + \mathrm{o}_p(n)$. The other components have at most one cycle: the largest tree components have order $\alpha \log(n) + \mathrm{o}_p(\log(n))$, where

$$\alpha^{-1} = -\log(\beta) + \beta - 1 \quad \text{and} \quad \beta = \lambda\big(1 - \theta(\lambda)\big)/q.$$

The remaining components (whose number is bounded) are unicyclic. The number of edges is $\lambda(1 + (q-1)\theta(\lambda)^2)n/(2q) + \mathrm{o}_p(n)$.

The results in (Bollobás et al., 1996a) also include a large deviations result for the number of components in $RCG(n, \lambda/n, q)$ (this result was new even for $q = 1$).

5.3. Random triangle model

It is natural to consider modifications of the idea of the random cluster model, where instead of biasing by $q^{c(V,A)}$ one biases by some other quantity instead. A first step in this direction has recently been taken in (Jonasson, 1999). Here the factor $q^{c(V,A)}$ is replaced by a factor $q^{t(V,A)}$ where $t(V, A)$ denotes the number of triangles in the graph $G(V, A)$, so that we now have

$$\mu_t(A) = \frac{p^{|A|}(1-p)^{|E|-|A|}q^{t(V,A)}}{\sum_A p^{|A|}(1-p)^{|E|-|A|}q^{t(V,A)}}.$$

This is called the random triangle model. If $q \geqslant 1$, this favours graphs with many triangles. One motivation for the work was the fact that social networks exhibit a large degree of transitivity (a friend of a friend of mine is more likely to be a friend of mine than someone plucked at random from the population). Note that in this model, unlike the random cluster model, whether or not an edge is present will depend only on edges incident with it.

Again the distribution arises as the stationary distribution of a suitable Markov chain: this time, if the set of edges already present is Y, one chooses $e \in E$, and lets the corresponding edge be present with probability

$$\frac{pq^{\Delta(Y,e)}}{pq^{\Delta(Y,e)} + 1 - p},$$

where $\Delta(Y, e)$ is the number of triangles in $Y \cup \{e\}$ of which e is an edge.

In (Jonasson, 1999), it is shown that, in the mean-field case $G = K_n$, then if $n^{-2} \leqslant p(n) \leqslant n^{-\alpha}$ for some $\alpha > 0$ and $q = 1 + h_1(n)/n$ where $\log(n) = o(h_1(n))$, then the tendency for friends of friends to become friends takes off explosively: **whp**, the graph is K_n. However, if $q = 1 + h_2(n)/n$ where $h_2(n) = o(\log(n))$, and $p = O(n^{-\alpha})$ for some $\alpha > 2/3$, the random triangle model behaves **whp** exactly like the $G(n, p)$ model. Some results are also obtained for the intermediate case: crudely speaking, if $p = an^{-2/3}$ and $q = 1 + h_3(n)/n$ where $h_3(n) = o(\log(n))$, the random triangle graph is approximately a $G(n, p)$ plus a $Po(a^3 h(n)/6)$ number of disjoint, uniformly spread out triangles. The idea is also used in (Jonasson and Häggström, 1999) to discuss a percolation problem on the two-dimensional triangular lattice.

6. Random randomly coloured graphs

We first briefly describe properties of RRC graphs in general.

6.1. Properties of RRC graphs

In (Penman, 2003a) various results are presented on the probability of trees and cycles in models $\Gamma(n, k, \mathbf{s}, P)$. Linear algebra arguments are use to show, for example, that cycles of even length are always at least as likely to arise as classically, and the conjecture that any tree is at least as likely to arise as classically is shown to be true as a result of work of Sidorenko. However, in certain models, odd cycles can be less likely to occur than classically. However, this is not true in $G_{p,q}$ with $p > q$, where one can show that the joint probability of any two sets of edges arising is always at least as large as the product of the individual probabilities of these two sets arising. The proof in (Biggins and Penman, 2003b) is, perhaps surprisingly, not based on the FKG inequality.

Penman (2003a) also shows that in any RRC model, the number of edges has a Poisson limit (in the sparse case) and a normal limit (in the dense case). However in some respects the number of edges exhibits different behaviour from classically: Biggins and Penman (2003a) uses the Gärtner–Ellis theorem to obtain the large deviations behaviour. One novelty is that the probability of a large deviation is, for some of the range, only exponentially small in the square root of the number of edges, rather than the full number (this is because a large deviation in the number of vertices of a given colour, which happens with this smaller probability, more or less automatically causes a large deviation in the number of edges).

In the particular case of $G_{p,q}$, more precise results are of course possible. For example, the maximum degree is **whp**

$$\Delta = \frac{p+q}{2} n + \sqrt{p(1-p) + q(1-q)} \sqrt{n \log(n)}$$

(see Penman, 2003b, where further results on connectedness, connectivity, etc. in $G_{p,q}$ are proven). Note that the formula implies that the maximum degree is maximised (for fixed $(p+q)/2$) when $p = q$.

6.2. Applications of RRC graphs

It is clear that $\Gamma(n, k, \mathbf{s}, P)$ corresponds to a multitype Reed–Frost model, where, before the process starts at all, we assign one of k types randomly to each individual, and then states of the process are given by having, for each $1 \leqslant l \leqslant k$, numbers i_l of infectives of type l and r_l of removed individuals of type l (so that the number of susceptibles of type l, which we will here denote z_l to avoid confusion with the s_l, the probability that a vertex receives colour l, is $n_l - i_l - r_l$) and then the system changes states with probabilities given by

$$z_l(t+1) \sim \mathrm{Bin}\left(z_l(t), \prod_{j=1}^{k}(1-p_{lj})^{i_j(t)}\right)$$

with

$$i_l(t+1) = z_l(t) - z_l(t+1) \quad \text{and} \quad i_l(0) = m_l, \quad z_l(0) = n_l.$$

A technical problem is that we cannot insist in advance that we have some number m_l of initial type l infectives, since in principle the colouring process may not give us that number; however, we may well be able to get round this problem, either by having the number of initial infectives of each type to depend on the number of vertices of that type, or (if the number of initial infectives is to be small compared with n) by using the fact that with overwhelming probability more than that number of type l vertices will arise. For example, the spread of a sexually transmitted disease in a largely heterosexual population who mix freely will be better modelled by a graph where there are two roughly equally sized classes and edges between the classes are more likely than edges within either class.

In the example about ESSs, the distribution of clique sizes in the new models may well differ from that in the classical model. For example, Penman (2003b) shows that in $G_{p,q}$ ($0 < p, q < 1$ constant) the largest clique size is asymptotically $2\log_{1/p}(n)$ for $p > q$ (that is, the largest clique in the red or blue vertices is in control) but for $p < q$ the size is bounded above by $\log_{1/\sqrt{pq}}(n)$ so is smaller than classically (by the AM-GM inequality). We can also get more radically different behaviour in general RRC models: for example, if all the edges between vertices of one particular colour arise (with probability 1), we will get a clique whose order is linear in n despite the fact that in $G(n, \alpha)$ the largest clique size would usually be a multiple of $\log(n)$. Whereas in $G(n, \alpha)$ **whp** all cliques are of order between $(1-\varepsilon)\log_{1/p}(n)$ and $(2+\varepsilon)\log_{1/p}(n)$, we can in the RRC case get cliques of orders concentrated around several different multiples of $\log(n)$.

A substantial challenge which remains for RRC graphs is to develop a good theory of graph processes which stand in the same relation to them as the ordinary graph process does to $G(n, \alpha)$.

6.3. The subgraph problem in RIC graphs

We now consider the threshold for the subgraph problem in $G(n, m, p)$. Recall $m = \lfloor n^\beta \rfloor$. In the Erdős–Rényi model $d^*(H) = \max_{L \leqslant H} d(L)$ was critical in determining the threshold for the graph to contain a copy of H **whp**. In this situation, the critical quantity is quite different: it is, very roughly, the maximum, over subsets S of $V(H)$, of the minimum number of complete subgraphs of G required to cover all edges in the subgraph induced by S. Moreover, we can also say when there cease to be **induced** copies of H.

THEOREM 6.1. *Let H be a fixed graph. A clique cover $C = \{C_1, C_2, \ldots, C_k\}$ of H is a collection of complete subgraphs of H such that every edge of H is in at least one of these subgraphs. C is reducible if, for some $C \in C$, all edges induced by C are contained in the union of the edges induced by the other members of C: otherwise it is irreducible. For such C, let $|C| = k$, and $\sum C$ be the sum of the orders of the complete graphs in C. Also, for $S \subseteq V(H)$, let*

$$C[S] = \{C_i \cap S: |C_i \cap S| \geqslant 1\} \quad and \quad C'[S] = \{C_i \cap S: |C_i \cap S| \geqslant 2\}.$$

Define then, for C a clique cover of H and $\emptyset \neq S \subseteq V(G)$,

$$\tau(H, C, S) = n^{-|S|/\sum C[S]} m^{-|C[S]|/\sum C[S]},$$
$$\tau'(H, C, S) = n^{-|S|/\sum C'[S]} m^{-|C'[S]|/\sum C'[S]},$$
$$\tau_1(H) = \max_S \min_C \{\tau(H, C, S), \tau'(H, C, S)\},$$

where if $C'[S] = \emptyset$, we set the corresponding τ' term to be zero. Then, letting $H \leqslant G$ mean that H is an induced subgraph of G, we have:

(1) *If $\lim_{n \to \infty} mp^2 = 0$, then*

$$\lim_{n \to \infty} P\{H \leqslant G(n, m, p)\} = \begin{cases} 0, & \text{if } p/\tau_1(H) \to 0, \\ 1, & \text{if } p/\tau_1(H) \to \infty. \end{cases}$$

(2) *Suppose $\varepsilon \leqslant mp^2 \leqslant 1/\varepsilon$. Then*

$$\lim_{n \to \infty} P\{H \leqslant G(n, m, p)\} = 1.$$

(3) *Suppose $mp^2 \to \infty$, $\omega(n)$ is a function tending to infinity arbitrarily slowly, and $p = \sqrt{(\log(n) + \omega(n))/(md^*(H))}$. Then*

$$\lim_{n \to \infty} P\{H \leqslant G(n, m, p)\} = \begin{cases} 1, & \text{if } \omega(n) \to -\infty, \\ 0, & \text{if } \omega(n) \to \infty. \end{cases}$$

This is Theorem 3 of (Karoński et al., 1999). The derivation of the following corollary to it is not trivial.

COROLLARY 6.2. *For fixed h,*

$$\tau_1(K_h) = \begin{cases} n^{-1}m^{-1/h}, & \text{if } \beta \leqslant 2h/(h-1), \\ n^{-1/(h-1)}m^{-1/2}, & \text{if } \beta \geqslant 2h/(h-1). \end{cases}$$

(The change of regimes as β passes a certain point is very typical of the thresholds obtained in this model.)

In the light of this result, it is natural to ask when particular subgraphs arise in $G(n, m, p)$ and in $G(n, \alpha)$. The table in (Fill et al., 2000), p. 163, shows that thresholds can be greater or smaller than classically. However, for large values of m, the large number of colours dilutes the correlation structure sufficiently that the model is very similar to the classical one. Here is a precise result in this direction.

THEOREM 6.3. *Let $m = \lfloor n^\beta \rfloor$, where $\beta > 6$. Suppose $G(n, m, p)$ is not* **whp** *edgeless or complete; that is, that*

$$\lim_{n \to \infty} p(n) n \sqrt{m} = \infty \quad \text{and} \quad \lim_{n \to \infty} \frac{p(n)\sqrt{m}}{\sqrt{2 \log(n)}} = 0.$$

Then, letting $\mathcal{L}(G_{n,m,p})$ be the law of $G(n, m, p)$ and $\mathcal{L}(G_\alpha)$ be that of G_α, where $\alpha = 1 - (1 - p^2)^m$, we have

$$\lim_{n \to \infty} d_{TV}\big(\mathcal{L}(G(n, m, p)), \mathcal{L}(G_\alpha)\big) = 0.$$

The result is not true in general for $\beta > 3$, but there is some evidence that the early parts of the evolution are very similar for such β. If $\beta < 3$, even the probability of a triangle differs in the RIC graph and the corresponding $G(n, p)$.

One observation worth making is that in $G(n, m, p)$ edges are more likely to arise conditional on others being present, rather than less likely (so, for example, complete graphs of fixed order k appear earlier in $G(n, n^{1/2}, p)$ than in $G(n, \alpha)$).

THEOREM 6.4. *The probability that any set of m labelled edges arises in a random intersection graph is always at least as large as classically.*

The proof is a consequence of facts about dissociated indicator variables: see, for example, (Barbour et al., 1990), Chapter 2.

6.4. Gate matrix layout problems

We refer to (Karoński et al., 1999) for more detailed discussion of this. In a Gate Matrix Layout (GML) design, we have a finite collection of gates (which are conductors or transistor gates) which we think of as vertical lines, and a collection of nets (transistor diffusions) which associate with each other, which we think of as horizontal lines. Any transistor with a given input must be placed over the given horizontal line, and all transistors associated in the same net must be on a common horizontal track. There may be more than one net in the same track provided the nets do not overlap.

This can always be achieved by having a separate track for each net, but this would be highly wasteful. Thus we want non-overlapping nets to share tracks. The question is than what ordering of the gate lines allows us to minimise the number of tracks required to lay out the circuit.

Mathematically, this is equivalent to having a binary $n \times m$ matrix M with as many rows as there are gates and as many columns as there are nets, with $m_{ij} = 1$ if and only if there is a transistor in net j which must be placed on gate i, and then seeking the column permutation of M which places the nets of 1s densely in their individual rows. Such a gate matrix M corresponds to an intersection graph G (independent of the order of the columns) whose vertices are the set of nets and whose edges are between those nets incident on a common gate.

The GML problem is hard to compute in general. We instead consider the following easier problem: given an $n \times m$ gate matrix M as above, and a fixed k, is there a permutation of the rows of M such that the layout is possible in at most k tracks? If there is such a permutation, we call M a yes instance: otherwise we call M a no instance.

It is a surprising result of Fellows and Langston that this problem is easy – indeed, soluble in time $O(n^2)$. Their proof uses the Robertson–Seymour theory of forbidden minors: a good introduction to that highly technical piece of graph theory is Chapter 12 of (Diestel, 1997). This shows (non-constructively) that there is a finite list of minimal (in a suitably careful sense) obstructions to k-GML. For $k = 1, 2, 3$, there are exactly 1, 2 and 110 respectively such minimal obstructions, known explicitly, but for $k = 4$ there are reasons to expect in excess of 120,000,000 obstructions, and the number seems to grow rapidly with k.

We now randomize the problem by saying that each element of the $n \times m$ matrix is 1 with probability p and 0 with probability $1 - p$, independently of all other elements. It turns out that there is a threshold value of p above which the matrix almost certainly is a no instance for k-GML, and below which it **whp** is a yes instance.

It is not hard to show that the obstruction of smallest order to to k-GML is K_{k+1}. Thus, the threshold for getting a K_k in a RIC graph in Corollary 6.2 already gives that we will **whp** need at least k tracks if

$$\lim_{n \to \infty} \frac{p}{f(n, m)} = \infty,$$

where

$$f(n, m) = \begin{cases} n^{-1} m^{-1/k}, & \text{if } \alpha \leqslant 2k/(k-1), \\ n^{-1/(k-1)} m^{-1/2}, & \text{if } \alpha \geqslant 2k/(k-1). \end{cases}$$

However, it is also believed that certain trees on $t_k = (5.3^{k-1} - 1)/2$ are the largest obstructions to k-GML. If this were true, it would mean that the threshold for k-GML is

$$p_k * (n) = \begin{cases} m^{-1/(k+1)}n^{-1} & \text{for } \alpha \leqslant (t_k - 2)(k+1)/((t_k - 1)(k-1)), \\ m^{-1/2}n^{-t_k/(2(t_k-1))} & \text{for } \alpha > (t_k - 2)(k+1)/((t_k - 1)(k-1)). \end{cases}$$

For results in this direction, see (Karoński and Szymkowiak, 2001) where it is shown that, for the 3-GML problem, the threshold is

$$p_3 * (n) = \begin{cases} m^{-1/4}n^{-1} & \text{for } \alpha \leqslant 10/21, \\ m^{-1/2}n^{-11/21} & \text{for } \alpha > 10/21 \end{cases}$$

(in agreement with the above). The preprint (Szymkowiak, 2002) shows that every tree obstruction to k-GML has t_k vertices: thus the remaining challenge is to show that these are the obstructions of largest order.

6.5. Proteins and cluster analysis

Proteins are polymers which play key roles as enzymes in almost all biological processes. The folding of the polymer creates various **domains** capable of interacting with domains on other proteins (as well as with other molecules) A natural way to represent the possible interactions within a set of proteins is a graph whose vertices are the proteins with edges indicating that the two proteins can interact. Extensive tables of such information exist: see, for example, http://pronet.doubletwist.com. A natural random model has each protein receiving a random subset of some set S of possible domains, each protein's set being independent of that of all the other proteins. If two proteins are assumed to interact if they have at least one domain in common, the result is clearly an RIC graph.

A possible extension is to allow two proteins to interact if and only if they have between t and s domains in common, or more generally to say that there is some fixed graph H whose vertex set is the $2^{|S|}$ possible sets of domains and that two proteins interact if and only if there is an edge between the two corresponding vertices of H. (This suggestion is a personal communication of R. Cannings.) One could also allow the interaction (if the two proteins have common domains) to happen with some probability other than 1. All these extensions can be modelled by RRC graphs. (Some results for deterministic intersection graphs on the case where t common elements are required follow from known results: see Prisner, 2001 for some references.)

Similar problems arise in cluster analysis, where we want to partition a set of objects into subsets, the objects in each subset being similar. Since we do not know which subgroup each object belongs to (or indeed how many subgroups there are), we commonly measure various characteristics of the objects (if there are m such characteristics, we thus generate an m-vector) and declare that two objects are in the same class if the m-vectors are close in a suitable sense. In the case where these m

characteristics are independent binary variables, and the proximity criterion is that they have the same value in at least one of the m places, RIC graphs are clearly relevant.

In practice, some of the measured characteristics are likely to be correlated with each other. Perhaps the most tractable looking step beyond independence is the following (still probably not very realistic for applications). Think of the intersection graph as a bipartite graph, with vertex classes $A = \{1, 2, \ldots, m\}$ and $V = \{1, 2, \ldots, n\}$, with an edge between $i \in A$ and $v \in V$ if and only if $i \in S_v$. Note that in this model of random bipartite graphs, each edge is independent of each other. J. Jaworski has introduced a model of random bipartite graphs where each vertex $v \in V$, independently of the others, chooses its degree $d(v)$ first (obtaining degree i with probability p_i) and then chooses a set of $d(v)$ vertices to be adjacent to uniformly and at random from all $d(v)$-element subsets of A, so that the edges are no longer independent of each other. This situation can still be modelled by an RRC graph, and may be tractable.

7. Other models of random graphs

7.1. Introduction

This section contains rather briefer discussion of a number of other models of random graphs. The choice of which topics are discussed here briefly and which in more detail previously is inevitably rather crude and subjective.

7.2. Restricted graph processes

Here we modify the graph process by saying that at each stage the next edge to be added is chosen uniformly and at random from the remaining absent edges, except that certain restrictions exist on the graph to be formed. For example, we could insist that every vertex degree is at most $f(n)$, the so-called f-graph process, or we could insist that the graph contains no copy of one or more graphs. This process is repeated until no more edges can be added without breaking the rules. Typical questions are: what does a typical such graph look like? Is the distribution obtained the uniform distribution on all maximal graphs satisfying the restrictions?

We first consider f-graph processes. In the case when $f(n)$ is a constant f, Ruciński and Wormald (1992) shows that **whp** this process results in a f-regular graph for nf even, and a graph with exactly one vertex of degree $f - 1$ and all others of degree f otherwise. The proof uses the differential equation method. Ruciński and Wormald (1997) discusses results for the case $f = 2$ (when the graph must consist entirely of paths and cycles), showing, for example, that the number of cycles of length l is asymptotically Poisson with mean a certain definite integral: for $l = 3$ this mean is $0.1887\ldots$. Note that this shows that the outcome of this process is different from the uniform distribution on 2-regular graphs, where the corresponding mean number of triangles is $1/6 < 0.1887$. Ruciński and Wormald (2002) shows that if $f \geqslant 3$, the resulting graph is **whp** connected: they conjecture it is in fact **whp** f-connected (for nf even).

It is natural to ask for how long in its evolution the f-graph process is essentially the same as the Erdős–Rényi graph process: (Balińska et al., 1991) contains some (often numerical) discussion on this.

A slightly different aspect is discussed in (Aldous, 1998). We start with $G(n, t/n)$ but freeze the size of components when it exceeds a threshold $w(n)$ for which $\lim_{n\to\infty} w(n) = \infty$ but $\lim_{n\to\infty} w(n)/n = 0$. Let $\mathcal{C}$ be the size of the component containing a given vertex and Z be the time at which that component first exceeds the threshold. (We think of t as time.) Then he conjectures that (as $n \to \infty$)

(1) $\mathcal{C}$ has the law of the a branching process in which the number of offspring of each individual is Poisson with mean 1, conditional on the non-extinction of the branching process;
(2) Z has density x^{-2} for $1 \leqslant x \leqslant \infty$ (and zero elsewhere);
(3) $\mathcal{C}$ and Z are independent.

We now turn to results where certain subgraphs, or classes of subgraphs, are prohibited. Aldous (1990) discusses a graph process where no cycle is allowed to arise: of course this ends up yielding a tree. He gives an explicit expression for the asymptotic degree distribution: in particular, this shows that the final distribution on trees is not the uniform distribution.

Erdős et al. (1995) gave results about the typical behaviour of the maximal graphs with the restrictions that the graph be triangle free or bipartite, and Bollobás and Riordan (2000) gives results for C_4-free and K_4-free graphs. Osthus and Taraz (2001) gives bounds which hold **whp** when the forbidden graph is strictly 2-balanced (a class which includes complete graphs and cycles).

Sometimes more detailed information can be obtained about the class of graphs for which the result fails: for example, the following result from (Prömel et al., 2001) about the structure of random bipartite graphs. Such a graph is **whp** triangle-free, by an earlier result of Erdős and Rothschild. Those which are not triangle free can **whp** be made triangle free by the removal of one vertex. Those which cannot be made triangle free by the removal of one vertex can **whp** be made triangle-free by the removal of two vertices, and so forth. Similar results are proved in (Prömel et al., 2001) linking r-partite graphs with K_r-free graphs.

7.3. Directed graphs

An obvious model of random directed graphs $D(n, p)$ has vertex set $\{1, 2, \ldots, n\}$ and a directed edge $i \to j$ with probability p, independently of all other pairs. We are interested in strongly connected components of $D(n, p)$, where one can get from any vertex to any other by going along directed edges in the legitimate direction. Łuczak and Karp independently observed the existence of a phase transition in this model: if $pn = c < 1$ all components are cycles whose length is bounded in probability, but if $pn > 1$ there is a giant component of size at least $\beta(c)n$ and some small components. Results on the existence of a giant component in a model where with probability r/n there is an arc $i \to j$ of the graph, with probability s/n there is an arc $j \to i$, with probability t/n there both these arcs are present and with probability $1 - (s + r + t)/n$ there is no directed edge linking the two, are described in (Łuczak, 1994).

Generalising the notion of RRC graphs, we can define a model of tournaments $\tau(n, k, \mathbf{s}, P)$ of random tournaments, where given n labelled vertices and k colours, vertices are coloured independently, receiving colour i with probability s_i, and the probability that an edge between a vertex of colour i and one of colour j goes $i \to j$ is p_{ij} (so $p_{ij} = 1 - p_{ji}$ $\forall i, j$; in particular each $p_{ii} = 1/2$). Thus in any $\tau(n, k, \mathbf{s}, P)$, the probability $\sum_{i,j=1}^{k} s_i p_{ij} s_j$ that the edge between i and j goes $i \to j$ is still $1/2$ so differences from the classical model reflect correlation structure; it may be helpful to think of the tournament as having groups of players of roughly equal strength, but with some groups being at an advantage or disadvantage against other groups. Penman (2003a) gives results on when given cycles are more or less likely than classically, showing, for example, that it can depend on the length of the cycle modulo 4. Some of these results apply to more general directed graphs.

A principal application of directed graphs is to food webs and phylogeny in ecology: a recent survey of random theory here is (Cohen, 1995).

7.4. Random interval graphs

These were introduced in (Scheinerman, 1988). A graph G is said to be an interval graph if one can attach intervals $I_v = [a_v, b_v] \subseteq \mathbf{R}$ to each vertex v in such a way that v and w are adjacent if and only if $I_v \cap I_w \neq \emptyset$. This is quite a strong condition, and interval graphs have good properties: for example, they are perfect (and so finding their clique number and chromatic number is easy). They often arise in applications.

It is natural thus to consider models of random interval graphs. One obvious approach is to generate $2n$ random variables $X_1, Y_1, X_2, Y_2, \ldots, X_n, Y_n$ which are independent and uniformly distributed on $[0, 1]$, and consider the intersection graph on $\{1, 2, \ldots, n\}$ with $I_i = [X_i, Y_i]$ (or $[Y_i, X_i]$ if $X_i > Y_i$). This turns out to be equivalent to selecting n intervals all of whose endpoints are distinct elements of $\{1, 2, \ldots, 2n\}$ and putting the uniform distribution on that space. Scheinerman (1988) proves that almost all graphs in this model have $n^2/3 + \mathrm{o}(n^2)$ edges.

The degrees take a much wider range of values in this model: for example, with probability **exactly** $2/3$ the maximum degree is n, and **whp** it is $\geqslant n - \omega(n)$ for $\omega(n)$ tending to infinity arbitrarily slowly. On the other hand the minimum degree δ satisfies

$$\lim_{n \to \infty} P\{\delta < k\sqrt{n}\} = 1 - \mathrm{e}^{-k^2/2}$$

which implies that the mean of δ is asymptotically $\sqrt{\pi n/2}$.

Almost all interval graphs are Hamiltonian. The clique number (and hence, as they are perfect, the chromatic number) is **whp** $n/2 + \mathrm{o}(n)$. The independence number is **whp** $2\sqrt{n/\pi} + \mathrm{o}(\sqrt{n})$: this is a sharpening from (Fernandez de la Vega, 2000) of results of (Scheinerman, 1988). Scheinerman (1990) gives results on the evolution of random graphs in this framework.

Another model of random interval graphs arises when one considers the vertices of the graph to be customers in some queueing system, and the interval attached to each vertex is the time interval they spend in the system. Some results on this are given in (Pippinger, 1998): of course in this situation even the number of vertices is random.

The term random interval graph is also sometimes used for a class of graphs we discuss in Section 4.4.

7.5. Random subgraphs of the cube, and generalisations

In this subsection only, the convention that n is the number of vertices is temporarily abandoned. The n-dimensional cube Q^n is the graph whose vertices are the subsets of $\{1, \ldots, n\}$, two such vertices being adjacent if and only if their symmetric difference has size 1. Clearly Q^n is an n-regular n-connected graph of diameter n. A random subgraph of it is obtained by keeping each edge with probability p, independent of all other edges. One can also add the possible edges in one by one as a graph process. In (Bollobás et al., 1996b), it is shown that **whp** the hitting time for minimum degree k is equal to the hitting time of vertex connectivity k (in fact a stronger result is proven). This generalises a result in the earlier paper (Bollobás et al., 1995), where the diameter is also discussed. The proofs depend on properties of the giant component of the graph.

The total weight of the minimum spanning tree if each edge of Q_n independently is assigned a uniform weight is discussed in (Penrose, 1998b).

Generalisations are considered in (Reidys, 1997) and (Reidys, 2000). The first of these deals with the generalized n-cube over an alphabet with α elements. The vertices are n-tuples $(x_1, \ldots, x_n)$, where $x_i \in \{1, 2, \ldots, \alpha\}$, for each $1 \leqslant i \leqslant n$. Two vertices are adjacent if they differ in exactly one coordinate. Each vertex of the graph is selected independently with probability $p(n)$, and then the subgraph h induced by the selected vertices is considered. The first theorem shows that for $p(n) = c \log(n)/n$ there exists a unique largest component which contains almost all vertices of H and that the size of the second largest component is $Cn/\log(n)$ where $C > 0$. A further theorem gives information about connectivity and numbers of independent paths between vertices when $p(n)$ is a positive constant. The results are apparently of interest in the study of neutral models of RNA secondary structures in evolutionary biology.

For the second generalisation, recall that if G is a finitely generated group and S is a symmetric generating set with $e_G \notin S$, then $X(G, S)$ is the graph with vertex set G and an edge between x and y if and only if $xy^{-1} \in S$. Its structure depends on both G and S. The n-cube is an example, taking $G = \mathbf{F}_2^n$ and S to be the standard basis. The result of (Reidys, 2000) concerns the largest component of a random subgraph of certain Cayley graphs of a rather restricted class of groups of prime-power order. Let (G_n) be a sequence of such groups, and suppose S_n, the generating set of G_n, is of the form $T_n \cup T_n^{-1}$ where T_n is a minimal generating set of G. Set $s_n = |S_n|$ for convenience: this does not depend on the choice of minimal generating set. We then form a random induced subgraph Γ_n of $X(G_n, S_n)$ by selecting each element of G_n with probability $p(n)$, independently of each other vertex, and considering the subgraph of $X(G_n, S_n)$ induced by the selected elements. The result is that there exists a constant $c > 0$ such that if we take

$$p(n) = c \frac{\log(s_n)}{s_n}$$

then **whp** the largest component of Γ_n contains almost all vertices of Γ_n.

7.6 Small world graphs

These are studied in (Watts, 1999). They attempt to model the rule of thumb that in many situations only about six links are required to connect any two individuals. This first came to prominence when a psychologist asked random people in Kansas to send a parcel, through acquaintances, to random people in Massachusetts: the median path length was six or seven. Much of this work is motivated by attempts to study the world-wide web as a graph. Mathematically, we want to generate graphs with fairly high clustering (i.e., tendency for the adjacency relation to be transitive) and very short paths between individuals. One of the models used is to take an n-cycle, and add edges between every vertex and its k nearest neighbours on either side. Then consider consecutively each edge of the graph, choosing a random vertex of the graph to be one endpoint with probability β and leaving the edge where it is with probability $1 - \beta$. Extensive simulations suggest that $\beta \simeq 0.1$ is one way to create the desired effect. The subject is beginning to develop at a more mathematical level: see the two recent surveys (Cooper and Frieze, 2002) and (Bollobás and Riordan, 2002) and references cited therein for details of this rapidly expanding area. Such models are likely also to be useful in practice: for example, Markov chain Monte Carlo methods have been used to study the spread of disease on such graphs. Barbour and Reinert (2001) discusses a variant of this idea.

7.7. Random hypergraphs

A hypergraph is an analogue of a graph where there is a set of vertices as before, but now the hyperedges are, as opposed to being subsets of $V(G)$ of size 2, more general subsets of $V(G)$. This is clearly a very general notion, and often in practice one has to restrict to special cases, e.g., regular hypergraphs where all edges have the same cardinality. Many of the key definitions in graph theory go over to hypergraphs: for example, vertices are adjacent if they both occur in some hyperedge, and the degree of a vertex is the number of vertices it is adjacent to.

One obvious model of random hypergraphs is $G(n, p_1, \ldots, p_k)$ where the vertex set is $\{1, 2, \ldots, n\}$ and subsets containing i elements, $2 \leqslant i \leqslant k$ (here k does not depend on n) occur as edges with probability $p_i(n)$, independently for each such subset. For example, it was shown in (Schmidt-Pruzan and Shamir, 1985) that if $c = \sum_{i=2}^{k}(i-1)p_i < 1$ then all components are of size $O(\log(n))$, but if $c > 1$ there is a giant component of order linear in n. (Note that again the giant component arises when the expected degree is 1.) (Karoński and Łuczak, 1996) contains more on the phase transition, and other results about such random hypergraphs.

In the case $p_i = 0$ for $i \neq r$ so that the hypergraph is r-uniform, Krivelevich and Sudakov (1998) considers, for each $1 \leqslant \gamma \leqslant (r-1)$, the smallest number χ_γ of sets V_i satisfying $|V_i \cap e| \leqslant \gamma$ for all hyperedges e into which we can partition the n vertices. These numbers are clearly a family of generalisations of the chromatic number. They show that there exists $d_0(\gamma)$ such that if

$$d_0(\gamma) \leqslant d(\gamma, n, p) = \gamma \binom{r-1}{\gamma} \binom{n-1}{r-1} p = \mathrm{o}(n^\gamma),$$

then **whp**

$$1 \leqslant \left(\frac{(\gamma+1)\log(d(\gamma,n,p))}{d(\gamma,n,p)}\right)^{1/\gamma} \chi_\gamma \leqslant \left(1 + \frac{1}{1+\log^{0.1}(d(\gamma,n,p))}\right)^{1/\gamma}.$$

7.8. Random planar graphs

Recall that a graph is **planar** if it can be drawn in the plane, with points representing vertices and straight lines representing edges, and no two edges crossing each other. Most introductions to graph theory discuss their properties. Many graphs in practice are planar (e.g., road networks, with vertices at their junctions): thus it is natural to ask about models of random planar graphs. An obvious first step towards understanding them is to sample from them (at least approximately) uniformly and at random.

In (Denise et al., 1996) the authors define an ergodic time-reversible Markov chain whose state space is the set of all planar subgraphs of any given graph G and whose stationary distribution is the uniform distribution on this state space. It is not known whether or not this chain is rapidly mixing. Nevertheless, use of the program suggested various conjectures on random planar graphs, for example, that they are **whp** not 2-connected.

Recently there has been some theoretical progress on these conjectures: for example, Osthus, Prömel and Taraz (2002) gives results and references to other recent papers. Amongst the recent results are the facts that

$$0 < \lim_{n\to\infty} P\{G \text{ is connected}\} \leqslant P\{G \text{ has an isolated vertex}\} < 1$$

and that, letting $\mathcal{P}_n$ denote the space of labelled planar graphs on n vertices and $\mathcal{P}_n^k$ be the number of k-connected labelled planar graphs on n vertices,

$$n!26.18^{n+o(n)} \leqslant |\mathcal{P}_n^2| \leqslant |\mathcal{P}_n| \leqslant n!37.3^{n+o(n)}$$

though the techniques do not at present seem to yield the conjecture that the graphs are **whp** not 2-connected. It is also known that the number of edges is between $13n/7$ and $2.56n$ **whp**: it is also shown that the number of edges of a typical graph in $\mathcal{P}_n^2$ is at most $2.69n$ **whp** and of a typical graph in $\mathcal{P}_n^3$ is at most $2.992n$ **whp**.

7.9. Random graphs with prescribed degrees: regular graphs

It is in general hard to say much about random graphs with a given degree sequence. Let $\underline{d} = \{d_1, d_2, \ldots, d_n\}$ be the set of degrees of vertices of a graph on n vertices, and let $G(\underline{d})$ denote a graph chosen uniformly at random among all graphs on n vertices with this degree set. Let the number of vertices of degree i be $\lambda_i n + o(n)$ for each $i \geqslant 0$, and set $Q = \sum_i i(i-2)\lambda_i$. In (Molloy and Reed, 1995) it is shown that if $Q > 0$ there is a giant component whose size is about cn for a suitable non-zero constant c, but if $Q < 0$ then **whp** all components are small. Molloy and Reed (1998) gives more detailed information about the size and degree set of the giant component when it exists.

The simplest case is models of regular labelled graphs, when all vertices have degree d: we call the space of all such graphs, with uniform distribution, $\mathcal{G}_{n,d}$. Of course this is only non-empty if nd is even, which we assume in the rest of this subsection. (Wormald, 1999) is a recent survey of this topic, where (for example) it is shown that if X_i is the number of i-cycles in $\mathcal{G}_{n,d}$, with $d \geq 3$ fixed, then for fixed k $X_3, \ldots, X_k$ are asymptotically independent Poisson random variables with $\mathbf{E}(X_i) = (d-1)^i/(2i)$, but on the other hand if F is a graph with more edges than vertices, a graph in $\mathcal{G}_{n,d}$ **whp** contains no graph isomorphic to F. Moreover, **whp** G has a Hamilton cycle (and hence a perfect matching when n is even. In fact Kim and Wormald (2001) show that it has $\lfloor d/2 \rfloor$ edge-disjoint Hamiltonian cycles, plus a perfect matching if d is odd). Also it has $\kappa = d$ **whp**. It is also conjectured that the independence number is about $\beta(d)n$ for some function $\beta(d)$ (not yet very fully understood): the chromatic number is also not well understood in this case.

Many of the results have long been known to extend to the case of $d(n)$ growing slowly with n (and in fact it appears that the independence and chromatic numbers are easier to handle when $d(n)$ is growing with n). Recent work in (Krivelevich et al., 2002) proved the conjectures that the graph is Hamiltonian with $\kappa = \delta$ **whp** for values of d greater than $\sqrt{n}\log(n)$, and the recent paper (Cooper et al., 2002a) fills the remaining gaps: thus for $3 \leq d(n) \leq n-4$, the graph is Hamiltonian and has $\delta = \kappa$ **whp**. Another preprint (Cooper et al., 2002b) gives the asymptotic behaviour **whp** of independence numbers and chromatic numbers, except for $d \geq 0.9n$ or d a small constant.

Wormald (1999) also discusses simulation procedures which often improve on those in (Tinhofer, 1993).

7.10. Unlabelled graphs

Very often in practice we are only concerned with the graph up to isomorphism and thus want results for unlabelled graphs rather than labelled ones. Unfortunately such results are harder to get, as counting the graphs is difficult. Unless the graph has very few edges or very many edges, the number of unlabelled graphs with $\wp$ is asymptotically the number of labelled graphs having $\wp$ divided by $n!$: see Chapter 13 of (Bollobás, 1985) for the precise statement of this result, which depends on the fact that **whp** a random graph has trivial automorphism group. Unlabelled graphs are surprisingly difficult to deal with: for example, they do not have a phase transition (see Łuczak, 1994).

7.11. Random trees

There are over 100 papers in MathReviews mentioning random trees in their title, many of them by J. W. Moon and co-authors. The results are hard to summarise, as the trees may be rooted (have a distinguished vertex) or unrooted, be labelled or unlabelled, and (especially) be grown in different ways depending on the application. Thus we restrict ourselves to describing some areas of application and a few results.

Many problems involving trees have natural interpretations in terms of search problems in computer science: for example, Flajolet (1988) discusses random tree models that occur in the average case analysis of symbol manipulation algorithms, compiling, comparison-based searching and sorting, digital retrieval techniques, file systems and communication protocols. For a more esoteric application, relating the distribution of the number of leaves (vertices of degree 1) to the reconstruction of the family trees of manuscripts (stemma), see (Najock and Heyde, 1982). In (Pittel, 1990) the growth in continuous time of polymers is modelled by a certain random forest: it is shown that the largest tree in the forest has size whose order of magnitude is $\log(n)$, $n^{2/3}$ or n according as a certain parameter is less than, equal to, or greater than, a certain critical value.

One common theme is the use of results from the theory of branching processes, which are known to be strongly linked to so-called simply generated trees. For example, Biggins and Grey (1997) uses branching random walk to study the height H_n of a random unlabelled tree on n vertices grown in a certain way.

Drmota and Gittenberger (1999) shows that, if the unlabelled trees are selected uniformly at random, the number of vertices of fixed degree k is asymptotically normal with mean $\mu_k n$ and variance $\sigma_k^2 n$ for positive μ_k, σ_k. Similar results are given for plane trees, labelled trees and forests.

If one samples from the n^{n-2} labelled trees on n vertices uniformly at random, Moon (1968) shows that the maximum degree D satisfies

$$E(D) \sim \frac{\log n}{\log \log n}.$$

Sharper results for labelled trees were obtained later in (Carr et al., 1994). Meir and Moon (1973) gives results on the independence number of a random tree.

7.12. Graphs on infinite vertex sets

These may be of less modelling interest, but we describe them briefly anyway. Consider a random graph on vertex set **N** with each possible edge present with probability p, independently of all others. A striking result of Erdős and Rényi is that, for $0 < p < 1$, the graph produced is with probability 1 isomorphic to one particular graph R, whose properties (interesting in pure mathematics) are surveyed in (Cameron, 1996). A more general random graph on **N** where each edge is known to arise with probability $\geq p$ but nothing at all is known about the correlations between different edges is studied in (Fremlin and Talagrand, 1985) where existence of subgraphs of various types is considered. There appears to have been little if any consideration of random graphs whose vertex set is uncountable.

In percolation theory one works on a suitable infinite lattice, and says that each edge is open with a certain probability and otherwise closed, independently of all other edges. Typical questions are: is there an infinite connected component all of whose edges are open? If so, is it unique? This subject initially arose from questions of whether water will penetrate to the centre of a porous body (the pores being those edges which are open, essentially). This important subject has a rather different flavour from finite random graph theory: for an introduction to it, we refer to (Grimmett, 1999).

7.13. Covers of graphs

This is an idea introduced recently in (Amit et al., 2001, 2002; Amit and Linial, 2002). The idea is to take some particular graph G (the base graph), and then construct a cover

$\widetilde{G}$ such that, for each vertex $v \in V(G)$ there are n vertices over it in $\widetilde{G}$, and, if v and w are adjacent in G, there is a random perfect matching in the bipartite graph whose two vertex classes are the set of vertices above v and the set of vertices above w. The idea is that these graphs are intermediate between being random and having properties determined by G. The primary interest is in asymptotic behaviour as $n \to \infty$. The base graph will be assumed connected throughout.

The lifts are not always typically connected. For example, if the base graph G is a tree, the lift is easily seen to consist of n disjoint copies of T: and if G is a cycle, one can show that the probability a lift is connected is $1/n$. However, if the minimum degree of G is at least 3, an n-lift of G is connected **whp**. They also show that **whp** we have

$$\chi(\widetilde{G}) \geqslant \sqrt{\frac{\chi(G)}{\log(\chi(G))}}$$

and conjecture that in fact **whp**

$$\chi(\widetilde{G}) \geqslant \Omega\left(\frac{\chi(G)}{\log(\chi(G))}\right)$$

which would be best possible if true. They also address the more complex question of when the graph has a perfect matching, and relate the expanding properties of the graph (which crudely speaking means how many vertices there are adjacent to at least one vertex in a given set: such questions arise naturally in the design of networks) to those of the base graph. There are some indications that various properties in this model may not obey 0–1 laws.

7.14. Expected running time of algorithms

It is by now fairly well-known that many problems about graphs (does it have a Hamiltonian cycle? What is its chromatic number? Does it have an independent set of order k?) are not easy in general. However, these results are for the worst possible case, and for many problems it is known that the proportion of cases which are not easy tends to zero as n goes to infinity (often very fast, e.g., exponentially in n). If one takes the **expected** running time of the algorithm as the graph ranges over all possible inputs, according to some distribution – that is, the expected time in some model of random graphs – many problems become soluble in polynomial time (and many already soluble in polynomial time are soluble in much smaller polynomial time). However there are some problems for which even the typical instances are hard. For more information on this, we refer to the two recent surveys (Frieze and McDiarmid, 1997) and (Frieze and Reed, 1998). For example, in (Frieze and McDiarmid, 1997) an algorithm is described which decides whether or not a $G(n, p)$ for $p > 0$ constant is Hamiltonian in linear (in n) expected time, and a different algorithm is given which finds a Hamiltonian cycle in $G(n, 1/2)$ with probability $1 - o(2^{-n})$ and has run time $O(n^3 \log(n))$. On the other hand, the only efficient colouring algorithms for random graphs at present require about twice the number of colours which are really needed.

Many problems remain unsolved, for example, finding algorithms for clique numbers and chromatic numbers whose expected runtime is polynomial in the number of vertices of G when the input is a random graph from $G(n, 1/2)$. For a recent overview of algorithmic problems related to colouring random graphs, see (Krivelevich, 2002).

For results on algorithms in geometric models, see (Steele, 1990).

References

Aldous, D. (1990). A random tree model associated with random graphs. *Random Structures Algorithms* **1**, 383–402.
Aldous, D. (1998). Tree-valued Markov chains and Poisson–Galton–Watson distributions. In *Microsurveys in Discrete Probability* (Eds. D. Aldous and J. Propp). DIMACS Ser. Discrete Math. Theoret. Comp. Sci., Vol. 41.
Alon, N., and J. H. Spencer (1992). *The Probabilistic Method*, 2nd ed. With an appendix by Paul Erdős. Wiley, New York. (A revised edition appeared in 2000.)
Alon, N., M. Krivelevich and B. Sudakov (1998). Finding a large hidden clique in a random graph. *Random Structures Algorithms* **13**, 457–466.
Amit, A., N. Linial, J. Matoušek and E. Rozenman (2001). Random lifts of graphs. In *Proceedings of the 12th annual ACM-SIAM Symposium on Discrete Algorithms*, pp. 883–894. SIAM, Philadelphia.
Amit, A., N. Linial and J. Matoušek (2002). Random lifts of graphs; Independence and chromatic number. *Random Structures Algorithms* **20**, 1–22.
Amit, A. and N. Linial (2002). Random graph coverings I. General theory and graph connectivity. *Combinatorica* **22**, 1–18.
Andersson, H. and T. Britton (2000). *Stochastic Epidemic Models and Their Statistical Analysis*. Springer Lecture Notes in Statist., Vol. 151.
Appel, M. J. B and R. P. Russo (1997a). The maximum vertex degree of a graph on uniform points in $[0, 1]^d$. *Adv. Appl. Probab.* **29**, 567–581.
Appel, M. J. B. and R. P. Russo (1997b). The minimum vertex degree of a graph on uniform points in $[0, 1]_d$. *Adv. Appl. Probab.* **29**, 582–594.
Balińska, K., E. Godehardt and L. V. Quintas (1991). When is the f-graph process the Erdős–Rényi graph process? *Sci. Ser. A. Math. (N.S.)*, 15–19.
Barbour, A. D., L. Holst and S. Janson (1990). *Poisson Approximation*. Oxford Mathematical Monographs, Vol. 2. Clarendon Press, Oxford.
Barbour, A. D. and D. Mollison (1987). Random graphs and epidemics. In *Lecture Notes in Biomathematics* (Eds. Lefevre and Gabriel). Vol. 89. Springer.
Barbour, A. D. and G. Reinert (2001). Small worlds. *Random Structures Algorithms* **19**, 54–74.
Biggins, J. D. and D. R. Grey (1997). A note on the growth of random trees. *Statist. Probab. Lett.* **32**, 339–342.
Biggins, J. D. and D. B. Penman (2003a). Large deviations in random randomly coloured graphs. To appear.
Biggins, J. D. and D. B. Penman (2003b). FKG-like inequalities in random randomly coloured graphs. To appear.
Bollobás, B. (1985). *Random Graphs*. Academic Press, London. (A second revised edition was published by Cambridge University Press as Vol. 73 of the Cambridge Studies in Advanced Mathematics in 2001.)
Bollobás, B. (1986). *Combinatorics*. Cambridge University Press.
Bollobás, B. (1998). *Modern Graph Theory*. Graduate Texts in Mathematics, Vol. 184. Springer-Verlag, New York.
Bollobás, B. and A. Frieze (1985). On matchings and Hamilton cycles in random graphs. *Ann. Discrete Math.* **28**, 23–46.
Bollobás, B., G. R. Grimmett and S. Janson (1996). The random-cluster model on the complete graph. *Probab. Theory Related Fields* **104**, 283–317.
Bollobás, B., Y. Kohayakawa and T. Łuczak (1995). Connectivity properties of random subgraphs of the cube. *Random Structures Algorithms* **6**, 221–230.

Bollobás, B., Y. Kohayakawa and T. Łuczak (1994). On the diameter and radius of random subgraphs of the cube. *Random Structures Algorithms* **5**, 627–648.

Bollobás, B. and O. Riordan (2000). Constrained graph processes. *Electron. J. Combin.* **7**, Research Paper 18, 20 pp. (electronic).

Bollobás, B. and O. Riordan (2002). Mathematical results on scale-free graphs. In *Handbook of Graphs and Networks* (Eds. S. Bornholdt and H. Schuster). Wiley-VCH, Berlin.

Bollobás, B. and A. G. Thomason (1987). Threshold functions. *Combinatorica* **7**, 35–38.

Bollobás, B. and A. G. Thomason (2000). The structure of hereditary properties and colourings of random graphs. *Combinatorica* **20**, 173–202.

Bourgain, J. and G. Kalai (1999). Threshold intervals under group symmetries. In *Convex Geometric Analysis*, pp. 59–63 (Berkeley, CA, 1996). MSRI Publications, Vol. 34. Cambridge University Press.

Cameron, P. J. (1996).The random graph. In *The Mathematics of Paul Erdős*, pp. 333–351. Algorithms Combinatorics, Vol. 14.

Cameron, P. J. (1999). First-order logic. In *Graph Connections: Relationships Between Graph Theory and Other Parts of Mathematics* (Eds. R. J. Wilson and L. Beineke). Oxford University Press.

Cannings, C. (1997). *Evolutionarily Stable Strategies*. Encycl. Maths. Supplement, Vol. 1. Kluwer Academic Publishers, Dordrecht.

Carr, R., W. Goh and E. Schmutz (1994). The maximum degree in a random tree and related problems. *Random Structures Algorithms* **5**, 13–24.

Cohen, J. E. (1995). Random graphs in ecology. In *Topics in Contemporary Probability and its Applications*, pp. 233–260. Probab. Stochastics Ser. CRC, Boca Raton, FL.

Cooper, C. and A. Frieze (2002). On a model of random web graphs. In *Proceedings of ESA2001*. Version at http://www.math.cmu.edu/~af1p/papers.html.

Cooper, C., A. Frieze and B. Reed (2002a) Random regular graphs of non-constant degree: connectivity and Hamiltonicity. *Combin. Probab. Comput.* **11**, 249–261.

Cooper, C., A. Frieze, B. Reed and O. Riordan (2002b) Random regular graphs of non-constant degree; independence and chromatic number. *Combin. Probab. Comput.* **11**, 323–341.

Denise, A., M. Vasconcellos and D. Welsh (1996). The random planar graph. *Congr. Numer.* **113**, 61–79.

Diaz, J., J. Petit, M. D. Penrose and M. Serna (2000). Convergence theorems for some layout measures on random lattice and random geometric graphs. *Combin. Probab. Comput.* **9**, 489–511.

Diestel, R. (1997). *Graph Theory*. Graduate Texts in Math., Vol. 173. Springer, Berlin.

Drmota, M. and B. Gittenberger (1999). The distribution of nodes of given degree in random trees. *J. Graph Theory* **31**, 227–253.

Erdős, P., S. Suen and P. Winkler (1995). On the size of a random maximal graph. *Random Structures Algorithms* **6**, 309–318.

Fill, J. A., E. R. Scheinerman and K. B. Singer-Cohen (2000). On random intersection graphs when $m = \omega(n)$. *Random Structures Algorithms* **16**, 156–176.

Fernandez de la Vega, W. (2000). The independence number of random interval graphs. In *Algorithms and Complexity*, pp. 59–62 (Rome, 2000). Lecture Notes in Comput. Sci., Vol. 1767. Springer, Berlin.

Flajolet, P. (1988). Random tree models in the analysis of algorithms. *Performance '87*, pp. 171–187 (Brussels, 1987). North-Holland, Amsterdam.

Fremlin, D. H. and M. Talagrand (1985). Subgraphs of random graphs. *Trans. Amer. Math. Soc.* **291**, 551–582.

Friedgut, E. (with an appendis by J. Bourgain) (1999). Sharp thresholds and the k-SAT problem. *J. Amer. Math. Soc.* **12**, 1017–1054.

Frieze, A. and C. McDiarmid (1997). Algorithmic theory of random graphs. *Random Structures Algorithms* **10**, 5–42.

Frieze, A. and B. Reed (1998). Probabilistic analysis of algorithms. In *Probabilistic Methods for Algorithmic Discrete Mathematics*, pp. 36–92. Algorithms Combinatorics, Vol. 16. Springer, Berlin.

Garey, M. R. and D. S. Johnson (1979). *Computers and Intractability. A Guide to the Theory of NP-completeness*. A Series of Books in the Mathematical Sciences. W. H. Freeman, San Francisco, CA.

Godehardt, E. (1990a). Connectivity of random graphs of small order and statistical testing. In *Random Graphs '87*, pp. 61–72 (Poznan, 1987). Wiley, Chichester.

Godehardt, E. (1990b). *Graphs as Structural Models. The Application of Graphs and Multigraphs in Cluster Analysis*, 2nd ed. Advances in System Analysis, Vol. 4. Friedr. Vieweg und Sohn, Braunschweig.

Godehardt, E. and B. Harris (1997). Asymptotic properties of random interval graphs and their use in cluster analysis. In *Probabilistic Methods in Discrete Mathematics*, pp. 19–30 (Petrozavodsk, 1996). VSP, Utrecht.

Godehardt, E. and J. Jaworski (1996). On the connectivity of a random interval graph. *Random Structures Algorithms* **9**, 137–161.

Grimmett, G. R. (1995). The stochastic random-cluster process and the uniqueness of random-cluster measures. *Ann. Probab.* **23**, 1461–1510.

Grimmett, G. R. (1999). *Percolation*, 2nd ed. Grundlehren Math. Wiss., B. 321. Springer-Verlag, Berlin.

Ivchenko, G. I. (1975). Unequally probable random graphs. *Trudy. Mosk. Inst. Electr. Mas.* **44**, 34–66 (Russian).

Janson, S., T. Łuczak and A.Ruciński (2000). *Random Graphs*. Wiley, New York.

Janson, S., D. E. Knuth, T. Łuczak and B. Pittel (1994), The birth of the giant component. *Random Structures Algorithms* **4**, 233–258.

Jonasson, J. (1999). The random triangle model. *J. Appl. Probab.* **36**, 852–867.

Jonasson, J. and O. Häggström (1999). Phase transition in the random triangle model. *J. Appl. Probab.* **36**, 1101–1115.

Juhász, F. (1991). The asymptotic behaviour of Fiedler's algebraic connectivity. *Discrete Math.* **96**, 59–63.

Karoński, M. (1995). Random graphs. Chapter in *Handbook of Combinatorics*, pp. 351–380 (Eds. R. Graham, M. Grotschel and L. Lovasz). North-Holland.

Karoński, M. and T. Łuczak (1996). Random hypergraphs. In *Combinatorics, Paul Erdős is Eighty*, Vol. 2, pp. 283–293 (Keszthely, 1993). Bolyai Soc. Math. Stud., Vol. 2. Janos Bolyai Math. Soc., Budapest.

Karoński, M., E. Scheinerman and K. B. Singer-Cohen (1999). On random intersection graphs: the subgraph problem. *Combin. Probab. Comput.* **8**, 131–159.

Karoński, M. and M. Szymkowiak (2001). On a solution of a randomized three tracks version of the Gate Matrix Layout Problem. *Discrete Math.* **236**, 179–189.

Kauffman, S. (1993). *The Origins of Order*. Oxford University Press.

Kim, J. H. and N. C. Wormald (2001). Random matchings which induce Hamilton cycles and Hamiltonian decompositions of random regular graphs. *J. Combin. Theory Ser. B* **81**, 20–44.

Kolchin, V. F. (1999). *Random Graphs*. Encyclopedia of Mathematics and its Applications, Vol. 53. Cambridge University Press, Cambridge.

Kovalenko, I. N. (1975). On the structure of random directed graphs. *Theory Probab. Math. Statist.* **6**, 83–92. (Russian).

Krivelevich, M. (2002). Coloring random graphs – an algorithmic perspective. In *Proceedings of the 2nd Colloquium on Mathematics and Computer Science* (MathInfo'2002), pp. 175–195 (Eds. B. Chauvin et al.). Birkhäuser, Basel.

Krivelevich, M. and B. Sudakov (1998). The chromatic numbers of random hypergraphs. *Random Structures Algorithms* **21**, 381–403.

Krivelevich, M., B. Sudakov, V. H. Vu and N. Wormald (2002). Random regular graphs of high degree. *Random Structures Algorithms* **18**, 346–363.

Łuczak, T. (1994). Phase transition phenomena in random discrete structures. *Discrete Math.* **136**, 225–242.

Maehara, H. (1990). On the intersection graph of random arcs on the circle. In *Random Graphs '87*, pp. 159–173 (Poznan, 1987). Wiley, Chichester.

McColm, G. L. (1999). First order zero-one laws for random graphs on the circle. *Random Structures Algorithms* **14**, 239–266.

McColm, G. (2003). Threshold functions for random graphs on a line segment. Submitted: version at http://www.math.usf.edu/~mccolm/research/Research.html

McDiarmid, C. (1989). On the method of bounded differences. In *Surveys in Combinatorics*, pp. 148–188. London Math. Soc. Lecture Note Series, Vol. 141.

McDiarmid, C. (1998). Concentration. Chapter in *Probabilistic Methods for Discrete Mathematics* (Eds. M. Habib, C. McDiarmid, J. Ramirez and B. Reed). Springer-Verlag.

McDiarmid, C. (1990). On the chromatic number of random graphs. *Random Structures Algorithms* **1**, 435–442.

McKay, B. D. (2001). Tables of combinatorial data. Available at http://cs.anu.edu/people/bdm/data.

Meir, A. and J. W. Moon (1973). The expected node-independence number of random trees. *Indag. Math.* **35**, 335–341.

Molloy, M. and B. Reed (1995). A critical point for random graphs with a given degree sequence. *Random Structures Algorithms* **6**, 161–179.

Molloy, M. and B. Reed (1998). The size of the giant component of a random graph with a given degree sequence. *Combin. Probab. Comput.* **7**, 295–305.

Moon, J. W. (1968). On the maximum degree in a random tree. *Michigan Math. J.* **15**, 429–432.

Moon, J. W. (1988). On the bipartition numbers of random trees. In *Eleventh British Combinatorial Conference*, pp. 3–10 (London, 1987). Ars Combin., Vol. 25.

Najock, D. and C. C. Heyde (1982). On the number of terminal vertices in certain random trees with an application to stemma construction in philology. *J. Appl. Probab.* **19**, 675–680.

O'Connell, N. (1998). Some large deviation results for sparse random graphs. *Probab. Theory Related Fields* **110**, 277–285.

Osthus, D. and A. Taraz (2001). Random maximal H-free graphs. *Random Structures Algorithms* **18**, 61–82.

Osthus, D., H.-J. Prömel and A. Taraz (2002). On random planar graphs, the number of planar graphs and their triangulations. *J. Combin. Theory Ser. B*. To appear.

Palmer, E. M. (1985). *Graphical Evolution. An Introduction to the Theory of Random Graphs*. Wiley, Chichester.

Penman, D. B. (1998). Random graphs with correlation structure. Ph.D. thesis, University of Sheffield.

Penman, D. B. (2003a). Trees and cycles in random randomly coloured graphs. To appear.

Penman, D. B. (2003b). The $G_{p,q}$ model of random graphs. To appear.

Penrose, M. D. (1997). The longest edge of the random minimal spanning tree. *Ann. Appl. Probab.* **7**, 340–361.

Penrose, M. D. (1998a). Extremes for the minimal spanning tree on normally distributed points. *Adv. Appl. Probab.* **30**, 628–639.

Penrose, M. D. (1998b). Random minimal spanning tree and percolation on the N-cube. *Random Structures Algorithms* **12**, 63–82.

Penrose, M. D. (1999). On k-connectivity for a geometric random graph. *Random Structures Algorithms* **15**, 145–164.

Penrose, M. D. (2000). Vertex ordering and partitioning problems for random spatial graphs. *Ann. Appl. Probab.* **10**, 517–538.

Penrose, M. D. (2003). *Random Geometric Graphs*. Oxford University Press. To appear.

Penrose, M. D. and N. Henze (1999). On the multivariate runs test. *Ann. Statist.* **27**, 290–298.

Pippinger, N. (1998). Random interval graphs. *Random Structures Algorithms* **12**, 361–380.

Pittel, B., W. A. Woyczynski and J. A. Mann (1990). Random tree-type partitions as a model for acyclic polymerization: Holtsmark 3/2-stable distribution of the supercritical gel. *Ann. Probab.* **18**, 319–341.

Prisner, E. (2001). Electronic notes on intersection graphs. Available at http://www.math.uni-hamburg.de/spag/gd/mitarbeiter/prisner/prisner.html.

Prömel, H.-J., T. Schickinger and A. Steger (2001). On the structure of clique-free graphs. *Random Structures Algorithms* **19**, 37–53.

Reidys, C. M. (1997). Random induced subgraphs of generalized n-cubes. *Adv. Appl. Math.* **19**, 360–377.

Reidys, C. M. (2000). Random subgraphs of Cayley graphs over p-groups. *Europ. J. Combin.* **21**, 1057–1066.

Royle, G. F. (2001). Electronic tables of graphs of small order. Available at http://www.cs.uwa.edu.au/~gordon/remote/data.html.

Ruciński, A. (1992). Proving normality in combinatorics. In *Random Graphs*, Vol. 2, pp. 215–231 (Poznan, 1989). Wiley, New York.

Ruciński, A. and N. C. Wormald (1992). Random graph processes with degree restrictions. *Combin. Probab. Comput.* **1**, 169–180.

Ruciński, A. and N. C. Wormald (1997). Random graph processes with maximum degree 2. *Ann. Appl. Probab.* **7**, 183–200.

Ruciński, A. and N. Wormald (2002). Connectedness of graphs generated by a random d-process. *J. Austral. Math. Soc.* **72**, 67–86.

Scheinerman, E. R. (1988). Random interval graphs. *Combinatorica* **8**, 357–371.

Scheinerman, E. R. (1990). An evolution of interval graphs. *Discrete Math.* **82**, 287–302.

Schmidt-Pruzan, J. and E. Shamir (1985). Component structure in the evolution of random hypergraphs. *Combinatorica* **5**, 81–94.

Shepp, L. A. (1989). Connectedness of certain random graphs. *Israel J. Math.* **67**, 23–33.

Stark, D. (2001). Compound Poisson approximation for subgraph counts in random graphs. *Random Structures Algorithms* **18**, 39–60.

Steele, J. M. (1990). Probabilistic and worst-case analysis of classical problems of combinatorial optimisation in Euclidean space. *Math. Oper. Res.* **15**, 749–770.

Szymkowiak, M. (2002). Tree obstructions for the k-Gate Matrix Layout Problem. Preprint, Technical University of Poznan.

Thomason, A. G. (1988). A graph property not satisfying a "zero-one law". *Europ. J. Combin.* **9**, 517–521.

Tinhofer, G. (1993). Verfahrensmuster zur Erzeugung von Zufallsgraphen. [Examples of procedures for the generation of random graphs]. In *Konstruktive Anwendungen von Algebra und Kombinatorik*, pp. 127–169 (Bayreuth, 1991). Bayreuth. Math. Schr., Vol. 43.

Vickers, G. T. and C. Cannings (1988). Patterns of ESSs I, II. *J. Theoret. Biol.* **132**, 387–408, 409–420.

Watts, D. J. (1999). *Small Worlds: The Dynamics of Networks between Order and Randomness.* Princeton University Press, Princeton, NJ.

Welsh, D. J. A. (1994). The random cluster process. *Discrete Math.* **136**, 373–390.

Wormald, N. C. (1995). Differential equations for random processes and random graphs. *Ann. Appl. Probab.* **5**, 1217–1235.

Wormald, N. C. (1999). Models of random regular graphs. In *Surveys in Combinatorics*, pp. 239–298 (Eds. J. D. Lamb and D. A. Preece). London Math. Soc. Lecture Note Series, Vol. 267. Cambridge University Press, Cambridge.

Locally Self-Similar Processes and their Wavelet Analysis

Joseph E. Cavanaugh, Yazhen Wang and J. Wade Davis

1. Introduction

A stochastic process $Y(t)$ is defined as *self-similar* with self-similarity parameter H if for any positive stretching factor c, the distribution of the rescaled and reindexed process $c^{-H}Y(ct)$ is equivalent to that of the original process $Y(t)$. This means for any sequence of time points $t_1, \ldots, t_n$ and any positive constant c, the collections $\{c^{-H}Y(ct_1), \ldots, c^{-H}Y(ct_n)\}$ and $\{Y(t_1), \ldots, Y(t_n)\}$ are governed by the same probability law. As a consequence, the qualitative features of a sample path of a self-similar process are invariant to magnification or shrinkage, so that the path will retain the same general appearance regardless of the distance from which it is observed.

Although self-similar processes were first introduced in a theoretical context by Kolmogorov (1941), statisticians were made aware of the practical applicability of such processes through the work of B. B. Mandelbrot (Mandelbrot and van Ness, 1968; Mandelbrot and Wallis, 1968, 1969). Self-similarity is a pervasive characteristic in naturally occurring phenomena. As a result, self-similar processes have been used to successfully model data arising in a variety of different scientific fields, including hydrology, geophysics, medicine, genetics, economics, and computer science. Recent applications include Buldyrev et al. (1993), Lelend et al. (1994), Ossadnik et al. (1994), Percival and Guttorp (1994), Peng et al. (1992, 1995a, 1995b), and Willinger et al. (1998).

The dynamics of a self-similar process $Y(t)$ are principally dictated by the value of the self-similarity parameter or scaling exponent H. For practical applications, it may be assumed that $Y(t)$ has finite second moments, and that its associated increment process $X(t) = Y(t) - Y(t-1)$ is stationary. Under these assumptions, H may be taken to lie over the interval $(0, 1)$, and the value of H may be used in describing the autocorrelation structure of the increment sequence $X(t), X(t+1), \ldots$ (see Beran, 1994, pp. 52–53). For $H \in (1/2, 1)$, $X(t)$ is characterized by serial correlations that decay slowly, and therefore exhibits long-range dependence or long memory. For $H \in (0, 1/2)$, $X(t)$ is characterized by serial correlations that decay rapidly and sum to zero. For $H = 1/2$, $X(t)$ is serially uncorrelated. The estimation of H as a constant has been extensively studied, predominantly in the context of long memory where it is assumed that $H \in (1/2, 1)$. A partial list of relevant references includes Geweke and Porter-Hudak (1983),

Taylor and Taylor (1991), Constantine and Hall (1994), Chen et al. (1995), Robinson (1995), Taqqu et al. (1995), Abry and Sellan (1996), Comte (1996), McCoy and Walden (1996), Hall et al. (1997), Kent and Wood (1997), Moulines and Soulier (1999), Abry et al. (2000), Bardet et al. (2000), Abry et al. (2001), and Bardet et al. (2001).

In modeling applications, treating the self-similarity parameter H as a constant implies that the self-similar features of the underlying phenomenon persist over time. However, many phenomena exhibit self-similar patterns that change as the phenomenon itself evolves: this characteristic may be embodied in the associated data by sections of irregular roughness. To adequately model such data, the class of self-similar processes must be expanded to allow the scaling exponent to vary as a function of time. Moreover, a procedure must be available that provides a statistical characterization of the exponent's progression.

In what follows, we consider a class of processes that are locally (as opposed to globally) self-similar. The defining property of this class, based on the covariance function, relates the local autocorrelation of the process to the value of $H(t)$. We provide two examples of stochastic processes that are locally self-similar. One of these, generalized fractional Brownian motion, is explored in detail. We then propose and describe a procedure based on wavelets for constructing an estimate of the time-varying scaling exponent of a locally self-similar process. We establish a consistency result for this estimate. We investigate the effectiveness of the procedure in a simulation study, and demonstrate its applicability in several practical analyses.

Techniques for estimating a constant self-similarity parameter H are often based on log-linear regression (e.g., Geweke and Porter-Hudak, 1983; Taylor and Taylor, 1991; Constantine and Hall, 1994). Such methods frequently exploit an approximate log-linear relationship between either the spectrum of $X(t)$ or the variogram of $Y(t)$ and the time index t, using least-squares regression to obtain the estimate of H. With a locally self-similar process $Y(t)$, $H(t)$ is a function of t; as a result, the associated increment process $X(t)$ is non-stationary. To estimate $H(t)$ locally, we take advantage of an approximate local log-linear relationship between the square of the wavelet transformation of $Y(t)$ and the scale for the transformation. Local least-squares regression is used to obtain the estimate of $H(t)$. The basic procedure is detailed in Wang et al. (2001) and was first presented by Wang et al. (1997). The method may be viewed as an extension of previously proposed log-linear regression techniques for the estimation of constant H.

Replacing the role of the spectrum or variogram with that of the wavelet transformation is logical for the problem at hand. Since wavelets are ideal for characterizing changes in scale content over time, they can accommodate the time-varying dynamics of $Y(t)$ induced by changes in $H(t)$. Our reliance on wavelets is further motivated by the computational efficiency of the discrete wavelet transformation as well as the successful application of wavelets in problems dealing with both non-stationarity and self-similarity (e.g., Farge et al., 1993; Hwang and Mallat, 1994). In fact, for globally self-similar processes, log-linear regression methods that utilize the wavelet transformation have been recently proposed for the estimation of constant H (e.g., Abry et al., 2000, 2001; Bardet et al., 2000). An early inspiration for these methods as well as for our procedure is provided by an illustration in Daubechies (1992, pp. 301–303) for estimating the Hölderian exponent.

The exposition in this chapter is organized as follows. Section 2 defines and briefly discusses locally self-similar processes. Section 3 provides an in-depth exploration of the properties of generalized fractional Brownian motion. Section 4 presents the procedure for the estimation of $H(t)$ via the continuous wavelet transformation. Section 5 outlines the implementation of the procedure to data collected in discrete time via the discrete wavelet transformation. Section 6 presents simulations designed to check the effectiveness of the procedure, and Section 7 features three practical applications. Section 8 concludes.

2. Locally self-similar processes

Let $Y(t)$ represent a mean-zero stochastic process with covariance function

$$\Sigma_t(u_1, u_2) = \text{Cov}\{Y(t+u_1), Y(t+u_2)\} \tag{2.1}$$

and variogram

$$\Upsilon_t(u_1, u_2) = \text{Var}\{Y(t+u_1) - Y(t+u_2)\}. \tag{2.2}$$

The process $Y(t)$ is said to be *locally self-similar* if

$$\Sigma_t(u_1, u_2) = \frac{1}{2}\{\Sigma_t(u_1, u_1) + \Sigma_t(u_2, u_2)\} - C(t)|u_1 - u_2|^{2H(t)}\{1 + o(1)\},$$
$$\text{as } |u_1| + |u_2| \to 0, \tag{2.3}$$

where $C(t)$ is a non-negative function of t, and $H(t)$ represents the local scaling exponent function, or scaling function for short. Note that (2.3) suggests

$$\Upsilon_t(u_1, u_2) = 2C(t)|u_1 - u_2|^{2H(t)}\{1 + o(1)\}, \quad \text{as } |u_1| + |u_2| \to 0. \tag{2.4}$$

Since

$$\Sigma_t(u_1, u_1) + \Sigma_t(u_2, u_2) = -2\Sigma_t(u_1, u_2) = \Upsilon_t(u_1, u_2),$$

the relations (2.3) and (2.4) are equivalent.

Definition (2.3) relates the local autocorrelation of $Y(t)$ to the value of $H(t)$. With $C(t) > 0$, for fixed $|u_1 - u_2|$ near 0, the covariance between $Y(t+u_1)$ and $Y(t+u_2)$ becomes smaller as $H(t)$ moves from 1 towards 0. As a consequence, sections of sample paths where $H(t)$ is near 0 should appear more rough and erratic than sections where $H(t)$ is near 1.

Let $X(t) = Y(t) - Y(t-1)$ denote the increment process of $Y(t)$, with covariance function

$$\Psi_t(u_1, u_2) = \text{Cov}\{X(t+u_1), X(t+u_2)\}.$$

The *Wigner–Ville distribution* of $X(t)$, which serves as a local version of the power spectrum, is defined as

$$g_t(\lambda) = \frac{1}{2\pi} \int \Psi_t(\tau/2, -\tau/2) e^{-i\lambda\tau} \, d\tau.$$

It can be shown that

$$g_t(\lambda) = K(t)\lambda^{1-2H(t)}\{1 + o(1)\}, \quad \text{as } \lambda \to 0, \tag{2.5}$$

where $K(t)$ is a non-negative function of t. When $H(t) \in (1/2, 1)$, (2.5) implies that $g_t(\lambda)$ is unbounded at $\lambda = 0$. This spectral characteristic is indicative of long-range dependence.

If the increment process $X(t)$ is stationary, $g_t(\lambda)$ is independent of t and reduces to the ordinary power spectrum of $X(t)$. In this instance, $K(t)$ and $H(t)$ are constant, and (2.5) becomes

$$g(\lambda) = K\lambda^{1-2H}\{1 + o(1)\}, \quad \text{as } \lambda \to 0. \tag{2.6}$$

Estimation of constant H in (2.6) has been considered by many authors, mostly in the context of long-memory time series where $H \in (1/2, 1)$.

We now present and discuss examples of processes that exhibit locally self-similar behavior. First, consider the process defined for real $t \geq 0$ by the stochastic integral

$$Y(t) = \int_{-\infty}^{0} \left\{ (t-u)^{H(t)-1/2} - (-u)^{H(t)-1/2} \right\} dW(u)$$

$$+ \int_{0}^{t} (t-u)^{H(t)-1/2} \, dW(u), \tag{2.7}$$

where $W(t)$ is standard Brownian motion, and $H(t) \in (0, 1)$ represents the scaling function. This process is an extension of fractional Brownian motion, fBm, that allows for the self-similarity parameter to vary over time. For this reason, the process is called *generalized fractional Brownian motion*, gfBm.

In the following section, we explore many of the important theoretical and conceptual properties of gfBm. We provide simple, explicit forms for its variance function, covariance function, and variogram. We show that gfBm may be locally approximated by fBm, and establish that gfBm obeys the law of the iterated logarithm. Finally, we derive the fractal, Hausdorff, and local dimensions of gfBm sample paths.

The increment process of fBm is called *fractional Gaussian noise*, fGn. fGn is stationary with a spectrum that satisfies (2.6). The increment process of gfBm is non-stationary (unless $H(t)$ is constant) with a spectrum that satisfies (2.5). Since gfBm relaxes many homogeneous restrictions of fBm, the process may be used to model many natural phenomena with time-varying dynamics that cannot be accommodated by either standard Brownian motion or fBm (cf. Mandelbrot, 1983; Gonçalvès and Flandrin, 1993; Flandrin and Gonçalvès, 1994).

Next, consider the process defined for non-negative integer t as

$$\Phi(B)(1-B)^{H(t)-1/2}X(t) = \Theta(B)\varepsilon(t), \qquad (2.8)$$

where B is a backshift operator defined by $BX(t) = X(t-1)$, $\Phi(B)$ and $\Theta(B)$ are polynomials in B having characteristic roots outside the unit circle, $\varepsilon(t)$ is Gaussian white noise, and $H(t) \in (0, 1)$ represents the scaling function. This process is an extension of a fractional autoregressive integrated moving-average, or fARIMA, process. Again, the extension allows for the self-similarity parameter to evolve over time. We refer to this process as a *generalized* fARIMA or gfARIMA process.

A fARIMA process is stationary with a spectrum that satisfies (2.6). A gfARIMA process is non-stationary (unless $H(t)$ is constant) with a spectrum that satisfies (2.5). It can be shown that the normalized partial sums of a fARIMA process have the same limiting distribution as that of a globally self-similar process (see Beran, 1994, pp. 48–50). Analogously, it can be shown that the normalized partial sums of a gfARIMA process have the same limiting distribution as that of a locally self-similar process. Thus, a fARIMA process may be regarded as the increment process for a globally self-similar process, and a gfARIMA process may be regarded as the increment process for a locally self-similar process.

fARIMA processes have been extensively studied in the context of long-memory time series (see Beran, 1994). gfARIMA processes have even greater potential for widespread applicability, since they provide for the modeling of non-stationary time series that exhibit irregular patterns of roughness.

A special case of (2.8) is defined by

$$(1-B)^{H(t)-1/2}X(t) = \varepsilon(t). \qquad (2.9)$$

We refer to such a process as *generalized fractionally integrated noise*, gfin, since it represents an extension of fractionally integrated noise, fin.

Many of the fundamental properties of the two aforementioned locally self-similar processes have yet to be established. In the next section, we investigate the first of these processes, gfBm. We exhibit the form of its variance function, covariance function, and variogram. We show that gfBm may be locally approximated by fBm, and establish that gfBm obeys the law of the iterated logarithm. We also derive the fractal, Hausdorff, and local dimensions of gfBm sample paths.

3. Generalized fractional Brownian motion

Fractional Brownian motion has been used in modeling a host of natural phenomena from a wide range of scientific fields (e.g., Mandelbrot and van Ness, 1968; Mandelbrot, 1983; Peitgen et al., 1992; Wornell and Oppenheim, 1992). Such a process may be defined by (2.7) provided that $H(t) \equiv H$ is regarded as constant. As with any self-similar process, the process dynamics are largely characterized by the value of $H \in (0, 1)$. When $H = 1/2$, (2.7) reduces to ordinary Brownian motion. When $H > 1/2$, (2.7) involves fractional integration, which is a smoothing operation. In this case, the process is said to be *persistent*, yielding sample paths that appear trend-like, exhibiting gradual, sustained ripples. When $H < 1/2$, (2.7) involves fractional differentiation, which is an unsmoothing operation. In this case, the process is said to be *anti-persistent*, yielding sample paths that appear noisy, exhibiting jagged, erratic shifts.

Like Brownian motion, fractional Brownian motion has many properties that are homogeneous in nature, including stationary increments, global self-similarity, and constant local dimensions. Such properties prohibit fBm from being used to describe complicated phenomena characterized by different modes of regularity. For instance, in modeling landscapes via random fractals and $1/f$ signals, processes that allow for varying patterns of roughness are desirable (cf. Mandelbrot, 1983; Wang, 1997). In this section, we explore the properties of generalized fractional Brownian motion and demonstrate that the process has the capability to describe the dynamics of complex phenomena.

We begin in Section 3.1 by exhibiting the form of the covariance function for gfBm, which then leads to expressions for the variance function and the variogram. In Section 3.2, we investigate local properties of gfBm. We show that the process may be locally approximated by fBm, and establish that it obeys the law of the iterated logarithm. Finally, in Section 3.3, we derive the fractal, Hausdorff, and local dimensions of its sample paths.

3.1. Covariance function, variance function, variogram

In the definition of gfBm provided in Section 2, we assume that the process $Y(t)$ has mean zero. Thus, the covariance between $Y(t)$ and $Y(s)$ is given by $E\{Y(t)Y(s)\}$ and the variance of $Y(t) - Y(s)$ by $E[\{Y(t) - Y(s)\}^2]$. In this section, we will denote the covariance function and variogram respectively by writing

$$R(s,t) = E\{Y(t)Y(s)\} \quad \text{and} \quad V(s,t) = E[\{Y(t) - Y(s)\}^2].$$

Thus, in the notation of (2.1) and (2.2), we have $\Sigma_t(u_1, u_2) = R(t + u_1, t + u_2)$ and $\Upsilon_t(u_1, u_2) = V(t + u_1, t + u_2)$.

We note that in the definition of gfBm from Section 2, a constant starting value $Y(0)$ may be added to the right-hand side of (2.7). Without loss of generality, we have assumed this constant is zero.

The following theorem provides an explicit form for the covariance function of gfBm. All proofs appear in the Appendix.

THEOREM 1.

$$\begin{aligned}R(s,t) &= t^{H(s)+H(t)}\left[\{H(s)+H(t)\}^{-1} + \int_0^\infty f_H(u;s,t)\,du\right]\Big/2 \\ &\quad + s^{H(s)+H(t)}\left[\{H(s)+H(t)\}^{-1} + \int_0^\infty f_H(u;t,s)\,du\right]\Big/2 \\ &\quad - |t-s|^{H(s)+H(t)}\left[\{H(s)+H(t)\}^{-1} + \int_0^\infty f_H(u;s\wedge t, s\vee t)\,du\right]\Big/2,\end{aligned}$$

where

$$f_H(u; s, t) = u^{H(s)+H(t)-1} + (1+u)^{H(s)+H(t)-1} \\ - 2u^{H(s)-1/2}(1+u)^{H(t)-1/2}.$$

From Theorem 1, formulas immediately follow for the variance function and the variogram of gfBm.

COROLLARY 1.

$$\mathrm{Var}\{Y(t)\} = t^{2H(t)}\left[\{2H(t)\}^{-1} + \int_0^\infty f_H(u; t, t)\,du\right],$$

$$V(t, s) = |t-s|^{H(s)+H(t)}\left[\{H(s)+H(t)\}^{-1} + \int_0^\infty f_H(u; s \wedge t, s \vee t)\,du\right] \\ + t^{2H(t)}\left[\{2H(t)\}^{-1} + \int_0^\infty f_H(u; t, t)\,du\right] \\ - t^{H(s)+H(t)}\left[\{H(s)+H(t)\}^{-1} + \int_0^\infty f_H(u; s, t)\,du\right] \\ + s^{2H(s)}\left[\{2H(s)\}^{-1} + \int_0^\infty f_H(u; s, s)\,du\right] \\ - s^{H(s)+H(t)}\left[\{H(s)+H(t)\}^{-1} + \int_0^\infty f_H(u; t, s)\,du\right].$$

REMARK 1.1. If $H(t) \equiv H$ is constant, $Y(t)$ is fBm with self-similarity parameter H. The formula in Theorem 1 reduces to the exact covariance function of fBm. Indeed, since

$$f_H(u; s, t) = u^{2H-1} + (1+u)^{2H-1} - 2u^{H-1/2}(1+u)^{H-1/2} \\ = \{u^{H-1/2} - (1+u)^{H-1/2}\}^2,$$

we have

$$R(s, t) = \left(t^{2H} + s^{2H} - |t-s|^{2H}\right) \\ \times \left[(2H)^{-1} + \int_0^\infty \{u^{H-1/2} - (1+u)^{H-1/2}\}^2\,du\right]\bigg/2.$$

REMARK 1.2. If $H(s) + H(t) = 1$, then

$$H(s) - 1/2 = 1/2 - H(t) = \{H(s) - H(t)\}/2,$$

and

$$f_H(u; s, t) = 2 - 2\left(\frac{u}{1+u}\right)^{\frac{H(s)-H(t)}{2}}.$$

The formula for $R(s, t)$ simplifies to

$$R(s, t) = (s \wedge t)\left[1 + \int_0^\infty \left\{2 - \left(\frac{u}{1+u}\right)^{\frac{H(s)-H(t)}{2}} - \left(\frac{u}{1+u}\right)^{\frac{H(t)-H(s)}{2}}\right\} du\right].$$

As an example of an instance where the preceding formula would be applicable, suppose $H(t) = \alpha$ for $t \in (0, 1/3]$ and $H(t) = 1 - \alpha$ for $t \in [2/3, 1)$. Then over the intervals $(0, 1/3]$ and $[2/3, 1)$, $Y(t)$ corresponds to fBm with respective self-similarity parameters α and $1 - \alpha$, and for $s \in (0, 1/3]$ and $t \in [2/3, 1)$,

$$R(s, t) = s\left[1 + \int_0^\infty \left\{2 - \left(\frac{u}{1+u}\right)^{\alpha-1/2} - \left(\frac{u}{1+u}\right)^{1/2-\alpha}\right\} du\right].$$

For $\alpha \neq 1/2$, α and $1 - \alpha$ fall on opposite sides of $1/2$, so $Y(t)$ is persistent over one of the intervals ($H > 1/2$) and anti-persistent over the other ($H < 1/2$). Moreover, the preceding covariance function provides the covariance between the persistent and anti-persistent sections of the process.

3.2. Local properties

Consider a fixed time point t_*. The following theorem shows for small δ, the variance of $Y(t_* + \delta) - Y(t_*)$ is approximately the same as that for an analogous difference corresponding to fBm with self-similarity parameter $H(t_*)$. The theorem further establishes that $Y(t)$ may be locally approximated at t_* by fBm.

THEOREM 2. *Suppose $H(t)$ is twice continuously differentiable at t_*.*

(1) $$V(t_* + \delta, t_*) = |\delta|^{2H(t_*)}\left[\{2H(t_*)\}^{-1} + \int_0^\infty f_H(u; t_*, t_*) du\right] + O(\delta^2).$$

(2) *Let $\widetilde{Y}(t)$ be fBm with parameter $H(t_*)$, i.e.,*

$$\widetilde{Y}(t) = \int_{-\infty}^0 \left\{(t-u)^{H(t_*)-1/2} - (-u)^{H(t_*)-1/2}\right\} dW(u)$$

$$+ \int_0^t (t-u)^{H(t_*)-1/2} dW(u).$$

Then,

$$\mathrm{Var}\{Y(t_* + \delta) - \widetilde{Y}(t_* + \delta)\} = O(\delta^2).$$

REMARK 2.1. If $H(t_*) < 1/2$, under the assumption that $H(t)$ is once continuously differentiable, the proof of Theorem 2 can be modified to show that the results of the theorem hold with error terms of order δ.

REMARK 2.2. Note that the first result of Theorem 2 reflects the local self-similarity of gfBm, since it clearly suggests the variogram property (2.4).

The following theorem establishes the law of the iterated logarithm for gfBm.

THEOREM 3. *Let*

$$M(t_*, \delta) = \sup\{|Y(s) - Y(t)|;\ s, t \in (t_* - \delta, t_* + \delta)\}.$$

Then there exist constants c and C depending only on $H(t_*)$ such that

$$\liminf_{\delta \to 0} \frac{M(t_*, \delta)}{(\delta/\log|\log \delta|)^{H(t_*)}} = c, \qquad \limsup_{\delta \to 0} \frac{M(t_*, \delta)}{\delta^{H(t_*)}/(\log|\log \delta|)^{1/2}} = C.$$

3.3. Fractal, Hausdorff, and local dimensions

Fundamental characteristics of the behavior of a stochastic process are quantified via the fractal, Hausdorff, and local dimensions of its sample paths. In what follows, we show that for gfBm, these dimensions are solely determined by $H(t)$. The results reinforce the assertion that the dynamics of a locally self-similar process $Y(t)$ are principally dictated by the scaling function $H(t)$.

We begin by providing general definitions for the fractal, Hausdorff, and local dimensions based on the graph of a function $g(x)$. The first two of these definitions will involve the graph of the function over an interval $[a, b] \subset [0, \infty)$, denoted by $G = \{(x, g(x));\ x \in [a, b]\} \subset \mathbb{R}^2$.

Let $d(z, \varepsilon)$ represent a disk centered at $z \in \mathbb{R}^2$ with radius $\varepsilon > 0$, and let $N_\varepsilon(G)$ be the minimum number of disks $d(z, \varepsilon)$ that cover G. The *fractal dimension* (or *box dimension*) of $g(x)$ on $[a, b]$ is defined as

$$\Delta_{[a,b]}(g) = \lim_{\varepsilon \to 0} \frac{\log N_\varepsilon(G)}{|\log \varepsilon|}. \tag{3.1}$$

Next, let $\{d_i\}$ denote a collection of open disks with diameter bounded by $\varepsilon > 0$ chosen so that the union of the disks covers G. Let the α-*dimensional outer measure* be defined as

$$S_\alpha(G) = \lim_{\varepsilon \to 0}\left[\inf\left\{\sum_i (\text{diameter } d_i)^\alpha\right\}\right],$$

where the infimum is taken over all such collections $\{d_i\}$, and the sum is taken over all disks in a collection. The *Hausdorff dimension* of $g(x)$ on $[a, b]$ is defined as

$$D_{[a,b]}(g) = \inf\{\alpha\colon S_\alpha(G) = 0\} = \sup\{\alpha\colon S_\alpha(G) = \infty\}. \tag{3.2}$$

(See Mandelbrot, 1983; Adler, 1981; Barnsley, 1993; Tricot, 1995.)

Finally, the *local dimensions* of $g(x)$ at a point x are defined in terms of (3.1) and (3.2) as

$$\Delta_x(g) = \lim_{\delta \to 0} \Delta_{[x-\delta, x+\delta]}(g), \qquad D_x(g) = \lim_{\delta \to 0} D_{[x-\delta, x+\delta]}(g).$$

The following two theorems provide the fractal and Hausdorff dimensions for the sample paths of gfBm.

THEOREM 4. *Let $H_{ab} = \min\{H(t); t \in [a, b]\}$. If $H_{ab} < 1/2$, suppose $H(t)$ is once continuously differentiable on $[a, b]$; if $H_{ab} \geq 1/2$, suppose $H(t)$ is twice continuously differentiable on $[a, b]$. Then the fractal and Hausdorff dimensions of the sample paths of $Y(t)$ are given by*

$$\Delta_{[a,b]}(Y) = D_{[a,b]}(Y) = 2 - H_{ab}.$$

REMARK 4.1. Theorem 4 confirms the intuition that the fractal and Hausdorff dimensions of gfBm are dominated by the roughest segments of its sample paths.

Our final theorem provides the local dimensions for the sample paths of gfBm. The result is obtained by applying Theorem 4 to the interval $[a, b] = [t - \delta, t + \delta]$ and letting $\delta \to 0$.

THEOREM 5. *If $H(t) < 1/2$, suppose $H(t)$ is once continuously differentiable at t; if $H_{ab} \geq 1/2$, suppose $H(t)$ is twice continuously differentiable at t. Then the local dimensions of the sample paths of $Y(t)$ are given by*

$$\Delta_t(Y) = D_t(Y) = 2 - H(t).$$

4. Estimating the scaling function

The results of Section 3.3 demonstrate the importance of the role of the scaling function $H(t)$ in characterizing the behavior of a locally self-similar process. In this section, we outline the procedure used to estimate this function.

Our development is based on the application of the continuous wavelet transformation to a locally self-similar process $Y(t)$ defined in continuous time. Of course, in practice, $Y(t)$ will be measured over a collection of discrete time points, yielding a finite sample of measurements. Thus, in Section 5, we discuss the implementation of the procedure to a finite series via the discrete wavelet transformation.

To begin, let ψ denote the mother wavelet. Assume ψ has at least two vanishing moments and is at least twice continuously differentiable. Let $TY(a, t)$ denote the wavelet transformation of the locally self-similar process $Y(t)$ corresponding to the scale a and the location t. We may then write

$$TY(a, t) = a^{-1/2} \int \psi\left(\frac{u-t}{a}\right) Y(u)\, du = a^{1/2} \int \psi(x) Y(t + ax)\, dx. \quad (4.1)$$

Using (4.1), (2.3), and the first vanishing moment of ψ, we obtain

$$E\{|TY(a,t)|^2\} = \frac{1}{a}\iint \psi\left(\frac{u-t}{a}\right)\psi\left(\frac{v-t}{a}\right)E\{Y(u)Y(v)\}\,du\,dv$$

$$= a\iint \psi(x)\psi(y)E\{Y(t+ax)Y(t+ay)\}\,dx\,dy$$

$$= a\iint \psi(x)\psi(y)\Sigma_t(ax,ay)\,dx\,dy$$

$$\sim a\iint \psi(x)\psi(y)\left[\frac{1}{2}\{\Sigma_t(ax,ax)+\Sigma_t(ay,ay)\}\right.$$

$$\left. - C(t)|ax-ay|^{2H(t)}\right]dx\,dy$$

$$= C_1 a^{1+2H(t)}, \quad \text{as } a \to 0, \tag{4.2}$$

where

$$C_1 = -C(t)\iint |x-y|^{2H(t)}\psi(x)\psi(y)\,dx\,dy.$$

Now let

$$y_t(a) = \log\{|TY(a,t)|^2\},$$
$$C_2 = E\left(\log[|TY(a,t)|^2/E\{|TY(a,t)|^2\}]\right),$$

and

$$\varepsilon_t(a) = \log[|TY(a,t)|^2/E\{|TY(a,t)|^2\}] - C_2.$$

Then clearly,

$$y_t(a) = C_2 + \log[E\{|TY(a,t)|^2\}] + \varepsilon_t(a). \tag{4.3}$$

Note that (4.2) and (4.3) imply the approximate regression model

$$y_t(a) \approx c + \{2H(t)+1\}\log a + \varepsilon_t(a), \quad \text{for small scale } a, \tag{4.4}$$

where $c = \log C_1 + C_2$.

Assuming that the error terms in model (4.4) are at least approximately uncorrelated, the model suggests the use of ordinary least squares to estimate $H(t)$. The general procedure is outlined as follows.

(1) Select a sequence of small scales $a_1 > \cdots > a_k$, say $a_j = 2^{-j}$ where $j = 1, \ldots, k$.
(2) Define a set of bivariate data (x_j, y_j), $j = 1, \ldots, k$, by setting

$$x_j = \log a_j \quad \text{and} \quad y_j = y_t(a_j) \quad \text{for each } j.$$

(3) Evaluate the least-squares estimate of $H(t)$ in (4.4) via

$$\widehat{H}(t) = \left\{ \frac{\sum (x_j - \bar{x})(y_j - \bar{y})}{\sum (x_j - \bar{x})^2} - 1 \right\} \Big/ 2, \qquad (4.5)$$

where $\bar{x} = \sum x_j / k$, $\bar{y} = \sum y_j / k$.

In the Appendix, we establish the following consistency result for the estimator $\widehat{H}(t)$.

THEOREM 6. *Suppose that Y(t) is Gaussian with a covariance function that satisfies* (2.3). *Then as* $k \to \infty$, $\widehat{H}(t)$ *converges in probability to* $H(t)$.

Intuitively, the magnitude of $TY(a, t)$ reflects the content of scale a present in the process at time t. The approximate regression model (4.4) implies that at time t, the degree to which the magnitude of $TY(a, t)$ decreases with a is governed by $H(t)$. According to the model, the larger the value of $H(t)$, the greater the rate at which the scale content decays with the corresponding scale. Since fine scale content should be more prevalent at times where a sample path appears rough than at times where a path appears smooth, larger values of $H(t)$ should therefore be associated with smoother sample paths.

Having conceptualized the procedure in terms of the continuous wavelet transformation, we now discuss its implementation to a finite series via the discrete wavelet transformation.

5. Implementation of the estimation procedure

In practice, we generally observe a process $Y(t)$ at a set of discrete time points. For convenience, we will assume that the points are scaled to lie over the interval $[0, 1)$. We will also assume that the points are equally spaced, and that the sample size is a power of two. Thus, we may index our measurements with $t_i = (i - 1)/n$, where $i = 1, \ldots, n = 2^J$.

If the sample size is not a power of two, the series may be extended to achieve such a sample size by repeating the final data value, and the results may be adjusted accordingly. (See the application in Section 7.1 for an example.)

Let the vector $Y = [Y(t_1), \ldots, Y(t_n)]'$ represent our sample of n measurements on $Y(t)$. To implement the estimation procedure, the discrete wavelet transformation or DWT will be applied to the vector Y, and an estimate of $H(t)$ will be derived from the transformed data.

The DWT of Y is a one-to-one transformation which reconfigures the information in Y so that two objectives are served: (1) large-scale behavior (e.g., trends, long-term cycles) may be easily delineated from small-scale behavior (e.g., local irregularities, noise); (2) the evolution of scale content may be monitored over time. The discrete Fourier transformation or DFT serves the first objective yet not the second, since the DFT is formulated under the assumption that frequencies persist over time.

The DWT may be written in the form WY, where W is an $n \times n$ orthogonal matrix that depends on both the wavelet and the boundary adjustment (Cohen et al., 1993a; Daubechies, 1994). Fast algorithms with complexity of order n are available for performing both the DWT and the inversion of the DWT that results in the reconstruction of the original data. (Such algorithms are even less computationally intensive than the fast Fourier transformation, which has complexity of order $n \log_2 n$.) Thus, the DWT provides an efficient, information-preserving mechanism for the characterization of scale content over time.

To label the $n = 2^J$ coefficients of the DWT, we index $n - 1$ of the coefficients dyadically:

$$y_{j,k}, \quad k = 0, \ldots, 2^j - 1, \; j = 0, \ldots, J - 1.$$

We then label the remaining coefficient $y_{-1,0}$. The coefficient $y_{j,k}$ is referred to as the DWT of Y at level j and location $k2^{-j}$. For $k = 0, \ldots, 2^j - 1$ and $j = 0, \ldots, J - 1$, the quantity $y_{j,k}$ may be viewed as an approximation to $TY(2^{-j}, k2^{-j})$, the continuous wavelet transformation $TY(a, t)$ evaluated at scale $a = 2^{-j}$ and location $t = k2^{-j}$ (Donoho and Johnstone, 1994). Intuitively, the magnitude of $y_{j,k}$ reflects the content of scale $a = 2^{-j}$ in that portion of the series which lies over the interval $[k2^{-j}, (k+1)2^{-j})$.

In what follows, we outline the implementation of the estimation procedure for $H(t)$ in a series of five steps.

1. Perform the DWT on the series Y to obtain the DWT coefficients

$$\{y_{j,k} \mid k = 0, \ldots, 2^j - 1; \; j = 0, \ldots, J - 1\}.$$

2. Partition the sampling interval $[0, 1)$ into 2^l nonoverlapping subintervals of equal length, where l is an integer chosen such that $0 \leq l \leq (J - 1)$. The 2^l subintervals are of the form

$$I_m = [m2^{-l}, (m+1)2^{-l}), \quad m = 0, \ldots, 2^l - 1.$$

3. Choose an integer $J' \leq (J - 1)$ such that $2^{-J'}$ represents the largest scale to be used in the procedure. All DWT coefficients $y_{j,k}$ where $J' \leq j \leq (J - 1)$ will be used in estimating the $H(t)$ curve.

4. For each I_m, $m = 0, \ldots, 2^l - 1$, pool together values of $y_{j,k}$ to be used in constructing the local estimate for $H(t)$. Use the $y_{j,k}$ corresponding to levels j such that $J' \leq j \leq (J - 1)$, and to locations $k2^{-j}$ such that the associated interval $[k2^{-j}, (k+1)2^{-j})$ overlaps with I_m. Specifically, for each $m = 0, \ldots, 2^l - 1$, perform the following steps.

a. Define the bivariate collection of data

$$\{X_m, Y_m\} = \{(\log(2^{-j}), \log(|y_{j,k}|^2)) \mid [k2^{-j}, (k+1)2^{-j}) \cap I_m \neq \emptyset;$$
$$0 \leq k \leq 2^j - 1, \; J' \leq j \leq (J - 1)\}.$$

b. Fit an ordinary least-squares line to $\{X_m, Y_m\}$, treating the X_m as the regressor measurements and the Y_m as the response measurements.

c. Solve for the estimate $\widehat{H}(t)$ by adjusting the estimate of the slope in the least-squares fit by first subtracting 1 and then dividing by 2, as indicated in (4.5).

One may envision each $\widehat{H}(t)$ as estimating the average value of the scaling function $H(t)$ over the corresponding subinterval I_m. The appropriate time index for the $\widehat{H}(t)$ associated with I_m might be regarded as the midpoint of I_m, namely $2^{-l-1} \times (2m+1)$.

5. Construct a curve from the collection of local estimates $\widehat{H}(t)$ by employing a smoothing algorithm, such as local polynomial smoothing. This curve then serves to approximate the shape of the scaling function $H(t)$.

As mentioned in the Introduction, if a globally self-similar process $Y(t)$ has stationary increments and finite second moments, H may be taken to lie over the interval $(0, 1)$. Values of H outside of this range lead to processes that are not of practical interest. However, in applying our estimation procedure for $H(t)$ to real data, estimates of $H(t)$ that fall below 0 or above 1 will occasionally occur. Such estimates may arise for a variety of reasons. For instance, if the procedure is applied to the first difference of a locally self-similar process, estimates of $H(t)$ between -1 and 0 generally result; if the procedure is applied to the partial sums of a locally self-similar process, estimates of $H(t)$ between $+1$ and $+2$ generally result. Thus, estimates of $H(t)$ outside the interval $(0, 1)$ may imply that the underlying series (or subseries) must be suitably differenced or accumulated in order to obtain a series amenable to our algorithm (i.e., one that may be regarded as locally self-similar with locally stationary increments and finite second moments). Of course, such estimates may also occur if the underlying series (or subseries) reflects the dynamics of a process that lies outside the realm of locally self-similar processes. This issue, however, is beyond the scope of the present paper.

6. Simulations

We test the performance of our algorithm in twelve simulation sets: six based on the gfBm process (2.7) and six based on the partial sums of the gfIn process (2.9). We simulate the processes so that the time index t is confined to the interval $[0, 1)$.

To simulate realizations of the gfBm process (2.7), consider setting

$$W_n(t_i) = \frac{1}{\sqrt{n}} \sum_{k=1}^{i} \varepsilon(t_k), \quad t_i = \frac{(i-1)}{n}, \quad i = 1, \ldots, n, \tag{6.1}$$

where the $\varepsilon(t_k)$ are variates of a Gaussian white noise process. For large n, the values of $W_n(t_i)$ may be treated as realizations of Brownian motion, since the normalized partial sum in (6.1) approximates the stochastic integral that defines such a process. The second of the two stochastic integrals that comprise (2.7) may then be approximated by its discretized analogue

$$Y_{2,n}(t_i) = \sum_{k=1}^{i}(t_i - t_k)^{H(t_i)-1/2}\{W_n(t_k) - W_n(t_{k-1})\} \quad \{W(t_0) \equiv 0\}$$

$$= \frac{1}{\sqrt{n}}\sum_{k=1}^{i}(t_i - t_k)^{H(t_i)-1/2}\varepsilon(t_k), \quad t_i = \frac{(i-1)}{n}, \, i = 1, \ldots, n.$$

Similarly, the first of the two stochastic integrals in (2.7) can be approximated by a sum of the form

$$Y_{1,N}(t_i) = \frac{1}{\sqrt{N}}\sum_{k=-N}^{0}\{(t_i - u_k)^{H(t_i)-1/2} - (-u_k)^{H(t_i)-1/2}\}\varepsilon(u_k),$$

$$u_k = \left(\frac{k}{N}\right)K, \; k = 0, -1, \ldots, -N, \; t_i = \frac{(i-1)}{n}, \, i = 1, \ldots, n,$$

where the $\varepsilon(u_k)$ are variates of a Gaussian white noise process, and the integer $N > 0$ and constant $K > 0$ are chosen such that $(u_{-N}, u_{-N+1}], (u_{-N+1}, u_{-N+2}], \ldots, (u_{-1}, u_0]$ provides a sufficiently fine partition of a suitably large interval $(-K, 0]$. The sequence

$$Y_n(t_i) = Y_{1,N}(t_i) + Y_{2,n}(t_i), \quad t_i = \frac{(i-1)}{n}, \, i = 1, \ldots, n,$$

may then be treated as realizations of the process (2.7).

We remark that the contribution of $Y_{1,N}(t_i)$ to $Y_n(t_i)$ is negligible for the purpose at hand. This contribution is therefore omitted in generating our sample paths. We also note that certain components in the sums that comprise $Y_{1,N}(t_i)$ and $Y_{2,n}(t_i)$ may be of the form 0^a where $a \leq 0$. Components of the form 0^0 can be treated as 1, and those of the form 0^a where $a < 0$ can be discarded (i.e., treated as 0).

To simulate realizations of the gfin process defined by (2.9), we use the fact that (2.9) may be represented as an infinite moving average of the form

$$X(t_i) = \sum_{k=0}^{\infty} a(t_i, k)\varepsilon(t_{i-k}), \tag{6.2}$$

where the $\varepsilon(t_{i-k})$ are variates of a Gaussian white noise process, and the $a(t_i, k)$ are defined by

$$a(t_i, k) = \frac{\Gamma(k + \{H(t_i) - 1/2\})}{\Gamma(k+1)\Gamma(H(t_i) - 1/2)} \tag{6.3}$$

(see Beran, 1994, p. 65). (Here, $\Gamma(\cdot)$ denotes the gamma function.) The infinite sum in (6.2) may be approximated by the finite sum

$$X_n(t_i) = \sum_{k=0}^{N} a(t_i,k)\varepsilon(t_{i-k}), \quad t_i = \frac{(i-1)}{n}, \quad i=1,\ldots,n, \quad (6.4)$$

provided that N is chosen to be a suitably large integer. The sequence (6.4) may then be treated as realizations of the process (6.2), or equivalently, (2.9). Accordingly, for large n, the normalized partial sums

$$Y_n(t_i) = \frac{1}{\sqrt{n}} \sum_{k=1}^{i} X_n(t_k), \quad t_i = \frac{(i-1)}{n}, \quad i=1,\ldots,n,$$

may be regarded as realizations of a locally self-similar process with a covariance function of the form (2.3). These sums are therefore amenable to our algorithm.

In computing the coefficients $a(t_i,k)$, we note that evaluating $\Gamma(k)$ for large k may lead to overflow. This problem can be avoided by using the log gamma function to compute $\log a(t_i,k)$ and exponentiating. We also note that when $H(t_i) = 1/2$, formula (6.3) for the $a(t_i,k)$ involves $\Gamma(0)$. In this instance, the $a(t_i,k)$ should be assigned the value 1 when $k=0$ and 0 otherwise.

We generate six samples of the gfBm process (2.7) and six of the gfIn process (2.9) using the following six scaling functions:

$H_1(t) = 0.7$ (constant),

$H_2(t) = 0.2 + 0.5t$ (linear),

$H_3(t) = 3.0(t - 0.5)^2 + 0.1$ (quadratic),

$H_4(t) = 4.0\{(t - 0.5)^3 + 0.125\}$ (cubic),

$H_5(t) = 0.25\{\log(10.0t + 0.5) + 1.0\}$ (logarithmic),

$H_6(t) = 0.01\exp(4.0t) + 0.25$ (exponential).

In each instance, we consider a sample of size $n = 2^{12} = 4096$ ($J = 12$).

In the estimation algorithm, we use the least asymmetric wavelet with a width of 8 (Daubechies, 1992) and the interval boundary conditions (Cohen et al., 1993b). We partition the sampling interval $[0,1)$ into $128 = 2^7$ ($l = 7$) subintervals, which results in 128 local estimates of $H(t)$ corresponding to the time indices $t = 1/256, 3/256, \ldots, 255/256$. In pooling values of $y_{j,k}$ for each estimate, we use the coefficients corresponding to the scales $2^{-4}, 2^{-5}, \ldots, 2^{-11}$ (i.e., $J' = 4$). We smooth the estimates using local polynomial smoothing.

For the simulation sets based on the gfBm processes, the sample paths and the estimated curves for $H(t)$ are displayed in Figures 1 through 12; for the sets based on the gfIn processes, the partial sums for the sample paths and the estimated curves for $H(t)$ are displayed in Figures 13 through 24.

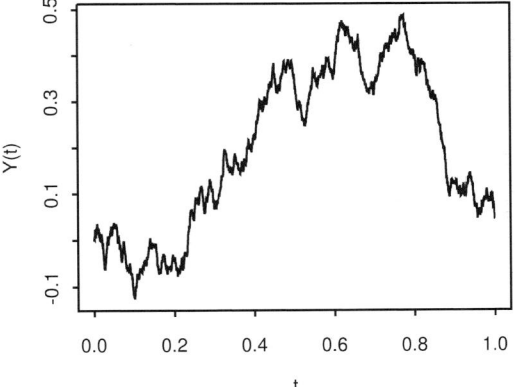

Fig. 1. Sample path of gfBm with scaling function $H_1(t)$ ($n = 4096$).

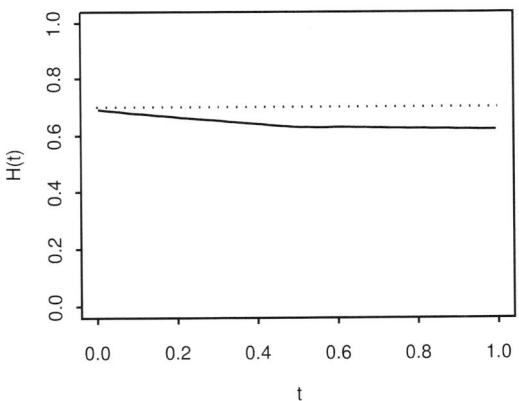

Fig. 2. Estimate of $H_1(t)$ for gfBm sample path in Figure 1.

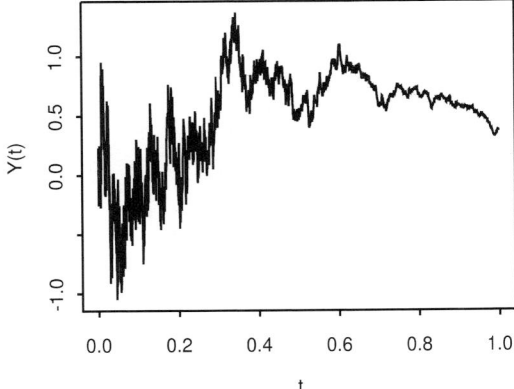

Fig. 3. Sample path of gfBm with scaling function $H_2(t)$ ($n = 4096$).

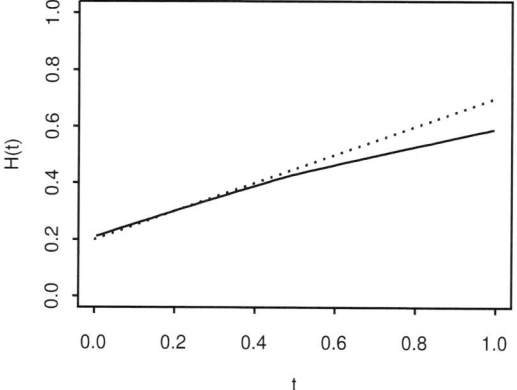

Fig. 4. Estimate of $H_2(t)$ for gfBm sample path in Figure 3.

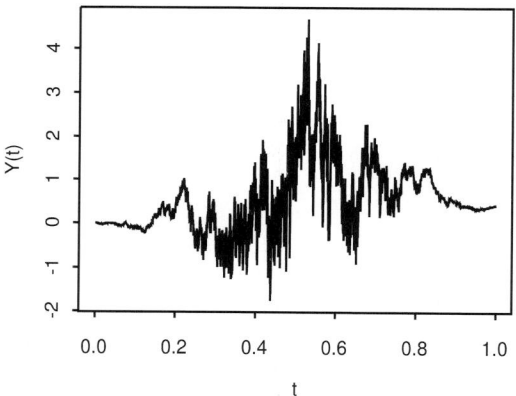

Fig. 5. Sample path of gfBm with scaling function $H_3(t)$ ($n = 4096$).

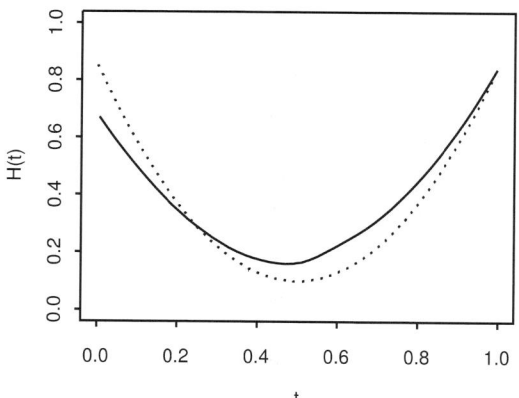

Fig. 6. Estimate of $H_3(t)$ for gfBm sample path in Figure 5.

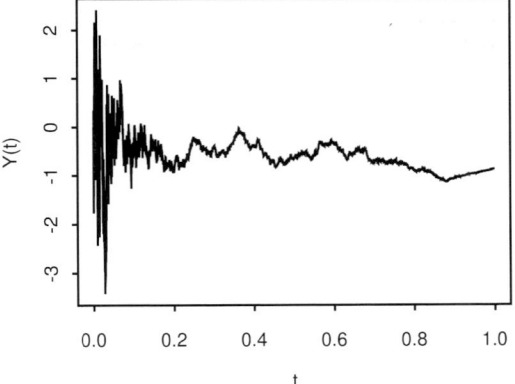

Fig. 7. Sample path of gfBm with scaling function $H_4(t)$ ($n = 4096$).

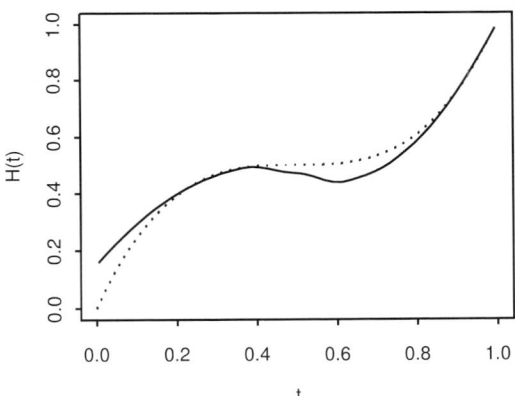

Fig. 8. Estimate of $H_4(t)$ for gfBm sample path in Figure 7.

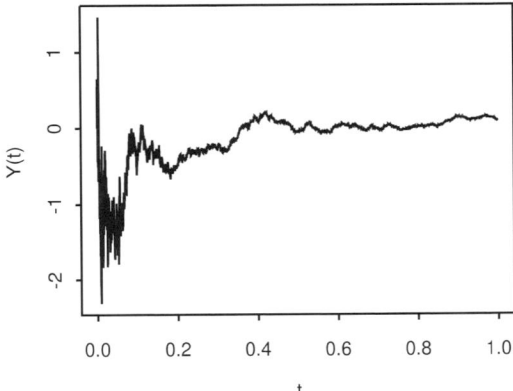

Fig. 9. Sample path of gfBm with scaling function $H_5(t)$ ($n = 4096$).

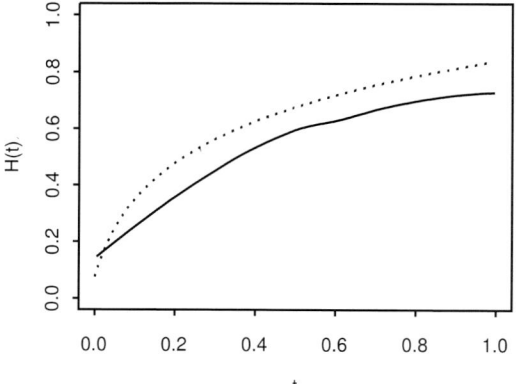

Fig. 10. Estimate of $H_5(t)$ for gfBm sample path in Figure 9.

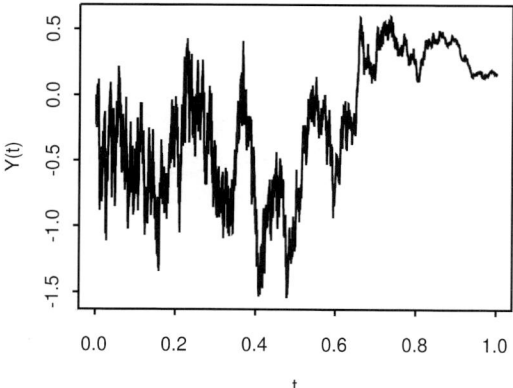

Fig. 11. Sample path of gfBm with scaling function $H_6(t)$ ($n = 4096$).

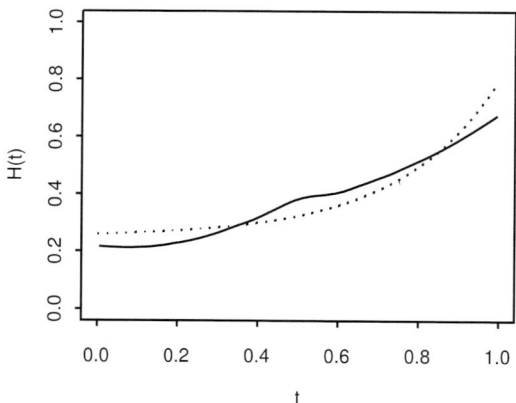

Fig. 12. Estimate of $H_6(t)$ for gfBm sample path in Figure 11.

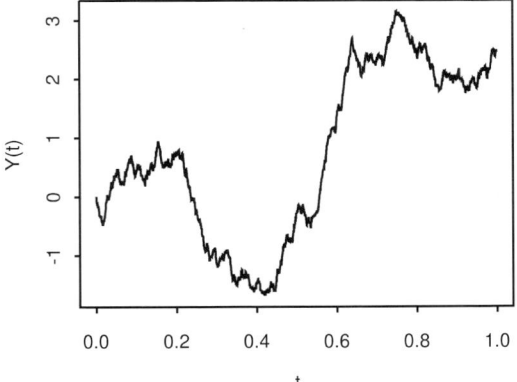

Fig. 13. Partial sums for sample path of gfin with scaling function $H_1(t)$ ($n = 4096$).

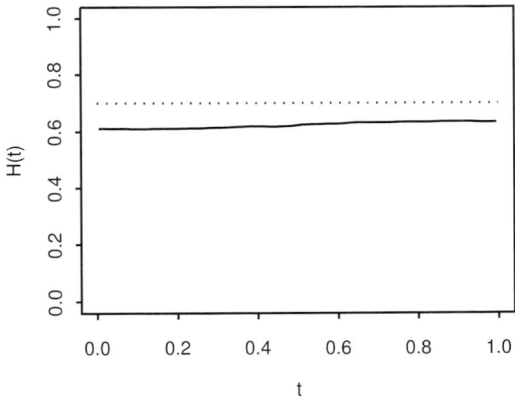

Fig. 14. Estimate of $H_1(t)$ for gfin sample path in Figure 13.

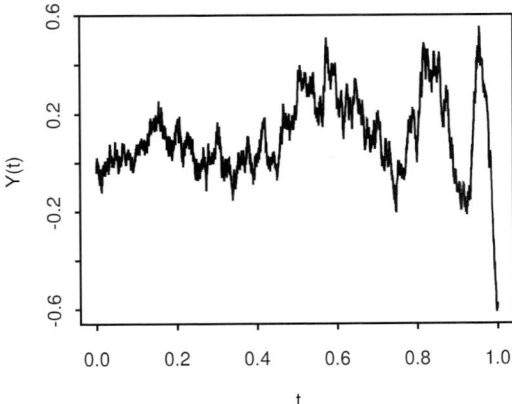

Fig. 15. Partial sums for sample path of gfin with scaling function $H_2(t)$ ($n = 4096$).

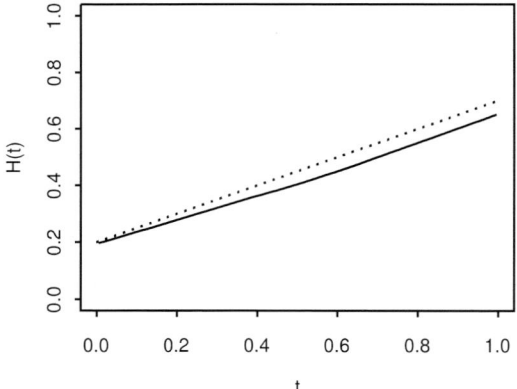

Fig. 16. Estimate of $H_2(t)$ for gfin sample path in Figure 15.

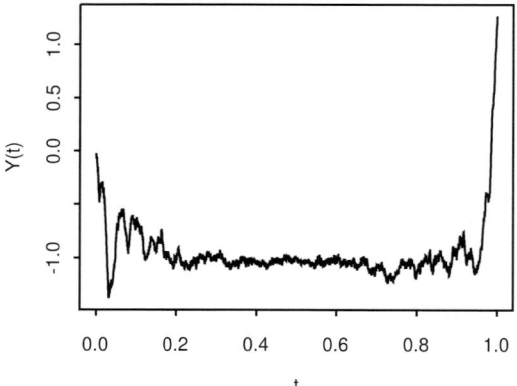

Fig. 17. Partial sums for sample path of gfin with scaling function $H_3(t)$ ($n = 4096$).

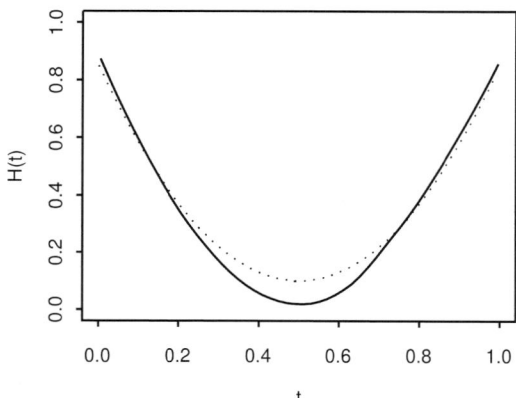

Fig. 18. Estimate of $H_3(t)$ for gfin sample path in Figure 17.

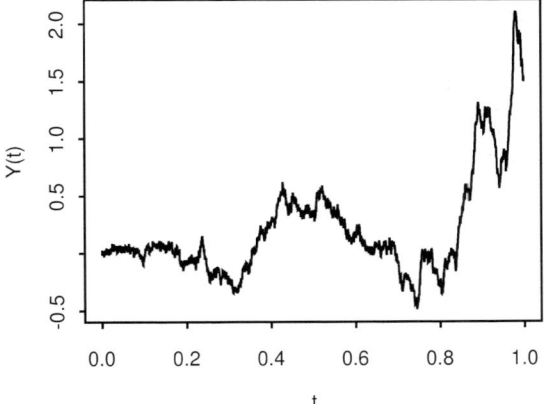

Fig. 19. Partial sums for sample path of gfin with scaling function $H_4(t)$ ($n = 4096$).

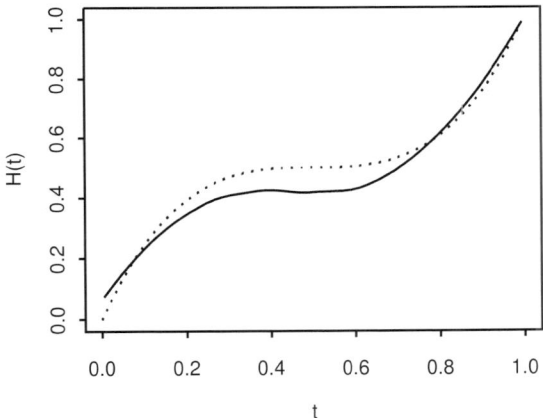

Fig. 20. Estimate of $H_4(t)$ for gfin sample path in Figure 19.

Each estimated curve effectively approximates the general shape of the corresponding scaling function $H(t)$. Note that the curves reflect a minor amount of negative bias in the $\widehat{H}(t)$: this phenomenon is also apparent in the simulation results reported by Wang et al. (2001). Not surprisingly, the number of subintervals used in the sampling interval partition (i.e., the choice of l) affects the bias: the larger the number of subintervals, the smaller the bias. Of course, employing a finer partition also inflates the variability of the estimated curve. However, the impact of this increased variation can be partly controlled by the manner in which the estimate is smoothed.

The encouraging results obtained in our simulations suggest that our estimation procedure should result in an effective characterization of $H(t)$ in large-sample settings. We now apply the procedure to three real data sets: two from hydrology and one from computer science.

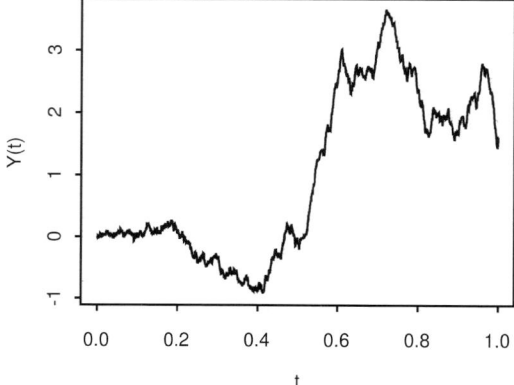

Fig. 21. Partial sums for sample path of gfin with scaling function $H_5(t)$ ($n = 4096$).

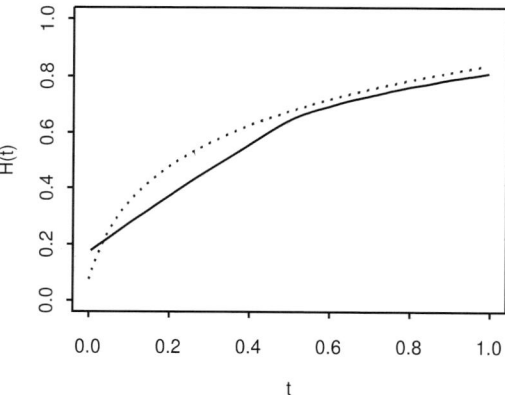

Fig. 22. Estimate of $H_5(t)$ for gfin sample path in Figure 21.

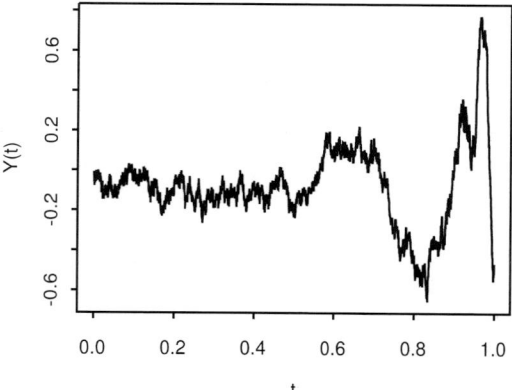

Fig. 23. Partial sums for sample path of gfin with scaling function $H_6(t)$ ($n = 4096$).

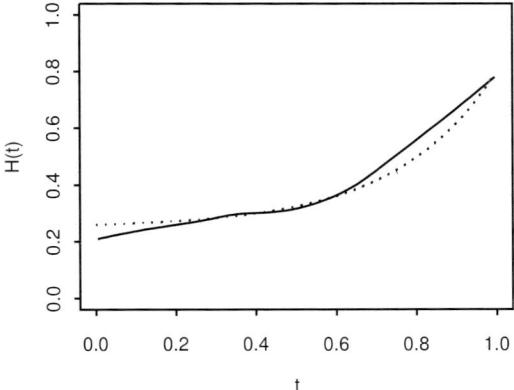

Fig. 24. Estimate of $H_6(t)$ for gfin sample path in Figure 23.

7. Applications

7.1. Vertical ocean shear measurements

Percival and Guttorp (1994) analyze a set of vertical ocean shear measurements. The data for the measurements are collected by dropping a probe into the ocean that records the water velocity every 0.1 meter as it descends. The "time" index is depth (in meters). The shear measurements (in 1/seconds) are obtained by taking a first difference of the velocity readings over 10 meter intervals, and then applying a low-pass filter to the differenced readings.

Vertical shear measurements display characteristics typical of self-similar processes; in particular, the increments of such a series often exhibit long-memory behavior. The data considered by Percival and Guttorp consist of 6875 values collected from a depth of 350.0 meters down to 1037.4 meters. The authors analyze 4096 of the values (chosen from the middle of the series) using wavelets and the Allan variance. Their justification for selecting the central part of the sample for their analysis is that "this subseries can be regarded as a portion of one realization of a process whose first backward difference is a stationary process" (p. 334). In other words, this part of the sample can be regarded as a sample path of a globally self-similar process with stationary increments.

We analyze the entire sample under the premise that the series can be treated as a realization of a locally self-similar process. Our goal is to estimate the scaling function $H(t)$.

Rather than redefining the time index so that the sampling interval is $[0, 1)$, we retain the values of the original index (in meters). Since the size of the series is not a power of two, we extend its length to $2^{13} = 8192$ ($J = 13$) by repeating the last value in the series 1317 times. We then apply our estimation algorithm by partitioning the sampling interval into $32 = 2^5$ ($l = 5$) subintervals, which results in 32 local estimates of $H(t)$. Only the first 27 estimates are relevant, since the remaining estimates pertain to the augmented portion of the series. We smooth these 27 estimates using local polynomial smoothing.

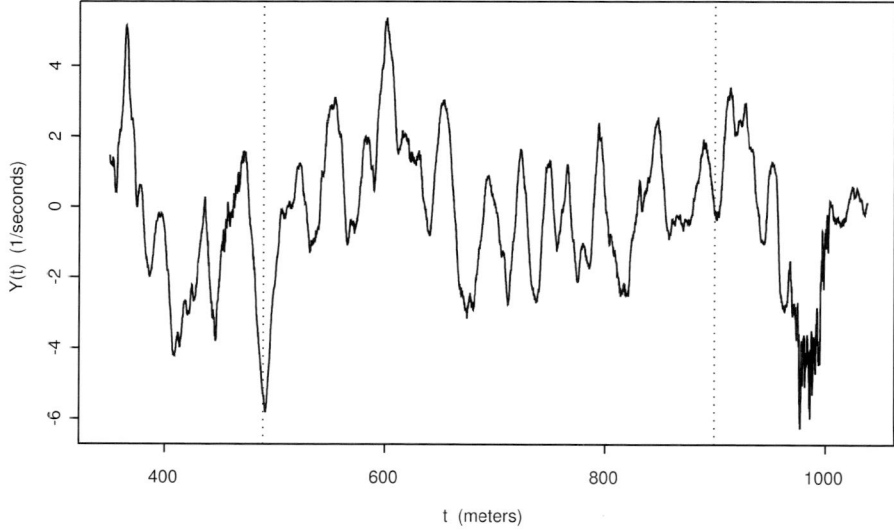

Fig. 25. Vertical ocean shear series ($n = 6875$).

In implementing our procedure, we employ the least asymmetric wavelet with a width of 8 and the interval boundary conditions. We use the DWT coefficients corresponding to the scales $2^{-4}, 2^{-5}, \ldots, 2^{-12}$ (i.e., $J' = 4$).

The series is plotted in Figure 25, and the smoothed estimated curve for $H(t)$ in Figure 26. In both figures, the range of t corresponding to the portion of the sample analyzed by Percival and Guttorp is delineated with dotted lines. We note that the $H(t)$ curve provides strong evidence that the self-similarity parameter is not constant. The curve varies over the range from 0.65 to 1.00, which reinforces the notion that the increments of the series exhibit long-memory behavior. However, the shape of the curve indicates that the nature of the long-range dependence changes as the depth increases. To model $H(t)$ as a constant over the entire sampling interval would be to ignore the local aspect of self-similarity that appears to characterize the series.

The estimated curve for $H(t)$ appears somewhat quadratic. The shape of the curve is similar to that featured in Wang et al. (2001), where the series is analyzed using the same estimation algorithm, albeit with a different choice for the wavelet. Note that the curve appears reasonably flat over the middle section between the dotted lines. This supports the argument of Percival and Guttorp that the central section of the series may be regarded as globally self-similar with stationary increments. However, when taken as a whole, the series exhibits depth-varying self-similarity patterns that can only be characterized by modeling the scaling exponent as a function of t and by estimating $H(t)$ accordingly.

7.2. Yearly minimum water levels for the Nile River

In the study of long-range dependence, one of the most widely analyzed time series is a collection of yearly minimum water levels for the Nile River. The 660 levels were

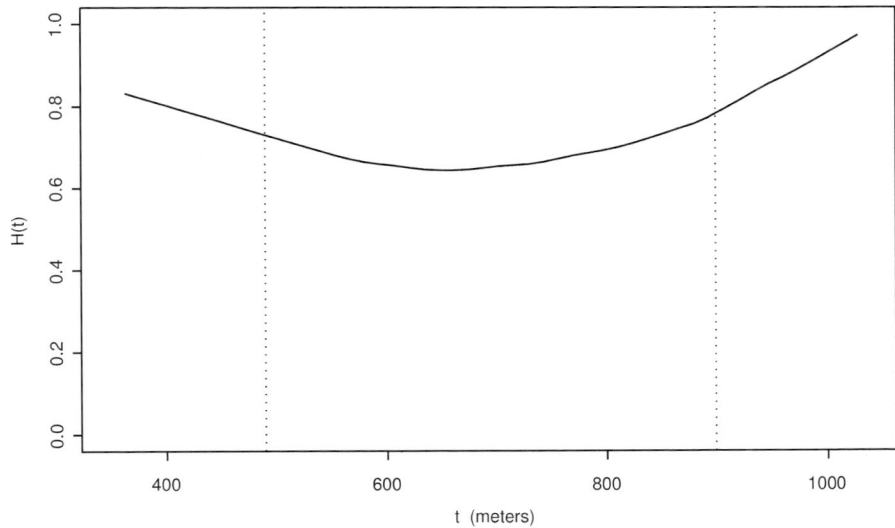

Fig. 26. Estimate of $H(t)$ for vertical ocean shear series.

recorded from 622 A.D. to 1281 A.D. at the Roda Gauge near Cairo, Egypt. Percival and Walden (2000, pp. 190–193) and Beran (1994, p. 22) discuss the data and provide interesting perspectives on its history.

Beran (1994, pp. 117–118) initially analyzes the series using two models: fGn (the increment process of fBm), and fin. Using a discrete version of Whittle's estimator, he constructs point estimates as well as 95% confidence intervals for the global self-similarity parameter H. The results are summarized below.

Model	Point estimate	95% Confidence interval
fGn	0.84	(0.79, 0.89)
fin	0.90	(0.84, 0.96)

Beran (1994, p. 206-207) presents a second analysis of the series based on a suspected change point in the self-similarity parameter around 722 A.D. He partitions the initial 600 observations of the series into 6 sections consisting of 100 observations each. The Whittle estimates for each of these 6 sections are featured below.

Time interval	(622, 721)	(722, 821)	(822, 921)	(922, 1021)	(1022, 1121)	(1122, 1221)
Point estimate	0.54	0.85	0.86	0.83	0.84	0.93

Treating the series as a realization of the increment process of a locally self-similar process, we analyze the initial $2^9 = 512$ ($J = 9$) measurements. The objective of our

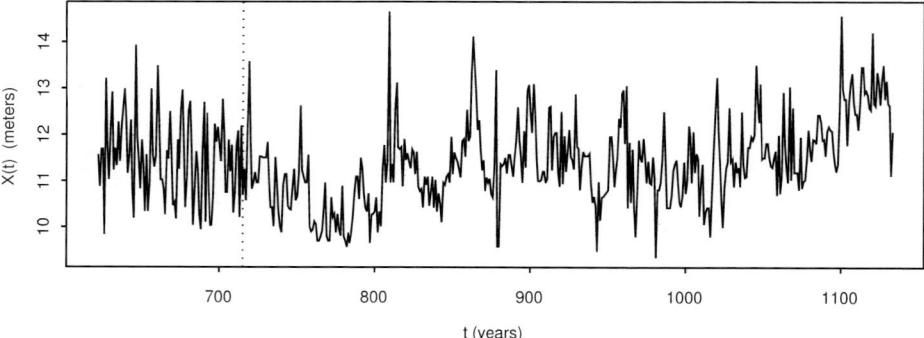

Fig. 27. Nile River water level series ($n = 512$).

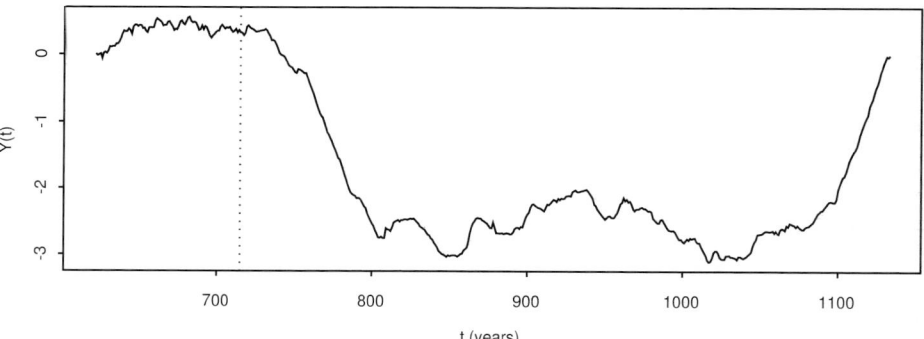

Fig. 28. Partial sums for Nile River water level series ($n = 512$).

analysis is to estimate $H(t)$ over two sections of the series: the points collected prior to 722 A.D., and the points collected from 722 onwards. We will then compare our results to those obtained by Beran.

To make the series amenable to our algorithm, the mean of the observations is subtracted, and the normalized partial sums of the resulting data are computed. The original series is plotted in Figure 27 and the partial sums in Figure 28. The time point of interest, 722 A.D., is demarcated with a dotted line.

Rather than redefining the time index so that the sampling interval is [0, 1), we retain the values of the original index (in years). We apply our algorithm by partitioning the sampling interval into $16 = 2^4$ ($l = 4$) subintervals, which results in 16 local estimates of $H(t)$. Of these 16 estimates, the first 3 apply to the time period prior to 722 A.D., and the remaining 13 apply to the period from 722 onwards. Using local polynomial smoothing, we smooth the first 3 estimates and the last 13 estimates separately, allowing for a discontinuity in the $H(t)$ curve at the time point 722 A.D.

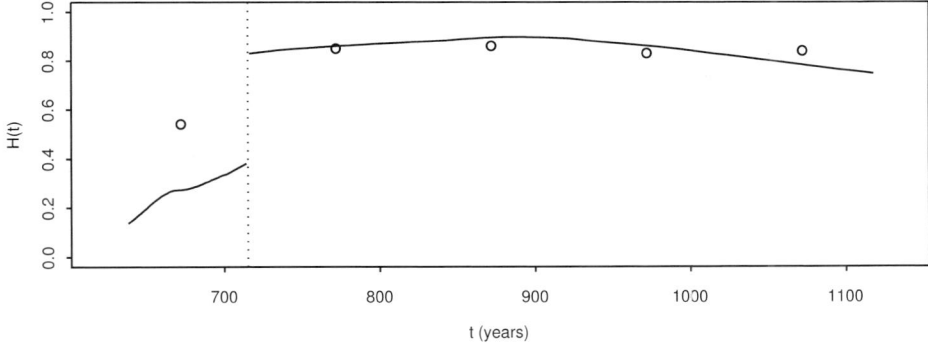

Fig. 29. Estimate of $H(t)$ for Nile River water level series.

In implementing our procedure, we employ the least asymmetric wavelet with a width of 12 and the interval boundary conditions. We use the DWT coefficients corresponding to the scales $2^{-5}, 2^{-6}, \ldots, 2^{-8}$ (i.e., $J' = 5$).

The smoothed estimated curve for $H(t)$ is featured in Figure 29, again with the time point 722 A.D. highlighted with a dotted line. The first 5 of Beran's 6 Whittle estimates are superimposed on the plot with dots. Clearly, our results support Beran's assertion that a change point in $H(t)$ occurs around 722 A.D.

Over the section of the series extending from 722 A.D. onwards, our results seem quite consistent with Beran's Whittle estimates: the $H(t)$ curve appears roughly level and oscillates around 0.85. In fact, the mean of the 13 local estimates of $H(t)$ is 0.85; curiously, this value lies within both of Beran's confidence intervals for the global estimation of H.

For the section of the series collected prior to 722 A.D., our results are somewhat different from Beran's. Here, the $H(t)$ curve rises abruptly from about 0.1 to about 0.4, and the mean of the 3 local estimates of $H(t)$ is 0.25. These results are indicative of anti-persistent behavior. On the other hand, Beran's Whittle estimate for this section is 0.54, which suggests that the observations are characteristic of white noise.

As a point of interest, Percival and Walden (2000, pp. 326–327, 386–388) discuss a suspected shift in the variance of the series around 715 A.D. They speculate that this shift is due to a change in the method used to measure the minimum water level. A test is conducted to determine if a change point occurs at 715 A.D., and the test shows significance. Based on the final 512 points in the series, a maximum likelihood estimate is constructed for a long-memory parameter δ, which is related to the self-similarity parameter H by $H = \delta + 0.5$. The result is $\hat{\delta} = 0.4452$ (p. 286), which corresponds to the estimate $\widehat{H} = 0.9452$. Thus, for the latter section of the series, the analysis of Percival and Walden indicates long-range dependence which is slightly stronger than that reflected by either our analysis or the analyses of Beran.

7.3. Ethernet network traffic measurements

Our final analysis considers a series comprised of $2^{18} = 262{,}144$ Ethernet local area network (LAN) traffic measurements collected over a one-hour period at the Bellcore Morris Research and Engineering Center. Each observation represents the number of packets sent over the Ethernet during a 10 millisecond period. A larger version of the series, featuring 360,000 observations spanning the entire hour of the experiment, is analyzed by Leland et al. (1994). The data were collected during a period of "normal" traffic load.

As explained by Leland et al., self-similarity is present in Ethernet LAN traffic data due to the absence of a natural length of a "burst." At various time scales, ranging from a few milliseconds to minutes to hours, traffic bursts are present that are similar in structure. Thus, a series of traffic measurements has the same basic appearance whether it is viewed over very short time scales (e.g., milliseconds), or over much longer scales (e.g., hundreds of seconds).

Analyzing the series of length 360,000, Leland et al. recover the self-similarity parameter H in three ways: by using the slope of a variance-time curve, by using the slope of an R/S plot, and by using a maximum likelihood estimator based on the periodogram. The results are as follows.

Method	Point estimate	95% Confidence interval
Slope of variance-time curve	0.87	NA
Slope of R/S plot	0.90	NA
Periodogram-based MLE	0.90	(0.85, 0.95)

We consider an analysis of the series of length $2^{18} = 262{,}144$ that treats the series as a realization of the increment process of a locally self-similar process. Our objective is to estimate $H(t)$, and to interpret our results in light of the analysis of Leland et al.

To make the series amenable to our algorithm, the mean of the observations is subtracted, and the normalized partial sums of the resulting data are computed. The first 1000 observations of the original series (covering a 10 second interval) are plotted in Figure 30; the entire set of partial sums (covering roughly 2600 seconds) is plotted in Figure 31.

Rather than redefining the time index so that the sampling interval is $[0, 1)$, we retain the values of the original index (in 10 millisecond increments). We apply our algorithm by partitioning the sampling interval into $1024 = 2^{10}$ ($l = 10$) subintervals, which results in 1024 local estimates of $H(t)$. We smooth these estimates using local polynomial smoothing.

In implementing our procedure, we employ the least asymmetric wavelet with a width of 8 and the interval boundary conditions. We use the DWT coefficients corresponding to the scales $2^{-4}, 2^{-5}, \ldots, 2^{-17}$ (i.e., $J' = 4$).

The smoothed estimated curve for $H(t)$ is featured in Figure 32. The limits of the aforementioned confidence interval reported by Leland et al. are represented in the

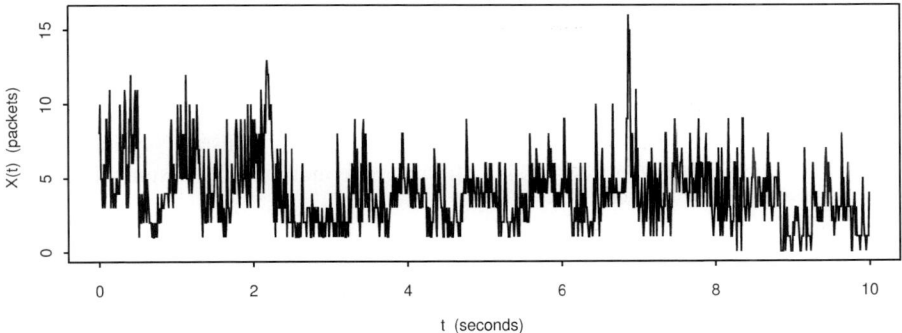

Fig. 30. Ethernet network traffic series (first 1000 observations).

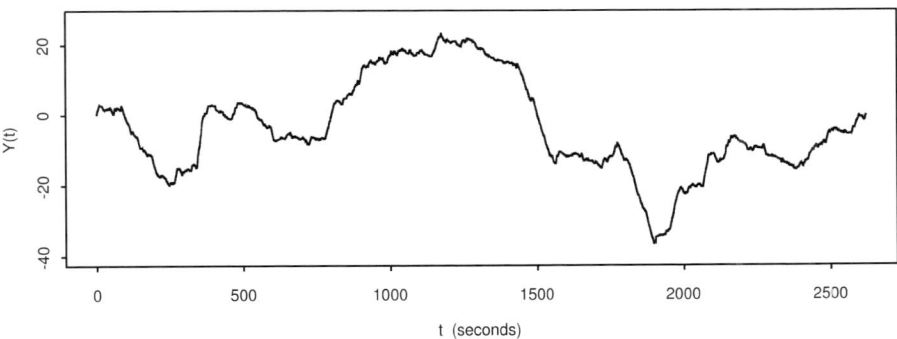

Fig. 31. Partial sums for Ethernet network traffic series ($n = 262,144$).

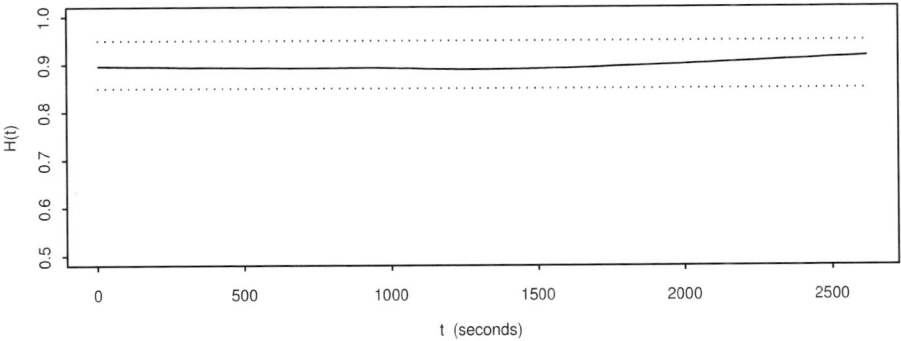

Fig. 32. Estimate of $H(t)$ for Ethernet network traffic series.

figure by dotted lines. Note that the curve for $H(t)$ is somewhat flat, and lies entirely within the confidence interval limits. Both our analysis and that of Leland et al. indicate that the series exhibits strong long-range dependence.

Our analysis suggests that in applications where the self-similarity parameter is roughly constant, our procedure should produce an estimated $H(t)$ curve which accurately reflects this characteristic. The simulation results based on the constant scaling function $H_1(t)$ reinforce this assertion. Thus, in analyzing data where it is not apparent whether the self-similar behavior remains globally constant or changes over time, our procedure provides a promising alternative to methods that assume the former.

8. Conclusion

Many naturally occurring phenomena produce data that exhibits self-similar behavior which evolves as the phenomenon progresses. Adequate modeling of such data requires the consideration of locally self-similar processes with time-varying scaling exponents.

In the preceding chapter, we have considered two examples of locally self-similar processes: generalized fractional Brownian motion, gfBm, and the limit of the normalized partial sums of generalized fractional ARIMA processes, gfARIMA. We have investigated several key theoretical properties of gfBm. Additionally, we have proposed, discussed, and investigated an algorithm for estimating the time-varying scaling function $H(t)$ of a locally self-similar process. Our algorithm exploits the time-scale localization facility of wavelets to produce a consistent estimator of $H(t)$ at a given time point t. Our simulation results, which feature both gfBm and gfARIMA processes, indicate that our method provides an accurate reflection of the progression of $H(t)$. Moreover, our applications illustrate that our method can be used to quantify time-dependent self-similarity patterns that arise in actual spatial and temporal series.

Appendix

In what follows, we present proofs for Theorems 1, 2, 3, and 4 (from Section 3), and for Theorem 6 (from Section 4). In our proofs, we use K to denote constants that may vary in specification from one usage to the next. In the proof of Theorem 6, these constants may depend on t but not on the scales a.

PROOF OF THEOREM 1. Assume $s \leqslant t$. Direct calculations establish

$$
\begin{aligned}
R(s,t) &= \int_{-\infty}^{0} \left\{ (t-u)^{H(t)-1/2} - (-u)^{H(t)-1/2} \right\} \\
&\quad \times \left\{ (s-u)^{H(s)-1/2} - (-u)^{H(s)-1/2} \right\} du \\
&\quad + \int_{0}^{s} (t-u)^{H(t)-1/2} (s-u)^{H(s)-1/2} du \\
&= \int_{0}^{\infty} \left\{ (t+u)^{H(t)-1/2} - u^{H(t)-1/2} \right\} \left\{ (s+u)^{H(s)-1/2} - u^{H(s)-1/2} \right\} du \\
&\quad + \int_{-s}^{0} (t+u)^{H(t)-1/2} (s+u)^{H(s)-1/2} du \\
&= \lim_{M \to \infty} A_M,
\end{aligned}
$$

where

$$A_M = \int_0^M \{(t+u)^{H(t)-1/2} - u^{H(t)-1/2}\}\{(s+u)^{H(s)-1/2} - u^{H(s)-1/2}\} du$$

$$+ \int_{-s}^0 (t+u)^{H(t)-1/2}(s+u)^{H(s)-1/2} du$$

$$= \int_{-s}^M (t+u)^{H(t)-1/2}(s+u)^{H(s)-1/2} du$$

$$- \int_0^M (t+u)^{H(t)-1/2} u^{H(s)-1/2} du$$

$$- \int_0^M u^{H(t)-1/2}(s+u)^{H(s)-1/2} du + \int_0^M u^{H(s)+H(t)-1} du$$

$$= \int_0^{s+M} (t-s+u)^{H(t)-1/2} u^{H(s)-1/2} du$$

$$- \int_0^M (t+u)^{H(t)-1/2} u^{H(s)-1/2} du$$

$$- \int_0^M u^{H(t)-1/2}(s+u)^{H(s)-1/2} du + M^{H(s)-H(t)}/\{H(s)+H(t)\}.$$

To make the integrals in A_M converge as $M \to \infty$, we must first subtract $u^{H(s)+H(t)-1}$ and $(1+u)^{H(s)+H(t)-1}$ from each of the integrands, and then perform a change of variables in the integrals. We obtain

$$A_M = \int_0^{s+M} \{(t-s+u)^{H(t)-1/2} u^{H(s)-1/2} - u^{H(s)+H(t)-1}/2$$

$$- (t-s+u)^{H(s)+H(t)-1}/2\} du$$

$$- \int_0^M \{(t+u)^{H(t)-1/2} u^{H(s)-1/2} - u^{H(s)+H(t)-1}/2$$

$$- (t+u)^{H(s)+H(t)-1}/2\} du$$

$$- \int_0^M \{u^{H(t)-1/2}(s+u)^{H(s)-1/2} - u^{H(s)+H(t)-1}/2$$

$$- (s+u)^{H(s)+H(t)-1}/2\} du$$

$$+ \{t^{H(s)+H(t)} + s^{H(s)+H(t)} - (t-s)^{H(s)+H(t)}\}/\{2H(s)+2H(t)\}$$

$$= -\left\{(t-s)^{H(s)+H(t)} \int_0^{(s+M)/(t-s)} f_H(u;s,t) du\right\}/2$$

$$+ \left\{t^{H(s)+H(t)} \int_0^{M/t} f_H(u;s,t) du\right\}/2$$

$$+ \left\{ s^{H(s)+H(t)} \int_0^{M/s} f_H(u; t, s) \, du \right\} \Big/ 2$$
$$+ \left\{ t^{H(s)+H(t)} + s^{H(s)+H(t)} - (t-s)^{H(s)+H(t)} \right\} \Big/ \{2H(s) + 2H(t)\}.$$

The theorem is then established by letting $M \to \infty$. □

PROOF OF THEOREM 2. (1) We consider $\delta \downarrow 0$ only. By Corollary 1, we have

$$V(t_* + \delta, t_*)$$
$$= \delta^{H(t_*)+H(t_*+\delta)} \left[\{H(t_*) + H(t_* + \delta)\}^{-1} + \int_0^\infty f_H(u; t_*, t_* + \delta) \, du \right]$$
$$+ (t_* + \delta)^{2H(t_*+\delta)} \left[\{2H(t_* + \delta)\}^{-1} + \int_0^\infty f_H(u; t_* + \delta, t_* + \delta) \, du \right]$$
$$- (t_* + \delta)^{H(t_*)+H(t_*+\delta)}$$
$$\times \left[\{H(t_*) + H(t_* + \delta)\}^{-1} + \int_0^\infty f_H(u; t_*, t_* + \delta) \, du \right]$$
$$+ t_*^{2H(t_*)} \left[\{2H(t_*)\}^{-1} + \int_0^\infty f_H(u; t_*, t_*) \, du \right]$$
$$- t_*^{H(t_*)+H(t_*+\delta)} \left[\{H(t_*) + H(t_* + \delta)\}^{-1} + \int_0^\infty f_H(u; t_* + \delta, t_*) \, du \right].$$

Note that $\int_0^\infty f_H(u; s, t) \, du$ is finite and has continuous derivatives with respect to s and t. Also, we can argue that

$$x^{H(t+\delta)} - x^{H(t)}$$
$$= \delta H'(t) x^{H(t)} \log x + \delta^2 x^{H(t^*)} \left[\{H'(t^*) \log x\}^2 + H''(t^*) \log x \right],$$

where t^* is between t_* and $t_* + \delta$, and that

$$H'(t) x^{H(t)} \log x, \qquad x^{H(t^*)} \left[\{H'(t^*) \log x\}^2 + H''(t^*) \log x \right]$$

are both bounded. Therefore, we have

$$V(t_* + \delta, t_*)$$
$$= \delta^{2H(t_*)} \left[\{2H(t_*)\}^{-1} + \int_0^\infty f_H(u; t_*, t_*) \, du \right] + O(\delta^2)$$
$$+ t_*^{2H(t_*)} \int_0^\infty \{ f_H(u; t_* + \delta, t_* + \delta) - f_H(u; t_*, t_* + \delta) + f_H(u; t_*, t_*)$$
$$- f_H(u; t_* + \delta, t_*) \} \, du$$

$$+ t_*^{2H(t_*)}\big[\{2H(t_*+\delta)\}^{-1} - 2\{H(t_*)+H(t_*+\delta)\}^{-1} + \{2H(t_*)\}^{-1}\big]$$

$$+ \delta H'(t_*) t_*^{2H(t_*)} \log t_* \int_0^\infty \{2f_H(u; t_*+\delta, t_*+\delta) - f_H(u; t_*, t_*+\delta)$$

$$- f_H(u; t_*+\delta, t_*)\}\, du$$

$$+ \delta H'(t_*) t_*^{2H(t_*)} \log t_* \big[2\{2H(t_*+\delta)\}^{-1} - 2\{H(t_*)+H(t_*+\delta)\}^{-1}\big]$$

$$+ 2\delta H(t_*) t_*^{2H(t_*)-1} \int_0^\infty \{f_H(u; t_*+\delta, t_*+\delta) - f_H(u; t_*, t_*+\delta)\}\, du$$

$$+ 2\delta H(t_*) t_*^{2H(t_*)-1} \big[\{2H(t_*+\delta)\}^{-1} - \{H(t_*)+H(t_*+\delta)\}^{-1}\big]$$

$$= \delta^{2H(t_*)} \bigg[\{2H(t_*)\}^{-1} + \int_0^\infty f_H(u; t_*, t_*)\, du\bigg] + O(\delta^2).$$

(2) By the arguments in the proof of Theorem 1, we can establish that

$$E\{Y(t_*+\delta)\widetilde{Y}(t_*+\delta)\}$$

$$= (t_*+\delta)^{H(t_*)+H(t_*+\delta)} \bigg(\{H(t_*)+H(t_*+\delta)\}^{-1}$$

$$+ \int_0^\infty \big[\{f_H(u; t_*, t_*+\delta) + f_H(u; t_*+\delta, t_*)\}/2\big]\, du\bigg).$$

Using Corollary 1 and the arguments from part (1), we can then prove $\mathrm{Var}\{Y(t_*+\delta) - \widetilde{Y}(t_*+\delta)\}$ is of order δ^2. □

PROOF OF THEOREM 3. Since $\widetilde{Y}(t)$ is fBm with self-similarity parameter $H(t_*)$, by the law of the iterated logarithm for fBm (Oodaira, 1972; Csáki, 1980; Goodman and Kuelbs, 1991; Monrad and Rootzén, 1995), there exist constants c and C depending only on $H(t_*)$ such that

$$\liminf_{\delta \to 0} \sup\left\{\frac{|\widetilde{Y}(s)-\widetilde{Y}(t)|}{(\delta/\log|\log\delta|)^{H(t_*)}};\ s,t \in (t_*-\delta, t_*+\delta)\right\} = c,$$

$$\limsup_{\delta \to 0} \sup\left\{\frac{|\widetilde{Y}(s)-\widetilde{Y}(t)|}{\delta^{H(t_*)}/(\log|\log\delta|)^{1/2}};\ s,t \in (t_*-\delta, t_*+\delta)\right\} = C.$$

Let $Z(t) = Y(t) - \widetilde{Y}(t)$, and

$$m(t_*, \delta) = \sup\{|Z(s)-Z(t)|;\ s,t \in (t_*-\delta, t_*+\delta)\}.$$

Then it is sufficient to prove

$$\lim_{\delta \to 0} \frac{m(t_*, \delta)}{\delta^{H(t_*)}/\log|\log\delta|} = 0. \tag{A.1}$$

By Theorem 2.1 of Monrad and Rootzén (1995) and by Theorem 2, for $2H(t_*) < \gamma < 2$, we have that

$$P\{m(t_*, \delta) > \varepsilon\} \leq 1 - \exp(-K\delta\varepsilon^{-2/\gamma}).$$

Let $0 < \zeta < \gamma/2 - H(t_*)$, and for $k = 1, 2, \ldots$, set

$$\delta_k = k^{1/[\{H(t_*)+\zeta\}/\gamma - 1/2]} \quad \text{and} \quad \varepsilon_k = \delta_k^{H(t_*)+\zeta}.$$

Then,

$$\sum_k P\{m(t_*, \delta_k) > \delta_k^{H(t_*)+\zeta}\} \leq \sum_k (1 - \exp[-K\delta_k^{1-2\{H(t_*)+\zeta\}/\gamma}])$$

$$\sim \sum_k K\delta_k^{1-2\{H(t_*)+\zeta\}/\gamma}$$

$$= K \sum_k k^{-2} < \infty.$$

By the Borel–Cantelli lemma, it follows that $m(t_*, \delta_k) \leq \delta_k^{H(t_*)+\zeta}$. Furthermore, for $\delta_{k+1} < \delta \leq t_k$,

$$m(t_*, \delta) \leq m(t_*, \delta_k) \leq \delta_k^{H(t_*)+\zeta} \leq \delta^{H(t_*)+\zeta}(\delta_k/\delta_{k+1})^{H(t_*)+\zeta}.$$

Thus, for small δ, $m(t_*, \delta) \leq 2\delta^{H(t_*)+\zeta}$. This establishes (A.1). □

PROOF OF THEOREM 4. Since $D_{[a,b]}(Y) \leq \Delta_{[a,b]}(Y)$, we need to show $D_{[a,b]}(Y) \geq 2 - H_{ab}$ and $\Delta_{[a,b]}(Y) \leq 2 - H_{ab}$.
(1) $\Delta_{[a,b]}(Y) \leq 2 - H_{ab}$. It is sufficient to show that for any $\gamma < H_{ab}$, we have $\Delta_{[a,b]}(Y) \leq 2 - \gamma$.

Suppose that for any $\gamma < H_{ab}$,

$$P[\sup\{|Y(t) - Y(s)|/|s-t|^\gamma; s, t \in [a,b]\} < \infty] = 1. \tag{A.2}$$

We will then have $\Delta_{[a,b]}(Y) \leq 2 - \gamma$ (Adler, 1981, Chapter 8; Tricot, 1995, Chapter 12). Now for $t - s$ bounded below from zero, with probability one, $|Y(t) - Y(s)|/|s-t|^\gamma < \infty$, so we need only show (A.2) for small $t - s$.

Theorem 2 implies that there exists a positive constant K such that for $s, t \in [a, b]$ and $t - s \to 0$,

$$V(t, s) \leq K|t-s|^{2H_{ab}}.$$

Let

$$m(\delta) = \sup\{|Y(t) - Y(s)|; s, t \in [a, b], |t - s| \leq \delta\}.$$

By Theorem 2.1 of Monrad and Rootzén (1995), we have

$$P\{m(\delta) \geq \varepsilon\} \leq 1 - \exp(-K\delta\varepsilon^{-1/H_{ab}}).$$

For $k = 1, 2, \ldots$, set $\delta_k = k^{2/(\gamma/H_{ab}-1)}$ and $\varepsilon_k = \delta_k^\gamma$. Then

$$\sum_k P\{m(\delta_k) \geq \delta_k^\gamma\} \leq \sum_k \{1 - \exp(-K\delta_k\varepsilon_k^{-1/H_{ab}})\}$$

$$\sim \sum_k K \delta_k^{1-\gamma/H_{ab}}$$

$$= K \sum_k k^{-2} < \infty.$$

By the Borel–Cantelli lemma, it follows that $m(\delta_k)\delta_k^\gamma \leq 1$. Furthermore, for $\delta_{k+1} < \delta \leq t_k$,

$$\frac{m(\delta)}{\delta^\gamma} \leq \frac{m(\delta_k)}{\delta_k^\gamma} \left(\frac{\delta_k}{\delta_{k+1}}\right)^\gamma.$$

Since $\delta_k/\delta_{k+1} \to 1$, we have $m(\delta)/\delta^\gamma \leq 2$. Finally,

$$|Y(t) - Y(s)|/|t-s|^\gamma \leq m(|t-s|)/|t-s|^\gamma.$$

This establishes (A.2).

(2) $D_{[a,b]}(Y) \geq 2 - H_{ab}$. It it sufficient to show that for any $\gamma > H_{ab}$, we have $D_{[a,b]}(Y) \geq 2 - \gamma$.

Suppose $t_* \in [a, b]$ is such that $H_{ab} = H(t_*)$. There exists an $\varepsilon > 0$ such that for $t \in [t_* - \varepsilon, t_* + \varepsilon]$, we have $\gamma > H(t) + \varepsilon$. Since $D_{[a,b]}(Y) \geq D_{[t_*-\varepsilon, t_*+\varepsilon]}(Y)$, to prove $D_{[a,b]}(Y) \geq 2 - \gamma$, it suffices to show that $D_{[t_*-\varepsilon, t_*+\varepsilon]}(Y) \geq 2 - \gamma$, which is implied by

$$\int_{t_*-\varepsilon}^{t_*+\varepsilon} \int_{t_*-\varepsilon}^{t_*+\varepsilon} E\{|Y(t) - Y(s)|^2 + |s-t|^2\}^{\gamma/2-1} \, ds \, dt < \infty.$$

(See Adler, 1981, Chapter 8, Lemma 8.2.4.)

Note that $Y(t) - Y(s)$ follows a normal distribution with mean zero. If $\phi(y)$ denotes the probability density function of the standard normal distribution, we then have

$$\int_{t_*-\varepsilon}^{t_*+\varepsilon} \int_{t_*-\varepsilon}^{t_*+\varepsilon} E\{|Y(t) - Y(s)|^2 + |s-t|^2\}^{\gamma/2-1} \, ds \, dt$$

$$= \int_{t_*-\varepsilon}^{t_*+\varepsilon} \int_{t_*-\varepsilon}^{t_*+\varepsilon} \int \{y^2 V(s,t) + |s-t|^2\}^{\gamma/2-1} \phi(y) \, dy \, ds \, dt$$

$$\leq K \int_{t_*-\varepsilon}^{t_*+\varepsilon} \int_{-2\varepsilon}^{2\varepsilon} \int \{y^2 V(t+u,t) + u^2\}^{\gamma/2-1} \, dy \, du \, dt.$$

Now the integral over $[-2\varepsilon, 2\varepsilon]$ with respect to u need only be considered over $[-\delta, \delta]$ for small $0 < \delta \leqslant 2\varepsilon$. By Theorem 2, we have that for small u, $V(t+u,t) \sim C(t)|u|^{2H(t)}$, with $C(t)$ bounded below from zero and above from infinity. Thus it is sufficient to establish the finiteness of

$$\int_{t_*-\varepsilon}^{t_*+\varepsilon} \int_{-\delta}^{\delta} \int \{y^2|u|^{2H(t)} + u^2\}^{\gamma/2-1} \, dy \, du \, dt.$$

This integral in turn is bounded by

$$8\delta \int_{t_*-\varepsilon}^{t_*+\varepsilon} \int_0^{\delta} \int_0^{\infty} \{y^2 u^{2H(t)} + u^2\}^{\gamma/2-1} \, dy \, du \, dt$$

$$= 8\delta \int_{t_*-\varepsilon}^{t_*+\varepsilon} \int_0^{\delta} u^{H(t)(\gamma-2)} \int_0^{\infty} \{y^2 + u^{2-2H(t)}\}^{\gamma/2-1} \, dy \, du \, dt$$

$$\leqslant K \int_{t_*-\varepsilon}^{t_*+\varepsilon} \int_0^{\delta} u^{H(t)(\gamma-2)} u^{\{1-H(t)\}(\gamma-1)} \, du \, dt$$

$$= K \int_{t_*-\varepsilon}^{t_*+\varepsilon} \int_0^{\delta} u^{\gamma - H(t) - 1} \, du \, dt$$

$$\leqslant K \int_{t_*-\varepsilon}^{t_*+\varepsilon} \int_0^{\delta} u^{\varepsilon-1} \, du \, dt$$

$$= K \varepsilon \delta^{\varepsilon} < \infty. \qquad \square$$

PROOF OF THEOREM 6. The theorem can be easily established by proving

$$E\{\widehat{H}(t)\} \to H(t) \quad \text{and} \quad \text{Var}\{\widehat{H}(t)\} \to 0. \tag{A.3}$$

The result then follows by Chebyshev's inequality.

Recall from Section 4 that the bivariate data values which lead to $\widehat{H}(t)$ are denoted by $x_j = \log(2^{-j})$ and $y_j = y_t(2^{-j})$ for $j = 1, \ldots, k$, where $y_t(a) = \log\{|TY(a,t)|^2\}$. Thus, $\widehat{H}(t)$ is based on a bivariate sample of size k. We will prove (A.3) by demonstrating that as $k \to \infty$,

$$E\{\widehat{H}(t)\} = H(t) + o(k^{-1}), \qquad \text{Var}\{\widehat{H}(t)\} = O(k^{-3}). \tag{A.4}$$

To ease notation, we suppress the time index t in the specifications of $y_t(a)$, $H(t)$, and $\widehat{H}(t)$, and write these objects simply as $y(a)$, H, and $\widehat{H}$, respectively.

Direct computations show that

$$\sum_{j=1}^{k}(x_j - \bar{x})^2 = (\log 2)^2 \left\{\sum_{j=1}^{k} j^2 - k(k+1)^2/4\right\}$$

$$= (\log 2)^2 k(k+1)(k-1)/12 \sim k^3, \tag{A.5}$$

and

$$\sum_{j=1}^{k} |x_j - \bar{x}| = (\log 2) \sum_{j=1}^{k} |j - (k+1)/2| \sim k^2. \tag{A.6}$$

By (4.2), we have as $a \to 0$,

$$E\{|TY(a,t)|^2\} = C_1 a^{2H+1}\{1 + o(1)\},$$

and thus

$$\log[E\{|TY(a,t)|^2\}] = \log C_1 + (2H+1)\log a + o(1).$$

The preceding relation and (4.3) together imply

$$E(y_j) = c + (2H+1)x_j + o(1), \tag{A.7}$$

where $c = \log C_1 + C_2$. From (4.5) along with (A.5), (A.6), and (A.7), we obtain

$$E(\widehat{H}) = \frac{\sum (x_j - \bar{x}) E(y_j - \bar{y})}{2 \sum (x_j - \bar{x})^2} - \frac{1}{2}$$

$$= H + o\left(\frac{\sum |x_j - \bar{x}|}{\sum (x_j - \bar{x})^2}\right)$$

$$= H + o(k^{-1}).$$

This establishes the first result of (A.4). We now derive the second result.

Utilizing (4.1), the covariance property (2.3), and the first vanishing moment of the mother wavelet ψ, we have

$$\mathrm{Cov}\{TY(a_1,t), TY(a_2,t)\}$$

$$= \frac{1}{(a_1 a_2)^{1/2}} \iint \psi\left(\frac{u-t}{a_1}\right) \psi\left(\frac{v-t}{a_2}\right) E\{Y(u)Y(v)\}\, du\, dv$$

$$= (a_1 a_2)^{1/2} \iint \psi(x)\psi(y) E\{Y(t+a_1 x)Y(t+a_2 y)\}\, dx\, dy$$

$$= (a_1 a_2)^{1/2} \iint \psi(x)\psi(y) \Sigma_t(a_1 x, a_2 y)\, dx\, dy$$

$$\sim (a_1 a_2)^{1/2} \iint \psi(x)\psi(y) \Big[\frac{1}{2}\{\Sigma_t(a_1 x, a_1 x) + \Sigma_t(a_2 y, a_2 y)\}$$

$$- C(t)|a_1 x - a_2 y|^{2H}\Big] dx\, dy$$

$$= -C(t)(a_1 a_2)^{1/2} \iint \psi(x)\psi(y)|a_1 x - a_2 y|^{2H} \, dx \, dy$$

$$= -C(t)a_1^{2H+1/2} a_2^{1/2} \iint \psi(u + ya_2/a_1)\psi(y)|u|^{2H} \, du \, dy,$$

as $a_1, a_2 \to 0$. (A.8)

Now assume the mother wavelet ψ has M vanishing moments and is M times continuously differentiable where $M \geq 2$. We take an Mth-order expansion of $\psi(u + ya_2/a_1)$ about u to establish that as $(a_2/a_1) \to 0$,

$$\psi(u + ya_2/a_1) = \psi(u) + y\psi'(u)(a_2/a_1) + \cdots$$
$$+ (1/M!)y^M \psi^{(M)}(u)(a_2/a_1)^M \{1 + o(1)\}.$$

This expansion allows us to show

$$\iint \psi(u + ya_2/a_1)\psi(y)|u|^{2H} \, du \, dy \sim K(a_2/a_1)^M, \quad \text{as } (a_2/a_1) \to 0.$$

Using the preceding in (A.8), we see that as $a_1, a_2 \to 0$ with $(a_2/a_1) \to 0$,

$$\text{Cov}\{TY(a_1, t), TY(a_2, t)\} \sim K a_1^{2H+1/2} a_2^{1/2} (a_2/a_1)^M.$$

This result implies that for small a_1, a_2 chosen such that $a_2 < a_1$,

$$\left|\text{Corr}\{TY(a_1, t), TY(a_2, t)\}\right| \leq K(a_2/a_1)^{M-H}. \tag{A.9}$$

Now for a bivariate normal pair of random variables (W, Z) with $E(W) = E(Z) = 0$, it can be shown that

$$\left|\text{Corr}\{\log(W^2), \log(Z^2)\}\right| \leq [\text{Corr}(W, Z)]^2.$$

It therefore follows from (A.9) that for small $a_2 < a_1$,

$$\left|\text{Corr}\{y(a_1), y(a_2)\}\right| \leq [\text{Corr}\{TY(a_1, t), TY(a_2, t)\}]^2$$
$$\leq K(a_2/a_1)^{2M-2H}. \tag{A.10}$$

Furthermore, if W is normal with variance σ^2, the variance of $\log(W^2)$ does not depend on σ^2, as can be seen by writing $\log(W^2) = \log\{(W/\sigma)^2\} + \log \sigma^2$. Utilizing this fact along with (A.10), we have that for small $a_2 < a_1$,

$$\left|\text{Cov}\{y(a_1), y(a_2)\}\right| \leq K(a_2/a_1)^{2M-2H}.$$

From the preceding, we have

$$\left|\text{Cov}(y_i, y_j)\right| \leq K 2^{|i-j|(2H-2M)}. \tag{A.11}$$

With (A.11), we can write

$$\left|\sum_{ij} x_i x_j \mathrm{Cov}(y_i, y_j)\right| \leqslant K \sum_{ij} ij 2^{|i-j|(2H-2M)}$$

$$\leqslant K \sum_{i \leqslant j} ij 2^{(j-i)(2H-2M)}$$

$$= K \sum_{i=1}^{k} \left[i 2^{-i(2H-2M)} \left\{ \sum_{j=i}^{k} j 2^{j(2H-2M)} \right\} \right]$$

$$\leqslant K \sum_{i=1}^{k} [i 2^{-i(2H-2M)} \{i 2^{i(2H-2M)}\}]$$

$$= K \sum_{i=1}^{k} i^2 \sim k^3. \tag{A.12}$$

Similarly, we can show

$$\bar{x}\left|\sum_{ij} x_i \mathrm{Cov}(y_i, y_j)\right| \sim k^3, \quad (\bar{x})^2 \left|\sum_{ij} \mathrm{Cov}(y_i, y_j)\right| \sim k^3. \tag{A.13}$$

By (A.12) and (A.13), we have

$$\sum_{ij} (x_i - \bar{x})(x_j - \bar{x}) \mathrm{Cov}(y_i, y_j)$$

$$= \sum_{ij} x_i x_j \mathrm{Cov}(y_i, y_j) - \bar{x} \sum_{ij} (x_i + x_j) \mathrm{Cov}(y_i, y_j)$$

$$+ (\bar{x})^2 \sum_{ij} \mathrm{Cov}(y_i, y_j) \sim k^3.$$

Finally, using the preceding along with (4.5) and (A.5), we have

$$\mathrm{Var}(\widehat{H}) = \frac{\sum_{ij}(x_i - \bar{x})(x_j - \bar{x}) \mathrm{Cov}(y_i, y_j)}{4\{\sum(x_i - \bar{x})^2\}^2} \sim k^{-3}. \qquad \square$$

References

Abry, P., P. Flandrin, M. S. Taqqu and D. Veitch (2000). Wavelets for the analysis, estimation and synthesis of scaling data. In *Self-Similar Network Traffic and Performance Evaluation* (Eds. K. Park and W. Willinger). Wiley, New York.

Abry, P., P. Flandrin, M. S. Taqqu and D. Veitch (2001). Self-similarity and long-range dependence through the wavelet lens. In *Long-Range Dependence: Theory and Applications* (Eds. P. Doukhan, G. Oppenheim and M. S. Taqqu). Birkhäuser, Boston.

Abry, P. and F. Sellan (1996). The wavelet-based synthesis for fractional Brownian motion proposed by F. Sellan and Y. Meyer: remarks and fast implementation. *Appl. Comput. Harmonic Anal.* **3**, 377–383.

Adler, R. L. (1981). *The Geometry of Random Fields.* Wiley, New York.

Bardet, J.-M., G. Lang, E. Moulines and P. Soulier (2000). Wavelet estimator of long-range dependent processes. *Statistical Inference for Stochastic Processes* **3**, 85–99.

Bardet, J.-M., G. Lang, G. Oppenheim, A. Philippe, S. Stoev and M. S. Taqqu (2001). Semi-parametric estimation of the long-range dependence parameter: a survey. In *Long-Range Dependence: Theory and Applications* (Eds. P. Doukhan, G. Oppenheim and M. S. Taqqu). Birkhäuser, Boston.

Barnsley, M. F. (1993). *Fractals Everywhere*, 2nd ed. Academic Press, Boston.

Beran, J. (1994). *Statistics for Long Memory Processes.* Chapman and Hall, New York.

Buldyrev, S. V., A. L. Goldberger, S. Havlin, C.-K. Peng, H. E. Stanley, M. H. R. Stanley and M. Simons (1993). Fractal landscapes and molecular evolution: modeling the myosin heavy chain gene family. *Biophys. J.* **65**, 2673–2679.

Chen, G., P. Hall and D. S. Poskitt (1995). Periodogram-based estimators of fractal properties. *Ann. Statist.* **23**, 1684–1711.

Cohen, A., I. Daubechies, B. Jawerth and P. Vail (1993a). Multiresolution analysis, wavelets and fast algorithms on an interval. *C. R. Acad. Sci. Ser. I Math.* **316**, 417–421.

Cohen, A., I. Daubechies and P. Vail (1993b). Wavelets on the interval and fast wavelet transforms. *Appl. Comput. Harmonic Anal.* **1**, 54–81.

Comte, F. (1996). Simulation and estimation of long memory continuous time models. *J. Time Series Anal.* **17**, 19–36.

Constantine, A. G. and P. Hall (1994). Characterizing surface smoothness via estimation of effective fractal dimension. *J. Roy. Statist. Soc. B* **56**, 97–113.

Csáki, E. (1980). A relation between Chung's and Strassen's laws of the iterated logarithm. *Z. Wahrsch. Verw. Gebiete* **54**, 287–301.

Daubechies, I. (1992). *Ten Lectures on Wavelets.* CBMS-NSF Regional Conference Series in Applied Mathematics. SIAM, Philadelphia.

Daubechies, I. (1994). Two recent results on wavelets: wavelet bases for the interval, and biorthogonal wavelets diagonalizing the derivative operator. In *Recent Advances in Wavelet Analysis* (Eds. L. L. Schumaker and G. Webb). Academic Press, Boston.

Doncho, D. L. and I. M. Johnstone (1994). Ideal spatial adaptation by wavelet shrinkage. *Biometrika* **81**, 425–455.

Farge, M., C. R. Hunt and J. C. Vassilicos (1993). *Wavelets, Fractals and Fourier Transformations.* Clarendon Press, Oxford.

Flandrin, P. and P. Gonçalvès (1994). From wavelets to time-scale energy distributions. In *Recent Advances in Wavelet Analysis* (Eds. L. L. Schumaker and G. Webb). Academic Press, Boston.

Geweke, J. and S. Porter-Hudak (1983). The estimation and application of long-memory time series models. *J. Time Series Anal.* **4**, 221–237.

Gonçalvès, P. and P. Flandrin (1993). Bilinear time-scale analysis applied to local scaling exponents estimation. In *Progress in Wavelet Analysis and Applications* (Eds. Y. Meyer and S. Roques). Frontieres, Paris.

Goodman, V. and J. Kuelbs (1991). Rates of clustering for some Gaussian self-similar processes. *Probab. Theory Related Fields* **88**, 47–75.

Hall, P., H. L. Koul and B. A. Turlach (1997). Note on convergence rates of semiparametric estimators of dependence index. *Ann. Statist.* **25**, 1725–1739.

Hwang, W. L. and S. Mallat (1994). Characterization of self-similar multifractals with wavelet maxima. *Appl. Comput. Harmonic Anal.* **4**, 316–328.

Kent, J. T. and A. T. A. Wood (1997). Estimating the fractal dimension of a locally self-similar Gaussian process by using increments. *J. Roy. Statist. Soc. B* **59**, 679–700.

Kolmogorov, A. N. (1941). Local structure of turbulence in an incompressible viscous fluid at very large Reynolds numbers. Translation in *Turbulence: Classic Papers on Statistical Theory* (1961) (Eds. S. K. Friedlander and L. Topper). Interscience Publishers Inc., New York.

Leland, W. E., M. S. Taqqu, W. Willinger and D. V. Wilson (1994). On the self-similar nature of Ethernet traffic (extended version). *IEEE/ACM Transactions on Networking* **2**, 1–15.

Mandelbrot, B. B. (1983). *The Fractal Geometry of Nature* (Updated and augmented edition). Freeman, New York.

Mandelbrot, B. B. and J. W. van Ness (1968). Fractional Brownian motions, fractional noises and applications. *SIAM Review* **10**, 422–437.

Mandelbrot, B. B. and J. R. Wallis (1968). Noah, Joseph and operational hydrology. *Water Resources Research* **4**, 909–918.

Mandelbrot, B. B. and J. R. Wallis (1969). Computer experiments with fractional Gaussian noises. *Water Resources Research* **5**, 228–267.

McCoy, E. J. and A. T. Walden (1996). Wavelet analysis and synthesis of stationary long-memory processes. *J. Comput. Graphical Statist.* **5**, 26–56.

Monrad, D. and H. Rootzén (1995). Small values of Gaussian processes and functional laws of the iterated logarithm. *Probab. Theory Related Fields* **101**, 173–192.

Moulines, E. and P. Soulier (1999). Broadband log-periodogram regression of time series with long-range dependence. *Ann. Statist.* **27**, 1415–1439.

Oodaira, H. (1972). On Strassen's version of the law of the iterated logarithm for Gaussian processes. *Z. Wahrsch. Verw. Gebiete* **21**, 289–299.

Ossadnik, S. M., S. V. Buldyrev, A. L. Goldberger, S. Havlin, R. N. Mantegna, C.-K. Peng, M. Simons and H. E. Stanley (1994). Correlation approach to identify coding regions in DNA sequences. *Biophys. J.* **67**, 64–70.

Peitgen, H. O., H. Jurgens and D. Saupe (1992). *Chaos and Fractals: New Frontiers of Science*. Springer-Verlag, New York.

Peng, C.-K., S. V. Buldyrev, A. L. Goldberger, S. Havlin, F. Sciortino, M. Simons and H. E. Stanley (1992). Long-range correlation in nucleotide sequences. *Nature* **356**, 168–170.

Peng, C.-K., J. M. Hausdorff, J. E. Mietus, S. Havlin, H. E. Stanley and A. L. Goldberger (1995a). Fractals in physiological control: from heartbeat to gait. In *Lévy Flights and Related Phenomena in Physics, Proceedings of the 1994 International Conference on Lévy Flights* (Eds. M. F. Shlesinger, G. M. Zaslavsky and U. Frisch). Springer-Verlag, Berlin.

Peng, C.-K., S. Havlin, H. E. Stanley and A. L. Goldberger (1995b). Quantification of scaling exponents and crossover phenomena in nonstationary heartbeat time series. *Chaos* **5**, 82–87.

Percival, D. B. and P. Guttorp (1994). Long-memory processes, the Allan variance and wavelets. In *Wavelets in Geophysics* (Eds. E. Foufoula-Georgiou and P. Kumar). Academic Press, New York.

Percival, D. B. and A. T. Walden (2000). *Wavelet Methods for Time Series Analysis*. Cambridge University Press, Cambridge.

Robinson, P. (1995). Gaussian semiparametric estimation of long range dependence. *Ann. Statist.* **23**, 1630–1661.

Taqqu, M. S., V. Teverovsky and W. Willinger (1995). Estimators for long-range dependence: an empirical study. *Fractals* **3**, 785–798.

Taylor, C. C. and S. J. Taylor (1991). Estimating the dimension of a fractal. *J. Roy. Statist. Soc. B* **53**, 353–364.

Tricot, C. (1995). *Curves and Fractal Dimension*. Springer-Verlag, New York.

Wang, Y. (1997). Fractal function estimation via wavelet shrinkage. *J. Roy. Statist. Soc. B* **59**, 603–613.

Wang, Y., J. E. Cavanaugh and C. Song (1997). Self-similarity index estimation via wavelets for locally self-similar processes. Contributed presentation, International Workshop on Wavelets in Statistics, Duke University, Durham, North Carolina.

Wang, Y., J. E. Cavanaugh and C. Song (2001). Self-similarity index estimation via wavelets for locally self-similar processes. *J. Statist. Plann. Inference* **99**, 91–110.

Willinger, W., V. Paxson and M. S. Taqqu (1998). Self-similarity and heavy tails: structural modeling of network traffic. In *A Practical Guide to Heavy Tails: Statistical Techniques and Applications* (Eds. R. J. Adler, R. E. Feldman and M. S. Taqqu). Birkhäuser, Boston.

Wornell, G. W. and A. V. Oppenheim (1992). Estimation of fractal signals from noisy measurements using wavelets. *IEEE Trans. Signal Process.* **40**, 611–623.

Stochastic Models for DNA Replication

Richard Cowan

1. Variation in biology

A statistician or probabilist entering into the biological domain may do so at a variety of levels. Most statisticians, when consulted for help by biologists, find themselves dealing at the *animal or plant level* – the animal (or plant) is the experimental unit. Since measurements usually display considerable variability from animal to animal, the consulting statistician finds ready application for the well-developed repertoire of statistical analyses – analysis of variance, regression, survival analysis, and so on – in interpreting the experiment. Quite often in our experiences at this level, the statistician is not greatly involved with the biology, beyond what the data tell him/her about these experimental units.

At *the organ or tissue level*, where a part of the animal or plant is studied, there are often similar considerations. For example, a sample of leaves from a single plant may be subjected to experiment. Once again, the standard statistical techniques might be applied, with the leaf being the experimental unit.

One might wish to study the animal or plant at its *cellular level*. Or, one may be working with an organism, such as a bacterium or yeast, which is unicellular and so the study of the "animal" necessarily means observation of the "cell". If we are to take observations on a single cell, either a microscope or a "flow cytometer" is typically involved. For example, we might observe a culture of cells on a petri dish and, by meticulous observation under the microscope, record the individual generation time of each cell (Powell, 1953, 1958; Cowan and Staudte, 1986) or the cell's size at division (Staudte et al., 1984). Or we might perhaps observe the cells individually as they flow, in a dilute solution, past some electronic sensor which measures (say) the amount of fluorescence emitted by each cell, the cell population having been previously labelled by a fluorescent dye which intercollates with the cell's DNA. This standard technique of "flow cytometry" is able to stage the cell's progression through its division cycle by effectively measuring the amount of DNA it currently has.

Such observations of individual cells can show considerable variability; for example, the generation times of the bacterial cells in Powell's culture had a coefficient of variation in the range 15–25%.

So it is possible to make the sort of observations on individual cells which would draw the statistician into biology at the cellular level. Very often, however, measurements are not done individually but are instead made on a large population of cells. For example, a biologist may measure the optical density of the millions of cells in a flask by using an appropriate scanner and may have the readings for 20 flasks, say. The flask now becomes the experimental unit and so the statistician's exposure to cellular biology may turn out again to be minimal. Certainly the inter-cellular variation is not being measured and is lost in the aggregate nature of the observation.

A great deal of the literature in biology is now at the *molecular level* and, at this level, observation of *the individual* is a rarity indeed. Generally, our powers of observation do not allow us to look at the individual molecule. We know from physical chemistry that these individuals display a great deal of randomness – in their movement, their speed, their orientation and energy levels, their electron status, and, for the larger molecules, their isomeric form, shape and configuration. Yet we do not see as a rule, except in the occasional electron micrograph, direct observation of this variation for individual molecules.

Our observations usually involve aggregates, where the impact of *Avogadro's number* and various *laws of large numbers* average out molecular variation. Hence one might think that there is little of stochastic interest in biology at the molecular level and little need for the skills of the statistician or probabilist. In this article, by focussing on a range of problems associated with the replication of DNA, I aim to show otherwise.

2. Stochastic chemistry

Firstly, however, I shall review the small literature which deals with these aggregation effects in systems with a large number of reacting molecules. This is the literature (surveyed by McQuarrie, 1967) which treats chemical reactions as stochastic processes.

Consider the reversible chemical reaction $A + B \rightleftharpoons AB$. This has been modelled as a Markov process with a state described by the triple (a_t, b_t, n_t). Here a_t and b_t denote the numbers of free A and B molecules respectively at time t and n_t is the number of AB molecules at t. Indeed the state can be described merely by n_t since, due to the conservation laws for the A and B molecules, $a_t + n_t = a_0 + n_0$ and $b_t + n_t = b_0 + n_0$. Introducing $p_n(t) := \mathbb{P}\{n_t = n\}$ and using the standard methods of Markovian process analysis, combined with *first-order chemical kinetics* for the breakdown of AB and *law of mass action* assumptions for the combining of free A with free B, we may write a family of differential equations. For $0 < n < n_0 + \min(a_0, b_0)$,

$$p'_n(t) = \rho(n+1)p_{n+1}(t) - \left[\rho n + \theta(a_0 + n_0 - n)(b_0 + n_0 - n)\right]p_n(t)$$
$$+ \theta(a_0 + n_0 - n + 1)(b_0 + n_0 - n + 1)p_{n-1}(t), \tag{1}$$

where ρ is the *infinitesimal rate* of breakdown of an AB molecule and θ is the infinitesimal rate of aggregation of every pair of free A and B. When $n = n_0 + \min(c_0, b_0)$ or $n = 0$, the first or third term respectively is omitted from (1).

The model leading to (1) is not unique from a chemical viewpoint; for example, the first-order-kinetics assumption for breakdown of AB into its products is one which attributes the breakdown solely to energy received from collision of an AB with either the side of the flask or the H_2O molecules which form the solution and not to the energy of collisions with other AB, A or B molecules. If these latter collisions were important, then the equation should contain breakdown rates which involve products of various molecular abundances. The model has also been criticised for the glib way it handles the combining events $A + B \to AB$, focussing as it does merely on abundance of the free A and B without regard for their spatial distribution (Clifford, Green and Pilling, 1987).

The stochastic approach leading to (1) is nevertheless a considerable refinement over the deterministic approach which is prevalent among chemists. The chemists' approach writes equations in terms of *chemical concentrations*, $[A]_t$, $[B]_t$ and $[AB]_t$ assumed to be continuous and differentiable functions of t. Whilst this assumption is strictly incorrect, since concentration is the number of molecules (an integer at all times) divided by the volume, its adoption leads to the differential equation

$$\frac{d[AB]_t}{dt} = -\rho[AB]_t + \theta[A]_t[B]_t$$
$$= -\rho[AB]_t + \theta([A]_0 + [AB]_0 - [AB]_t)([B]_0 + [AB]_0 - [AB]_t). \quad (2)$$

If required, the transient solution of this Riccati equation (2) is readily found, though in practice this does not seem to interest chemists greatly. Rather, they focus on the equilibrium state trivially found by equating the time derivative to zero.

By contrast, the transient solution of the stochastic model (1) is not easy to find except in the irreversible case where $\rho = 0$ (McQuarrie, 1967; Ishida, 1964). Darvey, Ninham and Staff (1966) have, however, given explicit results for the equilibrium probability distribution when the reaction is reversible (that is, $\rho > 0$) in terms of Hypergeometric functions.

Ishida's calculations for the irreversible case have revealed that $E(n_t)$ from the stochastic theory is virtually the same as $[AB]_t$ from the deterministic theory, except in the final stages of the reaction when nearly all of the less abundant of A or B has been converted. A minor, but physically insignificant, difference exists at this late stage. His formulae also show that, in the canonical case where $n_0 = 0$ and $a_0 = b_0$, the actual process n_t has very low variation about its mean, until this later stage (when a_t and b_t have become small).

The attitude expressed in the probability literature of that time (McQuarrie, 1967) was one of disappointment. Was the stochastic approach worthwhile, given its complexity and the identical practical conclusions to that of the simpler deterministic theory? Moreover Avogadro's number suggested that most reactions of interest to chemists involve massive numbers of molecules. Similar discouragement arose from the study of other reactions, for example, the stochastic study of the simple conversion of one substance, S say, to another, P say, through the action of an enzymic catalyst, E. This 'Michaelis-Menten' reaction, $S + E \rightleftharpoons ES \to P + E$, was studied stochastically by Heyde and Heyde (1971). With large numbers of molecules involved, stochastic variation is low and unobservable, though it might be reasoned that the time for the last few S to be converted to P could have sizeable stochastic variation relative to its mean (albeit unobservable in a practical sense).

Some change of thinking by mathematicians on the role of probability in chemical reactions came with the work of Hall (1983) and Dunstan and Reynolds (1981). Hall considered the $A + B \rightleftharpoons AB$ reaction when the ratio θ/ρ is either very small or very large. If the ratio is very small, then there will be very few AB molecules at equilibrium. If it is very large, then the less abundant of A and B will be a rarity at equilibrium. He showed that either the *Poisson distribution* or a so-called *modified Bessel distribution* approximates well the number at equilibrium of this less abundant free molecule (which of the two distributions being determined by subtleties on the numbers of the *more* abundant of A or B). Hall did not consider small numbers of molecules; rather his results were asymptotic as the numbers of molecules involved became infinite. The interesting results came because the ratio θ/ρ of reaction rates simultaneously became extreme as molecular abundance became infinite.

Hall (1983) also made precise an interesting observation made by Dunstan and Reynolds (1981), who had demonstrated earlier that a *normal* distribution could approximate the complicated equilibrium distribution of Darvey, Ninham and Staff (1966) for the reaction $A + B \rightleftharpoons AB$. Dunstan and Reynolds have one numerical example, whereby $a_0 = b_0 = 10^9$, $n_0 = 0$ and $\theta/\rho = 6^{-1} \times 10^{-14}$. They calculate that, at large times t, $n_t \sim N(1667, 1667)$ approximately, in this example. Noting a coefficient of variation of about 2.5%, they warn that a deterministic analysis may be somewhat inaccurate for molecular numbers below 10^9.

I believe that the warning of Dunstan and Reynolds needs qualification. The non-negligible coefficient of variation is a reflection, as in Hall's work, of reaction imbalance; their ratio θ/ρ is very small mathematically – only about 1667 molecules of bound AB exist at equilibrium compared to about 2×10^9 of free molecules. Had they chosen a ratio equal to 1, they would have concluded that $n_t \sim N(0.999968 \times 10^9, 126^2)$ approximately; the coefficient of variation would then have been truly negligible. Thus the warning based merely on initial numbers of A and B is unsound – the reaction imbalance is a major factor. In their defense, reversible reactions in practice are often energetically disposed to be rather one-directional and, in those cases, their warning may be apposite.

Hall neatly ties up the relationship between the modified Bessel, Poisson and Normal distributions by showing that the modified Bessel family is approximated by the Poisson or Normal distributions for some of its parameter values. So the theories for the population size of the rarer molecules of a reaction are unified by Hall.

Thus the work of Hall, Dunstan and Reynolds adds considerable insight to the earlier theories of stochastic chemistry. We might summarise that non-negligible stochastic effects for molecular-abundance quantities like n_t occur in these simple chemical reactions:

- whenever numbers of all molecular types are small;
- whenever we look at the rarer molecules in reversible reactions which are close to being irreversible;

- toward the completion of irreversible reactions when numbers of some molecules are becoming small.

3. Exponentially distributed waiting times

Later in this article, we need, for our model building, assumptions about the distribution of waiting time for one particular site on the DNA molecule to bind an enzyme (from a pool of identical enzymes which are attracted to that site). We have found that the exponential distribution is a convenient choice because of its mathematical simplicity. Is this an appropriate choice, however? Let us explore the issue of waiting times with reference to our learning example $A + B \rightleftharpoons AB$.

Models in chemistry of a Markovian nature, such as the one discussed above for the reversible reaction $A + B \rightleftharpoons AB$, have inbuilt consequences about the exponential distribution. Firstly, the time that a given AB molecule stays in a bound form is exponentially distributed with p.d.f. $\rho e^{-\rho x}$. Secondly, the time from t until the next event (either a joining of a free A and B or a disintegration of an AB) is, given n_t, exponentially distributed by model assumption. Unconditional upon n_t, this time is distributed as a mixture of exponentials, having p.d.f.

$$\sum_{n=0}^{\infty} \mathbb{P}\{n_t = n\} \lambda_n e^{-\lambda_n x},$$

where $\lambda_n = \rho n + \theta(a_0 + n_0 - n)(b_0 + n_0 - n)$. It must be noted, however, that unless a_0, b_0 and n_0 are *all* quite small, the *random variables* n_t and λ_{n_t} will have negligible coefficient of variation and so the mixed exponential will differ little in practical terms from a pure exponential distribution as a model for the *next event* waiting time.

Of more interest though, for some of the problems we shall be considering later, is the time T until a given free A molecule first binds to a B molecule. Here envisage the particular A as having a label which identifies it, but any B suffices.

Once again there are heuristic arguments which show that T is approximately exponentially distributed provided there are a reasonably large number of B molecules *and* provided that the chemical reaction is reversible. Our arguments draw on ideas from the theory of stationary point processes.

To see how the exponential arises in the simplest setting, consider the stationary point process formed by the epochs when our labelled A commences to bind with the ith B molecule in the system. This A will repeatably bind with this particular B when the reaction is reversible, staying bound for a while, and after many liaisons with other B's will find the ith B again. We assert that the expected number of "binding commencements" with that particular B per unit time, say λ, is finite and positive.

If b is the total number of B molecules in the system, either free or bound, then there will be b such point processes. These are not independent, but any dependence is weak and gets weaker as b is made larger.

Thus, heuristically at least, we see that the point process formed by the superposition of these b point processes will, for large b, be approximately a Poisson point process.

This process is none other than the point process of all "binding commencements" of the labelled A molecule. Thus the times between such binding epochs will be exponentially distributed approximately.

4. The biological cell

Stochastic interest in molecular biology can be high, because many important molecules in the cell are there in small numbers. DNA is the prime example. An *E. Coli* bacterial cell[1] has just one DNA molecule.[2] On this single DNA molecule, there is just one site which is receptive to the molecular complexes which initiate the replication of the DNA. This site is called *the origin* of replication. The initiating complexes themselves are in greater abundance, a few hundred perhaps (though there is no good data on the number); they are manufactured by the cell for a short period of each generation time.

One speculates that there may be some random delay as this unique site waits for a complex to bind and that this is part of the explanation as to why this early phase of the cell generation time, prior to the commencement of DNA replication, is variable in duration (Smith and Martin, 1973). The overall generation times of bacterial cells are quite variable, with coefficients of variation in the range 15–25% (Powell, 1953, 1958), and it has been found that the greatest source of the variation is in the waiting period until this critical binding event which initiates DNA replication.

There are, of course, some molecules in the cell which are present in massive numbers, perhaps even in Avogadrian abundance. Amongst these are the *nucleotides*, *amino acids* and *fatty acids*, the basic building blocks for the polymers which the cell manufactures. These polymers are the DNA itself, numerous RNA's, the proteins/enzymes and the cellular membranes.

There also exist some less abundant macro-molecules which possess multiple binding sites for some enzyme or protein. So the abundance of these sites arises not from an abundance of the macro-molecules, but from the multiplicity of the sites. For example, there is but one membrane (a geometrically-regular polymer of fatty acids) which encloses the cell, yet it has a vast number of receptors which attract other molecules (as part of the import and export of matter between the cell and its external environment). So these receptors form an abundant class of binding sites.

A second, and perhaps greater element of interest for stochastic modellers, is that the binding sites on some of the macro-molecules are arranged spatially in a quite random way. DNA is a prime example. A strand of this "double-stranded" polymer is a seemingly random sequence of its building blocks, the nucleotide "bases" *adenine, cytosine, guanine* and *thiamadine*, called A, C, G and T – for example, ...T–G–T–C–A–A–G–A–T–G–C–.... (The other strand is determined from the first by simple matching rules, A opposite T and C opposite G, and the two strands are held together in the famous Crick–Watson helical structure.) Other molecules, enzymes usually, can

---

[1] growing at normal temperatures and with normal nutrient, so that its generation time is about one hour on average (Mitchison, 1971, p. 102)

[2] We ignore here the DNA contained within the plasmids and phages, biological entities which reside within the bacterium.

be attracted to specific "letter" patterns on one strand; these patterns obviously occur at variable distances from each other.

For example, there is an enzyme (called Hpa I) which has a chemical affinity with any site on the DNA where a sequence G–T–T–A–A–C occurs. The local geometry and electron densities caused by these "letters" in sequence create the affinity. Yet, because of the seemingly random sequence of letters along the whole DNA molecule, locations of the specific pattern G–T–T–A–A–C appear to be random. For patterns such as this, with sufficient rarity due to its 6 "bases", any map of the sites (known as a *restriction map*) gives the impression of a Poisson point process. Indeed Poisson limit theorems of the type discussed in Barbour et al. (1992) and by Chryssaphinou and Papastavridis (1988) provide some mathematical substance to a claim that, despite the discrete nature of the problem, the locations of Hpa I sites form a Poisson point process on the continuum which is the linear DNA molecule.

Some biologists may raise a few objections to the statement that DNA is a random molecule, at least in the regions of that molecule which have a definite role to play in directing and regulating the nature of the animal or plant. Yet, a few hours spent reading Chapter 1 of the standard text, Alberts et al. (1983), shows a molecular evolution of DNA driven by random polymerisation and random mutation, with a gradual emergence of regions of some functional importance to the organism. These "functional regions", genes or regulatory sequences, tended to stop mutating at some point in evolution. Or rather, the random changes to the DNA polymer still occurred but produced an organism which did not survive to reproduce itself.

Even then, some random change was (and is) still allowed in these "functional" zones. For example, a gene can accept considerable mutation (at the third position of each *codon*) without affecting in any way the blueprint of life contained in the gene.[3]

Thus, the view that the DNA molecule is a random sequence of bases (along one of its two strands) prevails, provided we recognise that some regions are still randomly evolving and other regions are now fixed, with the result of past random change now "frozen" in time.[4]

5. A glib mathematical abstraction

Consider an interval $[0, L]$ as a section of our DNA molecule, with a point process of sites (which are perhaps of some importance because they attract some enzyme). We do not assume that the point process is Poisson, yet it soon becomes one under the following rather-glib model of mutation.

Because of mutation the sites may die (i.e., change into a form which has lost its role). This occurs randomly and is like a "thinning" of the point process. There is, however, the genesis of new sites through mutation. These are equally likely to occur anywhere

---

[3] A codon is a unit, comprising three bases, which provides information to the cell. The translation of that information for the cell's benefit is based on a genetic code that has many redundancies. Some random mutation of the DNA is allowed without changing the information that the cell receives.

[4] to borrow from Francis Crick who described the genetic code itself as a *frozen accident*.

in $[0, L]$ and independently so. Thus this spatial birth and death process quickly gives rise to a Poisson point process.

We assume throughout the sequel of our paper that sites which interest us form a Poisson process along the DNA. Our first example of this is with sites known as "replication origins".

6. The spatial pattern of replication origins

We have mentioned above that a bacterial cell has only one *origin of replication* on its sole DNA molecule. On the DNA of higher organisms, however, there are a large number of "origins" on each of the many DNA molecules that the cell possesses. These origins are randomly scattered along the linear molecule. Due to random evolution of the extremely long DNA molecule, perhaps along the lines of our glib model above, these specific sites can be modelled well by a Poisson process along the DNA molecule.

The replication of DNA is an interesting process (Kornberg, 1980). Each origin is recognised, after some random delay time, by an enzyme-complex which then binds to the site. The complex immediately initiates, via its influence on other enzymes, a bi-directional movement along the DNA. At both moving 'frontiers', the double-stranded structure unwinds from its helical structure and separates into two single strands. Because of the strand separation, the frontiers are called *forks* and the region of separated DNA between the pair of forks is known as an "eye" or "replicon". Replication of the single-strands in the "eye" then takes place (see later for details of the actual replication mechanism).

When a fork meets a fork from another eye, both forks stop and the enzymes involved in unwinding/separating fall off the DNA. The two eyes simply merge into one larger eye.

Beautifully clear photographs of this process, taken under the electron microscope, appear in Blumenthal, Kriegstein and Hogness (1974) and Kriegstein and Hogness (1973). We draw a typical picture below (Figure 1). One can see various segments of DNA within the photographic field and numerous fully-visible *eyes* where the DNA has separated into two strands. The authors used the fruit fly, *Drosophila melanogaster*, in the period prior to the so-called *mid-blastula transition*, for their studies. Generation times are extremely short, around 10 minutes, during the first 12 generations of the embryo before the "transition" and DNA replication is complete within about 3–4 minutes. (Later in the fly's development, generation times of cells lengthen markedly, as does the time taken for DNA replication; it is thought that a much smaller proportion of origins are activated in this later period of development.)

The lengths of the eyes extant at the time the photo is taken are quite variable, as are the distances between eyes. These are evidenced by histograms compiled from Blumenthal et al. (1974).

There are some difficulties in the interpretation of these histograms. One problem, of which the authors were well aware, was that the eyes were observed from molecules at different stages of the process. To combat this problem, they also published data on the eye-to-eye distances for molecular segments grouped according to the proportion of

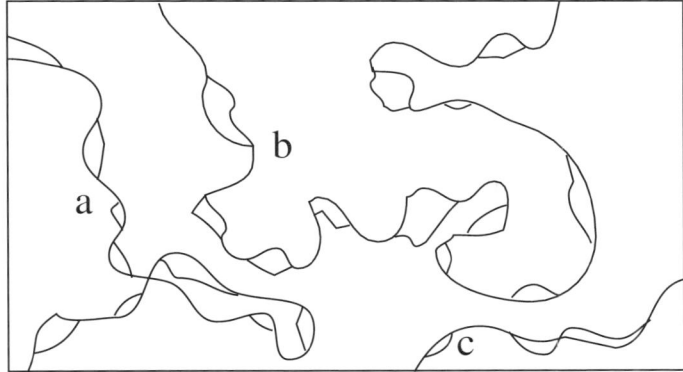

Fig. 1. The type of micrograph which shows the *eyes* of replicating DNA. Three DNA segments are seen in the rectangular field: a has 8 fully-visible eyes and one eye truncated by the boundary; b has 12 fully-visible eyes; c has two, one of which crosses over itself.

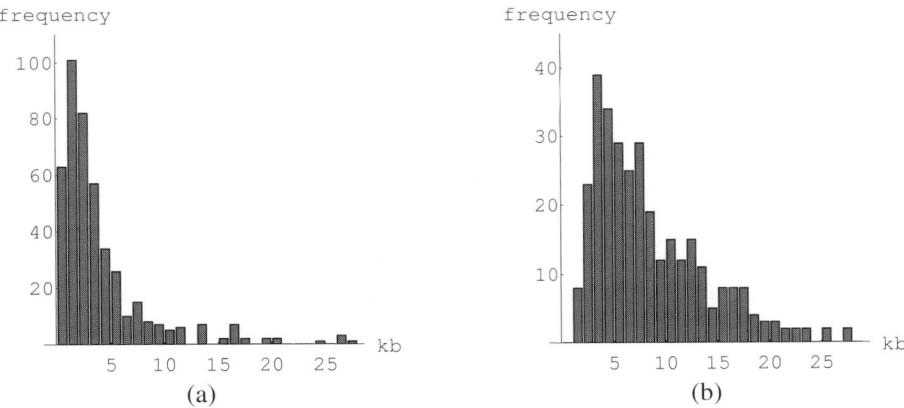

Fig. 2. (a) The histogram of eye length for 439 fully visible eyes; (b) the histogram of eye-to-eye distance as measured along the DNA from centre to centre of fully-visible adjacent eyes.

the molecule which has separated into two strands (a good indication of the stage in the process). These also show great variability.

There are other difficulties of interpretation due to the sampling protocols used and calculations employed by the authors (for example, in the treatment of eyes truncated by the edge of the photo and in the use of the full eyes and eye-to-eye distances only from segments with at least two full eyes). The authors do not adequately deal with the truncation and size-biassing effects encountered. The main message we draw now, however, is simply that the entities such as eye length and eye-to-eye distance show considerable variability; this provides the stochastic interest for the probabilist.

In Cowan, Chiu and Holst (1995), a stochastic model of this "strand-separation" process was posed. The Poisson process of origins was given an intensity γ. These sites must wait an i.i.d. exponentially distributed time V with mean $1/\mu$ until the approach of the enzyme complex, the times being independent of the Poisson process. There is a reasonable sized pool of enzyme complexes so, as we discussed above, an exponential distribution makes a plausible assumption.

If, in the waiting for an enzyme complex, a site has not been passed over by a moving fork initiated elsewhere, molecular binding of the complex takes place and the two bidirectional forks commence at the site. The forks move at constant speed r in each direction. The assumption of constant speed is reasonable and accords with experimental evidence. In reality, one might expect that the time to unwind/separate the two strands a distance of just *one* base might indeed be a random variable, but thousands of these small steps are involved in rapid succession, so variability is not observed in the total time to unwind (say) a thousand bases, variability being nullified by the strong law of large numbers. For *Drosophila*, Blumenthal et al. estimate fork speeds of 2,600 bases per minute.

Let T_L be the time when a section of length L, represented by $[0, L]$, is completely separated into single strands. In the paper, the authors found a useful asymptotic result for the distribution of T_L as $L \to \infty$. Their methods drew upon mathematical results by Janson (1983) and Hall (1988) on coverage processes.

We now look at the Cowan–Chiu–Holst model in some detail.

7. The time to separation of a long DNA molecule

Janson (1983) and Hall (1988) considered a coverage problem which relates to the model of Cowan, Chiu and Holst. Line-segments of random i.i.d. lengths are placed with left-ends (or centers) at the points of a stationary Poisson process. Janson and Hall considered the probability that these (possibly overlapping) segments cover the interval $[0, y]$.

Our situation fits into this framework. Consider the Poisson point process of all origins which have been *approached* by an enzyme-complex by time t. This includes those origins approached but already 'passed over' at the time of enzyme approach, so this Poisson process has intensity $\gamma(1 - e^{-\mu t})$. If at the epoch of approach to each origin a line-segment starts to grow at rate $v = 2r$, then by time t each point has acquired a line-segment with generic length U such that

$$\mathbb{P}\{U \leqslant x\} = \mathbb{P}\{v(t - V) \leqslant x | V \leqslant t\}$$
$$= \frac{e^{-\mu(t-x/v)} - e^{-\mu t}}{1 - e^{-\mu t}}, \quad 0 < x \leqslant vt, \tag{3}$$

since

$$\mathbb{P}\{V \leqslant x\} = 1 - e^{-\mu x}, \quad x > 0. \tag{4}$$

In the real process, passed-over origins do not initiate the growth of a line-segment, but the event that the interval $[0, y]$ is covered has the same probability whether or not we allow such initiations (and it is convenient to allow it in this analysis so that we can draw on the ideas of Janson and Hall).

Janson (1983, Lemma 2) and Hall (1988, Theorem 2.5) each prove a limit theorem of the following character. If their Poisson process intensity, λ say, tends to infinity and the mean segment-length a tends to 0 such that

$$\frac{e^{a\lambda}}{\lambda} \to e^u, \tag{5}$$

where $-\infty < u < \infty$, and if a number of other conditions are imposed on the *distribution* of segment-length and its mode of change as a changes, then

$$\mathbb{P}\{[0, y] \text{ is covered}\} \to e^{-ye^{-u}}.$$

These theorems do not have an immediate application to the problem, because it turns out that one of the 'other conditions' of Janson or Hall does not hold in this case. Nevertheless their work helped in the proof of the following theorem concerning the time T_L until the interval $[0, L]$ is covered.

THEOREM 1. *For each real number u,*

$$\mathbb{P}\left\{v\gamma T_L - \log(\gamma L) - \frac{v\gamma}{\mu} \leq u\right\} \to e^{-e^{-u}} \quad \text{as } L \to \infty.$$

The result involves a limit as $L \to \infty$. At each stage in this limiting process we may rescale the real axis by a factor L, so that $[0, L]$ becomes $[0, 1]$. This means that at time t, the mean rescaled segment length, denoted by a, is

$$a = \frac{v}{L}\left(\frac{t}{1 - e^{-\mu t}} - \frac{1}{\mu}\right),$$

whilst the initiated origins form a Poisson process with intensity

$$\lambda = \frac{v\gamma}{a}(1 - e^{-\mu t})\left(\frac{t}{1 - e^{-\mu t}} - \frac{1}{\mu}\right).$$

As $L \to \infty$, it is clear that $a \to 0$ and $\lambda \to \infty$, but for fixed t there does not exist a number $u \in \mathbb{R}$ such that (5) holds. It is necessary that t changes with L. Suppose, for $c_1 > 0$,

$$t = c_0 + c_1 \log L.$$

Then (5) holds, provided $c_1 = 1/\gamma v$ and $c_0 = 1/\mu + (u + \log \gamma)/\gamma v$. Suppose therefore that for some $u \in \mathbb{R}$

$$t = \frac{\log(\gamma L) + u}{\gamma v} + \frac{1}{\mu}. \tag{6}$$

Expressing a and λ as functions of t (eliminating L) we have

$$a_t = \frac{f(t)}{t}\left(\frac{t}{1-e^{-\mu t}} - \frac{1}{\mu}\right) \xrightarrow{t\to\infty} 0,$$

$$\lambda_t = \left(1-e^{-\mu t}\right)\frac{\gamma v t}{f(t)} \xrightarrow{t\to\infty} \infty,$$

where

$$f(t) = \gamma v t \exp\left(u + \frac{\gamma v}{\mu} - \gamma v t\right) \xrightarrow{t\to\infty} 0.$$

With relationship (6) between t and L, (5) holds as $L \to \infty$ and hence as $t \to \infty$. This suggests the application of the theorems of Janson or Hall, yet there is a difficulty. From (3), (4) and (6) the segment-lengths after rescaling have distribution function H_t given by

$$H_t(x) = \begin{cases} 1 - \dfrac{1-\exp[-\mu t\{1-x/f(t)\}]}{1-e^{-\mu t}}, & x < f(t), \\ 1, & x \geqslant f(t). \end{cases}$$

$H_t(x)$ changes with t in a different manner, however, from that which is required in Janson or Hall. They require that H_t satisfies $H_t(x) = H_1(a_1 x/a_t)$ for all $t > 0$. Let $p_t(y)$ be the probability that $[0, y]$ is covered at time t. From Janson (1983, Lemma 1) or Hall (1988, Theorem 2.6) we have that (in the current notation)

$$\begin{aligned}\pi_t(s) &\equiv \int_0^\infty e^{-sy} p_t(y)\,dy \\ &= \frac{1}{s} - \frac{1}{s^2 \int_0^\infty \exp[-sy + \lambda_t \int_y^\infty \{1 - H_t(x)\}\,dx]\,dy} \\ &= \frac{1}{s} - \frac{1}{s^2 \int_0^\infty \exp[-sy + I_t(y)]\,dy}, \quad \text{say.} \end{aligned} \quad (7)$$

In our context, the essence of Janson–Hall limit theory is a proof that if $H_t(x) = H_1(a_1 x/a_t)$ for $t > 0$, then as $t \to \infty$ (and hence $\lambda_t \to \infty$, $a_t \to 0$ with $\exp(a_t\lambda_t)/\lambda_t \to e^u$),

$$\int_0^\infty \exp[-sy + I_t(y)]\,dy \to \frac{1+se^u}{s}. \tag{8}$$

Cowan, Chui and Holst then prove that (8) holds, despite the absence of one of the Janson–Hall conditions. So, from (7),

$$\begin{aligned}\pi_t(s) &\to \frac{1}{s} - \frac{1}{s(1+se^u)} \\ &= \frac{e^u}{1+se^u}.\end{aligned}$$

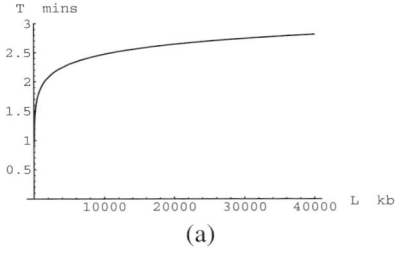 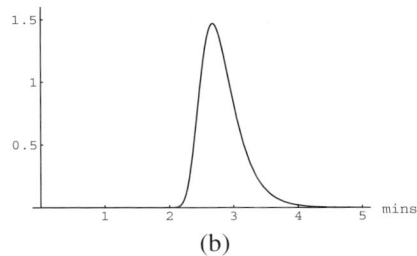

Fig. 3. (a) The slow logarithmic growth of the mean of T_L for lengths of DNA (in kb) up to the longest chromosomal arm of the fruit fly; (b) the approximate p.d.f. of T_L for $L = 40{,}000$ kb. In both plots, $v = 8$ kb/min, $1/\gamma = 2$ kb and $1/\mu = 12$ secs.

Thus $p_t(y) \to e^{-ye^{-u}}$. Now our original interval $[0, L]$ has been scaled to be $[0, 1]$, so as $L \to \infty$,

$$\mathbb{P}\left\{v\gamma T_L - \log(\gamma L) - \frac{v\gamma}{\mu} \leq u\right\} = \mathbb{P}\{T_L \leq t\} = p_t(1) \to e^{-e^{-u}},$$

which proves the theorem. Note that the distribution function on the RHS has mean 0.577 (Euler's constant) and variance $\pi^2/6$.

REMARK 1. A consequence of the result above is that

$$\frac{v\gamma T_L}{\log(\gamma L)} \to 1 \quad \text{in probability as } L \to \infty.$$

Thus for extremely large L, T_L is near to the constant $\log(\gamma L)/v\gamma$. Interestingly, for all the stochasticity in the model, the result is close to a constant for very large L. $\mathbb{E}T_L$ grows slowly like $\log L$ (see Fig. 3(a)) but $\mathbb{V}\text{ar}T_L$ settles down to $\pi^2/(6v^2\gamma^2)$.

The longest chromosome in *Drosophila melanogaster* is 62,000 kb. Allowing 40,000 for the longer of the two arms on either side of its centromere, we see below in Figure 3(b) a sample plot of the approximate p.d.f. of T_L. Note that the variance is still appreciable for $L = 40{,}000$ kb and the chosen parameter values.

REMARK 2. Cowan, Chiu and Holst took care not to refer to L as the length of the DNA molecule, merely that $[0, L]$ was a long section of the molecule. The DNA was modelled by the whole real line. So their model is not totally accurate since DNA has finite length, albeit very long relative to $1/\gamma$. Arguably, as they remark, we should treat L as the DNA length and study a process where origins outside $[0, L]$ cannot be initiated. Let T_L^* be the completion time for this new process. Usually $T_L^* = T_L$, but their difference (if any) is governed by the proportion, in our studied process, of $[0, L]$ covered by frontiers which are initiated outside $[0, L]$. This proportion at time t is bounded above by vt/L, a term which goes to zero as t and L become large in the manner of (6).

REMARK 3. The work of Cowan, Chiu and Holst relates to "a DNA model" studied by Vanderbei and Shepp (1988) and, in a context unrelated to DNA, by Quine and Robinson (1990, 1992). In the former model of Vanderbei and Shepp, there are no specified sites of initiation; instead initiation may occur randomly and indiscriminantly at any location that has not yet been 'passed over' by a fork frontier, that is, according to a Poisson process which is stationary in space-time. This model is not appropriate for DNA replication, because (as we have seen) origins of replication are defined by the specific local stereo-chemistry of the DNA; the approach of the enzyme-complex is not indiscriminant. So the model of Vanderbei and Shepp, though interesting, is not applicable in DNA studies. (Nevertheless, Cowan, Chiu and Holst also applied their method to this related model. As a result, they were able to prove rigorously the results suggested by Vanderbei and Schepp.)

REMARK 4. Holst, Quine and Robinson (1996) have addressed the distribution of T_L for finite L. They have found the Laplace transform of a related quantity and, in principle, this is invertible and the exact distribution can be found. They have not done the inversion, but as this chapter was going to press, Chiu and Yin (2000) have found the inverse Laplace transform. They present an explicit, though complicated, expression for the distribution of T_L. See these papers for details and generalisations.

8. The proportion of origins initiated

It is clear that some origins are not initiated during the process of strand separation described above. In the Cowan–Chiu–Holst model, it is possible to find, at least for large L, the probability that a given origin O is initiated and not passed over by the fork of an eye initiated elsewhere.

Let us condition on the time S until O is first approached by an enzyme complex. Suppose $S = s$. Let $q(s)$ be the probability that a fork from the right of O passes over O before time s. Note that the only origins lying to the right of O which can initiate moving forks in time to reach O by time s, are those within distance rs. Suppose there are $k > 0$ of these. Given k, the locations $U_1, U_2, \ldots, U_k$ of these k origins are independently distributed, uniformly on $[0, rs]$.

The times $S_1, S_2, \ldots, S_k$ at which these k origins are approached by enzyme are independent and identically distributed with p.d.f. $\gamma e^{-\gamma t}$. Thus, for the ith such origin, the chance that it is not approached earlier enough for a frontier to reach O by time s is

$$\mathbb{P}\left\{S_i > s - \frac{U_i}{r}\right\} = \int_0^{rs} e^{-\gamma(s-u/r)} \frac{du}{rs}$$

$$= \frac{1}{\gamma s}(1 - e^{-\gamma s}).$$

The origin O is not passed over from the right by time s if and only if all k such sites are approached too late, that is with probability (given k and s) of

$$\left[\frac{1}{\gamma s}(1 - e^{-\gamma s})\right]^k.$$

Therefore, after incorporating the case $k = 0$ as well,

$$q(s) = \sum_{k=0}^{\infty} \frac{(\mu r s)^k e^{-\mu r s}}{k!} \frac{(1 - e^{-\gamma s})^k}{(\gamma s)^k} = \exp\left[\frac{\mu r}{\gamma}(1 - \gamma s - e^{-\gamma s})\right].$$

Clearly, the chance that O is not passed over by time s from either the right or left is $[q(s)]^2$. Therefore, unconditional upon s, the probability that origin O is not passed over, and so initiates its own eye, is, with v replacing $2r$,

$$\int_0^\infty \gamma e^{-\gamma s} [q(s)]^2 \, ds = \int_0^\infty e^{-w} \exp\left[\frac{\mu v}{\gamma}(1 - w - e^{-w})\right] dw. \quad (9)$$

This result can be found in Chiu (1992) and, in a quite unrelated context, in the work of McPeek and Speed (1995). There is also a generalised formula for the case where the space-time positions of "enzyme approaches to sites" form a more general space-time Poisson process (Chiu and Quine, 1997).

9. Mean eye lengths and eye-to-eye distances

Blumenthal et al. collect statistics on mean eye lengths, c say, and mean eye-to-eye distances, d say (see their Figure 6). We do not reproduce their numbers, because, as mentioned above, we question their statistical protocol. We shall, however, attempt to emulate the broad character of their Figure 6, if not an exact fit.

Their figure plots the two means against the proportion of the molecule which has separated into eyes. We shall therefore try to relate these entities theoretically in the Cowan–Chiu–Holst model. We are able to do so at various times t, assuming a very long DNA molecule (indeed infinite, as we model the DNA molecule by the real line).

Consider again the preamble to Eq. (3). At time t, the origins which have been approached by an enzyme complex form a point process of intensity $\lambda_t = \gamma(1 - e^{-\mu t})$. The distribution function of "line segment" which has grown at these sites is given by (3). The mean length is

$$a_t = \int_0^{vt} \left[1 - \frac{e^{-\mu(t-x/v)} - e^{-\mu t}}{1 - e^{-\mu t}}\right] dx$$

$$= \frac{vt}{1 - e^{-\mu t}} - \frac{v}{\mu}. \quad (10)$$

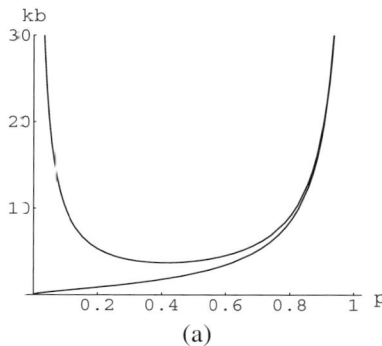 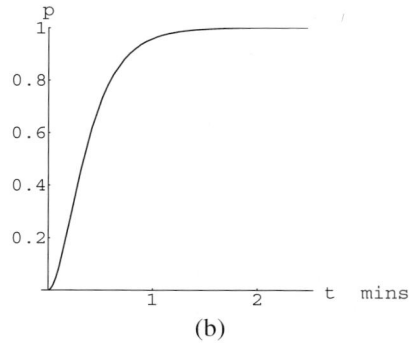

Fig. 4. (a) The upper curve shows the mean eye-to-eye distance d as a function of p, the proportion of a long DNA molecule which has separated. The lower curve shows the mean eye length c as a function of p. (b) p as a function of t. In both plots, $v = 8$ kb/min, $1/\gamma = 2$ kb and $1/\mu = 12$ secs.

The union of all these line segments is a realisation of the Boolean process of intervals on the line. Coverage results for such a process are summarised in Hall (1988, p. 89) in terms of the mean value a of line-segment lengths and the intensity λ of nuclei for these segments (Hall uses α instead of a). We have shown these entities above as a function of t.

Thus we can cite Hall's results. Let p_t be the proportion of the line covered by line segments at time t. This is given [Hall's (3.4)] as

$$p_t = 1 - e^{-a_t \lambda_t}. \tag{11}$$

The eye length (Hall's clump size, Theorem 2.2) has mean

$$c_t = \frac{e^{a_t \lambda_t} - 1}{\lambda_t} \tag{12}$$

whilst the gap of DNA separating two eyes can easily be shown to have mean $1/(a_t \lambda_t)$. Thus the mean eye-to-eye distance is

$$d_t = \frac{1}{a_t \lambda_t} + \frac{e^{a_t \lambda_t} - 1}{\lambda_t}. \tag{13}$$

We seek to relate both mean eye length c and mean eye-to-eye distance d to p. This is achieved by eliminating t from the equations above, since it is the common dependence on t which creates the sought relationship. Elimination of t is not possible by simple algebraic manipulation.

We can, however, prepare "parametric plots" using a package such as *Mathematica*; typical parametric plots are shown in Figure 4(a). Figure 4(b) plots p_t against t. The parameter values chosen for Figure 4 mean that fork speed is 4 kb/min, origins are on average 2 kb apart and the mean time until an origin is first approached by an enzyme complex is 12 seconds. From (9), I have calculated that 13.19% of origins are actually initiated when these parameter values are used.

My Figure 4(a) is of the same general character as Figure 6 in Blumenthal et al. There is not an exact numerical fit between the two figures and I have not attempted to find one. Some explorations suggest that the model will not capture the flatness of the plateau for d evident in their figure. I am not in a position to know if this is due to a failure of the Cowan–Chiu–Holst model or to inappropriate statistical calculations in the paper of Blumenthal et al. or to inapplicability of plots such as my Figure 4(b) when the DNA segments analysed are not extremely long.

I leave these as *open* issues and move on to other problems associated with DNA replication.

10. What is happening inside the eyes?

We have indicated that the separation of the DNA into 2 single strands allows these strands to be copied. The single strand acts as a template for the building of the copy; for example, an A on the template attracts a T to bind opposite, and vice versa, whilst C and G have similar complementarity. Thus, due to this base-pairing phenomenon, the copy formed is identical to the strand that was formerly bound to the template before the separation into the "eye" form.

This copying does not, however, rely on the attraction of A to T and C to G by chemical diffusion alone. It is enzyme mediated; an enzyme binds to the DNA and rapidly finds, from the neighbouring solution, the right match to place directly opposite each base in the template. The enzyme acts as a catalyst and makes the binding quick and efficient.

How does this start? In the most popular theory of DNA replication (Kornberg, 1980; Huberman and Horwitz, 1973; Denhardt and Faust, 1986), two distinct enzymic mechanisms are postulated. The first involves the same enzyme which separates the DNA. As this enzyme moves along the double-stranded molecule, separating the DNA and pushing the fork further away from the eye's "origin", its rear portion remains in contact with one of the two single strands behind it. This rear portion attracts the right base to match each base on this particular single strand, immediately converting the strand back into a double-stranded molecule. If the eye's origin is called O, then one strand is immediately replicated between O and the right-moving fork whilst the other strand is immediately replicated between O and the left-moving fork (there being a location O on both strands after separation occurs). Figure 5(a) shows this process.

Note, however, that half of the DNA is not replicated by this enzymic method, since the rear portion of the enzyme retains contact with only one strand, the so-called *leading* strand of that replicating fork. The strand that it loses contact with, the so-called *lagging* strand, does not enjoy the replicating assistance of this "separating" enzyme (and so *lags* somewhat in its replication).

Nature has devised another enzymic mechanism to replicate the lagging strand. Scattered randomly along a single strand of DNA are very short sequences of "letters" – specific letters but the sequence is not currently known. These letter patterns attract short

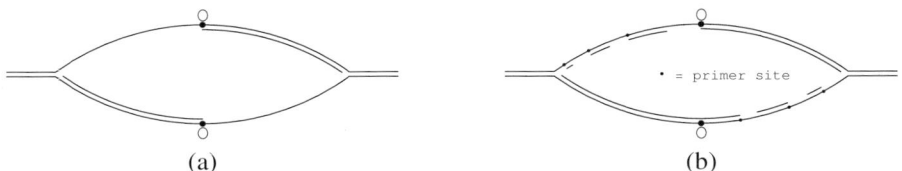

Fig. 5. (a) An eye and the effect on the leading strand of replication by the same enzyme that separates the two strands. (b) The discontinuous replication on the lagging strand from primer sites, with $b/r = 0.6$.

fragments of RNA known as *primers*; thus the sequence is called a *primer site*. When the primer lodges on the primer site, by random diffusion and chemical attraction, another enzyme known as *DNA polymerase* also lodges and starts the process of building the DNA copy on the single strand (by acting as the catalyst for base pair attraction and moving along the DNA, creating the correct base pairing as it goes).

Noteworthy, however, is that this enzyme can only function in the "moving catalyst" role in a given direction, namely away from the fork toward O. Thus, if we could look at the eye centered at O (before any of the complications that arise when eyes merge), we would see a pattern as in Figure 5(b). (Note that the magnification in micrographs such as Figure 1 is insufficient to see this detail.)

As Figure 5(b) shows, there are newly-copied 'islands' on the lagging strand adjacent to the fork (i.e., fragments of newly-created double-stranded DNA set on the single-stranded material). The first such fragment formed on (say) the right-hand side of O in the figure grows leftwards and soon reaches O, where it joins the large section of double-stranded DNA already in place to the left of O. Eventually each fragment on the right-hand side of O joins to the fragment on its left, and so becomes connected to the origin O in due course. In contrast to our 'islands', we refer to the collection of joined fragments connected to 0 as the 'mainland'. Any fragment which becomes connected to the mainland is no longer called a 'fragment'.

If the process is interrupted at time t, and the DNA heated so that all strands separate, short fragments of single-stranded DNA can be observed (our islands); these have been named Okazaki fragments after Okazaki et al. (1968) who observed these fragments in prokaryotic cells. (Huberman and Horwitz (1973) later established the existence of such fragments in eukaryotic cells.)

Of course the mean length of Okazaki fragments does not equal the mean inter-primer-site distance. Some Okazaki fragments will comprise just one inter-site spacing which is not completely copied and others will be the amalgam of numerous copied and partly copied inter-site spacings. The investigation of the relationship between mean length of Okazaki fragments, our 'islands', and the mean inter-site distance, μ say, was the major aim of a paper by Cowan and Chiu (1994). They formulated a stochastic model of fragment formation and, using data on mean lengths of Okazaki fragments, showed that the unknown μ can be estimated.

I give an account of their model, together with some useful recent analysis by Piau (2000).

11. The Cowan–Chiu model of fragment formation

We focus on the lagging strand to the right of O in Figure 5(b). Measured along this strand, let X_1 be the distance between O and the first primer site and X_i the distance between the $(i-1)$st and the ith sites ($i = 2, 3, \ldots$). It is readily shown that for almost any plausible model of DNA structure, these inter-site distances are approximately exponentially distributed, so we model the sites by a stationary Poisson process of intensity $1/\mu$.

Denote the fork speed by r and define $Y_i = X_i/r$. Thus $\{\sum_{i=1}^n Y_i\}_{n=1}^\infty$ is the Poisson process of times at which the right-moving fork passes the primer sites of the lagging strand. Site i becomes exposed to primers at time Y_i and, if there is no delay,[5] the fragment at this ith site starts immediately and grows leftward at rate b. This temporal Poisson process of primer sites "emerging" from the fork has intensity $\lambda = r/\mu$.

We assume that, when the fragment grows to reach the previous primer site, there is a time δ for the complete joining of the fragment to its neighbouring fragment (which may by this time have joined the mainland). Let D_t denote the distance between the fork and the nearest end of the mainland at time t, whilst P_t denotes the cumulative length of all copying up until time t. Also define L_t as the total length of extant Okazaki fragments at time t, whilst N_t is the number of such fragments at time t. Note that N_t decreases by one whenever two fragments join or when a fragment joins to the mainland, but increases when a primer site emerges from the fork. Since the length of the mainland equals $P_t - L_t$ and also equals $rt - D_t$, we have the identity

$$L_t = P_t - (rt - D_t). \tag{14}$$

The mean Okazaki-fragment length is effectively equal to $\mathbb{E}(L_t)/\mathbb{E}(N_t)$. To see this, note that experiments which try to estimate the mean length of Okazaki fragments are performed with cultures containing a massive number of cells. So there are large numbers of active forks at any time in the experiment (especially so in the eukaryote case where each cell has thousands of active forks). If fork i has variates L_t^i and N_t^i, the observed mean is $\sum_i L_t^i / \sum N_t^i$. The forks are independent and their number large, so the strong law of large numbers ensures that this ratio effectively equals $\mathbb{E}(L_t)/\mathbb{E}(N_t)$.

12. The expectations of N_t and P_t: renewal equations

We find $\mathbb{E}(N_t)$ first, using an argument which conditions on Y_1 and exploits the regenerative situation at time Y_1. Denoting $\mathbb{E}(N_t)$ by $n(t)$, $\mathbb{E}(N_t|Y_1 = y)$ by $n(t|y)$ and for brevity $b/(b+r)$ by a, one finds that $n(t|y) = 0$ for $y \geq t$. So for $t \geq 0$,

$$n(t) = \int_0^t n(t|y)\lambda e^{-\lambda y}\, dy.$$

---

[5] We assume no delay. This is for mathematical convenience but also fairly realistic. If there is a delay it should be random in duration, presumably exponentially distributed, but with a very small mean due to the obvious abundance of the primers.

When $t \geqslant \delta$, we have the breakdown

$$n(t) = \int_0^{a(t-\delta)} n(t-y)\lambda e^{-\lambda y}\,dy + \int_{a(t-\delta)}^t \big(1+n(t-y)\big)\lambda e^{-\lambda y}\,dy$$

$$= e^{-\lambda a(t-\delta)} - e^{-\lambda t} + \int_0^t n(t-y)\lambda e^{-\lambda y}\,dy,$$

whilst for $0 \leqslant t < \delta$,

$$n(t) = \int_0^t \big(1+n(t-y)\big)\lambda e^{-\lambda y}\,dy$$

$$= 1 - e^{-\lambda t} + \int_0^t n(t-y)\lambda e^{-\lambda y}\,dy.$$

Therefore, for all $t \geqslant 0$

$$n(t) = m(t) + \int_0^t n(t-y)\lambda e^{-\lambda y}\,dy,$$

where $m(t) = \exp(-\lambda a[t-\delta]_+) - e^{-\lambda t}$ and $[x]_+ := \max(x,0)$. Applying standard methods for this typical renewal equation, we find the solution

$$n(t) = m(t) + \lambda \int_0^t m(u)\,du. \tag{15}$$

So

$$n(t) = \lambda t \quad t < \delta$$

$$= \lambda \delta + \frac{r}{b}\big(1 - e^{-\lambda a(t-\delta)}\big), \quad t \geqslant \delta. \tag{16}$$

Therefore

$$n := \lim_{t \to \infty} \mathbb{E}(N_t) = \lambda\delta + \frac{r}{b}. \tag{17}$$

This result can also be derived by Little's formula, under quite general point-process models for the primer sites. Consider "births" of Okazaki fragments, at rate λ, when primer sites emerge from the fork and "deaths" when the fragments join to the previous fragment. Since μ equals r/λ, ν, the mean time between birth and death, is $\delta + r/(\lambda b)$. So by Little's formula, the expected number alive at equilibrium is $\nu\lambda$, or $\lambda\delta + r/b$.

As an aside, we note that similar regenerative arguments can be used to find higher moments of N_t. We merely note for the record that

$$\mathbb{E}(N_t^2) \to \frac{(b+2r+2b\lambda\delta)e^{\lambda a\delta}}{ab} - \frac{2(b+r)^3 e^{\lambda(2-a)a\delta}}{b^2(b+2r)} + 1, \quad t \to \infty.$$

A similar regenerative argument also yields an integral equation for $\mathbb{E}(P_t)$. Condition on $Y_1 = y$. If $0 < y \leq at$, the copying at the first primer site has progressed ry. If $at < y \leq t$, the copying has progressed $b(t - y)$. Defining $p(t) := \mathbb{E}(P_t)$ and $p(t|y) := \mathbb{E}(P_t|Y_1 = y)$, we have that for $t \geq 0$,

$$p(t) = \int_0^t p(t|y)\lambda e^{-\lambda y}\, dy$$

$$= \int_0^{at} (ry + p(t - y))\lambda e^{-\lambda y}\, dy + \int_{at}^t (b(t - y) + p(t - y))\lambda e^{-\lambda y}\, dy$$

$$= q(t) + \int_0^t p(t - y)\lambda e^{-\lambda y}\, dy,$$

where $q(t) = [r + be^{-\lambda t} - (r + b)e^{-\lambda at}]/\lambda$. It follows from the form of (15) that for $t \geq 0$,

$$p(t) = rt - \frac{r}{\lambda a}(1 - e^{-\lambda at}).$$

As $t \to \infty$, $rt - \mathbb{E}(P_t) \to r/\lambda a$. Thus if $\ell := \lim_{t \to \infty} \mathbb{E}(L_t)$ and $d := \lim_{t \to \infty} \mathbb{E}(D_t)$, then from (14),

$$\ell = d - r/\lambda a. \tag{18}$$

13. The expectation of D_t: the quasi-renewal equation

To find an equation for $\mathbb{E}(D_t)$, again condition on $Y_1 = y$. Consider $t \geq \delta$. If $y > a(t - \delta)$, then the copying started at the first primer site has not reached the mainland, so the distance between the fork and the mainland is rt. If $0 < y \leq a(t - \delta)$, it has reached the mainland and so the said expected distance is equal to $\mathbb{E}(D_{t-y})$. Define $d(t) := \mathbb{E}(D_t)$ and $d(t|y) := \mathbb{E}(D_t|Y_1 = y)$. Clearly $d(t) = rt$ for $t < \delta$, whilst for $t \geq \delta$

$$d(t) = \int_0^\infty d(t|y)\lambda e^{-\lambda y}\, dy$$

$$= \int_0^{a(t-\delta)} d(t - y)\lambda e^{-\lambda y}\, dy + \int_{a(t-\delta)}^\infty rt\lambda e^{-\lambda y}\, dy$$

$$= rte^{-\lambda a(t-\delta)} + \int_0^{a(t-\delta)} d(t - y)\lambda e^{-\lambda y}\, dy.$$

Define $d_\delta(t) := d(\delta + t)$ for $t \geq 0$ and $\delta \geq 0$. Then for $t \geq 0$, $\delta \geq 0$

$$d_\delta(t) = r(t + \delta)e^{-\lambda at} + \int_0^{at} d_\delta(t - y)\lambda e^{-\lambda y}\, dy. \tag{19}$$

This is not a renewal equation since $a < 1$. For reasons which will become clear, we define a new function g_δ by $g_\delta(t) := d_\delta(t)/r$. When (19) is suitably modified, to yield (20), we note that g_δ depends on r and b only via a, which itself depends on r and b solely via the ratio b/r.

$$g_\delta(t) = (t+\delta)e^{-\lambda at} + \int_0^{at} g_\delta(t-y)\lambda e^{-\lambda y}\,dy. \tag{20}$$

It is a simple matter to show, from (20), that

$$g_\delta(t) = g_0(t) + \delta.$$

Thus we may focus our attention on the case $\delta = 0$, whereby (20) looks like

$$g_0(t) = te^{-\lambda at} + \int_0^{at} g_0(t-y)\lambda e^{-\lambda y}\,dy. \tag{21}$$

Let us call equations of this type, *quasi-renewal equations*.

Cowan and Chiu were unable to find an explicit analytic solution to Eq. (21) and adopted a numerical approach. They presented graphs of g_0 versus t for a range of b/r, with λ set at 1.

They remark that the limit of g_0 as $t \to \infty$ is important to the theory, since $\mathbb{E}(D_t)/r \to d/r = d_\delta(\infty)/r = g_\delta(\infty) = g_0(\infty) + \delta$. They were, however, only able to give numerical calculations for $g_0(\infty)$.

Recently, Piau (2000) has made a study of the quasi-renewal equation, deriving explicit expressions for the limiting value of the solution as $t \to \infty$. In particular, he studied the equation

$$g(t) = h(t) + \int_0^{at} g(t-y)\lambda e^{-\lambda y}\,dy,$$

where $a \in (0, 1)$. He showed that, if L is the Laplace transform operator, defined for $x \geq 0$ by

$$Lg(x) := \int_0^1 e^{-xt} g(t)\,dt, \qquad Lh(x) := \int_0^1 e^{-xt} h(t)\,dt$$

and $c := 1 - a$ (Piau actually used the symbol b which clashes with our notation, hence our choice of something different), then

$$Lg(x) = \sum_{m \geq 0} (-1)^m Lh(A^m(x))\left(1 + \frac{A^m(x)}{\lambda}\right) \prod_{k=0}^m \frac{\lambda}{A^k(x)},$$

where A is the affine function $A(x) := (x + \lambda a)/c$ and A^n is the nth iterate of A, that is, $A^n(x) = -\lambda + (x + \lambda)/c^n$. Piau also showed that, as $t \to \infty$,

$$g(t) - h(t) \to \lambda \sum_{n \geq 0} (-1)^n Lh(x_n) \frac{P_n(c)}{c^n},$$

where $x_n := A^n(0) = \lambda(c^{-n} - 1)$, $P_0(c) := 1$ and, for $n \geq 1$,

$$P_n(c) := \prod_{k=1}^{n} \frac{c^k}{1 - c^k} = \prod_{k=1}^{n} \frac{\lambda}{x_k}.$$

Applying his results to (20), he derived the satisfying analytic result

$$g_0(\infty) = \frac{1}{\lambda(1-a)} \sum_{n \geq 1} \left[1 - \prod_{k \geq n} [1 - (1-a)^k] \right]. \tag{22}$$

Piau's calculations with (22) give numerical answers close to those of Cowan and Chiu, albeit slightly higher.

14. Relationship between fragment length and primer-site spacing

A major concern is the ratio at equilibrium, R say, of mean Okazaki-fragment length to mean inter-primer-site distance, that is,

$$R = \frac{\ell/n}{r/\lambda} = \frac{\lambda g_0(\infty) + \lambda \delta - 1/a}{\lambda \delta + r/b} \tag{23}$$

from (17) and (18). Note that R depends on b and r solely via b/r. Figure 6 shows a plot of R for a range of $\alpha := b/r$, with λ set at 1 and $\delta = 0$.

When $\lambda \neq 1$, Cowan and Chiu show that $n(t, \alpha, \lambda, \delta) = n(\lambda t, \alpha, 1, \lambda \delta)$, where the expression is now written as a function of four variables. Likewise, they show that $\lambda g_\delta(t, \alpha, \lambda) = g_{\lambda\delta}(\lambda t, \alpha, 1)$ for any $\delta \geq 0$. Eq. (23) shows that R is a function of α, δ and λ, namely

$$R(\alpha, \delta, \lambda) = \frac{\lambda g_0(\infty, \alpha, \lambda) + \lambda\delta - (1+\alpha)/\alpha}{\lambda\delta + 1/\alpha}.$$

So one can clearly see that, for all λ,

$$R(\alpha, \delta, \lambda) = R(\alpha, \lambda\delta, 1).$$

Thus Figure 6, based on $\delta = 0$ and $\lambda = 1$, has applicability for general λ since $R(\alpha, 0, \lambda) = R(\alpha, 0, 1)$. For $\delta \neq 0$, the use of (23) combined with Figure 6 easily yields the answers needed.

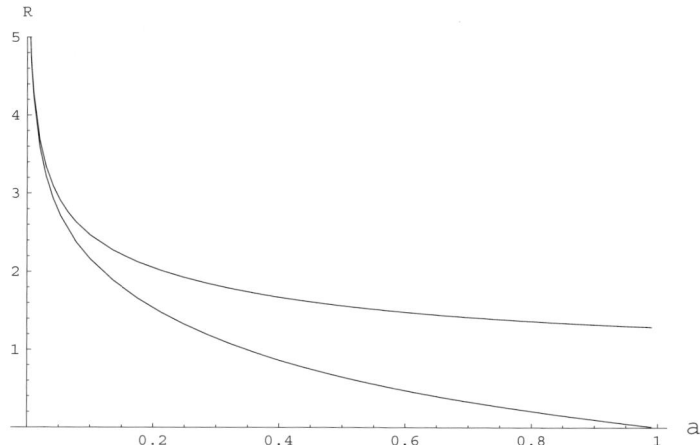

Fig. 6. The upper curve shows R, the ratio of mean Okazaki-fragment length to mean inter-primer-site distance, plotted against $\alpha := b/r$. The lower curve shows R^*, an entity which arises in the competing model (see Sections 16–19).

15. Estimated spacing between primer sites

The pattern of 'bases' which form a primer site is not known. It would be useful to know the expected distance between primer sites as this would give a guide to the number of invariant 'bases' in the primer site. For example, if the site had 4 invariant bases (e.g., *GCTA* or *GXAXGT* where $X =$ anything), then the approximate average distance between sites is $4^4 = 256$ bases.

The Cowan–Chiu theory, combined with data on the average length of Okazaki fragments, provides an estimate of the mean inter-primer-site distance and hence some information on the primer site pattern. Experimental data give mean Okazaki fragment length of 150 bases in eukaryotes and 1500 bases in prokaryotes (Kornberg, 1980).

Good data on the parameters b, r and δ are not available, but it is known roughly that b/r is about 0.6 for eukaryotes and perhaps 100 times less than this for prokaryotes. From (23) and Piau's formula (22), assuming $\delta = 0$ in the absence of better experimental knowledge, R is about 1.48 for eukaryotes and 4.77 for prokaryotes. Thus the mean inter-primer site distance is respectively $150/1.48 = 101$ and $1500/4.77 = 314$.

The data on which these calculations are based are very rough, but this calculation already suggests that, if these species have a common primer-site structure, then it might have 4 invariant bases and be spaced on average every 256 bases. The near contenders, 3 or 5 invariant bases, would lead to spacings of about 64 or 1024 bases respectively.

16. The competing theory

The discussion of Okazaki fragments cannot be closed without reference to an alternative theory considered by some biologists. We have stated, in building the model above, that the DNA copying on the *leading* strand is continuous due to the role of the unwinding/separating enzyme as a catalyst in the copying of the leading strand.

There have been suggestions, however, that replication of the *leading* strand is also done in discontinuous pieces and that the unwinding/separating enzyme plays no role at all in copying. Some experiments point to this (Callan, 1973; Kowalski and Denhardt, 1982). The putative mechanism is essentially similar to that described for the lagging strand, namely the lodging of primers at specific primer sites on the DNA and the subsequent copying by the moving *polymerase* enzyme. The only difference is that, because of the strict directionality of the polymerase movement along the DNA, the polymerase on the leading strand is moving toward the fork (in contrast to its movement away from the fork on the lagging strand). There would be a constraint, of course, that the speed b of the polymerase *must* be no greater than the fork speed r or else the polymerase would be bumping into the fork enzyme. Figure 7 illustrates this competing theory.

17. Notations for the competing theory

We now present an analysis of the Cowan–Chiu model, modified to take into account this competing theory. We follow the unpublished report of Cowan and Chiu (1992) in calling the new model, "Model B", and the earlier model, "Model A".

Let us focus on the lower strand of Figure 7. Okazaki fragments are present in both the leading strand (of the fork to the left of O) and the lagging strand (of the fork to the right of O). In model A, the 'mainland' commences at time zero because it is assumed that copying action starts immediately on the leading strand at O. In model B, we need to define what is meant by 'mainland', since polymerase activity is not initiated on the leading strand until after the leading-strand fork passes the first primer site on the leading strand (call this the *leading site*).

Let $X_1, X_2, \ldots$ be defined as before for the lagging strand and define $X_1^*, X_2^*, \ldots$ as the equivalent quantities on the leading strand. Thus the leading site is distance X_1^* from O. A polymerase lodges on this site at time Y_1^*, where $Y_i^* := X_i^*/r$. Prior to this time, we define the mainland to be solely the leading site itself. At later times, the

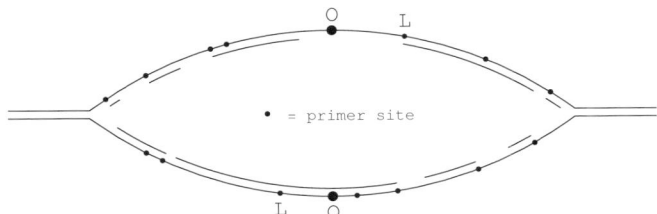

Fig. 7. The discontinuous-replication model for both the leading and lagging strands. Here $b/r = 0.6$. There are 3 "islands" and one "mainland" on the lower strand, the mainland being the piece which covers the "leading site" marked L. On the upper-strand there are 4 "islands" and the "mainland" which (at this early stage in the eye's development) is the piece with the "leading site" L at its left-hand end.

mainland is the contiguous section of double-stranded DNA which either *covers* the leading site or has one end at the leading site. Thus D_t equals $rt + X_1^*$ when $t < Y_1^*$ and D_t^*, the distance from the leading-strand fork to the mainland, is $rt - X_1^*$ for this same range of t.

Since one polymerase which lodges on the lagging strand can pass through O and travel on the leading strand, we need to be precise in our definitions of variables such as N_t, L_t and P_t. An Okazaki fragment is defined to be a *lagging*-strand fragment (resp. *leading*-strand fragment) if its polymerase lodged on the lagging strand (resp. leading strand). So, for example, P_t is the cumulative distance travelled by polymerases which lodge on the lagging strand. P_t^* is similarly defined for the leading strand, as are N_t^*, L_t^* and the functions n^*, p^*, ...

18. Analysis of the lagging strand for Model B

Lagging-strand properties are changed slightly from Model A. The methodology is simple and illustrated by $E(N_t)$. Condition on $\{Y_1^* = z\}$ and let $n(t|z)$ be $E(N_t|Y_1^* = z)$, different in meaning to that ascribed to $n(t|y)$ in Section 12. Then

$$n(t) = \int_0^\infty n(t|z) \lambda e^{-\lambda z} \, dz \tag{24}$$

and, by further conditioning on $\{Y_1 = y\}$,

$$n(t|z) = \int_0^t \left[1 + n(t - y|0)\right] \lambda e^{-\lambda y} \, dy \quad t < \delta + z/\alpha$$

$$= \int_0^{a(t-\delta-z/\alpha)} n(t - y|0) \lambda e^{-\lambda y} \, dy$$

$$+ \int_{a(t-\delta-z/\alpha)}^t \left[1 + n(t - y|0)\right] \lambda e^{-\lambda y} \, dy$$

when $t \geq \delta + z/\alpha$. Recall $\alpha = b/r$. Note that $n(\cdot|0)$ is known, from Model A, as the n in (16). So $n(t|z)$ can be calculated directly, and (24) used to show that, for Model B,

$$n(t) = \lambda t \quad 0 \leq t < \delta$$

$$= \lambda \delta + \frac{1}{\alpha}\left(1 - e^{-\lambda \alpha (t - \delta)}\right), \quad t \geq \delta,$$

$$\mathbb{E}(N) = \lambda \delta + \frac{1}{\alpha}. \tag{25}$$

For finite t, this result differs from the equivalent result for Model A, but asymptotically, there is no difference. A similar methodology shows that

$$\mathbb{E}(P_t) = rt - \frac{r}{\lambda \alpha}\left(1 - e^{-\lambda \alpha t}\right), \quad t \geq 0.$$

As $t \to \infty$, $rt - \mathbb{E}(P_t) \to r/(\lambda\alpha)$. It can also be readily calculated that, although $d(t)$ differs between Models A and B for finite t, $d(\infty)$ is the same in both cases (essentially the stochastic process D_t is equal under both models except for a short transient difference).

Due to our convention that the leading site is the separation point between lagging and leading strands, (14) no longer holds. It is replaced by

$$L_t = P_t - rt + D_t - X_1^*.$$

Yet, as $t \to \infty$, we find that (18) still holds. So

$$\mathbb{E}(L) = rg_0(\infty) + r\delta - \frac{r}{\lambda a}, \tag{26}$$

where $g_0(\infty)$ is the same as in Model A. The average length of lagging-strand Okazaki fragments is $\mathbb{E}(L)/\mathbb{E}(N)$, easily calculated from (26) and (25).

19. Analysis of the leading strand for Model B

The properties of the leading strand can be derived using a similar methodology. We firstly suppose $\{Y_1^* = 0\}$ and analyse the leading strand, using regenerative arguments. Then we condition on $\{Y_1^* = z\}$, find simple linkages with the results under $\{Y_1^* = 0\}$, and finally integrate over z. Details are omitted. We merely report that

$$n^*(t) = n(t) - 1 + e^{-\lambda t}, \quad t \geq 0,$$
$$\mathbb{E}(N^*) := \lim_{t \to \infty} \mathbb{E}(N_t^*) = \mathbb{E}(N) - 1,$$
$$p^*(t) = p(t). \tag{27}$$

Thus, as $t \to \infty$, $rt - \mathbb{E}(P_t^*) \to r/(\lambda\alpha)$, the same as on the lagging strand. We can also show that, if $g_0^*(t)$ is defined as $\mathbb{E}(D_t^*)/r$ for $t \geq 0$ when $\delta = 0$ and if $g_0^*(t|z)$ denotes $\mathbb{E}(D_t^*|Y_1^* = z)$, then $g_0^*(\cdot|0)$ satisfies

$$g_0^*(t|0) = (1-\alpha)te^{-\lambda\alpha t} + \int_0^{\alpha t} g_0^*(t-u|0)\lambda e^{-\lambda u} \, du. \tag{28}$$

Since $g_0^*(t|z) = g_0^*(t-z|0)$ for $t \geq z$, the solution of (28) also provides the solution for $g_0^*(\cdot|z)$. Recall $\alpha := b/r \leq 1$, so (28) is not a renewal equation except in one special case, $\alpha = 1$, where the solution of (28) is trivially $g_0^*(t|0) = 0$. For $\alpha < 1$, (28) is a quasi-renewal equation which can be analysed using Piau's methods. As before, it is only the limit $g_0^*(\infty|0)$ which is important, and moreover, $g_0^*(\infty|0) = g_0^*(\infty)$. Piau's method's show that

$$g_0^*(\infty) = \sum_{n \geq 1}\left[1 - \prod_{k \geq n}[1 + (1-\alpha)^k]\right].$$

For the leading strand, it is easy to see that

$$L_t^* = P_t^* - rt + D_t^* + X_1^*.$$

So, as $t \to \infty$,

$$\mathbb{E}(L^*) = \mathbb{E}(D^*) - \frac{r}{\lambda\alpha} + \frac{r}{\lambda}$$
$$= rg_0^*(\infty) + r\delta - \frac{r}{\lambda}\left(e^{-\lambda\alpha\delta} - \frac{1}{\alpha}\right). \tag{29}$$

The average length, R^* say, of leading-strand Okazaki fragment is $\mathbb{E}(L^*)/\mathbb{E}(N^*)$, easily derived from (29), (27) and (25) as:

$$R^* = \frac{\lambda g_0^*(\infty) + \lambda\delta + e^{-\lambda\alpha\delta} - 1/\alpha}{\lambda\delta + 1/\alpha - 1}.$$

In Figure 6, R^* is plotted against $\alpha := b/r$ for $\delta = 0$ and $\lambda = 1$. The plot captures the intuitive result that, as $b \to r$, the mean length of the Okazaki fragments on the leading strand tends to zero. It is easy to show that, in an analogous manner to the lagging-strand, $R^*(\alpha, \delta, \lambda) = R^*(\alpha, \lambda\delta, 1)$.

Finally, experiments which measure average length of extant Okazaki fragments involve averaging over all such fragments, be they on the leading or lagging strand. It is a simple matter to calculate this average and reach an overall R figure:

$$R_{\text{overall}} = \frac{\lambda[E(L) + E(L^*)]}{r[E(N) + E(N^*)]}.$$

I find model B to be extremely plausible since nature needs only to devise one method, namely the one involving the DNA polymerase and primers in order to achieve replication. It is not necessary to endow the unwinding/separating enzyme with the added role of copying the leading strand and one wonders why nature would evolve a redundancy of this sort. Further biological research will, no doubt, point to the correct model, A or B.

20. Concluding remarks

In this chapter I have considered a model for the formation of *eyes* in a molecule of DNA when it is undergoing replication. The sites where these eyes initiate, called *replication origins* are modelled by a Poisson process along the DNA and the times at which certain enzyme complexes first approach a given site have the exponential distribution (in line with our general discussion on the role of exponential distributions in chemical reactions). These two stochastic ingredients to the model provide the interest for the probabilist. Results on the time to replicate a long section of DNA and on the proportion of origins actually initiated are presented.

Secondly, I considered a model for the actual replication within the *eye*, identifying different mechanisms for the so-called *leading* and *lagging* strands. The model for the lagging strand succeeds in capturing the formation of Okazaki fragments. We are able to present results on their mean length. The same results are presented for an alternative model favoured by some biologists.

In presenting the material, I have been concious of drawing on the known biological facts and the snippets of relevant data to assist the estimation of our models' parameters. These data are not precise but their informal use enables us to cast some light on a few biological questions.

For the probabilist, my aim has been to describe the way that stochastic effects arise in molecular biology, in part due to the small numbers of some molecules (or molecular sites) – leading to greater stochastic variation in the chemistry – but mainly due to the random spatial organisation of binding sites on the long DNA molecule.

I hope that interest is aroused in this branch of *applied probability modelling*.

References

Alberts, B., D. Bray, J. Lewis, M. Raff, K. Roberts and J. D. Watson (1983). *Molecular Biology of the Cell.* Garland, New York.

Barbour, A. D., L. Holst and S. Janson (1992). *Poisson Approximation.* Oxford University Press, Oxford.

Blumenthal, A. B., H. J. Kriegstein and D. S. Hogness (1974). The units of DNA replication in *Drosophila melanogaster* chromosomes. *Cold Spring Harbor Symposium on Quantitative Biology* **38**, 205–213.

Callan, H. G. (1973). DNA replication in the chromosomes of eukaryotes. *Cold Spring Harbor Symposium on Quantitative Biology* **38**, 195–203.

Chiu, S. N. (1992). M.Phil. thesis, University of Hong Kong.

Chiu, S. N. and M. P. Quine (1997). Central limit theory for the number of seeds in a growth model in R^d with inhomogeneous Poisson arrivals. *Ann. Appl. Probab.* **7**, 802–814.

Chiu, S. N. and C. C. Yin (2000). The time of completion of a linear birth-growth model. Preprint received Dec. 1999.

Chryssaphinou, O. and S. Papastavridis (1988). A limit theorem for the number of non-overlapping occurrences of a pattern in a sequence of independent trials. *J. Appl. Probab.* **25**, 428–431.

Clifford, P., N. J. B. Green and M. J. Pilling (1987). Statistical models of chemical kinetics in liquids. *J. Roy. Statist. Soc. Ser. B* **49**, 266–300.

Cowan, R. and S. N. Chiu (1992). Mathematics of DNA replicating forks. *Research Report 28*, University of Hong Kong.

Cowan, R. and S. N. Chiu (1994). A stochastic model of fragment formation when DNA replicates. *J. Appl. Probab.* **31**, 301–308.

Cowan, R., S. N. Chiu and L. Holst (1995). A limit theorem for the replication time of a DNA molecule. *J. Appl. Probab.* **32**, 296–303.

Cowan, R. and R. Staudte (1986). The bifurcating autoregression model in cell lineage studies. *Biometrics* **42**, 769–783.

Darvey, I. G., B. W. Ninham and P. J. Staff (1966). Stochastic models for second-order chemical kinetics. The equilibrium state. *J. Chem. Phys.* **45**, 2145–2155.

Denhardt, D.T. and E. A. Faust (1986). Eukaryotic DNA replication. *BioEssay* **2**, 148–154.

Dunstan, F. D. J. and J. F. Reynolds (1981). Normal approximation for distributions arising in the stochastic approach to chemical reaction kinetics. *J. Appl. Probab.* **18**, 263–267.

Hall, P. (1983). On the roles of the Bessel and Poisson distributions in chemical kinetics. *J. Appl. Probab.* **20**, 585–599.

Hall, P. (1988). *Introduction to the Theory of Coverage Processes.* Wiley, New York.

Heyde, C. C. and E. Heyde (1971). Stochastic fluctuations in a one substrate one product enzyme system: are they ever relevant? *J. Appl. Probab.* **30**, 395–404.

Holst, L., M. P. Quine and J. Robinson (1996). A general stochastic model for nucleation and linear growth. *Ann. Appl. Probab.* **6**, 903–921.

Huberman, J A. and H. Horwitz (1973). Discontinuous DNA synthesis in mammalian cells. *Cold Spring Harbour Symposium on Quantitative Biology* **38**, 233–238.

Ishida, K. (1964). Stochastic model for bimolecular reaction. *J. Chem. Phys.* **41**, 2472–2478.

Janson, S. (1983). Random coverings of the circle with arcs of random lengths. In *Essays in honour of Carl-Gustav Esseen* (Eds. A. Gut and L. Holst). Dept. Math. Uppsala Univ.

Kornberg, A. (1980). *DNA Replication.* Freeman, San Francisco.

Kowalski, J. and D. T. Denhardt (1982). Adenovirus DNA replication *in vivo*. Properties of short DNA molecules extracted from infected cells. *Biochem. Biophys. Acta* **698**, 260–270.

Kriegstein, H. J. and D. S. Hogness (1973). The mechanism of DNA replication in *Drosophila* chromosomes: structure of replication forks and evidence for bidirectionality. *Proc. Nat. Acad. Sci.* **71**(1), 135–139.

McPeek, M. S. and T. P. Speed (1995). Modeling interference in genetic recombination. *Genetics* **139**, 1031–1044.

McQuarrie, D. A. (1967). Stochastic approach to chemical kinetics. *J. Appl. Probab.* **4**, 413–478.

Mitchison, J. M. (1971). *The Biology of the Cell Cycle.* Cambridge University Press, Cambridge.

Okazaki, R., T. Okazaki, K. Sakabe, K. Sugimoto, R. Kainuma, A. Sugimo and N. Iwatsuki (1968). *In vivo* mechanism of DNA chain growth. *Cold Spring Harbour Symposium on Quantitative Biology* **33**, 129–143.

Piau, D. (2000). Quasi-renewal estimates. *J. Appl. Probab.* **37**, 269–275.

Powell, E. O. (1953). Some features of the generation times of individual bacteria. *Biometrika* **42**, 16–44.

Powell, E. O. (1958). An outline of the pattern of bacterial generation times. *J. Gen. Microbiol.* **18**, 382–417.

Quine, M. P. and J. Robinson (1990). A linear random growth model. *J. Appl. Probab.* **27**, 499–509.

Quine, M. P. and J. Robinson (1992). Estimation for a linear growth model. *Statist. Probab. Lett.* **15**, 293–297.

Smith, J. A. and L. Martin (1973). Do cells cycle? *Proc. Nat. Acad. Sci. USA* **70**, 1263–1267.

Staudte, R., M. Guiguet and M. Collyn d'Hooghe (1984). Additive models for dependent cell populations. *J. Theor. Biol.* **109**, 127–146.

Vanderbei, R. J. and L. A. Shepp (1988). A probabilistic model for the time to unravel a strand of DNA. *Stochastic Models* **4**, 299–314.

An Empirical Process with Applications to Testing the Exponential and Geometric Models

J. A. Ferreira

We study an empirical lack-of-memory process defined on the basis of the Lau–Rao characterization of the exponential and geometric distributions. Using the Komlós–Major–Tusnády approximation theorem for the 'classical' empirical process, we prove some convergence results for the lack-of-memory process, including its convergence to a Gaussian process. We also study certain integral statistics, defined as integrals of the lack-of-memory process with respect to the empirical distribution function, and apply them to the problem of testing goodness-of-fit of the exponential and geometric models. Besides giving results on their performance, we point out the connections between these and other well-known statistics.

1. Introduction

Let μ be a σ-finite measure on $[0, +\infty)$ such that $\mu\{0\} < 1$, and f a non-negative, Borel measurable, locally integrable (with respect to Lebesgue measure) function defined on $[0, +\infty)$, not identically zero almost everywhere and satisfying the functional equation

$$f(x) = \int_{[0,\infty)} f(x+y)\mu(dy) \quad \text{for almost all } x \geq 0. \tag{1}$$

(Here and elsewhere, 'local integrability', 'almost everywhere' and 'almost all' without further qualification refer to Lebesgue measure.) Then either μ is arithmetic with some span $\lambda > 0$ and

$$f(x+n\lambda) = f(x)b^n, \quad n = 0, 1, \ldots, \text{ for almost all } x \geq 0,$$

for some $b > 0$ such that $\sum_{n=0}^{\infty} b^n \mu(\{n\lambda\}) = 1$, or μ is non-arithmetic and

$$f(x) \propto e^{\eta x}, \quad \text{for almost all } x \geq 0,$$

for some real η such that $\int_{[0,\infty)} e^{\eta y} \mu(dy) = 1$.

This is the Lau–Rao theorem, a result that plays a major role in the characterization of the exponential, geometric and related distributions. If we put $f = \overline{F}$ in (1), with $\overline{F} = 1 - F$ and F a probability distribution function on $[0, +\infty)$, then the theorem says that F is *essentially* exponential or geometric. More precisely, if F is assumed arithmetic with the same span λ as μ whenever μ is arithmetic, and non-arithmetic whenever μ is non-arithmetic, then F is either the geometric distribution on $\{\lambda, 2\lambda, \ldots\}$, or the exponential distribution, or a mixture of one of these and the degenerate distribution at the origin.

This characterization can be used to define a family of goodness-of-fit statistics for the exponential and geometric models. By replacing $f = \overline{F}$ and μ in (1) by sample counterparts, the idea is to construct a sample function measuring the difference between the two sides of the equation, and then to use certain functionals of this sample function as goodness-of-fit statistics. To be specific, let $X_1, X_2, \ldots$ be a sequence of independent, non-negative random variables with distribution function F, and denote by F_n the empirical distribution function associated with the sample $X_1, \ldots, X_n$: $F_n(x) = n^{-1} \sum_{i=1}^{n} I_{[X_i \leq x]}$, $x \in \mathbb{R}$, where I_A is the indicator function of the event A. For each n, let μ_n be a σ-finite measure on $[0, +\infty)$ depending on $X_1, \ldots, X_n$, and define a random function or *empirical process* Z_n by

$$Z_n(x) = \overline{F}_n(x) - \int \overline{F}_n(x+y) \mu_n(dy), \quad x \geq 0. \tag{2}$$

The properties of F_n, together with the Lau–Rao theorem, suggest that when F is (essentially) exponential or geometric the process Z_n should behave in a symmetric fashion around zero, and that such pattern should occur only when F is one of those distributions. If one has in mind the theory of empirical processes – the theory concerned with the convergence of $F_n - F$ and related random functions –, one is even led to conjecture that, when suitably normalized, Z_n converges in distribution to a Gaussian process with covariance and mean value functions depending on F and on the type of measures μ_n chosen. These intuitions being correct, it then makes sense to define statistics of Z_n that are sensitive to departures from the exponential and geometric distributions, and to use them for testing goodness-of-fit.

Eq. (1) was named 'integrated Cauchy functional equation' by Lau and Rao, who first proved the theorem in 1982. The result builds on a theorem of Marsaglia and Tubilla (1975), who gave a solution to a generalized form of the Cauchy equation, and on a lemma of Shanbhag (1977), who essentially formulated and solved (1) with μ arithmetic. For generalizations and applications of the theorem see Rao and Shanbhag (1994).

Because the characterization associated with (1) is a generalized form of the lack-of-memory property, sometimes called the 'integrated lack-of-memory property', we refer to Z_n and to any of its linear normalizations as an *empirical integrated lack-of-memory process*.

What we do in this chapter is to introduce a certain empirical lack-of-memory process Z_n, determine its asymptotic distribution, and apply the results to derive goodness-of-fit tests for the exponential and geometric distributions. Our program is as follows.

The empirical process is defined in Section 2. We give examples of what the measures or integrators μ_n may be and derive some useful formulae.

In Section 3 we point out the connection between our process and the so-called *total time on test empirical process* and *cumulative total time on test statistic*. The history behind these concepts is a very long one, going back to the 50's, and accordingly this is a rather long section. Although our account is not exhaustive and not every important reference is included, we explain how many of the methods and arguments used in the areas of goodness-of-fit and statistical reliability for arriving at the total time on test statistic have their roots in the integral equation (1). Apart from providing a unified view of several apparently unrelated ideas, our discussion clarifies the reason as to why goodness-of-fit procedures based on the total time on test empirical process work.

The asymptotic behaviour of the lack-of-memory process is studied in Section 4. We obtain a Glivenko–Cantelli-type consistency result, and derive the limiting (Gaussian) distribution of the normalized version of the process. Our results are based on the important strong approximation theorem of Komlós, Major and Tusnády (1975), and employ essentially the techniques of Csörgő and Révész (1981), Csörgő, Csörgő and Horváth (1986) and Shorack and Wellner (1986).

Section 5 contains the complete statement of the preceding results and several examples, followed by a comparison with the work of Csörgő, Csörgő and Horváth (1986) on the total time on test empirical process.

The applications to goodness-of-fit testing of the exponential and geometric distributions appear in Section 6. Specifically, we derive the asymptotic distributions of *integral statistics* of the type $\int Z_n(x) \, dF_n(x)$ and $\int Z_n^2(x) \, dF_n(x)$, propose goodness-of-fit tests based on them, and provide estimates of critical values and variances needed for their practical application. Instead of including the results of our power studies, we have decided to report only the most important and general conclusions on the performance of the statistics, and this we do at the end of Section 6.

One question that arises naturally from the point of view of applications is how should the measures μ_n be chosen in order that the integral statistics have good power against a given alternative or type of alternatives. The results mentioned in Section 6 suggest four choices of μ_n in the context of testing the exponential distribution, and the problem is taken up in Section 7, where we propose choosing μ_n in terms of relative efficiency, a well-known measure for comparing the performance of different statistics which we use to define an optimality criterion. The main conclusion is that, given μ_n, it is possible in principle to find a family of alternatives to the exponential distribution such that the test based on $\int Z_n(x) \, dF_n(x)$ represents the optimal choice, and this allows us to complement our comments on the performance of the statistics made in Section 6.

At a strictly practical level, the statistics studied here seem to represent a small improvement relative to the classical statistics for testing the exponential distribution, and a considerable improvement relative to the standard chi-square test for the geometric distribution.

2. The empirical integrated lack-of-memory process

Throughout the paper we let $\mathcal{F}$ be the class of distribution functions concentrated on $[0, +\infty)$, non-degenerate at zero, and with finite first moment, and denote its generic element by F. We let $\mathcal{F}_0$ be either the class of exponential distributions or the class of geometric distributions, and let $F_0(\cdot; \alpha)$ denote its generic element:

$$F_0(x; \alpha) = 1 - e^{-\alpha x}, \quad x \geq 0, \; \alpha > 0,$$

or

$$F_0(x; \alpha) = 1 - (1-\alpha)^{[x]}, \quad x \geq 0, \; 0 < \alpha < 1.$$

For ease of notation we often write $F_0(\cdot)$ or F_0 for $F_0(\cdot; \alpha)$.

We consider sequences $X_1, X_2, \ldots$ of independent random variables with distribution function F (any element in $\mathcal{F}$), and to each of these we associate the parameter $\theta := 1/E[X_i]$, which is finite and > 0 because of our definition of $\mathcal{F}$. When $F = F_0 \in \mathcal{F}_0$, we have $\theta \equiv \alpha$, hence $F_0(\cdot, \alpha) \equiv F_0(\cdot, \theta)$, and the maximum likelihood estimator of θ based on the sample $X_1, \ldots, X_n$ is $\theta_n = 1/\overline{X}_n$. For the more general F we also take $\theta_n = \overline{X}_n^{-1}$ in $[\overline{X}_n > 0]$, but define it arbitrarily in $[\overline{X}_n = 0]$. In this fashion, $\theta_n \to \theta$ with probability 1 irrespective of which F one has in mind.

Finally, we introduce a family of non-negative, non-decreasing, right-continuous functions indexed by θ', $M(x, \theta')$, $x \in \mathbb{R}$, $\theta' \in \Theta$, defined to be zero for $x < 0$, and satisfying $\sum_{n=0}^{\infty}(1-\theta')^n[M(n\lambda) - M((n-1)\lambda)] = 1$ if M has arithmetic support with span λ and $\int e^{-\theta' y} M(dy, \theta') = 1$ if M has non-arithmetic support, to generate examples of the measure μ of (1). The *index set* Θ is required to contain all possible values of θ_n (hence a neighbourhood of θ); thus Θ is in fact determined by the distribution function F under consideration, and it also depends on the specification of θ_n on the set $[\overline{X}_n = 0]$. The incomplete specification of θ_n presents no problem because it is the interior of Θ that matters, and on the other hand it also permits more flexibility in the definition of M. For example, if the X_i arise from a binomial distribution with proportion parameter $p \in (0, 1)$ and we set, say, $\theta_n = 1$ (resp. $\theta_n = 0$) on $[\overline{X}_n = 0]$, we require each M to have an index set Θ containing $[1, \infty)$ (resp. $\{0\} \cup [1, \infty)$).

We often specify M on $[0, \infty)$ only, being implicit that $M = 0$ on $(-\infty, 0)$. Sometimes neither θ nor θ' will be important, and then we write $M(x)$ for $M(x, \theta)$ or $M(x, \theta')$.

Replacing the measure μ of (1) by the Lebesgue–Stieltjes measure generated by M represents no loss of generality because it turns out, once the solutions to (1) have been identified, that μ must be bounded on bounded sets (this is easy to see), hence that it must be a Lebesgue–Stieltjes measure, and consequently that μ is determined by a non-decreasing, right-continuous function that is unique up to an additive constant; taking M non-negative amounts to a normalization, made possible here because μ is finite on $(-\infty, 0)$.

Sometimes only two cases will be contemplated within $\mathcal{F}$: F continuous, and F concentrated on $\{1, 2, \ldots\}$. For simplicity we will often refer to the 'continuous case' and 'discrete case', in order to distinguish between the two situations. In general, no restrictions are imposed on the support of M, but while dealing with testing problems it is natural to require that M be either continuous or concentrated on the positive integers. Thus, *whenever we talk about the 'continuous case', we always assume that M is also continuous, and whenever we talk about the 'discrete case' we always assume that M has $\{0, 1, 2, \ldots\}$ as support.*

Taking $M_n(x) = M(x, \theta_n)$ as the sample equivalent of $M(x, \theta)$, we define the *empirical integrated lack-of-memory process Z_n* by

$$Z_n(x) = \overline{F}_n x) - \int \overline{F}_n(x+y) \, dM_n(y), \quad x \geqslant 0, \qquad (3)$$

and its *normalized version Z_n^** by

$$Z_n^*(x) = n^{1/2}[Z_n(x) - Z(x)], \quad x \geqslant 0, \qquad (4)$$

where

$$Z(x) = \overline{F}(x) - \int \overline{F}(x+y) M(dy, \theta), \quad x \geqslant 0. \qquad (5)$$

The definition of Z_n presents no problems: for each n, $\overline{F}_n$ has bounded support, so the integral $\int \overline{F}_n(x+y) \, dM_n(y)$ is finite even when M is unbounded. On the other hand, the definition of Z_n^* requires that $\int \overline{F}(x+y) M(dy, \theta)$ be finite for all $x \geqslant 0$, or $\int \overline{F}(y) M(dy, \theta) < \infty$, which is equivalent to $\int M(x-, \theta) \, dF(x) < \infty$. This condition on the underlying F is thus necessary for deriving approximations to Z_n. Of course, the condition holds for bounded M; but it also holds, for example, if F has finite first moment and M is replaced by Lebesgue measure on $[0, \infty)$.

EXAMPLE 2.1 (*Possible choices of M*). Although M can be quite arbitrary, it is convenient to introduce specific examples from the start. In the continuous case we may consider M defined by

$$M(x, \theta') = \theta' x, \quad x \geqslant 0, \; \theta' \in \Theta,$$

or by

$$M(x, \theta') = 2(1 - e^{-\theta' x}), \quad x \geqslant 0, \; \theta' \in \Theta,$$

or, as a generalization of the latter case, by

$$M(x, \theta') = \int_0^x (2\theta')^r u^{r-1} e^{-\theta' u} \, du / \Gamma(r), \quad x \geqslant 0, \; \theta' \in \Theta, \; r > 1.$$

These are multiples of Lebesgue, exponential and gamma measures on $[0, \infty)$.

Analogous choices in the discrete case are the multiples of counting, geometric and negative binomial measures on $\{0, 1, 2, \ldots\}$; thus we can define M by

$$M(x, \theta') = \theta'[x + 1], \quad x \geq 0, \; \theta' \in \Theta, \quad \text{or by}$$

$$M(x, \theta') = (2 - \theta')\bigl(1 - (1 - \theta')^{[x+1]}\bigr), \quad x \geq 0, \; \theta' \in \Theta, \quad \text{or by}$$

$$M(x, \theta') = \bigl[1 - (1 - \theta')^2\bigr]^r \sum_{j=0}^{[x]} (1 - \theta')^j \frac{\Gamma(r + j)}{\Gamma(r) j!}, \quad x \geq 0, \; \theta' \in \Theta, \; r > 1$$

(the third definition being again a generalization of the second).

When calculating integral statistics such as $\int Z_n \, dF_n$ and $\int Z_n^2 \, dF_n$, we need to evaluate Z_n at the order statistics $X_{1:n}, \ldots, X_{n:n}$ or at $0, 1, 2, \ldots$. To derive formulae for Z_n at these points we observe that

$$\int \overline{F}_n(x + y) \, dM_n(y) = n^{-1} \sum_{j=1}^{n} \int I_{[0, X_j - x)}(y) \, dM_n(y)$$

$$= n^{-1} \sum_{j=1}^{n} M_n\bigl((X_j - x)-\bigr), \quad x \geq 0. \tag{6}$$

Hence in the continuous case

$$Z_n(X_{i:n}) = (n - i)/n - n^{-1} \sum_{j=i+1}^{n} M_n(X_{j:n} - X_{i:n}), \quad i = 0, 1, \ldots, n, \tag{7}$$

where we put $X_{0:n} := 0$, and in the discrete case

$$Z_n(i) = \overline{F}_n(i) - \sum_{j=i+1}^{\infty} M_n(j - i - 1)\bigl[F_n(j) - F_n(j - 1)\bigr], \quad i = 0, 1, \ldots. \tag{8}$$

3. Connection with certain test statistics and empirical processes

Consider the continuous case with $M(dx, \theta') = \theta' \, dx$. Then $dM_n(x) = dx/\overline{X}_n$, and (6) gives

$$Z_n(X_{i:n}) = \frac{n - i}{n} - \sum_{j=i+1}^{n} (n - j + 1)(X_{j:n} - X_{j-1:n})/(n\overline{X}_n),$$

$$i = 0, \ldots, n.$$

Using the fact that $\sum_{j=1}^{n}(n-j+1)(X_{j:n}-X_{j-1:n})=n\overline{X}_n$, and introducing the notation

$$U_{i:n-1}=T_i/T_n, \quad T_i=\sum_{j=1}^{i}X_j^*, \quad X_i^*=(n-i+1)(X_{i:n}-X_{i-1:n}),$$

we can write

$$\begin{aligned}Z_n(X_{i:n})&=\left\{\sum_{j=1}^{i}(n-j+1)(X_{j:n}-X_{j-1:n})/(n\overline{X}_n)\right\}-i/n\\ &=\left\{\sum_{j=1}^{i}X_j^*\bigg/\sum_{j=1}^{n}X_j^*\right\}-i/n\\ &=T_i/T_n-i/n=U_{i:n-1}-i/n, \quad i=0,\ldots,n.\end{aligned} \qquad (9)$$

With this choice of M we have

$$\int Z_n\,dF_n=\frac{1}{n}\left\{\sum_{i=1}^{n-1}T_i/T_n\right\}-\frac{n-1}{2n}=\frac{1}{n}\sum_{i=1}^{n-1}U_{i:n-1}-\frac{n-1}{2n}. \qquad (10)$$

This statistic and the transformation

$$(X_1,\ldots,X_n)\to\left(X_1^*,\ldots,X_n^*\right)\to(U_{1:n-1},\ldots,U_{n:n-1})$$

that lead to it have a long history in the statistical literature. The transformation $(X_1,\ldots,X_n)\to(X_1^*,\ldots,X_n^*)$ was introduced by Sukhatme (1937), who was the first to observe that if $X_1,\ldots,X_n$ are independent exponential variables then $X_1^*,\ldots,X_n^*$ – the so-called *normalized spacings* – are also independent and exponential.

The use of the *total time on test statistics* T_i and of the $U_{i:n-1}$ in life-testing problems goes back at least to Epstein and Sobel (1953) and Epstein (1960). The latter proposed test statistics based on the partial sums $\sum_{i=1}^{k}T_i$ ($k=1,\ldots,n$), of which (10) could be seen as a special case. If we regard the X_i as lifetimes of items being tested, then the formula $T_i=X_{1:n}+\cdots+X_{i:n}+(n-i)X_{i:n}$ expresses the 'i-th total time' as the sum of lifetimes of the i components that have failed by time $X_{i:n}$ and of those $n-i$ still in operation – whence the name of the T_i. Lewis (1965), inspired by the work of Durbin (1961), proposed using (10) for testing the Poisson process. Subsequently, the statistic was rediscovered and studied several times; in the statistical reliability area it became known as the *cumulative total time on test statistic*.

Partly, the ubiquity of the total time on test statistics in the goodness-of-fit context is due to a simple distributional property: if $X_1,\ldots,X_n$ are independent exponential variables, then the joint distribution of $U_{1:n-1},\ldots,U_{n-1:n-1}$ is that of the order statistics associated with a random sample of $n-1$ standard uniform random variables. This property suggests that in order to test exponentiality one may transform the original sample $(X_1, \ldots, X_n)$ into $(U_{1:n-1}, \ldots, U_{n-1:n-1})$, and then perform a uniformity test on the $U_{i:n-1}$, using for example the Kolmogorov–Smirnov statistic.

Hogg (1964) and Seshadri, Csörgő and Stephens (1969) were among the first to note that this distributional property, or some form of it, characterizes the exponential distribution (see Dufour, Maag and Eeden (1984) for the precise statement and proof, and Rao and Shanbhag (1997) for extensions), and to propose testing the exponential distribution using the $U_{i:n-1}$. Hogg's work has been somewhat forgotten (perhaps because it was published only as an abstract in the *Annals of Mathematical Statistics*), but he was probably the first to systematically use characterization theorems in goodness-of-fit by replacing a hypothesis by a formally equivalent one (he also gave a test based on a characterization of Fisz (1958)).

Seshadri, Csörgő and Stephens (1969), Csörgő, Seshadri and Yalovsky (1975) and Stephens (1986b) have shown by simulation that testing the uniformity of the $U_{i:n-1}$ is indeed a powerful method of testing exponentiality of the original sample $(X_1, \ldots, X_n)$. In particular, uniformity tests based on the Anderson–Darling and Cramér–von Mises statistics are at least as powerful as the modified (i.e., with the scale parameter replaced by its estimate) exponentiality tests based on these same statistics – and these two are already amongst the best omnibus tests.

Besides the argument of good power for favouring this test procedure, often invoked was the convenient elimination of the scale parameter, which in this way did not have to be estimated; at the time, tables for testing composite hypotheses through estimation of the parameters were not always available (Lilliefors's modification of the Kolmogorov–Smirnov test was given in 1969, and Stephens's modifications of the Anderson–Darling and other statistics were made available only in the mid 70's). Of course, the maximum likelihood estimate of the scale parameter *is* involved in the $U_{i:n-1}$, and this is explicitly shown in the definition of Z_n.

The idea that *formally* equivalent hypotheses should lead to *statistically* equivalent procedures, or at least that this should be so under certain conditions, was pursued later by O'Reilly and Stephens (1982). Since it became clear that certain transformations based on characterizations would yield weak or even biased tests, other reasons were given as an explanation for the good performance of the transformation $(X_1, \ldots, X_n) \to (U_{1:n-1}, \ldots, U_{n:n-1})$. O'Reilly and Stephens (1982) suggested that 'invariance' (with respect to the original order of the observations) of the transformation combined with the characterization itself might be the important factor, and consequently proposed a method for deriving such transformations (and associated characterizations). Later on, the explanations for the success of the transformation became more empirical; the articles of Stephens (1986a) and Quesenberry (1986) show that the equivalence of hypotheses provided by the characterization ceased to be regarded as the decisive element.

EXAMPLE 3.1 (*The J transformation*). Consider the transformation $(X_1, \ldots, X_n) \to (\widetilde{U}_{1:n-1}, \ldots, \widetilde{U}_{n:n-1})$ defined by

$$\widetilde{U}_{i:n-1} = \sum_{j=1}^{i} X_i \bigg/ \sum_{j=1}^{n} X_j, \quad i = 1, \ldots, n-1.$$

If the X_i are independent and exponentially distributed, then the $\widetilde{U}_{i:n-1}$ are distributed as the order statistics of $n-1$ standard uniform random variables. Conversely, if the $\widetilde{U}_{i:n-1}$ are distributed as the order statistics of $n-1$ standard uniform variables then, under certain conditions, the X_i must be exponentially distributed (e.g., Seshadri, Csörgő and Stephens, 1969). Just as with the characterization result about the distributions of $(X_1, \ldots, X_n)$ and $(U_{1:n-1}, \ldots, U_{n:n-1})$, the formal equivalence between the distributional statements about $(X_1, \ldots, X_n)$ and $(\widetilde{U}_{1:n-1}, \ldots, \widetilde{U}_{n:n-1})$ suggests testing exponentiality of the X_i by performing a uniformity test on the $\widetilde{U}_{i:n-1}$. However, as Seshadri, Csörgő and Stephens (1969) noted, such a test can be biased against certain alternatives, namely against alternative distributions with coefficient of variation smaller than the coefficient of variation of the exponential distribution (which is 1). To understand this, we just need to note that the less the X_i vary, the more $\widetilde{U}_{i:n-1}$ looks like i/n (the expected value of the ith order statistic associated with a sample of $n-1$ standard uniforms); in the limit, if the X_i are equal to a non-zero constant (thus having a coefficient of variation equal to 0), we have $\widetilde{U}_{i:n-1} \equiv i/n$ – the transformed observations become the 'perfect' or 'expected' ordered uniform sample, and most testing procedures will not reject the hypothesis of exponentiality.

A way of avoiding the bias is to perform a two-tailed uniformity test – thus rejecting also values of the statistic that are 'too good to be true' –, but the resulting test is very poor in terms of power; furthermore, it is not at all obvious that such two-tailed test should be unbiased against all alternatives.

It is clear that the $\widetilde{U}_{i:n-1}$ are not invariant under permutations of the X_i. Therefore, it is conceivable that two persons could arrive at different conclusions by performing the same test on the same sample just because they were given the data in different orderings. By itself, this possibility already suggests that tests based on the $\widetilde{U}_{i:n-1}$ can not be very powerful.

Of course, the $\widetilde{U}_{i:n-1}$ can not be expressed in terms of our Z_n process (as the $U_{i:n-1}$ can, through (9)), nor in terms of the empirical distribution of the X_i, since the empirical distribution is invariant under permutations.

The transformation $(X_1, \ldots, X_n) \to (\widetilde{U}_{1:n-1}, \ldots, \widetilde{U}_{n:n-1})$ is usually referred to as the 'J transformation'. When the X_i have coefficient of variation smaller than one, so that, as described above, the transformed observations become 'too good to be true', it is said (in this context) that the $\widetilde{U}_{i:n-1}$ are 'superuniform' observations. For general comments and references on this and other transformations see the book by D'Agostino and Stephens (1986).

By proposing tests on transformed variables, Lewis's (1965) original intention was to increase power against certain alternatives, and his arguments – analogous to Durbin's (1961) – on why such tests should work were purely empirical, concerning the typical behaviour of the transformed sample values under certain alternatives (e.g., coefficient of variation greater than one). Cox and Lewis (1966) followed the same sort of arguments in connection with tests for the Poisson process.

That the transformation $(X_1, \ldots, X_n) \to (U_{1:n-1}, \ldots, U_{n:n-1})$ should work is apparent from the definition of Z_n, the consistency properties of F_n and θ_n, and the characterization result associated with (1). On the other hand, the more general transformation obtained with M other than Lebesgue measure – namely the mapping defined by (7) – should work as well, and this possesses no analogue of the above distributional identity. It therefore follows that the good performance of the transformation can only be attributable to the characterization of the exponential distribution in terms of the functional equation (1).

The regular appearance of the total time on test statistics in the statistical literature is also explained by the central position occupied by the exponential model in the reliability classification of distributions. Indeed, the T_i crop up in test statistics against special alternatives of reliability classes of distributions – increasing/decreasing failure rate, mean residual life, etc. When deriving these statistics the starting point is usually the definition of a discrepancy measure expressing the departure from a specified reliability property – constant or linear failure rate, increasing failure rate, etc. –; a statistic is then defined as a sample equivalent of the discrepancy measure.

Although we do not wish to go into details about reliability concepts, it is important to point out the connection between these and the functional equation (1), so that the appearance of the total time statistics in the reliability literature is explained. Let us therefore introduce only a few definitions and mention how different authors arrived independently at the cumulative total time on test and similar statistics.

A distribution function $F \in \mathcal{F}$ is said to have *increasing failure rate* if, for each $y > 0$, $x \to \overline{F}(x+y)/\overline{F}(x)$ is non-increasing in the set $\{t: \overline{F}(t) > 0\}$. F is said to have *decreasing mean residual life* if $x \to \varepsilon_F(x) := \int_x^\infty \overline{F}(y)\,dy/\overline{F}(x)$ decreases in the set $\{t: \overline{F}(t) > 0\}$. F is said to be *new better than used in expectation* if $\varepsilon_F(x) \leq \varepsilon_F(0-)$ for all $x \in \{t: \overline{F}(t) > 0\}$, that is, if

$$\overline{F}(x) \geq \int_0^\infty \overline{F}(x+y)\theta\,dy \quad \text{for all } x \geq 0. \tag{11}$$

It is well known and easy to see that if F has increasing failure rate then it has decreasing mean residual life, and that if F has decreasing mean residual life then it is new better than used in expectation.

Barlow (1968) was the first to propose the total time on test statistic (10) for testing the exponential distribution against increasing failure rate alternatives. In this context the null hypothesis is rejected for large values of (10).

Confronting the inequality (11) with the equality in (1) with $\mu(dy)$ replaced by $\theta\,dy$ one sees that (10) is in fact a statistic for testing the exponential distribution against new better than used in expectation alternatives, rather than against the smaller class of alternatives with increasing failure rate. This fact had been noted already by Hollander and Proschan (1975), who arrived at (10) starting from a discrepancy measure expressing deviations towards new better than used in expectation alternatives.

More recently, Ahmad (1992) rediscovered (10) as a test against decreasing mean residual life alternatives, and subsequently Kumazawa (1993) pointed out that Ahmad's test was in effect the test against new better than used in expectation alternatives previously derived by Hollander and Proschan (1975).

The repeated derivation of (10) is explained by the fact that the discrepancy measures used for arriving at it are all related to the inequality in (11). Even when decreasing mean residual life alternatives or increasing failure rate alternatives are aimed at, the discrepancy measures involve some sort of weighting (with respect to F), and this usually turns them into discrepancy measures for new better than used in expectation alternatives.

Of course, (10) can also be used for testing exponentiality against *decreasing* failure rate, *increasing* mean residual life, and new *worse* than used in expectation alternatives (the definitions of these three concepts are obtained by reversing the monotonicity and ordering requirements in the definitions just given); in these cases the null hypothesis is rejected for small values of (10).

The total time on test statistics T_i appear in connection with other goodness-of-fit statistics as well. Having in mind an exponentiality test against decreasing mean residual time alternatives, Hollander and Proschan (1975) started from a certain discrepancy measure and derived a statistic of the form

$$\overline{X}_n^{-1} \sum_{i=1}^n c_{i,n} X_{i:n}, \qquad (12)$$

where the $c_{i,n}$ are non-random coefficients. Statistics of this type *without the term $\overline{X}_n$* – that is, linear combinations of order statistics – are known as *L-estimates*, and play an important role in statistical estimation (Serfling, 1980, p. 262). Recalling the notation immediately preceding (9), we see that $\overline{X}_n^{-1} \sum_{i=1}^n c_{i,n} X_{i:n}^*$ is also a statistic of this form; in particular (see (9) and (10)), the cumulative total time on test statistic is of the type (12) (apart from a non-random term).

Klefsjö (1983), motivated by characterizations of life distributions in terms of the total time on test statistics, derived several of these 'scaled' L-estimates (12) for testing the exponential distribution against *ageing alternatives* (i.e., increasing failure rate, etc.); in particular, he arrived at the cumulative total time on test statistic by a new argument.

The use of scaled L-estimates for testing the exponential distribution goes back to the 60's, at least. Barlow and Proschan (1966) and Bickel and Docksum (1969), for instance, considered several statistics of the form (12) for testing exponentiality against alternatives with increasing failure rate average (a definition similar to those given above). It is also interesting to note that (12) includes Jackson's statistic (see D'Agostino and Stephens, 1986, p. 222), originally devised as a *regression* exponentiality test (as opposed to a test based on the empirical distribution function).

Again, the connection with the empirical lack-of-memory process lies on the fact that statistics like (12) can be viewed as functionals of Z_n with $M_n(x) = \theta_n x$. To see this, we define a (right-continuous) *weight function* $w_n : [0, 1] \to \mathbb{R}$ such that $w_n(i/n) = c_{i,n}^*$ ($i = 1, \ldots, n$) for a given double sequence $c_{i,n}^*$, and use the formulae in (9) to calculate the *weighted integral statistic*

$$\int Z_n(w_n \circ F_n) \, dF_n$$

$$= \sum_{i=1}^n \left\{ \frac{\theta_n}{n} \sum_{j=1}^i X_j^* - \frac{i}{n} \right\} w_n\left(\frac{i}{n}\right)$$

$$= \sum_{i=1}^{n} \left\{ \theta_n \sum_{j=1}^{i} X_j^* - i \right\} \frac{c_{i,n}^*}{n}$$

$$= \frac{\theta_n}{n} \sum_{i=1}^{n} X_{i:n} \left\{ \sum_{j=i+1}^{n} c_{j,n}^* + (n-i+1)c_{i,n}^* \right\} - \frac{1}{n} \sum_{i=1}^{n} i c_{i,n}^*$$

which, apart from the non-random location term, is of the form (12) with

$$c_{i,n} = n^{-1} \left\{ \sum_{j=i+1}^{n} c_{j,n}^* + (n-i+1)c_{i,n}^* \right\}, \quad i = 1, \ldots, n.$$

Conversely, any statistic of the form (12) can be written (apart from a non-random location term) as a linear function of a weighted integral statistic $\int Z_n(w_n \circ F_n) \, dF_n$, with w_n satisfying

$$w_n\left(\frac{i}{n}\right) = \frac{n}{n-i+1} \left\{ c_{i,n} - \frac{1}{n-i} \sum_{j=i+1}^{n} c_{j,n} \right\},$$

$i = 1, \ldots, n-1$, $w_n(1) = n c_{n,n}$.

L-estimates have been studied by several authors; under quite general conditions, their distributions are asymptotically normal (Serfling, 1980, Chapter 8). Thus, the asymptotic distributions of weighted integral statistics $\int Z_n(w_n \circ F_n) \, dF_n$ with $M(dx, \theta') = \theta' \, dx$ can in principle be determined from results already available. Our distributional results of Section 6 go in another direction: they concern the asymptotic distributions of integral statistics, such as $\int Z_n \, dF_n$ and $\int Z_n^2 \, dF_n$, with M taking a general form.

We now turn to the relation between Z_n and the *total time on test empirical process*. We start with a definition: Let $F \in \mathcal{F}$; the *total time on test transform* of F is the function defined by

$$H_F^{-1}(u) = \int_0^{F^{-1}(u)} \overline{F}(x) \, dx, \quad 0 \leq u < 1,$$

where $F^{-1}(u) = \inf\{x: F(x) \geq u\}$ is the quantile function of F. Its sample counterpart is $H_{F_r}^{-1}$:

$$H_{F_n}^{-1}(u) = \int_0^{F_n^{-1}(u)} \overline{F}_n(x) \, dx, \quad 0 \leq u < 1.$$

Since $F_n^{-1}(i/n) = X_{i:n}$, simple integration gives the formula $T_i = n H_{F_n}^{-1}(i/n)$. Thus T_i/n is the sample counterpart of H_F^{-1} at the points $u = i/n$. There is another definition of H_F^{-1} (e.g., Csörgő, Csörgő and Horváth, 1986, pp. 4–5), but we find the present one (which is adopted for example by Shorack and Wellner (1986, pp. 775–777)) more convenient.[1] For general information and references concerning properties and applications of H_F^{-1} see Bergman and Klefsjö (1988).

The widespread use of the T_i in statistical reliability theory and the lack of general asymptotic results for them led Csörgő, Csörgő and Horváth (1986) to study the *scaled total time on test empirical process* $\mathbf{T}_n$, a process related to Z_n^* and defined *for continuous and strictly increasing F* by

$$\mathbf{T}_n(u) = n^{1/2}\bigl(\theta_n H_{F_n}^{-1}(u) - \theta H_F^{-1}(u)\bigr), \quad 0 \leqslant u < 1. \tag{13}$$

To compare this with (4), we first note that

$$\theta_n H_{F_n}^{-1}(u) = 1 - \int_0^\infty \overline{F}_n\bigl(F_n^{-1}(u) + y\bigr)\theta_n \, dy,$$

$$\theta H_F^{-1}(u) = 1 - \int_0^\infty \overline{F}\bigl(F^{-1}(u) + y\bigr)\theta \, dy,$$

whence

$$\begin{aligned}
\mathbf{T}_n(u) &= n^{1/2}\left(\int_0^\infty \overline{F}\bigl(F^{-1}(u) + y\bigr)\theta \, dy - \int_0^\infty \overline{F}_n\bigl(F_n^{-1}(u) + y\bigr)\theta_n \, dy\right) \\
&= n^{1/2}\bigl(\overline{F}(F^{-1}(u)) - \overline{F}_n(F_n^{-1}(u)) + Z_n(F_n^{-1}(u)) - Z(F^{-1}(u))\bigr) \\
&= n^{1/2}\bigl(F_n(F_n^{-1}(u)) - u\bigr) + n^{1/2}\bigl(Z(F_n^{-1}(u)) - Z(F^{-1}(u))\bigr) \\
&\quad + Z_n^*(F_n^{-1}(u)),
\end{aligned} \tag{14}$$

where in the last passage we have used the identity $F(F^{-1}(u)) = u$, valid under the sole assumption that F is continuous (e.g., Billingsley, 1995, p. 197). Since

$$n^{1/2}\bigl(u - F_n(F_n^{-1}(u))\bigr) = n^{1/2}(u - i/n),$$

$$(i-1)/n < u \leqslant i/n \ (i = 1, \ldots, n),$$

(14) gives

$$\mathbf{T}_n(u) = n^{1/2}\bigl(Z(F_n^{-1}(u)) - Z(F^{-1}(u))\bigr)$$

$$+ Z_n^*(F_n^{-1}(u)) \quad \text{for } u = i/n, \tag{15}$$

---

[1] The difference lies in the alternative definition of F^{-1}, with $F(x) > u$ instead of $F(x) \geqslant u$, which makes it right-continuous instead of left-continuous.

and

$$\sup_{0 \leqslant u \leqslant 1} \left| T_n(u) - n^{1/2} \left(Z(F_n^{-1}(u)) - Z(F^{-1}(u)) \right) - Z_n^*(F_n^{-1}(u)) \right| = n^{-1/2}.$$

Thus, if F is exponential, Z is identically zero, and so $T_n = Z_n^* \circ F_n^{-1} + o(1)$. This shows on the one hand that T_n and $Z_n^* \circ F_n^{-1}$ have the same asymptotic distribution under the null hypothesis H_0: $F \in \mathcal{F}_0$, and on the other hand that T_n and Z_n^* measure the same discrepancies (even though at different points), leading to equivalent exponentiality tests. To illustrate this last statement, we note that, since by (15) T_n and $Z_n^* \circ F_n^{-1}$ coincide at $u = i/n$ when F is exponential, we can write, for example,

$$\int Z_n^* dF_n = n^{-1} \sum_{i=1}^n Z_n^*(X_{i:n}) = n^{-1} \sum_{i=1}^n (Z_n^* \circ F_n^{-1})(i/n)$$

$$= n^{-1} \sum_{i=1}^n T_n(i/n) = \int T_n \circ F_n \, dF_n,$$

which shows that the cumulative total time on test statistic (10) can be written both as a functional of Z_n^* and as a functional of T_n under the null hypothesis H_0: $F \in \mathcal{F}_0$. Similar identities hold for other types of integral statistics.

But the similarities between T_n and Z_n^* end here, because if F is not exponential the other term in (15), $n^{1/2}(Z(F_n^{-1}(u)) - Z(F^{-1}(u)))$, is not negligible.

Csörgő, Csörgő and Horváth (1986) also study the empirical processes associated with the *mean residual life function* and the *Lorenz curve*. These functions are also related to the integral on the right-hand side of (1) with $\mu(dy) = \theta \, dy$, but the corresponding empirical processes are not expressible in terms of Z_n^*, like T_n is through formula (15). Previous studies on the total time on test process had been restricted and superficial; see the Introduction of Csörgő, Csörgő and Horváth (1986) for earlier references.

It is legitimate to ask what is gained by considering Z_n^* instead of T_n. Apart from achieving greater generality (other integrators M can be used, F does not need to be continuous nor strictly increasing), Z_n^* uses a more natural and convenient scale – the real line rather than the inverse image of F_n –, and this, as we shall see later, makes the asymptotic theory simpler and less restrictive.

We have attempted to show in this section that many of the exponentiality statistics found in the literature can be seen as functionals of the empirical process Z_n, and hence that such statistics are, at bottom, rooted in the functional equation (1). This is not surprising, as the derivation of many of the tests is explicitly or implicitly based on characterization properties, and, as we have mentioned in the Introduction, the Lau–Rao theorem underlies much of the characterization theory of the exponential distribution. Finally, we note three implications of our work: first, apparently unrelated statistics are, after all, related; second, some tests thought to be based on 'alternative' methods are in fact tests based on the empirical distribution function; third, analogous statistics can be applied to testing the geometric model.

4. Asymptotic behaviour of the process

We now derive the main results of the chapter, which concern the limiting behaviour of the process Z_n^*. There are several conditions to be imposed on F and M; these will be introduced as the need arises, and then collected in Section 5 together with the complete statement of results.

We start by showing that, under certain conditions on M and F, $Z_n \to Z$ uniformly with probability 1, or, in other words, that $\sup_x |n^{-1/2} Z_n^*(x)| \to 0$ almost surely (a.s.). This consistency result shows that (under the relevant conditions) Z_n goes uniformly to zero with probability 1 if and only if the sequence $X_1, X_2, \ldots$ has common distribution function F satisfying Eq. (1) – and, with the obvious restrictions on F and M ($M(0) < 1$, M arithmetic or non-arithmetic, etc.) if and only if F is exponential, geometric, or a mixture of any of these with the distribution degenerate at zero.

We have

$$\sup_x |n^{-1/2} Z_n^*(x)|$$

$$= \sup_x |Z_n(x) - Z(x)|$$

$$\leqslant \sup_x |\overline{F}_n(x) - \overline{F}(x)|$$

$$+ \sup_x \left| \int \overline{F}(x+y) M(\mathrm{d}y, \theta) - \int \overline{F}_n(x+y) M(\mathrm{d}y, \theta) \right|$$

$$+ \sup_x \left| \int \overline{F}_n(x+y) M(\mathrm{d}y, \theta) - \int \overline{F}_n(x+y) \, \mathrm{d}M_n(y) \right|. \tag{16}$$

As $\Delta_n^{(1)} := \sup_x |\overline{F}(x) - \overline{F}_n(x)| \to 0$ with probability 1 by the Glivenko–Cantelli theorem, it is sufficient to prove that the second and third terms on the right-hand side of (16) go to zero with probability 1; denote these terms by $\Delta_n^{(2)}$ and $\Delta_n^{(3)}$.

Recall the condition already imposed on M and F: we require that $\int \overline{F}(y) M(\mathrm{d}y, \theta) < \infty$. Let $\varepsilon > 0$ be given. We can choose T so large that $\int_{(T,\infty)} \overline{F}(y) M(\mathrm{d}y, \theta) < \varepsilon/2$. Then

$$\Delta_n^{(2)} \leqslant \int_{(T,\infty)} \overline{F}_n(y) M(\mathrm{d}y, \theta) + \int_{(T,\infty)} \overline{F}(y) M(\mathrm{d}y, \theta) + \int_{[0,T]} \Delta_n^{(1)} M(\mathrm{d}y, \theta)$$

$$< n^{-1} \sum_{i=1}^n \int_{(T,\infty)} I_{(-\infty, X_i)}(y) M(\mathrm{d}y, \theta) + \varepsilon/2 + M(T, \theta) \Delta_n^{(1)}.$$

By the strong law of large numbers the first term here goes to $\int_{(T,\infty)} \overline{F}(y) M(\mathrm{d}y, \theta)$ with probability 1, and so we have $\limsup_n \Delta_n^{(2)} < \varepsilon$ a.s. Since ε is arbitrary, this proves $\Delta_n^{(2)} \to 0$ a.s.

Further assumptions are needed in order to prove the convergence of $\Delta_n^{(3)}$. Using (6) we see that

$$\Delta_n^{(3)} = \sup_x \left| n^{-1} \sum_{i=1}^{n} M\big((X_i - x)-, \theta_n\big) - M\big((X_i - x)-, \theta\big) \right|,$$

and in order to get an appropriate bound for this we assume that, for each t, $M(t, \cdot)$, as a function of θ, has continuous derivative $m(t, \cdot)$ on Θ^0, the interior of Θ. Then, for each t, we can apply a particular version of the mean value theorem (e.g., Ferguson, 1996, p. 20) to get the representation

$$M(t, \theta') - M(t, \theta) = (\theta' - \theta) \int_0^1 m\big(t, \theta + u(\theta' - \theta)\big) \, du, \quad \theta' \in \Theta^0, \quad (17)$$

which in turn yields

$$\sup_t \big| M(t, \theta') - M(t, \theta) \big| \leqslant |\theta - \theta'| \int_0^1 \sup_t \big| m\big(t, \theta + u(\theta' - \theta)\big) \big| \, du.$$

To get a further bound, assume there exists a function K such that $|m(t, \theta')| \leqslant K(t)$ for all θ' in a neighbourhood of θ and all real t, and with $\int \sup_x K((y-x)-) \, dF(y) < \infty$. Because of the a.s. convergence of θ_n to θ, this enables us to say that, with probability 1, $|m(t, \theta + u(\theta_n - \theta))| \leqslant K(t)$ for large enough n, $0 \leqslant u \leqslant 1$ and all real t.

With these assumptions, we thus have with probability 1

$$\Delta_n^{(3)} \leqslant |\theta_n - \theta| n^{-1} \sum_{i=1}^{n} \sup_x K\big((X_i - x)-\big)$$

for large enough n; and since the series here converges to $\int \sup_x K((y-x)-) \, dF(y)$ a.s., we have $\Delta_n^{(3)} \to 0$ a.s.

4.1. Empirical processes and the Komlós–Major–Tusnády theorem

Empirical process theory is concerned with the asymptotic properties of random functions related to F_n, and in particular of the *empirical process* Y_n, defined by $Y_n(x) = n^{1/2}[F_n(x) - F(x)]$, $x \in \mathbb{R}$. One of the central convergence results provided by the theory states that Y_n converges in distribution to the Gaussian process $Y = \{Y(t): t \in \mathbb{R}\}$ with $E[Y(t)] = 0$, $E[Y(s)Y(t)] = F(s)\overline{F}(t)$ ($s \leqslant t$).

Because we can write

$$Z_n^*(x) = \int Y_n(x+y) M(dy, \theta) - Y_n(x)$$

$$- \int \overline{F}_n(x+y) n^{1/2} \big(M_n(dy) - M(dy, \theta)\big), \quad (18)$$

the convergence result just mentioned and some preliminary calculations suggest that the asymptotic distribution of Z_n^* should be that of

$$W(x) = \int Y(x+y) M(\mathrm{d}y, \theta) - Y(x)$$
$$- \theta^2 \int \overline{F}(x+y) m(\mathrm{d}y, \theta) \int Y(t)\, \mathrm{d}t, \qquad (19)$$

a Gaussian process defined in terms of the Y process above, with zero means and covariance structure depending on F and M (recall the definition of m given just before formula (17)).

This is in fact the case, and in order to prove it we will make use of the Komlós–Major–Tusnády strong approximation theorem. This very important theorem, originally proved in Komlós, Major and Tusnády (1975), appears in Section 4.4 of Csörgő and Révész (1981).

Recall that a Brownian bridge $B = \{B(u), u \in [0, 1]\}$ is a continuous Gaussian process on $[0, 1]$ specified by $E[B(u)] = 0$, $E[B(u)B(v)] = u(1-v)$, $u \leqslant v$, and $B(0) = B(1) = 0$ a.s.; it can be defined in terms of the Wiener process $\mathbf{W}$ by $B(u) = \mathbf{W}(u) - u\mathbf{W}(1)$, $0 \leqslant u \leqslant 1$.

THEOREM 4.1 (Komlós–Major–Tusnády). *Let $U_1, U_2, \ldots$ be a sequence of independent $U(0, 1)$ random variables, and denote by $G_1, G_2, \ldots$ the corresponding sequence of empirical distribution functions. There exists a sequence $B_r = \{B_n(u), u \in [0, 1]\}$ of Brownian bridges, defined on the same probability space, such that with probability 1*

$$\sup_{0 \leqslant u \leqslant 1} \left| n^{1/2} [G_n(u) - u] - B_n(u) \right| = \mathrm{O}(n^{-1/2} \log n).$$

From the definition of $Y = \{Y(t), t \in \mathbb{R}\}$, we see that if F is the uniform distribution on $[0, 1]$ then the restriction of Y to $[0, 1]$ is a Brownian bridge; since in this case B_n has the same distribution as Y for every n, the convergence of Y_n to Y follows from Theorem 4.1.

We refer to the B_n of Theorem 4.1 as the *Komlós–Major–Tusnády sequence of Brownian bridges*.

A fundamental and well-known property of the quantile function F^{-1} is that if U is a $U(0, 1)$ random variable then $X = F^{-1}(U)$ has distribution function F. We want to use a version of Theorem 4.1 that applies to sequences of variables with any distribution function, and in order to do so we will assume in everything that follows that our sequences $X_1, X_2, \ldots$ arise in this way from uniform variables – that is, $X_i = F^{-1}(U_i)$ for uniform U_i. This is legitimate because we are only concerned with distributional properties of $Y_n = n^{1/2}(F_n - F)$, and these are the same irrespective of the way the X_i arise, as long as they are independent and have distribution function F.

Thus, as in the statement of the theorem, let G_n denote the empirical distribution function constructed with $U_1, \ldots, U_n$. It is possible to show, using the properties of F^{-1}, that $F_n(x) = G_n(F(x))$ for all x. Consequently, we have $Y_n(x) =$

$n^{1/2}[G_n(F(x)) - F(x)]$ for all x, and it follows from the Komlós–Major–Tusnády theorem that there exists a sequence $\{B_n(u), u \in [0,1]\}$ of Brownian bridges such that

$$\sup_x |Y_n(x) - B_n(F(x))| = O(n^{-1/2} \log n) \tag{20}$$

with probability 1. For every n, the Gaussian process $B_n \circ F := \{B_n(F(x)), x \in \mathbb{R}\}$ has the same distribution as the process Y; hence the theorem implies the convergence in distribution of Y_n to Y in the general case.

These facts are well known from empirical process theory.

Our problem now is to prove that, under appropriate conditions on F and M, the supremum in x of the differences between $Z_n^*(x)$ and the *approximating process*

$$W_n(x) = \int B_n(F(x+y)) M(dy, \theta) - B_n(F(x))$$

$$- \theta^2 \int \overline{F}(x+y) m(dy, \theta) \int B_n(F(t)) dt,$$

which is just (19) with Y replaced by the $B_n \circ F$ of (20), goes to zero in probability. If this is proven, then the fact that $B_n \circ F$ has for each n the same distribution as Y implies that the limiting distribution of Z_n^* is that of the process W in (19).

The plan is to prove successively that the remainder terms

$$\Delta_n^{(1)} := \sup_x \left| \int Y_n(x+y) M(dy, \theta) - \int B_n(F(x+y)) M(dy, \theta) \right|$$

and

$$\Delta_n^{(2)} := \sup_x \left| \int \overline{F}_n(x+y) n^{1/2}(M_n(dy) - M(dy, \theta)) \right.$$

$$\left. - \theta^2 \int \overline{F}(x+y) m(dy, \theta) \int B_n(F(t)) dt \right|,$$

tend to zero in probability. This is enough because (20) holds and

$$\sup_x |Z_n^*(x) - W_n(x)| \leq \Delta_n^{(1)} + \Delta_n^{(2)} + \sup_x |Y_n(x) - B_n(F(x))|.$$

Before proceeding we need to make sure that the integrals of the Brownian bridge appearing in W_n are well defined.

4.2. Integrals of the Brownian bridge

Let r denote the covariance function of $B \circ F$: $r(s,t) = E[B(F(s))B(F(t))] = F(s)\overline{F}(t)$, $s \leq t$. In what follows, M is any non-negative, non-decreasing right-continuous function concentrated on $[0, \infty)$, and F is any probability distribution function in $\mathcal{F}$.

Since the Brownian bridge is a measurable process, the Lebesgue integral $\int B \circ F \, dM$ exists if and only if the realization of B satisfies $\int |B \circ F| \, dM < \infty$, and it turns out that this condition is satisfied a.s. if and only if $E \int |B \circ F| \, dM < \infty$. The sufficiency is obvious, and the necessity follows by a lemma of Csörgő and Horváth (1993, pp. 314–316). Thus, noting that $E \int |B \circ F| \, dM = \int E|B(F(t))| \, dM(t) = \sqrt{2/\pi} \int F(t)^{1/2} \overline{F}(t)^{1/2} \, dM(t)$, we conclude that $\int B(F(t)) \, dM(t)$ exists a.s. if and only if $\int F(t)^{1/2} \overline{F}(t)^{1/2} \, dM(t) < \infty$, i.e., if and only if $\int \overline{F}(t)^{1/2} \, dM(t) < +\infty$.

Under this condition, then, we have $E \int B \circ F \, dM = 0$, and

$$V \int B(F(t)) \, dM(t) = \iint r(s,t) \, dM(s) \, dM(t).$$

The same applies with obvious modifications to similar integrals. Thus, $\int_{(T,+\infty)} B(F(x+t)) \, dM(t) \ (x \geq 0)$ exists a.s. if $\int F(t+x)^{1/2} \overline{F}(t+x)^{1/2} \, dM(t) < \infty$, and we note that its variance, which is $\int_{(T,+\infty)} \int_{(T,+\infty)} r(s+x, t+x) \, dM(t) \, dM(s)$, tends to zero as $T \to +\infty$.

We next show that the distributions of these integrals are normal.

Let μ be any measure on $[0, +\infty)$. By change of variables,

$$\int_{[F(x),1)} B(u) \mu F^{-1}(du)$$

$$= \int_{[F^{-1}(F(x)),\infty)} B(F(t)) \mu(dt)$$

$$= \int_{[x,\infty)} B(F(t)) \mu(dt) + \int_{[F^{-1}(F(x)),x)} B(F(t)) \mu(dt),$$

where μF^{-1} is the measure on $[0,1]$ determined on the half-open intervals $[u, v)$ by $\mu F^{-1}[u,v) = \mu[F^{-1}(u), F^{-1}(v))$. Since $F(t) = F(x)$ if $t \in [F^{-1}(F(x)), x)$, this gives

$$\int_{[F(x),1)} B(u) \mu F^{-1}(du)$$

$$= \int_{[x,\infty)} B(F(t)) \mu(dt) + B(F(x)) \mu[F^{-1}(F(x)), x).$$

Using this formula with μ determined by $\mu(s,t] = M(t) - M(s)$, $s < t$, one can write

$$\int B(F(x+t)) \, dM(t)$$

$$= \int_{[x,\infty)} B(F(t)) \mu(dt)$$

$$= \int_{[F(x),1)} B(u) \mu F^{-1}(du) - B(F(x)) \mu[F^{-1}(F(x)), x). \qquad (21)$$

Here, the integral on the right is a.s. a (proper or improper) Riemann–Stieltjes integral because B is continuous on $[0,1]$ and μF^{-1} is a measure on $[0,1]$; since it exists a.s., it is obtained as the limit of finite Gaussian sums converging a.s. to a normal random variable. Since $B(F(x))\mu[F^{-1}(F(x)),x)$ is a normal variable times a constant, $\int B(F(x+t))\,dM(t)$ is indeed normally distributed.

The *joint* normality of any finite set of integrals and variables $B \circ F(t)$ is similarly justified.

Csörgő, Csörgő and Horváth (1986, p. 35) noticed that it is possible to define $\int B(F(t))\,dt$ as an improper Riemann integral even when the corresponding Lebesgue integral does not exist, i.e., even when $\int \overline{F}(t)^{1/2}\,dt = +\infty$. In fact, they showed that for $\int B(F(t))\,dt$ to exist as a Riemann integral it is enough for F to have finite second moment: $\int t^2\,dF(t) < \infty$.

It is easy to see that $\int \overline{F}(t)^{1/2}\,dt < +\infty$ implies $\int t^2\,dF(t) < \infty$. On the other hand, F defined by $F(t) = 1 - e^2 t^{-2}\log(t)^{-p}$ ($t \geqslant e$) for $p \in (1,2)$ satisfies $\int t^2\,dF(t) < \infty$ but not $\int \overline{F}(t)^{1/2}\,dt < +\infty$. Thus the observation of Csörgő, Csörgő and Horváth (1986) means that one can indeed define $\int B(F(t))\,dt$ under a weaker condition than the one imposed above. [In any case, let us note that the finiteness of $\int F(t)^{1/2}\overline{F}(t)^{1/2}\,dt$ is not too far from a second moment assumption. For example, if $\int t^2\log(1+t)^{\alpha+1}\,dF(t) < \infty$ for some $\alpha > 0$, then $\int t^2\,dF(t) < \infty$ implies $\int \overline{F}(t)^{1/2}\,dt < +\infty$ (Hoeffding, 1973); also, if F has regularly varying tail – i.e., $\overline{F}(x) = x^\alpha L(x)$ for some α, with L satisfying $L(tx)/L(t) \to 1$ as $t \to +\infty$ for all $x > 0$ – then $\int t^2\,dF(t) < \infty$ implies $\int \overline{F}(t)^{1/2}\,dt < +\infty$ (Stigler, 1974).]

Following their arguments, one can show that integrals of the type $\int B \circ F\,dM$ also exist a.s. as 'improper' Lebesgue integrals for general M satisfying $\int M^2\,dF < \infty$ (which is a weaker condition than $\int \overline{F}^{1/2}\,dM < \infty$). Using integration by parts, the right-hand side of (21) can be written as an integral with respect to the Brownian bridge (which exists a.s. as a by-product of integration theory with respect to the Wiener process) plus a Gaussian function that can be shown to be a.s. finite by using the Birnbaum–Marshall inequality (see Csörgő, Csörgő and Horváth, 1986, p. 25).

With this extended definition of the integrals, the existence of the approximating process W_n requires only that $\int M(t)^2\,dF(t) < \infty$ and $\int t^2\,dF(t) < \infty$ hold for the pair (F, M). However, to prove that Z_n^* can be approximated by W_n we shall need (though on one occasion only; see the following subsection) the stronger assumption $\int \overline{F}^{1/2}\,dM < \infty$.

4.3. The first remainder term $\Delta_n^{(1)}$

We can always choose a sequence $\{u_n\}$ such that $u_n \to \infty$ with n, and $M(u_n)n^{-1/2}\log n \to 0$. [If M is bounded then any sequence tending to ∞ will do; otherwise, take for instance $u_n = \sup\{x: M(x) \leqslant n^{1/4}\}$.] With such a sequence, we have

$$\Delta_n^{(1)} \leqslant \sup_x \left| \int_{[0,u_n]} Y_n(x+y) - B_n(F(x+y))\,dM(y) \right|$$

$$+ \sup_x \left| \int_{u_n}^\infty Y_n(x+y)\,dM(y) \right|$$

$$+ \sup_x \left| \int_{u_n}^{\infty} B_n(F(x+y)) \, dM(y) \right|, \qquad (22)$$

where the first term is bounded by $\sup_x |Y_n(x) - B_n(F(x))| \, M(u_n)$ and hence tends to zero a.s. by (20) and by the choice of $\{u_n\}$.

Since for $\varepsilon > 0$

$$P\left[\int_{u_n}^{\infty} |Y_n(y)| \, dM(y) > \varepsilon \right] \leq \frac{1}{\varepsilon} E\left[\int_{u_n}^{\infty} |Y_n(y)| \, dM(y) \right]$$

$$\leq \frac{1}{\varepsilon} \sqrt{\frac{2}{\pi}} \int_{u_n}^{\infty} \overline{F}(y)^{1/2} \, dM(y),$$

our assumption $\int \overline{F}^{1/2} \, dM < \infty$ implies

$$\sup_x \left| \int_{u_n}^{\infty} Y_n(x+y) \, dM(y) \right| \leq \int_{u_n}^{\infty} |Y_n(y)| \, dM(y) \to_P 0.$$

That $\sup_x |\int_{u_n}^{\infty} B_n(F(x+y)) \, dM(y)| \to_P 0$ follows in the same way, or alternatively by assuming only $\int M(y)^2 \, dF(y) < \infty$, since $\sup_x |\int_{u_n}^{\infty} B_n(F(x+y)) \, dM(y)|$ has for all n the same distribution as $\sup_x |\int_{u_n}^{\infty} B(F(x+y)) \, dM(y)|$ and the latter converges to zero in probability as long as the improper integral exists.

Thus the second and third terms in (22) go to zero in probability, and so $\Delta_n^{(1)} \to_P 0$.

REMARK. This is the only place where the stronger assumption $\int \overline{F}^{1/2} \, dM < \infty$ is used. When $M(dt) \propto dt$, the methods of Csörgő, Csörgő and Horváth (1986, pp. 36–37) show that $\int M^2 \, dF < \infty$ is enough to conclude $\sup_x |\int_{u_n}^{\infty} Y_n(x+y) \, dM(y)| \to_P 0$, but we have not been able to extend their arguments for general M.

4.4. The second remainder term $\Delta_n^{(2)}$

Here we assume first $\sigma^2 := V[X] < \infty$, so that $n^{1/2}(\theta_n^{-1} - \theta^{-1})$ is asymptotically normal with mean zero and variance σ^2: $n^{1/2}(\theta_n^{-1} - \theta^{-1}) \to^d N(0, \sigma^2)$. Then $n^{1/2}((\theta_n - \theta)/\theta^2) + n^{1/2}(\theta_n^{-1} - \theta^{-1}) \to_P 0$, and $n^{1/2}(\theta_n - \theta) \to^d N(0, \sigma^2 \theta^4)$; this follows from Cramér's and Slutsky's theorems.

Next we assume that $\sup_x |m(x, \theta) - m(x, \theta')| \to 0$ as $\theta' \to \theta$, and that $m(y, \theta)$ is right continuous and of bounded variation as a function of y, so that it generates a signed measure concentrated on $[0, +\infty)$ and the integral $\int \overline{F}_n \, dm$ exists.

We start by showing that

$$\int \overline{F}_n(x+y) n^{1/2} (M_n(dy) - M(dy, \theta)) \quad \text{and}$$
$$n^{1/2}(\theta_n - \theta) \int \overline{F}_n(x+y) m(dy, \theta) \qquad (23)$$

have the same asymptotic distribution.

Using (6) and (17), we can see that

$$\left| \int \overline{F}_n(x+y) n^{1/2} \big(M_n(\mathrm{d}y) - M(\mathrm{d}y, \theta) \big) \right.$$
$$\left. - n^{1/2}(\theta_n - \theta) \int \overline{F}_n(x+y) m(\mathrm{d}y, \theta) \right|$$
$$= n^{-1/2} \left| \sum_{i=1}^{n} \big[M\big((X_i - x)-, \theta_n\big) - M\big((X_i - x)-, \theta\big) \big] \right.$$
$$\left. - (\theta_n - \theta) \sum_{i=1}^{n} m\big((X_i - x)-, \theta\big) \right|$$
$$= n^{1/2} |\theta_n - \theta| n^{-1} \left| \sum_{i=1}^{n} \left[\int_0^1 m\big((X_i - x) - \theta + u(\theta_n - \theta)\big) \mathrm{d}u \right. \right.$$
$$\left. \left. - m\big((X_i - x)-, \theta\big) \right] \right|$$
$$\leqslant n^{1/2} |\theta_n - \theta| \int_0^1 \sup_x \big| m\big(x, \theta + u(\theta_n - \theta)\big) - m(x, \theta) \big| \, \mathrm{d}u,$$
$$\theta_n \in \Theta^0. \tag{24}$$

Since

$$\sup_x \big| m\big(x, \theta + u(\theta_n - \theta)\big) - m(x, \theta) \big|$$
$$\leqslant \sup_{-1 \leqslant u \leqslant 1} \sup_x \big| m(x, \theta + u\rho) - m(x, \theta) \big| < \infty$$

if $0 < \rho \leqslant 1$ is small enough and $|\theta_n - \theta| < \rho$, the sequence of integrands in (24) is uniformly bounded in n as long as $\theta_n \to \theta$ ($n \to \infty$); and since these integrands converge to zero for each u, the bounded convergence theorem implies that the integral in (24) converges to zero as $\theta_n \to \theta$.

Finally, the a.s. convergence of θ_n to θ and the convergence in distribution of $n^{1/2}(\theta_n - \theta)$ to a normal random variable imply that (24) tends to zero in probability.

Now assume $\int \overline{F}(y) m(\mathrm{d}y, \theta)$ is finite. Then the sample functions in (23) have the same asymptotic distribution as $n^{1/2}(\theta_n - \theta) \int \overline{F}(x+y) m(\mathrm{d}y, \theta)$. [This follows by establishing an upper bound for the absolute difference between the integrals $\int \overline{F}(x+y) m(\mathrm{d}y, \theta)$ and $\int \overline{F}_n(x+y) m(\mathrm{d}y, \theta)$, essentially as we have done with the term $\Delta_n^{(2)}$ in (16).] Consequently, we can prove $\Delta_n^{(2)} \to_P 0$ by proving

$$\left| n^{1/2}(\theta_n - \theta) - \theta^2 \int B_n\big(F(t)\big) \mathrm{d}t \right| \to_P 0,$$

and this we do next.

The fact that $n^{1/2}(\theta_n - \theta)/\theta^2 + n^{1/2}(\theta_n^{-1} - \theta^{-1}) \to_P 0$ permits one to express $n^{1/2}(\theta_n - \theta)$ in terms of the Y_n process. Indeed,

$$n^{1/2}(\theta_n^{-1} - \theta^{-1}) = n^{1/2} n^{-1} \sum_{i=1}^{n} (X_i - \theta^{-1})$$

$$= n^{1/2} \int \overline{F}_n(t) \, dt - n^{1/2} \int \overline{F}(t) \, dt$$

$$= -\int n^{1/2} [F_n(t) - F(t)] \, dt - \int Y_n(t) \, dt,$$

and therefore

$$n^{1/2}(\theta_n - \theta) - \theta^2 \int Y_n(t) \, dt \to_P 0.$$

Thus in fact we aim to prove $|\int Y_n(t) \, dt - \int B_n(F(t)) \, dt| \to_P 0$.

For given $T > 0$,

$$\left| \int (Y_n(t) - B_n(F(t))) \, dt \right| \leq \left| \int_{[0,T]} (Y_n(t) - B_n(F(t))) \, dt \right|$$

$$+ \left| \int_T^\infty B_n(F(t)) \, dt \right| + \left| \int_T^\infty Y_n(t) \, dt \right|.$$

Here, the first term on the right is bounded by $T \sup_x |Y_n(x) - B_n(F(x))|$. The second is the modulus of a normal variable with mean zero and variance σ_T^2 that goes to zero as $T \to \infty$.

Without the absolute value bars, the third term is

$$\int_T^\infty n^{-1/2} \sum_{i=1}^{n} (\overline{F}(x) - I_{(x,\infty)}(X_i)) \, dx$$

$$= n^{-1/2} \sum_{i=1}^{n} \left(\int_T^\infty \overline{F}(x) \, dx - I_{(T,\infty)}(X_i)(X_i - T) \right).$$

To see that this is a sum of independent variables with mean zero and variance $\bar{\sigma}_T^2 = \int_T^\infty 2(x - T) \overline{F}(x) \, dx - (\int_T^\infty \overline{F}(x) \, dx)^2$, and hence asymptotically a $N(0, \bar{\sigma}_T^2)$ random variable, we just need to check that

$$E[I_{(T,\infty)}(X_i)(X_i - T)] = \int_T^\infty \overline{F}(x) \, dx,$$

$$E[I_{(T,\infty)}(X_i)(X_i - T)^2] = \int_T^\infty 2(x - T) \overline{F}(x) \, dx.$$

Now let $\delta, \varepsilon > 0$ be chosen arbitrarily, and let Φ denote the standard normal distribution function. Because $\sigma_T^2, \bar{\sigma}_T^2 \to 0$ as $T \to \infty$, we can choose T so large that $2[1 - \Phi(\varepsilon/(3\sigma_T))]$, $2[1 - \Phi(\varepsilon/(3\bar{\sigma}_T))] < \delta/4$, and, because of the Komlós–Major–Tusnády and central limit theorems, choose n_0 so large that $n \geq n_0$ implies

$$P\left[T \times \sup_x |Y_n(x) - B_n(F(x))| > \varepsilon/3\right] < \delta/4$$

and

$$P\left[\left|\int_T^\infty Y_n(t)\,dt\right| > \varepsilon/3\right] \leq 2[1 - \Phi(\varepsilon/(3\bar{\sigma}_T))] + \delta/4.$$

Then we have

$$P\left[\left|\int Y_n(t) - B_n(F(t))\,dt\right| > \varepsilon\right] \leq \delta$$

for $n \geq n_0$, which proves the result.

5. Statement of results; examples and comparisons

We now list the various conditions to be imposed on the pairs (F, M), and collect the results in Theorem 5.1.

- **C0** $\int M(x-, \theta)\,dF(x) < \infty$, and $\int e^{-\theta' y} M(dy, \theta') = 1$ in case M has non-arithmetic support, and $\sum_{n=0}^\infty (1-\theta')^n [M(n\lambda) - M(n\lambda-)] = 1$ in case M has arithmetic support with span λ.
- **C1** For each t, $M(t, \cdot)$ has continuous derivative $m(t, \cdot)$ on Θ^0 (the interior of Θ).
- **C2** There exists a function K such that $|m(t, \theta')| \leq K(t)$ for all θ' in a neighbourhood of θ and all real t, and such that $\int \sup_{x \geq 0} K((y-x))\,dF(y) < \infty$.
- **C3** $\int F(t)^{1/2} \overline{F}(t)^{1/2} M(dt, \theta) < \infty$.
- **C4** $m(\cdot, \theta)$ is right-continuous and of bounded variation, $\int \overline{F}(y) m(dy, \theta) < \infty$, and $\sup_x |m(x, \theta) - m(x, \theta')| \to 0$ as $\theta' \to \theta$.

THEOREM 5.1. *Let $X_1, X_2, \ldots$ be a sequence of non-negative independent random variables with common distribution function F and $0 < \theta = 1/E[X_i] < \infty$. Set $\theta_n = \overline{X}_n^{-1}$ if $\overline{X}_n^{-1} > 0$ and define θ_n arbitrarily if $\overline{X}_n^{-1} = 0$. Let Θ be a set containing the range of θ_n, and let $\{M(\cdot, \theta'): \theta' \in \Theta\}$ be a family of non-negative Lebesgue–Stieltjes measure functions (non-decreasing, right-continuous) concentrated on $[0, +\infty)$ and satisfying C0. Define Z_n, Z_n^* and Z as in (3)–(5).*

(i) *If C1 and C2 hold, then*

$$\sup_x |Z_n(x) - Z(x)| \to 0 \quad (n \to \infty) \text{ a.s.}$$

(ii) *If C3 and C4 hold and $V[X_i] < \infty$, there exists a sequence of approximating processes W_n defined on the same probability space in terms of the Komlós–Major–Tusnády sequence of Brownian bridges B_n by*

$$W_n(x) = \int B_n(F(x+y)) M(dy, \theta) - B_n(F(x))$$
$$- \theta^2 \int \overline{F}(x+y) m(dy, \theta) \int B_n(F(t)) dt$$

such that

$$\sup_x |Z_n^*(x) - W_n(x)| \to_P 0 \quad (n \to \infty);$$

the processes W_n are Gaussian and have the same distribution for each n.

The process W of (19) has of course the same distribution as W_n. If we define φ by

$$\varphi(x, \theta) = \theta^2 \int \overline{F}(x+s) m(ds, \theta), \quad x \geq 0, \ \theta \in \Theta, \tag{25}$$

the formula for W becomes

$$W(x) = \int Y(x+s) M(ds, \theta) - Y(x) - \varphi(x, \theta) \int Y(t) dt, \quad x \geq 0, \tag{26}$$

where Y is the process introduced in Section 4.1.

Note that $Y =^d B \circ F$, where B is a Brownian bridge. In the sequel we deal mainly with distributional properties, and for this reason work with W as if Y were defined by $Y = B \circ F$ (rather than by $Y =^d B \circ F$), which in fact amounts to dropping the subscript n in the definition of W_n.

Let us examine some basic properties of W. First, $E[W(x)] = 0$, and the covariance function of W can be formally calculated as

$$C[W(s), W(t)]$$
$$= r(s, t) + \iint r(s+x, t+y) dM(x) dM(y) - \int r(s+x, t) dM(x)$$
$$- \varphi(t, \theta) \iint r(s+x, y) dM(x) dy - \int r(s, t+y) dM(y)$$
$$+ \varphi(t, \theta) \int r(s, y) dy - \varphi(s, \theta) \iint r(x, t+y) dM(y) dx$$
$$+ \varphi(s, \theta) \int r(x, t) dx + \varphi(s, \theta) \varphi(s, \theta) \iint r(x, y) dx dy, \quad s \leq t,$$

which is a rather intricate expression. Explicit calculations have only been made for the discrete and continuous cases,[2] but these will not be given here apart from the following *special case of $F = F_0$ exponential*:

$$C[W(s), W(t)] = \theta^{-2}\varphi_0(s, \theta)\varphi_0(t, \theta)$$
$$+ \int_t^\infty M(x-s)(M(x-t) - 1) \, dF_0(x)$$
$$- \varphi_0(s, \theta) \int_t^\infty x(M(x-t) - 1) \, dF_0(x)$$
$$- \varphi_0(t, \theta) \int_s^\infty x(M(x-s) - 1) \, dF_0(x), \quad s \leq t, \quad (27)$$

where φ_0 is defined as in (25) with F_0 in place of F.

The path properties of W are determined by those of Y and by the pair (F, M). Since $B(1) = 0$ a.s. and $Y = B \circ F$, we see that $\lim_{x \to +\infty} Y(x) = B(1) = 0$ a.s. and hence $\lim_{x \to +\infty} W(x) = 0$ a.s. Since both F and M are right-continuous and B is a.s. continuous, W is also a.s. right-continuous; if F is continuous, then W is also a.s. continuous.

5.1. Examples

We provide only a couple of specific examples. The objective is to illustrate the verification of the conditions C1–C4 and to partly justify our use of some of the pairs (F, M).

EXAMPLE 5.1 ($M(dx, \theta')$ *multiple of Lebesgue measure*). This is a simple case to check. C1 holds with $\Theta^0 = (0, +\infty)$, $m(t, \theta') = t$, t, $\theta' \geq 0$, and so C2 and C4 hold if F has finite first moment. If F has only finite second moment then C3 may not hold, but by the remark at the end of Section 4.3 part (ii) of the theorem is still true in this case. Thus in this case Theorem 5.1 applies to all $F \in \mathcal{F}$.

The calculation of (27) with $Mr(ds, \theta') = \theta' ds$ shows that if F is exponential then W has covariance function equal to r, so that W is identical to the Y process. Thus

$$\int Y(\cdot + s)M(ds, \theta) - Y(\cdot) - \varphi(\theta, \cdot) \int Y(t) \, dt =^d Y(\cdot) \quad (28)$$

if F is exponential and $M(ds, \theta) = \theta \, ds$. Equivalently, $W \circ F^{-1}$ is a Brownian bridge if F is exponential and $M(ds, \theta') = \theta' \, ds$. Csörgő, Csörgő and Horváth (1986) conjectured that, when $M(ds, \theta') = \theta' \, ds$, (28) holds only if F is exponential. One might also conjecture about (28) in the general case. This is a problem that requires studying the equation $C[W(s), W(t)] = F(s)\overline{F}(t)$, $s \leq t$. A similar conjecture has been solved by Deheuvels (1982).

---

[2] Recall that in the continuous case both F and M are continuous, and in the discrete case F is concentrated on $\{1, 2, \ldots\}$ and M on $\{0, 1, 2, \ldots\}$.

An extension of $M(x, \theta') = \theta' \, dx$ that springs to mind is $M(dx, \theta') \propto \theta'^p px^{p-1} \, dx$, with $p \geqslant 1$. This may satisfy C1, C2 and C3, but not C4, since $\sup_x |m(x, \theta) - m(x, \theta')| = +\infty$ for $\theta, \theta' > 0$ unless $p = 1$. Thus in this case part (i) of Theorem 5.1 may apply, but not part (ii). But $M(x, \theta') = \theta' \, dx$ is not the only unbounded measure to which Theorem 5.1 applies:

EXAMPLE 5.2 (*Other unbounded measures*). Suppose $m(x, \cdot)$ has continuous derivative $D_2 m(x, \cdot)$ on Θ^0 for each x, and that $\sup_x |D_2 m(x, \theta')| < \infty$ for θ' in a neighbourhood of θ. By the mean value theorem,

$$\sup_x |m(x, \theta) - m(x, \theta')|$$
$$\leqslant |\theta - \theta'| \int_0^1 \sup_{x \geqslant 0} |D_2 m(x, \theta + u(\theta' - \theta))| \, du, \qquad (29)$$

for θ' close enough to θ, and if the right-hand side here tends to zero as $\theta' \to \theta$, the second part of C4 is satisfied.

As a specific example, we take M defined by $M(x, \theta') = \kappa \log(1 + \theta' x)$, $x, \theta' \geqslant 0$ (κ satisfying $\int e^{-\theta x} M(dx, \theta) = 1$). We have

$$m(x, \theta') = \kappa x / (1 + \theta' x),$$
$$D_2 m(x, \theta') = -\kappa (x/(1 + \theta' x))^2, \quad x, \theta' > 0.$$

Here, $|D_2 m(x, \theta')|$ increases in x and satisfies $|D_2 m(x, \theta')| \uparrow \kappa \theta'^2$ as $x \to \infty$, so $\sup_x |D_2 m(x, \theta')| \leqslant \kappa \theta'^2$ for all $\theta' > 0$. Using this in (29), we see that

$$\sup_x |m(x, \theta) - m(x, \theta')| \leqslant |\theta - \theta'| \int_0^1 \kappa (\theta + u(\theta' - \theta))^2 \, du$$
$$= (\kappa/3) |\theta^3 - \theta'^3| \to 0$$

as $\theta' \to \theta$, and so the last part of C4 is satisfied. It can be seen that all the rest (including C3) is satisfied with any $F \in \mathcal{F}$.

EXAMPLE 5.3 ($M(dx, \theta')$ *multiple of counting measure on* $\mathbb{N}_0$). As in Example 5.1, it is easy to see that if we take M defined by $M(x, \theta') = \theta'[x + 1]$, $x \geqslant 0$, then C1, C2 and C4 are verified with any $F \in \mathcal{F}$; on the other hand, this time we need to assume C3. The calculation of the covariance function with this M and with $F = F_0$ geometric gives

$$C[W(s), W(t)] = (1 - (1-\theta)^{[s]})(1-\theta)^{[t]+1} = F_0(s) \overline{F}_0(t+1), \quad s \leqslant t,$$

revealing that in this case W is like a discretized version of the process Y of (28).

The verification of the conditions for the other choices of M considered in Example 2.1 is similar.

Many other choices of M are possible, of course. Multiples of probability distribution functions concentrated on $(0, +\infty)$ can provide convenient examples, for which C3 is automatically satisfied. Also, once a few functions are available one can create mixtures of them to get more examples.

In the continuous case we have only considered *scale invariant* examples of M, that is, functions of the type $M(x, \theta') = N(\theta' x)$, $x \geqslant 0$, $\theta' \in \Theta$, where $N : [0, \infty) \to [0, \infty)$ is some non-decreasing, right-continuous function. Such a choice seems natural and makes certain functionals of the limiting process scale invariant.

5.2. Comparison with the results of Csörgő, Csörgő and Horváth

Let us compare Theorem 5.1 with the results of Csörgő, Csörgő and Horváth on the process T_n defined by (13). Suppose F is continuous, strictly increasing and has finite mean. Under these assumptions, Csörgő, Csörgő and Horváth (1986, pp. 49–50, 60) prove the consistency result corresponding to (i) of Theorem 5.1: $n^{-1/2} T_n \to 0$ uniformly with probability 1.

In order to produce the statement analogous to (ii) we need to introduce a definition not directly of interest to us, that of an *O'Reilly weight function*: this is a continuous function q defined on $[0, 1]$, strictly positive and increasing on $(0, 1/2)$, symmetric around $1/2$, and such that

$$\int_0^{1/2} t^{-1} e^{-\varepsilon \frac{q^2(t)}{t}} \, dt < \infty \quad \text{for all } \varepsilon > 0.$$

[The significance of such type of function has to do with the rate at which the Brownian bridge approaches zero in the tails; see Csörgő, Csörgő and Horváth (1986, p. 27).]

Now make the additional assumptions that F has finite second moment and a density f that is strictly positive on the set $\{x : 0 < F(x) < 1\}$, and further that

$$\sup_{0 < u < 1} \frac{q(u)(1-u)}{f(F^{-1}(u))} < \infty \tag{30}$$

for some O'Reilly weight function q. Under these assumptions Csörgő, Csörgő and Horváth (1986, pp. 50, 61) prove that T_n can be approximated uniformly in probability by a sequence of processes (defined in terms of the Komlós–Major–Tusnády sequence of Brownian bridges) with the same distribution as the Gaussian process T defined for $0 \leqslant u \leqslant 1$ by

$$T(u) = -\theta \int_0^u B(v) \, dF^{-1}(v) - \theta \frac{(1-u)}{f(F^{-1}(u))} B(u)$$

$$+ \theta^2 H_F^{-1}(u) \int_0^1 B(v) \, dF^{-1}(v), \tag{31}$$

where B is a Brownian bridge and H_F^{-1} is the total time on test transform; in particular, T_n converges in distribution to T.

The moment conditions are the same as ours. The difference lies in the continuity, strict monotonicity and differentiability assumptions, and in the existence of q satisfying (30).

Condition (30) excludes those F for which $f(x)^2 = O(F(x))$ as $x \downarrow 0$. To prove it, we require a result on the tail behaviour of O'Reilly functions (see Csörgő, Csörgő and Horváth, 1986, p. 31): if q is an O'Reilly function, then $q(u)u^{-1/2} \to +\infty$ as $u \downarrow 0$. Using this with $u = F(x)$, we see that $q(F(x))F(x)^{-1/2} \times (F(x)/f(x)^2)^{1/2}\overline{F}(x) \to +\infty$ as $x \downarrow 0$, so that (30) does not hold for such F. A specific example is furnished by F with density of the form $f(x) = x^{\gamma-1}e^{-x}/\Gamma(\gamma)$, $x > 0$, where $\gamma \geqslant 2$ (gamma densities with shape parameter $\geqslant 2$).

The extra assumptions of Csörgő et al. are there because T_n uses the inverse image of F_n rather than the real line as its domain of definition, whence our claim at the end of Section 3 that considering Z_n^* instead of T_n makes the theory easier and less restrictive. To see this, we need to relate the limiting processes W and T in the same way as we have related Z_n^* and T_n around formula (15).

Take $M(ds, \theta') = \theta' ds$ and F under the assumptions just stated. By putting $x = F^{-1}(u)$ in (26), changing variables and performing some algebra, we can see that

$$W \circ F^{-1}(u) = T(u) - B(u) + \theta \frac{(1-u)}{f(F^{-1}(u))} B(u). \tag{32}$$

If F is the exponential distribution, $\theta(1-u) = f(F^{-1}(u))$, and (32) gives $T = W \circ F^{-1}$, which is analogous to the relation $T_n = Z_n^* \circ F_n^{-1} + o(1)$ obtained after (15). Thus, when F is exponential, the limiting process T is just W evaluated at the inverse image of F, in the same way that T_n is essentially Z_n^* evaluated at the inverse image of F_n. When F is not exponential, (32) says that $W \circ F^{-1}(u)$ and $T(u)$ differ by

$$-B(u) + \theta \frac{(1-u)}{f(F^{-1}(u))} B(u), \tag{33}$$

a term resulting from the 'non-negligible' term $n^{1/2}(Z \circ F_n^{-1}(u) - Z \circ F^{-1}(u))$ of (14) and (15); as seen by the proofs of Csörgő et al. (see their formula (37) on p. 51 and the statement preceding it), it is the convergence of the latter to the former that demands the additional assumptions. A 'weak' and 'pointwise' way of illustrating this fact (we could not really prove the convergence without reproducing their proofs) is by means of the *quantile process*; this is the process Q_n defined by

$$Q_n(u) = n^{1/2}(F_n^{-1}(u) - F^{-1}(u)), \quad 0 < u < 1.$$

If we invoke a well-known theorem saying that, under the present assumptions on F, the asymptotic distribution of $f \circ F^{-1}(u)Q_n(u)$ is that of $B(u)$ (a Brownian bridge) for each $u \in (0, 1)$, and the property that $F_n^{-1}(u) \to F^{-1}(u)$ a.s. if $0 < u < 1$,[3] and if we

---

[3] Reiss (1989, p. 19, pp. 108–109) gives elementary proofs of these two results.

write

$$n^{1/2}(Z \circ F_n^{-1}(u) - Z \circ F^{-1}(u))$$
$$= f(F^{-1}(u)) Q_n(u) \left(\frac{\theta(1-u)}{f(F^{-1}(u))} - 1 \right) + R_n(u),$$

with

$$R_n(u) = Q_n(u) \left(\frac{Z \circ F_n^{-1} - Z \circ F^{-1}(u)}{F_n^{-1}(u) - F^{-1}(u)} - \frac{dZ(x)}{dx}\bigg|_{x=F^{-1}(u)} \right),$$

we can conclude that $R_n(u) \to_P 0$ and

$$n^{1/2}(Z \circ F_n^{-1}(u) - Z \circ F^{-1}(u)) \to^d -B(u) + \theta \frac{(1-u)}{f(F^{-1}(u))} B(u),$$
$$0 < u < 1,$$

which suggests that $n^{1/2}(Z \circ F_n^{-1}(u) - Z \circ F^{-1}(u))$ yields indeed (33). This argument also indicates that the appearance of F_n^{-1} in T_n requires the existence of $F' = f$, since even a weak and pointwise statement (the convergence in distribution for fixed u just given) seems to require the consideration of Q_n, whose convergence in turn requires $F' = f$ – a look at Csörgő and Horváth (1993, Chapter 6) reveals that all the results available for Q_n assume at least the existence of f.

6. Integral statistics. Applications to testing

In the remainder of the chapter we will be concerned with the problem of testing the hypothesis H_0: $F \in \mathcal{F}_0$ against H_1: $F \in \mathcal{F} - \mathcal{F}_0$ using integral statistics, and especially *linear integral test statistics*

$$I_n^{(1)}(F_0, M) = \int n^{1/2} Z_n(x) \, dF_n(x). \tag{34}$$

It is assumed from now on that we are either in the continuous case or in the discrete case. Formula (34) can thus be written in a more explicit form: noting (7) and (8), we see that (34) becomes

$$I_n^{(1)}(F_0, M) = n^{-1/2}(n-1)/2 - n^{-3/2} \sum_{i=1}^{n-1} \sum_{j=i+1}^{n} M_n(X_{j:n} - X_{i:n}) \tag{35}$$

in the continuous case, and

$$I_n^{(1)}(F_0, M) = n^{1/2} \sum_{i=1}^{\infty} \overline{F}_n(i) [F_n(i) - F_n(i-1)]$$

$$-\sum_{i=1}^{\infty}\sum_{j=i+1}^{\infty} M_n(j-i-1)$$
$$\times \left[F_n(j)-F_n(j-1)\right]\left[F_n(i)-F_n(i-1)\right] \quad (36)$$

in the discrete case.

The reason why (35) and (36) should, respectively, detect departures from the exponential or geometric distributions, has to do with the behaviour of Z_n under alternative hypotheses: if F is neither exponential nor geometric, one may expect Z_n to become too large or too small 'on average', and thus, since (34) is an average with respect to F_n, one may also expect $I_n^{(1)}(F_0, M)$ to become too large or too small. On the other hand, there could be alternatives for which Z_n would become too large but in such a way that integrating with respect to F_n would, on average, cancel out the discrepancies and bring $I_n^{(1)}(F_0, M)$ close to zero. We shall see later that such alternatives exist indeed, and that the tests based on (34) are not always consistent. In spite of this shortcoming, $I_n^{(1)}(F_0, M)$ is a very convenient statistic. Besides having good power against a wide range of alternatives, it has an asymptotic normal distribution under both the null and alternative hypotheses, which makes it easy to use and also to study.

Consistent tests based on the Z_n process can always be obtained by considering *quadratic integral statistics* of the form

$$I_n^{(2)}(F_0, M) = \int n Z_n^2(x)\, dF_n(x). \quad (37)$$

The corresponding asymptotic theory is not so simple; we shall obtain the asymptotic distribution of $I_n^{(2)}(F_0, M)$ only under the null hypothesis and, even in this case, not explicitly.

Besides (34) and (37), one could also consider $\int n^{1/2}|Z_n(x)|\,dF_n(x)$, or weighted versions of any of the integral statistics, such as $\int n^{1/2}Z_n(x)w(x)\,dF_n(x)$, $\int nZ_n(x)^2 w(x)\,dF_n(x)$, etc., where w is some function. The corresponding asymptotic results amount to simple generalizations of those stated below and will not be explicitly mentioned.

Before considering questions related to consistency of the tests based on (34) and (35), and to how the tests are actually defined and carried out, we shall first obtain approximations to the distributions of $I_n^{(1)}(F_0, M)$, $I_n^{(2)}(F_0, M)$ and calculate certain variances.

With the purpose of determining approximations to the distribution of $I_n^{(1)}(F_0, M)$ under the alternative hypothesis we introduce the more general integral statistics

$$I_n^{(1)}(F, M) = n^{1/2}\left[\int Z_n(x)\,dF_n(x) - \int Z(x)\,dF(x)\right], \quad (38)$$

of which (34) is a special case corresponding to those F such that $\int Z(x)\,dF(x) = 0$.

6.1. Asymptotic distributions of the integral statistics

$I_n^{(1)}(F, M)$ can be written in terms of the empirical process Y_n and the integrated lack-of-memory process Z_n^* as

$$I_n^{(1)}(F, M) = \int Z_n^*(x) \, dF_n(x) + \int Z(x) \, dY_n(x). \tag{39}$$

We assume throughout the conditions stated in Theorem 5.1. By C0, $\int \overline{F}(y) M(dy, \theta) < \infty$, which implies that Z is bounded, and also of bounded variation, so that we can define integrals with respect to the corresponding signed measure. In fact, it will now be convenient to write the second integral in (39) as an integral with respect to Z. Using Fubini's theorem we see that (39) can be written as

$$I_n^{(1)}(F, M) = \int Z_n^*(x) \, dF_n(x) - \int Y_n(x-) \, dZ(x).$$

As might be expected, the weak limit of $I_n^{(1)}(F, M)$ is obtained by replacing F_n, Z_n^* and Y_n in this expression by F, W and Y, respectively:

PROPOSITION 6.1. *Under the conditions of Theorem 5.1, the asymptotic distribution of $I_n^{(1)}(F, M)$ in the continuous and discrete cases is the same as that of the normal random variable*

$$I^{(1)}(F, M) = \int W(x) \, dF(x) - \int Y(x-) \, dZ(x). \tag{40}$$

PROOF. The proof consists of showing that $I_n^{(1)}(F, M)$ is approximated in probability by $\int W_n(x) \, dF(x) - \int B_n(F(x-)) \, dZ(x)$, where again B_n is the Komlós–Major–Tusnády sequence and W_n the sequence of Theorem 5.1.

We have

$$\left| I_n^{(1)}(F, M) - \int W_n \, dF + \int B_n(F(x-)) \, dZ(x) \right|$$

$$\leq \left| \int Z_n^* \, dF_n - \int W_n \, dF \right| + \left| \int Y_n(x-) - B_n(F(x-)) \, dZ(x) \right|,$$

where the second term tends to zero a.s. because

$$\left| \int Y_n(x-) - B_n(F(x-)) \, d\overline{F}(x) \right.$$

$$\left. - \int [Y_n(x-) - B_n(F(x-))] \, d \int \overline{F}(x+y) \, dM(y) \right|$$

$$\leq - \int \left| Y_n(x-) - B_n(F(x-)) \right| d\overline{F}(x)$$

$$-\int |Y_n(x-) - B_n(F(x-))|\,\mathrm{d}\int \overline{F}(x+y)\,\mathrm{d}M(y) \to 0$$

with probability 1. Since

$$\left|\int Z_n^* \,\mathrm{d}F_n - \int W_n \,\mathrm{d}F\right| \leqslant \int |Z_n^* - W_n|\,\mathrm{d}F_n + \left|\int W_n\,\mathrm{d}F_n - \int W_n\,\mathrm{d}F\right|$$

$$\leqslant \sup_x |Z_n^*(x) - W_n(x)| + \left|\int W_n\,\mathrm{d}F_n - \int W_n\,\mathrm{d}F\right|$$

and $\sup_x |Z_n^*(x) - W_n(x)| \to_P 0$, to finish the proof we need to show that

$$\left|\int W_n\,\mathrm{d}F_n - \int W_n\,\mathrm{d}F\right| \to_P 0. \tag{41}$$

Consider first the discrete case. For $T > 0$,

$$\int_0^T W_n\,\mathrm{d}F_n - \int_0^T W_n\,\mathrm{d}F$$

$$= \int_0^T [\overline{F}_n(x-) - \overline{F}(x-)]\,\mathrm{d}W_n(x) + W_n(T)[\overline{F}(T) - \overline{F}_n(T)],$$

and therefore

$$\left|\int W_n\,\mathrm{d}F_n - \int W_n\,\mathrm{d}F\right|$$

$$\leqslant \left|\int_0^T \overline{F}_n(x-) - \overline{F}(x-)\,\mathrm{d}W_n(x)\right| + |W_n(T)||F_n(T) - F(T)|$$

$$+ \left|\int_T^\infty W_n\,\mathrm{d}F_n\right| + \left|\int_T^\infty W_n\,\mathrm{d}F\right|$$

$$\leqslant \sup_x |F_n(x) - F(x)|(V_{W_n}(T) + |W_n(T)|)$$

$$+ \sup_x |W_n(x)|(\overline{F}_n(T) + \overline{F}(T)), \tag{42}$$

where $V_{W_n}(T) = \sum_{i=1}^T |W_n(i) - W_n(i-1)|$ (the total variation of W_n over $[0, T]$).

Take arbitrary $\varepsilon, \delta > 0$. Since W_n is a.s. bounded and has the same distribution for each n, we can choose K_ε such that $P[\sup_x |W_n(x)| > K_\varepsilon] \leqslant \varepsilon/2$, then choose T so large that $\delta/(2K_\varepsilon) > 2\overline{F}(T)$, and finally choose $K_{\varepsilon,T}$ such that $P[V_{W_n}(T) + |W_n(T)| > K_{\varepsilon,T}] \leqslant \varepsilon/2$ (possible because W_n is a.s. of bounded variation on $[0, T]$). By (42), then,

$$P\left[\left|\int W_n\,\mathrm{d}F_n - \int W_n\,\mathrm{d}F\right| > \delta\right]$$

$$\leqslant \varepsilon/2 + P\Big[\sup_x |F_n(x) - F(x)| > \delta/(2K_{\varepsilon,T})\Big]$$
$$+ \varepsilon/2 + P\big[\overline{F}_n(T) - \overline{F}(T) > \delta/(2K_\varepsilon) - 2\overline{F}(T)\big],$$

so $\limsup_n P[|\int W_n \, dF_n - \int W_n \, dF| > \delta] \leqslant \varepsilon$, proving (41) in the discrete case.

Let again $\varepsilon, \delta > 0$ be given, and choose K_ε satisfying $P[\sup_x |W_n(x)| > K_\varepsilon] \leqslant \varepsilon/4$ and T so large that $\delta/(2K_\varepsilon) > 2\overline{F}(T)$. In the continuous case W_n is a.s. uniformly continuous on any bounded interval $[0, T]$. Therefore, given arbitrary $T, \delta, \varepsilon > 0$, there exist numbers $0 = \alpha_0 < \alpha_1 < \cdots < \alpha_m = T$ such that the sets

$$A_n = \Big\{ \sup_{\alpha_{i-1} \leqslant x \leqslant \alpha_i} W_n(x) - \inf_{\alpha_{i-1} \leqslant x \leqslant \alpha_i} W_n(x) \leqslant \delta/4, \ i = 1, \ldots, m \Big\}$$

satisfy $P(A_n) \geqslant 1 - \varepsilon/5$ for all n.

Since the convergence of F_n to F is uniform a.s., we have with probability 1

$$\big|F_n(\alpha_i) - F_n(\alpha_{i-1}) - \big(F(\alpha_i) - F(\alpha_{i-1})\big)\big| \leqslant \delta/(4m K_\varepsilon), \quad i = 1, \ldots, m, \tag{43}$$

for sufficiently large n. Now every point of A_n for which (43) holds also satisfies

$$\Big|\int_0^T W_n \, dF_n - \int_0^T W_n \, dF\Big|$$
$$\leqslant \delta/4 + \sup_x |W_n(x)|$$
$$\times \sum_{i=1}^m \big|F_n(\alpha_i) - F_n(\alpha_{i-1}) - \big(F(\alpha_i) - F(\alpha_{i-1})\big)\big|$$
$$\leqslant \delta/4 + \sup_x |W_n(x)| \delta/(4K_\varepsilon). \tag{44}$$

For each n, let A'_n be the subset of A_n for which (43) (hence (44)) holds, so that we have $P(A'_n) \geqslant 1 - \varepsilon/4$ for large enough n. By taking n so large that $P[\overline{F}_n(T) - \overline{F}(T) > \delta/(2K_\varepsilon) - 2\overline{F}(T)] < \varepsilon/4$, we then have

$$P\Big[\Big|\int W_n \, dF_n - \int W_n \, dF\Big| > \delta\Big]$$
$$\leqslant P\Big(\Big[\Big|\int_0^T W_n \, dF_n - \int_0^T W_n \, dF\Big| > \delta/2\Big] \cap A'_n\Big)$$
$$+ P\big(A'^C_n\big) + P\Big[\sup_x |W_n(x)|\big(\overline{F}_n(T) + \overline{F}(T)\big) > \delta/2\Big]$$
$$\leqslant P\Big[\sup_x |W_n(x)|\delta/(4K_\varepsilon) > \delta/4\Big] + \varepsilon/4$$

$$+\varepsilon/4 + P\bigl[\overline{F}_n(T) - \overline{F}(T) > \delta/(2K_\varepsilon) - 2\overline{F}(T)\bigr] < \varepsilon,$$

which proves (41) in the continuous case. $\square$

Now let $f : \mathbb{R} \to \mathbb{R}$ be a function satisfying

$$f(x) \to 0 \quad \text{as } x \to 0, \quad \text{and}$$
$$|f(x) - f(y)| \leq |x - y| g(x, y) \quad \text{for all } x, y, \tag{45}$$

where g is a continuous real function defined for all (x, y).

Assume also that f is such that $f \circ Z$ has bounded variation, and define $\widetilde{Z}$, $\widetilde{Z}_n^*$ and $\widetilde{W}_n$ as

$$\widetilde{Z} = f \circ Z, \qquad \widetilde{Z}_n^* = f \circ \bigl(n^{1/2}(Z_n - Z)\bigr) = f \circ Z_n^*, \qquad \widetilde{W}_n = f \circ W_n.$$

Because of the properties of f, these new functions share with Z, Z_n^* and W_n all those properties that are used in the proof of Proposition 6.1. More precisely, $\widetilde{Z}(x) \to 0$ as $x \to \infty$, and $\widetilde{Z}$ is the same type of function as Z, being bounded and of bounded variation on bounded intervals; the realizations of $\widetilde{Z}_n$ are of the same type as those of Z_n (continuous functions of unbounded variation on certain intervals, or step functions of bounded variation); since $\sup_x g(Z_n^*(x), W_n(x))$ is bounded in probability, we have by (45) that

$$\sup_x |\widetilde{Z}_n^*(x) - \widetilde{W}_n(x)| \leq \sup_x |Z_n^*(x) - W_n(x)| \sup_x g(Z_n^*(x), W_n(x)) \to_P 0.$$

This means that if f satisfies the above conditions then Proposition 6.1 holds with Z, Z_n^* and W_n replaced by $\widetilde{Z}$, $\widetilde{Z}_n^*$ and $\widetilde{W}_n$, respectively:

COROLLARY. *Let f be a function satisfying (45) and such that $f \circ Z$ is of bounded variation. Under the conditions of Theorem 5.1, the asymptotic distribution of*

$$I_n^{(f)}(F, M) := \int f \circ Z_n^*(x) \, dF_n(x) + \int f \circ Z(x) \, dY_n(x)$$
$$= \int f \circ Z_n^*(x) \, dF_n(x) - \int Y_n(x-) \, df \circ Z(x)$$

in the continuous and discrete cases is the same as that of the random variable

$$I^{(f)}(F, M) = \int f \circ W(x) \, dF(x) - \int Y(x-) \, df \circ Z(x). \tag{46}$$

Since the functions $x \to x^2$ and $x \to |x|$ satisfy (45) and Z^2 and $|Z|$ are of bounded variation on bounded intervals, the corollary gives us in particular the asymptotic distributions of $I_n^{(2)}(F_0, M)$ and $\int n^{1/2}|Z_n| \, dF_n$ under the null hypothesis $H_0 : F \in \mathcal{F}_0$.

6.2. Testing

We have just seen that in the discrete and continuous cases, and under the conditions of Theorem 5.1, we have

$$I_n^{(1)}(F, M) \to^d N(0, \sigma_{F,M}^2),$$

where $\sigma_{F,M}^2 := V[I^{(1)}(F, M)]$ and $I^{(1)}(F, M)$ is the variable of (40), and, for $F \in \mathcal{F}_0$,

$$I_n^{(2)}(F_0, M) \to^d Q(\theta, M),$$

where $Q(\theta, M)$ denotes the distribution of $\int W(x)^2 \, dF(x)$, a special case of (46).

Under H_0: $F \in \mathcal{F}_0$, both the variance $\sigma_{\theta,M}^2 := \sigma_{F,M}^2 = V[I_n^{(1)}(F_0, M)]$ and the distribution of $Q(\theta, M)$ depend on M and possibly on the unknown parameter θ.

If in the continuous case we take M scale invariant in the sense of Section 5.1, then $I_n^{(1)}(F_0, M)$ and $I^{(1)}(F_0, M)$ are scale invariant and $\sigma_M^2 := \sigma_{\theta,M}^2$ is independent of θ.[4] In this case, *a two-sided test of (approximate) size γ based on $I_n^{(1)}(F_0, M)$ of H_0: $F \in \mathcal{F}_0$ (that F is exponential) consists of rejecting the null hypothesis if $I_n^{(1)}(F_0, M)$ falls outside the interval* $(-z_{1-\gamma/2}\sqrt{\sigma_M^2}, z_{1-\gamma/2}\sqrt{\sigma_M^2})$, where as usual $z_p = \Phi^{-1}(p)$ ($0 < p < 1$) and Φ is the standard normal distribution function.

More generally, if $\sigma_{\theta,M}^2$ is a continuous function of θ, then $\sigma_{\theta_n,M}^2 \to_P \sigma_{\theta,M}^2$, and so

$$I_n^{(1)}(F, M)/\sqrt{\sigma_{\theta_n,M}^2} \to^d N(0, 1).$$

In this case, *a two-sided test of (approximate) size γ based on $I_n^{(1)}(F_0, M)$ of H_0: $F \in \mathcal{F}_0$ (that F is exponential or geometric, depending on the case under consideration), consists of rejecting the null hypothesis if $I_n^{(1)}(F_0, M)$ falls outside the interval* $(-z_{1-\gamma/2}\sqrt{\sigma_{\theta_n,M}^2}, z_{1-\gamma/2}\sqrt{\sigma_{\theta_n,M}^2})$.

This test rule needs a rectification in the discrete case, because there is positive probability that $\theta_n = 1$, which causes $\sigma_{\theta_n,M}^2$ to be zero (see (47) below). Thus, in the event that $\theta_n = 1$, we make the convention that the test rule rejects the null hypothesis that F is geometric. This makes sense, since $\theta_n = 1$ suggests a distribution degenerate at 1 rather than a geometric distribution as we have defined it.

Although the distribution of $Q(\theta, M)$ generally depends on θ, if M is scale invariant then $I_n^{(2)}(F_0, M)$ is scale invariant and $Q(M) := Q(\theta, M)$ is independent of θ. In this case, then, *a one-sided test of (approximate) size γ based on $I_n^{(2)}(F_0, M)$ of H_0: $F \in \mathcal{F}_0$ (that F is exponential) consists of rejecting the null hypothesis if $I_n^{(2)}(F_0, M)$ exceeds the quantile of probability $1 - \gamma$ of the distribution of $Q(M)$*.

---

[4] The subscripts in $\sigma_{F,M}^2$, $\sigma_{M,\theta}^2$, σ_M^2, etc., patently specify the context in which the variances are being used; in particular, the omission of F indicates that the variance is calculated under H_0.

In the remaining situations $I_n^{(2)}(F_0, M)$ cannot be used for testing (unless θ is known).

Although the above tests can be carried out with many choices of the function M, we have limited ourselves to investigating the following examples: M multiple of the Lebesgue measure function on $[0, +\infty)$, M multiple of the gamma measure function with shape parameter $r = 1, 2, 3$, M defined by $M(x, \theta') = \kappa \log(1 + \theta' x)$, $x \geqslant 0$ (Example 5.2), in the continuous case, and M multiple of counting measure on $\mathbb{N}_0$ in the discrete case. We shall refer to these in the abbreviated form $M \propto Leb$, $M \propto Ga(r)$ ($r = 1, 2, 3$), $M \propto Log$ and $M \propto Count$, respectively.

In the case of $M \propto Log$ the proportionality constant κ can not be calculated exactly; using the properties of the gamma and digamma functions and the tables for these functions in Abramowitz and Stegun (1965), we find that $\kappa^{-1} \approx 0.596347362$ with an error $< e \times 2 \times 10^{-10}$, which is the estimate used in what follows.

In order to perform the tests we require σ_M^2, $\sigma_{\theta_n, M}^2$ and the appropriate quantiles of $Q(M)$.

The latter will generally have to be estimated, because we have no explicit expression for the distribution of $Q(M)$. If $M \propto Leb$, then (Example 5.1) $W \circ F^{-1}$ restricted to $[0, 1]$ is a Brownian bridge, and therefore the distribution of $Q(M)$ reduces to the asymptotic distribution of the Cramér–von Mises statistic, which has been tabulated by Stephens (1986a, p. 105), and whose distribution function is known (e.g., Csörgő and Révész (1981, p. 43)). For other choices of M, the quantiles of the distribution of $Q(M)$ can be estimated via simulation. Table 1 gives estimates and approximate 95% confidence intervals of the 90% quantiles of the distribution of $I_n^{(2)}(F_0, M)$ for sample sizes of $n = 20, 50, 100$ and $M \propto Leb$, $M \propto Log$, $M \propto Ga(r)$, $r = 1, 2, 3$ (each estimate is based on 100,000 simulation runs). Our experience indicates that the critical values estimated with samples of 100 are a good approximation to the asymptotic ones, so for most practical purposes they can also be used with sample sizes larger than 100.

σ_M^2 and $\sigma_{\theta_n, M}^2$ are generally difficult to calculate exactly. Calculating the first requires integrating the covariance function in (27) with respect to the two-fold product of the

Table 1
Estimates and 95% confidence intervals of the 90% quantiles of the distribution of the quadratic integral statistic based on samples of size $n = 20, 50, 100$

			M		
n	$\propto Leb$	$\propto Log$	$\propto Ga(1)$	$\propto Ga(2)$	$\propto Ga(3)$
20	0.3316	0.2596	0.1465	0.4043	0.9392
	(0.329,0.334)	(0.258,0.262)	(0.146,147)	(0.401,0.407)	(0.931,0.947)
50	0.3399	0.2552	0.1492	0.4082	0.9492
	(0.337,0.342)	(0.253,0.257)	(0.148,0.150)	(0.405,0.412)	(0.942,0.959)
100	0.3429	0.2538	0.1509	0.4097	0.9593
	(0.340,0.346)	(0.252,0.256)	(0.150,0.152)	(0.407,0.413)	(0.951,0.969)

exponential measure. Excruciating calculations lead to the formula

$$\sigma_M^2 = \int_0^\infty \left[\int_0^x (M(x-t, 1) - 1)e^{-t} \, dt \right.$$
$$\left. - \frac{(x-1)}{2} \int_0^\infty t N'(t) e^{-t} \, dt \right]^2 e^{-x} \, dx, \qquad (47)$$

which naturally is independent of θ. Note that this is the formula for $V[I^{(1)}(F_0, M)]$ in the continuous case, with M scale invariant and F exponential.

For $M \propto Leb$ and $M \propto Ga(r)$, $r = 1, 2, 3$, formula (47) gives us the exact values of σ_M^2 as $1/12$, $5/432$, $69/972$ and $8957/34992$, respectively. However, except for simple cases such as these, (47) is not really useful for calculating σ_M^2. In most cases one would need numerical methods to evaluate the integrals involved, and it seems that in general the best is to estimate σ_M^2 through simulation. This is what we have done in the case of $M \propto Log$: we have estimated $V[I_n^{(1)}(F_0, M)]$ for several sample sizes n ranging from 50 to 500, and, by assessing the stability of these estimates, observed that a good estimate of the asymptotic variance σ_M^2 is 0.04839, which we refer to as the 'pseudo-asymptotic variance'.

Calculating $\sigma_{\theta_n, M}^2$ is easy if $\sigma_{\theta, M}^2$ is known, but we have not calculated a formula for the latter in the general case. In the discrete case, with $M \propto Count$, we have

$$\sigma_{\theta, M}^2 = \frac{\theta^3(1-\theta)^2(1+(1-\theta)^2)}{(1-(1-\theta)^2)(1-(1-\theta)^3)(1-(1-\theta)^4)}, \qquad 0 < \theta < 1, \qquad (48)$$

a formula obtained by integrating the covariance function of Example 5.3 with respect to the two-fold product of the geometric measure. In this situation $\sigma_{\theta_n, M}^2$ is obtained by simply replacing θ by θ_n in (48).

For small sample sizes n the distribution of $I_n^{(1)}(F_0, M)$ may differ considerably from a normal distribution, in which case it is advisable to perform the tests using critical values estimated from finite samples rather than the asymptotic ones. Table 2 gives estimates of the 5% and 95% quantiles of the distribution of $I_n^{(1)}(F_0, M)$ for a

Table 2
Estimates and 95% confidence intervals of the 5% (first row) and 95% (second row) quantiles of the distribution of the linear integral statistic based on samples of size 20

		M		
$\propto Leb$	$\propto Log$	$\propto Ga(1)$	$\propto Ga(2)$	$\propto Ga(3)$
−0.46503	−0.40912	−0.18913	−0.44255	−0.82976
(−0.470, 0.461)	(−0.412, −0.406)	(−0.190, −0.188)	(−0.445, −0.440)	(−0.835, −0.824)
0.45921	0.32494	0.18743	0.45375	0.83155
(0.456, 0.463)	(0.323, 0.328)	(0.185, 0.190)	(0.449, 0.458)	(0.826, 0.838)

Table 3
Point estimates and 90% confidence intervals of actual significance levels of the test based on $I_n^{(1)}(F_0, M)/\sqrt{\sigma_{\theta_n,M}^2}$, $M \propto$ Count, with nominal significance level 0.10, for $n = 20, 50, 100, 200, 350$ and $\alpha = 0.15, 0.25, 0.50, 0.75, 0.85$

n	$\alpha \equiv \theta$				
	0.15	0.25	0.50	0.75	0.85
20	0.0856 (0.081,0.090)	0.0815 (0.077,0.086)	0.0762 (0.072,0.081)	0.0531 (0.049,0.057)	0.0808 (0.076,0.085)
50	0.0902 (0.085,0.095)	0.0916 (0.087,0.096)	0.0864 (0.082,0.091)	0.0830 (0.078,0.088)	0.0603 (0.056,0.064)
100	0.0981 (0.093,0.103)	0.1020 (0.097,0.107)	0.0957 (0.091,0.101)	0.0929 (0.088,0.098)	0.0915 (0.087,0.096)
200	0.0978 (0.093,0.103)	0.0995 (0.095,0.104)	0.0968 (0.092,0.102)	0.0992 (0.094,0.104)	0.0935 (0.089,0.098)
350	0.0987 (0.094,0.104)	0.0999 (0.095,0.105)	0.0972 (0.092,0.102)	0.0968 (0.092,0.102)	0.0958 (0.091,0.101)

sample size of $n = 20$ and $M \propto$ Leb, $M \propto$ Log, $M \propto$ Ga(r), $r = 1, 2, 3$ (below each estimate is an approximate 95% confidence interval). For larger sample sizes ($n \geq 30$) the asymptotic critical values (the 'pseudo-asymptotic' in the case of $M \propto$ Log) seem sufficiently accurate for practical purposes.

When $F = F_0$ is geometric, the convergence of $I_n^{(1)}(F_0, M)/\sqrt{\sigma_{\theta_n,M}^2}$ to a standard normal variable is fast for values of $\alpha \equiv \theta < 0.50$, but considerably slower for values of $\alpha > 0.50$, especially for α near 1. In view of this, one may think of estimating critical values for small sample sizes and for values of $\alpha > 0.50$. Instead of this, we have adopted the simpler approach of estimating the actual size of a test of nominal size 0.10 based on the asymptotic critical values ($\approx \pm 1.6449$), when the parameter takes values $\alpha = 0.15, 0.25, 0.50, 0.75, 0.85$. In this way, given the estimate of θ, one will be able in most practical situations to judge how close is the actual to the nominal significance level. Estimates of actual test sizes for sample sizes of $n = 20, 50, 100, 200, 350$ appear in Table 3 accompanied by approximate 90% confidence intervals. We can see from the table that the use of asymptotic critical values will not affect much the nominal 10% size of the test for a good range of values of θ. In the worst scenario considered one can expect an actual size of about 5%. On the other hand, for $n \geq 100$ the critical values provide a good approximation in all cases.

6.3. Consistency and performance of the statistics

We have mentioned at the beginning of this section that there are alternatives for which the linear integral statistics are not consistent. This is apparent from the definition of $I_n^{(1)}(F, M)$ in (36), which tells us that the statistic $I_n^{(1)}(F_0, M)$ is sensitive to departures from the hypothesis that $\int Z \, dF = 0$ rather than to departures from the hypothesis that

$F \in \mathcal{F}_0$. As hinted before, there could be alternatives for which $Z > 0$ on some proper subset of the support of F, $Z \leqslant 0$ on its complement, and $\int Z \, dF = 0$, in which case $I_n^{(1)}(F_0, M)$ would simply fluctuate around zero no matter how large the sample size, and the statistic would be inconsistent. The following example shows that such alternatives exist.

EXAMPLE 6.1 (*Distribution Functions F for which $Z \neq 0$ and $\int Z \, dF = 0$*). Consider the family of beta distributions $Be(\alpha, \beta)$ with density $f = F'$ defined by $f(x) = x^{\alpha-1}(1-x)^{\beta-1}/B(\alpha, \beta)$, $0 < x < 1$, $\alpha, \beta > 0$. Spurrier (1984, p. 1645) noticed that the cumulative total time on test statistic (10) – a multiple of (34) with $M \propto Leb$ – is inconsistent against $Be(1/2, 1)$ alternatives. In fact, one can show that this statistic is inconsistent for several $Be(\alpha, \beta)$ alternatives by finding solutions (α, β) of the equation $\int Z \, dF = 0$, which is equivalent to $\theta \iint \overline{F}(x+y) \, dy \, dF(x) = 1/2$. Besides $(1/2, 1)$, we have for instance $((\sqrt{13}-1)/4, 2)$ as an exact solution and $(0.73, 3)$ as an approximate solution to $\int Z \, dF = 0$, and in principle one can find many more.

We note, however, that in spite of these choices of (α, β) giving examples of F against which (34) is inconsistent, the models have little resemblance to the exponential distribution.

Thus it is clear that the linear integral statistics should be used for testing H_0: $\int Z \, dF = 0$ versus H_1: $\int Z \, dF \neq 0$ (rather than H_0: $F \in \mathcal{F}_0$ versus H_1: $F \in \mathcal{F} - \mathcal{F}_0$). For this testing problem the linear integral statistics are indeed consistent, as expected: in the continuous case, for example, if F satisfies $\int Z \, dF = K > 0$ then, since $\int Z_n^* \, dF_n$ is a.s. bounded, we have

$$I_n^{(1)}(F_0, M) = \int n^{1/2} Z_n \, dF_n = \int Z_n^* \, dF_n + n^{1/2} \int Z \, dF_n \to +\infty \quad \text{a.s.}$$

For the more general test of H_0: $F \in \mathcal{F}_0$ versus H_1: $F \in \mathcal{F} - \mathcal{F}_0$, one can employ the quadratic integral statistics (in the continuous case), which can be shown to be consistent against all alternatives $F \in \mathcal{F} - \mathcal{F}_0$.

In spite of not being consistent against all alternatives to the exponential distribution, $I_n^{(1)}(F_0, M)$ with $M \propto Leb$ has been extensively used in the literature for testing exponentiality, and was shown, for instance by Klefsjö (1983, pp. 920–921) and Stephens (1986b, pp. 452–454), to have good power properties. By 'good' we mean that the power of the total time on test statistic is usually as good as, and sometimes better, than the power of the Anderson–Darling (A^2) and Cramér–von Mises (ω^2) statistics, which have been confirmed as being amongst the most powerful omnibus statistics (not only for testing the exponential distribution but also the normal, Cauchy, logistic and others; see Stephens (1974) and Stephens (1986, p. 166)) and will be used here as the standard of comparison.

Besides confirming the good power properties of the total time on test statistic against Weibull, gamma and generalized Pareto alternatives, we have also assessed the performance of the integral statistics for the other choices of M via simulation. The main realization of our power study is that a measure function M having its mass relatively well spread over $[0, +\infty)$, such as $M \propto Leb$ and $M \propto Ga(3)$, generally yields powerful tests based on both linear and quadratic integral statistics against 'plausible' alternatives to the exponential model. Even though $M \propto Leb$ seems to represent overall the best choice against the most typical alternatives, it turns out that for alternative models that resemble the exponential distribution but have a bounded support (e.g., generalized Pareto) or a lighter tail $M \propto Ga(3)$ is preferable to $M \propto Leb$. On the other hand, with of $M \propto Ga(r)$ ($r = 1, 2$) and $M \propto Log$ the integral statistics are generally quite bad against the above alternatives. However, it will be seen in Section 7 that even $M \propto Ga(1)$ and $M \propto Log$ can be very powerful against certain alternatives.

The generalized Pareto distributions with light (resp. heavy) tails are examples of alternatives against which the linear integral statistic with $M \propto Ga(3)$ (resp. $M \propto Leb$) is superior to the classical quadratic statistics A^2 and ω^2. However, the performance of our statistics is generally comparable to that of A^2 and ω^2, and whenever there is an improvement relative to these it is through the linear rather than through the quadratic integral statistics. These conclusions make it clear that the integral statistics will be of practical value only in specialized contexts, where supposedly exponential data is available on a regular basis; for superficial or more occasional analyses it should be enough to calculate the classical quadratic statistics.

Our recommendation is that $I_n^{(1)}(F_0, M)$ be used with $M \propto Leb$, $M \propto Ga(3)$, $M \propto Ga(1)$ and $M \propto Log$, in conjunction with A^2 and ω^2. These choices of M should allow the efficient detection of certain types of discrepancies to which the classical statistics are not very sensitive, while the use of the latter should compensate for the possible inconsistency of $I_n^{(1)}(F_0, M)$.

As to the performance of $I_n^{(1)}(F_0, M)/\sqrt{\sigma_{\theta_n, M}^2}$ with $M \propto Count$ in testing the geometric distribution, our simulation studies have shown that against various alternatives of negative binomial, (shifted-) Poisson and logarithmic distributions the integral statistic is considerably superior to the chi-square statistic (around 40% more powerful for samples of size 100, for instance). These results allow us to conclude that the linear integral statistic can be truly useful in data analyses involving the geometric distribution. Since, as in the continuous case, there should be examples of F against which our test will be inconsistent, it is advisable in practice to calculate the chi-square statistic whenever the integral statistic does not reject the null hypothesis and at the same time the data cast doubt on the geometric assumption (e.g., because of too large or small a variance).

With the choice of $M \propto Count$ the integral statistic can be seen as a discrete analogue of the total time on test statistic. Other choices of M are possible, but we have not studied them.

7. Asymptotic efficiency in the continuous case

We shall complement our study of the integral statistics by looking at their asymptotic efficiency. We consider only the linear statistics in the continuous case and impose some 'regularity' conditions on the families of alternatives. After defining an optimality criterion in terms of asymptotic efficiency, we show how, given an integrator M, it is possible in principle to find a family of alternatives to the exponential distribution against which the integral statistic based on M represents an optimal choice.

We consider the problem of testing the exponential distribution against alternatives $F_\gamma(\cdot\,;\alpha) \in \mathcal{F}$ with density function $f_\gamma(\cdot\,;\alpha)$ approaching the exponential density as γ approaches zero: $f_\gamma(\cdot\,;\alpha) \to f_0(\cdot\,;\alpha) := F_0'(\cdot\,;\alpha)$ as $\gamma \to 0$. The main objective will be to compare the performance of the integral statistics $I_n^{(1)}(F_0, M)$ with the performance of certain *efficient statistics* which in some sense are optimal for this testing problem.

Throughout, we take M scale invariant, i.e., $M(x, \theta') = N(\theta'x)$, $x \geq 0$, $\theta' \in \Theta$, for some non-negative, non-decreasing function $N : [0, \infty) \to [0, \infty)$, and assume that M satisfies the conditions of Theorem 5.1; in particular, N is required to be continuously differentiable on $(0, \infty)$, and $m(x, \theta') = x N'(x\theta')$, $x > 0$. Since in this case M is determined by N, we shall often refer to the latter in place of the former; for instance, we will write $I_n^{(1)}(F_0, N)$ in place of $I_n^{(1)}(F_0, M)$, $N \propto Leb$ instead of $M \propto Leb$, $Z(x, N)$ and $Z_n(x, N)$ for the functions in (3) and (5), σ_N^2 for the variance in (47), etc. Also, we may need to distinguish the F being used for calculating some quantities; so we may write $Z_\gamma(x, N)$ for (3) with $F = F_\gamma$, $\sigma_N^2(\gamma)$ for $\sigma_{F_\gamma, N}^2 = V[I^{(1)}(F_\gamma, N)]$, etc.

The parameter α is a scale parameter for each member $f_\gamma(\cdot\,;\alpha)$; specifically, we assume that $f_\gamma(x; \alpha) = \alpha\, g_\gamma(\alpha x)$, $x \geq 0$, for a family of real functions g_γ defined on $[0, \infty)$. Because M is scale invariant, it follows that all test statistics, distributions and quantities used below are independent of α. For this reason we take $\alpha = 1$ in much of what follows, and use F_γ, f_γ, etc., in place of $F_\gamma(\cdot\,; 1)$, $f_\gamma(\cdot\,; 1)$, etc.

It will be important to indicate the distribution of the random variables involved at each time. Following Hall and Mathiason (1990) and other authors, we use P_γ for probabilities, E_γ for expectation, $\to^d [\gamma]$ for convergence in distribution, $\to_P [\gamma]$ for convergence in probability, or 'under P_γ' for any of these, when the random variables under consideration have distribution function F_γ.

The alternatives F_γ need only be defined for γ in a (right- or left-) neighbourhood of the origin. It is assumed that for values of γ in such a neighbourhood the conditions of Theorem 5.1 are fulfilled relatively to the N under consideration. For instance, if $N \propto Leb$ or if $N \propto Ga(r)$, it is enough for F_γ to possess finite second moment for sufficiently small γ.

In a word, given a pair (F_γ, N), we are assuming enough to be able to state

$$\frac{I_n^{(1)}(F_\gamma, N)}{\sigma_N(\gamma)} = \frac{n^{1/2}[\int Z_n(x, N)\, dF_n(x) - \mu_N(\gamma)]}{\sigma_N(\gamma)} \to^d [\gamma] N(0, 1) \qquad (49)$$

for each γ, where $\mu_N(\gamma) := \int Z_\gamma(x, N)\, dF_\gamma(x)$.

We need to impose additional assumptions on F_γ. These consist, essentially, of the regularity conditions imposed on statistical models in order to derive large sample estimation and testing procedures. We are guided by Hall and Mathiason (1990, p. 80) and Ferguson (1996, pp. 107–108, 119–121).

Throughout, 'for all γ' means for all values of γ for which the alternative family is defined, and similarly for any other parameters. If the set of permissible values of γ is a (half-) closed interval, then continuity, differentiability, etc., refer to one-sided versions (e.g., right continuity, etc.).

Define ℓ by $\ell(\gamma, \alpha; x) = \log f_\gamma(x; \alpha)$ whenever $f_\gamma(x; \alpha) \neq 0$ and as zero otherwise. We assume that $\ell(\cdot, \cdot; x)$ has, for all x, continuous second-order partial derivatives. Define the *score* s by $s(x; \gamma, \alpha) \equiv [s_1(x; \gamma, \alpha), s_2(x; \gamma, \alpha)] = [\partial \ell(\gamma, \alpha; x)/\partial \gamma, \partial \ell(\gamma, \alpha; x)/\partial \alpha]$. We assume that $\int s(x; \gamma, \alpha) F_\gamma(dx; \alpha) = (0, 0)$. Denote by $B(\gamma, \alpha)$ the *Fisher information* – the covariance matrix with diagonal elements $\int s_i(x; \gamma, \alpha)^2 F_\gamma(dx; \alpha)$, $i = 1, 2$, and off-diagonal element $\int s_1(x; \gamma, \alpha) s_2(x; \gamma, \alpha) F_\gamma(dx; \alpha)$. We assume that $B(\gamma, \alpha)$ is non-singular and that it equals the expectation of the symmetric of the Hessian matrix of $\ell(\cdot, \cdot; x)$, which we denote by $H(\cdot, \cdot; x)$; that is, we assume that $B(\gamma, \alpha) = -\int H(\gamma, \alpha; x) F_\gamma(dx; \alpha)$. We also assume that $H(\gamma, \alpha; x)$ is bounded in absolute value by an integrable (with respect to $F_\gamma(dx; \alpha)$) function of x for all α and for all γ in a neighbourhood of zero.

These assumptions imply 'Assumption 4' of Hall and Mathiason (1990, p. 80), and justify the application of some of their results, which we shall need later.

Finally, we define the *score for* γ by $v_\gamma(x) = \partial \log f_\gamma(x)/\partial \gamma \equiv s_1(x; \gamma, 1)$, and the *Fisher information for* γ by $I(\gamma) = \int v_\gamma(x)^2 \, dF_\gamma(x)$ (the first diagonal element of $B(\gamma, 1)$).

Note that our differentiability assumptions imply the existence and continuity of $\partial f_\gamma(x; \alpha)/\partial \alpha = g_\gamma(\alpha x) + \alpha x g'_\gamma(\alpha x)$, $\partial^2 f_\gamma(x; \alpha)/\partial \alpha^2 = 2x g'_\gamma(\alpha x) + \alpha x^2 g''_\gamma(\alpha x)$, $\partial^2 f_\gamma(x; \alpha)/\partial \gamma \partial \alpha = \partial g_\gamma(\alpha x)/\partial \gamma + \alpha x \partial g'_\gamma(\alpha x)/\partial \gamma$, where g'_γ, g''_γ denote derivatives of $g_\gamma(\cdot)$; therefore, $f_\gamma(x)$ is twice continuously differentiable in x.

7.1. Asymptotic efficiency

Let $T^{(1)} \equiv T_n^{(1)} \equiv T_n^{(1)}(\gamma)$ and $T^{(2)} \equiv T_n^{(2)} \equiv T_n^{(2)}(\gamma)$ be two statistics based on a random sample $(X_1, \ldots, X_n)$ and satisfying

$$\frac{n^{1/2}(T_n^{(i)}(\gamma) - \mu_i(\gamma))}{\sigma_i(\gamma)} \to^d [\gamma] N(0, 1)$$

for certain functions μ_i, σ_i, with $\mu_i(0) = 0$ and $\sigma_i^2(0) = \lim_{\gamma \to 0} \sigma_i^2(\gamma)$, $i = 1, 2$. One way of comparing the performances of $T^{(1)}$ and $T^{(2)}$ for testing H_0: $\gamma = 0$ versus H_1: $\gamma > 0$ ($\gamma < 0$) is to calculate the *Pitman asymptotic efficiency of* $T^{(1)}$ *relative to* $T^{(2)}$, the quantity defined by

$$e(T^{(1)}, T^{(2)}) = (\mu'_1(0)/\sigma_1(0))^2 / (\mu'_2(0)/\sigma_2(0))^2 \tag{50}$$

whenever $\mu'_i(0)$ (the derivative of μ_i at 0) exists and is $\neq 0$ for $i = 1, 2$. This is a measure of the relative efficacy of $T^{(1)}$ and $T^{(2)}$ for rejecting 'local' alternatives with 'large' samples. To make this interpretation precise, and to state under which conditions such interpretation is valid, we follow Serfling (1980, pp. 50–52, 314–325) and Hall and Mathiason (1990, pp. 89–90).

Suppose, without loss of generality, that $\mu'_i(0) > 0$ ($i = 1, 2$), so that the tests of (asymptotic) size ε ($0 < \varepsilon < 1$) based on $T^{(1)}$ and $T^{(2)}$ consist of rejecting H_1: $\gamma > 0$ if $n^{1/2} T_n^{(i)}(0)/\sigma_i(0) > z_{1-\varepsilon}$ ($i = 1, 2$), where as usual $z_p = \Phi^{-1}(p)$ ($0 < p < 1$) and Φ is the standard normal distribution function.

Make the assumption that for a sequence of *local alternatives* of the form $H_{1,n}$: $\gamma = \gamma_n := hn^{-1/2}$, where h is an arbitrary positive number, we have for all x

$$P_{\gamma_n}\left(\frac{n^{1/2}T_n^{(i)}(0)}{\sigma_i(0)} \leqslant x\right)$$
$$= P_{\gamma_n}\left(\frac{n^{1/2}(T_n^{(i)}(0) - \mu_i(\gamma_n))}{\sigma_i(\gamma_n)} \leqslant x\frac{\sigma_i(0)}{\sigma_i(\gamma_n)} - h\frac{\gamma_n^{-1}\mu_i(\gamma_n)}{\sigma_i(\gamma_n)}\right)$$
$$\to \Phi(x - hc_i) \quad \text{as } n \to \infty \tag{51}$$

for constants c_i, $i = 1, 2$. Then it follows (directly from (51)) that $c_i = \mu_i'(0)/\sigma_i(0)$, and that the power of the test based on $T^{(i)}$ against the sequence $H_{1,n}$: $\gamma = \gamma_n$, which we denote by $\delta_i(\gamma_n)$, satisfies

$$\delta_i(\gamma_n) = P_{\gamma_n}\left(n^{1/2}T_n^{(i)}(0)/\sigma_i(0) > z_{1-\varepsilon}\right)$$
$$\to 1 - \Phi\left(z_{1-\varepsilon} - h\mu_i'(0)/\sigma_i(0)\right) =: \delta_i \tag{52}$$

if and only if

$$n^{1/2}\mu_i(\gamma_n)/\sigma_i(\gamma_n) \to \Phi^{-1}(1-\varepsilon) - \Phi^{-1}(1-\delta_i), \tag{53}$$

where, necessarily, $\varepsilon < \delta_i < 1$.

Consider now using $T^{(1)}$ and $T^{(2)}$ with two possibly different sequences of sample sizes $n_1 \equiv n_1(n)$, $n_2 \equiv n_2(n)$, satisfying $n_1, n_2 \to \infty$ as $n \to \infty$, for testing $\gamma = 0$ against the two sequences of alternatives $\gamma \equiv \gamma_{n_1} = h_1 n_1^{-1/2}$ and $\gamma \equiv \gamma_{n_2} = h_2 n_2^{-1/2}$, respectively. The *Pitman asymptotic efficiency* of $T^{(1)}$ relative to $T^{(2)}$ is defined as the limit of the ratio n_2/n_1, with n_1 and n_2 constrained to have equal asymptotic power against asymptotically equal alternatives. Informally, $e(T^{(1)}, T^{(2)}) = e_0 < 1$ (say) means that for $T^{(2)}$ to perform comparably to $T^{(1)}$ against 'close' alternatives, the sample size required using $T^{(2)}$ is roughly only $e_0\%$ of the sample size required using $T^{(1)}$, and in this sense $T^{(2)}$ is more efficient than $T^{(1)}$. Specifically, $e(T^{(1)}, T^{(2)})$ is the limit of n_2/n_1 with n_1 and n_2 satisfying $\gamma_{n_1} \sim \gamma_{n_2}$ (i.e., $h_1 n_1^{-1/2} \sim h_2 n_2^{-1/2}$) and $\delta_1(\gamma_{n_1}) \to \delta$, $\delta_2(\gamma_{n_2}) \to \delta$ for some $\delta \in (\varepsilon, 1)$. This last requirement, which is possible by virtue of the equivalence between (52) and (53), enables the determination of $e(T^{(1)}, T^{(2)})$, since by (53) $\delta_1(\gamma_{n_1})$ and $\delta_2(\gamma_{n_2})$ tend to the same limit if and only if

$$n_1^{1/2}\mu_1(\gamma_{n_1})/\sigma_1(\gamma_{n_1}) \sim n_2^{1/2}\mu_2(\gamma_{n_2})/\sigma_2(\gamma_{n_2}),$$

or

$$(n_2/n_1)^{1/2} \to \left(\mu_1'(0)/\sigma_1(0)\right)/\left(\mu_2'(0)/\sigma_2(0)\right),$$

which gives us formula (50) for $e(T^{(1)}, T^{(2)})$.

The same derivation applies to alternatives $\gamma < 0$. A more general concept of relative efficiency (involving higher derivatives) applies also to the case in which both $\mu'_1(0)$ and $\mu'_2(0)$ equal zero. If, however, $\mu'_1(0) = 0$ and $\mu'_2(0) \neq 0$, the present definition is still meaningful: $n_2/n_1 \to 0$ tells us that $T^{(1)}$ requires infinitely more data than $T^{(2)}$.

Given $T \equiv T_n \equiv T_n(\gamma)$ such that $n^{1/2}(T_n(\gamma) - \mu(\gamma))/\sigma(\gamma) \to^d [\gamma]N(0,1)$, with $\mu(0) = 0$, $\mu'(0) \neq 0$, $\sigma^2(0) = \lim_{\gamma \to 0} \sigma^2(\gamma)$, and satisfying the condition analogous to (51), we are usually interested in calculating $(\mu'(0)/\sigma(0))^2$, which can then be used for comparisons with other statistics. We write $\epsilon(T) = (\mu'(0)/\sigma(0))^2$, and refer to this simply as the *asymptotic efficiency* of T.

7.2. Asymptotic efficiency of the integral statistics

We first prove that $I_n^{(1)}(F_0, N)$ verifies the condition analogous to (51). We shall prove this indirectly by appealing to a version of 'Le Cam's third lemma' given by Hall and Mathiason (1990), which implies that for the condition in question to hold it is enough that, for each γ, the asymptotic joint distribution of $I_n^{(1)}(F_\gamma, N)$ (see (49)) and the score statistic under P_γ be bivariate normal.

The *score statistic for γ* based on $(X_1, \ldots, X_n)$, which we denote by $S_n(\gamma)$, is defined by

$$S_n(\gamma) = n^{-1/2} \sum_{i=1}^n v_\gamma(X_i) = n^{1/2} \int v_\gamma(x)\, dF_n(x),$$

and Le Cam's third lemma (Hall and Mathiason, 1990, p. 81) implies that if, for each γ,

$$[I_n^{(1)}(F_\gamma, N), S_n(\gamma)] \to^d [\gamma]N(0, \Sigma(\gamma)), \tag{54}$$

where the covariance matrix $\Sigma(\gamma)$ has off-diagonal elements equal to $\Sigma_{IS}(\gamma)$, then, for $\gamma_n = hn^{-1/2}$,

$$\lim_{n \to \infty} P_{\gamma_n}\left(n^{1/2}I_n^{(1)}(F_0, N)/\sigma_N(0) \leq x\right) = \Phi\left(x - h\Sigma_{IS}(0)/\sigma_N(0)\right),$$

which is the condition analogous to (51) with $\mu'_N(0) = \Sigma_{IS}(0)$.

Thus, if we prove (54), the discussion in the previous subsection gives us $(\Sigma_{IS}(0)/\sigma_N(0))^2$ as the asymptotic efficiency of $I_n^{(1)}(F_{0,N})$.

We can prove (54) by showing that every linear combination $sI_n^{(1)}(F_\gamma, N) + tS_n(\gamma)$ is asymptotically normal. Since, by the proof of Proposition 6.1, $I_n^{(1)}(F_\gamma, N)$ can be approximated in probability by a Gaussian sequence, it remains to show that $S_n(\gamma)$ can also be approximated in probability by a Gaussian sequence defined in terms of the same sequence of Brownian bridges used for approximating $I_n^{(1)}(F_\gamma, N)$. It is sufficient to prove that

$$\Delta_n := \left| S_n(\gamma) + \int B_n(F_\gamma(x))\, dv_\gamma(x) \right| \to_P [\gamma]0,$$

where the B_n are the Komlós–Major–Tusnády sequence of Brownian bridges.

The random integral appearing here is the Riemann integral $\int B_n(F_\gamma(x))v'_\gamma(x)\,dx$, which exists a.s. under the present assumption that $I(\gamma)$ is finite. This follows by the arguments mentioned at the end of Section 4.2. The rest is proved by arguments of the type already used: integration by parts and the formula $\int v_\gamma\,dF_\gamma = 0$ give

$$S_n(\gamma) = -\int n^{1/2}(F_n(x) - F_\gamma(x))\,dv_\gamma(x) = -\int n^{1/2}Y_n(x)\,dv_\gamma(x)$$

(Y_n is the process of Section 4.1), and so we have, for arbitrary $T > 0$,

$$\Delta_n \leqslant \sup_x |Y_n(x) - B_n(F_\gamma(x))| \sup_{0 \leqslant x \leqslant T} |v'_\gamma(x)| + \Delta_n^{(1)} + \Delta_n^{(2)},$$

with $\Delta_n^{(1)} := |\int_T^\infty B_n \circ F_\gamma(x) v'_\gamma(x)\,dx|$, $\Delta_n^{(2)} := |\int_T^\infty Y_n(x)v'_\gamma(x)\,dx|$. The first term converges in probability to zero under P_γ, and

$$\limsup_{n\to\infty} P_\gamma[\Delta_n^{(i)} > \delta] \leqslant \delta^{-2} E_\gamma(\Delta_n^{(i)})^2 = \delta^{-2} E_\gamma(\Delta_1^{(i)})^2,$$

for all $\delta > 0$ and $i = 1, 2$, which can be made arbitrarily small if T is taken large enough.

Now in order to find $\epsilon(\mathrm{I}_n^{(1)}(F_0, N))$ we need $\mu'_N(0) = \Sigma_{\mathrm{IS}}(0)$ and $\sigma_N^2 \equiv \sigma_N^2(0)$. The latter has already been given in (47). Noting that $m(u, 1) = uN'(u)$ and setting

$$\kappa_N = \int_0^\infty y N'(y)\,dF_0(y), \tag{55}$$

we can write (47) as

$$\sigma_N^2 = \int_0^\infty \left[\int_0^x N(x-t)\,dF_0(t) - F_0(x) - \frac{(x-1)}{2}\kappa_N\right]^2 dF_0(x). \tag{56}$$

A formula for $\mu'_N(0)$ is given next, together with an upper bound for $\epsilon(\mathrm{I}_n^{(1)}(F_0, N))$:

PROPOSITION 7.1. *We have*

$$\mu'_N(0) = \Sigma_{\mathrm{IS}}(0)$$
$$= -\int_0^\infty \left(\int_0^x N(x-t)\,dF_0(t) - F_0(x) - \frac{x}{2}\kappa_N\right) v_0(x)\,dF_0(x), \tag{57}$$

and

$$\epsilon(\mathrm{I}_n^{(1)}(F_0, N)) = \mu'_N(0)^2/\sigma_N^2(0) \leqslant I(0). \tag{58}$$

PROOF. From (54) we know that $\Sigma_{IS}(0)$ is the covariance between $I^{(1)}(F_0, N)$ – the variable of Proposition 6.1 with F_0 in place of F – and $S(0) := -\int B \circ F_0 \, dv_0$ – the weak limit of $S_n(\gamma)$ when $\gamma = 0$. Taking $\alpha \equiv \theta = 1$ without loss of generality and using (55) plus integration by parts, we have, from the definition of W given after Theorem 5.1,

$$I^{(1)}(F_0, N) = \int W(x) \, dF_0(x)$$

$$= \iint B \circ F_0(x+y) \, dN(y) \, dF_0(x)$$

$$- \int B \circ F_0 \, dF_0 - \frac{\kappa_N}{2} \int B \circ F_0(y) \, dy.$$

Thus, in order to find $C[I^{(1)}(F_0, N), -\int B \circ F_0 \, dv_0] = -E[I^{(1)}(F_0, N) \int B \circ F_0 \, dv_0]$ we need to calculate and then add the three following terms:

$$-\iiiint E[B \circ F_0(x+y) B \circ F_0(z)] \, dv_0(z) \, dN(y) \, dF_0(x),$$

$$\iint E[B \circ F_0(x) B \circ F_0(y)] \, dv_0(y) \, dF_0(x),$$

$$(\kappa_N/2) \iint E[B \circ F_0(x) B \circ F_0(y)] \, dv_0(y) \, dx.$$

Calculating first the integrals with respect to v_0 by turning them into integrals of v_0 with respect to dF_0, and using $\int v_0 \, dF_0 = 0$, it can be seen that the first term is successively equal to

$$-\int_0^\infty \Bigg[\int_0^\infty \int_{x+y}^\infty F_0(x+y) \overline{F_0}(z) \, dv_0(z) \, dN(y)$$

$$+ \int_0^\infty \int_0^{x+y} \overline{F_0}(x+y) F_0(z) \, dv_0(z) \, dN(y) \Bigg] dF_0(x)$$

$$= -\int_0^\infty \Bigg[\int_0^\infty \bigg(F_0(x+y) \int_{x+y}^\infty v_0(z) \, dF_0(z)$$

$$- \overline{F_0}(x+y) \int_0^{x+y} v_0(z) \, dF_0(z) \bigg) dN(y) \Bigg] dF_0(x)$$

$$= -\int_0^\infty \Bigg[\int_0^\infty \int_{x+y}^\infty v_0(z) \, dF_0(z) \, dN(y) \Bigg] dF_0(x)$$

$$= -\int_0^\infty \Bigg[\int_x^\infty \int_0^{z-x} v_0(z) \, dN(y) \, dF_0(z) \Bigg] dF_0(x)$$

$$= -\int_0^\infty \int_x^\infty N(z-x) v_0(z) \, dF_0(z) \, dF_0(x)$$

$$= -\int_0^\infty \left[\int_0^x N(x-t)\,dF_0(t)\right]v_0(x)\,dF_0(x).$$

Through similar steps one can see that the second and third terms are

$$\int_0^\infty v_0(x)F_0(x)\,dF_0(x) \quad \text{and} \quad (\kappa_N/2)\int_0^\infty xv_0(x)\,dF_0(x),$$

respectively. Addition of the three expressions gives (57).

Finally, we can add $0 = -(\kappa_N/2)\int_0^\infty v_0\,dF_0$ to the right-hand side of (57) to get

$$\mu_N'(0) = -\int_0^\infty \Psi_N(x)v_0(x)\,dF_0(x),$$

where $\Psi_N(x)$ is the term whose square appears as the integrand in (56); thus

$$\epsilon\bigl(I_n^{(1)}(F_0, N)\bigr) = \left(\int_0^\infty \Psi_N(x)v_0(x)\,dF_0(x)\right)^2 \bigg/ \int_0^\infty \Psi_N(x)^2\,dF_0(x)$$

$$\leq \int v_0(x)^2\,dF_0(x) = I(0) \tag{59}$$

by the Cauchy–Schwartz inequality. □

7.3. Efficient statistics. An optimality criterion

We have seen how the concept of asymptotic efficiency provides a way of assessing the power of a given statistic for testing the composite hypothesis H_0: $F(\cdot) = F_0(\cdot; \alpha)$ (some α) against one-sided alternatives of the form H_1: $F(\cdot) = F_\gamma(\cdot; \alpha)$, $\gamma > 0 (< 0)$, and we have calculated and established an upper bound for the asymptotic efficiency of $I_n^{(1)}(F_0, N)$. Now we come to the problem of finding a universal standard with which we can compare $\epsilon(I_n^{(1)}(F_0, N))$ – that is, of defining an optimality criterion for the asymptotic efficiency of test statistics.

There exists a class of *efficient statistics*, which are optimal for our testing problem, in the sense that (i) the power of tests of a given (asymptotic) size $\varepsilon \in (0, 1)$ based on them is asymptotically greater than or equal to the power of any other test of (asymptotic) size ε against sequences of alternatives $H_{1,n}$: $\gamma = \gamma_n := hn^{-1/2}$, $h > 0$ (< 0), and (ii) that their asymptotic efficiency (which is the same for each of these efficient statistics) is greater than or equal to the asymptotic efficiency of any other statistic. Furthermore, the two senses (i) and (ii) in which this 'local optimality' holds are essentially equivalent, because the power functions of tests based on efficient statistics converge to (and are bounded above by) $1 - \Phi(z_{1-\varepsilon} - ch)$ ($h > 0$) or $\Phi(z_\varepsilon - ch)$ ($h < 0$), where c is the square root of their asymptotic relative efficiency.

The class of efficient statistics includes the likelihood ratio statistic, the Wald statistic, the score or Rao statistic, and the 'C-alpha' statistics of Neyman, among others. These are the 'classical' statistics used for testing hypotheses about one parameter (γ, in our case) in the presence of nuisance parameters (α, in our case). Many authors have, in one form or another, suggested their optimality, but it seems that precise definitions and statements and rigorous proofs have never been given until very recently (the work of A. Wald on his test is an exception). The best-known textbooks in statistics and research articles in general always refer to the score statistic (for instance) in an ambiguous way: the terms used in the derivation and justification of the statistic are usually informal or imprecise, and the optimality justifications (if any) are usually based on other sources, which, according to our experience, turn out to contain no rigorous proofs or justifications. Rayner and Best (1989), in their discussion of optimality properties of the classical tests, puzzled over this indeterminacy, and so did Choi, Hall and Schick (1996), who finally put the problem in the right terms. We shall base ourselves on this reference and also on the related articles of Hall and Mathiason (1990) and Basawa (2001), which contain the necessary background.

Complete statements of the optimality properties mentioned above can be found in Theorem 1 and Corollary 1 of Choi et al. (1996, p. 848) (property (i)) and Hall and Mathiason (1990, pp. 89–90) (property (ii)). Property (i) is expressed by saying that tests based on efficient statistics are *asymptotically uniformly most powerful*. The definition of efficient statistic is given on p. 846 of Choi, Hall and Schick (1996). The fact that the classic statistics are efficient and asymptotically equivalent is proved in Section 4 of Hall and Mathiason (1990) (see also Section 7 of Choi, Hall and Schick (1996)). More restrictive optimality properties are proved in Hall and Mathiason (1990) and Basawa (2001).

Efficient statistics also provide *asymptotically uniformly most powerful unbiased tests*: that is, *two-sided* tests (with asymptotic size ε) based on efficient statistics have the greatest limiting power against sequences of alternatives $H_{1,n}$: $\gamma = hn^{-1/2}, h \neq 0$, among all asymptotically unbiased tests (with asymptotic size ε), and again the limiting power is a function of the asymptotic efficiency; see Theorem 2 and Corollary 2 of Choi, Hall and Schick (1996). Our integral statistics will generally belong to the class of asymptotically unbiased tests: for instance, if $N \propto Leb$ and F_γ is the Weibull family (with shape parameter $\gamma + 1$), then $\mu_N(\gamma) \neq 0$ if $\gamma \neq 0$, and consequently (by (54)) the limiting power of the two-sided test based on $I_n^{(1)}(F_0, N)$ against the sequence of alternatives $H_{1,n}$: $\gamma = hn^{-1/2}, h \neq 0$, is greater than or equal to the size of the test, which is what is meant by 'asymptotically unbiased test'.

Given that efficient statistics share optimal power properties, and given the connection between asymptotic power and asymptotic efficiency, it is natural to take the asymptotic efficiency of efficient statistics as the standard by which one measures the relative efficiency of any other statistic. Thus, following Hall and Mathiason (1990, p. 90), we define the *relative efficiency* of a test statistic T as its asymptotic efficiency divided by the asymptotic efficiency of efficient statistics, and denote it by $\epsilon^*(T)$. Property (ii) above states that $\epsilon^*(T) \leqslant 1$.

We say that a given test statistic T is *optimal* for the testing problem H_0: $F(\cdot) = F_0(\cdot; \alpha)$ (α unspecified) versus H_1: $F(\cdot) = F_\gamma(\cdot; \alpha), \gamma > 0, (<0, \neq 0)$ if $\epsilon^*(T) = 1$. If

$T^{(1)}$ and $T^{(2)}$ are two competing statistics for this problem, we say that $T^{(2)}$ is *better* (or *more efficient*) than $T^{(1)}$ if $\epsilon^*(T^{(1)}) \leq \epsilon^*(T^{(2)})$.

Hall and Mathiason (1990) (see the definition of 'effective score' on p. 82, and their Proposition 6 on p. 90) show that the asymptotic efficiency of any efficient statistic is given by

$$B_{11}(0, \alpha) - B_{12}(0, \alpha)^2 B_{22}(0, \alpha)^{-1}, \tag{60}$$

where B_{ij} designate the elements of B. Thus, the relative efficiency $\epsilon^*(T)$ equals $\epsilon(T)$ divided by (60). Some calculations show that in our context

$$B_{11}(0, \alpha) = \int_0^\infty v_0(x)^2 \, dF_0(x) = I(0), \qquad B_{22}(0, \alpha) = \alpha^{-2},$$

$$B_{12}(0, \alpha) = \alpha^{-1} \int_0^\infty x v_0(x) \, dF_0(x),$$

so that (60) equals

$$I(0) - \left(\int_0^\infty x v_0(x) \, dF_0(x) \right)^2, \tag{61}$$

a formula used later in computations of relative efficiency.

Since, by the above mentioned property (ii), the asymptotic efficiency of any statistic can not exceed (60), we have

$$\epsilon\left(I_n^{(1)}(F_0, N)\right) \leq I(0) - \left(\int_0^\infty x v_0(x) \, dF_0(x) \right)^2, \tag{62}$$

an inequality that is sharper than (58).

7.4. *Optimality of the integral statistics. Some numerical examples*

Finding the function or functions N maximizing $\epsilon(I_n^{(1)}(F_0, N))$ for a given family of alternatives seems a difficult problem. Being able to exhibit an optimal N for a given family F_γ could be of intrinsic interest and probably illuminating from a statistical point of view, but on the other hand it might not be of great practical consequence; for if one seeks protection against a specific alternative to the exponential model (which is seldom the case in applications), then it is simply more sensible to resort to an efficient statistic than to $I_n(F_0, N)$ with the optimal N.

A related question is the following: given N, is there a family of alternatives F_γ with which the equality in (62) is achieved? In other words: given N, is there a family of alternatives F_γ for which $\epsilon^*(I_n^{(1)}(F_0, N)) = 1$? A positive answer to this has some practical value because it may give us an idea of the kind of alternatives against which a choice of N will be most efficient. For instance, if we were able to exhibit an example of a heavy-tailed family of alternatives for which N is the optimal choice, then we should be able to trust N as a good choice against heavy-tailed 'close' alternatives in general. Unfortunately, this also seems to be a difficult problem.

A restricted form of this question, however, is quite easy to answer (because the Cauchy–Schwartz inequality is available): given N, is there a family of alternatives F_γ with which the equality in (58) is achieved? In other words: given N, is there a family of alternatives F_γ satisfying $\int x v_0(x) \, dF_0(x) = 0$ for which $\epsilon^*(I_n^{(1)}(F_0, N)) = 1$?

PROPOSITION 7.2. *Let F_γ be given. There exists a function N with which equality in (58) is achieved if and only if*

(i) $v_0(0) \neq 0, \quad v_0(0) \neq \pm\infty,$

(ii) $\displaystyle\int_0^\infty x v_0(x) \, dF_0(x) = 0, \quad \text{and}$

(iii) $\displaystyle\frac{v_0'(x) + v_0''(x)}{v_0(0) + v_0'(0)} \leq -\frac{v_0(0)}{v_0(0) + v_0'(0)}, \quad x \geq 0,$

in which case N is given by

$$N(x) = -\frac{v_0(0)}{v_0(0) + v_0'(0)} \left[x + \frac{v_0(x) + v_0'(x)}{v_0(0)} \right] + 1, \quad x \geq 0. \tag{63}$$

PROOF. We know that the equality in (59) is achieved if and only if $\Psi_M \propto v_0$. Thus, assume there exists $C_N \neq 0$ such that

$$\frac{(x-1)}{2} \kappa_N - \int_0^x N(x-t) \, dF_0(t) + F_0(x) = C_N v_0(x), \quad x \geq 0. \tag{64}$$

Putting $x = 0$ in (64) we see that for a solution N to exist it is necessary that $v_0(0) = -\kappa_N/(2C_N)$, which implies (i).

Writing the integral on the left as $e^{-x} \int_0^x N(u) e^u \, du$, we see that (64) is equivalent to

$$\frac{(x-1)}{2} \kappa_N e^x - \int_0^x N(t) e^t \, dt + e^x - 1 = C_N v_0(x) e^x, \quad x \geq 0. \tag{65}$$

Since N is continuously differentiable on $(0, \infty)$, the left-hand side of (65) is twice continuously differentiable. Differentiating both sides of (65) gives

$$N(x) = \frac{x}{2} \kappa_N + 1 - C_N (v_0(x) + v_0'(x)), \quad x \geq 0,$$

and using $v_0(0) = -\kappa_N/(2C_N)$ we conclude that any solution N must be of the form

$$N(x) = \frac{\kappa_N}{2} \left[x + \frac{v_0(x)}{v_0(0)} + \frac{v_0'(x)}{v_0(0)} \right] + 1, \quad x \geq 0. \tag{66}$$

Recalling $\int v_0 \, dF_0 = 0$, one can check that (66) satisfies the normalizing condition $\int e^{-x} \, dN(x) = 1$. N must also satisfy $\kappa_N = \int y N'(y) \, dF_0(y)$, and integration by parts shows that this condition is equivalent to $\kappa_N = -1 + \int y N(y) \, dF_0(y)$, or

$$\kappa_N = \kappa_N + \frac{\kappa_N}{v_0(0)} \int_0^\infty x v_0(x) e^{-x} \, dx,$$

implying (ii). Alternatively, (ii) follows directly from the comparison of (58) and (62).

Since we require $N(0) = 0$, v_0 must satisfy $(\kappa_N/2)[1 + v_0'(0)/v_0(0)] + 1 = 0$; imposing this condition on (66) gives (63).

Finally, for N to be non-decreasing we must have $N' \geq 0$, whence (iii). □

The formula for N in (63) suggests that many optimal integral statistics will be multiples of the total time on test statistic (i.e., $I_n^{(1)}(F_0, N)$ with $N \propto \text{Leb}$) plus additional terms that are determined by the particular families of alternatives. The total time on test statistic has, in this sense, a central role within the class of integral statistics, and can be regarded as a first order approximation to them.

The restriction to families F_γ satisfying $\int x v_0(x) \, dF_0(x) = 0$ has some importance because it excludes interesting alternatives, such as the Weibull, generalized Pareto and gamma models; for these, we know that there is no N such that $\epsilon(I_n^{(1)}(F_0, N)) = I(0)$, but the question of whether $\epsilon^*(I_n^{(1)}(F_0, N)) = 1$ for some N has not been answered.

In any case, we shall at least be able to exhibit a few examples of alternatives against which the integral statistics are optimal. Thus, to take up the idea expressed just before Proposition 7.2, we shall turn the problem round and determine v_0 in terms of N, and then, for fixed N, we can try to construct an alternative F_γ from v_0; if we succeed to do so, we end up with a pair (F_γ, N) for which the equality is achieved in (58), so that $I_n^{(1)}(F_0, N)$ is optimal for testing the exponential distribution against alternatives F_γ.

Solving (65) for v_0, we arrive at

$$v_0(x) = \frac{2v_0(0)}{\kappa_N} \left[\int_0^x N(x-t) \, dF_0(t) - \frac{\kappa_N(x-1)}{2} - F_0(x) \right], \quad x \geq 0. \tag{67}$$

This can be seen to satisfy conditions (i)–(iii) of Proposition 7.2 as well as $\int v_0(x) \, dF_0(x) = 0$. Since $v_0(0)$ is an arbitrary constant, $v_0(x)$ is in fact a constant ($\neq 0$ and independent of N) times the term in brackets on the right-hand side of (67).

In order to construct an F_γ corresponding to the v_0 of (67) we restrict ourselves to alternatives of the form

$$F_\gamma(x) = 1 - \exp\{-x + \gamma R(x)\}, \quad x \geq 0, \tag{68}$$

where R is a function to be determined. (Note that each v_0 can in principle arise from two different types of F_γ.)

The v_0 corresponding to (68) is $R - R'$, and so (noting that the integration constant in the equation $R - R' = v_0$ must be zero for F_γ to be continuous at the origin) we conclude that

$$R(x) \propto -e^x \int_0^x e^{-u} v_0(u)\, du, \quad x \geq 0.$$

If we now take v_0 as in (67) and perform some algebra, letting the proportionality constant be absorbed into the parameter γ in (68), we find that the general form we seek for R is

$$R(x) = -\frac{\kappa_N}{2}x - 1 + \frac{e^{-x} + e^x}{2}$$

$$+ \frac{1}{2}\int_0^x N(x-t)(e^{-t} - e^t)\, dt, \quad x \geq 0. \tag{69}$$

Whether (68) with this R is really a probability distribution function for γ in a (right- or left-) neighbourhood of the origin depends on the particular N. We examine only the simplest examples, for which (69) can be explicitly calculated.

EXAMPLE 7.1 (*Optimality of $N \propto Leb$*). Taking $N \propto Leb$ in (69), the family F_γ of (68) becomes

$$F_\gamma(x) = 1 - \exp\{-x + \gamma[(x/2) + e^{-x} - 1]\}, \quad x \geq 0, \tag{70}$$

which for $\gamma \in [-2, 2)$ is indeed a probability distribution function. Thus $N \propto Leb$ is optimal for testing the exponential distribution against 'local' alternatives from (70).

Working in the opposite direction, just for illustration, one can check that

$$v_0(x) = (3/2) - (x/2) - 2e^{-x}, \quad x \geq 0,$$
$$v_0(0) = -1/2, \quad v'(0) = 3/2,$$

(i)–(iii) of Proposition 7.2 are satisfied, and (63) gives us $N \propto Leb$.

Two examples of the density f_γ corresponding to (70) are represented in Figure 1, together with the standard exponential density. For $\gamma > 0$ ($\gamma < 0$), (70) has a tail that is heavier (lighter) than that of the exponential model. Overall, f_γ bears some resemblance with the Weibull densities (though there are important differences in the lower tail).

EXAMPLE 7.2 (*Optimality of $N \propto Ga(r)$*). Taking $N \propto Ga(r)$ in (69) gives

$$R(x) = -\frac{r}{4}x - 1 + \frac{e^{-x} + e^x}{2} + N(x) - \frac{N(2x)}{2^{r+1}}e^x - \frac{1}{2}\frac{e^{-x}(2x)^r}{\Gamma(r+1)}, \quad x \geq 0.$$

We consider only two examples. With $r = 1$, the F_γ corresponding to this R is

$$F_\gamma(x) = 1 - \exp\{-x + \gamma[1 - e^{-x} - xe^{-x} - (x/4)]\}, \quad x \geq 0, \tag{71}$$

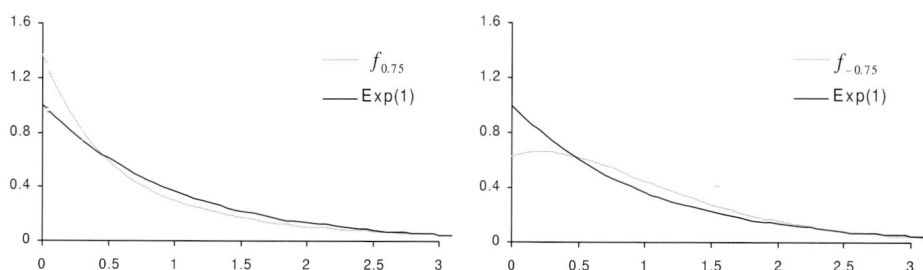

Fig. 1. Graphical comparison between the standard exponential density (Exp(1)) and the density f_γ corresponding to the model (70) with $\gamma = -0.75, 0.75$.

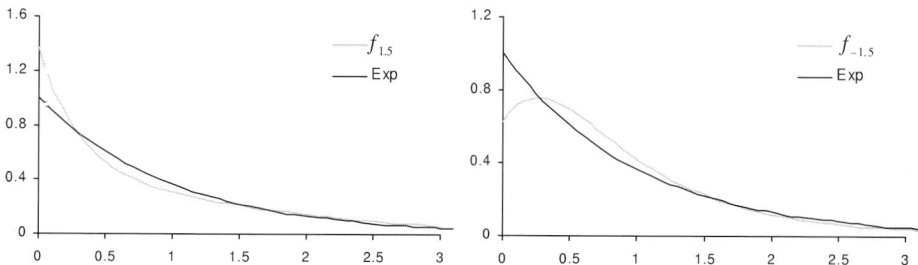

Fig. 2. Graphical comparison between the standard exponential density (Exp) and the density f_γ corresponding to the model (71) with $\gamma = -1.5, 1.5$.

which is a probability distribution function for $\gamma \in (-4, 4e/(4-e)]$.

Figure 2 shows graphical representations of the density f_γ of (71) for $\gamma = -1.5, 1.5$. For $\gamma > 0$ ($\gamma < 0$), the tail of (71) is lighter (heavier) than the exponential tail. There exists a considerable qualitative difference between these two densities and those of the preceding example: the first points where their curves meet the curve of the exponential density are roughly 0.25 and 0.5, respectively. In this respect, (70) is much closer to a Weibull model than (71). This is in agreement with our power study, which has shown that $I_n^{(1)}(F_0, N)$ has a poor performance in testing the exponential against Weibull alternatives with $N \propto Ga(1)$, and a good performance with $N \propto Leb$. On the other hand, it also suggests that against alternatives resembling (71) $I_n^{(1)}(F_0, N)$ with $N \propto Leb$ may represent a poor choice.

With $r = 3$ we get (68) with R given by

$$R(x) = (x^2 + (x/4) - 1) + e^{-x}((1/2) - 8x - 4x^2 - (4x^3/3)) + (e^x/2), \quad x \geqslant 0. \tag{72}$$

With this R, (68) is a distribution function only for values of γ in a left-neighbourhood of zero that is approximately $[-0.129, 0]$. Figure 3 compares the corresponding densities $f_{-0.02}$ and $f_{-0.05}$ with the standard exponential density. The overall shape of

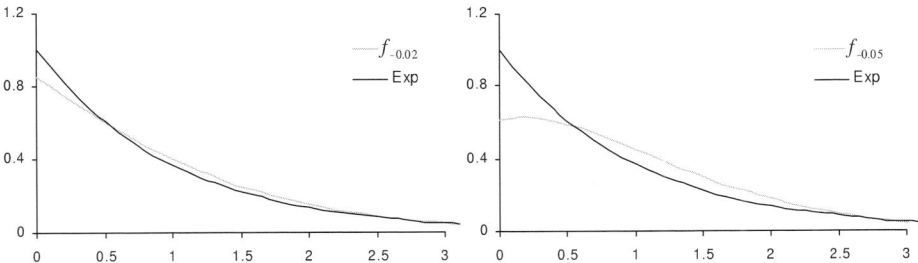

Fig. 3. Graphical comparison between the standard exponential density (Exp) and the density f_γ corresponding to the model (68) with R as given by (72) with $\gamma = -0.02, -0.05$.

$f_{-0.05}$ is quite similar to the density of Figure 1, but its tail is much lighter. In fact, it is interesting to note that the model corresponding to (72) has an extremely light tail for all permissible values of γ. To illustrate this point, we observe that $f_{-0.02}(11) \cong 8.04^{-264}$, while $f_0(11) \cong 5.67^{-5}$.

This observation is in agreement with the particularly good performance of $I_n^{(1)}(F_0, N)$ with $N \propto Ga(3)$ against generalized Pareto alternatives with bounded support.

The *Makeham* and *failure rate* distributions are defined by (68) with $R(x) = x + e^{-x} - 1$ and $R(x) = x^2/2$, respectively. Both of these fail to satisfy condition (ii) of Proposition 7.2; thus, not all models of the form (68) allow a choice of N yielding $\epsilon(I_n^{(1)}(F_0, N)) = I(0)$. Note that the expression of the Makeham distribution resembles (70); the two models, however, can be considerably different.

The preceding examples show that, given N, one can in principle find F_γ against which $I_n^{(1)}(F_0, N)$ represents an optimal test statistic, the only obstacle being the explicit calculation of (69). This is interesting and may be of some practical value (e.g., in targeting particular types of deviations from exponentiality), but it is important to provide a more global assessment of the integral statistics.

With this in mind, we have calculated the asymptotic efficiency and the relative efficiency of $I_n^{(1)}(F_0, N)$ with $N \propto Leb$, $N \propto Ga(3)$ and $N \propto Ga(1)$, for testing exponentiality against several alternatives, including the linear failure rate and Makeham distributions just introduced, and also the models obtained in Examples 8.1 and 8.2. The results appear in Table 4, and consist of asymptotic efficiency values and percentage of relative efficiency (asymptotic efficiency divided by the asymptotic efficiency of the efficient statistics). We should note that some of the results of Table 4 had already been presented by Lewis (1965), Hollander and Proschan (1975) and Klefsjö (1983).

[We have used exact, numerical and simulation methods to obtain the figures in the table. The asymptotic efficiency values in the last column were calculated as in (61), and the others from the formulae given in Proposition 7.1. Some of the integrals involved can be expressed in terms of gamma, digamma and trigamma functions, and for these we have used the tables of Abramowitz and Stegun (1965). Others are more

Table 4

Asymptotic efficiency and relative efficiency of the integral statistics $I_n^{(1)}(F_0, N)$ with $N \propto Leb$, $N \propto Ga(3)$ and $N \propto Ga(1)$, for testing exponentiality against various families of alternatives. The relative efficiency (%, in bold) equals the asymptotic efficiency (below the percentages) divided by the (optimal) asymptotic efficiency of the efficient statistics, which is given in the last column

Alternative	$N \propto Leb$	$N \propto Ga(3)$	$N \propto Ga(1)$	Efficient statistics
Weibull	**88%**	**75%**	**31%**	**100%**
	1.440	1.239	0.508	1.644
Generalized Pareto	**75%**	**55%**	**0%**	**100%**
	0.748	0.548	0	1
Gamma	**70%**	**60%**	**43%**	**100%**
	0.449	0.388	0.279	0.645
Makeham	**100%**	**92%**	**20%**	**100%**
	0.0833	0.0767	0.0166	0.0833
Linear failure rate	**75%**	**55%**	**0%**	**100%**
	0.748	0.548	0	1
Model (70)	**100%**	**93%**	**17%**	**100%**
	0.0833	0.0771	0.0139	0.0833
Model (68) with R as in (72)	**93%**	**100%**	**24%**	**100%**
	0.237	0.256	0.0606	0.256
Model (71)	**15%**	**17%**	**100%**	**100%**
	0.0023	0.0027	0.00157	0.0157

complicated, and we have used simulation for calculating them. This was done by estimating appropriate moments of exponential variables. The required precision varied with the magnitude of quantities being estimated, but the figures should be correct up to the decimals shown. Note that the results for generalized Pareto and linear failure rate are identical; this is because the v_0 function is the same for both models.]

Table 4 tells us that the total time on test statistic $- I_n^{(1)}(F_0, N)$ with $N \propto Leb$ – has a good performance for most of the alternatives considered; its efficiency is never less than 70% of the optimal efficiency against all alternatives except for model (71). $I_n^{(1)}(F_0, N)$ with $N \propto Ga(3)$ is also quite good, but generally inferior. However, as one might expect, these two statistics perform very poorly against alternatives of the type (71). That $N \propto Ga(1)$ represents a poor choice for all alternatives except for model (71) (against which it is optimal) was expected from the results of our power study.

Let us make some power comparisons in testing against model (71) with $\gamma = 1.5$. We take a sample size of $n = 100$ and compare power estimates (obtained by simulation) of both linear and quadratic integral statistics with power estimates of the Anderson–Darling and Cramér–von Mises statistics. The power of $I_n^{(1)}(F_0, N)$ with $N \propto Ga(1)$ is ≈ 0.45, while $I_n^{(2)}(F_0, N)$ with $N \propto Ga(1)$ has power of ≈ 0.35. A^2 and ω^2 have power estimates of 0.34 and 0.30, respectively. Thus, the integral quadratic and the standard quadratic statistics perform similarly, but the linear statistic (with the appropriate N) is considerably better.

The linear statistics with $N \propto Leb$ and $N \propto Ga(3)$ have an approximate power of 0.17 and 0.21, respectively, which is quite bad. The quadratic statistics with $N \propto Leb$ and $N \propto Ga(3)$ have power of ≈ 0.24 and ≈ 0.28, respectively, which represents an improvement but does not justify their use in place of the standard quadratic statistics.

This example illustrates our earlier statement about the linear integral statistics being generally more useful than the quadratic integral statistics: it is the former class rather than the latter that provides significant improvements in power whenever such improvements are possible, while in other cases the quadratic integral statistics are at most comparable to the classical quadratic statistics.

Finally, we reiterate the recommendations made in Section 6.3: as a general rule for testing exponentiality, we suggest using the linear integral statistics with $N \propto Leb$, $N \propto Ga(1)$, $N \propto Ga(3)$ and $N \propto Log$ together with the Anderson–Darling and Cramér–von Mises statistics. We deliberately exclude $N \propto Ga(2)$ because at least one of $N \propto Ga(1)$ and $N \propto Ga(3)$ should be able to replace it. Although other examples of N can in principle be used, these four already represent some variety.

Acknowledgements

This work is based on part of my PhD thesis, written at the University of Sheffield, UK; I am very grateful to D.N. Shanbhag, my supervisor, for all his comments and suggestions, and to Fundação para a Ciência e Tecnologia (the Portuguese Science Foundation) for funding my research through the Sub-programa Ciência e Tecnologia do 2º Quadro Comunitário de Apoio.

References

Abramowitz, M. and I. A. Stegun (1965). *Handbook of Mathematical Functions*. Dover.

Ahmad, I. A. (1992). A new test for mean residual life times. *Biometrika* **79**, 416–419.

Barlow, R. (1968). Likelihood ratio tests for restricted families of probability distributions. *Ann. Math. Statist.* **39**, 547–560.

Barlow, R. and F. Proschan (1966). Inequalities for linear combinations of order statistics from restricted families. *Ann. Math. Statist.* **37**, 1574–1592.

Basawa, I. (2001). Inference in stochastic processes. In *Handbook of Statistics, Vol. 19: Stochastic Processes: Theory and Methods*, pp. 55–77 (Eds. C. R. Rao and D. N. Shanbhag). Elsevier.

Bergman, B. and B. Klefsjö (1988). Total time on test transform. In *Encyclopedia of Statistical Sciences*, Vol. 9 (Eds. S. Kotz and N. Jonhson). Wiley, New York.

Bickel, P. J. and K. A. Docksum (1969). Tests for monotone failure rate based on normalized spacings. *Ann. Math. Statist.* **40**, 1216–1235.

Billingsley, P. (1995). *Probability and Measure*, 3rd ed. Wiley, New York.

Choi, S., W. J. Hall and A. Schick (1996). Asymptotically uniformly most powerful tests in parametric and semiparametric models. *Ann. Statist.* **24**(2), 841–861.

Cox, D. R. and P. A. W. Lewis (1966). *The Statistical Analysis of Series of Events*. Methuen, London.

Csörgő, M., S. Csörgő and L. Horváth (1986). An asymptotic theory for reliability and concentration processes. In *Lecture Notes in Statist.*, Vol. 33. Springer-Verlag.

Csörgő, M. and L. Horváth (1993). *Weighted Approximations in Probability and Statistics.* Wiley, New York.
Csörgő, M. and P. Révész (1981). *Strong Approximations in Probability and Statistics.* Academic Press, New York.
Csörgő, M., V. Seshadri and M. Yalovsky (1975). Applications of characterizations in the area of goodness-of-fit. In *Statistical Distributions in Scientific Work*, Vol. 2, pp. 79–90 (Eds. G. P. Patil, S. Kotz and J. K. Ord). D. Reidel.
D'Agostino, R. and M. Stephens (1986). *Goodness of Fit Techniques.* Marcel Dekker.
Deheuvels, P. (1982). Invariance of Wiener processes and of Brownian bridges by integral transforms and applications. *Stochastic Process. Appl.* **13**, 311–318.
Dufour, R., U. Maag and C. van. Eeden (1984). Correcting a proof of a characterization of the exponential distribution. *J. Roy. Statist. Soc.* **46**(2), 238–241.
Durbin, J. (1961). Some methods of constructing exact tests. *Biometrika* **48**, 41–55.
Epstein, B. (1960). Tests for the validity of the assumption that the underlying distribution of life is exponential, parts 1 and 2. *Technometrics* **2**, 83–101, 167–183.
Epstein, B. and M. Sobel (1953). Life testing. *J. Amer. Statist. Assoc.* **48**, 486–502.
Ferguson, T. S. (1996). *A Course in Large Sample Theory.* Chapman and Hall.
Fisz, M. (1958). Characterization of some probability distributions. *Skand. Aktuarietidskr.* **41**, 65–70.
Hall, W. J. and D. J. Mathiason (1990). On large-sample estimation and testing in parametric models. *Internat. Statist. Rev.* **58**, 77–97.
Hoeffding, W. (1973). On the centering of a simple linear rank statistic. *Ann. Statist.* **1**, 54–66.
Hogg, R. (1964). Applications of the characterizations of distributions to tests of fit, randomness and independence. *Ann. Math. Statist. (Abstract)* **35**, 1837.
Hollander, M. and F. Proschan (1975). Tests for the mean residual life. *Biometrika* **62**(3), 585–593.
Klefsjö, B. (1983). Some tests against ageing based on the total time on test transform. *Comm. Statist. Theory Methods* **12**(8), 907–927.
Komlós, J., P. Major and G. Tusnády (1975). An approximation of partial sums of independent and identically distributed r.v.'s and the sample DF. I. *Wahrsch. Verw. Gebiete* **32**, 111–131.
Kumazawa, Y. (1993). Comments on 'A new test for mean residual life times' by Ahmad. *Biometrika* **80**(2), 473–474.
Lewis, P. (1965). Some results on tests for Poisson processes. *Biometrika* **52**, 67–77.
Marsaglia, G. and A. Tubilla (1975). A note on the "Lack of memory" property of the exponential distribution. *Ann. Probab.* **3**(2), 353–354.
O'Reilly, F. and M. Stephens (1982). Characterizations and goodness-of-fit tests. *J. Roy. Statist. Soc.* **44**, 353–360.
Quesenberry, C. (1986). Transformation methods in goodness-of-fit. In *Goodness of Fit Techniques*, pp. 235–277 (Eds. R. D'Agostino and M. Stephens). Marcel Dekker.
Rao, C. and D. Shanbhag (1994). *Choquet-Deny Type Functional Equations with Applications to Stochastic Models.* Wiley, Chichester.
Rao, C. and D. Shanbhag (1997). Extensions of a characterization of an exponential distribution based on a censored ordered sample. In *Advances in the Theory and Practice of Statistics: A Volume in Honor of S. Kotz*, pp. 431–440 (Eds. N. Johnson and N. Balakrisnhan). Wiley.
Reiss, R. (1989). *Approximate Distributions of Order Statistics.* Springer-Verlag, New York.
Serfling, R. (1980). *Approximation Theorems of Mathematical Statistics.* Wiley, New York.
Seshadri, V., M. Csörgő and M. Stephens (1969). Tests for the exponential distribution using Kolmogorov-type statistics. *J. Roy. Statist. Soc.* **31**, 499–509.
Shanbhag, D. (1977). An extension of the Rao–Rubin characterization of the Poisson distribution. *J. Appl. Probab.* **14**, 640–646.
Shorack, G. R. and J. A. Wellner (1986). *Empirical Processes with Applications in Statistics.* Wiley, New York.
Spurrier, J. (1984). An overview of tests for exponentiality. *Comm. Statist. Theory Methods* **13**(13), 1635–1654.
Stephens, M. (1974). EDF statistics for goodness of fit and some comparisons. *J. Amer. Statist. Assoc.* **69**, 730–737.

Stephens, M. (1986a). Tests based on EDF statistics. In *Goodness of Fit Techniques*, pp. 95–193 (Eds. R. D'Agostino and M. Stephens). Marcel Dekker.

Stephens, M. (1986b). Tests for the exponential distribution. In *Goodness of Fit Techniques*, pp. 421–459 (Eds. R. D'Agostino and M. Stephens). Marcel Dekker.

Sukhatme, P. (1937). Tests of significance for samples of the χ^2-population with two degrees of freedom. *Ann. of Eugenics* **8**, 52–56.

D. N. Shanbhag and C. R. Rao, eds., *Handbook of Statistics, Vol. 21*
© 2003 Elsevier Science B.V. All rights reserved

Patterns in Sequences of Random Events

J. Gani

This chapter records three encounters with patterns in sequences of random events; it is not intended to be a review of the field, but concentrates rather on those areas which arose almost accidentally in the course of my research.

My first encounter with the topic was through the use of tables of random numbers, while my second resulted from a study of the theory of runs. My third and most recent encounter followed some editorial work on a short note of Siegel (1997). In it, he produced a neat proof that in a sequence of length n whose elements may be 0 or 1, the number of sequences avoiding the pattern 11 was the Fibonacci number $F(n+2)$, a result derivable from the work of Guibas and Odlyzko (1981), and also known to Rényi (1984).

Intrigued by this result, I read more widely in the field. I was eventually led to the recognition that there existed a Markov chain formulation for the case of general patterns; these arise when consecutive events from an alphabet of k letters themselves form a Markov chain.

1. Early encounters: random numbers and the theory of runs

My earliest encounter with patterns in sequences of random events occurred in 1954, when I first began to use Kendall and Babington Smiths's tables of *Random Sampling Numbers* (1939b), described in their Royal Statistical Society papers (1938, 1939a). These tables, consisting of 100,000 digits grouped in twos and fours, and in 100 separate thousands had been generated by a very rudimentary machine in which a circular disc divided into 10 equal sections numbered $0, 1, \ldots, 9$, was rotated rapidly and stopped randomly at some number which was then recorded.

Shortly after the publication of Kendall and Babington Smith's first paper (1938), the tables of Fisher and Yates (1938) appeared. The random numbers in these were obtained from the 15–19th digits of A. J. Thompson's tables of logarithms in *Logarithmica Britannica*, adjusted to correct for a surplus of sixes.

The use of random numbers in electronic computers resulted in the need to generate these in a completely deterministic manner. Methods for doing this are described briefly in Abramowitz and Stegun (1964, pp. 949–950), and at somewhat greater length in

Ripley (1987, Chapter 2), among other works. Such numbers, referred to as pseudo-random numbers, are now readily generated by computers. Tabulated random numbers, such as those compiled by the Rand Corporation (1955) have usually been subjected to a series of statistical tests to ensure their quality.

To satisfy the criterion of randomness, it is not enough for the digits $0, 1, \ldots, 9$, to occur singly with equal frequencies. Kendall and Babington Smith (1938) outlined four basic tests that need to be satisfied to ensure adequate randomness, so that certain patterns arise in the sequences of random numbers with prescribed frequencies. Ripley (1987, pp. 43–45) has also listed several tests for the independence and uniformity of the output of a pseudo-random number generator.

My second encounter with patterns in sequences of random events also took place in 1954, when I was studying the first edition of Feller's Volume 1 (1950). In it I came across the theory of runs of independent events, such as Bernoulli trials $X_i, i = 1, \ldots, n$, with $P\{X_i = 1\} = p$, $P\{X_i = 0\} = q$, $0 < p < 1$, $p + q = 1$, resulting in sequences containing runs of consecutive 1's or 0's.

The theory of such runs dates back to the end of the 19th century. In his classical paper, Mood (1940) traces the history of early results to the works of Bruns (1906), von Bortkiewicz (1917) and von Mises (1921). Ising (1925), in his note on the theory of ferromagnetism, first derived the number of ways of obtaining a given total number of runs, without regard to their length, in a sequence of length n consisting of n_1 +'s and n_2 −'s (or 1's and 0's), where $n_1 + n_2 = n$. Assuming that every possible sequence arrangement was equiprobable, Wald and Wolfowitz (1940) later proved that the distribution of this number was asymptotically normal.

Mood (1940) treated the problem of runs purely as a combinatorial one. He not only derived the distribution and moments of runs in a sequence of independent events having two possible outcomes, but extended his results to the case where there were $k > 2$ possible outcomes. He then discussed the asymptotic distributions of the total number of runs of a particular outcome in both the binomial case $k = 2$, and the multinomial case $k > 2$.

Feller's (1968) entirely different approach is based on the theory of recurrent events. In the third edition of his book in 1968, he defines a success run of Bernoulli trials of length r very specifically, to ensure the applicability of the theory. A success run of length r is a non-overlapping uninterrupted succession of exactly r 1's. Using this definition, the sequence [111]10[111][111]0, of length $n = 12$, contains three non-overlapping runs of length $r = 3$, as shown in the brackets. If an overlapping definition of runs were used, then the sequence would contain 6 success runs. It is clearly of importance to define the types of runs very precisely; for the various definitions of runs, the reader may refer to Schwager (1983).

It may be instructive to quote a simple result from Feller (1968, pp. 322–326) on success runs, in order to illustrate the theory of recurrent events in this area. Consider a sequence of independent Bernoulli trials X_i, $i = 1, \ldots, n$, with $P\{X_i = 1\} = p$, $P\{X_i = 0\} = q$, $0 < p < 1$, $p + q = 1$, and let E be the event of a success run of length r in the sequence. For n at least equal to r, we write f_n for the probability that E occurs for the first time at the nth trial, and u_n for the probability that E occurs at the nth trial, not necessarily for the first time. Now the probability that the last r trials $X_{n-r+1}, \ldots, X_n$

are successes is p^r, and we can see that this can be achieved if the event E occurs at one of the trials $n - r + 1, \ldots, n$, and is then followed by $r - 1, r - 2, \ldots, 0$ subsequent successes, with the respective probabilities $u_{n-r+1} p^{r-1}, \ldots, u_n$. Since these possibilities are mutually exclusive, it follows that

$$p^r = u_n + u_{n-1} p + u_{n-2} p^2 + \cdots + u_{n-r+1} p^{r-1}. \tag{1.1}$$

Clearly $u_1 = u_2 = \cdots = u_{r-1} = 0$, and u_0 is set equal to 1. Multiplying (1.1) by s^n, and summing over $n = r, r+1, r+2, \ldots$, we finally obtain the generating function (gf)

$$U(s) = \sum_{n=0}^{\infty} u_n s^n = \frac{1 - s + q p^r s^{r+1}}{(1-s)(1 - p^r s^r)}. \tag{1.2}$$

From the theory of recurrent events, the gf of the first recurrence times of the event E at the nth trial is given by

$$F(s) = 1 - U(s)^{-1} = \frac{p^r s^r (1 - ps)}{1 - s(1 - q p^r s^r)}. \tag{1.3}$$

The mean and variance of the recurrence times of runs of length r are derived from (1.3) as

$$\mu = \frac{1 - p^r}{q p^r}, \quad \sigma^2 = \frac{1}{(q p^r)^2} - \frac{2r+1}{q p^r} - \frac{p}{q^2} \tag{1.4}$$

these being the indices commonly used to characterize the distribution.

Feller (1968, p. 341) lists as Exercise 25 the proof of the Poisson distribution of long runs, originally derived by von Mises (1921). This was later generalized by Rajarshi (1974) for success runs in a two state Markov chain. Feller (1968) also considered more general cases such as that of the event E where $E = \{E_1 \text{ or } E\}$, with $E_1 = \{\text{a success run of length } r\}$, while $E_2 = \{\text{a failure run of length } k\}$ in a sequence of n independent Bernoulli trials. Denoting by $U_1(s)$, $U_2(s)$ the gfs of the events E_1 and E_2 respectively, and by $U(s)$ the gf of E, with $u_0 = 1$, Feller derives the gf of the first recurrence times as

$$F(s) = \frac{(1 - ps) p^r s^r (1 - q^k s^k) + (1 - qs) q^k s^k (1 - p^r s^r)}{1 - s + qs(ps)^r + ps(qs)^k - p^r q^k s^{r+k}}, \tag{1.5}$$

with the mean recurrence time

$$\mu = \frac{(1 - p^r)(1 - q^r)}{q p^r + p q^k - p^r q^k}. \tag{1.6}$$

A somewhat more complicated pattern of 0's and 1's in the Bernoulli trials was proposed by Leslie (1967). He was motivated by the firing of cone cells in the retina of the eye, where an input of photons results in output signals along the optic nerve. A cone will fire in response to a sequence of k photons (1's) separated by not more than a time g (0's), this event being denoted as $E(k, g)$. For example, in the following sequence of length $n = 44$,

$$00[10101][1011]0[111]010100[10101]0[1101][1101]100[1011]00$$

$E(3, 1)$ occurs 7 times, as indicated by the brackets.

Using a direct method, as well as an extension of Feller's principles, Leslie (1967) obtained the gf $F(s)$ of first recurrence times of the event $E(k, g)$ as

$$F(s) = \frac{psQ(s)^{k-1}}{1 - qs - pq^{g+1}s^{g+2}\left(\frac{1-Q(s)^{k-1}}{1-Q(s)}\right)}, \tag{1.7}$$

where $Q(s) = ps(1 - (qs)^{g+1})/(1 - qs)$. The mean and variance of the distribution of first recurrence times are given by

$$\mu = \frac{1 - (1 - q^{g+1})^k}{pq^{g+1}(1 - q^{g+1})^{k-1}},$$

$$\sigma^2 = \frac{\mu}{q^{g+1}}\left[\frac{2}{W(1)} - (2g+1)q^{g+1}\right] - \mu^2 \tag{1.8}$$

$$- \frac{2}{q^{g+1}}\left[\frac{W'(1)}{W(1)} + \frac{q}{p^2}(1 - q^g - gpq^g)\right],$$

where $W(s) = psQ(s)^{k-1}$. Tables of probabilities, as well as means and variances of the recurrence times are provided for various values of the probability p of a 1 in a Bernoulli trial.

Some generalizations of the theory of recurrent events to regenerative phenomena and fluctuations, as well as the limiting behaviour of the number of runs in a sequence of length n of such phenomena were considered by Heyde (1967) and Imhof (1974). We now consider some work on sequences which avoid certain repetitions, using matrix methods.

2. Sequences of events with repetitions

My third and most recent encounter with patterns in sequences of random events arose quite accidentally in the course of my editorial duties in 1996. Siegel (1997) submitted a note to *The Mathematical Scientist* containing a neat proof that the number of sequences consisting of either 0's or 1's which avoided the repetition 11 was the Fibonacci number $F(n + 2)$. This result, which was proved by Rényi (1984), is recorded in Vajda's (1989) book; it is also derivable from the general formulae of Guibas and Odlyzko (1981), as we shall see later. It does not, however, seem to be widely known by statisticians, so it may be worthwhile sketching Siegel's proof.

Let us write $t_n(2)$ for the number of sequences of length n whose elements can be either 0 and 1, which avoid the repetition 11. Let $a_n(2)$ be the number of sequences among these which avoid 11, but end with a single 1. Then it is readily seen that

$$
\begin{aligned}
t_n(2) &= 2t_{n-1}(2) - a_{n-1}(2), \\
a_n(2) &= t_{n-1}(2) - a_{n-1}(2),
\end{aligned} \quad n = 1, 2, \ldots, \tag{2.1}
$$

with $t_0(2) = 1$, $a_0(2) = 0$. The matrix in this set of recursive equations is

$$
\begin{bmatrix} 2 & -1 \\ 1 & -1 \end{bmatrix} \tag{2.2}
$$

which has the eigenvalues

$$
\lambda_1(2) = (1 + \sqrt{5})/2, \qquad \lambda_2(2) = (1 - \sqrt{5})/2. \tag{2.3}
$$

From these, one obtains the value of $t_n(2)$ as

$$
t_n(2) = \frac{1}{2^{n+1}} \left(1 + \frac{3}{\sqrt{5}} \right)(1 + \sqrt{5})^n + \frac{1}{2^{n+1}} \left(1 - \frac{3}{\sqrt{5}} \right)(1 - \sqrt{5})^n \tag{2.4}
$$

which is promptly verifiable as $F(n + 2)$. If we choose to consider a sequence consisting of n independent Bernoulli trials with equiprobable outcomes 0 and 1, then the probability of a sequence of length n which avoids the repetition 11 is $t_n(2)/2^n$. The question which arose in my mind was whether this recursive equations method could be extended to the case of a set of n independent trials with k equiprobable mutually exclusive outcomes $A_1, A_2, \ldots, A_k$, which avoided an m-repetition of A_i. It turned out that one could frame the problem in terms similar to the case $k = 2$, as follows.

Let $t_{m,n}(k)$ be the number of sequences of n trials with k equiprobable outcomes avoiding the m-repetition $A_i A_i \ldots A_i$, and let $a_{j,n}(k)$ be the number of sequences of n trials avoiding this m-repetition which end with $j < m$ A_i's. Then with $t_{m,0}(k) = 1$, and $a_{j,0}(k) = 0$ for $j = 1, 2, \ldots, m - 1$, we have for values of $n = 1, 2, \ldots$, the set of recursive equations

$$
\begin{aligned}
t_{m,n}(k) &= k t_{m,n-1}(k) - a_{m-1,n-1}(k), \\
a_{j,n}(k) &= a_{j-1,n-1}(k), \quad j = 2, \ldots, m - 1, \\
a_{1,n}(k) &= t_{m,n-1}(k) - a_{m-1,n-1}(k) - \cdots - a_{1,n-1}(k).
\end{aligned} \tag{2.5}
$$

The relevant matrix is now

$$A_m = \begin{bmatrix} k & -1 & 0 & 0 & \cdots & 0 \\ 0 & 0 & 1 & 0 & \cdots & 0 \\ \vdots & \vdots & \vdots & \vdots & \vdots & \vdots \\ 0 & 0 & 0 & 0 & \cdots & 1 \\ 1 & -1 & -1 & -1 & \cdots & -1 \end{bmatrix}$$

whose eigenvalues λ_j, $j = 1, \ldots, m$ are the roots of the equation

$$|A_m - I| = (-1)^m \left\{ \lambda^m + \sum_{i=0}^{m-1} \lambda^i (1-k) \right\} = 0. \tag{2.6}$$

One can now express $t_{m,n}(k)$ as

$$t_{m,n}(k) = \sum_{j=1}^{m} \alpha_j \lambda_j^n, \tag{2.7}$$

where the α_j are derived from the initial equations for $t_{m,0}(k)$ and $a_{j,0}(k)$. For further details, the reader is referred to Gani (1998). Once again, if one considers a sequence consisting of n independent trials resulting in one of k mutually exclusive equiprobable events, the probability that it will avoid an m-repetition $A_i A_i \ldots A_i$ will be $t_n(k)/k^n$.

Shortly after, I discussed similar problems with Albrecht Irle, a visitor from the University of Kiel, Germany. We agreed to collaborate on a paper (Gani and Irle, 1999) which summarized what we had learned about patterns in sequences of random events, both independent and forming a Markov chain. The following sections briefly outline the literature which we found of interest, as well as our contribution to it.

3. Strings and string overlaps

In a fundamental paper published in 1981, Guibas and Odlyzko (1981) used combinatorial methods to derive some general results on strings (particular patterns of letters from an alphabet), string overlaps and pattern matching. Consider an alphabet of size q, so that all strings consist of sequences composed of characters from this alphabet.

As an example, Guibas and Odlyzko (1981) examine the patterns X and Y from an alphabet $\{0, 1\}$ of size $q = 2$, where $X = 101001$, and $Y = 10010$. An important quantity in their exposition is the correlation of X and Y, denoted by XY; this is itself a string over $\{0, 1\}$ of the same length as X; it is determined by shifting Y rightward, and noting when the overlapping segments of X and Y are identical (or not) by recording a 1 (or a 0). We see that on shifting Y to the right 5 consecutive times, we obtain the correlation $XY = 001001$:

$$
\begin{array}{ll}
X = 101001 & \\
Y = 10010 & 0 \\
\rightarrow\ \ 10010 & 0 \\
10010 & 1 \\
10010 & 0 \\
10010 & 0 \\
10010 & 1
\end{array}
$$

Note, however, that $YX = 00010$ so that in general XY is not equal to YX. The correlation may also be expressed in terms of the polynomial $XY_z = z^3 + 1$ in the variable z; this is a simple method of encoding the correlation sequence.

Following precisely the notation used by Guibas and Odlyzko (1981), let $\{A, B, \ldots, T\}$ be a reduced set of patterns of decreasing lengths, so that for any two of these, one is never a substring of the other. Denote by

$$f(n) = f(A, B, \ldots, T; n) \tag{3.1}$$

the number of strings of length n using the given alphabet that do not contain any of $A, B, \ldots, T$. The corresponding gf, written in negative powers $z^{-j}, 0 \leqslant j < \infty$, is given by

$$F(z) = \sum_{n=0}^{\infty} f(n) z^{-n}. \tag{3.2}$$

We denote the number of strings of length n ending with H that do not contain any of $A, B, \ldots, T$, except for that single appearance of H at the end by $f_H(n)$, and write their gf as

$$F_H(z) = \sum_{n=0}^{\infty} f_H(n) z^{-n}. \tag{3.3}$$

Similarly, define $F_A(z), F_B(z), \ldots, F_T(z)$ as the gfs of $f_A(n), f_B(n), \ldots, f_T(n)$ the respective number of strings of length n which do not contain the patterns $A, B, \ldots, T$, except for the appearance of $A, B, \ldots, T$ at the end of the string.

In their main theorem, Guibas and Odlyzko (1981) prove that these gfs satisfy the nonsingular set of equations

$$(z - q) F(z) + z F_A(z) + z F_B(z) + \cdots + z F_T(z) = z,$$

$$F(z) - z A A_z F_A(z) - z B A_z F_B(z) - \cdots - z T A_z F_T(z) = 0,$$

$$\cdots\cdots \qquad \cdots\cdots \qquad \cdots\cdots$$

$$F(z) - z A T_z F_A(z) - z B T_z F_B(z) - \cdots - z T T_z F_T(z) = 0.$$

From these they derive the results

$$F(z) = \frac{zAA_z}{1+(z-q)AA_z}, \quad F_A(z) = \frac{1}{1+(z-q)AA_z}, \tag{3.4}$$

as well as the gfs $F_B(z), \ldots, F_T(z)$, from which one can find $f(n)$ and $f_H(n)$ readily. We now show that $F(z)$ will give the Rényi (1984) and Siegel (1997) result directly.

For $q = 2$, we see that if $A = \{11\}$ is to be avoided, then we calculate the correlation $AA = 11$ as follows:

$$\begin{array}{ll} A = 11 & \\ A = 11 & 1 \\ \rightarrow 11 & 1 \end{array}$$

with $A_z = z + 1$. It follows from (3.5) that

$$F(z) = \frac{z(z+1)}{1+(z-2)(z+1)} = \frac{1+z^{-1}}{1-(z^{-1}+z^{-2})}. \tag{3.5}$$

On expanding the denominator, we find

$$\begin{aligned} F(z) &= (1+z^{-1})\bigl[1 + (z^{-1}+z^{-2}) + (z^{-2}+2z^{-3}+z^{-4}) \\ &\quad + (z^{-3}+3z^{-4}+3z^{-5}+z^{-6}) \\ &\quad + (z^{-4}+4z^{-5}+6z^{-6}+4z^{-7}+z^{-8}) + (z^{-5}+\cdots) + \cdots\bigr] \\ &= (1+z^{-1})\bigl[1+z^{-1}+2z^{-2}+3z^{-3}+5z^{-4}+\cdots\bigr] \\ &= 1+2z^{-1}+3z^{-2}+5z^{-3}+8z^{-4}+\cdots. \end{aligned} \tag{3.6}$$

We see that the coefficients of z^{-n} for the values of $n \geq 2$, which are of interest to us, are indeed the Fibonacci numbers $F(n+2)$.

If the events occurring as the string is built up are assumed to be equiprobable, then from (3.1), the probability of strings not containing any of $A, B, \ldots, T$, is $f(n)/2^n$, but these results can be extended to the case where the probabilities of occurrence of the alphabet letters are unequal. The authors go on to discuss nontransitive games, optimal strategies in a coin tossing game, worst case behaviour of pattern matching algorithms, and the comparison of the number of strings which do not contain given patterns.

Guibas and Odlyzko (1981) attribute an earlier result of type (3.5) for the case of a single excluded pattern to Solov'ev (1966). Their methods are powerful, and resolve a number of important problems in sequences of independent events. A study of similar problems using alternative renewal theory methods was carried out by Breen, Waterman and Zhang (1985). They were motivated by the action of restrictive enzymes on DNA sequences, constructed with the four letter alphabet $\{A, T, G, C\}$. They obtained the gfs of renewal probabilities for several possible patterns, and from these derived the gfs of first occurrence of certain patterns, their specific interest being the distribution of trials between their occurrence. They also indicated how their results could be derived from those of Guibas and Odlyzko (1981).

We now proceed to review other developments in the area, before presenting some general methods which deal with independent events of unequal probabilities, as well as dependent events forming a Markov chain.

4. Further classical and martingale methods

Further research into patterns arising in sequences of random events, using classical methods continued in the 1980s. Blom (1982) explored the mean number of random digits required in sequences of independent equiprobable events until a particular pattern emerged. He noted that Solov'ev (1966) and Nielsen (1973) had already considered the problem using somewhat different techniques. One of Blom's main results concerns the pattern S of length k_0 with overlapping subsequences $S_1, S_2, \ldots, S_v$ of respective lengths $k_1 > k_2 > \cdots > k_v$, where such sequences appear both at the beginning and end of S. The mean time $E(X_S)$ until S occurs is shown to be

$$E(X_S) = \sum_{i=0}^{v} N^{k_i}. \tag{4.1}$$

If, for example, $S = (010)$, then $k_0 = 3$, for which there is a single overlapping sequence $S_1 = \{0\}$ of length $k_1 = 1$; then $E(X_S) = 2^3 + 2^1 = 10$. The result (4.1) was later generalized by Blom (1983) to sequences of mutually exclusive events with unequal probabilities. In Blom and Thorburn (1982), the authors consider several possible stopping patterns $S_1, S_2, \ldots, S_n$, all of the same length. They begin by discussing the probability distribution and gf of the time until a single pattern S occurs for the first time, and then proceed to extend this to the probability distribution and gf of the time until one of $S_1, S_2, \ldots, S_n$ is reached. They conclude with a discussion of the order statistics of the respective waiting times. Blom and Sandell (1994) have also considered the number of sequences of equiprobable Bernoulli trials ending in a particular pattern, or more generally containing a fixed number of such patterns before the last one.

Other investigations were concerned with asymptotic results; among these is Karnin's (1983) work on the time T until a particular pattern of equiprobable Bernoulli trials is repeated. Suppose the pattern consists of a string of m consecutive identical outcomes, such as 1111111 where $m = 7$. Karnin proves that as $m \to \infty$,

$$P\{T > z\sqrt{2^{m+1}}\} \to \exp(-z^2/2), \quad 0 < z < \infty, \tag{4.2}$$

and $E(T) \to \sqrt{2^m \pi}$. The motivation in this case was an encryption scheme. The problem had previously been solved by Zubkov and Mikhailov (1974) using some rather complicated methods, which Karnin simplified in his elementary proof.

In a somewhat different direction, Chryssaphinou and Papastavridis (1988) investigated the number T_n of non-overlapping occurrences of a particular pattern A_n in a sequence of n independent events with unequal probabilities $p_1, p_2, \ldots, p_q$ drawn from an alphabet of size q. Consider a sequence of patterns $\{A_n\}$ of lengths k_n, where $k \to \infty$ as $n \to \infty$, and let the mean recurrence time of non-overlapping A_n for fixed n be

$$\mu_n = \sum_i \frac{q_{i,n}}{P\{A_n\}}, \tag{4.3}$$

where $q_{i,n}$ is the conditional probability that A_n occurs with overlap i, after a given A_n. Then the authors prove that if

$$n/\mu_n \to \lambda > 0, \quad \text{as } n \to \infty,$$

the distribution of T_n tends to a Poisson distribution with the parameter λ.

The early 1980s also witnessed the application of martingale methods to problems of patterns in sequences of independent events. For example, Li (1980) made use of Doob's fundamental theorem on martingale stopping times to derive his results. He considered the discrete i.i.d. random variables $Z_1, Z_2, \ldots$, with $A_0, A_1, \ldots, A_n$ being finite sequences of possible values of the $\{Z_i\}$. For example, let $A = \{a_1 a_2 \ldots a_m\}$, $B = \{b_1 b_2 \ldots b_n\}$ be two patterns; then for every pair (i, j) of integers, write

$$\delta_{ij} = \begin{cases} 1/P\{Z = b_j\}, & \text{if } 1 < i < m, 1 < j < n, \text{ and } a_i = b_j, \\ 0, & \text{if } a_i \neq b_j \end{cases} \tag{4.4}$$

so that $A * B = \delta_{11}\delta_{22} \cdots \delta_{mm} + \delta_{21}\delta_{32} \cdots \delta_{mm-1} + \cdots + \delta_{m1}$.

Given the starting sequence A_0, let p_i be the probability that A_i precedes the remaining sequences in a realization of the process $Z_1, Z_2, \ldots$. Then for every i, and a sequence with the stopping time N when any A_j occurs,

$$\sum_{j=1}^n p_j A_j * A_i = E(N) + A_0 * A_i. \tag{4.5}$$

These equalities can be expressed in matrix form as

$$\begin{bmatrix} 0 & 1 & 1 & \cdots & 1 \\ -1 & A_1 * A_1 & A_2 * A_1 & \cdots & A_n * A_1 \\ -1 & A_1 * A_2 & A_2 * A_2 & \cdots & A_n * A_2 \\ \vdots & \vdots & \vdots & \vdots & \vdots \\ -1 & A_1 * A_n & A_2 * A_n & \cdots & A_n * A_n \end{bmatrix} \begin{bmatrix} E(N) \\ p_1 \\ p_2 \\ \vdots \\ p_n \end{bmatrix} = \begin{bmatrix} 1 \\ A_0 * A_1 \\ A_0 * A_2 \\ \vdots \\ A_0 * A_n \end{bmatrix}. \tag{4.6}$$

Li also quotes the famous J. H. Conway leading number algorithm, mentioned as a private communication in Guibas and Odlyzko (1981). Li presents it as a corollary in the following form. If A and B are not connected sequences of each other, then the odds that B precedes A in a realization of the process $Z_1, Z_2, \ldots$, are

$$\frac{A * A - A * B}{B * B - B * A}. \tag{4.7}$$

In a subsequent paper, Gerber and Li (1981) show how these methods can be extended to a process $Z_1, Z_2, \ldots$, which forms a Markov chain with stationary transition probabilities. Both the gf and the expectations of the time until a particular pattern occurs are derived. Benevento (1984) has also generalized Li's (1980) earlier results for sequences of events from an ergodic Markov chain.

Research on patterns in sequences of random events continued during the 1990s. Godbole (1991) in a paper reviewing the field, considered a sequence of Bernoulli trials $X_1, X_2, \ldots, X_n$, each with $P\{X_i = 1\} = p$, in which $N_{n,k}$ denotes the number of non-overlapping success runs of length $k \geqslant 2$. Using the Stein–Chen method, he obtained a total variation upper bound for the rate of convergence of $N_{n,k}$ to a Poisson random variable with parameter λ under the standard condition that $np^k \to \lambda$ as $n \to \infty$. He also extended his methods to cover the occurrence of word patterns from larger alphabets, as well as to the case of the two-state Markov chain.

Pekoz (1996) also used Stein–Chen techniques to bound errors for a geometric approximation to the distribution of the number of failures before a first success in a set of dependent trials. He used his result to approximate the waiting time distribution until particular patterns occur in coin tossing, and for visits to a rare set of events in a stationary Markov chain. His error bounds were sharper than those obtainable using ordinary Poisson approximations.

The concept of runs on a ring or a circle dates back to the paper of Barton and David (1958). Makri and Philippou (1994) were able to derive the distribution of $C_{n,k}$ the number of success runs of length k on a circle, in n Bernoulli trials, each with probability p of success. Note that in this case, it is possible to form additional runs by combining successes at the beginning and end of sequences.

A strong motivating factor in the study of patterns has been the set of problems which arises in the sequencing of DNA strands. Spurred by these, Alexander (1996) investigated strings formed with letters chosen independently with given probabilities from a finite alphabet. His concern was the length of the shortest common string S containing each of a set of given substrings $S_1, S_2, \ldots, S_n$, of respective lengths $m_1, m_2, \ldots, m_n$, in a consecutive order. The strings may overlap, so that the shortest common string will then have a length M_n which is smaller than $m_1 + m_2 + \cdots + m_n$. An asymptotic expression is derived for the savings V_n obtained by compression, this being defined by

$$V_n = \sum_{i=1}^{n} m_i - M_n, \tag{4.8}$$

the difference between the sum of the lengths of all the strings, and the shortest length. Alexander (1996) then extends his method to the case where the strings are independent realizations of an ergodic stationary process with positive entropy.

5. Markov chain techniques

Our brief account of patterns arising in sequences of random events indicates that several methods can be used to analyze these patterns. We have outlined methods based on direct combinatorial considerations, on the theory of recurrent events, on matrix relations summarizing sets of difference equations, on the combinatorics of strings and their overlaps, on martingale techniques, and also derived approximations relying on asymptotics. It would be helpful if one could fall back on some unifying principle in the study of such patterns, not only in the more usual case of a sequence of independent events, but also for events forming a Markov chain. Such a unifying principle is provided by the use of Markov chain techniques.

From the mid-1980s, workers in the theory of reliability such as Fu (1986), Fu and Hu (1987), and Chao and Fu (1991) had been using various Markov chain methods to analyze large series systems. Chao and Fu (1991), for example, studied the finite Markov chain $Y_1, Y_2, \ldots, Y_n$, for which the associated X_i are 1 if the component Y_i works, and 0 if it does not. The reliability of the system is a function of the X_i, and the authors proved that under certain regularity conditions, this reliability tends to a constant, often with an exponential form. While this was only indirectly related to the occurrence of patterns, the techniques used foreshadowed later developments.

In molecular biology, a paper by Biggins and Cannings (1987a), motivated by the cutting of DNA sequences using restriction enzymes, had been based on Markov renewal processes; it studied the occurrence of specific sequences of states in a Markov chain. The authors related the process to the Type 1 counter process, and derived results for the generating function of the time to next occurrence of a particular pattern in both non-overlapping and partially overlapping sequences. Biggins (1987) later simplified some of these results. Biggins and Cannings (1987b) also applied their technique to obtain the stationary probability of first reaching a particular pattern, as well as the mean restriction fragment lengths when DNA is digested using a restriction enzyme.

The basic techniques for a Markov chain treatment of patterns in sequences of n independent Bernoulli trials were set out in a paper by Fu and Koutras (1994), and extended in Fu (1996) and Fu, Lou and Wang (1999). Fu and Koutras (1994) explained in clear terms the method of construction of embedded Markov chains to trace the occurrence of particular patterns of 1's and 0's, and used these to derive the distributions of $E_{n,k}$, $G_{n,k}$, $N_{n,k}$, $M_{n,k}$, and L_n. These are respectively the numbers of runs of size exactly k as defined by Mood (1940), the number of runs of size k or greater, the number of non-overlapping runs of size k as defined by Feller (1968), the number of runs of overlapping consecutive k successes, and the size of the longest run. The subsequent development of Gani and Irle (1999) which concentrates on the case of several states, themselves forming a Markov chain, will be presented later. For the moment, we mention briefly some recent work by Koutras and colleagues, which makes use of the embedded Markov chain technique.

Koutras, Papadopoulos and Papastavridis (1995) generalized the result of Makri and Philippou (1994) for $C_{n,k}$, the number of runs of size k on a circle, to the case of n i.i.d (as well as non-i.i.d.) random variables. They went on to establish a Poisson limit theorem for $C_{n,k}$ as $n \to \infty$. Koutras (applied the embedded Markov chain technique to the study of a class of reliability structures, among them the k-out-of-n and consecutive k-out-of-n: F systems. Finally, in a recent paper, Koutras and Alexandrou (1997) examined the sequence of trials $X_1, X_2, \ldots$, with three possible outcomes A, B, and C, occurring on a line or a circle. Their interest was in the waiting time until an A-run of length k or a B-run of length r occurred. They were able to derive the gf of the waiting times in both the linear and circular cases, and also to obtain closed approximations to the distributions using the Stein–Chen method.

Let us now try to present a general account of a Markov chain method for dealing with sequences of events, extending the techniques of Fu and Koutras (1994). The main difference is that in Gani and Irle (1999) these techniques are applicable to the case where $X_1, X_2, \ldots, X_n$, are the results of trials from a set of k mutually exclusive events $A_1, A_2, \ldots, A_k$ which form a Markov chain. Let q_i be the initial probability that $P\{X_i = A_i\}$, and $p_{ij} = P\{X_r = A_j | X_{r-1} = A_i\}$. Then suppose that the pattern in which we are interested is $A_{i_1} A_{i_2} \ldots A_{i_m}$; we construct the augmented Markov chain consisting of the original states $A_1, A_2, \ldots, A_k$ and add the states $A_{i_1} A_{i_2}$, $A_{i_1} A_{i_2} A_{i_3}, \ldots, A_{i_1} \ldots A_{i_m}$.

We now need to place the transition probabilities in their appropriate positions; suppose, for example, that $A_{i_1} A_{i_2}$ is $A_1 A_2$, then $P\{A_1 \to A_1 A_2\} = p_{12}$, while $P\{A_1 \to A_2\}$ now becomes 0, and all other transitions probabilities in the original Markov chain matrix retain their places. Likewise if the transition $X_{r-1} \to X_r$, involves passage from one of the augmented states $A_{i_1} \ldots A_{i_{j-1}}$ to the next largest state $A_{i_1} \ldots A_{i_{j-1}} A_{i_j}$, $j \leqslant m$, it follows that the transition probability $p_{A_{i_{j-1}} A_{i_j}}$ should record this. If, $X_r \neq A_{i_j} = A_1$ say, then one must move backward along the augmented matrix until an overlap occurs, or one ends with the state A_1, and allocate the probability $p_{A_{i_{j-1}} A_1}$ to that position.

To illustrate the procedure, we give a simple example. Consider an alphabet of two states 0 and 1 forming a Markov chain with initial probabilities $q_0 = P\{0\}$ and $q_1 = P\{1\}$, and transition probability matrix

$$\begin{bmatrix} p_{00} & p_{01} \\ p_{10} & p_{11} \end{bmatrix}. \tag{5.1}$$

Suppose the particular pattern we are interested in is 011; we construct an augmented matrix with states 0, 1, 01, 011, having the transition probability matrix

$$\left[\begin{array}{cc|cc} p_{00} & 0 & p_{01} & 0 \\ p_{10} & p_{11} & 0 & 0 \\ p_{10} & 0 & 0 & p_{11} \\ 0 & 0 & 0 & 1 \end{array}\right] = \begin{bmatrix} P & E \\ 0 & 1 \end{bmatrix}, \tag{5.2}$$

where the p_{ij} are placed in the positions indicated, according to the principles outlined earlier.

We see that the probability that the waiting time $T = t$, when the state 011 is reached for the first time, is given by

$$P\{T = t\} = [q_0 \ q_1 \ 0] P^{t-2} E, \quad t > 2, \tag{5.3}$$

while the gf of T is

$$f_T(s) = [q_0\ q_1\ 0][I - Ps]^{-1} E s^2. \tag{5.4}$$

From this, the mean and variance of the waiting time T can readily be obtained. If one wishes to consider the waiting time until either 011 or 001 is reached, one must now construct an augmented matrix with states 0, 1, 00, 01, 001, 011, where the last two states are absorbing. Using the same principles for the allocation of the transition probabilities (5.1), one can readily derive the probability of the waiting time T until either 001 or 011 is reached for the first time. Thus, a Markov chain characterization of the pattern problem tends to simplify the methods of analyzing it.

There remain a variety of interesting problems in the field, motivated by its mathematical challenges and by practical issues in both reliability theory and DNA sequencing. I hope that this account of my own encounters with problems of patterns in sequences of random events will inspire further research on this fascinating topic.

References

Abramowitz, M. and I. A. Stegun (1964). *Handbook of Mathematical Functions* (6th printing, 1967). National Bureau of Standards, Washington, DC.

Alexander, K. S. (1996). Shortest common superstrings of random strings. *J. Appl. Probab.* **33**, 1112–1126.

Barton, D. E. and F. N. David (1958). Runs in a ring. *Biometrika* **45**, 572–578.

Benevento, R. V. (1984). The occurrence of sequence patterns in ergodic Markov chains. *Stochastic Process. Appl.* **11**, 369–373.

Biggins, J. D. (1987). A note on repeated sequences in Markov chains. *Adv. Appl. Probab.* **19**, 739–742.

Biggins, J. D. and C. Cannings (1987a). Markov renewal processes, counters and repeated sequences in Markov chains. *Adv. Appl. Probab.* **19**, 521–545.

Biggins, J. D. and C. Cannings (1987b). Formulas for mean restriction-fragment lengths and related quantities. *Amer. J. Hum. Genetics* **39**, 258–265.

Blom, G. (1982). On the mean number of random digits until a given sequence occurs. *J. Appl. Probab.* **19**, 136–142.

Blom, G. (1983). The mean waiting time to a repetition. *Adv. Appl. Probab.* **15**, 216–218.

Blom, G. and D. Sandell (1994). Patterns in binary sequences. *Math. Scientist* **19**, 93–102.

Blom, G. and D. Thorburn (1982). How many random digits are required until given sequences are obtained. *J. Appl. Probab.* **19**, 518–531.

Breen, S., M. S. Waterman and N. Zhang (1985). Renewal theory for several patterns. *J. Appl. Probab.* **22**, 228–234.

Bruns, H. (1906). *Wahrscheinlichkeitsrechnung und Kollektivmasslehre*. Leipzig.

Chao, M. T. and J. C. Fu (1991). The reliability of large series systems under Markov structure. *Adv. Appl. Probab.* **23**, 894–908.

Chryssaphinou, O. and S. Papastavridis (1988). A limit theorem for the number of non-overlapping occurrences of a pattern in a sequence of independent trials. *J. Appl. Probab.* **25**, 428–431.

Feller, W. (1950). *An Introduction to Probability Theory and its Applications*, Vol. 1 (3rd ed., 1968). Wiley, New York.

Fisher, R. A. and F. Yates (1938). *Statistical Tables for Biological, Agricultural and Medical Research* (Revised editions in 1942, 1948, 1953, 1957, 1963). Oliver and Boyd, Edinburgh.

Fu, J. C. (1986). Reliability of consecutive-k-out-of-n: F systems with $(k-1)$-step Markov dependence. *IEEE Trans. Reliability* **R35**, 602–606.

Fu, J. C. (1996). Distribution theory of runs and patterns associated with a sequence of multi-state trials. *Statist. Sinica* **6**, 957–974.

Fu, J. C. and B. Hu (1987). On reliability of a large consecutive-k-out-of-n: F system with $(k-1)$-step Markov dependence. *IEEE Trans. Reliability* **R36**, 75–77.

Fu, J. C. and M. V. Koutras (1994). Distribution theory of runs: a Markov chain approach. *J. Amer. Statist. Assoc.* **89**, 1050–1058.

Fu, J. C., W. Y. W. Lou and Y.-J. Wang (1999). On the exact distributions of Eulerian and Simon Newcomb numbers associated with random permutations. *Statist. Probab. Lett.* **42**, 115–125.

Gani, J. (1998). On sequences of events with repetitions. *Stochastic Models* **14**, 265–271.

Gani, J. and A. Irle (1999). On patterns in sequences of random events. *Monatsh. Math.* **127**, 295–309.

Gerber, H. U. and S.-Y. R. Li (1981). The occurrence of sequence patterns in repeated experiments and hitting times in a Markov chain. *Stochastic Process. Appl.* **11**, 101–108.

Godbole, A. (1991). Poisson approximations for runs and patterns of rare events. *Adv. Appl. Probab.* **23**, 851–865.

Guibas, L. J. and A. M. Odlyzko (1981). String overlaps, pattern matching and nontransitive games. *J. Combin. Theory A* **30**, 183–208.

Heyde, C. C. (1967). Some local limit results in fluctuation theory. *J. Austral. Math. Soc.* **7**, 455–464.

Imhof, J. P. (1974). Runs of discrete time regenerative phenomena. *J. Appl. Probab.* **11**, 588–593.

Ising, E. (1925). Beitrag zur Theorie des Ferromagnetismus. *Z. Physik* **31**, 253–258.

Karnin, E. D. (1983). The first repetition of a pattern in a symmetric Bernoulli sequence. *J. Appl. Probab.* **20**, 413–418.

Kendall, M. G. and B. Babington Smith (1938). Randomness and random sampling numbers. *J. Roy. Statist. Soc.* **101**, 147–166.

Kendall, M. G. and B. Babington Smith (1939a). Second paper on random sampling numbers. *J. Roy. Statist. Soc. Supp.* **6**, 51–61.

Kendall, M. G. and B. Babington Smith (1939b). *Random Sampling Numbers*. Tracts for Computers No. 24, Dept. of Statistics, University College London.

Koutras, M. V. (1996). On a Markov chain approach for the study of reliability structures. *J. Appl. Probab.* **33**, 357–367.

Koutras, M. V., G. K. Papadopoulos and S. G. Papastavridis (1995). Runs on a circle. *J. Appl. Probab.* **32**, 396–404.

Koutras, M. V. and V. A. Alexandrou (1997). Sooner waiting time problems in a sequence of trinary trials. *J. Appl. Probab.* **34**, 593–609.

Leslie, R. T. (1967). Recurrent composite events. *J. Appl. Probab.* **4**, 34–61.

Li, S.-Y. R. (1980). A martingale approach to the study of occurrence of sequence patterns in repeated experiments. *Ann. Probab.* **8**, 1171–1176.

Makri, F. S. and A. N. Philippou (1994). Binomial distributions of order k on the circle. In *Runs and Patterns in Probability: Selected Papers*, pp. 65–81 (Eds. A. Godbole and S. G. Papastavridis). Kluwer, Dordrecht.

Mood, A. M. (1940). The distribution theory of runs. *Ann. Math. Statist.* **11**, 367–392.

Nielsen, P. T. (1973). On the expected duration of a search for a fixed pattern in random data. *IEEE Trans. Inform. Theory* **19**, 702–704.

Pekoz, E. A. (1996). Stein's method for geometric approximation. *J. Appl. Probab.* **33**, 707–713.

Rand Corporation (1955). *A Million Random Digits with 100,000 Normal Deviates*. The Free Press, Glencoe, IL.

Rajarshi, M. B. (1974). Success runs in a two-state Markov chain. *J. Appl. Probab.* **11**, 190–192.

Rényi, A. (1984). Variation on a theme of Fibonacci. In *A Diary on Information Theory*. Akademiai Kiado, Budapest.

Ripley, B. D. (1987). *Stochastic Simulation*. Wiley, New York.

Schwager, S. J. (1983). Run probabilities in sequences of Markov-dependent trials. *J. Amer. Statist. Assoc.* **78**, 165–178.

Siegel, D. (1997). Bernoulli meets Fibonacci. *Math. Scientist* **22**, 122–124.

Solov'ev, A. D. (1966). A combinatorial identity and its application to the problem concerning the first occurrence of a rare event. *Theory Probab. Appl.* **11**, 276–282.

Vajda, S. (1989). *Fibonacci and Lucas Numbers, and the Golden Section*. Halsted Press, Chichester.

von Bortkiewicz, L. J. (1917). *Die Iterationen.* Berlin.
von Mises, R. (1921). Das Problem der Iterationen. *Z. Angew. Math. Mech.* **1**, 298–307.
Wald, A. and J. Wolfowitz (1940). On a test whether two samples are from the same population. *Ann. Math. Statist.* **11**, 147–162.
Zubkov, A. M. and V. G. Mikhailov (1974). Limit distribution of random variables associated with long duplications in a sequence of independent trials. *Theory Probab. Appl.* **19**, 172–179.

Stochastic Models in Telecommunications for Optimal Design, Control and Performance Evaluation

N. Gautam

In this chapter stochastic models for high-speed networks viz. traffic models, buffer content process models, LAN models, etc. are considered. The stochastic models are used to obtain end-to-end Quality-of-Service (QoS) measures that the network must guarantee users in the future. The QoS measures or performance measures can be used in optimal design problems and admission control problems. Other aspects such as TCP, routing, leaky buckets, wireless networks, etc. are also considered.

1. Introduction

One of the greatest success stories for stochastic models in engineering is in the field of telecommunications. Discrete and fluid queueing models have played a major role in the development of computer and communication networks. There are several branches of telecommunications that use stochastic models, however, in this paper the main focus is in networking systems and other stochastic systems that aide high-performance networking. A few other applications are briefly mentioned later in this introduction.

There are several interesting scenarios in the Internet and the emerging next generation networks (such as Internet2, NGI, etc.) where stochastic modeling is applicable. The future networks will carry a wide variety of traffic (Data, Voice, Video, etc.) and the users will demand very high-quality from the networks. Therefore it is very important to consider certain performance issues known as Quality-of-Service (QoS). There are four well-known end-to-end QoS measures, viz., loss probability, delay, delay-jitter and bandwidth. They are briefly described as follows:

When messages flow from a source to a destination (end-to-end) through a network, parts of a message or the whole message may be dropped due to unavailable resources (buffer capacity) to store the messages. The probability of delivering a message with some data loss is termed as loss probability. The time between the source sending a message and the destination receiving it is called latency or delay. Typically real-time or multimedia traffic (such as live video conference) can tolerate some loss but have very stringent delay requirements. However data traffic such as emails, fax, file transfers, etc. can tolerate some delay but almost zero loss. The other QoS measures are delay-jitter (which is a measure of the variation in the delay) and bandwidth (which is the rate at which messages are processed).

The message flow (will be called traffic henceforth) and the network conditions are extremely stochastic in nature. Given the growth in the Internet as well as users demanding QoS for their applications, it is important to be able to predict the QoS measures as they will have to be guaranteed to the users. Also the QoS measures can be used for **optimal design** and **admission control** of the networks. Some of the main **design** aspects include buffer sizes, link capacities, network parameters, traffic shaping parameters, etc. While exercising **admission control**, the network either rejects an incoming request for connection or accepts it (and provides the required QoS).

As mentioned earlier, the main concentration of this paper will be on high-speed telecommunication networks. Other applications of stochastic processes in communications include coding theory, signal processing, image processing, pattern recognition, speech recognition, etc. Stochastic models using hidden Markov processes (for a thorough exposition of hidden Markov models, see Rabiner, 1989), hidden semi-Markov models, Markov decision processes, etc. are used in signal processing, image processing, pattern recognition and speech recognition.

Broadly there are three types of telecommunication networks – telephony (telephone networks for voice calls, fax, and also dial-up connections), cable-TV networks (cable, web-TV, etc.), and high-speed networks such as the Internet. This paper focuses on high-speed networks with the motivation that in the very near future, Internet telephony, video-on-demand, networked homes, multimedia applications, etc. will possibly replace telephone and cable-TV networks due to low cost. However, unless the performance of high-speed networks improves greatly this will not be possible!

2. Traffic models

Traffic flowing through the networks can be classified into several types. Two of the most common traffic types are ethernet packets/frames and ATM cells. Depending on the network segment, all messages are broken down into either packets or cells. The length or size of an Ethernet packet ranges anywhere from 60 bytes to 1500 bytes and generally follows a bimodal distribution. The length of ATM cells is fixed at 53 bytes. Therefore the network traffic comprises of millions and billions of these little packets or cells!

One of the most important tasks before evaluating the performance of telecommunication networks is to fit appropriate models for traffic to capture their stochastic nature. Data can be obtained by using "sniffers" on the network and analyzing a "dump" of all the packets or cells that were generated during the time the sniffer was used. The information that can be obtained about each packet or cell by sniffing include: its arrival time, its source, its destination, its length, its type, etc. To fit traffic models, only the time of arrival and packet size are sufficient.

2.1. Hierarchical networks

Telecommunication networks are typically hierarchical in nature. Generally, traffic can be classified into four levels:

- Application Level: The traffic generated by an application, say, http or telnet or ftp which can vary significantly based on the protocols they follow.
- Source Level: Each workstation or computer can be thought of as a source that generates traffic. This traffic comprises of the traffic generated by different applications that are running on the source. Therefore the traffic that flows on a link that exits the computer is a mixture of the different applications. The process of mixing is known as multiplexing.
- Aggregate Level: Several computer, printers, etc are connected together to form a local area network (LAN). The traffic on a LAN pipe is the aggregated traffic that is multiplexed from all the sources.
- Backbone Level: The LANs are connected together by means of a backbone (say, the Internet backbone), and this forms the Metropolitan Arean Networks (MANs) or the Wide Area Networks (WANs). The traffic on a MAN/WAN pipe is the combination of the traffic from several LANs.

Appropriate traffic models can be used depending on the levels being considered. Some frequently used stochastic models for traffic flow are explained in the next section. Although, different researchers prefer to use different traffic models, the models can be broadly classified into two parts, discrete models and fluid models. In the discrete model each packet or cell is assumed to be a discrete entity that can be of varying sizes. In the fluid models it is assumed that the packets or cells are packed so close to each other that the traffic flow can be assumed to be a fluid flowing across a pipe, maybe at different rates.

2.2. Fluid-flow traffic models

In the fluid-flow models it is assumed that traffic is in the form of fluid which flows through a pipe at different rates at different times. For example, fluid flows at rate $r(1)$ bytes per second for a random amount of time t_1, then flows at rate $r(2)$ bytes per second for a random amount of time t_2, and so on. This behaviour can be captured as a discrete stochastic process that jumps from one state to another whenever the traffic flow rate changes. This can be formalized as a stochastic process $\{Z(t), t \geq 0\}$ that is in state $Z(t)$ at time t. Fluid flows in the pipe at rate $r(Z(t))$ at time t. Researchers have used different models for the $\{Z(t), t \geq 0\}$ process that are summarized in the following description:

2.2.1. DTMC environmental processes

Consider a DTMC where a transition occurs every θ seconds. When the DTMC is in state i, $\theta r(i)$ bytes flow through the pipe. Let $P = [p_{ij}]$ be the transition probability matrix, where p_{ij} is the probability that the DTMC goes from state i to state j in one-step. For an irreducible, aperiodic and positive recurrent DTMC, let π be the steady-state distribution such that $\pi = \pi P$ and $\pi \mathbf{a} = 1$ with $\mathbf{a}$ being a column vector of ones.

2.2.2. CTMC environmental processes

Let $\{Z(t), t \geq 0\}$ be an irreducible, finite state CTMC with generator matrix Q. When the CTMC is in state i, traffic flows at rate $r(i)$. Let $R = \text{diag}[r_{ii}]$, where $r_{ii} = r(i)$. Let p be the stationary distribution of the CTMC such that $pQ = 0$ and $p\mathbf{a} = 1$ with $\mathbf{a}$ being a column vector of ones.

2.2.3. Alternating renewal environmental processes

This is sometimes known as the on-off traffic where traffic either flows at the maximum link (or pipe) capacity r bytes per second or no traffic flows. The up times or on times (when traffic flows through the pipe) are distributed according to a general CDF $U(\cdot)$. The down or off times (when traffic does not flow through the pipe) are distributed according to a general CDF $D(\cdot)$.

2.2.4. SMP environmental processes

Consider a Semi-Markov Process (SMP) $\{Z(t), t \geq 0\}$ on state space $\{1, 2, \ldots, \ell\}$. Fluid is generated at rate $r(i)$ at time t when the SMP is in state $Z(t) = i$. Let S_n denote the time of the nth jump epoch in the SMP with $S_0 = 0$. Define Z_n as the state of the SMP immediately after the nth jump, i.e.,

$$Z_n = Z(S_n+).$$

Let

$$G_{ij}(x) = P\{S_1 \leq x; Z_1 = j | Z_0 = i\}. \tag{1}$$

The kernel of the SMP is

$$G(x) = \left[G_{ij}(x)\right]_{i,j=1,\ldots,\ell}.$$

Note that $\{Z_n, n \geq 0\}$ is a discrete time Markov chain (DTMC) which is embedded in the SMP. Assume that this DTMC is irreducible and recurrent with transition probability matrix

$$P = G(\infty).$$

Let

$$G_i(x) = P\{S_1 \leq x | Z_0 = i\} = \sum_{j=1}^{\ell} G_{ij}(x)$$

and the expected time the SMP spends in state i be

$$\tau_i = E(S_1 | Z_0 = i).$$

Let
$$\pi_i = \lim_{n\to\infty} P\{Z_n = i\}$$

be the stationary distribution of the DTMC $\{Z_n, n \geq 0\}$. It is given by the unique non-negative solution to

$$\pi = \pi P \quad \text{and} \quad \sum_i \pi_i = 1.$$

The stationary distribution of the SMP is given by

$$p_i = \lim_{t\to\infty} P\{Z(t) = i\} = \frac{\pi_i \tau_i}{\sum_{m=1}^{\ell} \pi_m \tau_m}. \tag{2}$$

2.2.5. MRGP environmental processes

Consider a regular Markov regenerative process $\{Z(t), t \geq 0\}$. For the definition of MRGP see Cinlar (1975), who calls it a semi-regenerative process. Also see Heyman and Sobel (1982) and Kulkarni (1995). Let $\{(Y_n, S_n), n \geq 0\}$ be an embedded Markov renewal sequence in the MRGP. Assume that $\{Y_n, n \geq 0\}$ has a finite state-space $\{1, 2, \ldots, \ell\}$. From the definition of the Markov renewal sequences,

$$P\{Y_{n+1} = j, S_{n+1} - S_n \leq x | Y_n = i, S_n, \ldots, Y_0, S_0\}$$
$$= P\{Y_1 = j, S_1 \leq x | Y_0 = i\} \tag{3}$$

for all $x \geq 0$, and $i, j = 1, 2, \ldots, \ell$. Furthermore, given the history $\{Z(t), 0 \leq t \leq S_n\}$ and $\{(Y_k, S_k), 0 \leq k \leq n\}$, the future of the $Z(t)$ process, viz., $\{Z(t), t \geq S_n\}$, depends on the past only via Y_n. Assume that the MRGP is regular, i.e., $S_n \to \infty$ with probability 1 as $n \to \infty$. It is clear from Eq. (3) that $\{Y_n, n \geq 0\}$ is a discrete time Markov chain (DTMC). Assume it is irreducible. Let $\mu_k = E[S_1 | Y_0 = k]$, $\pi = (\pi_k)$ is a positive solution to $\pi = \pi G(\infty)$, and,

$$\alpha_{kj} = E(\text{time spent by the } Z(t) \text{ process in state } j \text{ during } [0, S_1) | Y_0 = k).$$

Then,

$$p_j = \lim_{t\to\infty} P\{Z(t) = j\} = \frac{\sum_k \pi_k \alpha_{kj}}{\sum_k \pi_k \mu_k}.$$

2.3. Discrete traffic models

In the discrete traffic models it is assumed that the packet flow is in the form of discrete entities (like cars going on a highway) through the pipes. Therefore it is important to characterize the arrival process and the size of the packets. Some typical arrival processes, frequently used by researchers are explained below. The packet size distribution is usually a bimodal empirical distribution and is not discussed here.

2.3.1. Poisson process
If the interarrival distribution of the packets can be modeled using an independent and identically distributed exponential distribution, then the packet arrival process can be characterized as a Poisson process with parameter λ, where λ is the mean arrival rate (in terms of number of packets per unit time).

2.3.2. MMPP (Markov-modulated Poisson process)
Whenever a new connection is established through a pipe or an existing connection is terminated, the mean packet flow rate (number of packets per unit time) increases or decreases respectively. To account for this behavior, the traffic arrival process is modeled as an MMPP. Consider a CTMC $\{Y(t), t \geqslant 0\}$ with generator matrix Q. When the CTMC is in state i, packets flow according to a Poisson process with mean rate $\lambda(i)$. This captures the change in mean arrival rates effectively.

2.3.3. BMAP (batch Markovian arrival process)
The batch Markovian arrival process (BMAP) is explained in detail in Lucantoni (1991). The BMAP is a generalization of the Markovian arrival process (MAP) which was introduced by Lucantoni et al. (1990). A special case of the MAP are the phase type renewal processes and the MMPP. The following definition and properties of BMAP is reproduced from Lucantoni et al. (1993).

Consider a series of $m \times m$ matrices D_k, $k \geqslant 0$, such that D_0 has negative diagonal elements and nonnegative off-diagonal elements and for $k \geqslant 1$, D_k are nonnegative. Define an irreducible infinitesimal generator D such that

$$D = \sum_{k=0}^{\infty} D_k.$$

To assure that arrivals will occur assume that $D \neq D_0$.

Consider a two-dimensional Markov process $\{N(t), J(t), t \geqslant 0\}$ on the state space $\{(i, j): i \geqslant 0, 1 \leqslant j \leqslant m\}$ with an infinitesimal generator Q given by

$$Q = \begin{bmatrix} D_0 & D_1 & D_2 & D_3 & \dots \\ & D_0 & D_1 & D_2 & \dots \\ & & D_0 & D_1 & \dots \\ & & & D_0 & \dots \\ & & & & \dots \end{bmatrix}.$$

Here, $N(t)$ counts the number of arrivals in time t and $J(t)$ represents a state or phase. For example, a transition from state (i, j) to state $(i+k, l)$, $k \geqslant 1$, $1 \leqslant j, l \leqslant m$ denotes a batch arrival of size k and thus the batch size can depend on j and l. The matrix D_0 is nonsingular and the sojourn time in the set of states $\{(i, j): 1 \leqslant j \leqslant m\}$ is finite w.p. 1. Thus the arrival process does not terminate.

Let π denote the stationary probability vector of the Markov process with generator D such that

$$\pi D = 0, \qquad \pi e = 1.$$

The mean arrival rate of the process is hence

$$\lambda = \pi \sum_{k=1}^{\infty} k D_k e = \pi d.$$

One can think of D_0 as governing transitions in the phase process which do not generate arrivals and D_k as the rate of arrivals of size k (with the appropriate phase change). As a simple example, for Poisson arrivals with mean arrival rate λ, $m = 1$, $D_0 = -\lambda$, $D_1 = \lambda$ and $D_k = 0$ for all $k \geqslant 2$.

2.3.4. Fractals or self-similar arrival process

Willinger et al. (1996) describe the latest developments and advances in using self-similar traffic for performance modeling of high-speed telecommunication networks. Here, the notations and descriptions follow Willinger and Paxson (1998).

Experimental traces of traffic processes exhibit high spatial variability and long-range dependence (autocorrelations with a power law decay). Heavy-tailed distributions (such as Pareto distributions) with infinite variance are used to model the extreme spacial variability. Typical probability distributions $[F(\cdot)]$ are of the form

$$1 - F(x) = \kappa_1 x^{-\beta},$$

where κ_1 is a positive (finite) constant independent of x and the tail index β is such that $0 < \beta < 2$. A fractional Gaussian noise is used to model the fractal or long-range dependent or self-similar behavior. A covariance-stationary Gaussian process $X = (X_k: k \geqslant 1)$ is called a fractional Gaussian noise with Hurst parameter $H \in [0.5, 1)$ if the autocorrelation between X_n and X_{n+k}, $k \geqslant 0$, is given by

$$\mathrm{cor}(X_n, X_{n+k}) = 0.5\{(k+1)^{2H} - 2(k)^{2H} + (k-1)^{2H}\}.$$

The Hurst parameter H quantifies the strength of the fractal scaling.

A discrete-time, covariance-stationary, zero-mean stochastic process $X = (X_k: k \geqslant 1)$ is called exactly self-similar or fractal with scaling parameter $H \in [0.5, 1)$ if for all levels of aggregation (or resolution), $m \geqslant 1$,

$$X^{(m)} = m^{H-1} X,$$

where the aggregated processes $X^{(m)}$ are defined by

$$X^{(m)}(k) = \frac{X_{(m-1)k+1} + \cdots + X_{km}}{m}, \quad k \geqslant 1.$$

For an exactly self-similar process with scaling parameter H,

$$\mathrm{Var}\, X^{(m)} = \kappa_1 m^{2H-2}.$$

2.3.5. Fractional Brownian motion vs. Levy processes

The following is directly adapted from Konstantopoulos and Lin (1996). A Levy motion $\{Z_t\}$ is a process with stationary independent increments, and its marginal distribution is a stable random variable (invariant under affine transformations). It is a self-similar process with Hurst constant $1/\alpha$: $\{Z_{tx}, t \geq 0\} \stackrel{\mathcal{D}}{=} \{x^{1/\alpha} Z_t, t \geq 0\}$. Here, $\{Z_t\}$ is a zero mean process and Z_1 has finite β-moments for any $0 < \beta < \alpha$. Since $\alpha > 1$, $\{Z_t\}$ is a martingale, in fact an L^β martingale for any $1 < \beta < \alpha$. Note that the fractional Brownian motion is not even a semi-martingale. Konstantopoulos and Lin (1996) consider multiplexing of multiple number of sessions such that the sessions arrive according to a Poisson process with mean rate λ and each session remains active for a random time T with a heavy-tailed distribution

$$1 - F(x) = P\{T > x\} \sim \kappa x^{-\alpha}, \quad \text{as } x \to \infty$$

for a fixed $1 < \alpha < 2$. Also, when each session is active, traffic is generated at the rate of 1 per unit time. Konstantopoulos and Lin (1996) show that (i) in the limit the arrival processes converge to a Levy motion with stable non-Gaussian independent increments, and, (ii) the autocorrelation function is asymptotically that of a fractional Brownian motion and not a Levy motion!

3. Network performance using traffic models

When an application (also called source) sends a message to a destination, the message traverses several nodes (also called hops or network interfaces) before reaching the destination. The messages are stored in buffers at the nodes briefly before proceeding on to the next hop. At these nodes, traffic from other applications are either superposed along with or split from this application's traffic. In this section the traffic models considered in Section 2 will be used to model buffer content processes and evaluate performance measures.

3.1. Single class single buffer content processes

To begin the analysis, first concentrate on the simplest model: a network with a single node, a single class of traffic, hence a single buffer. The main aim is to obtain the probability distribution of the buffer contents in the long-run given a traffic model of the input to the buffer and the buffer emptying scheme.

3.1.1. Effective bandwidths of all the traffic model types

First consider the source generating traffic into the buffer. The concept of effective bandwidth of traffic generated by a source or traffic stream flowing through a telecommunication pipe or link is explained. Let $A(t)$ be the total amount of traffic (fluid or discrete) generated by a source or flowing through a pipe over time $(0, t]$. For the following analysis consider a fluid model. Note that it is straightforward to perform similar analysis for discrete models as well. Consider a stochastic process $\{Z(t), t \geq 0\}$

that models the traffic flow. Also let $r(Z(t))$ be the rate at which the traffic flows at time t. Then

$$A(t) = \int_0^t r(Z(u))\,du. \tag{4}$$

The *asymptotic log moment generating function* (ALMGF) of the traffic is defined as

$$h(v) = \lim_{t\to\infty} \frac{1}{t} \log E\{\exp(vA(t))\}. \tag{5}$$

Using Eq. (5) one can show that $h(v)$ is an increasing, convex function of v and for all $v > 0$,

$$r^{\text{mean}} \leqslant h'(v) \leqslant r^{\text{peak}}, \tag{6}$$

where

$$r^{\text{mean}} = E(r(Z(\infty))),$$

$$r^{\text{peak}} = \sup_z \{r(z)\},$$

and $h'(v)$ denotes the derivative of $h(v)$ with respect to v.

The *Effective Bandwidth* of the traffic is defined as

$$eb(v) = h(v)/v. \tag{7}$$

It can be shown that $eb(v)$ is an increasing function of v and

$$r^{\text{mean}} \leqslant eb(v) \leqslant r^{\text{peak}}$$

and

$$\lim_{v\to 0} eb(v) = r^{\text{mean}} \quad \text{and} \quad \lim_{v\to\infty} eb(v) = r^{\text{peak}}.$$

It is not easy to calculate the effective bandwidths using Eq. (7). However, when $\{Z(t),\ t \geqslant 0\}$ is a Continuous Time Markov Chain (CTMC), or a regenerative process, or a Markov Regenerative Process (MRGP), one can compute the effective bandwidths more easily. The methods are illustrated briefly.

1. **CTMC source**: Elwalid and Mitra (1993), and Kesidis et al. (1993), use eigenvalue techniques to show how to compute the effective bandwidths of sources that are modulated by CTMCs as follows. Let $\{Z(t),\ t \geqslant 0\}$ be an irreducible, finite state CTMC with generator matrix Q. When the CTMC is in state i, the source generates

fluid at rate $r(i)$. Let $R = \text{diag}[r_{ii}]$, where $r_{ii} = r(i)$. Let $e(M)$ denote the largest real-eigenvalue of a square matrix M. Then,

$$h(v) = e(Q + vR). \tag{8}$$

2. **MRGP source**: Kulkarni (1997) and Gautam (1995) show how to compute the effective bandwidths of sources that are modulated by Markov Regenerative Processes and Regenerative Processes. Let $\{Z(t), t \geq 0\}$ be an m-state Markov Regenerative Process (MRGP). Let $\{(Y_n, S_n), n \geq 0\}$ be an embedded Markov renewal sequence in the MRGP. Assume that $\{Y_n, n \geq 0\}$ is an irreducible Discrete Time Markov Chain (DTMC) with a finite state-space $\{1, 2, \ldots, m\}$. Let

$$F_1 = \int_0^{S_1} r(Z(t)) \, dt$$

be the total fluid generated by the source during $[0, S_1]$. Define

$$\Lambda_{ij}(u, v) = E\{e^{-uS_1 + vF_1}; Y_1 = j | Y_0 = i\}, \tag{9}$$

for $i, j = 1, 2, \ldots, m$ and $-\infty < u, v < \infty$. Let

$$\Lambda(u, v) = [\Lambda_{ij}(u, v)]$$

be an $m \times m$ matrix. Let $e(\Lambda(u, v))$ be the largest real-positive eigenvalue of $\Lambda(u, v)$. Define

$$e^*(v) = \sup_{\{u > 0: \, e(\Lambda(u,v)) < \infty\}} \{e(\Lambda(u, v))\}$$

and

$$u^*(v) = \inf\{u > 0: \, e(\Lambda(u, v)) < \infty\}.$$

Then for a given v,
(a) if $e^*(v) \geq 1$ (see Figure 1), $h(v)$ is a unique solution to $e(\Lambda(h(v), v)) = 1$,
(b) if $e^*(v) < 1$ (see Figure 2), $h(v) = u^*(v)$.
3. **Regenerative source**: Assume that $\{Z(t), t \geq 0\}$ is a regenerative process with regeneration epochs $\{S_n, n \geq 0\}$, with $S_0 = 0$. Let

$$F_1 = \int_0^{S_1} r(Z(t)) \, dt$$

be the total fluid generated by the source during $[0, S_1]$. Define

$$\Lambda(u, v) = E\{e^{-uS_1 + vF_1}\}. \tag{10}$$

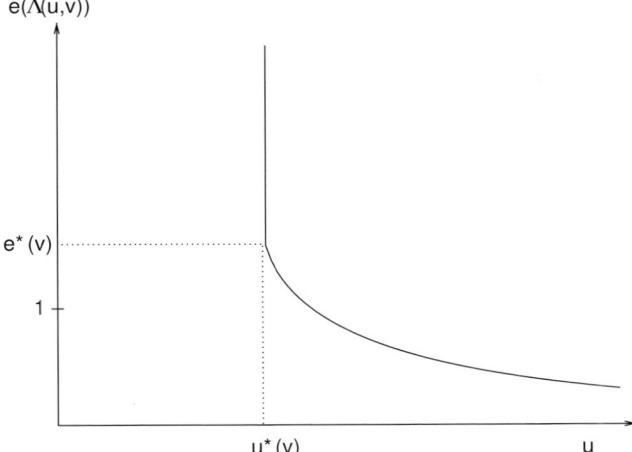

Fig. 1. $e(\Lambda(u,v))$ vs u.

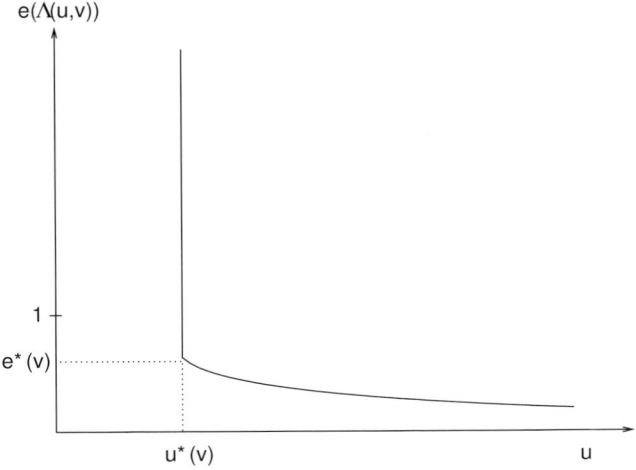

Fig. 2. $e(\Lambda(u,v))$ vs u.

Since $\Lambda(u,v)$ is a scalar, $e(\Lambda(u,v)) = \Lambda(u,v)$. Following the technique in the MRGP case above, define

$$e^*(v) = \sup_{\{u>0:\ e(\Lambda(u,v))<\infty\}} \{e(\Lambda(u,v))\}$$

and

$$u^*(v) = \inf\{u > 0:\ e(\Lambda(u,v)) < \infty\}.$$

Table 1
Effective bandwidth of input sources

Source type	Effective bandwidth
Constant arrival rate of R cells/second	R
Poisson source with intensity R cells/second	$\frac{R(e^\delta - 1)}{\delta}$
Irreducible and aperiodic *discrete-time Markov source* with transition probability matrix $\mathbf{P}$, rate matrix $\psi = \text{diag}(\psi_1, \ldots, \psi_m)$ where m is the size of the state space of of the DTMC, ψ_i is the number of cells that arrive when in state i, and $\rho(\mathbf{A})$ is the spectral radius of matrix $\mathbf{A}$	$\frac{R \log[\rho(e^{\delta\psi}\mathbf{P})]}{\delta}$
Markov Modulated Poisson Process (MMPP) with intensity $\psi = \text{diag}(\psi_i)$, where ψ is a function of a CTMC with infinitesimal generator $\mathbf{Q}$, and $\mu(\mathbf{A})$ is the largest eigenvalue of the matrix $\mathbf{A}$	$\frac{\mu(\mathbf{Q}+(e^\delta - 1)\psi)}{\delta}$

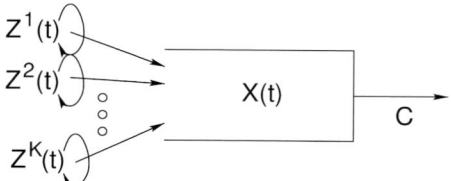

Fig. 3. Single buffer fluid model.

For a given v,
(a) if $e^*(v) \geqslant 1$, $h(v)$ is a unique solution to $e(\Lambda(h(v), v)) = 1$,
(b) if $e^*(v) < 1$, $h(v) = u^*(v)$.

Table 1 summarizes the effective bandwidths of some discrete ATM traffic in the forms of cells (see Krishnan et al. (1997) for the calculation of effective bandwidths for traffic modeled by fractional Brownian motion).

3.1.2. *Approximate methods and bounds (via martingales)*

Consider a single buffer that admits a single-class traffic from K independent sources, each driven by a random environment process $\{Z^k(t), t \geqslant 0\}$ (see Figure 3). Note that $Z^k(t)$ can be thought of as the state of the kth input source ($k = 1, 2, \ldots, K$) at time t. When source k is in state $Z^k(t)$, it generates fluid at rate $r^k(Z^k(t))$ into the buffer. Let $X(t)$ be the amount of fluid in the buffer at time t. The buffer has infinite capacity and is serviced by a channel of constant rate c. The dynamics of the buffer-content process $\{X(t),\ t \geqslant 0\}$ is described by

$$\frac{dX(t)}{dt} = \begin{cases} \sum_{k=1}^{K} r^k(Z^k(t)) - c, & \text{if } X(t) > 0, \\ \{\sum_{k=1}^{K} r^k(Z^k(t)) - c\}^+, & \text{if } X(t) = 0, \end{cases} \quad (11)$$

where $\{x\}^+ = \max(x, 0)$. The solution is given by (see Kulkarni and Rolski, 1994)

$$X(t) = \sup_{0 \leqslant u \leqslant t} \left(Y(t), \int_u^t \left(\sum_{k=1}^{K} r^k(Z^k(s)) - c \right) ds \right),$$

where

$$Y(t) = X(0) + \int_0^t \left(\sum_{k=1}^{K} r^k(Z^k(s)) - c \right) ds.$$

It has been shown in Kulkarni and Rolski (1994) that the buffer content process $\{X(t),\ t \geqslant 0\}$ is stable if

$$\sum_{k=1}^{K} E\{r^k(Z^k(\infty))\} < c, \quad (12)$$

in which case $X(t) \to X$ in distribution with

$$X = \sup_{u \leqslant 0} \int_u^0 \left(\sum_{k=1}^{K} r^k(Z^k(s)) - c \right) ds. \quad (13)$$

Exact expressions for the buffer content distribution $P\{X > x\}$ can be obtained only for special environment processes like CTMCs.

Let $h_k(v)$ and $eb_k(v)$ be the ALMGF and effective bandwidths of source k respectively. In Kesidis et al. (1993), it is shown that

$$\lim_{B \to \infty} P(X > B) e^{B\theta} \to \omega, \quad (14)$$

for some positive finite constant ω, where θ is the solution to

$$\sum_{k=1}^{K} \frac{h_k(\theta)}{\theta} = c. \quad (15)$$

Therefore the effective bandwidth approximation states that one can show that in the long-run, the buffer content distribution $P(X > B) \approx e^{-B\theta}$ (neglect the effect of ω by setting it to 1).

CDE approximation: It can be shown that the effective-bandwidth approximation is very conservative for most engineering applications, mainly because the statistical multiplexing gains are not taken advantage of. In this subsection the use of Chernoff Dominant Eigenvalue (CDE) approximation (see Elwalid et al., 1995 and Elwalid and Mitra, 1995) to further fine tune the effective-bandwidth analysis is explained.

Consider the model in Figure 3. The CDE approximation for the tail probability is given by

$$P\{X > x\} = \lim_{t \to \infty} P\{X(t) > B\} \approx L e^{-\theta B}, \tag{16}$$

where L is the fraction of the fluid that would be lost if there was no buffer and θ is as in Eq. (15). Note that L is an estimate of ω in Eq. (14).

Mathematically, L can be written as

$$L = \lim_{t \to \infty} \frac{\int_0^t \{[\sum_{k=1}^K r^k(Z^k(t))] - c\}^+ dt}{\int_0^t \{\sum_{k=1}^K r^k(Z^k(t))\} dt}. \tag{17}$$

Note that L is a function of c and the parameters of each of the K sources. Typically it may not be computationally simple to calculate L exactly in many applications. Hence Elwalid et al. (1995) suggest a method of estimating L by using Chernoff's theorem.

The input sources are characterized by a function $m_k(w)$, which is similar to the ALMGF ($h_k(v)$), and is defined as

$$m_k(w) = \lim_{t \to \infty} \log E\{\exp(wr^k(Z^k(t)))\}. \tag{18}$$

Let

$$s^* = \sup_{w \geq 0} \left\{ cw - \sum_{k=1}^K m_k(w) \right\}$$

and w^* be obtained by solving

$$\sum_{k=1}^K m'_k(w^*) = c.$$

Then the Chernoff estimate of L as given in Elwalid et al. (1995) and Elwalid and Mitra (1995) is

$$L \approx \frac{\exp(-s^*)}{w^* \sigma(w^*) \sqrt{2\pi}}, \tag{19}$$

where

$$\sigma^2(w^*) = \sum_{k=1}^K m''_k(w^*).$$

The main problem in the above analysis is computing $m_k(w)$. If $\{Z^k(t),\ t \geq 0\}$ can be modeled as a stationary and ergodic process with state space $\mathcal{S}$ and stationary probability vector, π, then

$$m_k(w) = \log\left\{\sum_{j\in\mathcal{S}} \pi_k^j e^{wr^k(j)}\right\}. \tag{20}$$

SMP bounds: Consider the case when $\{Z^k(t),\ t \geq 0\}$ ($k = 1, 2, \ldots, K$) are independent semi-Markov processes (SMPs) with state space $\mathcal{S}_k = \{1, 2, \ldots, \ell_k\}$ and kernel $G^k(x) = [G_{ij}^k(x)]$. The expected time the kth SMP spends in state i is τ_i^k. The stationary distribution of the kth SMP $\{Z^k(t), t \geq 0\}$ is p^k, where

$$p_i^k = \lim_{t\to\infty} P\{Z^k(t) = i\}.$$

First the computation of $eb_k(v)$ is described. Let $\widetilde{G}_{ij}(s)$ be the Laplace–Stieltjes transform (LST) of $G_{ij}(x)$. For a given $v > 0$, define

$$\chi_{ij}^k(v, u) = \widetilde{G}_{ij}^k\bigl(-v(r_k(i) - u)\bigr),$$

$$\chi^k(v, u) = [\chi_{ij}^k(v, u)].$$

Then $eb_k(v)$ is given by the smallest positive number such that the Perron–Frobenius eigenvalue of $\chi^k(v, eb_k(v))$ is one. Let η be a solution to Eq. (15), and denote $\Phi(\eta) = \chi(\eta, eb_k(\eta))$. Let h^k be the left eigenvector of $\Phi(\eta)$ corresponding to the eigenvalue 1, i.e.,

$$h^k = h^k \Phi^k(\eta).$$

Also,

$$P^k(i, j) = [G^k(\infty)]_{ij}. \tag{21}$$

Also define

$$H^k = \sum_{i=1}^{\ell_k} \frac{h_i^k}{\eta(r_k(i) - eb_k(\eta))} \left(\sum_{j=1}^{\ell_k} (\phi_{ij}^k(\eta)) - 1\right), \tag{22}$$

$$\Psi_{\min}^k(i, j) = \inf_x \left\{ \frac{h_i^k e^{-\eta(r_k(i) - eb_k(\eta))x} \int_x^\infty e^{\eta(r_k(i) - eb_k(\eta))y}\, \mathrm{d}G_{ij}^k(y)}{\frac{p_i^k}{\tau_i^k} \int_x^\infty \mathrm{d}G_{ij}^k(y)} \right\}, \tag{23}$$

and

$$\Psi_{\max}^k(i,j) = \sup_x \left\{ \frac{h_i^k e^{-\eta(r_k(i)-eb_k(\eta))x} \int_x^\infty e^{\eta(r_k(i)-eb_k(\eta))y} \, dG_{ij}^k(y)}{\frac{p_i^k}{\tau_i^k} \int_x^\infty dG_{ij}^k(y)} \right\}. \quad (24)$$

From Gautam et al. (1999),

$$C_* e^{-\eta x} \leqslant P(X > x) \leqslant C^* e^{-\eta x}, \quad x \geqslant 0, \quad (25)$$

where

$$C^* = \frac{\prod_{k=1}^K H^k}{\min_{\mathcal{A}} \prod_{k=1}^K \Psi_{\min}^k(i_k, j_k)}, \quad C_* = \frac{\prod_{k=1}^K H^k}{\max_{\mathcal{A}} \prod_{k=1}^K \Psi_{\max}^k(i_k, j_k)},$$

$$\mathcal{A} = \left\{ (i_1, j_1), (i_2, j_2), \ldots, (i_K, j_K) \colon i_k, j_k \in \mathcal{S}_k, \sum_{k=1}^K r_k(i_k) > c \text{ and} \right.$$

$$\left. \forall k, \ P^k(i_k, j_k) > 0 \right\}. \quad (26)$$

Computation of $\Psi_{\max}$ and $\Psi_{\min}$: Consider a nonnegative random variable Y with distribution $G_{ij}(x)/G_{ij}(\infty)$ and density

$$g_{ij}(x) = \frac{dG_{ij}(x)}{dx} \frac{1}{G_{ij}(\infty)}.$$

The failure rate function of Y is defined by

$$\lambda_{ij}(x) = \frac{g_{ij}(x)}{1 - \frac{G_{ij}(x)}{G_{ij}(\infty)}}. \quad (27)$$

Y is said to be an increasing failure rate (IFR) random variable if

$$\lambda_{ij}(x) \uparrow x$$

and Y is said to be a decreasing failure rate (DFR) random variable if

$$\lambda_{ij}(x) \downarrow x.$$

It is possible to obtain closed form algebraic expressions for $\Psi_{\max}(i,j)$ and $\Psi_{\min}(i,j)$ in Eqs. (24) and (23) respectively if a random variable Y with distribution

$G_{ij}(x)/G_{ij}(\infty)$ is an IFR or DFR random variable. The following notation is used to compute $\Psi_{\max}(i, j)$ and $\Psi_{\min}(i, j)$ in those cases. Let x^* and x_* be such that

$$x^* = \arg\sup_x \left\{ \frac{h_i \int_x^\infty e^{\eta(r_i-c)y} \, dG_{ij}(y)}{\frac{p_i}{\tau_i} e^{\eta(r_i-c)x} \int_x^\infty dG_{ij}(y)} \right\} \tag{28}$$

and

$$x_* = \arg\inf_x \left\{ \frac{h_i \int_x^\infty e^{\eta(r_i-c)y} \, dG_{ij}(y)}{\frac{p_i}{\tau_i} e^{\eta(r_i-c)x} \int_x^\infty dG_{ij}(y)} \right\}. \tag{29}$$

If Y is IFR or DFR, then $\Psi_{\max}(i, j)$ and $\Psi_{\min}(i, j)$ in Eqs. (24) and (23) respectively occur at x values given by the following table

	IFR		DFR	
	$r_i > c$	$r_i \leqslant c$	$r_i > c$	$r_i \leqslant c$
x^*	0	∞	∞	0
$\Psi_{\max}(i,j)$	$\frac{\phi_{ij}(-\eta(r_i-c))\tau_i h_i}{p_{ij} p_i}$	$\frac{\tau_i h_i \lambda_{ij}(\infty)}{p_i(\lambda_{ij}(\infty)-\eta(r_i-c))}$	$\frac{\tau_i h_i \lambda_{ij}(\infty)}{p_i(\lambda_{ij}(\infty)-\eta(r_i-c))}$	$\frac{\tilde\phi_{ij}(-\eta(r_i-c))\tau_i h_i}{p_{ij} p_i}$
x_*	∞	0	0	∞
$\Psi_{\min}(i,j)$	$\frac{\tau_i h_i \lambda_{ij}(\infty)}{p_i(\lambda_{ij}(\infty)-\eta(r_i-c))}$	$\frac{\tilde\phi_{ij}(-\eta(r_i-c))\tau_i h_i}{p_{ij} p_i}$	$\frac{\tilde\phi_{ij}(-\eta(r_i-c))\tau_i h_i}{p_{ij} p_i}$	$\frac{\tau_i h_i \lambda_{ij}(\infty)}{p_i(\lambda_{ij}(\infty)-\eta(r_i-c))}$

where

$$\lambda_{ij}(\infty) = \lim_{x \to \infty} \lambda_{ij}(x).$$

3.2. Multiple class (single) node models with multiplexing

In this section the single class model results obtained in Section 3.1 to solve scenarios in multi-class nodes by making suitable transformations. Consider the model of a multi-class node illustrated in Figure 4. The node consists of N input buffers, one for each class of traffic. The input to buffer j ($j = 1, \ldots, N$), is from the K_j sources of class j. The ith source of class j is driven by an independent random environment process $Z_{ij} = \{Z_{ij}(t), t \geqslant 0\}$ for $i = 1, 2, \ldots, K_j$. At time t, source i of type j generates fluid at rate $r_{ij}(Z_{ij}(t))$. Let $X_j(t)$ be the amount of fluid in buffer j at time t. All the classes of fluids are served by a single channel of constant capacity c, using a specified service scheduling policy. Three policies are studied here: timed round robin (polling) policy, static priority service policy, and, generalized processor sharing (GPS) policy.

Assume that all N buffers are of infinite capacity. If B_j is the actual size of buffer j ($j = 1, 2, \ldots, N$), then

$$\lim_{t \to \infty} P\{X_j(t) > B_j\} = P\{X_j > B_j\}$$

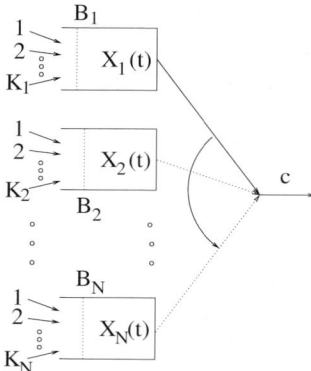

Fig. 4. The multi-class node model.

is the steady-state approximation of the overflow probability from buffer j. Let ε_j be the cell loss probability target for class j traffic ($j = 1, 2, \ldots, N$). The Quality of Service (QoS) criterion for loss probability that need to be satisfied class j traffic is

$$\lim_{t \to \infty} P\{X_j(t) > B_j\} = P\{X_j > B_j\} < \varepsilon_j. \tag{30}$$

Note that although bounds can be obtained for the delay, explicit expressions for delay QoS is a research issue that needs to be addressed. Also, delay-jitter QoS measures are research problems to be studied.

The three service scheduling policies, timed round robin policy, static priority service policy, and, GPS policy are discussed. Note that the effective bandwidth and the SMP bounds analysis for the multiclass model is not a trivial extension of that of the single class model. The output channel capacity for each buffer is not a constant in the multiclass node model. Therefore the model requires a careful transformation that results in a constant output channel capacity model for each of the buffers. From the transformed models, $P\{X_j > B_j\}$ needs to be computed.

3.2.1 Timed round robin (polling)

Consider the multi-class node model described in Section 3.2 and illustrated in Figure 4. All classes of fluids are multiplexed using a *Timed Round Robin* service scheduling policy which is described as follows. The scheduler allocates the entire output capacity c to each of the N buffers in a cyclic fashion. In each cycle, buffer j gets the entire capacity for an interval of length τ_j. Note that during this interval, buffer j could be empty. Hence the scheduler is not work conserving.

Let t_{so} be the total switch-over time during an entire cycle. Assume that t_{so} does not change with time. The *cycle time* T is defined as the amount of time the scheduler takes to complete a cycle, and is given by

$$T = t_{so} + \sum_{j=1}^{N} \tau_j. \tag{31}$$

First assume that all buffers are of infinite capacity. The dynamics of the buffer-content process $\{X_j(t), t \geq 0\}$ is described by

$$\frac{dX_j(t)}{dt} = \begin{cases} \sum_{i=1}^{K_j} r_{ij}(Z_{ij}(t)) - c, \\ \quad \text{if } X(t) > 0 \text{ and scheduler serving buffer } j, \\ \{\sum_{i=1}^{K_j} r_{ij}(Z_{ij}(t)) - c\}^+, \\ \quad \text{if } X(t) = 0 \text{ and scheduler serving buffer } j, \\ \sum_{i=1}^{K_j} r_{ij}(Z_{ij}(t)), \\ \quad \text{if scheduler not serving buffer } j. \end{cases} \tag{32}$$

Assume that the following stability condition is satisfied for buffer j ($j = 1, \ldots, N$)

$$\sum_{i=1}^{K_j} E\{r_{ij}(Z_{ij}(\infty))\} < c \frac{\tau_j}{T}. \tag{33}$$

Effective bandwidth analysis: If given $\tau_1, \tau_2, \ldots, \tau_N$ and t_{so} are given, then the buffer contents of a given buffer (say j) and its dynamics do not depend on the parameters of any other buffer (say $i \neq j$). Therefore, it is convenient to analyze each buffer separately. Buffer j can be modeled as a single-buffer-fluid model with variable output capacity and input from K_j different sources, such that source i of class j is modulated by an environmental process $\{Z_{ij}(t), t \geq 0\}$. The output capacity alternates between c (for τ_j units of time) and 0 (for $T - \tau_j$ units of time).

Note that the effective-bandwidth approximation and the SMP bounds assume that the output channel capacity is a constant. Therefore to utilize those techniques, one needs to first transform the model into an appropriate one with a constant output channel capacity as follows:

Consider a single-buffer-fluid model for buffer j with a constant output channel capacity c whose input is generated by the original K_j sources and a fictitious compensating source. The compensating source is such that it stays on for a deterministic amount of time $T - \tau_j$ and off for a deterministic amount of time τ_j. When the compensating source is on, it generates fluid at rate c and when it is off it generates fluid at rate 0. Note that the compensating source is independent of the original K_j sources. Clearly, the dynamics of the buffer-content process (of buffer j) in Eq. (32) remain unchanged for this transformed single-buffer-fluid model with $K_j + 1$

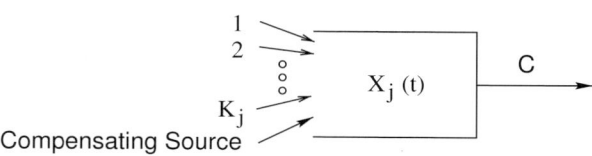

Fig. 5. Transformed buffer j model.

input sources (including the compensating source) and constant output capacity c. Refer to Figure 5 for an illustration of the transformed model for buffer j.

Using the effective bandwidth computations in Kulkarni (1997), one can show that the effective bandwidth of the compensating source described above is given by

$$eb_j^s(v) = \frac{c(T - \tau_j)}{T}. \tag{34}$$

Note that the effective bandwidth of this deterministic source is indeed its mean traffic generation rate. Let the effective bandwidth of source i ($i = 1, 2, \ldots, K_j$) of class j be $eb_{ij}(v)$. Therefore $P(X_j > B_j) \approx e^{-B_j \eta_j}$, where η_j (using Eq. (15)) is obtained by solving

$$\sum_{i=1}^{K_j} eb_{ij}(\eta_j) + c \frac{(T - \tau_j)}{T} = c. \tag{35}$$

The loss probability QoS criteria for all the classes of traffic are satisfied if for all $j = 1, 2, \ldots, N$,

$$e^{-B_j \eta_j} < \varepsilon_j. \tag{36}$$

Hence Eq. (35) indicates that the QoS guarantee using the effective-bandwidth approximation technique depends only on the ratio τ_j/T and not the individual values of τ_j or T. Consider two instances, one with large τ_j and T and the other with small τ_j and T, such that the ratio τ_j/T is the same in both instances. The effective bandwidth approximation implies that the loss probability will be less than ε_j in both instances. This goes against intuition. It is theoretically valid since the effective-bandwidth analysis assumes that $B_j \to \infty$. However in practice, this cannot be valid due to finite buffers.

Therefore the effective-bandwidth approximation technique fails for moderate to large sized buffers and works only for extremely large sized buffers! The Chernoff dominant eigenvalue approximation (see Elwalid and Mitra, 1995) also faces the same problem. The SMP bounds below resolve this issue.

Semi-Markov process (SMP) bounds analysis: Consider the transformed model of buffer j ($j = 1, 2, \ldots, N$) illustrated in Figure 5. Assume that the $\{Z_{ij}(t), t \geq 0\}$ processes ($i = 1, 2, \ldots, K_j$) are semi-Markov processes. Therefore there are $K_j + 1$

independent sources modulated by SMPs (including the compensating source) that generate traffic into buffer j whose the output capacity is a constant c.

For the SMP bounds analysis for buffer j, follow the single-class traffic analysis in Section 3.1.2 for a buffer with input generated by independent semi-Markovian sources multiplexed together. Let η_j be the smallest positive solution to Eq. (35).

Using Eqs. (22), (23) and (24), one can obtain H^{ij}, $\Psi_{\min}^{ij}$ and $\Psi_{\max}^{ij}$ respectively for source i ($i = 1, 2, \ldots, K_j$) of class j. The corresponding expressions H^{sj}, $\Psi_{\min}^{sj}$ and $\Psi_{\max}^{sj}$ for the jth compensating source are

$$H^{sj} = \frac{1 - \exp(-\eta_j c \frac{T-\tau_j}{T} \tau_j)}{\eta_j c} \left[\frac{T^2}{(T-\tau_j)\tau_j} \right], \tag{37}$$

$$\Psi_{\min}^{sj} = \begin{bmatrix} 0 & T \exp(-\eta_j c \frac{T-\tau_j}{T} \tau_j) \\ T \exp(-\eta_j c \frac{T-\tau_j}{T} \tau_j) & 0 \end{bmatrix}, \tag{38}$$

$$\Psi_{\max}^{sj} = \begin{bmatrix} 0 & T \\ T & 0 \end{bmatrix}. \tag{39}$$

Letting $s = K_j + 1$, the bounds on the limiting distribution of the buffer content process $\{X_j(t), t \geq 0\}$ as

$$C_{j*} e^{-\eta_j x} \leq P(X_j > x) \leq C_j^* e^{-\eta_j x},$$

where, η_j is from Eq. (35),

$$C_j^* = \frac{\prod_{k=1}^{K_j+1} H^{kj}}{\min_{\mathcal{A}^j} \prod_{k=1}^{K_j+1} \Psi_{\min}^{kj}(l_k, m_k)}, \tag{40}$$

$$C_{*j} = \frac{\prod_{k=1}^{K_j+1} H^{kj}}{\max_{\mathcal{A}^j} \prod_{k=1}^{K_j+1} \Psi_{\max}^{kj}(l_k, m_k)}, \tag{41}$$

and

$$\mathcal{A}^j = \left\{ (l_1, m_1), (l_2, m_2), \ldots, (l_{K_j+1}, m_{K_j+1}) : \\ l_k, m_k \in \mathcal{S}_k, \sum_{k=1}^{K_j+1} r_{kj}(l_k) > c \text{ and } \forall k, \, P^{kj}(l_k, m_k) > 0 \right\}. \tag{42}$$

The QoS criteria for all the classes of traffic are satisfied if, for $j = 1, 2, \ldots, N$,

$$C_j^* e^{-\eta_j B_j} < \varepsilon_j. \tag{43}$$

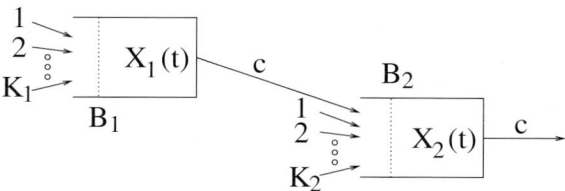

Fig. 6. The transformed model.

Clearly, H^{sj} and $\Psi_{\min}^{sj}$ are functions of τ_j, T and τ_j/T. Hence, C_j^* is a function of both τ_j and T and not simply of the ratio τ_j/T.

3.2.2. Static priority service policy

In this section, consider a *Static Priority Service Policy* (for the model in Section 3.2 and illustrated in Figure 4) to multiplex the multi-class traffic which operates as follows. Under this service policy, traffic of class j has higher service priority over traffic of class i, if $i > j$. The scheduler serves the traffic of class j only if there is no fluid of higher priority in the buffers. Thus all the available channel capacity (a maximum of c) is assigned for the class-1 fluid and the leftover channel capacity (if any) that class-1 does not need, to class-2 fluid. Any leftover channel capacity that class-1 and class-2 do not need, is assigned to class-3 fluid, and so on.

For a comprehensive study on effective bandwidths with priorities, see Berger and Whitt (1998a) and (1998b). See Gautam and Kulkarni (1997) for the study of effective bandwidth and CDE approximations for the static priority case. Here a two-class traffic case is explained, although the analysis can be extended to more than 2 classes. The K_j class-j sources, $j = 1, 2$, are independent and identical on-off sources with exponential on and off times, on-time parameter α_j, off-time parameter β_j and on-time rate r_j.

Consider the transformed model in Figure 6. The sample paths of the buffer content processes $\{X_1(t), t \geq 0\}$ and $\{X_2(t), t \geq 0\}$ in this model are identical to those in the original model in Figure 4 (for $N = 2$). This observation is made in Elwalid and Mitra (1995) and is immensely useful in the analysis. Note that the output from buffer 1 can be modeled as an SMP. Hence, the input to buffers 1 and 2 can be modeled as ones with multiplexing independent SMP sources.

Buffer 1: If $K_1 \leq c/r_1$, then $P\{X_1 > B_1\} = 0$, since buffer 1 will always be empty. Now for the case $K_1 > c/r_1$, the steady-state distribution of the buffer content process is bounded as

$$C_{*1} e^{-\eta_1 B_1} \leq P\{X_1 > B_1\} \leq C_1^* e^{-\eta_1 B_1},$$

where

$$\eta_1 = \frac{K_1(c\alpha_1 + c\beta_1 - K_1\beta_1 r_1)}{c(K_1 r_1 - c)},$$

$$C_1^* = \frac{\left(\frac{K_1 r_1}{K_1 r_1 - c} \frac{\alpha_1}{\alpha_1 + \beta_1}\right)^{K_1}}{\left(\frac{c\alpha_1}{\beta_1(K_1 r_1 - c)}\right)^{\lceil \frac{c}{r_1} \rceil}},$$

and

$$C_{*1} = \left(\frac{K_1 r_1 \beta_1}{c(\alpha_1 + \beta_1)}\right)^{K_1}. \tag{44}$$

Buffer 2. First model the K_2 exponential on-off sources as a single (K_2+1)-state SMP with the states denoting the number of priority-2 sources that are on and then derive expressions for H^1, $\Psi_{\max}^1(i,j)$ and $\Psi_{\min}^1(i,j)$ as defined in Eqs. (22), (23) and (24). In Kulkarni and Gautam (1997) it is shown that the output process from buffer 1 can be modeled as an SMP. The corresponding expressions H^2, $\Psi_{\max}^2(i,j)$ and $\Psi_{\min}^2(i,j)$ for the SMP model of the output from buffer 1 can be derived. Therefore one can analyze the input to buffer 2 as traffic from two sources (output from buffer 1 and the (K_2+1)-state SMP), each modulated by an SMP.

Begin by obtaining η_2. Note that η_2 solves either

$$K_1 eb_1(\eta_2) + K_2 eb_2(\eta_2) = c \quad \text{and} \quad \eta_2 \leqslant v^* \tag{45}$$

or

$$\frac{v^*}{\eta_2} K_1 eb_1(v^*) + K_2 eb_2(\eta_2) = \frac{cv^*}{\eta_2} \quad \text{and} \quad \eta_2 > v^*, \tag{46}$$

where

$$v^* = \frac{\beta_1}{r_1}\left(\sqrt{\frac{c\alpha_1}{\beta_1(K_1 r_1 - c)}} - 1\right) + \frac{\alpha_1}{r_1}\left(1 - \sqrt{\frac{\beta_1(K_1 r_1 - c)}{c\alpha_1}}\right),$$

and for $j = 1, 2$

$$eb_j(v) = \frac{r_j v - \alpha_j - \beta_j + \sqrt{(r_j v - \alpha_j - \beta_j)^2 + 4\beta_j r_j v}}{2v}. \tag{47}$$

Therefore using the expressions for H^1, $\Psi_{\max}^1(i,j)$, $\Psi_{\min}^1(i,j)$, H^2, $\Psi_{\max}^2(i,j)$ and $\Psi_{\min}^2(i,j)$ we have

$$C_2^* = \frac{H^1 H^2}{\min_{(i_1,j_1),(i_2,j_2):\,\min\{i_1 r_1, c\} + i_2 r_2 > c,\, p_{i_1 j_1} > 0,\, p_{i_2 j_2} > 0} \Psi_{\min}^1(i_1, j_1) \Psi_{\min}^2(i_2, j_2)}$$

and

$$C_{*2} = \frac{H^1 H^2}{\max_{(i_1,j_1),(i_2,j_2):\,\min\{i_1 r_1, c\} + i_2 r_2 > c,\, p_{i_1 j_1} > 0,\, p_{i_2 j_2} > 0} \Psi_{\max}^1(i_1, j_1) \Psi_{\max}^2(i_2, j_2)}.$$

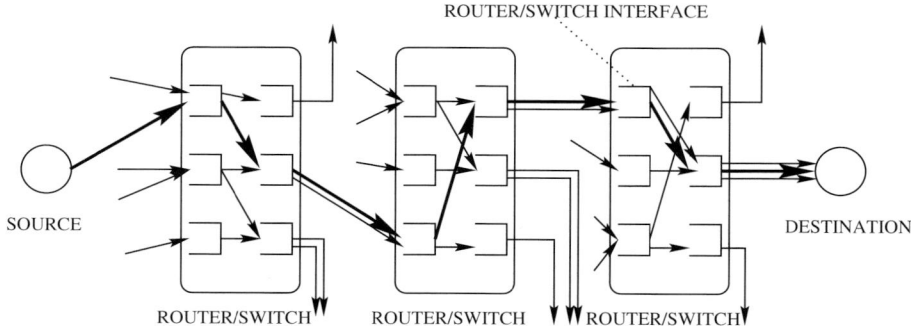

Fig. 7. Example scenario.

3.2.3. Generalized processor sharing (GPS)

Consider the multi-class node model described in Section 3.2 and illustrated in Figure 4. All classes of fluids are multiplexed using a *generalized processor sharing* service scheduling policy which is described in the following manner. Consider preassigned numbers $\phi_1, \phi_2, \ldots, \phi_N$ for each of the N classes numbers such that if all the input buffers have traffic entering, the scheduler allocates output capacity c in the ratio $\phi_1 : \phi_2 : \ldots : \phi_N$ to each of the N buffers. If one or more of the buffers are empty and no traffic enters those buffers, then the capacity c is divided in the ratio of ϕ_j's of the remaining buffers.

The discrete version of the GPS is called the Packetized General Processor Sharing (PGPS). The PGPS service policy is based on the generalized processor sharing approach explained in Parekh and Gallager (1993). The Quality-of-Service aspects, effective bandwidths, admission control, etc. for the GPS and PGPS have been addressed in detail in de Veciana et al. (1995) and de Veciana and Kesidis (1994). The PGPS is also known as "weighted fair queueing" in the literature.

3.3. Network of nodes models with single and multiple class models

The scenario for this section is depicted in Figure 7. Consider data flowing from a source to a destination through routers and switches in a high speed network handling multiple classes of traffic. The different classes of traffic are differentiated according to the priorities they receive. For example, critical information could receive very high priority, latency sensitive applications moderately high priority, and latency insensitive applications such as email very low priority. When packets belonging to different classes or priorities arrive at a router or switch interface, the packets are flushed out of the interface buffer using a given scheduling mechanism (such as static priority, round-robin polling, generalized processor sharing, etc.). Traffic belonging to all classes are multiplexed when they leave a router or switch interface. However when the traffic stream encounters another router or switch interface downstream, it gets demultiplexed into its original classes. The objective of the network is to provide guaranteed end-to-end QoS.

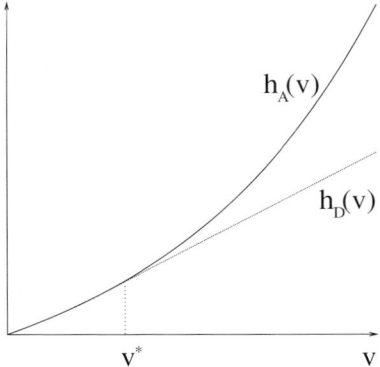

Fig. 8. $h_A(v)$ and $h_D(v)$ vs v.

For the single class case, the following is the analysis for determining the effective bandwidth of the output traffic from a buffer. The output traffic from a buffer acts as input traffic for a downstream node in a network. Typically, it may not be possible to characterize some output processes as tractable stochastic processes and compute the effective bandwidths. Consider a network with nodes in tandem where the output from one buffer acts as input to another buffer. Clearly, the output cannot be characterized by simple processes like a CTMC. Although it can be shown to be a regenerative source, the characterization is intractable.

Chang and Thomas (1995), Chang and Zajic (1995) and de Veciana et al. (1994), derive the effective bandwidths of the output of a node in terms of the input source. Let $A(t)$ be the total input to the buffer over $(0, t]$ and $D(t)$ be the total output from the buffer of a single-buffer-fluid model over $(0, t]$. The capacity of the output channel is c. Analogous to Eq. (5), define the ALMGF of the output as

$$h_D(v) = \lim_{t \to \infty} \frac{1}{t} \log E\{\exp(vD(t))\}. \tag{48}$$

If $h_A(v)$ is the ALMGF of the input to a buffer then the ALMGF of the output (of capacity c) from the buffer is given in terms of $h_A(v)$ by

$$h_D(v) = \begin{cases} h_A(v), & \text{if } 0 \leqslant v \leqslant v^*, \\ h_A(v^*) - cv^* + cv, & \text{if } v > v^*, \end{cases} \tag{49}$$

where v^* is obtained by solving for v in the equation,

$$\frac{d}{dv}[h_A(v)] = c.$$

Figure 8 illustrates the relationship between $h_A(v)$ and $h_D(v)$. The QoS problem for different types of single-class networks are studied by de Veciana et al. (1994) and (1995) and Gautam (1995). Essentially the sum of the effective bandwidths of all the traffic sources into a buffer should be compared to the output capacity. Also since the route taken by a traffic stream is known, it is easy to use the effective bandwidth of the output from a node to derive the effective bandwidth of a downstream node.

One of the important research issues is to obtain end-to-end performance measures in multi-class networks. Since the multiple classes of traffic use a common service scheduling mechanism that would result in a low volume of a particular class and high volume of traffic of another class. Intuitively the multiple classes of traffic are negatively correlated. This could potentially be used to obtain conservative estimates of QoS measures by exploiting stochastic monotonicity properties. Using the results in Puhalskii and Whitt (1999) for functional large deviations principle for waiting and departure processes, it is possible to obtain the required performance measures.

4. LAN (multiaccess communication) models

4.1. Slotted and unslotted Aloha

One of the foremost multiaccess communication protocol is the Aloha, developed in the University of Hawaii. The following stochastic models for slotted and unslotted Aloha are adapted from Kulkarni (1995). Assume that there are N users at geographically diverse locations that transmit messages (in the form of packets) via satellites. In the **slotted Aloha** version, it is assumed that the clocks of all users are synchronized. Therefore at time slots $n = 1, 2, 3, \ldots$, each user, independent of other users, transmits a packet with probability p. If more than one user transmits a packet during a given slot, then a collision results between all the packets and the resulting message is garbled. All the users involved in a collision retransmit at the beginning of a slot with probability r, however, if a user has a message to retransmit no new messages are transmitted by this user. If a user has a packet to retransmitted, then this users is termed a "backlogged" user. Let X_n be the number of backlogged users at the beginning of the nth slot. Clearly there will be $N - X_n$ unbacklogged users at the beginning of the nth slot. It can be shown that the process $\{X_n, n \geq 0\}$ is a DTMC (as shown in Kulkarni, 1995) with transition probability matrix that can be derived using:

$$P\{X_{n+1} = i - 1 | X_n = i\} = (1 - p)^{N-i} i r (1 - r)^{i-1},$$

$$P\{X_{n+1} = i + 1 | X_n = i\} = (N - i) p (1 - p)^{N-i-1} \left(1 - (1 - r)^i\right),$$

$$P\{X_{n+1} = i + j | X_n = i\} = \binom{N - i}{j} p^j (1 - p)^{N-i-j}, \quad 2 \leq j \leq N - i,$$

$$P\{X_{n+1} = i | X_n = i\} = (N - i) p (1 - p)^{N-i-1} (1 - r)^i$$
$$+ (1 - p)^{N-i} \left(1 - i r (1 - r)^{i-1}\right).$$

There are several modifications to the slotted Aloha that have eventually resulted in efficient satellite communications.

One of the **unslotted Aloha** versions considers a system where each user, when not backlogged, generates messages according to a Poisson process. Message transmission times are according to an exponential distribution. A collision results when a user attempts to transmit while another user is transmitting. If a collision results, all transmissions are terminated instantaneously. All messages involved in a collision wait for an exponential time before attempting to retransmit. If $X(t)$ denotes the number of backlogged messages at time t and $Y(t)$ a binary variable that denotes whether or not a message is under transmission at time t. It is possible to model $\{(X(t), Y(t)), t \geqslant 0\}$ as a CTMC. Using the steady-state probability distributions of the DTMC for the slotted Aloha and the CTMC for the unslotted Aloha, performance measures such as throughput (expected number of successful transmissions per unit time), satellite utilization, expected number of backlogged messages, expected delay in successfully transmitting a message, etc. can be computed. Using the performance measures it is possible to derive optimal designs for the Aloha systems.

4.2. Ethernet models

The most popular local area network (LAN) is the Ethernet (see Walrand and Varaiya, 1996). The popularity is due to the high performance and low cost. The protocol used in Ethernet is CSMA/CD (Carrier Sense Multiple Access with Collision Detection). The following is a simplified model of an Ethernet where there are a number of identical nodes (say N) connected onto a common cable. A significant portion of the following description is adapted from Bertsekas and Gallager (1992).

When one node transmits a packet (and the others are silent), all the other nodes hear that packet. In addition, a node can listen to the cable before transmitting (i.e., conceptually, 0, 1, and idle can be distinguished). Finally, because of the physical properties of the cable, it is possible for a node to listen to the cable while transmitting. Thus, if two nodes start to transmit almost simultaneously, they will shortly detect a collision in process and both cease transmitting. This technique is called CSMA/CD. On the other hand, if one node starts transmitting and no other node starts before the first node's signal has propagated throughout the cable, the first node is guaranteed to finish its packet without collision. Thus, the first portion of a packet can be viewed as making a reservation for the rest of the packet.

Slotted CSMA/CD in Ethernet, DTMC model: For analytic purposes, it is easier to visualize Ethernet in terms of megaslots and minislots. The minislots are of duration β, which denotes the time required for a signal to propagate from one end of the cable to the other and to be detected. If the nodes are all synchronized into minislots of this duration, and if only one node transmits in a minislot, all other nodes will detect the transmission and not use subsequent minislots until the entire packet is completed. If more than one node transmits in a minislot, each transmitting node will detect the condition by the end of the minislot and cease transmitting. Packets or messages are backlogged for retransmission when there is a collision.

To model the system, first assume that the minislots could be of 3 types – no transmission (idle), one transmission (one) and many transmissions (many). An *idle-minislot* is followed by another *idle-minislot* if none of the users (backlogged or non-backlogged) decide to transmit. An *idle-minislot* is followed by a *many-minislot* if many of the users (backlogged or non-backlogged) decide to transmit which results in a

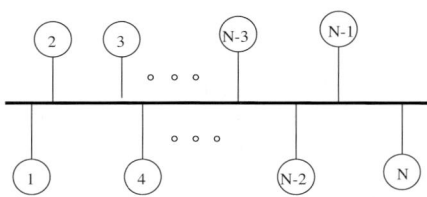

Fig. 9. A simplified model for Ethernets.

collision and messages are backlogged. An *idle-minislot* is followed by a *one-minislot* if exactly one of the users (backlogged or non-backlogged) decides to transmit. A *one-minislot* is always followed by a megaslot (when the entire packet is transmitted without collision). A *many-minislot* is always followed by an *idle-minislot*. All minislots (idle, one and many) are of duration β. The megaslot is of random duration with mean α (here it is assumed that megaslots are of constant duration α) during which only one message is transmitted successfully. A megaslot is always followed by an *idle-minislot*. It is assumed that at the end of an *idle-minislot*, each backlogged user will attempt to retransmit with probability r and each non-backlogged user will attempt to transmit with probability p. The system can be modeled as a DTMC where the state of the DTMC at the end of a slot is the number of backlogged users and the type of slot (idle, one, many, mega).

Unslotted CSMA/CD in Ethernet, CTMC model: Messages (packets) are generated by each of the N nodes according to Poisson processes. As soon as a message (packet) is generated, the node attempts to transmit it onto the cable. If the node detects the transmission of another packet during the attempt, it withdraws the attempt and this packet is backlogged. There are also backlogged packets whenever a collision occurs. If the node does not detect the transmission of another packet and there are no other packets starting to transmit then this packet begins transmission starting with an initial phase. If during the initial phase (analogous to a minislot) there are no collisions then the packet is transmitted successfully in the final phase. Assume that during the initial phase of transmission, none of the other nodes can detect the packet being transmitted whereas during the final phase all nodes can detect packet transmission. Also assume that collisions that occur during the initial phase can be immediately detected! Note that the initial and final phases are each exponentially distributed. All backlogged packets wait for a random amount of time (distributed exponentially with mean $1/\theta$) before retransmission. The system can be modeled as a CTMC where the state of the CTMC at the end of a slot is the number of backlogged users and the type of slot (idle, initial and final).

For both the DTMC and the CTMC models, performance measures such as throughput (expected number of successful transmissions per unit time), cable utilization, expected number of backlogged messages, expected delay in successfully transmitting a message, etc. can be computed. Using the performance measures it is possible to evaluate optimal designs for the Ethernet systems. See Bertsekas and Gallager (1992) for modified versions and approximations.

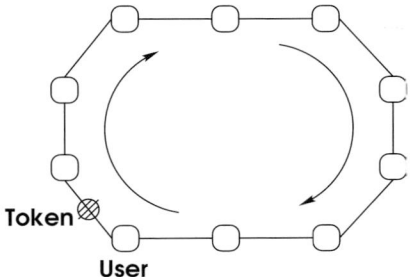

Fig. 10. Model of a token ring LAN.

4.3. Token rings

Besides Ethernet, the other commonly used LAN architecture and protocol is the Token Ring, developed by IBM. Here a simplistic model for a token ring is explained and is based on the description in Roy (1990). Consider N independent and identical users that are arranged logically in the form of a ring (see Figure 10).

Unlike the Ethernet model where all the users are allowed to transmit simultaneously which could potentially result in collisions, the token ring scheme is such that at a time there is at most one user generating a message over the cable or ring (thus there are no collisions). A designated user generates a "free" token into the ring. This token traverses the ring in a given direction. When a user with a message to transmit receives the free token, the user holds on to the token (now called "busy" token) and transmits the message onto the ring or cable. There are two basic types of implementation, the exhaustive and the gated service. In the exhaustive service a user that receives a free token transmits packets until there are no packets to transmit, however in the gated service case, only the packets that arrived prior to receiving the free token are transmitted and the packets arriving during the transmission will be transmitted during the next free token arrival to the user! Once the user completes transmission, the busy token is converted into a free token and passed along the ring.

To model the system as a CTMC, one could assume that the packets are generated according to a Poisson process, the length of the packets are exponentially distributed, and, the propagation time (including latency at the user) is also exponentially distributed. Since all the users are identical, a CTMC of the form $\{(X(t), Y(t), t \geq 0\}$ modeled where $X(t)$ is the number of messages in the network and $Y(t)$ is the status of the token (free or busy) at time t. Using the steady state distribution of the CTMC, performance measures such as throughput, delay, and blocking probability (if the users have finite buffers) can be computed.

Realistically speaking users may belong to multiple classes and are not necessarily identical. Also, the exponential distribution may not be the most appropriate. Several researchers have addressed these shortcomings (see Chae and Nilsson, 1991 for the performance analysis of a prioritized token ring with reservation model) and there are other interesting problems to be addressed in the future.

5. Other topics and models

5.1. TCP and flow control

When a message needs to be sent from a source to a destination, it is broken down into small packets and transported from the source to the destination. The protocols responsible for this transport of packets over networks are user datagram protocol (UDP) and transmission control protocol (TCP). Certain applications (typically real-time) use UDP where the destination does not acknowledge the receipt of packets to the source. Therefore in UDP the source does not know if the message sent, firstly reached the destination, and if it did, whether it reached without any losses.

On the other hand, TCP is an acknowledgement (ACK) based protocol. Every packet that reaches the destination is acknowledged. Therefore TCP is useful for applications that cannot tolerate losses, at the same time can tolerate slow transmission. There has been tremendous amount of research in the area of speeding up TCP, and also modeling it for different networks such as ATM, wireless, etc. A simplified version of TCP is explained below. Readers are encouraged to refer to Jacobson (1988) and (1990), Romanow and Floyd (1995), Stevens (1994), etc. for a detailed description.

Instead of waiting for an ACK for every packet sent, the source sends n packets to the destination before receiving an ACK. These n packets constitute the window with n as the window size. The window size is not a constant throughout the connection. If the connection is across a network with low congestion then the window size gradually increases up to a prenegotiated maximum $W_{\max}$. However if congestion develops and packets are lost, TCP backs off the packet generation by reducing the window size to half. The backing off is performed using a timer so that if the source does not receive an ACK before a certain time (this is also variable, in fact it depends on the connection and round trip time between source and destination for packets), the source retransmits with a smaller window.

Marsan et al. (1999) develop an approximate CTMC model for performance analysis of TCP connections in high-speed ATM networks. The modeling takes into account the slow start (initially the window size increases slowly), fast recovery, and congestion-avoidance (window size reduction) strategies commonly used in TCP. The sparse but regular structure of the infinitesimal generator matrix is taken advantage of in the analysis.

Misra and Ott (1999) analyze the stationary behavior of the TCP congestion window. Most of the earlier analysis assumed that the loss probability is constant with respect to window size. With the development of Random Early Detection (RED) it is important to consider the loss probability that varies with respect to the window size (as the window size increases, the loss probability increases in a stochastic sense). A Markov process that is further approximated as a continuous time, continuous state space system is modeled. The stationary distribution of the process is analyzed.

Kumar (1998) studies the performance of various TCP versions such as TCP-OldTahoe (uses timeout recovery), TCP-Tahoe (uses fast retransmit), TCP-Reno (uses fast retransmit and fast but conservative recovery), and, TCP-NewReno (uses fast retransmit and fast recovery). A stochastic model (Markov renewal reward process) is used to study the throughput performance of the different TCP versions in the presence of random loss on a wireless link. The main results include the following: TCP-Reno performs no better or worse that TCP-Tahoe for large packet loss probability. TCP-NewReno is a considerable improvement over TCP-Tahoe.

Baccelli and Bonald (1997) consider window flow control in lossless packet-switched networks (essentially applicable to TCP). However the window is assumed to be static and all packets follow the same route between a source and destination. There is also exogenous traffic along the route. General stochastic processes that are stationary and ergodic are used to model input processes. The stability of the system is evaluated and performance measures such as bounds on the maximal throughput are obtained.

5.2. Routing

Routing in the current version of the Internet uses best effort schemes and does not use any congestion avoidance mechanisms. The routers in the Internet use a learning process to develop a routing table. Based on the routing table an incoming packet is delivered to the appropriate neighboring router. This procedure continues from the source to the destination. The final source-destination route is usually the minimal hop path. In the case of a breakdown, the routers reconfigure their routing tables appropriately. The benefits to the best effort scheme are: easy implementation, fast learning (or recovery after breakdown), and fairness.

However in circuit-switched networks where the network topology is known and the number of circuits n_i of each link i between a pair of switches is given. Assume that the rate of call requests per unit time between every pair of source and destination is given. Consider a source S and a destination D between which there are R possible routes. Any message between S and D is sent across route j with probability $p_j(SD)$. Therefore when a new connection needs to be established between S and D, a random number is generated and the resulting route is selected. However if that route is not free, the call for connection is blocked or rejected. This is the static routing policy which could potentially result in a large number of rejected calls. Optimal values of $p_j(SD)$ (for $j = 1, \ldots, R$) are selected based on minimizing the total cost.

Gibbens et al. (1988) developed the dynamic alternate routing (DAR) strategy where the (stochastic) k-shortest paths are obtained between all sources and destinations at all points of time. Clearly the paths vary dynamically. Then calls between a source and destination are routed through the current shortest path or current second shortest path. If both are full then the calls are rejected. In Gibbens et al. (1993) some of the consequences for dynamic routing schemes for dual- and multi-parented networks (where a call can enter or leave the network in two or more points) are considered. Bounds are obtained for optimal dynamic routing strategies. The robustness is also illustrated. Gibbens and Kelly (1995) use stochastic analysis of dynamic routing for classical mathematical programming (optimization) to design networks. The methods used are network flow optimization and Markov decision processes for bounds on dynamic routing strategies. In Dasylva and Srikant (1999), non-trivial lower bounds on the lost revenue under any routing scheme in a multi-class loss network are obtained.

The bound is used to obtain linear programs which give bounds for sparsely connected networks with multiple classes and alternate routing.

One of the key factors that will enable networks to provide guaranteed QoS is the concept of QoS routing. Suppose an application desires the following QoS measures between the source and destination: maximum loss rate ε, maximum delay δ and maximum jitter ρ. Then the QoS routing problem is to send the message from the source to destination by the least expensive route such that the loss rate is less than ε, delay is less than δ and jitter is less than ρ, each with probability, say 0.9999. This is an important problem to be addressed and is being actively pursued. See Apostolopoulos et al. (1999) for a description of QoS routing, cost function, comparisons to best-effort routing, etc.

5.3. Leaky bucket policing

5.3.1. Description

The proliferation of the Internet and its excessive congestion has led researchers working on emerging high-speed telecommunication networks to develop tools to police and control the traffic at the user or source end. These policing mechanisms need to not only ensure that the telecommunication network traffic generated by the sources are kept below a negotiated threshold but also ensure that the users receive the Quality of Service (QoS) that they have been promised. One such policing mechanism is the leaky bucket (see Cidon and Gopal, 1998; Gu et al., 1995; Gün et al., 1994; Vamvakos and Anantharam, 1998; Butto et al., 1991; Callegati et al., 1993; Holtsinger and Perros, 1992; Sohraby and Sidi, 1994; Wu and Mark, 1992; Yin and Hluckyj, 1993).

A leaky bucket is essentially a credit management mechanism that controls the traffic entering the network. A single or a series of leaky buckets can be used to optimally regulate the source traffic in communication networks (see Anantharam and Konstantopoulos, 1999). In the recent literature a few researchers have proposed models to optimally select leaky bucket parameters (see Anantharam and Konstantopoulos, 1993; Raha et al., 1995; Elwalid and Mitra, 1997; Gorinsky et al., 1997; de Veciana, 1994, etc.).

Here stochastic fluid-flow models are used to describe the traffic flow, following the large literature using fluid-flow models for communication systems (see Anick et al., 1982; Elwalid and Mitra, 1991, etc.). Chen and Yao (1992, 1993), Ott and Shanthikumar (1997), Harrison (1985), Chen and Mandelbaum (1991), etc., demonstrate how to convert any discrete arrival system into a fluid-flow system and apply the fluid-flow model results.

"Leaky Bucket" is a control mechanism for admitting data into a network. It consists of a data buffer and a token pool as shown in Figure 11. Use a fluid-flow leaky bucket model assuming that the data traffic and tokens can be modeled as fluids. Tokens are generated continuously at a fixed rate γ into the token pool of size M. The new tokens are discarded if the token pool is full. External data traffic enters the data buffer (of size B_D) from a source modulated by an environmental process $\{Z(t), t \geqslant 0\}$. Traffic is generated by this source at rate $r(Z(t))$ at time t.

If there are tokens in the token pool, the incoming fluid takes an equal amount of tokens and enters the network. Broadly, there are two types of leaky buckets,

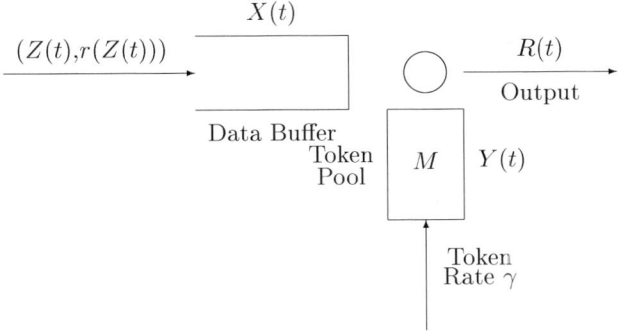

Fig. 11. Fluid model of a leaky bucket.

the buffered and the unbuffered leaky buckets. If the token pool is empty then two alternative implementations are considered:

- Buffered Leaky Bucket: the packets wait in the infinite capacity data buffer ($B_D = \infty$) for tokens to arrive,
- Unbuffered Leaky Bucket: there is no data buffer ($B_D = 0$) for the packets and any packet that does not find a token enters the network carrying a "violation" tag. Later such violation traffic can be dropped if congestion develops.

The buffered and unbuffered leaky bucket models are described in the following sections, and their respective output processes are studied.

5.3.2. Buffered leaky buckets

The output from a buffered leaky bucket acts as an input to a downstream network node. Hence, in this section the output from the leaky bucket is characterized. Refer to Figure 11. Let $X(t)$ be the amount of traffic in the data buffer at time t. Let $Y(t)$ be the amount of tokens in the token pool at time t ($Y(t) \leqslant M$). Note that fluid starts accumulating in the data buffer ($X(t) > 0$) only when the token pool is empty ($Y(t) = 0$). As long as tokens are available ($Y(t) > 0$), fluid does not wait at the data buffer ($X(t) = 0$). Therefore $X(t)Y(t) = 0$, for all t. Clearly, when the token pool is not empty ($Y(t) > 0$), the output from the leaky bucket is at rate $r(Z(t))$ at time t and when the token pool is empty, the output from the leaky bucket is at rate γ. Hence the output rate from the leaky bucket at time t, $R(t)$, is given by

$$R(t) = \begin{cases} \gamma, & \text{if } Y(t) = 0, \\ r(Z(t)), & \text{if } Y(t) > 0. \end{cases} \tag{50}$$

Define a process $\{W(t), t \geqslant 0\}$ (see Anantharam and Konstantopoulos, 1994) as

$$W(t) = X(t) + M - Y(t). \tag{51}$$

First characterize the $\{W(t), t \geq 0\}$ process. The dynamics of the $X(t)$ and the $Y(t)$ processes are given by

$$\frac{dX(t)}{dt} = \begin{cases} r(Z(t)) - \gamma, & \text{if } X(t) > 0, \\ 0, & \text{if } X(t) = 0, \end{cases} \tag{52}$$

$$\frac{dY(t)}{dt} = \begin{cases} \gamma - r(Z(t)), & \text{if } 0 < Y(t) < M, \\ -\{r(Z(t)) - \gamma\}^+, & \text{if } Y(t) = M, \\ 0, & \text{if } Y(t) = 0. \end{cases} \tag{53}$$

From Eq. (51),

$$W(t) > M \Rightarrow X(t) > 0 \text{ and } Y(t) = 0,$$
$$0 < W(t) \leq M \Rightarrow X(t) = 0 \text{ and } 0 < Y(t) < M,$$
$$W(t) = 0 \Rightarrow X(t) = 0 \text{ and } Y(t) = M.$$

Therefore,

$$\frac{dW(t)}{dt} = \frac{dX(t)}{dt} - \frac{dY(t)}{dt}$$
$$= \begin{cases} r(Z(t)) - \gamma, & \text{if } X(t) > 0 \text{ and } Y(t) = 0, \\ r(Z(t)) - \gamma, & \text{if } X(t) = 0 \text{ and } 0 < Y(t) < M, \\ \{r(Z(t)) - \gamma\}^+, & \text{if } X(t) = 0 \text{ and } Y(t) = M, \end{cases}$$
$$= \begin{cases} r(Z(t)) - \gamma, & \text{if } W(t) > 0, \\ \{r(Z(t)) - \gamma\}^+, & \text{if } W(t) = 0. \end{cases} \tag{54}$$

Thus the dynamics of the $W(t)$ process are identical to those of the buffer-content process of an infinite-sized buffer with output capacity γ and input rate $r(Z(t))$ at time t. Therefore to obtain the properties of the $W(t)$ process, for example its probability distribution, all one needs to do is look up the vast literature on the buffer-content process of an infinite sized buffer with output capacity γ and input rate $r(Z(t))$. The structure of the $\{W(t), t \geq 0\}$ process is exploited in the analysis that follow.

Sample paths of $Z(t)$, $X(t)$, $Y(t)$, and $W(t)$ are shown in Figure 12. Define the first passage time V (see Figure 12) as

$$V = \inf\{t > 0: W(t) = 0 | W(0) = 0, W(0+) > 0\}. \tag{55}$$

Let $\Theta(V)$ be the total amount of traffic output from the leaky bucket in time V. During the time interval $(0, V)$, $W(t) > 0$ and token pool is non-full. Hence the tokens enter the token pool at rate γ during the time interval $(0, V)$. Since the token pool is full at times 0 and V, the total number of tokens removed from the pool over $(0, V)$ must be the same as the total number of tokens that entered the pool over $(0, V)$. Hence

$$\Theta(V) = \gamma V. \tag{56}$$

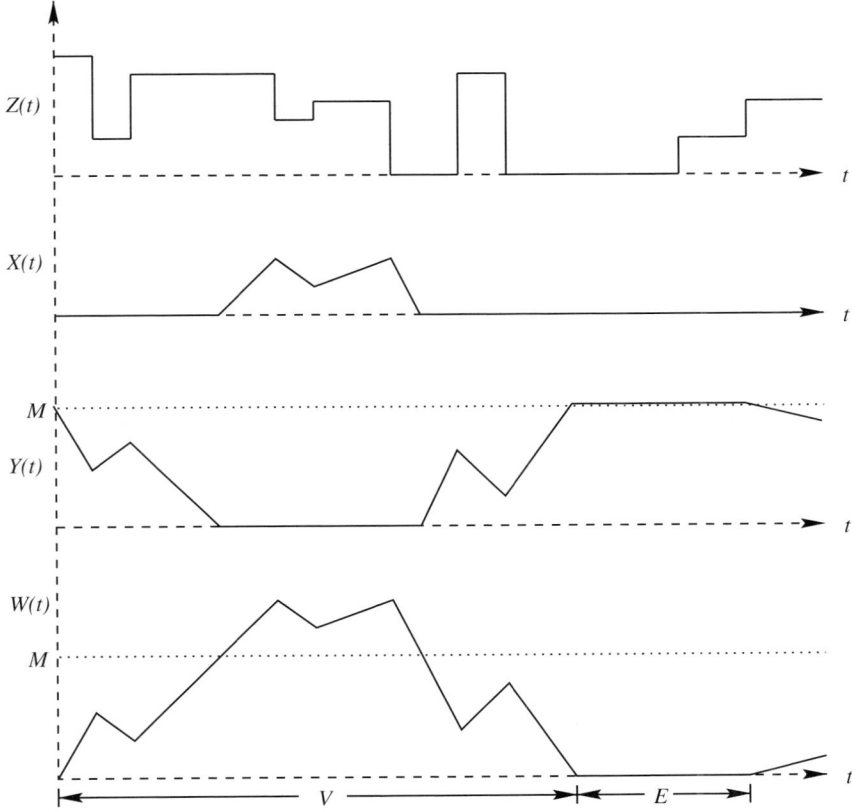

Fig. 12. $Z(t)$, $X(t)$, $Y(t)$, and $W(t)$ for buffered leaky buckets.

Define $A(t)$ as the total fluid arrival into the leaky bucket from the source in time t. Also, let $O(t)$ be the total fluid output from the leaky bucket in time t. Using the result in Eq. (56), the following result states the effective bandwidth of the output of the leaky bucket when $\{Z(t), t \geqslant 0\}$ is a semi-Markov process (SMP). Note that de Veciana (1994) derives a similar expression for the effective bandwidth of the output of the leaky bucket for a discrete traffic model.

Let $\{Z(t), t \geqslant 0\}$ be an SMP on a finite state space $\mathcal{S}$. Let $O(t)$ be the total output from the leaky bucket over $[0, t]$. The effective bandwidth of the output of the leaky bucket

$$eb_O(v) = \lim_{t \to \infty} \frac{1}{vt} \log E\{\exp(vO(t))\}$$

is given in terms of the effective bandwidth of the input, $eb_A(v)$, as

$$eb_O(v) = \begin{cases} eb_A(v), & \text{if } 0 \leqslant v \leqslant v^*, \\ \frac{v^*}{v}eb_A(v^*) - \gamma\frac{v^*}{v} + \gamma, & \text{if } v > v^*, \end{cases} \tag{57}$$

where v^* is obtained by solving

$$\frac{d}{dv^*}[v^* eb_A(v^*)] = 0$$

and

$$eb_A(v) = \lim_{t\to\infty} \frac{1}{vt} \log E\left\{\exp\left(v \int_0^t r(Z(t))\,dt\right)\right\}.$$

For a proof of the above result refer to Gautam (2002).

Therefore, given the effective bandwidth of the input traffic to the leaky bucket, it is easy to obtain the effective bandwidth of the output traffic from the leaky bucket by simply replacing the leaky bucket by a single infinite capacity buffer with capacity γ and measuring the output effective bandwidth of this infinite capacity buffer in terms of its input. When the environmental processes of the input traffic can be modeled as Continuous time Markov Chains, Semi-Markov Processes, Markov Regenerative Processes (MRGP) or regenerative processes, etc., one can compute their effective bandwidths using the results shown in Elwalid and Mitra (1993), Kesidis et al. (1993), Kulkarni (1997), etc.

5.3.3. Unbuffered leaky buckets

For the unbuffered leaky bucket, consider only the case when the environmental process governing the fluid input from a source, $\{Z(t),\ t \geqslant 0\}$, is a 2-state on-off process ($Z(t) = 0$ or 1, which implies whether the source is off or on respectively at time t). Therefore the fluid input is from a general on-off source with on time distribution $U(\cdot)$ (with mean τ_U) and off time distribution $D(\cdot)$ (with mean τ_D). When the source is on it generates traffic at rate r and at rate 0 when off. Therefore $r(Z(t)) = r\,Z(t)$.

In this unbuffered leaky bucket case, a packet that arrives at the leaky bucket is sent into the network with a "violation" tag if no tokens are available at the time of its arrival. The emphasis will be on the untagged packets as the tagged ones would be dropped in the event of a congestion. Note that $X(t) = 0$ for all t in this unbuffered leaky bucket case. The sample path of $W(t)$ is shown in Figure 13. Since there is no data buffer, $W(t) = M - Y(t)$ and $W(t)$ ranges from 0 to M. Note that $W(t)$ process is identical to a buffer content process of a fluid queue with on-off input, constant output with rate γ, and, a finite buffer of size M.

To obtain the effective bandwidth of the output process, the output rate from the leaky bucket is $R(t)$ at time t and is given by

$$R(t) = \begin{cases} \gamma, & \text{if } W(t) = M, \\ r, & \text{if } W(t) < M \text{ and } Z(t) = 1, \\ 0, & \text{if } W(t) < M \text{ and } Z(t) = 0. \end{cases} \tag{58}$$

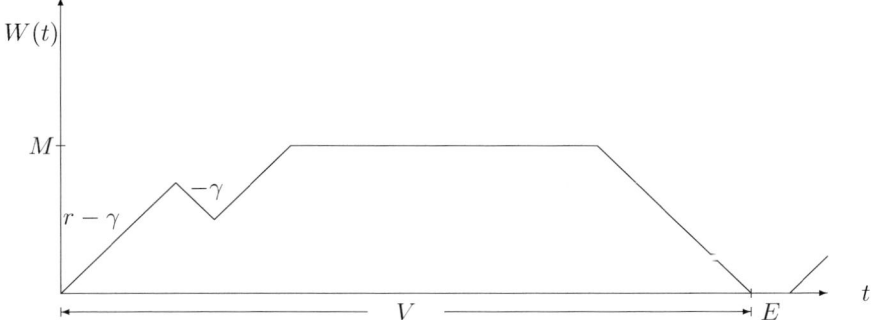

Fig. 13. $W(t)$ process for unbuffered leaky bucket.

Let V be as in Eq. (55). Then Eq. (56) remains valid in the unbuffered case. Hence the effective bandwidth of the output process from the unbuffered leaky bucket is equivalent to that of of the output process from a single finite buffer (of size M, with general on-off source input and output capacity γ. However, the effective bandwidth of the output cannot be easily written in terms of that of the input due to the fluid loss (as a result of untagged traffic) at the input buffer.

Closed-form algebraic expressions for $eb_O(v)$ are intractable even when the sources are exponential on-off sources. For general on-off sources, an approximation is developed in Gautam (2002).

5.4. Wireless network models

One of hottest research topics in telecommunications is wireless communications technology and a survey paper would certainly be incomplete without describing some of the on-going research work in mobile communications. However, the field is relatively new and most of the techniques are not well-established. Therefore only a brief summary of some of the current papers in the area of stochastic models in wireless networks are presented here. Almost all the forementioned traffic models, performance analysis, flow control, congestion control, etc. do not make any assumptions about whether the networks are at least partially wireless or not. It is to be noted that mobile communications where the users (sources and destinations) are mobile are called wireless communication here. Since the sources and destinations are not static an important problem is to locate the users to send and receive messages.

A theoretical framework for the study of mobility tracking based on user (or for that matter service or host) location probability distributions are provided in Rose and Yates (1997). Using stochastic ordering and information theory, quantitative comparisons of various mobility schemes are demonstrated and insights are obtained into the mobility tracking problem over a wide range of mobility characteristics.

Awduche et al. (1996) describe location management issues that involve tracking components that maintain dynamic data on the locations of mobile stations through a distributed database. The main focus is on a search component that prescribes the manner in which the wireless network is to be paged so as to determine the location of mobile stations whose whereabouts are unknown. The methods used are based on search theory where a stochastic sequential framework that systematically determines the location of mobile stations situated within a group of cells. Search algorithms are hence developed.

A Poisson-arrival location model (PALM) was introduced in Massey and Whitt (1993) in which customers arrive according to a nonhomogeneous Poisson process and move independently through a general location state space according to a location stochastic process. That was extended to a version of PALM to study communicating mobiles on a highway. Leung et al. (1994) stress the need for combining teletraffic theory and vehicular traffic theory. Their numerical results indicate that both the time-dependent behavior and the mobility of vehicles play important roles in determining the system performance.

5.5. *Other topics*

There have been several important topics that have been left out of this exposition. Some of the topics are listed below:

- One of the most critical factor that will enable QoS provisioning in high-speed networks is pricing. F.P. Kelly and colleagues have developed some optimal pricing models (see Kelly, 1996 for an example).
- ATM switch design and router design involve significant amount of stochastic modeling, particularly queueing. All the multiclass scheduling policies (polling, static priority, waited fair queueing, etc.) can be implemented on the currently available switches and routers.
- All the models considered here were unicast where traffic flows from a single source to a single destination. There are interesting stochastic models for multicasting (single source and a few destinations like an Internet classroom with students globally located) and for broadcasting (single source and all nodes as destinations) applications.
- Congestion control aspects at the packet level have not been addressed. The most common scheme is to have small buffers and if the buffer overflows, newly arriving packets are dropped. Modifications of that model include dropping from the beginning of the queue and RED (Random Early Detection: where packets from a non-full buffer are dropped with probability $p(n)$ if there are n packets in the buffer).
- Several scenarios in telecommunication networks (such as client-server systems) can be modeled as Queueing Networks. Walrand (1988) provides several applications of Queueing Networks in Telecommunications.

Acknowledgements

The author is grateful to Prof. C. R. Rao and Prof. D. N. Shanbhag for providing the opportunity to contribute to the Handbook of Statistics. The author would also like to thank the anonymous reviewers.

References

Anantharam, V. and T. Konstantopoulos (1993). An optimal flow control scheme that regulates the burstiness of traffic subject to delay constraints. In *Proceedings of the 32nd IEEE Conference on Decision and Control*, Vol. 4, pp. 3606–3610.

Anantharam, V. and T. Konstantopoulos (1994). Optimality and interchangeability of leaky buckets. In *32nd Allerton Conference, Monticello, IL*, pp. 235–244.

Anantharam, V. and T. Konstantopoulos (1999). A methodology for the design of optimal traffic shapers in communication networks. *IEEE Trans. Automat. Control* **44**(3), 583–586.

Anick, D., D. Mitra and M. M. Sondhi (1982). Stochastic theory of a data handling system with multiple sources. *Bell System Tech. J.* **61**, 1871–1894.

Apostolopoulos, G., R. Guerin, S. Kamat, A. Orda and S. K. Tripathi (1999). Intradomain QoS routing in IP networks: a feasibility and cost/benefit analysis. *IEEE Network* **3**(5), 42–54.

Awduche, D. O., A. Ganz and A. Gaylord (1996). An optimal search strategy for mobile stations in wireless networks. In *5th IEEE Intl. Conf. on Universal Personal Communications*.

Baccelli, F. and R. Bonald (1997). On the stable behaviors of window flow control. In *Proceedings of the 36th IEEE Conference on Decision and Control*.

Berger, A. W. and W. Whitt (1998a). Effective bandwidths with priorities. *IEEE/ACM Trans. on Networking* **6**(4).

Berger, A. W. and W. Whitt (1998b). Extending the effective bandwidth concept to network with priority classes. *IEEE Comm. Magazine*.

Bertsekas, D. P. and R. G. Gallager (1992). *Data Networks*, 2nd ed. Prentice-Hall, Englewood Cliffs, NJ.

Butto, M., E. Cavallero and A. Tonietti (1991). Effectiveness of the leaky bucket policing mechanism in ATM networks. *IEEE J. Sel. Areas Commun.* **SAC-9**, 335–342.

Callegati, F., G. Corazza and C. Raffaelli (1993). On the dimensioning of the leaky bucket policing mechanism for multiplexer congestion avoidance. In *IEEE International Conf. on Information Engg.*, Vol. 2, pp. 617–21.

Chen, H. and D. D. Yao (1992). A fluid model for systems with random disruptions. *Oper. Res.* **40**(Suppl. 2), S239–S247.

Chen, H. and D. D. Yao (1993). Dynamic scheduling of a multiclass fluid network. *Oper. Res.* **41**(6), 1104–1115.

Chang, C. S. and J. A. Thomas (1995). Effective bandwidth in high-speed digital networks. *IEEE J. Sel. Areas Commun.* **13**(6), 1091–1100.

Chang, C. S. and T. Zajik (1995). Effective bandwidths of departure processes from queues with time varying capacities. In *INFOCOM'95*, pp. 1001–1009.

Chen, H. and A. Mandelbaum (1991). Discrete flow networks: bottleneck analysis and fluid approximations. *Math. Oper. Res.* **16**(2), 408–446.

Chae, K. and A. A. Nilsson (1991). Performance analysis of prioritized token ring. In *Proc. of the 34th Midwest Symposium on Circuits and Systems*, Vol. 1, pp. 447–450.

Cidon, I. and I. S. Gopal (1998). Paris: an approach to integrated high-speed private networks. *Int. J. of Digital and Analog Cabled Systems* **1**(2).

Cinlar, E. (1975). *Introduction to Stochastic Processes*. Prentice-Hall, Engelwood Cliffs, NJ.

Dasylva, A. and R. Srikant (1999). Bounds on the performance of admission control and routing policies for general topology networks with multiple call classes. In *IEEE INFOCOM-99*, pp. 505–512.

de Veciana, G. (1994). Leaky buckets and optimal self-tuning rate control. In *GLOBECOM'94*, pp. 1207–1211.

de Veciana, G., C. Courcoubetis and J. Walrand (1994). Decoupling bandwidths for networks: a decomposition approach to resource management. In *INFOCOM'94*, pp. 466–473.

de Veciana, G., G. Kesidis and J. Walrand (1995). Resource management in wide-area ATM networks using effective bandwidths. *IEEE J. Sel. Areas Commun.* **13**(6), 1081–1090.

de Veciana, G. and G. Kesidis (1994). Bandwidth allocation for multiple qualities of service using generalized processor sharing. In *IEEE GLOBECOM-94*, pp. 1550–1554.

Elwalid, A. I. and D. Mitra (1991). Analysis and design of rate-based congestion control of high speed networks, Part I: Stochastic fluid models, access regulation. *Queueing Systems Theory Appl.* **9**, 29–64.

Elwalid, A. I. and D. Mitra (1993). Effective bandwidth of general Markovian traffic sources and admission control of high-speed networks. *IEEE/ACM Trans. on Networking* **1**(3), 329–343.

Elwalid, A. I. and D. Mitra (1995). Analysis, approximations and admission control of a multi-service multiplexing system with priorities. In *INFOCOM'95*, pp. 463–472.

Elwalid, A. I. and D. Mitra (1997). Traffic shaping at a network node: theory, optimum design, admission control. In *INFOCOM'97*, pp. 444–454.

Elwalid, A. I., D. Heyman, T. V. Lakshman, D. Mitra and A. Weiss (1995). Fundamental bounds and approximations for ATM multiplexers with applications to video teleconferencing. *IEEE J. Sel. Areas Commun.* **13**(6), 1004–1016.

Gautam, N. (1995). Effective bandwidth methodologies for network management. Master of Science Expository Paper, Dept. of Operations Research, University of North Carolina, Chapel Hill, NC 27599.

Gautam, N. (2002). Buffered and unbuffered leaky bucket policing: guaranteeing QoS, design and admission control. *Telecommunication Systems* **21**(1), 35–63.

Gautam, N., V. G. Kulkarni, Z. Palmowski and T. Rolski (1999). Bounds for fluid models driven by semi-Markov inputs. *Probab. Engrg. Inform. Sci.* **13**, 429–475.

Gibbens, R. J. and P. J. Hunt (1991). Effective bandwidths for the multi-type UAS channel. *Queueing Systems Theory Appl.* **9**, 17–28.

Gibbens, R. J., F. P. Kelly and P. B. Key (1988). Dynamic alternative routing – modelling and behaviour. In *Proc. 12th Int. Teletraffic Congr.*, Turin, Italy.

Gibbens, R. J., F. P. Kelly and S. R. E. Turner (1993). Dynamic routing in multiparented network. *IEEE/ACM Trans. on Networking* **1**(2), 261–270.

Gibbens, R. J. and F. P. Kelly (1995). Network programming methods for loss networks. *IEEE JSAC* **13**(7), 1189–1198.

Gorinsky, S., S. Baruah and A. Stoyen (1997). Boosting the network performance via traffic reshaping. In *Proceedings of the Sixth International Conference on Computer Communications and Networks*, pp. 285–290.

Gu, X., K. Sohraby and D. R. Vaman (1995). *Control and Performance in Packet, Circuit and ATM Networks*. Kluwer Academic Publishers, Boston.

Gün, L., V. G. Kulkarni and A. Narayanan (1994). Bandwidth allocation and access control in high-speed networks. *Ann. Oper. Res.* **49**, 161–183.

Harrison, J. M. (1985). *Brownian Motion and Stochastic Flow Systems*. Wiley.

Heyman, D. P. and M. J. Sobel (1982). *Stochastic Models in Operations Research*, Vol. 1. McGraw-Hill, New York.

Holtsinger, D. and H. Perros (1992). Performance analysis of leaky bucket policing mechanisms. In *Proc. Tricomm '92*, Raleigh, NC.

Jacobson, V. (1988). Congestion avoidance and control. In *ACM SIGCOMM-88*, Stanford, CA.

Jacobson, V. (1990). Berkeley TCP evolution from 4.3-tahoe to 4.3-reno. In *18th IETF*, Vancouver, BC.

Kelly, F. P. (1996). *Charging and Accounting for Bursty Connections*. Internet Economics, MIT Press.

Kesidis, G., J. Walrand and C. S. Chang (1993). Effective bandwidths for multiclass Markov fluids and other ATM sources. *IEEE/ACM Trans. on Networking* **1**(4), 424–428.

Konstantopoulos, T. and S. Lin (1996). High variability versus long-range dependence for network performance. In *Proceedings of the 35th IEEE Conference on Decision and Control, Kobe, Japan*, pp. 1354–1359.

Krishnan, K. R., A. L. Neidhardt and A. Erramilli (1997). Scaling analysis in traffic management of self-similar processes. In *Proc. 15th Intl. Teletraffic Congress*, pp. 1087–1096. Washington, DC.

Kulkarni, V. G. (1995). *Modeling and Analysis of Stochastic Systems*. Texts in Statistical Science Series. Chapman and Hall, Ltd., London.

Kulkarni, V. G. (1997). Effective bandwidths for Markov regenerative sources. *Queueing Systems Theory Appl.* **24**.

Kulkarni, V. G. and N. Gautam (1996). Leaky buckets: sizing and admission control. *Proceedings of the 35th IEEE Conference on Decision and Control*, Kobe, Japan.

Kulkarni, V. G. and N. Gautam (1997). Admission control of multi-class traffic with service priorities in high-speed networks. *Queueing Systems Theory Appl.* **27**, 79–97.

Kulkarni, V. G. and T. Rolski (1994). Fluid model driven by an Ornstein–Ühlenbeck process. *Probab. Engrg. Inform. Sci.* **8**, 403–417.

A. Kumar (1998). Comparative performance analysis of versions of TCP in a local network with a lossy link. *IEEE/ACM Trans. on Networking* **6**(4), 485–498.

Leung, K. K., W. A. Massey and W. Whitt (1994). Traffic models for wireless communication networks. In *INFOCOM-94*, pp. 1029–1037.

Lucantoni, D. M., K. S. Meier-Hellstern and M. F. Neuts (1990). A single-server queue with server vacations and a class of non-renewal arrival processes. *Adv. Appl. Probab.* **22**, 676–705.

Lucantoni, D. M. (1991). New results for the single server queue with a batch Markovian arrival process. *Stochastic Models* **7**, 1–46.

Lucantoni, D. M., G. L. Choudhury and W. Whitt (1993). Computing transient distributions in general single-server queues. In *GLOBECOM '93*, Vol. 2, pp. 1045–1050.

Marsan, M. A., E. de Souza e Silva, R. L. Gigno and M. Meo (1999). A Markovian model for TCP over ATM. *Telecommunication Systems* **12**(4), 341–368.

Massey, W. A. and W. Whitt (1993). Network of infinite server queues with nonstationary Poisson input. *Queueing Systems Theory Appl.* **13**.

Misra, A., and T. J. Ott (1999). The window distribution of idealized TCP congestion avoidance with variable packet loss. In *IEEE INFOCOM-99*, pp. 1564–1572.

Narayanan, A. and V. G. Kulkarni (1996). First passage times in fluid models with an application to two-priority fluid systems. In *Proceedings of the IEEE International Computer Performance and Dependability Symposium*.

Palmowski, Z. and T. Rolski (1996). The superposition of alternating on-off flows and a fluid model. *Report no 82*, June 1996, Mathematical Institute, Wrocław University.

Ott, T. J. and J. G. Shanthikumar (1997). Discrete storage processes and their Poisson flow and fluid flow approximations. *Queueing Systems Theory Appl.* **24**, 101–136.

Parekh, A. K. and R. G. Gallager (1993). A generalized processor sharing approach to flow control in integrated services networks: the single node case. *IEEE/ACM Trans. on Networking* **1**(3), 344–357.

Puhalskii, A. A. and W. Whitt (1999). Functional large deviation principle for waiting and departure processes. *Probab. Engrg. Inform. Sci.* **12**, 479–507.

Rabiner, L. R. (1989). A tutorial on hidden Markov models and selected applications in speech recognition. *Proc. IEEE* **77**(2).

Raha, A., S. Kamat and W. Zhao (1995). Using traffic regulation to meet end-to-end deadlines in ATM LANs. In *Proceedings of the International Conference on Network Protocols*, pp. 152–159.

Romanow, A. and S. Floyd (1995). Dynamics of TCP traffic over ATM networks. *IEEE JSAC* **13**(4) 633–641.

Rose, C. and R. Yates (1997). Location uncertainty in mobile networks: a theoretical framework. *IEEE Commun. Magazine* **35**(2), 94–101.

Roy, R. R. (1990). Continuous time Markov chain model for token ring local area network. In *IEEE Aerospace Applications Conference*, pp. 247–251.

Sohraby, K. and M. Sidi (1994). On the performance of bursty and modulated sources subject to leaky bucket rate-based access control schemes. *IEEE Trans. Commun.* **42**(2–4).

Stevens, W. (1994). *TCP/IP Illustrated*, Vols. 1 and 2. Addison-Wesley.

Vamvakos, V. and V. Anantharam (1998). On the departure process of a leaky bucket system with long-range dependent input traffic. *Queueing Systems Theory Appl.* **28**(1–3).

Walrand, J. (1988). *An Introduction to Queueing Networks*. Prentice-Hall, Englewood Cliffs, NJ.

Walrand, J. and P. P. Varaiya (1996). *High-Performance Communication Networks*. Morgan Kaufmann Publishers, Inc.

Willinger, W. and V. Paxson (1998). Where mathematics meets the Internet. *Notices Amer. Math. Soc.* **45**(8), 961–970.

Willinger, W., M. S. Taqqu and A. Erramilli (1996). A bibliographical guide to self-similar traffic and performance modeling for modern high-speed networks. In *Stochastic Networks: Theory and Applications*, pp. 339–366 (Eds. F. P. Kelly, S. Zachary and I. Ziedins). Royal Statistical Lecture Note Series, Vol. 4. Clarendon Press, Oxford, UK.

Wu, G. and J. W. Mark (1992). Discrete time analysis of leaky bucket congestion control. In *Proc. of ICC '92*.
Yin, N. and M. G. Hluckyj (1993). Analysis of the leaky bucket algorithm for on-off data sources. *J. High Speed Networks* **2**(1), 81–98.

… 8

Stochastic Processes in Epidemic Modelling and Simulation

David Greenhalgh

1. Introduction

This article aims to discuss the uses of stochastic processes in epidemiology, mathematical modelling and the simulation of epidemics. This is a very large area and it will not be possible to cover the entire literature on the topic. Instead, we aim to survey the main areas. We start by looking at some classical applications of stochastic processes in epidemic theory, namely the Chain Binomial, followed by the simple and general stochastic epidemic models. In Section 4 we consider the application of stochastic processes in spatial epidemic modelling, particularly percolation processes, simulation modelling and diffusion processes. One of the major motivations for studying epidemic models is to make predictions about control strategies such as vaccination of susceptible individuals or removal of infected ones, so Section 5 discusses stochastic models for the control of epidemics. Section 6 looks at applications of stochastic modelling for several specific diseases, including HIV/AIDS. This is followed by a description of some uses of stochastic methods in parameter estimation and hypothesis testing. A brief summary concludes our survey.

2. Chain binomial models

2.1. Introduction

One of the earliest uses of stochastic processes in epidemic theory was due to McKendrick (1926), who formulated what were essentially stochastic models for various problems in medicine, including the spread of epidemics. He derived equations for the moments and the duration of epidemics, with applications to malaria and influenza. However, this approach was not followed up and the next advance was in a different direction, the Chain Binomial epidemic model.

In this section we shall examine Chain Binomial epidemic models. First of all we shall discuss the classical Greenwood and Reed–Frost Chain Binomial models and an earlier version of the Reed–Frost model due to En'ko. Then we look at fitting these

2.2. Basic chain binomial models

Consider a disease such as measles with a fixed incubation period followed by a short infectious period. New cases of the disease occur in generations, separated by the incubation period. For such a disease spreading amongst a small group of people, as in a household, a Chain Binomial model is suitable. The simplest such model was due to Greenwood (1931), used to model the 1926 measles outbreak in St. Pancras, London, UK. The population was divided into infectious, infected and susceptible individuals. In each generation, each susceptible who had met at least one infected person became infected with probability $p = 1 - q$. All such infections were independent and this probability did not depend on the number of infectious people whom the susceptible individual had met. In 1928 Frost gave two lectures at Harvard in which he explained a discrete time model for the spread of infection in a susceptible population. The first lecture was published only recently (Frost, 1976). In the stochastic analogue of the more complicated Reed–Frost model (Wilson and Burke, 1942; Abbey, 1952; Maia, 1952; Elveback and Varma, 1965), in each generation a susceptible was infected independently by each infected person whom he had met. Each of these people infected him independently with probability $p = 1 - q$ and every susceptible behaved independently. Thus the more infectious people a susceptible met, the greater his or her chances of becoming infected. R_t denoted the number of susceptibles and S_t the number of infected people in the tth generation. For both models (R_t, S_t) formed a Markov Chain and it was straightforward to write down the transition probabilities of the process in each case.

Dietz (1988) has described the relatively unknown work of the Russian physician En'ko (1889) as the first Chain Binomial model. En'ko's epidemic model took the following iterative form: C_t denoted the expected number of infectious individuals, S_t the expected number of susceptible individuals and N_t the total size of the population at the end of time interval t. Then

$$C_{t+1} = S_t \left(1 - \left(1 - \frac{C_t}{N_{t-1}}\right)^{kN_t}\right),$$

$$S_{t+1} = S_t \left(1 - \frac{C_t}{N_{t-1}}\right)^{kN_t},$$

and

$$N_{t+1} = N_t - C_t.$$

The parameter k was the number of contacts of a susceptible individual during a single time interval. Dietz showed that En'ko's model was equivalent to the mean of the Reed–Frost Chain Binomial model. En'ko then proceeded to fit his model to epidemiological data from boarding schools in St. Petersburg and discussed the difficulties of applying it to scarlet fever data. En'ko's paper was translated into English by Dietz and republished in 1989 (En'ko, 1989). The model assumed a fixed number of contacts during a given time interval whereas the Reed–Frost model did not specify the number of contacts. Dietz and Schenzle (1985) discussed possible ways to generalise the En'ko–Reed–Frost model by choosing different distributions for the number of contacts, such as the negative binomial distribution. In their recent book on stochastic modelling of epidemics, Daley and Gani (1999) also gave an outline of the basic Greenwood and Reed–Frost Chain Binomial epidemic models and discussed further theoretical properties and extensions.

2.3. Applications of chain binomial models

Although Chain Binomial models were originally motivated by measles they have been applied to other diseases. Lidwell and Sommerville (1951), Heasman (discussion of Bailey, 1955) and Heasman and Reid (1961) discussed the application of Chain Binomial models to the common cold. Hope-Simpson and Sutherland (1954) obtained ambiguous results when attempting to use Chain Binomial models to describe the spread of influenza. Chain Binomial models were also used by Hope-Simpson (1952) to investigate the infectiousness of chicken-pox and mumps, as well as measles.

Heasman and Reid pointed out that it was not sufficient to consider only the total number of cases in order to discriminate between different models, one also needed to consider the frequencies of each type of epidemic chain, in other words the number of cases in each generation. They fitted only the Reed–Frost model. This has been further studied by Becker (1981) and Schenzle (1982). Becker also attempted to fit the Greenwood model to this data, but showed it to be inadequate. He then took the probabilities p_i that a given susceptible was infected when exposed to i infectives as parameters to be estimated (thus generalising the Greenwood and Reed–Frost models). Later Becker (1989) discussed further modification of the model to take into account heterogeneity in infectiousness of individuals, so that individuals not only had different levels of infectiousness, but also remained infectious for long periods of time. Schenzle (1982) presented three more models which assumed that either the chain probabilities had a beta distribution (for the Reed–Frost and the Greenwood model) or that they had two discrete values with probabilities π and $1 - \pi$ (for the Reed–Frost model). They all gave satisfactory fits. Some of these and other studies on the common cold dataset were discussed by Dietz and Schenzle (1985).

Bailey (1975) examined the use of the Greenwood and Reed–Frost Chain Binomial models to predict the spread of epidemics amongst households and illustrated this with data from measles outbreaks which occurred at Providence, Rhode Island, USA, 1929–1934. These data were provided by Wilson et al. (1939). Bailey discussed maximum likelihood estimation of the Chain Binomial model parameters from the observed chains of infection occurring in households of various sizes. Daley and Gani (1999) reviewed this and other work on fitting Chain Binomial models to epidemic data. They concluded that the overall evidence suggested that models which involved some heterogeneity in infectiousness of individuals fitted the data better.

Gani and Jerwood (1971) provided an account of the stochastic version of the Reed–Frost model in terms of Markov Chain methods. Lefèvre and Picard (1990) obtained and studied the final size distribution of the Reed–Frost epidemic model. A simple martingale argument and a non-standard family of polynomials were used. Picard and Lefèvre (1990) generalised these results to include an extension of the Reed–Frost model and the distribution of the final size as well as severity of the epidemic. The latter was interpreted as the total personal time units of infection during the course of the epidemic (in other words the area under the epidemic curve).

Dayananda and Hogarth (1977) studied a control model for immunisation of susceptibles using the Greenwood and Reed–Frost Chain Binomial epidemic models. They used dynamic programming methods for both deterministic and stochastic versions of these models, and obtained results for the situation where the probability of contact between two individuals was either very large or very small. In a later paper, Dayananda and Hogarth (1978), they presented a study of the Reed–Frost epidemic model with control by isolation of infectives, and a combination of immunisation and isolation. Lefèvre (1979, 1981) generalised and corrected this work. Optimal control for approximate Chain Binomial epidemics was examined when control action was affected by treatments that immunised a random number of susceptibles which depended on the treatment intensity level. Dynamic programming was used to determine the optimal operating strategy and to find conditions under which bang-bang policies were optimal. Having completed our survey of Chain Binomial models and their applications, in the next section we shall look at the use of stochastic processes to model the spread of epidemics in continuous time.

3. The simple and general stochastic epidemic models

3.1. Introduction

Section 2 has examined both basic Chain Binomial models and their applications, in particular fitting Chain Binomial models to data and modelling certain control strategies. These models were essentially in discrete time. In this section we shall consider stochastic epidemic processes in continuous time. We start off by examining epidemics over a relatively short time-scale, without the additional complications of births and deaths in the population. We consider the simple stochastic epidemic, with no removals, and the general stochastic epidemic with removals. We compare deterministic and stochastic epidemic models and outline some threshold theorems. Then we examine generalisations of results for different disease transmission terms and epidemics spreading in non-homogeneous populations, in particular epidemics spreading amongst groups of households. We next study stochastic recurrent epidemic models, which consider the epidemic over a longer time-scale and include the stochastic effects of births and deaths in the population. The final section discusses some specific applications of stochastic recurrent epidemic models, particularly to measles and polio.

3.2. The simple non-recurrent stochastic epidemic model

Bailey (1975) gave a comprehensive survey of mathematical models for infectious diseases as the field stood in 1975. He outlined the simple stochastic epidemic model where individuals were either susceptible or infected and once in the latter state remained infectious and infected indefinitely. There were $n+1$ individuals in total. If $X(t)$ was the number of susceptibles and $Y(t)$ the number of infecteds at time t then $X(t) + Y(t) = n + 1$. The chance of contact between any two specified individuals in an interval of time of length Δt was $\beta \Delta t + o(\Delta t)$. Bailey investigated the differential equations for $p_r(\tau)$, defined as the probability that there were still r susceptibles remaining uninfected at time τ. These equations were solved using the Laplace Transform method, but expressions for $p_r(\tau)$ were complicated and the work involved in evaluating them was very laborious. An alternative approach was to derive the partial differential equation satisfied by the probability generating function $P(x, \tau)$. This approach led to Haskey's formula for the stochastic mean (Haskey, 1954),

$$\mu'_1(\tau) = \sum_{j=1} \frac{n!}{(n-j)!(j-1)!} \left((n - 2j + 1)^2 \tau + 2 - (n - 2j + 1) \sum_{u=j}^{n-j} u^{-1} \right)$$
$$\times e^{-j(n-j+1)\tau},$$

where the upper limit of the main summation was $\frac{1}{2}n$ for n even; and $\frac{1}{2}(n+1)$ with the introduction of the factor $1/2$ into the term given by $j = \frac{1}{2}(n+1)$ for n odd. Renshaw (1991) has compared the mean of the deterministic and stochastic simple epidemic models. Daley and Gani (1999) also discussed the simple stochastic epidemic using slightly different methods and described probability generating function techniques to obtain information about the distribution of the duration time of the epidemic.

3.3. The general non-recurrent stochastic epidemic model

Bailey also discussed briefly the stochastic general epidemic model. In the deterministic general epidemic model infectious individuals were removed at a constant rate γ. In the corresponding stochastic model each infectious individual was removed after a time which had an exponential distribution with parameter γ. A partial differential equation for the probability generating function of $(X_t, Y_t; t \geq 0)$ the number of susceptibles and infectives in this model was given. $p_{r,s}(t)$ denoted the probability that $X_t = r$ susceptibles and $Y_t = s$ infectives at time t, given that $X_0 = n$ and $Y_0 = a$ at time $t = 0$, the change of time-scale $\tau = \beta t$ was made and the probability generating function

$$P(z, w, \tau) = \sum_{r=0}^{\infty} \sum_{s=0}^{\infty} p_{r,s}(\tau) z^r w^s$$

was introduced. This satisfied the partial differential equation

$$\frac{\partial P}{\partial \tau} = (w^2 - zw) \frac{\partial^2 P}{\partial z \partial w} + \rho(1-w) \frac{\partial P}{\partial w},$$

with initial condition $P(z, w, 0) = z^n w^a$, where $\rho = (\gamma/\beta)$ was the relative removal rate. A full analysis was not attempted by Bailey himself, but an expression for the probability distribution of the final size of the epidemic was reported. P_w denoted the probability of an epidemic of total size w (not counting the initial a infectives). Then

$$P_w = \frac{n!\rho^{a+w}}{(n-w)!(n+\rho)(n+\rho-1)\cdots(n+\rho-w)}$$
$$\times \sum_\alpha (n+\rho)^{-\alpha_0}(n+\rho-1)^{-\alpha_1}\cdots(n+\rho-w)^{-\alpha_w},$$

where the summation was over all compositions of $a+w-1$ into $w+1$ parts such that

$$0 \leqslant \sum_{j=0}^{i} \alpha_j \leqslant a+i-1,$$

for $0 \leqslant i \leqslant w-1$ and $1 \leqslant \alpha_w \leqslant a+w-1$. If n was large, difficulties could arise from the partitional nature of the summation, as it could be difficult to ensure that all terms had been included.

In a review of early mathematical models for HIV (Human Immunodeficiency Virus) and AIDS (Acquired Immune Deficiency Syndrome) Isham (1988) compared the simple deterministic and stochastic epidemics with removals. $X(t)$ and $Y(t)$ denoted respectively the number of susceptibles and infectives at time t and in any small time interval $[t, t+\Delta t)$ each infective had a probability $\alpha\Delta t + o(\Delta t)$ of spreading the infection to each susceptible. Then

$$X(t+\Delta t) = \begin{cases} X(t) & \text{with probability } 1-\alpha X(t)Y(t)\Delta t + o(\Delta t), \\ X(t)-1 & \text{with probability } \alpha X(t)Y(t)\Delta t + o(\Delta t). \end{cases}$$

For the simple stochastic epidemic model

$$\frac{d}{dt}E[X(t)] = -\alpha E[X(t)(n-X(t))]$$
$$= -\alpha E[X(t)](n - E[X(t)]) + \alpha \operatorname{var}[X(t)],$$

whereas for the simple deterministic epidemic model,

$$\frac{d}{dt}X(t) = -\alpha X(t)(n-X(t)).$$

Thus the deterministic model did not give the mean of the stochastic model. It could be shown that if $X(0)$ was sufficiently large then the approximation was good (Chapter 5 of Bailey, 1975).

In general a deterministic model gave only an approximation to the stochastic mean, and in some circumstances variation between realisations of the stochastic epidemic could be such that knowledge of the behaviour of their average would be of limited use or even misleading. In particular deterministic models gave no insight into phenomena such as the probability of epidemic fade-out, and as we shall see in Section 3.8, they could give a misleading picture of the minimum population size necessary to maintain endemic fluctuations. Specific problems were likely to arise with small populations, or at the beginning of the epidemic where it was possible that the infection would die out rather than develop into a true epidemic. Isham also pointed out that several different stochastic models could all correspond to the same deterministic approximation.

3.4. Whittle's and Williams' threshold theorems

The threshold theorem of Kermack and McKendrick (1927) stated that for the general deterministic epidemic model, the relative removal rate $\rho = (\gamma/\beta)$ was a threshold value for the initial susceptible population size. If a small number of infected individuals entered a population of size n where everyone was initially susceptible, then an epidemic would occur if and only if $n > \rho$. If n just exceeded the threshold value, so $n = \rho + \varepsilon$, where ε was small, then the resulting epidemic was of size approximately 2ε and reduced the susceptible population size to $\rho - \varepsilon$ as time t became large.

Whittle's famous (1955) note followed on from Bailey's paper (Bailey, 1953). In it Whittle derived a stochastic threshold theorem which corresponded to Kermack and McKendrick's deterministic result. $X(t)$ and $Y(t)$ denoted the number of susceptible and infectious individuals respectively at time t. Then in a small time interval $[t, t + \Delta t)$

$$\Pr\big((X, Y) \to (X-1, Y+1)\big) = \beta X Y \Delta t + o(\Delta t) \tag{3.1}$$

and

$$\Pr\big((X, Y) \to (X, Y-1)\big) = \gamma Y \Delta t + o(\Delta t). \tag{3.2}$$

At the start of the epidemic when the number of susceptibles could reasonably be approximated by n, the population size, and the number of infectives was a, Eqs. (3.1) and (3.2) could be compared to a simple birth and death process with a constant birth rate $n\beta$ and a constant death rate γ. Whittle said that an epidemic of intensity i took place if a proportion less than or equal to i of the n initial susceptibles was eventually infected. π_i denoted the probability that an epidemic of intensity i did not occur.

Whittle's Stochastic Threshold Theorem stated that:

(i)   If $\rho < n(1-i)$, $\left(\dfrac{\rho}{n}\right)^a \leqslant \pi_i \leqslant \left(\dfrac{\rho}{n(1-i)}\right)^a$;

(ii)   If $n(1-i) \leqslant \rho < n$, $\left(\dfrac{\rho}{n}\right)^a \leqslant \pi_i \leqslant 1$;

(iii)   If $n \leqslant \rho$, $\pi_i = 1$.

Whittle's Theorem could be interpreted as stating that if $\rho \geqslant n$, then there was zero probability of an epidemic exceeding any pre-assigned intensity i; while if $\rho < n$ then the probability of an epidemic was approximately $1 - (\rho/n)^a$ for small i. So if at the start of the epidemic there were fewer susceptibles than the threshold value, then the disease would never affect more than an arbitrarily small proportion of the population and the distribution of the final epidemic size would be J-shaped on $[0, n]$. If, however, at the start of the epidemic there were more susceptible people than the threshold value then a major epidemic infecting a substantial number of people would occur with probability $1 - (\rho/n)^a$. In this case the distribution of the final size of the epidemic would be U-shaped on $[0, n]$. The deterministic threshold theorem provided less insight than the stochastic threshold theorem as it averaged out this second case.

In the general stochastic epidemic with n initial susceptibles and a initial infecteds with infection rate β and removal rate γ, $T_{n,a}(\beta, \gamma)$ was written to denote the total size of the epidemic, in other words the total number of individuals infected during the course of the outbreak. Rajarshi (1981) generalised a result of Williams (1971) to show that

$$\lim_{n \to \infty} \Pr\bigl(T_{n,a}(n^{-1}\beta, \gamma) = k\bigr)$$
$$= \frac{(2k + a - 1)!a}{k!(k + a)!} \frac{\theta^{k+a}}{(1 + \theta)^{2k+a}}, \quad k = 0, 1, 2, \ldots, \tag{3.3}$$

where $\theta = n^{-1}(\gamma/\beta)$ was the relative removal rate per initial susceptible. By summing the right-hand side of (3.3) over $k = 0, 1, 2, \ldots$ it could be seen that in the limit as $n \to \infty$, the probability $P_a(\theta)$ of an epidemic of finite size was given by

$$P_a(\theta) = \bigl(\min(\theta, 1)\bigr)^a. \tag{3.4}$$

This constituted Williams' threshold theorem for the general epidemic, namely that there was zero probability of a 'true epidemic' if $\theta \geq 1$, whilst if $\theta < 1$ a 'true epidemic' occurred with probability $1 - \theta^a$. A 'true epidemic' was one in which in the limit as n tended to infinity, infinitely many susceptibles became infected.

Williams (1971) proved (3.3) for the special case $a = 1$, by using a set of recurrence relations for $\Pr(T_{n,1}(\beta, \gamma) = k)$, $k = 0, 1, 2, \ldots, n$. His proof was simplified by Bailey (1975). Rajarshi (1981) applied the reflection principle to the embedded random walk of the process $\{(X(t), Y(t)); t \geq 0\}$ to yield an alternative proof. Watson (1980) used a random time-scale transformation which gave a proof of Eq. (3.4) and a normal approximation to the size of a true epidemic. By relating the total size of the general stochastic epidemic to that of a randomised Reed–Frost Chain Binomial epidemic, von Bahr and Martin-Löf (1980) used harmonic functions for the Markov Chain of the Reed–Frost epidemic to obtain a proof of Eq. (3.3) together with a normal approximation for the size of the epidemic.

In a recent book, Diekmann and Heesterbeek (1999) also discussed some of these issues for the stochastic Kermack–McKendrick model. This book is very intelligently written and interwoven with numerous exercises for the interested reader. They discussed the stochastic basis of the deterministic Kermack–McKendrick model and showed how these ideas could be used to give a precise set of conditions that guaranteed that the deterministic model was a good approximation of a stochastic one. The analysis was based on ideas of J. A. J. Metz.

In the deterministic Kermack–McKendrick model, the proportion of susceptibles at the end of the outbreak was denoted by $s(\infty)$, which was a root of the equation,

$$\ln(s(\infty)) = R_0(s(\infty) - 1).$$

When $R_0 < 1$, the relevant root was $s(\infty) = 1$, this meant that the introduction of the disease did not lead to a major epidemic outbreak. When $R_0 > 1$, there existed a unique root between zero and one, which was the relevant one. Diekmann and Heesterbeek described how the distribution of the final epidemic size in a finite population of size N approached the corresponding value for the infinite size population model as N tended to infinity. Later in the book they extended the formula to determine the extinction probability in a population which was partially vaccinated with an imperfect vaccine which reduced susceptibility and infectivity of an individual (but not necessarily to zero), while leaving the time course of the infectivity unaltered.

Daley and Gani (1999) discussed some of the topics which we have mentioned here, in particular the general stochastic epidemic, its total size distribution, and Whittle's Stochastic Threshold Theorem. They also described the general stochastic epidemic in a stratified population where the intergroup infection rates between each pair of groups and the removal rates for each group were different.

In their analysis of common cold data Heasman and Reid (1961) also fitted the stochastic version of the Kermack–McKendrick model, as well as Chain Binomial models. the former model gave a better fit than the Reed–Frost model for the epidemics which occurred in families with crowded homes, defined as those with at most three rooms. The Kermack–McKendrick model has been shown by Becker (1980) to be a special case of a general Chain Binomial model. However this general model provided only a small improvement to the Reed–Frost model as judged by comparing the log-likelihood functions.

3.5. Generalisations of results

Ball (1983) has outlined a method of constructing a sequence of general stochastic epidemics, indexed by the initial number of susceptibles N, from a time-homogeneous birth-and-death process. The construction was used to show strong convergence of the general stochastic epidemic to a birth-and-death process over any finite interval $[0, t]$, and almost sure convergence of the total size of the general stochastic epidemic to that of a birth-and-death process. The latter result furnished a new proof of the Threshold Theorem of Williams (1971). The methods were quite general and in the remainder of the paper, similar models were developed for a wide variety of epidemic models including Chain Binomial, host vector and geographical models.

An SI (or simple) epidemic model is one in which an individual starts off susceptible, at some stage becomes infected (and infectious) and from then on remains permanently infected (and infectious). Similarly to the SI model, in an SIS model an individual starts off susceptible, at some point becomes infected and infectious, but now after an infectious period (which is usually relatively short) returns to the susceptible class. An SIR model is as an SIS model except that at the end of the infectious period the individual enters a third class of recovered, non-infectious, permanently immune individuals. An SIRS model is like an SIR model, but the disease-induced immunity is only temporary and at the end of this period of immunity the individual becomes susceptible to the infection again.

Jacquez and O'Neill (1991) have compared threshold results for the deterministic and stochastic versions of the homogeneously mixing SI epidemic model with a constant recruitment rate, death due to the disease, a background death rate and a disease transmission rate $\beta c XY/(X+Y)$. This transmission term was motivated by mathematical models for the spread of HIV and AIDS. They found that for the deterministic model, if a small number of infectives was introduced into an equilibrium population consisting entirely of susceptibles, then an epidemic would take off if and only if $R_0 > 1$. Here R_0 was the basic reproduction number, defined as the average number of secondary infections produced by a single new infected individual entering the equilibrium situation where all individuals were susceptible.

For the stochastic model it was found that the expected number of infected individuals tended to zero as time became large if $R_0 \leqslant 1$. If $R_0 > 1$ then simulations showed that there was a finite probability, strictly less than one, that an epidemic would take off, resulting in a quasi-equilibrium, but also a non-zero probability that the epidemic would become extinct, and the probability of extinction decreased as R_0 increased above one. Similar analytical results were obtained for the corresponding SIS, SIR and SIRS epidemic models, but simulation was not attempted for these models.

Ball and O'Neill (1993) considered a modification of the general epidemic model in a closed, homogeneously mixing population, in which new infections occurred at rate $\beta xy/(x+y)$, where x and y were respectively the number of susceptible and infectious individuals at time t and β was the constant per capita contact rate. This transmission term was again motivated by AIDS modelling. If z was the number of removed individuals at time t, then the deterministic version of the modified epidemic was

$$\frac{dx}{dt} = -\frac{\beta xy}{x+y},$$
$$\frac{dy}{dt} = \frac{\beta xy}{x+y} - \gamma y,$$

and

$$\frac{dz}{dt} = \gamma y,$$

with initial conditions $x(0) = n$, $y(0) = a$ and $z(0) = 0$. This model had an exact closed form solution. The spread of infection was faster and more severe in the modified deterministic epidemic than in the general epidemic.

For the corresponding stochastic model, $p_{r,s}(t)$ was used to denote the probability that there were exactly r susceptibles and s infectives at time t. As before, the time-scale transformation $\tau = \beta t$ was made and ρ was defined as (γ/β), the relative removal rate.

Then the probabilities $p_{r,s}(t)$ satisfied the differential-difference equations

$$\frac{d}{d\tau} p_{r,s} = \frac{(r+1)(s-1)}{(r+s)} p_{r+1,s-1} + \rho(s+1) p_{r,s+1}$$
$$- s\left(\frac{r}{r+s} + \rho\right) p_{r,s}, \quad 0 \leqslant r \leqslant n, \ 0 \leqslant s \leqslant a,$$

with initial conditions $p_{r,s}(0) = 0$, $0 \leqslant r \leqslant n$, $0 \leqslant s \leqslant a$, $(r,s) \neq (n,a)$ and $p_{n,a}(0) = 1$. Here the convention adopted was that $p_{r,s} \equiv 0$ unless $0 \leqslant r \leqslant n$ and $0 \leqslant s \leqslant a$.

Ball and O'Neill derived a (rather complex) explicit solution of these equations. They then examined the probability distribution of the total epidemic size. If $P_w^{(n)}$ denoted the probability of an epidemic of total size w (the notation here expressed explicitly the dependence of $P_w^{(n)}$ on n, the initial number of susceptibles) then $P_w^{(n)} = \rho f_{n-w,1}^{(n)}$ where the $f_{r,s}^{(n)}$ satisfied the recurrence relations

$$(r+1)(s-1) f_{r+1,s-1}^{(n)} + \rho(r+s)(s+1) f_{r,s+1}^{(n)} - s(r+\rho(r+s)) f_{r,s}^{(n)} = 0,$$
$$0 \leqslant r \leqslant n, \ 0 \leqslant s \leqslant a, \ (r,s) \neq (n,a),$$

and

$$a\left(\frac{n}{a+n} + \rho\right) f_{n,a}^{(n)} = 1.$$

Ball and O'Neill also compared the final size distributions of the modified general stochastic epidemic and the general stochastic epidemic numerically for specific parameter values. They outlined proofs of Williams' and Whittle's threshold theorems for the modified general stochastic epidemic. If π_i denoted the probability of an epidemic which infected a fraction not larger than i of the population, as before, Ball and O'Neill showed that for sufficiently large n we had that approximately

$$\left(\min(\rho(1-i), 1)\right)^a \leqslant \pi_i \leqslant \left(\min\left(\frac{\rho(n+a)}{n(1-i)}, 1\right)\right)^a.$$

Thus for large n if $\rho \geqslant 1$ then there was zero probability of an epidemic exceeding any fixed intensity $i > 0$; whilst if $\rho < 1$ the probability of a major epidemic was approximately $1 - \rho^a$ for small i. This was Whittle's threshold theorem for Ball and O'Neill's modified general stochastic epidemic model.

Ball and O'Neill also demonstrated that in the limit as n became large

$$P_w^{(n)} \to P_w = \frac{(2w-a+1)!a}{w!(w+a)!} \frac{\rho^{w+a}}{(1+\rho)^{2w+a}}, \quad w = 0, 1, 2, \ldots,$$

the same value as for the general stochastic epidemic (Williams, 1971). This was the total size of a birth-and-death process with birth rate one, death rate ρ and initial population size a (Ball, 1983). Thus in the limit as n tended to infinity, the probability of a finite epidemic was given by $(\min(\rho, 1))^a$, so a major epidemic could occur if and only if $\rho < 1$. This was Williams' threshold theorem for Ball and O'Neill's model. Finally Ball and O'Neill considered the effect of introducing variability in susceptibility into their epidemic model.

3.6. Non-recurrent epidemics amongst groups of individuals

Metz (1978) examined the general stochastic epidemic model when there were several classes of infectives. More recently, Ball (1996) discussed a very general model for the spread of an epidemic among a population of m groups or households each of size n. The asymptotic situation in which the number of households m tended to infinity whilst the household size, n, remained fixed was analysed. For large m the process of infected households could be approximated by a branching process. A coupling argument was used to make this approximation precise as m tended to infinity, thus allowing a threshold theorem to be developed, and stated as follows:

THEOREM 3.1 (Ball, 1996). *Let R be the number of between group contacts resulting from a typical single household epidemic, $R_T = E[R]$ and $f(s) = E(s^R)$, the probability generating function of R. Then as $m \to \infty$:*

(i) *a global epidemic occurs with non-zero probability if and only if $R_T > 1$;*
(ii) *the probability of a global epidemic is $1 - p$ where p is the smallest root of $f(s) = s$ in $[0, 1]$;*
(iii) *the probability generating function, $h(s)$ say, of the limiting total number of groups infected, $T(\infty) + 1$, satisfies $h(s) = sf(h(s))$. Further*

$$P(T(\infty) + 1 = k) = k^{-1} P(R_1 + R_2 + \cdots + R_k = k), \quad k = 1, 2, 3, \ldots,$$

where $R_1, R_2, \ldots$ are independent and identically distributed copies of R.

(i) still held if there were a initial infectives where $a < \infty$. If these a initial infectives were in different groups then the probability of a global epidemic was $1 - p^a$ and (iii) could be modified appropriately. If some groups contained more than one initial infective then the situation was more complicated.

Specialisation to the case where infective individuals made infectious contacts within their own group with rate λ_W and outside their own group with rate λ_B was also briefly considered. This model was analysed in more detail by Ball, Mollison and Scalia-Tomba (1997). For fixed group sizes but with the number of groups increasing without limit, the threshold values R_0 in the deterministic and stochastic versions of the model differed, but they converged as the fixed group size increased to infinity. Becker and Dietz (1995) considered a similar model using different methods, although some of the results are the same. Becker and Dietz have also looked at optimal vaccination strategies.

3.7. Stochastic recurrent epidemic models

Bailey (1975) has also discussed the stochastic general recurrent epidemic model. This was similar to the stochastic general epidemic model except that new susceptibles were now recruited into the model at rate μ. Again $p_{r,s}(t)$ denoted the probability that there were exactly r susceptibles and s infectives at time t, given n susceptibles and a infectives at time $t = 0$, and if

$$\Pi(z, w, t) = \sum_{r=0}^{\infty} \sum_{s=0}^{\infty} p_{r,s}(t) z^r w^s,$$

was the probability generating function then $\Pi(z, w, t)$ satisfied the partial differential equation

$$\frac{\partial \Pi}{\partial t} = \beta(w^2 - zw) \frac{\partial^2 \Pi}{\partial z \partial w} + \gamma(1-w) \frac{\partial \Pi}{\partial w} + \mu(z-1)\Pi,$$

subject to $\Pi(z, w, 0) = z^n w^a$. Again a full stochastic analysis of this equation was not attempted.

Bailey also considered a modified stochastic model where new infectives were introduced at a relatively small rate $\varepsilon > 0$. With this model, he deduced a formula for the average recurrence time $\overline{R}$ starting from a state near the unique deterministic endemic equilibrium (with $\varepsilon = 0$) with $x_0 = (\gamma/\beta)$ susceptibles and $y_0 = (\mu/\gamma)$ infectives, and ending with zero infectives as

$$\overline{R} \sim \frac{(2\pi(x_0 + y_0))^{1/2}}{\gamma y_0} \exp\left(\frac{(y_0 + (x_0/y_0))^2}{2(x_0 + y_0)}\right),$$

Bartlett (1957, 1960a, 1960b). These results held in the limit as ε went to zero, which was the case of the general stochastic recurrent epidemic model. A similar type of result has been proved by Ridler-Rowe (1967) who showed that the expected duration time $\overline{T}$ from the start of the process, with a infectives and n susceptibles to the point at which the infectives first became extinct was given by

$$\overline{T} \sim \gamma^{-1} \log_e(a+n) \quad \text{as } n \to \infty.$$

The difference in the forms of these two results was presumably due to the fact that Bartlett was concerned with appropriate passage times to zero starting from a state near to the equilibrium point, whereas Ridler-Rowe considered starting from a point for which the number of susceptibles (though not necessarily the number of infectives) was very large and a long way from the equilibrium point.

Diekmann and Heesterbeek (1999) also investigated the relationship between critical community size and epidemic fade-out. They concluded that there was not so much a critical community size, but rather a critical relationship between population size and the ratio of the two timescales involved (that of demography and that of disease transmission). When both $\gamma/(\sqrt{N}\mu)$ and $\beta/(\sqrt{N}\mu)$ were very small, they expected a single outbreak, while when they were both very large they expected an endemic situation. As a numerical example for measles in a moderately developed country they estimated the critical population size of the order of 1.65 million. They quoted from Nasell (1999) the following approximation formula for $\bar{t}_e$, the expected time to extinction under critical conditions:

$$\bar{t}_e \approx \frac{(R_0 - 1)N}{2\left(\frac{\gamma}{\mu}\right)^2} \frac{1}{\mu}.$$

In their book, Anderson and May (1991) discussed a discrete-time version of the recurrent epidemic model. The total number of susceptible, infected and immune individuals at time t were denoted by X_t, Y_t and Z_t respectively. The constant total population size was $N = X_t + Y_t + Z_t$. The model advanced in timesteps of duration D, where D was the average time interval between an individual acquiring infection and passing it on to the next person. As usual R_0 was the basic reproduction number. Susceptible individuals were lost by infection and gained by new births. Then the discrete time dynamic equations which described the spread of the disease were

$$X_{t+D} = X_t - Y_{t+D} + B, \tag{3.5}$$

and

$$Y_{t+D} = \left(\frac{R_0 X_t}{N}\right). \tag{3.6}$$

A routine linearised stability analysis showed that the system of Eqs. (3.5) and (3.6) was neutrally stable.

Incorporating 'demographic stochasticity' into these models, in other words using stochastic processes to describe the birth, death and infection of individuals, could tip this model into sustained oscillations of outbreaks of disease (Section 7.6.1 of Bailey, 1975; Anderson and May, 1991). This happened provided that the population was large enough to avoid stochastic extinction of infection and the indefinitely maintained disease cycles had a period of approximately

$$T \approx 2\pi \left(\frac{LD}{R_0 - 1}\right)^{1/2} \approx 2\pi (AD)^{1/2},$$

for the model with no incubation period, or approximately

$$T \approx 2\pi \left(\frac{L(D + D')}{R_0 - 1}\right)^{1/2} \approx 2\pi \left(A(D + D')\right)^{1/2},$$

if an incubation period of length D' was included. Here A denoted the average age at infection, which measured the time taken for the number of susceptibles to be replenished by birth processes. The inter-epidemic period thus scaled as $A^{1/2}$. For smaller population sizes there was a stochastic 'fade-out', and repeated epidemics depended on introductions from outside the population. The inter-epidemic period now tended to scale as A (Anderson and May, 1991).

Bartlett (1957, 1960a) performed a similar analysis. He began with a discrete time model similar to Eqs. (3.5) and (3.6) above, modified by replacing the infection probability $(R_0 Y_t/N)$ in Eq. (3.6) by the expression

$$1 - \left(1 - \frac{R_0}{N}\right)^{Y_t}.$$

This gave a more accurate account of the Binomial infection process (it corresponded to the Reed–Frost model) and reduced to $(R_0 Y_t/N)$ when $(R_0 Y_t/N)$ was very small. Bartlett showed that further incorporation of stochastic birth and death processes into the model had the effect of producing sustained oscillations, unless the population was too small to perpetuate the infection, which then exhibited 'fade-out'. By considering small stochastic departures from the equilibrium and obtaining analytic expressions for the variances and covariances of these small stochastic departures, Bailey (1975) proved that in the absence of fade-out, the oscillations would be indefinitely maintained. This was in sharp contrast to the damped convergence to the endemic equilibrium predicted by the deterministic model.

3.8. Specific applications of recurrent stochastic epidemic models

We now look at some specific applications of recurrent stochastic epidemic models. Bartlett (1953) used stochastic simulations to mimic the behaviour of recurrent epidemics of measles in a boarding school, with parameters chosen to be in reasonable conformity with the observed data. A stochastic simulation was set up and run for thirteen years' data. During this time six major epidemic outbreaks occurred and the average inter-epidemic period was 125 weeks. As expected, a type of stochastic stability was observed, with later epidemics being roughly the same size as earlier ones. There were also four minor outbreaks during the simulated period. The simulation results agreed with both the theoretical predictions of stochastic models and the qualitative character of observed measles data, although it was not possible to fit the model to actual detailed statistical data.

A more ambitious study, also specifically related to the spread of measles was reported by Bartlett (1957, 1960b, 1961). It suggested that the critical community size to prevent fade-out should be around 250,000 individuals, provided that the initial number of susceptibles was somewhere near the quasi-equilibrium value. Smaller critical sizes applied to non-isolated districts because of the immigration of new infectives. In an analysis of measles data in the UK Bartlett observed fade-out in towns of population size less than approximately 250,000 and no fade-out in towns with greater than this number of people. A similar figure was obtained by Bartlett (1960a) in his analysis of United States measles data. Black (1966) has further refined and confirmed Bartlett's results in a detailed study of measles endemicity in nineteen island communities. Bartlett's critical community size was smaller than that estimated by Diekmann and Heesterbeek. The reason for this could have been that Diekmann and Heesterbeek were estimating only a rough order of magnitude and also did not have a similar restriction on the initial number of susceptibles. Also their estimate assumed that the population was at a steady state, and thus ignored the oscillatory character of the dynamics which was included in Bartlett's simulations. Dietz and Schenzle (1985) stated that demographic stochasticity alone could not cause the observed seasonal fluctuations in epidemic data, although they did consider stochastic instability of endemic states to be important.

Anderson and May (1986) have used Monte Carlo simulation methods with parameters based on the spread of measles in Iceland to perform a similar analysis. In one simulation the community size was set at 250,000 and in the other it was set at 100,000. In the smaller community stochastic extinction occurred and the recurrent epidemics were triggered off by the immigration of infectives, whereas in the larger community the epidemic sustained itself. The inter-epidemic interval was roughly 2–3 years in the larger community and 4 years in the smaller one. The deterministic model predicted a much lower threshold for extinction of 6,500 individuals.

Murray and Cliff (1975) studied a stochastic version of the classic Hamer–Soper deterministic model for measles (Hamer, 1906; Soper, 1929). The model was extended to analyse patterns of disease in a spatial model for ten contiguous sub-areas centered on the city of Bristol. Following a description of the nature of the data, the Hamer–Soper model and a stochastic equivalent were outlined for a single areal unit and some of the properties of the stochastic model illustrated on a small dataset. A small rate of immigration of susceptibles and infectives into the model was necessary to sustain the oscillations. The multi-region model was then described and fitted to notification data of measles cases in the ten sub-area system, and the results of the analysis outlined. A modification of the model to give a better fit to the incidence of measles in Bristol itself was also discussed.

Eichner et al. (1996) considered the critical community size necessary to eradicate polio using a stochastic model. If the infection was able to persist, the stochastic fluctuations could lead to oscillations with an average duration of 3.45 years. This was slightly less than the 3.67 years predicted by the deterministic model (Anderson and May, 1991). Eichner et al. discussed the differences in the spread of polio between developing countries and developed (or industrialised) countries. There were two key epidemiological differences between these types of countries. First, average life expectancy would be higher in the industrialised country. Second, hygiene standards would also be better. As usual R_0 represented the expected number of secondary infections generated by a single, newly infected individual who entered a population at demographic equilibrium where disease was not initially present. In Eichner et al.'s model $R_0 = \beta/(\mu+\gamma) \approx \beta/\gamma$, where β was the rate of contacts which were sufficiently close for virus transmission, $(1/\gamma)$ the average infectious period and μ the common per capita birth and death rate. Hence in the industrialised country, β (and hence R_0) would be lower due to improved hygiene standards. The basic reproduction number was assumed to be 12 for a developing country and 5 for an industrialised one. (These numbers were derived from published age-specific antibody data.) In a hypothetical developing country where the life expectancy was relatively short (45 years) and the hygiene standards were relatively poor (as measured by setting $R_0 = 12$) a population size of around 100,000 was needed for the disease to persist. In contrast, in a developed country where the life expectancy was 75 years and hygiene standards were better (reducing R_0 to 5), a population of 500,000 inhabitants was needed. Life expectancy was a major factor in determining whether or not the disease would persist, and proved to be more important than hygiene standards (as measured by R_0). The average duration of persistence increased in an approximately exponential manner as the population size or the basic reproductive number was increased, and it decreased sharply with increasing life expectancy.

Next Eichner et al. discussed populations where there was geographical heterogeneity in the spread of the disease, and the majority of the population was divided into distinct social groups with a small minority mixing homogeneously with all groups. The ability of the infection to persist decreased with an increasing number of subpopulations. If the number of subpopulations was kept constant but the fraction of contacts with individuals of other subpopulations increased, virus persistence improved. It was believed that seasonal variation in polio virus transmission rates was practically important and would increase the minimum population size needed for persistence of the disease (Schenzle and Dietz, 1987). We now proceed to examine spatial epidemic models.

4. Spatial models

4.1. Introduction

This section considers models which remove the homogeneous mixing assumption by incorporating some sort of spatial structure. We start by outlining percolation process models, originally motivated by problems in fluid dynamics. This is followed by a review of spatial stochastic simulation models. The next subsection discusses two specific spatial stochastic simulation studies, one by Sattenspiel (1987) for hepatitis A triggered by children in day-care centers, and one by White and Harris (1995a) for the spread of bovine tuberculosis in South-West England. The section concludes with an analysis of how the spread of epidemics can be modelled by diffusion processes, which regard disease as travelling waves of infection.

4.2. Percolation processes

It is difficult to formulate a model to describe how a disease will spread in two dimensions. One way of doing this is by using percolation processes which were originally motivated by problems in fluid dynamics. Consider a fluid flowing through a porous medium which consists of 'sites' joined together by 'bonds'. If these 'bonds' are directed then the fluid flows through them in only one direction, while if they are undirected then the fluid flows through them in both directions. We say that the fluid 'wets' any site that it reaches.

Percolation theory started from two basic types of problems. In a bond percolation problem each of the bonds was blocked independently with probability p and the fluid could not flow through any blocked bond. In a site percolation problem each of the sites was blocked independently with probability p and the fluid could not flow through or wet any blocked site. Suppose that an orchard consisted of trees which were planted at the vertices of a square lattice. It was assumed that the disease started in one corner of the orchard and that once a tree became infected it would independently infect each of its four immediate neighbours with probability p. This could be modelled as a bond percolation problem. Percolation processes would tell us how far the disease would spread and how many sites could be expected to be infected. Such models could also help to describe how diseases spread among groups of people.

In an excellent review of percolation theory Shante and Kirkpatrick (1971) considered such a medium with a regular geometric pattern (which we shall call a lattice). They defined the percolation probability $P(p)$ to be the probability that infection would spread from a single tree to infinitely many trees, or equivalently the probability that a major epidemic outbreak would occur. They defined the critical probability

$$p_c = \sup\{p \ni P(p) = 0\}.$$

Thus if $p < p_c$ then the disease would spread only locally, whereas if $p > p_c$ then an epidemic of infinite size might occur. Hammersley (1961) showed that the critical probability was less for the bond percolation problem than for the site percolation problem. Sykes and Essam (1963, 1964) estimated critical probabilities for square, triangular, diamond and honeycomb lattices using a connection between the bond and site percolation problems. A tree-like lattice was any lattice which contained some point from which any random walk was self-avoiding. The percolation problem has been solved exactly for certain of these tree-like lattices (Fisher and Essam, 1961).

Morgan and Welsh (1965) considered a disease spreading among people who were situated at the vertices of a two-dimensional square lattice. They assumed that this lattice was directed and that the disease could spread only in the direction of the positive x and y axes. One person was initially infected and this person was taken to be at the origin. If an infected person was in contact with a susceptible, then the susceptible would catch the disease after a random infection time which had the exponential distribution with parameter λ. The maximum number of steps that the disease had taken from the origin at time t was defined to be N_t and $E[N_t]$ was denoted by $M(t; \lambda)$. Then $M(t; \lambda)$ was finite for all values of t and λ and moreover

$$M(t; \lambda) \equiv M(t\lambda; 1).$$

They therefore defined

$$M(t) \equiv M(t; \lambda),$$

and showed that if the times t_1 and t_2 were both strictly positive, then

$$M(t_1 + t_2) \geqslant M(t_1) + M(t_2).$$

Thus the disease would travel faster at later times. They denoted by r_t the number of people at time t who were the furthest distance from the person who was originally infected. They used the notation

$$R(t; \lambda) \equiv E[r_t].$$

Morgan and Welsh conjectured two results. The first implied that the distance which the disease had spread at time t grew no faster than t. The second conjecture was that

$$\frac{d}{dt} M(t; \lambda) = 2\lambda R(t; \lambda).$$

Hammersley (1966) generalised much of the above work. He assumed that individuals in the population were situated on a much more general graph and allowed the infection time to have a more general distribution than that above. In the same paper he proved most of the two conjectures of Morgan and Welsh. Diekmann and Heesterbeek (1999) discussed stochastic models for the spread of epidemics on random networks or graphs and derived formulae for the basic reproduction number and the probability of extinction.

4.3. Results from percolation process models

We shall now turn to models which tell us about the spread of the disease in more general terms, such as those describing the rough shape of the infected area and how fast the disease was travelling. Richardson (1973) has considered a more general version of Morgan and Welsh's model. His simulations suggested that as time progressed, the shape of the infected area changed from roughly a circle to roughly a diamond. Thus a long time after the start of the epidemic, the boundary of the infected area would move with the same speed in each direction. Similar results were true if the shape of the infected area at time t for fixed t was examined as the infection rate λ increased.

Mollison (1972) considered a spatial epidemic model where the people were situated in a single straight line with σ people at each integer point. Each infected person made contacts independently after random times which occurred independently as a Poisson process of rate $\alpha\sigma$. The destination of a given contact had a symmetrical probability distribution $v(u)$ where u was the distance of the contact from the source. He found two types of distinct behaviour: if the contact distribution $v(u)$ had exponentially bounded tails then the epidemic advanced in a steady wave; but if the contact distribution had infinite variance then the epidemic sometimes advanced in great leaps forwards.

In a later paper, Mollison (1977) examined these types of behaviour theoretically. θ was defined to be a unit vector in a particular direction and S_t the distance to the furthest infected person in that direction. The front velocity $\gamma_t(\theta)$ was the average speed at which the distance to the furthest infected person in direction θ increased at time t. Provided that it was certain that the epidemic would not die out, and under certain extra weak conditions, $\gamma_t(\theta)$ would tend in probability to a constant value $\gamma(\theta)$ as time progressed.

In fact by imposing extra conditions, this result could be strengthened to give almost sure convergence. Moreover, if a body Γ was defined by

$$\Gamma \equiv \{\mathbf{s} \text{ s.t. } \mathbf{s}.\boldsymbol{\theta} \leqslant \gamma(\boldsymbol{\theta})\},$$

then Γ was a convex body, with H_t denoting the convex hull of the set of infected individuals at time t. Then the area represented by (H_t/t) tended to Γ as time progressed. Thus a long time after the start of the epidemic, the infected area would have roughly the same shape as Γ and the edge of this area would be moving in every direction with a constant speed.

Cox and Durrett (1988) proved that the spatial 'nearest-neighbour' epidemic with removals grew linearly, had an asymptotic velocity in the strong law sense, and an asymptotic shape on the set of non-extinction. Simulations of McKendrick (1992) confirmed that the velocity c_0', say, was significantly lower than the velocity c_0 of its linearisation, the nearest neighbour birth and death process.

4.4. Simulation models

For the above epidemic models we found that it was difficult to obtain useful theoretical results. We shall now consider computer models for the simulation of epidemics. This work described the overall spatial pattern of infectives and how it changed as time progressed. Bartlett (1957) examined the consequences of including a spatial effect in a study of a recurrent epidemic by considering a 6×6 grid of sites, this was the largest layout that could be handled computationally at the time. Infectives moved from their own site to any neighbouring site with a common boundary. Although the threshold value for sustained oscillations was shown to be relatively unaffected by this spatial factor he demonstrated that persistence of the infection in just a few sites was sufficient to trigger a new epidemic. We briefly mention an epidemic simulation model due to Bartlett (1960b) which used Monte Carlo methods to simulate a measles epidemic in two dimensions. He found that a typical realisation showed a series of periodic epidemic outbreaks as predicted by the Hamer–Soper model (Hamer, 1906; Soper, 1929). Bartlett (1961) used a recurrent general epidemic model to simulate the spread of measles over a lattice. He studied the time to extinction of the disease and whether there were recurrent epidemic outbreaks, as well as the effect of seasonal differences on the probabilities of extinction.

The next model which we describe was due to Bailey (1967). He supposed that a disease spread at times which were separated by a fixed interval. He considered an 11×11 square lattice. At the start there was just one infected individual at the centre of the grid and at each other point there was one susceptible individual. Each time that the disease spread each infected individual independently infected each of his eight nearest neighbours independently with probability p. The disease spread away from the centre until the boundary was reached. A discrete time approach was used with results being obtained both for the model which corresponded to the simple epidemic without removals (where infectives continued to spread the disease indefinitely) and the general epidemic with removals (where infectives may stop spreading the disease).

Bailey found that the disease spread across the grid with roughly constant speed until it got near the boundary. To begin with the disease spread very quickly, but this rate of growth tailed off as the boundaries were approached. This behaviour was similar to the simple deterministic epidemic, although the manner in which the disease spread was very different. Bailey also observed a threshold effect. When p was small the number of people eventually infected tended to be very small, but when p was large the number of people infected was very large. The threshold value for p dividing these two types of behaviour was between 0.28 and 0.32.

Kendall (1957, discussion of Bartlett, 1957) demonstrated that a spatial threshold existed. He showed that a pandemic affecting every part of the region would occur if and only if the initial population density of susceptibles σ exceeded the non-spatial threshold $\rho = (\gamma/\beta)$. If a pandemic did occur then its severity θ was the unique root of the equation

$$\theta = 1 - \exp\left(-\frac{\sigma\theta}{\rho}\right). \tag{4.1}$$

This meant that the proportion of individuals who eventually became infected would be at least θ in any part of the region, no matter how far from the original source of infection. This spatial pandemic threshold result therefore corresponded to the non-spatial result of Kermack and McKendrick (1927). Moreover it could be shown that the solution to Eq. (4.1) was also the severity of the non-spatial epidemic. This was a rather surprising result; namely that the inclusion of a spatial effect did not affect the threshold value.

4.5. Simulation studies for hepatitis A and bovine tuberculosis

Sattenspiel (1987) studied a spatial stochastic simulation model for the spread of hepatitis A amongst children (in the USA day-care centers are important foci for hepatitis A transmission). The probability of transmission of the disease from an infective to a susceptible child depended on two factors, the geographical distance between the homes of the children, and whether they attended the same day-care facility. The probability of transmission as a result of geographical proximity was determined by a function that was inversely proportional to the square of the distance between children. The probability of transmission due to day-care attendance was zero for a child who attended no day-care facility, while for children at day-care centers, it was a function of the proportion of infective children in the facility. Two simulation models were run, the first among 200 children on a 100×100 grid, the second among 500 children on a 500×500 grid. Each epidemic was started with a single child infected. For six different initially infected children, each simulation was run 25 times and the average total number of cases was plotted against time. The results clearly illustrated that day-care centers were an extremely important factor both in starting the epidemic, and in increasing the severity and duration of community-wide epidemics.

White and Harris (1995a) described the use of a spatial stochastic model to examine the dynamics of bovine tuberculosis in badger populations. The model was simulated on a 10×10 grid-cell basis with each grid-cell potentially occupied by one badger social group; immigration into and emigration from the main grid were incorporated. Population regulation was assumed to occur at the group level through density-dependent fecundity and cub mortality, and the model could be run for various disease-free equilibrium group sizes (determined by the carrying capacity of the environment). The model worked on a three monthly basis. Three classes of individual (adults, yearlings and cubs) and three classes of infection (susceptible, infected but not infectious and infectious) were recognised.

Bovine tuberculosis was shown to persist in badger populations for long periods of time, even in those with a disease-free equilibrium population size of only four adults and yearlings. However, with standard rates of intergroup infection and movement, disease became endemic only in populations with a disease-free equilibrium group size greater than six adults and yearlings. In the endemic situation the prevalence of disease ranged between 11–22% depending on the combination of inter- and intra-group infection probabilities. Endemic infection within the homogeneous environment of the grid was characterised by a high degree of heterogeneity. Patches of infection were spatio-temporally unstable but shifted in location relatively slowly.

Spread of the disease from a point source of infection, with standard rates of intergroup movement and infection, only occurred to any marked extent in populations with disease-free equilibrium group sizes of eight or more adults or yearlings. Increasing the intergroup infection probability had a significant effect on increasing the probability and rate of spread, and considerably lowered the threshold group size for the spread of the disease from a point source to about four adults and yearlings. However, increasing the rates of intergroup movement reduced the probability of spread of the disease except at the lowest group sizes. When both intergroup infection and movements were increased, the effect of the increased infection in advancing spread was offset to some degree by increased movements. Perturbation to the badger population such as could be caused by the control operations could therefore increase the probability of persistence or spread of infection.

Recall that the work of Mollison (1977) showed that we could often regard an epidemic as travelling forwards in great waves of infection. In the next subsection we shall examine some analytical models for the spread of a disease where the epidemic is regarded as a wave travelling in one or two dimensions.

4.6. Diffusion processes

First we focus on the problems involved in defining the wave velocity exactly. Mollison (1977) has pointed out that to do this we must consider deterministic and stochastic epidemic models separately. The simplest epidemic models were those where the disease was spreading amongst people arranged in a single straight line. Then both x, the number of susceptibles, and y, the number of infectives, were functions of both position s and time t. Mollison looked for travelling wave solutions which described the spread of the disease of the form

$$x(s,t) = x(s - ct)$$

and

$$y(s, t) = y(s - ct),$$

where c was a constant called the wave velocity. A typical result was that epidemic waves were possible only above a certain minimum velocity. This was a threshold theorem for waves. Recall the stochastic model due to Mollison (1972) described in Section 4.3. The people amongst whom the disease was spreading were at the integer-valued points on a straight line. The symmetrical contact probability distribution $v(u)$ gave the destination of a given contact from the source. Mollison (1977) described a deterministic version of this model. He showed that if the contact distribution was a double exponential distribution

$$v(s) = \frac{1}{\sigma\sqrt{2}} \exp\left(-\frac{|s|}{\sqrt{2}}\sigma\right),$$

then an epidemic wave was possible for each velocity greater than or equal to a certain minimum velocity called the critical velocity. The smallest velocity that waves could travel at was

$$c_V = \inf\left(\frac{\Psi(\theta)}{\theta} : \theta > 0\right),$$

where $\Psi(\theta)$ was the moment generating function of the contact distribution $v(u)$. If $\Psi(\theta)$ diverged for all positive values of θ, then the velocity of the epidemic would increase without limit as time progressed. It was shown that if the contact distribution had exponentially bounded tails then an epidemic wave was possible at any velocity greater than or equal to the critical velocity, but epidemic waves were not possible below this velocity (Atkinson and Reuter, 1976; Brown and Carr, 1977). For stochastic epidemic models the situation was more difficult. There were two different sorts of wave velocity. We have already met the front velocity at time t. The expected instantaneous velocity in direction θ at time t was defined to be the rate of change of the expected distance to the furthest infected person in that direction.

Kendall (1957) investigated a two-dimensional version of the Kermack–McKendrick model discussed earlier. Under plausible assumptions about the behaviour of the number of people removed at a given point, he showed that introducing a small extra number of infected people would cause a pandemic exactly when the average population density, σ, exceeded a certain threshold value, ρ. He then gave an expression for the severity of that pandemic. We discussed Kendall's paper previously in Section 4.4. Later Kendall (1965) investigated a one-dimensional version of this model in greater detail. Kendall's Threshold Theorem for waves stated that epidemic waves were not possible unless the average population density σ exceeded a certain threshold value, ρ. If this condition was satisfied then the critical velocity was $v\sqrt{2(\sigma - \rho)/\sigma}$ and epidemic waves were possible at all velocities above this critical value. Here v was a constant independent of ρ and σ.

Finally we mention a stochastic epidemic model due to Bartlett (1960b). Again the epidemic spread among people arranged in a two-dimensional plane with average density n susceptibles per unit area. Bartlett now supposed that the infected individuals did not move around and the contact distribution $v(\mathbf{r})$ was given by the formula

$$v(\mathbf{r}) = \frac{\lambda}{2\pi\sigma^2} \exp\left(-\frac{|\mathbf{r}|^2}{2\sigma^2}\right),$$

where σ was small. This meant that the disease could spread only locally. This was a situation similar to the percolation process models which were discussed earlier. Bartlett found that if $n\lambda < \mu$, then it was not possible to define a wave of infection but if $n\lambda > \mu$ then epidemic waves were possible. In the latter case the final shape of the infected area was roughly circular and its boundary would then move with speed $\sqrt{2n\lambda\sigma^2(n\lambda - \mu)}$. All of these threshold results were relevant to the control of an epidemic. If we could reduce the density of susceptible individuals per unit area to below the threshold value by immunising susceptibles then a major epidemic outbreak would not occur.

Mollison (1991) described the use of linear deterministic models for examining the spread of epidemics, and discussed their advantages and limitations. He paid particular attention to spatial epidemic models and the speed of spatial epidemic spread. The main advantages of these models were that their assumptions were relatively transparent and easy to analyse, yet they generally gave the same velocity as more complex linear stochastic and nonlinear deterministic models. He felt that their simplicity allowed greater freedom to choose a biologically realistic model, and greatly facilitated examination of the dependence of conclusions on model components, as well as how these were incorporated into the model and fitted from data. A variety of examples were discussed including both dispersion and diffusion models. As the data available to fit the model was often limited, Mollison believed that it was more convincing for a model to be rejected on the basis of the data rather than be fitted to the sparse data available. Linear models provided only an upper bound on the velocity of nonlinear stochastic models, and were almost wholly inadequate when it came to modelling more complex aspects such as the onset of endemicity and endemic patterns. We now continue with a discussion of ways to control epidemics.

5. Stochastic models for control of epidemics

5.1. Introduction

One of the main aims in studying mathematical models for the spread of epidemics is to make predictions about how best they can be controlled, whether by vaccination, chemotherapy or other means. Although there has been a great deal of work done on deterministic models for epidemic control, there has been relatively little work on stochastic models. In this section we shall survey some of this work, paying particular attention to spatial stochastic simulation models.

5.2. General control models

Control methods in the deterministic and stochastic Kermack–McKendrick models have been considered by Abakuks (1973, 1974). He assumed that x, the number of infectives, and y, the number of susceptibles, and thus the state of the epidemic, represented by (x, y) were always known. Each infection incurred unit cost, and the epidemic could be controlled by removing infected people at a cost k each. For both models it was possible to draw a curve in the xy-plane such that if the point (x, y) lay above this curve then the best policy was not to remove anyone, whereas if (x, y) lay below this curve then it was best to remove all of the infectives and so stop the epidemic. In the later paper Abakuks (1974) considered trying to control the epidemic by immunisation and obtained similar results. Neymann and Scott (1964) had earlier considered a version of the Chain Binomial epidemic model in which the disease spread in a two-dimensional plane. The number of individuals infected by a single person depended on where that person was in the habitat, and the model allowed an infectious person to travel from where he became infected to where he became infectious. The effect of an immunisation program was also briefly discussed.

Watson (1972) studied a stochastic heterogeneously mixing epidemic model by using simulations and analytic approximations. Cane and McNamee (1982) analysed versions of the simple and general epidemic models with variable contact rates, so that the population was divided into n groups and the contact rate between group i and group j had the form $\beta_{ij} = \lambda_i \lambda_j$, $i, j = 1, 2, \ldots, n$, where $\lambda_1, \lambda_2, \ldots, \lambda_n > 0$. This assumption is usually called proportional mixing. Cane and McNamee obtained theoretical results for the deterministic versions of these models, but also performed simulations for the stochastic models. They found that, compared with the usual homogeneous mixing assumption, on average their model predicted a more rapid initial spread of infection and a slower final spread. It also predicted that, for a given removal rate, an epidemic could develop under the heterogeneous mixing assumption among a smaller group of susceptibles.

Greenhalgh (1986c) considered the stochastic version of Cane and McNamee's model where the epidemic was controlled both by vaccination of susceptible individuals and removal of infected ones. For control by immunisation only, the optimal policy to maximise the expected number of people immunised at a fixed time T was always to vaccinate the susceptible individual with the largest value of λ_i, the "most susceptible" individual remaining. This policy also minimised the expected number of people infected at time T. For control of the epidemic by removing infected individuals, it was shown that the optimal policy to maximise the expected number of individuals removed at time T was characterised by a series of switching times, alternately applying no removal effort and full removal effort. Such an objective function might be appropriate for an epidemic such as German measles or mumps, which spread through a class of schoolchildren of a similar age. Here immunisation might often be inconvenient or expensive, and the effect of catching the disease could often be much worse in adult life than in childhood. If the disease spread through a school in a short time (so that age effects could be ignored), then under those circumstances it might be reasonable to try to maximise the expected number of people removed. Some possible extensions of the results were briefly discussed.

Greenhalgh (1988) has also discussed further control of an epidemic by the removal and isolation of infected individuals, so as to maximise the expected number of individuals removed at some terminal time. A homogeneously mixing deterministic epidemic model was analysed using the maximum principle. It was found that the optimal policy was to wait until a switching time and then attempt to remove as many infected people as possible. For Cane and McNamee's heterogeneously mixing stochastic model it was found that, with the same objective, the optimal policy was to wait until a switching time $\xi(T, S_0)$, which depended on the terminal time T and the initial state S_0 of the epidemic and then apply full removal effort to the least infectious person. For the stochastic homogeneously mixing case, the relationship between the switching times, the starting state of the epidemic and the terminal time was explored.

Daley and Gani (1999) discussed an immunisation model to control an epidemic which spread amongst a stratified population. The inter-group contact rates β_{ij} had two values, a contact rate β_H which corresponded to two individuals in the same group and a second contact rate β_C which corresponded to individuals in different groups. There were m groups and for $i = 1, 2, \ldots, m$, group i contained N_i individuals. Daley and Gani derived an expression for R_0 and concluded that a maximal reduction in R_0 was obtained for a given overall reduction in the total number of susceptibles when $\max_{1 \leq i \leq m} N_i$ was reduced maximally subject to this given total reduction.

Becker (1972) has used stochastic models to investigate whether the US government's smallpox vaccination program should be stopped. He balanced the expected number of deaths due to a smallpox vaccination campaign against the costs and risks in deaths of an outbreak of smallpox. He used data on the incidence of smallpox in America to show that purely on these grounds smallpox vaccination should be stopped. His arguments were even more convincing if the detrimental side-effects of the vaccine which were not fatal were taken into account. Greenhalgh (1987) has used methods similar to Becker's in a simple theoretical stochastic model, to calculate the probability of a major outbreak of a disease. His model was appropriate for a disease spreading in a large population subject to control by immunisation. In the absence of immunisation the disease was assumed to spread by each infective infecting at random some of a fixed number of susceptibles. Recurrence relations were derived for the probability of a major epidemic outbreak. These were solved using analytical methods, and the resulting solutions were then explored numerically. Van Druten et al. (1986) have reported simulation studies of eradication of Congenital Rubella Syndrome (CRS) in the Netherlands by vaccination. Simulations with both deterministic and stochastic models were performed and found to give similar results. Dietz and Schenzle (1985) stated that there was great potential for further use of the tools of stochastic differential equations and population processes in random environments to modelling epidemics.

5.3. Spatial control models

Ball (1981) also tackled the problem of control of an epidemic spreading over a large area. He used a percolation process epidemic model to describe the spread of fox rabies. He considered the effect of reducing the size of a group of foxes in a control zone around the infected area and deduced a threshold effect which depended on the fox group size.

White and Harris (1995b) used a spatial stochastic simulation model to compare the efficacy of different badger control policies, and to determine the theoretical requirements for the control of endemic bovine tuberculosis in South-West England. Culling strategies for controlling the endemic disease were compared with strategies employing a yet to be developed oral vaccine. A comparative assessment was made of the efficacy of previous and proposed culling-based strategies employed by the UK Ministry of Agriculture, Fisheries and Food for the control of localised disease, and the potential for an oral-based vaccination strategy examined. For endemic bovine tuberculosis, to achieve a reasonable probability ($p > 0.7$) of successful control with a strategy involving a single culling operation, a very high proportion ($> 90\%$) of the cattle had to be culled. Single vaccination would not be successful in combating endemic disease. However strategies which involved repeated annual vaccination had a very high probability of eradicating endemic disease, even with a very low (40–50%) annual vaccination efficiency.

White and Harris (1995b) examined the following strategies which have been used to control bovine tuberculosis. From August 1975 to June 1982 the gassing strategy was used. This consisted of gassing badgers by pumping hydrogen cyanide powder into setts (the badgers' burrows). From August 1982 cage trapping replaced gassing. Trapping was concentrated around setts, and badger groups were removed in a centrifugal manner until a 'clean ring' of uninfected social groups had been removed (the clean ring strategy). From April 1986 an 'interim' strategy was introduced, where trapping was confined to those parts of the farm where cattle were believed to have contracted the disease from badgers, or to the whole farm if it was not possible to be more precise. In November 1994 a new approach was introduced. Part of this approach was a 'live test' strategy where a blood test for diagnosing tuberculosis in live badgers was used in the surrounding land to see whether any badgers around the original farm were infected with bovine tuberculosis. All animals caught in the first week were tested for bovine tuberculosis. If at least one infected animal was found, trapping continued at that sett for a further three weeks. All badgers trapped at positive setts, except lactating sows, were killed.

The most successful culling-based strategies were the gassing and clean ring strategies. Compared with no control at all, the interim strategy offered benefits of a lower probability of disease spread and persistence only in populations with low disease-free equilibrium group sizes or low initial prevalences of infection. In all other instances the benefits were negligible. A live test strategy offered an improvement over the interim strategy but would not be as effective as either gassing or clean ring strategies. In addition it was likely to necessitate the culling of approximately four times as many badgers each year as the interim strategy, and the proportion of those killed that were infected would be approximately half that under the interim strategy. The efficacy of a strategy involving annually repeated oral vaccination of badgers within a similar area to that covered by the live test depended on the efficiency of vaccination. A vaccination efficiency of 20–60% represented an overall improvement in efficacy over the interim strategy, being equivalent to the live test strategy. However only vaccination efficiencies of 60–80% or greater achieved results similar to the gassing strategy, and none were as successful as the clean ring strategy. The authors concluded that reactive strategies based on culling or vaccination would not solve the problem of bovine tuberculosis in badgers. Proactive strategies directed in those areas with a recent history of bovine tuberculosis in badgers should be considered as an alternative short-term measure. The only strategy likely to eradicate bovine tuberculosis from badger populations in the long term was the use of repeated vaccination in pro-active control operations in areas with a history of badger bovine tuberculosis.

In a series of papers Greenhalgh (1986a, 1986b, 1989) discussed the control of an epidemic spreading in one or two dimensions by using vaccination. Greenhalgh (1986a) considered a simple stochastic model with one or several epidemics spreading in one dimension, controlled by vaccination. He used dynamic programming methods to show that the optimal policy was to treat the population which was immediately at risk, and if there was more than one epidemic then it was optimal to treat the fastest moving epidemics first. In a second paper Greenhalgh (1989) extended these analytical results to the spread of an epidemic on a two-dimensional square lattice. Again the optimal policy was always to immunise one of the populations immediately at risk from the disease. Greenhalgh (1986b) discussed a two-dimensional Monte Carlo simulation model of the spread of a disease on a square lattice and considered the question of attempting to stop the disease by immunising a small or a large ring around an infected area. His results showed that to minimise the expected number of people ever to catch the disease, the policies which always treated one of the susceptible populations immediately at risk were better than the policies which attempted to immunise a ring around the infected area. For trying to minimise the probability of a major epidemic outbreak, it was found that there was a 'threshold' effect. If the infection rate was below a certain threshold value, then it appeared that by choosing an arbitrarily large ring to immunise, the probability of the disease breaking through the ring before it was complete could be made arbitrarily close to zero. On the other hand if the infection rate was above this threshold value, then the optimal policy was to try and stop the disease as soon as possible, that was by using a 'nearest-neighbour' policy.

So far, this article has concentrated mainly on theoretical advances in the use of stochastic processes in epidemic modelling, although we have also mentioned several very practical examples which were based on these theoretical ideas. The next section discusses some more practical uses of stochastic modelling for describing the spread and control of infectious diseases.

6. Specific applications

6.1. Introduction

This section describes some more specific applications of stochastic modelling. We start off by considering some models for parasitic diseases of humans, and follow this with applications of stochastic processes in veterinary modelling. However, we concentrate on stochastic modelling in HIV and AIDS epidemiology, particularly the spread of HIV and AIDS amongst injecting intravenous drug users, as this is an area with which the author is particularly familiar.

6.2. Parasitic diseases

A very practical example of the use of stochastic simulation models is the computer modelling of the spread and control of onchocerciasis or river blindness using the program ONCHOSIM (Plaisier et al., 1990, 1991a, 1991b; Habbema et al., 1992). This was developed in collaboration with the Onchocerciasis Control Program (OCP) in West Africa, and used as a tool in the planning and evaluation of control programs such as larvicide application and chemotherapy with ivermectin. The model comprised a detailed description of the life history of the parasite *Onchocerca Volvulus*, its reproductive lifespan, and its transmission from person to person by Similium flies. Two simulation techniques were mixed in the program. Stochastic microsimulation was used to calculate the life events of individual persons and inhabitant parasites, while the dynamics of the Similium population and the development of the parasite in flies were simulated deterministically. The model outputs included both the microfilarial load and the prevalence of river blindness in the population, and real data was used to validate the model.

Stochastic applied probability models of superinfection and acquired immunity to multiple parasite strains were considered by Milligan and Downham (1996). These models were suitable for diseases such as malaria and trypanosomiasis. The prevalence of infection, the number of infections per individual and the mean duration of infection increased rapidly in young individuals, but decreased in older individuals as immunity was acquired. The mean interval between successive infections varied with age. The models explained observed cross-sectional patterns of age prevalence and longitudinal patterns in which individuals typically continued to become infected as they aged, albeit with decreasing frequency. In these models the time spent infected depended on parasite diversity, as well as the inoculation and recovery rates. Disease control could cause an increase in the number of infections and the prevalence of infection in older individuals, and in the average prevalence in the community, even when strain-specific immunity was life-long. Diekmann and Heesterbeek (1999) also outlined a stochastic model for a parasitic disease where hosts were differentiated by their parasite load.

6.3. Veterinary modelling

Another area where stochastic modelling has been used to model the spread of infectious diseases is in veterinary epidemiology. Thrusfield and Gettinby (1984) have surveyed the use of deterministic, stochastic, Monte Carlo simulation and matrix and network modelling approaches in veterinary epidemiology, with illustrations by suitable examples of each type of approach. De Jong (1995) discussed the use of the stochastic recurrent general epidemic model and Whittle's stochastic threshold theorem in veterinary epidemiology. De Jong et al. (1996) have used a stochastic Monte Carlo simulation model for the spread of bovine respiratory syncytial virus (BRSV) among cattle. Parameters were estimated from monthly data on the prevalence of antibodies against BRSV in sera from all cattle in six dairy herds. The objective of the model was to decide whether conjectured transmission of BRSV amongst seropositive cattle could explain the regular cyclic behaviour of BRSV in dairy herds. With the parameter values estimated, they concluded that persistence of BRSV by transmission amongst seropositive cattle would be accompanied by frequent extinctions (once every one to fifty years). They suggested that transmission amongst seropositive cattle was not a plausible mechanism for the persistence of BRSV in dairy herds.

6.4. HIV and AIDS

AIDS was first discovered in 1981 when young men in the USA with similar symptoms sought medical attention; these symptoms included pneumocystis carinii, a form of pneumonia and Kaposi's sarcoma, a skin tumour (CDC, 1981). Since then AIDS has developed into a world-wide pandemic. In some parts of the world, particularly sub-Saharan Africa, heterosexual AIDS transmission is a major problem, but in the developed world, particularly the USA and the UK, the transmission of HIV and AIDS is largely confined to high risk groups, particularly homosexual men and injecting intravenous drug users. The increase in the number of cases of HIV and AIDS has been accompanied by a vast number of academic papers on the subject. Most of this work has concentrated on the sexual transmission of HIV. This review will mainly focus on transmission of HIV among drug users, as it is the area where the author is most familiar with the use of stochastic processes in HIV modelling. However first of all we shall survey some work on estimation of parameters for HIV and AIDS modelling, followed by work on stochastic models in sexual transmission of HIV and AIDS.

6.5. Estimation of the AIDS incubation period

Longini et al. (1989) used a staged Markov model to estimate the distribution and mean length of the AIDS incubation period. The model divided the infected period into four progressive stages: infected but antibody negative; antibody positive but asymptomatic; pre-AIDS symptoms, abnormal haematological indicator or both, and clinical AIDS. The time in each stage was assumed to follow an exponential distribution with a different mean for each stage, these were taken to be independent. Maximum likelihood methods were used to estimate the mean and median waiting times in each stage, using data from a cohort of 603 HIV infected individuals. The mean AIDS incubation period from initial infection to development of clinical AIDS was 9.8 years with a 95% confidence interval (8.4, 11.2) years. The paper also considered the estimated density function of the AIDS incubation period, and the estimated survival functions for each stage of infection. Longini et al. (1996) surveyed further work in this area, which included more sophisticated staged Markov models, and used a larger cohort of HIV infected individuals and a greater number of stages in the model.

Daley and Gani (1999) discussed extrapolation forecasting for AIDS, the estimation of the AIDS incubation period and the use of the back-calculation method. The AIDS incubation period was estimated using a two- or three-parameter Weibull distribution and the gamma density function. The parameters were estimated by maximum likelihood. Daley and Gani described how Solomon et al. (1991) used AIDS data from Australia and the back calculation method to reconstruct HIV prevalence and hence made short-term predictions of future new AIDS diagnoses in Australia.

6.6. Sexual transmission of HIV/AIDS

Isham (1988) has reviewed mathematical modelling of the spread of HIV and AIDS up to 1987. Blanchard et al. (1990) introduced modelling of an AIDS epidemic on a random graph. Cairns (1991) analysed a model for the spread of AIDS in a homosexual population. The population was divided into four groups, each having a different mean number of sexual partners per month. In the uninfected population μ_P and σ_P were respectively the mean and standard deviation of the number of sexual partners that an individual had per month. The heterogeneity coefficient $S = \sigma_P/\mu_P$ was a measure of the relative variability in partner change rates. Cairns identified a minimal set of primary components that dictated the dynamics of the model: the initial growth rate θ, the basic reproduction number R_0 and the heterogeneity coefficient S. He then showed that it was sufficient to maximise the likelihood over the three primary components; further maximisation over the remaining secondary parameters did not produce a significant improvement in fit or affect the projection of the epidemic. This method also permitted construction of confidence limits for the projected incidence curve, which allowed the quantification of the uncertainties associated with such model fitting procedures. The method was tested on simulation data to analyse how the accuracy of estimates and projections changed as more data was gained.

Van Druten et al. (1994) discussed the application of Markov Chain theory to homosexual role behaviour (meaning the role, or roles, if any, normally taken in anal homosexual intercourse) and the spread of HIV. Change in role behaviour in an Amsterdam study of 748 gay men for the period 1984–1988 was examined. Van Druten et al. demonstrated that Markov Chain theory could be used to predict role behaviour over time and that this could to a large extent explain the evolution of HIV in role-specific subgroups. Data on role behaviour was collected at eight specific points in time and the Markov Chain transition matrix $T(i)$ between two successive time points i and $i+1$, $i = 1, 2, \ldots, 7$ estimated. It was shown that the initial vector of role behaviour together with the initial transmission matrix $T(1)$ could predict reasonably well the proportions of the cohort having different types of role behaviour at subsequent time points. Different types of behaviour were also shown to carry different risks of HIV infection.

Kaplan (1990) has examined two commonly made hypotheses concerning the sexual transmission of HIV. The first was that an uninfected person in repeated acts of risky sex with an infected partner, stood a constant probability of infection with each sex act (independently of other sex acts), while the second assumed that an uninfected person stood a constant probability of infection over the course of a relationship, without regard to the specific number of sex acts that occurred.

The sexual transmission of HIV from infectious to susceptible partners on a per sex act (or per partnership) basis, could be described by a random variable N, the number of sex acts (or sex partners) necessary for HIV transmission. Bernoulli type models of HIV transmission assumed that

$$\Pr(N \leqslant s) = 1 - (1-p)^s, \quad s = 0, 1, 2, \ldots, \quad 0 \leqslant p \leqslant 1,$$

where p was the per sexual contact (per partner) HIV transmission probability, and s was the number of sex acts (sex partners) for the individual in question. If HIV transmission was modelled on a per partner basis, then the transmission probability p was the product of the likelihood that a selected partner was infected and the conditional probability of HIV transmission given the selection of an infected partner. The observed data was of the form (s_i, x_i), $i = 1, 2, \ldots, n$, where S_i referred to the number of sex acts (sex partners) for the ith person in a sample of n people, while X_i was an indicator variable equal to 1 if HIV was transmitted (i.e., if the ith individual seroconverted) and 0 otherwise. The log-likelihood function for p could be written

$$\mathcal{L}(p) \equiv \sum_{i=1}^{n} \left[x_i \log(1 - (1-p)^{s_i}) + (1 - x_i) s_i \log(1 - p) \right]. \tag{6.1}$$

Assuming that p was small, Kaplan showed that its maximum likelihood estimator $\hat{p}$ was approximately

$$\frac{\sum_{i=1}^{n} x_i}{\sum_{i=1}^{n} (x_i + (1 - x_i) s_i)},$$

with approximate variance

$$\frac{p}{\sum_{i=1}^{n} s_i (1 + p(1-p)^{s_i - 2})}.$$

Maximum likelihood estimates of p, and associated 95% confidence intervals were thus obtained for two datasets, one per sex act for heterosexuals ($\hat{p} = 0.00014$, 95% confidence interval (0.0006, 0.0022)) and one per partnership among homosexuals ($\hat{p} = 0.051$, 95% confidence interval (0.022, 0.08)).

To test the appropriateness of the Bernoulli process described above, Kaplan considered a non-parametric alternative. The data was re-ordered so that $s_1 \leq s_2 \leq \cdots \leq s_n$ and P_i was defined to be $\Pr(N \leq S_i)$. This implied that $P_1 \leq P_2 \leq \cdots \leq P_n$. Hence the non-parametric likelihood estimators $\widehat{P_i}$ had to solve the constrained maximisation problem:

$$\text{Maximise} \quad \sum_{i=1}^{n} \left[x_i \log P_i + (1 - x_i) \log(1 - P_i) \right] \tag{6.2}$$

$$\text{subject to} \quad 0 \leq P_1 \leq P_2 \leq \cdots \leq P_n.$$

This was an isotonic regression problem, and the values of $\widehat{P_i}$ could be found by the pool adjacent violators algorithm (Ayer et al., 1955). Statistical testing of the Bernoulli process model against the non-parametric alternative was performed using simulation to evaluate the p-values. The null hypothesis of the Bernoulli process model was rejected for the per sex act dataset, but accepted for the per partnership dataset, indicating that while HIV infectivity could be modelled as a Bernoulli process with a constant infection probability per partner, the model might not be so appropriate at the level of sexual contacts. Some probabilistic interpretations of these results were also discussed.

Daley and Gani (1999) discussed mathematical modelling of an education campaign to prevent the spread of an AIDS epidemic, based on a STOP-AIDS campaign launched by the Swiss AIDS Foundation in 1987. An optimal policy was derived which balanced the costs of the campaign against the medical costs saved because of the prevention of new infections due to the campaign. Later in the same chapter they discussed controlling the spread of HIV in a prison, by screening and quarantine of prisoners. They considered both an isolated prison model and a prison-city model in which prisoners entered into the prison from the city in which it was situated and on release went out of the prison back into the city. For the isolated prison model an optimal policy was derived which balanced the costs of screening and quarantine against the medical costs saved due to the prevention of new HIV infections.

6.7. HIV and AIDS amongst drug users

We now turn to models describing the spread of HIV and AIDS among injecting, intravenous drug users (IVDUs). Mackintosh and Stewart (1979) described a mathematical model for the spread of heroin-use in a population, and used it to examine possible control policies. The model had close similarities with the Reed–Frost epidemic model.

To our knowledge the first model to explore the spread of HIV and AIDS amongst injecting intraveneous drug users was the pioneering work of Kaplan (1989). We summarise this model here, even though it was deterministic, as it formed the basis for several stochastic epidemic modelling applications. Kaplan considered a population of homogeneously mixing drug users, and moreover assumed that the population was of constant size, so that drug users who left it, whether infected or not, were immediately replaced by new susceptible drug users. $\pi(t)$ denoted the fraction of drug users who were infected at time t and $\beta(t)$ the fraction of needles infected at time t. Then the basic equations determining the progress of the epidemic were

$$\frac{d\beta}{dt} = \lambda\gamma\pi(t) - \lambda\gamma\beta(t)\big(1 - \big(1 - \pi(t)\big)(1 - \theta)\big),$$

and

$$\frac{d\pi}{dt} = \big(1 - \pi(t)\big)\lambda\beta(t)\alpha - \mu\pi(t).$$

Here λ was the rate per individual of visiting shooting galleries; γ was the gallery ratio, defined as $\gamma = (N/n)$, where N was the constant number of needles within the various shooting galleries and n the constant number of addicts. θ was the probability that an uninfected IVDU flushed an infected needle (in other words rid the needle of infected blood) on a single injection; and μ the total rate at which infected IVDUs left the sharing, injecting population. This included IVDUs who ceased sharing because they developed AIDS and those who ceased sharing for other reasons. α was the probability that a susceptible IVDU was infected following a single injection with an

HIV infected needle. Massad et al. (1994) outlined a probability model for calculation of the probability δ that an uninfected needle was infected after a single injection by an infected user. Kaplan's models and the simulation models described below assumed that $\delta = 1$.

Kaplan (1994, 1995) and Kaplan and Heimer (1994) used stochastic processes to estimate the parameters of their model, and to evaluate needle exchange schemes. Kaplan postulated a two-stage Markov process to capture the essence of needle infection dynamics over a short time-scale. When a needle was uninfected, it was assumed that transition to the infected state occurred at rate λ, while infected needles returned to the uninfected state at rate μ.

$\pi(t)$ was defined as the probability that a needle was HIV infected t time units following its introduction to the population. It was shown that

$$\frac{d\pi}{dt} = \lambda(1 - \pi(t)) - \mu\pi(t),$$

which implied that

$$\pi(t) = \frac{\lambda}{\lambda + \mu}\left(1 - e^{-(\lambda+\mu)t}\right) + \pi_0 e^{-(\lambda+\mu)t},$$

where π_0 was the probability that the needle was initially infected. Kaplan then assumed that the length of time needles remained in the population following introduction was given by an exponential distribution with mean τ. $\bar{\pi}(\tau)$ was defined as the probability that a randomly selected needle was infected. Kaplan showed that

$$\bar{\pi}(\tau) = \frac{\pi_0 + \lambda\tau}{1 + (\lambda + \mu)\tau}.$$

The mean circulation time τ could be affected by needle exchange. Needle exchanges aimed to reduce the transmission of HIV and AIDS by exchanging used needles for clean ones. The trial New Haven needle exchange studied by Kaplan operated essentially on a one for one exchange basis. τ was a function of the exchange rate ν of needles per program client per unit time. A reasonable performance measure for the impact of a needle exchange program was therefore the relative reduction in the proportion of infected needles

$$1 - \frac{\bar{\pi}(\tau(\nu))}{\bar{\pi}(\tau(0))}.$$

Kaplan used this result to evaluate the trial needle exchange program in New Haven, Connecticut, USA. The parameters of the model were estimated by maximum likelihood methods.

Kaplan (1995) also outlined a maximum likelihood method to estimate the HIV incidence rate amongst IVDUs (the rate of new infections per unit time). The data used consisted of the results of tests for the HIV virus on a collection of sequences of needles, each sequence collected over different timepoints from each of a group of intravenous drug injectors. The analysis was based on a change point method. Kaplan's results indicated that transmission had been significantly slowed among program participants in the trial New Haven needle exchange program. These results were instrumental in changing the Connecticut State legislation on needle exchange programs.

Gani and Yakowitz (1993) considered a random allocation model for the spread of HIV and AIDS by needle sharing among a fixed size group of intravenous drug users who are friends or relatives (sometimes called buddy-users). A Markov Chain approach was used to track the increase in infection in a stable group of such intravenous drug users when some of them were HIV infected, and some numerical examples given. The model was modified to allow for the replacement of infectives in the group, with the group size remaining constant. Daley and Gani (1999) outlined extensions of the Greenwood and Reed–Frost Chain Binomial models by incorporating replacement of emigrating infected individuals by immigration of a mixture of new susceptible and other infected individuals. This was relevant to the model for HIV and AIDS amongst IVDUs discussed by Gani and Yakowitz. Daley and Gani calculated the group threshold sizes necessary for endemicity.

Yakowitz (1994) has described a stochastic simulation model for the spread of HIV and AIDS among a large population of drug users. This work was strongly influenced by the deterministic model of Kaplan (1989). In the simulation there were a large number of small local groups of drug users, and a large residual population (called the population at large, or PAL). On a local scale, the spread of HIV among a small group of sharing drug users was modelled as in Gani and Yakowitz (1993) except that the new recruits into the group came from the PAL, and the departures from the group went into the PAL. The PAL was supplied with a sequence of newcomers, all susceptible, and the members of the PAL ceased to inject drugs at a category-dependent rate for reasons such as death, illness, a successful treatment program and so on. Simulations of the model were explored. Yakowitz found that for a single group there was considerable variability in the number of susceptibles over time, and consequently felt that classical deterministic approximations would not be reliable. However, if a large number of groups were considered simultaneously, Yakowitz found that the epidemic sampling variability of the total number of susceptibles and infectives was relatively small, even though the states of the individual groups varied considerably. An analytical justification for this was discussed. At the quasi-equilibrium, about 40% of the population was found to be infected which was in rough agreement with observed levels of HIV amongst IVDUs. Yakowitz then derived the Markov transition law for the evolution of the group-generated HIV epidemic.

Yakowitz performed simulations for the large-scale epidemic model. These simulations showed the regularity expected as limit theorems took hold, and he concluded that here deterministic approximations would indeed be reliable. For a large number of simulations for low infection rates, either the susceptible population eventually disappeared or grew in an unbounded fashion, which depended on the recruitment rate and infectivity. No steady-state appeared possible for prevalence or susceptibles.

Hay (1999) developed stochastic simulations and analysis based on Kaplan's basic model. Partial differential equations for the joint probability generating function

$\pi(s, z; t)$ of the number of infected addicts and the number of infected needles at time t were derived, although these turned out to be intractable. Stochastic simulations based on Kaplan's model showed considerable variability, but tended to converge to a quasi-equilibrium, although there was a relatively small probability that they would die out. Later, more realistic features such as the decline in infectivity of HIV over time in unused needles, and variable HIV infectivity over the course of an addict's infection were included. The parameter α, the probability that a susceptible addict was infected from a single injection with an infectious needle, was estimated from data from the Scottish Centre for Infection and Environmental Health, Glasgow, UK.

Peterson et al. (1990) outlined a complex stochastic Monte Carlo simulation model for HIV/AIDS in an intravenous drug user community. This model incorporated a model of the social networks describing the pattern of equipment sharing by drug users. The disease progression of an infected individual was modelled from susceptible, to acute infection, asymptomatic, pre-AIDS symptoms, clinical AIDS and finally death. The model assumed that the levels of infectiousness for the four HIV infected states were in the ratio $5 : 1 : 3 : 5$. Similarly the progression of drug use was modelled from 'monthly' to 'weekly' to 'daily' user, with possible transitions to and from 'being in jail and not using' and 'being in the community and not using' and to 'non-HIV related death'. A sensitivity analysis was performed to examine the dependence of the results on the unknown parameter α, the probability that a susceptible drug user became infected after a single injection with an infectious needle. The implications of random sharing versus structured sharing were also explored.

Tan and Tang (1993) studied a stochastic model for the spread of HIV involving both intravenous drug use and homosexual contact. There were three interacting subpopulations: IVDUs, homosexual men, and homosexual men who were also IVDUs. The disease progression was modelled as five sequential infective stages with variable infectivity for the different stages. The model included variable levels of sexual activity and needle sharing, immigration, emigration and mixing patterns. They examined four scenarios to compare deterministic and stochastic models with a given mixing pattern and random mixing. They concluded that stochastic variation in the number of different partners per unit time seemed to have little effect on the spread of the disease, provided that the expected number per unit time was reasonably large. However the deterministic model did not correspond well to the mean of the realisations of the stochastic model. We have previously noted a similar lack of agreement for simpler deterministic and stochastic epidemic models.

Atkinson (1996) has investigated the use of a discrete event simulation package, the General Purpose Simulation System (GPSS), to develop a model of HIV transfer in a hypothetical cohort of IVDUs under conditions resembling those found in a high volume shooting gallery. (Shooting galleries are places where addicts meet to share injection equipment.) Following the introduction of an index case, the HIV infection rate in this cohort was followed over five simulated years. The model kept track of the infected state of an infected user, but all infected users had the same infectivity. The model was then used to consider the effects of systematic variation in the frequency of injection and needle-cleaning behaviour.

Kretzschmar and Wiessing (1998) have studied the spread of HIV in social networks of injecting drug users. A stochastic simulation model was used to describe a network of long-term buddy relationships in a population of drug users. HIV transmission took place when injection equipment was borrowed. The probability of transmission depended on the duration of infection. It was very high for a short period immediately after infection, then low for a long intermediate period, but started to rise again in a period immediately before the individual developed clinical AIDS symptoms. Two surveys amongst drug users in The Netherlands were used to estimate the behavioural parameters of the model. The aim of the study was to investigate the effect of different prevention strategies.

They found evidence of a threshold effect. Beneath a certain threshold sharing frequency, the disease never took off, while above this threshold there was a large stochastic variation in prevalence. This is consistent with the threshold effects for other stochastic models which we have considered. After reduction of risk behaviour, HIV prevalence decreased very slowly. Reduction of sharing with strangers was by far the most effective way to decrease long-term disease prevalence. Also, prevention focused on new IVDUs was a very effective strategy. Because newly infected IVDUs were very infectious for 6–8 weeks after initial infection, most transmission of infection took place then. For this reason prevention strategies based on HIV testing did not appear to be very effective.

In the next section we turn to a slightly different area which we have also touched on briefly here, namely the use of stochastic processes and stochastic methods in estimating parameters (and also hypothesis testing) for epidemic models.

7. Stochastic processes in parameter estimation and hypothesis testing

7.1. Introduction

In this section we examine the use of stochastic processes in estimating parameters for epidemic models and for hypothesis testing. A large portion of this section discusses work contained in Becker's (1989) book. First of all we discuss tests for homogeneity in susceptibility, or the presence of acquired immunity in a population. Then we continue with the use of stochastic processes in parameter estimation, particularly estimation of the disease transmission function and the joint distribution of the incubation and infectious periods, including some work by Bailey and co-workers. Next we examine applications of martingale methods. Finally we consider two applications of these estimation methods, first to statistical inference in veterinary epidemiology, and second to vaccination programs for hepatitis A in Bulgaria.

7.2. Tests for homogeneity in susceptibility or acquired immunity

Becker (1989) discussed the use of stochastic processes in estimating the parameters of epidemic models. We have already briefly mentioned Becker's work on the estimation of parameters for simple Chain Binomial epidemic models. He later considered analyses of data for a series of r independent epidemics (labelled $i = 1, 2, \ldots, r$) observed spreading among a community of n people. For $i, j = 1, 2, \ldots, r$, $i \neq j$, there were N_i observed cases in epidemic i and $N_{i,j}$ individuals who were observed cases in both epidemics i and j. The Mantel–Haenszel-type statistic

$$\frac{\sum_{i=1}^{r-1} N_{i,i+1} - \sum_{i=1}^{r-1} \frac{N_i N_{i+1}}{n}}{\left\{ \sum_{i=1}^{r-1} \frac{N_i N_{i+1}(n-N_i)(n-N_{i+1})}{n^2(n-1)} \right\}^{1/2}}$$

could be used to test for evidence of acquired immunity for a homogeneous community, or alternatively varying susceptibility for a disease with no acquired immunity. Small values of this statistic indicated evidence of acquired immunity for a homogeneous community, while large values indicated varying susceptibility for a disease with no acquired immunity. Non-significant values of the test statistic suggested no acquired immunity in a homogeneous community, or that acquired immunity had tended to make susceptibility more homogeneous in a community which was originally non-homogeneous. Some other χ^2-based tests for homogeneity were discussed and illustrated with an application to common cold data.

7.3. Parameter estimation

Continuous time removal data was first used to estimate an infection rate, and a removal rate by Bailey and Thomas (1971). They used maximum likelihood estimation for the general epidemic model. The method was described, with a correction, by Bailey (1975, Section 6.8.3). Further work on estimation of latent, infectious and incubation periods was also described by Bailey (1975, Chapter 15). Maximum likelihood methods were used based on data for measles, where there were households of only two individuals, one of whom was the index case. This was later extended to larger households. The data were provided by Dr. R. E. Hope-Simpson of the Cirencester Public Health Laboratory Service (see Hope-Simpson, 1952; Hope-Simpson and Sutherland, 1954).

The first hypothesis which was examined was to have a variable latent period with variance v, but a very short infectious period immediately followed by recognisable symptoms was still retained. However the data did not support this hypothesis. The next hypothesis discussed was a latent period which was normally distributed, followed by an extended but constant period of infectiousness. This was in much better agreement with the data (Bailey, 1954, 1955, 1956a, 1956b, 1956c). For the measles data maximum likelihood estimates of the mean of the latent period and of the constant infectious period using a scoring method were 8.58 ± 0.32 days and 6.57 ± 0.76 days respectively. Various modifications were described including extensions of the methods to data from households containing exactly three individuals using the Greenwood Chain Binomial model. These gave estimates of the mean latent period as 7.63 ± 0.50 days and the constant infectious period as 7.05 ± 1.13 days.

Next computerised applications of the method were considered. These maximised the log-likelihood function directly rather than using maximum likelihood scoring methods. The measles data were re-analysed and the results were very similar. The methods were also applied to infectious hepatitis data (Bailey and Alff-Steinberger, 1970). They concluded that a good description of the data was obtained by assuming a mean latent period of about 15.5 days, with a standard deviation of about 4 days, followed by an infectious period of 21.5 days. Bailey (1975) discussed extension of the methods to larger groups than households of three. The procedure was slightly heuristic and was illustrated by an application to the Abakaliki smallpox data analysed by Bailey and Thomas (1971). This involved a total of thirty cases in a population of 120 individuals at risk in a closed community of Abakaliki in South-Eastern Nigeria.

Becker (1989) also discussed maximum likelihood methods for parameter estimation. First he considered estimation of the transmission function. He supposed that at time t there were $X(t)$ susceptibles and $Y(t)$ infectious individuals in the community. The rate of spread of disease was given by

$$\Pr(X(t+\Delta t) = x - 1 | X(t) = x, Y(t) = y) = h(x, y)\Delta t + o(\Delta t),$$

and

$$\Pr(X(t+\Delta t) = x | X(t) = x, Y(t) = y) = 1 - h(x, y)\Delta t + o(\Delta t),$$

where the correction term $o(\Delta t)$ became negligible for small Δt. Here h was a function of x and y, but could also depend on other epidemiologically important factors. Normally the form of $h(x, y)$ was specified as a monotone increasing function of both x and y with a few unknown parameters. For example, in the general epidemic model

$$h(x, y) = \beta xy, \quad x, y = 0, 1, 2, \ldots,$$

where β was an unknown parameter which had to be estimated. Another alternative was

$$h(x, y) = \begin{cases} \beta xy, & x \geq 0, \ y > 0, \\ 0, & x \geq 0, \ y = 0, \end{cases}$$

which corresponded to the Greenwood Chain Binomial epidemic model. After becoming infected, individuals independently passed through a latent period of duration S, followed by an infectious period of duration G and then became permanently immune. Identification of the joint distribution of S and G via the joint density function, $f_{S,G}$, completed the specification of the model.

Becker supposed that the epidemic process was observed completely over $[0, \tau]$ where τ was the duration of the epidemic. He also assumed that the successive times of the K infectious contacts in $(0, \tau]$ were $T_1, T_2, \ldots, T_K$. For $i = 1, 2, \ldots, K$, s_i and g_i denoted respectively the latent and infectious period of the ith individual. The likelihood function corresponding to the observations $K = k$ and (s_i, g_i, t_i), $i = 1, 2, \ldots, k$, where $0 < t_1 \leq t_2 \leq \cdots \leq t_k < \tau$ was derived. This was a function of the parameters contained in $f_{S,G}$ and h and could be used to make inferences about them.

It was rarely possible to observe the epidemic process in sufficient detail to construct this likelihood function. However, the likelihood which corresponded to a partially observed infection process could be very complex. Becker discussed the form of this likelihood function when the times of infectious contacts were not observed and the latent period was a (known or unknown) constant μ_S, and also when removals only were observed and both the latent and infectious periods were (known or unknown) constants μ_S and μ_G. Assuming that both μ_S and μ_G were known for the disease under consideration, Becker outlined generalised linear model (GLIM) approaches to estimate the unknown function h and its parameters. The method was then applied to the Abakaliki smallpox data and a larger dataset consisting of respiratory disease data collected in the island of Tristan da Cunha, October to November 1967.

7.4. Martingale methods

Becker then turned to the application of martingale methods in epidemic parameter estimation. The material was formulated in continuous time, but the methods derived did not necessarily require continuous observation of the epidemic. Indeed some methods needed only the eventual sizes of outbreaks. For example, the problem of estimating the infection rate for a simple epidemic model was considered. In a community of size $x(0) - y(0)$ there were initially $x(0)$ susceptibles and $y(0)$ infectives. $N(t)$ denoted the number of individuals infected during $[0, t)$. Individuals who were infected immediately became infectious and remained so for the duration of the epidemic. The process

$$M(t) = N(t) - \int_0^t \beta X(\xi) Y(\xi) \, d\xi$$

was a zero mean martingale. If the infection process was observed continuously over $[0, T]$ one could then obtain an estimating equation for β by evaluating M at time T, and equating it to its mean, which was zero. Hence the estimate for β was

$$\hat{\beta} = \frac{N(T)}{\int_0^T X(\xi) Y(\xi) \, d\xi}$$

(also the maximum likelihood estimate of β). The standard error of this estimate was

$$\text{s.e.}(\hat{\beta}) = \frac{N(T)^{1/2}}{\int_0^T X(\xi) Y(\xi) \, d\xi}.$$

Then it was supposed that the state of the epidemic was observed only at time T. Becker defined

$$J(\xi) = I[X(\xi) > 0],$$

where $I[E]$ was the standard indicator function of the event E, which took the value of one if E occurred and zero otherwise, and

$$B(\xi) = \begin{cases} 0, & \text{if } J(x^-) \text{ was zero,} \\ \frac{J(\xi^-)}{X(\xi^-) Y(\xi^-)}, & \text{otherwise.} \end{cases}$$

A method similar to the case where the whole epidemic was observable gave

$$\hat{\beta} = \frac{\int_0^T B(\xi)\,dN(\xi)}{\int_0^T J(\xi)\,d\xi}.$$

In the computation of the integral, $\int_0^T B(\xi)\,dN(\xi)$, the values of X and Y were inserted just prior to each of the jump times. Thus

$$\int_0^T B(\xi)\,dN(\xi) = \frac{1}{x(0)y(0)} + \frac{1}{(x(0)-1)(y(0)+1)} + \cdots$$
$$+ \frac{1}{(x(T)+1)(y(T)-1)}$$

and $\int_0^T J(\xi)\,d\xi = T$, provided that if $x(T) = 0$, T was taken to be $\inf\{t \geq 0: x(t) = 0\}$. The standard error of $\hat{\beta}$ was

$$\text{s.e.}(\hat{\beta}) = \frac{[\int_0^T B^2(\xi)\,dN(\xi)]^{1/2}}{\int_0^T J(\xi)\,d\xi},$$

where

$$\int_0^T B^2(\xi)\,dN(\xi) = \frac{1}{x(0)^2 y(0)^2} + \frac{1}{(x(0)+1)^2(y(0)-1)^2} + \cdots$$
$$+ \frac{1}{(x(T)+1)^2(y(T)-1)^2}.$$

In either situation confidence intervals could be constructed and hypothesis tests performed, justified by the use of the Central Limit Theorem for Martingales.

Becker next considered the stochastic general epidemic model and inference about $\theta = (\beta/\gamma)$, which measured the potential that an infective had for infecting a given susceptible, and was closely related to the threshold value $R_0 = N\theta$ for the general epidemic. Here N was the total population size. Becker's estimate for θ was

$$\hat{\theta} = \frac{\int_0^{T_N} \overline{B}(\xi)\,dN(\xi)}{z(T_N)}, \tag{7.1}$$

where T_N was the time when the infection process ended, in other words the earliest time when either every infected individual had been removed or there were no susceptible individuals left. $z(T_N)$ was the number of removed people at time T_N and $\overline{B}(\xi) = J(\xi-)/X(\xi-)$. Also

$$\int_0^{T_N} \overline{B}(\xi)\,dN(\xi) = \frac{1}{x(0)} + \frac{1}{x(0)-1} + \cdots + \frac{1}{x(T_N)+1}.$$

The standard error was

$$\text{s.e.}(\hat{\theta}) = \frac{\left(\int_0^{T_N} \overline{B}^2(\xi)\, dN(\xi) + \hat{\theta}^2 z(T_N)\right)^{1/2}}{z(T_N)}. \tag{7.2}$$

There was a minor technical problem when $x(T_N) = 0$, in that $z(T_N)$ was not observable, but this was easily resolved. Becker then outlined how the infection potential and its standard error could be estimated if there were several independent outbreaks. The methods were applied to data on outbreaks of smallpox in Abakaliki, Nigeria and on outbreaks of the common cold.

The next topic considered was the application of these methods to epidemics in a series of households, if there were two different rates β_b and β_w for transmission between and within households respectively. He discussed inferences about the two transmission potentials $\theta_b = (\beta_b/\gamma)$ and $\theta_w = (\beta_w/\gamma)$. This method was then applied to data on an epidemic of the common cold on Tristan da Cunha between August 1964 and April 1968. Next estimation of a time dependent infection rate $\beta(t)$ was looked at, again illustrated by estimation from the Abakaliki smallpox data. Becker then investigated methods of inference for large populations, using the idea that if the total number of infectives remained small, then the growth of an epidemic could be approximated by a branching process. He derived estimators for the mean of this branching process. These were applied to find the threshold value from data for a 1956 outbreak of variola minor in a district of São Paulo, Brazil. This was followed by a description of a general method for analysis of data on the size of outbreaks in households which permitted the possibility of infection from outside the household. The method was illustrated with investigation of data from an Asian flu epidemic and an epidemic of respiratory disease in Tristan da Cunha.

7.5. Statistical inference in veterinary epidemiology

De Jong and Kimman (1994) reported experiments on the transmission of Aujesky's disease (sometimes called pseudorabies virus) among vaccinated and unvaccinated pigs. Becker's martingale formula was used to estimate the transmission potential θ for the two groups, and for the combined group, under the null hypothesis that vaccination had no effect on θ. Given θ, the probabilities of the distribution of the size of the final epidemic outbreak were calculated. From these probabilities, the combined probability of at least the observed difference in outcome was computed and used to test the null hypothesis, which was rejected. Hence the vaccine successfully reduced disease transmission. The values of the reproduction number R in the unvaccinated and vaccinated groups were 10.0 and 0.5 respectively.

A second study (Bouma, Jong and Kimman, 1995) was performed using similar methods to compare disease transmission terms βXY and $\beta XY/N$ which Bouma et al. called pseudo and true mass action respectively. Experimental groups of either ten or forty vaccinated pigs per group were housed at equal density. Half of each group were vaccinated and the value of the reproduction number R estimated from the groups of ten pigs using methods similar to those of de Jong and Kimman (1994). Then the final size distributions were calculated for the outbreak in the experiment with the group of forty pigs, using the true and pseudo mass action models. If the probability of an observed (or more extreme) outcome was below 5% ($p < 0.05$) in either model then that model was considered false. The results suggested that the true mass action term $\beta XY/N$ was the correct one.

Bouma et al. (1996) used statistical methods similar to those of de Jong and Kimman (1994) and Bouma et al. (1995) to investigate whether two pseudorabies strains which differed in virulence, also differed in transmission amongst inoculated pigs. The conclusions were that although vaccinated pigs showed a significant difference in virus excretion for the two strains, virus transmission of the two strains did not differ. Mulder et al. (1995) used similar methods to quantify the transmission of genetically engineered pseudorabies virus, as did also Stegeman et al. (1995) to compare pseudorabies virus transmission in pigs vaccinated once and twice. The incidence of pseudorabies virus infections was significantly higher ($p = 0.039$) in the group vaccinated once (38%) than in the group vaccinated twice (10%). The reproduction number was also significantly higher ($p = 0.025$) amongst pigs vaccinated once ($R = 3.4$) than amongst pigs that had had two vaccinations ($R = 1.5$).

7.6. Application to hepatitis A in Bulgaria

Keiding (1991) discussed a framework for understanding disease incidence as a hazard rate and disease prevalence as a probability in the parameter space; he related observable quantities to these via a statistical model. The paper focused on statistical analysis of epidemiological cross-sectional data using methods from modern non-parametric continuous time survival analysis, which included random censoring and truncation models and estimation under monotonicity constraints. The exposition was illustrated by examples on age at menopause, diabetes incidence from a case study in Denmark, estimation of the reproduction number for infectious diseases with lifelong immunity applied to hepatitis A in Bulgaria, incidence of the chronic skin disease pustulosis palmo-plantaris before and after menopause, and incidence and prevalence of Myasthenia Gravis in Denmark.

The application to hepatitis A in Bulgaria was discussed in more detail by Keiding (1990). It was supposed that age-stratified serological data was available, collected before mass vaccination of the population. $\lambda_0(a)$ denoted the age-dependent force of infection at equilibrium in the absence of vaccination, $\mu(a)$ was the age-dependent background death rate (there was no appreciable mortality from hepatitis A) and $\phi(a)$ was the proposed steady state age-dependent vaccination rate at age a.

To estimate $\lambda_0(a)$ the data available was serological data from n individuals in the form $(A_1, \delta_1), (A_2, \delta_2), \ldots, (A_n, \delta_n)$ where A_i was the current age of individual i and

$$\delta_i = \begin{cases} 1, & \text{if individual } i \text{ was seropositive,} \\ 0, & \text{if individual } i \text{ was seronegative.} \end{cases}$$

Without loss of generality $A_1, A_2, \ldots, A_n$ were taken to be in ascending order. Keiding proceeded by inferring the distribution function $G(y) = 1 - e^{-\Lambda_0(y)}$ where $\Lambda_0(y) = \int_0^y \lambda_0(\xi)\,d\xi$.

A function $H : \{0, 1, 2, \ldots, n\} \to \mathcal{R}$ was defined by

$$H(i) = \sum_{j=1}^{i} \delta_j,$$

then the nonparametric maximum likelihood estimate $\widehat{G}$ of G was given by letting $\widehat{G}(A_i)$ be the left continuous derivative of the convex minorant of H at i, and completing the definition of the step function $\widehat{G}$ by left continuity.

The kernel smoothing techniques of Ramlau-Hansen (1983) could then be used to estimate λ_0 as follows. K was a symmetric probability density on $[-1, 1]$ and b was the bandwidth. Keiding took $K(\xi) = \frac{3}{4}(1 - \xi^2)$ with appropriate treatment of the tail problems which arose for ages near the boundary of the age range. Then

$$\hat{\lambda}_0(a) = \frac{1}{b} \int_{a-b}^{a+b} K\left(\frac{\xi - a}{b}\right) \frac{d\widehat{G}(\xi)}{1 - \widehat{G}(\xi^-)},$$

and similarly a kernel estimate of the density g of G was

$$\hat{g}(a) = \frac{1}{b} \int_{a-b}^{a+b} K\left(\frac{\xi - a}{b}\right) d\widehat{G}(\xi).$$

The resulting estimate of the reproduction number R_ϕ under the proportional mixing assumption was

$$R_\phi = \frac{\int_0^\infty e^{-M(a) - \Phi(a)} \hat{\lambda}_0^2(a) \, da}{\int_0^\infty e^{-M(a)} \hat{\lambda}_0(a) \hat{g}(a) \, da}.$$

Here $\Phi(a) = \int_0^a \phi(\xi) \, d\xi$ and $M(a) = \int_0^a \mu(\xi) \, d\xi$. If a steady-state vaccination campaign ϕ was applied then the disease was expected to die out if $R_\phi < 1$, but take off if $R_\phi > 1$. This result could be used to evaluate vaccination programs such as vaccination of fixed proportions of individuals at a given age or two given ages. Greenhalgh and Dietz (1994) have discussed this further with more general mixing assumptions.

8. Summary and conclusions

In this chapter we have examined the use of stochastic processes in epidemic modelling. We started off by looking at some classical epidemic models, the discrete time Chain Binomial models and the continuous time simple and general stochastic epidemic models. For the latter we compared deterministic and stochastic models and surveyed threshold theorems as well as discussing both recurrent and non-recurrent epidemic models. We then considered stochastic models for the spatial spread of a disease:

in particular percolation processes, stochastic spatial simulation models and diffusion processes. Next we discussed practical applications of stochastic models to epidemic control programs such as immunisation of susceptible individuals and removal of infected ones. This was followed by an exposition of some applications of stochastic modelling to specific diseases, with particular attention paid to HIV and AIDS. The final section examined a slightly different area, namely the use of stochastic theory to estimate parameters and test hypotheses for epidemic models.

For the sake of completeness we should also mention the volumes of papers edited by Grenfell and Dobson (1995), Mollison (1995) and Isham and Medley (1996), all of which arose from a six month research program on Epidemic Models at the Isaac Newton Institute at Cambridge University, Cambridge, UK in 1993. These provided many illustrations of problems in both deterministic and stochastic epidemic modelling. The book by Grenfell and Dobson emphasised parasites and pathogens in animal populations which ranged from insects through to mammals, although significant sections also dealt with plant pathogens. A full range of infectious disease agents was considered ranging from microparasites (viruses, bacteria, protozoa and prions) through to macroparasites (helminths and arthropods). Mollison's book surveyed the state (in 1993) of epidemic modelling in relation to understanding, prediction and evaluation of epidemics and implementation of disease control strategies. The book of Isham and Medley concentrated on mathematical models for infectious human diseases with particular emphasis on model structure and real data. The book contained a breadth of different quantitative approaches used in understanding the patterns of infectious diseases in human populations, and the design of control strategies to lessen their effect.

Stochastic modelling in epidemiology covers a very wide area, and we cannot claim to have covered all of it. Indeed the balance of this article is probably biased by the personal research interests and experience of the author. Nevertheless we have outlined the basic classical stochastic epidemic models. The work discussed here can be loosely divided into three areas. First the theoretical models, which aimed to give generic answers to general questions about the spread of the disease, for example, the threshold theorems of Williams and Whittle and the theoretical work on the spread of percolation processes. Such work tried to give a broad qualitative insight into features of the spread of the disease, rather than a precise description of exactly how and where the epidemic will spread. The second area consisted of detailed stochastic simulation models to give precise answers to specific questions about a particular disease, for example, the work of White and Harris on the spread and control of bovine tuberculosis in South-West England, or some of the stochastic simulations of HIV and AIDS among injecting intravenous drug users. The third area was the use of stochastic processes to estimate parameters; we have described mainly results outlined in the book by Becker (1989), but some of Kaplan's work also fell into this category. Of course there was some overlap between these three areas and advances in one area may support developments in the others.

Acknowledgement

Some of the material describing literature on the spread of HIV among intravenous drug users was derived from the Ph.D. thesis of Gordon Hay and we are grateful for permission to use this. We are also grateful to an anonymous referee for his comments on the manuscript.

References

Abakuks, A. (1973). An optimal immunisation policy for an epidemic. *J. Appl. Probab.* **10**, 247–262.
Abakuks, A. (1974). Optimal immunisation policies for epidemics. *Adv. Appl. Probab.* **6**, 494–511.
Abbey, H. (1952). An examination of the Reed–Frost theory of epidemics. *Human Biol.* **24**, 201–233.
Anderson, R.M. and R. M. May (1986). The invasion, persistence and spread of infectious diseases within animal and plant communities. *Phil. Trans. Roy. Soc. B* **314**, 533–570.
Anderson, R.M. and R. M. May (1991). *Infectious Diseases of Humans*. Oxford University Press, Oxford, UK.
Atkinson, C. and G. E. H. Reuter (1976). Deterministic epidemic waves. *Math. Proc. Cambridge Philos. Soc.* **80**, 315–330.
Atkinson, J. (1996). A simulation model of the dynamics of HIV transmission in intravenous drug users. *Comp. Biomed. Res.* **29**, 338–349.
Ayer, M., H. D. Brunk, G. M. Ewing, W. T. Reid and E. Silverman (1955). An empirical distribution function for sampling with incomplete information. *Ann. Math. Statist.* **26**, 641–647.
Bailey, N. T. J. (1953). The total size of a general stochastic epidemic. *Biometrika* **40**, 177–185.
Bailey, N. T. J. (1954). A statistical method of estimating the periods of incubation and infection of an infectious disease. *Nature* **174**, 139–140.
Bailey, N. T. J. (1955). Some problems in the statistical analysis of epidemic data. *J. Roy. Statist. Soc. Ser. B* **17**, 35–58.
Bailey, N. T. J. (1956a). On estimating the latent and infectious periods of measles. I. Families with two susceptibles only. *Biometrika* **43**, 15–22.
Bailey, N. T. J. (1956b). On estimating the latent and infectious periods of measles. II. Families with three or more susceptibles. *Biometrika* **43**, 322–331.
Bailey, N. T. J. (1956c). Significance tests for a variable chance of infection in Chain Binomial theory. *Biometrika* **43**, 332–336.
Bailey, N. T. J. (1967). The simulation of stochastic epidemics in two dimensions. In *Proceedings of the Fifth Berkeley Symposium on Mathematical Statistics and Probability*, Vol. 4, pp. 237–257. Berkeley and Los Angeles, University of California, California, USA.
Bailey, N. T. J. (1975). *The Mathematical Theory of Infectious Diseases and Its Applications*, 2nd ed. Griffin, London, UK.
Bailey, N. T. J. and C. Alff-Steinberger (1970). Improvements in the estimation of latent and infectious periods of a contagious disease. *Biometrika* **57**, 141–153.
Bailey, N. T. J. and A. S. Thomas (1971). The estimation of parameters from population data on the general stochastic epidemic. *Theor. Pop. Biol.* **2**, 53–70.
Ball, F. G. (1981). Some statistical problems in the epidemiology of fox rabies. Ph.D. thesis, Department of Biomathematics, University of Oxford, Oxford, UK.
Ball, F. G. (1983). The threshold behaviour of epidemic models. *J. Appl. Probab.* **20**, 227–241.
Ball, F. G. (1996). Threshold behaviour in stochastic epidemics amongst households. In *Athens Conference on Applied Probability and Time Series, Vol. I: Applied Probability*, pp. 253–266 (Eds. C. Heyde, R. Pyke and T. Rachev). Lecture Notes in Statist., Vol. 114.
Ball, F. G., D. Mollison and G. Scalia-Tomba (1997). Epidemics with two levels of mixing. *Ann. Appl. Probab.* **7**, 46–89.

Ball, F. G. and P. O'Neill (1993). A modification of the general stochastic epidemic motivated by AIDS modelling. *Adv. Appl. Probab.* **25**, 39–62.

Bartlett, M. S. (1953). Stochastic processes or the statistics of change. *Appl. Statist.* **2**, 44–64.

Bartlett, M. S. (1957). Measles periodicity and community size. *J. Roy. Statist. Soc. Ser. A* **120**, 48–70.

Bartlett, M. S. (1960a). The critical community size for measles in the United States. *J. Roy. Statist. Soc. Ser. A* **123**, 37–44.

Bartlett, M. S. (1960b). *Stochastic Population Models in Ecology and Epidemiology*. Methuen, London, UK.

Bartlett, M. S. (1961). Monte Carlo studies in ecology and epidemiology. In *Proceedings of the Fourth Berkeley Symposium on Mathematical Statistics and Probability*, Vol. 4, pp. 39–55. Berkeley and Los Angeles, University of California Press, California, USA.

Becker, N. G. (1972). Vaccination programs for rare infectious diseases. *Biometrika* **59**, 443–453.

Becker, N. G. (1980). An epidemic chain model. *Biometrics* **36**, 249–254.

Becker, N.G. (1981). A general Chain Binomial model for infectious diseases. *Biometrics* **37**, 251–258.

Becker, N. G. (1989). *Analysis of Infectious Disease Data*. Chapman and Hall, London.

Becker, N. G. and K. Dietz (1995). The effect of the household distribution on transmission and control of highly infectious diseases. *Math. Biosci.* **127**, 207–219.

Black, F. L. (1966). Measles endemicity in insular populations: critical community size and its evolutionary implications. *J. Theor. Biol.* **11**, 207–211.

Blanchard, P., G. F. Bolz and T. Kruger (1990). Modelling AIDS epidemics or any venereal disease on random graphs. *Lect. Notes Biomath.* **86**, 104–117.

Bouma, A., M. C. M. de Jong and T. G. Kimman (1995). Transmission of pseudorabies virus within pig populations is independent of the size of the population. *Prev. Vet. Med.* **23**, 163–172.

Bouma, A., M. C. M. de Jong and T. G. Kimman (1996). Transmission of two pseudorabies virus strains that differ in virulence and virus excretion in groups of vaccinated pigs. *Amer. J. Vet. Res.* **57**, 43–47.

Brown, K. J. and J. Carr (1977). Deterministic epidemic waves of critical velocity. *Math. Proc. Cambridge Philos. Soc.* **81**, 431–433.

Cairns, A. J. G. (1991). Model fitting and prediction of the AIDS epidemic. *Math. Biosci.* **107**, 451–489.

Cane, V.R. and R. McNamee (1982). The spread of infection in a heterogeneous population, *J. Appl. Probab.* **19A**, 173–184.

CDC (1981). Pneumocystis pneumonia – Los Angeles. *Morbidity and Mortality Weekly Report* **30**, 250–252.

Cox, J. T. and R. Durrett (1988). Limit theorems for the spread of epidemics and forest fires. *Stochastic Process. Appl.* **30**, 171–191.

Daley, D. J. and J. Gani (1999). *Epidemic Modelling: An Introduction*. Cambridge Studies in Mathematical Biology. Cambridge University Press, Cambridge, UK.

Dayananda, P. W. A. and W. L. Hogarth (1977). Control for some approximations to Chain Binomial epidemic models. *Math. Biosci.* **35**, 151–163.

Dayananda, P. W. A. and W. L. Hogarth (1978). Optimal health programs of immunization and isolation for some approximations to Chain Binomial epidemic models. *Math. Biosci.* **41**, 241–251.

De Jong, M. C. M. (1995). Mathematical modelling in veterinary epidemiology: why model building is important. *Prev. Vet. Epidemiol.* **25**, 183–193.

De Jong, M. C. M. and T. G. Kimman (1994). Experimental quantification of vaccine-induced reduction in virus transmission. *Vaccine* **12**, 761–766.

De Jong, M. C. M., W. H. M. Van der Poel, J. A. Kramps, A. Brand and J. T. Van Oirschot (1996). Quantitative investigation of population persistence and recurrent outbreaks of bovine respiratory syncytial virus on dairy farms. *Amer. J. Vet. Res.* **57**, 628–633.

Diekmann, O. and J. A. P. Heesterbeek (1999). *Mathematical Epidemiology of Infectious Diseases: Model Building, Analysis and Interpretation*. Wiley, New York, USA.

Dietz, K. (1988). The first epidemic model: a historical note on P. D. En'ko. *Austral. J. Statist.* **30A**, 56–65.

Dietz, K. and D. Schenzle (1985). Mathematical models for infectious disease statistics. In *A Celebration of Statistics: Centenary Volume of the International Statistical Institute*, pp. 167–204 (Eds. A. C. Atkinson and S. E. Fienberg). Springer-Verlag, New York, USA.

Eichner, M., K. P. Hadeler and K. Dietz (1996). Stochastic models for the eradication of poliomyelitis: minimum population size for virus persistence. In *Models for Infectious Human Diseases: Their Structure and Relation to Data*, pp. 315–327 (Eds. V. Isham and G. F. Medley). Publications of the Isaac Newton Institute, Cambridge University Press, Cambridge, UK.

En'ko, P. D. (1889). On the course of epidemics of some infectious diseases. *Vrach.* (St. Petersburg, Russia) **x**, 1008–1010; 1039–1042; 1061–1063.

En'ko, P. D. (1989). On the course of epidemics of some infectious diseases. *Int. J. Epidemiol.* **18**, 749–755.

Elveback, L. and A. Varma (1965). Simulation of mathematical models for public health problems. *Public Health Report* **80**, 1067–1076.

Fisher, M. E. and J. W. Essam (1961). Some cluster size and percolation problems. *J. Math. Phys.* **2**, 609–619.

Frost, W. H. (1976). Some conceptions of epidemics in general. *Amer. J. Epidemiol.* **103**, 141–151.

Gani, J. and D. Jerwood (1971). Markov Chain methods in Chain Binomial epidemic models. *Biometrics* **27**, 591–604.

Gani, J. and S. Yakowitz (1993). Modelling the spread of HIV among intravenous drug users. *IMA J. Math. Appl. Med. Biol.* **10**, 51–65.

Greenhalgh, D. (1986a). Simple models for control of epidemics. *Math. Modelling* **7**, 753–763.

Greenhalgh, D. (1986b). Optimal control of an epidemic by ring vaccination. *Stochastic Models* **2**, 339–364.

Greenhalgh, D. (1986c). Control of an epidemic spreading in a heterogeneously mixing population. *Math. Biosci.* **80**, 23–45.

Greenhalgh, D. (1987). Solution of recurrence relations with applications in epidemic control. *Int. J. Systems Sci.* **18**, 57–74.

Greenhalgh, D. (1988). Some results on optimal control applied to epidemics. *Math. Biosci.* **88**, 125–158.

Greenhalgh, D. (1989). Simple two dimensional models for the spread of a disease. *Stochastic Models* **5**, 131–159.

Greenhalgh, D. and K. Dietz (1994). Some bounds on estimates for reproductive ratios derived from the age-specific force of infection. *Math. Biosci.* **124**, 9–57.

Greenwood, M. (1931). On the statistical measure of infectiousness. *Camb. J. Hyg.* **31**, 336–351.

Grenfell, B. T. and A. P. Dobson (Eds.) (1995). *Ecology of Infectious Diseases in Natural Populations*. Publications of the Isaac Newton Institute, Cambridge University Press, Cambridge, UK.

Habbema, J. D. F., E. S. Alley, A. P. Plaisier, G. J. Van Oortmarssen and J. H. F. Remme (1992). Epidemiological modelling for onchocerciasis control. *Parasitology Today* **8**, 99–103.

Hamer, W. H. (1906). Epidemic disease in England. *Lancet* **1**, 733–739.

Hammersley, J. M. (1961). Comparison of atom and bond percolation processes. *J. Math. Phys.* **2**, 728–733.

Hammersley, J. M. (1966). First passage percolation. *J. Roy. Statist. Soc. Ser. B* **28**, 491–496.

Haskey, H. W. (1954). A general expression for the mean in a simple stochastic epidemic. *Biometrika* **41**, 272–275.

Hay, G. (1999). Modelling the spread of HIV/AIDS amongst injecting drug users. Ph.D. thesis, Department of Statistics and Modelling Science, University of Strathclyde, Glasgow, UK.

Heasman, M. A. and D. D. Reid (1961). Theory and observation in family epidemics of the common cold. *Brit. J. Prev. Soc. Med.* **15**, 12–16.

Hope-Simpson, R. E. (1952). Infectiousness of communicable diseases in the household. *Lancet* **2**, 755–760.

Hope-Simpson, R. E. and I. Sutherland (1954). Does influenza spread within the household? *Lancet* **1**, 721–726.

Isham, V. (1988). Mathematical modelling of the transmission dynamics of HIV infection and AIDS: a review. *J. Roy. Statist. Soc. Ser. A* **151**, 5–49.

Isham, V. and G. F. Medley (Eds.) (1996). *Models for Infectious Human Diseases: Their Structure and Relation to Data*. Publications of the Isaac Newton Institute, Cambridge University Press, Cambridge, UK.

Jacquez, J. A. and P. O'Neill (1991). Reproduction numbers and thresholds in stochastic epidemic models. I. Homogeneous populations. *Math. Biosci.* **107**, 161–186.

Kaplan, E. H. (1989). Needles that kill: modelling human immunodeficiency virus via shared drug injection equipment in shooting galleries. *Rev. Infect. Dis.* **11**, 289–298.

Kaplan, E. H. (1990). Modelling HIV infectivity: must sex acts be counted? *J. Acquired Immune Deficiency Syndromes* **3**, 55–61.

Kaplan, E. H. (1994). A method for evaluating needle exchange programs. *Statistics in Medicine* **13**, 2179–2187.

Kaplan, E. H. (1995). Probability models of needle exchange. *Oper. Res.* **43**, 558–569.
Kaplan, E. H. and R. Heimer (1994). A circulation theory of needle exchange. *AIDS* **8**, 567–574.
Keiding, N. (1990). Non-parametric estimation of Dietz and Schenzle's transmission potential for age-dependent infections. *Research Report 90/2*, Statistical Research Unit, University of Copenhagen, Copenhagen, Denmark.
Keiding, N. (1991). Age-specific incidence and prevalence: a statistical perspective. *J. Roy. Statist. Soc. Ser. A* **154**, 371–412.
Kendall, D. G. (1957). La propogation d'une epidemie au d'un bruit dans une population limitée. In *Publications of the Institute of Statistics, University of Paris*, Vol. 6, pp. 307–311. Paris, France.
Kendall, D. G. (1965). Mathematical models for the spread of infection. In *Mathematics and Computer Science in Biology and Medicine*, pp. 213–215. H.M.S.O., London, UK.
Kermack, W. O. and A. G. McKendrick (1927). Contributions to the mathematical theory of epidemics. Part I. *Proc. Roy. Soc. Ser. A* **115**, 700–721.
Kretzschmar, M. and L. G. Wiessing (1998). Modelling the spread of HIV in social networks of injecting drug users. *AIDS* **12**, 801–811.
Lefèvre, C. (1979). Generalization of a control model for approximate Chain Binomial epidemics. *Math. Biosci.* **45**, 21–35.
Lefèvre, C. (1981). Optimal immunization and isolation policies for the Greenwood Chain Binomial epidemic model. *Biometrical J.* **23**, 55–67.
Lefèvre, C. and P. Picard (1990). A non-standard family of polynomials and the final size distribution of Reed–Frost epidemic processes. *Adv. Appl. Probab.* **22**, 25–48.
Lidwell, O. M. and T. Sommerville (1951). Observations on the incidence and distribution of the common cold in a rural community during 1948 and 1949. *Camb. J. Hyg.* **49**, 365–381.
Longini, I. M., W. Scott Clark, R. H. Byers, J. W. Ward, W. W., Darrow, G. F. Lemp and H. W. Hethcote (1989). Statistical analysis of the stages of HIV infection using a Markov model, *Statistics in Medicine* **8**, 831–843.
Longini, I. M., W. Scott Clark, G. A. Satten, R. H. Byers and J. M. Karon (1996). Staged Markov models based on $CD4^+$ T-lymphocytes for the natural history of HIV infection. In *Models for Infectious Human Diseases: Their Structure and Relation to Data*, pp. 439–459 (Eds. V. Isham and G. F. Medley). Publications of the Isaac Newton Institute, Cambridge University Press, Cambridge, UK.
Mackintosh, D. R. and G. T. Stewart (1979). A mathematical model of a heroin epidemic. *J. Epidemiol. Comm. Health* **33**, 299–304.
Maia J. de O. C. (1952). Some mathematical developments in the epidemic theory formulated by Reed and Frost. *Human Biol.* **24**, 167–200.
Massad, E., F. A. B. Coutinho, H. M. Yang, H. B. de Carvalho, F. Mesquita and M. N. Burattini (1994). The basic reproduction ratio of HIV amongst intravenous drug users. *Math. Biosci.* **123**, 227–247.
McKendrick, A. G. (1926). Applications of mathematics to medical problems. *Proc. Edinburgh Math. Soc.* **14**, 98–130.
McKendrick, I. J. (1992). The spatial modelling of epidemics: velocity and pattern. Ph.D. thesis, Department of Actuarial Mathematics and Statistics, Heriot-Watt University, Edinburgh, UK.
Metz, J. A. J. (1978). The epidemic in a closed population with all susceptibles equally vulnerable: some results for large susceptible populations and small initial infections. *Acta Biotheoretica* **27**, 75–123.
Milligan, P. J. M. and D. Downham (1996). Models of superinfection and acquired immunity to multiple parasite strains. *J. Appl. Probab.* **33**, 915–932.
Mollison, D. (1972). The rate of spatial propogation of simple epidemics. In *Proceedings of the Sixth Berkeley Symposium on Mathematical Statistics and Probability*, Vol. 3, pp. 579–614. Berkeley and Los Angeles, University of California, California, USA.
Mollison, D. (1977). Spatial contact models for ecological and epidemic spread. *J. Roy. Statist. Soc. Ser. B* **39**, 283–326.
Mollison, D. (1991). Dependence of epidemic and population velocities on basic parameters. *Math. Biosci.* **107**, 255–287.
Mollison, D. (Ed.) (1995). *Epidemic Models: Their Structure and Relation to Data*. Publications of the Isaac Newton Institute, Cambridge University Press, Cambridge, UK.

Morgan, R. W. and D. J. A. Welsh (1965). A two dimensional Poisson growth process. *J. Roy. Statist. Soc. Ser. B* **27**, 497–504.

Mulder, W. A. M., M. C. M. de Jong, J. Priem, A. Bouma, J. M. A. Pol and T. G. Kimman (1995). Experimental quantification of transmission of genetically engineered pseudo-rabies virus. *Vaccine* **13**, 1763–1769.

Murray, G. D. and A. D. Cliff (1975). A stochastic model for measles epidemics in a multi-region setting. *Inst. Br. Geog.* **2**, 158–174.

Nasell, I. (1999). On the time to extinction in recurrent epidemics. *J. Roy. Statist. Soc. Ser. B* **61**, 309–330.

Neymann, J. and E. Scott (1964). A stochastic model of epidemics. In *Stochastic Models in Biology and Medicine*, pp. 45–83. University of Wisconsin Press, Madison, Wisconsin, USA.

Peterson, D., K. Willard, M. Altmann, L. Gatewood and G. Davidson (1990). Monte Carlo simulation of HIV transmission in an intravenous drug user community. *J. Acquired Immune Deficiency Syndromes* **3**, 1086–1095.

Picard, P. and C. Lefèvre (1990). A unified analysis of the final size and severity distribution in collective Reed–Frost epidemic processes. *Adv. Appl. Probab.* **22**, 269–294.

Plaisier, A. P., G. J. Van Oortmarssen, J. D. F. Habbema, J. H. F. Remme and E. S. Alley (1990). ONCHOSIM: a model and computer simulation program for the transmission and control of onchocerciasis. *Computer Methods and Programs in Biomedicine* **31**, 43–56.

Plaisier, A. P., G. J. Van Oortmarssen, J. H. F. Remme, E. S. Alley and J. D. F. Habbema (1991). The risk and dynamics of onchocerciasis recrudescence after cessation of vector control. *Bull. W.H.O.* **69**, 169–178.

Plaisier, A. P., G. J. Van Oortmarssen, J. H. F. Remme and J. D. F. Habbema (1991). The reproductive lifespan of *Onchocerca Volvulus* in West African savanna. *Acta Tropica* **48**, 271–284.

Rajarshi, M. B. (1981). Simpler proofs of two threshold theorems for a general stochastic epidemic. *J. Appl. Probab.* **18**, 721–724.

Ramlau-Hansen, H. (1983). Smoothing counting process intensities by means of kernel functions. *Ann. Statist.* **11**, 453–466.

Renshaw, E. (1991). *Modelling Biological Populations in Space and Time*. Cambridge University Press, Cambridge, UK.

Richardson, D. (1973). Random growth in a tessellation. *Proc. Cambridge Philos. Soc.* **74**, 515–528.

Ridler-Rowe, C. J. (1967). On a stochastic model of an epidemic. *J. Appl. Probab.* **4**, 19–33.

Sattenspiel, L. (1987). Epidemics in non-randomly mixing populations: a simulation. *Amer. J. Phys. Anthropology* **73**, 251–265.

Schenzle, D. (1982). Problems in drawing epidemiological inferences by fitting epidemic models to lumped data. *Biometrics* **38**, 843–847.

Schenzle, D. and K. Dietz (1987). Critical population sizes for endemic virus transmission. In *Räumliche Persistenz und Diffusion von Krankenheiten*, pp. 31–42 (Eds. W. Fricke and E. Hinz). Selbstverlag des Geographischen Instituts der Universität Heidelberg, Heidelberg, Germany.

Shante, V. K. S. and S. Kirkpatrick (1971). An introduction to percolation theory. *Adv. Phys.* **20**, 325–357.

Solomon, P. J., R. G. Attewell, E. B. Freeman and S. R. Wilson (1991). AIDS in Australia: reconstructing the epidemic from 1980 to 1990 and predicting future trends in HIV disease. National Centre for Epidemiology and Public Health, Working Paper 29, Australian National University, Canberra, Australia.

Soper, H. E. (1929). Interpretation of periodicity in disease prevalence. *J. Roy. Statist. Soc. Ser. A* **92**, 34–73.

Stegeman, A., A. Van Nes, M. C. M. de Jong and F. W. M. M. Bolder (1995). Assessment of the effectiveness of vaccination against pseudorabies in finishing pigs. *Amer. J. Vet. Res.* **56**, 573–578.

Sykes, M.F. and J. W. Essam (1963). Some exact critical percolation probabilities for bond and site problems in two dimensions. *Phys. Rev. Lett.* **10**, 3–4.

Sykes, M. F. and J. W. Essam (1964). Exact critical probabilities for site and bond problems in two dimensions. *J. Math. Phys.* **5**, 1117–1127.

Tan, W. Y. and S. C. Tang (1993). A stochastic model of the HIV epidemic involving both sexual contact and IV drug use. *Math. Comp. Modelling* **17**, 31–57.

Thrusfield, M. V. and G. Gettinby (1984). An introduction to techniques of veterinary modelling. In *Proceedings of the Society for Veterinary Epidemiology and Preventive Medicine, 10–11 July 1984*, pp. 114–138 (Ed. M. V. Thrusfield). University of Edinburgh, Edinburgh, UK (Also published separately, ISBN 0 948073 01 2).

Van Druten, J. A. J., T. de Boo and A. D. Plantinga (1986). Measles, mumps and rubella: control by vaccination. In *Proceedings of the 19'th IABS Congress on Use and Standardization of Combined Vaccines, Amsterdam, The Netherlands*, Vol. 65, pp. 53–63 (Ed. S. Karger). Develop. Biol. Standard, Basel, Switzerland.

Van Druten, J. A. M., F. J. P. Van Griensven and J. C. M. Hendriks (1994). Homosexual role behaviour and the spread of HIV. In *Modelling the AIDS Epidemic: Planning, Policy and Prediction*, pp. 515–532 (Eds. E. H. Kaplan and M. L. Brandeau). Raven Press, New York, USA.

Von Bahr and A. Martin-Löf (1980). Threshold limit theorems for some epidemic processes. *Adv. Appl. Probab.* **12**, 319–349.

Watson, R. K. (1972). On an epidemic in a stratified population. *J. Appl. Probab.* **9**, 659–666.

Watson, R. K. (1980). On the size distribution for some epidemic models. *J. Appl. Probab.* **17**, 912–921.

White, P. C. L. and S. Harris (1995a). Bovine tuberculosis in badger (Meles meles) populations in southwest England: the use of a spatial stochastic simulation model to understand the dynamics of the disease. *Phil. Trans. Roy. Soc. London B* **349**, 391–413.

White, P. C. L. and S. Harris (1995b). Bovine tuberculosis in badger (Meles meles) populations: an assessment of past, present and possible future control strategies using simulation modelling. *Phil. Trans. Roy. Soc. London B* **349**, 415–432.

Whittle, P. (1955). The outcome of a stochastic epidemic – a note on Bailey's paper. *Biometrika* **42**, 116–122.

Williams, T. (1971). An algebraic proof of the threshold theorem for the general stochastic epidemic (abstract). *Adv. Appl. Probab.* **3**, 223.

Wilson, E. B., C. Bennett, M. Allen and J. Worcester (1939). Measles and scarlet fever in Providence, Rhode Island, 1929–34 with respect to age and size of family. *Proc. Amer. Phil. Soc.* **80**, 357–476.

Wilson, E. B. and M. H. Burke (1942). The epidemic curve. *Proc. Nat. Acad. Sci. Washington* **28**, 361–367.

Yakowitz, S. J. (1994). From a microcosmic IDU model to a macrocosmic HIV epidemic. In *Modelling the AIDS Epidemic, Planning, Policy and Prediction*, pp. 365–383 (Eds. E. H. Kaplan and M. L. Brandeau). Raven Press, New York, USA.

Empirical Estimators Based on MCMC Data

Priscilla E. Greenwood and Wolfgang Wefelmeyer

1. Introduction

Suppose we want to calculate the expectation of a function f under a distribution π on some space E. If E is of high dimension, or if π is defined indirectly, it may be difficult to calculate the expectation $\pi f = E_\pi f = \int \pi(dx) f(x)$ analytically or even by numerical integration. (The notation πf will be used throughout the paper.) The classical Monte Carlo method generates i.i.d. realizations $X^0, \ldots, X^n$ from π, and approximates πf by the *empirical estimator*

$$E_n f = \frac{1}{n} \sum_{i=1}^n f(X^i).$$

If f is π-integrable, the estimator is strongly consistent. If f is π-square-integrable, the estimator is asymptotically normal with variance $\pi(f - \pi f)^2$. Often, however, this Monte Carlo method is difficult to implement. One reason is that high-dimensional distributions are hard to simulate. Additional difficulties arise when π is defined indirectly, as in many Bayesian modeling situations, or known only up to a normalizing constant, as is usually the case for random fields.

Markov chain Monte Carlo methods (MCMC) generate realizations $X^0, \ldots, X^n$ of a Markov chain with π as invariant law. (Here and in the following, by Markov chain we mean a discrete-time Markov process with arbitrary state space, not a continuous-time Markov process with discrete state space.) Again, the empirical estimator $E_n f$ can be used to approximate πf, and we have an explicit expression for the asymptotic variance of the estimator from the ergodic theory for Markov chains; see Section 2.

Choice of an MCMC method amounts to choice of a transition distribution Q from the large family of those with invariant law π. One important criterion is the speed with which the law of the Markov chain converges to π. This problem is well-studied, together with the associated question of how long the sampler must run until the observations are satisfactorily close to stationarity. Recent references are (Schervish and Carlin, 1992; Chan, 1993; Frigessi et al., 1993; Tierney, 1994; Meyn and Tweedie, 1994; Ingrassia, 1994; Roberts and Polson, 1994; Athreya, Doss and Sethuraman, 1996; Rosenthal, 1995; Mengersen and Tweedie, 1996; Roberts and Tweedie, 1996,

1999, 2000; Johnson, 1996; Roberts and Sahu, 1997; Kira and Ji, 1997; Robert, 1998; Diaconis and Saloff-Coste, 1998; Jerrum, 1998; Roberts and Rosenthal, 1998; Jarner and Roberts, 2002). The initial observations from this "burn-in" period are usually discarded.

At this point the transition distribution, Q, used in the sampler may be changed to one which is optimized according to a different criterion. Now the simulated data will be used to estimate πf using either the empirical estimator or possibly an improved estimator which exploits some property of the sampler. It is reasonable to judge the sampler by the asymptotic variance of the empirical estimator. This criterion is utilized by Peskun (1973), Frigessi, Hwang and Younes (1992), Green and Han (1992), Liu, Wong and Kong (1994, 1995), Clifford and Nicholls (1995), Liu (1996), Fishman (1996), and Mira and Tierney (1999). This survey will be about efforts to choose a sampler and an estimator of πf where one starts from an already (approximately) stationary initial distribution. For a short overview see Wefelmeyer (1998).

MCMC methods originated with the study of interacting particle systems (Metropolis et al., 1953). More recently, MCMC methods have been applied extensively to image analysis, starting with the Gibbs sampler of Grenander (1983) and Geman and Geman (1984), and to Bayesian statistics (Smith and Roberts, 1993), spatial statistics (Besag and Green, 1993; Graham, 1994), expert systems (Pearl, 1987; Spiegelhalter et al., 1993), incomplete data problems (Tanner and Wong, 1987), and hierarchical models (Gelfand et al., 1990).

The algorithm of Metropolis, Metropolis, Rosenbluth, Teller and Teller (1953) and its generalization by Hastings (1970) construct MCMC samplers as follows. Let $K(x, dy)$ be a *candidate* transition distribution on E. Write $\varepsilon_x(dy)$ for the one-point probability measure with mass at x. Find a function $\alpha(x, y)$ with values in $[0, 1]$ such that

$$Q(x, dy) = K(x, dy)\alpha(x, y) + \varepsilon_x(dy) \int Q(x, dz)(1 - \alpha(x, z))$$

is in *detailed balance* with π,

$$\pi(dx)Q(x, dy) = \pi(dy)Q(y, dx). \tag{1.1}$$

This implies that π is the invariant law of Q, and the chain is *reversible* under the stationary distribution. Assume, for simplicity, that $K(x, dy)$ has density $k(x, y)$, except perhaps for an atom at $x = y$. We refer to Tierney (1998) for a more general discussion. The *Metropolis algorithm* takes $k(x, y)$ symmetric, and

$$\alpha(x, y) = \begin{cases} \min\{1, \frac{\pi(y)}{\pi(x)}\}, & \pi(x)k(x, y) > 0, \\ 1, & \pi(x)k(x, y) = 0. \end{cases}$$

The *Hastings algorithm* does not assume k to be symmetric, and takes

$$\alpha(x, y) = \begin{cases} \min\{1, \frac{\pi(y)k(y,x)}{\pi(x)k(x,y)}\}, & \pi(x)k(x, y) > 0, \\ 1, & \pi(x)k(x, y) = 0. \end{cases}$$

The *independence Hastings algorithm* uses independent candidate realizations, $k(x, y) = k(y)$,

$$\alpha(x, y) = \begin{cases} \min\{1, \frac{\pi(y)k(x)}{\pi(x)k(y)}\}, & \pi(x)k(y) > 0, \\ 1, & \pi(x)k(y) = 0. \end{cases}$$

The algorithms accept the proposal from $K(x, dy)$ with probability $\alpha(x, y)$. If the proposal is rejected, the same position is retained by the chain and the next transition is considered. Tierney and Mira (1999) show that performance is improved if, upon rejection, instead of moving on to the next transition, another attempt to move is made by proposing a new candidate, generated from a different distribution, which is allowed to depend on the previously rejected value. This idea of delaying the rejection and adapting the proposal distribution is generalized to a more flexible class of methods in Green and Mira (1999). These methods apply in particular to settings in which the dimension varies. Optimal scaling of $K(x, dy)$ for high-dimensional state space E is discussed in Gelman, Roberts and Gilks (1996) and Roberts, Gelman and Gilks (1997).

Auxiliary variable algorithms (also called *substitution sampler* or *data augmentation*) consider π as the marginal of an appropriate distribution. For notational convenience, we write π_1 and E_1 for the distribution and state space of interest. Introduce a new state space E_2 and a distribution $\pi(dx_1, dx_2)$ on $E = E_1 \times E_2$, with first marginal $\pi_1(dx_1)$. The distribution π can be factored into marginal and conditional distributions in two different ways:

$$\pi(dx) = \pi_1(dx_1) p_2(x_1, dx_2) = p_1(x_2, dx_1) \pi_2(dx_2).$$

The *auxiliary variable algorithm* is the Markov chain with transition distribution

$$Q(x_1, dy_1) = \int p_2(x_1, dx_2) p_1(x_2, dy_1). \tag{1.2}$$

The algorithm of Swendsen and Wang (1987), see also Edwards and Sokal (1988), is a data augmentation algorithm. The monograph of Tanner (1996) has a chapter on data augmentation. We refer also to Besag and Green (1993), Higdon (1998) and Mira and Tierney (1999).

Another example of an auxiliary variable method is the slice sampler. See Neal (2000), Damien, Wakefield and Walker (1999), Roberts and Rosenthal (1999) and Fishman (1999). The underlying idea is that, as in ordinary rejection sampling, one can simulate from a distribution by simulating uniformly from under its density. For the *simple slice sampler*, we write again π_1 and E_1 for the distribution and state space of interest. We assume that $\pi_1(dx_1)$ has density proportional to $f(x_1)$, and choose a factorization

$$f(x_1) = f_1(x_1) f_2(x_1),$$

where $\sup_{x_1} f_2(x_1) = 1$. We take $E_2 = (0, \infty)$ and $p_2(x_1, dx_2)$ the uniform distribution on $(0, f_2(x_1))$, and introduce the joint distribution

$$\pi(dx) = \pi_1(dx_1) p_2(x_1, dx_2).$$

The conditional distribution $p_1(x_2, dx_1)$ of x_1 given x_2 has density proportional to

$$f_1(x_1) 1_{(f_2(x_1) > x_2)}(x_1).$$

The transition distribution of the simple slice sampler is now defined by (1.2).

The *Gibbs sampler* requires that the state space E is a product space, say $E = E_1 \times \cdots \times E_k$. For each $j = 1, \ldots, k$, we can express $x \in E$ by separating out the jth component: $x = (x_j, x_{-j})$, where x_{-j} is obtained from x by omitting the jth component x_j. Factor π in k different ways,

$$\pi(dx) = \pi_{-j}(dx_{-j}) p_j(x_{-j}, dx_j), \quad j = 1, \ldots, k, \tag{1.3}$$

with $p_j(x_{-j}, dx_j)$ the one-dimensional conditional distribution under π of x_j given x_{-j}, and $\pi_{-j}(dx_{-j})$ the $(k-1)$-dimensional marginal distribution of x_{-j}. Gibbs samplers successively use the transition distributions

$$Q_j(x, dy) = p_j(x_{-j}, dy_j) \varepsilon_{x_{-j}}(dy_{-j})$$

which change only the jth component of x.

The Gibbs sampler with *deterministic* (and *cyclic*) sweep applies Q_j cyclically according to the numbering $j = 1, \ldots, k$ of the components. The transition distribution at time $i = (q-1)k + j$ is Q_j. The chain is neither homogeneous nor reversible. For $j = 1, \ldots, k-1$, the realization $X^{(q-1)k+j}$ is determined by $X^{(q-1)k}$ and X^{qk} as $(X^{qk}_{\leq j}, X^{(q-1)k}_{>j})$, where $x_{\leq j} = (x_1, \ldots, x_j)$, $x_{>j} = (x_{j+1}, \ldots, x_k)$. Hence nothing is lost if we observe only the chain $X^{(q-1)k}$, $q = 1, 2, \ldots$. By "Gibbs sampler" one often means this subchain of full sweeps, with transition distribution

$$Q_{(d)}(x, dy) = Q_1 \cdots Q_k(x, dy) = \prod_{j=1}^{k} p_j(y_{<j}, x_{>j}, dy_j). \tag{1.4}$$

The subscript (d) stands for *deterministic*. The auxiliary variable method may be viewed as the marginal of a two-step full sweep Gibbs sampler, $k = 2$.

For the Gibbs sampler with *random* sweep (with equal probabilities), each index j is picked according to the uniform distribution on $1, \ldots, k$, independently at successive time steps. The transition distribution of the corresponding Markov chain at each time is

$$Q_{(r)}(x, dy) = \frac{1}{k} \sum_{j=1}^{k} Q_j(x, dy) = \frac{1}{k} \sum_{j=1}^{k} p_j(x_{-j}, dy_j) \varepsilon_{x_{-j}}(dy_{-j}).$$

The subscript (r) stands for *random*. The chain is reversible.

Sections 2 and 3 recall probabilistic and statistical results for general Markov chains. The asymptotic variance of the empirical estimator $E_n f$ is described in Section 2. Section 3 determines a lower bound for the asymptotic variance of estimators for πf when the observations come from a Markov chain model. Section 4 shows, for reversible Markov chains, that for arbitrary f the asymptotic variance of $E_n f$ is reduced if f is replaced by an appropriate conditional expectation, a form of Rao–Blackwellization of the empirical estimator. Section 5 applies Section 3 to Gibbs samplers with deterministic and random sweep and compares the asymptotic variances of the empirical estimator obtained using these samplers. The asymptotic variance under deterministic sweep is about half that under random sweep. Section 6 applies Section 4 to Gibbs samplers and gives lower bounds for the asymptotic variance of estimators for πf. The information bounds coincide for continuous π. If the components of π are not strongly dependent, the empirical estimator is close to efficient under any deterministic sweep. In Sections 7 and 8 we consider Gibbs samplers for random fields on a square lattice and exploit local interactions and symmetries of the random field to improve empirical estimators.

For an introduction to MCMC methods, with emphasis on convergence diagnostics, we refer to the monographs by Robert (1996), Gamerman (1997) and Robert and Casella (1999). For a review with applications to probabilistic inference see Neal (1993). MCMC methods for Gibbs fields are described in Brémaud (1999). Applications to Bayesian statistics are discussed in Besag, Green, Higdon and Mengersen (1995) and in the monograph by Gilks, Richardson and Spiegelhalter (1996).

Preprints on MCMC methods can be downloaded via the MCMC Preprint Service of the Statistical Laboratory at the University of Cambridge. The BUGS software for various samplers is developed jointly by the Biostatistics Unit of the Medical Research Council in Cambridge and by the Imperial College School of Medicine at St Mary's in London; see Spiegelhalter, Thomas and Best (1996). Christian Robert has a web page on convergence diagnostics; see also Mengersen, Robert and Guihenneuc-Jouyaux (1999). The homepages are:

http://www.statslab.cam.ac.uk/~mcmc/,
http://www.mrc-bsu.cam.ac.uk/bugs/welcome.shtml,
http://www.crest.fr/pageperso/ls/robert/robert.htm.

2. The asymptotic variance of empirical estimators for Markov chains

In this section we consider observations $X^0, \ldots, X^n$ from an Markov chain on an arbitrary state space E, with transition distribution $Q(x, dy)$. See Tierney (1996) for an introduction to Markov chains on general state spaces. We give conditions under which the empirical estimator

$$E_n f = \frac{1}{n} \sum_{i=1}^{n} f(X^i)$$

is asymptotically normal, and describe its asymptotic variance in various ways. Applications to MCMC methods will be given in later sections.

As usual, we write

$$\pi \otimes Q(dx, dy) = \pi(dx)Q(x, dy).$$

For functions $f(x)$ and $h(x, y)$, we write

$$Q_x f = Q(x, f) = \int Q(x, dy) f(y),$$

$$Q_x h = Q(x, h) = \int Q(x, dy) h(x, y),$$

$$\pi \otimes Qh = \int \pi(dx) \int Q(x, dy) h(x, y).$$

Then $(Qf)(x) = Q_x f$ defines an operator on $L_2(\pi)$. We assume that the chain is positive Harris recurrent, with invariant distribution $\pi(dx)$, and that it is V-*uniformly ergodic*, i.e., $V: E \to [1, \infty)$ and

$$\sup_x \sup_{|v| \leqslant V} \frac{|Q_x^r v - \pi v|}{V(x)} \to 0 \quad \text{for } r \to \infty.$$

We refer to Meyn and Tweedie (1993) for these concepts. Under these assumptions, if $f^2 \leqslant V$, the empirical estimator $E_n f$ is asymptotically normal. The asymptotic variance is described as follows. Introduce the *potential U* by

$$U_x f = \sum_{r=0}^{\infty} Q_x^r f \quad \text{if } \pi f = 0. \tag{2.1}$$

Define the operator A by centering U conditionally given x,

$$A f(x, y) = U_y(f - \pi f) - Q_x U(f - \pi f) = \sum_{r=0}^{\infty} (Q_y^r f - Q_x^{r+1} f). \tag{2.2}$$

The empirical estimator admits a *martingale approximation*

$$n^{1/2}(E_n f - \pi f) = n^{-1/2} \sum_{i=1}^{n} A f(X^{i-1}, X^i) + o_P(1). \tag{2.3}$$

The approximation is due to Gordin (1969). Write

$$H = \{h(x, y): h \in L_2(\pi \otimes Q), Q_x h = 0 \text{ for } x \in E\}. \tag{2.4}$$

Under the stationary distribution of the chain, $h(X^{i-1}, X^i)$ is a martingale increment. For $h \in H$ we have a martingale central limit theorem,

$$n^{-1/2} \sum_{i=1}^{n} h(X^{i-1}, X^i) \Rightarrow (\pi \otimes Qh^2)^{1/2} \cdot N, \qquad (2.5)$$

where N is a standard normal random variable, and convergence is in distribution. See, e.g., Meyn and Tweedie (1993, Chapter 17). Note that Af is in H. Hence the term $Af(X^{i-1}, X^i)$ is a martingale increment. From the martingale approximation (2.3) and the central limit theorem (2.5) it follows that the empirical estimator $E_n f$ is asymptotically normal with variance

$$\pi \otimes Q(Af)^2 = \pi \big(U(f - \pi f)\big)^2 - \pi \big(QU(f - \pi f)\big)^2$$

$$= \pi(f - \pi f)^2 + 2\sum_{r=1}^{\infty} \pi\big((f - \pi f) \cdot Q^r(f - \pi f)\big). \qquad (2.6)$$

Suppose that the Markov chain is reversible. This means that Q is in detailed balance (1.1) with π. Then the asymptotic variance of the empirical estimator can be written in the following way; compare Mira and Geyer (1999). By Theorem 2.1 of Roberts and Rosenthal (1997) and the V-uniform ergodicity of Q, the transition distribution Q is $L_2(\pi)$-*geometrically ergodic*: there are constants $\rho < 1$ and $C < \infty$ such that

$$\sup_{\pi f^2 \leqslant 1} \pi (Q^r f - \pi f)^2 \leqslant C\rho^r.$$

Further, detailed balance is equivalent to selfadjointness of Q as an operator on $L_2(\pi)$,

$$\pi(f \cdot Qg) = \pi(Qf \cdot g) \quad \text{for } f, g \in L_2(\pi). \qquad (2.7)$$

Write $I(x, dy) = \varepsilon_x(dy)$ for the identity kernel. The *spectrum* σ of Q is the set of λ such that $\lambda I - Q$ is not invertible. It is a nonempty closed subset of $[-1, 1]$. Let

$$L_{2,0}(\pi) = \{f \in L_2(\pi): \pi f = 0\}.$$

By the spectral theorem, there is a unique *spectral measure* M on Borel sets of σ such that

$$Qf = \int \lambda M(d\lambda) f \quad \text{for } f \in L_{2,0}(\pi).$$

See, e.g., Conway (1990, Theorem IX.2.2). Introduce

$$M_f(d\lambda) = \pi(f \cdot M(d\lambda)f) \quad \text{for } f \in L_{2,0}(\pi). \qquad (2.8)$$

For $f \in L_2(\pi)$, the asymptotic variance of the empirical estimator $E_n f$ can be written as

$$\int_{-1}^{1} \frac{1+\lambda}{1-\lambda} M_{f-\pi f}(d\lambda); \tag{2.9}$$

see Kipnis and Varadhan (1986). We refer to Mira and Geyer (1999) for an exposition of these results.

As noted in the Introduction, it is reasonable to judge an MCMC sampler by the asymptotic variance of empirical estimators. Let $P(x, dy)$ and $Q(x, dy)$ be transition distributions of reversible Markov chains with common invariant distribution $\pi(dx)$. Write $v(f, P)$ for the asymptotic variance of the empirical estimator $E_n f$ if the Markov chain is generated by P. Mira and Geyer (1999) say that P is *at least as efficient as Q* if

$$v(f, P) \leq v(f, Q) \quad \text{for all } f \in L_2(\pi).$$

They show that this *efficiency ordering* is equivalent to *covariance ordering*,

$$\pi(f \cdot Pf) \leq \pi(f \cdot Qf) \quad \text{for all } f \in L_{2,0}(\pi).$$

Tierney (1998) says that *P dominates Q off the diagonal* if

$$P(x, B) \geq Q(x, B) \quad \text{for } B \text{ not containing } x, \text{ and for } \pi\text{-a.a. } x.$$

He proves that in this case P is at least as efficient as Q, and $Q - P$ is positive definite on $L_2(\pi)$. Domination off the diagonal is a strong ordering. If the probability of staying in x is zero under Q, then we must have $P(x, \cdot) = Q(x, \cdot)$. Domination off the diagonal was introduced by Peskun (1973) for discrete state space; he also proved that the property implies efficiency ordering. Mira and Tierney (1999) show that, given any independence Metropolis algorithm, it is possible to construct a slice sampler that dominates it off the diagonal.

Suppose that the state space E is finite, with N elements. Let $1 = \lambda_1$ and $1 > \lambda_2 \geq \cdots \geq \lambda_N$ be the eigenvalues of Q, and $e_1 = 1, e_2, \ldots, e_N$ the corresponding eigenvectors, with $\pi e_r = 1$ for all r. For all f,

$$\pi\big((f - \pi f) \cdot e_1\big) = \pi(f - \pi f) = 0$$

and

$$f - \pi f = \sum_{r=2}^{N} \pi\big((f - \pi f) \cdot e_r\big) e_r.$$

If the chain is reversible, we have

$$M_f(d\lambda) = \sum_{r=2}^{N} \varepsilon_{\lambda_r}(d\lambda)\big(\pi(f \cdot e_r)\big)^2 \quad \text{for } f \text{ with } \pi f = 0, \tag{2.10}$$

and the asymptotic variance (2.9) of $E_n f$ is

$$\sum_{r=2}^{N} \frac{1+\lambda_r}{1-\lambda_r}\big(\pi((f-\pi f) \cdot e_r)\big)^2. \tag{2.11}$$

We refer to Frigessi, Hwang and Younes (1992) and Green and Han (1992).

A large spectral gap $1 - \lambda_2$ entails a fast rate of convergence of the Markov chain to stationarity. On the other hand, the asymptotic variance (2.11) of the empirical estimator involves all eigenvalues and is small if $\lambda_2, \ldots, \lambda_N$ are small and negative.

Suppose P and Q are reversible transition matrices with common invariant probability vector π and eigenvalues $1 = \lambda_{1P} > \lambda_{2P} \geq \cdots \geq \lambda_{NP}$ and $1 = \lambda_{1Q} > \lambda_{2Q} \geq \cdots \geq \lambda_{NQ}$. Mira and Geyer (1999) note that if $Q - P$ is positive, then $\lambda_{rP} \leq \lambda_{rQ}$ for all r. This follows from the Courant–Fisher minimax representation

$$\lambda_{r+1,Q} = \min_{g_1,\ldots,g_r} \max_{\substack{f:\, \pi(f\cdot g_s)=0 \\ s=1,\ldots,r}} \frac{\pi(f \cdot Qf)}{\pi(f \cdot f)},$$

where the minimum is taken over all vectors $g_1, \ldots, g_r$. For this representation see, e.g., Horn and Johnson (1985, Theorem 4.2.11).

For Monte Carlo methods based on i.i.d. realizations $X^0, \ldots, X^n$ from π, a well-known variance reduction method consists in generating *antithetic variables* $Y^0, \ldots, Y^n$ with the same distribution as $X^0, \ldots, X^n$ but negative correlation between $f(X^i)$ and $f(Y^i)$. Then the empirical estimator $\frac{1}{2n}\sum_{i=1}^{n}(f(X^i) + f(Y^i))$ has smaller asymptotic variance than the usual empirical estimator based on $2n$ realizations. Similar results hold for MCMC data; see in particular Frigessi, Gåsemyr and Rue (2000) for Gibbs samplers.

3. Efficient estimation for Markov chain models

In this section we determine a lower bound for the asymptotic variance of estimators for πf when the observations $X^0, \ldots, X^n$ come from a Markov chain on an arbitrary state space E. For a review of efficient estimation of functionals on Markov chain models we refer to Wefelmeyer (1999). The variance bound is based on a nonparametric version of Hájek's (1970) convolution theorem. It requires the model to be locally asymptotically normal in the following sense.

Let Θ be a possibly infinite-dimensional set, the *parameter space*. A Markov chain model is described by a family Q_ϑ, $\vartheta \in \Theta$, of transition distributions on the state space E. Fix $\vartheta \in \Theta$ such that the Markov chain corresponding to $Q = Q_\vartheta$ is positive Harris recurrent with invariant distribution $\pi = \pi_\vartheta$. Assume that Θ is smooth in the following sense. There are a linear space M, the *tangent space* of Θ at ϑ, and a linear map $D: M \to H$, with H defined in (2.4), and for each $m \in M$ there is a sequence ϑ_{nm} such that $Q_{nm} = Q_{\vartheta_{nm}}$ is *Hellinger differentiable* with derivative Dm,

$$\int Q(x, dy) \left(\left(\frac{dQ_{nm}}{dQ}(x, y) \right)^{1/2} - 1 - \frac{1}{2} n^{-1/2} Dm(x, y) \right)^2$$
$$\leqslant n^{-1} r_n(x), \tag{3.1}$$

where r_n decreases to 0 pointwise and is π-integrable for large n. This version of Hellinger differentiability is due to Höpfner, Jacod and Ladelli (1990).

Write P_n and P_{nm} for the joint distribution of $X^0, \ldots, X^n$ under Q and Q_{nm}, respectively. As in Höpfner (1993) we have a nonparametric version of *local asymptotic normality* for the likelihood ratio. For $m \in M$,

$$\log \frac{dP_{nm}}{dP_n} = n^{-1/2} \sum_{i=1}^{n} Dm(X^{i-1}, X^i) - \frac{1}{2} \pi \otimes Q(Dm)^2 + o_{P_n}(1),$$

$$n^{-1/2} \sum_{i=1}^{n} Dm(X^{i-1}, X^i) \Rightarrow \left(\pi \otimes Q(Dm)^2 \right)^{1/2} \cdot N \quad \text{under } P_n, \tag{3.2}$$

where N is a standard normal random variable. The last result is just the central limit theorem (2.5). Local asymptotic normality for Markov chains was first proved by Roussas (1965) for parametric models, and by Penev (1991) for nonparametric models.

The norm $\pi \otimes Q(Dm)^2$ induces an inner product $\pi \otimes Q(Dm \cdot Dm')$ on M. Consider $\pi_\vartheta f$ as a functional of ϑ. The functional is *differentiable* at ϑ with *gradient* g if $g \in H$ and

$$n^{1/2}(\pi_{nm} f - \pi f) \to \pi \otimes Q(Dm \cdot g) \quad \text{for } m \in M. \tag{3.3}$$

The *canonical gradient* $g_0 = Dm_0$ is the projection of g onto DM. The function m_0 is uniquely determined by

$$n^{1/2}(\pi_{nm} f - \pi f) \to \pi \otimes Q(Dm \cdot Dm_0) \quad \text{for } m \in M. \tag{3.4}$$

The canonical gradient is not always easy to calculate. Sometimes it is easier to find another gradient first (which may, in turn, be canonical in a larger model). One can then try to project that gradient into the tangent space. One such gradient is the function Af defined in (2.2), as follows from the perturbation expansion of Kartashov (1985a, 1985b) and (1996, Section 4.2), and using $QDm = 0$: for $m \in M$,

$$n^{1/2}(\pi_{nm} f - \pi f) \to \int \pi(dx) Q(x, dy) Dm(x, y) U_y f$$
$$= \pi \otimes Q(Dm \cdot Af). \tag{3.5}$$

Another approach to calculating the canonical gradient is possible when the tangent space M comes equipped with some inner product, as is usually the case. This approach is used in Section 6 for the Gibbs sampler. We denote the inner product by $(m, m')_M$. It may then be possible to find the gradient m_M with respect to this inner product,

$$n^{1/2}(\pi_{nm} f - \pi f) \to (m, m_M)_M \quad \text{for } m \in M.$$

Comparing with (3.4), we see that the canonical gradient Dm_0 is now determined by

$$(m, m_M)_M = \pi \otimes Q(Dm \cdot Dm_0) \quad \text{for } m \in M.$$

If D has an adjoint $D^* : H \to M$, we have

$$\pi \otimes Q(Dm \cdot Dm') = (m, D^*Dm')_M \quad \text{for } m, m' \in M.$$

Hence, if D^*D has an inverse, the canonical gradient is Dm_0 with

$$m_0 = (D^*D)^{-1} m_M.$$

In fact, one can avoid calculating D^*. It suffices to find an operator C such that

$$\pi \otimes Q(Dm \cdot Dm') = (m, Cm')_M \quad \text{for } m, m' \in M.$$

Then the canonical gradient is Dm_0 with $m_0 = C^{-1} m_M$. It may happen that C^{-1} is difficult to determine but that C can be written as a perturbation of the identity operator, say $C = I - B$. If B is not too large, C^{-1} may then be written as the von Neumann series $C^{-1} = \sum_{r=0}^{\infty} B^r$, and $m_0 = \sum_{r=0}^{\infty} B^r m_M$.

Efficient estimators for πf are characterized as follows. Call an estimator T_n *regular* for πf with *limit* L if

$$n^{1/2}(T_n - \pi_{nm} f) \Rightarrow L \quad \text{under } P_{nh} \text{ for } m \in M.$$

Call T_n *asymptotically linear* with *influence function* h if $h \in H$ and

$$n^{1/2}(T_n - \pi f) = n^{-1/2} \sum_{i=1}^{n} h(X^{i-1}, X^i) + o_{P_n}(1).$$

By a result of LeCam, see Bickel, Klaassen, Ritov and Wellner (1998, Section A.9), an asymptotically linear estimator is regular if and only if its influence function is a gradient. The martingale approximation (2.3) says that the empirical estimator $E_n f$ is asymptotically linear with influence function Af. Since Af is a gradient by (3.5), the empirical estimator is regular.

The convolution theorem of Hájek (1970) in the version of Pfanzagl and Wefelmeyer (1982, Theorem 9.3.1), or see Bickel, Klaassen, Ritov and Wellner (1998, p. 63,

Theorem 2), applied now for Markov chains, says that if T_n is regular with limit L, then

$$\left(n^{-1/2}\sum_{i=1}^{n}g_0(X^{i-1},X^i),\,n^{1/2}(T_n-\pi f)-n^{-1/2}\sum_{i=1}^{n}g_0(X^{i-1},X^i)\right)$$
$$\Rightarrow \left((\pi\otimes Qg_0^2)^{1/2}\cdot N,\,M\right)\quad\text{under }P_n, \tag{3.6}$$

with M independent of N, and g_0 the canonical gradient. In particular,

$$L = (\pi\otimes Qg_0^2)^{1/2}\cdot N + M\quad\text{in distribution.}$$

For every $a > 0$ we have $P(-a < cN < a) \geqslant P(-a < cN + M < a)$. This justifies calling T_n *efficient* if

$$L = (\pi\otimes Qg_0^2)^{1/2}\cdot N\quad\text{in distribution.}$$

It follows from (3.6) that T_n is efficient if and only if it is asymptotically linear with influence function equal to the canonical gradient,

$$n^{1/2}(T_n - \pi f) = n^{-1/2}\sum_{i=1}^{n}g_0(X^{i-1},X^i) + o_{P_n}(1).$$

An efficient estimator is asymptotically normal with variance $\pi\otimes Qg_0^2$. We call this variance the *asymptotic variance bound*. In Section 6 we calculate the asymptotic variance bound for Gibbs samplers with random and deterministic sweep.

4. Improving empirical estimators by conditioning

To begin let $X^1,\ldots,X^n$ be *independent* and identically distributed as π, and let f be a π-square-integrable function. The empirical estimator $E_n f = \frac{1}{n}\sum_{i=1}^{n}f(X^i)$ for the expectation πf is asymptotically normal with variance $\pi(f - \pi f)^2$. Now replace $f(X^i)$ by a conditional expectation $E_\pi(f(X^i)|h(X^i))$, where h is some function. Then the *Rao–Blackwellized* empirical estimator

$$E_n E_\pi(f|h) = \frac{1}{n}\sum_{i=1}^{n}E_\pi\bigl(f(X^i)|h(X^i)\bigr) \tag{4.1}$$

has asymptotic variance $\pi(E_\pi(f|h) - \pi f)^2$. The Rao–Blackwell theorem says that it is smaller than the asymptotic variance of $E_n f$,

$$\pi\bigl(E_\pi(f|h) - \pi f\bigr)^2 \leqslant \pi(f - \pi f)^2. \tag{4.2}$$

Of course, the "estimator" $E_n E_\pi(f|h)$ can be used only if $E_\pi(f|h)$ does not depend on π, e.g., when h is sufficient.

Recently, there has been considerable interest in developing versions of the Rao–Blackwell theorem in the context of stochastic simulation, and for Markov chain Monte Carlo (MCMC) in particular. See Casella and Robert (1996) and the references cited therein. Early references are Kalos and Whitlock (1986, Section 4.2) and Pearl (1987). See also Neal (1993, Section 6.3). Gelfand and Smith (1990, 1991) consider i.i.d. runs of a Gibbs sampler. In the empirical estimator based on the final value of each run, they replace f by a conditional expectation under π. For long runs, the final values are distributed approximately according to π, so the classical Rao–Blackwell theorem (4.2) implies that the variance is reduced. Single runs of Markov chains are studied in the following references. Liu, Wong and Kong (1994) consider auxiliary variable algorithms of the form $Q(x_1, dy_1) = \int p_2(x_1, x_2) p_1(x_2, dy_1)$ and the Rao–Blackwellized empirical estimator

$$E_n p_1 f = \frac{1}{n} \sum_{i=1}^{n} \int p_1(X_2^i, dy_1) f(y_1).$$

They prove that the variance is always reduced. Casella and Robert (1996) propose some types of Rao–Blackwellization for the Metropolis–Hastings algorithm. Their approach is to integrate out some or all of the uniform random variables involved.

Let $X^0, \ldots, X^n$ be realizations of an arbitrary Markov chain with transition distribution $Q(x, dy)$ and invariant distribution $\pi(dx)$. Assume that the Markov chain is positive Harris recurrent and V-uniformly ergodic, and that $f^2 \leq V$. Then the Rao–Blackwellized empirical estimator $E_n E_\pi(f|h)$ is asymptotically normal, and by (2.6) its asymptotic variance is

$$\pi\left(E_\pi(f|h) - \pi f\right)^2 + 2 \sum_{r=1}^{\infty} \pi\left((E_\pi(f|h) - \pi f) \cdot Q^r (E_\pi(f|h) - \pi f)\right).$$

Geyer (1995) gives necessary and sufficient conditions for $E_n E_\pi(f|h)$ to have smaller asymptotic variance than $E_n f$ for all f simultaneously, and points out that these conditions are unlikely to be satisfied in practice.

McKeague and Wefelmeyer (2000) suggest a different version of Rao–Blackwellization. Rather than conditioning $f(X^i)$ on a function $h(X^i)$, they condition on the previous value of the chain. The function $f(x)$ in the empirical estimator $E_n f$ is replaced by $Q(x, f) = E(f(X^i)|X^{i-1} = x)$. Their Rao–Blackwellized empirical estimator is therefore

$$E_n Q f = \frac{1}{n} \sum_{i=1}^{n} Q(X^i, f).$$

Rao–Blackwellization can be repeated, leading to the estimator

$$E_n Q^k f = \frac{1}{n} \sum_{i=1}^{n} Q^k(X^i, f).$$

For *reversible* chains, this Rao–Blackwellization reduces the asymptotic variance simultaneously for all f. Schmeiser and Chen (1991) prove this result for the Hit-and-Run algorithm proposed by Belisle, Romeijn and Smith (1993). The result does not hold, in general, for non-reversible chains.

THEOREM 1. *Let $X^0, \ldots, X^n$ be realizations of a Markov chain which is positive Harris recurrent, V-uniformly ergodic and reversible. For $f \in L_2(\pi)$, the asymptotic variance of $E_n Q^k f$ is less than that of $E_n f$, and the variance reduction is*

$$\sum_{j=0}^{k-1} \pi\big((I+Q)Q^j(f-\pi f)\big)^2.$$

The asymptotic variance of $E_n Q^k f$ tends to zero as k goes to infinity.

The proof is simple. By (2.6) and selfadjointness (2.7), the asymptotic variance of $E_n Q^k f$ is

$$\pi\bigg(Q^k(f-\pi f) \cdot \bigg(I + 2\sum_{r=1}^{\infty} Q^r\bigg) Q^k(f-\pi f)\bigg)$$

$$= \pi\big((f-\pi f) \cdot Q^{2k}(f-\pi f)\big)$$

$$+ 2 \sum_{r=2k+1}^{\infty} \pi\big((f-\pi f) \cdot Q^r(f-\pi f)\big). \tag{4.3}$$

The asymptotic variance (4.3) of $E_n Q^k f$ is obtained from the asymptotic variance (2.6) of $E_n f$ by omitting the term $\pi(f-\pi f)^2$, the terms of order $r = 1, \ldots, 2k-1$, and half the term of order $2k$. This implies the second part of Theorem 1.

The difference between (2.6) and (4.3) can be written as

$$\pi\bigg((f-\pi f) \cdot \sum_{j=0}^{k-1}(I+Q)^2 Q^{2j}(f-\pi f)\bigg).$$

This implies the first part of Theorem 1.

Theorem 1 can also be expressed in terms of the spectral measure. By (2.9) and (2.8), the asymptotic variance of $E_n Q^k f$ is

$$\int_{-1}^{1} \frac{1+\lambda}{1-\lambda} \lambda^{2k} \pi\big((f-\pi f) \cdot E(d\lambda)(f-\pi f)\big),$$

and the variance reduction over $E_n f$ is

$$\int_{-1}^{1} \frac{1+\lambda}{1-\lambda} (1-\lambda^{2k}) \pi\big((f-\pi f) \cdot E(d\lambda)(f-\pi f)\big).$$

Suppose that the state space E is finite, with N elements. Let $1 = \lambda_1 > \lambda_2 \geqslant \cdots \geqslant \lambda_N$ be the eigenvalues of Q, and $e_1 = 1, e_2, \ldots, e_N$ the corresponding eigenvectors, with $\pi e_r = 1$. Then the variance reduction is

$$\sum_{r=2}^{N} \frac{1+\lambda_r}{1-\lambda_r}\left(1-\lambda_r^{2k}\right)\left(\pi\left((f-\pi f) \cdot e_r\right)\right)^2.$$

McKeague and Wefelmeyer (2000) illustrate Theorem 1 with simulations for the Ising model and the Gibbs sampler, and for f the rth nearest neighbor correlation.

5. Asymptotic variance of empirical estimators for Gibbs samplers

Let $E = E_1 \times \cdots \times E_k$ be a product of measurable spaces, with product σ-field, and let $\pi(dx)$ be a distribution on E. Gibbs samplers successively use the transition distributions $Q_j(x, dy) = p_j(x_{-j}, dy_j)\varepsilon_{x_{-j}}(dy_{-j})$, with $p_j(x_{-j}, dx_j)$ the one-dimensional conditional distribution under π of x_j given x_{-j}, introduced in (1.3). For a function $f(x)$ we have by definition of Q_j,

$$Q_j(x, f) = \int Q_j(x, dy) f(y) = \int p_j(x_{-j}, dx_j) f(x_{-j}, x_j) = p_j(x_{-j}, f).$$

In particular, $Q_j(x, dy)$ does not depend on x_j. Hence Q_j is *idempotent*,

$$Q_j^2 = Q_j. \tag{5.1}$$

This means that Q_j is a *projection operator* on $L_2(\pi)$. Indeed, we can write $L_2(\pi)$ as the orthogonal sum of two subspaces, one consisting of functions f with $Q_j f = 0$, the other of functions $f(x)$ not depending on x_j, and Q_j is the projection on the second subspace along the first. Therefore,

$$\pi(f \cdot Q_j f') = \pi(Q_j f \cdot Q_j f') \quad \text{for } f, f' \in L_2(\pi). \tag{5.2}$$

Relation (5.2) implies that Q_j, as an operator on $L_2(\pi)$, is *positive*,

$$\pi(f \cdot Q_j f) = \pi(Q_j f \cdot Q_j f) \geqslant 0 \quad \text{for } f \in L_2(\pi), \tag{5.3}$$

and *selfadjoint*,

$$\pi(f \cdot Q_j f') = \pi(Q_j f \cdot Q_j f') = \pi(Q_j f \cdot f') \quad \text{for } f, f' \in L_2(\pi). \tag{5.4}$$

The last relation is seen to be equivalent to detailed balance,

$$\pi(dx) Q_j(x, dy) = \pi(dy) Q_j(y, dx). \tag{5.5}$$

This, in turn, implies again that Q_j has invariant law π.

The Gibbs sampler for π with deterministic (and cyclic) sweep has transition distribution Q_j at time $i = (q-1)k + j$; the subchain of full sweeps has transition distribution $Q_1 \cdots Q_k$. Let $X^0, \ldots, X^n$ be realizations from the Gibbs sampler with deterministic sweep, with n a multiple of k, say $n = pk$. We want to estimate the expectation πf of a function f. The most common estimator for πf is the empirical estimator based on the *subchain* $X^k, \ldots, X^{pk}$,

$$E_n^k f = \frac{1}{p} \sum_{q=1}^p f(X^{qk}).$$

The empirical estimator based on the *full* chain $X^1, \ldots, X^n$ is

$$E_n f = \frac{1}{n} \sum_{i=1}^n f(X^i) = \frac{1}{k} \sum_{j=1}^k E_n^j f$$

with

$$E_n^j f = \frac{1}{p} \sum_{q=1}^p f_{\leqslant j}(X^{(q-1)k}, X^{qk})$$

and

$$f_{\leqslant j}(x, y) = f(y_{\leqslant j}, x_{> j}).$$

The estimator $E_n^j f$ is based on the subchain $X^j, X^{k+j}, \ldots, X^{(p-1)k+j}$.

To fix things, by *asymptotic distribution* of an estimator T_n we will mean the asymptotic distribution of $n^{1/2}(T_n - \pi f)$, even though standardizing by $p^{1/2}$ rather than $n^{1/2}$ is more common for the empirical estimator $E_n^j f$.

Greenwood, McKeague and Wefelmeyer (1998) calculate the asymptotic variance of $E_n^j f$ and $E_n f$, using the form (2.6) of the asymptotic variance in the central limit theorem for Markov chains.

THEOREM 2. *Assume that the Gibbs sampler for π with deterministic sweep is positive Harris recurrent and the subchains are V-uniformly ergodic, and that $f^2 \leqslant V$. Then the empirical estimator $E_n^j f$ is asymptotically normal with variance*

$$\sigma_j^2 = k\pi(f - \pi f)^2 + 2k \sum_{r=1}^\infty \pi\big((f - \pi f) \cdot p_j^{\text{cycl } rk}(f - \pi f)\big),$$

where $p_j^{\text{cycl } r} = p_j p_{j+1} \cdots p_k p_1 p_2 \cdots$ *with r terms.*

THEOREM 3. *Under the assumptions of Theorem 2, the empirical estimator $E_n f$ is asymptotically normal with variance*

$$\sigma_d^2 = \pi(f - \pi f)^2 + 2\sum_{r=1}^{\infty} \frac{1}{k} \sum_{j=1}^{k} \pi\big((f - \pi f) \cdot p_j^{\text{cycl } r}(f - \pi f)\big).$$

Because the empirical estimator $E_n^k f$ based on the subchain of full sweeps is often used in practice, we have included the description of its asymptotic variance in Theorem 2. However, we do not recommend this estimator; the simulations in Figure 1 below show that $E_n^k f$ can be considerably worse than $E_n f$. This is true even when π has only two components; see Greenwood, McKeague and Wefelmeyer (1996).

THEOREM 4. *Assume that the Gibbs sampler for π with random sweep is positive Harris recurrent and V-uniformly ergodic, and that $f \in L_2(\pi)$. Then the empirical estimator $E_n f$ is asymptotically normal with variance*

$$\sigma_{(r)}^2 = \pi(f - \pi f)^2 + 2\sum_{r=1}^{\infty} \pi\big((f - \pi f) \cdot Q_{(r)}^r(f - \pi f)\big)$$

$$= \pi(f - \pi f)^2$$
$$+ 2\sum_{r=1}^{\infty} \frac{1}{(k-1)^r} \sum_{\substack{j_1,\ldots,j_r=1 \\ j_i \neq j_{i+1}}}^{k} \pi\big((f - \pi f) \cdot p_{j_1} \cdots p_{j_r}(f - \pi f)\big).$$

The second summation in $\sigma_{(r)}^2$ contains $k(k-1)^{r-1}$ terms, each being an r-order autocovariance of the form $\pi((f - \pi f) \cdot p_{j_1} \cdots p_{j_r}(f - \pi f))$. From Theorem 3, the r-order term in $\sigma_{(d)}^2$ is an average of k of these r-order autocovariances, those of the form $\pi((f - \pi f) \cdot p_j^{\text{cycl } rk}(f - \pi f))$. One might expect that $\sigma_{(r)}^2 \approx (k/(k-1))\sigma_{(d)}^2$, or that $\sigma_{(r)}^2$ is slightly larger than $\sigma_{(d)}^2$. Such a result holds if one considers a random sweep without repetition, see Fishman (1996, Theorem 8). Greenwood, McKeague and Wefelmeyer (1996) argue, however, that $\sigma_{(r)}^2$ can be up to twice as large as $\sigma_{(d)}^2$, even if k is large. This is also seen in simulations; see Figure 1. The reason is that the higher-order terms in $\sigma_{(r)}^2$ can decay more slowly than those in $\sigma_{(d)}^2$. This is easily seen in the special case of *independent* components. Then $\pi((f - \pi f) \cdot p_j^{\text{cycl } rk}(f - \pi f))$ vanishes for $s \geqslant k$, because integration of $f(x)$ cyclically over k components gives πf. However, $\pi((f - \pi f) \cdot p_{j_1} \cdots p_{j_s}(f - \pi f))$ vanishes only if all k components are present among $j_1, \ldots, j_s$. Also, if some of the j_i are equal, fewer than r components are integrated out, so $\pi((f - \pi f) \cdot p_{j_1} \cdots p_{j_r}(f - \pi f))$ is larger than any $\pi((f - \pi f) \cdot p_j^{\text{cycl } rk}(f - \pi f))$ "covering" $j_1, \ldots, j_r$.

6. Asymptotic variance bounds for Gibbs samplers

In Section 5 we have determined the asymptotic variance of the empirical estimator for an expectation πf under the Gibbs sampler with random and deterministic sweep.

In this section we consider another criterion by which MCMC methods can be judged: How much information about πf is contained in the simulated values $X^0, \ldots, X^n$, given the knowledge that a particular sampler was used to generate them? In particular: what fraction of the information is exploited by the empirical estimator? It is assumed that no information about π itself is made available to the statistician, apart from the link between π and the transition distribution of the observed Markov chain. Of course, π is known in principle, and part of that knowledge can sometimes be exploited to improve upon the empirical estimator. An example is Rao–Blackwellization, see Section 4. If π is a random field on a lattice and the interactions between the sites are known to be local, improved estimators are described in Section 7. Symmetries of π are exploited Section 8.

We keep the setting of Section 5. Consider first the Gibbs sampler with *deterministic* sweep. The subchain of full sweeps has transition distribution $Q_{(d)} = Q_1 \cdots Q_k$, see (1.4). It is parametrized by π. To determine the information bound of (regular) estimators of πf, we must prove that the model is locally asymptotically normal (3.2). A perturbation of π is of the form

$$\pi_{nk}(\mathrm{d}x) = \pi(\mathrm{d}x)\big(1 + n^{-1/2}k(x)\big),$$

with k in

$$M = \{m(x)\colon m \text{ measurable, bounded, } \pi m = 0\}. \tag{6.1}$$

Write $p_{j,nm}(x_{-j}, \mathrm{d}x_j)$ for the one-dimensional conditional distribution under $\pi_{nm}(\mathrm{d}x)$ of x_j given x_{-j}. The effect of the perturbation of π on p_j is easily obtained as follows (Greenwood, McKeague and Wefelmeyer, 1998, Lemma 1): uniformly in x,

$$p_{j,nm}(x_{-j}, \mathrm{d}x_j) = p_j(x_{-j}, \mathrm{d}x_j)\big(1 + n^{-1/2}m_j(x) + \mathrm{O}(n^{-1})\big)$$

with $m_j(x) = m(x) - Q_j(x, m) = m(x) - p_j(x_{-j}, m)$.

Write $Q_{(d)nm}$ for the transition distribution (1.4) of the Gibbs sampler for π_{nm} with deterministic sweep,

$$Q_{(d)nm}(x, \mathrm{d}y) = (Q_{1,nm} \cdots Q_{k,nm})(x, \mathrm{d}y) = \prod_{j=1}^{k} p_{j,nm}(y_{<j}, x_{>j}, \mathrm{d}y_j).$$

It follows easily that $Q_{(d)nm}$ is obtained by perturbing $Q_{(d)}$ as follows: uniformly in x and y,

$$Q_{(d)nm}(x, \mathrm{d}y) = Q_{(d)}(x, \mathrm{d}y)\big(1 + n^{-1/2}D_{(d)}m(x, y) + \mathrm{O}(n^{-1})\big) \tag{6.2}$$

with

$$D_{(d)}m(x, y) = \sum_{j=1}^{k} m_j(y_{\leq j}, x_{>j}). \tag{6.3}$$

Relation (6.2) implies that $Q_{(d)nm}$ is Hellinger differentiable (3.1) with derivative $D_{(d)}m$.

Write $P_{(d)n}$ for the joint distribution of $X^0, X^k, \ldots, X^{pk}$ if π is true, and $P_{(d)nm}$ if π_{nm} is true. If the Gibbs sampler for π with deterministic sweep is positive Harris recurrent, we obtain local asymptotic normality (3.2) of the form

$$\frac{\log dP_{(d)nm}}{dP_{(d)n}} = n^{-1/2} \sum_{q=1}^{p} D_{(d)}m\bigl(X^{(q-1)k}, X^{qk}\bigr) - \frac{1}{2}\pi \otimes Q(D_{(d)}m)^2$$
$$+ o_{P_{(d)n}}(1). \tag{6.4}$$

The desired minimal asymptotic variance of regular estimators of πf is the squared length of the gradient of πf. To determine this gradient, we note that by definition of π_{nm}, and since $\pi m = 0$,

$$n^{1/2}(\pi_{nm}f - \pi f) = \pi m f = \pi\bigl(m \cdot (f - \pi f)\bigr) \quad \text{for } m \in M.$$

Comparing this relation with definition (3.4) of the gradient, we see that the canonical gradient is $g_{(d)} = D_{(d)}m_{(d)}$ with $m_{(d)}$ fulfilling

$$\pi \otimes Q(D_{(d)}m \cdot D_{(d)}m_{(d)}) = \pi\bigl(m \cdot (f - \pi f)\bigr) \quad \text{for } m \in M.$$

The left side can be written

$$\pi\bigl(m \cdot (I - Q_{(r)})m_{(d)}\bigr);$$

see Greenwood, McKeague and Wefelmeyer (1998, Lemma 2). Surprisingly, this inner product for *deterministic* sweep involves the transition distribution for *random* sweep. Let $\|\ \|_2$ be the operator norm on $L_2(\pi)$, defined by $\|Q\|_2^2 = \sup_{\pi f^2 \leqslant 1} \pi(Qf)^2$ for an operator Q on $L_2(\pi)$. If $\|Q_{(r)}^t\|_2 < 1$ for some t, we obtain

$$m_{(d)} = (I - Q_{(r)})^{-1}(f - \pi f).$$

Hence the asymptotic variance bound is

$$\pi\bigl((f - \pi f) \cdot (I - Q_{(r)})^{-1}(f - \pi f)\bigr).$$

After some calculation we arrive at the following result; see Greenwood, McKeague and Wefelmeyer (1998, Theorem 1).

THEOREM 5. *Let $f \in L_2(\pi)$, and assume that $\|Q_{(r)}^t\|_2 < 1$ for some t. For the Gibbs sampler with deterministic sweep, the asymptotic variance bound is*

$$B_{(d)} = \pi\bigl((f - \pi f) \cdot (I - Q_{(r)})^{-1}(f - \pi f)\bigr)$$

$$= \pi(f - \pi f)^2$$
$$+ \sum_{r=1}^{\infty} \frac{1}{(k-1)^r} \sum_{\substack{j_1,\dots,j_r=1 \\ j_i \neq j_{i+1}}}^{k} \pi\big((f - \pi f) \cdot p_{j_1} \cdots p_{j_r}(f - \pi f)\big).$$

Note that $B_{(d)}$ does not depend on the order of the deterministic sweep; it only depends on π and f.

The result for *random* sweep is more involved. Write $Q_{(r)nm}$ for the transition distribution of the Gibbs sampler with random sweep for π_{nm},

$$Q_{(r)nm}(x, dy) = \frac{1}{k} \sum_{j=1}^{k} Q_{j,nm}(x, dy) = \frac{1}{k} \sum_{j=1}^{k} p_{j,nm}(x_{-j}, dy_j) \varepsilon_{x_{-j}}(dy_{-j}).$$

The perturbation of $Q_{(r)}$ now involves the probabilities of not changing the value when updating a component. The reason is that the transition distribution $Q_j(x, dy)$ is supported by the line through x parallel to the jth coordinate axis, $\{y: y_{-j} = x_{-j}\}$. Hence the support of $Q_{(r)}(x, dy)$ is contained in the union of the k lines. The supports of the $Q_j(x, dy)$ are disjoint except for the point x, which may be charged by some or all of them. Therefore, to calculate the $Q_{(r)}(x, dy)$-density of $Q_{(r)nm}(x, dy)$, we must treat x separately. We assume that the σ-field on each E_j contains the one-point sets, which will be the case in all applications.

Greenwood, McKeague and Wefelmeyer (1998, Lemma 4) show that for $m \in M$, and uniformly in x and y,

$$Q_{(r)nm}(x, dy) = Q_{(r)}(x, dy)\big(1 + n^{-1/2} D_{(r)}m(x, y) + O(n^{-1})\big) \tag{6.5}$$

with

$$D_{(r)}m(x, y) = \sum_{j=1}^{k} (D_j m)(x, y),$$

$$(D_j m)(x, y) = \left(1(y_{-j} = x_{-j}) - \left(1 - \frac{r_j(x)}{r(x)}\right) 1(y = x)\right) m_j(x_{-j}, y_j),$$

$$r_j(x) = p_j(x_{-j}, \{x_j\}), \qquad r(x) = \sum_{j=1}^{k} r_j(x).$$

Relation (6.5) implies that $Q_{(r)nm}$ is Hellinger differentiable (3.1) with derivative $D_{(r)}m$. Write $P_{(r)n}$ for the joint distribution of $X^0, X^k, \dots, X^{pk}$ if π is true, and $P_{(r)nm}$ if π_{nm} is true. If the Gibbs sampler for π with random sweep is positive Harris recurrent, we obtain local asymptotic normality (3.2) of the form

$$\frac{\log dP_{(r)nm}}{dP_{(r)n}} = n^{-1/2} \sum_{i=1}^{n} (K_{(r)}m)(X^{i-1}, X^i) - \frac{1}{2}\pi \otimes Q(D_{(r)}m)^2$$
$$+ o_{P_{(r)n}}(1). \tag{6.6}$$

Similarly as for deterministic sweep, the canonical gradient is $g_{(r)} = D_{(r)} m_{(r)}$ with $m_{(r)}$ fulfilling

$$\pi \otimes Q(D_{(r)} m \cdot D_{(r)} m_{(r)}) = \pi \left(m \cdot (f - \pi f) \right) \quad \text{for } m \in M.$$

By Greenwood, McKeague and Wefelmeyer (1998, Lemma 5), the left side can be written

$$\pi \left(m \cdot (I - Q_{(r)} + S) m_{(r)} \right)$$

with

$$Sm = \frac{1}{k} \sum_{i,j=1}^{k} Q_i (R_{ij} Q_j m),$$

$$R_{ij}(x) = \delta_{ij} r_j(x) - r_i(x) r_j(x) / r(x).$$

If $\|(Q_{(r)} - S)^t\|_2 < 1$ for some t, we obtain

$$m_{(r)} = (I - Q_{(r)} + S)^{-1} (f - \pi f).$$

After some calculation, we arrive at the following result, Greenwood, McKeague and Wefelmeyer (1998, Theorem 2).

THEOREM 6. *Let $f \in L_2(\pi)$, and assume that $\|(Q_{(r)} - S)^t\|_2 < 1$ for some t. For the Gibbs sampler with random sweep, the asymptotic variance bound is*

$$B_{(r)} = \pi \left((f - \pi f) \cdot (I - Q_{(r)} + S)^{-1} (f - \pi f) \right)$$

$$= \pi (f - \pi f)^2 + \sum_{r=1}^{\infty} \pi \left((f - \pi f) \cdot (Q_{(r)} - S)^r (f - \pi f) \right).$$

Both S and $I - Q_{(r)}$ are positive operators on $L_2(\pi)$. Write $K \geqslant L$ if $K - L$ is positive. Then $I - Q_{(r)} + S \geqslant I - Q_{(r)}$ and therefore $(I - Q_{(r)} + S)^{-1} \leqslant (I - Q_{(r)})^{-1}$. Thus the variance bound is no larger for random sweep than for deterministic sweep: $B_{(r)} \leqslant B_{(d)}$.

Suppose that π is continuous in the sense that it is absolutely continuous with respect to the product of its marginals, and the marginals have no atoms. Then $p_j(x_{-j}, dx_j)$ has no atoms for $\pi_{-j}(dx_{-j})$-a.s. x_{-j}. Hence $r_j(x) = p_j(x_{-j}, \{x_j\}) = 0$ for π-a.a. x, and therefore $R_{ij}(x) = 0$ for π-a.a. x, and the operator S reduces to 0. This implies that information bound for random sweep coincides with the information bound for deterministic sweep: $B_{(r)} = B_{(d)}$.

The last term in the asymptotic variance bound $B_{(d)}$ for deterministic sweep appears with a factor 2 in the asymptotic variance $\sigma_{(r)}^2$ of the empirical estimator for random sweep. In most applications, the leading term $\pi(f - \pi f)^2$ of the variances is relatively

k=10

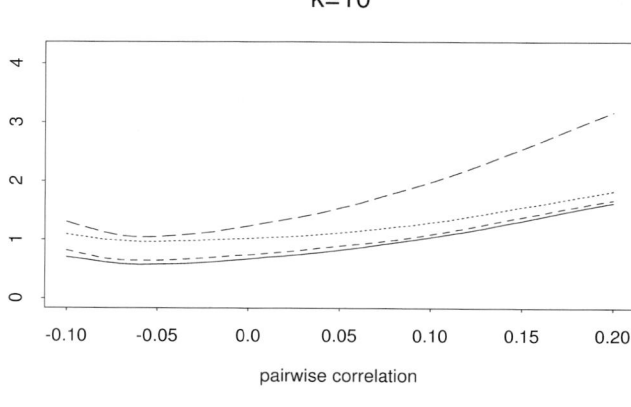

k=20

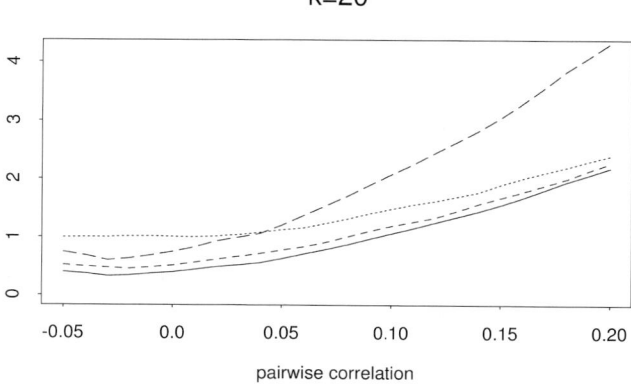

Fig. 1. Exchangeable k-dimensional normal example. The information bounds for both random and deterministic sweep (solid line) and the asymptotic variances (in units of $k\pi(f - \pi f)^2$) of the usual empirical estimator $E_n^k f$ under deterministic sweep (dotted line), and the full chain estimator $E_n f$ under deterministic sweep (short dashed line) and random sweep (long dashed line).

small. Then $\sigma_{(r)}^2$ is nearly twice as large as $B_{(d)}$. When π is continuous, we have $B_{(d)} = B_{(r)}$. Hence the efficiency of the empirical estimator for random sweep is close to 50%.

As mentioned in Section 5, the asymptotic variance of the empirical estimator is about twice as large for random sweep as for deterministic sweep. This implies that the empirical estimator for deterministic sweep is close to efficient.

We illustrate the results with an exchangeable k-dimensional multivariate normal distribution π in which each component has zero mean and unit variance, and all the pairwise correlations are identical. This example has been widely used in the literature for studying convergence rates of Gibbs samplers, see, e.g., Raftery and Lewis (1992, Example 3) and Roberts and Sahu (1997). The function f is taken to be the indicator that the random field exceeds a unit threshold: $f(x) = 1(\max_j x_j > 1)$. The results for 10 and 20 dimensions are shown in Figure 1.

7. Improving empirical estimators for random fields with local interactions

To begin let π be a distribution on a K-dimensional space $E = E_1 \times \cdots \times E_K$, and let $X^1, \ldots, X^n$ be *independent* and identically distributed as π. Let f be a π-square-integrable function. In the nonparametric setting, with nothing known about π, the empirical estimator $E_n f = \frac{1}{n} \sum_{i=1}^n f(X^i)$ is efficient; see Bickel, Klaassen, Ritov and Wellner (1998, Section 3.3). If the components of π are known to be independent, $\pi = \pi_1 \otimes \cdots \otimes \pi_K$, then $E_n f$ is no longer efficient, and a better estimator of πf is the generalized von Mises statistic

$$M_n f = \frac{1}{n^K} \sum_{i_1, \ldots, i_K = 1}^n f(X_1^{i_1}, \ldots, X_K^{i_K}).$$

Since it is the expectation of f under the product of the marginal empiricals, $M_n f$ is again efficient if nothing is known about the components $\pi_1, \ldots, \pi_K$; see Levit (1974) and Koshevnik and Levit (1976).

Note that the terms $(X_1^{i_1}, \ldots, X_K^{i_K})$ have law π: they are obtained by mixing the components from the different i.i.d. copies $X^i = (X_1^i, \ldots, X_K^i)$. In other words, the von Mises statistic is obtained by replacing values of the components by values with different time indices. This works because there are no interactions either among the K components or among values with different time indices.

Greenwood, McKeague and Wefelmeyer (1999) extend the idea behind the von Mises statistic to samplers on random fields with local interactions. For simplicity, we restrict attention to nearest neighbor random fields and the Gibbs sampler with a specific sweep. Let $S = \{0, \ldots, k-1\}^d$ be a square lattice of dimension d. For simplicity, take k to be even. The lattice has $K = k^d$ sites. Let π be the law of a random field on E^S. As in (1.3), factor π in K different ways,

$$\pi(dx) = \pi_{-s}(dx_{-s}) p_s(x_{-s}, dx_s), \quad s \in S,$$

where x_{-s} is obtained from x by omitting x_s. The one-dimensional conditional distributions $p_s(x_{-s}, dx_s)$ are called the *local characteristics* of the random field.

A Gibbs sampler with deterministic sweep is based on some ordering $s_1, \ldots, s_K$ of the sites. Let $n = qK$. The subchain $X^0, X^K, \ldots, X^{qK}$ of full sweeps, see (1.4), has transition distribution

$$Q(x, dy) = \prod_s p_s(y_{<s}, x_{>s}, dy_s),$$

where $x_{<s}$ is the subconfiguration of all sites that come before site s. As in Section 5, the usual estimator for πf is the empirical estimator based on the subchain,

$$E_n^K f = \frac{1}{p} \sum_{q=1}^p f(X^{qK}).$$

For each s, the partially updated configuration $(X_{\leq s}^{qK}, X_{>s}^{(q-1)K})$ also has stationary law π, and further empirical estimators are

$$E_n^s f = \frac{1}{p} \sum_{q=1}^{p} f(X_{\leq s}^{qK}, X_{>s}^{(q-1)K}).$$

The empirical estimator based on the full chain $X^1, \ldots, X^n$ is

$$E_n f = \frac{1}{n} \sum_{i=1}^{n} f(X^i) = \frac{1}{K} \sum_{s} E_n^s f.$$

The set of nearest neighbors of a site s is $\partial s = \{t: |t - s| = 1\}$, with $|t - s| = \sum_j |t_j - s_j|$. We use a free boundary, in which case the boundary sites have fewer than $2d$ neighbors. We assume *nearest neighbor interactions*,

$$p_s(x_{-s}, dx_s) = p_s(x_{\partial s}, dx_s), \tag{7.1}$$

i.e., the local characteristics at site s depend only on the nearest neighbors of s.

A widely used updating scheme for nearest neighbor models *respects the checkerboard pattern* of the lattice in the sense that it updates first the sites with, say, even parity and then those with odd parity. See, e.g., Heermann and Burkitt (1992). The corresponding Gibbs sampler updates a single site s using the local characteristic $p_s(x_{\partial s}, dx_s)$. Therefore, all even, or all odd, sites can be updated simultaneously, and the sampler can be written as a two-step Gibbs sampler. Write a configuration $x = (y_e, y_o)$, where y_e and y_o are the subconfigurations of x on the even and odd sites, respectively. The subchain of full sweeps has transition distribution

$$Q(y_e, y_o, d(z_e, z_o)) = Q_e(y_o, dz_e) Q_o(z_e, dz_o) \tag{7.2}$$

with

$$Q_e(y_o, dy_e) = \prod_{s \text{ even}} p_s(y_{o, \partial s}, dy_s),$$

$$Q_o(y_e, dy_s) = \prod_{s \text{ odd}} p_s(y_{e, \partial s}, dy_s).$$

Let $X^0 = (Y^0, Y^1)$ be an initial configuration. The Gibbs sampler based on this updating scheme first creates a subconfiguration Y^2 on the even sites, then a subconfiguration Y^3 on the odd sites, and so on. Here, rather than counting the update of a complete configuration as a time step, we define a *full time step* to be the update of an even or an odd subconfiguration. This means that the output of the Gibbs sampler is $Y^0, Y^1, Y^2, \ldots$, and the sequence of complete configurations, or full sweeps, is given by $X^0 = (Y^0, Y^1)$, $X^K = (Y^2, Y^3), \ldots$.

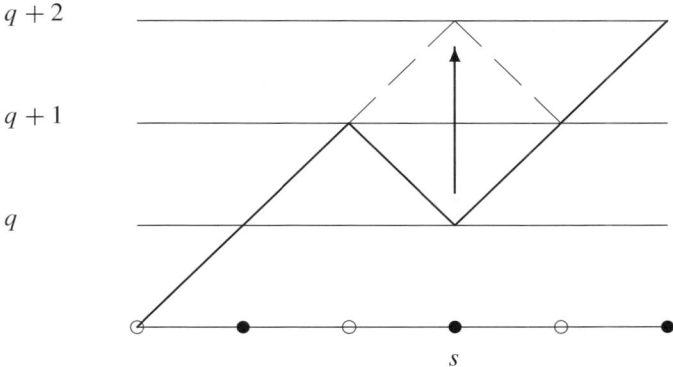

Fig. 2. An admissible move at site s.

To motivate the construction of our estimators, we assume for now that the initial configuration X^0 is distributed according to the stationary law π. Then the Gibbs sampler Markov chain $X^0, X^1, \ldots$ is stationary. Now suppose that we replace a component X_s^{qK} of the configuration X^{qK} by a future value $X_s^{(q+j)K}$. Which replacements leave the joint law of the configuration unchanged? We have already seen in $E_n^s f$ an example of such replacements for the general case with possibly non-local interactions – we replaced the values X_t^{qK} by $X_t^{(q+1)K}$ for $t \leqslant s$. We will see that for nearest neighbor models more general replacements are possible.

It is convenient to describe such replacements by an *update function* $I : S \to \{0, 1, \ldots\}$, with $I(s)$ even for s even, and odd otherwise. An update function I describes a new configuration $Z^I = (Y_s^{I(s)})_{s \in S}$ in terms of the observed chain $Y^0, Y^1, Y^2, \ldots$ by specifying, for each site s, the time index $I(s)$ of the value $Y_s^{I(s)}$ going into this configuration. For example, the initial configuration is $X^0 = (Y^0, Y^1) = Z^{I^0}$, where

$$I^0(s) = \begin{cases} 0, & s \text{ even}, \\ 1, & s \text{ odd}, \end{cases} \tag{7.3}$$

and X^{qK} is obtained from X^0 by shifting I^0 to yield $X^{qK} = Z^{I^0+2q}$ for $q = 1, 2, \ldots$. We say that an update function is *admissible* if its values at any two neighboring sites differ by 1. A *move* picks a site s, then replaces $I(s)$ by $I(s) + 2$, leaving I unchanged otherwise. A move is *admissible* if it preserves admissibility of the update function, see Figure 2. Note that an admissible move can be made at site s if and only if

$$I(t) = I(s) + 1 \quad \text{for all } t \in \partial s.$$

Note also that I^0 is an admissible update function, and that all admissible update functions are built up by applying finitely many admissible moves to I^0.

The following theorem of Greenwood, McKeague and Wefelmeyer (1999) shows that for an admissible update function I, the process Z^{I+2q}, $q = 0, 1, \ldots$, is distributed as output from another Gibbs sampler for π. The sweep of this new Gibbs sampler is *ordered* by I in the sense that it first updates the sites on the lowest level of I, then proceeds upwards layer by layer. In general, the new sweep does not respect the checkerboard pattern.

THEOREM 7. *Suppose π has nearest neighbor interactions and X^i, $i = 0, 1, \ldots$, is generated by a Gibbs sampler for π whose updating respects the checkerboard pattern of the lattice. If I is an admissible update function, then Z^{I+2q}, $q = 0, 1, \ldots$, is distributed as a full sweep Gibbs sampler for π having sweep ordered by I.*

The idea of the proof is simple: note that an update at a site s is obtained by adding 2 to the current value of the update function at that site. Thus, the configuration Z^{I+2q} is obtained from Z by applying Gibbs sampler updates site by site in the order of the sweep associated with I. A more formal version of this argument is as follows. Let I_s denote the update function obtained from I by applying the moves at the sites before s in the order of the new sweep. If $I_s(s) = q$, then since I_s is admissible, $I_s(t) = q + 1$ for $t \in \partial s$. The move at s replaces $I_s(s) = q$ by $q + 2$. Recall that $Y_s^{q+2} = Z_s^{I+2}$ was generated using the conditional law $p_s(Y_{\partial s}^{q+1}, dx_s)$ which equals $p_s(Z_{-s}^{I_s}, dx_s)$. Hence Z^{I+2} is obtained from Z^I using the Gibbs sampler with the new sweep.

Call an admissible update function I a update function if it uses part of the initial configuration X^0, i.e., if $\min I$ equals 0 or 1. Each update function I gives an estimator for πf,

$$E_n^I f = \frac{1}{n-h+2} \sum_{q=0}^{n-h+1} f(Z^{I+2q}).$$

Here h is the *height* of I, i.e., the number of full time steps, or half sweeps, it straddles. This means that $\max I$ equals $2h - 2$ or $2h - 1$.

The asymptotic variance of $E_n^I f$ can be substantially different from that of the usual empirical estimator $E_n f$. Estimators with reduced variance might be obtained by averaging over some family $\mathcal{I}$ of update functions:

$$E_n^{\mathcal{I}} f = \frac{1}{|\mathcal{I}|} \sum_{I \in \mathcal{I}} E_n^I f,$$

where $|\mathcal{I}|$ denotes the cardinality of $\mathcal{I}$.

Averaging over update functions can be interpreted as symmetrizing $E_n f$, as in a generalized von Mises statistic. In general we expect such estimators to have smaller variance for larger families of update functions. However, there is a trade-off in terms of computational cost: for high update functions we would need to store more configurations, and for large families we would need to evaluate f more frequently, which could be critical in large lattices or when $f(x)$ is expensive to compute.

If the random field is arbitrary, with not necessarily local interactions, we can only use the update functions involved in the empirical estimators $E_n^s f$. For s even the update function I^s is

$$I^s(t) = \begin{cases} 0, & t > s, t \text{ even,} \\ 2, & t \leqslant s, t \text{ even,} \\ 1, & t \text{ odd.} \end{cases}$$

The update function for s odd is similar:

$$I^s(t) = \begin{cases} 1, & t > s, t \text{ odd,} \\ 3, & t \leqslant s, t \text{ odd,} \\ 2, & t \text{ even.} \end{cases}$$

To make use of the nearest neighbor assumption, we must go beyond the update functions just described. For large lattices it may not be computationally feasible to use all update functions. If one uses only a few update functions, they should be well spaced to reduce correlation between different $E_n^I f$. The higher the update functions we allow, the better we can space them. However, high update functions require more storage: to calculate $E_n^I f$ for a update function I of height h, we must store h configurations at a time.

Simulations in Greenwood, McKeague and Wefelmeyer (1999) for the Ising model and the Gibbs and Metropolis samplers, and for f the rth nearest neighbor correlation, show that the variance reduction can be considerable, even if only a few of the update functions are used.

8. Exploiting symmetries of random fields

Let $X^0, \ldots, X^n$ be observations from an arbitrary Markov chain with transition distribution $Q(x, dy)$. Assume that the chain is positive Harris recurrent, with invariant distribution $\pi(dx)$, and V-uniformly ergodic. Let T be a measurable transformation on E which leaves π invariant and has a measurable inverse. For $f^2 \leqslant V$, we obtain a new consistent and asymptotically normal empirical estimator for πf,

$$E_n(f \circ T) = \frac{1}{n} \sum_{i=1}^{n} f(TX^i).$$

The same is true for any power T^j of T.

Suppose now that the transition distribution Q is invariant under T in the sense that $Q(x, dy) = Q(Tx, T dy)$. This forces π to be invariant under T. Also, for the stationary chain, $(X^0, X^1, \ldots, X^n)$ is distributed as $(TX^0, \ldots, TX^n)$. Hence $E_n(f \circ T^j)$ has the same asymptotic variance as $E_n f$. Better estimators, in the sense of asymptotic variance, can be obtained by linear combinations of $E_n(f \circ T^j)$. The following result of Greenwood, McKeague and Wefelmeyer (1996) shows how best to use linear combinations of such estimators if the powers form a finite cyclic group.

PROPOSITION 1. *Let Q be invariant under a transformation T with $T^m = T^0$ for some $m \geq 2$. Then the best linear combination of $E_n(f \circ T^j)$, $j = 0, \ldots, m-1$, is the average,*

$$\overline{E}_n f = \frac{1}{m} \sum_{j=0}^{m-1} E_n(f \circ T^j).$$

The proof is simple. Observe that the pair $E_n(f \circ T^k)$, $E_n(f \circ T^j)$ is the pair $E_n f$, $E_n(f \circ T^{(j-k) \bmod m})$ evaluated with the chain $X^0, X^1, \ldots$ replaced by the chain $T^k X^0, T^k X^1, \ldots$. Hence the asymptotic covariances of the two pairs agree. Therefore, the asymptotic covariance matrix of $E_n(f \circ T^j)$, $j = 0, \ldots, m-1$, is circulant. In particular, it has equal row sums. If the covariance matrix is nonsingular, Proposition 1 follows from Greenwood, McKeague and Wefelmeyer (1996, Lemma 2). Proposition 1 also follows from the observation that if Σ is positive semidefinite with equal row sums, then $b'\Sigma b$ is minimized over vectors b with $\sum_{j=1}^m b_j = 1$ by $b_j = 1/m$ for all j. To see this, let $e_1, \ldots, e_m$ be an orthonormal basis of eigenvectors of Σ with nonnegative eigenvalues $\mu_1, \ldots, \mu_m$. Assume w.l.g. that $e_1 = (m^{-1/2}, \ldots, m^{-1/2})$. Write $b = \sum_{j=1}^m \lambda_j e_j$. Then $\lambda_1 = m^{-1/2}$ and

$$b'\Sigma b = \sum_{j=1}^m \lambda_j^2 \mu_j = \mu_1/m + \sum_{j=2}^m \lambda_j^2 \mu_j,$$

which is minimized by $\lambda_j = 0$ for $j = 2, \ldots, m$.

Greenwood, McKeague and Wefelmeyer (1996) apply Proposition 1 to the two-step Gibbs sampler. Let π be a distribution on a two-dimensional space $E = E_1 \times E_2$. Let $p_1(x_2, dx_1)$ be the conditional distribution of x_1 given x_2, and $p_2(x_1, dx_2)$ the conditional distribution of x_2 given x_1. Let $n = 2p$, and let $X^0, X^1, \ldots, X^{2p}$ be observations from the two-step Gibbs sampler with deterministic sweep. The subchain $X^0, X^2, \ldots, X^{2p}$ of full sweeps has transition distribution

$$Q(x, dy) = p_1(x_2, dy_1) p_2(y_1, dy_2).$$

Call a transformation T on $E_1 \times E_2$ *parallel* if it is a direct product $T(x_1, x_2) = (T_{11}x_1, T_{22}x_2)$, and *transverse* if $T(x_1, x_2) = (T_{21}x_2, T_{12}x_1)$. Note that the composition of two transverse transformations is parallel, and the composition of a parallel with a transverse is transverse. If T is parallel and leaves π invariant, then

$$p_1(x_2, dx_1) = p_1(T_{22}x_2, T_{11} dx_1),$$
$$p_2(x_1, dx_2) = p_2(T_{11}x_1, T_{22} dx_2). \tag{8.1}$$

It follows that under the stationary law of the sampler,

$$(X^0, X^1, \ldots, X^{2p}) = (TX^0, TX^1, \ldots, TX^{2p}) \quad \text{in distribution.}$$

If T is transverse and leaves π invariant, then

$$p_1(x_2, dx_1) = p_2(T_{21}x_2, T_{12}\,dx_1),$$
$$p_2(x_1, dx_2) = p_1(T_{12}x_1, T_{21}\,dx_2). \quad (8.2)$$

It is easy to see that the transformed time-reversed chain has the same law as the original chain:

$$(X^0, X^1, \ldots, X^{2p}) = (TX^{2p}, TX^{2p-1}, \ldots, TX^0) \quad \text{in distribution.}$$

As in Section 5, the empirical estimator $E_n f = \frac{1}{n}\sum_{i=1}^n f(X^i)$ can be written as the average $\frac{1}{2}(E_n^1 f + E_n^2 f)$ of the empirical estimators based on the two subchains,

$$E_n^1 f = \frac{1}{p}\sum_{q=1}^p f(X^{2q-1}), \qquad E_n^2 f = \frac{1}{p}\sum_{q=1}^p f(X^{2q}).$$

Greenwood, McKeague and Wefelmeyer (1996, Theorem 2) use Proposition 1 to show the following result.

THEOREM 8. *Assume that the two-step Gibbs sampler for π is positive Harris recurrent, and that the two subchains are V-uniformly ergodic. Let π be invariant under a parallel or transverse transformation T with $T^m = T^0$, and let $f^2 \leqslant V$. Then the empirical estimators $E_n^1(f \circ T^j)$, $E_n^2(f \circ T^j)$, $j = 0, \ldots, m-1$, have equal asymptotic variances, and the best linear combination is the average,*

$$\overline{E}_n f = \frac{1}{m}\sum_{j=0}^{m-1} E_n(f \circ T^j).$$

Theorem 8 can be generalized to more than one transformation, as long as the transformations commute; see Greenwood, McKeague and Wefelmeyer (1996, Theorem 3).

Theorem 8 can be applied to nearest neighbor random fields. For simplicity, let

$$S = \{0, \ldots, k_1 - 1\} \times \{0, \ldots, k_2 - 1\},$$

where k_1 and k_2 are even. Let $\pi(dx)$ be the law of a random field on E^S with nearest neighbor interactions (7.1). As in Section 7, number first the even and then the odd sites, respecting the checkerboard pattern of the lattice. Then the Gibbs sampler with deterministic sweep can be written as a two-step Gibbs sampler, with full sweep transition distribution (7.2),

$$Q(y_e, y_o, d(z_e, z_o)) = Q_e(y_o, dz_e)Q_o(z_e, dz_o),$$

where y_e and y_o are the subconfigurations of x on the even and odd sites, respectively.

Define addition on S by $(s+t)_1 = s_1 + t_1 \mod k_1$, $(s+t)_2 = s_2 + t_2 \mod k_2$. For $t \in S$, the translation of S by t is defined as $T_t s = s - t$. This induces a translation on E^S by $(T_t x)_s = x_{T_t^{-1} s} = x_{s+t}$. Translations on E^S by an even or odd number of sites are parallel or transverse transformations, respectively. For a *horizontal* translation, think of the lattice as wrapped around a cylinder so that the vertical boundaries meet. The neighbors of each site $s = (s_1, s_2)$ along the vertical boundary now include $(s_1 \pm 1, s_2)$ with addition mod k_1.

A horizontal translation by an *even* number of sites is $T = T_{(p,0)}$, with p even. This translation takes even into even sites and odd into odd and is a parallel transformation in the sense of Section 4. Suppose that k_1 is a multiple of p, say $k_1 = mp$. Suppose that π is invariant under T. Then it is also invariant under powers $T^j = T_{(jp,0)}$, $j = 0, \ldots, m-1$. These transformations form a cyclic group. Theorem 8 implies that the empirical estimators

$$E_n^1(f \circ T_{(jp,0)}), \qquad E_n^2(f \circ T_{(jp,0)}), \quad j = 1, \ldots, m-1,$$

have equal asymptotic variances, and the best linear combination is the average.

A horizontal translation by an *odd* number of sites is $T = T_{(p,0)}$, with p odd. This translation takes even into odd sites and odd into even and is a transverse transformation. Even powers of T are parallel. Suppose that k_1 is a multiple of p, say $k_1 = mp$. Suppose that π is invariant under T. As above, the best linear combination of the corresponding empirical estimators is the average.

Horizontal and vertical translations commute. Hence the best linear combination of the corresponding empirical estimators is again the average. Greenwood, McKeague and Wefelmeyer (1996) present simulations for the Ising model without and with external field, and for f the nearest neighbor correlation.

Acknowledgement

Work supported by NSERC, Canada. We thank Antonietta Mira, Radford Neal and Gareth Roberts for their comments, and the referee for his careful reading of the manuscript and his many corrections and suggestions.

References

Athreya, K. B., H. Doss and J. Sethuraman (1996). On the convergence of the Markov chain simulation method. *Ann. Statist.* **24**, 69–100.

Belisle, C. J., P. Romeijn and R. L. Smith (1993). Hit-and-run algorithms for generating multivariate distributions. *Math. Oper. Res.* **18**, 255–266.

Besag, J. and P. Green (1993). Spatial statistics and Bayesian computation. *J. Roy. Statist. Soc. Ser. B* **55**, 25–37.

Besag, J., P. Green, D. Higdon and K. Mengersen (1995). Bayesian computation and stochastic systems (with discussion). *Statist. Sci.* **10**, 3–66.

Bickel, P. J., C. A. J. Klaassen, Y. Ritov and J. A. Wellner (1998). *Efficient and Adaptive Estimation for Semiparametric Models*. Springer, New York.

Brémaud, P. (1999). *Markov Chains. Gibbs Fields, Monte Carlo Simulation, and Queues.* Texts in Appl. Math., Vol. 31. Springer, New York.
Casella, G. and C. P. Robert (1996). Rao–Blackwellization of sampling schemes. *Biometrika* **83**, 81–94.
Chan, K. S. (1993). Asymptotic behavior of the Gibbs sampler. *J. Amer. Statist. Assoc.* **88**, 320–326.
Clifford, P. and G. Nicholls (1995). A Metropolis sampler for polygonal image reconstruction. *Technical Report*, Department of Statistics, Oxford University.
Conway, J. B. (1990). *A Course in Functional Analysis*, 2nd ed. Graduate Texts in Mathematics, Vol. 96. Springer, New York.
Damien, P., J. Wakefield and S. Walker (1999). Gibbs sampling for Bayesian non-conjugate and hierarchical models by using auxiliary variables. *J. Roy. Statist. Soc. Ser. B* **61**, 331–344.
Diaconis, P. and L. Saloff-Coste (1998). What do we know about the Metropolis algorithm? *J. Comput. Syst. Sci.* **57**, 20–36.
Edwards, R. G. and A. D. Sokal (1988). Generalization of the Fortuin–Kasteleyn–Swendsen–Wang representation and Monte Carlo algorithm. *Phys. Rev. Lett.* **38**, 2009–2012.
Fishman, G. (1996). Coordinate selection rules for Gibbs sampling. *Ann. Appl. Probab.* **6**, 444–465.
Fishman, G. (1999). An analysis of Swendsen–Wang and related sampling methods. *J. Roy. Statist. Soc. Ser. B* **61**, 623–641.
Frigessi, A., J. Gåsemyr and H. Rue (2000). Antithetic coupling of two Gibbs sampler chains. *Ann. Statist.* **28**.
Frigessi, A., C.-R. Hwang, S. J. Sheu and P. Di Stefano (1993). Convergence rates of the Gibbs sampler, the Metropolis algorithm, and other single-site updating dynamics. *J. Roy. Statist. Soc. Ser. B* **55**, 205–220.
Frigessi, A., C.-R. Hwang and L. Younes (1992). Optimal spectral structure of reversible stochastic matrices, Monte Carlo methods and the simulation of Markov random fields. *Ann. Appl. Probab.* **2**, 610–628.
Gamerman, D. (1997). *Markov Chain Monte Carlo. Stochastic Simulation for Bayesian Inference.* Chapman and Hall, London.
Gelfand, A. E., S. E. Hills, A. Racine-Poon and A. F. M. Smith (1990). Illustration of Bayesian inference in normal data models using Gibbs sampling. *J. Amer. Statist. Assoc.* **85**, 972–985.
Gelfand, A. E. and A. F. M. Smith (1990). Sampling-based approaches to calculating marginal densities. *J. Amer. Statist. Assoc.* **85**, 398–409.
Gelfand, A. E. and A. F. M. Smith (1991). Gibbs sampling for marginal posterior expectations. *Comm. Statist. Theory Methods* **20**, 1747–1766.
Gelman, A., G. O. Roberts and W. R. Gilks (1996). Efficient Metropolis jumping rules. In *Bayesian Statistics*, Vol. 5, pp. 599–608 (Eds. J. M. Bernardo, J. O. Berger, A. P. Dawid and A. F. M. Smith). Oxford University Press.
Geman, S. and D. Geman (1984). Stochastic relaxation, Gibbs distributions, and the Bayesian restoration of images. *IEEE Trans. Pattern Anal. Mach. Intell.* **6**, 721–741.
Geyer, C. J. (1992). Practical Markov chain Monte Carlo. *Statist. Sci.* **7**, 473–483.
Geyer, C. J. (1995). Conditioning in Markov chain Monte Carlo. *J. Comput. Graph. Statist.* **4**, 148–154.
Gilks, W. R., S. Richardson and D. J. Spiegelhalter (Eds.) (1996). *Introducing Markov Chain Monte Carlo.* Chapman and Hall, London.
Gordin, M. I. (1969). The central limit theorem for stationary processes. *Soviet Math. Dokl.* **10**, 1174–1176.
Graham, J. (1994). Monte Carlo Markov chain likelihood ratio test and Wald test for binary spatial lattice data. *Technical Report*, Department of Statistics, North Carolina State University.
Green, P. J. and X.-L. Han (1992). Metropolis methods, Gaussian proposals and antithetic variables. In *Stochastic Models, Statistical Methods, and Algorithms in Image Analysis*, pp. 142–164 (Eds. P. Barone, A. Frigessi and M. Piccioni). Lecture Notes in Statist., Vol. 74. Springer, Berlin.
Green, P. J. and A. Mira (1999). Delaying rejection in Metropolis–Hastings algorithms with reversible jumps. *Technical Report*, Department of Mathematics, University of Bristol.
Greenwood, P. E., I. W. McKeague and W. Wefelmeyer (1996). Outperforming the Gibbs sampler empirical estimator for nearest neighbor random fields. *Ann. Statist.* **24**, 1433–1456.
Greenwood, P. E., I. W. McKeague and W. Wefelmeyer (1998). Information bounds for Gibbs samplers. *Ann. Statist.* **26**, 2128–2156.
Greenwood, P. E., I. W. McKeague and W. Wefelmeyer (1999). Von Mises type statistics for single site updated local interaction random fields. *Statistica Sinica* **9**, 699–712.

Grenander, U. (1983). *Tutorial in Pattern Theory*. Lecture Notes, Division of Applied Mathematics, Brown University.

Hájek, J. (1970). A characterization of limiting distributions of regular estimates. *Z. Wahrsch. Verw. Gebiete* **14**, 323–330.

Hastings, W. K. (1970). Monte Carlo sampling methods using Markov chains and their applications. *Biometrika* **57**, 97–109. Reprinted with introduction in Kotz and Johnson (1997).

Heermann, D. W. and A. N. Burkitt (1992). Parallel algorithms for statistical physics problems. In *The Monte Carlo Method in Condensed Matter Physics*, pp. 53–74 (Ed. K. Binder). Springer, Berlin.

Higdon, D. M. (1998). Auxiliary variable methods for Markov chain Monte Carlo with applications. *J. Amer. Statist. Assoc.* **93**, 585–595.

Höpfner, R. (1993). On statistics of Markov step processes: representation of log-likelihood ratio processes in filtered local models. *Probab. Theory Related Fields* **94**, 375–398.

Höpfner, R., J. Jacod and L. Ladelli (1990). Local asymptotic normality and mixed normality for Markov statistical models. *Probab. Theory Related Fields* **86**, 105–129.

Horn, R. A. and C. R. Johnson (1985). *Matrix Analysis*. Cambridge University Press.

Ingrassia, S. (1994). On the rate of convergence of the Metropolis algorithm and Gibbs sampler by geometric bounds. *Ann. Appl. Probab.* **4**, 347–389.

Jarner, S. F. and G. Roberts (2002). Polynomial convergence rates of Markov chains. *Ann. Appl. Probab.* **12**, 224–247.

Jerrum, M. (1998). Mathematical foundations of the Markov chain Monte Carlo method. In *Probabilistic Methods for Algorithmic Discrete Mathematics*, pp. 116–165 (Eds. M. Habib, C. McDiarmid, J. Ramirez-Alfonsin and B. Reed). Algorithms and Combinatorics, Vol. 16. Springer, New York.

Johnson, V. E. (1996). Studying convergence of Markov chain Monte Carlo algorithms using coupled sample paths. *J. Amer. Statist. Assoc.* **91**, 154–166.

Kartashov, N. V. (1985a). Criteria for uniform ergodicity and strong stability of Markov chains with a common phase space. *Theory Probab. Math. Statist.* **30**, 71–89.

Kartashov, N. V. (1985b). Inequalities in theorems of ergodicity and stability for Markov chains with common phase space. I. *Theory Probab. Appl.* **30**, 247–259.

Kartashov, N. V. (1996). *Strong Stable Markov Chains*. VSP, Utrecht.

Kalos, M. H. and P. A. Whitlock (1986). *Monte Carlo Methods. Vol. 1. Basics*. Wiley, New York.

Kipnis, C. and S. R. S. Varadhan (1986). Central limit theorem for additive functionals of reversible Markov processes and applications to simple exclusions. *Comm. Math. Phys.* **104**, 1–19.

Kira, E. and C. Ji (1997). Rates of convergence for the Gibbs sampler. *Markov Process. Related Fields* **3**, 89–102.

Koshevnik, Y. A. and B. Y. Levit (1976). On a non-parametric analogue of the information matrix. *Theory Probab. Appl.* **21**, 738–753.

Kotz, S. and N. L. Johnson (Eds.) (1997). *Breakthroughs in Statistics*, Vol. III. Springer, New York.

Levit, B. Y. (1974). On optimality of some statistical estimates. In *Proceedings of the Prague Symposium on Asymptotic Statistics*, Vol. 2, pp. 215–238 (Ed. J. Hájek). Charles University, Prague.

Liu, J. S. (1996). Peskun's theorem and a modified discrete-state Gibbs sampler. *Biometrika* **83**, 681–682.

Liu, J. S., W. H. Wong and A. Kong (1994). Covariance structure of the Gibbs sampler with applications to the comparisons of estimators and augmentation schemes. *Biometrika* **81**, 27–40.

Liu, J. S., W. H. Wong and A. Kong (1995). Covariance structure and convergence rate of the Gibbs sampler with various scans. *J. Roy. Statist. Soc. Ser. B* **57**, 157–169.

McEachern, S. N. and L. M. Berliner (1994). Subsampling the Gibbs sampler. *Amer. Statist.* **48**, 188–190.

McKeague, I. W. and W. Wefelmeyer (2000). Markov chain Monte Carlo and Rao–Blackwellization. *J. Statist. Plann. Inference* **85**, 171–182.

Mengersen, K. L., C. P. Robert and C. Guihenneuc-Jouyaux (1999). MCMC convergence diagnostics: a "reviewww". In *Bayesian Statistics*, Vol. 6, pp. 415–440 (Eds. J. M. Bernardo, J. O. Berger, A. P. Dawid and A. F. M. Smith). Oxford University Press.

Mengersen, K. L. and R. L. Tweedie (1996). Rates of convergence of the Hastings and Metropolis algorithms. *Ann. Statist.* **24**, 101–121.

Metropolis, N., A. W. Rosenbluth, M. N. Rosenbluth, A. H. Teller and E. Teller (1953). Equation of state calculations by fast computing machines. *J. Chem. Phys.* **21**, 1087–1092. Reprinted with introduction in Kotz and Johnson (1997).

Meyn, S. P. and R. L. Tweedie (1993). *Markov Chains and Stochastic Stability*. Springer, London.

Meyn, S. P. and R. L. Tweedie (1994). Computable bounds for geometric convergence rates of Markov chains. *Ann. Appl. Probab.* **4**, 981–1011.

Mira, A. and C. J. Geyer (1999). Ordering Monte Carlo Markov chains. *Technical Report*, School of Statistics, University of Minnesota.

Mira, A. and L. Tierney (1999). On the use of auxiliary variables in Markov chain Monte Carlo sampling. *Technical Report*, Medical Informatics Laboratory, University of Pavia. To appear in *Scand. J. Statist.*

Neal, R. M. (1993). Probabilistic inference using Markov chain Monte Carlo methods. *Technical Report*, Department of Statistics, University of Toronto. http://www.cs.toronto.edu/~radford/.

Neal, R. M. (2000). Slice sampling. *Technical Report*, Department of Statistics, University of Toronto. To appear in *Ann. Statist.*

Pearl, J. (1987). Evidential reasoning using stochastic simulation. *Artificial Intelligence* **32**, 245–257.

Penev, S. (1991). Efficient estimation of the stationary distribution for exponentially ergodic Markov chains. *J. Statist. Plann. Inference* **27**, 105–123.

Peskun, P. H. (1973). Optimum Monte Carlo sampling using Markov chains. *Biometrika* **60**, 607–612.

Pfanzagl, J. and W. Wefelmeyer (1982). *Contributions to a General Asymptotic Statistical Theory*. Lecture Note in Statistics, Vol. 13. Springer, New York.

Raftery, A. E. and S. M. Lewis (1992). How many iterations in the Gibbs sampler? In *Bayesian Statistics*, Vol. 4, pp. 763–773 (Eds. J. M. Bernardo, J. O. Berger, A. P. Dawid and A. F. M. Smith). Oxford University Press.

Robert, C. P. (1996). *Méthodes de Monte Carlo par Chaînes des Markov*. Economica, Paris.

Robert, C. P. (Ed.) (1998). *Discretization and MCMC. Convergence Assessment*. Lecture Notes in Statistics, Vol. 135. Springer, New York.

Robert, C. P. and G. Casella (1999). *Monte Carlo Statistical Methods*. Springer, New York.

Roberts, G. O., A. Gelman and W. R. Gilks (1997). Weak convergence and optimal scaling of random walk Metropolis algorithms. *Ann. Appl. Probab.* **7**, 110–120.

Roberts, G. O. and N. G. Polson (1994). On the geometric convergence of the Gibbs sampler. *J. Roy. Statist. Soc. Ser. B* **56**, 377–384.

Roberts, G. O. and J. S. Rosenthal (1997). Geometric ergodicity and hybrid Markov chains. *Electron. Comm. Probab.* **2**, 13–25. http://www.math.washington.edu/~ejpecp/.

Roberts, G. O. and J. S. Rosenthal (1998). Markov-chain Monte Carlo: some practical implications of theoretical results (with discussion). *Canad. J. Statist.* **26**, 5–31.

Roberts, G. O. and J. S. Rosenthal (1999). Convergence of slice sampler Markov chains. *J. Roy. Statist. Soc. Ser. B* **61**, 643–660.

Roberts, G. O. and S. K. Sahu (1997). Updating schemes, correlation structure, blocking and parameterisation for the Gibbs sampler. *J. Roy. Statist. Soc. Ser. B* **59**, 291–317.

Roberts, G. O. and R. L. Tweedie (1996). Geometric convergence and central limit theorems for multivariate Hastings and Metropolis algorithms. *Biometrika* **83**, 95–110.

Roberts, G. O. and R. L. Tweedie (1999). Bounds on regeneration times and convergence rates for Markov chains. *Stochastic Process. Appl.* **80**, 211–229.

Roberts, G. O. and R. L. Tweedie (2000). Rates of convergence of stochastically monotone and continuous time Markov models. *J. Appl. Probab.* **37**, 359–373.

Rosenthal, J. S. (1995). Minorization conditions and convergence rates for Markov chain Monte Carlo. *J. Amer. Statist. Assoc.* **90**, 558–566.

Roussas, G. G. (1965). Asymptotic inference in Markov processes. *Ann. Math. Statist.* **36**, 987–992.

Schervish, M. J. and B. P. Carlin (1992). On the convergence of successive substitution sampling. *J. Comput. Graph. Statist.* **1**, 111–127.

Schmeiser, B. and M. H. Chen (1991). On random-direction Monte Carlo sampling for evaluating multidimensional integrals. *Technical Report*, Department of Statistics, Purdue University.

Smith, A. F. M. and G. O. Roberts (1993). Bayesian computation via the Gibbs sampler and related Markov chain Monte Carlo methods. *J. Roy. Statist. Soc. Ser. B* **55**, 3–23.

Spiegelhalter, D. J., A. P. Dawid, S. L. Lauritzen and R. G. Cowell (1993). Bayesian analysis in expert systems (with discussion). *Statist. Sci.* **8**, 219–283.

Spiegelhalter, D. J., A. Thomas and N. G. Best (1996). Computation on Bayesian graphical models. In *Bayesian Statistics*, Vol. 5, pp. 407–426 (Eds. J. M. Bernardo, J. O. Berger, A. P. Dawid and A. F. M. Smith). Oxford University Press.

Swendsen, R. H. and J.-S. Wang (1987). Nonuniversal critical dynamics in Monte Carlo simulations. *Phys. Rev. Lett.* **58**, 86–88.

Tanner, M. A. (1996). *Tools for Statistical Inference. Methods for the Exploration of Posterior Distributions and Likelihood Functions*, 3rd ed. Springer, New York.

Tanner, M. A. and W. H. Wong (1987). The calculation of posterior distributions by data augmentation. *J. Amer. Statist. Assoc.* **82**, 528–540.

Tierney, L. (1994). Markov chains for exploring posterior distributions (with discussion). *Ann. Statist.* **22**, 1701–1762.

Tierney, L. (1996). Introduction to general state-space Markov chain theory. In *Introducing Markov Chain Monte Carlo*, pp. 59–74 (Eds. W. R. Gilks, S. Richardson and D. J. Spiegelhalter). Chapman and Hall, London.

Tierney, L. (1998). A note on Metropolis–Hastings kernels for general state spaces. *Ann. Appl. Probab.* **8**, 1–9.

Tierney, L. and A. Mira (1999). Some adaptive Monte Carlo methods for Bayesian inference. *Statistics in Medicine* **18**, 2507–2515.

Wefelmeyer, W. (1998). Judging MCMC estimators by their asymptotic variance. In *Prague Stochastics '98*, Vol. 2, pp. 591–596 (Eds. M. Hušková, P. Lachout and J. A. Víšek). Union of Czech Mathematicians and Physicists, Prague.

Wefelmeyer, W. (1999). Efficient estimation in Markov chain models: an introduction. In *Asymptotics, Nonparametrics, and Time Series*, pp. 427–459 (Ed. S. Ghosh). Dekker, New York.

Fractals and the Modelling of Self-Similarity

B. M. Hambly

Fractals provide a geometric framework for the modelling of self-similarity and have become widely used in recent years in a broad range of applications. In this chapter we restrict the discussion to the modelling of environments which are self-similar and we provide a limited review of models which have natural fractal structure. The aim is to demonstrate a range of mathematical ideas which can facilitate the use of fractals in modelling. We shall see that modelling with fractals raises interesting mathematical and statistical questions as well as offering new insights into the phenomena being modelled.

1. Introduction

In the 1970s Benoit Mandelbröt published the first version of an essay that developed into 'The fractal geometry of nature' (Mandelbrot, 1982) in which a new approach to modelling natural structures was set out. This view begins with the idea that the geometry of regular bodies is quite inappropriate when we consider the natural world. Many structures exhibit features over a range of length scales and are not necessarily well represented by smooth geometries. Indeed such everyday objects as trees, clouds, coastlines and mountain ranges are highly complex and can have quite different appearance depending on the scale at which they are viewed. Despite the obvious differences, all these examples have a common feature, there is some form of self-similarity in appearance over a range of length scales.

From this starting point we can ask what is the natural way to analyse sets which have structure over many length scales and this leads us to some concrete mathematical questions. In fact some of these questions had already been considered from the pure mathematical perspective in the field of geometric measure theory. The construction of self-similar sets was used to produce counter examples to intuitive mathematical ideas concerning the properties of curves and surfaces. In this setting such sets were seen as pathological, only of interest for their strangeness but this changed rapidly with the recognition that self-similarity was actually an essential feature in many natural systems. The benefit of the earlier mathematical development was the existence of a range of mathematical tools available to study sets with this type of complex structure.

The word fractal was introduced by Mandelbrot to describe these objects but it is a word without rigorous mathematical definition. The term encompasses sets with quite different features but all will have some aspect of self-similarity over a range of scales. Essentially fractals provide a basis for describing geometrically systems which satisfy certain scaling properties. In many instances, where there is some sort of scale invariance, we can expect to find fractal structure.

The main emphasis of this paper will be on the tools available to the modeller who wishes to use fractals. Once we have decided that a fractal model may provide the right framework in which to work we have to understand how that affects the quantities we are interested in. For purely geometric questions we want to be able to find an adequate description of the fractal set that allows us to compare it with other sets and captures the key features of the geometry of the fractal. There have been many lively and wide ranging discussions of the geometry of fractals, see (Barnsley, 1993; Falconer, 1990, 1997; Mattila, 1995) for the mathematical end of the spectrum and (Feder, 1988; Turcotte, 1997) for the physical viewpoint.

We will begin with a discussion of geometric modelling and provide an introduction to the idea of fractal dimension and the construction of various classes of sets with a range of self-similarity properties. The fractal dimension provides only one parameter to describe a set and we will extend this idea to multifractals; sets which have local variation in their dimension. Other features of fractals which may provide a more detailed description, such as lacunarity are difficult to quantify and we will just consider dimensional descriptions of fractal sets. The constructions we describe lead naturally to the application, image compression, and we discuss the use of similarity transformations to obtain efficient encodings of images (Barnsley, 1993).

Once we have established that a fractal model may be appropriate, and fine tuned it to our problem, we may be interested in how that choice of model affects the problem we are concerned with. This can, for instance, mean trying to understand how the fractal geometry affects physical processes within the fractal. The examples that we focus on are transport problems – imagine that the set represents some heterogeneous medium and we are interested in the behaviour of waves or diffusion within the medium. For some discussion in this direction see (Barlow, 1998; Ben-Avraham and Havlin, 2000; Kigami, 2001). Other examples that we will not discuss here include, reactions on fractal surfaces (Ben-Avraham and Havlin, 2000), erosion, drainage and fragmentation in geology (Turcotte, 1997) and fractal antennae (Dekking et al., 1999).

In time series we are often confronted with observations from an underlying diffusion process. The continuous trajectories of such processes naturally have fractal structure, though the possible dimensions of the trajectories are restricted. However recent interest in trying to understand complex large scale data sets which arise in teletraffic modelling or finance suggest modelling using processes which have flexible dimensional properties. Examples include fractional Brownian motion, which for a certain parameter range, exhibits long range dependence, or Lévy processes. Note that the fractals that we consider arise from self-similar constructions or random processes and we will not address fractals which arise as attractors in chaotic dynamical systems.

The outline of this chapter is as follows. To begin with we will describe the geometry of fractals and discuss fractal models that are generated via the iterated function system approach expounded, for instance, in (Barnsley, 1993). These ideas lead naturally to techniques for simulating fractals and could be used to produce the wonderful pictures of fractals as found in (Peitgen, Jürgens and Saupe, 1992) which have brought fractals to the attention of the general public. However we will not discuss the details of image generation here.

The next section will consider briefly the fractal properties of stochastic processes. We will not discuss this in any depth but instead observe some of the different fractal characteristics of various processes which may be useful when selecting models. The emphasis will be on one-dimensional processes and the properties of the surfaces of random fields will not be discussed.

Finally we will consider the dynamic properties of fractals and look at the modelling of diffusion on such sets. We give a brief discussion of the mathematical issues and then describe a class of fractals where precise results can be obtained. The problem that we will eventually focus on is the modelling of the diffusion of gas through soil. In this setting a fractal model may be appropriate and we will consider the implications of working with such a model.

The discussion in this paper will only touch on a tiny proportion of all the work that there is in this field and even in the restricted areas we do consider, this will be a far from comprehensive survey. This area is still developing rapidly as self-similarity can be found in a wide variety of settings and new tools are required to answer the questions thrown up by new applications. The analysis of fractal models provides many new and interesting challenges in the mathematics, the statistics and in the applications.

2. Fractal geometry

We begin by discussing some of the basic ideas in fractal geometry. For more in depth accounts see (Falconer, 1990, 1997; Mattila, 1995). There is no generally accepted definition of a fractal but most people can agree whether any particular example should be called a fractal or not. The desirable features for a fractal set are

(1) some form of self-similarity,
(2) recursive structure,
(3) structure at all length scales.

A typical feature of a fractal is that it has a non-integer dimension. There are many notions of dimension for fractals and we briefly introduce the most important for what we will discuss here. The fractal dimension, in essence, describes the power law scaling between the size of the set and the length scale at which it is measured. Through out this paper we will only consider fractal subsets of Euclidean space, $\mathbb{R}^n$.

2.1. Fractal dimension

A natural mathematical definition of dimension is that of Hausdorff dimension. This allows us to define the notion of dimension for any set in $\mathbb{R}^n$ and provides us with a natural measure (a 'volume') as well (compare with $\mathbb{R}^n$ where we have the n-dimensional Lebesgue measure). We begin by introducing the Hausdorff s-dimensional measure of a set F as a limit. Firstly define

$$\mathcal{H}^s_\delta(F) = \inf\left\{\sum_i \text{diam}(U_i)^s;\ \{U_i\} \text{ is a } \delta\text{-cover of the set } F\right\},$$

where a δ-cover is a family of sets $\{U_i\}$ such that $F \subset \bigcup_i U_i$ and $\text{diam}(U_i) < \delta$, where $\text{diam}(U) = \sup\{|x - y|: x, y \in U\}$ is the diameter of the set U. Now as $\delta \to 0$ the sequence $\mathcal{H}^s_\delta(F)$ is decreasing and hence we can define the s-dimensional Hausdorff measure of F as the limit,

$$\mathcal{H}^s(F) = \lim_{\delta \to 0} \mathcal{H}^s_\delta(F).$$

The quantity $\mathcal{H}^s(F)$, considered as a function in s, is seen to take the value infinity for small s and then to drop to 0 after one particular value. We think of this as the value of the dimension at which there could be a meaningful notion of 'volume' for the set. The point at which it makes the transition is defined to be the Hausdorff dimension of the set. The reader can easily verify that this accords with the usual notion of dimension when considering sets with non-empty interior in n-dimensions. Thus

$$\dim_H(F) = \inf\{s:\ \mathcal{H}^s(F) = 0\} = \sup\{s:\ \mathcal{H}^s(F) = \infty\}.$$

At the Hausdorff dimension $\dim_H(F)$ we can consider the $\dim_H$-dimensional Hausdorff measure of the set F. It is possible that $\mathcal{H}^{d_f}(F) = 0, \infty$ or it may have a value in between. In the case of self-similar sets, introduced in Section 2.4, it takes a finite non-zero value but for random fractals such as random self-similar sets, or those that arise in the path properties of stochastic processes, it is often the case that the d_f-dimensional Hausdorff measure is almost surely 0. In this case it is possible to introduce a finer notion of dimension by choosing a different gauge function in the definition of Hausdorff dimension. That is we consider the Hausdorff measure of the set F in the gauge h, defined as

$$\mathcal{H}^h(F) = \lim_{\delta \to 0} \inf\left\{\sum_i h(\text{diam}(U_i));\ \{U_i\} \text{ is a } \delta\text{-cover of the set } F\right\}.$$

The usual s-dimensional Hausdorff measure is found by setting $h(r) = r^s$. In the case of random fractals it is typical to find that the Hausdorff measure constructed with a gauge function such as $h(r) = r^s(\log|\log(r)|)^\alpha$ is positive and finite. Such a function is often called the exact Hausdorff measure function for the fractal F.

The Hausdorff dimension provides a robust mathematical formulation of the concept of dimension, however it is somewhat unwieldy to use in practice. The problem of estimating the dimension of a given fractal involves determining the power law relationship that exists between the 'volume' of the fractal and the length scale of measurement. One way of viewing our usual concept of n-dimensions is that an n-dimensional cube can be covered by 2^n cubes of side $1/2$. By analogy, we hope that for a fractal F, of fractal dimension d_f, it should be the case that, if we cover the set with balls of small radius ε, the number required should scale with the fractal dimension, ε^{-d_f} as $\varepsilon \to 0$.

We formalise this as the box counting dimension for the fractal. Let $N_\varepsilon(F)$ denote the minimum number of balls (or cubes, the exact type of set is not important) of radius ε required to cover F, that is $F \subset \bigcup_i B(x_i, \varepsilon)$. Thus as $\varepsilon \to 0$, the counting function $N_\varepsilon(F)$ increases at a certain rate. As there may be oscillations in the counting function we define the upper box counting dimension $\dim_{UB}(F)$ and the lower box counting dimension $\dim_{LB}(F)$ as

$$\dim_{UB}(F) = \limsup_{\varepsilon \to 0} \frac{\log N_\varepsilon(F)}{-\log \varepsilon},$$

$$\dim_{LB}(F) = \liminf_{\varepsilon \to 0} \frac{\log N_\varepsilon(F)}{-\log \varepsilon}.$$

The box counting dimension is defined to be the common value if the upper and lower dimensions agree. Note that there is no natural measure associated with this dimension and it may be possible for countable sets to have non-zero box counting dimension.

We also briefly mention an idea of dimension which is complementary to the Hausdorff dimension – that of packing dimension. The Hausdorff dimension of F was derived by constructing a measure from the consideration of optimal covers of F. The packing dimension of F is constructed from a packing measure obtained by considering optimal packings of F with disjoint sets. This is slightly more complicated to set up as the packing measure, obtained by taking the obvious analogue of the Hausdorff measure, is only an outer measure. However a straightforward modification ensures, one can define packing dimension and measure as for Hausdorff measure and dimension, see (Falconer, 1990) for details.

These notions of dimension are related through the fundamental inequalities

$$\dim_H(F) \leqslant \dim_{LB}(F) \leqslant \dim_P(F),$$

and

$$\dim_H(F) \leqslant \dim_{LB}(F) \leqslant \dim_{UB}(F).$$

Note that there is no general relation between the upper box and packing dimensions.

For our purposes, if the values of the dimensions of a set F are all the same we will usually just refer to d_f, the fractal dimension of F. In modelling it is not possible to distinguish these notions of dimension as they are asymptotic formulae and hence any practical notion of fractal dimension is based on box counting dimension. Any real system will also have a lower cut off, a limit beyond which we cannot measure, and hence a set is considered fractal if there is a linear relationship between $\log N_\varepsilon(F)$ and

$-\log \varepsilon$ over a sufficiently large range of scales. In this setting the dimension is the slope of the best fit line.

In statistical physics many models are set on lattices and then claims are made about the fractal dimension of random subsets of the lattice. In particular the percolation model or the Ising model, when the parameter is in the neighbourhood of the phase transition. In this setting a mathematical approach to the definition of fractal dimension for subsets of $\mathbb{Z}^n$ is needed and can be found in (Barlow and Taylor, 1992). Here the fundamental idea is to reverse the picture and consider the scaling in the set over larger and larger windows.

An alternative approach to box dimension is to consider the Minkowski content of the fractal set. If we have a fractal $F \subset \mathbb{R}^n$, then take the δ-parallel set F_δ defined by

$$F_\delta = \{x \in \mathbb{R}^n : |x - y| \leqslant \delta \text{ for some } y \in F\}. \tag{2.1}$$

As we let $\delta \to 0$ we can determine the rate of change of the volume of F_δ. If F is d_f dimensional we should find that $\text{vol}(F_\delta)\delta^{d_f - n} \to M(F)$ as $\delta \to 0$. The quantity $M(F)$ is the Minkowski content of the set F. Thus in general we can define upper and lower Minkowski dimensions of the set F as

$$\dim_{UM}(F) = n - \liminf_{\delta \to 0} \frac{\log \text{vol}(F_\delta)}{-\log \delta},$$

$$\dim_{LM}(F) = n - \limsup_{\delta \to 0} \frac{\log \text{vol}(F_\delta)}{-\log \delta}.$$

It is easy to show that this is the same as the upper and lower box dimensions for the set F. As before the Minkowski dimension is equal to the box dimension when the limits agree.

There are many other dimensions which are commonly referred to in the fractal literature. Indeed, if it is found that there is a scaling relation between two quantities, then the exponent is often called a dimension. Here we have been focused entirely on the geometry of the set and a way to provide an index reflecting the scaling in the set. In the next subsection we will discuss the idea of multifractal analysis which gives a more detailed description of the geometry of a fractal.

The notion of dimension can be extended to define the dimension of a measure. Let μ be a measure in $\mathbb{R}^n$ which can be thought of roughly as a mass distribution in n-dimensional space. We define its dimension as the dimension of the largest set in the support of the measure,

$$\dim_H(\mu) = \sup\{\dim_H(A) : A \subset \text{supp}(\mu)\}.$$

Any fractal set can be viewed as a measure where we think of the measure as a mass distribution which is spread 'evenly' across the fractal set. It is easy to make this precise in the examples that we construct below (see Falconer, 1990).

2.2. Multifractal analysis

In the previous section, when discussing the box counting dimension, we implicitly assumed that the fractal F was homogeneous. We could interpret this statement as saying that for each point $x \in F$, the measure of the intersection of F with the ball of radius r at $x \in F$ should scale according to the dimension, $\mu(B(x,r) \cap F) \asymp r^{d_f}$ as $r \to 0$ (where we write $\asymp$ to denote bounded above and below by constants), with exponent independent of x. However, it may be the case that the set is inhomogeneous, in that the scaling at a point x may vary as x varies across the fractal F. This suggests a notion of the local dimension of a measure at a point $x \in F$ as the scaling exponent obtained for the measure of a ball of radius r at x in the limit as $r \to 0$.

It may be the case that these local dimensions vary across the fractal F allowing it to be decomposed into subsets each with the same local dimension. The curve obtained by calculating the dimension of the set containing all the points with a particular local dimension is called the multifractal spectrum of the fractal. In mathematical terms we define the set of points where the measure has local dimension α as

$$J_\alpha = \left\{ x \in F \colon \lim_{r \to 0} \frac{\log(\mu(B_r(x)))}{\log(r)} \text{ exists and equals } \alpha \right\}.$$

The fine multifractal spectrum is then defined to be the Hausdorff dimension of these sets as a function of α,

$$f_f(\alpha) = \dim_H(J_\alpha).$$

We note that typically the sets J_α are dense in the fractal and hence box-counting dimensions will not discriminate between the different J_α.

There is a coarse multifractal spectrum that can be obtained through a box counting approach. Firstly we consider the lattice $\varepsilon \mathbb{Z}^n$, the n-dimensional integer lattice scaled to have side length ε. Let $C_\varepsilon(x)$ denote the cube, side length ε, centered at position x in the lattice. We now count the cubes of sufficiently large measure by setting

$$N_\varepsilon(\alpha) = \left| \left\{ C_\varepsilon(x) \colon \mu(C_\varepsilon(x)) \geq \varepsilon^\alpha \right\} \right|,$$

where, for a finite set, $|A|$ denotes the number of elements in A. The coarse spectrum is then defined to be, when it exists, the following double limit,

$$f_C(\alpha) = \lim_{\delta \to 0} \lim_{\varepsilon \to 0} \frac{\log(N_\varepsilon(\alpha + \delta) - N_\varepsilon(\alpha - \delta))}{-\log \varepsilon}. \tag{2.2}$$

This spectrum is a global one in that it involves all the sets which scale with a particular exponent value and does not have the local interpretation of the fine spectrum. In particular $f_C(\alpha)$ is not the box dimension of the set of x with a local dimension α. If the limit as $\varepsilon \to 0$ does not exist in (2.2) we can consider upper and lower coarse multifractal spectra for the measure μ. As in the case of the dimensions of the fractal itself there is a relationship between the spectra. For a finite measure in $\mathbb{R}^n$,

$$f_f(\alpha) \leq \underline{f}_C(\alpha) \leq \bar{f}_C(\alpha).$$

There are many cases where these spectra agree, in particular for a wide class of random self-similar graph directed measures (Olsen, 1996). The natural spectrum to compute in practice is the coarse spectrum and in order to do this we can use the so called multifractal formalism.

Multifractal analysis was developed initially in the physics literature and has been made precise in certain mathematical settings (Olsen, 1995). Early ideas are apparent in Mandelbrot (1974) in the study of turbulence but more formal arguments appeared in Frisch and Parisi (1985) and then in Halsey et al. (1986) where the 'multifractal formalism' was first developed. This is a technique for calculating the multifractal spectrum which is useful in practice as it exploits a remarkable connection between the local structure of the set and certain global quantities. Those familiar with the theory of large deviations will recognise the probabilistic content.

Let $M_\varepsilon(q) = \sum_{C_\varepsilon} \mu(C_\varepsilon)^q$ denote the moment sums where the sets C_ε are cubes of side ε which cover the set F. For each $q \in \mathbb{R}$ we can find the scaling in the moment measures as

$$\beta(q) = \lim_{\varepsilon \to 0} \frac{\log M_\varepsilon(q)}{-\log \varepsilon}.$$

If the limit does not exist we can define upper and lower functions $\underline{\beta}, \bar{\beta}$. We can relate the moment measures to the counting function which appeared in our definition of the coarse multifractal spectrum, as for each $q > 0$

$$M_\varepsilon(q) = \sum_{C_\varepsilon} \mu(C_\varepsilon)^q \asymp \int_0^\infty \varepsilon^{\alpha q} N_\varepsilon(\alpha) \, d\alpha.$$

A slight modification gives a similar estimate for the case when $q < 0$. Using the coarse multifractal spectrum, $N_\varepsilon(\alpha) \asymp \varepsilon^{-f_C(\alpha)}$, we have, for small ε,

$$\varepsilon^{-\beta(q)} \asymp M_\varepsilon(q) \asymp \varepsilon^{\inf_\alpha \{\alpha q - f_C(\alpha)\}}, \quad \forall q. \tag{2.3}$$

If the spectrum is concave, (2.3) shows that the multifractal spectrum f_C and the function $-\beta$ are Legendre conjugates. This yields an expression for the coarse multifractal spectrum

$$f_C(\alpha) = \inf_{-\infty < q < \infty} \{\alpha q + \beta(q)\}.$$

Note that the Hausdorff dimension of the measure can be read off from the multifractal spectrum as it corresponds to $\dim_H(\mu) = \sup_\alpha f(\alpha)$, which occurs when $q = 0$. For the value $q = 1$ the corresponding $f(\alpha)$ is called the information dimension.

In some simple examples it is quite easy to determine this spectrum. For instance the case of a Bernoulli measure on [0, 1]. In this example we start with the unit interval and divide it successively into dyadic pieces and assign mass p to the left side and $q = 1 - p$ to the right. The resulting measure on the unit interval, obtained as the limit of this procedure, is singular with respect to Lebesgue measure for $p \neq 1/2$ and has an explicit non-trivial multifractal spectrum in the interval $[-\log p/\log 2, -\log q/\log 2]$ for $p \neq 1/2$. We remark that there are natural sets where the formalism breaks down, for example the occupation measure of a stable subordinator (Hu and Taylor, 1997).

2.3. Projections, products and intersections

Imagine that we are interested in a particular fractal set but can only observe it through some lower-dimensional transect. For instance we have a rough surface but can only observe it along a line, or a three-dimensional soil sample which has been cross sectioned. What is the relationship between the dimension of the original structure and the dimension of the intersection with a lower-dimensional subset? Or we may have several fractals and be interested in a structure arising as a product of fractals with different dimensions.

These questions have been answered theoretically and can feed in to any practical aspects of fractal modelling. We will just state the main theorems and leave the reader to determine when these are useful. For deeper discussion see (Falconer, 1990; Mattila, 1995) and for statistical applications see (Davies and Hall, 1999).

Firstly, for projections we denote by proj_{S_k} the orthogonal projection onto the k-dimensional hyperplane S_k.

THEOREM 2.1. *Let $F \subset \mathbb{R}^n$ be a fractal and H_k a k-dimensional hyperplane.*

(1) *If $F \subset \mathbb{R}^n$, then if $\dim_H(F) \leqslant k$, then $\dim_H(\text{proj}_{H_k}(F)) = \dim_H(F)$ for almost all H_k.*
(2) *If $F \subset \mathbb{R}^n$, then if $\dim_H(F) > k$, then $\dim_H(\text{proj}_{H_k}(F)) = k$ for almost all H_k.*

For a product of two fractals we have the following.

THEOREM 2.2. *If the Hausdorff and box dimensions of the fractals $E, F \subset \mathbb{R}^n$ are the same, then $\dim(E \times F) = \dim(E) + \dim(F)$.*

The section or slice theorem allows us to relate the dimensions of cross sections to the dimension of the original set.

THEOREM 2.3. *Let $F \subset \mathbb{R}^n$ and let L_k denote a k-dimensional hyperplane passing through the origin, then*

$$\dim_H(F \cap L_k) = \dim_H(F) - (n - k),$$

for almost every L_k with respect to $(n - k)$-dimensional Lebesgue measure.

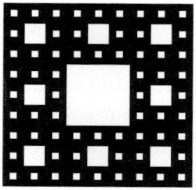

Fig. 1. The Sierpinski gasket and carpet.

Thus in general we are safe in assuming that if we cut a fractal in $\mathbb{R}^n$ with an $(n-1)$-dimensional hyperplane, the section through the fractal has dimension 1 less than the fractal itself. The only problem could be if we were unlucky in the choice of direction of the section. For example consider the Sierpinski carpet in $\mathbb{R}^2$, shown on the right side of Figure 1. If we chose a vertical line with x coordinate given by a point on the standard middle third Cantor set, the slice will have dimension one. If we choose the vertical line with $x = 1/2$, the slice will be a standard Cantor set with dimension $\log 2/\log 3$. However the intersection of a vertical line, at a point chosen uniformly at random from the unit interval, with the carpet will be a subset of dimension $\log(8/3)/\log(3)$.

So far we have considered projection and intersection theorems for the Hausdorff dimension. For the multifractal spectrum such results are more difficult to determine. If we examine a cross section of our fractal we can ask about the relationship between f_{cs}, the multifractal spectrum of the cross section by L_k, a k-dimensional hyperplane in $\mathbb{R}^n$, and the spectrum f for the fractal itself. In (Olsen, 2000) a multifractal slice theorem shows that under some conditions

$$f_{cs}(\alpha) = f(\alpha + n - k) - (n - k),$$

for almost every L_k with respect to $(n-k)$-dimensional Lebesgue measure.

2.4. Constructing fractals

We consider some straightforward mathematical constructions of fractals.

2.4.1. Self-similar sets
A very natural way to specify a fractal is as the fixed point for a family of contraction maps. In this setting there is a rigorous mathematical justification for the dimension formulas first determined by Mandelbrot (1982). We begin by recalling that a contraction map ϕ on a subset A of $\mathbb{R}^n$ satisfies

$$|\phi(x) - \phi(y)| \leqslant L|x - y|, \quad \forall x, y \in A,$$

where $0 < L < 1$ is the Lipschitz constant of the map ϕ. If there is equality we will call the map a similitude.

Let $\mathcal{A}$ denote the set of all non-empty compact subsets of a closed set Ω. This can be made into a metric space by introducing the Hausdorff metric $d_H(U, V)$, defined by

$$d_H(U, V) = \inf\{\delta: U \subset V_\delta, V \subset U_\delta\}, \quad U, V \in \mathcal{A},$$

where U_δ is defined in (2.1).

A family of contraction maps $\{\phi_i, i = 1, \ldots, m\}$ is called an *iterated function system*. It defines a map Φ on $\mathcal{A}$ as $\Phi(A) = \bigcup_{i=1}^{m} \phi_i(A)$ for $A \in \mathcal{A}$. The fundamental result, which follows from the fact that Φ is a contraction map with respect to the Hausdorff metric, is the following.

THEOREM 2.4. *There exists a unique non-empty compact set F which satisfies*

$$F = \bigcup_{i=1}^{m} \phi_i(F). \tag{2.4}$$

Let $A \in \mathcal{A}$ be such that $\Phi(A) \subset A$ and define $\Phi^k(A) = \Phi \circ \cdots \circ \Phi(A)$, then

$$F = \bigcap_{k=0}^{\infty} \Phi^k(A).$$

If the iterated function system consists of similitudes and there is no significant overlap between components (they only intersect on their boundaries), the dimension of the limit fractal can be explicitly calculated. This non-overlap of components can be made rigorous in the 'open set condition' (OSC) which is said to hold if there is a non-empty, bounded, open set V such that the $\phi_i(V)$ are disjoint and $\bigcup_{i=1}^{m} \phi_i(V) \subset V$. Two examples of fractals satisfying the OSC are shown in Figure 1.

The following result was established by Hutchinson (1981).

THEOREM 2.5. *Let $\{\phi_i, i = 1, \ldots, m\}$ be a family of similitudes with contraction factors L_i. Suppose that the OSC holds, then if $F = \bigcup_{i=1}^{m} \phi_i(F)$, then $\dim_H(F) = \dim_B(F) = s$ where $\sum_{i=1}^{m} L_i^s = 1$. The Hausdorff s-dimensional measure of the set F is strictly positive and finite.*

Let $\mathcal{S} = \{1, \ldots, m\}$ be an alphabet and define $I_n = \bigcup_j \mathcal{S}^j$ to be the set of sequences from $\mathcal{S}$ up to length n. The set of infinite sequences is denoted by I. This space of sequences describes the fractal as each letter of $\mathcal{S}$ corresponds to a similitude and we can denote the successive application of n similitudes as a word in I_n. In dynamical terms the fractal is encoded by a one-sided shift space, where the shift map produces a new sequence by removing the first term of the old one. Thus we can also think about an m-ary tree as describing the fractal and the shift map as moving down a branch. Any point in the fractal set can be represented as the limit of a sequence of subsets of F and we can identify the fractal with the topological boundary of the tree arising from the shift space.

A word of warning is that if the maps ψ_i are affine transformations and not similitudes the situation can be more complicated. The limit fractal exists but the dimension results of Theorem 2.5 do not necessarily hold. Indeed it is possible to find self-affine fractals which have different box and Hausdorff dimensions. However a theorem of Falconer (1990), shows that the generic situation is that the two notions of dimension agree and the dimension can be computed as the radius of convergence of a certain power series.

The fine multifractal spectrum for a Bernoulli measure (in which a mass p_i is associated with each contraction map and $\sum_{i=1}^{m} p_i = 1$) on a self-similar set can be found from the multifractal formalism. Let

$$\Phi(q, \beta) = \sum_i p_i^q L_i^\beta.$$

This is used to define for each value of q a value of $\beta(q)$ such that $\Phi(q, \beta(q)) = 1$. The multifractal spectrum of the Bernoulli measure is then given as the Legendre transform of the function β, that is

$$f(\alpha) = \inf_{-\infty < q < \infty} \{\alpha q + \beta(q)\}.$$

We now consider two extensions of the class of self-similar fractals which remove the exact self-similarity and hence allow a greater degree of flexibility in the sets which can be modelled.

2.4.2. Graph directed constructions

The class of graph directed constructions was defined in (Mauldin and Williams, 1988) and provides a vector valued extension of the class of self-similar sets. Let $J = \{J_j\}_{j \in S}$ be a set of compact and connected subsets of $\mathbb{R}^n$, indexed by a finite set S. Let S be the vertex set for a directed graph $G = (S, E)$, with edge set E. For each $i \in S$, let $E_i = \{e(i, j) \in E : j \in S\}$ and assume $|E_i| \geq 1$. For each edge $e(i, j) \in E$, let $\phi_e : \mathbb{R}^n \to \mathbb{R}^n$ be an L_e-similitude with $L_e < 1$. Also for each $i \in S$, assume

$$\bigcup_{e(i,j) \in E_i} \phi_{e(i,j)}(J_j) \subset J_i.$$

The fractal K is then the unique vector of compact sets $(K_1, \ldots, K_n)$ which satisfies

$$K_i = \bigcup_{e(i,j) \in E_i} \phi_{e(i,j)}(K_j).$$

This result is Theorem 1 of (Mauldin and Williams, 1988) and is proved by demonstrating that the map $\Phi = \bigcup_e \phi_e$ is a contraction on the set of compact subsets of J in the Hausdorff metric. The graph G is called the construction graph for K and some properties of the graph are reflected in the fractal. An example of a graph directed fractal is shown in Figure 2.

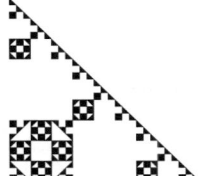

Fig. 2. A graph directed fractal.

A point in this type of fractal is no longer described via a branch of a tree but instead can be described using the paths in the graph. Let E^n be the set of length n paths in the graph G. Note that E^n is not the full set of n-sequences of elements of $\mathcal{S}$, but consists of the allowable sequences $\sigma = (e(i_0, i_1), \ldots, e(i_{n-1}, i_n))$ for which $e(i_j, i_{j+1}) \in E$ for $j = 0, \ldots, n-1$. In the language of dynamical systems E^∞ is a Markov subshift of the full shift space generated by $\mathcal{S}$.

A formula can be found for the Hausdorff dimension of a graph directed fractal under the obvious extension of the open set condition which ensures that the overlap between sets is sufficiently small. To find the dimension of K we consider the non-negative $|\mathcal{S}| \times |\mathcal{S}|$-matrix $\mathbf{H}^s$ with entries $h_{ij}^s = \sum_{e(i,j) \in E} L_e^s$. Let $\Phi(s)$ be the principal eigenvalue of the matrix $\mathbf{H}^s$; Φ is strictly monotone decreasing and continuous in s.

THEOREM 2.6 (Mauldin and Williams, 1988, Theorem 3). *If K satisfies the OSC and if G is strongly connected, then the common Hausdorff dimension of the K_i, is given by $\dim_H(K_i) = \alpha = \Phi^{-1}(1)$ and $0 < \mathcal{H}^\alpha(K_i) < \infty$ for all $i \in \mathcal{S}$.*

If the graph G is not strongly connected we write $SC(G)$ for the set of strongly connected components of G. The Hausdorff dimension is then given by $\dim_H(K) = \max\{\alpha_F \colon F \in SC(G)\}$. The Hausdorff measure is finite if and only if $\{F \in SC(G) \colon \alpha = \alpha_F\}$ consists of pairwise incomparable elements.

2.4.3. Random recursive fractals

We can define a class of random recursive fractals in which there is a set of families of similitudes and each is applied at random. This will generate fractals which have only statistical self-similarity in that there are subsets of the fractal that are identically distributed copies of the whole set. These fractals were first considered in (Falconer, 1986; Mauldin and Williams, 1986; Graf, 1987) and have been generalised to include random recursive measures in (Hutchinson and Rüschendorff, 1999). Subsequently their local geometry and multifractal structure have been explored (Arbeiter and Patzschke, 1996; Hambly and Jones, 2003). An example based on two types of Sierpinski gasket is shown in Figure 3.

We describe our random fractals with an address space which is a random tree. Let $I_n = \bigcup_{k=0}^n \mathbb{N}^k$ and let $I = \bigcup_k I_k$ be the space of arbitrary length sequences. We will write $\mathbf{i}, \mathbf{j}$ for concatenation of sequences. For a point $\mathbf{i} \in I$ denote by $|\mathbf{i}|$ its length and, if $|\mathbf{i}| \geq n$, let $\mathbf{i}|_n$ be the sequence of length n such that $\mathbf{i} = \mathbf{i}|_n, \mathbf{k}$ for a sequence $\mathbf{k}$.

Fig. 3. A random recursive Sierpinski gasket.

Let $\phi^a = \{\phi_1^a, \ldots, \phi_M^a\}$ for $a \in A$ denote a set of contracting maps in $\mathbb{R}^n$ which each satisfy the open set condition. We let

$$U_{\mathbf{i}} = (M(\mathbf{i}), L_1(\mathbf{i}), \ldots, L_{M(\mathbf{i})}(\mathbf{i})), \quad \mathbf{i} \in I,$$

be independent and identically distributed $(\mathbb{N}, (0, 1)^{\mathbb{N}})$-valued random variables, where the L_i are the Lipschitz coefficients of the maps in the family, and consider the associated probability measure $\mathbb{P}$ as lying on the set A of possible families of maps.

A random tree T is a subset of I such that

(i) $\emptyset \in T$ the root of the tree;
(ii) $\mathbf{i} \in T$ implies $\mathbf{i}|_k \in T$ for all $k < |\mathbf{i}|$;
(iii) $\mathbf{i} \in T$ implies $\mathbf{i}1, \mathbf{i}2, \ldots, \mathbf{i}M(\mathbf{i}) \in T$, and $\mathbf{i}(M(\mathbf{i})+1), \mathbf{i}(M(\mathbf{i})+2), \ldots \notin T$.

Level n of the tree will be denoted $T_n = T \cap \mathbb{N}^n$. We will write $(\Omega, \mathbb{P})$ for the natural probability space associated with these trees. That is, a sample point $\omega \in \Omega$ will denote a random tree T with the associated random variables $\{U_{\mathbf{i}}: \mathbf{i} \in T\}$, and the probability measure, $\mathbb{P}$, is as defined previously. If we project this measure onto its first coordinate it is the offspring distribution for a Galton–Watson branching process.

Let $E = E_\emptyset$ and set $E_{\mathbf{i}}, \mathbf{i} \in T_n$, geometrically similar to E, to be

$$E_{\mathbf{i}} = \phi_{\mathbf{i}}(E) = \phi_{\mathbf{i}|_1}^{U_{\mathbf{i}|_1}}\left(\cdots\left(\phi_{\mathbf{i}|_n}^{U_{\mathbf{i}|_n}}(E)\right)\right).$$

We regard $\mathbf{i}$ as the address of the set $E_{\mathbf{i}}$ and identify the points in the fractal with the topological boundary of the tree. A random recursive fractal can then be defined by

$$F^\omega = \bigcap_{n=1}^\infty \bigcup_{\mathbf{i} \in T_n(\omega)} E_{\mathbf{i}}.$$

The Hausdorff dimension of the set F^ω can be found by applying the results of (Falconer, 1986; Mauldin and Williams, 1986; Graf, 1987) and is given by

$$d_f(F^\omega) = \inf\left\{\alpha: \mathbb{E}\left(\sum_{i=1}^{M_\emptyset} L_i^{-\alpha}(\emptyset)\right) = 1\right\}, \quad \text{for a.e. } \omega \in \Omega. \tag{2.5}$$

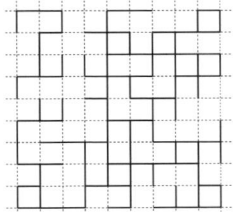

Fig. 4. A percolation configuration in part of $\mathbb{Z}^2$ for $p = 1/2$.

We note that the Hausdorff d_f-dimensional measure is 0 in general, the exact Hausdorff measure function was found in (Graf, Mauldin and Williams, 1988). For finer results on the local structure of these random sets see (Hambly and Jones, 2003).

We remark that this class and the previous can be put together and the resulting graph directed random recursive multifractal measures are considered in (Olsen, 1996).

2.5. Fractals in physics

We have been concerned so far with mathematical constructions of fractal sets. These may serve as models for natural structures and provide a firm mathematical foundation for studying the dynamic properties of fractals. However fractals also arise naturally in physical systems when they approach a phase transition. The rigorous mathematical description of this behaviour presents a significant challenge and so we will content ourselves with some discussion.

A basic model of statistical physics is the percolation model. For bond percolation on the n-dimensional integer lattice $\mathbb{Z}^n$ each edge of the lattice is assigned one of two possible values, either a 0 denoting that it is closed or a 1 denoting that it is open. If the edges are assigned their values independently at random with probability p, we can consider the structure of the random set formed by the open edges (see Figure 4 for an example of such a set). This provides a natural model for a random medium and hence we would like to know when 'percolation' occurs, that is, there is an infinite open connected cluster in the lattice, and fluid injected at the origin could move an arbitrary distance through the medium. For an account of the mathematical theory of percolation see (Grimmett, 1999).

This is a classical model which exhibits phase transition, in that there exists a critical probability p_c such that for $p < p_c$ there are only finite clusters, but for $p > p_c$ there is a positive probability that the origin is connected to infinity. For the two-dimensional square lattice $p_c = 1/2$ but in higher dimensions the exact value is not known. One mathematical issue is that, at the critical probability, $p = p_c$, the probability of the existence of an infinite cluster is 0 in 2 dimensions, and conjectured to be so in 3 dimensions. This has led to various attempts to rigorously define the 'incipient infinite cluster', an infinite cluster of open edges at the critical probability. In (Kesten, 1986a) a critical percolation cluster is obtained by conditioning on the event that in the supercritical regime, the infinite cluster contains the origin and then letting $p \downarrow p_c$. This should provide a natural discrete fractal but as yet there is no mathematical proof for the value of the fractal dimension which is conjectured from physical arguments to be 91/48 in 2 dimensions. However recent work has established the dimension of the scaling limit of the outer hull of two-dimensional percolation clusters to be 4/3 (Werner, 2001).

Although the mathematics is not yet rigorous much is known about this model. It is clear that as the probability tends towards the critical probability the open connected clusters have fractal like properties. It is conjectured that many of the natural quantities of interest, such as the correlation length (the typical length between two points in a finite cluster) will behave like $|p - p_c|^\xi$ as $p \to p_c$ for some exponent ξ. This has given rise to a multitude of exponents which describe the asymptotic scaling of features of the model (Ben-Avraham and Havlin, 2000). A deep and still as yet unsolved mathematical question is to demonstrate the existence of these quantities. Once this is assumed there are various scaling relations that can be deduced which relate the exponents to each other. In the two dimensional setting physics techniques allow exact values to be calculated under an assumption of conformal invariance. Recent mathematical progress providing rigorous treatments for some models can be found in (Werner, 2001).

2.6. Graphs of functions

Graphs of functions can display self-similarity and are another way of generating fractals. In order to capture the fractal structure we introduce the Hölder exponents for a function which describe the irregularity at each point in the graph. Let $f : [0, 1] \to \mathbb{R}$. The local Hölder exponent of f at a point x is

$$H_x^{(l)}(f) = \sup\left\{\alpha \leqslant 1 : \lim_{r \to 0} \sup_{t,s \in B(x,r)} \frac{|f(t) - f(s)|}{|t - s|^\alpha} < \infty\right\}.$$

There is also a notion of pointwise Hölder exponent for f,

$$H_x^{(p)}(f) = \sup\left\{\alpha \leqslant 1 : \lim_{r \to 0} \sup_{t,s \in B(x,r)} \frac{|f(t) - f(s)|}{r^\alpha} < \infty\right\}.$$

Note that the exponent describes the order of differentiability and will be 1 for differentiable functions. We always have $H_x^{(l)}(f) \leqslant H_x^{(p)}(f)$. Extensions to higher dimensions in space and time are considered in many places. We note that the Hausdorff dimension of the graph of the function is bounded above by $2 - \inf_{x \in [0,1]} H_x^{(p)}(f)$.

The local Hölder exponent is similar to the local dimension of a measure and, in the same way, there is a multifractal spectrum for these exponents. A technique to approach the computation of this spectrum is to use wavelets. Wavelets provide an analogue of Fourier series where the basis functions are compactly supported. They are obtained from a suitable mother wavelet by scaling and translation and provide a basis of compactly supported functions for $L^2(\mathbb{R})$. The property of compact support allows the irregularities in a function to be exactly localised in the frequency domain. This suggests that if our function has multifractal structure in its Hölder exponents, these should be revealed in the behaviour of the wavelet transform. We begin by recalling the definition of the wavelet transform. Let g be a wavelet (see Jaffard, Meyer and Ryan, 2001 for more on wavelets) and define

$$W(\lambda, x) = \frac{1}{\lambda} \int_{-\infty}^{\infty} f(y) \bar{g}\left(\frac{y-x}{\lambda}\right) dy.$$

If the function f scales locally with a particular exponent α, then this will be seen in the local decay of this transform with λ. The fundamental idea is that if the local dimension of a point is a, then it will be reflected in the wavelet transform $W(\lambda, r) = O(\lambda^a)$. Naturally this is not true in complete generality but under mild regularity conditions on the function.

The spectrum can then be obtained using a Legendre transform approach and, under further conditions, is given (naively) by

$$f(\alpha) = \inf_{-\infty < q < \infty} \{q\alpha - \tau(q) + 1\},$$

where

$$\tau(q) = \lim_{\lambda \to 0} \frac{\log \int_{-\infty}^{\infty} |W(\lambda, r)|^q \, dr}{\log \lambda}.$$

Note that if $W(\lambda, r) = 0$ we will not be able to obtain estimates for $q < 0$ and hence a more practical way to estimate this is to use the wavelet-maxima method. Let

$$\theta(q) = \lim_{\lambda \to 0} \frac{\log \sum_l \sup_{r=l(\lambda)} |W(\lambda, r)|^q}{\log \lambda},$$

where l is a line of local maxima of the wavelet transform in the neighbourhood of λ, then $f(\alpha) = \inf_q \{q\alpha - \theta(q)\}$. For some discussion of when the spectrum can be obtained this way see (Jaffard, 1997a, 1997b). If the function is self-similar, then this method yields the correct spectrum and there is a formula involving the scalings of the separate pieces.

2.7. Images and the collage theorem

Fractal image compression has received quite a bit of attention as it appears to provide a technique for greatly reducing the amount of information that is required in order to store an image. The idea is to look at the image and determine self-similar pieces, that is subsets of the image which can be mapped via similitudes to other parts. Then, by encoding only the transformations which define the similarities, it is possible to store the image very efficiently and reconstruct it rapidly. The difficulty is naturally how to find the transformations for a real image.

The collage theorem tells us that such an approach will be successful as any compact set in $\mathbb{R}^n$ can be approximated arbitrarily closely in the Hausdorff metric by the invariant set determined by a family of contraction maps.

THEOREM 2.7. *Let $\Phi = \{\phi_i; i = 1, \ldots, m\}$ be a family of contractions on $\mathbb{R}^n$ with Lipschitz constants bounded above by $L < 1$. Let $E \subset \mathbb{R}^n$ be a non-empty compact set and let F be the fixed point for the map Φ, then*

$$d_H(E, F) \leqslant \frac{1}{1-L} d_H\left(E, \bigcup_{i=1}^{m} \phi_i(E)\right).$$

Also given $\delta > 0$ we can find Φ such that $d_H(E, F) < \delta$.

The problem, of course, is that this theorem does not indicate how many contractions may be required and a significant problem is to choose the contractions in some minimal way, yet keep the approximation close. This is a very difficult inverse problem and some discussion of the various approaches can be found in (Fisher, 1995). The usual approach is to consider a class of possible maps and to try and find the best way of using them; a computationally intensive search problem. There are some images that will suit a fractal compression approach, for instance, those which already have a greater amount of self-similarity that the compression technique can exploit.

2.8. Simulation

The constructions in Section 2.4 lead easily to techniques for simulation of geometric fractals. A simple algorithm, often called the chaos game, for the generation of self-affine sets is to take some initial point and then apply the maps according to a probability distribution on the basic alphabet. Let $x_0 \in \mathbb{R}^n$ and take $\{\xi_i\}_{i=1}^{N}$, a sequence of independent and identically distributed random variables on $\mathcal{S}$. Now set $x_i = \phi_{\xi_i}(x_{i-1})$ for $i = 1, 2, \ldots, N$ and plot the points. The trajectory of the point, as it moves around from cell to cell, will gradually fill out the fractal set. The probability distribution on $\mathcal{S}$ will give a picture of the corresponding Bernoulli mass distribution and can be used to enhance the picture. A great many iterations are required to get a good picture of the fractal (see Morato and Siri, 2001 for an optimal algorithm of this type).

The same approach can be applied to graph directed fractals using cellular automata. By also incorporating the random recursive fractals a wide range of sets can be generated and the resulting pictures are demonstrated and discussed in (Barnsley, 1993; Peitgen, Jürgens and Saupe, 1992).

3. Fractals and stochastic processes

Now that we have completed a survey of the geometry of fractal sets we come to the connection with stochastic processes. These provide another approach to generating fractals as the paths of many continuous time stochastic processes are fractal. The canonical noise process is Brownian motion but recently there has been interest in processes which display long range dependence, or self-similarity and have a range of dimensional properties. We will only give a brief discussion of some classes of processes which have different fractal structure.

3.1. Brownian motion

Brownian motion is one of the fundamental processes of probability and has many deep connections with analysis. It originally arose from modelling applications – it was Brown who first observed the irregular movement of pollen grains on the surface of water when viewed under a microscope.

Let $\{B_t;\ t \in \mathbb{R}_+\}$ denote an n-dimensional Brownian motion. That is

(1) The increment $B_{t+s} - B_s$ is independent of B_s.
(2) The increment $B_{t+s} - B_s$ has a multivariate normal distribution with mean 0, covariance matrix tI.
(3) The path $t \to B_t$ is continuous with $B_0 = 0$.

The existence of a process with these properties on a probability space $(\Omega, \mathcal{F}, \mathbb{P})$ is non-trivial and was first demonstrated by Wiener who showed the existence of the probability measure $\mathbb{P}$ on the space of continuous functions. One reason for its great usefulness comes from Donsker's theorem, that rescaling finite variance random walks will lead to Brownian motion.

We will now consider the one dimensional case. The fractal properties are immediately apparent as, although the paths of Brownian motion are continuous, they are nowhere differentiable. The Hölder continuity properties are given by Lévy's modulus of continuity

$$\lim_{h \to 0} \sup_{\substack{0 \leqslant s < t \leqslant 1 \\ t-s<h}} \frac{|B_t - B_s|}{\sqrt{|t-s|\log|t-s|}} = 1, \quad \mathbb{P}\text{-a.s.}$$

and the extreme behaviour of the path is seen in the law of the iterated logarithm,

$$\limsup_{t \to \infty} \frac{B_t}{\sqrt{2t \log \log t}} = 1, \quad \mathbb{P}\text{-a.s.}$$

The modulus of continuity result shows that both Hölder exponents are $1/2$ for almost all time points t for Brownian motion.

There are a number of interesting fractal sets associated with the Brownian path, for a review see (Taylor, 1986). We just mention a couple. Brownian motion in one dimension is recurrent, returning to zero infinitely often and accumulating local time, but not real time at zero. The dimension of the zero set is $1/2$ almost surely. For a recent fine multifractal analysis of the occupation measure of the Brownian path see (Dembo et al., 2000a, 2000b). The dimension of the graph $\{(t, B_t): t \in [0, 1]\}$ can be shown to be $3/2$ almost surely, which shows that despite the ubiquitous nature of Brownian motion as a noise function it has quite restricted dimensional properties.

Brownian motion can be extended in several directions to provide stochastic processes with paths of variable dimension. By considering processes with stationary and independent increments we arrive at Lévy processes. By considering Gaussian processes with different correlation structure we can obtain fractional Brownian motions. We will discuss these two cases in the one-dimensional setting.

3.2. Lévy processes

A Lévy process is a stochastic process which has stationary and independent increments. Thus Brownian motion is a special case, where the process is continuous and the increments normally distributed. A Lévy process can be decomposed into three parts, a Brownian motion with drift, a compound Poisson process with jumps of size greater than one and a pure jump martingale with jumps of size less than 1. The Lévy process must have an infinitely divisible law for its increments and this allows us to write its characteristic function, using the Lévy–Khintchine formula, as

$$E\left(e^{i\theta X_t}\right) = e^{-t\Psi(\theta)},$$

where

$$\Psi(\theta) = -i\mu\theta + \frac{1}{2}\sigma^2\theta^2 + \int_{-\infty}^{\infty}\left(1 - e^{i\theta x} + i\theta x I_{\{|x|<1\}}\right)\Pi(dx).$$

In this formula, μ, σ are constants, the mean and variance of the diffusive part and the jumps are specified by a jump measure Π, which requires an integrability condition, that $\int_{\mathbb{R}}(1 \wedge |x|^2)\Pi(dx) < \infty$. For an extended discussion of Lévy processes see (Bertoin, 1996).

As the Lévy process can have a diffusive component and it can also be a compound Poisson process, the paths are less rough than Brownian motion and the dimension of the graph of the path lies between 1 and 3/2. The class of stable processes provides a set of symmetric self-similar Lévy processes with variable dimension. They arise from taking the limits of normalised sums of random variables in which the random variables have infinite variance and their characteristic exponent is given by $\Psi(\theta) = c|\theta|^\alpha$ for $0 < \alpha \leqslant 2$. It is easy to see from this the self-similarity property satisfied by the stable process, $X_t = aX_{t/a^\alpha}$. The dimensional properties are that its zero set is of dimension $\max\{1 - 1/\alpha, 0\}$ and its graph has dimension $\max\{1, 2 - 1/\alpha\}$. Note that for $\alpha < 1$, the stable process is a pure jump process.

3.3. Fractional Brownian motion

Brownian motion is an excellent model for noise in randomly evolving systems. However not all noise has independent increments. It is clear that there are many situations, such as in network data traffic, where long range dependence occurs in practice and understanding models which display features of this kind is becoming more important.

A widely used model in this area is fractional Brownian motion, which is a Gaussian process but it is not Markov in general. This feature means that it can display long range dependence and is potentially useful in situations where the time behaviour depends on the past history of the process. The process can be defined as a Gaussian process with mean zero and covariance function

$$c(s,t) = \frac{1}{2}\left(|t-s|^{2H} - |t|^{2H} - |s|^{2H}\right).$$

The parameter H is called the Hurst parameter and varies between 0 and 1. It is the Hölder exponent for almost all time points. For $H \in (1/2, 1)$, the path is smoother than Brownian motion and exhibits long range dependence. When $H \in (0, 1/2)$ it is rougher than Brownian motion. The case where $H = 1/2$ is Brownian motion itself. For fractional Brownian motion there is a relationship between the covariance index and the dimensionality of the path in that $\dim_H(\{(t, X_t); \ t \in [0, 1]\}) = 2 - H$. Conditions for this to hold for more general Gaussian processes with correlation controlled by H can be found in (Hall and Roy, 1994).

In order to extend the flexibility of the dimensional exponents available, a recent paper (Ayache and Lévy Vehel, 1999) sets up multifractional Brownian motion, in which the Hölder exponents can vary over time and indeed can even be an irregular function.

It has proved important in practice to be able to simulate realizations of long paths of fractional Brownian motion. For a range of papers on issues relating to statistics and simulation of both stable processes and fBm (Adler et al., 1998). An approach to simulating fractional Brownian motion can be found in (Paxson, 1997). One area of modelling where this process has been used is in telecommunications and network traffic where there are self-similar and long range dependent features. For modelling the price processes of financial assets for the purposes of derivative pricing, though, due to the fact that the associated financial market admits arbitrage, it does not appear to be a good model without some modification.

A random field can be generated using stochastic processes with more than one time parameter. These surfaces can have complicated geometry. Examples include the Brownian sheet or Lévy's multiparameter Brownian motion and we refer the reader to (Adler, 1981) and the large literature which has developed subsequently for more information.

4. Dynamic fractal models

In this section we consider partial differential equations on fractals and show how the Laplace operator can be constructed on classes of geometric fractals as introduced in Section 2.4. This allows us to define diffusion in fractal media and we will show how this theory could be used in a potential modelling application.

The first question is how do we define a partial differential equation, such as the diffusion equation, which requires second-order differentiation in space, on a set which has no smooth structure? In this section we will develop some of the mathematical ideas required to discuss this problem and assess the robustness of the solution.

There is a substantial physics literature which tackles similar questions, primarily in the context of the percolation model. In that setting some of the mathematical issues that we discuss are not regarded as relevant as any physical system will have a lower cut off. The approach is to model the world as a discrete lattice and study diffusion through the large scale behaviour of random walks to deduce global effects when the microscopic lattice structure is fractal. In the same way that differential equations provide continuous versions of difference equations which may be easier to solve via calculus, the differential equations on true fractals will provide good approximations for use in studying these physical models.

4.1. The percolation model

The original ideas concerning modelling of the transport properties of disordered systems were due to de Gennes. He proposed the 'ant in the labyrinth' (random walks on percolation clusters) as a technique for understanding diffusion in disordered media. This motivated a long and detailed study of the properties of random walks on percolation clusters near criticality. It is now a basic model for the study of transport problems in disordered systems in the physics literature and is discussed at length in (Ben-Avraham and Havlin, 2000). The rigorous analysis of the dynamic behaviour of the percolation model is difficult. For a mathematical result in this direction (Kesten, 1986b) was able to demonstrate sub-diffusivity; a random walk on the incipient infinite percolation cluster in two dimensions does not have mean square displacement which scales linearly with time. In two dimensions the values of the dynamic critical exponents can be computed using physical arguments and conformal invariance, in a similar way to the geometric exponents of percolation clusters. Our mathematical results are broadly consistent with these results and hence, for the purpose of modelling, the physical arguments give weight to the use of this approach in considering disordered systems with self-similar features.

The key ideas that arose from the physics are that there are two dynamic exponents which describe the behaviour of the fractal. The first is the walk dimension, d_w, which determines the time to distance scaling for the random walker on the fractal. The second is the spectral or fracton dimension, d_s, which describes the asymptotic scaling in the eigenvalues for the associated Laplace operator. One further exponent that we mention here is the resistance scaling exponent, d_r, which determines the scaling of resistance to distance. These are related by two key formulae, the first is called the Einstein relation,

$$d_w = d_f + d_r, \tag{4.1}$$

$$\frac{d_s}{2} = \frac{d_f}{d_w}. \tag{4.2}$$

We will give rigorous definitions of these dimensions and a discussion of these relations later.

4.2. Connected geometric fractals

We now turn our attention to the geometric fractals constructed in Section 2.4 and develop some mathematical tools for studying diffusion problems in such media. We will only consider self-similar sets here, the extension to graph directed sets is straightforward though that to general random recursive fractals is more difficult. Firstly, in order to consider partial differential equations on fractals, we require a Laplace operator. This is a local operator and therefore it is essential that our set is connected.

The class of connected fractals can be divided into two different varieties. The finitely ramified case, where any connected subset of the fractal can be disconnected from the rest of the set by the removal of a finite number of points. The basic example being the Sierpinski gasket. The other variety are infinitely ramified fractals in which the dimension of the boundary of the self-similar components is greater than 0. The canonical example of an infinitely ramified fractal is the Sierpinski carpet, see Figure 1.

We will mainly discuss finitely ramified fractals. Recall that a self-similar set F is defined by a family of contraction maps $\{\phi_i : i \in S\}$ and we denote by I_n the address space of length n sequences. For $\sigma \in I_n$, we write $\phi_\sigma = \phi_{w_1,\ldots,w_n} = \phi_{w_1} \circ \cdots \circ \phi_{w_n}$. For $\sigma \in I_n$, we define $K_\sigma = \phi_\sigma(F)$, and call such sets n-cells. The set F is the 0-cell. Let F_0 be the smallest set such that for all $n \in \mathbb{N}$, when $a, b \in I_n$, and $a \neq b$, $K_a \cap K_b = \phi_a(F_0) \cap \phi_b(F_0)$. This is the set of basic ramification points and is the effective boundary of the fractal. It is finitely ramified if the cardinality of this set, $|F_0|$, is finite.

By using the ramification points we can define a sequence of graph approximations to the fractal. The key feature is that any diffusing particle on the fractal must pass through these points as it moves about the fractal. For $n \geq 1$, we define $F_n = \bigcup_{\sigma \in I_n} \phi_\sigma(F_0)$. Note that $\{F_n\}_{n \in \mathbb{N}}$ is an increasing sequence. Let $F_n = \bigcup_{i \in S} \phi_i(F_{n-1})$ and $F_\infty = \bigcup_{n=0}^\infty F_n$. The fractal F can then be recovered as the closure of this set of ramification points.

In order to prove existence theorems it is usual to assume some sort of symmetry in the fractal. A very strong assumption in this direction is the nesting axiom of Lindstrom (1990). This was weakened slightly in (Fitzsimmons, Hambly and Kumagai, 1994) to affine nested fractals but is still highly restrictive. The definition of an affine nested fractal is as follows.

DEFINITION 4.1. The set F is an affine nested fractal if $\{\phi_1, \ldots, \phi_m\}$ satisfy the OSC and:

(A1) (*Connectivity*) For any 1-cells K and K', there is a sequence $\{K_i : i = 0, \ldots, n\}$ of 1-cells such that $K_0 = K$, $K_n = K'$ and $K_{i-1} \cap K_i \neq \emptyset$, $i = 1, \ldots, n$.
(A2) (*Symmetry*) If $x, y \in F_0$, then reflection in the hyperplane $H_{xy} = \{z : |z - x| = |z - y|\}$ maps F_n to itself.
(A3) (*Nesting*) If $\{i_1, \ldots, i_n\}, \{j_1, \ldots, j_n\}$ are distinct sequences, then

$$\phi_{i_1,\ldots,i_n}(F) \cap \phi_{j_1,\ldots,j_n}(F) = \phi_{i_1,\ldots,i_n}(F_0) \cap \phi_{j_1,\ldots,j_n}(F_0).$$

The symmetry assumption is the most restrictive and the sets that fall into this category are those based on polygons in $\mathbb{R}^2$ or n-dimensional tetrahedra. A general class of exactly self-similar finitely ramified fractals, containing these nested fractals, are the p.c.f. self-similar sets as defined by Kigami (1993, 2001). This framework defines the fractals as self sufficient metric spaces, avoiding the embedding into some Euclidean space and sets up Laplace operators on fractals abstractly. However there are no general existence theorems in this setting.

We can extend our fractals from compact sets to unbounded ones by inverting one of the contraction maps used for constructing the fractal and then iterating it. It is straightforward to extend the results we obtain to the unbounded fractal setting.

4.3. The Laplace operator on a fractal

We now turn to a concrete mathematical formulation of the question that will interest us in this section. Our aim is to consider the diffusion equation on F. That is to find a function $u(x,t): F \times \mathbb{R}_+ \to \mathbb{R}$ which satisfies the following Cauchy problem

$$\frac{\partial u}{\partial t} = \Delta_F u,$$
$$u(x,0) = u_0(x). \tag{4.3}$$

The first question is how to make sense of Δ_F, the Laplace operator on F. As the underlying set has no smooth structure it is not a priori clear how to define such a mathematical object. However there is now quite a substantial mathematical literature on the existence, uniqueness and properties of the Laplace operator on fractals, see (Barlow, 1998; Kigami, 2001).

At this stage much of the mathematical work has focused on quite restricted classes of fractals with the most general form of this construction outlined in (Kigami, 1995). Here we will discuss this construction problem for self-similar sets and only briefly mention the key questions of existence and uniqueness. Existence for us will mean that there is a corresponding non-degenerate diffusion process which can move throughout the fractal. The most general criteria available for existence and uniqueness of diffusions on finitely ramified fractals can be found in (Sabot, 1997).

The first approach to constructing a Laplace operator on a fractal was to define it as the generator of a diffusion process. The canonical diffusion is Brownian motion and thus a notion of Brownian motion on a fractal is required. This was first achieved mathematically in three independent papers by Kusuoka, Goldstein and Barlow and Perkins (see Barlow, 1998). It is very natural to approach this problem by taking a sequence of random walks on the lattice approximations to the fractal, F_n and rescaling them to obtain a process on the fractal itself. We take the approach of (Kigami, 2001) using electrical networks but we will always think of our fractals as embedded in some Euclidean space.

The connection between random walks and electrical networks as detailed in (Doyle and Snell, 1984) shows that an equivalent way to think of the lattice approximation is as a resistor network. We define a *resistor network* to be a finite graph $G = (V, C)$, containing a set of vertices V and a conductivity matrix C which assigns to each pair of vertices (i, j) a conductivity c_{ij}. The corresponding Markov chain is reversible and moves as follows. It jumps out of state i at rate $c_i = \sum_j c_{ij}$ and moves to state j with probability $p_{ij} = c_{ij}/c_i$. We define the diagonal term $c_{ii} = -c_i$, so that C is the generator of a Markov chain with reversing measure $\mu_i = c_i / \sum_{i \in V} c_i$. Let $\ell(V) = \{f | V \to \mathbb{R}\}$, then for a function $f \in \ell(V)$, define the discrete Laplace operator as

$$\Delta_C f(x) = \sum_{(x,y) \in E} (f(y) - f(x)) \frac{c_{xy}}{\mu_x}.$$

A Dirichlet form, a symmetric bilinear quadratic form, can be constructed on G. For the discrete Laplacian given above we define the Dirichlet form as

$$\mathcal{E}_C(f, g) = -\sum_{x \in V} \Delta_C f(x) g(x) \mu(x)$$
$$= -f^T C g$$
$$= \frac{1}{2} \sum_{x,y} (f(x) - f(y))(g(x) - g(y)) c_{xy}, \quad f, g \in \ell(V). \qquad (4.4)$$

There is a one-to-one correspondence between Dirichlet forms, conductance matrices and Markov chains on finite graphs. With some technical conditions, this can be extended to continuous state spaces, to show the equivalence of regular Dirichlet forms to transition semigroups, their generators and corresponding strong Markov processes, (Fukushima, Oshima and Takeda, 1994).

A Dirichlet form $\mathcal{E}_C$ is irreducible if $\mathcal{E}_C(u, u) \geq 0$ for all $u \in \ell(V)$ with equality iff u is constant. We will be interested in producing irreducible Dirichlet forms for our fractals as these correspond to non-degenerate processes. We note that this Dirichlet form does not contain information about the measure; it only enters when we write the form using an inner product and hence, in our setting, any measure equivalent to μ can be used.

If (V_1, C_1) and (V_2, C_2) are two resistor networks, $V_1 \subset V_2$, and

$$\mathcal{E}_{C_1}(u, u) = \min\{\mathcal{E}_{C_2}(v, v) : v \in \ell(V_2), \, v|_{V_1} = u\} \qquad (4.5)$$

we say they are *compatible*, and write $(V_1, C_1) \leq (V_2, C_2)$. The network (V_1, C_1) is called the trace of (V_2, C_2) on V_1. The corresponding Markov chain on V_1 is obtained from the chain on V_2, by watching it in the local time on V_1.

In order to construct a Laplace operator on the fractal, as in (Kigami, 1993), we need to find a compatible sequence of resistor networks on lattice approximations to the fractal. For a self-similar set defined by $\{\phi_1, \ldots, \phi_N\}$, we need a discrete Laplace operator which is the fixed point of a renormalization map. We define the resistivity of a cell by a choice of a positive finite weight r_i, $i \in V$. Let D denote a conductivity matrix and consider (F_0, D) as a resistor network. Letting $\mathcal{E}_0$ be the associated Dirichlet form, we define the Dirichlet form on F_1 as

$$\mathcal{E}_1(u, u) = \sum_{i=1}^{N} r_i^{-1} \cdot \mathcal{E}_0(u \circ \phi_i, u \circ \phi_i). \qquad (4.6)$$

The weights r_i allow us to build in a self-similar conductivity structure to each component of the set, giving a 'fractured' resistance to the fractal medium. Typically, for a homogeneous structure, we take $r_i = 1$ for all i.

We now go on to define a simple renormalization map. Firstly we note that the set of all Dirichlet forms on the graph F_0 is a cone, $\mathbb{D}$; in fact it is isomorphic to the positive orthant in $\mathbb{R}^{\frac{1}{2}|F_0|(|F_0|-1)}$. The renormalization map $\Lambda : \mathbb{D} \to \mathbb{D}$ is parameterised by r and is defined by taking the trace (4.5) of $\mathcal{E}_1$ on F_0. Equivalently we can regard Λ as acting on the generators of the chains, D. If we write the generator for the chain on F_1 in terms of transitions from F_0 and $F_1 \setminus F_0$,

$$H = \begin{bmatrix} T & J^T \\ J & X \end{bmatrix},$$

then $\Lambda(D) = T - J^T X^{-1} J$.

In order to prove existence of the Laplace operator on a self-similar set we will assume that the map Λ, with a given fixed set of r's, has an eigenform (non-degenerate fixed-point) D, i.e., that there exists a $\lambda > 0$ such that $\Lambda(D) = \lambda^{-1} D$ and D corresponds to an irreducible Dirichlet form on F_0.

The classical example is the Sierpinski gasket where, if we take $r_i = 1$, $i = 1, 2, 3$, then $\lambda = 5/3$ with $D(i, j) = 1$ for $i \neq j$, provides a suitable fixed point. For nested fractals existence was proved in (Lindstrøm, 1990) and uniqueness in (Sabot, 1997). It has also been shown that even if the initial conductivity matrix is not the fixed point the iterates of the renormalization map will typically converge to the fixed point (Metz, 1996). This even occurs when the initial conditions are random (Kumagai and Kusuoka, 1996).

From now on we will assume that we have a fixed D and r and define the conductivity of cell i by $\rho_i = \lambda/r_i$. We will let $\mathcal{E}_0$ be a Dirichlet form on F_0 corresponding to this fixed-point D. For $\sigma \in I_n$, let $\rho_\sigma = \prod_{i=1}^n \rho_{\sigma_i}$. For $n \in \mathbb{N}$, define

$$\mathcal{E}_n(f, f) = \sum_{\sigma \in I_n} \rho_\sigma \mathcal{E}_0(f|_{\phi_\sigma(F_0)} \circ \phi_\sigma, f|_{\phi_\sigma(F_0)} \circ \phi_\sigma),$$

and let H_n be the matrices defined by

$$\mathcal{E}_n(f, g) = -f^T H_n g.$$

As is easily seen, $(F_n, H_n) \leqslant (F_m, H_m)$ whenever $n \leqslant m$. This ensures that the sequence of Dirichlet forms is monotone increasing and allows us to construct the limit form.

We can define the limiting Dirichlet form on the fractal, $(\mathcal{E}, \mathcal{F})$ on $L^2(F, \mu)$ for any Borel measure μ, by

$$\mathcal{E}(u, u) = \lim_{n \to \infty} \mathcal{E}_n(u, u), \quad \forall u \in \mathcal{F},$$

$$\mathcal{F} = \left\{ u \in L^2(F, \mu) : \sup_n \mathcal{E}_n(u, u) < \infty \right\}.$$

The key mathematical result ensures that all the objects which we hope exist do.

THEOREM 4.2. (a) *The pair $(\mathcal{E}, \mathcal{F})$ is a regular, local Dirichlet form on $L^2(F, \mu)$.*

(b) *Associated with the Dirichlet form $(\mathcal{E}, \mathcal{F})$ on $L^2(F, \mu)$ there is a probability space $(\Omega, \mathcal{F}, \mathbb{P})$ supporting a non-degenerate diffusion process W on F which is reversible with respect to the measure μ.*

(c) *The Dirichlet Laplacian is recovered as* $\mathcal{E}(f, g) = (-\Delta_F f, g)_\mu$ *for* $g \in \mathcal{F}, g = 0$ *on* ∂F.

PROOF. The proof follows as in (Kigami, 1993, 2001; Barlow, 1998). □

There is a particular natural choice of measure, dependent on the r_i and even though we have not shown uniqueness for our process we will call this diffusion Brownian motion and the associated generator the Laplacian. For the Sierpinski gasket natural measure is the $\log 3/\log 2$-dimensional Hausdorff measure.

We can define the effective resistance between two points $x, y \in F$, using the trace of the Dirichlet form

$$r(x, y) = \left(\inf\{\mathcal{E}(u, u): u \in \mathcal{F}, u(x) = 1, u(y) = 0\}\right)^{-1}.$$

If the harmonic structure is regular, in that $\rho_i > 1$ for $i = 1, \ldots, N$, the effective resistance determines a metric on the fractal. In this regular case let $\mu_i = \rho_i^{-S}$, where S satisfies $\sum_{i=1}^{N} \rho^{-S} = 1$, then the Hausdorff dimension of the fractal F, in the effective resistance metric, is given by $\dim_{H,R}(K) = S$.

For the Sierpinski carpet, the rigorous approach is given in (Barlow and Bass, 1999). In this setting the fractal is approximated by a sequence of domains in $\mathbb{R}^n$ and the associated sequence of reflecting Brownian motions is shown to converge weakly. For an approach via graph approximation see (Kusuoka and Zhou, 1992). The key to the construction is a Harnack inequality which ensures the existence of a suitable time scaling for the approximating processes. As yet rigorous work for higher-dimensional carpets requires spatial symmetry in order to prove this Harnack inequality.

4.4. The dynamic exponents

The key dynamic exponents are the spectral dimension d_s, which is the effective geometric dimension, and the walk dimension, the effective order of the Laplace operator. We briefly give their definitions and explain the relationships between them.

To define the spectral dimension we consider the Dirichlet eigenvalues associated with our Laplace operator, i.e., the $\lambda \in \mathbb{R}$ such that there exists a $u \in \mathcal{D}(\Delta_F)$, satisfying

$$\Delta_F u = -\lambda u, \quad \text{in } F,$$
$$u(x) = 0, \quad x \in \partial F. \tag{4.7}$$

The boundary of the fractal $\partial F = F_0$, is the finite set of basic ramification points. It is also possible to define Neumann eigenvalues, which requires a notion of normal derivative at the boundary, see (Kigami, 2001). For symmetric fractals such as the Sierpinski gasket there are eigenfunctions that are strictly localised, in that they are zero outside some compact subset of the fractal. This is quite different from $\mathbb{R}^n$ and is similar to the behaviour of disordered media.

As the eigenvalues form an increasing sequence whose only accumulation point is infinity, we can define the eigenvalue counting function as

$$N(\lambda) = \left|\{\lambda_i: \lambda_i \leq \lambda, \lambda_i \text{ solves } (4.7)\}\right|.$$

The spectral dimension d_s is defined from the asymptotics of $N(\lambda)$ as

$$d_s := 2 \lim_{\lambda \to \infty} \frac{\log N(\lambda)}{\log \lambda}.$$

More precise results concerning the asymptotics of the eigenvalue counting function can be found in (Kigami and Lapidus, 1993). For finitely ramified fractals with regular harmonic structure, the spectral dimension is always less than two. It is possible for infinitely ramified fractals, such as Sierpinski carpets in dimensions greater than 2, to have spectral dimension greater than 2. If a fractal has spectral dimension less than 2, the process has similar properties to a process living in a space of dimension less than 2. For instance, on an unbounded version of the fractal, the diffusion process is point recurrent.

For modelling purposes we can take as an assumption about diffusions on fractals that there is an exponent d_w which describes the scaling of the time to distance on the fractal. The physicist would define this via the mean square displacement of a random walker on the fractal. That is, if we consider an unbounded fractal lattice and look at the large scale behaviour of a diffusing particle started from x with position X_t at time t, we expect that

$$E^x |X_t - x|^2 \asymp t^{2/d_w}, \quad \text{as } t \to \infty.$$

Thus $d_w = 2$ for diffusion in $\mathbb{R}^n$. However in an extended fractal there is anomalous diffusion, due to the presence of obstacles at all scales. The diffusion is said to be sub-diffusive in that it moves more slowly than diffusion in Euclidean space and the exponent $d_w > 2$.

The relationships between the exponents are as follows. As our Laplace operator has compact resolvent we can write down a spectral decomposition for the transition kernel,

$$p_t(x, y) = \sum_{i=1}^{\infty} e^{-\lambda_i t} \phi_i(x) \phi_i(y),$$

where $\{\phi_i\}$ is an orthonormal set of eigenfunctions in $L^2(F, \mu)$ corresponding to the eigenvalues $\{\lambda_i\}$. It is easy to see from this that

$$\int_0^{\infty} e^{-st} N(ds) = \int_F p_t(x, x) \mu(dx). \tag{4.8}$$

By using the definition of the eigenvalue counting function and a Tauberian theorem for the left-hand side of (4.8), and the homogeneity of F for the right-hand side, we have that for $x \in F$, $p_t(x, x) \asymp t^{-d_s/2}$ as $t \to 0$.

Alternatively, as the diffusion speed is governed by d_w, we expect that the diffusion is quite likely to be in $B(x, ct^{1/d_w})$, the ball of radius ct^{1/d_w} about x, at time t for some c, and hence there is a constant c' such that

$$c' \asymp P^x\left(X_t \in B(x, ct^{1/d_w})\right) = \int_{B(x, ct^{1/d_w})} p_t(x, y) \mu(dy).$$

By homogeneity there is a further constant c'' such that $c'' p_t(x,x) \leqslant p_t(x,y) \leqslant p_t(x,x)$ for $y \in B(x, ct^{1/d_w})$ and hence

$$p_t(x,x) \asymp \mu\big(B(x, ct^{1/d_w})\big)^{-1} \asymp (t^{1/d_w})^{-d_f}, \quad t \to 0. \tag{4.9}$$

Thus, combing with the computation after (4.8), we have the relationship (4.2) for spatially homogeneous fractals.

Finally the Einstein relation (4.1) follows from its discrete counter part on graphs, first proved in (Chandra et al., 1996). Let $T_x = \inf\{t: X_t = x\}$ denote the hitting time of x, then

$$E^x T_y + E^y T_x = R_C(x,y) M(G),$$

where $M(G)$ is the total mass of all the edges in the graph G and $R_C(x,y)$ is the effective resistance from x to y. By translating this result into the fractal setting we have (4.1) provided that there is homogeneity in the set so that $E^x T_y$ is of the same order as $E^y T_x$.

For the finitely ramified fractals with spectral dimension less than 2, we can combine the dynamic and static exponents in that $d_s = 2S/(S+1)$, where S is the Hausdorff dimension in the effective resistance metric. Other relationships, such as $d_s \leqslant d_f$ and $2 \leqslant d_w \leqslant d_f + 1$ can be derived from this fact.

A multifractal approach to the dynamic exponents is also possible as they can be given local interpretations. The heat kernel estimate from (4.9) suggests a definition of the local spectral dimension as

$$d_s(x) = \lim_{t \to 0} 2 \frac{\log p_t(x,x)}{\log t}.$$

Let $T_A = \inf\{t: X_t \notin A\}$, we can define a local walk dimension in terms of these exit times

$$d_w(x) = \lim_{r \to 0} \frac{\log(E^x(T_{B(x,r)}))}{\log(r)}.$$

By analogy with multifractal analysis for measures it is possible to discuss the dimensions of sets which have a particular local spectral or walk dimension, see (Hambly, Kigami and Kumagai, 2002) for details.

4.5. Transition density estimates

There is now a well developed machinery for proving estimates on the transition density of the diffusion process on the fractal. These estimates only provide uniform bounds in terms of arbitrary constants and there are no general results concerning how close these constants may be. Only in the case of the Sierpinski gasket is it known from large deviation estimates that the size of the oscillations we may expect is small (Ben Arous and Kumagai, 2000). For a recent summary and discussion of the transition kernel on fractals from a physics perspective see (Schulzky et al., 2000). As the fractal will be too rough to support a nice differentiable probability density function, in the physics literature, an averaged version is considered.

In certain types of problems, for instance, eigenvalue asymptotics, it is known that oscillations in the counting function may occur if there is symmetry in the fractal (Kigami and Lapidus, 1993; Kigami, 2001), and this is a consequence of the existence of localized eigenfunctions for the Laplace operator, but this will not be the case for the oscillations in the heat kernel. Indeed once we move to random fractals it is clear that the local geometry can be quite different and it will not be possible to have uniform estimates on the heat kernel. What we can hope for is that the size of the fluctuations is small relative to the time and length scales that we are interested in, and there is some ergodicity which averages these fluctuations at larger scales, leading to a sensible model for the diffusion on the fractal. For evidence of this see (Hambly, 2000; Hambly and Kumagai, 2001).

We now consider the scaling properties of our diffusion process and give a heuristic argument for the form of the transition density. As the set has some self-similarity this will be manifest in the diffusion process. Let P^x denote the probability measure for the process $\{X_t, t \geqslant 0\}$ started at x. By scaling $X_t = l X_{t l^{-d_w}}$ in distribution, for some scale factor l, and hence

$$P^x(X_t \in B_r(x)) = P^{x/l}(X_{t l^{d_w}} \in B_{r/l}(x/l)), \quad \forall x \in F, \ r > 0.$$

If there is a fundamental solution to the diffusion equation (4.3) (i.e., a solution with a delta function initial condition), this will be the transition density for the associated diffusion process. In sufficiently regular fractals this can be established and we see that $p_t(x, y)$ must satisfy a scaling relation of the form

$$p_t(x, y) = l^{d_f} p_{t l^{d_w}}(lx, ly).$$

By considering the functional form of any solution which satisfies this relationship we have

$$p_t(x, y) = t^{-d_f/d_w} f\left(\frac{|x - y|^{d_w}}{t}\right).$$

For short time asymptotics the form of the function f can be established and in a variety of settings is shown to be $f(a) = \exp(-a^{1/(d_w - 1)})$. The form of the metric on the fractal must also be considered and, if the shortest paths in the fractal are themselves fractal, we have to use the shortest path metric. This is then related to the Euclidean metric through what is often called a chemical exponent d_c, in that the shortest path metric d satisfies,

$$d(x, y) \asymp |x - y|^{d_c}.$$

The effective resistance metric will also scale with the Euclidean metric through the resistance exponent,

$$r(x, y) \asymp |x - y|^{d_r}.$$

Finally we summarise the transition density estimates for the short time asymptotics in a suitable metric d on a compact symmetric fractal subset of some Euclidean space.

THEOREM 4.3. *There exists a jointly continuous transition density $p_t(x, y)$ and there are positive and finite constants c_1, c_2, c_3, c_4 such that*

$$c_1 t^{-d_s/2} \exp\left(-c_2 \left(\frac{d(x, y)^{d_w}}{t}\right)^{1/(d_w-1)}\right)$$
$$\leqslant p_t(x, y)$$
$$\leqslant c_3 t^{-d_s/2} \exp\left(-c_4 \left(\frac{d(x, y)^{d_w}}{t}\right)^{1/(d_w-1)}\right), \quad \forall x, y \in F, \ 0 < t < 1.$$

Following classical work on elliptic operators in divergence form estimates of the form above are called Aronson type estimates on the heat kernel. As we consider fractals with fewer symmetries or operators with uneven r, these estimates do break down. The natural metric is a shortest path metric built from the effective resistance metric. If there is insufficient symmetry in the set, then there may exist directions in the fractal which have a different form for the estimate (Hambly and Kumagai, 1999). The derivation of the results in (Hambly and Kumagai, 1999) shows that if we are interested in studying diffusion over a certain time and space scale, then we only need to take into account structure down to a certain depth in the fractal. This is another important point in favour of these models and will be useful for simulation purposes.

From this result we can bound the solution to the Cauchy problem (4.3) that we initially formulated in that

$$u(x, t) = \int_F p_t(x, y) u_0(y) \mu(\mathrm{d}y), \quad \forall x \in F, \ 0 < t < 1.$$

It is also then easy to derive the expected crossing time, and mean square displacement results. There are constants such that

$$c_5 r^{d_w} \leqslant E^x(T_{B(x,r)}) \leqslant c_6 r^{d_w}, \quad 0 < r < 1, \tag{4.10}$$
$$c_7 t^{2/d_w} \leqslant E^x |X_t - x|^2 \leqslant c_8 t^{2/d_w}, \quad 0 < t < 1. \tag{4.11}$$

We note that if the fractals are extended to infinity, the estimates of Theorem 4.3 and (4.10), (4.11) hold for long times and distances as well as short ones.

4.6. *A soil problem*

Consider the diffusion of gas in soil. It has typically been assumed that soil is a heterogeneous medium which can be modelled by a random field for the diffusion coefficient. This is a homogenisation assumption, in that the microscopic irregularities provide some effective diffusion constant for the medium. However there is some evidence that certain types of soil show self-similarity over a range of length scales and this suggests that a fractal model may be appropriate and reveal the true effects of the heterogeneity in the soil structure. We give here an analysis of a soil problem which shows how modelling with fractals can be used to make qualitative statements which differ from what would be deduced from a classical approach. The use of fractals in modelling soil is reviewed in (Perfect and Kay, 1995).

Soil contains bacteria which converts nitrates, which are used as fertilisers, into nitrogen. There is a chemical pathway for the breakdown of the nitrate into various products until it is finally emitted as nitrogen. However, if this process occurs in the absence of oxygen, the full reduction to nitrogen is not achieved and instead nitrous oxide is produced. This is a green house gas and it is of interest to know the extent to which this occurs (Smith, 1997).

Thus we need to understand what conditions in the soil lead to the absence of oxygen. In a soil aggregate, which we assume has a uniform distribution of bacteria, the oxygen concentration is depleted as we move further from the surface. This can mean that at some distance from the surface the oxygen concentration falls to 0. At this stage the centre of the aggregate becomes anaerobic and the conditions for the production of nitrous oxide will prevail.

The classical approach to this problem is to regard the soil aggregate as spherical, of radius R, and assume that it is homogeneous with a diffusion coefficient D which is much smaller than the diffusion coefficient in air. Thus the oxygen concentration $u(x)$ at a point x will satisfy a Poisson equation in the aggregate $A \subset \mathbb{R}^3$,

$$D\Delta u = Q \quad \text{in } A,$$
$$u|_{\partial C} = C \quad \text{on } \partial A,$$

where Q is the rate of absorption of oxygen in the aggregate and C is the concentration of oxygen outside the aggregate (Smith, 1980).

This can be solved by reducing it to a one-dimensional equation using the assumed spherical symmetry of A. We need one further boundary condition, which is that there will be no flux at the point r_0 where the aggregate becomes anaerobic $\frac{\partial u}{\partial r}(r_0) = 0$.

Solving this, where the aggregate has $r_0 > 0$ gives

$$u(r) = \begin{cases} 0, & 0 \leqslant r \leqslant r_0, \\ \frac{Q}{6D}r^2 + \frac{Qr_0^3}{3Dr} + C - \frac{Q}{6DR}(R^3 - 2r_0^3), & r_0 < r \leqslant R, \end{cases}$$

and we see that, if the radius, R, of the aggregate is greater than a critical size, r_c (the value of R such that $u(r_0) = 0$ and $r_0 = 0$), an anaerobic zone will form. The critical radius is then given by $r_c = \sqrt{6DC/Q}$.

Now there is evidence that some soils appear fractal over a range of length scales and thus we may see what the effect of such a model would have on the results obtained above. We begin by assuming that the soil aggregate is fractal. Thus the diffusion through the aggregate is a fractal diffusion and therefore it scales differently to diffusion in air. In order to solve the Poisson equation above we can use the probabilistic approach. If we assume that there is a diffusion process X, then, from the associated martingale problem, we know that

$$M_t = f(X_t) - f(X_0) - \int_0^t \Delta_F f(X_s)\,ds,$$

is a martingale for $f \in \mathcal{D}(\Delta_F)$. If let $f = u$, the solution to the Poisson equation (which is in $\mathcal{D}(\Delta_F)$), and use the optional stopping theorem with $T_A = \inf\{t: X_t \in \partial A\}$, then

$$0 = E^x\left(u(X_{T_A})\right) - u(x) - E^x \int_0^{T_A} \frac{Q}{D}\,ds,$$

where now D is the effective sub-diffusion speed in the fractal. Rearranging shows that

$$u(x) = C - \frac{Q}{D} E^x T_A.$$

Letting A be the ball of radius r and using the expected crossing time estimates for fractals (4.10) we can immediately deduce that the dependence of the value of r_c, such that $u(r_c) = 0$, on the parameters is given by

$$r_c \asymp (DC/Q)^{1/d_w}.$$

The difference between the upper and lower constants arises from the short time asymptotics of the mean square displacement.

Thus we can make a qualitative statement; if the walk dimension of the soil aggregates is substantially bigger than 2 we would expect to see that anaerobic zones were larger, and more aggregates contained them, than in the classical case. Of course we know that our estimates are less accurate as more randomness is included in the model but, to leading order, this is what we expect. This is a simplistic analysis of a complicated problem as there are typically effects of water in the soil aggregate which need to be considered. Nevertheless it reveals that there could be important differences if a fractal model were used.

5. Further applications and conclusion

The work in the area of fractals is vast and a simple search in the library will reveal the diversity of modelling applications. In the past fractals have often been used as a purely descriptive tool. The claim that a particular system is fractal stems from the observation that there is power law scaling of some feature. In many instances this provides little motivation for the name fractal and explanatory power is not necessarily gained from the observation. However, recent developments in this area are helping to gain an understanding of the consequences of choosing a fractal model for other features of interest within the system.

In this chapter I have tried to show how mathematics will allow an expanded use of fractals in more than a purely descriptive way. If fractal models are to be truly effective then they must offer insights into the underlying behaviour of the physical system. An example is the recent recognition that fractal antennae are surprisingly effective. In order to understand this phenomenon it is important to understand the physical system, to solve the appropriate set of equations for the fractal, and then to consider how to optimise the design of the antenna. All these issues require an understanding of fractal dynamics.

There is also a need to investigate the statistical problems that arise in fractal models. There is now a substantial statistical literature on the estimation of fractal dimension, however there is nothing as yet on parameter estimation in the dynamic fractal setting. How should we estimate the walk and spectral dimensions and their multifractal structure in practice? How much detail do we need to simulate in order to obtain good approximations to a diffusion on a fractal? These and many other questions give scope for work in the further development of fractal models for self-similar structures.

References

Adler, R. (1981). *The Geometry of Random Fields*. Wiley, New York.
Adler, R. et al. (1998). *A Practical Guide to Heavy Tails. Statistical Techniques and Applications* (Eds. R. J. Adler, R. Epstein-Feldman and M. S. Taqqu). Birkhäuser, Boston.
Arbeiter, M. and N. Patzschke (1996). Random self-similar multifractals. *Math. Nachr.* **181**, 5–42.
Ayache, A. and J. Lévy Vehel (1999). Generalized multifractional Brownian motion: definition and preliminary results. In *Fractals: Theory and Applications in Engineering*, pp. 17–32. Springer-Verlag, Berlin.
Barlow, M.T. (1998). *Diffusions on Fractals*. Lectures on Probability Theory and Statistics. École d'été de probabilités de St Flour XXV. Springer-Verlag, Berlin.
Barlow, M. T. and R. F. Bass (1999). Brownian motion and harmonic analysis on Sierpinski carpets. *Canad. J. Math.* **51**, 673–744.
Barlow, M. T. and S. J. Taylor (1992). Defining fractal subsets of Z^d. *Proc. London Math. Soc.* **64**, 125–152.
Barnsley, M. F. (1993). *Fractals Everywhere*. Academic Press, Boston.
Ben Arous, G. and T. Kumagai (2000). Large deviations of Brownian motion on the Sierpinski gasket. *Stochastic Process. Appl.* **85**, 225–235.
Ben-Avraham, D. and S. Havlin (2000). *Diffusion and Reactions in Fractals and Disordered Systems*. Cambridge University Press, Cambridge.
Bertoin, J. (1996). *Lévy Processes*. Cambridge University Press, Cambridge.
Chandra, A. K., P. Raghavan, W. L. Ruzzo, R. Smolensky and P. Tiwari (1996). The electrical resistance of a graph captures its commute and cover times. *Comput. Complexity* **6**, 312–340.
Davies, S. and P. Hall (1999). Fractal analysis of surface roughness by using spatial data. *J. Roy. Statist. Soc. Ser. B* **61**, 3–37.
Dekking, M. et al. (1999). *Fractals: Theory and Applications in Engineering* (Eds. M. Dekking, J. Lévy Vehel, E. Lutton and C. Tricot). Springer-Verlag, London.
Dembo, A., Y. Peres, J. Rosen and O. Zeitouni (2000a). Thick points for spatial Brownian motion: multifractal analysis of occupation measure. *Ann. Probab.* **28**, 1–35.
Dembo, A., Y. Peres, J. Rosen and O. Zeitouni (2000b). Thin points for Brownian motion. *Ann. Inst. H. Poincaré Probab. Statist.* **36**, 749–774.
Doyle, P. G. and L. J. Snell (1984). *Random Walks and Electrical Networks*. Math. Assoc. Amer. Washington.
Falconer, K. J. (1986) Random fractals. *Math. Proc. Cambridge Philos. Soc.* **100**, 559–582.
Falconer, K. J. (1990). *Fractal Geometry, Mathematical Foundations and Applications*. Wiley, New York.

Falconer, K. J. (1997). *Techniques in Fractal Geometry*. Wiley, New York.
Feder, J. (1988). *Fractals*. Plenum Press, New York.
Fisher, Y. (1995). *Fractal Image Compression. Theory and Application* (Ed. Y. Fisher). Springer-Verlag, New York.
Fitzsimmons, P. J., B. M. Hambly and T. Kumagai (1994). Transition density estimates for Brownian motion on affine nested fractals. *Comm. Math. Phys.* **165**, 595–620.
Frisch, U. and G. Parisi (1985). Fully developed turbulence and intermittency. In *Proc. E. Fermi Int. Summer School in Physics*, pp. 84–88. North-Holland.
Fukushima, M., Y. Oshima and M. Takeda (1994). *Dirichlet Forms and Symmetric Markov Processes*. De Gruyter, Berlin.
Graf, S. (1987). Statistically self-similar fractals. *Probab. Theory Related Fields* **74**, 357–392.
Graf, S., R. D. Mauldin and S. C. Williams (1988). The exact Hausdorff dimension in random recursive constructions, *Mem. Amer. Math. Soc.* **381**.
Grimmett, G. (1999). *Percolation*. Springer-Verlag, Berlin.
Hall, P. and R. Roy (1994). On the relationship between fractal dimension and fractal index for stationary stochastic processes. *Ann. Appl. Probab.* **4**, 241–253.
Halsey, T. C., M. H. Jensen, L. P. Kadanoff, I. Procaccia and B. I. Shraiman (1986). Fractal measures and their singularities – the characterization of strange sets. *Phys. Rev. A* **33**, 1141–1151.
Hambly, B. M. (2000). On the asymptotics of the eigenvalue counting function for random recursive Sierpinski gaskets. *Probab. Theory Related Fields* **117**, 221–247.
Hambly, B. M. and O. D. Jones (2003). Thick and thin points for random recursive fractals. To appear in *Adv. Appl. Probab.*
Hambly, B. M., J. Kigami and T. Kumagai (2002). Multifractal formalisms for the spectral and walk dimensions. *Math. Proc. Cambridge Philos. Soc.* **132**, 555–571.
Hambly, B. M. and T. Kumagai (1999). Transition density estimates for diffusion processes on post critically finite self-similar fractals. *Proc. London. Math. Soc.* **78**, 431–458.
Hambly, B. M. and T. Kumagai (2001). Fluctuation in the transition density for Brownian motion on a random recursive fractal. *Stochastic Process. Appl.* **92**, 61–85.
Hu, X. and S. J. Taylor (1997). The multifractal structure of stable occupation measure. *Stochastic Process. Appl.* **66**, 283–299.
Hutchinson, J. E. (1981). Fractals and self-similarity. *Indiana Univ. Math. J.* **30**, 713–747.
Hutchinson, J. E. and L. Rüschendorff (2000). Random fractals and probability metrics. *Adv. Appl. Probab.* **32**, 925–947.
Jaffard, S., Y. Meyer and R. D. Ryan (2001). *Wavelets. Tools for Science and Technology*. SIAM, Philadelphia.
Jaffard, S. (1997a). Multifractal formalism for functions I: Results valid for all functions. *SIAM J. Math. Anal.* **28**, 944–970.
Jaffard, S. (1997b). Multifractal formalism for functions II: Self-similar functions. *SIAM J. Math. Anal.* **28**, 971–998.
Kesten, H. (1986a). The incipient infinite cluster in two-dimensional percolation. *Probab. Theory Related Fields* **73**, 369–394.
Kesten, H. (1986b). Subdiffusive behavior of random walk on a random cluster. *Ann. Inst. H. Poincaré Probab. Statist.* **22**, 425–487.
Kigami, J. (1993). Harmonic calculus on P. C.F. self-similar sets. *Trans. Amer. Math. Soc.* **335**, 721–755.
Kigami, J. (1995). Harmonic calculus on limits of networks and its application to dendrites. *J. Funct. Anal.* **128**, 48–86.
Kigami, J. (2001). *Analysis on Fractals*. Cambridge University Press, Cambridge.
Kigami, J. and M. L. Lapidus (1993). Weyl's problem for the spectral distribution of the Laplacian on P.C.F. self-similar fractals. *Comm. Math. Phys.* **158**, 93–125.
Kumagai, T. and S. Kusuoka (1996). Homogenization on nested fractals. *Probab. Theory Related Fields* **104**, 375–398.
Kusuoka, S. and X. Y. Zhou (1992). Dirichlet forms on fractals: Poincaré constant and resistance. *Probab. Theory Related Fields* **93**, 169–196.
Lindstrøm, T. (1990). Brownian motion on nested fractals. *Mem. Amer. Math. Soc.* **420**.

Mandelbrot, B. B. (1974). Intermittent turbulence in self-similar cascades: divergence of high moments and dimension of the carrier. *J. Fluid Mech.* **62**, 331.

Mandelbrot, B. B. (1982). *The Fractal Geometry of Nature.* Freeman, San Fransisco.

Mattila, P. (1995). *Geometry of Sets and Measures in Euclidean Spaces. Fractals and Rectifiability.* Cambridge University Press, Cambridge.

Mauldin R. D. and S. C. Williams (1986). Random recursive constructions: asymptotic geometric and topological properties. *Trans. Amer. Math. Soc.* **295**, 325–346.

Mauldin, R. D. and S. C. Williams (1988). Hausdorff dimension in graph directed constructions. *Trans. Amer. Math. Soc.* **309**, 811–829.

Metz, V. (1996). Renormalization contracts on nested fractals. *J. Reine Angew. Math.* **480**, 161–175.

Morato, L. M. and P. Siri (2001). A stochastic algorithm to compute optimal probabilities in the chaos game. *Adv. Appl. Probab.* **33**, 423–436.

Paxson, V. (1997). Fast, approximate synthesis of fractional Gaussian noise for generating self-similar network traffic. *Comp. Comm. Rev.* **27**, 5–18.

Peitgen, H.-O., H. Jürgens and D. Saupe (1992). *Chaos and Fractals. New Frontiers of Science.* Springer-Verlag. New York.

Perfect, E. and B. D. Kay (1995). Applications of fractals in soil and tillage research: a review. *Soil Till. Res.* **36**, 1–20.

Olsen, L. (1995). A multifractal formalism. *Adv. Math.* **116**, 82–196.

Olsen, L. (1996). *Multifractal Spectrum for Graph Directed Random Recursive Self-Similar Measures.* Longman, London.

Olsen, L. (2000). Multifractal geometry. In *Fractal Geometry and Stochastics II*, pp. 3–37 (Eds. C. Bandt, S. Graf and M. Zähle). Progress in Probability, Vol 46.

Sabot, C. (1997). Existence and uniqueness of diffusions on finitely ramified self-similar fractals. *Ann. Sci. Ecole Norm. Sup.* **30**, 605–673.

Schulzky, C., C. Essex, M. Davison, A. Franz and K. H. Hoffmann (2000). The similarity group and anomalous diffusion equations. *J. Phys. A Math. Gen.* **33**, 5501–5511.

Smith, K. A. (1980). A model of the extent of anaerobic zones in aggregated soils, and its potential application to estimates of denitrification. *J. Soil Sci.* **31**, 263–277.

Smith, K. A. (1997). The potential for feedback effects induced by global warming on emissions of nitrous oxide by soils. *Global Change Bio.* **3**, 327–338.

Taylor, S. J. (1986). The measure theory of random fractals. *Math. Proc. Cambridge Philos. Soc.* **100**, 383–406.

Turcotte, D. L. (1997). *Fractals and Chaos in Geology and Geophysics.* Cambridge University Press, Cambridge.

Werner, W. (2001). Critical exponents, conformal invariance and planar Brownian motion. In *Proceedings of the 3rd Europ. Congress of Mathematics*, pp. 87–103. Birkhäuser.

Numerical Methods in Queueing Theory

D. Heyman

1. Introduction

Queueing theory relies heavily on Markov chains and Laplace transforms to obtain analytical results. Markov chains frequently describe the dependency in queueing processes. Laplace transforms are often convenient for solving differential equations and describing sums of i.i.d. random variables that occur in solving queueing models. Probability generating functions are used for similar reasons. Consequently, this entry describes numerical methods for analyzing Markov chains and for inverting Laplace transforms and probability generating functions.

A numerical method that is used extensively is not considered – simulation. Random number generation, experimental design and output analysis raise substantial statistical issues. The numerical methods that are used to address these issues are different from the methods described here.

The numerical solution of a queueing model consists of more than obtaining a tractable algorithm. Computer roundoff errors can cause mathematically correct algorithms to produce inaccurate numerical solutions, so numerical accuracy of an algorithm has to be understood.

1.1. Examples

The numerical methods we shall consider appear when analyzing the M/G/1 queue. This queueing model has Poisson arrivals at rate λ. There is a single server that takes a random time with density function g to serve a customer. Customers wait as long as necessary to receive service and are served in order of arrival. Let $X(t)$ be the number of customers present (in queue and in service) at time t, X_n be the number present just after the nth service completion, and W_n be the waiting time of the nth customer served.

D. Kendall showed that X_n, $n = 1, 2, \ldots$, is a Markov chain. The transition matrix is

$$P = \begin{pmatrix} a_0 & a_1 & a_2 & a_3 & \cdots \\ a_0 & a_1 & a_2 & a_3 & \cdots \\ 0 & a_0 & a_1 & a_2 & \cdots \\ 0 & 0 & a_0 & a_1 & \cdots \\ 0 & 0 & 0 & a_0 & \cdots \\ \cdots & \cdots & \cdots & \cdots & \cdots \end{pmatrix}, \qquad (1)$$

where $a_k = \int_0^\infty e^{-\lambda t} \frac{(\lambda t)^k}{k!} g(t)\,dt$. A closed form equation for

$$\sum_{j=1}^\infty z^j \int_0^\infty \mathbf{P}[X(t) = j | X_0 = i] e^{-st}\,dt \qquad (2)$$

is available (Cohen, 1969, Eq. 4.49). The transform inversion techniques described in Section 2 allow numerical inversion of this double transform.

Let μ^{-1} be the mean service time, $\rho = \lambda/\mu$ and $\pi_{ij}(n) = \mathbf{P}[X_n = j | X_0 = i]$. When $\rho < 1$ the following limits exist and are independent of i.

$$\pi_j = \lim_{n\to\infty} \pi_{ij}(n) = \lim_{n\to\infty} \mathbf{P}[X_n = j] \qquad (3)$$

and

$$W(t) = \lim_{n\to\infty} \mathbf{P}[W_n \leq t], \quad t \geq 0. \qquad (4)$$

Moreover, $\pi_0 = W(0) = 1 - \rho$ and W has density for $t > 0$ (say w) because the service times do. Let $\tilde{w}(s) = \int_0^\infty e^{-st} w(t)\,dt$; then

$$\tilde{w}(s) = (1-\rho)\frac{s\tilde{g}(s)}{s - \lambda + \lambda\tilde{g}(s)}, \quad \Re[s] \geq 0, \qquad (5)$$

where $\tilde{g}$ is the Laplace transform of g and $\Re[s]$ is the real part of s. Let $\hat{\pi}(z) = \sum_{j=0}^\infty z^j \pi_j$ with $|z| \leq 1$; then

$$\hat{\pi}(z) = (1-\rho)\frac{(z-1)\tilde{g}(\lambda - \lambda z)}{z - \lambda + \lambda\tilde{g}(\lambda - \lambda z)}, \quad |z| \leq 1. \qquad (6)$$

Another important operating characteristic is the *busy period*, which starts when a customer arrives at an empty system and ends the next time the system is empty. Let b be the density function of a busy period and $\tilde{b}$ be its Laplace transform. Then $\tilde{b}$ is the smallest positive solution of

$$\tilde{b}(s) = \tilde{g}(s + \lambda - \lambda\tilde{b}(s)). \qquad (7)$$

There are known formal solutions of (5) and (7); these are infinite series and have not been found suitable for numerical computation. Numerical inversion of the Laplace transforms is preferable. The $\{\pi_j\}$ can be computed from the balance equations of the Markov chain governed by P, but numerical inversion of the probability generating function in (6) is typically more convenient because it avoids calculating the $\{a_k\}$.

When the customers are partitioned into two classes, one of which is given priority over the other, numerical transform inversion is practical and no potentially computable time-domain expressions are known because the Laplace transform of the "effective service-times" has $\tilde{b}$ as part of the argument of $\tilde{g}$. The equations were derived by

D. P. Gaver and may be found in (Heyman and Sobel, 1982, Section 11-5). Similar issues occur when the customers are served "shortest service-time first" (Takács, 1964).

Finite Markov chains are used to model queues with limited (or no) waiting space. When there are K waiting positions, the embedded Markov chain with M/G/1 assumptions is the $(K+1) \times (K+1)$ Northwest corner of the matrix P in (1) with the last column augmented so that the row sums are one. The steady-state probabilities can be obtained by numerically solving the balance equations. The special structure of this transition matrix (upper-Hessenberg or skip-free-to-the-right) permits a solution by recursion.

The following example is more typical of those models that are solved by the numerical method described in Section 4.2. There are two queues, each with s servers and b waiting positions. A customer that tries to enter queue 1 when all of its waiting positions are full will attempt to join the second queue. Customers that try to join the second queue when all of its waiting positions are full are lost. Assume that the arrival processes are Poisson and the service times at all servers are exponential. Let λ_i be the arrival rate (including lost customers but not including overflow customers) and μ_i be the service rate at queue i. Let $N_i(t)$ be the number of customers in queue i at time t. The bivariate process $\{N_1(t), N_2(t), t \geq 0\}$ is Markov. Let Q be the generator; Q has the block-partitioned form

$$Q = \begin{pmatrix} B_0 & C_0 & & & & & \\ A_1 & B_1 & C_1 & & & & \\ & A_2 & B_2 & C_2 & & & \\ & \vdots & \vdots & \vdots & \vdots & & \vdots \\ & & & & A_{b-1} & B_{b-1} & C_{b-1} \\ & & & & & A_b & B_b \end{pmatrix}. \tag{8}$$

The blocks are

$$A_k = \mu_1 \min(k, s) I, \quad k = 1, 2, \ldots, b, \tag{9}$$

$$C_k = \lambda_1 I, \quad k = 0, 1, 2, \ldots, b-1, \tag{10}$$

where I is the $(b+1) \times (b+1)$ identity matrix. For $k = 0, 1, \ldots, b-1$ and $i = 1, 2, \ldots, b$,

$$B_k(i, i-1) = \mu_2 \min(i, s), \quad B_k(i-1, i) = \lambda_2 \tag{11}$$

and is zero otherwise. Finally,

$$B_b(i, i-1) = \mu_2 \min(i, s), \quad B_b(i-1, i) = \lambda_1 + \lambda_2. \tag{12}$$

The stationary distribution of Q can be used to calculate performance measures such as the probability that a customer is lost and the probability that a customer arriving at queue 1 is served in queue 2.

2. Numerical inversion of Laplace transforms

The numerical inversion of Laplace transforms has long been thought to be difficult, but that is not correct for probability distributions. Several programs for inverting Laplace transforms were available at Bell Laboratories in the late 1960's. We describe algorithms for inverting Laplace transforms and probability generating functions (z-transforms) based on Abate and Whitt (1992a, 1992b); the former contains an extensive history and literature review. The expositions by Abate and Whitt contain many probabilistic insights that are lost in the abbreviated treatment given here.

Let f be a non-negative function on the positive real line, and let $\tilde{f}$ be the Laplace transform of f:

$$\tilde{f}(s) = \int_0^\infty e^{-st} f(t) \, dt, \tag{13}$$

where s is a complex variable. The standard inversion formula is the Bromwich contour integral

$$f(t) = \frac{1}{2\pi i} \int_{a-i\infty}^{a+i\infty} e^{st} \tilde{f}(s) \, ds, \quad i = \sqrt{-1}, \tag{14}$$

where a is chosen so that $\tilde{f}(s)$ has no singularities on or to the right of the vertical line $s = a$ in the complex plane. ($a = 0$ suffices when f is a bounded continuous probability density.) Writing $s = a + iu$ and using the identity from complex variable theory

$$e^s = e^a \big[\cos(u) + i \sin(u)\big] \tag{15}$$

the right side of (14) yields the following string of equations ($\Im[z]$ denotes the imaginary part of z):

$$f(t) = \frac{1}{2\pi} \int_{-\infty}^{\infty} \big[\cos(ut) + i \sin(ut)\big] \tilde{f}(a + iu) \, du$$

$$= \frac{e^{at}}{2\pi} \int_{-\infty}^{\infty} \big\{\Re[\tilde{f}(a+iu)] \cos(ut) - \Im[\tilde{f}(a+iu) \sin(ut)]\big\} \, du$$

$$= \frac{2e^{at}}{\pi} \int_0^{\infty} \Re[\tilde{f}(a+iu)] \cos(ut) \, du. \tag{16}$$

The integral in (16) is evaluated symbolically with the trapezoidal rule. There is a known error bound for the discretization error introduced. Let f_h be the value of the integral when a step size of h is used. Then,

$$f(t) \approx f_h(t) = \frac{h e^{at}}{\pi} \Re(\tilde{f})(a) + \frac{2h e^{at}}{\pi} \sum_{k=1}^{\infty} \Re(\tilde{f})(a + ikh) \cos(kht). \tag{17}$$

Taking $h = \pi/(2t)$ and $a = A/(2t)$ yields

$$f_h(t) = \frac{e^{A/2}}{\Re(\tilde{f})}\left(\frac{A}{2t}\right) + \frac{e^{A/2}}{t}\sum_{k=1}^{\infty}(-1)^k \Re(\tilde{f})\left(\frac{A+2k\pi i}{2t}\right), \tag{18}$$

The Poisson summation formula (see, e.g., Franklin, 1964, p. 514) can be used to show that the discretization error $f(t) - f_h(t) = e_d$ (say) is given by

$$e_d = \sum_{k=1}^{\infty} e^{-kA} f\big[(2k+1)t\big]. \tag{19}$$

When f is either a distribution function or a complementary distribution function, $f(t) \leqslant 1$ for all t, so

$$e_d \leqslant \frac{e^{-A}}{1 - e^{-A}}, \tag{20}$$

which is approximately e^{-A} when e^{-A} is small. Thus, when $A = \gamma \log 10$ the discretization error is no larger than $10^{-\gamma}$; $A = 19.1$ achieves a 10^{-7} discretization error. This is the only place a probabilistic property of f is used.

The first n terms in (18) are summed explicitly and the next m terms are summed via Euler summation (Franklin, 1964). This introduces a truncation error that can be *estimated* and reduced by increasing n. Abate and Whitt recommend taking $m = 11$ and $n = 15$; m and n are increased when the error estimate is larger than desired. Let $E(m, n, t)$ be the Euler sum as described above. Then

$$E(m, n, t) = \sum_{k=0}^{m} m\binom{m}{k} 2^{-m} s_{n+k}(t), \tag{21}$$

where

$$s_n(t) = \frac{e^{A/2}}{2t}\Re(\tilde{f})\left(\frac{A}{2t}\right) + \frac{e^{A/2}}{t}\sum_{k=1}^{n} a_k(t) \tag{22}$$

and

$$a_k(t) = (-1)^k \Re(\tilde{f})\left(\frac{A+2k\pi i}{2t}\right). \tag{23}$$

Eqs. (18), (21), (22) and (23) are embodied in an algorithm called EULER (Abate and Whitt, 1992a), which is shown in Figure 1. The Laplace transform variable s is written as $s = x + iy$ and $\tilde{f}(s)$ is written as $\tilde{f}(x, y)$.

1. INPUT t

2. Set Parameters:
   A = 19.1, n = 15, m = 11, U = exp(A/2)/t, c(1) = 1

3. Set Variables: x = A/(2t), h = π/t

4. Sum First n Terms of (18):
   sum = $\Re \tilde{f}(x, 0)$: For k = 1, 2, ..., n

   (a) y = kh
   (b) sum $\leftarrow$ sum $+(-1)^k \tilde{f}(x, y)$
   (c) end for

5. Sum Next m terms of (3) by Euler Summation:
   Compute Constants:

   (a) For k = 2, 3, ..., m + 1
   (b) c(k + 1) = c(k)(m − k + 1)/k
   (c) end for

6. Euler Summation:
   s(1) = sum

   (a) For k = 1, 2, ..., m + 1
   (b) j = n + k
   (c) y = jh
   (d) s(k + 1) = s(k) + $(-1)^j \tilde{f}(x, y)$
   (e) end for

7. ans = 0

   (a) For k = 1, 2, ..., m + 1
   (b) ans $\leftarrow$ ans $+c(k)s(k + 1)$
   (c) end for

8. ans $\leftarrow$ ans $\times (U/2^m)$

9. OUTPUT: f(t) = ans

Fig. 1. Algorithm EULER.

The Fourier analysis, trapezoidal rule integration and Poisson summation techniques can also be applied to generating functions (Abate and Whitt, 1992b). Let

$$\hat{q}(z) = \sum_{k=0}^{\infty} z^k q_k, \tag{24}$$

where z is complex. When q is a probability mass function, we have the convenient bound $|q_k| \leqslant 1$ which is important for error analysis but is otherwise inessential. With this bound, $\hat{q}$ is analytic inside the unit circle. Assume that $\hat{q}$ can be computed for any z inside the region of convergence. Then

$$q_0 = \hat{q}(0). \tag{25}$$

The inversion formula is

$$q_k = \frac{1}{2kr^k}\left\{G(r) + G(-r) + 2\sum_{j=1}^{k-1}(-1)^j\Re[G(re^{\pi ji/k})]\right\}, \quad k = 1, 2, \ldots, \tag{26}$$

where $0 < r < 1$. The error in (26) is at most $r^{2k}/(1 - r^{2k})$, which is close to the numerator when the numerator is small.

The ideas behind the inversion algorithms described above have been extended to produce algorithms to invert multi-dimensional Laplace transforms, including lattice functions (Choudhury, Lucantoni and Whitt, 1994). In particular, transforms such as

$$P(z, s) = \sum_{j=0}^{\infty} z^j \int_0^{\infty} e^{-st} p_j(t)\, dt \tag{27}$$

can be inverted when each p_j is a probability mass function.

3. The ubiquity of Markov chains

Markov chains are ubiquitous in the numerical solution of queueing problems because there is a lot known about numerical solution of Markov chains and because many queueing problems can be described by a Markov chain. The reason for the latter is that even when interarrival or service times are not exponentially distributed, they can be approximated by sums and mixtures of exponential distributions. This idea started with Erlang's method of stages (Heyman and Sobel, 1982, p. 298) and has been extended in recent years. Neuts (1975, 1981) introduced the *phase-type* distributions, which are first-passage-times of an absorbing continuous-time Markov chain. In addition, some non-Markovian queueing processes have an embedded Markov chain, such as the M/G/1 queue.

Another way Markov chains arise is that non-Poisson arrival processes are often modeled with a *Markov-modulated Poisson Process* (*MMPP*) in which the state of an ergodic continuous-time Markov chain determines the rate of a Poisson stream of events. Each event can be the arrival of a batch of customers. The most general process of this type is the *batch Markovian arrival process* (*BMAP*), which is described in Section 6.

It can be a difficult task to turn a Markovian model as described above into a computer program to construct the transition matrix. Berson, de Souza de Silva and Muntz (1991) describe a way to generate the transition matrix from a description of the system dynamics.

4. Finite Markov chains

4.1. Transient probabilities

For a continuous-time Markov chain with generator Q, let $\Pi_{ij}(t) = \mathbf{P}[X(t) = j | X(0) = i]$, and $\Pi(t) = (\Pi_{ij}(t))$. The balance equations are

$$\dot{\Pi} = \Pi Q, \quad t > 0, \tag{28}$$

where the dot denotes differentiation and $\Pi(0) = I$. When the structure of Q permits, a transform of the transient probabilities can be obtained and numerical transform inversion applied, as illustrated by the transient probabilities of the M/G/1 queue described in Section 1. The solution of (28) is

$$\Pi(t) = e^{Qt}, \quad t \geq 0, \tag{29}$$

and for a finite number of states, one can "just compute the matrix exponential" and the solution is obtained. Computing a matrix exponential accurately is not easy. There are at least 19 dubious ways (Moler and Van Loan, 1978); see Golub and Van Loan (1989) for a good way. One could apply numerical differential equation solvers to (28). These methods may display numerical inaccuracies caused by rapidly decaying components of the transient solution (i.e., *stiffness*). See (Stewart, 1994, Section 8.4) for an introduction to these methods.

The most commonly used method is *Jensen's algorithm* (Grassmann, 1991; Stewart, 1994, Section 8.2) which is based on the uniformization procedure for continuous-time Markov chains. The uniformization procedure (Heyman and Sobel, 1982, Section 8-7) replaces the generator Q with the transition matrix $P = I + Q/q_{\max}$, where $q_{\max} = \max_i \{-q_{ii}\}$. Uniformization is used repeatedly in the sequel; it provides both computational and analytic results. Let $N(t)$ be the the number of transitions that occur by time t; it has a Poisson distribution with mean $q_{\max} t$ and

$$\Pi_{ij}(t) = \sum_{k=0}^{\infty} P^k \mathbf{P}[N(t) = k]. \tag{30}$$

Numerical implementations of (30) replace the infinite sum with a finite sum with upper limit K say. The truncation error is

$$\sum_{k=K+1}^{\infty} P^k \mathbf{P}[N(t) = i] \leq \sum_{k=K+1}^{\infty} \mathbf{P}[N(t) = i] = \mathbf{P}[N(t) > K] \tag{31}$$

which is kept below some number $\varepsilon > 0$ when K is large enough so that

$$\sum_{k=0}^{K} \frac{(q_{\max} t)^k}{k!} \geq (1 - \varepsilon) e^{q_{\max} t}. \tag{32}$$

1. Calculate K from (32)
   (a) Set K = 0, $\xi = 1$, $\sigma = 1$, $\eta = (1 - \varepsilon)e^{q_{max}t}$.
   (b) While $\sigma < \eta$ do
      i. K ← K + 1, $\xi \leftarrow \xi \times \frac{q_{max}t}{K}$.
2. Calculate $p(t)$
   (a) p ← p(0), y = p(0)
   (b) For $k = 1, 2, \ldots, K$ do
      i. y ← yP × $\frac{q_{max}t}{k}$
      ii. p ← p + y
   (c) p(t) = $e^{-q_{max}t}$p

Fig. 2. Jensen's algorithm.

Since there are no subtractions in (30) it should be resistant to computer round-off errors (Grassmann, 1983), so the computed probabilities should differ from the exact probabilities by no more than ε. From (32) we see that for ε fixed, the computational effort grows exponentially with the product $q_{max}t$. To mitigate this effect, one can divide the interval $(0, t]$ into subintervals and solve a smaller problem on each subinterval. For $0 = t_0 < t_1 < t_2 < \cdots < t_n = t$, compute $\tau_i = t_i - t_{i-1}$ and

$$\Pi(t_i) = \Pi(t_{i-1}) \sum_{k=0}^{K_i} P^k e^{-q_{max}\tau_i} \frac{(q_{max}\tau_i)^k}{k!}, \quad i = 1, 2, \ldots, n. \tag{33}$$

Figure 2 shows Jensen's algorithm (Stewart, 1994, p. 413) for computing the vector $p(t) = p(0)\Pi(t)$ where $p(0)$ is given vector of initial probabilities.

For each time point (t_i) that Jensen's algorithm evaluates, the number of multiplications (which is the major computational expense) is $K(s + s_p)$ where s is the number of states in the Markov chain and s_p is the number of positive entries in P. The vector $p(0)$, the matrix P and two work vectors of length s need to be stored.

4.2. Steady-state probabilities

Since transient probabilities are often difficult to compute, and since steady-state probabilities, when they exist, describe long-run averages, steady-state probabilities are often computed. The statement of the problem is this. The states of the Markov chain are numbered from 1 to N, and the transition matrix is $P = (p_{ij})$. We assume P is ergodic and seek a solution of

$$\pi = \pi P, \tag{34}$$

with

$$\pi \geq 0 \quad \text{and} \quad \sum_{i=1}^{N} \pi_i = 1, \tag{35}$$

where π is a row vector.

There are two general classes of algorithms for this problem. *Direct* algorithms compute an exact solution in exact arithmetic (no roundoff errors) and the number of arithmetic operations can be computed from the description of the problem. An *iterative* algorithm produces a sequence of approximations that is expected to converge to the exact solution. (Some algorithms appear to converge when applied to problems that don't satisfy the conditions that have been needed to prove convergence.)

Gaussian elimination can be applied to (34) and (35); it will require $N^3/3$ (plus terms in N^2 and N) floating point operations. (A floating point operation, or *flop*, is the arithmetic required to do $a_i \leftarrow a_i + b_j \times c_k$, which is an addition, a multiplication and some subscripting.) A variant of Gaussian elimination designed for this system of equations is described in Section 4.2.

Since $\pi(0) \lim_{k \to \infty} P^k = \pi$ for any initial probability vector $\pi(0)$, forming successive powers of P is a convergent algorithm which is called the *power method*. Since computing $P \times P$ requires N^3 flops in general, it cannot require fewer flops than Gaussian elimination unless the structure of P can be exploited. The rate of convergence of the power method is controlled by the eigenvalue of P with second largest modulus, and can be very slow. This is particularly noticed with nearly completely decomposable Markov chains (Courtois, 1977). (A nearly completely decomposable Markov chain has a block partitioning such that the blocks along the diagonal have row sums that are close to one. These blocks describe local changes of state which occur much more frequently than transitions to states in other blocks. These Markov chains occur, for example, with MMPP arrivals when many changes in the queue length occur between changes in the arrival rate. Algorithms have been devised that exploit this property (Stewart, 1994, Chapter 6).)

Jacobi and Gauss–Seidel iterations are based on computing the iterations $\pi(n+1) \leftarrow \pi(n)P$ in an effective way. Successive over-relaxation (SOR) methods attempt to speed convergence by averaging successive iterates:

$$\pi^{\text{new}} \leftarrow \pi(n)P \quad \text{and}$$
$$\pi(n+1) \leftarrow \omega \pi^{\text{new}} + (1-\omega)\pi(n), \quad 0 < \omega < 2. \tag{36}$$

The choice of ω determines the rate of convergence, and selecting good values is currently more of an art than a science. Seelen (1992) had success in using successive over-relaxation on queueing problems. He developed ways to adapt ω as the computations proceed.

Since premultiplying a matrix by a vector takes N^2 flops, these iterations should be used when P is sparse or so large that it cannot be stored all at once. Greenberg and Vanderbei (1991) find that these iterations require approximately N iterations to converge, so Gaussian elimination will require less computation unless P is sparse and Gaussian elimination cannot take advantage of this.

A direct algorithm

The direct method of choice for solving the steady-state balance equations is the GTH algorithm, which was introduced in Grassmann, Taksar and Heyman (1985). It is a variant of Gaussian elimination that accurately computes the stationary vector of an irreducible, finite stochastic matrix. The accuracy is achieved by avoiding subtractions, which are a main source of computer round-off errors. Empirical evidence of its accuracy is given in Heyman (1987), and analytic evidence in O'Cinneide (1993). If there are transient states in P, the algorithm will either compute their steady-state probabilities as zero or attempt a division by zero. The latter can be used to discard transient states (Heyman, 1987).

The GTH algorithm also finds the steady-state probabilities of a continuous-time Markov chain with an irreducible generator. This follows directly by uniformizing the continuous-time process and treating it as a discrete-time Markov chain, and is described in more detail after the description of the algorithm.

The probabilistic reasoning behind the GTH algorithm is this. Suppose state N is invisible, so the sequence of transitions $i \to N \to j$ appears to be a transition from i directly to j. With this supposition, the Markov chain with N states would appear to be a Markov chain with $N - 1$ states, and the transition probabilities would appear to be

$$p_{ij} + p_{iN}(1 + p_{NN} + p_{NN}^2 + \cdots)p_{Nj} = \frac{p_{iN}p_{Nj}}{1 - p_{NN}}, \quad i, j < N, \qquad (37)$$

the terms in parentheses account for the instantaneous transitions from state N back to itself. This operation is called *state reduction*. Algebraically, (37) is solving the Nth equation in (34) for π_N and substituting the results in the other equations. The state reduction step can be repeated until there is only one state left in the Markov chain; since there is exactly one redundant equation in (34) when P is irreducible, the result of all the state reduction steps is the equation

$$\pi_1 = \pi_1. \qquad (38)$$

Any positive solution of (34) can be normalized to sum to one to satisfy (35), so we pick $\pi_1 = 1$. The back-substitution steps reverse the state reduction steps, computing π_i from π_j, $j < i$. The normalization step makes the probabilities sum to one. We display the GTH algorithm in Figure 3. The notation $a \leftarrow b$ means b replaces a.

For continuous-time Markov chains, the uniformization procedure allows the algorithm to be applied to the transition matrix $P = I + Q/q_{\max}$. The normalization step cancels the effect of carrying $q_{\max}$ through the state reduction step, and since the diagonal terms are never used, one can apply the algorithm directly to the generator Q.

The state reduction step overwrites P so the algorithm has no storage requirements; the matrix P has to be stored if it is to be used in future calculations. The back-substitution and normalization steps can be combined so overflow of TOT or underflow of π_j are avoided. In Step 1(a), S_n should be tested to see if it is zero to machine precision. If it is, then (to machine precision) P is such that states $n + 1, n + 2, \ldots, N$ cannot be reached from state n.

The algorithm requires $N^3/3$ additions and an equal number of multiplications, plus terms of order N^2 and order N. The following important special structure can be exploited to reduce the number of computations. P is called *banded* when either

1. (State Reduction) For n = N, N − 1, . . . , 2, do the following:
    (a) Let $S_n = \sum_{j=1}^{n-1} p_{nj}$.
    (b) Let $p_{in} \leftarrow p_{in}/S_n$, $i < n$.
    (c) Let $p_{ij} \leftarrow p_{ij} + p_{in}p_{nj}$, $i, j < n$.
2. (Initialization) Initialize TOT = 1 and $\pi_1 = 1$.
3. (Back-substitution) For j = 2, 3, . . . , N do the following:
    (a) Let $\pi_j = p_{1j} + \sum_{k=2}^{j-1} \pi_k p_{kj}$.
    (b) Let TOT $\leftarrow$ TOT + π_j.
4. (Normalization) Let $\pi_j \leftarrow \pi_j/\text{TOT}$, j = 1, 2, . . . , N.

Fig. 3. GTH algorithm.

$p_{i,i+k} = 0$ whenever $k > g$ or $p_{i,i-k} = 0$ whenever $k > h$ for some positive numbers g and h (with appropriate modifications at the boundaries – when $i + k > N$ or $i - k < 0$, the condition is valid). This means that the chain cannot jump too far from the current state. Two examples are the M/M/1/k queue and the Engset model with several arrival classes (Cooper, 1981, Section 3.7). The overflow model in Section 1 with generator (8) has a particular banded structure called *block tri-diagonal*. There is no need to perform the calculations in Steps 1 and 2 that are outside of the band.

5. Infinite Markov chains

Markov chains with an infinite number of states often are more analytically tractable than the same model with a finite number of states because a boundary condition that is in the finite state version of the model is absent from the infinite state version. The M/G/1 queue is an example. Structured infinite-state Markov chains may also exhibit this property, which can be exploited in numerical algorithms. This work started in the mid 1960's (Wallace and Rosenberg, 1966; Evans, 1967) and gained impetus a decade later when computing became easier.

5.1. The main paradigms

We assume that the Markov chains under study are irreducible and persistent, so that they possess a unique stationary distribution. For discrete-time chains, we do not require aperiodicity; if this property is present, the stationary distribution is also a limiting distribution. Transition matrices and generators are expressed in block partition form, P (or Q) $= (A_{ij})$ say, where A_{ij} is a matrix (not necessarily square). This corresponds to a state space of ordered pairs, (i, k) say, where i runs from zero to infinity and is called the *level* and k has a finite support and is called the *phase*. The interpretation of A_{ij} is model specific, but the following example shows the general idea. Suppose the arrivals form a 2-state MMPP, the service times are exponential, and there is a single server. Let λ_k be the arrival rate in phase k and q_k be the rate the MMPP leaves phase k.

Let μ be the service rate. Then i and j represent the number of customers present and phase respectively, and the generator has

$$A_{i,i-1} = \begin{pmatrix} \mu & 0 \\ 0 & \mu \end{pmatrix}, \qquad A_{i,i} = \begin{pmatrix} y_1 & q_1 \\ q_2 & y_2 \end{pmatrix},$$

$$A_{i,i+1} = \begin{pmatrix} \lambda_1 & 0 \\ 0 & \lambda_2 \end{pmatrix}, \tag{39}$$

where y_1 and y_2 make the row sums equal zero.

The structure that leads to explicit solutions is repeating rows, except possibly for the first row and first column to allow for different boundary conditions. Specifically, the structure is a block version of a Toeplitz matrix,

$$\begin{pmatrix} B_0 & B_1 & B_2 & B_3 & \cdots \\ C_1 & A_0 & A_1 & A_2 & \cdots \\ C_2 & A_{-1} & A_0 & A_1 & \cdots \\ C_3 & A_{-2} & A_{-1} & A_0 & \cdots \\ \cdots & \cdots & \cdots & \cdots & \cdots \end{pmatrix}. \tag{40}$$

A formal representation of the stationary distribution of (40) as a transition matrix can be obtained (Grassmann and Heyman, 1990), and an algorithm for its computation developed (Grassmann and Heyman, 1993). When the matrices to the left of A_{-1} are all zero (*block upper-Hessenberg* or *skip-free-to-the-left in blocks*) or the matrices to the right of A_1 are all zero (*block lower-Hessenberg* or *skip-free-to-the-right in blocks*) an explicit solution can be obtained. The former is called the $M/G/1$ *paradigm* (Neuts, 1989) because it's the block form of the Markov chain used to analyze the M/G/1 queue that is shown in (1). The latter is called the $GI/M/1$ *paradigm* (Neuts, 1981) for an analogous reason. When the chain is skip-free in both directions it is called a *quasi-birth-and-death* (*QBD*) *process* because it's the block form of the generator of a birth-and-death process. We now examine algorithms for these three special cases.

5.2. The $GI/M/1$ paradigm

We change the subscripting used in (40) to make the subsequent equations neater. The specific matrix we study here is

$$\begin{pmatrix} C_0 & H_0 & 0 & 0 & \cdots \\ C_1 & H_1 & H_0 & 0 & \cdots \\ C_2 & H_2 & H_1 & H_0 & \cdots \\ C_3 & H_3 & H_2 & H_1 & \cdots \\ \cdots & \cdots & \cdots & \cdots & \cdots \end{pmatrix}. \tag{41}$$

Let (41) be a transition matrix, let $\pi = (\pi_0, \pi_1, \pi_2, \ldots)$ be the stationary vector partitioned conformally (π_i is associated with the row that starts with C_i). In the scalar GI/M/1 queue, the (scalar) stationary probabilities satisfy $\pi_{i+1} = r\pi_i$ where r is the root of $x = \sum_0^\infty x^i H_i$ inside the unit interval. Thus, $\pi_i = (1-r)r^i$, which is the geometric distribution. Suppose an analogous condition applies to (41); specifically, suppose there is a matrix R such that

$$\pi_{i+1} = \pi_i R \Leftrightarrow \pi_i = \pi_0 R^i, \quad i \geq 0. \tag{42}$$

This is called a *matrix-geometric distribution*. The balance equation for π_0 is

$$\pi_0 = \pi_0 \sum_0^\infty R^k C_k. \tag{43}$$

The balance equation for π_1 is

$$\pi_1 = \sum_0^\infty \pi_k H_k = \pi_0 \sum_0^\infty R^k H_k. \tag{44}$$

From (44), (42) can be valid only if

$$R = \sum_0^\infty R^k H_k. \tag{45}$$

Substituting (45) in the balance equation for π_2 yields

$$\pi_2 = \sum_0^\infty \pi_{k+1} H_k = \pi_0 R \sum_0^\infty R^k H_k = \pi_0 R^2. \tag{46}$$

Proceeding in this way yields (42). In order for

$$\sum_0^\infty \pi_k \mathbf{1} = 1 \tag{47}$$

(**1** is a column vector of ones), we must have $\lim_{k \to \infty} R^k = 0$, which implies that $(I - R)^{-1}$ exists and all of the eigenvalues of R are inside the unit disk. These results are formally proven in (Neuts, 1981, Chapter 1) and displayed in Theorem 1.2.1 there, along with the additional results that the matrix $\sum_0^\infty R^k C_k$ appearing in (43) is stochastic and that R is the minimal non-negative solution of

$$X = \sum_0^\infty X^k H_k. \tag{48}$$

The matrix R can be found from (48) by successive substitution starting with the initial solution $X = 0$. This produces an increasing sequence of matrices converging to

R (Neuts, 1981, Lemma 1.2.3). Various other algorithms for computing R have been proposed; Latouche (1993) compares them. Once R is computed, (43) is solved, and then π_0 is normalized by (47). The state probabilities then follow from (42).

From the matrix-geometric solution for the state probabilities, one can compute that the mean number of customers present in the steady-state is

$$L = \sum_0^\infty k\pi_k \mathbf{1} = \pi_0 \left(\sum_0^\infty k R^k \right) \mathbf{1} = \pi_0 (I - R)^{-1} \mathbf{1}. \tag{49}$$

The matrix R has a probabilistic meaning that can be deduced from (45). The skip-free-to-the-left structure implies that when there is a sequence of transitions such that the levels go monotonically from $n+k$ to n ($k > 1$), when the levels decrease, they decrease by one. Since the rows repeat, the probabilistic laws governing these transitions do not depend on n. Let r_{ij} be the mean number of times the chain enters state $(n+1, j)$ before returning to level n when starting from state (n, i), and $R = (r_{ij})$. The repeating rows implies that the (i, j)th element of R^k is the mean number of visits to state $(n+k, j)$ before returning to level n when starting in state (n, i), and conditioning on the first transition will show that R satisfies (45).

In computations, all H_k with k greater than some finite n are zero. The usual stopping criterion for iterating (48) is that the difference between successive iterates is smaller than a given tolerance. This criterion doesn't guarantee that the computed R is within a known, or even small, distance of the true R.

These results for a transition matrix formally carry over to the generator of a continuous-time Markov chain by uniformization (Neuts, 1981, Section 1.7). The numerical values are not the same, but (43) with the left side replaced by zero, (47) and (48) are valid.

5.3. The M/G/1 paradigm

The block matrix analog of the embedded Markov chain of the M/G/1 queue is

$$\begin{pmatrix} B_0 & B_1 & B_2 & B_3 & \cdots \\ A_0 & A_1 & A_2 & A_3 & \cdots \\ 0 & A_0 & A_1 & A_2 & \cdots \\ 0 & 0 & A_0 & A_1 & \cdots \\ \cdots & \cdots & \cdots & \cdots & \cdots \end{pmatrix}. \tag{50}$$

Permitting the first row to differ from the second row allows one to model various boundary conditions, such as letting the first customer served in a busy period to have a different service-time distribution from all other customers, and introduces only minor notational changes in the analysis. As above, let (50) be an ergodic transition matrix, and $\pi = (\pi_0, \pi_1, \pi_2, \ldots)$ be the stationary vector partitioned conformally.

The skip-free-to-the-left property leads to an analog of (45). Let g_{ij} be the probability that when starting in state $(n+1, i)$, $n > 0$, the state entered when the Markov chain next reaches level n is (n, j), and $G = (g_{ij})$. Then G is the minimal non-negative solution to

$$G = \sum_0^\infty A_k G^k \qquad (51)$$

and G is stochastic (Neuts, 1989, Chapter 2). The solution of (51) can be obtained by successive iterations. Experience has shown that convergence is most rapid when the initial trial solution is a stochastic matrix (Lucantoni, 1993).

The matrix G doesn't suffice to compute the steady-state distribution for the M/G/1 paradigm as the matrix R did for the GI/M/1 paradigm. The reason is the following. Temporarily use scalar states and let P be a irreducible transition matrix with positive-recurrent states. Then

$$I - P = (A - I)(B - S), \qquad (52)$$

where A is strictly upper-triangular, B is strictly lower-triangular, and S is diagonal with the upper-left element equal to zero and all other diagonal elements positive; (Grassmann, 1993) for a finite number of states and (Heyman, 1995) for infinitely many states. This is the analog of the LU factorization that arises when solving systems of linear equations. To obtain π, one first solves for a row vector α that satisfies $\alpha(B - S) = 0$. Since the top row of $B - S$ is zero, $\alpha = (d, 0, 0, \ldots)$ where d is any number is a solution. Then one solves

$$\pi(A - I) = \alpha \qquad (53)$$

for π. Choosing $d = -1$ yields $\pi_0 = 1$ and since $A - I$ is upper-triangular (53) is solved recursively, yielding the unnormalized solution

$$\pi_j = \sum_{i=0}^{j-1} \pi_i a_{ij}, \quad \pi_0 = 1, \; A = (a_{ij}). \qquad (54)$$

In the the M/G/1 paradigm, the matrix G determines the matrix B in the factorization. In the GI/M/1 paradigm, the matrix R determines the matrix A in the factorization and (54) yields the matrix-geometric solution.

The requisite formulas for obtaining π for the M/G/1 paradigm were derived by Ramaswami (1988). For $\pi = (\pi_0, \pi_1, \ldots)$ partitioned conformally with the transition matrix,

$$\pi_j + \left[\pi_0 B_j(G) + \sum_{i=1}^{j-1} \pi_i A_{j+1-i}(G) \right] [I - A_1(G)]^{-1}, \quad j \geq 1, \qquad (55)$$

where

$$B_j(G) = \sum_{i=j}^{\infty} B_i G^{i-j} \quad \text{and} \quad A_j(G) = \sum_{i=j}^{\infty} A_i G^{i-j}, \quad j \geqslant 0. \tag{56}$$

One can obtain the generating function of π just as in the scalar M/G/1 queue. Define the generating functions

$$\hat{\pi}(z) = \sum_{k=0}^{\infty} z^k \pi_k, \quad \widehat{A}(z) = \sum_{k=0}^{\infty} z^k A_k \quad \text{and} \quad \widehat{B}(z) = \sum_{k=0}^{\infty} z^k B_k \tag{57}$$

for $|z| \leqslant 1$. Write the balance equations in block form, multiply the kth equation by z^k and sum, to obtain

$$\hat{\pi}(z) = \pi_0 \widehat{B}(z) + \pi_1 \widehat{A}(z) + z\pi_2 \widehat{A}(z) + z^2 \pi_3 \widehat{A}(z) + \cdots \tag{58}$$

whence

$$\hat{\pi}(z)\bigl[zI - \widehat{A}(z)\bigr] = \pi_0 \bigl[z\widehat{B}(z) - \widehat{A}(z)\bigr]. \tag{59}$$

Obtaining π_0 is beyond the scope of this presentation; the formula and derivation are given by (Neuts, 1989, Theorem 3.2.1). This generalizes the Pollaczek–Khinchine formula for the scalar case. The difficulty in using (59) in numerical work is the computation of the matrices $\widehat{A}(z)$ and $\widehat{B}(z)$. The next section describes a special case in which this computation is not difficult. The special case is broad enough to include a wide variety of useful models.

6. The BMAP/G/1 queue

The batch Markovian arrival process (*BMAP*) was introduced by Neuts (1979). The BMAP/G/1 queue is a special case of the M/G/1 paradigm whose special structure allows the same performance measures to be computed as can be computed for the traditional M/G/1 queue. The BMAP includes a wide variety of non-Poisson processes that have been found useful in many applied studies, so this queueing model is frequently used. The first detailed solution of the BMAP/G/1 queue was given by Ramaswami (1980). Further results that make the computations tractable were obtained by Lucantoni (1991). This presentation follows Lucantoni (1993).

A BMAP is built on an underlying continuous-time Markov chain on a finite state space; these are the phases of the BMAP. Let D be the generator of this Markov process. When the phase process makes a transition to state i, a random number of arrivals occur, possibly zero. Let λ_i be the transition rate in phase-state i and $p_i(k, j)$ be the probability that when state i is entered, k arrivals occur and the next phase state will be j. Define the matrices D_k by $(D_0)_{ii} = -\lambda_i$, $1 \leqslant i \leqslant m < \infty$, $(D_k)_{ij} = \lambda_i p_i(k, j)$, $i \neq j$ and

$k \geq 0$. Then $D = \sum_0^\infty D_k$. Let δ be the stationary distribution of the generator D and $\bar{\lambda}$ be the arrival rate of the BMAP. Then

$$\bar{\lambda} = \delta \sum_{k=0}^{\infty} k D_k \mathbf{1}. \tag{60}$$

The matrix generating function

$$\widehat{D}(z) = \sum_{k=0}^{\infty} z^k D_k, \quad |z| \leq 1, \tag{61}$$

is a key quantity.

Let $N(t)$ be the number of arrivals in $(0, t]$ and $J(t)$ be the phase at time t. The process $\{N(t), J(t); t \geq 0\}$ is a continuous-time Markov chain with generator

$$\begin{pmatrix} D_0 & D_1 & D_2 & D_3 & \cdots \\ 0 & D_0 & D_1 & D_2 & \cdots \\ 0 & 0 & D_0 & D_1 & \cdots \\ 0 & 0 & 0 & D_0 & \cdots \\ \cdots & \cdots & \cdots & \cdots & \cdots \end{pmatrix}, \tag{62}$$

where the cumulative number of arrivals is the level. Let $P_{ij}(n, t) = \mathbf{P}[N(t) = n, J(t) = j | N(0) = 0, J(0) = i]$, $P(n, t) = (P_{ij}(n, t))$, and

$$\widehat{P}(z, t) = \sum_{n=0}^{\infty} z^n P(n, t) = e^{D(z)t}, \quad |z| \leq 1, \ t \geq 0. \tag{63}$$

It is this exponential form for $\widehat{P}$ which makes the analysis of the BMAP/G/1 queue tractable; it's analogous to the exponential form of the probability generating function of a Poisson process and permits a useful explicit formula for $\widehat{A}(z)$. Let H be the distribution of the service times, then

$$\widehat{A}(z) = \int_0^\infty \sum_{n=0}^{\infty} z^n P(n, t) \, dH(t) = \int_0^\infty \widehat{P}(z, t) \, dH(t)$$

$$= \int_0^\infty e^{D(z)} \, dH(t) = \widetilde{H}[-D(z)]. \tag{64}$$

When H has a rational Laplace–Stieltjes transform, say

$$\widetilde{H}(s) = \frac{\sum_0^m a_k s^k}{\sum_0^n b_k s^k}, \quad m \leq n, \tag{65}$$

then

$$\widehat{A}(z) = \left(\sum_0^n b_k \widehat{D}^k(z)\right)^{-1} \left(\sum_0^m a_k \widehat{D}^k(z)\right), \qquad (66)$$

which requires evaluating two matrix polynomials and one matrix inverse.

Among the special cases of a BMAP are renewal processes with Erlang, hyper-exponential and phase-type distributions, and MMPPs. An MMPP with underlying phase transition process with generator Q and arrival rates λ_i is the BMAP with $\Lambda = \text{diag}(\lambda_1, \ldots, \lambda_m)$, $D_0 = Q - \Lambda$, $D_1 = \Lambda$, and $D_k = 0$ for $k > 1$.

For BMAP arrivals, one can obtain

$$B_n = -D^{-1} \sum_{k=0}^n D_{k+1} A_{n-k} \quad \text{and}$$

$$\widehat{B}(z) = -D_0^{-1}[\widehat{D}(z) - D_0]\widehat{A}(z). \qquad (67)$$

Substituting (67) into (59) yields

$$\hat{\pi}(z)[zI - \widehat{A}(z)] = -\pi_0 D_0^{-1} \widehat{D}(z)\widehat{A}(z). \qquad (68)$$

To evaluate π_0, first obtain the matrix G, and the steady-state distribution of G, γ say. Let $\bar{\mu}^{-1}$ be the mean service time; the traffic intensity of the queue is $\rho = \bar{\lambda}/\bar{\mu}$, and $\rho < 1$ is the ergodicity condition. Then (Lucantoni, 1991, Eq. (54))

$$\pi_0 = (1 - \rho)\gamma(-D_0)/\bar{\lambda}. \qquad (69)$$

Taken together, (64), (68) and (69) provide an explicit solution for the probability generating function of the steady-state probabilities.

The Laplace transforms of the waiting-time distribution and transient state probabilities can be obtained by generalizing the analysis of the scalar case. Since the arrivals are not a pure Poisson process and may come in batches, care must be taken to distinguish probabilities at customer arrival epochs, at customer departure epochs and at an arbitrary time. Numerical values are then obtained by numerical transform inversion, so both of the general methods described in this article are used together. Special cases such as arrivals occuring one at a time, as a renewal process with a phase-type inter-arrival time distribution, service times having a phase-type distribution, and arrivals forming a 2-state MMPP produce more explicit formulas (Lucantoni, 1991).

7. The quasi birth-and-death process

The quasi-birth-and-death process (QBD) is skip-free in both directions, so it enjoys the special properties of the $GI/M/1$ and $M/G/1$ paradigms, but it's more than just a special case. By making the blocks large enough, infinite if necessary, the QBD can describe the general case of repeating rows. Latouche and Ramaswami (1999) develop the theory in this fashion. They also show how to exploit the special structure of the blocks to obtain efficient algorithms for the GI/M/1 and M/G/1 paradigms. The discussion here is limited to describing how the special structure of the QBD leads to a very fast algorithm for the state probabilities. The QBD is used frequently in applied studies because in prospective studies one often only has estimates of the mean arrival rate and service time, so exponential assumptions are made.

The specific matrix we study here is

$$\begin{pmatrix} C_0 & H_0 & 0 & 0 & \ldots \\ C_1 & H_1 & H_0 & 0 & \ldots \\ 0 & H_2 & H_1 & H_0 & \ldots \\ 0 & 0 & H_2 & H_1 & \ldots \\ \ldots & \ldots & \ldots & \ldots & \ldots \end{pmatrix}. \tag{70}$$

We interpret this as a transition matrix; continuous-time Markov chains are analyzed by uniformization. From (45) and (51) we have

$$R = H_0 + R H_1 + R^2 H_2 \tag{71}$$

and

$$G = H_0 + H_1 G + H_2 + H_2 G^2. \tag{72}$$

Let U_{ij} be the probability that starting in level $n > 0$ and phase i, the next visit to level n starts in phase j and there are no visits to level $n - 1$ in the interim. The repeating rows property implies that this probability does not depend on n. Then U is substochastic and is the minimal non-negative solution of

$$U = H_1 + H_0 (I + U + U^2 + \cdots) H_2 = H_1 + H_0 (I - U)^{-1} H_2. \tag{73}$$

These fundamental quantities are related by

$$G = (I - U)^{-1} H_2, \tag{74}$$

$$R = H_0 (I - U)^{-1} \tag{75}$$

and

$$U = H_1 + H_0 G = H_1 + R H_2. \tag{76}$$

From the discussion of the GI/M/1 paradigm we know we can compute the steady-state probabilities from R. Rather than compute R from (71), one uses a rapidly converging recursion for G, then computes U from G using (76), and then computes

1. (Initialize)
    (a) Set i = 0.
    (b) Compute $B_0 = (I - H_1)^{-1} H_0$.
    (c) Compute $B_2 = (I - H_1)^{-1} H_2$.
    (d) Set $G = B_2$ and $P = B_0$.
2. (Compute G) While $\min_i \sum_j G_{ij} < 1 - \varepsilon$ do the following:
    (a) $i \leftarrow i + 1$.
    (b) Compute $H_1^* = B_0 B_2$, $H_0^* = B_0^2$ and $H_2^* = B_2^2$.
    (c) Compute $B_0 = (I - H_1^*)^{-1} H_0^*$ and $B_2 = (I - H_1^*)^{-1} H_2^*$.
    (d) $G \leftarrow G + P B_2$ and $P \leftarrow P B_0$.
3. (Termination)
    (a) Compute $U = H_1 + H_0 G$.
    (b) Compute $R = H_0 + (I - U)^{-1}$.

Fig. 4. Logarithmic reduction algorithm.

R from U using (75). The new recursion for G was developed by Latouche and Ramaswami (1993). They consider censored versions of the Markov chain that visits only even numbered states. This yields a sequence of substochastic matrices G_n that converge monotonically to the stochastic matrix G;

$$G_n = \sum_{k=0}^{n} \left(\prod_{i=0}^{k-1} B_0^{(i)} \right) B_2^{(k)}, \quad G = \lim_{n \to \infty} G_n, \tag{77}$$

where

$$B_i^{(0)} = (I - H_1)^{-1} H_i, \quad i = 0, 2, \tag{78}$$

and

$$B_i^{(k+1)} = \left(I - B_0^{(k)} B_2^{(k)} - B_2^{(k)} B_0^{(k)} \right)^{-1} \left(B_i^{(k)} \right)^2 \tag{79}$$

for $i = 0, 2$ and $k = 0, 1, 2, \ldots$. (An empty product is defined to be I.) Since G_n is substochastic, a convenient convergence criterion is to terminate the iterations when all of the row sums are larger than $1 - \varepsilon$ for some specified small number $\varepsilon > 0$. The algorithm is displayed in Figure 4.

When there are m rows in each block, an iteration of this algorithm requires $(25/3)m^3 + O(m)$ flops. Computational experience has shown that this algorithm converges very rapidly and that all the computations can be done accurately.

References

Abate, J. and W. Whitt (1992a). The Fourier-series method for inverting transforms of probability distributions. *Queueing Systems* **10**, 5–88.

Abate, J. and W. Whitt (1992b). Numerical inversion of probability generating functions. *Oper. Res. Lett.* **12**, 245–251.

Berson, S., E. de Souza de Silva and R. R. Muntz (1991). A methodology for the specification and generation of Markov models. In Stewart (1991), pp. 11–36.

Choudhury, G. L., D. M. Lucantoni and W. Whitt (1994). Multidimensional transform inversion with application to the transient M/G/1 queue. *Ann. Appl. Probab.* **4**, 719–740.

Cohen, J. W. (1969). *The Single Server Queue*. North-Holland, Amsterdam.

Cooper, R. B. (1981). *Introduction to Queueing Theory*, 2nd ed. North-Holland, New York.

Courtois, P.-J. (1977). *Decomposability*. Academic Press.

Evans, R. V. (1967). Geometric distribution in some two dimensional queueing systems. *Oper. Res.* **5**.

Franklin, P. (1964). *A Treatise on Advanced Calculus*. Dover, New York.

Golub, G. H. and C. F. Van Loan (1989). *Matrix Computations*, 2nd ed. Johns Hopkins University Press, Baltimore.

Grassmann, W. K. (1983). Rounding errors. *Technical report*, Operations Research Department, Stanford University.

Grassmann, W. K. (1991). Finding transient solutions in Markovian event systems. In Stewart (1991), pp. 357–371.

Grassmann, W. K. (1993). Means and variances in Markov reward systems. In Meyer and Plemmons (1993).

Grassmann, W. K. and D. P. Heyman (1990). Equilibrium distribution of block-structured Markov chains with repeating rows. *J. Appl. Probab.* **27**, 557–576.

Grassmann, W. K. and D. P. Heyman (1993). Computation of steady-state probabilities for infinite-state Markov chains with repeating rows. *ORSA J. Comput.* **5**, 292–303.

Grassmann, W. K., M. I. Taksar and D. P. Heyman (1985). Regenerative analysis and steady-state distributions for Markov chains. *Oper. Res.* **33**, 1107–1116.

Greenberg, A. and R. Vanderbei (1991). Quicker convergence for iterative numerical solutions to stochastic problems: probabilistic interpretation, ordering heuristics, and parallel processing. *Probab. Engrg. Inform. Sci.* **4**, 493–521.

Heyman, D. P. (1987). Further comparisons of some direct methods for computing stations distributions of Markov chains. *SIAM J. Alg. and Disc. Meth.* **8**, 52–60.

Heyman, D. P. (1995). A decomposition theorem for infinite stochastic matrices. *J. Appl. Probab.* 893–901.

Heyman, D. P. and M. J. Sobel (1982). *Stochastic Models in Operations Research*, Vol. 1. McGraw-Hill, New York.

Latouche, G. (1993). Algorithms for infinite Markov chains with repeating columns. In Meyer and Plemmons (1993).

Latouche, G. and V. Ramaswami (1993). A logarithmic reduction algorithm for the quasi-birth-and-death process. *J. Appl. Probab.* **30**, 650–674.

Latouche, G. and V. Ramaswami (1999). *Introduction to Matrix Analytic Methods in Stochastic Modeling*. SIAM.

Lucantoni, D. (1991). New results on the single server queue with a batch Markovian arrival process. *Stochastic Models* **7**, 1–46.

Lucantoni, D. (1993). The BMAP/G/1 queue: a tutorial. In *Models and Techniques for Performance Evaluation of Computer and Communication Systems*, pp. 330–358 (Eds. L. Donatiello and R. Nelson). Springer-Verlag.

Meyer, C. D. and R. J. Plemmons (Eds.) (1993). *Linear Algebra, Markov Chains and Queueing Models*. Springer-Verlag.

Moler, C. and C. Van Loan (1978). Nineteen dubious ways to compute the exponential of a matrix. *SIAM Rev.* **20**, 801–836.

Neuts, M. F. (1975). Probability distributions of phase type. In *Liber Amicorium Prof. Emeritus H. Florin*, pp. 173–206. University of Louvin.

Neuts, M. F. (1979). A versatile Markovian point process. *J. Appl. Probab.* **16**, 764–779.
Neuts, M. F. (1981). *Matrix-Geometric Solutions in Stochastic Models*. The Johns Hopkins University Press.
Neuts, M. F. (1989). *Structured Stochastic Matrices of M/G/1 Type and their Applications*. Marcel Dekker.
O'Cinneide, C. A. (1993). Error analysis of a variant of Gaussian elimination for steady-state distributions of Markov chains. *Numer. Math.* **65**, 109–120.
Ramaswami, V. (1980). The N/G/1 queue and its detailed analysis. *Adv. Appl. Probab.* **12**, 222–261.
Ramaswami, V. (1988). A stable recursion for the steady-state vector in Markov chains of M/G/1 type. *Stochastic Models* **4**, 193–188.
Seelen, L. P. (1992). An algorithm for Ph/Ph/c queues. *Eur. J. Oper. Res.* **23**, 118–127.
Stewart, W. J. (Ed.) (1991). *Numerical Solution of Markov Chains*. Marcel Dekker.
Stewart, W. J. (1994). *Introduction to the Numerical Solution of Markov Chains*. Princeton University Press, Princeton.
Takács, L. (1964). Priority queues. *Oper. Res.* **12**, 63–74.
Wallace, V. L. and R. S. Rosenberg (1966). Markovian models and numerical analysis of computer system behavior. In *Proc. AFIPS Spring Joint Computer Conference*, Vol. 28. AFIPS Press, New Jersey.

Applications of Markov Chains to the Distribution Theory of Runs and Patterns

M. V. Koutras

1. Introduction

The concept of run and pattern arises in a quite natural way in every application involving experimental trials with two or more possible outcomes in each trial. An ecologist studies the spread of a disease by identifying specific patterns of infected/non-infected plants in a transect through a field (Pielou, 1962, 1963a, 1963b, 1977). For a mechanical engineer performing a start-up test for a new machine, it is reasonable to couch his decision (accepting the machine or rejecting it) on the number of consecutive successful or unsuccessful attempted start-ups (Hahn and Gage, 1983; Viveros and Balakrishnan, 1993; Balakrishnan et al., 1995, 1997). In the context of reliability, the same model leads to the well known consecutive-k-out-of-n: F system and its variations (for a review refer to Chao et al., 1995; see also Feder, 1974). An educational psychologist evaluates his subject's learning capability or his method's efficiency by examining patterns of level achievements of students exposed to the learning process (Grant, 1946, 1947; Bogartz, 1965). Many quality control plans base the acceptance/rejection of the sample lot on the occurrence of prolonged sequences of successive working/failed components (Wolfowitz, 1943; Kitagawa and Seguchi, 1956; Balakrishnan et al., 1993). Finally, another interesting application of the concept of runs comes from the area of non-parametric runs tests (Gibbons, 1971; Gibbons and Chakraborti, 1992; Koutras and Alexandrou, 1997a; Koutras et al., 1994, 1995). In this case, the interest focuses on the conditional distribution of runs or equivalently, on runs defined in a sequence of outcomes of pre-specified composition (Barton and David, 1957; David and Barton, 1962; Schuster, 1991, 1994).

We have already mentioned runs and patterns without having defined them formally. This is not surprising, since the term run and pattern is used in the field of Probability and Statistics in almost the same way as it is used in common language. Thus, if a sequence of multistate trials is considered (sequence of trials with two or more possible outcomes in each trial) a *run* of a certain type of elements is an uninterrupted sequence of such elements. By the term *pattern* $\mathcal{E}$ a specific string or family of strings with given composition is understood. For example, in the sequence 2111123133312333 we have one run of 1's of length 4, 2 runs of 3's of length 3, two occurrences of the pattern $\mathcal{E} = 123$, three occurrences of the composite pattern $\mathcal{E} = 12$ or 13, etc.

Run and pattern problems have attracted the attention of probabilists and statisticians as far back as the 18th century. The origin of problems pertaining to success runs in sequences of binary trials begins with De Moivre (1756). The classical framework for a *fixed length* run-related problem is mentioned in Feller (1968). A sequence of n Bernoulli trials is observed, with the possible outcomes labelled as success (S) or failure (F), and the number of *non-overlapping* and recurrent success runs of length k is counted (k is a fixed positive integer). A common notation for this statistic is $N_{n,k}$. There are several other alternative ways of counting runs. Ling (1988) suggested the *overlapping* scheme where a string of $m \geqslant k$ successes bordered at each end by failures or by the beginning or by the end of the sequence, accounts for $m - k + 1$ success runs. The number of overlapping success runs of length k in a sequence of n Bernoulli trials will be denoted by $M_{n,k}$. Of great statistical importance is also the number $G_{n,k}$ of success runs of length at least k (cf. Mood, 1940 or Gibbons, 1971). Finally the success runs of length exactly k will be denoted by $E_{n,k}$ (see Mood, 1940). To make the distinction between the four enumeration schemes more transparent we mention by way of example that in the sequence of 20 outcomes

$$SFSSSSSSFSSSFSSSSFSS$$

we have

$$N_{n,2} = 7, \quad M_{n,2} = 11, \quad G_{n,2} = 4, \quad E_{n,2} = 1,$$
$$N_{n,3} = 4, \quad M_{n,3} = 7, \quad G_{n,3} = 3, \quad E_{n,3} = 1.$$

The distributions of the aforementioned random variables (for fixed k) have been termed in the statistical literature as *binomial distributions of order k* and have been extensively studied by Philippou (1984), Philippou and Makri (1986), Aki and Hirano (1988), Godbole (1990, 1991, 1992), Hirano et al. (1991), Hirano and Aki (1993), etc. Clearly, the binomial distributions of order $k = 1$ coincide to the usual binomial law, a fact justifying the *order k* nomenclature. For a detailed and systematic exposition of the distribution theory of runs and scans (a special case of patterns) the interested reader may wish to consult the monograph by Balakrishnan and Koutras (2002).

Recently, Fu and Koutras (1994) developed a unified method for capturing the exact distribution of the number of runs of specified length by employing a Markov chain embedding technique. Koutras and Alexandrou (1995) refined the method and expressed these distributions in terms of multidimensional *binomial type* probability vectors. Fu (1996) extended the original method to cover the case of arbitrary patterns (instead of runs) whereas Koutras (1997a, 1997b) treated several waiting time problems within this framework; see also Chadjiconstantinidis, Antzoulakos and Koutras (2000). Finally Alexandrou (1997), Doi and Yamamoto (1998) and Han and Aki (1999) considered the case of multivariate run related distributions and offered simple solutions to the problem by exploiting proper extensions of the Markov chain embedding technique.

The present paper is mainly of an expository nature, its aim being to present the theoretical aspects of the Markov chain embedding techniques and elucidate the way this approach can be used to derive the exact distributions of the number of success runs and patterns in sequences of trials. In Section 2 we introduce the general Markov chain embedding method and present the main result on the evaluation of the exact distribution of enumerating random variables. In Section 3 we deal with specific run related distributions and illustrate how the exact distribution of the number of patterns can be established as well. Section 4 deals with the most interesting subclass of the Markov chain embeddable family, the variables of binomial type or shortly *MVB*; the distribution of these variables can be calculated by the aid of binomial type (triangular) vector recurrence relations, a fact justifying the nomenclature used. In Section 4 we present the main tools offered by the *MVB* approach for obtaining the exact distribution (probability mass function and generating functions) and proceed to Section 5 where waiting time problems associated with *MVB*'s are treated. In Section 6 we apply the *MVB* methodology to success runs and patterns problems. Finally Sections 7 and 8 treat in some detail the cases of multivariate enumerating variables within the framework of the Markov chain embedding technique.

2. The Markov chain embedding technique

A typical method for handling success run problems pertaining to repeated trials each of which admits two possible outcomes S or F (with probabilities p, q respectively), is to establish a proper discrete Markov chain which keeps track of the current success run length (see, e.g., Taylor and Karlin, 1984). Let $\{Y_t, t \geq 0\}$ be a stochastic process over $\{0, 1, 2, \ldots\}$ obtained by labeling the present state of the process by the length of the success run under way, i.e.,

- if the tth trial resulted in a failure, then $Y_t = 0$,
- if the preceding $i + 1$ trials in order resulted in $FS\ldots S$, the state variable Y_t would carry the label i.

Manifestly, the process forms a Markov chain and its transition probability matrix has the form

$$\begin{bmatrix} q & p & 0 & 0 & \cdot & \cdot & \cdot \\ q & 0 & p & 0 & \cdot & \cdot & \cdot \\ q & 0 & 0 & p & \cdot & \cdot & \cdot \\ \cdot & \cdot & \cdot & \cdot & \cdot & & \\ \cdot & \cdot & \cdot & \cdot & & \cdot & \\ \cdot & \cdot & \cdot & \cdot & & & \cdot \end{bmatrix}.$$

Accumulating states $k, k+1, \ldots$ in an absorbing state, labeled as "state k", we deduce a finite Markov chain over the state space $\{0, 1, \ldots, k\}$ with transition probability matrix

$$M = \begin{bmatrix} q & p & 0 & \cdot & 0 & 0 \\ q & 0 & p & \cdot & 0 & 0 \\ \cdot & \cdot & \cdot & \cdot & \cdot & \cdot \\ q & 0 & 0 & \cdot & 0 & p \\ 0 & 0 & 0 & \cdot & 0 & 1 \end{bmatrix}_{(k+1) \times (k+1)} \tag{1}$$

It is now clear that the probability distribution function of the random variables $N_{n,k}$, $M_{n,k}$, $G_{n,k}$, $E_{n,k}$ at $x = 0$ can be expressed as

$$\Pr(N_{n,k} = 0) = \Pr(M_{n,k} = 0) = \Pr(G_{n,k} = 0) = \Pr(E_{n,k} = 0)$$
$$= 1 - \Pr(Y_n = k) = 1 - \mathbf{e}_1 M^n \mathbf{e}'_{k+1}, \tag{2}$$

where $\mathbf{e}_i$, $i = 1, 2, \ldots, k+1$, are the unit (row) vectors of the space $\mathbb{R}^{k+1}$.

The preceding analysis and formula (2) has been used by Chao and Fu (1989, 1991), Fu and Lou (1991) and Koutras (1996) for the study of reliability structures and by Feder (1974) for modelling certain large assemblies with serially connected fasteners. For a further application in start-up demonstration tests, the interested reader may refer to Viveros and Balakrishnan (1993).

Recently, Fu and Koutras (1994) developed a method for capturing the whole distribution (i.e., not only at $x = 0$) of run-related statistics by exploiting a proper Markov chain embedding technique. As a matter of fact, the approach taken there is much more general and can practically cover almost all cases where the interest focuses on the distribution of an enumeration variable in a sequence of binary (or multistate) trials.

The Markov chain approach possesses great potential and simplifies substantially the solution of the problem it is applied to. As will be elucidated later, a great advantage of this approach is that, by some trivial adjustments in the form of the transition probability matrix, it remains valid even for cases where the assumption of i.i.d. trials is relaxed.

We shall first introduce the notion of a *Markov chain embeddable variable*, in a way similar to the one used by Fu and Koutras (1994) and proceed next to the examination of the most commonly used success run statistics by this approach.

Let X_n (n a non-negative integer) be a non-negative finite integer-valued random variable and denote by $l_n = \max\{x \colon \Pr(X_n = x) > 0\}$ its upper end point.

DEFINITION 2.1. The random variable X_n will be called a Markov chain embeddable variable if

(a) there exists a Markov chain $\{Y_t \colon t \geq 0\}$ defined on a state space $\Omega = \{\alpha_1, \alpha_2, \ldots\}$ which can be partitioned as

$$\Omega = \bigcup_{x \geq 0} C_x.$$

(b) the probability mass function of X_n can be captured by considering the projection of the probability space of Y_n onto C_x, i.e.,

$$\Pr(X_n = x) = \Pr(Y_n \in C_x), \quad x = 0, 1, \ldots, l_n.$$

Let us denote by Λ_t the one-step transition probability matrix of the Markov chain ($\{Y_t \colon t \geq 0\}, \Omega$) and by $\mathbf{e}_i$ the unit (row) vectors having 1 at the ith coordinate and 0 elsewhere. Then the exact distribution of X_n can be derived by the aid of the next theorem whose proof is an immediate consequence of the well known Chapman–Kolmogorov equations.

THEOREM 2.1. *If X_n is a Markov chain embeddable variable with transition probability matrix Λ_t then*

$$\Pr(X_n = x) = \pi_0 \left(\prod_{t=1}^{n} \Lambda_t \right) \sum_{i:\, \alpha_i \in C_x} \mathbf{e}'_i, \quad x = 0, 1, \ldots, l_n, \qquad (3)$$

where

$$\pi_0 = \big(\Pr(Y_0 = \alpha_1),\ \Pr(Y_0 = \alpha_2),\ \Pr(Y_0 = \alpha_3), \ldots\big)$$

is the vector of initial probabilities of the Markov chain $\{Y_t: t \geq 0\}$.

In order to keep the previous (and several forthcoming results) valid for $n = 0$, the following convention will be used: $\prod_{t=a}^{b} \Lambda_t = I$ if $a > b$. Evidently, if $\Lambda_t = \Lambda$ for all $t \geq 1$ (homogeneous Markov chain), the exact distribution of X_n takes on the next form, which exhibits a clear resemblance to formula (2)

$$P(X_n = x) = \pi_0 \Lambda^n \sum_{i:\, \alpha_i \in C_x} \mathbf{e}'_i, \quad x = 0, 1, \ldots, l_n.$$

In view of Theorem 2.1, in order to establish the exact distribution of a Markov chain embeddable variable, one needs to identify

(a) a state space Ω and a proper partition $\{C_x,\ x \geq 0\}$ of it;
(b) the transition probability matrix associated with the embedded variable;
(c) the initial probability vector π_0.

The next three sections will serve as an illustration of how one can obtain the exact distributions of runs and patterns by the aid of the Markov chain embedding technique.

3. Success runs and pattern distributions

Let $Z_1, Z_2, \ldots, Z_n$ be a sequence of Bernoulli trials with success (S) probabilities $0 < p < 1$ and failure (F) probabilities $q = 1 - p$. A typical element of the state space Ω will be represented by a 2-tuple (x, i). The state variable Y_t, $t \geq 1$, will take the value $(x, 0)$ if in the sequence of outcomes up to the tth trial, x success runs of length k have been registered and the last outcome of the sequence is F, i.e., $Z_t = 0$. Roughly speaking, the rest state labels (x, i), $i \geq 1$, indicate that, at time t, x occurrences of success runs of length k have already been observed and there is under way an additional success run of length $i < k$.

3.1. Non-overlapping success runs

In order to embed $N_{n,k}$, we observe first that $l_n = \max\{x\colon \Pr(N_{n,k} = x) > 0\} = [n/k]$, and partition the state space $\Omega = \{(x; i)\colon x = 0, 1, \ldots, l_n \text{ and } i = 0, 1, \ldots, k-1\}$ as

$$\Omega = \bigcup_{0 \leqslant x \leqslant l_n} C_x, \quad C_x = \{(x, i)\colon i = 0, 1, \ldots, k-1\}.$$

Define next Y_t, $t \geqslant 1$ as follows: $Y_t = (x, i)$ if and only if in the sequence of outcomes leading to the tth trial, say $SFSSF \ldots F \overbrace{SS \ldots S}^{m}$, there exist x non-overlapping success runs and m trailing successes with $i = m \pmod{k}$. For $t = 0$ set $Y_t = (0, 0)$, and therefore $\pi_0 = \mathbf{e}_1$. In order to obtain a stochastic transition probability matrix we label the last state $(l_n, k-1)$ as an absorbing state.

By the above definitions, there are only three feasible types of transitions for $\{Y_t, t \geqslant 0\}$ (excluding the transitions to the absorbing state):

- from state (x, i) to state $(x, 0)$, for $0 \leqslant x \leqslant l_n$ and $0 \leqslant i \leqslant k-1$;
- from state (x, i) to state $(x, i+1)$, for $0 \leqslant x \leqslant l_n$ and $0 \leqslant i \leqslant k-2$;
- from state $(x, k-1)$ to state $(x+1, 0)$, for $0 \leqslant x \leqslant l_n - 1$.

The first transition is associated with the occurrence of a failure at the tth trial while the other two with the occurrence of a success. For the second transition the sequence leading to the $(t-1)$th trial ends up with $i \leqslant k-2$ trailing successes (which have not been accounted for in a registered success run of length k); for the third type, there are $k-1$ trailing successes and the outcome of the tth trial results in the formulation of an additional success run of length k.

Summarizing the aforementioned transitions, the only non-zero entries of the transition probability matrix of $\{Y_t, t \geqslant 0\}$ are given by

$$\begin{aligned}
\Pr(Y_t = (x, 0) | Y_t = (x, i)) &= q, & 0 \leqslant x \leqslant l_n,\ 0 \leqslant i \leqslant k-1, \\
\Pr(Y_t = (x, i+1) | Y_t = (x, i)) &= p, & 0 \leqslant x \leqslant l_n,\ 0 \leqslant i \leqslant k-2, \\
\Pr(Y_t = (x+1, 0) | Y_t = (x, k-1)) &= p, & 0 \leqslant x \leqslant l_n - 1, \\
\Pr(Y_t = (x, k-1) | Y_t = (x, k-1)) &= 1, & x = l_n.
\end{aligned} \quad (4)$$

By way of example consider the case $n = 7$, $k = 3$. Then $l_n = [7/3] = 2$ and $\Omega = C_0 \cup C_1 \cup C_2$ where

$$C_0 = \{(0, 0), (0, 1), (0, 2)\},$$
$$C_1 = \{(1, 0), (1, 1), (1, 2)\},$$
$$C_2 = \{(2, 0), (2, 1), (2, 2)\}.$$

The transition probability matrix is now given by

$$\Lambda = \begin{bmatrix} (0,0) & (0,1) & (0,2) & (1,0) & (1,1) & (1,2) & (2,0) & (2,1) & (2,2) \\ q & p & 0 & & & & & & \\ q & 0 & p & & & & & & \\ q & 0 & 0 & p & & & & & \\ \hline & & & q & p & 0 & & & \\ & & & q & 0 & p & & & \\ & & & q & 0 & 0 & p & & \\ \hline & & & & & & q & p & 0 \\ & & & & & & q & 0 & p \\ & & & & & & 0 & 0 & 1 \end{bmatrix} \tag{5}$$

and the probability mass function of $N_{7,3}$ can be easily captured by the formula

$$\Pr(N_{7,3} = x) = \mathbf{e}_1 \Lambda^7 \sum_{i=3x+1}^{3x+3} \mathbf{e}'_i, \quad x = 0, 1, 2,$$

where $\mathbf{e}_i$ are the unit (row) vectors of $\mathbb{R}^9$.

For the general case, it is not difficult to verify that the transition probability matrix Λ can be written as a bidiagonal block matrix of the form

$$\Lambda = \begin{bmatrix} A & B & & & \\ & A & B & & \\ & & \cdot & \cdot & \\ & & & \cdot & \cdot \\ & & & & A & B \\ & & & & & A^* \end{bmatrix}, \tag{6}$$

where

$$A = \begin{bmatrix} q & p & 0 & \cdot & 0 \\ q & 0 & p & \cdot & 0 \\ \cdot & \cdot & \cdot & \cdot & \cdot \\ q & 0 & 0 & \cdot & p \\ q & 0 & 0 & \cdot & 0 \end{bmatrix}_{k \times k}, \quad B = \begin{bmatrix} 0 & 0 & 0 & \cdot & 0 \\ 0 & 0 & 0 & \cdot & 0 \\ \cdot & \cdot & \cdot & \cdot & \cdot \\ 0 & 0 & 0 & \cdot & 0 \\ p & 0 & 0 & \cdot & 0 \end{bmatrix}_{k \times k}, \tag{7}$$

Matrix A^* is of the same form as A with its last row having been replaced by $(0, \ldots, 0, 1)$. Clearly Λ's dimension is $\sum_{x=0}^{l_n} |C_x| = (l_n + 1) \cdot k$.

It is worth mentioning that transition probability matrices (5) and (6) have been given in their full form, in the sense that, all possible states (x, i) are listed no matter whether they can be reached by the state variable Y_t or not. For example, in the special case $n = 7$, $k = 3$ considered earlier, it is evident that Y_t cannot reach state $(2, 2)$. We could therefore eliminate it from the state space Ω, turn state $(2, 1)$ to an absorbing state and work with an 8×8 transition probability matrix instead of the 9×9 matrix Λ. Such shortcuts are quite useful especially when we are interested in numerical calculations and the values of n, k are large.

3.2. Overlapping success runs

The embedding of the random variable $M_{n,k} = \sum_{j=1}^{n-k+1}(\prod_{i=j}^{j+k-1} Z_i)$ which enumerates the overlapping success runs in a sequence of n Bernoulli trials $Z_1, Z_2, \ldots, Z_n$ can be achieved as follows: observe first that $l_n = \max\{x\colon \Pr(M_{n,k} = x) > 0\} = n - k + 1$, introduce the state space

$$\Omega = \{(x, i)\colon x = 0, 1, \ldots, l_n \text{ and } i = 0, 1, \ldots, k - 1\}$$
$$\cup \{(x, -1)\colon x = 0, 1, \ldots, l_n\}$$

and partition it as

$$\Omega = \bigcup_{0 \leqslant x \leqslant l_n} C_x, \quad C_x = \{(x, i)\colon i = -1, 0, \ldots, k - 1\}.$$

Consider next a typical sequence of outcomes, say $SFSSF\ldots F\overbrace{SS\ldots S}^{m}$ leading up to the tth trial, and define

- $Y_t = (x, m)$ if $m \leqslant k - 1$ and x overlapping success runs of length k have been counted so far;
- $Y_t = (x, -1)$ if $m \geqslant k$ and x overlapping success runs of length k have been counted.

State $(0, -1)$, which has no meaning in terms of the latter description, could be ruled out; however, in order to keep the cardinalities of all C'_xs equal (which will be proved useful later on), we shall retain it with the understanding that this does not cause any complication to the process, since $(0, -1)$ is in fact inaccessible.

It is now easy to verify that $\{Y_t, t \geqslant 0\}$ is a Markov chain over Ω with transition probability matrix of the form (6) where

$$A = \begin{bmatrix} q & p & 0 & \cdot & 0 & 0 \\ q & 0 & p & \cdot & 0 & 0 \\ \cdot & \cdot & \cdot & \cdot & \cdot & \cdot \\ q & 0 & 0 & \cdot & p & 0 \\ q & 0 & 0 & \cdot & 0 & 0 \\ q & 0 & 0 & \cdot & 0 & 0 \end{bmatrix}_{(k+1)\times(k+1)},$$

$$B = \begin{bmatrix} & & & & & 0 \\ & & & & & 0 \\ & & \mathbf{O}_{k\times k} & & & \cdot \\ & & & & & 0 \\ & & & & & p \\ 0 & 0 & 0 & \cdot & 0 & p \end{bmatrix}_{(k+1)\times(k+1)}$$

(8)

and A^* is a duplicate of A with its last row replaced by $(0, \ldots, 0, 1)$. It should be stressed that the first block of matrices A and B corresponds to states (x, i): $i = 0, \ldots, k - 1$ while the last row and column to states $(x, -1)$.

3.3. Success runs of length at least k

The random variable of interest is now

$$G_{n,k} = \sum_{j=1}^{n-k+1} (1 - Z_{j-1}) \left(\prod_{i=j}^{j+k-1} Z_i \right)$$

(convention: $Z_0 = 0$) and its upper end point is $l_n = \max\{x: \Pr(G_{n,k} = x) > 0\} = [(n+1)/(k+1)]$. The state space Ω and the partition used for $M_{n,k}$ can also be used for embedding $G_{n,k}$. Nevertheless, the states $(x, -1)$ have a quite different interpretation (and a different impact on the enumerating procedure) than the one employed in $M_{n,k}$. To become more precise, $(x, -1)$ will be serving here as *waiting state* in the sense that, when the process is in this state, $m \geq k$ trailing successes (S) have been observed and we are waiting for the occurrence of a failure (F) so that a complete success run of length at least k is registered. The formal definition of the state variable Y_t, $t \geq 0$, is as follows: let $SFSFF \ldots F \overbrace{SS \ldots S}^{m}$ be a typical sequence of outcomes for the segment $Z_1, Z_2, \ldots, Z_t$ of the binary sequence $Z_1, Z_2, \ldots, Z_n$; then we set

- $Y_t = (x, i)$, $0 \leq i \leq k - 1$, if there exist exactly $x \geq 0$ success runs of length at least k before the last $m + 1$ outcomes and $i = m \leq k - 1$;
- $Y_t = (x, -1)$, $x \geq 1$, if there exist $x - 1$ success runs of length at least k before the last $m + 1$ outcomes and $m \geq k$.

As far as state $(0, -1)$ is concerned, a comment similar to the one made in Section 3.2 applies here as well.

The aforementioned definitions establish a finite Markov chain on Ω with transition probability matrix of the form (6) where

$$A = \begin{bmatrix} q & p & 0 & \cdot & 0 & 0 \\ q & 0 & p & \cdot & 0 & 0 \\ \cdot & \cdot & \cdot & \cdot & \cdot & \cdot \\ q & 0 & 0 & \cdot & p & 0 \\ q & 0 & 0 & \cdot & 0 & 0 \\ q & 0 & 0 & \cdot & 0 & p \end{bmatrix}_{(k+1) \times (k+1)},$$

$$B = \begin{bmatrix} & & & & 0 & \\ & & & & 0 & \\ & \mathbf{0}_{k \times k} & & & \cdot & \\ & & & & 0 & \\ & & & & p & \\ 0 & 0 & 0 & \cdot & 0 & 0 \end{bmatrix}_{(k+1) \times (k+1)}.$$

(9)

Matrix A^* is the same with the one used for the non-overlapping and overlapping cases.

3.4. Success runs of length exactly k

For the study of the distribution of the number

$$E_{n,k} = \sum_{j=1}^{n-k+1} (1 - Z_{j-1}) \left(\prod_{i=j}^{j+k-1} Z_j \right) (1 - Z_{j+k})$$

(convention: $Z_0 = Z_{n+1} = 1$) of success runs of length exactly k, we may use the state space

$$\Omega = \{(x, i): x = 0, 1, \ldots, l_n \text{ and } i = 0, 1, \ldots, k - 1\}$$
$$\cup \{(x, -1): x = 0, 1, \ldots, l_n\} \cup \{(x, -2): x = 0, 1, \ldots, l_n\},$$

where $l_n = [(n + 1)/(k + 1)]$. States (x, i), $x = 0, 1, \ldots, l_n$, $i = 0, 1, \ldots, k - 1$, are defined as in the previous cases. The reason for introducing the additional classes of states $(x, -1)$, $(x, -2)$ arises from the need to establish a mechanism which will cancel the counting of a success run whose length exceeds k and wait for the termination of the series of successes before the success run formulation counter restarts.

The exact meaning of the additional states $(x, -1)$, $(x, -2)$ is the following: let $SFSSF \ldots F \overbrace{SS \ldots S}^{m}$ be a typical sequence ending with the tth trial. Then

- $Y_t = (x, -1)$, $x = 0, 1, \ldots, l_n$, if $m > k$ and x success runs of length exactly k have appeared before the last $m + 1$ outcomes;
- $Y_t = (x, -2)$, $x = 0, 1, \ldots, l_n$, if $m = k$ and exactly x success runs of length exactly k have appeared before.

Note again that state $(0, -2)$ is unreachable by the process and could be ruled out if a reduction of the transition probability matrix is desirable. The (non-zero) transition probabilities for the Markov chain established above are given by

$$\Pr(Y_t = (x, i + 1) | Y_{t-1} = (x, i)) = p \quad \text{for } 0 \leq x \leq l_n,\ 0 \leq i \leq k - 2,$$
$$\Pr(Y_t = (x + 1, -2) | Y_{t-1} = (x, k - 1)) = p \quad \text{for } 1 \leq x \leq l_n - 1,$$
$$\Pr(Y_t = (x - 1, -1) | Y_{t-1} = (x, -2)) = p \quad \text{for } 1 \leq x \leq l_n,$$
$$\Pr(Y_t = (x, -1) | Y_{t-1} = (x, -1)) = p \quad \text{for } 0 < x \leq l_n,$$
$$\Pr(Y_t = (x, 0) | Y_{t-1} = (x, i)) = q \quad \text{for } 0 \leq x \leq l_n.$$

3.5. Distribution of the number of patterns

We shall now consider a sequence of n multistate trials $Z_1, Z_2, \ldots, Z_n$ with $m \geq 2$ possible outcomes each. For typographical convenience let us label the m outcomes as $1, 2, \ldots, m$ and assume that the trials are i.i.d., i.e., they are independent and have the common distribution

$$p_j = \Pr(Z_i = j), \quad j = 1, 2, \ldots, m \text{ (for all } i = 1, 2, \ldots, n).$$

Under this set up, it is clear that we can define many types of runs of fixed length, for example runs of 1's of length k_1, runs of 2's of length k_2, etc. More generally we may define fixed length blocks with pre-specified composition which we shall term as *patterns* or *events* and denote by $\mathcal{E}$. A *single pattern* will consist of one specific string of symbols while a *composite* or *multiple pattern* will consist of at least two different strings of symbols. For example, $\mathcal{E} = 123221$ is a single pattern while $\mathcal{E} = 12$ or 312 is a composite pattern. Manifestly a success run of length k is a special case of single pattern.

In the sequel we shall present a method for computing the distribution of the number of occurrences of an event $\mathcal{E}$ (single or composite) in n multistate trials $Z_1, Z_2, \ldots, Z_n$. The type of counting used hereafter will be the non-overlapping one, although other enumeration schemes could be analyzed as well by performing some trivial adjustments to the methodology presented in this section.

To start with, let us consider first the case of a single pattern $\mathcal{E} = i_1 i_2 \ldots i_k$ of length k and denote by $\mathcal{E}^{(j)}$ the initial segment of $\mathcal{E}$ which includes the j first symbols of it ($j = 1, 2, \ldots, k-1$). The Markovian structure for the problem under inspection can be easily revealed if we employ once again states of the form (x, j), and define Y_t, $t \geq 0$, as follows: $Y_t = (x, j)$, $t \geq 1$, $j \geq 1$, if in the subsequence $Z_1, Z_2, \ldots, Z_t$ the pattern $\mathcal{E}$ has been observed x times and the ending block of $Z_1, Z_2, \ldots, Z_t$ coincides with $\mathcal{E}^{(j)}$. If the ending block does not match any of $\mathcal{E}^{(j)}$ then we set $Y_t = (x, 0)$. To complete the definition of the process, we set $Y_0 = (0, 0)$.

Clearly, if $Y_{t-1} = (x, k-1)$ and the last symbol of $\mathcal{E}$ is i_k then $Y_t = (x+1, 0)$ with probability p_{i_k}. On the other hand, if $Y_{t-1} = (x, j)$, $j \geq 0$, then $Y_t = (x, j')$ with probability $\sum p_i$ where the summation is performed over all i's which guarantee that the ending block $\mathcal{E}^{(j)}$ is turned (by the inclusion of i at the end) to an ending block of the form $\mathcal{E}^{(j')}$.

To make these notions more transparent, let us consider the special case $m = 3$ and the pattern $\mathcal{E} = 2112$ of length $k = 4$. Then $\mathcal{E}^{(1)} = 2$, $\mathcal{E}^{(2)} = 21$, $\mathcal{E}^{(3)} = 211$ and an appropriate state space for establishing a Markov chain embedding approach is offered by

$$\Omega = \bigcup_{x \geq 0} C_x, \quad C_x = \{(x, 0), (x, 1), (x, 2), (x, 3)\}.$$

By way of example we mention that, if the next $n = 10$ outcomes had been observed

3211221131,

then

$$
\begin{aligned}
&Y_1 = (0, 0), &&Y_2 = (0, 1), &&Y_3 = (0, 2), \\
&Y_4 = (0, 3), &&Y_5 = (1, 0), &&Y_6 = (1, 1), \\
&Y_7 = (1, 2), &&Y_8 = (1, 3), &&Y_9 = (1, 0), \\
&Y_{10} = (1, 0).
\end{aligned}
$$

The one step transition probabilities are given by

$$\Pr(Y_t = (x+1, 0) | Y_{t-1} = (x, 3)) = p_2,$$

$$\Pr(Y_t = (x, j') | Y_{t-1} = (x, 3)) = \begin{cases} 0, & \text{if } j' \neq 0, \\ 1 - p_2, & \text{if } j' = 0, \end{cases}$$

$$\Pr(Y_t = (x, j') | Y_{t-1} = (x, 0)) = \begin{cases} 1 - p_2, & \text{if } j' = 0, \\ p_2, & \text{if } j' = 1, \\ 0, & \text{if } j' = 2 \text{ or } 3, \end{cases}$$

$$\Pr(Y_t = (x, j') | Y_{t-1} = (x, 1)) = \begin{cases} p_{3-j'}, & \text{if } j' = 0, 1, 2, \\ 0, & \text{if } j' = 3 \end{cases}$$

and

$$\Pr(Y_t = (x, j') | Y_{t-1} = (x, 2)) = \begin{cases} p_1, & \text{if } j' = 3, \\ p_2, & \text{if } j' = 1, \\ p_3, & \text{if } j' = 0, \\ 0, & \text{if } j' = 2. \end{cases}$$

Summarizing these results in a transition probability matrix, we deduce (for $n = 10$)

$$M = \begin{bmatrix} A & B & \\ & A & B \\ & & A^* \end{bmatrix}, \tag{10}$$

where

$$A = \begin{bmatrix} p_1 + p_3 & p_2 & 0 & 0 \\ p_3 & p_2 & p_1 & 0 \\ p_3 & p_2 & 0 & p_1 \\ p_1 + p_3 & 0 & 0 & 0 \end{bmatrix}, \quad B = \begin{bmatrix} 0 & 0 & 0 & 0 \\ 0 & 0 & 0 & 0 \\ 0 & 0 & 0 & 0 \\ p_2 & 0 & 0 & 0 \end{bmatrix} \tag{11}$$

and A^* is of the same form as A, with its last row replaced by $(0, 0, 0, 1)$ (evidently $A + B$ is a stochastic matrix). The exact distribution of the number of occurrences of the pattern in the $n = 10$ multistate trials can be easily derived by formula (3) of Theorem 2.1 ($l_n = l_{10} = 2$, $\pi_0 = \mathbf{e}_1$ and $\Lambda_t = M$ for $t = 1, 2, \ldots, 10$).

The decomposition of a single pattern $\mathcal{E}$ to the sequence of subpatterns $\mathcal{E}^{(j)}$, $j = 1, 2, \ldots, k - 1$, and the subsequent initiation of the Markov chain embedding scheme as described above, has been termed by Fu (1996) as *forward and backward principle*.

The same principle can be effortlessly extended to cover the case of composite patterns. As an illustration, let us consider again $m = 3$ and take $n = 5$ and $\mathcal{E} = 12$ or 31. Then the multiple pattern $\mathcal{E}$ can be decomposed in $\mathcal{E}^{(1)} = 1$, $\mathcal{E}^{(2)} = 3$ and setting $Y_t = (x, j)$ if in the subsequence $Z_1, Z_2, \ldots, Z_t$ the pattern $\mathcal{E}$ has been observed x times and the ending block of $Z_1, Z_2, \ldots, Z_t$ is $\mathcal{E}^{(j)}$ ($Y_t = (x, 0)$ means that the ending block is neither $\mathcal{E}^{(1)}$ nor $\mathcal{E}^{(2)}$) we may easily deduce a transition probability matrix of the form (10) with

$$A = \begin{bmatrix} p_2 & p_1 & p_3 \\ 0 & p_1 & p_3 \\ p_2 & 0 & p_3 \end{bmatrix}, \qquad B = \begin{bmatrix} 0 & 0 & 0 \\ p_2 & 0 & 0 \\ p_1 & 0 & 0 \end{bmatrix}. \tag{12}$$

Theorem 2.1 can then be used to obtain the exact distribution of the total number of occurrences of the pattern 12 or 31 in the sequence of trinary outcomes $Z_1, Z_2, \ldots, Z_5$.

In closing, we mention that a proper modification of the Markov chain definition used before could produce the appropriate environment to accommodate other enumeration schemes, for example overlapping counting or enumeration in the sense of the "at least" principle. The details are left to the reader.

4. Markov chain embeddable variables of binomial type

From the previous sections it is clear that for the majority of enumerating random variables the transition probability matrix can be viewed as a bidiagonal blocked matrix with non-zero blocks appearing only on the main diagonal and the diagonal next to it (cf. (6)). Therefore, considering appropriate *probability vectors* describing the overall state formulation of the Markovian process at time t (i.e., when the segment $Z_1, Z_2, \ldots, Z_t$ is under inspection) we would naturally be led to certain multidimensional triangular recurrence relations. Motivated by this observation, Koutras and Alexandrou (1995) proceeded to the introduction of a significant subclass of the family of Markov chain embeddable variables which offers a computationally efficient framework for tackling problems related to the distribution of enumerating variables in sequences of binary or multistate trials.

To start with, let us first observe that without loss of generality we may assume that the state subspaces C_x, $x = 0, 1, \ldots$, have the same cardinality $s = |C_x|$; if this is not true we can merely incorporate into the C_x's, with $|C_x| < \max_{x \geq 0} |C_x|$, a number of hypothetical states, inaccessible to the process, so that its behavior is not affected.

DEFINITION 4.1. A non-negative random variable X_n will be called a *Markov chain Embeddable variable of Binomial type* (MVB) if

(a) there exists a Markov chain $\{Y_t, t \geq 0\}$ defined on a discrete state space Ω which can be partitioned as

$$\Omega = \bigcup_{x \geq 0} C_x, \quad C_x = \{c_{x,0}, c_{x,1}, \ldots, c_{x,s-1}\}.$$

(b) $\Pr(Y_t \in C_v | Y_{t-1} \in C_x) = 0$ for all $v \neq x, x+1$ and $t \geq 1$.
(c) The event $X_n = x$ is equivalent to $Y_n \in C_x$, i.e.,

$$\Pr(X_n = x) = \Pr(Y_n \in C_x).$$

From now on we shall be using index t for the steps of the Markov chain (Y_{t-1} will offer the information needed to proceed from the subsequence $Z_1, Z_2, \ldots, Z_{t-1}$ to $Z_1, Z_2, \ldots, Z_{t-1}, Z_t$). Index n will be reserved for the final stage where the whole sequence $Z_1, Z_2, \ldots, Z_n$ has been unfolded. The value of Y_n at this stage can uniquely determine the distribution of X_n.

Roughly speaking, an MVB is a subfamily of Markov chain embeddable variables which is characterized by the following property: the state subclasses C_x, $x \geqslant 0$, can be ordered in such a way that once the chain enters C_x, the one step transitions could lead either to the same state subclass or to the next one.

The two types of transition described above give rise to the next two $s \times s$ transition probability matrices

$$A_t(x) = \left(\Pr(Y_t = c_{x,j} | Y_{t-1} = c_{x,i})\right)_{s \times s},$$
$$B_t(x) = \left(\Pr(Y_t = c_{x+1,j} | Y_{t-1} = c_{x,i})\right)_{s \times s}, \quad t \geqslant 1.$$

The first one describes the transitions between states of the same state subclass C_x (*within states* one step transition matrix), while the second describes the transitions between states of two consecutive state subclasses C_x and C_{x+1} (*between states* one step transition matrix). Clearly condition (b) of Definition 4.1 guarantees that matrix $A_t(x) + B_t(x)$ is stochastic.

Denoting by

$$\boldsymbol{\pi}_x = \left(\Pr(Y_0 = c_{x,0}), \Pr(Y_0 = c_{x,1}), \ldots, \Pr(Y_0 = c_{x,s-1})\right), \quad x \geqslant 0,$$

the (row) vector of initial probabilities, we may state the following theorem.

THEOREM 4.1. *For every MVB X_n the sequence of vectors*

$$\mathbf{f}_t(x) = \left(\Pr(Y_t = c_{x,0}), \Pr(Y_t = c_{x,1}), \ldots, \Pr(Y_t = c_{x,s-1})\right),$$

$$0 \leqslant x \leqslant l_n, \ 1 \leqslant t \leqslant n,$$

satisfies the recurrence relations

$$\mathbf{f}_t(0) = \mathbf{f}_{t-1}(0) A_t(0),$$
$$\mathbf{f}_t(x) = \mathbf{f}_{t-1}(x) A_t(x) + \mathbf{f}_{t-1}(x-1) B_t(x-1), \quad 1 \leqslant x \leqslant l_n,$$

$$t = 1, 2, \ldots, n,$$

with initial conditions $\mathbf{f}_0(x) = \boldsymbol{\pi}_x$, $0 \leqslant x \leqslant l_n$. The probability distribution of X_n is given by

$$\Pr(X_n = x) = \mathbf{f}_n(x) \mathbf{1}', \quad x = 0, 1, \ldots, l_n,$$

where $\mathbf{1} = (1, 1, \ldots, 1) \in \mathbb{R}^s$.

The proof is an immediate consequence of the Chapman–Kolmogorov equations and Definition 4.1.

It is sufficient for our purposes (and facilitates the establishment of more compact formulae as well) to assume that $\pi_x = 0$, $x \geq 1$ and $\pi_0 \mathbf{1}' = 1$; this convention, which is actually equivalent to the condition $\Pr(X_0 = 0) = 1$, will be taken for granted in the sequel without further explicit reference to it.

If X_n enumerates certain patterns in a sequence of i.i.d. trials, the within states and between states one step transition probability matrices usually do not depend on x and t, i.e., $A_t(x) = A$, $B_t(x) = B$ for all x and t. In this case the random variable X_n will be called a *homogeneous MVB*, and the recurrence relations of Theorem 4.1 reduce to

$$\mathbf{f}_t(0) = \mathbf{f}_{t-1}(0)A,$$
$$\mathbf{f}_t(x) = \mathbf{f}_{t-1}(x)A + \mathbf{f}_{t-1}(x-1)B, \quad 1 \leq x \leq l_n, \qquad t = 1, 2, \ldots, n. \quad (13)$$

These recurrences lead to the establishment of a neat formula for the generating function of the vectors $\mathbf{f}_t(x)$, $x = 0, 1, \ldots, l_n$, i.e.,

$$\boldsymbol{\phi}_t(z) = \sum_{x=0}^{l_t} \mathbf{f}_t(x) z^x.$$

To this end, multiply both sides of the second recurrence by z^x, sum up for all $x = 1, 2, \ldots, l_t$ and add the first recurrence to deduce

$$\boldsymbol{\phi}_t(z) = \left(\sum_{x=0}^{l_t} \mathbf{f}_{t-1}(x) z^x \right) A + z \left(\sum_{x=0}^{l_t - 1} \mathbf{f}_{t-1}(x) z^x \right) B. \quad (14)$$

Due to condition (b) of Definition 4.1 we shall have either $l_t = l_{t-1}$ or $l_t = l_{t-1} + 1$. In the first case formula (14) reads

$$\boldsymbol{\phi}_t(z) = \boldsymbol{\phi}_{t-1}(z)(A + zB) - z^{l_{t-1}+1} \mathbf{f}_{t-1}(l_{t-1}) B$$

while in the second

$$\boldsymbol{\phi}_t(z) = \boldsymbol{\phi}_{t-1}(z)(A + zB) + z^{l_t} \mathbf{f}_{t-1}(l_t) A.$$

It is not difficult to verify that the last terms of the right-hand side of the above identities vanish (taking into account that $\boldsymbol{\phi}_t(1)\mathbf{1}' = \boldsymbol{\phi}_{t-1}(1)\mathbf{1}' = 1$, $(A + B)\mathbf{1}' = 1$); therefore

$$\boldsymbol{\phi}_t(z) = \boldsymbol{\phi}_{t-1}(z)(A + zB), \quad t \geq 1,$$

and since

$$\boldsymbol{\phi}_0(z) = \sum_{x=0}^{l_n} \mathbf{f}_0(x) z^x = \sum_{x=0}^{l_n} \pi_x z^x = \pi_0$$

we may state the following theorem.

THEOREM 4.2. *The probability generating function*

$$\phi_n(z) = \sum_{x=0}^{l_n} \Pr(X_n = x) z^x$$

of a homogeneous MVB X_n can be expressed as

$$\phi_n(z) = \boldsymbol{\phi}_n(z) \mathbf{1}' = \boldsymbol{\pi}_0 (A + zB)^n \mathbf{1}'. \tag{15}$$

Multiplying both sides of (15) by w^n and summing up for all $n = 0, 1, \ldots$ we derive the next expression for the double generating function of $\Pr(X_n = x)$, $x = 0, 1, \ldots, l_n$, $n = 0, 1, \ldots,$

$$\Phi(z, w) = \sum_{n=0}^{\infty} \sum_{x=0}^{l_n} \Pr(X_n = x) z^x w^n = \boldsymbol{\pi}_0 \big[I - w(A + zB)\big]^{-1} \mathbf{1}'. \tag{16}$$

Formulae (15) and (16) offer two efficient tools for the evaluation of the generating functions of homogeneous *MVB*'s X_n.

The nomenclature "Binomial Type" used to describe the *MVB* can be justified by the apparent similarity of (13) to the well known recurrence relations satisfied by the binomial distribution law $b(n, p; x) = \binom{n}{x} p^x q^{n-x}$, namely

$$b(t, p; 0) = b(t - 1, p; 0) q,$$
$$b(t, p; x) = b(t - 1, p; x) q + b(t - 1, p; x - 1) p, \quad \begin{aligned} t &= 1, 2, \ldots, n. \\ x &= 1, 2, \ldots, n, \end{aligned}$$

An analogous similarity exists between (15), (16) and the following expressions for the probability generating function and double generating function of $b(n, p; x)$,

$$\sum_{x=0}^{n} b(n, p; x) z^x = (q + zp)^n,$$

$$\sum_{n=0}^{\infty} \sum_{x=0}^{n} b(n, p; x) z^x w^n = \big[1 - (q + zp) w\big]^{-1}.$$

5. Waiting time distributions associated with *MVB*'s

Let $Z_1, Z_2, \ldots$ be a sequence of i.i.d. binary or multistate trials and X_n be a random variable counting the number of occurrences of a single (or composite) event $\mathcal{E}$ among $Z_1, Z_2, \ldots, Z_n$. We shall denote by T_r, $r \geq 1$, the waiting time for the rth occurrence of the event enumerated by X_n, that is $T_r = n$ if and only if $X_n = r$ and $X_{n-1} = r - 1$.

If we assume that X_n is a homogeneous MVB and retain the notation introduced in the previous section we may write

$$\Pr(T_r = n) = \Pr(Y_n \in C_r \text{ and } Y_{n-1} \in C_{r-1})$$
$$= \sum_{i=0}^{s-1} \Pr(Y_n \in C_r | Y_{n-1} = c_{r-1,i}) \Pr(Y_{n-1} = c_{r-1,i})$$
$$= \sum_{i=1}^{s} \beta_i \mathbf{f}_{n-1}(r-1) \mathbf{e}'_i,$$

where $\beta_i = \sum_{j=1}^{s} \mathbf{e}_i B \mathbf{e}'_j = \mathbf{e}_i B \mathbf{1}'$, $1 \leq i \leq s$ and $\mathbf{e}_i = (0, \ldots, 1, \ldots, 0)$ denotes the ith unit (row) vector of $\mathbb{R}^s$. Hence

$$\Pr(T_r = n) = \mathbf{f}_{n-1}(r-1) B \mathbf{1}' \tag{17}$$

thereby providing an efficient tool for evaluating the exact distribution of T_r. To this end it suffices to initiate the recursive scheme (13) to get $\mathbf{f}_{n-1}(r-1)$, compute the row sums β_i of B, $i = 1, 2, \ldots, s$, and finally substitute the results in (17).

Should we wish to obtain a compact formula for the double generating function

$$H(z, w) = \sum_{n=1}^{\infty} \sum_{r=1}^{l_n} \Pr(T_r = n) w^n z^r$$

it suffices to multiply (17) by $w^n z^r$ and sum up for all $n = 0, 1, \ldots$ and $r = 0, 1, \ldots, l_n$ to deduce

$$H(z, w) = wz \left(\sum_{n=1}^{\infty} \sum_{r=1}^{l_n} \mathbf{f}_n(r) w^n z^r \right) B \mathbf{1}'.$$

Taking into account the identity

$$\sum_{n=1}^{\infty} \sum_{r=1}^{l_n} \mathbf{f}_n(r) w^r z^n = \pi_0 \big[I - w(A + zB) \big]^{-1}$$

we obtain the next theorem.

THEOREM 5.1. *The double generating function of the waiting time T_r for the rth occurrence of an event associated with a homogeneous MVB is given by*

$$H(z, w) = wz\pi_0 \sum_{i=1}^{s} \beta_i \big[I - w(A + zB) \big]^{-1} \mathbf{e}'_i$$
$$= wz\pi_0 \big[I - w(A + zB) \big]^{-1} B \mathbf{1}',$$

where $\beta_i = \mathbf{e}_i B \mathbf{1}'$, $1 \leq i \leq s$.

Writing the matrix $I - w(A + zB)$ as

$$I - w(A + zB) = (I - wA)\left[I - zw(I - wA)^{-1}B\right]$$

and expanding the inverse of the bracketed term as

$$\left[I - zw(I - wA)^{-1}B\right]^{-1} = \sum_{j=0}^{\infty}\left[(I - wA)^{-1}B\right]^{j}(zw)^{j},$$

we may express $H(z, w)$ as

$$H(z, w) = \pi_0 \sum_{r=1}^{\infty} w^r \left[(I - wA)^{-1}B\right]^{r-1}(I - wA)^{-1}B\mathbf{1}'w^r z^r.$$

This reveals that the probability generating function of T_r admits the following form

$$\sum_{n=1}^{\infty} \Pr(T_r = n)w^n = w^r \pi_0 \left[(I - wA)^{-1}B\right]^r \mathbf{1}', \quad r \geqslant 1. \tag{18}$$

For $r = 1$ we may obtain the following formula for the probability generating function of the waiting time distribution for the first occurrence

$$H(w) = \sum_{n=1}^{\infty} \Pr(T_1 = n)w^n = w\pi_0(I - wA)^{-1}B\mathbf{1}'. \tag{19}$$

Expanding $(I - wA)^{-1}$ in a power series we get

$$\Pr(T_1 = n) = \pi_0 A^{n-1} B\mathbf{1}',$$

which is an exact parallel of the well known formula for the ordinary geometric distribution. Similarly, if $AB = BA$, it can be easily verified that formula (18) leads to the next analogue of the Pascal distribution

$$\Pr(T_r = n) = \pi_0 \binom{n-1}{r-1} A^{n-r} B^r \mathbf{1}', \quad n \geqslant r.$$

6. The number of runs and patterns as members of the *MVB* family

As already mentioned in Sections 2 and 3 many success run and pattern enumerating random variables lead to bidiagonal transition probability matrices of the form (6). Hence, they belong to the family of *MVB*'s, this membership conferring a number of instrumental formulae for the evaluation of their exact distributions and probability generating functions.

We shall now proceed to review a few typical results for success runs and patterns distributions which can be established with ease from the general formulae developed in Sections 4 and 5.

6.1. Success runs

Let us start with the case of the number $N_{n,k}$ of success runs of length k in a sequence of a fixed number of binary trials $Z_1, Z_2, \ldots, Z_n$. It is rather straightforward to show that $N_{n,k}$ is an *MVB*, with transition probability matrices as given in (7). Hence, the exact distribution of $N_{n,k}$ could be easily calculated by implementing the recurrence scheme (13) with initial conditions $\mathbf{f}_0(0) = \mathbf{e}_1$, $\mathbf{f}_0(x) = \mathbf{0}$, $x \geq 1$, and then applying the formula

$$\Pr(X_n = x) = \mathbf{f}_n(x)\mathbf{1}', \quad x = 0, 1, \ldots, [n/k].$$

Formula (15) may be used to deduce a recursive scheme for the probability generating function

$$\phi_n(z) = \sum_{x=0}^{[n/k]} \Pr(X_n = x) z^x.$$

More specifically, for $n = k$, a thorough inspection of the matrix $(A + zB)^k$ reveals that

$$\phi_k(z) = \mathbf{e}_1 (A + zB)^k \mathbf{1}' = (1 - p^k) + p^k z. \qquad (20)$$

On the other hand, for $n > k$ we may write

$$\phi_n(z) = \mathbf{e}_1 (A + zB)(A + zB)^{n-1} \mathbf{1}',$$

which reduces to

$$\phi_n(z) = q\phi_{n-1}(z) + p\mathbf{e}_2 (A + zB)^{n-1} \mathbf{1}'$$
$$= q\phi_{n-1}(z) + qp\phi_{n-2}(z) + p^2 \mathbf{e}_3 (A + zB)^{n-2} \mathbf{1}.$$

Repeating a similar procedure till $(A + zB)^{n-k}$ shows up, we deduce

$$\phi_n(z) = \sum_{i=0}^{k-1} p^i q \phi_{n-i-1}(z) + p^k z \phi_{n-k}(z), \quad n > k. \qquad (21)$$

The double generating function

$$\Phi(z, w) = \sum_{n=0}^{\infty} \sum_{x=0}^{[n/k]} \Pr(N_{n,k} = x) z^x w^n = \sum_{n=0}^{\infty} \phi_n(z) w^n$$

could be easily evaluated by multiplying (21) by w^n and summing up for all $n > k$ (formula (20) will also be proved useful to fill in the term corresponding to $n = k$).

An alternative (direct) method for evaluating $\Phi(z, w)$ is offered by (16). Recalling A and B from (7) we have

$$I - w(A + zB) = \begin{bmatrix} 1-qw & -pw & 0 & \cdot & 0 & 0 \\ -qw & 1 & -pw & \cdot & 0 & 0 \\ -qw & 0 & 1 & \cdot & 0 & 0 \\ \cdot & & & & \cdot & \cdot \\ -qw & 0 & 0 & \cdot & 1 & -pw \\ -(q+zp)w & 0 & 0 & \cdot & 0 & 1 \end{bmatrix}.$$

Straightforward calculations reveal that

$$|I - w(A + zB)| = (1 - pw)^{-1}(1 - w + qp^k w^{k+1}) - zp^k w^k$$

and that the sum of the entries of the first row of $[I - w(A + zB)]^{-1}$ equals

$$(1 - (pw)^k)(1 - pw)^{-1}|I - w(A + zB)|^{-1}.$$

Accordingly, (16) yields

$$\Phi(z, w) = \mathbf{e}_1 [I - w(A + zB)]^{-1} \mathbf{1}'$$

$$= \frac{1 - (pw)^k}{1 - w + qp^k w^{k+1} - (1 - pw)zp^k w^k}. \tag{22}$$

If one is interested in the distribution of the waiting time for the rth (non-overlapping) appearance of a success run of length k, he could apply Theorem 5.1 for

$$s = k, \qquad \pi_0 = \mathbf{e}_1, \qquad \beta_i = \mathbf{e}_i B \mathbf{1}' = \begin{cases} 0, & \text{if } 1 \leq i < k-1, \\ p, & \text{if } i = k, \end{cases}$$

and after some routine algebraic calculations the following expression ensues

$$H(z, w) = \frac{z(pw)^k}{1 - qw \sum_{i=0}^{k-1}(pw)^i - z(pw)^k}. \tag{23}$$

The probability generating function $H(w)$ of the waiting time for the first occurrence of a success run of length k can be easily calculated either by the obvious formula

$$H(w) = \left[\frac{1}{z} H(z, w)\right]_{z=0},$$

or by a direct application of (19) as

$$H(w) = \frac{(pw)^k}{1 - qw \sum_{i=0}^{k-1}(pw)^i} = \frac{(1-pw)(pw)^k}{1 - w + qp^k w^{k+1}}. \tag{24}$$

The aforementioned formulae for the distribution of $N_{n,k}$ have been derived by several authors using different approaches. The recursive scheme (21) seems to have appeared for the first time in Aki and Hirano (1988), while (22) and (24) are mentioned in Feller (1968). For additional formulae pertaining to Markov dependent trials the interested reader may refer to Antzoulakos (1999), Aki and Hirano (1993), Hirano and Aki (1993), Mohanty (1994), Uchida and Aki (1995), etc.

The evaluation of the generating functions of $M_{n,k}$ and $G_{n,k}$ by the aid of the formulae developed in Sections 3 and 4 depends on essentially the same considerations as were used in obtaining the generating functions for $N_{n,k}$. Therefore we do not deem it necessary to go through them in full detail. Many results within this framework may be found in Koutras and Alexandrou (1995) and Koutras (1997b).

In closing, we mention that the Markov chain embedding approach established in Section 2 for the number $E_{n,k}$ of success runs of length exactly k, does not lead to an *MVB* (the form of the associated transition probability matrix is not (blocked) bidiagonal). To overcome this hurdle, Han and Aki (1999) introduced the class of *Markov chain embeddable variables of returnable type* by replacing condition (b) of Definition 4.1 by

$$\Pr(Y_t \in C_v | Y_{t-1} \in C_x) = 0 \quad \text{for all } v \neq x-1, x, x+1 \text{ and } t \geq 1.$$

Under this condition, besides the within states one step transition matrix $A_t(x)$, two between states matrices show up: matrix

$$B_t(x) = \left(\Pr(Y_t = C_{x+1,j} | Y_{t-1} = C_{x,i})\right)_{s \times s}$$

which controls the upwards transitions, and matrix

$$C_t(x) = \left(\Pr(Y_t = C_{x-1,j} | Y_{t-1} = C_{x,i})\right)_{s \times s}$$

which controls the downwards transitions. The recurrence relations of Theorem 4.1 now read

$$\mathbf{f}_t(x) = \mathbf{f}_{t-1}(x) A_t(x) + \mathbf{f}_{t-1}(x-1) B_t(x-1) + \mathbf{f}_{t-1}(x+1) C_t(x+1),$$

and in the special case where $A_t(x) = A$, $B_t(x) = B$, $C_t(x) = C$ for all x and t, the probability generating function

$$\phi_n(z) = \sum_{x=0}^{l_n} \Pr(X_n = x) z^x$$

takes the form $\phi_n(z) = \pi_0(A + zB + z^{-1}C)^n \mathbf{1}'$. For more details on the class of Markov chain embeddable variables of returnable type the reader may consult Han and Aki (1999).

6.2. Patterns

The machinery developed for *MVB*'s can also be effectively used for the study of the exact distribution of random variables enumerating patterns in sequences of multistate trials. Recalling the notations introduced in Section 2, we may state that if the transition probability matrix M associated with the Markov chain embedding procedure for the variable X_n is blocked bidiagonal, then the respective probability vectors will satisfy a triangular recurrence relation, and therefore X_n would be a *MVB*.

Such a situation arises in both cases studied in Section 3. For example, as illustrated in Section 3.5, the application of the forward and backward principle for the pattern $\mathcal{E} = 2112$ leads to a transition probability matrix of the form

$$M = \begin{bmatrix} A & B & & & \\ & A & B & & \\ & & A & B & \\ & & & \ddots & \end{bmatrix}$$

with A and B given by (11). As a consequence, introducing the sequence of vectors

$$\mathbf{f}_t(x) = \big(\Pr(Y_t = (x, 0)),\ \Pr(Y_t = (x, 1)),\ \Pr(Y_t = (x, 2)),\ \Pr(Y_t = (x, 3))\big)$$

we may write the following set of recurrence relations

$$\mathbf{f}_t(0) = \mathbf{f}_{t-1}(0)A,$$
$$\mathbf{f}_t(x) = \mathbf{f}_{t-1}(x)A + \mathbf{f}_{t-1}(x-1)B, \quad 1 \leqslant x \leqslant [n/4],$$
$$t = 1, 2, \ldots, n.$$

The exact distribution of the number X_n of occurrences of $\mathcal{E} = 2112$ in n trials can then be expressed as

$$\Pr(X_n = n) = \mathbf{f}_n(x)\mathbf{1}', \quad x = 0, 1, \ldots, [n/4].$$

Moreover, the double generating function

$$\Phi(z, w) = \sum_{n=0}^{\infty} \sum_{x=0}^{[n/4]} \Pr(X_n = x) z^x w^n$$

can be evaluated by (16) as

$$\Phi(z, w) = \mathbf{e}_1 \big[I - w(A + zB)\big]^{-1} \mathbf{1}'$$

$$= \frac{1 + p_1^2 p_2 w^3}{1 - w + p_1^2 p_2 w^3 - p_1^2 p_2 w^4 (1 - p_2 + p_2 z)}.$$

If one is interested in the waiting time T_r for the rth occurrence of the pattern $\mathcal{E} = 2112$ in an (infinite) sequence of trials $Z_1, Z_2, \ldots$, one could employ Theorem 5.1 to derive the double generating function

$$H(z, w) = \sum_{n=0}^{\infty} \sum_{r=0}^{[n/4]} \Pr(T_r = n) w^n z^r$$

as

$$H(z, w) = wz\mathbf{e}_1 \sum_{i=1}^{4} \beta_i [I - w(A + zB)]^{-1} \mathbf{e}'_i$$

$$= wz\mathbf{e}_1 [I - w(A + zB)]^{-1} B\mathbf{1}',$$

where

$$\beta_i = \mathbf{e}_i B\mathbf{1}' = \begin{cases} 0 & \text{for } i = 1, 2, 3, \\ p_2 & \text{for } i = 4. \end{cases}$$

Direct algebraic calculations on the inverse of the matrix

$$I - w(A + zB) = \begin{bmatrix} 1 - w(p_1 + p_2) & -wp_2 & 0 & 0 \\ -wp_3 & 1 - wp_2 & -wp_1 & 0 \\ -wp_3 & -wp_2 & 1 & -wp_1 \\ -w[(p_1 + p_3) + zp_2] & 0 & 0 & 1 \end{bmatrix}$$

reveal that

$$H(z, w) = \frac{p_1^2 p_2^2 w^4 z}{1 - w + p_1^2 p_2 w^3 - p_1^2 p_2 w^4 (1 - p_2 + p_2 z)}. \tag{25}$$

As a by-product we can get the generating function for the first occurrence of $\mathcal{E} = 2112$ as

$$E[w^{T_1}] = \left[\frac{1}{z} H(z, w) \right]_{z=0} = w\mathbf{e}_1 (I - wA)^{-1} B\mathbf{1}'$$

$$= \frac{p_1^2 p_2^2 w^4}{1 - w + p_1^2 p_2 w^3 - p_1^2 p_2 w^4 (1 - p_2)}.$$

Expanding (25) with respect to z and picking out the coefficient of z^r we may derive the generating function of T_r as

$$E[w^{T_r}] = \left(\frac{p_1^2 p_2^2 w^4}{1 - w + p_1^2 p_2 w^3 - p_1^2 p_2 w^4 (1 - p_2)}\right)^r = \{E[w^{T_1}]\}^r,$$

a formula that can be easily interpreted by taking into account the fact that the pattern occurrences form a recurrent event (since the enumeration is performed by the non-overlapping scheme).

Results similar to the one stated above, can be easily extracted for the composite pattern $\mathcal{E} = 12$ or 31 by exploiting the *MVB* methodology. Practically, this can be done for any pattern (single or composite), the only practical restriction being the size of calculations needed to invert the resulting matrix when the forward and backward principle produces large matrices A and B.

7. Multivariate *MVB* distributions

The notion of *MVB* can be easily adjusted to a multivariate framework. Such an extension offers efficient tools for capturing the exact joint distribution of enumerating random variables defined on sequences of multistate trials.

For typographical convenience, we shall first present in some detail the case of bivariate *MVB*'s, and discuss in brief how this approach could be extended to the general multivariate setup.

Let $X_n^{(1)}$, $X_n^{(2)}$ be discrete random variables with upper end points $l_n^{(1)}$, $l_n^{(2)}$ respectively.

DEFINITION 7.1. The bivariate random variable $(X_n^{(1)}, X_n^{(2)})$ defined on a subset L_n of $\{0, 1, \ldots, l_n^{(1)}\} \times \{0, 1, \ldots, l_n^{(2)}\}$ will be called *Markov chain Embeddable Variable of Trinomial type* (*MVT*) if

(a) there exists a Markov chain $\{Y_t, t \geq 0\}$ defined on a discrete state space Ω which can be partitioned as

$$\Omega = \bigcup_{x_1, x_2 \geq 0} C_{x_1, x_2}, \quad C_{x_1, x_2} = \{c_{x_1, x_2; 0}, c_{x_1, x_2; 1}, \ldots, c_{x_1, x_2; s-1}\},$$

(b) $\Pr(Y_t = c_{v_1, v_2; j} | Y_{t-1} = c_{x_1, x_2; i}) = 0$ for $(v_1, v_2) \neq (x_1, x_2), (x_1 + 1, x_2)$ and $(x_1, x_2 + 1)$,
(c) the event $\{X_n^{(1)} = x_1, X_n^{(2)} = x_2\}$ is equivalent to $Y_n \in C_{x_1, x_2}$, i.e.,

$$\Pr(X_n^{(1)} = x_1, X_n^{(2)} = x_2) = \Pr(Y_n \in C_{x_1, x_2}), \quad (x_1, x_2) \in L_n.$$

The three different types of transitions allowed by the above definition give birth to the next three families of $s \times s$ matrices

- *within states* one step transition probability matrices

$$A_t(x_1, x_2) = \left(\Pr(Y_t = c_{x_1,x_2;j}|Y_{t-1} = c_{x_1,x_2;i})\right)_{s \times s}, \tag{26}$$

- *between states* one step transition probability matrices

$$B_t^{(1)}(x_1, x_2) = \left(\Pr(Y_t = c_{x_1+1,x_2;j}|Y_{t-1} = c_{x_1,x_2;i})\right)_{s \times s},$$
$$B_t^{(2)}(x_1, x_2) = \left(\Pr(Y_t = c_{x_1,x_2+1;j}|Y_{t-1} = c_{x_1,x_2;i})\right)_{s \times s}. \tag{27}$$

Manifestly the sum $A_t(x_1, x_2) + B_t^{(1)}(x_1, x_2) + B_t^{(2)}(x_1, x_2)$ is a stochastic matrix for all $t = 1, 2, \ldots$ and x_1, x_2. Denoting by

$$\pi_{x_1,x_2} = \left(\Pr(Y_0 = c_{x_1,x_2;0}), \Pr(Y_0 = c_{x_1,x_2;1}), \ldots, \Pr(Y_0 = c_{x_1,x_2;s-1})\right)$$

the (row) vector of initial probabilities, we may state the following theorem, which can be immediately verified by the Chapman–Kolmogorov equations.

THEOREM 7.1. *The sequence of vectors*

$$\mathbf{f}_t(x_1, x_2) = \left(\Pr(Y_t = c_{x_1,x_2;0}), \Pr(Y_t = c_{x_1,x_2;1}), \ldots, \Pr(Y_t = c_{x_1,x_2;s-1})\right),$$

$0 \leq t \leq n$, $(x_1, x_2) \in L_n$, *satisfies the recurrence relations*

$$\mathbf{f}_t(0, 0) = \mathbf{f}_{t-1}(0, 0) A_t(0, 0),$$

$$\mathbf{f}_t(0, x_2) = \mathbf{f}_{t-1}(0, x_2) A_t(0, x_2) + \mathbf{f}_{t-1}(0, x_2 - 1) B_t^{(2)}(0, x_2 - 1),$$
$$1 \leq x_2 \leq l_n^{(2)},$$

$$\mathbf{f}_t(x_1, 0) = \mathbf{f}_{t-1}(x_1, 0) A_t(x_1, 0) + \mathbf{f}_{t-1}(x_1 - 1, 0) B_t^{(1)}(x_1 - 1, 0),$$
$$1 \leq x_1 \leq l_n^{(1)},$$

$$\mathbf{f}_t(x_1, x_2) = \mathbf{f}_{t-1}(x_1, x_2) A_t(x_1, x_2) + \mathbf{f}_{t-1}(x_1 - 1, x_2) B_t^{(1)}(x_1 - 1, x_2)$$
$$+ \mathbf{f}_{t-1}(x_1, x_2 - 1) B_t^{(2)}(x_1, x_2 - 1),$$
$$1 \leq x_1 \leq l_n^{(1)} \text{ and } 1 \leq x_2 \leq l_n^{(2)}, \tag{28}$$

for $t = 1, 2, \ldots, n$, *with initial conditions* $\mathbf{f}_0(x_1, x_2) = \pi_{x_1,x_2}$ *for* $(x_1, x_2) \in L_n$. *Furthermore, the probability distribution function of the bivariate MVT* $(X_n^{(1)}, X_n^{(2)})$ *is given by*

$$P\left(X_n^{(1)} = x_1, X_n^{(2)} = x_2\right) = \mathbf{f}_n(x_1, x_2) \mathbf{1}', \quad (x_1, x_2) \in L_n. \tag{29}$$

It is worth noting that the set of recurrences mentioned above exhibits a clear resemblance to the set of recurrences satisfied by the classical trinomial distribution. This fact justifies the nomenclature "trinomial type" used for $(X_n^{(1)}, X_n^{(2)})$.

Let us next assume that the within states and between states transition probability matrices do not depend on x_1, x_2 and t, a case which is quite common when dealing with sequences of i.i.d. trials. We shall refer to this case by the term "homogeneous" MVT and denote the matrices $A_t(x_1, x_2)$, $B_t^{(1)}(x_1, x_2)$ and $B_t^{(2)}(x_1, x_2)$ by A, B_1, and B_2 respectively.

After some lengthy but rather straightforward algebraic manipulations (exploiting the recurrences of Theorem 7.1) we may easily verify that the vector generating function

$$\boldsymbol{\phi}_t(y, z) = \sum_{(x_1,x_2) \in L} \mathbf{f}_t(x_1, x_2) y^{x_1} z^{x_2}$$

satisfies the recursive relation

$$\boldsymbol{\phi}_t(y, z) = \boldsymbol{\phi}_{t-1}(y, z)(A + yB_1 + zB_2), \quad t \geq 1$$

(for a detailed proof, see Alexandrou, 1997). Thus we may immediately obtain the following results which are quite pleasing in their simplicity

THEOREM 7.2. *The joint probability generating function*

$$\phi_n(y, z) = \sum_{(x_1,x_2) \in L_n} \Pr(X_n^{(1)} = x_1, X_n^{(2)} = x_2) y^{x_1} z^{x_2}$$

and double generating function

$$\phi(y, z, w) = \sum_{n=0}^{\infty} \phi_n(y, z) w^n$$

of a homogeneous MVT $(X_n^{(1)}, X_n^{(2)})$ *can be expressed as*

$$\phi_n(y, z) = \boldsymbol{\phi}_0(y, z)(A + yB_1 + zB_2)^n \mathbf{1}',$$

$$\phi(y, z, w) = \boldsymbol{\phi}_0(y, z)\big[I - w(A + yB_1 + zB_2)\big]^{-1} \mathbf{1}',$$

respectively, where

$$\boldsymbol{\phi}_0(y, z) = \sum_{(x_1,x_2) \in L_0} \boldsymbol{\pi}_{x_1,x_2} y^{x_1} z^{x_2}.$$

A by-product of Theorem 7.2 is the following result which offers a manageable expression for the generating functions of the first- and second-order moments of the bivariate random variable $(X_n^{(1)}, X_n^{(2)})$.

THEOREM 7.3. *Let* $(X_n^{(1)}, X_n^{(2)})$ *be a homogeneous MVT with* $\boldsymbol{\phi}_0(y, z) = \boldsymbol{\phi}_0$ *(independent of y, z). Then the generating functions of* $E(X_n^{(i)})$, $i = 1, 2$, *and* $E(X_n^{(1)} X_n^{(2)})$ *can be expressed as*

$$\sum_{n=1}^{\infty} E(X_n^{(i)}) w^n = \frac{w}{1-w} \boldsymbol{\phi}_0 (I - w\Delta)^{-1} B_i \mathbf{1}', \quad i = 1, 2,$$

$$\sum_{n=1}^{\infty} E(X_n^{(1)} X_n^{(2)}) w^n = \frac{w^2}{1-w} \boldsymbol{\phi}_0 (I - w\Delta)^{-1}$$
$$\times \left[B_2 (I - w\Delta)^{-1} B_1 + B_1 (I - w\Delta)^{-1} B_2 \right] \mathbf{1}',$$

where $\Delta = A + B_1 + B_2$ (w *belongs to an appropriate neighborhood of zero*).

A detailed proof of Theorem 7.3 can be found in Alexandrou (1997); see also Han and Aki (1999).

So far we have discussed the case of bivariate enumerating variables which can be studied by the use of appropriate finite Markov chains. The extension of the above definitions to the case of m-variate random variables $\mathbf{X} = (X_1^{(n)}, X_2^{(n)}, \ldots, X_m^{(n)})$ can be accomplished in a rather routine manner. To this end we shall need $m + 1$ transition probability matrices, say $A_t(\mathbf{x})$ (within states) and $B_t^{(j)}(\mathbf{x})$, $j = 1, 2, \ldots, m$ (between states), and the evaluation of the joint distribution of $\mathbf{X}$ will be given by

$$\Pr(\mathbf{X} = \mathbf{x}) = \mathbf{f}_n(\mathbf{x}) \cdot \mathbf{1}',$$

where $\mathbf{f}_n(\mathbf{x})$ are sequences of vectors satisfying the recurrence relations

$$\mathbf{f}_t(\mathbf{x}) = \mathbf{f}_{t-1}(\mathbf{x}) A_t(\mathbf{x}) + \sum_{j=1}^{m} \mathbf{f}_{t-1}(\mathbf{x} - \mathbf{e}_j) B_t(\mathbf{x} - \mathbf{e}_j)$$

(convention: $\mathbf{f}_t(x) = 0$ and $B_t(\mathbf{x}) = 0$ if $x_i < 0$ for at least one $i = 1, 2, \ldots, m$).

It is not difficult to verify that the arguments exploited in Theorems 7.2 and 7.3 can be properly adjusted to provide formulae for the joint probability functions and generating functions of first- and second-order moments of $\mathbf{X}$. For a detailed study of this class of variables, which have been termed *Markov chain Embeddable variables of Multinomial type*, the reader may consult Han and Aki (1999).

8. Multivariate success runs distributions

Let $Z_1, Z_2, \ldots$ be a sequence of i.i.d. trials with three possible outcomes in each trial, labelled as F (failure, $Z_i = 0$), S_1 (type I success, $Z_i = 1$) and S_2 (type II success,

$Z_i = 2$). Assume also that

$$p_1 = \Pr(Z_i = 1), \qquad p_2 = \Pr(Z_i = 2), \qquad q = \Pr(Z_i = 0) = 1 - p_1 - p_2$$

for all $i = 1, 2, \ldots$ and let $k \leqslant r$ be two positive integers. For an uninterrupted sequence of k consecutive S_1's or r consecutive S_2's we shall be using the term *type I success run* of length k or *type II success run* of length r respectively. We shall now proceed to a detailed study of the joint distribution of the number of runs of each type for the three enumeration schemes (non-overlapping, overlapping and at least) by taking advantage of the methodology described in Section 7. It is worth mentioning that recently, Doi and Yamamoto (1998) proceeded to a detailed study of multivariate success runs distributions by exploiting the original Markov chain approach of Fu and Koutras (1994). The technique used here not only offers more efficient schemes for the computation of the exact probability mass function (due to the substantially smaller dimensions of the transition probability matrices involved), but facilitates the evaluation of the respective probability generating functions as well.

8.1. Non-overlapping success runs

Let $N_{n,k}^{(1)}$ be the number of non-overlapping type I success runs of length k and $N_{n,r}^{(2)}$ the number of non-overlapping type II success runs of length r in a sequence of a fixed number of trials $Z_1, Z_2, \ldots, Z_n$. Then the upper end points of $N_{n,k}^{(1)}, N_{n,r}^{(2)}$ are $l_n^{(1)} = [n/k], l_n^{(2)} = [n/r]$ and the bivariate random variable $(N_{n,k}^{(1)}, N_{n,r}^{(2)})$ can be treated as *MVT* by applying the following procedure: let Ω be the state space $\Omega = \bigcup_{x_1, x_2 \geqslant 0} C_{x_1, x_2}$ where

$$C_{x_1,x_2} = \{(x_1, x_2; 0, 0)\} \cup \{(x_1, x_2; m, 1): 1 \leqslant m \leqslant k - 1\}$$
$$\cup \{(x_1, x_2; m, 2): 1 \leqslant m \leqslant r - 1\}, \quad x_1, x_2 \geqslant 0.$$

Assume next that in the first t trials $Z_1, Z_2, \ldots, Z_t$ there have been spotted x_1 non-overlapping type I success runs of length k and x_2 non-overlapping type II success runs of length r. Finally, let c be the number of trailing identical symbols in the observed sequence, i.e., $Z_t = Z_{t-1} = \cdots = Z_{t-c+1} \neq Z_{t-c}$ (convention: $Z_j = 0$ for $j \leqslant 0$). Then a Markov chain that meets the requirements of Definition 7.1 can be established by assigning values to $Y_t, t \geqslant 1$, as follows:

- $Y_t = (x_1, x_2; c \pmod{k}, 1)$ if and only if $Z_t = 1$ and $c \pmod{k} \neq 0$;
- $Y_t = (x_1, x_2; c \pmod{r}, 2)$ if and only if $Z_t = 2$ and $c \pmod{r} \neq 0$;
- $Y_t = (x_1, x_2; 0, 0)$ in all other cases, i.e., if $Z_t = 0$ or $(Z_t = 1$ and $c \pmod{k} = 0)$ or $(Z_t = 2$ and $c \pmod{r} = 0)$.

(Convention: $Y_0 = (0, 0; 0, 0)$.) It can be easily verified that, the corresponding within states transition probability matrices have the form

$A_t(x_1, x_2) = A$

$$= \begin{bmatrix} q & p_1 & 0 & \cdots & 0 & 0 & p_2 & 0 & \cdots & 0 & 0 \\ q & 0 & p_1 & \cdots & 0 & 0 & p_2 & 0 & \cdots & 0 & 0 \\ q & 0 & 0 & \cdots & 0 & 0 & p_2 & 0 & \cdots & 0 & 0 \\ \vdots & \vdots & \vdots & \ddots & \vdots & \vdots & \vdots & \vdots & \ddots & \vdots & \vdots \\ q & 0 & 0 & \cdots & 0 & p_1 & p_2 & 0 & \cdots & 0 & 0 \\ q & 0 & 0 & \cdots & 0 & 0 & p_2 & 0 & \cdots & 0 & 0 \\ q & p_1 & 0 & \cdots & 0 & 0 & 0 & p_2 & \cdots & 0 & 0 \\ q & p_1 & 0 & \cdots & 0 & 0 & 0 & 0 & \cdots & 0 & 0 \\ \vdots & \vdots & \vdots & \ddots & \vdots & \vdots & \vdots & \vdots & \ddots & \vdots & \vdots \\ q & p_1 & 0 & \cdots & 0 & 0 & 0 & 0 & \cdots & 0 & p_2 \\ q & p_1 & 0 & \cdots & 0 & 0 & 0 & 0 & \cdots & 0 & 0 \end{bmatrix}_{(k+r-1) \times (k+r-1)}.$$

The between states transition probability matrices $B_t^{(1)}(x_1, x_2) = B_1$ have all their entries zero except for entry $(k, 1)$ which is p_1, whereas $B_t^{(2)}(x_1, x_2) = B_2$ have all their entries zero except for entry $(k+r-1, 1)$ which is p_2.

Applying Theorem 7.1 we may obtain a set of recurrence relations which facilitates the evaluation of the probability mass function of the bivariate random variable $(N_{n,k}^{(1)}, N_{n,r}^{(2)})$. Furthermore, exploiting Theorem 7.2 we may deduce the generating function

$$\Phi(y, z; w) = \sum_{n=0}^{\infty} \phi_n(y, z) w^n$$

$$= \sum_{n=0}^{\infty} \sum_{x_1=0}^{l_n^{(1)}} \sum_{x_2=0}^{l_n^{(2)}} \Pr(N_{n,k}^{(1)} = x_1, N_{n,r}^{(2)} = x_2) y^{x_1} z^{x_2} w^n$$

as

$$\Phi(y, z; w) = \frac{P(w)}{Q(y, z, w)},$$

where $P(w) = (1 - (p_1 w)^k)(1 - (p_2 w)^r)$ and

$$Q(y, z, w) = 1 - w + (p_1 w)^k (1 - p_1) w$$
$$+ (p_2 w)^r (1 - p_2) w - (p_1 w)^k (p_2 w)^r (1 + qw)$$
$$- \left[y (p_1 w)^k (1 - p_1 w)(1 - (p_2 w)^r) \right.$$
$$+ \left. z (p_2 w)^r (1 - p_2 w)(1 - (p_1 w)^k) \right].$$

The last formula for $\Phi(y,z;w)$ can be used to obtain recurrences for the joint probability generating function

$$\phi_n(y,z) = \sum_{x_1=0}^{[n/k]} \sum_{x_2=0}^{[n/r]} \Pr(N_{n,k}^{(1)} = x_1, N_{n,r}^{(2)} = x_2) y^{x_1} z^{x_2}.$$

To be more precise, for $n > k + r$ we find

$$\begin{aligned}\phi_n(y,z) &= \phi_{n-1}(y,z) - p_1^k(1-p_1)\phi_{n-k-1}(y,z) - p_2^r(1-p_2)\phi_{n-r-1}(y,z) \\ &\quad + p_1^k p_2^r[\phi_{n-k-r}(y,z) + q\phi_{n-k-r-1}(y,z)] \\ &\quad + yp_1^k[\phi_{n-k}(y,z) - p_1\phi_{n-k-1}(y,z) - p_2^r\phi_{n-k-r}(y,z) \\ &\quad + p_2^r p_1\phi_{n-k-r-1}(y,z)] \\ &\quad + zp_2^r[\phi_{n-r}(y,z) - p_2\phi_{n-r-1}(y,z) - p_1^k\phi_{n-k-r}(y,z) \\ &\quad + p_1^k p_2\phi_{n-k-r-1}(y,z)],\end{aligned}$$

with initial conditions $\phi_n(y,z) = 1$, for $0 \leqslant n < k$; likewise, for $k \leqslant n \leqslant k+r$ we deduce

$$\begin{aligned}\phi_n(y,z) &= \phi_{n-1}(y,z) - p_1^k(1-p_1)\phi_{n-k-1}(y,z) - p_2^r(1-p_2)\phi_{n-r-1}(y,z) \\ &\quad + p_1^k p_2^r[\phi_{n-k-r}(y,z) + q\phi_{n-k-r-1}(y,z)] \\ &\quad + yp_1^k[\phi_{n-k}(y,z) - p_1\phi_{n-k-1}(y,z) - p_2^r\phi_{n-k-r}(y,z) \\ &\quad + p_2^r p_1\phi_{n-k-r-1}(y,z)] \\ &\quad + zp_2^r[\phi_{n-r}(y,z) - p_2\phi_{n-r-1}(y,z) - p_1^k\phi_{n-k-r}(y,z) \\ &\quad + p_1^k p_2\phi_{n-k-r-1}(y,z)] + s_n,\end{aligned}$$

where $s_k = -p_1^k$, $s_r = -p_2^r$, $s_{k+r} = p_1^k p_2^r$ and $s_n = 0$ for $n \notin \{k, r, k+r\}$. By convention we set $\phi_n(y,z) = 0$ for all $n < 0$.

Applying Theorem 7.3 we may readily derive the following expression for the generating function of $E(N_{n,k}^{(1)} N_{n,r}^{(2)})$

$$M_{1,2}^N(w) = \sum_{n=1}^{\infty} E(N_{n,k}^{(1)} N_{n,r}^{(2)}) w^n = \frac{2(p_1 w)^k (p_2 w)^r}{(1-w)^3 a(w) b(w)},$$

where

$$a(w) = \frac{1-(p_1 w)^k}{1-p_1 w} \quad \text{and} \quad b(w) = \frac{1-(p_2 w)^r}{1-p_2 w}.$$

The exact distribution of the bivariate random variable $(X_n^{(1)}, X_n^{(2)})$ can be used to tackle the class of sooner-later waiting time problems. In the classical *sooner waiting time* problem the interest lies in the number of Bernoulli trials (i.e., $q = 0$) needed until the first occurrence of a fixed length run of either type. In the *later waiting time* problem the sequence of trials terminates upon the occurrence of at least one type I success run and at least one type II success run. The sooner problem was first introduced in Feller (1968). For recent advances on both sooner and later waiting time problems (and extensions to non-i.i.d. trials) the interested reader may consult Ebneshahrashoob and Sobel (1990), Ling and Low (1993), Aki (1992, 1997), Aki and Hirano (1993), Balakrishnan et al. (1993), Aki et al. (1996), Antzoulakos and Philippou (1996), Koutras and Alexandrou (1997b) among others. It is clear that the sooner and later random variables W_{sooner}, W_{later} can be expressed in terms of the probability vectors $\mathbf{f}_t(x_1, x_2)$ of Theorem 7.1 (associated to $(N_{n,k}^{(1)}, N_{n,r}^{(2)})$) as follows

$$\Pr(W_{\text{sooner}} = n) = p_1 \mathbf{f}_{n-1}(0, 0) \mathbf{e}_k' + p_2 \mathbf{f}_{n-1}(0, 0) \mathbf{e}_{k+r-1}',$$

$$\Pr(W_{\text{later}} = n) = p_1 \sum_{x_2=1}^{l_n^{(2)}} \mathbf{f}_{n-1}(0, x_2) \mathbf{e}_k' + p_2 \sum_{x_1=1}^{l_n^{(1)}} \mathbf{f}_{n-1}(x_1, 0) \mathbf{e}_{k+r-1}'.$$

8.2. Overlapping success runs

Let $M_n^{(1)}, M_n^{(2)}$ denote the number of overlapping type I, type II success runs of length k, r respectively, in the sequence $Z_1, Z_2, \ldots, Z_n$. Then $l_n^{(1)} = n - k + 1$, $l_n^{(2)} = n - r + 1$ and we may introduce the state space $\Omega = \bigcup_{x_1, x_2 \geq 0} C_{x_1, x_2}$ where C_{x_1, x_2} are as in the non-overlapping case with the addition of two substates labelled as $(x_1, x_2; -1, 1)$ and $(x_1, x_2; -1, 2)$. The assignment of values in Y_t, $t \geq 1$, can now be performed as follows (x_1, x_2 denote the number of overlapping type I and type II success runs of length k and r respectively, in the sequence $Z_1, Z_2, \ldots, Z_t$, while $c \geq 1$ is again the number of trailing identical symbols):

- $Y_t = (x_1, x_2; c, 1)$ if and only if $Z_t = 1$ and $c \leq k - 1$,
- $Y_t = (x_1, x_2; c, 2)$ if and only if $Z_t = 2$ and $c \leq r - 1$,
- $Y_t = (x_1, x_2; -1, 1)$ if and only if $Z_t = 1$ and $c \geq k$,
- $Y_t = (x_1, x_2; -1, 2)$ if and only if $Z_t = 2$ and $c \geq r$,
- $Y_t = (x_1, x_2; 0, 0)$ if and only if $Z_t = 0$.

Manifestly, for all $x_1, x_2 \geq 0$, states $(0, x_2; -1, 1)$ and $(x_1, 0; -1, 2)$ are dummy states and cannot be reached by the chain. The only reason we have kept them in our state space is in order to increase C_{0, x_2}'s and $C_{x_1, 0}$'s cardinality to $k + r + 1$.

Under this setup, $(M_{n,k}^{(1)}, M_{n,r}^{(2)})$ turns out to be a *MVT*, its within states transition probability matrices given by

$$A_t(x_1, x_2) = A$$

$$= \begin{bmatrix} q & p_1 & 0 & \cdots & 0 & 0 & 0 & p_2 & 0 & \cdots & 0 & 0 & 0 \\ q & 0 & p_1 & \cdots & 0 & 0 & 0 & p_2 & 0 & \cdots & 0 & 0 & 0 \\ q & 0 & 0 & \cdots & 0 & 0 & 0 & p_2 & 0 & \cdots & 0 & 0 & 0 \\ \vdots & \vdots & \vdots & \ddots & \vdots & \vdots & \vdots & \vdots & \vdots & \ddots & \vdots & \vdots & \vdots \\ q & 0 & 0 & \cdots & 0 & p_1 & 0 & p_2 & 0 & \cdots & 0 & 0 & 0 \\ q & 0 & 0 & \cdots & 0 & 0 & 0 & p_2 & 0 & \cdots & 0 & 0 & 0 \\ q & 0 & 0 & \cdots & 0 & 0 & 0 & p_2 & 0 & \cdots & 0 & 0 & 0 \\ q & p_1 & 0 & \cdots & 0 & 0 & 0 & 0 & p_2 & \cdots & 0 & 0 & 0 \\ q & p_1 & 0 & \cdots & 0 & 0 & 0 & 0 & 0 & \cdots & 0 & 0 & 0 \\ \vdots & \vdots & \vdots & \ddots & \vdots & \vdots & \vdots & \vdots & \vdots & \ddots & \vdots & \vdots & \vdots \\ q & p_1 & 0 & \cdots & 0 & 0 & 0 & 0 & 0 & \cdots & 0 & p_2 & 0 \\ q & p_1 & 0 & \cdots & 0 & 0 & 0 & 0 & 0 & \cdots & 0 & 0 & 0 \\ q & p_1 & 0 & \cdots & 0 & 0 & 0 & 0 & 0 & \cdots & 0 & 0 & 0 \end{bmatrix}_{(k+r+1) \times (k+r+1)}$$

The corresponding between states transition probability matrices $B_t^{(1)}(x_1, x_2) = B_1$ have their entries zero except $(k, k+1)$, $(k+1, k+1)$ which are p_1, while $B_t^{(2)}(x_1, x_2) = B_2$ have all their entries zero except $(k+r, k+r+1)$, $(k+r+1, k+r+1)$ which equal p_2.

Making use of Theorem 7.1 we may deduce recurrent relations for the probability mass function of the bivariate random variable $(M_n^{(1)}, M_n^{(2)})$. Furthermore, by virtue of Theorem 7.2 the generating function

$$\Phi(y, z; w) = \sum_{n=0}^{\infty} \sum_{x_1=0}^{n-k+1} \sum_{x_2=0}^{n-r+1} \Pr(M_n^{(1)} = x_1, M_n^{(2)} = x_2) y^{x_1} z^{x_2} w^n$$

can be expressed as

$$\Phi(y, z; w) = \frac{P(y, z, w)}{Q(y, z, w)},$$

where

$$P(y, z, w) = C(y, z, w) + p_1 w \left[1 - p_2 w z - (p_2 w)^r (1-z)\right]$$
$$\times \left[1 - p_1 w y - (p_1 w)^{k-1}(1-y)\right]$$
$$+ p_2 w \left[1 - p_1 w y - (p_1 w)^k (1-y)\right]$$
$$\times \left[1 - p_2 w z - (p_2 w)^{r-1}(1-z)\right],$$

$$C(y, z, w) = (1 - p_1 w y)(1 - p_2 w z)(1 - p_1 w)(1 - p_2 w)$$
$$- (p_1 w)(p_2 w)\left[1 - p_1 w y - (p_1 w)^{k-1}(1-y)\right]$$
$$\times \left[1 - p_2 w z - (p_2 w)^{r-1}(1-z)\right],$$

$$Q(y, z, t) = C(t, z, w) - qwP(y, z, w).$$

Exploiting Theorem 7.3 we may deduce

$$M_{1,2}^M(w) = \sum_{n=1}^{\infty} E(M_{n,k}^{(1)} M_{n,r}^{(2)}) w^n = \frac{2(p_1 w)^k (p_2 w)^r}{(1-w)^3}.$$

Moreover, multiplying both sides by $(1-w)^3$ and picking out the coefficients of w^n, $n = 1, 2, \ldots$, we get

$$E(M_{n,k}^{(1)} M_{n,r}^{(2)}) = (n - k - r + 1)(n - k - r + 2) p_1^k p_2^r, \quad n \geq k + r.$$

By virtue of $E(M_{n,k}^{(1)}) = (n - k + 1) p_1^k$ and $E(M_{n,r}^{(2)}) = (n - r + 1) p_2^r$, we may also write

$$\text{Cov}(M_{n,k}^{(1)}, M_{n,r}^{(2)})$$
$$= \left[kr + (k-1)^2 + (r-1)^2 - n(k+r-1) - 1 \right] p_1^k p_2^r, \quad \text{for } n \geq k + r.$$

In the special case of binary trials ($q = 0$) and runs of equal length ($k = r$), the above formulae for $E(M_{n,k}^{(1)} M_{n,r}^{(2)})$ and $\text{Cov}(M_{n,k}^{(1)}, M_{n,r}^{(2)})$ coincide with the ones derived by Ling and Tai (1990).

In closing, we mention that from the obvious expression

$$M_{1,2}^N(w) = M_{1,2}^M(w) / \big(a(w) b(w)\big),$$

we may obtain the following formula for the evaluation of $E(N_{n,k}^{(1)} N_{n,r}^{(2)})$ through $E(M_{n,k}^{(1)} M_{n,r}^{(2)})$

$$E(N_{n,k}^{(1)} N_{n,r}^{(2)}) = \alpha_n - (p_1 + p_2) \alpha_{n-1} + p_1 p_2 \alpha_{n-2},$$

where $\alpha_n = \sum_{j=0}^{[n/k]-1} \sum_{i=0}^{[(n-(j+1)k)/r]-1} E(M_{n-jk-ir,k}^{(1)} M_{n-jk-ir,r}^{(2)}) p_1^{jk} p_2^{ir}.$

8.3. Success runs of length at least k, r

Let $G_{n,k}^{(1)}$ denote the number of type I success runs of length at least k and $G_{n,r}^{(2)}$ the number of type II success runs of length at least r. Then $l_n^{(1)} = [(n+1)/(k+1)]$, $l_n^{(2)} = [(n+1)/(r+1)]$, and the introduction of an appropriate Markov chain $\{Y_t : t \geq 0\}$ can be achieved in exactly the same way as for the overlapping case; of course (x_1, x_2) will now denote the number of type I success runs of length at least k and type II success runs of length at least r in the first t trials.

Once more, the substates $(0, x_2; -1, 1)$ and $(x_1, 0; -1, 2)$ for $x_1, x_2 \geq 0$ are hypothetical substates (inaccessible by the chain); nevertheless, we need them in order to have $|C_{0,x_2}| = |C_{x_1,0}| = k + r + 1$. Manifestly, the requirements of Definition 7.1

are met by $(G_{n,k}^{(1)}, G_{n,r}^{(2)})$ which is therefore a *MVT*. The transition probability matrices (of dimension $k + r + 1$) are similar to the overlapping case with some trivial modifications. To be more precise, the $(k+1, k+1)$ and $(k+r+1, k+r+1)$ entries of $A_t(x_1, x_2) = A$ are now p_1 and p_2 respectively (instead of 0), the $(k+1, k+1)$ entry of $B_t^{(1)}(x_1, x_2) = B_1$ is zero (instead of p_1), and finally the $(k+r+1, k+r+1)$ entry of $B_t^{(2)}(x_1, x_2) = B_2$ is zero (instead of p_2).

A direct application of Theorem 7.2 reveals that the generating function

$$\Phi(y, z; w) = \sum_{n=0}^{\infty} \sum_{x_1=0}^{[(n+1)/(k+1)]} \sum_{x_2=0}^{[(n+1)/(r+1)]} \Pr(G_{n,k}^{(1)} = x_1, G_{n,k}^{(2)} = x_2) y^{x_1} z^{x_2} w^n$$

can be expressed as

$$\Phi(y, z; w) = \frac{P(y, z, w)}{Q(y, z, w)},$$

where

$$P(y, z, w) = \left[1 - (p_1 w)^k (1 - y)\right]\left[1 - (p_2 w)^r (1 - z)\right],$$

$$Q(y, z, w) = 1 - w + (p_1 w)^k (1 - p_1) w (1 - y) + (p_2 w)^r (1 - p_2) w (1 - z)$$
$$\quad - (p_1 w)^k (p_2 w)^r (1 + qw)(1 - y)(1 - z).$$

Applying Theorem 7.3 we find

$$M_{1,2}^G(w) = \frac{2(p_1 w)^k (p_2 w)^r (1 - p_1 w)(1 - p_2 w)}{(1 - w)^3},$$

and taking into account that $M_{1,2}^G(w) = M_{1,2}^M(w)(1 - p_1 w)(1 - p_2 w)$, we deduce

$$E(G_{n,k}^{(1)} G_{n,r}^{(2)}) = \beta_n - (p_1 + p_2)\beta_{n-1} + p_1 p_2 \beta_{n-2},$$

where $\beta_n = E(M_{n,k}^{(1)} M_{n,r}^{(2)})$.

8.4. Calculations

In Tables 1, 2 and 3, we provide some numerical results deduced by direct application of the recurrence relations of Theorem 7.1. More specifically the exact distributions of the bivariate random variables $(N_{n,k}^{(1)}, N_{n,r}^{(2)})$, $(M_{n,k}^{(1)}, M_{n,r}^{(2)})$ and $(G_{n,k}^{(1)}, G_{n,r}^{(2)})$ are given for the case $(n = 5, k = r = 2)$ and the exact distribution of $(M_{n,k}^{(1)}, M_{n,r}^{(2)})$ for $n = 15, k = 5, r = 3$. Three different choices were considered for the success and failure probabilities $\Pr(Z_t = 1), \Pr(Z_t = 2), \Pr(Z_t = 0) = 1 - \Pr(Z_t = 1) - \Pr(Z_t = 2), t = 1, 2, \ldots, n$, namely

Table 1
Exact distribution of bivariate binomial distributions for $n = 5, k = r = 2$

(x_1, x_2)	$p_t = (1+t)^{-1}$, $p_t^* = (1+2t)^{-1}$			$p_t = 0.5 - 2^{-(t+1)}$, $p_t^* = 0.5 + 2^{-(t+2)}$			$p_t = 0.2$, $p_t^* = 0.4$		
	$N^{(1)}_{5,2}, N^{(2)}_{5,2}$	$G^{(1)}_{5,2}, G^{(2)}_{5,2}$	$M^{(1)}_{5,2}, M^{(2)}_{5,2}$	$N^{(1)}_{5,2}, N^{(2)}_{5,2}$	$G^{(1)}_{5,2}, G^{(2)}_{5,2}$	$M^{(1)}_{5,2}, M^{(2)}_{5,2}$	$N^{(1)}_{5,2}, N^{(2)}_{5,2}$	$G^{(1)}_{5,2}, G^{(2)}_{5,2}$	$M^{(1)}_{5,2}, M^{(2)}_{5,2}$
(0,0)	0.64115	0.64115	0.64115	0.08808	0.08808	0.08808	0.46336	0.46336	0.46336
(0,1)	0.09452	0.09577	0.08353	0.28241	0.40805	0.15809	0.3456	0.38656	0.25088
(0,2)	0.00183	0.00058	0.01156	0.16879	0.04315	0.16746	0.05632	0.01536	0.11008
(0,3)			0.00115			0.07674			0.03072
(0,4)			0.00009			0.04890			0.01024
(1,0)	0.23846	0.24819	0.19361	0.18973	0.23689	0.12177	0.09984	0.10272	0.08448
(1,1)	0.01014	0.01014	0.00861	0.21185	0.21185	0.11074	0.03072	0.03072	0.02304
(1,2)			0.00056			0.05544			0.00512
(2,0)	0.01389	0.00417	0.04902	0.05914	0.01197	0.07992	0.00416	0.00128	0.01664
(2,1)			0.00097			0.04566			0.00256
(3,0)			0.00833			0.03785			0.00256
(4,0)			0.00139			0.00931			0.00032

Table 2
Exact distribution of bivariate binomial distributions for $n = 15, k = 5, r = 3$

(x_1, x_2)	$p_t = (1+t)^{-1}$ $p_t^* = (1+2t)^{-1}$		$p_t = 0.5 - 2^{-(t+1)}$ $p_t^* = 0.5 + 2^{-(t+2)}$		$p_t = 0.2$ $p_t^* = 0.4$	
	$N_{15,3}^{(1)}, N_{15,3}^{(2)}$	$G_{15,3}^{(1)}, G_{15,3}^{(2)}$	$N_{15,3}^{(1)}, N_{15,3}^{(2)}$	$G_{15,3}^{(1)}, G_{15,3}^{(2)}$	$N_{15,3}^{(1)}, N_{15,3}^{(2)}$	$G_{15,3}^{(1)}, G_{15,3}^{(2)}$
(0,0)	0.98341	0.98341	0.22181	0.22181	0.54578	0.54578
(0,1)	0.01476	0.01477	0.36108	0.41869	0.35972	0.38126
(0,2)	$2.5 \cdot 10^{-5}$	$1.6 \cdot 10^{-5}$	0.20534	0.18404	0.08342	0.06753
(0,3)	$9.6 \cdot 10^{-9}$	$2.6 \cdot 10^{-9}$	0.04770	0.01529	0.00792	0.00253
(0,4)	$8.3 \cdot 10^{-13}$	$1.6 \cdot 10^{-14}$	0.00389	$4.5 \cdot 10^{-5}$	0.00026	$3.6 \cdot 10^{-6}$
(0,5)	$5.2 \cdot 10^{-18}$		$4.8 \cdot 10^{-5}$		10^{-6}	
(1,0)	0.00179	0.00179	0.07116	0.07310	0.00207	0.00207
(1,1)	$3 \cdot 10^{-6}$	$3 \cdot 10^{-6}$	0.06606	0.07226	0.00074	0.00076
(1,2)	$8.9 \cdot 10^{-10}$	$5.9 \cdot 10^{-10}$	0.01638	0.01163	$6.9 \cdot 10^{-5}$	$4.6 \cdot 10^{-5}$
(1,3)	$3.6 \cdot 10^{-14}$	$3.7 \cdot 10^{-15}$	0.00084	$8.5 \cdot 10^{-5}$	$1.2 \cdot 10^{-6}$	10^{-7}
(2,0)	$6.3 \cdot 10^{-8}$	$3.5 \cdot 10^{-8}$	0.00437	0.00243	$1.4 \cdot 10^{-6}$	$9.8 \cdot 10^{-7}$
(2,1)	$1.8 \cdot 10^{-11}$	10^{-11}	0.00128	0.00059	$1.4 \cdot 10^{-7}$	$8.1 \cdot 10^{-8}$
(3,0)	$4.7 \cdot 10^{-14}$		$8.8 \cdot 10^{-6}$		$3.3 \cdot 10^{-11}$	

(a) $\Pr(Z_t = 1) = \frac{1}{1+t}$, $\Pr(Z_t = 2) = \frac{1}{1+2t}$,
(b) $\Pr(Z_t = 1) = \frac{1}{2} - \frac{1}{2^{t+1}}$, $\Pr(Z_t = 2) = \frac{1}{2} + \frac{1}{2^{t+2}}$,
(c) $\Pr(Z_t = 1) = 0.2$, $\Pr(Z_t = 2) = 0.4$.

Note that for the non-i.i.d. cases (i) and (ii) an appropriate adjustment of the transition probability matrices $A_t(x_1, x_2)$, $B_t^{(1)}(x_1, x_2)$, $B_t^{(2)}(x_1, x_2)$ is necessary before applying the recursive scheme (28). A more detailed numerical experimentation along with some additional theoretical results (pertaining to applications of the *MVT* approach to associated waiting time problems) can be found in Alexandrou (1997).

9. Alternative methods for exact distribution evaluation

As already mentioned in Section 1, the primary objective of the present work is to elucidate the use and usefulness of the Markov chain approach in the exact distribution computation of the number of run occurrences or more generally of the number of occurrences of specific patterns in sequences of binary/multistate trials. Needless to say, several alternative methods exist which can capture the exact distributions of the runs/patterns statistics mentioned in the previous sections. We shall briefly review some of them now. Since the purpose of this section is mainly focused to bringing to the reader's attention several bibliographical references, no technical details will be included herein.

Run-related problems have always attracted the attention of probabilists and statisticians since de Moivre's era. In his "Doctrine of Chance" (1756) he derived a recurrence relation for the evaluation of the probability that the first success run is

Table 3
Exact distribution of $(M^{(1)}_{15,5}, M^{(2)}_{15,3})$

(x_1, x_2)	$p_t = (1+t)^{-1}$ $p_t^* = (1+2t)^{-1}$ $M^{(1)}_{15,5}, M^{(2)}_{15,3}$	$p_t = 0.5 - 2^{-(t+1)}$ $p_t^* = 0.5 + 2^{-(t+2)}$ $M^{(1)}_{15,5}, M^{(2)}_{15,3}$	$p_t = 0.2$ $p_t^* = 0.4$ $M^{(1)}_{15,5}, M^{(2)}_{15,3}$
(0,0)	0.98341	0.22181	0.54578
(0,1)	0.01332	0.20051	0.23389
(0,2)	0.00134	0.16062	0.12054
(0,3)	0.00011	0.11106	0.05651
(0,4)	$8.5 \cdot 10^{-6}$	0.06902	0.02454
(0,5)	$5.5 \cdot 10^{-7}$	0.03897	0.00997
(0,6)	$3.2 \cdot 10^{-8}$	0.02035	0.00381
(0,7)	$1.6 \cdot 10^{-9}$	0.01013	0.00137
(0,8)	$7.8 \cdot 10^{-11}$	0.00431	0.00046
(0,9)	$3.4 \cdot 10^{-12}$	0.00189	0.00015
(0,10)	$1.3 \cdot 10^{-13}$	0.00086	$4.6 \cdot 10^{-5}$
(0,11)	$4.9 \cdot 10^{-15}$	0.00018	10^{-5}
(0,12)	$1.6 \cdot 10^{-16}$	$7.7 \cdot 10^{-5}$	$3.2 \cdot 10^{-6}$
(0,13)	$5.2 \cdot 10^{-18}$	$4.8 \cdot 10^{-5}$	10^{-6}
(1,0)	0.00155	0.03650	0.00167
(1,1)	$2.6 \cdot 10^{-6}$	0.02242	0.00042
(1,2)	$1.3 \cdot 10^{-7}$	0.01371	0.00016
(1,3)	$6.2 \cdot 10^{-9}$	0.00736	$5.8 \cdot 10^{-5}$
(1,4)	$2.6 \cdot 10^{-10}$	0.00334	$1.9 \cdot 10^{-5}$
(1,5)	10^{-11}	0.00136	$5.4 \cdot 10^{-6}$
(1,6)	$3.4 \cdot 10^{-13}$	0.00057	$1.4 \cdot 10^{-6}$
(1,7)	$9.9 \cdot 10^{-15}$	0.00011	$3 \cdot 10^{-7}$
(1,8)	$2.1 \cdot 10^{-16}$	$5.7 \cdot 10^{-5}$	$6.7 \cdot 10^{-8}$
(2,0)	0.00021	0.01948	0.00032
(2,1)	$2.4 \cdot 10^{-7}$	0.01044	$7.1 \cdot 10^{-5}$
(2,2)	$1.1 \cdot 10^{-8}$	0.00599	$2.6 \cdot 10^{-5}$
(2,3)	$4.8 \cdot 10^{-10}$	0.00288	$8.7 \cdot 10^{-6}$
(2,4)	$1.8 \cdot 10^{-11}$	0.00122	$2.6 \cdot 10^{-6}$
(2,5)	$6.3 \cdot 10^{-13}$	0.00052	$6.8 \cdot 10^{-7}$
(2,6)	$1.8 \cdot 10^{-14}$	0.00011	$1.5 \cdot 10^{-7}$
(2,7)	$4 \cdot 10^{-16}$	$5.7 \cdot 10^{-5}$	$3.3 \cdot 10^{-8}$
(3,0)	$2.6 \cdot 10^{-5}$	0.01006	$6 \cdot 10^{-5}$
(3,1)	$2.1 \cdot 10^{-8}$	0.00471	$1.1 \cdot 10^{-5}$
(3,2)	$8.7 \cdot 10^{-10}$	0.00242	$4 \cdot 10^{-6}$
(3,3)	$3.4 \cdot 10^{-11}$	0.00108	$1.2 \cdot 10^{-6}$
(3,4)	$1.1 \cdot 10^{-12}$	0.00047	$3.3 \cdot 10^{-7}$
(3,5)	$3.4 \cdot 10^{-14}$	0.00011	$7.5 \cdot 10^{-8}$
(3,6)	$7.6 \cdot 10^{-16}$	$5.7 \cdot 10^{-5}$	$1.6 \cdot 10^{-8}$
(4,0)	$2.9 \cdot 10^{-6}$	0.00506	$1.1 \cdot 10^{-5}$
(4,1)	$1.6 \cdot 10^{-9}$	0.00197	$1.8 \cdot 10^{-8}$
(4,2)	$6.2 \cdot 10^{-11}$	0.00094	$5.8 \cdot 10^{-7}$
(4,3)	$2.1 \cdot 10^{-12}$	0.00042	$1.6 \cdot 10^{-7}$
(4,4)	$6.4 \cdot 10^{-14}$	0.00011	$3.7 \cdot 10^{-8}$
(4,5)	$1.4 \cdot 10^{-15}$	$5.7 \cdot 10^{-5}$	$8.4 \cdot 10^{-9}$

Table 3
(Continued.)

(x_1, x_2)	$p_t = (1+t)^{-1}$, $p_t^* = (1+2t)^{-1}$	$p_t = 0.5 - 2^{-(t+1)}$, $p_t^* = 0.5 + 2^{-(t+2)}$	$p_t = 0.2$, $p_t^* = 0.4$
	$M_{15,5}^{(1)}, M_{15,3}^{(2)}$	$M_{15,5}^{(1)}, M_{15,3}^{(2)}$	$M_{15,5}^{(1)}, M_{15,3}^{(2)}$
(5,0)	$2.8 \cdot 10^{-7}$	0.00244	$2.1 \cdot 10^{-6}$
(5,1)	$1.1 \cdot 10^{-10}$	0.00079	$2.7 \cdot 10^{-7}$
(5,2)	$3.9 \cdot 10^{-12}$	0.00037	$7.8 \cdot 10^{-8}$
(5,3)	$1.2 \cdot 10^{-13}$	0.00011	$1.9 \cdot 10^{-8}$
(5,4)	$2.7 \cdot 10^{-15}$	$5.6 \cdot 10^{-5}$	$4.2 \cdot 10^{-9}$
(6,0)	$2.6 \cdot 10^{-8}$	0.00112	$3.7 \cdot 10^{-7}$
(6,1)	$7.3 \cdot 10^{-12}$	0.00032	$3.8 \cdot 10^{-8}$
(6,2)	$2.2 \cdot 10^{-13}$	0.00010	$9.4 \cdot 10^{-9}$
(6,3)	$5.1 \cdot 10^{-15}$	$5.5 \cdot 10^{-5}$	$2.1 \cdot 10^{-9}$
(7,0)	$2.1 \cdot 10^{-9}$	0.00050	$6.6 \cdot 10^{-8}$
(7,1)	$4 \cdot 10^{-13}$	0.00010	$4.7 \cdot 10^{-9}$
(7,2)	$9.5 \cdot 10^{-15}$	$5.3 \cdot 10^{-5}$	10^{-9}
(8,0)	$1.6 \cdot 10^{-10}$	0.00022	$1.1 \cdot 10^{-8}$
(8,1)	$1.7 \cdot 10^{-14}$	$4.9 \cdot 10^{-5}$	$5.2 \cdot 10^{-10}$
(9,0)	$1.1 \cdot 10^{-11}$	0.00010	$1.8 \cdot 10^{-9}$
(10,0)	$7.6 \cdot 10^{-13}$	$3.5 \cdot 10^{-5}$	$2.6 \cdot 10^{-10}$
(11,0)	$4.7 \cdot 10^{-14}$	$8.8 \cdot 10^{-6}$	$3.27 \cdot 10^{-11}$

completed at the nth trial in a sequence of Bernoulli trials. A concise account of the early history of attempts to find the distribution of the total number of runs in sequences of Bernoulli trials has been given in Mood (1940).

Feller (1968) treated several run-related problems as applications of renewal theory and recurrent events, thereof obtaining the generating functions of the associated distributions.

From early 80's an upsurge of research activity has been observed in the distribution theory of fixed length success runs. As a result, three different types of exact formulae have been established by several authors for the probability mass function of the run enumerating variables $N_{n,k}$, $M_{n,k}$ and $G_{n,k}$:

(a) recursive formulae (see, e.g., Aki and Hirano, 1988, Chryssaphinou et al., 1994, Ling, 1988);
(b) expressions involving multinomial coefficients and multiple summations over index sets determined by Diophantine equations (Hirano, 1986; Hirano et al., 1991; Philippou and Makri, 1986, etc.);
(c) expressions in terms of binomial coefficients and explicit single or double summations (Godbole, 1990, 1992; Koutras and Alexandrou, 1997a, etc.).

Similar formulae have been developed for more complicated run-related problems (for example the sooner-later waiting time problem) or for non-i.i.d. setups. For a detailed description of the recent advances in this area, the interested reader might wish to consult Balakrishnan and Koutras (2002) and the references quoted therein.

Until recently, there have not been published many general results pertaining to the exact distribution of the number of fixed length patterns in sequences of binary or multistate trials. To the best of the author's knowledge, apart from the "forward and backward principle" technique introduced by Fu (1996), no other universal procedure seems to be available for the exact computation of the whole distribution of the number of pattern occurrences. Nevertheless, several approaches have been suggested for the associated waiting time problem where one is interested in the first occurrence of the pattern or equivalently in the distribution of no pattern occurrences in a pre-specified number of trials.

As Feller (1968) indicated, when applying renewal and recurrent event theory, there is essentially no difference in the procedure of deriving generating functions for pattern waiting time problems as compared to the one used to establish generating functions for the respective run waiting time problems.

Guibas and Odlyzko (1980, 1981) employed a subtle combinatorial approach to study the general problem of enumerating the strings of a given length that contain no elements of a fixed set of patterns as substrings. In their approach, making crucial use of the key notion of "pattern correlation", they established a system of linear equations whose solution leads to the evaluation of the generating function of the quantities of interest. It is worth mentioning that these generating functions are rational and their special form can be profitably used to handle several important problems in asymptotic enumeration, as well as to investigate additional algebraic and combinatorial applications.

Breen et al. (1985), motivated by the action of restriction enzymes on DNA sequences, presented a similar generating function approach for a related pattern occurrence problem. Finally, Li (1980) and Gerber and Li (1981) employed martingale arguments to derive explicit expressions for the generating function of the number of trials it takes to observe a member of a finite set of patterns for the first time. For an overview of these results we refer to the monograph by Waterman (2000).

It is noteworthy that, there exists a very neat formula expressing the double generating function of the distribution of the number X_n of occurrences of patterns in a pre-specified number of multistate trials, in terms of the double generating function of the waiting time distribution for the rth occurrence of the patterns and vice versa (see Koutras, 1997a). If the non-overlapping enumeration scheme is in use, it is clear that the waiting time $T_r, r \geq 2$, for the rth occurrence of the pattern is the rth convolution of T_1 and therefore its probability generating function can be easily calculated (in the i.i.d. case) as the rth power of the probability generating function of T_1. Hence in the cases where one of the aforementioned techniques can be applied to capture the probability generating function of T_1, the double generating function of X_n is within reach and some additional algebraic manipulations could probably give rise to exact or effective recursive formulae for the computation of X_n's distribution. Unfortunately, for the other enumeration schemes, the transition from T_1 to T_r is not as smooth as in the non-overlapping case and the evaluation of T_r's single or double generating function by direct methods is in general a very tedious and intractable task.

In closing we mention that the problem of evaluating the exact probability of not observing a pattern (or any member of a family of patters) in a sequence of outcomes, has also been treated by Schwager (1983). His approach involves recursive techniques and can be adopted even for non-i.i.d. setups, e.g., when the probabilities of the outcomes can vary arbitrarily from trial to trial or exhibit a higher-order Markov dependence. However, no indication is given of how this technique could be extended to capture the whole distribution of the random variable X_n or T_r described above.

References

Aki, S. (1992). Waiting time problems for a sequence of discrete random variables. *Ann. Inst. Statist. Math.* **44**, 363–378.

Aki, S. (1997). On sooner and later problems between success and failure runs. In *Advances in Combinatorial Methods and Applications to Probability and Statistics*, pp. 385–400 (Ed. N. Balakrishnan). Birkhäuser, Boston.

Aki, S., N. Balakrishnan and S. G. Mohanty (1996). Sooner and later waiting time problems for success and failure runs in higher order Markov dependent trials. *Ann. Inst. Statist. Math.* **48**, 773–787.

Aki, S. and K. Hirano (1988). Some characteristics of the binomial distribution of order k and related distributions. In *Statistical Theory and Data Analysis II*, pp. 211–222. Elsevier, Amsterdam.

Aki, S. and K. Hirano (1993). Discrete distributions related to succession events in a two-state Markov chain. In *Statistical Science & Data Analysis*, pp. 467–474 (Eds. K. Matusita et al.). VSP.

Alexandrou, V. (1997). A study of enumerating random variables in sequences of trials by the aid of Markov chains, and applications. Ph.D. thesis, University of Athens, Greece (in Greek).

Antzoulakos, D. L. (1999). On waiting time problems associated with runs in Markov dependent trials. *Ann. Inst. Statist. Math.* **51**, 323–330.

Antzoulakos, D. L. and A. N. Philippou (1996). Derivation of the probability distribution functions for succession quota random variables. *Ann. Inst. Statist. Math.* **48**, 551–561.

Balakrishnan, N., K. Balasubramanian and R. Viveros (1993). On sampling inspection plans based on the theory of runs. *Math. Sci.* **18**, 113–126.

Balakrishnan, N., K. Balasubramanian and R. Viveros (1995). Start-up demonstration tests under correlation and corrective action. *Naval Res. Logistics* **42**, 1271–1276.

Balakrishnan, N. and M. V. Koutras (2002). *Runs and Scans with Applications*. Wiley, New York.

Balakrishnan, N., S. G. Mohanty and S. Aki (1997). Start-up demonstration tests under Markov dependence model with corrective actions. *Ann. Inst. Statist. Math.* **49**, 155–169.

Balasubramanian, K., R. Viveros and N. Balakrishnan (1993). Sooner and later waiting time problems for Markovian Bernoulli trials. *Statist. Probab. Lett.* **18**, 153–161.

Barton, D. E. and F. N. David (1957). Multiple runs. *Biometrika* **44**, 168–178.

Bogartz, R. (1965). The criterion method: some analyses and remarks. *Psychological Bulletin* **64**, 1–14.

Breen, S., M. S. Waterman and N. Zhang (1985). Renewal theory for several patterns. *J. Appl. Probab.* **22**, 228–234.

Chadjiconstantinidis, S., D. L. Antzoulakos and M. V. Koutras (2000). Joint distributions of successes, failures and patterns in enumeration problems. *Adv. Appl. Probab.* **32**, 866–884.

Chao, M. and J. Fu (1989). A limit theorem of certain repairable systems. *Ann. Inst. Statist. Math.* **41**, 809–818.

Chao, M. T. and J. C. Fu (1991). The reliability of large series system under a Markovian structure. *Adv. Appl. Probab.* **23**, 894–908.

Chao, M. T., J. C. Fu and M. V. Koutras (1995). Survey of the reliability studies of consecutive-k-out-of-n: F and related systems. *IEEE Trans. on Reliability* **44**, 120–127.

Chryssaphinou, O., S. Papastavridis and T. Tsapelas (1994). On the waiting time of appearance of given patterns. In *Runs and Patterns in Probability*, pp. 231–241 (Eds. A. P. Godbole and S. G. Papastavridis). Kluwer Academic Publishers, Netherlands.

David, F. N. and D. E. Barton (1962). *Combinatorial Chance*. Charles Griffins and Co., London.

De Moivre, A. (1756). *The Doctrine of Chance*, 3rd ed. Chelsea, New York.

Doi, M. and E. Yamamoto (1998). On the joint distribution of runs in a sequence of multi-state trials, *Statist. Probab. Lett.* **39**, 133–141.

Ebneshahrashoob, M. and M. Sobel (1990). Sooner and later waiting time problems for Bernoulli trials: frequency and run quotas. *Statist. Probab. Lett.* **9**, 5–11.

Feder, P. I. (1974). Problem solving: Markov chain method. *Industrial Engineering* **6**, 23–25.

Feller, W. (1968). *An Introduction to Probability Theory and its Applications*, Vol. I, 3rd ed. Wiley, New York.

Fu, J. (1996). Distribution theory of runs and patterns associated with a sequence of multistate trials. *Statistica Sinica* **6**, 957–964.

Fu, J. C. and M. V. Koutras (1994). Distribution theory of runs: a Markov chain approach. *J. Amer. Statist. Assoc.* **89**, 1050–1058.

Fu, J. C. and W. Y. Lou (1991). On reliabilities of certain large linearly connected engineering systems. *Statist. Probab. Lett.* **12**, 291–296.

Gerber, H. U. and S. R. Li (1981). The occurrence of sequence patterns in repeated experiments and hitting times in a Markov chain. *Stochastic Process. Appl.* **11**, 101–108.

Gibbons, J. D. (1971), *Nonparametric Statistical Inference*. McGraw-Hill, New York.

Gibbons, J. D. and S. Chakraborti (1992). *Nonparametric Statistical Inference*, 3rd ed. Marcel Dekker, New York.

Godbole, A. P. (1990). Specific formulae for some success run distributions. *Statist. Probab. Lett.* **10**, 119–124.

Godbole, A. P. (1991). Poisson approximations for runs and patterns of rare events, *Adv. Appl. Probab.* **23**, 851–865.

Godbole, A. P. (1992). The exact and asymptotic distribution of overlapping success runs. *Comm. Statist. Theory Methods* **21**, 953–967.

Grant, D. (1946). New statistical criteria for learning and problem solution in experiments involving repeated trials. *Psychological Bulletin* **43**, 272–282.

Grant, D. (1947). Additional tables of the probability of "runs" of correct responses in learning and problem solving. *Psychological Bulletin* **44**, 276–279.

Guibas, L. J. and A. M. Odlyzko (1980). Long repetitive patterns in random sequences. *Z. Wahrsch. Verw. Gebiete* **53**, 241–262.

Guibas, L. J. and A. M. Odlyzko (1981). String overlaps, pattern matching and nontransitive games. *J. Combin. Theory A* **30**, 183–208.

Hahn, G. J. and J. B. Gage (1983). Evaluation of a start-up demonstration test. *J. Quality Technology* **15**, 103–105.

Han, Q. and S. Aki (1999). Joint distributions of runs in a sequence of multi-state trials. *Ann. Inst. Statist. Math.* **51**, 419–447.

Hirano, K. (1986). Some properties of the distributions of order k. In *Fibonacci Numbers and their Applications*, pp. 43–53 (Eds. A. N. Philippou, A. F. Horadam and G. E. Bergum). D. Reidel Publishing Company.

Hirano, K. and S. Aki (1993). On number of occurrences of success runs of specified length in a two-state Markov chain, *Statistica Sinica* **3**, 313–320.

Hirano, K., S. Aki, N. Kashiwagi and H. Kuboki (1991). On Ling's binomial and negative binomial distributions of order k. *Statist. Probab. Lett.* **11**, 503–509.

Kitagawa, T. and T. Seguchi (1956). The combined use of runs in statistical quality controls. *Bull. Math. Statist.* **7**, 25–45.

Koutras, M. V. (1996). Reliability evaluation through Markov chain techniques. *J. Appl. Probab.* **33**, 357–367.

Koutras, M. V. (1997a). Waiting times and number of appearances of events in a sequence of discrete random variables. In *Advances in Combinatorial Methods and Applications to Probability and Statistics*, pp. 363–384 (Ed. N. Balakrishnan). Birkhäuser, Boston.

Koutras, M. V. (1997b). Waiting time distributions associated with runs of fixed length in two-state Markov chains. *Ann. Inst. Statist. Math.* **49**, 123–139.

Koutras, M. V. and V. A. Alexandrou (1995). Runs, scans and urn model distributions: a unified Markov chain approach. *Ann. Inst. Statist. Math.* **47**, 743–766.

Koutras, M. V. and V. A. Alexandrou (1997a). Non-parametric randomness tests based on success runs of fixed length. *Statist. Probab. Lett.* **32**, 393–404.

Koutras, M. V. and V. A. Alexandrou (1997b). Sooner waiting time problems in a sequence of trinary trials. *J. Appl. Probab.* **34**, 593–609.

Koutras, M. V., G. K. Papadopoulos and S. G. Papastavridis (1994). Circular overlapping success runs. In *Runs and Patterns in Probability*, pp. 287–305 (Eds. A. P. Godbole and S. G. Papastavridis). Kluwer Academic Publishers, Netherlands.

Koutras, M. V., G. K. Papadopoulos and S. G. Papastavridis (1995). Runs on a circle. *J. Appl. Probab.* **32**, 396–404.

Li, S.-Y. R. (1980). A martingale approach to the study of occurrence of sequence patterns in repeated experiments. *Ann. Probab.* **8**, 1171–1176.

Ling, K. (1988). On binomial distributions of order k. *Statist. Probab. Lett.* **6**, 247–250.

Ling, K. and T. Low (1993). On the soonest and latest waiting time distributions: succession quotas. *Comm. Statist.* **22**, 2207–2221.

Ling, K. and T. Tai (1990). On bivariate binomial distributions of order k. *Soochow J. Math.* **16**, 211–220.

Mohanty, S. G. (1994). Success runs of length-k in Markov dependent trials. *Ann. Inst. Statist. Math.* **46**, 777–796.

Mood, A. M. (1940). The distribution theory of runs. *Ann. Math. Statist.* **11**, 367–392.

Pielou, E. C. (1962). Runs of one species with respect to another in transects through plant populations. *Biometrics* **18**, 579–593.

Pielou, E. C. (1963a). Runs of healthy and diseased trees in transects through an infected forest. *Biometrics* **19**, 603–614.

Pielou, E. C. (1963b). The distribution of the diseased trees with respect to healthy ones in a patchily infected forest. *Biometrics* **19**, 450–459.

Pielou, E. C. (1977). *Mathematical Ecology*. Wiley, New York.

Philippou, A. N. (1984). The negative binomial distribution of order k and some of its properties. *Biometrical J.* **26**, 789–794.

Philippou, A. N. and F. Makri (1986). Successes, runs and longest runs. *Statist. Probab. Lett.* **4**, 211–215.

Schuster, E. F. (1991). Distribution theory of runs via exchangeable random variables. *Statist. Probab. Lett.* **11**, 379–386.

Schuster E. (1994). Exchangeability and recursion in the conditional distribution theory of number and length of runs. In *Runs and Patterns in Probability* (Eds. A. P. Godbole and S. G. Papastavridis). Kluwer Academic Publishers, Netherlands.

Schwager, S. (1983). Run probabilities in sequences of Markov dependent trials. *J. Amer. Statist. Assoc.* **78**, 168–175.

Taylor, H. M. and S. Karlin (1984). *An Introduction to Stochastic Modelling*. Academic Press.

Uchida, M. and S. Aki (1995). Sooner and later waiting time problems in a two-state Markov chain. *Ann. Inst. Statist. Math.* **47**, 415–433.

Viveros, R. and N. Balakrishnan (1993). Statisical inference from start-up demonstration test data. *J. Quality and Technology* **22**, 119–130.

Waterman. M. S. (2000). *Introduction to Computational Biology*. CRC Press, Boca Raton, FL.

Wolfowitz, J. (1943). On the theory of runs with some applications to quality control. *Ann. Math. Statist.* **14**, 280–288.

Modeling Image Analysis Problems Using Markov Random Fields

Stan Z. Li

1. Introduction

Modeling problems in this article are addressed mainly from the computational viewpoint. The primary concerns are how to define an objective function for the optimal solution for an image analysis problem and how to find the optimal solution. The reason for defining the solution in an *optimization* sense is due to various uncertainties in imaging processes. It may be difficult to find the perfect solution, so we usually look for an optimal one in the sense that an objective, into which constraints are encoded, is optimized.

Contextual constraints are ultimately necessary in the interpretation of visual information. A scene is understood in the spatial and visual context of the objects in it; the objects are recognized in the context of object features at a lower level representation; the object features are identified based on the context of primitives at an even lower level; and the primitives are extracted in the context of image pixels at the lowest level of abstraction. The use of contextual constraints is indispensable for a capable image analysis system.

Markov random field (MRF) theory provides a convenient and consistent way for modeling context dependent entities such as image pixels and correlated features. This is achieved through characterizing mutual influences among such entities using conditional MRF distributions. Rosanov (1967) presented the first comprehensive treatment of Gaussian MRF, looking into the problem of consistency among a set of conditional distributions. The practical use of MRF models is largely ascribed to a theorem stating the equivalence between MRFs and Gibbs distributions which was established by Hammersley and Clifford (1971) and further developed by Besag (1974). This is because the joint distribution is required in most applications but deriving the joint distribution from conditional distributions turns out to be very difficult for MRFs. The MRFs–Gibbs equivalence theorem points out that the joint distribution of an MRF is a Gibbs distribution, the latter taking a simple form. This gives us a not only mathematically sound but also mathematically tractable means for statistical image analysis (Grenander, 1983; Geman and Geman, 1984). From the computational perspective, the local property of MRFs leads to algorithms which can be implemented in a local and massively parallel manner. Furthermore, MRF theory provides a foundation for multi-resolution computation (Gidas, 1989).

For the above reasons, MRFs have been widely employed to solve image analysis problems at all levels. Most of the MRF models are for low level processing. These include image restoration and segmentation (Hassner and Slansky, 1980; Hansen and Elliott, 1982; Chellappa and Kashyap, 1982; Derin et al., 1984; Geman and Geman, 1984; Chellappa, 1985; Cohen and Cooper, 1987; Li, 1990, 1995b, 1998b), surface reconstruction (Barrow and Tenenbaum, 1981a; Grimson, 1981; Marroquin, 1985; Blake and Zisserman, 1987; Marroquin, Mitter and Poggio, 1987; Chou and Brown, 1990; Geiger and Girosi, 1991), edge detection (Mumford and Shah, 1985; Torre and Poggio, 1986; Geman et al., 1990), texture analysis (Chellappa and Kashyap, 1982; Cross and Jain, 1983; Elliott et al., 1984; Derin and Cole, 1986; Derin and Elliott, 1987) optical flow (Horn and Schunck, 1981; Hildreth, 1984; Koch, 1988; Shulman and Herve, 1989; Harris et al., 1990), shape from X (Barrow and Tenenbaum, 1981b, Ikeuchi and Horn, 1981), active contours (Kass, Witkin and Terzopoulos, 1987; Amini, Tehrani and Weymouth, 1988; Storvik, 1994), deformable templates (Mardia, Kent and Walder, 1991; Mardia, Hainsworth and Haddon, 1992; Jain, Zhong and Lakshmanan, 1996) data fusion (Clark and Yuille, 1990), visual integration, and perceptual organization (Aloimonos and Shulman, 1989; Szeliski, 1989). The use of MRFs in high level, such as for object matching and recognition, has also emerged in recent years (Modestino and Zhang, 1989; Cooper, 1990; Grenander and Chow, 1991; Baddeley and van Lieshout, 1992; Friedland and Rosenfeld, 1992; Kim and Yang, 1992; Cooper et al., 1993; Li, 1994, 1997, 1998a).

MRF theory tells us how to model a priori probabilities of contextual dependent patterns, such as textures and object features. A particular MRF model favors the class of patterns encoded by itself by associating them with larger probabilities than other pattern classes. MRF theory is often used in conjunction with statistical decision and estimation theories, so as to formulate objective functions in terms of established optimality principles. *Maximum a posteriori* (MAP) probability is one of the most popular statistical criteria for optimality and in fact, has been the most popular choice in MRF modeling for image analysis. MRFs and the MAP criterion together give rise to the MAP–MRF framework adopted in this book as well as in most other MRF works. This framework, advocated by Geman and Geman (1984) and others, enables us to develop algorithms for a variety of problems systematically using rational principles rather than relying on ad hoc heuristics.

An objective function is completely specified by its *form*, i.e., the parametric family, and the involved *parameters*. In the MAP–MRF framework, the objective is the joint posterior probability of the MRF labels. Its form and parameters are determined, in turn, according to the Bayes formula, by those of the joint prior distribution of the labels and the conditional probability of the observed data. "A particular MRF model" referred in the previous paragraph means a particular probability function (of patterns) specified by the functional form and the parameters. Two major parts of the MAP–MRF modeling is to derive the form of the posterior distribution and to determine the parameters in it, so as to completely define the posterior probability. Another important part is to design optimization algorithms for finding the maximum of the posterior distribution.

Several books exist on the subject of MRF modeling and its applications in image analysis. The two volumes edited by Mardia and Kanji (1993, 1994) and by Chellapa and Jain (1993) contain collections of research papers in the areas. The book by Winkler (1994) introduces mathematical principles and theories of MRFs, and Monte Carlo sampling methods for image analysis. The book by Guyon (1995) is a graduate level text. It is presented from the viewpoint of spatial modeling over lattices, using well-established theories from statistical mechanics and spectral analysis, with applications ranging from statistical mechanics to image analysis.

This chapter introduces fundamentals of MRF modeling in image analysis, important theoretical results, and modeling approaches. It is based on Chapter 1 of the book (Li, 2001) (of which the first edition appeared as Li, 1995a). The interested reader may refer to that book for a systematic and comprehensive presentation of the following issues essential in the subject: (1) *Formulations of MRF models*, for applications ranging from image restoration and reconstruction, edge and region segmentation, flow and motion, texture analysis and synthesis at lower level, to object matching and recognition at higher level. (2) *MRF parameter estimation*, which is needed for a complete formulation of any MRF model. (3) *Optimization techniques*, including various local and global optimization algorithms for finding solutions. See http://research.microsoft.com/~szli/MRF_Book/MRF_Book.html.

The rest of the article is organized as follows: Section 2 introduces image labeling and related notions in which MRF modeling is applied. Section 3 presents fundamental concepts and results of MRF theories. Section 4 describes classic and contemporary MRF models for image analysis. Section 5 presents a Bayesian framework for MRF modeling of image analysis problems and discusses related issues.

2. Image labeling

Many image analysis problems can be posed as labeling problems; the solution to a problem is represented by a set of labels assigned to image pixels or features. Labeling is also a natural representation for the study of MRFs (Besag, 1974).

2.1. Sites and labels

A *labeling problem* is specified in terms of a set of *sites* and a set of *labels*. Let $\mathcal{S}$ index a discrete set of m sites

$$\mathcal{S} = \{1, \ldots, m\} \tag{1}$$

in which $1, \ldots, m$ are indices. A site often represents a point or a region in a lattice space of image pixels, or an image feature such as a corner point, a line segment or a surface patch. A set of sites may be categorized in terms of their "regularity". Sites on a lattice are considered as spatially *regular*. A rectangular lattice for a 2D image of size $n \times n$ can be denoted by

$$\mathcal{S} = \{(i, j) \mid 1 \leqslant i, j \leqslant n\}. \tag{2}$$

Its elements correspond to the locations at which an image is sampled. Sites which do not present spatial regularity are considered as *irregular*. This is the usual case corresponding to features extracted from images at a more abstract level, such as the detected corners and lines.

We normally treat the sites in MRF models as un-ordered. For an $n \times n$ image, pixel (i, j) can be conveniently re-indexed by a single number k where k takes on values in $\{1, 2, \ldots, m\}$ with $m = n \times n$. This notation of single-number site index is used in this article also for images unless an elaboration is necessary. The inter-relationship between sites is maintained by a so-called *neighborhood system* (to be introduced later).

A label is an event that may happen to a site. Let $\mathcal{L}$ be a set of *labels*. A label set may be categorized as being continuous or discrete. In the continuous case, a label set may correspond to the real line $\mathbb{R}$ or a compact interval of it

$$\mathcal{L}_c = [X_l, X_h] \subset \mathbb{R}. \tag{3}$$

An example is the dynamic range for an analog pixel intensity. It is also possible that a continuous label takes a vector or matrix value, for example, $\mathcal{L}_c = \mathbb{R}^{a \times b}$ where a and b are dimensions.

In the discrete case, a label assumes a discrete value in a set of M labels

$$\mathcal{L}_d = \{\ell_1, \ldots, \ell_M\} \tag{4}$$

or simply

$$\mathcal{L}_d = \{1, \ldots, M\}. \tag{5}$$

In edge detection, for example, the label set is $\mathcal{L} = \{\text{edge, non-edge}\}$.

Besides the continuity, another essential property of a label set is the ordering of the labels. For example, elements in the continuous label set $\mathbb{R}$ (the real space) can be ordered by the relation "smaller than". When a discrete set, say $\{0, \ldots, 255\}$, represents the quantized values of intensities, it is an ordered set because for intensity values we have $0 < 1 < 2 < \cdots < 255$. When it denotes 256 different symbols such as texture types, it is considered to be un-ordered unless an artificial ordering is imposed.

For an ordered label set, a numerical (quantitative) measure of similarity between any two labels can usually be defined. For an unordered label set, a similarity measure is symbolic (qualitative), typically taking a value on "equal" or "non-equal". Label ordering and similarity not only categorize labeling problems but more importantly, affect our choices of labeling algorithms and hence the computational complexity.

2.2. The labeling problem

The labeling problem is to assign a label from the label set $\mathcal{L}$ to each of the sites in $\mathcal{S}$. Edge detection in an image, for example, is to assign a label f_i from the set $\mathcal{L} = \{\text{edge, non-edge}\}$ to site $i \in \mathcal{S}$ where elements in $\mathcal{S}$ index the image pixels. The set

$$f = \{f_1, \ldots, f_m\} \tag{6}$$

is called a *labeling* of the sites in $\mathcal{S}$ in terms of the labels in $\mathcal{L}$. When each site is assigned a unique label, $f_i = f(i)$ can be regarded as a function with domain $\mathcal{S}$ and image $\mathcal{L}$. Because the support of the function is the whole domain $\mathcal{S}$, it is a *mapping* from $\mathcal{S}$ to $\mathcal{L}$, that is,

$$f : \mathcal{S} \to \mathcal{L}. \tag{7}$$

Mappings with continuous and discrete label sets are demonstrated in Figure 1. A labeling is also called a *coloring* in mathematical programming.

In the terminology of random fields (cf. Section 3.2), a labeling is called a *configuration*. In image analysis, a configuration or labeling can correspond to an image, an edge map, an interpretation of image features in terms of object features, or a pose transformation, and so on.

When all the sites have the same label set $\mathcal{L}$, the set of all possible labelings, that is, the configuration space, is the following Cartesian product

$$\mathbb{F} = \underbrace{\mathcal{L} \times \mathcal{L} \times \cdots \times \mathcal{L}}_{m \text{ times}} = \mathcal{L}^m, \tag{8}$$

where m is the size of $\mathcal{S}$. In image restoration, for example, $\mathcal{L}$ contains admissible pixel values which are common to all pixel sites in $\mathcal{S}$ and $\mathbb{F}$ defines all admissible images. When $\mathcal{L} = \mathbb{R}$ is the real line, $\mathbb{F} = \mathbb{R}^m$ is the m-dimensional real space. When $\mathcal{L}$ is a discrete set, the size of $\mathbb{F}$ is combinatorial. For a problem with m sites and M labels, for example, there exist a total number of M^m possible configurations in $\mathbb{F}$.

In certain circumstances, admissible labels may not be common to all the sites. Consider, for example, feature based object matching. Supposing there are three types of features: points, lines and regions, then a constraint is that a certain type of image features can be labeled or interpreted in terms of the same type of model features. Therefore, the admissible label for any site is restricted to one of the three types. In an extreme case, every site i may have its own admissible set, $\mathcal{L}_i$, of labels and this gives the following configuration space

$$\mathbb{F} = \mathcal{L}_1 \times \mathcal{L}_2 \times \cdots \times \mathcal{L}_m. \tag{9}$$

This imposes constraints on the search for wanted configurations.

2.3. Labeling problems in image analysis

In terms of the regularity and the continuity, we may classify a labeling problem into one of the following four categories:

LP1: Regular sites with continuous labels.
LP2: Regular sites with discrete labels.
LP3: Irregular sites with discrete labels.
LP4: Irregular sites with continuous labels.

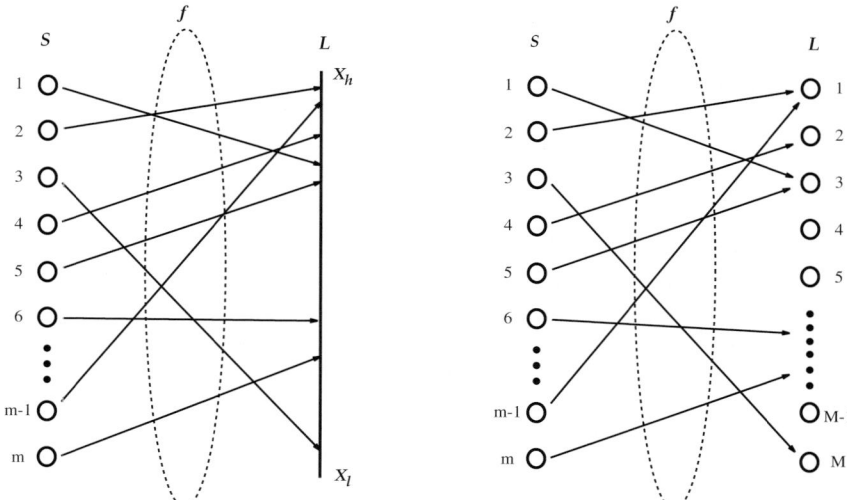

Fig. 1. A labeling of sites can be considered as a mapping from the set of sites $\mathcal{S}$ to the set of labels $\mathcal{L}$. The above shows mappings with continuous label set (left) and discrete label set (right).

The first two categories characterize low level processing performed on observed images and the other two do high level processing on extracted token features. The following describes some typical problems in terms of the four categories.

Restoration or smoothing of images having continuous pixel values is an LP1. The set $\mathcal{S}$ of sites corresponds to image pixels and the set $\mathcal{L}$ of labels is a real interval. The restoration is to estimate the true image signal from a degraded or noise-corrupted image.

Restoration of binary or multi-level images is an LP2. Similar to the continuous restoration, the aim is also to estimate the true image signal from the input image. The difference is that each pixel in the resulting image here assumes a discrete value and thus $\mathcal{L}$ in this case is a set of discrete labels.

Region segmentation is an LP2. It partitions an observation image into mutually exclusive regions, each of which has some uniform and homogeneous properties whose values are significantly different from those of the neighboring regions. The property can be, for example, grey tone, color or texture. Pixels within each region are assigned a unique label.

The prior assumption in the above problems is that the signal is smooth or piecewise smooth. This is complementary to the assumption of abrupt changes made for edge detection.

Edge detection is also an LP2. Each edge site, located between two neighboring pixels, is assigned a label in {edge, non-edge} if there is a significant difference between the two pixels. Continuous restoration with discontinuities can be viewed as a combination of LP1 and LP2.

Perceptual grouping (Lowe, 1985) is an LP3. The sites usually correspond to initially segmented features (points, lines and regions) which are irregularly arranged. The fragmentary features are to be organized into perceptually more significant features. Between each pair of the features is assigned a label in {connected, disconnected}, indicating whether the two features should be linked.

Feature-based object matching and recognition is an LP3. Each site indexes an image feature such as a point, a line segment or a region. Labels are discrete in nature and each of them indexes a model feature. The resulting configuration is a mapping from the image features to those of a model object.

Pose estimation from a set of point correspondences might be formulated as an LP4. A site is a given correspondence. A label represents an admissible (orthogonal, affine or perspective) transformation. A prior (unary) constraint is that the label of transformation itself must be orthogonal, affine or perspective. A mutual constraint is that the labels $f_1, \ldots, f_m$ should be close to each other to form a consistent transformation.

For a discrete labeling problem of m sites and M labels, there exist a total number of M^m possible labelings. For a continuous labeling problem, there are an infinite number of them. However, among all labelings, there are only a small number of them which are good solutions and may be just a few are optimal in terms of a criterion. How to define the optimal solution for a problem and how to find it are two important topics in the optimization approach to visual labeling.

2.4. Labeling with contextual constraints

The use of contextual information is ultimately indispensable in image understanding (Pavlidis, 1986). The use of contextual information in image analysis and pattern recognition dates back to (Chow, 1962; Abend, Harley and Kanal, 1965). In (Chow, 1962) character recognition is considered as a statistical decision problem. A nearest neighborhood dependence of pixels on an image lattice is obtained by going beyond the assumption of statistical independence. Information on the nearest neighborhood is used to calculate conditional probabilities. That system also includes parameter estimation from sample characters; recognition is done by using the estimated parameters. The work by Abend, Harley and Kanal (1965) is probably the earliest work using the Markov assumption for pattern recognition. There, a Markov mesh model is used to reduce the number of parameters required for the processing using contextual constraints. Fu and Yu (1980) use MRFs defined on an image lattice to develop a class of pattern classifiers for remote sensing image classification. Another development of context-based models is relaxation labeling (RL) (Rosenfeld, Hummel and Zucker, 1976). RL is a class of iterative procedures which use contextual constraints to reduce ambiguities in image analysis. A theory is given in (Haralick, 1983) to explain RL from a Bayes point of view.

In probability terms, contextual constraints may be expressed locally in terms of conditional probabilities $P(f_i | \{f_{i'}\})$, where $\{f_{i'}\}$ denotes the set of labels at the other sites $i' \neq i$, or globally as the joint probability $P(f)$. Because local information is more directly observed, it is normal that a global inference is made based on local properties.

In situations where labels are independent of one another (no context), the joint probability is the product of the local ones

$$P(f) = \prod_{i \in S} P(f_i). \tag{10}$$

The above implies conditional independence

$$P(f_i|\{f_{i'}\}) = P(f_i), \quad i' \neq i. \tag{11}$$

Therefore, a global labeling f can be computed by considering each label f_i locally. This is advantageous for problem solving.

In the presence of context, labels are mutually dependent. The simple relationships expressed in (10) and (11) do not hold any more. How to make a global inference using local information becomes a non-trivial task. Markov random field theory provides a mathematical foundation for solving this problem.

3. Markov random fields and Gibbs distributions

Markov random field theory is a branch of probability theory for analyzing the spatial or contextual dependencies of physical phenomena. It is used in visual labeling to establish probabilistic distributions of interacting labels. This section introduces notations and results relevant to image analysis.

3.1. Neighborhood system and cliques

The sites in S are related to one another via a neighborhood system. A neighborhood system for S is defined as

$$\mathcal{N} = \{\mathcal{N}_i \mid \forall i \in S\}, \tag{12}$$

where $\mathcal{N}_i$ is the set of sites neighboring i. The neighboring relationship has the following properties:

(1) a site is not neighboring to itself: $i \notin \mathcal{N}_i$;
(2) the neighboring relationship is mutual: $i \in \mathcal{N}_{i'} \Leftrightarrow i' \in \mathcal{N}_i$.

For a regular lattice S, the set of neighbors of i is defined as the set of sites within a radius of $\sqrt{r}$ from i

$$\mathcal{N}_i = \{i' \in S \mid [\text{dist}(\text{pixel}_{i'}, \text{pixel}_i)]^2 \leq r, \ i' \neq i\}, \tag{13}$$

where $\text{dist}(A, B)$ denotes the Euclidean distance between A and B and r takes an integer value. Note that sites at or near the boundaries have fewer neighbors.

In the first-order neighborhood system, also called the 4-neighborhood system, every (interior) site has four neighbors, as shown in Figure 2(a) where x denotes the

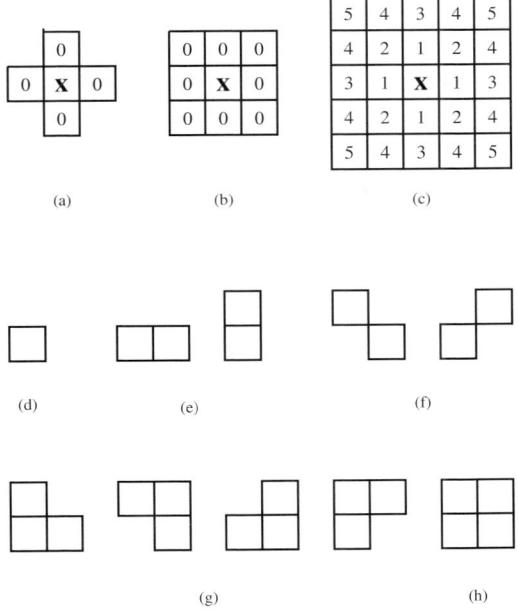

Fig. 2. Neighborhood and cliques on a lattice of regular sites.

considered site and 0's its neighbors. In the second-order neighborhood system, also called the 8-neighborhood system, there are eight neighbors for every (interior) site, as shown in Figure 2(b). The numbers $n = 1, \ldots, 5$ shown in Figure 2(c) indicate the outermost neighboring sites in the nth-order neighborhood system. The shape of a neighbor set may be described as the hull enclosing all the sites in the set.

When the ordering of the elements in $\mathcal{S}$ is specified, the neighbor set can be determined more explicitly. For example, when $\mathcal{S} = \{1, \ldots, m\}$ is an ordered set of sites and its elements index the pixels of a 1D image, an interior site $i \in \{2, \ldots, m-1\}$ has two nearest neighbors, $\mathcal{N}_i = \{i-1, i+1\}$, and a site at the boundaries (the two ends) has one neighbor each, $\mathcal{N}_1 = \{2\}$ and $\mathcal{N}_m = \{m-1\}$. When the sites in a regular rectangular lattice $\mathcal{S} = \{(i, j) \mid 1 \leqslant i, j \leqslant n\}$ correspond to the pixels of an $n \times n$ image in the 2D plane, an internal site (i, j) has four nearest neighbors as $\mathcal{N}_{i,j} = \{(i-1, j), (i+1, j), (i, j-1), (i, j+1)\}$, a site at a boundary has three and a site at the corners has two.

For an irregular $\mathcal{S}$, the neighbor set $\mathcal{N}_i$ of i is defined in the same way as (13) to comprise nearby sites within the radius of $\sqrt{r}$

$$\mathcal{N}_i = \{i' \in \mathcal{S} \mid [\mathrm{dist}(\mathrm{feature}_{i'}, \mathrm{feature}_i)]^2 \leqslant r, \; i' \neq i\}. \tag{14}$$

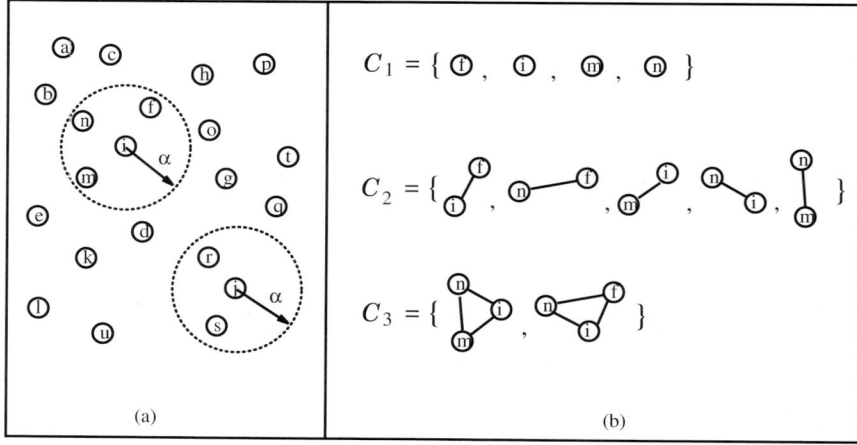

Fig. 3. Neighborhood and cliques on a set of irregular sites.

The dist(A, B) function needs to be defined appropriately for non-point features. Alternatively, the neighborhood may be defined by the Delaunay triangulation,[1] or its dual, the Voronoi polygons, of the sites (Besag, 1975). In general, the neighbor sets $\mathcal{N}_i$ for an irregular $\mathcal{S}$ have varying shapes and sizes. Irregular sites and their neighborhoods are illustrated in Figure 3(a). The neighborhood areas for sites i and j are marked by the dotted circles. The sizes of the two neighbor sets are $\#\mathcal{N}_i = 3$ and $\#\mathcal{N}_j = 2$.

The pair $(\mathcal{S}, \mathcal{N}) \stackrel{\triangle}{=} \mathcal{G}$ constitutes a graph in the usual sense; $\mathcal{S}$ contains the nodes and $\mathcal{N}$ determines the links between the nodes according to the neighboring relationship. A *clique* c for $(\mathcal{S}, \mathcal{N})$ is defined as a subset of sites in $\mathcal{S}$. It consists either of a single site $c = \{i\}$, or of a pair of neighboring sites $c = \{i, i'\}$, or of a triple of neighboring sites $c = \{i, i'. i''\}$, and so on. The collections of single-site, pair-site and triple-site cliques will be denoted by $\mathcal{C}_1$, $\mathcal{C}_2$ and $\mathcal{C}_3$, respectively, where

$$\mathcal{C}_1 = \{i \mid i \in \mathcal{S}\}, \tag{15}$$

$$\mathcal{C}_2 = \{\{i, i'\} \mid i' \in \mathcal{N}_i, \ i \in \mathcal{S}\} \tag{16}$$

and

$$\mathcal{C}_3 = \{\{i, i', i''\} \mid i, i', i'' \in \mathcal{S} \text{ are neighbors to one another}\}. \tag{17}$$

Note that the sites in a clique are *ordered*, and $\{i, i'\}$ is not the same clique as $\{i', i\}$, and so on. The collection of all cliques for $(\mathcal{S}, \mathcal{N})$ is

$$\mathcal{C} = \mathcal{C}_1 \cup \mathcal{C}_2 \cup \mathcal{C}_3 \cup \cdots, \tag{18}$$

---

[1] Algorithms for constructing a Delaunay triangulation in $(k \geqslant 2)$-dimensional space can be found in (Bowyer, 1981; Watson, 1981).

where "···" denotes possible sets of larger cliques.

The type of a clique for $(\mathcal{S}, \mathcal{N})$ of a regular lattice is determined by its size, shape and orientation. Figure 2(d)–(h) show clique types for the first- and second-order neighborhood systems for a lattice. The single-site and horizontal and vertical pair-site cliques in (d) and (e) are all those for the first-order neighborhood system (a). The clique types for the second-order neighborhood system (b) include not only those in (d) and (e) but also diagonal pair-site cliques (f) and triple-site (g) and quadruple-site (h) cliques. As the order of the neighborhood system increases, the number of cliques grow rapidly and so the involved computational expenses.

Cliques for irregular sites do not have fixed shapes as those for a regular lattice. Therefore, their types are essentially depicted by the number of involved sites. Consider the four sites f, i, m and n within the circle in Figure 3(a) in which m and n are supposed to be neighbors to each other and so are n and f. Then the single-site, pair-site and triple-site cliques associated with this set of sites are shown in Figure 3(b). The set $\{m, i, f\}$ does not form a clique because f and m are not neighbors.

3.2. Markov random fields

Let $F = \{F_1, \ldots, F_m\}$ be a family of random variables defined on the set $\mathcal{S}$, in which each random variable F_i takes a value f_i in $\mathcal{L}$. The family F is called a random field. We use the notation $F_i = f_i$ to denote the event that F_i takes the value f_i and the notation $(F_1 = f_1, \ldots, F_m = f_m)$ to denote the joint event. For simplicity, a joint event is abbreviated as $F = f$ where $f = \{f_1, \ldots, f_m\}$ is a *configuration* of F, corresponding to a realization of the field. For a discrete label set $\mathcal{L}$, the probability that random variable F_i takes the value f_i is denoted $P(F_i = f_i)$, abbreviated $P(f_i)$ unless there is a need to elaborate the expressions, and the joint probability is denoted $P(F = f) = P(F_1 = f_1, \ldots, F_m = f_m)$ and abbreviated $P(f)$. For a continuous $\mathcal{L}$, we have probability density functions (p.d.f.'s), $p(F_i = f_i)$ and $p(F = f)$.

F is said to be a Markov random field on $\mathcal{S}$ with respect to a neighborhood system $\mathcal{N}$ if and only if the following two conditions are satisfied:

$$P(f) > 0, \quad \forall f \in \mathbb{F} \quad \text{(positivity)}, \tag{19}$$

$$P(f_i | f_{\mathcal{S}-\{i\}}) = P(f_i | f_{\mathcal{N}_i}) \quad \text{(Markovianity)}, \tag{20}$$

where $\mathcal{S} - \{i\}$ is the set difference, $f_{\mathcal{S}-\{i\}}$ denotes the set of labels at the sites in $\mathcal{S} - \{i\}$ and

$$f_{\mathcal{N}_i} = \{f_{i'} | i' \in \mathcal{N}_i\} \tag{21}$$

stands for the set of labels at the sites neighboring i. The positivity is assumed for some technical reasons and can usually be satisfied in practice. For example, when the positivity condition is satisfied, the joint probability $P(f)$ of any random field is uniquely determined by its local conditional probabilities (Besag, 1974). The Markovianity depicts the local characteristics of F. In MRFs, only neighboring labels have direct interactions with each other. If we choose the largest neighborhood in which the neighbors of any sites include all other sites, then any F is an MRF with respect to such a neighborhood system.

An MRF can have other properties such as homogeneity and isotropy. It is said to be homogeneous if $P(f_i|f_{\mathcal{N}_i})$ is independent of the relative location of the site i in $\mathcal{S}$. So, for a homogeneous MRF, if $f_i = f_j$ and $f_{\mathcal{N}_i} = f_{\mathcal{N}_j}$, there will be $P(f_i|f_{\mathcal{N}_i}) = P(f_j|f_{\mathcal{N}_j})$ even if $i \neq j$. The isotropy will be illustrated in the next subsection with clique potentials.

In modeling some problems, we may need to use several *coupled* MRFs; each of the MRFs is defined on one set of sites, and the sites due to different MRFs are spatially interwoven. For example, in the related tasks of image restoration and edge detection, two MRFs, one for pixel values ($\{f_i\}$) and the other for edge values ($\{l_{i,i'}\}$), can be defined on the image lattice and its dual lattice, respectively. They are coupled to each other, e.g., via conditional probability $P(f_i|f_{i'}, l_{i,i'})$.

The concept of MRFs is a generalization of that of Markov processes (MPs) which are widely used in sequence analysis. An MP is defined on a domain of time rather than space. It is a sequence (chain) of random variables $\{\ldots, F_1, \ldots, F_m, \ldots\}$ defined on the time indices $\{\ldots, 1, \ldots, m, \ldots\}$. An nth-order unilateral MP satisfies

$$P(f_i|\ldots, f_{i-2}, f_{i-1}) = P(f_i|f_{i-1}, \ldots, f_{i-n}). \tag{22}$$

A bilateral or non-causal MP depends not only on the past but also on the future. An nth order bilateral MP satisfies

$$P(f_i|\ldots, f_{i-2}, f_{i-1}, f_{i+1}, f_{i+2}, \ldots)$$
$$= P(f_i|f_{i+n}, \ldots, f_{i+1}, f_{i-1}, \ldots, f_{i-n}). \tag{23}$$

It is generalized into MRFs when the time indices are considered as spatial indices.

There are two approaches for specifying an MRF, that in terms of the conditional probabilities $P(f_i|f_{\mathcal{N}_i})$ and that in terms of the joint probability $P(f)$. Besag (1974) argues for the joint probability approach in view of the disadvantages of the conditional probability approach: firstly, no obvious method is available for deducing the joint probability from the associated conditional probabilities. Secondly, the conditional probabilities themselves are subject to some non-obvious and highly restrictive consistency conditions. Thirdly, the natural specification of an equilibrium of statistical process is in terms of the joint probability rather than the conditional distribution of the variables. Fortunately, a theoretical result about the equivalence between Markov random fields and Gibbs distributions (Hammersley and Clifford, 1971; Besag, 1974) provides a mathematically tractable means of specifying the joint probability of an MRF.

3.3. Gibbs random fields

A set of random variables F is said to be a *Gibbs random field* (GRF) on $\mathcal{S}$ with respect to $\mathcal{N}$ if and only if its configurations obey a *Gibbs distribution*. A Gibbs distribution takes the following form

$$P(f) = Z^{-1} \times e^{-\frac{1}{T}U(f)}, \qquad (24)$$

where

$$Z = \sum_{f \in \mathbb{F}} e^{-\frac{1}{T}U(f)} \qquad (25)$$

is a normalizing constant called the *partition function*, T is a constant called the *temperature* which shall be assumed to be 1 unless otherwise stated, and $U(f)$ is the *energy function*. The energy

$$U(f) = \sum_{c \in \mathcal{C}} V_c(f) \qquad (26)$$

is a sum of *clique potentials* $V_c(f)$ over all possible cliques $\mathcal{C}$. $V_c(f)$ can be considered the cost of f as far as the clique c is concerned, and so its value depends on the local configuration, i.e., the f_i values for i involved in c. Obviously, the Gaussian distribution is a special member of this Gibbs distribution family.

A GRF is said to be homogeneous if $V_c(f)$ is independent of the relative position of the clique c in $\mathcal{S}$. It is said to be isotropic if V_c is independent of the orientation of c. It is considerably simpler to specify a GRF distribution if it is homogeneous or isotropic than one without such properties. The homogeneity is assumed in most MRF models for mathematical and computational convenience. The isotropy is a property of direction-independent blob-like regions.

To calculate a Gibbs distribution, it is necessary to evaluate the partition function Z which is the sum over all possible configurations in $\mathbb{F}$. Since there are a combinatorial number of elements in $\mathbb{F}$ for a discrete $\mathcal{L}$, as illustrated in Section 2.2, the evaluation is prohibitive even for problems of moderate sizes. Several approximation methods exist for solving this problem.

$P(f)$ measures the probability of the occurrence of a particular configuration, or "pattern", f. The more probable configurations are those with lower energies. The temperature T controls the sharpness of the distribution. When the temperature is high, all configurations tend to be equally distributed. Near the zero temperature, the distribution concentrates around the global energy minima. Given T and $U(f)$, we can generate a class of "patterns" by sampling the configuration space $\mathbb{F}$ according to $P(f)$.

For discrete labeling problems, a clique potential $V_c(f)$ can be specified by a number of values some times called *parameters*. For example, letting $f_c = (f_i, f_{i'}, f_{i''})$ be the local configuration on a triple-clique $c = \{i, i', i''\}$, f_c takes a finite number of states and therefore $V_c(f)$ takes a finite number of values. For continuous labeling problems, f_c

can vary continuously. In this case, $V_c(f)$ is a (possibly piecewise) continuous function of f_c.

Sometimes, it may be convenient to express the energy of a Gibbs distribution as the sum of several terms, each ascribed to cliques of a certain size, that is,

$$U(f) = \sum_{\{i\}\in\mathcal{C}_1} V_1(f_i) + \sum_{\{i,i'\}\in\mathcal{C}_2} V_2(f_i, f_{i'})$$
$$+ \sum_{\{i,i',i''\}\in\mathcal{C}_3} V_3(f_i, f_{i'}, f_{i''}) + \cdots. \qquad (27)$$

The above implies a homogeneous Gibbs distribution because V_1, V_2 and V_3 are independent of the locations of i, i' and i''. For non-homogeneous Gibbs distributions, the clique functions should be written as $V_1(i, f_i)$, $V_2(i, i', f_i, f_{i'})$, and so on.

An important special case is when only cliques of size up to two are considered. In this case, the energy can also be written as

$$U(f) = \sum_{i\in\mathcal{S}} V_1(f_i) + \sum_{i\in\mathcal{S}} \sum_{i'\in\mathcal{N}_i} V_2(f_i, f_{i'}). \qquad (28)$$

Note that in the second term on the RHS, $\{i, i'\}$ and $\{i', i\}$ are two distinct cliques in $\mathcal{C}_2$ because the sites in a clique are *ordered*. The conditional probability can be written as (letting $T = 1$)

$$P(f_i | f_{\mathcal{N}_i}) = \frac{e^{-[V_1(f_i) + \sum_{i'\in\mathcal{N}_i} V_2(f_i, f_{i'})]}}{\sum_{f_i\in\mathcal{L}} e^{-[V_1(f_i) + \sum_{i'\in\mathcal{N}_i} V_2(f_i, f_{i'})]}}. \qquad (29)$$

3.4. Markov–Gibbs equivalence

An MRF is characterized by its local property (the Markovianity) whereas a GRF is characterized by its global property (the Gibbs distribution). The Hammersley–Clifford theorem (Hammersley and Clifford, 1971) establishes the equivalence of these two types of properties. The theorem states that F is an MRF on $\mathcal{S}$ with respect to $\mathcal{N}$ if and only if F is a GRF on $\mathcal{S}$ with respect to $\mathcal{N}$. Many proofs of the theorem exist, e.g., in (Besag, 1974; Moussouris, 1974; Kindermann and Snell, 1980).

A proof that a GRF is an MRF is given as follows. Let $P(f)$ be a Gibbs distribution on $\mathcal{S}$ with respect to the neighborhood system $\mathcal{N}$. Consider the conditional probability

$$P(f_i | f_{\mathcal{S}-\{i\}}) = \frac{P(f_i, f_{\mathcal{S}-\{i\}})}{P(f_{\mathcal{S}-\{i\}})} = \frac{P(f)}{\sum_{f'_i\in\mathcal{L}} P(f')}, \qquad (30)$$

where $f' = \{f_1, \ldots, f_{i-1}, f_i', \ldots, f_m\}$ is any configuration which agrees with f at all sites except possibly i. Writing $P(f) = Z^{-1} \times e^{-\sum_{c \in \mathcal{C}} V_c(f)}$ out gives[2]

$$P(f_i | f_{\mathcal{S}-\{i\}}) = \frac{e^{-\sum_{c \in \mathcal{C}} V_c(f)}}{\sum_{f_i'} e^{-\sum_{c \in \mathcal{C}} V_c(f')}}. \tag{31}$$

Divide $\mathcal{C}$ into two set $\mathcal{A}$ and $\mathcal{B}$ with $\mathcal{A}$ consisting of cliques containing i and $\mathcal{B}$ cliques not containing i. Then the above can be written as

$$P(f_i | f_{\mathcal{S}-\{i\}}) = \frac{[e^{-\sum_{c \in \mathcal{A}} V_c(f)}][e^{-\sum_{c \in \mathcal{B}} V_c(f)}]}{\sum_{f_i'} \{[e^{-\sum_{c \in \mathcal{A}} V_c(f')}][e^{-\sum_{c \in \mathcal{B}} V_c(f')}]\}}. \tag{32}$$

Because $V_c(f) = V_c(f')$ for any clique c that does not contain i, $e^{-\sum_{c \in \mathcal{B}} V_c(f)}$ cancels from both the numerator and denominator. Therefore, this probability depends only on the potentials of the cliques containing i,

$$P(f_i | f_{\mathcal{S}-\{i\}}) = \frac{e^{-\sum_{c \in \mathcal{A}} V_c(f)}}{\sum_{f_i'} e^{-\sum_{c \in \mathcal{A}} V_c(f')}}. \tag{33}$$

That is, $P(f_i | f_{\mathcal{S}-\{i\}}) = P(f_i | f_{\mathcal{N}_i})$ This proves that a Gibbs random field is a Markov random field. The proof that an MRF is a GRF is much more involved; a result to be described in the next subsection, which is about the uniqueness of the GRF representation (Griffeath, 1976), provides such a proof.

The practical value of the theorem is that it provides a simple way of specifying the joint probability in which a priori knowledge or preference about interactions between labels can be encoded. We can specify the joint probability $P(F = f)$ by specifying the clique potential functions $V_c(f)$ and choosing appropriate potential functions for desired system behavior.

How to choose the forms and parameters of the potential functions for a proper encoding of constraints is a major topic in MRF modeling. The forms of the potential functions determine the form of the Gibbs distribution. When all the parameters involved in the potential functions are specified, the Gibbs distribution is completely defined.

To calculate the joint probability of an MRF, which is a Gibbs distribution, it is necessary to evaluate the partition function (25). Because it is the sum over a combinatorial number of configurations in $\mathbb{F}$, the computation is usually intractable. The explicit evaluation can be avoided in maximum-probability based MRF models when $U(f)$ contains no unknown parameters, as we will see subsequently. However, this is not true when the parameter estimation is also a part of the problem. In the latter case, the energy function $U(f) = U(f | \theta)$ is also a function of parameters θ and so is the partition function $Z = Z(\theta)$. The evaluation of $Z(\theta)$ is required. To circumvent the formidable difficulty therein, the joint probability is often approximated in practice.

---

[2] This also provides a formula for calculating the conditional probability $P(f_i | f_{\mathcal{N}_i}) = P(f_i | f_{\mathcal{S}-\{i\}})$ from potential functions.

3.5. Normalized and canonical forms

The choices of clique potential functions for a specific MRF are not unique; there may exist many equivalent choices which specify the same Gibbs distribution. However, there exists a unique normalized potential, called the *canonical potential*, for every MRF (Griffeath, 1976).

Let $\mathcal{L}$ be a countable label set. A clique potential function $V_c(f)$ is said to be *normalized* if $V_c(f) = 0$ whenever for some $i \in c$, f_i takes a particular value in $\mathcal{L}$. The particular value can be any element in $\mathcal{L}$, e.g., 0 in $\mathcal{L} = \{0, 1, \ldots, M\}$. Griffeath (1976) establishes the mathematical relationship between an MRF distribution $P(f)$ and the unique canonical representation of clique potentials V_c in the corresponding Gibbs distribution (Griffeath, 1976; Kindermann and Snell, 1980). The result is described below.

Let F be a random field on a finite set $\mathcal{S}$ with local characteristics $P(f_i | f_{\mathcal{S}-\{i\}}) = P(f_i | f_{\mathcal{N}_i})$. Then F is a Gibbs field with *canonical potential function* defined by the following:

$$V_c(f) = \begin{cases} 0, & c = \phi, \\ \sum_{b \subset c} (-1)^{|c-b|} \ln P(f^b), & c \neq \phi, \end{cases} \tag{34}$$

where ϕ denotes the empty set, $|c - b|$ is the number of elements in the set $c - b$ and

$$f_i^b = \begin{cases} f_i, & \text{if } i \in b, \\ 0, & \text{otherwise} \end{cases} \tag{35}$$

is the configuration which agrees with f on set b but assigns the value 0 to all sites outside of b. For nonempty c, the potential can also be obtained as

$$V_c(f) = \sum_{b \subset c} (-1)^{|c-b|} \ln P(f_i^b | f_{\mathcal{N}_i}^b), \tag{36}$$

where i is any element in b. Such canonical potential function is *unique* for the corresponding MRF. Using this result, the canonical $V_c(f)$ can be computed if $P(f)$ is known.

However, in MRF modeling using Gibbs distributions, $P(f)$ is defined after $V_c(f)$ is determined and therefore, it is difficult to compute the canonical $V_c(f)$ from $P(f)$ directly. Nonetheless, there is an indirect way: use a non-canonical representation to derive $P(f)$ and then canonicalize it using Griffeath's result to obtain the unique canonical representation.

The normalized potential functions appear to be immediately useful. For instance, for the sake of economy, one would use the minimal number of clique potentials or parameters to represent an MRF for a given neighborhood system. The concept of normalized potential functions can be used to reduce the number of nonzero clique parameters.

4. Useful MRF models

The following introduces some useful MRF models for modeling image properties such as regions and textures. We are interested in their conditional and joint distributions, and the corresponding energy functions. The interested reader may refer to Derin and Kelly (1989) for a systematic study and categorization of Markov random processes and fields in terms of what is called there strict-sense Markov and wide-sense Markov properties.

4.1. Auto-models

Contextual constraints on two labels are the lowest-order constraints to convey contextual information. They are widely used because of their simple form and low computational cost. They are encoded in the Gibbs energy as pair-site clique potentials. With clique potentials of up to two sites, the energy takes the form

$$U(f) = \sum_{i \in S} V_1(f_i) + \sum_{i \in S} \sum_{i' \in \mathcal{N}_i} V_2(f_i, f_{i'}), \tag{37}$$

where "$\sum_{i \in S}$" is equivalent to "$\sum_{\{i\} \in \mathcal{C}_1}$" and "$\sum_{i \in S} \sum_{i' \in \mathcal{N}_i}$" to "$\sum_{\{i,i'\} \in \mathcal{C}_2}$". The above is a special case of (27), which we call a second-order energy because it involves up to pair-site cliques. It the most frequently used form owing to the mentioned feature that it is the simplest in form but conveys contextual information. A specific GRF or MRF can be specified by proper selection of V_1's and V_2's. Some important such GRF models are described subsequently. Derin and Kelly (1989) present a systematic study and categorization of Markov random processes and fields in terms of what they call strict-sense Markov and wide-sense Markov properties.

When $V_1(f_i) = f_i G_i(f_i)$ and $V_2(f_i, f_{i'}) = \beta_{i,i'} f_i f_{i'}$, where $G_i(\cdot)$ are arbitrary functions and $\beta_{i,i'}$ are constants reflecting the pair-site interaction between i and i', the energy is

$$U(f) = \sum_{\{i\} \in \mathcal{C}_1} f_i G_i(f_i) + \sum_{\{i,i'\} \in \mathcal{C}_2} \beta_{i,i'} f_i f_{i'}. \tag{38}$$

The above is called *auto-models* (Besag, 1974). The auto-models can be further classified according to assumptions made about individual f_i's.

An auto-model is said to be an *auto-logistic* model, if the f_i's take on values in the discrete label set $\mathcal{L} = \{0, 1\}$ (or $\mathcal{L} = \{-1, +1\}$). The corresponding energy is of the following form

$$U(f) = \sum_{\{i\} \in \mathcal{C}_1} \alpha_i f_i + \sum_{\{i,i'\} \in \mathcal{C}_2} \beta_{i,i'} f_i f_{i'}, \tag{39}$$

where $\beta_{i,i'}$ can be viewed as the *interaction coefficients*. When $\mathcal{N}$ is the nearest neighborhood system on a lattice (4 nearest neighbors on a 2D lattice or 2 nearest neighbors on a 1D lattice), the auto-logistic model is reduced to the *Ising model*. The conditional probability for the auto-logistic model with $\mathcal{L} = \{0, 1\}$ is

$$P(f_i | f_{\mathcal{N}_i}) = \frac{e^{-(\alpha_i f_i + \sum_{i' \in \mathcal{N}_i} \beta_{i,i'} f_i f_{i'})}}{\sum_{f_i \in \{0,1\}} e^{-(\alpha_i f_i + \sum_{i' \in \mathcal{N}_i} \beta_{i,i'} f_i f_{i'})}}$$

$$= \frac{e^{-(\alpha_i f_i + \sum_{i' \in \mathcal{N}_i} \beta_{i,i'} f_i f_{i'})}}{1 + e^{-(\alpha_i + \sum_{i' \in \mathcal{N}_i} \beta_{i,i'} f_{i'})}}. \tag{40}$$

When the distribution is homogeneous, we have $\alpha_i = \alpha$ and $\beta_{i,i'} = \beta$, regardless of i and i'.

An auto-model is said to be an *auto-binomial* model if the f_i's take on values in $\{0, 1, \ldots, M-1\}$ and every f_i has a conditionally binomial distribution of $M-1$ trials and probability of success q

$$P(f_i | f_{\mathcal{N}_i}) = \binom{M-1}{f_i} q^{f_i} (1-q)^{M-1-f_i}, \tag{41}$$

where

$$q = \frac{e^{-(\alpha_i + \sum_{i' \in \mathcal{N}_i} \beta_{i,i'} f_{i'})}}{1 + e^{-(\alpha_i + \sum_{i' \in \mathcal{N}_i} \beta_{i,i'} f_{i'})}}. \tag{42}$$

The corresponding energy takes the following form

$$U(f) = -\sum_{\{i\} \in \mathcal{C}_1} \ln \binom{M-1}{f_i} - \sum_{\{i\} \in \mathcal{C}_1} \alpha_i f_i - \sum_{\{i,i'\} \in \mathcal{C}_2} \beta_{i,i'} f_i f_{i'}. \tag{43}$$

It reduces to the auto-logistic model when $M = 1$.

An auto-model is said to be an *auto-normal model*, also called a Gaussian MRF (Chellappa, 1985), if the label set $\mathcal{L}$ is the real line and the joint distribution is multivariate normal. Its conditional p.d.f. is

$$p(f_i | f_{\mathcal{N}_i}) = \frac{1}{\sqrt{2\pi\sigma^2}} e^{-\frac{1}{2\sigma^2} [f_i - \mu_i - \sum_{i' \in \mathcal{N}_i} \beta_{i,i'} (f_{i'} - \mu_{i'})]^2}. \tag{44}$$

It is a normal distribution with conditional mean

$$\text{mean}(f_i | f_{\mathcal{N}_i}) = \mu_i - \sum_{i' \in \mathcal{N}_i} \beta_{i,i'} (f_{i'} - \mu_{i'}) \tag{45}$$

and conditional variance

$$\text{var}(f_i | f_{\mathcal{N}_i}) = \sigma^2. \tag{46}$$

The joint probability is a Gibbs distribution

$$p(f) = \frac{\sqrt{\det(B)}}{\sqrt{(2\pi\sigma^2)^m}} e^{-\frac{(f-\mu)^T B(f-\mu)}{2\sigma^2}}, \qquad (47)$$

where f is viewed as a vector, μ is the $m \times 1$ vector of the conditional means, and $B = [b_{i,i'}]$ is the $m \times m$ *interaction matrix* whose elements are unity and off-diagonal element at (i, i') is $-\beta_{i,i'}$, i.e., $b_{i,i'} = \delta_{i,i'} - \beta_{i,i'}$ with $\beta_{i,i} = 0$. Therefore, the single-site and pair-site clique potential functions for the auto-normal model are

$$V_1(f_i) = (f_i - \mu_i)^2/2\sigma^2 \qquad (48)$$

and

$$V_2(f_i, f_{i'}) = \beta_{i,i'}(f_i - \mu_i)(f_{i'} - \mu_{i'})/2\sigma^2, \qquad (49)$$

respectively. A field of independent Gaussian noise is a special MRF whose Gibbs energy consists of only single-site clique potentials. Because all higher-order clique potentials are zero, there is no contextual interaction in the independent Gaussian noise. B is related to the covariance matrix Σ by $B = \Sigma^{-1}$. The necessary and sufficient condition for (47) to be a valid p.d.f. is that B is symmetric and positive definite.

A related but different model is the simultaneous auto-regression (SAR) model (Woods, 1972). Unlike the auto-normal model which is defined by the m conditional p.d.f.'s, this model is defined by a set of m simultaneous auto-regression equations

$$f_i = \mu_i + \sum \beta_{i,i'}(f_{i'} - \mu_{i'}) + e_i, \qquad (50)$$

where $e_i \sim N(0, \sigma^2)$ are independent Gaussian. It also generates the class of all multivariate normal distributions but with joint p.d.f. as

$$p(f) = \frac{\det(B)}{\sqrt{(2\pi\sigma^2)^m}} e^{-\frac{(f-\mu)^T B^T B(f-\mu)}{2\sigma^2}}, \qquad (51)$$

where B is defined as before. Any SAR model is an auto-normal model with the B matrix in (47) being $B = B_2 + B_2^T - B_2^T B_2$ where $B_2 = B_{\text{auto-regressive}}$. The reverse can also be done, though in a rather unnatural way via Cholesky decomposition (Ripley, 1981). Therefore, both models can have their p.d.f.'s in the form (47). However, for (51) to be a valid p.d.f., it requires only that $B_{\text{auto-regressive}}$ be non-singular.

4.2. Multi-level logistic model

The auto-logistic model can be generalized to *multi-level logistic* (MLL) model (Elliott et al., 1984; Derin and Cole, 1986; Derin and Elliott, 1987), also called Strauss process (Strauss, 1977) and generalized Ising model (Geman and Geman, 1984). There are M (> 2) discrete labels in the label set, $\mathcal{L} = \{1, \ldots, M\}$. In this type of models, a clique

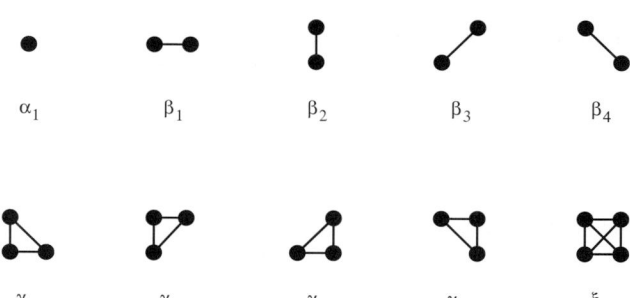

Fig. 4. Clique types and associated potential parameters for the second order neighborhood system. Sites are shown in dots and neighboring relationships in joining lines.

potential depends on the type c (related to size, shape and possibly orientation) of the clique and the local configuration $f_c \stackrel{\triangle}{=} \{f_i | i \in c\}$. For cliques containing more than one site ($\#c > 1$), the MLL clique potentials are defined by

$$V_c(f) = \begin{cases} \zeta_c, & \text{if all sites on c have the same label,} \\ -\zeta_c, & \text{otherwise,} \end{cases} \quad (52)$$

where ζ_c is the potential for type-c cliques; for single site cliques, they depend on the label assigned to the site

$$V_c(f) = V_c(f_i) = \alpha_I, \quad \text{if } f_i = I \in \mathcal{L}_d, \quad (53)$$

where α_I is the potential for label value I. Figure 4 shows the clique types and the associated parameters in the second-order (8-neighbor) neighborhood system.

Assume that an MLL model is of second-order as in (37), so that only α (for single-site cliques) and β (for pair-site cliques) parameters are non-zero. The potential function for pair-wise cliques is written as

$$V_2(f_i, f_{i'}) = \begin{cases} \beta_c, & \text{if sites on } \{i, i'\} = c \in \mathcal{C}_2 \text{ have the same label,} \\ -\beta_c, & \text{otherwise,} \end{cases} \quad (54)$$

where β_c is the β parameter for type-c cliques and $\mathcal{C}_2$ is set of pair-site cliques. For the 4-neighborhood system, there are four types of pair-wise cliques (cf. Figure 4) and so there can be four different β_c's. When the model is isotropic all the four take the same value. Owing to its simplicity, the pair-wise MLL model (54) has been widely used for modeling regions and textures (Elliott et al., 1984; Geman and Geman, 1984; Derin and Cole, 1986; Derin and Elliott, 1987; Murray and Buxton, 1987; Lakshmanan and Derin, 1989; Won and Derin, 1992).

When the MLL model is isotropic, it depicts blob-like regions. In this case, the conditional probability can be expressed as follows (Strauss, 1977)

$$P(f_i = I | f_{\mathcal{N}_i}) = \frac{e^{-\alpha_I - \beta n_i(I)}}{\sum_{I=1}^{M} e^{-\alpha_I - \beta n_i(I)}}, \qquad (55)$$

where $n_i(I)$ is the number of sites in $\mathcal{N}_i$ which are labeled I. It reduces to (40) when there are only two labels, 0 and 1. In contrast, an anisotropic model tends to generate texture-like patterns.

A hierarchical two-level Gibbs model has been proposed to represent both noise-contaminated and textured images (Derin and Cole, 1986; Derin and Elliott, 1987). The higher-level Gibbs distribution uses an isotropic random field, e.g., MLL, to characterize the blob-like region formation process. A lower-level Gibbs distribution describes the filling-in in each region. The filling-in may be independent noise or a type of texture, both of which can be characterized by Gibbs distributions. This provides a convenient approach for MAP–MRF modeling. In segmentation of noisy and textured image (Derin and Cole, 1986; Derin and Elliott, 1987; Lakshmanan and Derin, 1989; Hu and Fahmy, 1987; Won and Derin, 1992), for example, the higher-level determines the prior of f for the region process while the lower-level Gibbs contributes to the conditional probability of the data given f. Note that different levels of MRFs in the hierarchy can have different neighborhood systems.

4.3. The smoothness prior

A generic contextual constraint on this world is the *smoothness*. It assumes that physical properties in a neighborhood of space or in an interval of time present some coherence and generally do not change abruptly. For example, the surface of a table is flat, a meadow presents a texture of grass, and a temporal event does not change abruptly over a short period of time. Indeed, we can always find regularities of a physical phenomenon with respect to certain properties. Since its early applications (Grimson, 1981; Horn and Schunck, 1981; Ikeuchi and Horn, 1981) aimed to impose constraints, in addition to those from the data, on the computation of image properties, the smoothness prior has been one of the most popular prior assumptions in low level problems. It has been developed into a general framework, called regularization (Tikhonov and Arsenin, 1977; Poggio, Torre and Koch, 1985; Bertero, Poggio and Torre, 1988), for a variety of low level problems.

Smoothness constraints are often expressed as the prior probability or equivalently an energy term $U(f)$ measuring the extent to which the smoothness assumption is violated by f. There are two basic forms of such smoothness terms corresponding to situations with discrete and continuous labels, respectively.

Eqs. (52) and (54) of the MLL model with negative ζ and β coefficients provide a method for constructing smoothness terms for un-ordered, discrete labels. Whenever all labels f_c on a clique c take the same value, which means the solution f is locally smooth on c, they incur a negative clique potential (cost); otherwise, if they are not all the same, they incur a positive potential. Such an MLL model tends to give a smooth solution which prefers uniform labels.

For spatially (and also temporally in image sequence analysis) continuous MRFs, the smoothness prior often involves derivatives. This is the case with the analytical regularization. There, the potential at a point is in the form of $[f^{(n)}(x)]^2$. The order n determines the number of sites in the involved cliques; for example, $[f'(x)]^2$ where $n = 1$ corresponds to a pair-site smoothness potential. Different orders implies different class of smoothness.

Let us take continuous restoration or reconstruction of non-texture surfaces as an example. Let $f = \{f_1, \ldots, f_m\}$ be the sampling of an underlying "surface" $f(x)$ on $x \in [a, b]$ where the surface is one-dimensional for simplicity. The Gibbs distribution $P(f)$, or equivalently the energy $U(f)$, depends on the type of the surface f we expect to reconstruct. Assume that the surface is flat – a priori. A flat surface which has equation $f(x) = a_0$ should have zero first-order derivative, $f'(x) = 0$. Therefore, we may choose the prior energy as

$$U(f) = \int [f'(x)]^2 \, dx \qquad (56)$$

which is called a *string*. The energy takes the minimum value of zero only if f is absolutely flat or a positive value otherwise. Therefore, the surface which minimizes (56) alone has a constant height (grey value for an image).

In the discrete case where the surface is sampled at discrete points $a \leqslant x_i \leqslant b, i \in \mathcal{S}$, we use the first-order difference to approximate the first derivative and use a summation to approximate the integral; so the above energy becomes

$$U(f) = \sum_i [f_i - f_{i-1}]^2, \qquad (57)$$

where $f_i = f(x_i)$. Expressed as the sum of clique potentials, we have

$$U(f) = \sum_{c \in \mathcal{C}} V_c(f) = \sum_{i \in \mathcal{S}} \sum_{i' \in \mathcal{N}_i} V_2(f_i, f_{i'}), \qquad (58)$$

where $\mathcal{C} = \{(1, 2), (2, 1), (2, 3), \ldots, (m-2, m-1), (m, m-1), (m-1, m)\}$ consists of only pair-site cliques and

$$V_c(f) = V_2(f_i, f_{i'}) = \frac{1}{2}(f_i - f_{i'})^2. \qquad (59)$$

Its 2D equivalent is

$$\iint \{[f_x(x, y)]^2 + [f_y(x, y)]^2\} \, dx \, dy \qquad (60)$$

and is called a *membrane*.

Similarly, the prior energy $U(f)$ can be designed for planar or quadratic surfaces. A planar surface, $f(x) = a_0 + a_1 x$, has zero second-order derivative, $f''(x) = 0$. Therefore, the following may be chosen

$$U(f) = \int [f''(x)]^2 \, dx \qquad (61)$$

which is called a *rod*. The surface which minimizes (61) alone has a constant gradient. In the discrete case, we use the second-order difference to approximate the second-order derivative and the above energy becomes

$$U(f) = \sum_i [f_{i+1} - 2f_i + f_{i-1}]^2. \qquad (62)$$

For a quadratic surface, $f(x) = a_0 + a_1 x + a_2 x^2$, the third-order derivative is zero, $f'''(x) = 0$ and the prior energy may be

$$U(f) = \int [f'''(x)]^2 \, dx. \qquad (63)$$

The surface which minimizes the above energy alone has a constant curvature. In the discrete case, we use the third-order difference to approximate the second-order derivative and the above energy becomes

$$U(f) = \sum_i [f_{i+1} - 3f_i + 3f_{i-1} - f_{i-2}]^2. \qquad (64)$$

The above smoothness models can be extended to 2D. For example, the 2D equivalent of the rod, called a plate, comes in two varieties, the quadratic variation

$$\iint \{[f_{xx}(x, y)]^2 + 2[f_{xy}(x, y)]^2 + [f_{yy}(x, y)]^2\} \, dx \, dy \qquad (65)$$

and the squared Laplacian

$$\iint \{f_{xx}(x, y) + f_{yy}(x, y)\}^2 \, dx \, dy. \qquad (66)$$

The surface which minimizes one of the smoothness prior energy alone has either a constant grey level, a constant gradient or a constant curvature. This is undesirable because constraints from other sources such as the data are not used. Therefore, a smoothness term $U(f)$ is usually utilized in conjunction with other energy terms. In regularization, an energy consists of a smoothness term and a closeness term and the minimal solution is a compromise between the two constraints.

The encodings of the smoothness prior in terms of derivatives usually lead to *isotropic* potential functions. This is due to the assumption that the underlying surface is non-textured. *Anisotropic* priors have to be used for texture patterns. This can be done, for example, by choosing (37) with direction-dependent V_2's.

Care must be taken in the application of the smoothness assumption where *discontinuities* can be involved. The assumption of the uniform smoothness implies the smoothness *everywhere*. However, improper imposition of it can lead to undesirable, oversmoothed, solutions. This occurs when the uniform smoothness is violated, for example, at discontinuities where abrupt changes occur. It is necessary to take care of discontinuities when using smoothness priors. Therefore, how to apply the smoothness constraint while preserving discontinuities has been one of the most active research areas in image processing and low level vision, especially in connection of step and roof edge detection, motion and region segmentation. Discontinuities can be sensibly studied in terms of analytical properties (e.g., Blake and Zisserman, 1987; Li, 1995b, 1998b), from the probabilistic viewpoint (e.g., Geman and Geman, 1984; Marroquin, Mitter and Poggio, 1987), or based on robust statistics (e.g., Kashyap and Eom, 1988; Shulman and Herve, 1989). A comprehensive treatment of discontinuities and its relations with robust statistics can be found in (Li, 2001).

4.4. Hierarchical GRF model

A hierarchical two-level Gibbs model has been proposed to represent both noise-contaminated and textured images (Derin and Cole, 1986; Derin and Elliott, 1987). The higher-level Gibbs distribution uses an isotropic random field, e.g., MLL, to characterize the blob-like region formation process. A lower-level Gibbs distribution describes the filling-in in each region. The filling-in may be independent noise or a type of texture, both of which can be characterized by Gibbs distributions. This provides a convenient approach for MAP–MRF modeling. In segmentation of noisy and textured image (Derin and Cole, 1986; Derin and Elliott, 1987; Lakshmanan and Derin, 1989; Hu and Fahmy, 1987; Won and Derin, 1992), for example, the higher level determines the prior of f for the region process while the lower-level Gibbs contributes to the conditional probability of the data given f. Note that different levels of MRFs in the hierarchy can have different neighborhood systems.

Various hierarchical Gibbs models result according to what are chosen for the regions and for the filling-in's, respectively. For example, each region may be filled in by an auto-normal texture (Manjunath, Simchony and Chellappa, 1990; Won and Derin, 1992) or an auto-binomial texture (Hu and Fahmy, 1987); the MLL for the region formation may be substituted by another appropriate MRF.

A drawback of the hierarchical model is that the conditional probability $P(d_i | f_i = I)$ for regions given by $\{i \in S \mid f_i = I\}$ cannot always be written exactly. For example, when the lower level MRF is a texture modeled as an auto-normal field, its joint distribution over an irregularly shaped region is not known. This difficulty may be overcome by using approximate schemes such as pseudo-likelihood (Besag, 1975, 1977) (a proof of the consistency of the pseudo-likelihood estimate is given in (Geman and Graffigne, 1987)) or by using the eigen-analysis method (Wu and Leahy, 1993).

4.5. Multi-resolution methods

Multi-resolution methods provide a means for improving the convergence of iterative relaxation procedures (Hackbusch, 1985). It is shown by (Terzopoulos, 1986) that multi-resolution relaxation can be used to solve efficiently a number of low level vision problems. This class of techniques has been used for MRF computation especially in texture and motion analysis, e.g., (Konrad and Dubois, 1988; Barnard, 1989; Bouman and Liu, 1991; Kato, Berthod and Zerubia, 1993; Lakshmanan and Derin, 1993; Lakshmanan, Jain and Zhong, 1994; Bouman and Shapiro, 1994; Krishnamachari and Chellappa, 1997). It has been shown that multi-resolution techniques generally produce better results and requires less computation than single resolution algorithms.

An important issue in multi-resolution computation of MRFs is how to preserve the Markovianity and define consistent model descriptions at different resolutions. In general, the local Markovian property is not preserved at the coarse levels after a sub-sampling. In (Jeng, 1992), two theorems are given on a periodic sub-sampling of MRFs. One gives necessary and sufficient conditions for preserving the Markovianity and the other state that there is at least one sub-sampling scheme by which the Markovianity is preserved. A multi-resolution treatment is presented by Lakshmanan and Derin (1993) in which (possibly) non-Markov Gaussian fields are approximated by linear Gaussian MRFs. In (Heitz and Bouthemy, 1994), a consistent set of parameters are determined for objective functions at different resolutions; a nonlinear multi-resolution relaxation algorithm, which has fast convergence towards quasi-optimal solutions, is developed. A general transformation model is considered in (Perez and Heitz, 1994) as the "restriction" of an MRF, defined on a finite arbitrary non-directed graph, to a subset of its original sites; several results are derived for the preservation of the Markovianity which may be useful for designing consistent and tractable multi-resolution relaxation algorithms.

4.6. The FRAME model

The FRAME (Filter, Random Fields And Maximum Entropy) model, proposed by Zhu et al. (Zhu, Wu and Mumford, 1997, 1998; Zhu and Mumford, 1997), is a generalized MRF model which fuses the essence of filtering theory and MRF modeling through the maximum entropy principle. It is generalized in the following two aspects: (1) The FRAME model is defined in terms of statistics, i.e., potential functions, calculated from the output of a filter bank by which the image is filtered, instead of the cliques potentials of the image itself. Given an image (a realization of a MRF), the image is filtered by a bank of filters, giving a set of output images. Some statistics are then calculated from the output images. (2) The FRAME model provides a means of learning the model parameters from a set of samples (example images) representative of the MRF to be modeled. Besides, it also gives an algorithm for filter selection.

The joint distribution of the FRAME model is constrained in such a way that the model can reproduce the statistics of the example images. It is found by solving a constrained maximum entropy problem. Let $G^{(k)}$ ($k = 1, \ldots, K$) be a bank of K filters (such a Gabor filters), $f^{(k)} = G^{(k)} * f$ the output of filtering f by $G^{(k)}$, and $H^{(k)} \in \mathcal{L}^S$

(the $\mathcal{L}$ is assumed to be the same for all the K filter outputs) the histogram of $f^{(k)}$ defined by

$$H^{(k)}(I) = \frac{1}{|\mathcal{S}|} \sum_{i \in \mathcal{S}} \delta(I - f_i^{(k)}), \tag{67}$$

where $|\mathcal{S}|$ is the number of sites in S, and $\delta(t) = 1$ if $t = 0$ or 0 otherwise. For the filtered sample images, we denote the averaged histogram of the kth filter output by $\overline{H_{\text{samp}}^{(k)}}$ (averaged across all example images). Now, the joint distribution of the FRAME is defined below:

$$p(f) = \arg\max_p \left\{ -\int p(f) \log(p(f)) \, df \right\} \tag{68}$$

subject to $\overline{H_{p(f)}^{(k)}}(I) = \overline{H_{\text{samp}}^{(k)}}(I) \quad \forall k, \forall I,$ (69)

$$\int p(f) \, df = 1, \tag{70}$$

where

$$\overline{H_{p(f)}^{(k)}} = \int H^{(k)}(f) p(f) \, df \tag{71}$$

is the expectation of $H^{(k)}$ with respect to $p(f)$. By using Lagrange multipliers $\theta = \{\theta_I^{(k)}\}$ for the constraints of (69), we get the Lagrangian

$$L(p, \theta) = -\int p(f) \log(p(f)) \, df \tag{72}$$

$$+ \int_I \sum_k \theta_I^{(k)} \left\{ \int_f p(f) \sum_i \delta(I - f_i^{(k)}) \, df - |\mathcal{S}| \overline{H_{\text{samp}}^{(k)}}(I) \right\} dI$$

noting that the constraints are multiplied by the factor of $|\mathcal{S}|$. By setting $\partial L(p, \theta)/\partial p = 0$, the solution to the constrained optimization (ME) problem can be derived as (consider $p(f) = p(f | \theta)$ when θ is given)

$$p(f | \theta) = \frac{1}{Z(\theta)} e^{-\sum_{k=1}^K \sum_{i \in \mathcal{S}} \{\int \theta^{(k)}(I) \delta(I - f_i^{(k)}) \, dI\}}$$

$$= \frac{1}{Z(\theta)} e^{-\sum_{k=1}^K \sum_{i \in \mathcal{S}} \{\theta^{(k)}(f_i^{(k)})\}}, \tag{73}$$

where $\theta^{(k)}(\cdot)$ are the potential functions of the FRAME model and Z the normalizing factor.

In the discrete form, assume that $I^{(k)} = f_i^{(k)}$ is quantized into L discrete values $I_1^{(k)}, \ldots, I_L^{(k)}$. The solution in (73) can be written as

$$p(f|\theta) = \frac{1}{Z(\theta)} e^{-\sum_{k=1}^{K} \sum_{i \in S} \sum_{\ell=1}^{L} \{\theta_\ell^{(k)} \delta(I_i^{(k)} - f_i^{(k)})\}}$$
$$= \frac{1}{Z(\theta)} e^{-\sum_{k=1}^{K} \sum_{\ell=1}^{L} \theta_\ell^{(k)} H_\ell^{(k)}}$$
$$= \frac{1}{Z} e^{-\langle \theta, H \rangle}, \qquad (74)$$

where $\theta_\ell^{(k)} = \theta^{(k)}(I_\ell^{(k)})$, $H_\ell^{(k)} = H^{(k)}(I_\ell^{(k)})$, and $\langle \theta, H \rangle$ is the inner product of θ and H.

Whereas the traditional MRF model has to keep its neighborhood small to keep the model tractable and therefore unable to model patterns in which interaction in a large neighborhood is necessary, the FRAME model's neighborhood is implicitly determined by the filter windows which can be much larger than that of the traditional MRF model. The latter provides a means of modeling complicated high order patterns in a tractable way. Moreover, the FRAME model uses an accompanying learning procedure to estimate high order potential functions from the filter outputs. This makes the high order model tractable in formulation albeit expensive in computation. Two things are to be learned in the FRAME model: (1) the potential functions $\theta_I^{(k)}$ and (2) the types of filters $G^{(k)}$ to use. These are described in (Zhu, Wu and Mumford, 1998).

More recently, Zhu and his colleagues (Wu, Zhu and Liu, 2000) have established an equivalence between the FRAME model and another mathematical model of texture called Julesz ensembles (Julesz, 1962) when the size of the image lattice goes to infinity. On the other hand, they also proposes fast MCMC algorithms for sampling $p(f|\theta)$ which involves hundreds of parameters to estimate in a large neighborhood (Zhu and Liu, 2000).

5. The MAP–MRF framework

Optimization has been playing an essential and important role in image analysis. A problem can be formulated as optimizing a criteria, explicitly or implicitly, where given the data d, the solution f^* is defined as

$$f^* = \arg\min_f E(f|d). \qquad (75)$$

Here, E is the objective function of which the minimum, usually the global minimum, is defined as the optimal solution. In the MAP–MRF framework introduced below, $E(f|d)$ is the energy function in the posterior distribution.

The extensive use of optimization principles is due to various uncertainties in imaging processes. Noise and other degradation factors, such as caused by disturbances and quantization in sensing and signal processing, are sources of uncertainties. Different appearances and poses of objects, their mutual and self occlusion and possible shape deformation also cause ambiguities in visual interpretation. Under such circumstances, we can hardly obtain exact or perfect solutions and have to resort to inexact yet optimal solutions.

Bayes statistics is a theory of fundamental importance in estimation and decision making. According to this theory, when both the prior distribution and the likelihood function of a pattern are known, the best that can be estimated from these sources of knowledge is the Bayes labeling. The maximum a posterior (MAP) solution, as a special case in the Bayes framework, is sought in many image analysis algorithms.

The MAP–MRF framework is advocated by Geman and Geman (1984) and others (Geman and McClure, 1985; Derin and Elliott, 1987; Geman and Graffigne, 1987; Dubes and Jain, 1989; Besag, 1989; Szeliski, 1989; Geman and Gidas, 1991). Since the paper of Geman and Geman (1984), numerous problems have been formulated in this framework. This section reviews related concepts and derives involved probabilistic distributions and energies in MAP–MRF labeling. For more detailed materials on Bayes theory, the reader is referred to books like (Therrien, 1989).

In the following, the general optimization-based approach is introduced. It is followed by a description of the MAP–MRF framework where an image analysis problem, albeit at lower or high level, is formulated as an MAP estimation problem in which MRFs are used to model the prior distribution. Finally some issues in finding optimal solutions are discussed.

5.1. Optimization-based approach

In a pioneer vision system (Roberts, 1965), object identification and pose estimation are performed using the simplest least squares (LS) fitting. Nowadays, optimization is pervasive in all aspects of image analysis, including image restoration and reconstruction (Grimson, 1981; Terzopoulos, 1983; Geman and Geman, 1984; Leclerc, 1989; Hung, Cooper and Cernuschi-Frias, 1991; Li, 1998c), shape from shading (Ikeuchi and Horn, 1981), stereo, motion and optical flow (Ullman, 1979; Horn and Schunck, 1981; Hildreth, 1984; Murray and Buxton, 1987; Barnard, 1987), texture (Hassner and Slansky, 1980; Kashyap, Chellappa and Khotanzad, 1982; Cross and Jain, 1983), edge detection (Torre and Poggio, 1986; Tan, Gelfand and Delp, 1992), image segmentation (Silverman and Cooper, 1988; Li, 1990), perceptual grouping (Lowe, 1985; Mohan and Nevatia, 1989; Herault and Horaud, 1993), interpretation of line drawings (Leclerc and Fischler, 1992), object matching and recognition (Fischler and Elschlager, 1973; Davis, 1979; Shapiro and Haralick, 1981; Bhanu and Faugeras, 1984; Ben-Arie and Meiri, 1987; Modestino and Zhang, 1989; Nasrabadi, Li and Choo, 1990; Wells, 1991; Friedland and Rosenfeld, 1992; Li, 1992, 1994), and pose estimation (Haralick et al., 1989).

In all of the above cited examples, the solution is explicitly defined as an optimum of an objective function by which the goodness, or otherwise cost, of the solution is measured. Optimization may also be performed implicitly: the solution may optimize an objective function but in an implicit way which may or may not be realized. Hough transform (Hough, 1962; Duda and Hart, 1972; Ballard, 1981; Illingworth and Kittler, 1988) is a well-known technique for detecting lines and curves by looking at peaks of an accumulation function. It is later found to be equivalent to template matching (Stockman and Agrawala, 1977) and can be reformulated as a maximizer of some probabilities such as the likelihood (Haralick and Shapiro, 1992). Edge detection was performed using some simple operators like derivatives of Gaussian (Rosenfeld and Kak, 1976). The operators can be derived by using regularization principles in which an energy function is explicitly minimized (Poggio, Voorhees and Yuille, 1985).

We find it important to study image analysis problems from the viewpoint of optimization and to develop methodologies for optimization-based modeling. The following presents some discussions on the optimization-based approach.

5.2. Research issues

There are three basic issues in the optimization-based approach to image analysis: problem representation, objective function and optimization algorithms. There are two aspects of a representation: descriptive and computational. The former concerns how to represent image features and object shapes, which relates to photometry and geometry (Koenderink, 1990; Mundy and Zisserman, 1992; Kanatani, 1993) and is not an emphasis of this chapter. The latter concerns how to represent the solution, which relates to the choice of sites and label set for a labeling problem. For example, in image segmentation, we may use a chain of boundary locations to represent the solution; we may alternatively use a region map to do the same job. Comparatively speaking, however, the region map is a more natural representation for MRFs.

The second issue is how to formulate an objective function for the optimization. The objective function maps a solution to a real number measuring the quality of the solution in terms of some goodness or cost. The formulation determines how various constraints, which may be pixel properties like intensity and color and/or context like relations between pixels or object features, are encoded into the function. Because the optimal solution which is the optimum of the objective function, the formulation defines the optimal solution.

The third is how to optimize the objective, i.e., how to search for the optimal solution in the admissible space. Two major concerns are (1) the problem of local minima existing in non-convex functions and (2) the efficiency of algorithms in space and time. They are somewhat contradictory and currently there is no algorithms which guarantee the global solution with good efficiency.

These three issues are related to one another. In the first place, the scheme of representation influences the formulation of the objective function and the design of the search algorithm. On the other hand, the formulation of an objective function affects the search. For example, suppose two objective functions have the same point as the unique global optimum but one of them is convex whereas the other is not; obviously the convex one is much more desired because it provides convenience for the search.

In the following presentation, we will be mainly dealing with minimization problems. An objective function is in the form of an energy function and is to be minimized.

5.3. Role of energy functions

The role of an energy function is twofold: (1) as the quantitative measure of the global quality of the solution and (2) as a guide to the search for a minimal solution. As the quantitative cost measure, an energy function defines the minimal solution as its minimum, usually a global one. In this regard, it is important to formulate an energy function so that the "correct solution" is embedded as the minimum. We call this the correctness of the formulation.

To understand an optimization approach, one should not mix problems in formulation and those in search. Differentiating the two different kinds of problems helps debug the modeling. For example, if the output of an optimization procedure (assuming the implementation is correct) is not what is expected, there are two possible reasons: (1) the formulation of the objective function is not a correct one for modeling the reality and (2) the output is a low quality local minimum. Due to which one is the problem should be identified before the modeling can be improved.

The role of an energy function as a guide to the search may or may not be fully played. In real minimization, for example, when the energy function is smooth and convex w.r.t. its variables, global minimization is equivalent to local minimization and the gradient of the energy function provides sufficient information about where to search for the global solution. In this case, the role of guiding the search can be fully played. However, when the problem is non-convex, there is no general method which can efficiently utilize the energy function to guide the search. In this case, the role as the search-guide is underplayed.

In certain cases, it may be advantageous to consider the formulation of an energy function and the search simultaneously, in order to find the global minimum solution. This is to formulate the function appropriately to facilitate the search. Simulated annealing (Geman and Geman, 1984), graduated non-convexity (GNC) (Blake and Zisserman, 1987), and mean-field annealing (Peterson and Soderberg, 1989; Yullie, 1990; Geiger and Girosi, 1991) are examples in this regard. There, an energy function is deformed gradually from a convex form to its target form in the process of approximating the global solution using a local minimization algorithm.

Local minimization in real spaces is the most mature area in optimization and many formal approaches exist for solving it. This is not so for combinatorial and global minimization. In the latter cases, heuristics become an important and perhaps necessary element in practice. In the heuristic treatment of global minimization, rather restrictive assumptions are made. An example is the bounded model (Baird, 1985; Breuel, 1992). It assumes that a measurement error is upper-bounded by a certain threshold (within the threshold, the error may be assumed to be evenly distributed). Whether the assumption is valid depends on the threshold. It is absolutely true when the threshold is infinitely large. But in practice, the threshold is almost always set to a value which is less than that required to entirely validate the bounded-error assumption. The lower the value, the higher the efficiency is, but the less general the algorithm becomes.

In the hypothesis-verification approach, efficient algorithms are used to generate hypothetical solutions, such as Hough transform (Hough, 1962; Duda and Hart, 1972), interpretation tree search (Grimson and Lozano-Prez, 1987) and geometric hashing (Lamdan and Wolfson, 1988). The efficiency comes from the fast elimination of infeasible solutions, or pruning of the solution space, by taking advantage of heuristics. In this way, a relatively small number of solution candidates are picked up relatively quickly and are then verified or evaluated thoroughly, for example, by using an energy function. In this strategy, the energy function is used for the evaluation only, not as a guide to the search.

Note that the advantage of formal approaches is in the evaluation and the advantage of heuristic approaches is in the search. A good strategy for the overall design of a specialized system may be the following: use a heuristic algorithm to quickly find a small number of solution candidates and then evaluate the found candidates using an energy function derived formally to give the best solution.

5.4. Formulation of objective functions

In pattern recognition, there are two basic approaches to formulating an energy function: parametric and nonparametric. In the parametric approach, the types of underlying distributions are known and the distributions are parameterized by a few parameters. Therefore, the functional form of the energy can be obtained and the energy function is completely defined when the parameters are specified.

In the nonparametric approach, sometimes called distribution free approach, no assumptions about the distributions are made. There, a distribution is either estimated from the data or approximated by a pre-specified basis functions with several unknown parameters in it to be estimated. In the latter case, the pre-specified basis functions will determine the functional form of the energy.

Despite the terms parametric and nonparametric, both approaches are somewhat parametric in nature. This is because in any case, there are always parameters that must be determined to define the energy function.

The two most important aspects of an energy function are its form and the involved parameters. The form and parameters together define the energy function which in turn defines the minimal solution. The form depends on assumptions about the solution f and the observed data d. We express this using the notation $E(f|d)$. Denote the set of involved parameters by θ. With θ, the energy is expressed further as $E(f|d,\theta)$. In general, given the functional form for E, a different d or θ defines a different energy function, $E(f|d,\theta)$, w.r.t. f and hence a (possibly) different minimal solution f^*.

Since the parameters are part of the definition of the energy function $E(f|d,\theta)$, the minimal solution $f^* = \arg\min_f E(f|d)$ is not completely defined if the parameters are not specified even if the functional form is known. These parameters must be specified or estimated by some means. This is an important area of study in the MRF modeling.

5.5. Optimality criteria

In formal models, as opposed to heuristic ones, an energy function is formulated based on an established criterion. Because of inevitable uncertainties in imaging and vision processes, principles from statistics, probability and information theory are often used as the formal basis. When the knowledge about the data distribution is available but not about the prior information, the *maximum likelihood* (ML) criterion may be used, $f^* = \arg\max P(d|f)$. On the other hand, if only the prior information is available, the *maximum entropy* criterion may be chosen, $f^* = \arg\max -\sum_{i=1}^{m} P(f_i) \ln P(f_i)$. The maximum entropy criterion is simply taking this fact into account: configurations with higher entropy are more likely because nature can generate them in more ways (Jaynes, 1982).

When both the prior and likelihood distributions are known, the best result is achieved by that maximizes a Bayes criterion according to Bayes statistics (Therrien, 1989). Bayes statistics is a theory of fundamental importance in estimation and decision making. Although there have been philosophical and scientific controversies about their appropriateness in inference and decision making (see Clark and Yuille, 1990 for a short review), Bayes criteria, the MAP principle in particular, are the most popular ones in image analysis; in fact, MAP is the most popular criterion in optimization-based MRF modeling. The equivalence theorem of between Markov random fields and Gibbs distribution established in Section 3.4 provides a convenient way for specifying the joint prior probability, solving a difficult issue in MAP–MRF labeling.

In the principle of *minimum description length* (MDL) (Rissanen, 1978, 1983), the optimal solution to a problem is that needs the smallest set of vocabulary in a given language for explaining the input data. The MDL has close relationships to the statistical methods such as the ML and MAP (Rissanen, 1983). For example, if $P(f)$ is related to the description length and $P(d|f)$ related to the description error, then MDL is equivalent to MAP. However, it is a more natural and intuitive when prior probabilities are not well defined. The MDL has been used for image analysis problems at different levels such as segmentation (Leclerc, 1989; Pentland, 1990; Darrell, Sclaroff and Pentland, 1990; Dengler, 1991; Keeler, 1991) and object recognition (Breuel, 1993).

5.6. *Bayes estimation*

In Bayes estimation, a risk is minimized to obtain the optimal estimate. The Bayes risk of estimate f^* is defined as

$$R(f^*) = \int_{f \in \mathbb{F}} C(f^*, f) P(f|d) \, df, \tag{76}$$

where d is the observation, $C(f^*, f)$ is a cost function and $P(f|d)$ is the posterior distribution. First of all, we need to compute the posterior distribution from the prior and the likelihood. According to the Bayes rule, the posterior probability can be computed by using the following formulation

$$P(f|d) = \frac{p(d|f) P(f)}{p(d)}, \tag{77}$$

where $P(f)$ is the prior probability of labelings f, $p(d|f)$ is the conditional p.d.f. of the observations d, also called the likelihood function of f for d fixed, and $p(d)$ is the density of d which is a constant when d is given.

The cost function $C(f^*, f)$ determines the cost of estimate f when the truth is f^*. It is defined according to our preference. Two popular choices are (1) the quadratic cost function $C(f^*, f) = \|f^* - f\|^2$, and (2) the δ (0–1) cost function $C(f^*, f) = 0$ if $\|f^* - f\| \leq \delta$, where $\delta > 0$ is a small constant, or 1 otherwise.

With the quadratic cost function, the Bayesian risk is equivalent to the variance of the posterior, and the Bayesian estimate becomes the mean of the posterior

$$f^* = \int_{f \in \mathbb{F}} f P(f|d) \, df. \tag{78}$$

With the δ cost function with $\delta \to 0$, the Bayes risk is approximated by $R(f^*) = 1 - \kappa P(f|d)$ where κ is the volume of the space containing all points f for which $\|f^* - f\| \leq \delta$. Minimizing the above is equivalent to maximizing the posterior probability. Therefore, the minimal risk estimate is

$$f^* = \arg \max_{f \in \mathbb{F}} P(f|d) \tag{79}$$

which is known as the MAP estimate. Because $p(d)$ in (77) is a constant for a fixed d, the MAP estimate is found as $f^* = \arg \max_{f \in \mathbb{F}} \{p(d|f) P(f)\}$. Obviously, when the prior distribution, $P(f)$, is flat, the MAP is equivalent to the maximum likelihood.

5.7. MAP–MRF labeling

In the MAP–MRF labeling, $P(f|d)$ is the posterior distribution of an MRF. An important step in Bayes labeling of MRFs is to derive this distribution. Here we use a simple example to illustrate the formulation of a MAP–MRF labeling problem. The problem is to restore images from noisy data. Assuming that the image surfaces are flat, then the joint prior distribution of f is

$$P(f) = \frac{1}{Z} e^{-E(f)}, \tag{80}$$

where $E(f) = \sum_i \sum_{i' \in \{i-1, i+1\}} (f_i - f_{i'})^2$ is the *prior energy* for the type of surfaces. Assuming that the observation is the true surface height plus the independent Gaussian noise, $d_i = f_i + e_i$, where $e_i \sim N(\mu, \sigma^2)$, then the likelihood distribution is

$$p(d|f) = \frac{1}{\prod_{i=1}^{m} \sqrt{2\pi\sigma^2}} e^{-E(d|f)}, \tag{81}$$

where

$$E(d|f) = \sum_{i=1}^{m} (f_i - d_i)^2 / 2\sigma^2 \tag{82}$$

is the *likelihood energy*. Now the posterior probability is

$$P(f|d) \propto e^{-E(f|d)}, \tag{83}$$

where

$$E(f|d) = E(d|f) + E(f) = \sum_{i=1}^{m}(f_i - d_i)^2/2\sigma_i^2 + \sum_{i=1}^{m}(f_i - f_{i-1})^2 \qquad (84)$$

is the *posterior energy*. In this example, there is only one parameter in this simple example, σ_i. When it is determined, $E(f|d)$ is fully specified and the MAP–MRF solution is completely defined. The MAP estimate is the minimum of the posterior energy function, and hence Eq. (75).

5.8. Optimization methods

Energy minimization algorithms aim to find a minimum of an energy function $E(f)$ (or that of an objective function in parameter estimation). Design of minimization algorithms is an issue in general engineering. A minimum of a function is always said with respect to a neighborhood. In energy minimization based methods, the quality of a solution is measured by the minimized energy value, and the minimal solution defined in (75) is meant to be the global one, i.e., the one with the lowest possible energy value. Therefore, it is critical that the solution has the lowest possible energy value.

A minimization problem can be categorized as continuous or combinatorial according to whether the label set $\mathcal{L}$ is continuous or discrete. The basic structure of an algorithm depends mostly on the continuity of $\mathcal{L}$, due to the ordering of the elements in $\mathcal{L}$. A minimization can be further categorized as constrained or unconstrained according to whether additional constraints are imposed on the variables $\{f_1, \ldots, f_m\}$. There may be a combination of several minimization problems; for example, in boundary preserving surface reconstruction, the minimization is with respect to a continuous surface field and a discrete line process field (Marroquin, 1985; Blake and Zisserman, 1987). When a solution is represented by an image or another form of mutually constrained labels, it is generally impossible to express the solution in closed-form. An iterative minimization algorithm is usually used.

Local search is the basis for many minimization algorithms including global search algorithms. There are two broad classes of local search algorithms: deterministic and stochastic. A deterministic local search is performed as follows: given the current configuration $f \in \mathbb{F}$, a new configuration $f' \in \mathcal{N}(f)$ is generated; if f' leads to an improvement, $E(f') < E(f)$, f is replaced by f'. When $E(f)$ is lower bounded, the algorithm converges to a local minimum. In a stochastic local search, a new configuration f' is accepted according to a probabilistic rule; it is possible that f' is accepted even if $E(f') > E(f)$. Several local search algorithms are described in the following.

Generally speaking, *global optimization* problem is extremely hard except for a few special cases. Indeed, no efficient solutions exist which solve the general global optimization problem. The global optimization problem is much more difficult, for instance, than the related task of local optimization in which only the determination of local optima is desired. In local optimization, tests exist to determine if a point is a local optimum (assuming that the function to be optimized is twice differentiable). With such tests, a sequence of points converging to a local optimum can be constructed.

The relative difficulty of solving the global optimization problem on a computer as compared with the local optimization problem stems from the fact that, in general, the objective function may contain numerous local optima with corresponding function values varying significantly. Although an iterative algorithm for finding a local minimum (local method), such as Newton's method, may find one of these local minima efficiently, it is difficult to know whether this local solution is indeed a global solution.

One might suggest finding all of the local minima and choosing the smallest value(s) as the global minimum (minima). In general, however, the number of local minima may be quite large. Furthermore, due to the unpredictable global convergence behavior of local iterative methods (see Blake and Zisserman, 1987), one cannot determine whether all the local minima have been identified simply by varying the initial starting guesses.

When the energy function is non-convex (with continuous label set) or the problem is combinatorial (with discrete label set), finding a global minimum is a difficult problem. Whereas local minimization is relatively well studied, global minimization is still young area. Currently, there are no efficient algorithms which guarantee a global solution.

Global minimization requires to (i) find all (a finite number of) local minima and (ii) evaluate them. It needs to prove that there are no more local minima. Without efficient algorithm, this amounts to exhaustive search. In dealing with a complex system such as one in image analysis, one is facing one of the two choices: (i) to find the exact global minimum with intolerable expenses or (ii) to find some approximations to it with affordable cost. The second is usually the practical choice. In image analysis, two broad classes of iterative methods have been used as effective means for approximating the global solution: (i) annealing combined with local search and (ii) population-based methods. The former moves a single configuration to a solution whereas the latter maintains and operates on a population of configurations. The interested reader may refer to (Li, 2001) for various algorithms and techniques for local and global optimization.

Acknowledgement

The author would like to thank Prof. D. Shanbhag and the anonymous referee for their constructive comments and suggestions for improving the manuscript.

References

Abend, K., T. J. Harley, and L. N. Kanal (1965). Classification of binary random patterns. *IEEE Trans. Inform. Theory* **11** (4), 538–544.
Aloimonos, J. and D. Shulman (1989). *Integration of Visual Modules*. Academic Press, London, UK.
Amini, A., S. Tehrani and T. Weymouth (1988). Using dynamic programming for minimizing the energy of active contours in the presence of hard constraints. In *Proceedings of IEEE International Conference on Computer Vision*, pp. 95–99.
Baddeley, A. J. and M. N. M. van Lieshout (1992). Object recognition using Markov spatial processes. In *Proceedings of International Conference Pattern Recognition*, Vol. B, pp. 136–139.
Baird, H. S. (1985). *Model-Based Image Matching Using Location*. MIT Press, Cambridge, MA.

Ballard, D. H. (1981). Generalizing the Hough transform to detect arbitrary shapes. *Pattern Recognition* **13**(2), 111–122.
Barnard, S (1989). Stochastic stereo matching over scale. *Internat. J. Computer Vision* **3**, 17–32.
Barnard, S. T. (1987). Stereo matching by hierarchical, microcanonical annealing. In *Proceedings of International Joint Conference on Artificial Intelligence*, pp. 832–835.
Barrow, H. G. and J. M. Tenenbaum (1981a). Computational vision. *Proc. IEEE* **69**(5), 572–595.
Barrow, H. G. and J. M. Tenenbaum (1981b). Interpreting line drawings as three dimensional surfaces. *Artificial Intelligence* **17**, 75–117.
Ben-Arie, J. and A. Z. Meiri (1987). 3d objects recognition by optimal matching search of multinary relations graphs. *Computer Vision, Graphics and Image Processing* **37**, 345–361.
Bertero, M., T. A. Poggio and V. Torre (1988). Ill-posed problems in early vision. *Proc. IEEE* **76**(8), 869–889.
Besag, J. (1974). Spatial interaction and the statistical analysis of lattice systems (with discussions). *J. Roy. Statist. Soc. Ser. B* **36**, 192–236.
Besag, J. (1975). Statistical analysis of non-lattice data. *The Statistician* **24**(3), 179–195.
Besag, J. (1977). Efficiency of pseudo-likelihood estimation for simple Gaussian fields. *Biometrika* **64**, 616–618.
Besag, J. (1989). Towards Bayesian image analysis. *J. Appl. Statist.* **16**(3), 395–406.
Bhanu, B. and O. D. Faugeras (1984). Shape matching of two-dimensional objects. *IEEE Trans. on Pattern Anal. and Machine Intelligence* **6**(2), 137–155.
Blake, A. and A. Zisserman (1987). *Visual Reconstruction*. MIT Press, Cambridge, MA.
Bouman, C. and B. Liu (1991). Multiple resolution segmentation of texture segmentation. *IEEE Trans. on Pattern Anal. and Machine Intelligence* **13**(2), 99–113.
Bouman, C. and M. Shapiro (1994). A multiscale random field model for Bayesian segmentation. *IEEE Trans. on Image Process.* **3**(2), 162–177.
Bowyer, A. (1981). Computing Dirichlet tessellations. *Comput. J.* **24**, 162–166.
Breuel, T. M. (1992). Fast recognition using adaptive subdivision of transformation space. In *Proceedings of IEEE Computer Society Conference on Computer Vision and Pattern Recognition*, pp. 445–451.
Breuel, T. M. (1993). Higher-order statistics in visual object recognition. Memo #93-02, IDIAP, Martigny, Switzerland.
Chellappa, R. (1985). Two-dimensional discrete Gaussian Markov random field models for image processing. In *Progress in Pattern Recognition*, Vol. 2, pp. 79–112 (Eds. L. N. Kanal and A. Rosenfeld).
Chellappa, R. and A. Jain (Eds.) (1993). *Markov Random Fields: Theory and Applications*. Academic Press.
Chellappa, R. and R. L. Kashyap (1982). Digital image restoration using spatial interaction models. *IEEE Trans. on Acoustic, Speech and Signal Process.* **30**, 461–472.
Chou, P. B. and C. M. Brown (1990). The theory and practice of Bayesian image labeling. *Internat. J. Computer Vision* **4**, 185–210.
Chow, C. K. (1962). A recognition method using neighbor dependence. *IRE Trans. on Electronic Computer* **11**, 683–690.
Clark, J. J. and A. L. Yuille (1990). *Data Fusion for Sensory Information Processing Systems*. Kluwer Academic Publishers, Norwell, MA.
Cohen, F. S. and D. B. Cooper (1987). Simple parallel hierarchical and relaxation algorithms for segmenting noncasual Markovian random fields. *IEEE Trans. on Pattern Anal. and Machine Intelligence* **9**(2), 195–218.
Cooper, D. B., J. Subrahmonia, Y. P. Hung and B. Cernuschi-Frias (1993). The use of Markov random fields in estimating and recognizing object in 3D space. In *Markov Random Fields: Theory and Applications*, pp. 335–367 (Eds. R. Chellappa and A. Jain). Academic Press, Boston.
Cooper, P. R. (1990). Parallel structure recognition with uncertainty: coupled segmentation and matching. In *Proceedings of IEEE International Conference on Computer Vision*, pp. 287–290.
Cross, G. C. and A. K. Jain (1983). Markov random field texture models. *IEEE Trans. on Pattern Anal. and Machine Intelligence* **5**(1), 25–39.
Darrell, T., S. Sclaroff and A. Pentland (1990). Segmentation by minimal description. In *Proceedings of IEEE International Conference on Computer Vision*, pp. 112–116.
Davis, L. S. (1979). Shape matching using relaxation techniques. *IEEE Trans. on Pattern Anal. and Machine Intelligence* **1**(1), 60–72.

Dengler, J. (1991). Estimation of discontinuous displacement vector fields with the minimum description length criterion. In *Proceedings of IEEE Computer Society Conference on Computer Vision and Pattern Recognition*, pp. 276–282.

Derin, H. and W. S. Cole (1986). Segmentation of textured images using using Gibbs random fields. *Computer Vision, Graphics and Image Process.* **35**, 72–98.

Derin, H. and H. Elliott (1987). Modeling and segmentation of noisy and textured images using Gibbs random fields. *IEEE Trans. on Pattern Anal. and Machine Intelligence* **9**(1), 39–55.

Derin, H., H. Elliott, R. Cristi and D. Geman (1984). Bayes smoothing algorithms for segmentation of binary images modeled by Markov random fields. *IEEE Trans. on Pattern Anal. and Machine Intelligence* **6**(6), 707–720.

Derin, H. and P. A. Kelly (1989). Discrete-index Markov-type random fields. *Proc. IEEE* **77**(10), 1485–1510.

Dubes, R. C. and A. K. Jain (1989). Random field models in image analysis. *J. Appl. Statist.* **16**(2), 131–164.

Duda, R. O. and P. E. Hart (1972). Use of Hough transform to detect lines and curves in picture. *Comm. ACM* **15**(1), 11–15.

Elliott, H., H. Derin, R. Cristi and D. Geman (1984). Application of the Gibbs distribution to image segmentation. In *Proceedings of the International Conference on Acoustic, Speech and Signal Processing*, pp. 32.5.1–32.5.4. San Diego.

Fischler, M. and R. Elschlager (1973). The representation and matching of pictorial structures. *IEEE Trans. on Computers* **C-22**, 67–92.

Friedland, N. S. and A. Rosenfeld (1992). Compact object recognition using energy-function based optimization. *IEEE Trans. on Pattern Anal. and Machine Intelligence* **14**, 770–777.

Geiger, D. and F. Girosi (1991). Parallel and deterministic algorithms from MRF's: surface reconstruction. *IEEE Trans. on Pattern Anal. and Machine Intelligence* **13**(5), 401–412.

Geman, D., S. Geman, C. Graffigne and P. Dong (1990). Boundary detection by constrained optimization. *IEEE Trans. on Pattern Anal. and Machine Intelligence* **12**(7), 609–628.

Geman, D. and B. Gidas (1991). *Image Analysis and Computer Vision*, Chapter 2, pp. 9–36. National Academy Press.

Geman, S. and D. Geman (1984). Stochastic relaxation, Gibbs distribution and the Bayesian restoration of images. *IEEE Trans. on Pattern Anal. and Machine Intelligence* **6**(6), 721–741.

Geman, S. and C. Graffigne (1987). Markov random field image models and their applications to computer vision. In *Proceedings of the International Congress of Mathematicians: Berkeley, August 3–11, 1986*, pp. 1496–1517 (Ed. A. M. Gleason).

Geman, S. and D. McClure (1985). Bayesian image analysis: an application to single photon emission tomography. In *Proceedings of the Statistical Computing Section*, pp. 12–18. Washington, DC.

Gidas, B. (1989). A renormalization group approach to image processing problems. *IEEE Trans. on Pattern Anal. and Machine Intelligence* **11**, 164–180.

Grenander, U. (1983). *Tutorials in Pattern Synthesis*. Brown University, Division of Applied Mathematics.

Grenander, U. and D. M. K. Y. Chow (1991). *Hands: A Pattern Theoretic Study of Biological Shapes*. Springer-Verlag, New York.

Griffeath, D. (1976). Introduction to random fields. In *Denumerable Markov Chains*, 2nd ed. Chapter 12, pp. 425–458 (Eds. J. G. Kemeny, J. L. Snell and A. W. Knapp). Springer-Verlag, New York.

Grimson, W. E. L. (1981). *From Images to Surfaces: A Computational Study of the Human Early Visual System*. MIT Press, Cambridge, MA.

Grimson, W. E. L. and T. Lozano-Prez (1987). Localizing overlapping parts by searching the interpretation tree. *IEEE Trans. on Pattern Anal. and Machine Intelligence* **9**(4), 469–482.

Guyon, X. (1995). *Random Fields on a Network: Modeling, Statistics and Applications*. Probability and Its Applications. Springer-Verlag.

Hackbusch, W. (1985). *Multi-Grid Methods and Applications*. Springer-Verlag, Berlin.

Hammersley, J. M. and P. Clifford (1971). Markov field on finite graphs and lattices. Unpublished.

Hansen, F. R. and H. Elliott (1982). Image segmentation using simple Markov random field models. *Computer Graphics Image Process.* **20**, 101–132.

Haralick, R. M. (1983). Decision making in context. *IEEE Trans. on Pattern Anal. and Machine Intelligence* **5**(4), 417–428.

Haralick, R. M., H. Joo, C. Lee, X. Zhuang, V. Vaidya and M. Kim (1989). Pose estimation from corresponding point data. *IEEE Trans. Systems Man Cybernet.* **19**, 1426–1446.

Haralick, R. M. and L. G. Shapiro (1992). *Computer and Robot Vision*. Addison-Wesley, Reading, MA.

Harris, J. G., C. Koch, E. Staats and J. Lou (1990). Analog hardware for detecting discontinuities in early vision. *Internat. J. Computer Vision* **4**, 211–223.

Hassner, M. and J. Slansky (1980). The use of Markov random field as models of texture. *Computer Graphics Image Process.* **12**, 357–370.

Heitz, F. and P. Bouthemy (1994). Multiscale minimization of global energy functions in visual recovery problems. *CVGIP: Image Understanding* **59**(1), 125–134.

Herault, L. and R. Horaud (1993). Figure-ground discrimination: a combinatorial optimization approach. *IEEE Trans. on Pattern Anal. and Machine Intelligence* **15**, 899–914.

Hildreth, E. C. (1984). *The Measurement of Visual Motion*. MIT Press, Cambridge, MA.

Horn, B. K. P. and B. G. Schunck (1981). Determining optical flow. *Artificial Intelligence* **17**, 185–203.

Hough, P. V. C. (1962). A method and means for recognizing complex patterns. U.S. Patent No. 3,069,654.

Hu, R. and M. M. Fahmy (1987). Texture segmentation based on a hierarchical Markov random field model. *Signal Process.* **26**, 285–385.

Hung, Y. P., D. B. Cooper and B. Cernuschi-Frias (1991). Asymtotic Bayesian surface estimation using an image sequence. *Internat. J. Computer Vision* **6**(2), 105–132.

Ikeuchi, K. and B. K. P. Horn (1981). Numerical shape from shading and occluding boundaries. *Artificial Intelligence* **17**, 141–184.

Illingworth, J. and J. Kittler (1988). A survey of Hough transform. *Computer Vision, Graphics and Image Process.* **43**, 221–238.

Jain, A. K., Y. Zhong and S. Lakshmanan (1996). Object matching using deformable templates. *IEEE Trans. on Pattern Anal. and Machine Intelligence* **18**(3), 267–278.

Jaynes, E. (1982). On the rationale of maximum-entropy methods. *Proc. IEEE* **70**(9), 939–952.

Jeng, F. C. (1992). Subsampling of Markov random fields. *J. Visual Communication and Image Representation* **3**, 225–229.

Julesz, B. (1962). Visual pattern discrimination. *IRE Trans. Inform. Theory* **IT-8**, 84–92.

Kanatani, K. (1993). *Geometric Computation for Machine Vision*. Oxford University Press, New York.

Kashyap, R. L., R. Chellappa and A. Khotanzad (1982). Texture classification using features derived from random process models. *Pattern Recognition Lett.* **1**, 43–50.

Kashyap, R. L. and K. N. Eom (1988). Robust image modeling techniques with their applications. *IEEE Trans. on Acoustic, Speech and Signal Process.* **36**(8), 1313–1325.

Kass, M., A. Witkin and D. Terzopoulos (1987). Snakes: active contour models. In *Proceedings of IEEE International Conference on Computer Vision*, pp. 259–268.

Kato, Z., M. Berthod and J. Zerubia (1993). Multiscale Markov random field models for parallel image classification. In *Proceedings of IEEE International Conference on Computer Vision*, pp. 253–257.

Keeler, K. (1991). Map representations and coding based priors for segmentation. In *Proceedings of IEEE Computer Society Conference on Computer Vision and Pattern Recognition*, pp. 420–425.

Kim, I. Y. and H. S. Yang (1992). Efficient image understanding based on the Markov random field model and error backpropagation network. In *Proceedings of International Conference Pattern Recognition*, Vol. A, pp. 441–444.

Kindermann, R. and J. L. Snell (1980). *Markov Random Fields and Their Applications*. Amer. Math. Soc., Providence, RI.

Koch, C. (1988). Computing motion in the presence of discontinuities: algorithm and analog networks. In *Neural Computers*, pp. 101–110 (Eds. R. Eckmiller and C. C. D. Malsburg). NATO ASI Series, Vol. F41. Springer-Verlag.

Koenderink, J. J. (1990). *Solid Shape*. MIT Press.

Konrad, J. and E. Dubois (1988). Multigrid Bayesian estimation of image motion fields using stochastic relaxation. In *Proceedings of IEEE International Conference on Computer Vision*, pp. 354–362.

Krishnamachari, S. and R. Chellappa (1997). Multiresolution Gauss–Markov random field models for texture segmentation. *IEEE Trans. on Image Process.* **6**(2), 251–267.

Lakshmanan, S. and H. Derin (1989). Simultaneous parameter estimation and segmentation of Gibbs random fields using simulated annealing. *IEEE Trans. on Pattern Anal. and Machine Intelligence* **11**, 799–813.

Lakshmanan, S. and H. Derin (1993). Gaussian Markov random fields at multiple resolutions. In *Markov Random Fields: Theory and Applications*, pp. 131–157 (Eds. R. Chellappa and A. Jain). Academic Press, Boston.

Lakshmanan, S., A. K. Jain and Y. Zhong (1994). Multi-resolution image representation using Markov random fields. In *Proceedings of IEEE International Conference on Image Processing*, pp. 855–860.

Lamdan, Y. and H. Wolfson (1988). Geometric hashing: a general and efficient model-based recognition scheme. In *ICCV88*, pp. 238–249.

Leclerc, Y. G. (1989). Constructing simple stable descriptions for image partitioning. *Internat. J. Computer Vision* **3**, 73–102.

Leclerc, Y. G. and M. A. Fischler (1992). An optimization-based approach to the interpretation of single line drawings as 3D wire frames. *Internat. J. Computer Vision* **9**, 113–136.

Li, S. Z. (1990). Invariant surface segmentation through energy minimization with discontinuities. *Internat. J. Computer Vision* **5**(2), 161–194.

Li, S. Z. (1992). Towards 3D vision from range images: an optimization framework and parallel networks. *CVGIP: Image Understanding* **55**(3), 231–260.

Li, S. Z. (1994). A Markov random field model for object matching under contextual constraints. In *Proceedings of IEEE Computer Society Conference on Computer Vision and Pattern Recognition*, pp. 866–869, Seattle, Washington.

Li, S. Z. (1995a). *Markov Random Field Modeling in Computer Vision*. Springer-Verlag, New York.

Li, S. Z. (1995b). On discontinuity-adaptive smoothness priors in computer vision. *IEEE Trans. on Pattern Anal. and Machine Intelligence* **17**(6), 576–586.

Li, S. Z. (1997). Parameter estimation for optimal object recognition: theory and application. *Internat. J. Computer Vision* **21**(3), 207–222.

Li, S. Z. (1998a). Bayesian object matching. *J. Appl. Statist.* **25**(3), 425–443.

Li, S. Z. (1998b). Close-form solution and parameter selection for convex minimization based edge-preserving smoothing. *IEEE Trans. on Pattern Anal. and Machine Intelligence* **20**(9), 916–932.

Li, S. Z. (1998c). MAP image restoration and segmentation by constrained optimization. *IEEE Trans. on Image Process.* **7**(12), 1730–1735.

Li, S. Z. (2001). *Markov Random Field Modeling in Image Analysis*. Springer-Verlag, New York.

Lowe, D. G. (1985). *Perceptual Organization and Visual Recognition*. Kluwer Academic Publishers, Dordrecht.

Manjunath, B. S., T. Simchony and R. Chellappa (1990). Stochastic and deterministic networks for texture segmentation. *IEEE Trans. on Acoustic, Speech and Signal Process.* **38**, 1030–1049.

Mardia, K. V., T. J. Hainsworth and J. F. Haddon (1992). Deformable templates in image sequences. In *Proceedings of International Conference Pattern Recognition*, Vol. B, pp. 132–135.

Mardia, K. V. and G. K. Kanji (Eds.) (1993,1994). *Statistics and Images: 1 & 2*. Advances in Applied Statistics. Carfax.

Mardia, K. V., J. T. Kent and A. N. Walder (1991). Statistical shape models in image analysis. In *Proceedings of 23rd Symposium Interface*, pp. 550–575.

Marroquin, J. L. (1985). Probabilistic solution of inverse problems. *A. I. Lab. Tech. Report No. 860*, MIT, Cambridge, MA.

Marroquin, J. L., S. Mitter and T. Poggio (1987). Probabilistic solution of ill-posed problems in computational vision. *J. Amer. Statist. Assoc.* **82**(397), 76–89.

Modestino, J. W. and J. Zhang (1989). A Markov random field model-based approach to image interpretation. In *Proceedings the IEEE Computer Society Conference on Computer Vision and Pattern Recognition*, pp. 458–465.

Mohan, R. and R. Nevatia (1989). Using perceptual organization to extract 3-d structures. *IEEE Trans. on Pattern Anal. and Machine Intelligence* **11**, 1121–1139.

Moussouris, J. (1974). Gibbs and Markov systems with constraints. *J. Statist. Phys.* **10**, 11–33.

Mumford, D. and J. Shah (1985). Boundary detection by minimizing functionals: I. In *Proceedings the IEEE Computer Society Conference on Computer Vision and Pattern Recognition*, pp. 22–26. San Francisco, CA.

Mundy, J. L. and A. Zisserman (Eds.) (1992). *Geometric Invariants in Computer Vision*. MIT Press, Cambridge, MA.

Murray, D. and B. Buxton (1987). Scene segmentation from visual motion using global optimization. *IEEE Trans. on Pattern Anal. and Machine Intelligence* **8**, 220–228.

Nasrabadi, N., W. Li and C. Y. Choo (1990). Object recognition by a Hopfield neural network. In *Proceedings of Third International Conference on Computer Vision*, pp. 325–328. Osaka, Japan.

Pavlidis, T. (1986). A critical survey of image analysis methods. In *ICPR*, pp. 502–511.

Pentland, A. P. (1990). Automatic extraction of deformable part models. *Internat. J. Computer Vision* **4**, 107–126.

Perez, P. and F. Heitz (1994). Restriction of a Markov random field on a graph and multiresolution image analysis. *RR 2170*, INRIA, Sophia-Antipolis Cedex, France.

Peterson, C. and B. Soderberg (1989). A new method for mapping optimization problems onto neural networks. *Internat. J. Neural Systems* **1**(1), 3–22.

Poggio, T., V. Torre and C. Koch (1985). Computational vision and regularization theory. *Nature* **317**, 314–319.

Poggio, T., H. Voorhees and A. Yuille (1985). Regularizing edge detection. *A. I. Lab. Memo No. 773*, MIT, Cambridge, MA.

Ripley, B. D. (1981). *Spatial Statistics*. Wiley, New York.

Rissanen, J. (1978). Modeling by shortest data description. *Automatica* **14**, 465–471.

Rissanen, J. (1983). A universal prior for integers and estimation by minimal discription length. *Ann. Statist.* **11**(2), 416–431.

Roberts, L. G. (1965). Machine perception of three-dimensional solids. In *Optical and Electro-Optical Information Processing* (Ed. E. A. J. T. Tippett). MIT Press, Cambridge, MA.

Rosanov, Y. A. (1967). On Gaussian fields with given conditional distributions. *Theory Probab. Appl.* **XII**(3), 381–391.

Rosenfeld, A., R. Hummel and S. Zucker (1976). Scene labeling by relaxation operations. *IEEE Trans. Systems Man Cybernet.* **6**, 420–433.

Rosenfeld, A. and A. C. Kak (1976). *Digital Image Processing*. Academic Press, New York.

Shapiro, L. G. and R. M. Haralick (1981). Structural description and inexact matching. *IEEE Trans. on Pattern Anal. and Machine Intelligence* **3**, 504–519.

Shulman, D. and J. Herve (1989). Regularization of discontinuous flow fields. In *Proc. Workshop on Visual Motion*, pp. 81–86.

Silverman, J. F. and D. B. Cooper (1988). Bayesian clustering for unsupervised estimation of surface and texture models. *IEEE Trans. on Pattern Anal. and Machine Intelligence* **10**, 482–495.

Stockman, G. C. and A. K. Agrawala (1977). Equivalence of Hough curve detection to template matching. *Comm. ACM* **20**, 820–822.

Storvik, G. (1994). A Bayesian approach to dynamic contours through stochastic sampling and simulated annealing. *IEEE Trans. on Pattern Anal. and Machine Intelligence* **16**(10), 976–986.

Strauss, D. J. (1977). Clustering on colored lattice. *J. Appl. Probab.* **14**, 135–143.

Szeliski, R. (1989). *Bayesian Modeling of Uncertainty in Low-Level Vision*. Kluwer Academic Publishers, Dordrecht.

Tan, H. L., S. B. Gelfand and E. Delp (1992). A cost minimization approach to edge detection using simulated annealing. *IEEE Trans. on Pattern Anal. and Machine Intelligence* **14**, 3–18.

Terzopoulos, D. T. (1983). Multilevel computational process for visual surface reconstruction. *Computer Vision, Graphics and Image Process.* **24**, 52–96.

Terzopoulos, D. (1986). Image analysis using multigrid relaxation methods. *IEEE Trans. on Pattern Anal. and Machine Intelligence* **8**, 129–139.

Therrien. C. W. (1989). *Decision, Estimation, and Classification: An Introduction to Pattern Recognition and Related Topics*. Wiley, New York.

Tikhonov, A. N. and V. A. Arsenin (1977). *Solutions of Ill-Posed Problems*. Winston & Sons, Washington.

Torre, V. and T. Poggio (1986). On edge detection. *IEEE Trans. on Pattern Anal. and Machine Intelligence* **8**(2), 147–163.

Ullman, S. (1979). *The Interpolation of Visual Motion*. MIT Press, Cambridge, MA.

Watson, D. F. (1981). Computing the n-dimentional Delaunay tessellation with application to Voronoi polytopes. *Computer J.* **24**, 167–172.

Wells, W. M., III (1991). MAP model matching. In *Proceedings of IEEE Computer Society Conference on Computer Vision and Pattern Recognition*, pp. 486–492.

Winkler, G. (1994). *Image Analysis: Markov Fields and Dynamic Monte Carlo Methods*. Springer-Verlag.

Won, C. S. and H. Derin (1992). Unsupervised segmentation of noisy and textured images y using Markov random fields. *CVGIP: Graphics Model and Image Process.* **54**, 308–328.

Woods, J. W. (1972). Two-dimensional discrete Markovian fields. *IEEE Trans. on Inform. Theory* **18**, 232–240.

Wu, Y. N., S. C. Zhu and X. W. Liu (2000). Equivalence of Julesz ensemble and frame models. *Internat. J. Computer Vision* **38**(3), 245–261.

Wu, Z. and R. Leahy (1993). An approximation method of evaluating the joint likelihood for first-order GMRFs. *IEEE Trans. on Image Process.* **2**(4), 520–523.

Yullie, A. L. (1990). Generalized deformable models, statistical physics and matching problems. *Neural Comput.* **2**, 1–24.

Zhu, S. C. and X. W. Liu (2000). Learning in Gibbsian fields: how accurate and how fast can it be?. In *Proceedings of IEEE Computer Society Conference on Computer Vision and Pattern Recognition*, pp. 104–109. Hilton Head, NC, USA.

Zhu, S. C. and D. Mumford (1997). Prior learning and Gibbs reaction-diffusion. *IEEE Trans. on Pattern Anal. and Machine Intelligence* **19**(11).

Zhu, S. C., Y. Wu and D. Mumford (1997). Minimax entropy principle and its applications to texture modeling. *Neural Comput.* **9**(8).

Zhu, S. C., Y. N. Wu and D. Mumford (1998). FRAME: Filters, random field and maximum entropy: – towards a unified theory for texture modeling. *Internat. J. Computer Vision* **27**(2), 1–20.

An Introduction to Semi-Markov Processes with Application to Reliability

Nikolaos Limnios and Gheorghe Oprişan
Dedicated to Academician Vladimir S. Korolyuk on the occasion of his 75th birthday

In this chapter, we present first the basic properties of the semi-Markov processes with application to reliability. We give mainly probabilist results, but statistical inference is also present. Korolyuk's method of state space merging is presented. The state spaces of the studied processes include general, discrete and finite cases. More general processes are considered, as the non-homogeneous processes, the $(f\text{-}g)$-processes, the semi-Markov random walks and especially the K-dependent semi-Markov processes introduced here.

1. Introduction

The theory of Stochastic processes, which can be considered as an extension of Probability theory, allows the modeling of the evolution of systems along the time. It cannot be properly understood just as pure mathematics, separated from the body of experience and examples which have brought it to life.

The theory of stochastic processes entered in a period of intensive development, which is not finished yet, when the idea of Markov property was brought in. Not even a serious study of the renewal processes is possible without using the strong tool of Markov processes. The modern theory of Markov processes has its origins in the studies of A. A. Markov (1856–1922) on sequences of experiments "connected in a chain" and in the attempts to describe mathematically the physical phenomenon known as Brownian motion.

Later, there were many generalization (in fact all kinds of "weakenings" of the Markov property) of stochastic processes of Markov type. Some of them, like semi-Markov processes, has led to the new classes of stochastic processes and useful applications. The semi-Markov processes generalize the renewal processes as well as the Markov jump processes and have numerous applications in reliability, theory of dams, queuing theory, etc.

The semi-Markov process were introduced by Lévy (1954) and Smith (1955) simultaneously. Also, at the same time, Takács (1954) introduced essentially the same type of stochastic process and applied it to some problems in counter theory. The bases of the theory of the semi-Markov were set by Pyke (1961a, 1961b) and farther developed by Pyke and Schaufele (1964, 1967), Korolyuk (1976), Çinlar (1969a), Janssen and Limnios (1999) etc., and very soon the development of this theory together with its applications has got of vast proportions (Korolyuk and Turbin, 1982; Korolyuk and Shwitshchuk, 1995; Yackel, 1966; Moore and Pyke, 1968; Howard, 1964; O'Brien, 1974; Onicescu, Oprişan and Popescu, 1983; Iosifescu et al., 1984; Grigorescu and Oprişan, 1976; Osaki, 1985; Limnios and Oprişan, 1999, 2001; Korolyuk and Limnios, 2002), etc.

In many applied problems as reliability, queuing theory, etc. we use semi-Markov processes. The main advantage of semi-Markov processes is to allow non exponential distributions for transitions between states and to generalize several kind of stochastic processes. The most suited mathematical models for describing an operating system along the time are based on stochastic processes. The use of Markov processes is one of the most convenient solutions and the results are satisfactory in a lot of cases. However, a model using Markov processes is based on the assumption that the time spent by the system in each state is an exponentially distributed random variable. On the other hand, in practice there are cases where this assumption cannot be accepted. Hence it is necessary to elaborate methods for calculating the reliability indicators based on a semi-Markov evolution of the system.

An operating system is made up of components laid out in a certain configuration. The various possible combinations of failed and operating components define the states which the system is supposed to pass along time through. So, the system passes over from one state to another, as some components gradually fail and others are repaired. Some of the state are *down (refusal, failure) states* and other are *up (successful, functioning) states*. Here we follow reference (Limnios and Oprişan, 2001).

We give here the basic definitions and results concerning the semi-Markov processes in general state space case and some extensions concerning the non-homogeneous case, the $(f-g)$-processes, the semi-Markov random walk, the K-dependent semi-Markov processes that we introduce here, and Korolyuk's method of state space merging. Finaly, in the last section, we give some information on its application in reliability with statistical inference.

2. Semi-Markov kernel

The study of a semi-Markov process is closely related to a semi-Markov kernel in the same way as a Markov process is defined by its (probability) transition functions (see, e.g., Blumenthal and Getoor, 1968; Gihman and Skorohod, 1974).

Throughout this text $(E, \mathcal{E})$ will be a measurable space such that $\{x\} \in \mathcal{E}$ for all $x \in E$, and $(\mathbb{R}, \mathcal{B})$ and $(\mathbb{R}_+, \mathcal{B}_+)$ are the usual Borel spaces of the real line and half-real line respectively.

DEFINITION 2.1. A function $p(x, A)$, $x \in E$, $A \in \mathcal{E}$, is called a sub-Markov transition function (or sub-Markov kernel) on $(E, \mathcal{E})$ if:

(1) for each $x \in E$, $p(x, \cdot)$ is a measure on $\mathcal{E}$ such that $p(x, E) \leqslant 1$;
(2) for each $A \in \mathcal{E}$, $p(\cdot, A)$ is a Borel measurable function (i.e., $(\mathcal{E}, \mathcal{B})$-measurable).

If $p(x, E) = 1$ for all $x \in E$, then $p(x, A)$ is a Markov transition function (or a Markov kernel) on $(E, \mathcal{E})$.

If E is a finite or countable set and $\mathcal{E} = \mathcal{P}(E)$, then a sub-Markov transition function on $(E, \mathcal{E})$ is given by a sub-stochastic matrix $(p_{ij}; i, j \in E)$ and so $p(i, A) = \sum_{j \in A} p_{ij}$, $A \in \mathcal{E}$.

DEFINITION 2.2. A function $Q(x, A, t)$, $x \in E$, $t \in \mathbb{R}_+$, $A \in \mathcal{E}$, is called a semi-Markov kernel on $(E, \mathcal{E})$ if:

(1) $Q(x, A, \cdot)$, for all $x \in A$, $A \in \mathcal{E}$, is a non-decreasing, right continuous real function such that $Q(x, A, 0) = 0$;[1]
(2) $Q(\cdot, \cdot, t)$, for all $t \in \mathbb{R}_+$, is a sub-Markov kernel on $(E, \mathcal{E})$;
(3) $p(\cdot, \cdot) = Q(\cdot, \cdot, \infty)$ is a Markov kernel on $(E, \mathcal{E})$.

The following properties of a semi-Markov kernel are straightforward consequences of the above definitions

(1) $Q(x, \cdot, \cdot)$, for each $x \in E$, defines a probability measure on the σ-algebra $\mathcal{B}_+ \otimes \mathcal{E}$.
(2) For each $x \in E$, the function $H(x, \cdot) = Q(x, E, \cdot)$ is a distribution function such that $H(x, \cdot) = 0$.
(3) For each $t \in \mathbb{R}_+$, $A \in \mathcal{E}$, $Q(\cdot, A, t)$ is an $\mathcal{E}$-measurable function.

Due to the inequality $Q(x, A, t) \leqslant p(x, A)$, $x \in E$, $t \in \mathbb{R}_+$, $A \in \mathcal{E}$, the measure $Q(x, \cdot, t)$ is absolutely continuous with respect to the measure $p(x, \cdot)$, for each fixed $t \in \mathbb{R}_+$, $x \in E$ (i.e., $p(x, A) = 0$ implies $Q(x, A, t) = 0$). According to Radon–Nikodym theorem (see, e.g., Shiryaev, 1996), there is a real $\mathcal{E}$-measurable function $F(x, A, \cdot)$ such that

$$Q(x, A, t) = \int_A F(x, y, t) p(x, dy), \quad A \in \mathcal{E}. \tag{1}$$

It is not difficult to see that, for fixed $x, y \in E$, the function $F(x, y, \cdot)$ is non-decreasing. Hence $F(x, y, \cdot)$ can be choosen right continuous as is $Q(x, A, \cdot)$. Moreover, throughout we shall assume that $F(\cdot, \cdot, t)$, for fixed $t \in \mathbb{R}_+$, is $\mathcal{E} \otimes \mathcal{E}$-measurable.[2]

If Q_1 and Q_2 are two semi-Markov kernels on $(E, \mathcal{E})$, then their convolution, denoted by $Q_1 \star Q_2$, is defined by

$$(Q_1 \star Q_2)(x, A, t)$$
$$= \int_E \int_0^t Q_1(x, dy, du) Q_2(y, A, t - u), \quad x \in E, \ t \in \mathbb{R}_+, \ A \in \mathcal{E}. \tag{2}$$

---
[1] A non-decreasing, right continuous function $F : \mathbb{R} \to \mathbb{R}$ such that $F(0) = 0$, $F(+\infty) \leqslant 1$ is called *mass function*.
[2] This assumption is verified if E is finite or countable and $\mathcal{E} = \mathcal{P}(E)$.

The function $Q_1 \star Q_2$ is also a semi-Markov kernel. Note that generally $Q_1 \star Q_2 \neq Q_2 \star Q_1$.

We set by induction

$$Q^{(1)} = Q, \quad Q^{(2)} = Q \star Q, \quad \ldots, \quad Q^{(n)} = Q \star Q^{(n-1)} \tag{3}$$

and

$$Q^{(0)}(x, t, A) = \begin{cases} 0 & \text{if } t \leq 0, \\ \mathbb{1}_A(x), & \text{if } t > 0. \end{cases} \tag{4}$$

Using Fubini's theorem we see easily that

$$Q^{(m+n)} = Q^{(m)} \star Q^{(n)}. \tag{5}$$

3. Markov renewal processes (MRP)

On the measurable space $(E \times \mathbb{R}_+, \mathcal{E} \otimes \mathcal{B}_+)$ let $P((x, s), A \times [0, t])$ be the Markov transition function defined by

$$\begin{aligned} P((x, s), A \times [0, t]) &= Q(x, A, t - s), \\ (x, s) &\in E \times \mathbb{R}_+, \ A \times [0, t] \in \mathcal{E} \otimes \mathcal{B}_+. \end{aligned} \tag{6}$$

It is well known (see, e.g., Doob, 1953) that, for each $(x, s) \in E \times \mathbb{R}_+$, there is a probability space $(\Omega, \mathcal{F}, \mathbb{P}_{(x,s)})$ and a sequence of random variables $(J_n, S_n)_{n \in \mathbb{N}}$ such that

$$\mathbb{P}_{(x,s)}(J_0 = x, S_0 = s) = 1, \tag{7}$$

$$\begin{aligned} \mathbb{P}_{(x,s)}(J_{n+1} \in A, S_{n+1} &\leq t | \sigma(J_m, S_m, m \leq n)) \\ &= \mathbb{P}_{(x,s)}(J_{n+1} \in A, S_{n+1} \leq t | J_n, S_n) = Q(J_n, A, t - S_n) \end{aligned} \tag{8}$$

for all $n \in \mathbb{N}$, $t \in \mathbb{R}_+$, $A \in \mathcal{E}$.

Thus $(J_n, S_n)_{n \in \mathbb{N}}$ is a Markov process with the state space $(E \times \mathbb{R}_+, \mathcal{E} \otimes \mathcal{B}_+)$ and the transition probability function given by (6).

DEFINITION 3.1. The process $(J_n, S_n)_{n \in \mathbb{N}}$ is called the Markov renewal process (MRP) associated to the semi-Markov kernel Q.

It is not difficult to see that, for $n \in \mathbb{N}^*$, $(x, s) \in E \times \mathbb{R}_+$, $t \in \mathbb{R}_+$, $A \in \mathcal{E}$, we have

$$\begin{aligned} \mathbb{P}_{(x,s)}(S_n < S_{n+1}) &= \mathbb{E}_{(x,s)} \mathbb{P}_{(x,s)}(S_n < S_{n+1} | J_n, S_n) \\ &= \mathbb{E}_{(x,s)}\big[1 - H(J_n, \cdot - S_n) \circ S_n\big] \end{aligned}$$

$$= \mathbb{E}_{(x,s)}\big[1 - H(J_n, 0)\big]$$
$$= 1. \tag{9}$$

From (9) we deduce that, for all $n \in \mathbb{N}^*$, $x \in E$, $s \in \mathbb{R}_+$

$$0 < S_n < S_{n+1} < \infty, \quad \mathbb{P}_{(x,s)}\text{-a.s.} \tag{10}$$

For each $n \in \mathbb{N}$, we set

$$\mathcal{F}_n = \sigma\big((J_m, S_m), m \leqslant n\big), \qquad \mathcal{M}_n = \sigma(J_m, m \leqslant n)$$

and we note that we can suppose $\mathcal{F} = \sigma((J_n, S_n), n \in \mathbb{N})$. Obviously, we have $\mathcal{M}_n \subset \mathcal{F}_n \subset \mathcal{F}$, $n \in \mathbb{N}$.

Let us denote $X_0 = S_0$, $X_n = S_n - S_{n-1}$, $n \in \mathbb{N}^*$.

THEOREM 3.1. *For each $s \in \mathbb{R}_+$ the processes $(J_n, \mathcal{M}_n, \mathbb{P}_{(x,s)})$ and $((J_n, X_n), \mathcal{F}_n, \mathbb{P}_{(x,s)})$ are Markov chains with the state spaces $(E, \mathcal{E})$ and $(E \times \mathbb{R}_+, \mathcal{F} \otimes \mathcal{B}_+)$ respectively. Their transition probability functions are given by*

$$\mathbb{P}_{(x,s)}(J_{n+1} \in A | \mathcal{M}_n) = p(J_n, A), \quad x \in E, \ A \in \mathcal{E}, \tag{11}$$

$$\mathbb{P}_{(x,s)}(J_{n+1} \in A, X_{n+1} \leqslant t | \mathcal{F}_n) = Q(J_n, A, t),$$
$$x \in E, \ t \in \mathbb{R}_+, \ A \in \mathcal{E}. \tag{12}$$

PROOF. From (8) we have

$$\mathbb{P}_{(x,s)}(J_{n+1} \in A | \mathcal{M}_n) = \mathbb{E}_{(x,s)}\big[\mathbb{P}_{(x,s)}(J_{n+1} \in A, S_{n+1} < \infty | \mathcal{F}_n) | \mathcal{M}_n\big]$$
$$= \mathbb{E}_{(x,s)}\big[Q(J_n, A, \infty) | \mathcal{M}_n\big]$$
$$= Q(J_n, A, \infty)$$
$$= p(J_n, A)$$

and

$$\mathbb{P}_{(x,s)}(J_{n+1} \in A, X_{n+1} \leqslant t | \mathcal{F}_n) = \mathbb{P}_{(x,s)}(J_{n+1} \in A, S_{n+1} - S_n \leqslant t | J_n, S_n)$$
$$= Q(J_n, A, t + \cdot - S_n) \circ S_n$$
$$= Q(J_n, A, t). \qquad \square$$

The process $(J_n, X_n)_{n \in \mathbb{N}}$ is called $(J$–$X)$-process with state space E (see, e.g., Janssen and Limnios, 1999).

THEOREM 3.2. *For the process defined above, the following relations are fulfilled*:

$$\mathbb{P}_{(x,0)}(J_0 \in A_0, S_0 \leqslant t_0 - s, J_1 \in A_1, S_1 \leqslant t_1 - s, \ldots, J_n \in A_n, S_n \leqslant t_n - s)$$
$$= \mathbb{P}_{(x,s)}(J_0 \in A_0, S_0 \leqslant t_0, J_1 \in A_1, S_1 \leqslant t_1, \ldots, J_n \in A_n, S_n \leqslant t_n), \tag{13}$$

for all $n \in \mathbb{N}$, $x \in E$, $A_0, A_1, \ldots, A_n \in \mathcal{E}$, $t_0, \ldots, t_n \in \mathbb{R}_+$, $0 \leqslant s \leqslant \min\{t_0, \ldots, t_n\}$,

$$\mathbb{P}_{(x,s)}(X_n \leqslant t | \mathcal{M}_{n-1}) = H(J_{n-1}, t), \quad \mathbb{P}_{(x,s)}\text{-a.s.} \tag{14}$$

for all $n \in \mathbb{N}^*$, $s, t \in \mathbb{R}_+$, $x \in E$.

$$\mathbb{P}_{(x,s)}(X_n \leqslant t | \mathcal{M}_n) = F(J_{n-1}, J_n, t), \quad \mathbb{P}_{(x,s)}\text{-a.s.} \tag{15}$$

for all $n \in \mathbb{N}^*$, $s, t \in \mathbb{R}_+$, $x \in E$.

$$\mathbb{P}_{(x,s)}(S_n \leqslant t) = Q^{(n)}(x, E, t), \quad x \in E, \ s, t \in \mathbb{R}_+, \tag{16}$$

$$\mathbb{P}_{(x,s)}\big(X_{n_1} \leqslant t_1, \ldots, X_{n_k} \leqslant t_k | \sigma(J_n, n \in \mathbb{N})\big)$$

$$= \mathbb{P}_{(x,s)}\big(X_{n_1} \leqslant t_1, \ldots, X_{n_k} \leqslant t_k | \sigma(J_n, n \leqslant n_k)\big)$$

$$= \prod_{i=1}^{k} \mathbb{P}_{(x,s)}\big(X_{n_i} \leqslant t_i | \sigma(J_n, n \leqslant n_i)\big)$$

$$= \prod_{i=1}^{k} F(J_{n_i-1}, J_{n_i}, t_i), \quad \mathbb{P}_{(x,s)}\text{-a.s.} \tag{17}$$

for all $1 \leqslant n_1 < n_2 < \cdots < n_k$, $s, t_1, \ldots, t_k \in \mathbb{R}_+$, $x \in E$.

PROOF. For $x \notin A_0$ the equality (13) is obvious. Hence, we take $x \in A_0$. According to (6) we have

$$P\big((x, s), A \times [0, t]\big) = P\big((x, 0), A \times [0, t - s]\big), \quad s \leqslant t.$$

Consequently

$$\mathbb{P}_{(x,s)}(J_1 \in A_1, T_1 \leqslant t_1 - s, \ldots, J_n \in A_n, T_n \leqslant t_n - s)$$

$$= \int_{A_1 \times [0, t_1 - s]} P\big((x, 0), \mathrm{d}y_1 \times \mathrm{d}u_1\big)$$

$$\cdots \times \int_{A_{n-1} \times [0, t_{n-1} - s]} P\big((y_{n-2}, u_{n-2}), \mathrm{d}y_{n-1} \times \mathrm{d}u_{n-1}\big)$$

$$\times \int_{A_n \times [0, t_n - s]} P\big((x, s), \mathrm{d}y_1 \times \mathrm{d}u_n\big)$$

$$= \int_{A_1 \times [0, t_1 - s]} P\big((x, s), \mathrm{d}y_1 \times (\mathrm{d}u_n + s)\big)$$

$$\cdots \times \int_{A_{n-1} \times [0, t_1 - s]} P\big((x, s), \mathrm{d}y_{n-1} \times (\mathrm{d}u_n + s)\big)$$

$$\times \int_{A_n \times [0, t_n - s]} P\big((x, s), \mathrm{d}y_n \times (\mathrm{d}u_n + s)\big)$$

$$= \int_{A_1 \times [0, t_n]} P\big((x, s), \mathrm{d}y_1 \times \mathrm{d}u_n\big)$$

$$\cdots \times \int_{A_{n-1} \times [0, t_{n-1}]} \big((x, s), \mathrm{d}y_{n-1} \times \mathrm{d}u_{n-1}\big)$$

$$\times \int_{A_n \times [0, t_n]} P\big((x, s), \mathrm{d}y_1 \times \mathrm{d}u_n\big)$$

$$= \mathbb{P}_{(x,s)}(J_1 \in A_1, S_1 \leqslant t, \ldots, J_n \in A_n, S_n \leqslant t_n)$$

which proves (13).

The equalities (14) and (16) are consequences of (13) and respectively (12). The relation (15) derives from the definition of the function $F(x, y, t)$ (see (1)).

Now we shall prove that the σ-algebras $\sigma(X_k, k \leqslant n)$ and $\sigma((J_k, X_k), k > n)$ are $\mathbb{P}_{(x,s)}$-conditionally independent given $\mathcal{M}_n$.

Indeed, let $\varphi_1 : \mathbb{R}^{n+1} \to \mathbb{R}$ and $\varphi_2 : \mathbb{R}^\infty \to \mathbb{R}$ be measurable Borel functions. Using the Markov property of the process $(J_n, X_n)_{n \in \mathbb{N}}$ and (12), we have

$$\mathbb{E}_{(x,s)}\big[\varphi_1(X_0, \ldots, X_n)\varphi_2(J_{n+1}, \ldots)|\mathcal{M}_n\big]$$
$$= \mathbb{E}_{(x,s)}\big[\varphi_1(X_0, \ldots, X_n)\mathbb{E}_{(x,s)}\big(\varphi_2(J_{n+1}, \ldots)|\mathcal{M}_n\big)|\mathcal{M}_n\big]$$
$$= \mathbb{E}_{(x,s)}\big[\varphi_1(X_0, \ldots, X_n)|\mathcal{M}_n\big]\mathbb{E}_{(x,s)}\big[\varphi_2(J_{n+1}, \ldots)|\mathcal{M}_n\big] \qquad (18)$$

and the conditional independence is proved.

Now, we shall prove the first equality in (17). For this, let $\varphi : \mathbb{R}^n \to \mathbb{R}$, $\psi_1 : \mathbb{R}^{n+1} \to \mathbb{R}$, $\psi_2 : \mathbb{R}^\infty \to \mathbb{R}$ be measurable functions $A, B \in \mathcal{B}$, $\Omega_1 = \{\omega \in \Omega : \psi_1(J_0, J_1, \ldots, J_n) \in A\}$, $\Omega_2 = \{\omega \in \Omega : \psi_2(J_{n+1}, \ldots) \in B\}$.

Using the properties of the conditional expectation and (17), we have

$$\int_{\Omega_1 \cap \Omega_2} \mathbb{E}_{(x,s)}\big[\varphi(X_1, \ldots, X_n)|\mathcal{M}_n\big] \, \mathrm{d}\mathbb{P}_{(x,s)}$$

$$= \int_{\Omega_1} \mathbb{1}_{\Omega_2} \mathbb{E}_{(x,s)}\big[\varphi(X_1, \ldots, X_n)|\mathcal{M}_n\big] \, \mathrm{d}\mathbb{P}_{(x,s)}$$

$$= \int_{\Omega_1} \mathbb{E}_{(x,s)}\big[\mathbb{1}_{\Omega_2} \mathbb{E}_{(x,s)}\big(\varphi(X_1, \ldots, X_n)|\mathcal{M}_n\big)|\mathcal{M}_n\big] \, \mathrm{d}\mathbb{P}_{(x,s)}$$

$$= \int_{\Omega_1} \mathbb{E}_{(x,s)}\big[\big(\varphi(X_1, \ldots, X_n)|\mathcal{M}_n\big)\mathbb{E}_{(x,s)}[\mathbb{1}_{\Omega_2}|\mathcal{M}_n]\big] \, \mathrm{d}\mathbb{P}_{(x,s)}$$

$$= \int_{\Omega_1} \mathbb{E}_{(x,s)}\big(\mathbb{1}_{\Omega_2} \varphi(X_1, \ldots, X_n)|\mathcal{M}_n\big) \, \mathrm{d}\mathbb{P}_{(x,s)}$$

$$= \int_{\Omega_1} \mathbb{1}_{\Omega_2} \varphi(X_1, \ldots, X_n) \, \mathrm{d}\mathbb{P}_{(x,s)}$$

$$= \int_{\Omega_1 \cap \Omega_2} \varphi(X_1, \ldots, X_n) \, d\mathbb{P}_{(x,s)}.$$

So, we have proved that

$$\mathbb{E}_{(x,s)}\big[\varphi(X_1, \ldots, X_n) | \mathcal{M}_n\big]$$
$$= \mathbb{E}_{(x,s)}\big[\varphi(X_1, \ldots, X_n) | \mathcal{M}\big], \quad \mathbb{P}_{(x,s)}\text{-a.s.} \tag{19}$$

whose special case is the first equality (17).

The remained equalities in (17) can be deduced easily from the relation

$$\mathbb{P}_{(x,s)}(X_1 \leqslant \lambda_1, \ldots, X_n \leqslant \lambda_n | \mathcal{M}_n)$$
$$= \prod_{k=1}^{n} F(J_{k-1}, J_k, \lambda_k), \quad \mathbb{P}_{(x,s)}\text{-a.s.} \tag{20}$$

for all $n \in \mathbb{N}^*$, $x \in E$, $s, \lambda_1, \ldots, \lambda_n \in \mathbb{R}_+$.

To prove (20) we see that from (15) we have

$$\mathbb{P}_{(x,s)}(X_1 \leqslant \lambda_1, \ldots, X_n \leqslant \lambda_n | \mathcal{M}_n)$$
$$= \mathbb{E}_{(x,s)}\bigg[\mathbb{P}_{(x,s)}\bigg(\bigcap_{k=1}^{n}\{X_k \leqslant \lambda_k\} | \sigma(J_0, X_0, \ldots, J_{n-1}, X_{n-1}, J_n)\bigg) \bigg| \mathcal{M}_n\bigg]$$
$$= \mathbb{E}_{(x,s)}\big[\mathbb{1}_{\prod_{k=1}^{n-1}[0,\lambda_k]}(X_1, \ldots, X_n) \mathbb{P}_{(x,s)}$$
$$\times \big(X_k \leqslant \lambda_k | \sigma(J_0, X_0, \ldots, J_{n-1}, X_{n-1}, J_n)\big) | \mathcal{M}_n\big]$$
$$= F(J_{n-1}, J_n, \lambda_n) \big[\mathbb{P}_{(x,s)}(X_1 \leqslant \lambda_1, \ldots, X_{n-1} \leqslant \lambda_{n-1} | \mathcal{M}_n\big].$$

But

$$\mathbb{P}_{(x,s)}(X_1 \leqslant \lambda_1, \ldots, X_{n-1} \leqslant \lambda_{n-1} | \mathcal{M}_n)$$
$$= \mathbb{P}_{(x,s)}(X_1 \leqslant \lambda_1, \ldots, X_{n-1} \leqslant \lambda_{n-1} | \mathcal{M}_{n-1}), \quad \mathbb{P}_{(x,s)}\text{-a.s.}$$

by (19) and therefore the proof can be achieved by induction on n. □

For fixed $x \in E$ and $A \in \mathcal{E}$, the measure $Q(x, A, \cdot)$ on $(\mathbb{R}_+, \mathcal{B}_+)$, is absolute continuous with respect to the measure $H(x, \cdot)$. Hence there is (see, e.g., Shiryaev, 1996) a real, $\mathcal{E}$-measurable function $q(x, A, \cdot)$ such that

$$Q(x, A, t) = \int_0^t q(x, A, u) H(x, du), \quad t \in \mathbb{R}_+, \, x \in E, \, A \in \mathcal{E}. \tag{21}$$

Obviously, $q(x, \cdot, t)$ is a measure on $\mathcal{E}$ and, since $q(x, A, \cdot)$ is non-decreasing and right continuous, the function $q(\cdot, A, \cdot)$ is $\mathcal{E} \times \mathcal{E}$-measurable for each fixed $A \in \mathcal{E}$.

In fact, for all $y \in E$, $s \in \mathbb{R}_+$, we have

$$q(x, A, u - v)$$
$$= \mathbb{P}_{(x,s)}\big[J_{n+1} \in A | \sigma((J_k, S_k), 0 \leqslant k \leqslant n-1),$$
$$J_n = x, S_n = v, S_{n+1} = u\big]. \tag{22}$$

We set $\mathcal{G}_n = \sigma((J_k, X_k, X_{k+1}), 0 \leqslant k \leqslant n)$.

THEOREM 3.3. *For each $x \in E$, $s \in \mathbb{R}_+$, we have*

$$\mathbb{P}_{(x,s)}(J_{n+1} \in A, S_{n+1} \leqslant t_1, S_{n+2} \leqslant t_2 | \mathcal{G}_n)$$
$$= \mathbb{1}_{[0,t_1]}(S_{n+1}) \int_A H(y, t_2 - S_{n+1}) q(J_n, dy, S_{n+1} - S_n). \tag{23}$$

PROOF. According to the relation (12) and (22), we have

$$\mathbb{P}_{(x,s)}(J_{n+1} \in A, S_{n+1} \leqslant t_1, S_{n+2} \leqslant t_2 | \mathcal{G}_n)$$
$$= \mathbb{E}_{(x,s)}\big[\mathbb{P}_{(x,s)}(J_{n+1} \in A, S_{n+1} \leqslant t_1, S_{n+2} \leqslant t_2 | \mathcal{F}_{n+1}) | \mathcal{G}_n\big]$$
$$= \mathbb{E}_{(x,s)}\big[\mathbb{1}_A(J_{n+1}) \mathbb{1}_{[0,t_1]}(S_{n+1}) H(J_{n+1}, t_2 - S_{n+1}) | \mathcal{G}_n\big]$$
$$= \mathbb{1}_{[0,t_1]}(S_{n+1}) \int_A H(y, t_2 - S_{n+1}) q(J_n, dy, S_{n+1} - S_n)$$

and the proof is complete. □

Sometimes it is convenient to consider the special class of MRPs for which the function $F(x, y, t)$ does not depends on y. In this case, from (1), $F(x, y, t) = H(x, t)$ and

$$Q(x, A, t) = p(x, A) H(x, t) \tag{24}$$

for all $x, y \in E$, $t \in \mathbb{R}_+$, $A \in \mathcal{E}$.

This may be done without loss of generality by considering the semi-Markov kernel $Q^\sharp$ defined on $(E \times E, \mathcal{E} \times \mathcal{E})$ as follows.

$$Q^\sharp\big((x, y), A \times B, t\big) = \mathbb{1}_A(y) p(y, B) F(x, y, t),$$
$$(x, y) \in E \times E, \ t \in \mathbb{R}_+, \ A, B \in \mathcal{E}. \tag{25}$$

The functions related to this kernel are:

$$H^\sharp\big((x, y), t\big) = F(x, y, t), \quad (x, y) \in E \times E, \ t \in \mathbb{R}_+, \tag{26}$$
$$p^\sharp\big((x, y), A \times B\big) = \mathbb{1}_A(y) p(y, B), \quad (x, y) \in E \times E, \ A, B \in \mathcal{E}, \tag{27}$$

and therefore

$$Q^\sharp((x, y), A \times B, t) = p^\sharp((x, y), A \times B) H^\sharp((x, y), t). \tag{28}$$

THEOREM 3.4. *If we set $\mathcal{F}'_n = \sigma((J_{k+1}, S_k), k \leqslant n)$ then, for all $x \in E$, $s, t \in \mathbb{R}_+$, $A, B \in \mathcal{E}$, we have*

$$\mathbb{P}_{(x,s)}(J_{n+1} \in A, J_{n+2} \in B, S_{n+1} \leqslant t | \mathcal{F}'_n)$$
$$= \mathbb{1}_A(J_{n+1}) p(J_{n+1}, B) F(J_n, J_{n+1}, t - S_n). \tag{29}$$

PROOF. See (Limnios and Oprişan, 2001). □

We note that the MRP corresponding to the semi-Markov kernel $Q^\sharp$ is $(J'_n, S_n)_{n \in \mathbb{N}}$ where $J'_n = (J_n, J_{n+1})$.

THEOREM 3.5. *If we set $\mathcal{F}''_n = \sigma((J_k, S_{k+1}), k \leqslant n)$ then, for all $x \in E$, $s, t_1, t_2 \in \mathbb{R}_+$, $A \in \mathcal{E}$, we have*

$$\mathbb{P}_{(x,s)}(J_{n+1} \in A, S_{n+1} \leqslant t_1, S_{n+2} \leqslant t_2 | \mathcal{F}''_n)$$
$$= \mathbb{1}_{[0, t_1]}(S_{n+1}) \int_A H(y, t_2 - S_{n+1}) q(J_n, dy, S_{n+1} - S_n). \tag{30}$$

PROOF. See (Limnios and Oprişan, 2001). □

COROLLARY 3.1. *For $x \in E$, $s, t_1, t_2 \in \mathbb{R}_+$, $A \in \mathcal{E}$, we have*

$$\mathbb{P}_{(x,s)}(J_{n+1} \in A, X_{n+1} \leqslant t_1, X_{n+2} \leqslant t_2 | \mathcal{F}''_n)$$
$$= \mathbb{1}_{[0, t_1]}(X_{n+1}) \int_A H(y, t_2) q(J_n, dy, X_{n+1}).$$

PROOF. An obvious consequence of (12) and (30). □

REMARK 1. It is not difficult to see that the function

$$\Pi((x, s); A \times [0, t]) = \int_A H(y, t) q(x, dy, s),$$
$$(x, s) \in E \times \mathbb{R}_+, \ A \in \mathcal{E}, \ t \in \mathbb{R}_+,$$

defines a Markov transition function on $(E \times \mathbb{R}_+, \mathcal{E} \times \mathcal{B}_+)$ (according to Definition 1).

So, for each $(x, s) \in E \times \mathbb{R}_+$, there is a probability $\overline{\mathbb{P}}_{(x,s)}$ on $\mathcal{F}$, such that $((J_n, X_{n+1}), \mathcal{F}''_n, \overline{\mathbb{P}}_{(x,s)})$, $n \in \mathbb{N}$, is a Markov process with the state space $E \times \mathbb{R}_+$ and the Markov transition function $\Pi(\cdot\,;\,\cdot)$.

Sometimes one considers MRPs with possible finite lifetimes. This means a $\mathcal{F}$-measurable r.v. $\zeta : \Omega \to \overline{\mathbb{N}}$, which is the length of the trajectories of the MRP. This is the case when the transition function $p(\cdot, \cdot)$ on $(E, \mathcal{E})$ is sub-Markovian, i.e., $p(x, E) = Q(x, E, \infty) \leq 1$, $x \in E$. (See Definition 2.1.)

We shall consider a point Δ not in E. We write $E^\sharp = E \cup \{\Delta\}$ and let $\mathcal{E}^\sharp$ be the σ-algebra in $E^\sharp$ generated by $\mathcal{E}$. Note that $\{\Delta\} \in \mathcal{E}^\sharp$. Now we define a new semi-Markov kernel $Q^\sharp$ as follows

$$Q^\sharp(x, A, t) = \begin{cases} Q(x, A, t), & \text{if } x \in E, A \in \mathcal{E}, \\ 1 - p(x, E), & \text{if } x \in E, A = \{\Delta\}, \\ 0, & \text{if } x = \Delta, A \in \mathcal{E}, \\ 1, & \text{if } x = \Delta, \Delta \in A. \end{cases} \quad (31)$$

The corresponding MRP, denoted by $((J_n^\sharp, S_n^\sharp), \mathcal{F}_n^\sharp, \mathbb{P}_{(x,s)}^\sharp)_{n \in \mathbb{N}}$ makes transitions in E until it "dies", at which time it is transported to Δ where it remains forever. The lifetime of the process is

$$\zeta = \inf\{n \in \mathbb{N}^*, J_n^\sharp = \Delta\} \quad (32)$$

provided the set is not empty and $\zeta = \infty$ if it is empty.

Using the relations (2), (3) and (13) we have

$$\mathbb{P}_{(x,s)}^\sharp(\zeta > n) = \mathbb{P}_{(x,s)}^\sharp(J_n^\sharp \in E)$$
$$= Q^{(n)}(x, E, \infty)$$
$$= p^{\sharp(n)}(x, E), \quad x \in E, \ s \in \mathbb{R}_+, \quad (33)$$

where $p^{\sharp(n)}(x, A)$, $x \in E^\sharp$, $A \in \mathcal{E}^\sharp$ are the n-step transition probabilities of the Markov process $(J_n^\sharp)_{n \in \mathbb{N}}$.

Obviously, for all $n \in \mathbb{N}$, $\{\zeta > n\} \in \mathcal{F}_n^\sharp$ and

$$\mathbb{P}_{(x,s)}^\sharp(J_{n+1}^\sharp, S_{n+1}^\sharp \leq t | \mathcal{F}_n^\sharp) = Q(J_n^\sharp, A, t - S_n^\sharp) \quad (34)$$

almost everywhere on $\{\zeta > n\}$, for any $n \in \mathbb{N}$, $x \in E$, $s, t \in \mathbb{R}_+$, $A \in \mathcal{E}$.

4. Semi-Markov processes with an arbitrary state space

In the sequel, for $x \in E$, the probability $\mathbb{P}_{(x,0)}$ on $\mathcal{F}$, will be denoted by $\mathbb{P}_x$.

Let $Q(x, A, t)$, $x \in E$, $A \in \mathcal{E}$, $t \in \mathbb{R}_+$, be a semi-Markov kernel on $(E, \mathcal{E})$ and let $(J_n, S_n)_{n \in \mathbb{N}}$, $(J_n, X_n)_{n \in \mathbb{N}}$ be the associated MRP and $(J-X)$-process respectively (see Section 3).

DEFINITION 4.1. Let $(Z(t), t \in \mathbb{R}_+)$ be a stochastic process defined on a probability space $(\Omega, \mathcal{F}, \mathbb{P})$ with values in $(E, \mathcal{E})$. The process is called a jump process if for all

$\omega \in \Omega$ and all $t \in \mathbb{R}_+$ there exists a $\delta = \delta(t, \omega)$ such that $Z(t+h) = Z(t)$ for $0 \leqslant h < \delta$ equivalently, the trajectories are right-continuous in the discrete topology (on the state space).

If we set

$$N(t) = \begin{cases} 0, & \text{if } X_1 > t, \\ \sup\{n \in \mathbb{N}^* : X_1 + \cdots + X_n \leqslant t\}, & \text{if } X_1 < t \end{cases} \quad (35)$$

then, using relation (2) from Section 2, we can define the jumping process

$$Z(t) = Z_t = J_n, \quad \text{for } S_n \leqslant t + S_0 < S_{n+1}, \ t \in \mathbb{R}_+, n \in \mathbb{N}, \quad (36)$$

or, equivalently

$$Z_t = J_{N(t)}, \quad t \in \mathbb{R}_+. \quad (37)$$

The jump times are $S_1 - S_0, S_2 - S_0, \ldots$ and the intervals between jumps are $X_1, X_2, \ldots$ We note that the process $(Z_t, t \in \mathbb{R}_+)$ is observed after the random time S_0. (See Figure 1.)

DEFINITION 4.2. The stochastic process $(Z(t), t \in \mathbb{R}_+)$ defined above is called the semi-Markov process corresponding to the semi-Markov kernel Q.

The r.v. $S_\infty = \lim_{n \to \infty} S_n$ is called the *explosion time*. The semi-Markov kernel (as well as the corresponding Markov renewal and semi-Markov processes) is said to be *regular* if

$$S_\infty = \infty, \quad \mathbb{P}_{(x,s)}\text{-a.s. for all } x \in E, \ s \in \mathbb{R}_+. \quad (38)$$

According to the relation (13), we have

$$\mathbb{P}_{(x,s)}(N(t) \geqslant n) = \mathbb{P}_{(x,s)}(S_n \leqslant t+s) = \mathbb{P}_{(x,0)}(S_n \leqslant t)$$
$$= \mathbb{P}_{(x,0)}(N(t) \geqslant n) \quad (39)$$

for all $n \in \mathbb{N}, \ x \in E, \ s, t \in \mathbb{R}_+$.

Hence (39) is equivalent to each of the following relations

$$\lim_{n \to \infty} \mathbb{P}_{(x,s)}(N(t) \geqslant n) = 0, \quad \text{for all } x \in E, \ t \in \mathbb{R}_+, \quad (40)$$

$$\lim_{n \to \infty} \mathbb{P}_{(x,0)}(S_n \leqslant t) = 0, \quad \text{for all } x \in E, \ t \in \mathbb{R}_+. \quad (41)$$

THEOREM 4.1. *Each of the following conditions are necessary and sufficient for the semi-Markov process to be regular*:

$$\lim_{n \to \infty} Q^{(n)}(x, E, t) = 0, \quad \text{for all } x \in E, \ t \in \mathbb{R}_+, \quad (42)$$

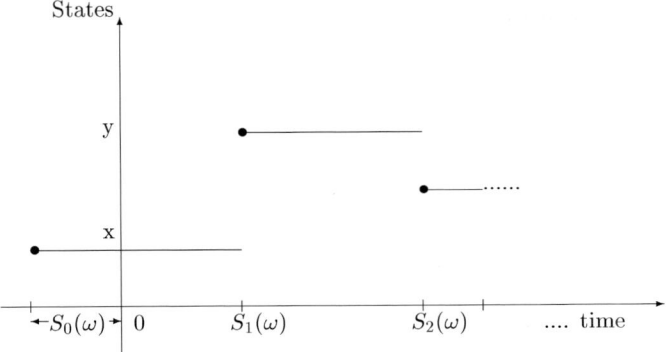

Fig. 1. System's trajectory ω.

where $Q^{(n)}$ is as defined in Section 2, relation (3)

$$\lim_{n\to\infty} \int_{E^\infty} \star_{j=0}^{n-1} F(y_j, y_{j+1}, t) \prod_{j=0}^{n-1} p(y_j, dy_{j+1}) = 0 \qquad (43)$$

for all $y_0 \in E$, $t \in \mathbb{R}_+$, where $F(y_j, y_{j+1}, t)$ is as defined in Section 2, relation (1).

PROOF. First, condition (42) is a consequence of (41) and (16).
To prove the second condition (43) we must prove first the relation

$$\mathbb{P}_{(x,0)}(S_n \leq t)$$

$$= \int_{E^\infty} \star_{j=0}^{n-1} F(y_j, y_{j+1}, t) \prod_{j=0}^{n-1} p(y_j, dy_{j+1}), \quad y_0 = x, t \in \mathbb{R}_+, \qquad (44)$$

which, in turn, is a consequence of relations (1) and (16). □

COROLLARY 4.1. *If there exists a mass function $F : \mathbb{R} \to \mathbb{R}$ such that*

$$F(x, y, t) \leq F(t), \quad \text{for all } x, y \in E, t \in \mathbb{R}_+, \qquad (45)$$

then the semi-Markov process is regular.

PROOF. Indeed, we have

$$\star_{j=0}^{n-1} F(y_j, y_{j+1}, t) \leq F^{(n)}(t) \qquad (46)$$

for all $y_j, y_{j+1} \in E, t \in \mathbb{R}_+, n \in \mathbb{N}^*$, where $F^{(n)}$ is the convolution n times of F by itself. On the other hand $\lim_{n\to\infty} F^{(n)}(t) = 0$, for all $t \in \mathbb{R}_+$ (see, e.g., Feller, 1966, Vol. II, Chapter 6(6)) and hence the second condition (43) of Theorem 3.1 is fulfilled. □

COROLLARY 4.2. *If E is a finite set and $\mathcal{E} = \mathcal{P}(E)$, then the semi-Markov process is regular.*

PROOF. The function $F(t) = \max_{x,y \in E} F(x, y, t)$, $t \in \mathbb{R}_+$, is a mass function which verifies the conditions of Corollary 1. □

A jump Markov process is a special case of a semi-Markov process whose semi-Markov kernel takes the particular form (see, e.g., Blumenthal and Getoor, 1968, p. 65; Korolyuk and Turbin, 1982, p. 16).

$$Q(x, A, t) = p(x, A)\bigl(1 - e^{\lambda(x)t}\bigr), \quad x \in E, A \in \mathcal{E}, t \in \mathbb{R}_+, \tag{47}$$

where $p(\cdot, \cdot)$ is a Markov kernel on $(E, \mathcal{E})$ and $\lambda : E \to \mathbb{R}_+$ is called the *rate of leaving the state x*.

5. Markov renewal equation

DEFINITION 5.1. An integral linear equation of the form

$$\varphi(x, t) = g(x, t) + \int_E \int_0^t Q(x, \mathrm{d}y, \mathrm{d}s)\varphi(y, t - s), \tag{48}$$

where Q is a semi-Markov kernel on $(E, \mathcal{E})$, g is a given function defined on $E \times \mathbb{R}_+$ and φ is the unknown function, is called a *Markov renewal equation*.

Using (2), Eq. (48) may be written

$$\varphi = g + Q \star \varphi. \tag{49}$$

We shall use the sign "~" to denote the Laplace transform, namely

$$\tilde{\varphi}(x, \lambda) = \int_0^\infty e^{-\lambda t} \varphi(x, t)\, \mathrm{d}t, \quad x \in E, \lambda > 0, \tag{50}$$

$$\tilde{Q}(x, A, \lambda) = \int_0^\infty e^{-\lambda t} Q(x, A, \mathrm{d}t), \quad x \in E, \lambda > 0, A \in \mathcal{E}. \tag{51}$$

We note that $\tilde{Q}(x, A, \lambda)$ is a sub-Markov kernel for each fixed $\lambda > 0$ (see Definition 1.1).

Taking the Laplace transform (see, e.g., Feller, 1966) in (48) we obtain the classical integral equation

$$\tilde{\varphi}(x, \lambda) = \tilde{g}(x, \lambda) + \int_E \tilde{Q}(x, \mathrm{d}y, \lambda)\tilde{\varphi}(y, \lambda). \tag{52}$$

The following function plays a prominent part in the Markov renewal theory

$$\psi(x, A, t) = \sum_{n \in \mathbb{N}} Q^{(n)}(x, A, t), \quad x \in E, \ A \in \mathcal{E}, \ t \in \mathbb{R}_+. \tag{53}$$

It is called the *Markov renewal function* corresponding to Q.

DEFINITION 5.2. The semi-Markov kernel Q (as well as the corresponding semi-Markov process) is called *normal* if $\psi(x, E, t) < \infty$ for all $x \in E$, $t \in \mathbb{R}_+$.

From (41), (16) and (53) it follows that a normal semi-Markov kernel is regular. We note that (45) is a sufficient condition for the normality of the semi-Markov process.

Throughout this section we will assume the semi-Markov process to be normal. From (53) and (3), we can deduce easily that the Markov renewal function ψ verifies the Markov renewal equation

$$\psi(x, A, t) = \mathbf{1}_A(x) + \int_E \int_0^t Q(x, \mathrm{d}y, \mathrm{d}s) \psi(y, A, t - s). \tag{54}$$

To discuss existence and uniqueness of solution, we need first some observation. So, let us denote by $\mathcal{V}$ the Banach space of all bounded real functions defined on E with the norm $\|u\| = \sup_{x \in E} |u(x)|$. If $\varphi(x, \cdot)$ is a function on $\mathbb{R}$ such that, for fixed $t \in \mathbb{R}$, $\varphi(\cdot, t)$ belongs to $\mathcal{V}$, then we write $\|\varphi\|$ for the function whose value at $t \in \mathbb{R}$ is the real number $\|\varphi(\cdot, t)\|$. Let $\mathcal{L}$ be the collection of all functions $\varphi(x, t)$ such that $\varphi(x, t) = 0$ for $t \leq 0$, the function $\|\varphi\|(\cdot)$ is bounded on finite intervals and, for each $x \in E$, $\varphi(x, \cdot)$ is Borel measurable. Further, we denote by $\mathcal{L}^+$ the set of non-negative functions in $\mathcal{L}$. Obviously, $\varphi \in \mathcal{L}$ iff φ^+ and φ^- belongs to $\mathcal{L}^+$.

The proofs of the following proposition can be found in (Çinlar, 1969a) or (Korolyuk and Turbin, 1976).

THEOREM 5.1 (Çinlar, 1969a). *For $g \in \mathcal{L}^+$, the Markov renewal equation (48) has a solution $\varphi \in \mathcal{L}$ iff $R \star g \in \mathcal{L}^+$. Any solution has the representation $\varphi = R \star g + f$, where f satisfies $f = Q \star f$, $f \in \mathcal{L}$.*

THEOREM 5.2 (Çinlar, 1969a). *If $\|H\|(t) < 1$ for some $t > 0$, then $R \star g$ exists and it is the unique solution of Eq. (48) for any $g \in \mathcal{L}$.*

COROLLARY 5.1. *If E is a finite set and $\mathcal{E} = \mathcal{P}(E)$, then $R \star g$ exists and is the unique solution of Eq. (48) for any $g \in \mathcal{L}$.*

PROOF. In this case $Q^{(n)} \to 0$ as $n \to \infty$ and $\psi(x, E, t) < \infty$ for any $x \in E$ and $t \geq 0$; thus the proof of the preceding theorem goes through. $\square$

Now, let us introduce two important functions for the characterization of the semi-Markov process. These functions verify Markov renewal equations. So, for $t \in \mathbb{R}_+$, $A \in$

$\mathcal{E}$, we set

$$p(x, A, t) = \mathbb{P}_x\big(Z(t) \in A\big), \quad x \in E,$$
$$G(x, A, t) = \mathbb{P}_x\big(\tau_{A^c} \leqslant t\big), \quad x \in A,$$

where τ_{A^c} is the first exit time from A, i.e.,

$$\tau_{A^c} = \inf\{t > S_1;\ Z(t) \in A^c\}, \quad A \in \mathcal{E},$$

with the convention $\inf \varnothing = \infty$.

THEOREM 5.3. *The functions $p(x, A, t)$ and $G(x, A, t)$ verify the following Markov renewal equations*

$$p(x, A, t) = \mathbf{1}_A(x)\big(1 - H(x, t)\big) + \int_E \int_0^t Q(x, dy, ds) p(x, A, t - s), \quad (55)$$

$$G(x, A, t) = Q\big(x, A^c, t\big) + \int_E \int_0^t Q(x, dy, ds) G(x, A, t - s). \quad (56)$$

PROOF. Using the Markov property we have

$$p(x, A, t) = \mathbb{P}_x\big(Z(t) \in A,\ S_1 \leqslant t\big) + \mathbb{P}_x\big(Z(t) \in A,\ S_1 > t\big)$$
$$= \mathbf{1}_A(x)\big[1 - H(x, t)\big] + \sum_{n \in \mathbb{N}^*} \mathbb{P}_x(J_n \in A,\ S_n \leqslant t < S_{n+1})$$
$$= \mathbf{1}_A(x)\big[1 - H(x, t)\big]$$
$$\quad + \sum_{n \in \mathbb{N}^*} \mathbb{E}_x\big[\mathbb{P}_x\big(\theta_1(J_{n-1} \in A,\ X_1 + \cdots + X_{n-1} \leqslant t - \cdot$$
$$\quad < X_1 + \cdots + X_n | \mathcal{F}_1)\big) \circ X_1\big]$$
$$= \mathbf{1}_A(x)\big[1 - H(x, t)\big]$$
$$\quad + \sum_{n \in \mathbb{N}^*} \mathbb{E}_x\big[\mathbb{P}_{J_1}(J_{n-1} \in A,\ S_{n-1} \leqslant t < S_n) \circ X_1\big]$$
$$= \mathbf{1}_A(x)\big[1 - H(x, t)\big] + \mathbb{E}_x\big[\mathbb{P}_{J_1}(Z(t - \cdot) \in A) \circ X_1\big]$$
$$= \mathbf{1}_A(x)\big[1 - H(x, t)\big] + \int_E \int_0^t Q(x, dy, ds) \mathbb{P}_y\big(Z(t - s) \in A\big)$$

and (55) is proved.

For (56) we proceed in a similar way as follows

$$G(x, A, t) = 1 - \mathbb{P}_x(\tau_{A^c} > t)$$

$$= 1 - \mathbb{P}_x\big(Z(u) \in A, \forall u \in [0,t]\big)$$
$$= 1 - \mathbb{P}_x\big(Z(u) \in A, \forall u \in [0,t], S_1 \leq t\big)$$
$$\quad - \mathbb{P}_x\big(Z(u) \in A, \forall u \in [0,t], S_1 > t\big)$$
$$= 1 - \sum_{n \in \mathbb{N}^*} \mathbb{P}_x\bigg(\bigcap_{k=0}^{n}\{J_k \in A\}, S_n \leq t < S_{n+1}\bigg) - \mathbb{P}_x(S_1 > t)$$
$$= H(x,t)$$
$$\quad - \sum_{n \in \mathbb{N}^*} \mathbb{E}_x\bigg[\mathbb{P}_x\bigg(\bigcap_{k=1}^{n}\{J_k \in A\}, X_2 + \cdots + X_n$$
$$\leq t - X_1 < X_1 + \cdots + X_{n+1}\big|\mathcal{F}_1\bigg)\bigg]$$
$$= H(x,t)$$
$$\quad - \sum_{n \in \mathbb{N}^*} \mathbb{E}_x\bigg[\mathbf{1}_A(J_1)\mathbb{P}_x\bigg(\theta_1\bigg(\bigcap_{k=0}^{n}\{J_k \in A\}, X_1 + \cdots + X_{n-1}$$
$$\leq t - \cdot < X_1 + \cdots + X_n\bigg)\big|\mathcal{F}_1\bigg) \circ X_1\bigg]$$
$$= H(x,t)$$
$$\quad - \sum_{n \in \mathbb{N}^*} \mathbb{E}_x\bigg[\mathbf{1}_A(J_1)\mathbb{P}_{J_1}\bigg(\bigcap_{k=0}^{n-1}\{J_k \in A\}, S_{n-1}$$
$$\leq t - \cdot < S_n\bigg) \circ X_1\bigg]$$
$$= H(x,t) - \mathbb{E}_x\big[\mathbf{1}_A(J_1)\mathbb{P}_{J_1}\big(Z(u) \in A\big), \forall u \in [0, t - \cdot] \circ X_1\big]$$
$$= H(x,t) - \int_A \int_0^t Q(x, dy, ds)\mathbb{P}_y(\tau_{A^c} > t - s)$$
$$= Q(x, A^c, t) + \int_A \int_0^t Q(x, dy, ds)G(x, A, t - s).$$

The proof is complete. $\square$

If we set

$$\mu(x, A) = \mathbb{E}_x[\tau_{A^c}] = \int_0^\infty t G(x, A, dt)$$

and

$$m(x) = \mathbb{E}_x[S_1] = \int_0^\infty t H(x, \mathrm{d}t)$$

for $x \in A$, $A \in \mathcal{E}$, then

$$\mu(x, A) = -\lim_{\lambda \downarrow 0} \frac{\mathrm{d}}{\mathrm{d}\lambda} \big[\widetilde{G}(x, A, \lambda)\big], \tag{57}$$

$$m(x) = -\lim_{\lambda \downarrow 0} \frac{\mathrm{d}}{\mathrm{d}\lambda} \big[\widetilde{H}(x, \lambda)\big] \tag{58}$$

and (56) yields

$$\mu(x, A) = m(x) + \int_A p(x, \mathrm{d}y) \mu(y, A). \tag{59}$$

We note that $\mu(x, A)$ is the mean time spent by the process in A before the jump outside A, if the process starts from $x \in A$.

6. The countable case

By "countable case" we mean that the state space E is finite or countable and that $\mathcal{E} = \mathcal{P}(E)$.

Let $E = \{1, 2, \ldots\}$ be the state space. A semi-Markov kernel on $(E, \mathcal{E})$ becomes a semi-Markov matrix and will be denoted by $Q_{ij}(t)$ (instead of $Q(i, \{j\}, t)$), $i, j \in E$, $t \in \mathbb{R}_+$. The related functions will be denoted as follows:

$$p(i, \{j\}) = p_{ij}, \quad i, j \in E,$$
$$H(i, t) = H_i(t), \quad i \in E, \ t \in \mathbb{R}_+,$$
$$F(i, j, t) = F_{ij}(t), \quad i \in E, \ t \in \mathbb{R}_+,$$
$$q(i, \{j\}, t) = q_{ij}(t), \quad i \in E, \ t \in \mathbb{R}_+.$$

The convolution of two semi-Markov kernels Q^1 and Q^2, defined by (2), is

$$(Q^1 * Q^2)_{ij}(t) = \sum_{k \in E} \int_0^t Q^1_{ik}(\mathrm{d}u) Q^2_{kj}(t - u), \quad i, j \in E, \ t \in \mathbb{R}_+. \tag{60}$$

Relation (1) becomes

$$Q_{ij}(t) = p_{ij} F_{ij}(t), \quad i, j \in E, \ t \in \mathbb{R}_+. \tag{61}$$

When $|E| = 1$, the sequence of random variables $(X_n)_{n \in \mathbb{N}}$ is a renewal sequence. Hence, a renewal process is a special case of $(J-X)$-process.

Let $\mathbf{p} = (p_{ij})_{i,j \in E}$ be a stochastic matrix. When the semi-Markov kernel is of the form

$$Q_{ij}(t) = \begin{cases} 0, & t \leq 0, \\ p_{ij}, & t > 0, \end{cases} \quad i, j \in E. \tag{62}$$

The corresponding $(J-X)$-process becomes a Markov chain with the state space E and the matrix of transition probabilities $\mathbf{p}$.

A special case of a MRP is that whose semi-Markov kernel takes the form (see Section 3)

$$Q_{ij}(t) = \begin{cases} 0, & t \leq 0, \\ p_{ij}(1 - e^{-\lambda_i t}), & t > 0, \end{cases} \tag{63}$$

for constants $\lambda_i > 0$ and $p_{ii} = 0$, $i \in E$. In this case we have a Markov process.

Often, the transitions of the process $(J_n)_{n \in \mathbb{N}}$ from a state to itself may be disturbing. To avoid that, one may work with the related MRP $(J'_n, S'_n)_{n \in \mathbb{N}}$ defined as follows. We shall consider that there are no absorbing states, i.e., states i such that $p_{ii} = 1$. The shift operators will be denoted by $\theta_k, k \in \mathbb{N}$ (i.e., $\theta_m h(J_0, S_0, J_1, S_1, \ldots) = h(J_m, S_m, J_{m+1}, S_{m+1}, \ldots)$). Now, we define a sequence $(\tau_n)_{n \in \mathbb{N}}$ of stopping times (with respect to the increasing family of σ-algebras $(\mathcal{F}_n)_{n \in \mathbb{N}}$) by recurrence

$$\tau_0 = 0, \quad \tau_1 = \min\{k: J_k \neq J_0\}, \quad \tau_{n+1} = \tau_n + \theta_{\tau_n} \tau_1. \tag{64}$$

These times are the "changing state" times of the process $(J_n)_{n \in \mathbb{N}}$. Now, let us consider the process $(J'_n, S'_n)_{n \in \mathbb{N}}$ where $J'_n = J_{\tau_n}$, $S'_n = S_{\tau_n}$, $n \in \mathbb{N}$.

This process is a MRP associated to the semi-Markov matrix $Q'_{ij}(t)$, $i, j \in E$, defined by

$$Q'_{ij}(t) = \begin{cases} \frac{Q_{ij}(t)}{1 - p_{ii}}, & \text{if } i \neq j, \\ 0, & \text{if } i = j. \end{cases} \tag{65}$$

For real stochastic systems the set E is finite and the transitions of the MRP from a state to another are determined by a finite number of random independent factors (e.g., the failure of a component.) Each of these factors is characterized by a random time of influence at the end of which the process changes the state. Usually, this random time depends on the "departure state". Therefore for each $i \in E$ we consider the family $\{\tau_{ik}, k \in E\}$ of random independent times (defined on a probability space $(\Omega, \mathcal{F}, \mathbb{P})$) of influence of the factors which determine the jump from i. We assume that the random variables τ_{ik}, $k \in E$, take values in $\overline{\mathbb{R}}_+$. The time spent in i is $\eta_i = \min_{k \in E} \tau_{ik}$.

THEOREM 6.1 (Korolyuk and Turbin, 1976). *Let $(J_n, S_n)_{n \in \mathbb{N}}$ be a MRP defined on a probability space $(\Omega, \mathcal{F}, \mathbb{P})$ with the state space $E \times \mathbb{R}_+$ (E is a finite set) and with the semi-Markov matrix $Q_{ij}(t)$, $i, j \in E$, $t \in \mathbb{R}_+$.*

For each $i \in E$ there exists a family of independent r.v. $\{\tau_{ik}, k \in E\}$, $\tau_{ik} : \Omega \to \overline{\mathbb{R}}_+$ with the distribution functions

$$A_{ik}(t) = \begin{cases} 0, & \text{if } t \leqslant 0, \\ 1 - \exp\left[-\int_0^t \frac{Q_{ik}(\mathrm{d}u)}{1-H_i(u)}\right], & \text{if } t > 0, \end{cases} \quad (66)$$

such that

$$Q_{ij}(t) = \int_0^t h_{ij}(u) A_{ij}(\mathrm{d}u), \quad i, j \in E, \quad (67)$$

where

$$h_{ij}(u) = \frac{1 - H_i(u)}{1 - A_{ij}(u)} = \mathbb{E}[I_{ij} | \tau_{ij} = u] \quad (68)$$

and I_{ij}, $i, j \in E$ is the indicator function of the event $\{\min_{k \in E} \tau_{ik} = \tau_{ij}\}$.

The proof of this theorem can be found in (Korolyuk and Turbin, 1976; Limnios and Oprişan, 2001).

REMARK 2. From (68) we obtain

$$\mathbb{E}[I_{ij}] = \int_0^\infty h_{ij}(u) A_{ij}(\mathrm{d}u) = Q_{ij}(\infty) = p_{ij}, \quad i, j \in E. \quad (69)$$

REMARK 3. Let us consider the mass functions $\lambda_{ij}(t) = \int_0^t \frac{Q_{ij}(\mathrm{d}u)}{1-H_i(u)}$, $i, j \in E, t > 0$. So, for $i, k \in E$,

$$A_{ik}(t) = \begin{cases} 0, & \text{if } t \leqslant 0, \\ 1 - \mathrm{e}^{-\lambda t}, & \text{if } t > 0, \end{cases}$$

and the functions $\lambda'_{ij}(t)$ are called *renewal (or failure) rates* provided that the derivatives $\lambda'_{ij}(t)$ exist.

From (68), we have

$$h_{ij}(u) = \frac{\exp[-\sum_{k \in E} \lambda_{ik}(u)]}{\exp[-\lambda_{ij}(u)]},$$

which, together with (67), gives

$$Q_{ij}(t) = \int_0^t \mathrm{e}^{-\lambda_i(u)} \lambda_{ij}(\mathrm{d}u), \quad (70)$$

where $\lambda_i(u) = \sum_{k \in E} \lambda_{ik}(u)$, $i \in E$.

The distribution functions of $\eta_i = \min_{k \in E} \tau_{ik}$, $i \in E$, are

$$H_i(t) = \mathbb{P}(\eta_i \leq t) = \begin{cases} 0, & \text{if } t \leq 0, \\ 1 - e^{-\lambda_i(t)}, & \text{if } t > 0. \end{cases} \tag{71}$$

The function $\lambda'_{ij}(t)$, $i \in E$, is the rate of leaving the state i. We note that $\lambda_i(\infty) = \infty$ and $\lambda_{ij}(0) = 0$ for all $i, j \in E$.

REMARK 4. Let $i \in E$. It is possible that, for some $j \in E$, $\tau_{ij} = \infty$ $\mathbb{P}$-a.s. For eliminating this we consider, for each $i \in E$, the set $E_i \subset E$ such that $\mathbb{P}(\tau_{ik} < \infty) > 0$ for all $k \in E_i$.

Consequently, $\eta_i = \min_{k \in E_i} \tau_{ik}$, $\lambda_i(t) = \sum_{k \in E_i} \lambda_{ik}(t)$.

Let us now define the MRP (J, S), with $S_0 = 0$. Then the counting process $(N(t), t \geq 0)$ associated to the point process $(S_n, n \geq 0)$, is defined, for each time $t \geq 0$, by

$$N(t) = \sup\{n : S_n \leq t\}.$$

Consider also the semi-Markov process $Z = (Z(t), t \in \mathbb{R}_+)$, defined by

$$Z(t) = J_{N(t)}. \tag{72}$$

Define also the transition function and the mean sojourn time in state i respectively

$$P_{ij}(t) = \mathbb{P}(Z(t) = j | Z(0) = i), \tag{73}$$

$$m_i = \int_0^\infty [1 - H_i(u)] \, du. \tag{74}$$

For $i, j, k \in E$, let $_k P_{ij}(t)$ denote the taboo transition probabilities, i.e.,

$$_k P_{ij}(t) = \mathbb{P}_i(Z(t) = j, \forall u \leq t, Z(u) \neq k). \tag{75}$$

Note that $_j P_{ij}(t) = \delta_{ij}[1 - H_j(t)]$. We also define in an obvious way the taboo distributions $_k G_{ij}(t)$.

The taboo functions defined above, verify the following equations:

$$P_{ij}(t) = {_k P_{ij}(t)} + \int_0^t P_{kj}(t-u) G_{ik}(du), \tag{76}$$

$$G_{ij}(t) = {_k G_{ij}(t)} + \int_0^t G_{kj}(t-u) {_j G_{ik}(du)}. \tag{77}$$

It is clear that for all $i, j \in E$, $Q_{ij}(t) \leq H_i(t)$ and thus by Radon–Nikodym theorem (see Section 3)

$$q_{ij}(x) := \frac{Q_{ij}(dx)}{H_i(dx)}$$

exists. Since

$$Q_{ij}(t) = \mathbb{P}(J_n = j, X_n \leq t | J_{n-1} = i)$$
$$= \int_0^t \mathbb{P}(J_n = j | J_{n-1} = i, X_n = x) H_i(\mathrm{d}x),$$

we have

$$q_{ij}(x) = \mathbb{P}(J_n = j | J_{n-1} = i, X_n = x) \quad \text{(a.s.)}.$$

Let $S_1^j, S_2^j, \ldots, S_n^j, \ldots$ be the recurrence times for a fixed state j. As the MRP has a stationary semi-Markov kernel and the strong Markov property, it is a renewal process. The r.v.s $S_{n+1}^j - S_n^j$, $n \geq 1$, are i.i.d. having a common distribution denoted by G_{jj}. The distribution of S_1^j is denoted by G_{ij} on $\{J_0 = i\}$. If j is the initial state, i.e., $\{J_0 = j\}$ (a.s.), then $S_1^j, S_{n+1}^j - S_n^j$, $n \geq 1$ are i.i.d. Let μ_{ij} be the first moment of G_{ij}.

The counting function of this renewal process is

$$N_j(t) = \sum_{n=0}^{\infty} \mathbf{1}_{\{J_n=j, S_n \leq t\}}$$

and

$$G_{ij}(t) := \mathbb{P}_i(S_1^j \leq t) = \mathbb{P}_i(N_j(t) > 0).$$

It is clear that, for every $t \geq 0$,

$$N(t) = \sum_j N_j(t).$$

The n-fold convolution of the semi-Markov kernel is

$$Q_{ij}^{(n)}(t) = \begin{cases} \sum_k \int_0^t Q_{ik}(\mathrm{d}u) Q_{kj}^{(n-1)}(t-u) = \sum_k \int_0^t Q_{ik}^{(n-1)}(\mathrm{d}u) Q_{kj}(t-u), \\ \quad \text{if } t > 0, \\ 0, \quad \text{if } t \leq 0, \end{cases}$$

and

$$Q_{ij}^{(0)}(t) = \delta_{ij} 1(t),$$
$$Q_{ij}^{(1)}(t) = Q_{ij}(t)$$

with $1(t) = 1$, if $t \geq 0$ and $= 0$, if $t < 0$. As in Section 2, we have

$$\mathbb{P}_i(J_n = j, S_n \leq t) = Q_{ij}^{(n)}(t) \tag{78}$$

and therefore a MRP is regular if and only if $\sum_j Q_{ij}^{(n)}(t) \to 0$, as $n \to \infty$, for all i.

For fixed states i and j, define the renewal function $\psi_{ij}(t) := \mathbb{E}_i[N_j(t)]$ of the renewal process $(S_n^j, n \geq 1)$ with counting function $N_j(t)$.

We have

$$\psi_{ij}(t) = \mathbb{E}_i[N_j(t)]$$

$$= \mathbb{E}_i\left[\sum_{n=0}^{\infty} \mathbf{1}_{\{J_n=j, S_n \leq t\}}\right]$$

$$= \sum_{n=0}^{\infty} \mathbb{P}_i(J_n = j, S_n \leq t)$$

$$= \sum_{n=0}^{\infty} Q_{ij}^{(n)}(t). \tag{79}$$

The matrix function $\psi(t) = (\psi_{ij}(t); i, j \in E)$ is called a *Markov renewal matrix*. We remind that the MRP is called *normal* if $\sum_j \psi_{ij}(t) < \infty$, for all $i \in E$ and all $t \geq 0$. It is obvious that a normal MRP is also regular.

Relation (79) in matrix form can be written as

$$\psi(t) = \sum_{n=0}^{\infty} Q^{(n)}(t). \tag{80}$$

It is easy to verify that the following useful equation holds

$$\psi_{ij}(t) = G_{ij}(t) + G_{ij} * \psi_{jj}(t). \tag{81}$$

7. Classification of states

Considering a normal semi-Markov process (or a MRP) over the countable state space E we will now give a classification of states where the part is played by the number of visits into states (see, e.g., Pyke and Schaufele, 1964; Çinlar, 1969a; Korolyuk and Turbin, 1976).

Let $(v_i, i \in E)$ be an invariant measure of $P = (p_{ij}, i, j \in E)$, i.e., $vP = v$.

DEFINITION 7.1. (1) States i and j are said to be *communicating* if $i = j$ or $G_{ij}(\infty)G_{ji}(\infty) > 0$. The communication is an equivalence relation.

(2) A state i is said to be *recurrent* if $G_{ii}(\infty) = 1$, otherwise it is *transient*.

(3) A recurrent state i is said to be *positive-recurrent* if $\mu_{ii} < \infty$ and *null-recurrent* if $\mu_{ii} = \infty$.

(4) State i is said to be *periodic* of period $c > 0$, if $G_{ii}(\cdot)$ is arithmetic, i.e., concentrated on $\{nc: n \in \mathbb{N}\}$. If not, it is called *aperiodic*.

DEFINITION 7.2. A MRP whose all states are:

(1) communicating is said to be *irreducible*;
(2) positive (or null) recurrent is said to be *positive (or null) recurrent*.

THEOREM 7.1. *Consider an irreducible positive recurrent MRP.*

(1) *For j fixed, the $(\mu_{ij}, i \in E)$ is the unique bounded solution of*

$$\mu_{ij} = m_i + \sum_{k \neq j} p_{ik} \mu_{kj}.$$

(2) *For all j, we have*

$$\mu_{jj} = \frac{1}{v_j} \sum_i v_i m_i.$$

PROOF. (1)

$$\begin{aligned}
\mathbb{E}_i[S_1^j] &= \mathbb{E}_i[S_1^j; J_1 = j] + \mathbb{E}_i[S_1^j; J_1 \neq j] \\
&= p_{ij} m_i + \sum_{k \neq j} \mathbb{E}_i[S_1^j; J_1 = k] \\
&= p_{ij} m_i + \sum_{k \neq j} \mathbb{E}_i[S_1 + \theta_{S_1} S_1^j | J_1 = k] p_{ik} \\
&= p_{ij} m_i + \sum_{k \neq j} (\mathbb{E}_i[S_1] + \mathbb{E}_k[S_1^j]) p_{ik} \\
&= m_i + \sum_{k \neq j} p_{ik} \mathbb{E}_k[S_1^j].
\end{aligned} \quad (82)$$

Consider now another bounded solution $(\lambda_{ij}, i \in E)$ and suppose that $p_{ij} > 0$. Then

$$\mu_{ij} - \lambda_{ij} = \sum_{k \neq j} p_{ik}(\mu_{kj} - \lambda_{kj}).$$

Let us denote $M = \sup_{k,l} |\mu_{kj} - \lambda_{kj}| < \infty$. We have

$$|\mu_{kj} - \lambda_{kj}| \leqslant \sum_{k \neq j} p_{ik} M = (1 - p_{ij}) M$$

from which we get

$$M \leqslant (1 - p_{ij}) M,$$

which is false, unless $M = 0$.

(2) From (1) we have

$$\sum_i v_i \mu_{ij} = \sum_i v_i m_i + \sum_i v_i \sum_{k \neq j} p_{ik} \mu_{kj}$$

$$= \sum_i v_i m_i + \sum_{k \neq j} \left(\sum_i v_i p_{ik} \right) \mu_{kj}$$

$$= \sum_i v_i m_i + \sum_{k \neq j} v_k \mu_{kj}$$

$$= \sum_i v_i m_i + \sum_k v_k \mu_{kj} - v_j \mu_{jj}. \tag{83}$$

This last equality proves the assertion. □

THEOREM 7.2. (1) *A MRP is irreducible if and only if its EMC is irreducible.*

(2) *A state i is recurrent (transient) for the MRP if and only if i is recurrent (transient) for the EMC.*

(3) *For an irreducible finite MRP, a state i is positive recurrent for the MRP if and only if i is recurrent for the EMC and if for all $j \in E$, $m_j < \infty$.*

(4) *If the EMC of an MRP is irreducible and recurrent, then all states are*

(a) *positive-recurrent if and only if $\sum_i v_i m_i < \infty$;*
(b) *null-recurrent if and only if $\sum_i v_i m_i = \infty$.*

Let μ_{ij} and μ_{ij}^* denote the mean hitting times of a state j, starting from the state i, of the SMP and the EMC respectively. When $j = i$, then μ_{ii} and μ_{ii}^* are the mean recurrence times of state i of the SMP and the EMC respectively. In other words, μ_{ij} is the mean of the distribution G_{ij}. And let η_{ij} denote the mean value of the distribution F_{ij}.

Consider the following set, for $n \in \mathbb{N}^*$,

$$\Delta_n^i = \{(i_0, i_1, \ldots, i_{n-1}, i_n) \in E^{n+1} \colon i_0 = i_n = i,\ i_k \neq i,\ 1 \leqslant k < n\}. \tag{84}$$

It is easy to verify that (see Korolyuk and Turbin, 1976):

$$\mu_{ii} = \sum_{n=1}^{\infty} \sum_{(i_0,i_1,\ldots,i_n) \in \Delta_n^i} \prod_{k=0}^{n-1} p_{i_k i_{k+1}} (\eta_{i_0 i_1} + \cdots + \eta_{i_{n-1} i_n}) \tag{85}$$

and

$$\mu_{ii}^* = \sum_{n=1}^{\infty} n \sum_{(i_0,i_1,\ldots,i_n) \in \Delta_n^i} \prod_{k=0}^{n-1} p_{i_k i_{k+1}}. \tag{86}$$

PROOF. (1) This follows by the equality of events: $\{S_1^j < \infty\} = \{\exists n\colon J_n = j\}$.

(2) The above equality is also verified on the event: $\{J_0 = i\}$ from which the result follows.

(3) From relations (85) and (86), we can write the following inequalities

$$\left(\min_{k,\ell \in E} \eta_{k\ell}\right)\mu_{ii}^* \leq \mu_{ii} \leq \left(\max_{k,\ell \in E} \eta_{k\ell}\right)\mu_{ii}^* \tag{87}$$

from which the conclusion follows.

(4) From the above Proposition 7.1(2) we have: $\mu_{jj} < \infty$ if and only if $\sum_i v_i m_i < \infty$. □

In the countable case the Markov renewal equation (MRE) (see Section 5) takes the form

$$\Theta_{ij}(t) = L_{ij}(t) + \sum_{k \in E} \int_0^t Q_{ik}(\mathrm{d}u)\Theta_{kj}(t-u), \tag{88}$$

where $L(t) = (L_{ij}(t); i, j \in E)$ is a known matrix valued function and $\Theta(t) = (\Theta_{ij}(t); i, j \in E)$ is an unknown matrix valued function. Both $L(t)$ and $\Theta(t)$ have their components are null on $(-\infty, 0)$.

It can be proved that, if the semi-Markov process is irreducible and positive recurrent, then the MRE (88) has a unique bounded solution of the form $\Theta(t) = \psi \star L(t)$ (see, e.g., Korolyuk and Turbin, 1976).

8. Asymptotic behavior

The following two propositions are straightforward applications of Blackwell renewal theorem and of the key renewal theorem from the Renewal theory (see, e.g., Feller, 1966). In fact, if $Z_0 = j$ (a.s.), then $(S_n^j, n \geq 0)$ is an ordinary renewal process, otherwise it is a delayed renewal process.

THEOREM 8.1 (Blackwell type theorem, Çinlar, 1969a). *If the MRP is recurrent, then, for all $i \in E$ and $c > 0$,*

$$\psi_{ii}(t) - \psi_{ii}(t-c) \to \frac{c}{\mu_{ii}}, \quad \text{as } t \to \infty.$$

THEOREM 8.2 (Key renewal type theorem, Çinlar, 1969a). *If i is an aperiodic recurrent state and h_i a direct Riemann integrable function, then*

$$\int_0^t \psi_{ji}(\mathrm{d}y)h_i(t-y) \to \frac{1}{\mu_{ii}} \int_0^\infty h_i(y)\,\mathrm{d}y, \quad \text{as } t \to \infty.$$

The next two propositions give the law of large numbers and the central limit theorem for the process $(X_n, n \geq 1)$ (see Grigorescu and Oprişan, 1976).

THEOREM 8.3 (Limit distribution, Howard, 1964). *If the MRP is positive recurrent, then*

$$\pi_j = \lim_{t \to \infty} P_{ij}(t) = \frac{v_j m_j}{vm},$$

where $v = (v_i)$ is an invariant measure of (J_n) and $m = (m_i)$.

THEOREM 8.4 (Law of large numbers).

$$\frac{1}{n} \sum_{i=1}^{n} X_i \to \mathbb{E}_\pi[X_1], \quad a.s., \text{ as } n \to \infty.$$

THEOREM 8.5 (Central limit theorem). *Under the following hypotheses: $\mathbb{E}_\pi[X_1] < \infty$ and $\sigma_i^2 = \mathbb{E}_i([Y_1^i - S_1^i \mathbb{E}_\alpha(X_1)]^2) < \infty$; where $Y_n^i = \sum_{j=S_{n-1}^i+1}^{S_n^i} X_j$, $S_0^i = 0, n = 1, 2, \ldots$, we have:*

$$\frac{\sqrt{n}}{\sqrt{\pi_i} \sigma_i} \left(\frac{X_1 + \cdots + X_n}{n} - \mathbb{E}_\pi[X_1] \right) \xrightarrow{\mathcal{L}} N(0, 1), \quad \text{as } n \to \infty.$$

THEOREM 8.6 (Çinlar, 1969a). *Consider an irreducible recurrent MRP and v a positive solution of $vP = v$. If h is a directly Riemann integrable function, then*

- *for the aperiodic case, we have, as $t \to \infty$*

$$\sum_{j \in E} \int_0^t \psi_{ij}(du) h_j(t-u) \to \frac{1}{vm} \int_0^\infty vh(u) \, du,$$

- *for the periodic case with period δ, we have*

$$\sum_{j \in E} \int_0^{x+n\delta} \psi_{ij}(du) h_j(t-u) \to \frac{\delta}{vm} \sum_{k=0}^{\infty} vh(x - \delta_{ij} + k\delta),$$

where δ_{ij} is the first jump point of the distribution G_{ij}.

Important results concerning the asymptotic behavior of a functional of a semi-Markov process can be found in (Pyke and Schaufele, 1964; Limnios and Oprişan, 1999).

9. Some recent approaches to semi-Markov processes

9.1. The (f, g)-processes

We shall present now a slight generalization of semi-Markov processes obtained by allowing more flexibility to the transition mechanism (see Hou, Liu and Zou, 1997; Iosifescu, 1999).

Consider a complete probability space $(\Omega, \mathcal{K}, \mathbb{P})$ and a random variable $\tau : \Omega \to \mathbb{R}_+$. For any $i, j \in \mathbb{N}^*$ consider a non-decreasing mapping $f(\cdot, i, j) : \mathbb{R}_+ \to \mathbb{R}_+$ and a measurable mapping $g(\cdot, i, j) : \mathbb{R}_+ \to \mathbb{R}_+$ such that:

(1) $f(0, i, j) = 0$, $i, j \in \mathbb{N}^*$;
(2) $\lim_{t \to \infty} S_i(t) = 1$ where $S_i(t) = \sum_{j \in \mathbb{N}^*} f(t, i, j)$, $t \in \mathbb{R}_+, i \in \mathbb{N}^*$;
(3) for any $i \in \mathbb{N}^*$ there exists $t_i > 0$ such that $S_i(t) < 1$ for all $t \in [0, t_i]$;
(4) $g(0, i, j) = \delta_{ij}$ (Kroneker's symbol), $i, j \in \mathbb{N}^*$;
(5) $\lim_{t \to \infty} g(t, i, j) = 0$, $i, j \in \mathbb{N}^*$;
(6) $\sum_{j \in \mathbb{N}^*} f(t, i, j) = 1 - S_i(t)$, $t \in \mathbb{R}_+, i \in \mathbb{N}^*$.

Note that it follows from (2) that setting $p_{ij} = \lim_{t \to \infty} f(t, i, j)$, $i, j \in \mathbb{N}^*$, the matrix $\mathbf{P} = (p_{ij})_{i, j \in \mathbb{N}^*}$ is a stochastic one.

We shall denote by $\mathbf{f} = (f(\cdot, i, j))_{i, j \in \mathbb{N}^*}$ and $\mathbf{g} = (g(\cdot, i, j))_{i, j \in \mathbb{N}^*}$ the matrix of functions satisfying properties (1)–(6) above.

DEFINITION 9.1. A right continuous $\mathbb{N}^*$-valued stochastic process $X = (X(t, \omega), 0 \leq t \leq \tau(\omega), \omega \in \Omega)$ is called an $(\mathbf{f}, \mathbf{g})$-*process* if there exists a sequence $(\tau_n)_{n \in \mathbb{N}}$ of stopping times for X satisfying $0 = \tau_0 < \tau_1 < \cdots$ and $\lim_{n \to \infty} \tau_n = \tau$, $\mathbb{P}$-a.s., for which the equations

$$\mathbb{P}\big(X(\tau_{n+1}) = j, \tau_{n+1} - \tau_n \leq t \mid X(\tau_m), 0 \leq m \leq n\big)$$
$$= \mathbb{P}\big(X(\tau_{n+1}) = j, \tau_{n+1} - \tau_n \leq t \mid X(\tau_n)\big) = f\big(t, X(\tau_n), j\big)$$

and

$$\mathbb{P}\big(X(\tau_n + t) = j, \tau_{n+1} - \tau_n > t \mid X(\tau_m), 0 \leq m \leq n\big)$$
$$= \mathbb{P}\big(X(\tau_n + t) = j, \tau_{n+1} - \tau_n > t \mid X(\tau_n)\big) = g\big(t, X(\tau_n), j\big)$$

hold $\mathbb{P}$-a.s. for all $t \in \mathbb{R}_+$ and $n, j \in \mathbb{N}^*$.

It follows from Definition 9.1 that $(X(\tau_n))_{n \in \mathbb{N}^*}$ is an $\mathbb{N}^*$-valued Markov chain with transition matrix $\mathbf{P}$. Moreover

$$p(t, i, j) = \mathbb{P}\big(X(t) = j \mid X(0) = i\big), \quad t \in \mathbb{R}_+, \ i, j \in \mathbb{N}^*,$$

is the conditional probability of the random event $(X(\tau_n + t) = j)$ given a jump to i occurs at time $\tau \in \mathbb{R}_+$.

Considering the first jump following τ we obtain the Markov renewal equation

$$p(t, i, j) = \sum_{k \in \mathbb{N}^*} \int_0^t f(du, i, k) \, p(t - u, k, j) + g(t, i, j) \tag{89}$$

for all $t \in \mathbb{R}_+$ and $n, j \in \mathbb{N}^*$.

REMARK 5. A semi-Markov process is a an $(\mathbf{f}, \mathbf{g})$-process where

$$g(t, i, j) = \delta_{ij}\left(1 - \sum_{j \in \mathbb{N}^*} f(t, i, j)\right), \quad t \in \mathbb{R}_+, \; i, j \in \mathbb{N}^*,$$

while $(\tau_n)_{n \in \mathbb{N}^*}$ is the very sequence of the jump times of the latter.

Beside, in the special case where X is the minimal Markov process associated with a conservative infinitesimal transition matrix and $(\tau_n)_{n \in \mathbb{N}^*}$ is the sequence of the jump times of the process, Eq. (89) are equivalent to Kolmogorov's system of backward differential equations.

Passing to Laplace transform and matrix notation, for $\lambda > 0$ and $i, j \in \mathbb{N}^*$, let us set

$$\pi_{ij}(\lambda) = \int_{\mathbb{R}_+} e^{\lambda t} p(t, i, j) \, dt,$$

$$\varphi_{ij}(\lambda) = \int_{\mathbb{R}_+} e^{\lambda t} f(dt, i, j),$$

$$\gamma_{ij}(\lambda) = \int_{\mathbb{R}_+} e^{\lambda t} g(t, i, j) \, dt$$

and

$$\mathbf{\Pi} = (\pi_{ij})_{i, j \in \mathbb{N}^*}, \qquad \mathbf{\Phi} = (\varphi_{ij})_{i, j \in \mathbb{N}^*}, \qquad \mathbf{\Gamma} = (\gamma_{ij})_{i, j \in \mathbb{N}^*}.$$

So, in terms of Laplace transforms, Eq. (89) becomes

$$\mathbf{\Pi} = \mathbf{\Phi} \mathbf{\Pi} + \mathbf{\Gamma}. \tag{90}$$

We give the following theorem concerning the existence and the uniqueness of the solutions of Eq. (89) whose proof can be found in (Iosifescu, 1999). We note that there exist similar results for a semi-Markov process or functionals of a semi-Markov process (see, e.g., Çinlar, 1969a; Limnios and Oprişan, 2000).

THEOREM 9.1. *Eq. (89) has a minimal solution whose matrix Laplace transform $\mathbf{\Pi}_{\min}$ is given by*

$$\mathbf{\Pi}_{\min} = (\mathbf{I} - \mathbf{\Phi})^{-1} \mathbf{\Gamma}.$$

The minimal solution is either stochastic or sub-stochastic. It is stochastic for all $t \in \mathbb{R}_+$ iff for some $\lambda > 0$ the only solution of the equation $\boldsymbol{\Phi}\boldsymbol{\xi} = \boldsymbol{\xi}$, where $\mathbf{0} \leqslant \boldsymbol{\xi} \leqslant \mathbf{1}$ (componentwise), is $\boldsymbol{\xi} = \mathbf{0}$.

COROLLARY 9.1. *The solution of Eq. (89) is unique iff the minimal solution is strictly stochastic.*

For other results, as asymptotic behavior, see (Iosifescu, 1999).

9.2. Non-homogeneous Markov renewal processes

There are different approaches of this concept (see, e.g., De Dominicis and Manca, 1984; Wajda, 1992) but we give here only some hints concerning non-homogeneous Markov renewal processes (or semi-Markov processes). So, following (De Dominicis and Manca, 1984) and (Janssen and Manca, 1998) with some modifications, we shall consider a probability space $(\Omega, \mathcal{K}, \mathbb{P})$ and the set $E = \{1, 2, \ldots, m\}$. Now we define the random variables $J_n : \Omega \to E$, representing the state of the nth transition, and $S_n : \Omega \to [0, \infty)$ representing the time of the nth transition. The associated non-homogeneous Markov kernel Q is defined by

$$Q_{ij}(s, t) = \mathbb{P}(J_{n+1} = j, S_{n+1} \leqslant t | J_n = i, S_n = s),$$
$$i, j \in E, s, t \in \mathbb{R}_+. \tag{91}$$

If

$$p_{ij}(s) = \lim_{t \to \infty} Q_{ij}(s, t) = \mathbb{P}(J_{n+1} = j, S_{n+1} \leqslant t | J_n = i, S_n = s), \quad i, j \in E,$$

then $P(s) = [p_{ij}(s)]$ is a non-homogeneous sub-Markov transition probability matrix for each $s \in \mathbb{R}_+$.

The conditional distribution of the times between jumps, given the left state and the moment of the last jump, are

$$H_i(s, t) = \sum_{j \neq i} Q_{ij}(s, t + s).$$

The non-homogeneous semi-Markov process can be defined in the same way as in the homogeneous case by setting $Z_t = J_{N(t)}$, $t \in \mathbb{R}_+$, where $N(t) = \max\{n : S_n \leqslant t\}$. Obviously we have

$$\mathbb{P}(Z_t = j) = \mathbb{P}(S_n \leqslant t < S_{n+1}, J_n = j), \quad j \in E.$$

If $\phi_{ij}(s, t) = \mathbb{P}(Z_t = j | Z_s = i)$, then the following Markov renewal type equation holds (see Janssen and Manca, 1998):

$$\phi_{ij}(s, t) = \delta_{ij}\bigl(1 - H_i(s, t)\bigr) + \sum_{\ell \in E} \int_0^\infty \phi_{\ell j}(\tau, t) Q_{i\ell}(s, \mathrm{d}\tau).$$

In (Janssen and Manca, 1998) numerical solutions of this equation are studied.

Another point of view for introducing a non-homogeneous semi-Markov process on an arbitrary state space $(E, \mathcal{E})$ is to consider the non-homogeneous Markov kernel. So, a function $Q(n, x, A, t), n \in \mathbb{N}^*, x \in E, t \in \mathbb{R}_+, A \in \mathcal{E}$, will be called a *non-homogeneous semi-Markov kernel* on $(E, \mathcal{E})$ if

(1) $Q(n, x, A, \cdot)$ is a mass function, for all $n \in \mathbb{N}^*, x \in A, A \in \mathcal{E}$;
(2) $Q(n, \cdot, \cdot, t), n \in \mathbb{N}^*$, is a non-homogeneous sub-Markov kernel on $(E, \mathcal{E})$, for all $t \in \mathbb{R}_+$;
(3) $p(n, \cdot, \cdot) = Q(n, \cdot, \cdot, \infty)$ is a non-homogeneous Markov kernel on $(E, \mathcal{E})$.

By usual considerations, similar to those presented in Section 1, we can prove the existence of a non-homogeneous Markov chain $(J_n, S_n), n \in \mathbb{N}$, on the state space $(E, \mathcal{E})$ such that

$$\mathbb{P}(J_{n+1} \in A, S_{n+1} \leqslant t | J_n = x, S_n = s) = Q(n, x, t - s, A),$$
$$n \in \mathbb{N}, x \in E, s, t \in \mathbb{R}_+, A \in \mathcal{E}.$$

The imbedded non-homogeneous Markov chain $(J_n, n \in \mathbb{N})$ has the probability transition functions

$$p(n, x, A) = \mathbb{P}(J_{n+1} \in A | J_n = x), \quad n \in \mathbb{N}, x \in E, A \in \mathcal{E}.$$

The intervals between jumps are $X_{n+1} = S_{n+1} - S_n, n \in \mathbb{N}$ and we have

$$\mathbb{P}(J_{n+1} \in A, S_{n+1} \leqslant t | J_n = x, S_n = s) = Q(n, x, t, A),$$
$$n \in \mathbb{N}, x \in E, s, t \in \mathbb{R}_+, A \in \mathcal{E}.$$

The processes (J_n, S_n) and $(J_n, X_n), n \in \mathbb{N}$, are called *non-homogeneous Markov renewal process* and *non-homogeneous J–X process*, respectively. The non-homogeneous semi-Markov process can be defined in the usual way.

9.3. Semi-Markov random walks

A semi-Markov random walk (SMRW) is determined by superposition of two independent renewal processes (see Bratiychuk, 1998; Korolyuk, 1999). More precisely, let $(X_n^{(i)}), (\eta_n^{(i)}), i = 1, 2, n = 1, 2, \ldots$, be four sequences of non-negative independent r.v. having the same distribution within each sequence. The random values $\eta_n^{(i)}$ are supposed to be integer-valued. The time intervals $(X_n^{(i)})$ have the same distribution function $F_i(x) = \mathbb{P}(X_n^{(i)} \leqslant t), i = 1, 2$, and the renewal processes generated they are given by

$$S_n^{(i)} = \sum_{k=1}^{n} X_n^{(i)} \quad \text{and} \quad N_i(t) = \max\{n \colon S_n^{(i)} \leqslant t\}, \quad t \geqslant 0, \ i = 1, 2.$$

The superposition of these two renewal processes is given by

$$N(t) = N_1(t) + N_2(t) \tag{92}$$

and the SMRW is defined by

$$\xi(t) = \sum_{k=1}^{N_1(t)} \eta_n^{(1)} - \sum_{k=1}^{N_2(t)} \eta_n^{(2)}. \tag{93}$$

Such walks constitute a quite natural model of $GI/GI/1$-type queueing systems if $(X_n^{(1)})$ are seen as inter-arrival times of the customer, with $\eta_n^{(1)}$ of them arriving at the moment $S_n^{(1)}$. The single server accepts the customers in groups of size $\eta_n^{(2)}$ and the nth group takes time $X_n^{(2)}$ to be served.

The superposition of two renewal processes (92) is completely determined by the Markov Renewal Process (J_n, S_n), $n \in \mathbb{N}$, which is given by the following stochastic relations (see Korolyuk, 1999)

$$J_{n+1} = (J_n - X_{n+1}^{(1)}) \mathbb{1}_{\{J_n > 0\}} + (J_n + X_{n+1}^{(2)}) \mathbb{1}_{\{J_n \leq 0\}},$$
$$S_n = S_n + (S_n \wedge X_{n+1}^{(1)}) \mathbb{1}_{\{J_n > 0\}} + (-S_n \wedge X_{n+1}^{(2)}) \mathbb{1}_{\{J_n \leq 0\}}.$$

The process (92) is a counting process of the MRP (J_n, S_n):

$$N(t) = \max\{n: S_n \leq t\}, \quad t \geq 0,$$

and the imbedded SMRW is defined by $\xi_n = \xi(S_n)$, $n \in \mathbb{N}$. The SMRW can be written $\xi(t) = \xi_{N(t)}$, $t \in \mathbb{R}_+$.

If we set $G_i(x) = \mathbb{P}(\eta_n^{(i)} \leq x)$, $x \in \mathbb{R}_+$, $i = 1, 2$, then the process (ξ_n, J_n), $n \in \mathbb{N}$, is a Markov chain with the transition probabilities

$$\mathbb{P}(\xi_{n+1} \in u + dv, J_{n+1} \in dy | \xi_n = u, J_n = x)$$
$$= \begin{cases} G_1(dv) F_1(x - dy), & x > 0, y > 0, \\ G_2(dv) F_1(x - dy), & x > 0, y < 0, \\ G_1(dv) F_2(dy - x), & x < 0, y > 0, \\ G_2(dv) F_2(dy - x), & x < 0, y < 0. \end{cases}$$

For the study of queueing systems of the types $GI/GI/1$ and $E/G/1/N$ see (Bratiychuk, 1998).

9.4. Korolyuk's state space merging method

Merging (also called lumping or aggregation or consolidation) methods were developed for Markov chains for studying conditions under which, given a Markov chain, $(J_n)_{n \in \mathbb{N}}$ say, the process $(f(J_n))_{n \in \mathbb{N}}$ is also a Markov chain. The application f is defined on the state space E of the initial Markov chain, with the range set V, which is the state space of $f(J_n)$. Generally, the set V is simpler than the set E, for example a finite set. Thus the study of the Markov chain $(f(J_n))_{n \in \mathbb{N}}$ is easier.

Here, the method proposed by Korolyuk and his co-authors is a more general merging method concerning semi-Markov processes in an asymptotic scheme (see Korolyuk and Shwitshchuk, 1995; Korolyuk and Turbin, 1993).

Let $(J_n, S_n)_{n \in \mathbb{N}}$ be a MRP on the state space $(E, \mathcal{E})$ and semi-Markov kernel $Q(x, A, t)$, $x \in E$, $A \in \mathcal{E}$, $t \in \mathbb{R}_+$. Consider now a partition of the initial state space E, say $(E_v, v \in V)$, i.e.,

$$E = \bigcup_{v \in V} E_v, \quad E_v \cap E_{v'} = \emptyset, \; v \neq v' \text{ and } E_v \in \mathcal{E}. \tag{94}$$

On the same space, consider an MRP, $(J_n^0, S_n^0)_{n \in \mathbb{N}}$ say, such that its EMC is uniformly ergodic on each subset E_v of the partition (94), with the invariant distribution $\nu_v(dx)$.

In fact, in the merging method we consider the following three MRPs:

- the initial MRP $(J_n, S_n)_{n \in \mathbb{N}}$ on the state space $(E, \mathcal{E})$;
- the supporting MRP $(J_n^0, S_n^0)_{n \in \mathbb{N}}$ on the state space $(E, \mathcal{E})$;
- the merged MRP $(\widehat{J}_n, \widehat{S}_n)_{n \in \mathbb{N}}$ on the state space $(V, \mathcal{V})$.

We assume that between the initial and the supporting MRP there is a "continuity" relation. The goal here is to study the initial MRP by means of the merged one which is close to it in some sense and in general much more simpler. The supporting ergodic MRP is an intermediate tool.

The heuristic state space merging algorithm is as follows:

The merged MRP is defined by the semi-Markov kernel

$$\widehat{Q}(v, B, t) = \widehat{P}(v, B)\left(1 - e^{-\Lambda(v)t}\right),$$

$v \in V$, $B \in \mathcal{V}$, where the transition probabilities of EMC are defined by

$$\widehat{P}(x, B) = \int_{E_v} \nu_v(dx) P(x, E_B),$$

where $E_B := \bigcup_{k \in B} E_k$ and $\Lambda(v)$ is the hazard rate of the sojourn time in state $v \in V$ of the merged process or, equivalently, the sojourn time of the initial process in the subset E_v

$$\Lambda(v) = \frac{q(v)}{M(v)}$$

with

$$q(v) := 1 - \int_{E_v} \nu_v(dx) P(x, E_v),$$

$$M(v) := \int_{E_v} \nu_v(dx) m(x),$$

and

$$m(x) := \int_0^\infty \overline{H}(x, dt).$$

The foundation of this algorithm is asymptotic. The initial MRP is considered as a perturbation of the supporting ergodic MRP. We will present here the general problem and theorem without proof. The interested reader can find a detailed proof and explanations in (Korolyuk and Turbin, 1976, 1982; Korolyuk and Shwitshchuk, 1995; Korolyuk, 1999).

Consider a family of MRPs $(J_n^\varepsilon, S_n^\varepsilon)_{n \in \mathbb{N}}$, $\varepsilon > 0$, with semi-Markov kernels

$$Q^\varepsilon(x, A, t) = P^\varepsilon(x, A) H(x, t)$$

and

$$P^\varepsilon(x, A) = P(x, A) - \varepsilon P_1(x, A),$$

where the transition kernel $P(x, A)$ of the non-perturbed supporting EMC is supposed to be uniformly ergodic and coherent with the partition (94), i.e.,

$$P(x, E_v) = \mathbf{1}_{E_v}(x), \quad v \in V.$$

And $P_1(x, A)$ is the perturbation kernel.

Consider the Markov processes defined as follows.

The counting process $N^\varepsilon(t)$ is defined by:

$$N^\varepsilon(t) := \sup\{n > 0: S_n^\varepsilon \leq t\}.$$

The initial semi-Markov process is:

$$Z^\varepsilon(t) := J_{N(t)}^\varepsilon$$

and the merged semi-Markov process is:

$$\widehat{Z}^\varepsilon(t) := v(Z^\varepsilon(t)).$$

Without loss of generality, we can suppose in the sequel that: $Q(x, B, t) = P(x, B) H(x, t)$.

The basic assumptions are the followings:

(1) The state space of the initial MRP is partitioned as in (94).
(2) The supporting EMC (J_n^0) has transition probabilities $P(x, B)$, $x \in E$, $B \in \mathcal{E}$. This transition probabilities are coherent with the above partition, i.e., $P(x, E_v) = \mathbf{1}_{E_v}(x)$.

(3) The supporting EMC (J_n^0) is uniformly ergodic in each class E_v, $v \in V$ and has stationary distributions $\nu_v(dx)$, $v \in V$.
(4) The stationary escape probabilities, $P^\varepsilon(x, B)$, $x \in E$, $B \in \mathcal{E}$, of the original EMC, are positive and sufficiently small, i.e.,

$$q_v := \int_{E_v} \nu_v(dx) P^\varepsilon(x, E \setminus E_v) > 0, \quad v \in V.$$

(5) The stationary lifetimes in the classes of states are uniformly bounded, i.e.,

$$0 < A \leqslant m_v := \int_{E_v} \nu_v(dx) m(x) \leqslant B < +\infty.$$

(6) We have, for all $x, y \in E$

$$\overline{H}(x, t) = \exp\left(-\int_0^t \lambda(x, u)\, du\right);$$

where the intensities $\lambda(x, u)$ are uniformly bounded on x, i.e.,

$$\sup_{x \in E} \lambda(x, u) \leqslant \lambda(u) < \infty, \quad \text{for all } u \in \mathbb{R}_+.$$

The last condition implies compactness of the processes $Z^\varepsilon(t/\varepsilon)$.

THEOREM 9.2 (Korolyuk and Turbin, 1976). *Under the above conditions* (1)–(6), *the merged process* $\widehat{Z}^\varepsilon(t) := v(Z^\varepsilon(t/\varepsilon))$ *converges, as* $\varepsilon \to 0$, *to the jump Markov process* $\widehat{Z}(t)$ *in the sense of convergence of finite-dimensional distributions.*

PROOF. See (Korolyuk, 1999). □

9.5. *K-dependent semi-Markov processes*

In the "standard" MRP the sojourn time in a state depends on the next state to be visited. In the generalization we are going to present the sojourn time in a state depends not only on the next state but also on the last K visited states.

Let $(\Omega, \mathcal{K}, \mathbb{P})$ be a probability space and consider the stochastic process $J_{-K+1}, \ldots, J_{-1}, J_0, J_1, \ldots$ with a countable state space E, defined on it; here $K \in \mathbb{N}^*$ is given. Moreover, we consider a non-decreasing sequence of real non-negative random variables $S_{-K+1} \leqslant \cdots \leqslant S_{-1} \leqslant S_0 \equiv 0 \leqslant S_1 \leqslant \cdots$ defined on $(\Omega, \mathcal{K}, \mathbb{P})$.

DEFINITION 9.2. The process $((J_{n-K+1}, J_{n-K+2}, \ldots, J_n), S_n)$, $n \in \mathbb{N}$, is called a *K-dependent MRP* (*K-MRP*) if, for all $j \in E$, $n \in \mathbb{N}$, and $t \in \mathbb{R}_+$, we have

$$\mathbb{P}(J_{n+1} = j, S_{n+1} - S_n \leqslant t | (J_q, S_q), q \leqslant n)$$
$$= \mathbb{P}(J_{n+1} = j, S_{n+1} - S_n \leqslant t | J_{n-K+1}, \ldots, J_n). \tag{95}$$

Obviously, an 1-dependent MRP is an ordinary MRP. The K-dependent semi-Markov kernel is defined by

$$Q_{i_1,i_2,\ldots,i_K,i_{K+1}}(t)$$
$$= \mathbb{P}(J_{n+1} = i_{K+1}, S_{n+1} - S_n \leqslant t | J_{n-K+1} = i_1, \ldots, J_n = i_K), \qquad (96)$$

$i_1, i_2, \ldots, i_K, i_{K+1} \in E$ and $Q_{i_1,i_2\ldots,i_K,i_{K+1}}(\cdot) \equiv 0$ if $i_m = i_r$, $1 \leqslant m, r \leqslant K+1$, $m \neq r$.

From (95) and (96), we see that $(J_n, n \in \mathbb{N})$ is a K-dependent Markov chain with transition probabilities

$$P_{i_1,i_2,\ldots,i_K,i_{K+1}} = Q_{i_1,i_2,\ldots,i_K,i_{K+1}}(\infty)$$
$$= \mathbb{P}(J_{n+1} = i_{K+1} | J_{n-K+1} = i_1, \ldots, J_n = i_K). \qquad (97)$$

The conditional distribution of the sojourn time in the state, given the next visited state and the last $K - 1$ visited states, is

$$F_{i_1,i_2,\ldots,i_{K-1},i,j}(t)$$
$$= \mathbb{P}(S_{n+1} - S_n \leqslant t | J_{n-K+1} = i_1, \ldots, J_{n-1} = i_{K-1}, J_n = i, J_{n+1} = j),$$
$$n \in \mathbb{N}, \qquad (98)$$

and we have

$$Q_{i_1,i_2,\ldots,i_K,i_{K+1}}(t) = P_{i_1,i_2,\ldots,i_K,i_{K+1}} F_{i_1,i_2,\ldots,i_K,i_{K+1}}(t). \qquad (99)$$

The unconditional distribution function of the sojourn time in state i_K is

$$H_{i_1,i_2,\ldots,i_K}(t) = \sum_{j \in E} Q_{i_1,i_2,\ldots,i_K,j}(t). \qquad (100)$$

Let us define the nth fold convolution of the semi-Markov kernel Q, for $i_1, i_2, \ldots, i_K$ pair wise different by

$$Q^{(0)}_{i_1,i_2,\ldots,i_K,i_{K+1}}(t) = \mathbb{1}_{\{i_K = i_{K+1}, t \geqslant 0\}}(t), \qquad (101)$$

$$Q^{(1)}_{i_1,i_2,\ldots,i_K,i_{K+1}}(t) = Q_{i_1,i_2,\ldots,i_K,i_{K+1}}(t), \qquad (102)$$

$$Q^{(n)}_{i_1,i_2,\ldots,i_K,i_{K+1}}(t) = \sum_{r \in E} \int_0^t Q_{i_1,i_2,\ldots,i_K,r}(ds) Q^{(n-1)}_{i_1,i_2,\ldots,i_K,r,i_{K+1}}(t-s),$$
$$n \geqslant 2. \qquad (103)$$

And we have

$$Q^{(n)}_{i_1,i_2,\ldots,i_K,j}(t) = \mathbb{P}(J_n = j, S_n \leqslant t | J_{-K+1} = i_1, \ldots, J_0 = i_K). \qquad (104)$$

Let $N_j(t)$, $j \in E$, be the number of visits to the state j by process J_n, i.e.,

$$N_j(t) = \sum_{n \in \mathbb{N}} \mathbb{1}_{\{J_n = j, S_n \leq t\}}. \tag{105}$$

We use the notation

$$\mathbb{P}_{i_1, i_2, \ldots, i_K}(\cdot) = \mathbb{P}(\cdot | J_{-K+1} = i_1, \ldots, J_0 = i_K)$$

$$\mathbb{E}_{i_1, i_2, \ldots, i_K}(\cdot) = \mathbb{E}(\cdot | J_{-K+1} = i_1, \ldots, J_0 = i_K)$$

and define

$$\psi_{i_1, i_2, \ldots, i_K, j}(t) = \mathbb{E}_{i_1, i_2, \ldots, i_K}[N_j(t)].$$

From (104) and (105) we can easily deduce that

$$\psi_{i_1, i_2, \ldots, i_K, j}(t) = \sum_{n \in \mathbb{N}} Q^{(n)}_{i_1, i_2, \ldots, i_K, j}(t). \tag{106}$$

The number of jumps in the interval $(0, t]$ is defined by $N(t) = \sup\{n : S_n \leq t\}$ and the E^K-valued process $(Z_t, t \in \mathbb{R}_+)$ defined by

$$Z_t = (Z_t^1, \ldots, Z_t^K) = (J_{N(t)-K+1}, \ldots, J_{N(t)})$$

will be called a *K-dependent semi-Markov process*.

If we set

$$P_{i_1, i_2, \ldots, i_K; j_1, j_2, \ldots, j_K}(t) = \mathbb{P}_{i_1, i_2, \ldots, i_K}[Z_t = (j_1, j_2, \ldots, j_K)]$$

then the following Markov renewal equation type is verified:

$$P_{i_1, i_2, \ldots, i_K; j_1, j_2, \ldots, j_K}(t) = \mathbb{1}_{\{i_1 = j_1, \ldots, i_K = j_K\}}[1 - H_{i_1, i_2, \ldots, i_K}(t)]$$
$$+ \sum_{r \in E} \int_0^t Q_{i_1, i_2, \ldots, i_K, r}(\mathrm{d}s) \, P_{i_2, \ldots, i_K, r; j_1, j_2, \ldots, j_K}(t-s).$$

10. Reliability modeling and estimation

10.1. Reliability related indicators modeling

Let E_0 and E_1 be the sets of up (successful) states and down (failure) states respectively. The *availability* of the system at time t is the probability that the system is in an up state in time t, i.e.,

$$A(t) = \sum_{j \in E_0} P_j(t) = \alpha(i) \sum_{j \in E_0} P_{ij}(t),$$

where $(\alpha(i), i \in E)$ is the initial distribution of the process (or of the EMC (J_n)). It can be obtain in the discrete case by solving (55).

Now, let $R_i(t)$ be the conditional probability that the first failure does not occurs before to time t, given that the process started from the state i. The vector $(R_i(t), i \in E)$ verifies the Markov renewal equation (see Section 4 and Iosifescu et al., 1984; Limnios and Oprişan, 1997b)

$$R_i(t) - \sum_{j \in E_0} \int_0^t R_j(t-s) Q_{ij}(ds) = 1 - H_i(t), \quad i \in E_0,$$

and the *reliability* of the system is given by

$$R(t) = \sum_{j \in E_0} \alpha(i) R_i(t).$$

In the same way, the maintainability vector $(M_i(t), i \in E_1)$, verifies the renewal equation

$$M_i(t) - \sum_{j \in E_0} \int_0^t M_j(t-s) Q_{ij}(ds) = 1 - H_i(t), \quad i \in E_1,$$

and the *maintainability* of the system is given by

$$M(t) = \sum_{j \in E_1} \alpha(i) M_i(t).$$

If μ_i is the conditional mean value of the operation period up to the first failure (i.e., up to the first entering into E_1), given that the process started from the state $i \in E_0$ at $t = 0$

$$\mu_i - \sum_{j \in E_1} p_{ij} \mu_j = \int_0^t [1 - H_i(t)] dt, \quad i \in E_0,$$

which is an algebraic system. The *mean time to failure* (MTTF), is given by

$$MTTF = \sum_{j \in E_1} \alpha(i) \mu_i(t).$$

The conditional mean value of time within the interval $[0, t]$ spent in the state j, given that the process started from the state i at $t = 0$, denoted by $U_{ij}(t)$ verifies the MRE (see, e.g., Iosifescu et al., 1984)

$$U_{ij}(t) - \sum_{k \in E} U_{kj}(t-s) Q_{ik}(ds) = \delta_{ij} \left[1 - H_i(t) + \int_0^t s\, H_i(ds) \right], \quad j \in E.$$

And the total successful time within the interval $[0, t]$ is

$$s(t) = \sum_{i \in E} \sum_{j \in E_0} U_{ij}(t).$$

The average value of the number of visits to the state j, within the interval $[0, t]$ is

$$\mathbb{E}[N_j(t)] = \sum_{i \in E} \psi_{ij}(t),$$

where $(\psi_{ij}(t), i, j \in E)$ is the Markov renewal matrix (see Section 6).

For a more general setting of reliability modeling see (Limnios and Oprişan, 2001).

10.2. Reliability estimation

We consider an observation of the system in the time interval $[0, t]$. The problem is to construct estimators for reliability function. For this, we use an estimator of the semi-Markov kernel and then we obtain estimator for reliability via the equation giving these quantity as function of the semi-Markov kernel.

The observation of a sample path, in the time interval $[0, t]$, of a semi-Markov process may be described as

$$\mathcal{H}_t = \{J_0, J_1, \ldots, J_{N(t)}, X_1, \ldots, X_{N(t)}\}. \tag{107}$$

Let us define the following empirical estimator of the semi-Markov kernel $Q_{ij}(t)$

$$\widehat{Q}_{ij}(x, t) := \frac{1}{N_i(t)} \sum_{k=1}^{N(t)} \mathbf{1}_{\{J_{k-1}=i, J_k=j, X_k \leq x\}}. \tag{108}$$

From this definition we obtain:

$$\widehat{Q}_{ij}(x, t) = \hat{p}_{ij}(t) \widehat{F}_{ij}(x, t), \tag{109}$$

where:

$$\hat{p}_{ij}(t) := \frac{N_{ij}(t)}{N_i(t)}, \tag{110}$$

$$\widehat{F}_{ij}(x, t) := \frac{1}{N_{ij}(t)} \sum_{k=1}^{N(t)} \mathbf{1}_{\{J_{k-1}=i, J_k=j, X_k \leq x\}} \tag{111}$$

are estimator of the transitions probabilities and state transition functions.

Define also

$$\widehat{H}_i(x, t) = \sum_{j \in E} \widehat{Q}_{ij}(x, t).$$

Then, the reliability estimator is defined by:

$$\widehat{R}(x,t) = \alpha_0 \big(I - \widehat{Q}_{00}(x,t)\big)^{(-1)} \star \big(I - \widehat{H}_0(x,t)\big)\mathbf{1}_r. \tag{112}$$

The following results give uniform strong consistency and asymptotic normality of the above estimator.

THEOREM 10.1 (Ouhbi and Limnios, 1999). *The reliability estimator obtained by the above relation is uniformly strongly consistent, i.e., for any fixed $L \in \mathbb{R}_+$,*

$$\sup_{0 \leqslant x < L} \big|\widehat{R}(x,t) - R(x)\big| \to 0, \quad (a.s.) \ t \to \infty.$$

THEOREM 10.2 (Ouhbi and Limnios, 1999). *The above reliability estimator is asymptotically normal, i.e., for all $x > 0$,*

$$t^{1/2}\big(\widehat{R}(x,t) - R(x)\big) \xrightarrow{d} N\big(0, \sigma^2(x)\big), \quad t \to \infty,$$

with

$$\sigma^2(x) = \sum_{i \in E_0} \sum_{j \in E} \mu_{ii} \left\{ \left(B_{ij}^0 \mathbf{1}_{(j \in E_0)} - \sum_{r \in E_0} \alpha(r)\psi_{ri}^0 \right)^2 * Q_{ij}(x) \right.$$

$$\left. - \left[\left(B_{ij}^0 \mathbf{1}_{(j \in E_0)} - \sum_{r \in E_0} \alpha(r)\psi_{ri}^0 \right) * Q_{ij}(x) \right]^2 \right\}$$

and

$$B_{ij} = \sum_E \sum_{E_0} \alpha(i) B_{nijk} * \big(I - \mathrm{diag}(Q(x)\mathbf{1}_{kk})\big),$$

$$B_{irkj}(x) = \sum_{n \geqslant 1} \sum_{\ell=1}^{n} Q_{ir}^{(\ell-1)} Q_{kj}^{(n-\ell)}(x).$$

For more detailed results on statistical estimation of reliability, see Ouhbi and Limnios (1997, 1999).

References

Billingsley, P. (1968). *Convergence of Probability Measures*. Wiley, New York.
Blumenthal, R. M. and R. K. Getoor (1968). *Markov Processes and Potential Theory*. Academic Press, New York.
Bratiychuk, M. S. (1998). Semi-Markov walks in queuing theory. In *Proceedings of the 2nd International Symposium on Semi-Markov Models: Theory and Applications*. Université de Technologie de Compiègne, December 9–11.

Breiman, L. (1968). *Probability*. Addison-Wesley, Reading, MA.

Çinlar, E. (1969a). Markov renewal theory. *Adv. Appl. Probab.* **1**, 123–187.

Çinlar, E. (1969b). On semi-Markov processes on arbitrary spaces. *Proc. Cambridge Philos. Soc.* **66**, 381–392.

De Dominicis, R. and R. Manca (1984). An algorithmic approach to non-homogeneous semi-Markov processes. *Comm. Statist. Simulation Comput.* **13**, 823–838.

Doob, J. L. (1953). *Stochastic Processes*. Wiley, New York.

Duma, I. and Gh. Oprişan (1983). Semi-Markov models for the analysis of channel coding. In *Progress in Electronics and Computer Science*, pp. 97–105. Ed. Acad., Bucharest (in Romanian).

Feller, W. (1966). *An Introduction to Probability Theory and its Applications*, Vols. 1 and 2. Wiley, New York.

Gihman, I. I. and A. B. Skorohod (1974). *Theory of Stochastic Processes*, Vols. 1, 2 and 3. Springer, Berlin.

Grigorescu, S. and Gh. Oprişan (1976). Limit theorems for J–X processes with a general state space. *Z. Wahrsch. Verv. Gebiete* **35**, 65–73.

Hou, Z., Z. Liu and J. Zou (1997). QNQL processes-(H,Q)-processes and their applications. *Chinese Sci. Bull.* **42**(11), 881–886.

Howard, R. A. (1964). Research in semi-Markovian decision structures. *J. Oper. Res. Soc. Japan* **6**(4), 114–124.

Iosifescu, M., S. Grigorescu, Gh. Oprişan and Gh. Popescu (1984). *Elements of Stochastic Modelling*. Ed. Tehnica, Bucharest (in Romanian).

Iosifescu, M. (1999). A Generalization of semi-Markov processes. In *Semi-Markov Models and Applications*. (Eds. J. Janssen and N. Limnios). Kluwer Academic Publishers, Boston.

Janssen, J. and N. Limnios (Eds.) (1999). *Semi-Markov Models and Applications*. Kluwer Academic Publishers, Boston.

Janssen, J. and R. Manca (1998). Numerical solution of non-homogeneous semi-Markov processes. In *Proceedings of the 2nd International Symposium on Semi-Markov Models: Theory and Applications*. Université de Technologie de Compiègne, December 9–11.

Korolyuk, V. S. (1999). Semi-Markov random walks. In *Semi-Markov Models and Applications*. (Eds. J. Janssen and N. Limnios). Kluwer Academic Publishers, Boston.

Korolyuk, V. S. and N. Limnios (2002). Poisson approximation of homogeneous stochastic additive functionals with semi-Markov switching. *Theory Probab. Math. Statist.* **64**, 75–84.

Korolyuk, V. S. and A. Shwitshchuk (1995). *Random Evolution for Semi-Markov Systems*. Kluwer Academic Publishers, Singapore.

Korolyuk, V. S. and A. F. Turbin (1976). *Semi-Markov Processes and their Applications*. Naukova Dumka, Kiev (in Russian).

Korolyuk, V. S. and A. F. Turbin (1982). *Markov Renewal Processes in Problems of Systems Reliability*. Naukova Dumka, Kiev (in Russian).

Korolyuk, V. S. and A. F. Turbin (1993). *Decomposition of Large Scale Systems*. Kluwer Academic Publishers, Singapore.

Kulkarni, V. G. (1995). *Modeling and Analysis of Stochastic Systems*. Chapman & Hall.

Lamperti, J. (1977). *Stochastic Processes*. Springer-Verlag, Berlin.

Lévy, P. (1954). Processus semi-markoviens. In *Proc. Int. Cong. Math.* (Amsterdam), pp. 416–426.

Limnios, N. and G. Oprişan (1997a). A general framework for reliability and performability analysis of semi-Markov systems. In *8th Intern. Conf. ASMDA*, Anacapri (Napoli), Italy.

Limnios, N. and G. Oprişan (1997b). Semi-Markov process to regard of their applications. *World Energy System J.* **1**(1), 64–75.

Limnios, N. and G. Oprişan (1999). The invariance principle for an additive functional of a semi-Markov process. *Rev. Roumaine Math. Pures Appl.* **44**(1), 75–83.

Limnios, N. and Gh. Oprişan (2000). An unified approach for reliability and performability, *Appl. Stochastic Models in Business and Industry* **15**(4), 353–368.

Limnios, N. and Gh. Oprişan (2001). *Semi-Markov Processes and Reliability*. Birkhäuser, Boston.

Moore, E. and R. Pyke (1968). Estimation of the transition distribution of a Markov renewal process. *Ann. Inst. Statist. Math.* **20**, 411–424.

O'Brien, G. L. (1974). Limit theorems for sums of chain-dependent processes. *J. Appl. Probab.* **11**, 582–587.

Onicescu, O., Gh. Oprişan and Gh. Popescu (1983). Renewal processes with complete connections. *Rev. Roumaine Math. Pures Appl.* **28**, 985–998.

Oprişan, Gh. (1976). On the J–X processes. *Rev. Roumaine Math. Pures Appl.* **21**, 717–724.

Osaki, S. (1985). *Stochastic System Reliability Modeling*. World Scientific, Singapore.

Osmann, C. and G. Haßlinger (1998). Modeling and analysis of channel coding in the presence of error bursts. In *Proceedings of the 2nd International Symposium on Semi-Markov Models: Theory and Applications*. Université de Technologie de Compiègne, December 9–11.

Ouhbi, B. and N. Limnios (1997). Estimation of kernels, availability and reliability functions of semi-Markov systems. In *Statistical and Probabilistic Models in Reliability*, pp. 113–124 (Eds. D. C. Ionescu and N. Limnios). Birkhäuser, Boston.

Ouhbi, B. and N. Limnios (1999). Nonparametric estimation for semi-Markov processes based on their hazard rate functions. *Statist. Inference for Stochastic Process.* **2**(2), 151–173.

Pyke, R. (1961a). Markov renewal processes: definitions and preliminary properties. *Ann. Math. Statist.* **32**, 1231–1242.

Pyke, R. (1961b). Markov renewal processes with finitely many states. *Ann. Math. Statist.* **32**, 1243–1259.

Pyke, R. and R. Schaufele (1964). Limit theorems for Markov renewal processes. *Ann. Math. Statist.* **35**, 1746–1764.

Pyke, R. and R. Schaufele (1967). The existence and uniqueness of stationary measures for Markov renewal processes. *Ann. Statist.* **37**, 1439–1462.

Resnick, S. (1994). *Adventures in Stochastic Processes*. Birkhäuser, Boston.

Silvestrov, D. S. (1980). *Semi-Markov Process with Discrete State Space*. Sovetskoe Radio, Moscow (in Russian).

Shiryaev, A. N. (1996), *Probability*, 2nd ed. Springer.

Smith, W. L. (1955). Regenerative stochastic processes. *Proc. Roy. Soc. Ser. A* **232**, 6–31.

Takacs, L. (1954). Some investigations concerning recurrent stochastic processes of a certain type. *Magyar Tud. Akad. Mat. Kutato Int. Kzl.* **3**, 115–128.

Vassiliou, P. C. G. and A. A. Papadopoulou (1992). Non-homogeneous semi-Markov systems and maintainability of the state sizes. *J. Appl. Probab.* **39**, 519–534.

Wajda, W. (1992). Uniformly strong ergodicity for non-homogeneous semi-Markov processes. *Demonstratio Math.* **25**, 755–764.

Yackel, J. (1966). Limit theorems for semi-Markov processes. *Trans. Amer. Math. Soc.* **123**, 402–424.

Departures and Related Characteristics in Queueing Models

M. Manoharan, M. H. Alamatsaz and D. N. Shanbhag

1. Introduction

The problem of identifiability is of paramount importance in all statistical methods and data analysis and it occurs in almost all fields where stochastic modelling is widely used. In many fields, the objective of the investigator's interest is not just the population or the probability distribution of an observable random variable but the physical structure or the model leading to the probability distribution. Identification problems arise when observations can be explained in terms of one of several available models. Any statistical procedure developed based on a stochastic model is meaningful only if the model is identifiable. The importance of identifiability in diverse areas such as reliability theory, survival analysis, econometrics, etc., is now well-known.

The major motivation for writing this article is to bring together relevant materials on identifiability problems in queueing theory as it occurs in diverse fields such as networks, telecommunications, etc. as well as to discuss some new results and observations on identifiability or characterisation of queueing systems not found elsewhere.

Departure processes are of considerable importance in the study of any queueing model. A natural question that arises is whether or not from the behaviour of the departures of a queue one can identify or gain some information about the unknown input process or service system. A problem of great interest in its initial development has been that of finding queueing systems with departure processes that are renewal processes. The original motivation for this problem was that such a process might itself form the arrival process of another queueing system, and hence, if this process happens to be a renewal process, the whole system would be easier to model. On the other hand for instance, in a packet oriented network, the departure processes of queues sometimes directly characterize the network performance. This is because the network congestion influences the communication quality through the arrival process of a packet stream at the destination terminal and also the arrival to the destination is sometimes directly given by the departure process of a queue.

It is important to have some knowledge of the structural aspects of distributions frequently encountered in the theory of queues. The properties such as unimodality and infinite divisibility could provide us an insight about the behavior of distributions. We shall have the investigation based on these properties. Several of the distributions in the standard queuing models, such as those of queue length, waiting times and inter departure times are of well known form like Poisson, exponential, geometric or mixtures of these. Structural aspects of these distributions are already well established in the existing literature. However there also exist cases in which one does not know much about the distribution in question and may have very few sketchy papers to learn about them.

Keilson (1971) and Roster (1980) discuss the unimodality properties of distributions occurring in some situations. Keilson establishes certain results connected with logconcavity and logconvexity and hence unimodality of passage time densities of diffusion and birth death processes whereas Roster asserts that the first passage time densities in state free Markov processes are unimodal. Here we shall study these properties relative to the stationary population size distribution $\{p_n\}_{n \geqslant 0}$ of the birth and death processes. The concept of unimodality and related topics are often the underlying assumptions in estimation theory and decision theory and are employed in many smoothing operations via convolutions, mixing and other suitable transformations. Further, they are of substantial importance in optimization theory and mathematical programming.

Many of the well known distributions belong to either or both of the classes of unimodal and infinitely divisible distributions. Unimodality and infinite divisibility are two of the most important concepts in distribution theory. These properties as well as related topics like strong unimodality, logconvexity, self decomposability, stability, etc. are of potential importance in both theoretical and practical problems in statistics and allied fields. Indeed, these properties determine certain structural aspects of probability distributions.

In this chapter, we study these properties with regard to the stationary population size distribution of birth and death processes and aim at some characterization results on certain queuing models. Section 2 gives a brief review of some characterizations of queueing systems based on departures. Section 3 deals with that based on infinite divisibility property. In this latter section, we characterize certain $M/M/s$, $1 \leqslant s \leqslant \infty$, queuing systems via infinite divisibility properties of their stationary queue length distributions. In Section 4, we discuss a birth and death process and some structural aspects of related characteristics.

2. Characterization/identifiability via output processes

There is vast literature on output processes from queues. Much of it is reviewed in Daley (1976). That review is updated later by Gnedenko et al. (1983, Chapter 11). The earliest work of which we are aware of this topic is (Doob, 1953, p. 405) who, in a study of the behaviour of the random translation of points, essentially proved that the stationary/steady state departure process in an $M/G/\infty$ queueing system is a Poisson process with the same rate as that of the arrival process. Later, Mirasol (1963) showed that if the input is a nonhomogeneous Poisson process with rate $\lambda(t)$ in an $M/G/\infty$

queueing system that is initially empty, then the output process is also a nonhomogeneous Poisson process but with rate $\lambda^\star(t) = \int_0^t \lambda(t-u)\,\mathrm{d}B(u)$, where B denotes the service time distribution function. A more satisfying argument to arrive at the result of Mirasol (or indeed a more general version of it) is that based on an appropriate planar Poisson process; for a scholarly account of Poisson processes on general spaces, see Kingman (1993). (See Shanbhag's, 1972 comment on Mirasol, 1963.) The main reason for the study of departure processes is that the departure from a queue may join another queue and hence be the input for the second queue. The important role of departure processes has mostly to do with queueing in tandem. However, in situations where only observable events of the system are the departures (e.g.: a Geiger counter), one may also utilise the output to find some information about the input processes or the servicing system. The possibility of deducing such information from the output are studied by Finch (1959), Shanbhag (1972), Shanbhag and Sharma (1972), Disney et al. (1973) and several others under identifiability or characterization. Burke (1956) (see also Burke, 1968) was the first to show that the steady state output process from an $M/M/s$ queue is a Poisson process.

If T_n is the nth service completion, then the sequence $\{T_n;\ n = 1, 2, \ldots\}$ is called the output process of the system. Let $\mathbf{X} = \{X(T_n),\ n = 1, 2, \ldots\}$ be the queue length process embedded at the points T_n and S_n be the service time of the nth served customer,

$$F(t) = P\{S_n \leq t\}, \quad t \geq 0,$$

and assume that F has a finite mean and variance. We will investigate the output processes in terms of the incremental process

$$\mathbf{I} = \{T_{n+1} - T_n,\ n = 1, 2, \ldots\}.$$

2.1. Output processes from $M/G/1/L$ $(L \leq \infty)$ queueing systems

Finch (1959) gave a nearly complete characterization of $M/G/1/L$ queue with renewal output processes and the gap was later filled in by Disney et al. (1973) by the following result.

THEOREM 2.1. *For the $M/G/1/L$ queue with $FCFS$ discipline, the process $\mathbf{I}$ is a renewal process if and only if the the queuelength process $\mathbf{X}$ is stationary and any of the following cases hold*:

(a) *The service times are identically 0*;
(b) $L = 0$, *for any G service process*;
(c) $L = 1$ *and* $G = D$;
(d) $L = \infty$ *and* $G = M$.

In these cases, respectively, the interdeparture distributions $P\{T_{n+1} - T_n \leq t\}$ are

(a) $1 - e^{-\lambda t},\ t \geq 0$;
(b) $(1 - e^{-\lambda t}) * F(t),\ t \geq 0$;
(c) $(1 - e^{-\lambda t}) * 1_d(t),\ t \geq 0$;
(d) $1 - e^{-\lambda t},\ t \geq 0$.

Here $*$ denotes the convolution operator and 1_d is the Dirac function, $1_d(t) = 0$ or 1 according as $t < d$ or $t \geqslant d$.

REMARK 2.1. Theorem 2.1 yields the finite-dimensional distributions of the departure process. From this result we can obtain the stationary joint distributions for any collection of increments and also for auto-covariances or auto-correlation functions. Disney and de Morais (1976) explore the auto-covariance functions for the output process increments for $M/E_k/1/L$ systems. They show for example, that for some settings of the parameters, the first order auto-covariance in the output process increments may be zero even though these increments are not independent.

Concerning characterization of a queueing system by its departure process, Finch (1959) showed that amongst $M/G/1/L$ ($1 \leqslant L \leqslant \infty$) systems with twice differentiable service time distribution function $B(\cdot)$, only $M/M/1/\infty$ has a stationary renewal output. King (1971) proved that a stationary $M/G/1/1$ queueing system has its successive interdeparture times D_n and D_{n+1} to be independent if, and only if, $G = D$ in which case the departure process is renewal. However, these results of Finch and King follow as corollaries of the result of Shanbhag and Sharma (1972). Shanbhag and Sharma remove Finch's prior condition of twice differentiability of the service time distribution $B(\cdot)$ and give a necessary and sufficient condition for the independence of successive interdeparture intervals for an $M/G/1/L$ ($L > 0$) queueing system.

Disney et al. (1973) consider the independence of D_n and D_{n+2} as well as that of D_n and D_{n+1} and prove that the only stationary $M/G/1/L$ ($1 \leqslant L \leqslant \infty$) systems, that have renewal departure processes are $M/M/1$ and $M/D/1/1$. Later this result was improved by Daley and Shanbhag (1975) who showed that the extra step of the independence of D_n and D_{n+2} is necessary. This paper considers a stationary $M/G/1/L$ queueing system with $1 \leqslant L \leqslant \infty$ and finds the distribution functions of the service time such that two successive interdeparture times are independent. For $L = 1$ and $L = \infty$, we have from Disney et al. (1973), Finch (1959) and King (1971) that $G = D$ and $G = M$ respectively. When $L = 2, 3, \ldots$ Daley and Shanbhag prove that for some constants $a > 0$, $k > 0$ and $\gamma > 0$ satisfying in particular $\gamma = e^{-\lambda a}$, $B(\cdot)$ has the left extremity a and density

$$b(x) = Ce^{-\lambda k(x-a)}\left(1 - e^{-\lambda(x-a)}\right)^{-(1-\gamma)k}, \quad x > a, \tag{1}$$

where C is the norming constant, and λ is the input rate.

Gerlach and Warmuth (1978) observed that the service time distribution given above is in fact the distribution of the random variable $-\{\log(Z\gamma)\}/\lambda$, where Z is a Beta-I random variable with parameter k and $1 - (1 - \gamma)k$. Clearly in view of a result of Shanbhag et al. (1977), the distribution in question is infinitely divisible, a property also possessed by systems with $L = 1$ and $L = \infty$. However it is interesting to note that in contrast with the case $L = 1$ and $L = \infty$, $b(\cdot)$ given above does not possess the strong unimodality property; indeed in this case $b(\cdot)$ is (shifted) logconvex and hence unimodal. (As a by-product of the logconvexity also, we get that a distribution with density as in (1) is infinitely divisible, see, for example, Steutel (1970).) This contrast may perhaps be related to the fact that in the case $L = 1$ and $L = \infty$, the interdeparture times D_n are renewal, while in the case when $L = 2, 3, \ldots$, this is not so. On the other hand, it is now clear that unimodality and infinite divisibility are necessary conditions on the service time distribution of a stationary $M/G/1/L$ ($1 \leqslant L \leqslant \infty$) queueing system for the independence of successive interdeparture times.

2.2. The effect of queue discipline and number of servers

In a steady state, a change of queue discipline does not affect the corresponding departure process unless it interferes with the servicing system. Shanbhag and Tambouratzis (1973) have established via an approach of Shanbhag (1966) that for an $M/G/s$ model with $L = 0$ and losses being considered as departures, the steady state output process is Poisson with the same parameter as the input process. Kelly (1976a) has studied an $M/G/1$ system with the preemption last come first served queue discipline. Kelly proved that the departures in this system in equilibrium also form a Poisson process and that the distribution of the number of customers in the system is geometric. Later the second assertion of Kelly's result was proved by Fakinos (1981) to be true more generally for a $GI/G/1$ system. A study of the effect of queue discipline on output process seems to be due originally to Muntz (1973). It has long been known as implied before from Doob (1953) and other sources (see, also, Kendall, 1964) that the steady state output process of an $M/G/\infty$ queue is a Poisson process (insensitivity property). It has been well realized that the result of Theorem 2.1 depends on the FCFS queue discipline, though mildly, and the number of servers by the following result of Disney and Konig (1985).

THEOREM 2.2. *In an $M/G/s$ queueing system, the steady state output process is a Poisson process for any G service time distribution if the system has any of the following four properties.*

(a) $s = \infty$.
(b) *The service discipline is processor-sharing.*
(c) *The service discipline is LCFS and preemptive resume discipline. (This holds for $L < \infty$ also, i.e. in $M/G/s/L$.)*
(d) $G = M$.

2.3. Output processes of $GI/G/s$ queueing systems

Daley (1974) showed that in a stationary pure loss $GI/G/1$ queueing system, the departure process is a renewal process if and only if the idle times and service times are independent. Berman and Westcott (1983) studied the output process of $GI/G/s$, $s < \infty$, without losses under $FCFS$ discipline and proposed a necessary condition for a renewal output. It is assumed that the queueing system possesses: (i) stability and stationarity, (ii) renewal arrival process and (iii) strong renewal property (conditional on there being an arrival at t, all future arrival times are independent of the bivariate process of arrivals and outputs prior to t). Then Berman and Westcott established the following interesting result:

THEOREM 2.3. *Under the aforementioned assumptions and some additional technical assumptions, a necessary condition for the queue to have a renewal output process is that the interarrival time and interdeparture times have the same distribution.*

By the above theorem, it appears that the class of stationary $GI/G/s$, $s < \infty$, queues with $FCFS$ discipline, the subclass having renewal output processes is somewhat restrictive. Theorem 2.1 also supports this view. The stationary $GI/D/s$ queue with an interarrival time almost surely greater than a service time is not included in Theorem 2.1, but the output process here is a renewal process. The further subclass that have Poisson output processes (with $FCFS$ discipline) is smaller still and possibly, under some mild constraints, identities stationary $M/M/s$ queues.

Finally, we point out that, a variation of the Daley's result mentioned above is provided by the following characterization of Rao and Shanbhag (1994).

THEOREM 2.4. *Let $\{D_n: n = 1, 2, \ldots\}$ be the stationary departure process of a $GI/G/1/0$ system with interarrival time distribution function A such that $A(0) = 0$ and service time distribution function B such that $B(0) = 0$ and $B(x) > 0$ for all $x > 0$. Then D_1 and D_2 (or equivalently any two successive D_n's) are independent if, and only if, the system is $M/G/1/0$.*

The 'if' part of the theorem is known and easy to prove. The 'only if' part follows as a consequence of the Lau–Rao theorem (Theorem 2.2.2 in Rao and Shanbhag, 1994).

REMARK 2.2. There still exist possibilities of arriving at several further characterization results in queueing theory via the Lau–Rao theorem referred to above.

2.4. Output processes from stationary birth and death queueing systems

So far the results for output processes have been for queues with, amongst other things, renewal arrivals (GI) and i.i.d. service times (G). We now discuss output processes from the class of stationary birth and death (B–D) queues.

For any B–D queue, **X** is a Markov process. Reich (1963, 1965) using reversibility arguments proved that constant birth rate is a sufficient condition for a stationary B–D queue with constant death rate to have a renewal output process. Simon and Disney (1984) proved the necessary part of this result and indicated results for several other systems as applications of a method of comparing Markov renewal processes and renewal processes. So, we have,

THEOREM 2.5. *Suppose we have a stationary B–D queue with FCFS discipline and $L = \infty$.*

(a) *If $\mu_n = \mu$ for all n, then $\lambda_n = \lambda$ is necessary and sufficient for the output process to be a renewal process. In this case, the output process is also a Poisson process. (See Theorem 2.1.)*
(b) *If $\mu_n \neq \mu$, then $\lambda_n = \lambda$ is a necessary condition for the output process to be a renewal process but it is not sufficient.*

REMARK 2.3. There are several generalizations or variations of these results. Vlach and Disney (1969) study the output process of a stationary $GI/G/1/L$ queue and show that the output process is once again a Markov renewal process. McNickle (1974) considers the number of output for the queue with Markov renewal arrival process and a Markov renewal service time process. In the study of stationary $GI/M/1/L$ queueing output processes, Laslett (1975) observed that the only $GI/M/1/L$ queues with a renewal output processes are $M/M/1$ and $M/M/1/0$. Also many studies of special cases now exist. See, e.g., Pack (1975, 1977, 1978) for details on queues with deterministic arrival or service time processes. Kelly (1976a, 1976b, 1979), Natvig (1977), Gerlach and Warmuth (1978), Foley (1982), Fakinos (1982), etc. treat some other specific models and/or variants.

Finally, we point out a more general result for a $GI/G/1/L$ system with a minor condition on service time. Laslett (1975) establishes that any stationary $GI/M/1/L$ queueing system with $0 < L < \infty$ cannot have a renewal output through a cumbersome and lengthy argument. The following result, which was first reported in Alamatsaz (1983), clearly subsumes the finding of Laslett and, as illustrated here, follows via a simple argument.

THEOREM 2.6. *In any $GI/G/1/L$ queueing system with $1 \leqslant L < \infty$ and interarrival time distribution A and service time distribution B such that $A(0) = 0$, $A(t)(1 - B(t)) > 0$ for some $t > 0$ and $B(t) > 0$ for all $t > 0$, it is impossible that the stationary departure process be renewal.*

PROOF. Let $\{D_n: n = 1, 2, \ldots\}$ denote the stationary departure process of the system considered.

Clearly, we have

$$P\{D_1 \leqslant x\} = \pi_0 I * B(x) + (1 - \pi_0) B(x), \quad 0 \leqslant x < \infty, \tag{2}$$

where $\{\pi_i\}$ is the stationary queue length distribution immediately after a departure and I is the distribution function of the limiting idle period (i.e., the limiting conditional distribution function of the residual interarrival time at a departure epoch given that the departing customer leaves the system empty). In view of our assumptions, it is clear that $0 < \pi_0 < 1$. It now follows that if $1 \leqslant L < \infty$, by virtue of Eq. (2), we have

$$P\{D_1 \leqslant x, D_2 \leqslant x, \ldots, D_{L+2} \leqslant x\} = o\big((B(x))^{L+2}\big)$$
$$= o\big((P\{D_1 \leqslant x\})^{L+2}\big), \tag{3}$$

as $x \to 0$. In the above expressions the notation 'o' is used to mean that $o(h(x))/h(x) \to 0$ as $x \to 0$ (even when $h(x)$ itself does not tend to zero as as $x \to 0$).

Consequently, we have that $\{D_n\}$ cannot be renewal. □

REMARK 2.4. A slightly extended version of the argument used in the proof of Theorem 2.6 implies under the assumptions of the theorem except that L could now be allowed to be infinite and that in this case we take the relative traffic intensity of the system to be less than 1, that the stationary departure process of the system cannot be renewal unless the stationary queue length distribution referred to in the proof of the theorem is geometric. Theorem 2.6 with "$GI/G/s/L$" in place of "$GI/G/1/L$" holds, if for $s > 1$ we have additionally (for example, as in the case when B is exponential) that the steady state probability that there are j residual service times at a departure epoch that lie in $(0, x]$ is of the order of $(B(x))^j$ as $x \to 0$ for each $j = 1, 2, \ldots, s - 1$ and for some (and hence all) $c \neq 1$, $B(x)$ and $B(cx)$ are of the same order as $x \to 0$; this latter result follows essentially via a minor variation of the existing proof of Theorem 2.6.

Many authors have considered characterization of queueing systems by second-order moment properties of the interdeparture times $\{D_n\}$. For a review on this we refer to Daley's (1976) work on departure processes. Ishikawa (1991) studied a stationary departure process of an $M/G/1/N$ queue using a Markov renewal process and obtained the joint density function of the k-successive departure intervals in order to examine the covariance of the departure process. More recently, departure process of a more general $M/G_r/1$ queueing model with server vacation under exhaustive service discipline has been considered by Tang (1994). This paper provides a new method for analysis of the departure process of the single server queue through the arguments of direct probability decomposition and renewal theory.

3. Characterization/identifiability via infinite divisibility property

Lukacs (1970) and van Harn (1978) discuss respectively the usual self-decomposability and lattice self-decomposability. One may also refer to the books of Feller (1971) and Loéve (1963) for the concept of infinite divisibility and/or self decomposability.

3.1. Characterization of $M/G/\infty$ queueing systems

Daley (1968) shows that amongst stationary $GI/M/1$ systems only the system $M/M/1$ has a renewal departure process. Shanbhag (1973) characterizes the $M/G/\infty$ queues amongst $M/G/s/L$ ($0 \leq L \leq \infty$) system via infinite divisibility of the stationary joint distribution of the number of arrivals and departures during any fixed interval. In view of the following theorem of Rao and Shanbhag (1994), an extended version of the result for $GI/G/s/L$ follows; somewhat different versions of the theorem given here appear in Haberland (1975) and Alamatsaz (1983).

THEOREM 3.1. *Let $\{N(t): t \geq 0\}$ denote a renewal process for which the underlying distribution F satisfies*

$$\limsup_{\delta \to 0+} \frac{F(\delta)}{\delta} < \infty. \tag{4}$$

Then for all t, the distribution of $N(t)$ is infinitely divisible if, and only if, $\{N(t)\}$ is a Poisson process.

PROOF. The 'if' part of the assertion is standard. We shall establish the 'only if' part.
Suppose that we have a renewal process and that $\limsup_{\delta \to 0+} \frac{F(\delta)}{\delta} = \lambda < \infty$. Then, for a sufficiently small δ, we have that if $t \in (0, \delta)$,

$$P\{N(t) = n\} \leq \frac{(\lambda + \frac{1}{2})^n t^n}{n!}, \quad n \in \mathbf{N}_0 \ (= \{0, 1, \ldots\}). \tag{5}$$

This follows because

$$\begin{aligned}
P\{N(t) = n\} &= F_n(t) - F_{n+1}(t) \\
&\leq F_n(t) \\
&= \int_{(0,t]} F(t - y) F_{n-1}(\mathrm{d}y), \quad n \in \{1, 2, \ldots\}, t \in (0, \infty), \tag{6}
\end{aligned}$$

where $F_m(t)$ for $m \geq 1$ denotes the m-fold convolution of F with itself and F_0 is the degenerate distribution function at zero. (5) clearly holds for $n = 0$ and 1. Assuming now that it holds for $n = 0, 1, \ldots, k$, where $k \geq 1$, we get for $n = k + 1$, from (6) and Fubini's theorem, that

$$\begin{aligned}
P\{N(t) = n\} &\leq \int_{(0,t]} \left(\lambda + \frac{1}{2}\right)(t - y) F_{n-1}(\mathrm{d}y) \\
&= \left(\lambda + \frac{1}{2}\right) \int_{(0,t]} \left(\int_{[y,t]} \mathrm{d}x\right) F_{n-1}(\mathrm{d}y) \\
&= \left(\lambda + \frac{1}{2}\right) \int_{(0,t]} F_{n-1}(x) \, \mathrm{d}x \\
&\leq \frac{(\lambda + \frac{1}{2})^n}{(n-1)!} \int_{(0,t]} x^{n-1} \, \mathrm{d}x \\
&= \frac{(\lambda + \frac{1}{2})^n t^n}{n!}, \quad t \in (0, \delta).
\end{aligned}$$

If $P\{N(t) \neq 0\} = 0$ for some $t \in (0, \infty)$, then it follows that the left extremity of F is positive, which in turn implies that there exists a point t at which $N(t)$ is a nondegenerate bounded random variable and hence a noninfinitely divisible random variable. Hence we have to assume $P\{N(t) \neq 0\} > 0$ for all $t \in (0, \infty)$. If it is then $N(t)$ is infinitely divisible for all t, we get that unless all $N(t)$ for $t \in (0, \delta)$ are Poisson, we have a contradiction (because the existence of a non-Poisson infinitely divisible $N(t)$ for some $t \in (0, \delta)$ implies that $E\{z^{N(t)}\}/\exp\{(\lambda + \frac{1}{2})z\} \to \infty$ as $z \to \infty$, contradicting the above inequality). It is easily seen that if $z \geq 1$ and k is a positive integer

$$E\left(z^{N(t)}\right) \leq z^{k-1} \left(E\left(z^{N(t/k)}\right)\right)^k,$$

which implies that if $N(t)$ is infinitely divisible for all t, then, as $N(t)$ for $t \in (0, \delta)$ are Poisson, we have $N(t)$ to be Poisson for all t (because, for a sufficiently large k, the ratio of the left-hand side of this last inequality tends to infinity otherwise). From Theorem 8.3.6 of Rao and Shanbhag (1994) or an elementary moment argument involving the moments of an exponential distribution, it follows that, since $N(t)$'s are all marginally Poisson, $\{N(t)\}$ is a Poisson process. Hence we have the theorem. □

COROLLARY 3.1. *In a $GI/G/s/L$ ($0 \leq L \leq \infty$) queueing system with interarrival time distribution A satisfying* $\limsup_{\delta \to 0+} \frac{A(\delta)}{\delta} < \infty$, *the stationary joint distribution of the number of arrivals and the number of departures, when exists, is infinitely divisible for every finite interval if, and only if, the queueing system is $M/G/\infty$.*

PROOF. The proof follows from Theorem 3.1 with modification as in Remark 8.3.11, (iii) of Rao and Shanbhag (1994), and the result of Shanbhag (1973). □

Also it is evident from the observation in Shanbhag (1973) and Theorem 3.1 that the following corollary holds:

COROLLARY 3.2. *In a $GI/G/s/L$ ($0 \leq L \leq \infty$) queueing system with interarrival time distribution A satisfying the restriction of the above theorem, the joint distribution of the number of arrivals and that of departures, during $(0, t)$, given that the queue length is initially zero, is inf. div. for all t if, and only if, the system is $M/G/\infty$.*

In connection with departure processes, as noted in the works of Daley (1972), Shanbhag (1973), Natvig (1975) and in the preceding results of the section, we see that the inf. div. property can be employed as a tool for characterizing certain queueing systems. We have already revealed that this property is a necessary condition for the service times to have the independence of successive interdeparture times in stationary $M/G_f1/L$, $1 \leq L \leq \infty$, queueing systems. Next we aim at characterizing the queueing system $M/M/s$, $1 \leq s \leq \infty$, via inf. div. property of their stationary queue length distribution $\{p_n\}$. We may recall that the distribution in question is 'geometric' when $s = 1$ and Poisson when $s = \infty$ which are both inf. div. More specifically, $\{p_n\}$ is self decomposable when $s = 1$ and stable when $s = \infty$. However, the inf. div. properties, if any, of $\{p_n\}$ for $1 < s < \infty$ seem to have not been considered seriously earlier. We have the following result.

THEOREM 3.2. *The stationary queue length distribution in an $M/M/s$ queueing system is infinitely divisible if and only if either $s = 1$ or $s = 2$ or $s = \infty$.*

PROOF. Since the distribution is Poisson for $s = \infty$ and geometric for $s = 1$ and in each of these cases the distribution is a standard infinitely divisible (or indeed self decomposable) distribution, it is sufficient if we show that for $s = 2$, the distribution is infinitely divisible and for $2 < s < \infty$ it is noninfinitely divisible.

Note that for $s = 2$, the distribution has the generating function

$$P(z) = P(0)\left\{1 + \left(\frac{\lambda}{2\mu}\right)z\right\} \bigg/ \left\{1 - \left(\frac{\lambda}{2\mu}\right)z\right\}$$

on the relevant domain of definition, it is clear that the distribution is infinitely divisible (but not self decomposable).

Now consider the case of $2 < s < \infty$. Recall that $\{p_n\}$ is given by

$$p_n = \begin{cases} \frac{p_0}{n!}\left(\frac{\lambda}{\mu}\right)^n, & 0 \leq n \leq s, \\ \frac{p_0}{s^{n-s}s!}\left(\frac{\lambda}{\mu}\right)^n, & n \geq s. \end{cases} \qquad (7)$$

Let $P(\cdot)$ denote the probability generating function of $\{p_n\}$. Following Feller (1968) we know that P is infinitely divisible if and only if it has the form $P(z) = ce^{Q(z)}$, where c is the norming constant and $Q(z) = \sum_{n=1}^{\infty} q_n z^n$, $q_n \geq 0 \ \forall n \geq 1$.

Here $c = p_0$ and thus

$$P(z) = p_0\left[1 + zq_1 + \frac{(zq_1)^2}{2!} + \cdots\right]\left[1 + z^2 q_2 + \frac{(z^2 q_2)^2}{2!} + \cdots\right]\cdots. \qquad (8)$$

On comparing the coefficients on both sides we have,

$$q_1 = \frac{\lambda}{\mu} \quad \text{and} \quad q_2 = q_3 = \cdots = q_s = 0.$$

Then we can write,

$$p_{s+1} = p_0\left[\frac{(\lambda/\mu)^{s+1}}{(s+1)!} + q_{s+1}\right].$$

Comparing this with (7) we get,

$$q_{s+1} = \frac{(\lambda/\mu)^{s+1}}{s(s+1)!}.$$

Hence (8) implies,

$$p_{s+2} \geq p_0\left[\frac{q_1^{s+2}}{(s+2)!} + q_1 q_{s+1}\right]$$

$$= p_0\left[\frac{(\lambda/\mu)^{s+2}(2s+2)}{s(s+2)!}\right].$$

In view of p_{s+2} given by (7), this means

$$\frac{1}{s^2 s!} \geq \frac{2(s+1)}{s(s+2)!},$$

which is impossible when $2 < s < \infty$. Therefore, $\{p_n\}$ cannot be infinitely divisible when $2 < s < \infty$. □

The following remarks are worth giving at this point; they provide us with interesting information of relevance to the findings of Theorems 3.1 and 3.2.

REMARK 3.1. In each of Theorem 3.1 and its modified version implied in Rao and Shanbhag (1994, Remark 8.3.11, (iii)), under the respective assumptions, the assertion that any $N(t)$ has an infinitely divisible distribution is equivalent to that it has a nonvanishing entire characteristic function. Incidentally, (in obvious notation) in the more general case of $F(0) = 0$, it is a minor exercise to see that each $N(t)$ has an entire characteristic function.

REMARK 3.2. Given any positive integer s, it is an easy exercise to see that there exists an infinitely divisible stationary distribution on $\{0, 1, \ldots\}$ such that for each integer $m < s$, the conditional distribution of the corresponding random variable X given that $X \geqslant m$ is neither geometric nor shifted geometric, and for the integer $m = s$ (and hence for each integer $m \geqslant s$), the corresponding conditional distribution is shifted geometric. (An example for $s = 1$ in this case is provided by Theorem 3.2; this corresponds to the stationary queue length distribution relative to $M/M/2$.)

4. Strong unimodality and other relevant properties

The book of Lukacs (1970) and of Dharmadhikari and Joag-Dev (1988) discuss the usual unimodality and discrete unimodality. The paper of Ibragimov (1956) and of Keilson and Gerber (1971) discuss logconcavity (strong unimodality) for absolutely continuous and for lattice distributions respectively. Also Karlin (1968) discusses various other aspects of logconcavity. Goldie (1967), Steutel (1970), Kingman (1972), Shanbhag and Westcott (1977) and van Harn (1978) deal with logconvexity or complete monotonicity, aspects of probability distributions and related infinitely divisibility problems implicitly or explicitly.

In what follows, we shall briefly review, amongst other things, some further results on birth and death processes addressing aspects of strong unimodality and other relevant properties.

The birth–death processes have several important applications to areas as diverse as queueing theory, reliability theory, bacteriology, ecology, inventory theory and so on. For example, in a Markovian queueing system the process $\{N(t)\}$ giving the number of customers in the system is a birth–death process. Insight into the structure of transition distributions and passage time distributions may be obtained through their calculation using the existing formulae, but these formulae usually involve infinite series of k-fold convolutions or impractical inversion of Laplace transforms. Interestingly, certain key passage time densities and sojourn time densities for such processes have a simple property of logconcavity or logconvexity and associated unimodality. These properties provide information on the character of distributions and they may also have value for the purposes of estimation and optimization.

We have for ergodic cases of interest for an $M/M/s$ queueing system:

(a) the passage time densities from emptiness to any level of system occupation N are unimodal and logconcave;
(b) the sojourn time densities at or above any level of occupation are logconvex. In particular the system occupation time density (or system busy period density) corresponding to sojourns on the set $n = 1, 2, \ldots$ are logconvex.

The above assertions still hold good for the model with balking (customers refusing to join too long lines) or reneging (departures of impatient customers) so long as the birth–death character of $N(t)$ remains intact.

Now we shall report a more general result due to Keilson (1971).

THEOREM 4.1. *Let $N(t)$ be a birth–death process governed by transition rates λ_n, μ_n with $\lambda_n > 0$, $n \geqslant 0$; $\mu_n > 0$, $n \geqslant 1$; $\mu_0 = 0$. Then*

(i) *the passage time densities from state $n = 0$ to any level are strongly unimodal and hence logconcave;*
(ii) *the random sojourn time on the set of states $M, M+1, \ldots$ initiated by arrival at state M from state $M - 1$ has a probability density which is superconvex (completely monotone). Also the sojourn time density on the set $0, 1, \ldots$ initiated by arrival at $M - 1$ from state M is superconvex and has the form of a finite sum of real exponentials with positive coefficients. In particular, the passage densities $S_{N,N+1}(t)$ (from state N to $N + 1$) and $S_{N,N-1}(t)$ are superconvex.*

REMARK 4.1. The infinite divisibility of passage time densities for birth–death processes and for a more general class of skip-free processes has been established by Miller (1967). The result for the birth–death processes follows now as a consequence of the complete monotonicity in view of the above theorem and a result of Steutel (1969) (given also in Steutel, 1970) on infinite divisibility of completely monotone densities.

REMARK 4.2. It is a straight forward exercise to see that the stationary queue length distribution for $M/M/s$ is strongly unimodal (logconcave). Moreover, it is obvious that given any arbitrary distribution with support $\{0, 1, \ldots, N\}$ or $\{0, 1, \ldots\}$, where N is an integer greater than or equal to 1, there exists a birth and death process with this as its stationary distribution; consequently, it follows that there exist several other cases of birth and death processes with stationary distributions that are either strongly unimodal or logconvex. (Clearly, the stationary distribution relative to a birth and death process is logconcave or logconvex if we have, in obvious notation assuming without loss of generality μ_n's to be nonzero for $n > 0$, $\{\frac{\lambda_n}{\mu_{n+1}}\}$ to be decreasing or increasing respectively; a formula for the stationary distribution and the criterion for its existence appear in Karlin (1966, p. 194).)

REMARK 4.3. As observed by Biggins and Shanbhag (1981), the wet period distribution relative to a certain discrete dam model (and hence that of the number of customers served during a busy period of an $M/G/1$ queueing system) or the distribution of the total progeny in a discrete branching process is (shifted) compound negative binomial. Also, in view of what is observed by Feller (1971) and others, it follows that the stationary waiting time distributions (under queue discipline "first come, first served") for $GI/G/1$ and $GI/M/s$ systems are compound geometric. While we are on the topic of compound geometric distributions, it is worth pointing out that essentially in view of what is seen by Kingman (1972) and others about Kaluza sequences, we have that every logconvex stationary distribution of a birth and death process is compound geometric; from Kingman (1972) or otherwise it is also clear that there exist compound geometric stationary distributions of birth and death processes that are not logconvex.

REMARK 4.4. Recently Fosam and Shanbhag (1997) have identified a class of distributions for which a certain regression property, which is an extended version of the Laha–Lukacs property (9.2.1) in Theorem 9.2.1 of Rao and Shanbhag (1994), holds. This class includes among others distributions that are related to those of busy periods of $M/G/1$ queueing systems and wet periods of certain dams. (Incidentally, there are typos in the cited paper: in (vi) of Theorem 1, '$(b - \delta)$' should have appeared just as 'b' and in Remark 9, in three places, '(14)' should have appeared as '(17)'.)

REMARK 4.5. There exist stationary queue length distributions relative to $GI/G/s$ queueing systems (or, in particular, to specialized versions of $D/G/s$ queueing systems) that are indecomposable.

References

Alamatsaz, M. H. (1983). On structural aspects of probability distributions and their mixtures with applications in queueing theory. Ph.D. thesis, University of Sheffield, UK.

Berman, M. and M. Westcott (1983). On queueing systems with renewal departure processes. *Adv. Appl. Probab.* **15**, 657–673.

Biggins, J. D. and D. N. Shanbhag (1981). Some decomposability problems in branching processes. *Math. Proc. Cambridge Philos. Soc.* **29**, 321–330.

Burke, P. J. (1956). The output of a queueing system. *Oper. Res.* **4**, 699–704.

Burke, P. J. (1968). The output process of a stationary $M/M/s$ queueing system. *Ann. Math. Statist.* **39**, 1144–1152.

Daley, D. J. (1968). The correlation structure of the output process of some single server queueing system. *Ann. Math. Statist.* **39**, 1007–1019.

Daley, D. J. (1972). A bivariate Poisson queuing process that is not infinitely divisible. *Proc. Cambridge Philos. Soc.* **72**, 449–450.

Daley, D. J. (1974). Characterizing pure loss $GI/G/1$ queues with renewal output. *Proc. Cambridge Philos. Soc.* **75**, 103–107.

Daley, D. J. (1976). Queueing output processes. *Adv. Appl. Probab.* **8**, 395–415.

Daley, D. J. and D. N. Shanbhag (1975). Independent inter departure times in $M/G/1/N$ queues. *J. Roy. Statist. Soc. B* **37**, 259–263.

Dharmadhikari, S. and K. Joag-Dev (1988). *Unimodality, Convexity and Applications*. Academic Press Inc., San Diego, CA.

Disney, R. L. and P. R. De-Morais (1976). Covariance properties for the departure processes of $M/Ek/1/N$ queues. *AIIE Trans.* **8**, 169–175.

Disney, R. L. and D. Konig (1985). Queueing networks: a survey of their random processes, *SIAM Rev.* **27**, 335–403.

Disney, R. L., R. L. Farrel and P. R. De-Morais (1973). A characterization of $M/G/1/N$ queue with renewal departure processes. *Mgmt. Sci.* **19**, 1222–1228.

Doob, J. L. (1953). *Stochastic Processes*. Wiley, New York.

Fakinos, D. (1981). The $GI/G/1$ queueing system with a particular queue discipline. *J. Roy. Statist. Soc. B* **43**, 190–196.

Fakinos, D. (1982). $M/G/k$ group-arrival, group departure loss system. *J. Appl. Probab.* **19**, 826–834.

Feller, W. (1968). *An Introduction to Probability Theory and its Applications*, Vol. I, 3rd ed. Wiley, New York.

Feller, W. (1971). *An Introduction to Probability Theory and its Applications*, Vol. II, 2nd ed. Wiley, New York.

Finch, P. D. (1959). The output process of the queueing system $M/G/1$. *J. Roy. Statist. Soc. B* **21**, 375–380.

Foley, R. D. (1982). The non-homogeneous $M/G/\infty$ queue. *Opsearch* **19**, 40–48.

Fosam, E. B. and D. N. Shanbhag (1997). An extended Laha–Lukacs result based on a regression property. *J. Statist. Plann. Inference* **63**, 173–186.

Gerlach, B. and W. Warmuth (1978). A comment to a paper: independent inter departures in $M/G/1/N$ queues by Daley and Shanbhag. *Electron. Informationserab. Kybernet* **14**(1–2), 49–51.

Gnedenko, B. V. and D. Konig et al. (1983). *Hand Book of Queueing Theory, Volume I: Fundamentals and Methods*. Academic Verlag, Berlin (in German); Elsevier–North-Holland, New York (in English).

Goldie, C. M. (1967). A class of infinitely divisible distributions. *Proc. Cambridge Philos. Soc.* **63**, 1141–1143.

Haberland, E. (1975). Infinite divisible recurrent point processes. *Math. Nachr.* **70**, 259–264.

Ibragimov, I. A. (1956). On the composition of unimodal distributions. *Theory Probab. Appl.* **1**, 255–260.

Ishikawa, A. (1991). On the joint distribution of the departure intervals in an $M/G/1/N$ queue. *J. Oper. Res. Soc. Japan* **34**, 422–435.

Karlin, S. (1966). *A First Course in Stochastic Processes*. Academic Press, New York.

Karlin, S. (1968). *Total Positivity*, Vol. I. Stanford University Press, California.

Keilson, J. (1971). Log-concavity and log-convexity in passage time densities of diffusion and birth–death processes. *J. Appl. Probab.* **8**, 391–398.

Keilson, J. and H. Gerber (1971). Some results for discrete unimodality. *J. Amer. Statist. Assoc.* **66**, 386–389.

Kelly, F. P. (1976a). The departure process from a queueing system. *Math. Proc. Cambridge Philos. Soc.* **80**, 283–285.

Kelly, F. P. (1976b). Networks of queues. *Adv. Appl. Probab.* **8**, 416–432.

Kelly, F. P. (1979). *Reversibility and Stochastic Networks*. Wiley, New York.

Kendall, D. G. (1964). Some recent work and further problems in the theory of queues. *Theory Probab. Appl.* **9**, 1–13.

King, R. A. (1971). The covariance structure of the departure process from $M/G/1$ queues with finite waiting lines. *J. Roy. Statist. Soc. B* **33**, 401–406.

Kingman, J. F. C. (1972). *Regenerative Phenomena*. Wiley, London.

Kingman, J. F. C. (1993). *Poisson Processes*. Clarendon Press, Oxford.

Laha, R. G. and E. Lukacs (1960). On a problem connected with quadratic regression. *Biometrika* **47**, 335–343.

Laslett, G. M. (1975). Characterizing the future capacity $GI/M/1$ queue with renewal output. *Mgmt. Sci.* **22**, 106–110.

Loéve, M. (1963). *Probability Theory*. Van Nostrand, New York.

Lukacs, E. (1970). *Characteristic Functions*, 2nd ed. Griffin, London.

Mcnickle, D. C. (1974). The number of departures from a semi-Markov queue. *J. Appl. Probab.* **11**, 825–828.

Miller, H. D. (1967). A note on passage time and infinitely divisible distributions. *J. Appl. Probab.* **4**, 401–405.

Mirasol, N. M. (1963). The output of an $M/G/\infty$ queueing system is Poisson. *Oper. Res.* **11**, 282–284.

Muntz, R. R. (1973). Poisson departure process and queueing networks. *IBM Research Report RC 4145*, IBM Thomas J. Watson Research center Yorktown Heights, New York.

Natvig, B. (1975). On the input and output processes for a general birth and death queuing model. *Adv. Appl. Probab.* **7**, 576–592.

Natvig, B. (1977). On the input and output processes for a general birth and death queueing model. *J. Appl. Probab.* **14**, 876–883.
Pack, D. C. (1975). The output of an $M/D/1$ queue. *Oper. Res.* **23**, 750–760.
Pack, D. C. (1977). The output of a $D/M/1$ queue. *SIAM J. Appl. Math.* **32**, 571–587.
Pack, D. C. (1978). The output of multi server queueing system. *Oper. Res.* **26**, 492–509.
Rao, C. R. and D. N. Shanbhag (1994). *Choquet–Deny Type Functional Equations with Applications to Stochastic Models*. Wiley, Chichester.
Reich, E. (1963). Note on queues in tandem. *Ann. Math. Statist.* **34**, 338–341.
Reich, E. (1965). Departure process (with discussion). In *Congestion Theory*, pp. 141–143 (Eds. W. L. Smith and W. E. Wilkinson). University of North Carolina Press, Chapel Hill.
Roster, U. (1980). Unimodality of first passage times for one dimensional strong Markov processes. *Ann. Probab.* **8**, 853–859.
Shanbhag, D. N. (1966). On infinite server queues with batch arrivals. *J. Appl. Probab.* **3**, 274–279.
Shanbhag, D. N. (1972). Letter to the editor. *J. Appl. Probab.* **9**, 470.
Shanbhag, D. N. (1973). Characterization for the queuing system $M/G/\infty$. *Proc. Cambridge Philos. Soc.* **74**, 141–143.
Shanbhag, D. N., D. D. F. Pestanna and M. Sreehari (1977). Some further results in infinite divisibility. *Math. Proc. Phil. Soc.* **82**, 289–295.
Shanbhag, D. N. and O. P. Sharma (1972). Characterizations based on the departure processes for the queueing systems $M/M/1$ and $M/G/1$. *Research Report 117/DNS and OPS 1*, Sheffield University.
Shanbhag, D. N. and D. G. Tambouratzis (1973). Erlang's formula and some results on the departure processes for a loss system. *J. Appl. Probab.* **10**, 233–240.
Shanbhag, D. N. and M. Westcott (1977). A note on infinitely divisible point processes. *J. Roy. Statist. Soc. B* **39**, 331–332.
Simon, B. and R. L. Disney (1984). Markov renewal processes and renewal processes, some conditions for equivalence. *J. New Zealand Oper. Res. Soc.* **12**, 19–29.
Steutel, F. W. (1969). Note on completely monotone densities. *Ann. Math. Statist.* **40**, 1130–1131.
Steutel, F. W. (1970). Preservation of infinite divisibility under mixing and related topics. *Mathematical Centre Tracts 33*. Mathematisch Centrum, Amsterdam.
Tang, Y. (1994). The departure process of the $M/G/1$ queueing model with server vacation and exhaustive service discipline. *J. Appl. Probab.* **31**, 1070–1082.
Van Harn, K. (1978). *Classifying Infinitely Divisible Distributions by Functional Equations*. Mathematisch Centrum, Amsterdam.
Vlach, T. L. and R. L. Disney (1969). The departure process from the $GI/G/1$ queue. *J. Appl. Probab.* **6**, 704–707.

Discrete Variate Time Series

Eddie McKenzie

1. Introduction

Modelling discrete variate time series is the most challenging and, as yet, least well developed of all areas of research in time series. The fact that variate values are integer renders most traditional representations of dependence either impossible or impractical. In the last two decades there have been a number of imaginative attempts to develop a suitable class of models. Our purpose here is to briefly review some of the most interesting and exciting of these.

Discrete variate time series occur in many contexts, often as counts of events, objects or individuals in consecutive intervals or at consecutive points in time. Some simple examples are the numbers of accidents in a manufacturing plant each month, the numbers of patients treated by a hospital's accident and emergency unit each hour, the numbers of fish caught in a particular area of sea each week, the numbers of busy lines in a telephone network noted every thirty minutes, and the numbers of lifts in a tall office building which are fully operational at the start of business each day. Such data may also arise from the discretization of continuous variate time series. An example of this is the reduction of daily rainfall volumes to a binary series of ones and zeros, i.e., wet and dry days. See, for example, Phatarfod and Srikanthan (1981).

Simulation is an important use for models of discrete variate time series, since the need to generate sequences of dependent random variates with a particular marginal distribution and correlation structure is common. Sometimes this is done for purely practical reasons, e.g., the attempt to simulate dam input more realistically, as in Phatarfod and Mardia (1973). Sometimes the reasons are more statistical, e.g., to assess effects of serial correlation on procedures, i.e., tests, estimators, etc., developed originally for independent data. Often the motivation can be both theoretical and practical. For example, such models are a useful source of often new and interesting discrete multivariate distributions. More generally, they may yield insight into the construction of such distributions. The work of Phatarfod and Mardia (1973) used a bivariate distribution introduced by Edwards and Gurland (1961) in studies of accident proneness. Both pairs of authors were less concerned with the underlying structure of the generating processes involved than with their joint distributions. However, two of the processes we shall see here, *viz* the Poisson and binomially distributed $AR(1)$ processes defined by (6) and (18), respectively, are versions of theirs.

In many cases, the discrete variates are large numbers and it makes sense to approximate them by continuous variates. Often, however, this is not possible, and it is necessary to find models for time series of relatively small integers. In addition, the observed series often display strong time-dependent trends or more general dependence on covariates. For example, four of the five examples noted above would be expected to exhibit the repeated patterns of seasonality. Since practical interest often lies in the nature of such trends and dependencies, much of the modelling of time series of counts was performed in the past using generalized linear models or a similar regression based approach, and treating any serial correlation as a nuisance which, optimistically, might be explained adequately by the fitted trends. See, for example, Stern and Coe (1984). In such cases, allowance would sometimes be made for the effects of unmodelled serial correlation, e.g., extra-Poisson variation might be necessary in a Poisson regression.

Despite the increasing recognition of the need to model and simulate such data, there were remarkably few models available until the late 1970's. In practice, Markov chains represented the only general class of models which might be suitable. They, however, tend to be overparametrized for most practical purposes, except in the two-state or binary case. In addition, their correlation structure is often limited for application and they are not easily generalized to alternative model forms. To some extent, higher order chains can help with these problems, but again only at the cost of large numbers of parameters.

The great upsurge of interest in simple linear time series models in the Gaussian case, and their corresponding success in practice, systematized and popularized by the work of Box and Jenkins (1970), led to attempts to extend that approach to the non-linear and non-Gaussian cases. The start of much of this is well catalogued in the book by Tong (1990). In particular, serious attempts were made to develop similar models for discrete random variates. This resulted in a considerable volume of work, outlined in the following sections, devoted to the construction of linear-like models for discrete time series which exhibit recognizable correlation structures. Simultaneously, work was also underway to extend generalized linear modelling concepts to discrete data in such a way that both trends and serial correlation are adequately modelled.

The rest of this review consists of several sections on models, or classes of models, presented roughly in the order in which they were first introduced. Thus, we begin with a brief section on Markov chains, concentrating mainly on methods designed to simplify and facilitate the use of higher-order chains in modelling discrete variate time series. This is followed by an introduction to the DARMA processes. These represent the first real attempt to define a class of models purely for discrete variate time series. Their properties and structure are based on the well known ARMA processes. The following section presents a wide range of models based on the idea of thinning. Both the practicality and the elegance of these models has been and is still being explored and much of this is reflected in the length of this section. Next, we consider regression type models. These are usually based on the concepts of generalized linear modelling but attempt to incorporate both trend and serial correlation adequately. Following a brief look at the state-space and Bayesian approaches, the review ends with a final section considering some future directions.

The area is still in its infancy and contains substantive and important practical problems to which we have only the beginnings of solutions. It also yields theoretical structures of considerable elegance and raises many questions which connect quite disparate areas of theory and practice. It is an exciting and rewarding area and what follows is only a very brief introduction and a superficial overview. Needless to say, although I have tried to cover the most important contributions and ideas, the relative lengths of the discussions on each topic reflect my own personal interests and should not be taken as a measure of their relative importance.

2. Markov chains

The theory and structure of Markov chains is well known in general and readily available elsewhere, e.g., Cox and Miller (1965). Further, we noted in the introduction that they are generally overparametrized and too limited in correlation structure for wide application to discrete variate time series. Consequently, we simply note here that there have been attempts to simplify the models, i.e., essentially to reparametrize them in terms of many fewer parameters. In particular, we discuss the work of Raftery (1985a), generalizing an earlier approach by Pegram (1980), on simplified forms for higher-order chains.

2.1. Raftery's MTD models

Raftery (1985a) introduced a class of models for higher-order Markov chains which simplify their structure and greatly reduce the number of parameters necessary. Consider a kth order Markov chain over a finite set of states denoted by $\{1, 2, \ldots, m\}$, and denote the transition probabilities by $\{p(s_0|s_1, s_2, \ldots, s_k)\}$. This model may require as many as $(m-1)m^k$ parameters. Raftery (1985a) simplifies the requirements greatly by proposing that

$$p(s_0|s_1, s_2, \ldots, s_k) = \sum_{j=1}^{k} \lambda_j q(s_0|s_j), \tag{1}$$

where $\sum_{j=1}^{k} \lambda_j = 1$ and $\{q(i|j): i = 1, 2, \ldots, m\}$ is a probability distribution for each value of $j = 1, 2, \ldots, m$. Note that the λ_j need not all be positive. This model has only $m(m-1) + (k-1)$ parameters. From its structure, the model appears most relevant for systems which tend to revert to previously occupied states. It appears to be much more general in form than an earlier attempt by Pegram (1980), and is somewhat simpler to use.

The models are autoregressive-like in structure in that each unit increase in the order of dependence requires only one more parameter. Raftery (1985a) also shows that the bivariate distributions of the process satisfy Yule–Walker-like equations. Raftery later calls these models and their extensions mixture transition distribution (MTD) models.

Note that, in estimation, it is necessary to ensure the non-negativity of the values defined by (1), i.e., that they are actually probabilities. This can be a difficult problem computationally, because of the nature and number of constraints implied. In a later paper, Raftery and Tavaré (1994), these computational problems are ameliorated by reducing the numbers of constraints.

Note, however, that if we also assume that the $\{\lambda_j\}$ are a probability distribution then this non-negativity problem vanishes. In fact, the estimated $\{\lambda_j\}$ are probability distributions in the examples of Raftery (1985a), but in the examples of Raftery and Tavaré (1994), this is no longer so, and the alternative estimation procedures are evaluated.

Problems of extending the MTD models to an infinite state space may be handled by defining the conditional probabilities $\{q(i|j): i = 1, 2, \ldots, m\}$ in terms of a finite number of parameters. The asymptotics for this case, i.e., $m \to \infty$, are validated by Adke and Deshmukh (1988).

It is clear that MTD models may be used as models for discrete variate time series with particular marginal distributions if required, and Raftery (1985b) introduces binomial and Poisson versions in particular.

3. The *DARMA* models

In a series of papers, Jacobs and Lewis (1978a, 1978b, 1978c) introduced a simple method for obtaining a stationary sequence of dependent random variables with a specified marginal distribution and correlation structure chosen independently. It was the first attempt at a general class of simple models for discrete variate time series. The models are based structurally on the well known autoregressive moving average processes and are referred to as DARMA models. A brief review of the simplest case reveals most of the salient features of this class of models.

The DAR(1) model
This is the first-oder autoregressive form and is given by

$$X_t = V_t X_{t-1} + (1 - V_t) Z_t, \tag{2}$$

where $\{V_t\}$ are i.i.d. binary r.v.s with $P(V_t = 1) = \alpha$ and $\{Z_t\}$ are i.i.d. with distribution given by π. If X_0 is also sampled from π then (2) generates a stationary process $\{X_t\}$ whose marginal distribution is π. The model defines the current observation to be a mixture of two independent r.v.s: it is either the last observation, with probability α, or another, independent, sample from the same distribution.

It is a very simple and very general model since π can be the distribution of any kind of r.v. and the correlation structure is independent of π. On the other hand, it is clear from (2) that dependence is realized by runs of constant values in the sample path, and the larger the value of α the longer the runs. Such behaviour is extremely unlikely for a continuous r.v. but may be of value for discrete or categorical ones. We concentrate on the discrete case here and assume that the sample space for π is a subset of the integers.

Note also that all *DARMA* models are constructed in this way, i.e., as mixtures of i.i.d. r.v.s, all sharing a common distribution, *viz* the marginal distribution. One of the effects of this is that all correlations are positive.

The autocorrelation function of $\{X_t\}$ as defined by (2) is given by $\rho_X(k) = \alpha^k$ for $k = 0, 1, \ldots$, and the conditional mean of X_t given X_{t-1} is linear in X_{t-1}. The conditional variance is quadratic in X_{t-1}. In addition, $\{X_t\}$ is a Markov chain with transition probability matrix given by $\alpha I + (1 - \alpha)Q$ where I is the identity matrix and Q is a matrix each of whose rows is the distribution π.

The model (2) extends to higher orders in an obvious way. The pth order model, $DAR(p)$, is given by

$$X_t = V_t X_{t-A_t} + (1 - V_t) Z_t, \qquad (3)$$

where $\{V_t\}$ and $\{Z_t\}$ are as before and $\{A_t\}$ are i.i.d. r.v.s defined on the set $\{1, 2, \ldots, p\}$ so that $P(A_t = k) = \phi_k$. Thus, the current value is either one of the last p observed values, chosen stochastically, or an independent choice Z_t.

The autocorrelation function satisfies the usual form of Yule–Walker equations for an $AR(p)$:

$$\rho_X(k) = \alpha \sum_{i=1}^{p} \phi_i \rho_X(k-i).$$

The DMA(q) model

This is the moving average model of order q and is given by $X_t = Z_{t-S_t}$ where $\{Z_t\}$ is as before and $\{S_t\}$ are i.i.d. r.v.s defined on $\{0, 1, \ldots, q\}$ with $P(S_t = k) = \theta_k$ for $k = 0, 1, \ldots, q$. The autocorrelation function of $\{X_t\}$ is given by

$$\rho_X(k) = \begin{cases} \sum_{i=0}^{q-k} \theta_i \theta_{i+k}, & k = 1, 2, \ldots, q, \\ 0, & k > q. \end{cases}$$

The DARMA(p, q+1) model

The mixed model is constructed by coupling the two component models, thus:

$$\begin{aligned} X_t &= U_t Y_{t-S_t} + (1 - U_t) Z_{t-q-1}, \\ Z_t &= V_t Z_{t-A_t} + (1 - V_t) Y_t, \end{aligned} \qquad (4)$$

where $\{Z_t\}$, $\{V_t\}$, $\{A_t\}$ and $\{S_t\}$ are as before and $\{U_t\}$ are i.i.d. binary with $P(U_t = 1) = \beta$.

In a later paper, Jacobs and Lewis (1983), this model is simplified into a single equation as the $NDARMA(p, q)$ model, thus:

$$X_t = V_t X_{t-A_t} + (1 - V_t) Z_{t-S_t}.$$

Comments

In the four referenced papers, the authors explore this family of models, deriving, in particular, autocorrelation functions, joint distributions, runs and asymptotic properties, some estimators and comparisons of their performance and a goodness of fit test for the marginal distribution. They also consider possible extensions to negative correlations.

The models are very simple and very general but by the very nature of their construction are somewhat unusual processes for practical application. It may be that they would be more suited to modelling dependent sequences of categorical observations, but this does not seem to have been attempted yet.

Some applications have appeared, mainly in the hydrological literature, which is perhaps not surprising given the models' representation of dependence as runs. These include Buishand (1978), Chang et al. (1984a, 1984b, 1987) and Delleur et al. (1989).

4. Models based on thinning

In a review paper in 1980, P. A. W. Lewis argued that it was time to seek general classes of useful time series models for non-Gaussian data. Such models, he suggested, should be simple and flexible in the following senses. In the case of stationarity, the model should be specified in terms of its marginal distribution and correlation structure. It should have few parameters and these should be interpretable, if possible, making the model easy to fit to data both formally and informally. Finally, it should be simple in form, i.e., in practice, it should be linear in structure if at all possible. In a series of papers from the late 1970's onwards, Lewis and his co-workers, in particular A. J. Lawrance, were successful in constructing a variety of time series models with these appealing properties for a number of positive continuous random variables. Most of these models were of the *ARMA* class with the next observation defined to be a mixture of linear combinations of independent random variables, with both random and scalar coefficients.

The simplest example is the *AR*(1) in which a sequence of variates $\{X_t\}$ is defined by

$$X_t = \alpha X_{t-1} + Z_t, \tag{5}$$

where $\{Z_t\}$ is a sequence of i.i.d. variates, and, in general, $|\alpha| \leqslant 1$, although when only non-negative r.v.'s are considered, as here, this becomes $0 \leqslant \alpha \leqslant 1$. Now, if $\{X_t\}$ in Eq. (5) are to be discrete r.v.'s defined on the non-negative integers, and α is to be chosen from (0, 1), it is clear that, in general, Z_t cannot be independent of X_{t-1}. The approach developed replaces the scalar multiplication by *binomial thinning* which is defined as follows. If N is a non-negative integer and $\alpha \in [0, 1]$ then $\alpha * N = \sum_{i=1}^{N} B_i(\alpha)$ where $\{B_i(\alpha)\}$ is a sequence of i.i.d. binary r.v.'s, independent of N, and for which $P(B_i(\alpha) = 1) = \alpha$. Notice that, conditional on N, $\alpha * N$ is a binomial r.v., the number of successes in N independent trials in each of which the probability of success is α.

Thus, the original *AR*(1) model of Eq. (5) is replaced by

$$X_t = \alpha * X_{t-1} + Z_t, \tag{6}$$

where $\{Z_t\}$ and α have the same properties as before, and the thinning operation of α on X_{t-1} is independent of Z_t.

The procedure of *thinning* is commoner in the study of point processes, but is also appropriate here as many discrete variate processes arise as aggregated point processes, i.e., counts of a point process in consecutive intervals of time. The application here as an alternative to scalar multiplication by a fraction arises naturally from the work of Steutel and Van Harn (1979) on self-decomposability for discrete r.v.'s. In particular, all distributions which are discrete self-decomposable in the sense of Steutel and Van Harn can be marginal distributions for the stationary solution to Eq. (6). This includes many of the commonest distributions on the non-negative integers, e.g., Poisson, geometric and negative binomial, but none defined on bounded sets, so that the binomial distribution is excluded and alternative model forms must be sought.

In order to generate the stationary sequence $\{X_t\}$ with a particular marginal distribution using (6) the innovation distribution, i.e., that of Z_t, must be specified first. By deriving probability generating functions (p.g.f.s) throughout (6), and using the stationarity of $\{X_t\}$, it is easily shown that the p.g.f. of Z_t is given by

$$P_Z(s) = E(s^Z) = P_X(s)/P_X(1-\alpha+\alpha s), \tag{7}$$

where P_X is the p.g.f. of X_t. It is more usual in this area to use the *alternate probability generating function* or a.p.g.f. defined by $G_X(s) = E((1-s)^X)$, in which case the a.p.g.f. of Z_t is given by

$$G_Z(s) = G_X(s)/G_X(\alpha s). \tag{8}$$

Thus, if a sequence of i.i.d. r.v.'s, $\{Z_t\}$, with a.p.g.f. $G_Z(s)$, is used as input to (6), and X_0 has a.p.g.f. $G_X(s)$, the output will be a dependent sequence of r.v.'s whose stationary marginal distribution also has a.p.g.f. $G_X(s)$.

The form of Eq. (6) is appealingly linear in appearance, and in fact such models share many of the properties of the linear $AR(1)$ as given by (5). In particular, the autocorrelation function of $\{X_t\}$ is given by $\rho_X(k) = \alpha^k$, for $k = 0, 1, \ldots$, and $E(X_{t+k}|X_t, X_{t-1}, \ldots)$ is a linear function of X_t. However, unlike (5) where the conditional variance is constant, in (6) $\text{Var}(X_{t+k}|X_t, X_{t-1}, \ldots)$ is also linear in X_t.

It is also worth noting that the form of (6) readily lends itself to interpretation in terms of a population evolving in continuous time. Since $\alpha * X_{t-1}$ is a thinned version of X_{t-1}, we may interpret X_t as being the sum of those who survive from time $(t-1)$ to t, i.e., $\alpha * X_{t-1}$, and those who arrive in the interval $(t-1, t)$ and survive until time t, i.e., Z_t. The connection with the simple death process with immigration is apparent.

The model given by (6) is a discrete variate version of the $AR(1)$ but the idea of binomial thinning has been extended to a variety of models, usually for particular marginal distributions. For convenience, we consider the models from the viewpoint of their intended marginal distributions.

4.1. Poisson marginals

The Poisson AR(1)

Suppose that the $AR(1)$ model given by (6) is to have as its marginal distribution the Poisson distribution of mean θ, i.e., $P(X_t = x) = e^{-\theta}\theta^x/x!$, for $x = 0, 1, \ldots$. The a.p.g.f. of this distribution is $G_X(s) = E((1-s)^X) = e^{-\theta s}$ and so the a.p.g.f. of the innovation distribution is $G_Z(s) = e^{-\theta s}/e^{-\theta\alpha s} = e^{-\theta(1-\alpha)s}$, which is the a.p.g.f. of a Poisson r.v. of mean $\theta(1-\alpha)$. Thus, given a sequence of i.i.d. Poisson r.v.s $\{Z_t\}$ of mean $\theta(1-\alpha)$ the model given by Eq. (6) yields a stationary sequence of Poisson r.v.s of mean θ and $AR(1)$-like properties.

The conditional moments have been discussed in general, but the conditional distribution is given by

$$p(y|x) = P(X_t = y | X_{t-1} = x)$$

$$= x!\exp\bigl((1-\alpha)\theta\bigr) \sum_{k=0}^{\min(y,x)} \frac{\alpha^k(1-\alpha)^{x+y-2k}\theta^{y-k}}{k!(x-k)!(y-k)!}, \quad y = 0, 1, \ldots.$$

The $AR(1)$ given by (6) is a Markov process in discrete time with transition probability matrix $[p(j|i)]$. This leads to a bivariate Poisson distribution for consecutive observations which is well known. It has joint a.p.g.f.

$$G_{X_t, X_{t-1}}(u, v) = \exp\bigl[-\theta(u + v - \alpha uv)\bigr]. \tag{9}$$

Since all the joint distributions of this Markov process are constructed from this bivariate one shown in (9), and it is symmetric, it follows that this Poisson $AR(1)$ is time-reversible.

The property that the innovation distribution is from the same family as the marginal distribution characterizes the Poisson distribution with respect to the model (6), as was shown by Al-Osh and Alzaid (1987a). The Poisson thus appears to play a similar role for the discrete $AR(1)$ model (6) to that of the normal distribution for the classical model (5). Indeed, when thinning is used to replace scalar multiplication in linear combinations of random variates, several of the characterizations of the normal distribution involving the properties of linear combinations of r.v.s have been shown to have Poisson analogues by McKenzie (1991).

As discussed before, the model given by (6) has a direct interpretation which, in this case, is a well known one. It is the $M/M/\infty$ queueing system observed at regularly spaced intervals of time. This, of course, does not restrict its application to only such situations, but it is reassuring that such a process does exist. In terms of usefulness, it is notable that the realizations of this $AR(1)$ are similar to sample paths generated from the usual linear model (5) using normal r.v.s. This is in stark contrast to the noted behaviour of the $DAR(1)$ model discussed already where positive correlation is realised by runs of constant value. The differences are displayed clearly in Figure 1 of McKenzie (1988b) in which this $AR(1)$ model is extended to other $ARMA$-type models which we now discuss.

The Poisson MA(1)
The first-order Poisson moving average model has equation

$$X_t = Z_t + \beta * Z_{t-1}, \tag{10}$$

where $\beta \in [0, 1]$ and $\{Z_t\}$ is a sequence of i.i.d. Poisson r.v.s of mean $\theta/(1+\beta)$. In this case, the model given by (10) generates a stationary process $\{X_t\}$ whose marginal distribution is Poisson of mean θ. Again, this is not a linear model, though it clearly has the characteristics of the $MA(1)$ process, i.e., independence of X_t and X_{t-k} for $k > 1$. In addition, the bivariate distribution of two consecutive observations in this process has the same form as for the $AR(1)$ and so the same form of conditional distributions and moments. Also, from McKenzie (1988b) it follows that the Poisson $MA(1)$ is time-reversible. Interestingly, it is also the $M/D/\infty$ queueing system sampled at equidistant points in time in such a way that the length of the sampling interval is less than the constant service time but greater than half of it.

The Poisson MA(q)
The Poisson $MA(1)$ extends to higher order in an obvious way:

$$X_t = Z_t + \beta_1 * Z_{t-1} + \cdots + \beta_q * Z_{t-q}, \tag{11}$$

where $\beta_i \in [0, 1]$ for each i, $\beta = 1 + \sum_{i=1}^{q} \beta_i$, and $\{Z_t\}$ are i.i.d. Poissons of mean θ/β. All thinning operations at time t are performed independently and are independent of all previous thinning operations. In this case, $\{X_t\}$ is a sequence of dependent Poissons of mean θ with $MA(q)$-type correlation structure. In fact, writing $\beta_0 = 1$,

$$\rho_X(k) = \begin{cases} \sum_{i=0}^{q-k} \beta_i \beta_{i+k} / \sum_{i=0}^{q} \beta_i, & k = 0, 1, \ldots, q, \\ 0, & k > q. \end{cases}$$

The Poisson $MA(q)$ can also be interpreted as a regularly observed continuous time processes. They are generalizations of the $M/D/\infty$ processes, e.g., for the $MA(2)$, service times are a mixture of two constant times.

The Poisson ARMA(1, q)
The mixed process is constructed by coupling the two component models. Thus, there are two equations:

$$Y_t = \alpha * Y_{t-1} + Z_t,$$

$$X_t = Y_{t-q} + \sum_{k=1}^{q} \beta_k * Z_{t+1-k}, \tag{12}$$

where, again, all thinning operations are independent. As before, choosing $\{Z_t\}$ as i.i.d. Poissons of mean $(1-\alpha)\theta$ and Y_0 as an independent Poisson of mean θ yields $\{Y_t\}$

as a Poisson $AR(1)$ of mean θ, and $\{X_t\}$ as a stationary process with Poisson marginal distribution of mean $(1+(1-\alpha)\sum_{k=1}^{q}\beta_k)\theta$. The process is structurally an $ARMA(1,q)$ with $\rho_X(k) = \alpha^{k-q}\rho_X(q)$ for $k \geqslant q$. Setting $\alpha = 0$ yields the Poisson $MA(q)$ and similarly setting all β_k to zero yields the Poisson $AR(1)$.

The joint distribution of n consecutive observations in the Poisson $ARMA(1,q)$ is given by the multivariate Poisson distribution of Teicher (1954) and some of the properties of the model can be derived from it. In particular, the process is not, in general, time-reversible, except for the $AR(1)$, $MA(1)$ and $MA(2)$ cases. On the other hand, the directional bivariate moments, $\text{Cov}(X_t^2, X_{t-k})$ and $\text{Cov}(X_t, X_{t-k}^2)$, are always equal because of the symmetry of the bivariate distributions. The relevance of this result is that such directional moments are often recommended for examination to assess real data for time-reversibility. In addition, as the mean of the Poisson $ARMA(1,q)$ process increases, these joint distributions tend to those of the corresponding Gaussian $ARMA(1,q)$ process.

4.2. The vector AR(1)

One approach to higher-order models is to derive them as components of a vector $AR(1)$ and we define such a process now. If α is a p-dimensional vector of probabilities whose sum does not exceed unity and X is a non-negative integer, we define $\alpha * X$ conditional on $X = x$ to be multinomial(x, α), i.e., the ith component of the p-vector $\alpha * X$ is the number of outcomes of type i in x independent and identical trials where the probability of such an outcome in any trial is the ith component of α. We refer to the operation of α on X in $\alpha * X$ as *multinomial thinning*.

Now, consider a vector $X = (X_1, X_2, \ldots, X_p)$ where each X_i is a non-negative integer and a matrix $A = [\alpha_1, \alpha_2, \ldots, \alpha_p]$ where each α_i is a vector of the same type as α, and write $A * X = \sum_{i=1}^{p} \alpha_i * X_i$, where each multinomial thinning is performed independently of all others. The stationary p-dimensional $AR(1)$ is now defined by

$$X_t = A * X_{t-1} + Z_t, \tag{13}$$

where $\{Z_t\}$ is a sequence of i.i.d. random vectors. Taking a.p.g.f.s throughout Eq. (13) yields

$$G_X(u) = G_X(A'u) G_Z(u), \tag{14}$$

so that, at least in theory, we can generate a sequence of dependent vectors $\{X_t\}$ from this model with a particular marginal distribution, provided we can find a suitable p-dimensional distribution for the innovation sequence $\{Z_t\}$.

As before, we can derive some useful properties for this general vector $AR(1)$ model (13). For example, the autocovariances $\Gamma(k) = \text{Cov}(X_t, X_{t-k}) = A^k \Gamma(0)$ and the conditional expectation of X_t given X_{t-k} is linear in the latter.

In order to specialize to Poisson components, we consider the case where $\{X_t\}$ is to be multiple Poisson, i.e., the components of the vector are independent Poisson r.v.s. Thus, X_t has a.p.g.f. $\exp(-\theta'u)$ where θ is the vector of means of the multiple Poisson.

Substituting this into the a.p.g.f. equation (14) reveals that Z_t must also be multiple Poisson with mean vector $(I - A)\theta$ in order for $\{X_t\}$ to be stationary.

Further,

$$G_{X_t,X_{t-1}}(u,v) = \exp[-\theta'u - \theta'v + u'\Gamma(1)'v]$$

and so $\{X_t\}$ is time-reversible if and only if $\Gamma(1)$ is symmetric, the same condition as for a Gaussian vector $AR(1)$.

The multiple Poisson $AR(1)$ can be extended to an $ARMA(1, q)$ in the same way as the univariate Poisson. Note also that any component of the multiple Poisson $AR(1)$ will in general have an $ARMA(p, p-1)$ structure. For example, the autocorrelation function of the jth component of the vector is given by

$$\rho_j(k) = \begin{cases} [A^k]_{jj}, & k = 1, 2, \ldots, p, \\ \sum_{i=1}^{p} \phi_i \rho_j(k-i), & k > p, \end{cases}$$

where $\phi_1, \phi_2, \ldots, \phi_p$ are the coefficients of the characteristic equation of the matrix A, i.e., $A^p = \sum_{i=1}^{p} \phi_i A^{p-i}$. The case $p = 2$ is discussed in detail in McKenzie (1988b). This same paper also briefly discusses the possibility of using more general multivariate Poisson distributions other than the multiple Poisson.

4.3. Compound Poisson models

Many of the standard distributions on the non-negative integers are either compound Poisson distributions or can be approximated by them. Thus, it is worth noting that all of the models discussed for the Poisson distribution here can be modified in a simple way to handle compound Poisson marginals. If X is a compound Poisson r.v. it may be written as $X = \sum_{i=1}^{N} Y_i$ where N is a Poisson r.v. of mean θ, say, and $\{Y_i\}$ are i.i.d. r.v.s. A generalized form of thinning may be defined as $\alpha \circ X = \sum_{i=1}^{N} B_i(\alpha) Y_i$ where $\{B_i(\alpha)\}$ are i.i.d. binary r.v.s with $P(B_i(\alpha) = 1) = \alpha$. Thus, an $AR(1)$ form suitable for compound Poisson marginals is

$$X_t = \alpha \circ X_{t-1} + Z_t,$$

where $\{Z_t\}$ are i.i.d. compound Poissons of the form $Z = \sum_{i=1}^{M} Y_i$ with $\{Y_i\}$ i.i.d. with the same distribution as before and M an independent Poisson of mean $(1 - \alpha)\theta$. Evidently, this process is constructed simply by marking the events in the original Poisson $AR(1)$ process.

If binomial thinning is replaced by this generalized form in all the Poisson models discussed above then they will support compound Poisson marginals.

4.4. Geometric marginals

Consider the geometric distribution with probability function $p_X(x) = (1 - \theta)\theta^x$ for $x = 0, 1, \ldots$, and a.p.g.f. $G_X(s) = (1 + \mu s)^{-1}$ where $\mu = \theta/(1 - \theta)$ is the mean of the distribution. If this is to be the marginal distribution of the discrete $AR(1)$ defined by (6) then the innovation distribution must have a.p.g.f. given by

$$G_Z(s) = (1 + \mu\alpha s)/(1 + \mu s) = \alpha + (1 - \alpha)\frac{1}{1 + \mu s},$$

i.e., $Z = 0$ with probability α and has the same geometric distribution as X with probability $(1 - \alpha)$. Thus, the discrete $AR(1)$ can be rewritten in this case in the form

$$X_t = \alpha * X_{t-1} + U_t W_t,$$

where $\{U_t\}$ are i.i.d. binary r.v.s with $P(U = 0) = \alpha$, independent of $\{W_t\}$ which are i.i.d. geometric of mean μ. If X_0 is also geometric of mean μ the $\{X_t\}$ will be a stationary $AR(1)$ with the same marginal distribution as X_0.

The joint a.p.g.f. of two consecutive observations in this process is given by

$$G_{X_t, X_{t-1}}(u, v) = \frac{1 + \alpha\mu u}{(1 + \mu u)(1 + \alpha\mu u + \mu v)}$$

and from its lack of symmetry we see that this process is not time-reversible. In fact, if α is near unity realizations exhibit sudden rises (when $U_t = 1$) and longer recessions (when $U_t = 0$).

It is an exact discrete analogue of the $EAR(1)$ process, an $AR(1)$ based on (5) but with negative exponential marginals, presented by Gaver and Lewis (1980). The reason for the relationship is interesting and revealing. It lies in two facts: (i) the equation in a.p.g.f.s associated with (6) is exactly the same as the equation in Laplace transforms (L.T.) associated with (5), and (ii) the a.p.g.f. of the geometric distribution of mean μ is identical to the Laplace transform of the negative exponential distribution with the same mean. Intuitively, we might view the geometric observations as counts in a homogeneous Poisson process of unit rate observed during consecutive intervals of time whose lengths are negative exponential random variables of mean μ.

As a consequence, any models with negative exponential marginals based on linear combinations of independent r.v.s with fractional and/or random binary coefficients can be duplicated for geometric marginals by replacing the scalar multiplication by binomial thinning. In fact, the $EAR(1)$ was extended to a $MA(1)$ by Lawrance and Lewis (1977), to an $ARMA(1, 1)$ by Jacobs and Lewis (1977) and to a general $ARMA(p, q)$ form by Lawrance and Lewis (1980). All have geometric analogues which are detailed in McKenzie (1986). Lawrance and Lewis (1981) developed a further form, a random coefficient autoregression, the $NEAR(1)$, and we show both it and its geometric analogue below to illustrate the closeness of the relationship. The $NEAR(1)$ is given by

$$Y_t = \beta U_t Y_{t-1} + [1 - V_t + (1 - \alpha)\beta V_t] E_t,$$

where $\{E_t\}$ is a sequence of negative exponentials of mean μ and $\{U_t\}$ and $\{V_t\}$ are independent sequences of i.i.d. binary r.v.s such that $P(U = 1) = \alpha$ and $P(V = 1) = \alpha\beta/[1 - (1 - \alpha)\beta]$. The geometric analogue is given by

$$X_t = (\beta U_t) * X_{t-1} + [1 - V_t + (1 - \alpha)\beta V_t] * W_t,$$

where $\{U_t\}$ and $\{V_t\}$ are exactly as before and $\{W_t\}$ are i.i.d. geometric of mean μ. Higher-order exponential forms have been developed and McKenzie (1985b) gives a geometric analogue of the *NEAR*(2) of Lawrance and Lewis (1985).

It is perhaps worth noting that higher-order *AR* models are not entirely satisfactory, both in the Poisson case above and here in the geometric one, when we try to construct them directly. Lack of suitable multivariate distributions means we have to resort to structures of mixtures like the *DARMA* models, i.e., an $AR(p)$ for the geometric would be of the form $X_t = \alpha_{N_t} * X_{t-N_t} + U_{t,N_t} W_t$ where $P(N_t = k) = a_k$ for $k = 1, 2, \ldots, p$, $\{W_t\}$ are i.i.d. geometric of mean μ and $\{U_{t,N}\}$ are i.i.d. binary with $P(U_{t,N} = 0) = \alpha_N$ for $N = 1, 2, \ldots, p$.

4.5. Negative binomial marginals

The negative binomial distribution has $p_X(x) = \Gamma(\beta + x)(1 - \theta)^\beta \theta^x / \Gamma(\beta) x!$ for $x = 0, 1, \ldots$, and corresponding a.p.g.f. $G_X(s) = (1 + \mu s)^{-\beta}$ where $\mu = \theta/(1 - \theta)$. Notice that the mean of this distribution is $\beta\mu$. We will say such a r.v. is $NB(\beta, \theta)$. Note that its a.p.g.f. is identical to the L.T. of a gamma density with shape and scale parameters β and $1/\theta$, respectively, i.e., $Ga(\beta, 1/\theta)$, with density $f(x) = \theta^{-\beta} x^{\beta-1} e^{x/\theta} / \Gamma(\beta)$ for $x > 0$. In fact, the $NB(\beta, \theta)$ and $Ga(\beta, 1/\theta)$ distributions are related in exactly the same way as the geometric, i.e., $NB(1, \theta)$, and negative exponential, i.e., $Ga(1, 1/\theta)$, distributions in the last section.

Gaver and Lewis (1980) derived the L.T. of the density of the innovation distribution for the $AR(1)$ based on (5) which is marginally $Ga(\beta, 1/\theta)$ and Lawrance (1982) derived a construction for the r.v. based on the Gamma distributed shot-noise process. The corresponding solutions in the $NB(\beta, \theta)$ case are that the innovation distribution has a.p.g.f. $G_Z(s) = \{\alpha + (1 - \alpha)\frac{1}{1+\mu s}\}^\beta$ and that Z_t may be constructed for each t using

$$Z = \sum_{i=1}^{N} (\alpha^{R_i}) * W_i, \tag{15}$$

where $\{R_i\}$ are i.i.d. uniform on $(0, 1)$, $\{W_i\}$ are i.i.d. $NB(1, \theta)$, and N is Poisson of mean $(-\beta \ln \alpha)$. Thus, if $\{Z_t\}$ are i.i.d. and given by (15) and X_0 is $NB(\beta, \theta)$ then (6) will generate a stationary $AR(1)$ which is marginally $NB(\beta, \theta)$.

The representation (15) suggests the interpretation of this $AR(1)$ as a regularly observed continuous time shot-noise process of the form

$$X(t) = \sum_{i=-N(\infty)}^{N(t)} (\alpha^{t-\tau_i}) * W_i,$$

where $\{W_i\}$ are i.i.d. as above, and $N(t)$ is a Poisson process with occurrence times $\{\tau_i\}$. There is more discussion of this model and other $AR(1)$-type models with negative binomial marginal distributions, including the linear birth–death–immigration process observed at regularly spaced times, in McKenzie (1987).

The complexity of the innovations processes for these models suggests that there may be simpler forms for an $AR(1)$ with a negative binomial marginal than simply (6). Lewis (1985) pursued this same idea for the $AR(1)$ given by (5) for Gamma distributed marginals and developed a random coefficient version for which the negative binomial form, given in McKenzie (1986), is

$$X_t = A_t * X_{t-1} + Z_t, \tag{16}$$

where $\{A_t\}$ are i.i.d. beta r.v.s with parameters $(\alpha\beta, (1-\alpha)\beta)$, i.e., A has density $f_A(x) = x^{\alpha\beta-1}(1-x)^{(1-\alpha)\beta-1}/B(\alpha\beta, (1-\alpha)\beta)$ for $0 \leqslant x \leqslant 1$. Further, $\{Z_t\}$ are i.i.d. $NB((1-\alpha)\beta, \theta)$ and if X_0 is $NB(\beta, \theta)$ then $\{X_t\}$ is a stationary $AR(1)$ with $NB(\beta, \theta)$ marginal distribution. The autocorrelation function of this process is $\rho_X(k) = \alpha^k$, for $k = 0, 1, \ldots$, and from the bivariate distribution of consecutive observations the process is time-reversible.

Notice that this random coefficient model (16) based on binomial thinning also provides a much simpler $AR(1)$ for the geometric, using $\beta = 1$, since the innovations are then negative binomial.

Notice too that this model (16) now has essentially the same structure as the Poisson $AR(1)$, i.e., the innovation distribution lies in the same family as the marginal distribution and the forms of the distribution which appear in the defining equation are additive, i.e., sum to another member of the family. The effect of these properties is that we can extend the $AR(1)$ model to a $MA(q)$ model and also an $ARMA(1, q)$ form, just as we did with the Poisson. There is little new there except in the properties of the individual processes, so we simply note the form of the $ARMA(1, q)$, thus:

$$Y_t = A_t * Y_{t-1} + Z_t,$$

$$X_t = Y_{t-q} + \sum_{k=1}^{q} B_{k,t} * Z_{t+1-k},$$

where all thinning operations are independent, the $\{A_t\}$ and the $\{Z_t\}$ are exactly as before, and the $\{B_{k,t}: k = 1, 2, \ldots, q\}$ are i.i.d. random vectors of independent beta distributed components so that $B_{k,t}$ have parameters $(\beta\alpha_k, \beta(1 - \alpha - \alpha_k))$. Thus, choosing $\{Z_t\}$ as i.i.d. $NB(\beta(1-\alpha), \theta)$ and Y_0 to be $NB(\beta, \theta)$ ensures that $\{Y_t\}$ is a stationary $AR(1)$ with $NB(\beta, \theta)$ marginal distribution, and $\{X_t\}$ is a stationary process whose marginal distribution is $NB(\beta(1+\sum_{k=1}^{q}\alpha_k), \theta)$. The process is structurally an $ARMA(1, q)$ with $\rho_X(k) = \alpha^{k-q}\rho_X(q)$ for $k \geqslant q$.

Further, in exactly the same way that we constructed a vector $AR(1)$ with multiple Poisson marginals using (13), we can derive a vector $AR(1)$ whose marginal distribution is multiple negative binomial, i.e., a vector of the form $(X_1, X_2, \ldots, X_p)$ where, for each i, X_i is $NB(\beta_i, \theta)$ and the X_is are independent. The form of the model is exactly the same, i.e., as in (13), but now the columns of the matrix A are Dirichlet random vectors which operate on the independent negative binomial components of X_t using multinomial thinning as before. More details on the derivation are available in Al-Osh and Alzaid (1993). Just as in the Poisson case, this model will extend to higher orders although, again, more benefit may be derived from considering the structure of univariate components of the vector $AR(1)$.

In addition, the random coefficient negative binomial $AR(1)$ given by (16) was extended by Böckenholt (1999) for use with the multivariate negative binomial, or negative multinomial, as its marginal distribution. The paper details an application to longitudinal multivariate count data from a panel study on personality and emotion.

4.6. Binomial marginals

We will write that the r.v. X is binomial(N, θ) if its probability function is given by $p_X(x) = \binom{N}{x} \theta^x (1-\theta)^{N-x}$ for $x = 0, 1, \ldots, N$. As noted, it is not discrete self-decomposable and so, in general, is not a suitable marginal distribution for the discrete $AR(1)$ defined by (6). However, Al-Osh and Alzaid (1993) presented an alternative form of discrete $AR(1)$ based on *hypergeometric thinning* which may be regarded as a 'natural' form for binomial marginals. When we wish the marginal distribution of $\{X_t\}$ to be binomial(N, θ) hypergeometric thinning is defined in the following way. The term in the discrete $AR(1)$, i.e., (6), corresponding to survivors from the count at time $(t-1)$, i.e., $\alpha * X_{t-1}$, is replaced by sampling M of the N trials (where $M < N$) without replacement. Thus, if the number of these survivors is $S_M(X_{t-1})$ then

$$P(S_M(X_{t-1}) = y | X_{t-1} = x) = \binom{x}{y}\binom{N-x}{M-y} / \binom{N}{M}.$$

It follows that if X_{t-1} is binomial(N, θ) then $S_M(X_{t-1})$ is binomial(M, θ). The corresponding $AR(1)$ can be written as

$$X_t = S_M(X_{t-1}) + Z_t, \tag{17}$$

where Z_t is binomial$(N - M, \theta)$. Thus, if $\{Z_t\}$ are i.i.d. binomial$(N - M, \theta)$ r.v.s then (17) will generate a stationary process $\{X_t\}$ of $AR(1)$-type whose marginal distribution is binomial(N, θ).

This version of the discrete $AR(1)$ appears very different from (6) but, in fact, it shares many of its important properties, when the marginal distribution is binomial. For example, the conditional mean of X_t given X_{t-1} is linear in X_{t-1}, and the autocorrelation function of $\{X_t\}$ is $\rho_X(k) = (M/N)^k$, for $k = 0, 1, \ldots$. The conditional variance of X_t given X_{t-1}, however, is quadratic in X_{t-1}.

The usefulness of the model (17) is somewhat limited by the fact that the range of values for autocorrelation depends on integer multiples of $1/N$, and N may not be very large. On the other hand, like (16), this model has essentially the same structure as the Poisson $AR(1)$ given by (6) and so can be extended in the same way as those two $AR(1)$ models. Thus, we can derive $MA(q)$ and $ARMA(1, q)$ models and a vector $AR(1)$ whose marginal distribution is a multiple binomial, i.e., the vector is of the form

$(X_1, X_2, \ldots, X_p)$ where, for each i, X_i is binomial(N_i, θ) and the X_is are independent. Details of this model are given in Al-Osh and Alzaid (1993).

An alternative binomial $AR(1)$ based on binomial thinning was provided by McKenzie (1985a). If the marginal distribution of $\{X_t\}$ is binomial(N, θ) and $\theta \leqslant 0.5$ then the model is

$$X_t = \alpha * X_{t-1} + \beta * (N - X_{t-1}), \tag{18}$$

where $\beta = (1 - \alpha)\theta/(1 - \theta)$, $0 < \alpha < 1$, and, as usual, all thinning operations are independent. If X_0 is binomial(N, θ) then this Eq. (18) generates a stationary $AR(1)$ process $\{X_t\}$ with a binomial(N, θ) marginal distribution.

If $\theta > 0.5$ we can recast the model (18) in terms of $Y_t = N - X_t$ and then $\beta = (1 - \alpha)(1 - \theta)/\theta$.

This model (18) is clearly a Markov chain but does not look like the other $AR(1)$ models we have examined. However, it does have $AR(1)$ structure, e.g., $\rho_X(k) = \rho^k$ where $\rho = (\alpha - \theta)/(1 - \theta)$ and both the conditional mean and variance of X_t given X_{t-1} are linear in X_{t-1}. Note also that, unlike the other $AR(1)$ models we have seen, the autocorrelation function of (18) can be both positive and negative. In fact, the full range of values is possible. In addition, the process $\{X_t\}$ is time-reversible.

As with the other $AR(1)$ models, (18) has a direct interpretation in terms of a continuous time Markov process observed at regularly spaced times. It is a model used in teletraffic to describe the number of busy lines when there are as many customers as lines (N) and demand arises as a Poisson process, call lengths are negative exponential and lost calls are cleared, i.e., waiting is not allowed. Evidently, the two components of (18) represent calls surviving the sampling interval and newly arrived ones.

Binary marginals

Binary r.v.s can be treated as binomial$(1, \theta)$ but then the model (17) is not helpful, and the model (18) can be rewritten as

$$X_t = A_t X_{t-1} + B_t(1 - X_{t-1}), \tag{19}$$

where $\{A_t\}$ and $\{B_t\}$ are independent sequences of i.i.d. binary r.v.s with $P(A_t = 1) = \alpha$ and $P(B_t = 1) = \beta$. Since the variate is binary we can replace thinning by multiplication by a binary r.v. The model (19) is a two-state Markov chain with marginal distribution $\{1 - \theta, \theta\}$ on $\{0, 1\}$ and parametrized by setting $P(X_t = 1 | X_{t-1} = 1) = \alpha$. In fact, whatever parametrization is used, the stationary binary $AR(1)$ is a two-state Markov chain and can be modelled as such.

Another model for a binary $AR(1)$ is due to Kanter (1975) and is given by

$$X_t = Z_t(X_{t-1} \oplus U_t) + (1 - Z_t)U_t, \tag{20}$$

where $\{Z_t\}$ and $\{U_t\}$ are independent sequences of i.i.d. binary r.v.s with $P(Z_t = 1) = p$ and $P(U_t = 1) = \theta(1-p)/(1-2p\theta)$, and $\oplus$ represents addition modulo 2. In this case, $\rho_X(k) = [(1 - 2\theta)p]^k$ for $k = 0, 1, \ldots$, which can again take the full range of values.

Of course, this process is still a two-state Markov chain but the model form (20) has the advantage that it has been extended to higher-order *AR*s by Kanter (1975) and to the full range of *ARMA* models by McKenzie (1981).

4.7. Models with unspecified marginals

We have noted in detail the results for three families of marginal distributions, *viz* Poisson, negative binomial and binomial, but there are a number of papers whose results refer to some of these processes in a more general context. These include Al-Osh and Alzaid (1987a, 1988), Aly and Bouzar (1994) and Alzaid and Al-Osh (1988, 1990, 1993). These authors (and some others) refer to these models based on thinning as *INAR*(1) and *INMA*(*q*), etc. I have not used those acronyms here since there are often a number of quite distinct models sharing each one. In addition, some of these papers deal with questions of inference and compare different approaches. The usual methods of estimation considered are conditional least squares (CLS), maximum likelihood (ML) and the generalized method of moments (GMM). Inference for these models is also considered in the papers by Brännäs (1994, 1995a, 1995b). In addition, Brännäs and Hall (1998) extended Al-Osh and Alzaid (1993), presenting four different forms of the *INMA*(*q*) model and comparing estimators, including a new GMM estimator based on the a.p.g.f. The procedures are illustrated by modelling the numbers of stocks of a small Swedish firm traded daily.

In all previous discussions of *ARMA*-like structure and behaviour there is one notable omission. There have been no 'natural' extensions of these models for particular marginals to higher order *AR*s. However, Alzaid and Al-Osh (1990) presented their *INAR*(*p*) where the current observation is a somewhat complex linear combination (using thinning) of the last p observations. It is given by

$$X_t = \sum_{i=1}^{p} \alpha_i * X_{t-i} + Z_t, \qquad (21)$$

where $\{Z_t\}$ are i.i.d. r.v.s as before, $\{\alpha_i\}$ are probabilities whose sum is less than unity and, conditional on X_t, the random vector $(\alpha_1 * X_t, \alpha_2 * X_t, \ldots, \alpha_p * X_t)$ has a multinomial distribution with parameters $(\alpha_1, \alpha_2, \ldots, \alpha_p, X_t)$. Notice that although this is multinomial thinning it takes place *through time*, since $\alpha_k * X_t$ is an explicit component of X_{t+k} and is evaluated at time $(t+k)$ for $k=1,2,\ldots$. These multinomial thinnings are independent of each other. The effect of this construction is that the correlation structure of the process mirrors that of the *ARMA*(*p*, *p* − 1) rather than the *AR*(*p*).

A timely paper by Du and Li (1991) presented a simpler and more general *AR*(*p*) model using thinning. It has the *AR*(*p*) form (21) but now each thinning operation $\alpha_k * X_{t-k}$ is performed independently, and at time t, and we require only that each of $\{\alpha_i\}$ is a probability. This process has the classical *AR*(*p*) correlation structure. They show that an unique stationary solution to Eq. (21) as defined by them exists if all the roots of the polynomial equation

$$\lambda^p - \sum_{i=1}^{p} \alpha_i \lambda^{p-i} = 0 \tag{22}$$

lie inside the unit circle. This is the same condition as for the classical linear $AR(p)$. They also show that the sample autocovariance and autocorrelation functions are strongly consistent as is conditional least squares.

This work was generalized by Gauthier and Latour (1994) and Latour (1997, 1998). They generalized the binomial thinning operator to the form $\alpha \odot X = \sum_{i=1}^{X} Y_i$ where X is a non-negative integer and $\{Y_i\}$ are i.i.d. r.v.s with mean α and finite variance. Thus, they present a model, the $GINAR(p)$, of the same form as Du and Li (1991) but in which binomial thinning $*$ is replaced by $\odot$. They examine the same properties, i.e., existence and uniqueness as well as estimation problems. In addition, they show that an equivalent condition to (22) for these models is the more easily checked

$$\sum_{i=1}^{p} \alpha_i < 1. \tag{23}$$

Latour (1997) also extends these models to a vector $AR(p)$, the $MGINAR(p)$, in which matrices operate on vectors using this generalized thinning operator. Thus, if A is an $(r \times r)$ matrix of such operators, $\{\alpha_{ij} \odot\}$, and $X' = (X_1, X_2, \ldots, X_r)$ then the ith component of the vector $A \odot X$ is given by $\sum_{j=1}^{r} \alpha_{ij} \odot X_j$. As usual, all thinning operations are independent. Dion et al. (1995) demonstrate the relationship of $GINAR(p)$ to multitype branching processes with immigration and show how this can be exploited to establish theoretical properties and some inference for the integer processes. They also indicate an extension to a $GINARMA(p,q)$ process in an analogous way, i.e.,

$$X_t = \sum_{i=1}^{p} \left(\sum_{k=1}^{X_{t-i}} U_{i,k,t} \right) + Z_t + \sum_{i=1}^{q} \left(\sum_{k=1}^{Z_{t-i}} V_{i,k,t} \right)$$

$$= \sum_{i=1}^{p} \alpha_i \odot X_{t-1} + Z_t + \sum_{i=1}^{q} \beta_i \odot Z_{t-i}.$$

We have commented on the similarity of structure of the richest models with the specific marginal distributions we considered, i.e., Poisson, negative binomial and geometric, and binomial and binary. In each case, the $AR(1)$ form was

$$X_t = S_t(X_{t-1}) + Z_t, \tag{24}$$

where $\{Z_t\}$ are i.i.d. r.v.s from the same family as the marginal distribution of $\{X_t\}$, and $S_t(X_{t-1})$ is a random operation performed on X_{t-1} which reduces it, e.g., binomial thinning or, in the binomial marginal case, hypergeometric thinning. Joe (1996) characterizes this form in the following construction. Let $\{F_\theta\}$ be the distribution function of an infinitely divisible convolution-closed parametric family (with $\theta > 0$) such that $F_{\theta_1} \otimes F_{\theta_2} = F_{\theta_1+\theta_2}$ where $\otimes$ is the convolution operator. If Z_{θ_1} and Z_{θ_2} are independent and such that Z_{θ_i} has distribution function F_{θ_i} then define $G_{\theta_1,\theta_2,z}$ as the distribution function of $Z_1 | Z_1 + Z_2 = z$. Let S be a random operator such that, given $X = x$, $S(X)$ has the distribution function $G_{\alpha\theta,(1-\alpha)\theta,x}$, and when $X \sim F_\theta$, $S(X) \sim F_{\alpha\theta}$. Then, if $\{Z_t\}$ are i.i.d. $F_{(1-\alpha)\theta}$, $X_0 \sim F_\theta$ and $\{S_t\}$ are independent replications of the operator S, the process $\{X_t\}$ generated by (24) is a stationary $AR(1)$-type process. Thus, any member of the convolution-closed, infinitely divisible class of distributions will lend itself to such a construction. On the other hand, the fact that the construction is possible for the binomial, which is not infinitely divisible, provided we use hypergeometric thinning, makes it clear that alternative forms for this equation and its extensions are possible. What matters is finding the correct form of thinning and the functional form under which the family of distributions is closed.

In the same paper, Joe (1996) developed an $AR(2)$ model. It is based on a 'natural' extension to the multivariate form of the univariate distributions identified for the $AR(1)$ and its extensions. These multivariate distributions are analogous in form to the multivariate Poisson distributions given by Teicher (1954). The results, though interesting, yield an $AR(2)$ which is very complex in form and whose conditional means, $E(X_t | X_{t-1}, X_{t-2})$ are non-linear in X_{t-1} and X_{t-2}. It would seem that solving the problems of finding the best 'natural' bivariate distributions for such processes still has some way to go.

4.8. Related models

Brännäs (1994) extended the work of Al-Osh and Alzaid (1987a) comparing estimators for $AR(1)$ with different marginals, and also presented an extension of the discrete $AR(1)$ for use with panel data of counts. Consider a panel of M cross-section units observed for T time periods and construct an M-dimensional vector $AR(1)$ for the counts in the form $X_t = A * X_{t-1} + Z_t$ for $t = 1, 2, \ldots, T$ where $A* = \text{diag}\{\alpha_1, \alpha_2, \ldots, \alpha_M\}$, $X'_t = (X_{1t}, X_{2t}, \ldots, X_{M,t})$ and $Z'_t = (Z_{1t}, Z_{2t}, \ldots, Z_{M,t})$. The $AR(1)$ coefficients $\{\alpha_i\}$ differ for different cross-section units and may even vary in time, if necessary, e.g., as functions of covariates. Dependence between units is generated solely by the form of the vector Z_t which is given by $Z_{i,t} = V_{i,t} + W_t$, where $\{V_{i,t}\}$ are independent r.v.s for all i and t, and $\{W_t\}$ are i.i.d. r.v.s. If the counts are Poisson, $\{V_{i,t}\}$ are independent Poissons of means $\{\lambda_i\}$ and $\{W_t\}$ are i.i.d. Poissons of mean $\{\delta\}$. This model was also discussed in Berglund and Brännäs (1996).

One attempt to find alternative structural forms for the $AR(1)$ process is represented by the work of Littlejohn (1992a, 1992b) on discrete minification processes. These are discrete analogues of the positive continuous variate minification processes of Lewis and McKenzie (1991). The defining equation for the continuous variate process is

$$X_t = \min(X_{t-1}, Z_t)/\rho, \qquad (25)$$

where $0 < \rho < 1$ and Z_t are i.i.d r.v.s. Now, $\{X_t\}$ exhibits the $AR(1)$ autocorrelation function $\rho_X(k) = \rho^k$ for $k = 0, 1, \ldots$, and a number of interesting properties including the fact that when $\{X_t\}$ is marginally negative exponential the process is the time reversed version of the $EAR(1)$ referred to earlier. Part of Littlejohn's interest lies in whether similar results are available for the geometric distribution as marginal of the discrete $AR(1)$ given by (6). However, the major practical problem is to find a suitable analogue for scalar division by a fraction as in (25). Littlejohn's solution (1992a) is the process

$$X_t = \rho \setminus \min(X_{t-1}, Z_t), \tag{26}$$

where Z_t are i.i.d. and satisfy $P(Z \geqslant n) = P(\rho * X \geqslant n)/P(X \geqslant n)$, and $\rho\setminus$ is a random operation, analogous to scalar division by a fraction and an inverse function for binomial thinning, defined for the distribution of $\{X_t\}$ by

$$P(\rho \setminus W = n | W = m) = \frac{\binom{n}{m}(1-\rho)^n p_X(n)}{\sum_{k=m}^{\infty} \binom{k}{m}(1-\rho)^k p_X(k)}$$

for $m = 0, 1, \ldots$; $n = m, m+1, \ldots$. Littlejohn (1992a) refers to this operator as X-re-composition. Note that it depends on the distribution of X. If $W = \min(X, Z)$ then $P(\rho \setminus W = i) = p_X(i)$.

4.9. Non-stationary models

In practice, most time series, whether discrete-valued or not, are non-stationary. Typically, changes in long term trend and periodic trends reflecting seasonality appear and must be modelled. These are not easily handled in discrete-variate time series models.

Consider the basic discrete $AR(1)$ (6) and taking expectations yields

$$\mu_t = \alpha \mu_{t-1} + \omega_t, \tag{27}$$

where $\mu_t = E(X_t)$ and $\omega_t = E(W_t)$. In the stationary case, we have $\omega_t = (1-\alpha)\mu_{t-1}$, but, in general, Eq. (27) will hold whatever the forms for the sequences of means, provided only that they are positive. Thus, we can endow $\{X_t\}$ with a time-dependent mean $\{\mu_t\}$, simply by manipulating the innovation means $\{\omega_t\}$. Suppose, for example, we wish $\{X_t\}$ to display a linear trend of the form $a + bt$ then setting $\omega_t = \alpha b + (1-\alpha)(a+bt)$ in (27) will ensure that $\{\mu_t\}$ has the correct form. This idea is discussed in McKenzie (1985a) where alternative forms such as seasonality and changes in level are considered. It appears to offer a reasonable approach to the problem of modelling deterministic trends in these models based on thinning, and it will extend to more general forms of model than the $AR(1)$ considered here.

On the other hand, the deterministic trend modelled and the correlation structure are not independent, and this may be a serious problem in practice. In the example above, the innovation means, $\{\omega_t\}$, must, of course, be positive. It is easy to show that this implies that $\alpha < 1 + b/(a + bt)$. Now, b may be negative over the range of data considered whereas $(a + bt)$ is positive so that α may have an upper bound strictly less than one. More generally, it is clear from (27) that we require that $\alpha < \mu_t/\mu_{t-1}$ for all t, and if $\{\mu_t\}$ is falling then this may create problems in finding a suitable model of this type.

This problem can occur for any type of deterministic trend and in all of the models we have discussed. For example, in the binary model given by (19) there is no explicit innovation process. Nevertheless, we can introduce time dependence into the probability θ_t, the mean of the process, by allowing either or both of the binary r.v.s A_t and B_t to be time dependent. This yields

$$\theta_t = \alpha_t \theta_{t-1} + \beta_t(1 - \theta_{t-1}), \tag{28}$$

where, in the stationary case, $\alpha_t = \alpha$ and $\beta_t = \beta = (1 - \alpha)\theta/(1 - \theta)$. As noted already, this model (19) is a two-state Markov chain with $P(X_t = 1|X_{t-1} = 1) = \alpha_t$ and $P(X_t = 1|X_{t-1} = 0) = \beta_t$. Since, in the stationary case, we model β as a function of α and the correlation is simply related to α it seems natural to induce the deterministic trend by making β_t time-dependent and allowing α to remain constant. Even here, however, we find that the correlation and trend are inter-dependent. Thus, to ensure that β_t is a probability, we find that α must satisfy

$$1 - \frac{1-\theta_t}{\theta_{t-1}} < \alpha < \frac{\theta_t}{\theta_{t-1}}$$

which, again, will be problematic if $\{\theta_t\}$ is decreasing.

On the other hand, it is possible that in some cases, at least, this problem can be avoided. For example, Azzalini (1994) has proposed a Markov chain model for binary data. Essentially, it has the same structure as (19) and so $\{\theta_t\}$ satisfies (28) above. He relates α_t and β_t by fixing the odds ratio $\psi = \alpha_t(1 - \beta_t)/(1 - \alpha_t)\beta_t$ either as a constant or as a function of covariates, and $\{\theta_t\}$ may be a function of time-dependent covariates also. In this formulation, however, it appears that the form of the deterministic components does not restrict the range of values available to the model parameters. In addition, he proposes extensions to repeated measures binary data and deals with the problems with missing data. He also argues that it may not matter if serial dependence is not accurately modelled, provided this does not affect the estimation of the regression parameters. This seems a little facile since often interest lies in testing the significance of these parameters and poor tests usually result from failure to deal adequately with serial correlation. On the other hand, his simulations suggest that estimated standard errors may be robust against mis-specification of the dependence structure.

There have been some attempts to make the standard models based in thinning more attractive for economic application by introducing covariates. For example, Brännäs (1995a, 1995b) models a time series of annual counts of Swedish paper and pulp mills using the Poisson $AR(1)$ in which the parameters are functions of covariates. Thus, the model takes the form $X_t = \alpha_t * X_{t-1} + Z_t$ where $\ln(\alpha_t/(1 - \alpha_t)) = \beta'Y_t$, Z_t is Poisson of mean $\lambda_t = e^{\gamma'Y_t}$ and Y_t is a p-vector of covariates at time t. He discusses estimation and prediction.

5. Regression models

An important approach to the incorporation of covariates was initiated by Zeger (1988) who was concerned with testing for a falling trend in monthly U.S. polio incidence data. He uses Poisson regression to model trend and seasonality explicitly, but recognizes the presence of residual serial correlation and the need to model it adequately to ensure the reliability of his test of the trend. His solution is to model the autocorrelation by means of an unobserved stationary process $\{\varepsilon_t\}$ with unit mean and autocovariance function $\text{Cov}(\varepsilon_t, \varepsilon_{t-k}) = \rho_\varepsilon(k)\sigma^2$ for $k = 0, 1, \ldots$. Conditional on ε_t, it is assumed that X_t is Poisson of mean and variance $\mu_t \varepsilon_t$ where μ_t is the deterministic component of the mean, i.e., $\ln \mu_t = \beta' Y_t$ where Y_t is a vector of covariates at time t. Thus, unconditionally,

$$E(X_t) = \mu_t, \qquad \text{Var}(X_t) = \mu_t + \sigma^2 \mu_t^2$$

and

$$\rho_X(k) = \frac{\rho_\varepsilon(k)}{[1 + (\mu_t \sigma^2)^{-1}]^{1/2}[1 + (\mu_{t-k}\sigma^2)^{-1}]^{1/2}}.$$

An estimating equation approach is adopted for this model and discussed in detail. As one of the motivating factors for such a model, Zeger notes that there is a complete separation of the marginal expectation μ_t, which is a function of the covariates, and the unknown parameters of the latent process $\{\varepsilon_t\}$ which define the generating mechanism for the serial correlation.

Numerically, the serial correlation in the polio series is relatively small and Zeger models $\{\varepsilon_t\}$ as though it were an $AR(1)$, i.e., it has autocorrelation function $\rho_\varepsilon(k) = \rho^k$. Campbell (1994) extended this approach to higher orders of dependence in modelling occurrences of sudden infant death syndrome and relating trends to environmental factors. Zeger's approach has been further developed by Brännäs and Johansson (1994), and extended to panel data by the same authors (1996). Questions of prediction and control were discussed by Brännäs (1995a).

The modelling of binary time series by means of Markov chains whose transition probabilities are functions of covariates and/or previously occupied states is discussed by Cox (1970) and is well illustrated in practice by, for example, Korn and Whittemore (1979) and Muenz and Rubinstein (1985). In the latter, the transition probabilities p_{00} and p_{10} are modelled by logistic regression on covariates and the approach is applied to longitudinal data studies in the health sciences. Missing data, time-dependent covariates and second-order chains are all discussed.

These ideas were generalized in a natural way by Fahrmeir and Kaufmann (1987). Although dealing with categorical time series, their models are suitable for multinomial (or binomial) time series. The conditional multinomial probabilities, i.e., conditional on the past history of the process, are modelled within the usual generalized linear model framework, e.g., as log odds ratios, a variety of models is considered and the properties of the maximum likelihood estimators discussed.

In a similar context, Zeger et al. (1985) note that modelling conditional distributions in this way may not yield simple or even useful expressions for the marginal properties which are often more relevant in such studies. They are also interested in modelling longitudinal data and assume that there is a time-independent vector of covariates Y_i associated with the the ith subject. The binary series for the ith subject $\{X_{it}\}$ is modelled as a stationary Markov chain with $\pi_i = P(X_{it} = 1|Y_i)$ given by logit $\pi_i = \beta' Y_i$, and all subjects share a common correlation structure given by $\text{corr}(X_{it}, X_{i,t-1}|Y_i) = \rho$. Consistent estimates of the regression parameters and their standard errors are obtained under mild assumptions about the true underlying dependence structure.

Liang and Zeger (1986) extend this idea of modelling the marginal structure via covariates (and treating dependence as a nuisance) to time-dependent covariates and other distributions using the formulations of generalized linear models. Thus, they do not elaborate multivariate marginal distributions for the sequence of observations but use a working generalized linear model to specify the form of the marginal distribution of X_{it}. Inference is performed using estimating equations which yield consistent estimates of the regression parameters and their variances under fairly mild assumptions about the form of the time dependence. The approach is closely related to quasi-likelihood. They note, however, that if the nature of the time dependence is important in the analysis then modelling the conditional distribution directly would be a better approach.

Zeger and Qaqish (1988) present several models based on the conditional distributions and these are also discussed in the context of longitudinal data in the book by Diggle et al. (1994). In the binary case, the model is an obvious extension of logistic regression. Thus,

$$\text{logit } p_t = \beta' Y_t + \sum_{i=1}^{q} \theta_i X_{t-i}, \tag{29}$$

where $p_t = P(X_t = 1|Y_t, X_{t-1}, X_{t-2}, \ldots)$. The dependence on $\{X_{t-i}: i = 1, 2, \ldots, q\}$ can be more complicated in form, and, of course, if all $\theta_i = 0$ then the usual logistic regression results. In addition, the individual θ_i have interpretations in terms of odds ratios.

The extension to more general count data is more problematic and several alternatives are considered for the case where, conditional on past observations and the vector of covariates Y_t, X_t is Poisson of mean μ_t. The direct analogue of (29) in the Poisson case, and taking $q = 1$ for simplicity, is

$$\ln \mu_t = \beta' Y_t + \theta X_{t-1}. \tag{30}$$

However, such a form has limited application since, unless $\theta < 0$, the conditional mean will grow exponentially with X_{t-1}. Thus, only negative association is possible here. In addition, the model's obvious interpretation is intuitively unappealing since it implies that $\exp(\beta' Y_t)$ is the mean when the previous observation was zero, whereas we would prefer that it approximate (at least) the unconditional mean. They suggest the more practical alternative

$$\ln \mu_t = \beta' Y_t + \theta \{\ln X^*_{t-1} - \beta' Y_{t-1}\}, \qquad (31)$$

where $X^*_{t-1} = \max(X_{t-1}, c)$ and $0 < c < 1$. Now, both positive and negative correlation are possible, depending on the sign of θ. An alternative form adds the constant c to all X_{t-1} and not simply zero values, thus:

$$\ln \mu_t = \beta' Y_t + \theta \{\ln(X_{t-1} + c) - \ln(\exp(\beta' Y_{t-1}) + c)\}. \qquad (32)$$

Such models as these arise naturally in the context of size-dependent branching processes and background theory from that area can be useful here.

Zeger and Qaqish (1988) refer to these models as *Markov regression models*. It is clear that they offer a useful class of models since they are flexible, readily understood, since they describe dependence on the past explicitly and in the same way as dependence on covariates, and can be fitted using available standard software.

Li (1994) attempts to extend these Markov regression models in a way that allows them to mirror moving average structure by incorporating past values of the sequence of conditional means. Thus, ignoring covariates, the $MA(1)$ form for binary data uses

$$\text{logit } \mu_t = \mu + \theta (X_{t-1} - \mu_{t-1}) \qquad (33)$$

and, for the Poisson, the $MA(1)$ form uses

$$\ln \mu_t = \mu + \theta \ln(X_{t-1}/\mu_{t-1}). \qquad (34)$$

As with the models of Zeger and Qaqish (1988), some allowance must be made for the possibility that $X_{t-1} = 0$. Higher-order models are considered.

5.1. Parameter driven models

Zeger's model (1988) for the polio data belongs to class of models which Cox (1981) describes as parameter-driven because, conditional on an unobserved parameter process, $\{\varepsilon_t\}$ in the above, the data are independent with their instantaneous marginal distribution determined by the current parameter value. A number of methods for time series of count data have been developed on this principle. For example, Keenan (1982) models a stationary binary time series $\{X_t\}$ by assuming the existence of an underlying stationary process $\{Z_t\}$ and a distribution function F such that $P(X_t = 1|Z_t) = F(Z_t)$. In particular, he considers the case when $\{X_t\}$ is marginally $N(0, \sigma^2)$ with a specified autocorrelation function and F is the distribution function of the $N(0, \tau^2)$ distribution. He derives some of the properties of such models including joint distributions, prediction and estimation of the underlying process.

In principle, such models are straightforward to construct. The data have a standard distributional form conditional on the parameter process which generates the correlation structure. Some applications, such as Zeger (1988) and Campbell (1994) appear to deal with the deterministic components and the serial correlation very well using a mixture of fairly standard models. Another successful class of parameter-driven models uses Markov chains as the parameter process. They are hidden Markov models and are described in detail in the book by MacDonald and Zucchini (1997) which also contains a large number of interesting applications of the approach. A simple example is provided by the Poisson-hidden Markov model. We assume that $\{S_t\}$ is an irreducible, homogeneous Markov chain on a set of states denoted by $\{1, 2, \ldots, m\}$ and that, conditional on $S_1 = s_1, S_2 = s_2, \ldots, S_n = s_n$ we have $X_1, X_2, \ldots, X_n$ are independent and X_t is Poisson of mean λ_{s_t}. Thus, there are m Poisson distributions with means $\{\lambda_k: k = 1, 2, \ldots, m\}$ and the process $\{X_t\}$ chooses its current marginal distribution from amongst them according to the state of the Markov chain at time t. There are m^2 parameters in such a model, the m means and $m(m-1)$ transition probabilities for the Markov chain. The number of states m is usually fairly small, sometimes as low as 2.

Evidently, this form of model can be used with any distribution in place of the Poisson. The book referenced above gives details on correlation and distributional structure and inference. The problem of model selection, in particular the number of states of the Markov chain, has yet to be solved satisfactorily, although the use of information criteria like BIC looks promising. Also dealt with are extensions to higher-order Markov chains, possibly using the mixture transition distribution models of Raftery and Tavaré (1994), multivariate processes and processes in which the parameters indicated by the state of the chain are functions of covariates, e.g., the regression coefficients are determined by the state of the chain. This last can also be modelled by allowing the transition probabilities of the Markov chain to be functions of the covariates.

6. State space and Bayesian models

In the linear, Gaussian structure, state-space models are parameter-driven models in which an unobserved state-vector θ_t evolves linearly through time and only a linear function of it is observed, usually in noise. Thus, there are two defining equations: an observation equation, and an evolution or system equation, typically of the form:

$$X_t = F_t \theta_t + \nu_t, \tag{35}$$

$$\theta_t = G_t \theta_{t-1} + \omega_t, \tag{36}$$

where $\{\nu_t\}$ and $\{\omega_t\}$ are independent sequences of i.i.d. r.v.s of zero mean and covariance matrices V_t and W_t, respectively. It is usually assumed that $\{F_t\}$, $\{G_t\}$, $\{V_t\}$ and $\{W_t\}$ are known. Evidently, there are all the usual difficulties in extending such a structure to discrete-valued data.

The usual approach, as described for example by Durbin and Koopman (2000), is to replace (35) by the relationship

$$p(x_t | \theta_t, x^*_{t-1}, y^*_t) = p(x_t | F_t \theta_t), \tag{37}$$

where $x_{t-1}^* = \{x_{t-1}, x_{t-2}, \ldots, x_1\}$, $y_t^* = \{y_t, y_{t-1}, \ldots, y_1\}$ are possible covariates, and F_t may be a function of either or both. It is usually assumed that the state-vector, θ_t, still satisfies (36) but now the innovation process $\{\omega_t\}$ may be non-Gaussian. In some models, the distribution is not specified, only the first two moments. This description, of course, is very broad, and may be used for non-Gaussian data more generally.

The most successful of earlier attempts used the Bayesian approach to statistical inference and we discuss it briefly now. The main proponents of Bayesian forecasting are West and Harrison who present the details of their approach in their book (1989). In the Gaussian case, their model is linear, as noted above, and is called the dynamic linear model. They extend this model to deal with non-Gaussian but exponential family data, developing their dynamic generalized linear model (GLM). As the name suggests, they use the basic approach of generalized linear modelling and so, in principle, can model binomial, Poisson, negative binomial and other discrete-valued series. Thus, the natural parameter η_t of the distribution is modelled *via* a link function $g(\cdot)$ as a linear function of the state vector θ_t, i.e., $g(\eta_t) = F_t \theta_t$, and θ_t evolves through time according to an equation of the usual form (36). The analysis of this model is Bayesian, but not fully so, in that the prior distribution for η_t is the conjugate prior for the exponential family but the parameter and distributional revisions derive from approximating the appropriate distributions by the behaviour of their first two moments. The work was originally presented in West, Harrison and Migon (1985) which also contains contributed discussion and additional examples.

This construction, using (37) for the exponential family of distributions, and (36) for the linear predictor η_t, is very appealing in the context of discrete variate time series, and forms the basic model for most other approaches in this area. It is often referred to as the dynamic GLM. The main differences between approaches tend to be computational and in the details of the estimation.

Singh and Roberts (1992), for example, define (36) in terms of only the first two moments, and estimate using a filtered and iterative weighted least squares algorithm. Kashiwagi and Yanagimoto (1992) build on the work of Kitagawa (1987) on non-normal state space models in which the the non-normal densities at each step of the Kalman filter are numerically derived. Their models, too, are related in structure to generalized linear modelling. For example, in the Poisson case, it is assumed that the series $\{X_t\}$ is Poisson of mean λ_t and that the natural parameter $\ln \lambda_t$ undergoes a Gaussian random walk, i.e., $\nabla \ln \lambda_t \sim N(0, \sigma^2)$ where $\nabla \ln \lambda_t = \ln \lambda_t - \ln \lambda_{t-1}$. If it is anticipated that the mean process is even smoother then the second differences may be used.

Fahrmeir (1992) uses the basic model of (37) and (36) with $\{\omega_t\}$ in (36) normally distributed. He refers to *this version* as a dynamic GLM, and estimates by maximising the posterior densities using a generalization of the extended Kalman filter and smoother.

Chan and Ledolter (1995) return to Zeger's (1988) polio data and model it as a dynamic GLM in which the latent process, $\{\theta_t\}$, is an *AR*(1). Explicitly, conditional on trend and seasonal covariates, $\{Y_t\}$, $(X_t | Y_t, \theta_t) \sim \text{Poisson}(a_t \mu_t)$, where $a_t = e^{\beta' Y_t}$ and $\ln \mu_t = \theta_t = \rho \theta_{t-1} + \omega_t$, where $\{\omega_t\}$ is Gaussian. Their approach is to treat the unobserved process $\{\theta_t\}$ as missing values and use a variant of the *EM*-algorithm (Dempster, Laird and Rubin, 1977) to maximize the likelihood. In their version of the algorithm, the E-step is performed indirectly using Monte Carlo simulation. The approach is direct and prediction and smoothing are straightforward. They note that diagnostics still need to be developed for this class of models, though Harvey and Koopman (1992) and Singh and Roberts (1992) are relevant.

The use of simulation, in particular Markov Chain Monte Carlo and the Gibbs sampler, in this class of models was considered by Carlin et al. (1992) and developed further by Carter and Kohn (1994) and Frühwirth-Schnatter (1994). It is illustrated in, for example, Cargnoni et al. (1997) in a longitudinal study dealing with collections of multinomial vectors, numbers of students going through the Italian secondary education system.

In the discussion following Lee and Nelder's (1996) paper on *hierarchical* GLMs, the authors, in response to commentary, discuss dynamic hierarchical GLMs and their relationship with the Kalman filter and smoother. Again, the basic model using (37) for the exponential family of distributions, and (36), is assumed. In their formulation, the filter is derived as an updating algorithm for maximizing the hierarchical likelihood.

The paper by Jørgensen et al. (1999) develops the dynamic GLM further by including covariates in two distinct ways, and so distinguishing between two kinds of covariate, and replacing (36) by an explicit Gamma Markov process. Conditioning on covariates $\{Y_t\}$, their observation process (37) is similar to that of Chan and Ledolter (1995), i.e., $(X_t|Y_t, \theta_t) \sim \text{Poisson}(a_t\theta_t)$ where $a_t = e^{\beta'Y_t}$, but now (36) is replaced by $(\theta_t|\theta_{t-1}) \sim \text{Ga}(\theta_{t-1}/\sigma^2, b_t\sigma^2)$. Thus, (36) still holds since $\theta_t = b_t\theta_{t-1} + \omega_t$ where ω_t is non-Gaussian. As noted, the authors distinguish two types of covariate: those with short-term effects appear in $\{Y_t\}$, while the long-term ones appear in $\{Z_t\}$, and $\ln b_t = \alpha'\nabla Z_t$ where $\nabla Z_t = Z_t - Z_{t-1}$. In an application to the polio data, $\{Y_t\}$ contains the seasonal covariates and $\{Z_t\}$ the trend. An unbiased estimating equation is derived from the EM-algorithm in which the E-step is replaced by the Kalman smoother. Various diagnostics are explored and several illustrative examples are presented.

Durbin and Koopman (2000) use the same dynamic GLM as before, i.e., (37) and (36), but now inference is based on the Kalman filter and smoother, simulation (but not MCMC), and Gaussian importance sampling. Using this approach, the authors show how to estimate all the quantities usually required in both the classical frequentest and Bayesian procedures. Avoiding MCMC means that there are no problems of convergence and most of the required quantities are readily computed. The Gaussian importance density needed for the simulation is constructed from linear Gaussian models which approximate the non-Gaussian one in the neighbourhood of the conditional mode of the state vector given the data.

The discussion following this paper accurately reflects the many different approaches to inference on what is, in many ways, the same model. It highlights the contribution of computational power made to solving this problem, and also the relative complexity of the solutions.

In one case, however, there are solutions which are relatively straightforward and intuitively appealing. It is the simplest case, but certainly one of the most useful, and it is given by (35) and (36) above with $F_t = 1$ and $G_t = 1$, i.e., the series has a mean θ_t which undergoes a random walk. This is one of the standard forms for modelling a series whose only parameter is its mean which may change through time. The form of the solution here when the noise processes are assumed to be stationary tends to a simple exponentially weighted moving average whose smoothing constant is a function of the ratio of the two noise variances. Smith (1979) extended this model to the exponential family of distributions, using an approach which is different from that of West et al. (1985). He noted that the observation equation (35) may be rewritten in this case in the form $E(X_t|\theta_t) = \theta_t$, and that, although this will extend to other distributions easily for this model, there is no simple way to generalize the random walk form of evolution (36). The problem lies in how to generalize the revision of the distribution of $(\theta_{t-1}|X^{t-1})$ to that of $(\theta_t|X^{t-1})$, where X^t is all the data available up to and including the observation at time t. His solution is to require that the density of the second distribution be proportional to the density of the first one raised to a fractional power, ω. This ensures that the revision has similar properties to the linear Gaussian case, i.e., the means are the same but the variance increases, and that the forecasts of the process are simple exponentially weighted moving averages. He presents details of the solution for a number of distributions including the Poisson, binomial and negative binomial.

Harvey and Fernandes (1989) extend this approach and Harvey presents the details and other examples in his book (1989). In the Poisson case, $(X_t|\mu_t)$ is Poisson of mean μ_t and it is assumed that μ_t has the conjugate prior for the Poisson, i.e., the gamma distribution. We will write that a r.v. has the Ga(a, b) distribution if it has density function $f(x) = b^a x^{a-1} e^{-bx}/\Gamma(a)$ for $x > 0$. Now, $(\mu_{t-1}|X^{t-1})$ is Ga(a_{t-1}, b_{t-1}) and, following Smith (1979), $(\mu_t|X^{t-1})$ will be Ga$(a_{t|t-1}, b_{t|t-1})$ where

$$a_{t|t-1} = \omega a_{t-1}, \qquad b_{t|t-1} = \omega b_{t-1} \tag{38}$$

and $0 \leqslant \omega \leqslant 1$. With the arrival of the next observation, X_t, the posterior distribution of μ_t conditional on X^t may be obtained as Ga(a_t, b_t) where $a_t = a_{t|t-1} + X_t$ and $b_t = b_{t|t-1} + 1$. Thus, we have $\hat{\mu}_t = E(\mu_t|X^t) = a_t/b_t$ and, using the established relationship between (a_t, b_t) and (a_{t-1}, b_{t-1}), i.e.,

$$a_t = \omega a_{t-1} + X_t, \qquad b_t = \omega b_{t-1} + 1 \tag{39}$$

we have

$$\hat{\mu}_t = \hat{\mu}_{t-1} + k_t(X_t - \hat{\mu}_{t-1}), \tag{40}$$

$$k_t = k_{t-1}/(k_{t-1} + \omega) \tag{41}$$

and, as $t \to \infty$, $k_t \to (1 - \omega)$ and $\hat{\mu}_t$ is obtained in (40) by exponential smoothing. In fact, Eqs. (40), (41) are the recursive equations for the estimator of the mean when discounted least squares is used with discount factor ω.

Harvey (1989) estimates the discount factor ω from historical data by maximizing the likelihood which is given by a product of the conditional probabilities $p(X_t|X^{t-1})$. These probabilities define the predictive distribution which, in this case, i.e., the Poisson with a gamma prior for its mean, is the negative binomial distribution, i.e., $(X_t|X^{t-1}) \sim$ NB$(a_{t|t-1}, (1 + b_{t|t-1})^{-1})$.

He also notes that although the stochastic mechanism underlying the transition from μ_{t-1} to μ_t at time $(t-1)$ is defined only implicitly by the behaviour of the parameters of the gamma distribution, it is possible to specify such a mechanism explicitly. In fact, we can write $\mu_t = \omega^{-1}\mu_{t-1}\beta_t$ where β_t is a beta$(\omega a_{t-1}, (1-\omega)a_{t-1})$ r.v. A beta(α, β) r.v. has density $f(x) = x^{\alpha-1}(1-x)^{\beta-1}/B(\alpha, \beta)$ where $0 < x < 1$. This process reflects the fact that in predicting μ_t from μ_{t-1} at time $(t-1)$ without any new information the location indicated does not change but our uncertainty about it increases.

The binomial distribution is treated in exactly the same way with analogous results. If X_t is binomial(n, p_t) and $(p_t|X^t) \sim$ beta(a_t, b_t) then Eqs. (38) hold and Eqs. (39) become

$$a_t = \omega a_{t-1} + X_t, \qquad b_t = \omega b_{t-1} + (n - X_t). \tag{42}$$

Now, $\hat{\mu}_t = E(np_t|X^t)$ and Eqs. (40) and (41) still hold. The predictive distribution in this case is the beta-binomial, i.e., $(X_t|X^{t-1})$ has probability function

$$p(x) = \binom{n}{x} \frac{B(a+x, b+n-x)}{B(a,b)},$$

where $a = a_{t|t-1}$ and $b = b_{t|t-1}$.

As with the Poisson case, it is possible to define a stochastic mechanism which effects the specified transition from p_{t-1} to p_t at time $(t-1)$. In this case, it is given by $p_t = B_a p_{t-1}/[B_a p_{t-1} + B_b(1 - p_{t-1})]$ where B_a is beta$(\omega a_{t-1}, (1-\omega)a_{t-1})$ and B_b is beta$(\omega b_{t-1}, (1-\omega)b_{t-1})$ and the two are independent.

The binary case is treated as the binomial with $n = 1$ and Harvey (1989) extends the procedure to the multinomial using the Dirichlet distribution as the conjugate prior.

The negative binomial may be treated almost exactly as the binomial. The conjugate prior for p_t in NB(β, p_t) is the beta distribution, and if we reparametrize a little so that $(p_t|X^t) \sim$ beta$(a_t, b_t + 1)$ then Eqs. (38) hold as before, and Eqs. (39) become

$$a_t = \omega a_{t-1} + X_t, \qquad b_t = \omega b_{t-1} + \beta. \tag{43}$$

Now, $\hat{\mu}_t = E(\beta p_t/(1-p_t)|X^t)$ and Eqs. (40) and (41) still hold. The predictive distribution in this case is the beta-Pascal, i.e., $(X_t|X^{t-1})$ has probability function

$$p(x) = \binom{\beta + x - 1}{x} \frac{B(a+x, b+1+\beta)}{B(a, b+1)},$$

where $a = a_{t|t-1}$ and $b = b_{t|t-1}$. The slightly different form for the prior here differs a little from Harvey (1989) but it ensures the existence of the prior mean and renders the analyses essentially as in the other cases. Again, an underlying stochastic mechanism can be described for the transition from p_{t-1} to p_t at time $(t-1)$. Alternatively, we may consider the transition of the odds, i.e., $r_t = p_t/(1-p_t)$, from r_{t-1} to r_t at time $(t-1)$. It is given by $r_t = B_a r_{t-1}/B_{b1}$ where B_a is the same as before,

$B_{b1} \sim \text{beta}(\omega b + 1, (1-\omega)b)$, B_b and B_{b1} are independent, and $a = a_{t-1}$ and $b = b_{t-1}$. Note that r_t has a beta distribution of the second kind with parameters $(a, b+1)$, i.e., it has density $f(x) = x^{a-1}/(1+x)^{a+b+1} B(a, b+1)$ for $x > 0$.

7. The future

This is a very rapidly changing area with no sign that the emergence of a uniformly best approach is imminent. Nevertheless, the state space representation using GLMs will clearly be an important component of data modelling in this area. The proliferation of models and the variety of different approaches to inference is no surprise once we move away from normality and linearity, and so one area whose development is essential is that of model identification and validation.

In addition, there will be a considerable expansion into two areas of application. The first is longitudinal count data, and development is already advanced there. Although we did not emphasize it, many of the models and procedures discussed in this review were constructed for just such data. The books by Diggle et al. (1994), Fahrmeir and Tutz (1994) and Cameron and Trivedi (1998) discuss particular cases and give brief reviews of current work in the area. The second area of expansion is the modelling of spatio-temporal discrete variate processes. Little has been done as yet, but approaches similar to those discussed here for time series have been used in modelling spatially distributed count data. See, for example, Diggle et al. (1998), where the count at a point Z_i is modelled as a GLM conditional on an unobserved Gaussian spatial stochastic process, $\{S(Z_i)\}$, and any covariates $\{Y_i\}$. The usual linear predictor is given by $\eta_i = S(Z_i) + \beta' Y_i$. The need to model and analyze spatio-temporal data is becoming commoner, especially in the natural sciences, and the computational power required much more readily available.

References

Adke, S. R. and S. R. Deshmukh (1988). Limit distribution of a high order Markov chain. *J. Roy. Statist. Soc. B* **50**, 105–108.

Al-Osh, M. A. and E.-E. A. A. Aly (1992). First order autoregressive time series with negative binomial and geometric marginals. *Comm. Statist. Theory Methods* **21**, 2483–2492.

Al-Osh, M. A. and A. A. Alzaid (1987a). First-order integer-valued autoregressive (INAR(1)) process. *J. Time Series Anal.* **8**, 261–275.

Al-Osh, M. A. and A. A. Alzaid (1987b). Binomial autoregressive moving average models. *Stochastic Models* **7**, 261–282.

Al-Osh, M. A. and A. A. Alzaid (1988). Integer-valued moving average (INMA) process. *Statist. Papers* **29**, 281–300.

Al-Osh, M. A. and A. A. Alzaid (1993). Some Gamma processes based on the Dirichlet-Gamma transformation. *Comm. Statist. Stochastic Models* **9**(1), 123–143.

Aly, E.-E. A. A. and N. Bouzar (1994). On some integer-valued autoregressive moving average models. *J. Multivariate Anal.* **50**, 132–151.

Alzaid, A. A. and M. A. Al-Osh (1988). First-order integer-valued autoregressive (INAR(1)) process: distributional and regression properties. *Statist. Neerl.* **42**, 53–61.

Alzaid, A. A. and M. A. Al-Osh (1990). An integer-valued pth-order autoregressive structure (INAR(p)) process. *J. Appl. Probab.* **27**, 314–324.

Alzaid, A. A. and M. A. Al-Osh (1993). Some autoregressive moving average processes with generalized Poisson marginal distributions. *Ann. Inst. Statist. Math.* **45**, 223–232.

Azzalini, A. (1994). Logistic regression for autocorrelated data with application to repeated measures. *Biometrika* **81**, 767–775.

Berglund, E. and K. Brännäs (1996). Entry and exit of plants: a study based on Swedish plant count data for municipalities. In *The Yearbook of the Finnish Statistical Society 1995*, pp. 95–111. Helsinki.

Böckenholt, U. (1999). Autoregressive negative multinomial regression models. *J. Amer. Statist. Assoc.* **94**, 757–765.

Box, G. E. P. and G. M. Jenkins (1970). *Time Series Analysis: Forecasting and Control*. Holden-Day, San Francisco.

Brännäs, K. (1994). Estimation and testing in integer-valued AR(1) models. *Umeå Economic Studies No. 335*.

Brännäs, K. (1995a). Prediction and control of a time series count data model. *Int. J. Forecasting* **11**, 263–270.

Brännäs, K. (1995b). Explanatory variables in the AR(1) count data model. *Umeå Economic Studies No. 381*.

Brännäs, K. and A. Hall (1998). Estimation in integer-valued moving average models. *Umeå Economic Studies No. 477*.

Brännäs, K. and P. Johansson (1994). Time series count data regression. *Comm. Statist. Theory Methods* **23**, 2907–2925.

Brännäs, K. and P. Johansson (1996). Panel data for regression of counts. *Statist. Papers* **37**, 191–213.

Buishand, T. A. (1978). The binary DARMA(1,1) process as a model for wet-dry sequences. *Technical Note 78-01*, Dept Mathematics, Statistics Division, Agricultural University, Wageningen, Netherlands.

Cameron, A. C. and P. K. Trivedi (1998). *Regression Analysis of Count Data*. Cambridge University Press.

Campbell, M. J. (1994). Time series regression for counts: an investigation of the relationship between sudden infant death syndrome and environmental temperature. *J. Roy. Statist. Soc. A* **157**, 191–208.

Cargnoni, C., P. Müller and M. West (1997). Bayesian forecasting of multinomial time series through conditionally Gaussian dynamic models. *J. Amer. Statist. Assoc.* **92**, 640–647.

Carlin, B. P., N. G. Polson and D. S. Stoffer (1992). A Monte Carlo approach to nonnormal and nonlinear state space modelling. *J. Amer. Statist. Assoc.* **87**, 493–500.

Carter, C. K. and R. Kohn (1994). On Gibbs sampling for state space models. *Biometrika* **81**, 541–553.

Chan, K. S. and J. Ledolter (1995). Monte Carlo EM estimation for time series models involving counts. *J. Amer. Statist. Assoc.* **90**, 242–252.

Chang, T. J., J. W. Delleur and M. L. Kavvas (1987). Application of discrete autoregressive moving average models for estimation of daily runoff. *J. Hydrol.* **91**, 119–135.

Chang, T. J., M. L. Kavvas and J. W. Delleur (1984a). Modelling of sequences of wet and dry days by binary discrete autoregressive moving average processes. *J. Clim. Appl. Meteor.* **23**, 1367–1378.

Chang, T. J., M. L. Kavvas and J. W. Delleur (1984b). Daily precipitation modelling by discrete autoregressive moving average processes. *Water Resour. Res.* **20**, 565–580.

Cox, D. R. and H. D. Miller (1965). *The Theory of Stochastic Processes*. Methuen, London.

Cox, D. R. (1970). *The Analysis of Binary Data*. Methuen, London.

Cox, D. R. (1981). Statistical analysis of time series: some recent developments. *Scand. J. Statist.* **8**, 93–115.

Delleur, J. W., T. J. Chang and M. L. Kavvas (1989). Simulation models of sequences of wet and dry days. *J. Irrig. and Drainage Engr.* **115**, 344–357.

Dempster, A. P., N. M. Laird and D. Rubin (1977). Maximum likelihood from incomplete data via the EM algorithm (with discussion). *J. Roy. Statist. Soc. B* **39**, 1–38.

Diggle, P. J., K.-Y. Liang and S. L. Zeger (1994). *Analysis of Longitudinal Data*. Oxford University Press, Oxford.

Diggle, P. J., J. A. Tawn and R. A. Moyeed (1998). Model-based geostatistics (with discussion). *Appl. Statist.* **47**, 299–350.

Dion, J.-P., G. Gauthier and A. Latour (1995). Branching processes with immigration and integer-valued time series. *SERDICA* **21**, 123–136.

Du, J.-G. and Y. Li (1991). The integer valued autoregressive (INAR(p)) model. *J. Time Series Anal.* **12**, 129–142.

Durbin, J. and S. J. Koopman (2000). Time series analysis of non-Gaussian observations based on state space models from both classical and Bayesian perspectives (with discussion). *J. Roy. Statist. Soc. B* **62**, 3–56.

Edwards, C. B. and J. Gurland (1961). A class of distributions applicable to accidents. *J. Amer. Statist. Assoc.* **56**, 503–517.

Fahrmeir, L. (1992). Posterior mode estimation by extended Kalman filtering for multivariate dynamic generalized linear models. *J. Amer. Statist. Assoc.* **87**, 501–509.

Fahrmeir, L. and H. Kaufmann (1987). Regression models for non-stationary categorical time series. *J. Time Series Anal.* **8**, 147–160.

Fahrmeir, L. and G. T. Tutz (1994). *Multivariate Statistical Modelling based on Generalized Linear Models.* Springer-Verlag, New York.

Frühwirth-Schnatter, S. (1994). Data augmentation and dynamic linear models. *J. Time Series Anal.* **15**, 183–202.

Gauthier, G. and A. Latour (1994). Convergence forte des estimateurs des paramètres d'un processus GENAR(p). *Ann. Sci. Math. Québec* **18**, 49–71.

Gaver, D. P. and P. A. W. Lewis (1980). First order autoregressive gamma sequences and point processes. *Adv. Appl. Probab.* **12**, 724–745.

Harvey, A. C. (1989). *Forecasting, Structural Time Series Models and the Kalman Filter.* Cambridge University Press, Cambridge.

Harvey, A. C. and C. Fernandes (1989). Time series models for count or qualitative observations. *J. Bus. Econ. Statist.* **7**, 407–422.

Harvey, A. C. and S. J. Koopman (1992). Diagnostic checking of unobserved-components time series models. *J. Bus. Econ. Statist.* **10**, 377–389.

Jacobs, P. A. and P. A. W. Lewis (1977). A mixed autoregressive-moving average exponential sequence and point process, EARMA(1,1). *Adv. Appl. Probab.* **9**, 87–104.

Jacobs, P. A. and P. A. W. Lewis (1978a). Discrete time series generated by mixtures I: correlational and runs properties. *J. Roy. Statist. Soc. B* **40**, 94–105.

Jacobs, P. A. and P. A. W. Lewis (1978b). Discrete time series generated by mixtures II: asymptotic properties. *J. Roy. Statist. Soc. B* **40**, 222–228.

Jacobs, P. A. and P. A. W. Lewis (1978c). Discrete time series generated by mixtures III: autoregressive processes (DAR(p)). *Naval Postgraduate School Technical Report NPS55Lw 73061A*.

Jacobs, P. A. and P. A. W. Lewis (1983). Stationary discrete autoregressive-moving average time series generated by mixtures. *J. Time Series Anal.* **4**, 19–36.

Joe, H. (1996). Time series models with univariate margins in the convolution-closed infinitely divisible class. *J. Appl. Probab.* **33**, 664–677.

Jørgensen, B., S. Lundbye-Christensen, P. X.-K. Song and L. Sun (1999). A state-space model for multivariate longitudinal count data. *Biometrika* **86**, 169–181.

Kanter, M. (1975). Autoregression for discrete processes mod 2. *J. Appl. Probab.* **12**, 371–375.

Kashiwagi, N. and T. Yanagimoto (1992). Smoothing serial count data through a state-space model. *Biometrics* **48**, 1187–1194.

Kitagawa, G. (1987). Non-Gaussian state space modelling for non-stationary time series (with discussion). *J. Amer. Statist. Assoc.* **82**, 1032–1063.

Keenan, D. M. (1982). A time series analysis of binary data. *J. Amer. Statist. Assoc.* **77**, 816–821.

Korn, E. L. and A. S. Whittemore (1979). Methods for analysing panel studies of acute health effects of air pollution. *Biometrics* **35**, 795–802.

Latour, A. (1997). The multivariate GINAR(p) process. *Adv. Appl. Probab.* **29**, 228–248.

Latour, A. (1998). Existence and stochastic structure of a non-negative integer-valued autoregressive process. *J. Time Series Anal.* **19**, 439–455.

Lawrance, A. J. (1982). The innovation distribution of a gamma distributed autoregressive process. *Scand. J. Statist.* **9**, 234–236.

Lawrance, A. J. and P. A. W. Lewis (1977). An exponential moving average sequence and point process, EMA(1). *J. Appl. Probab.* **14**, 98–113.

Lawrance, A. J. and P. A. W. Lewis (1980). The exponential autoregressive-moving average EARMA(p,q) process. *J. Roy. Statist. Soc. B* **42**, 150–161.

Lawrance, A. J. and P. A. W. Lewis (1981). A new autoregressive time series model in exponential variables (NEAR(1)). *Adv. Appl. Probab.* **13**, 826–845.

Lawrance, A. J. and P. A. W. Lewis (1985). Modelling and residual analysis of nonlinear autoregressive time series in exponential variables. *J. Roy. Statist. Soc. B* **47**, 165–183.

Lee, Y. and J. A. Nelder (1996). Hierarchical generalized linear models (with discussion). *J. Roy. Statist. Soc. B* **58**, 619–678.

Léon, L. F. and C.-L. Tsai (1998). Assessment of model adequacy for Markov regression time series models. *Biometrics* **54**, 1165–1175.

Lewis, P. A. W. (1980). Simple models for positive-valued and discrete-valued time-series with ARMA correlation structure. In *Multivariate Analysis*, Vol. IV, pp. 151–166 (Ed. P. R. Krishniah). North-Holland, Amsterdam.

Lewis, P. A. W. (1985). Some simple models for continuous variate time series. *Water Res. Bull.* **21**, 635–644.

Lewis, P. A. W. and E. McKenzie (1991). Minification processes and their transformations. *J. Appl. Probab.* **28**, 45–57.

Li, W. K. (1991). Testing model adequacy for some Markov regression models for time series. *Biometrika* **78**, 83–89.

Li, W. K. (1994). Time series models based on generalized linear models: some further results. *Biometrics* **50**, 506–511.

Liang, K.-Y. and S. L. Zeger (1986). Longitudinal data analysis using generalized linear models. *Biometrika* **73**, 13–22.

Littlejohn, R. P. (1992a). Discrete minification processes and reversibility. *J. Appl. Probab.* **29**, 82–91.

Littlejohn, R. P. (1992b). An operation which inverts Bernoulli multiplication and associated stationary reversible Markov processes. *J. Appl. Probab.* **29**, 234–238.

MacDonald, I. A. and W. Zucchini (1997). *Hidden Markov and Other Models for Discrete-valued Time Series*. Chapman & Hall, London.

McKenzie, E. (1981). Extending the correlation structure of exponential autoregressive moving average processes. *J. Appl. Probab.* **18**, 1–9.

McKenzie, E. (1985a). Some simple models for discrete variate time series. *Water Res. Bull.* **21**(4), 645–650.

McKenzie, E. (1985b). Contribution to the discussion of 'modelling and residual analysis of nonlinear autoregressive time-series' by A. J. Lawrance and P. A. W. Lewis. *J. Roy. Statist. Soc. B* **47**, 187–188.

McKenzie, E. (1986). Autoregressive-moving average processes with negative-binomial and geometric marginal distributions. *Adv. Appl. Probab.* **18**, 679–705.

McKenzie, E. (1987). Innovation distributions for gamma and negative binomial autoregressions. *Scand. J. Statist.* **14**, 79–85.

McKenzie, E. (1988a). The distributional structure of finite moving average processes. *J. Appl. Probab.* **25**, 313–321.

McKenzie, E. (1988b). Some ARMA models for dependent sequences of Poisson counts. *Adv. Appl. Probab.* **20**, 822–835.

McKenzie, E. (1991). Linear characterizations of the Poisson distribution. *Statist. Probab. Lett.* **11**, 459–461.

Muenz, L. R. and L. V. Rubinstein (1985). Markov models for covariate dependence of a binary sequence. *Biometrics* **41**, 91–101.

Pegram, G. G. S. (1980). An autoregressive model for multilag Markov chains. *J. Appl. Probab.* **17**, 350–362.

Phatarfod, R. M. and K. V. Mardia (1973). Some results for dams with Markovian inputs. *J. Appl. Probab.* **10**, 166–180.

Phatarfod, R. M. and R. Srikanthan (1981). Discretization in stochastic reservoir theory with Markovian inflows. *J. Hydrology* **52**, 199–218.

Raftery, A. E. (1985a). A model for high-order Markov chains. *J. Roy. Statist. Soc. B* **47**, 528–539.

Raftery, A. E. (1985b). A new model for discrete-valued time series: autocorrelations and extensions. *Rass. Met. Statist. Appl.* **3–4**, 149–162.

Raftery, A. E. and S. Tavaré (1994). Estimation and modelling repeated patterns in high order Markov chains with the mixture transition distribution model. *Appl. Statist.* **43**, 179–199.

Singh, A. C. and G. R. Roberts (1992). State space modelling of cross-classified series of counts. *Int. Statist. Rev.* **60**, 321–335.

Smith, J. Q. (1979). A generalization of the Bayesian steady forecasting model. *J. Roy. Statist. Soc. B* **41**, 375–387.

Stern, R. D. and R. Coe (1984). A model fitting analysis of daily rainfall data (with discussion). *J. Roy. Statist. Soc. A* **147**, 1–34.

Steutel, F. W. and K. van Harn (1979). Discrete analogues of self-decomposability and stability. *Ann. Probab.* **7**, 893–899.

Teicher, H. (1954). On the multivariate Poisson distribution. *Skand. Aktuarietidskr.* **37**, 1–9.

Tong, H. (1990). *Non-linear Time Series: A Dynamical System Approach.* Oxford University Press, Oxford.

West, M. and P. J. Harrison (1989). *Bayesian Forecasting and Dynamic Models.* Springer-Verlag, New York.

West, M., P. J. Harrison and H. S. Migon (1985). Dynamic generalized linear models and Bayesian forecasting (with discussion). *J. Amer. Statist. Assoc.* **80**, 73–97.

Zeger, S. L. (1988). A regression model for time series of counts. *Biometrika* **75**, 621–629.

Zeger, S. L., K.-Y. Liang and S. G. Self (1985). The analysis of binary longitudinal data with time-independent covariates. *Biometrika* **72**, 31–38.

Zeger, S. L. and B. Qaqish (1988). Markov regression models for time series: a quasi-likelihood approach. *Biometrics* **44**, 1019–1031.

Extreme Value Theory, Models and Simulation

Saralees Nadarajah

1. Introduction

Extreme value theory concerns the behavior of the extremes of a process or processes. The fundamentals of this probabilistic theory have been known since the beginning of the twentieth century. But the relevant statistical models emerged only much more recently. The aim of this chapter is to review the fundamentals, relevant statistical models and simulation schemes, and to note the various applications that the models have attracted. Most of the material presented is for independent and identically distributed observations. However, we shall survey the development of extreme value theory for nontrivial stochastic processes as well (Sections 8 and 15).

Univariate extremes are traditionally modeled by the annual maximum method. This method, discussed in Section 4 of the chapter, has been known in the literature since the 1920s. The theoretical motivation for it comes from the well known extremal types theorem: details of this limit theorem, its variants and corresponding rate of convergence results are given in Sections 2 and 3. A major weakness of the annual maximum method is that it utilizes only the largest-order statistic and thus its use wastes a lot of data. In Sections 5 and 6 we describe two modeling approaches based, respectively, on the generalized Pareto distribution and the joint distribution of the r-largest-order statistics that make use of more of the data.

Models for multivariate extremes are much more recent. The last decade, in particular, has seen the development of several models. In this chapter we describe nine models for bivariate extreme value distributions (Sections 13.1–13.9) and five models for multivariate extreme value distributions (Sections 14.1–14.5). We also provide tools for constructing further models (Section 14.6). Some of the models presented are motivated by specific applications: those in Sections 13.1, 13.5, 14.1 and 14.2 are motivated by spatial modeling of the extreme storms while those in Sections 13.6 and 14.5 are motivated, respectively, by directional modeling of the extreme wind speeds and modeling of sea storms along a coast. Most of the other models presented in the paper are theoretically motivated: the motivation comes from the various probabilistic characterizations given on the form of multivariate extreme value distributions (Sections 10 and 11).

Simulation is a useful tool for studying statistical properties of a given model. Simulation of univariate extremes, when the extreme is a single value such as the largest value of a process, is straightforward since commonly known models for it such as the generalized extreme value and the generalized Pareto models can be inverted to a closed form; thus, the inversion method can be applied. Simulation of multivariate extremes is less obvious and there has only been little work concerned with it. In this chapter we give details of work known in the literature: Sections 10.1.1 and 11.0.7 describe two comprehensive schemes for simulating bivariate extremes – one of them is based on approximating a limiting point process result while the other uses an exact probabilistic characterization; Sections 13.1 and 14.2, respectively, contain suggestions of schemes for a special class of bivariate and trivariate extreme value distributions – these schemes exploit the structure of dependence within the given class.

For recent authoritative reviews of extreme value theory we refer the reader to Deheuvels (1981), Smith (1985b, 1990), Leadbetter and Rootzén (1988), Galambos (1995, 1998) and Beirlant et al. (1998). For recent monographs on related topics see Leadbetter et al. (1983), Galambos (1987, 1995a), Resnick (1987), Castillo (1988), Pfeifer (1989), Berman (1992), Reiss (1993), Falk et al. (1994), Embrechts et al. (1997), Joe (1997), Reiss and Thomas (1997), Rolski et al. (1999) and Kotz and Nadarajah (2000).

2. Limit laws in univariate extremes and characterizations

Suppose $X_1, X_2, \ldots$ are independent and identically distributed (i.i.d.) random variables with common distribution function (d.f.) F. Let $M_n = \max\{X_1, \ldots, X_n\}$ denote the maximum of the first n random variables and let $w(F) = \sup\{x: F(x) < 1\}$ denote the upper end point of F. Since

$$\Pr(M_n \leqslant x) = \Pr(X_1 \leqslant x, \ldots, X_n \leqslant x) = F^n(x),$$

M_n converges almost surely to $w(F)$ whether it is finite or infinite. The limit theory in univariate extremes seeks norming constants $a_n > 0$, b_n and a nondegenerate d.f. G such that the d.f. of a normalized version of M_n converges to G, i.e.,

$$\Pr\left(\frac{M_n - b_n}{a_n} \leqslant x\right) = F^n(a_n x + b_n) \to G(x) \tag{1}$$

as $n \to \infty$. If this holds for suitable choices of a_n and b_n then we say that G is an extreme value d.f. and F is in the domain of attraction of G, written as $F \in D(G)$. We say further that two extreme value d.f.s G and G^* are of the same type if $G^*(x) = G(ax + b)$ for some $a > 0$, b and all x. The Extremal Types Theorem (Fisher and Tippett, 1928; Gnedenko, 1943; De Haan, 1970, 1976; Weissman, 1978) characterizes the limit d.f. G as of the type of one of the following three classes:

$$\text{I:} \quad \Lambda(x) = \exp\{-\exp(-x)\}, \quad x \in \Re;$$

II: $\Phi_\alpha(x) = \begin{cases} 0, & \text{if } x < 0, \\ \exp\{-x^{-\alpha}\}, & \text{if } x \geq 0 \end{cases}$

for some $\alpha > 0$; (2)

III: $\Psi_\alpha(x) = \begin{cases} \exp\{-(-x)^\alpha\}, & \text{if } x < 0, \\ 1, & \text{if } x \geq 0 \end{cases}$

for some $\alpha > 0$.

Thus, any extreme value distribution can be classified as one of Type I, II or III. The three types are often called the Gumbel, Fréchet and Weibull types, respectively. For alternative characterizations of the Type I, II and III distributions see Erdös and Rényi (1961), Dubey (1966), Ballerini (1987), Khan and Beg (1987), Nagaraja (1988) and Moothathu (1990).

Leadbetter et al. (1983) give a comprehensive account of necessary and sufficient conditions for $F \in D(G)$ and characterizations of a_n and b_n when G is one of the three extreme value d.f.s above. The necessary and sufficient conditions for the three d.f.s in (2) are:

I: $\exists \gamma(t) > 0$ s.t. $\lim_{t \uparrow w(F)} \dfrac{1 - F(t + x\gamma(t))}{1 - F(t)} = \exp(-x), \quad x \in \Re,$

II: $w(F) = \infty$ and $\lim_{t \uparrow \infty} \dfrac{1 - F(tx)}{1 - F(t)} = x^{-\alpha}, \quad x > 0,$ (3)

III: $w(F) < \infty$ and $\lim_{t \downarrow 0} \dfrac{1 - F(w(F) - tx)}{1 - F(w(F) - t)} = x^\alpha, \quad x > 0;$

see Galambos and Obretenov (1987) and Galambos and Xu (1991) for alternative conditions involving the hazard and residual life functions. The corresponding characterizations of a_n and b_n are:

I: $a_n = \gamma(F^\leftarrow(1 - n^{-1}))$ and $b_n = F^\leftarrow(1 - n^{-1}),$

II: $a_n = F^\leftarrow(1 - n^{-1})$ and $b_n = 0,$ (4)

III: $a_n = w(F) - F^\leftarrow(1 - n^{-1})$ and $b_n = w(F),$

where $F^\leftarrow$ denotes the inverse function of F. Note that only distributions with $w(F) = \infty$ (respectively, $w(F) < \infty$) can qualify for membership in $D(\Phi_\alpha)$ (respectively, $D(\Psi_\alpha)$).

An equivalent characterization of G is by means of the definition of max-stability: a d.f. G is max-stable if there exists $\alpha_n > 0$, β_n such that for each $n \geq 1$

$$G^n(x) = G(\alpha_n x + \beta_n).$$ (5)

It can be shown (Resnick, 1987, Proposition 5.9) that the class of extreme value d.f.s G is precisely the class of nondegenerate max-stable d.f.s. It is easily checked that α_n and

β_n for the three d.f.s in (2) are:

I: $\alpha_n = 1$ and $\beta_n = -\log n$,
II: $\alpha_n = n^{-1/\alpha}$ and $\beta_n = 0$, (6)
III: $\alpha_n = n^{1/\alpha}$ and $\beta_n = 0$.

Various methods have been developed to test whether a sequence of i.i.d. observations belong to the domain of attraction of one the three distributions (see Tiago de Oliveira and Gomes, 1984 and Marohn, 1998a, 1998b for the Gumbel type; Tiku and Singh, 1981 and Shapiro and Brain, 1987a, 1987b for the Weibull type; Galambos, 1982; Öztürk, 1986; Öztürk and Korukoğlu, 1988; Castillo, Galambos and Sarabia, 1989 and Hasofer and Wang, 1992 for general tests).

Anderson (1970, 1980) and Anderson et al. (1997) develop limit laws corresponding to (2) for discrete random variables. For instance, Anderson (1980) shows that if X_i are discrete random variables then under certain conditions either there exist $a_n > 0$ and b_n such that

$$\lim_{n \to \infty} a_n \Pr(M_n = a_n(x + o(1)) + b_n) = \exp\{-\exp(-x)\}$$

or there exist $a_n > 0$ such that

$$\lim_{n \to \infty} a_n \Pr(M_n = a_n(x + o(1))) = \exp(-x^{-\alpha}).$$

Other developments on the extremes of discrete random variables include Arnold and Villaseñor (1984) and Gordon et al. (1986) who consider sequences $X_1, \ldots, X_n$ of independent Bernoulli random variables and study the limiting behavior for a fixed $l \geq 0$ of the longest l-interrupted head run, i.e., the maximal value of $k - i$, where $0 \leq i < k \leq n$, and where there are exactly l values of $i < m < k$ such that $X_m = 0$ (tails).

Pancheva (1984) extends (2) for power normalization to obtain what is known as the p-max stable laws. Namely, she seeks constants $a_n > 0$ and $b_n > 0$ such that

$$\lim_{n \to \infty} \Pr\left\{ \left|\frac{M_n}{a_n}\right|^{1/b_n} \operatorname{sign}(M_n) \leq x \right\} = \lim_{n \to \infty} F^n\left(a_n |x|^{b_n} \operatorname{sign}(x)\right) = G(x)$$

for all $x \in \mathcal{C}(G)$, the set of continuity points of G, where $\operatorname{sign}(x) = -1, 0, 1$ according as $x < 0$, $x = 0$ or $x > 0$. Then it is shown that G must be of the same p-type (G and G^* are of the same p-type if there exists constants $a > 0$ and $b > 0$ such that $G(x) = G^*(a|x|^b \operatorname{sign}(x))$ for all x real) as one of the following six d.f.s

$$G(x) = \begin{cases} 0, & \text{if } x \leq 1, \\ \exp\{-(\log x)^{-\alpha}\}, & \text{if } x > 1 \end{cases}$$

for some $\alpha > 0$;

$$G(x) = \begin{cases} 0, & \text{if } x < 0, \\ \exp\{-(|\log x|)^\alpha\}, & \text{if } 0 \leq x < 1, \\ 1, & \text{if } x \geq 1 \end{cases}$$

for some $\alpha > 0$;

$$G(x) = \begin{cases} 0, & \text{if } x \leq -1, \\ \exp\{-(|\log(|x|)|)^{-\alpha}\}, & \text{if } -1 < x < 0, \\ 1, & \text{if } x \geq 0 \end{cases}$$

for some $\alpha > 0$;

$$G(x) = \begin{cases} \exp\{-(\log|x|)^\alpha\}, & \text{if } x < -1, \\ 1, & \text{if } x \geq -1 \end{cases}$$

for some $\alpha > 0$;

$$G(x) = \Phi_1(x), \quad -\infty < x < \infty;$$

$$G(x) = \Psi_1(x), \quad -\infty < x < \infty.$$

Necessary and sufficient conditions for F to belong to the domain of attraction of a p-max stable law are given in Mohan and Subramanya (1991) and Mohan and Ravi (1993).

Sweeting (1985) provides a variant of (2) that seeks norming constants $a_n > 0$ and b_n such that

$$f_n(x) = n a_n f(a_n x + b_n) F^{n-1}(a_n x + b_n) \to g(x) \tag{7}$$

as $n \to \infty$, where f_n is the p.d.f. of $(M_n - b_n)/a_n$ and g is the derivative of G. Letting $a(t) = F^{\leftarrow}(1 - 1/t)$ and $b(t) = tf(a(t))$, the following necessary and sufficient conditions are derived for (7) for distributions in the domain of attraction of the three types distributions:

I: $b(t)$ is slowly varying, i.e., $b(tx)/b(t) \to 1$ as $t \to \infty$,

II: $w(F) = \infty$ and $\lim_{t \to \infty} a(t)b(t) = \alpha$,

III: $w(F) < \infty$ and $\lim_{t \to \infty} \{w(F) - a(t)\}b(t) = \alpha$.

In related developments, Pickands (1968) derives conditions under which the various moments of $(M_n - b_n)/a_n$ converge to the corresponding moments of G. It is proved that this is indeed true for all d.f.s $F \in D(G)$ provided that the moments are finite for sufficiently large n. Pickands (1986) gives necessary and sufficient conditions for the first or second derivatives of the left-hand side of (1) to converge to the corresponding derivative of G. He shows that the second derivatives converge if and only if

$$\lim_{x \uparrow w(F)} \frac{d}{dx} \frac{1 - F(x)}{f(x)} = c$$

for some constant $c \in \Re$.

Another extension of (2) is to consider the limiting distribution of $\max(h(X_1), \ldots, h(X_n))$, after suitable normalization, when h is a continuous function defined on some domain. Dorea (1987) provides sufficient conditions for the existence of the limiting distribution as well as a characterization of the limiting distribution.

Recently Nasri-Roudsari (1996) extended (2) for generalized order statistics. Generalized order statistics have been defined by Kamps (1955) as follows: let $k > 0$, $m \in \Re$ be parameters such that $\gamma_r = k + (m+1)(n-r) > 0$ for all $r = 1, \ldots, n-1$, and let $\tilde{m} = (m, \ldots, m)$, a vector of length $n-1$, if $n \geq 2$, $\tilde{m} \in \Re$ arbitrary, if $n = 1$. If the random variables $U_1, \ldots, U_n$ possess a joint probability density function (p.d.f.) of the form

$$k \left(\prod_{i=1}^{n-1} \gamma_i\right)\left(\prod_{i=1}^{n-1}(1-u_i)^m\right)(1-u_n)^{k-1}$$

on the cone $0 \leq u_1 \leq \cdots \leq u_n \leq 1$ of $\Re^n$, then they are called generalized order statistics. Nasri-Roudsari's extension of (2) is that U_n approaches one of the following three d.f.s after suitable normalization:

I: $\quad \Lambda(x) = \dfrac{1}{\Gamma\left(\frac{k}{m+1}\right)} \Gamma\left(\dfrac{k}{m+1}, \exp\{-(m+1)x\}\right), \quad x \in \Re;$

II: $\quad \Phi_\alpha(x) = \begin{cases} 0, & \text{if } x < 0, \\ \dfrac{1}{\Gamma\left(\frac{k}{m+1}\right)} \Gamma\left(\dfrac{k}{m+1}, x^{-(m+1)\alpha}\right), & \text{if } x \geq 0 \end{cases}$

for some $\alpha > 0$; $\hfill (8)$

III: $\quad \Psi_\alpha(x) = \begin{cases} \dfrac{1}{\Gamma\left(\frac{k}{m+1}\right)} \Gamma\left(\dfrac{k}{m+1}, (-x)^{(m+1)\alpha}\right), & \text{if } x < 0, \\ 1, & \text{if } x \geq 0 \end{cases}$

for some $\alpha > 0$,

which reduce in the particular case $m = 0$ and $k = 1$ to (2). Here, $\Gamma(a, x)$ denotes the incomplete gamma function

$$\Gamma(a, x) = \int_x^\infty t^{a-1} \exp(-t) \, dt,$$

for $a > 0$, $x \geq 0$. Further work by Nasri-Roudsari and Cramer (1999) derives the rate of convergence corresponding to (8).

In practical applications of the limit law, (1), we assume that n is so large that

$$\Pr\left(\dfrac{M_n - b_n}{a_n} \leq x\right) \approx G(x),$$

which implies that

$$\Pr(M_n \leqslant x) \approx G\left(\frac{x - b_n}{a_n}\right) = G^*(x), \tag{9}$$

where G^* is of the same type as G. Thus, the three types distributions can be fitted directly to a series of observations of M_n.

Practical applications have been wide-ranging. Type I distributions have been applied to fire protection and insurance problems and the prediction of earthquake magnitudes (Ramachandran, 1982), to model extremely high temperatures (Brown and Katz, 1995), and to predict high return levels of wind speeds relevant for the design of civil engineering structures (Naess, 1998). Type II distributions have been applied to estimate probabilities of extreme occurrences in Germany's stock index (Broussard and Booth, 1998) and to predict the behavior of solar proton peak fluxes (Xapson, Summers and Barke, 1998). Type III distributions have been used to model failure strengths of load-sharing systems (Harlow, Smith and Taylor, 1983) and window glasses (Behr, Karson and Minor, 1991), for evaluating the magnitude of future earthquakes in the Pacific (Burton and Makropoulos, 1985), in Argentina (Osella, Sabbione and Cernadas, 1992), in Japan (Suzuki and Ozaka, 1994) and in the Indian subcontinent (Rao, Rao and Kaila, 1997), for partitioning and floorplanning problems (Sastry and Pi, 1991), to predict the diameter of crops for growth and yield modeling purposes (Kuru, Whyte and Woollons, 1992), for the analysis of corrosion failures of lead-sheathed cables at the Kennedy Space center (Lee, 1992), to predict the occurrence of geomagnetic storms (Silbergleit, 1996), and to estimate the occurrence probability of giant freak waves in the sea area around Japan (Yasuda and Mori, 1997).

3. Rates of convergence

The study of rate of convergence of (1) is important as it provides a means for improving statistical models such as (9) based on it. There have been two common ways to measure the rate of convergence of the distribution of the sample maximum. The first is to use the uniform metric between the d.f.s F and G,

$$d_n = \sup_{x \in \Re}\left|F^n(a_n x + b_n) - G(x)\right|,$$

and the second is to use the total variation metric,

$$D_n = \sup\left|\Pr\left(\frac{M_n - b_n}{a_n} \in B\right) - G(B)\right|,$$

where the supremum is taken over the Borel sets and $G(B)$ is the measure of B corresponding to the d.f. in (1). In this section we mention some recent estimates of d_n and D_n. The results depend heavily on the theory of regularly varying functions: that an ultimately positive function δ with domain $(0, \infty)$ is regularly varying with index $\alpha \in \Re$ (written $\delta \in RV_\alpha$) if, for $x > 0$,

$$\lim_{t \to \infty} \frac{\delta(tx)}{\delta(t)} = x^\alpha$$

(De Haan, 1970).

3.1. Smith's rate of convergence

The seminal paper of Smith (1982) gives the following estimates for d_n when F belongs to the domains of attraction of the Fréchet and Weibull distributions: suppose that $-\log F \in RV_{-\alpha}$ for some $\alpha > 0$, and that $L(x) := -x^{-\alpha} \log F(x)$ satisfies

$$\frac{L(tx)}{L(t)} = 1 + O(\delta(t)), \quad x > 0, \ t \to \infty,$$

for some positive function δ satisfying

$$Bx^{-\theta} \leq \frac{\delta(tx)}{\delta(t)} \leq C, \quad x \geq 1, \ t \geq t_0, \ B > 0, \ C > 0, \ t_0 > 0, \ \theta > 0.$$

Then, with $b_n = 0$, a_n defined by $-\log F(a_n) = 1/n$ and $r_n = \delta(a_n)$

$$d_n = \sup_{x \in \Re} |F^n(a_n x + b_n) - \Phi_\alpha(x)| = O(r_n)$$

as $n \to \infty$, and with $b_n = w(F)$, a_n defined by $-\log F(w(F) - a_n) = 1/n$ and $r_n = \delta(1/a_n)$

$$d_n = \sup_{x \in \Re} |F^n(a_n x + b_n) - \Psi_\alpha(x)| = O(r_n)$$

as $n \to \infty$. Zolotarev and Rachev (1985) use the theory of probability metrics to obtain the exact form of the O-terms above. When F belongs to the domain of attraction of the unit Fréchet distribution they show that

$$d_n \leq C \max(\rho_0, \rho_r, \rho_r^{1/(r-1)}) n^{1-r}, \quad 1 < r \leq 2,$$

where C is a constant depending only on r and

$$\rho_r = \sup_{x > 0} x^r |F(x) - \Phi_1(x)|,$$

the Kolmogorov weighted semimetric. Omey and Rachev (1988) extend the above estimate for $r > 2$ and consider asymptotic expansions for $F^n(a_n x + b_n)$.

3.2. Cohen's rate of convergence

Cohen (1982a) extends Smith's result for the Gumbel domain of attraction. It requires substantially more conditions: that there exists functions a, b, c, f defined on $[X, x_0)$, for some constant $X < x_0 \leqslant \infty$, and constants $c_1, K > 1, r, s$ such that

$$b(x) > 0, \quad X \leqslant x < x_0,$$
$$\text{or} \quad b(x) < 0, \quad X \leqslant x < x_0;$$
$$\{a(x) - 1\}/b(x) \to r \quad \text{as } x \uparrow x_0;$$
$$\{c(x) - c_1\}/b(x) \to s \quad \text{as } x \uparrow x_0;$$
$$f'(x) \text{ exists and } f'(x) = b(x) \quad \text{for } X \leqslant x < x_0;$$
$$xf(z) + z \uparrow x_0 \quad \text{uniformly in } |x| \leqslant -K \log|b(z)|, \text{ as } z \uparrow x_0;$$
$$b(xf(z) + z)/b(z) \to 1 \quad \text{uniformly in } |x| \leqslant -K \log|b(z)|, \text{ as } z \uparrow x_0.$$

Then, as $n \to \infty$,

$$d_n = \sup_{x \in \Re} |F^n(a_n x + b_n) - \Lambda(x)| = O(b(b_n)),$$

where b_n is defined by $-\log F(b_n) = 1/n$ and $a_n = f(b_n)$.

3.3. Balkema and de Haan's rate of convergence

More recently Balkema and de Haan (1990) gave the following alternative estimate for d_n when F belongs to the domain of attraction of the Gumbel distribution: suppose

$$\lim_{t \to w(F)} \frac{\frac{1-F(t+x\gamma(t))}{1-F(t)} - \exp(-x)}{\rho(t)} = \frac{x^2}{2} \exp(-x), \quad x \in \Re, \text{ locally uniformly,}$$

for γ as in (3) and for some function ρ of constant sign with $\rho(t) \neq 0$ for all t and $\lim_{t \to w(F)} \rho(t) = 0$. Then,

$$b_n = f(\log n), \qquad a_n = \frac{1 - F_1(b_n)}{1 - F(b_n)}$$

and

$$d_n = \sup_{x \in \Re} |F^n(a_n x + b_n) - \Lambda(x)| = \left| \frac{\{1 - F(b_n)\}\{1 - F_2(b_n)\}}{\{1 - F_1(b_n)\}^2} - 1 \right|,$$

where $F_0 := F$ and for $k = 1, 2, \ldots,$

$$1 - F_k(t) = \max\left\{0, \int_t^{w(F)} (1 - F_{k-1}(u)) \, du \right\}.$$

For other rate of convergence results for the Gumbel domain of attraction see Hall and Wellner (1979), Gomes (1984), Obretenov (1989) and Beirlant and Willekens (1990).

3.4. Rate of convergence for normal extremes

It is well known that the normal distribution belongs to the Gumbel domain of attraction. Hall (1979) shows that the rate of convergence, d_n, is at most $1/\log n$ when F is the standard normal d.f. In fact,

$$\frac{c}{\log n} < d_n < \frac{3}{\log n},$$

where c is a positive constant, and d_n cannot be improved upon to $o(1/\log n)$ by using a different choice of constants to replace a_n and b_n. However, Fisher and Tippett (1928) show empirically that the Type III limit, Ψ_α, is closer to $F^n(a_n x + b_n)$ than the Type I limit, Λ, when a_n and b_n are defined by

$$b_n \sqrt{2\pi} \exp(b_n^2/2) = n \quad \text{and} \quad a_n = 1/b_n,$$

respectively. Cohen (1982b) follows this up theoretically and shows the following. Let

$$P(x) = \begin{cases} \exp\{-(-A_n x + B_n)^{k_n}\}, & \text{if } A_n x < B_n, \\ 0, & \text{if } A_n x \geqslant B_n, A_n < 0, \\ 1, & \text{if } A_n x \geqslant B_n, A_n > 0, \end{cases}$$

where A_n, B_n and k_n satisfy

$$k_n \log(B_n - A_n b_n) = -a_n^2,$$
$$k_n A_n a_n / (B_n - A_n b_n) = 1 + a_n^2,$$
$$k_n (A_n a_n)^2 / (B_n - A_n b_n)^2 = a_n^2$$

and $A_n k_n > 0$ for all n. Then for all x, and for all $n \geqslant 200$,

$$|F^n(x) - P(x)| < 12.5796 a_n^4 < \frac{7.5596}{(\log n)^2}.$$

This gives a penultimate approximation for normal extremes that is of order $1/(\log n)^2$. Cohen also shows that the rate of convergence for this approximation cannot be improved upon to $o(1/(\log n)^2))$. In a related development, Hall (1980) gives the following bounds for $F^n(x)$:

$$\exp\{-z_n(x)[1 - x^{-2} + 3x^{-4} + z_n(x)/2(n-1)]\}$$
$$< F^n(x) < \exp[-z_n(x)(1 - x^{-2})],$$

for all $x \geq b_n$, where

$$z_n(x) = (1/\sqrt{2\pi})nx^{-1}\exp(-x^2/2).$$

Hall claims that this result gives very good approximations of $F^n(x)$ even for n as large as 10.

3.5. De Haan and Resnick's rate of convergence

Falk and Marohn (1993) show that d_n and D_n are determined by the rate of convergence in (3) and that the converse is also true. Here, we state a related result in De Haan and Resnick (1996) that gives the most recent estimates for d_n and D_n. Suppose that the p.d.f. f of F satisfies a second-order von Mises condition with parameters $\beta \in \Re$ and $\rho \leq 0$, i.e., that f is twice differentiable, f' is eventually positive and the function

$$\delta(x) = \frac{xf''(x)}{f'(x)} - \beta + 1$$

has constant sign near infinity and satisfies $\lim_{t \to \infty} \delta(t) = 0$ and $|\delta| \in RV_\rho$. Also define the function H_β by

$$H_\beta(x) = \begin{cases} \int_0^x \exp(\beta u) \int_0^u \exp(\rho s)\, ds\, du, & \text{if } \beta \geq 0, \\ -\int_x^\infty \exp(\beta u) \int_0^u \exp(\rho s)\, ds\, du, & \text{if } \beta < 0. \end{cases}$$

Then, the estimates are

$$\lim_{n \to \infty} \frac{d_n}{|\delta(n)|} = \sup_{x \in \Re} \exp\{-(1+\beta)x\}\exp\{-\exp(-x)\}|H_\beta(x)|$$

and

$$\lim_{n \to \infty} \frac{D_n}{|\delta(n)|} = \frac{1}{2}\int_{-\infty}^\infty \exp\{-(1+\beta)x\}\exp\{-\exp(-x)\}$$
$$\times |H_\beta'(x) - (1+\beta)H_\beta(x) + \exp(-x)H_\beta(x)|\, dx,$$

respectively.

3.6. Omey's rate of convergence

Rates of convergence for the variant (7) of (1) have also been studied: Omey (1988) seeks estimates of

$$\Delta_n = \sup |f_n(x) - g(x)|,$$

where f_n denotes the p.d.f. of $a_n^{-1} M_{n+1}$ with a_n defined by $-\log F(a_n) = 1/n$ and g is the derivative of G. Assume F has a bounded p.d.f. f. Let $L(x) := -x \log F(x)$ and $u(x) := x^2 f(x)$ and assume further that

$$\frac{L(xt)}{L(x)} = 1 + O(\delta(x)), \qquad \frac{u(xt)}{u(x)} = 1 + O(\delta(x)), \qquad x \to \infty,$$

for some positive functions δ satisfying

$$Bx^{-\theta} \leqslant \frac{\delta(tx)}{\delta(t)} \leqslant C, \quad x \geqslant 1, \; t \geqslant t_0, \; B > 0, \; C > 0, \; t_0 > 0, \; \theta > 0.$$

Then it shown that as $n \to \infty$

$$\Delta_n = O(\delta(a_n)) + O\left(\left|\frac{n+1}{a_n} u(a_n) - 1\right|\right).$$

4. Generalized extreme value (GEV) distribution

The three types of distributions introduced in (2) may be combined into the single distribution with d.f.

$$G(x) = \exp\left\{-\left(1 + \xi \frac{x-\mu}{\sigma}\right)^{-1/\xi}\right\} \qquad (10)$$

defined when $1 + \xi(x-\mu)/\sigma > 0$; $\mu \in \Re$, $\sigma > 0$ and $\xi \in \Re$. This distribution is known as the generalized extreme value distribution. We denote it by GEV(μ, σ, ξ). The range of values of μ, σ and ξ encompasses the distributions in each of the three types, e.g., GEV$(1, 1, 1/\alpha)$, $\alpha > 0$, and GEV$(-1, -1, -1/\alpha)$, $\alpha > 0$, are of the same type as Φ_α and Ψ_α, respectively. Thus, the Fréchet and Weibull types correspond to $\xi > 0$ ($\xi = 1/\alpha$) and $\xi < 0$ ($\xi = -1/\alpha$), respectively. The case $\xi = 0$ is interpreted as the limit $\xi \to 0$ and thus $G(x)$ reduces to the Gumbel type:

$$G(x) = \exp\left\{-\exp\left(-\frac{x-\mu}{\sigma}\right)\right\}, \qquad -\infty < x < \infty.$$

Evidently the value of ξ dictates the tail behavior of G, thus we refer to ξ as the shape parameter. We refer to μ and σ as the location and scale parameters, respectively. For a four parameter generalization of (10) see Scarf (1992).

The so-called Annual Maximum Method consists of fitting the Generalized Extreme Value distribution to a series of annual maximum data with n taken to be the number of i.i.d. events in a year, i.e., apply Eq. (9) with the right-hand side replaced by (10). Estimates of extreme quantiles of the annual maxima are then obtained by inverting Eq. (10):

$$G(x_T) = 1 - \frac{1}{T} \Rightarrow x_T = \mu - \frac{\sigma}{\xi}\left[1 - \{-\log(1 - 1/T)\}^{-\xi}\right]. \qquad (11)$$

In extreme value terminology, x_T is the return level associated with the return period T, and it is common to extrapolate the relationship (11) to obtain estimates of return levels considerably beyond the end of the data to which the model is fitted.

Various methods of estimation for fitting the Generalized Extreme Value distribution have been proposed. These include: least squares estimation (Maritz and Munro, 1967), direct estimation based on order statistics and records (Hill, 1975; Pickands, 1975; Davis and Resnick, 1984; Dekkers, Einmahl and De Haan, 1989; Berred, 1995; Qi, 1998), maximum likelihood estimation (Prescott and Walden, 1980), probability weighted moments (Hosking, Wallis and Wood, 1985), minimum risk point estimation (Mukhopadhyay and Ekwo, 1987), best linear unbiased estimation (Balakrishnan and Chan, 1992), Bayes estimation (Ashour and El-Adl, 1980; Lye, Hapuarachchi and Ryan, 1993), method of moments (Christopeit, 1994), robust bootstrap estimation (Seki and Yokoyama, 1996), and minimum distance estimation (Dietrich and Hüsler, 1996). The most serious competitors are maximum likelihood and probability weighted moments. Madsen et al. (1997), Dupuis and Field (1998) and Dupuis (1999b) compare and contrast the performance of these methods. There are a number of regular problems associated with ξ: when $\xi < -1$ the maximum likelihood estimates do not exist, when $-1 < \xi < -1/2$ they may have problems, and when $\xi > 1/2$ second and higher moments do not exist. See Smith (1985a, 1987) for details. The most recent method proposed by Castillo and Hadi (1997) circumvents these problems: it provides well defined estimates for all parameter values and performs well compared to any of the existing methods. In any case, experience with data suggests that the condition $-1/2 < \xi < 1/2$ is valid for most applications.

Also several measures and tests have been devised to assess goodness of fit of the distribution (Stephens, 1977; Tsujitani, Ohta and Kase, 1980; Kinnison, 1989; Auinger, 1990; Chowdhury, Stedinger and Lu, 1991; Aly and Shayib, 1992; Fill and Stedinger, 1995; De Waal, 1996; Zempléni, 1996).

It is often of interest to test whether the extreme values of a physical process are distributed according to a Type I extreme value distribution rather than one of Types II or III. This is equivalent to testing whether $\xi = 0$ in the Generalized Extreme Value distribution. Hosking (1984) compares thirteen tests of this hypothesis and finds evidence to suggest that a modified likelihood ratio test and a test due to van Montfort and Otten (1978) give the best overall performance.

Prescott and Walden (1980) derive the Fisher information matrix for the Generalized Extreme Value distribution. For a sample of n i.i.d. observations from (10), the elements of the matrix are:

$$E\left(-\frac{\partial^2 l}{\partial \mu^2}\right) = \frac{n}{\sigma^2} p,$$

$$E\left(-\frac{\partial^2 l}{\partial \sigma^2}\right) = \frac{n}{\sigma^2 \xi^2}\{1 - 2\Gamma(2+\xi) + p\},$$

$$E\left(-\frac{\partial^2 l}{\partial \xi^2}\right) = \frac{n}{\xi^2}\left\{\frac{\pi^2}{6} + \left(1 - \gamma + \frac{1}{\xi}\right)^2 - \frac{2q}{\xi} + \frac{p}{\xi^2}\right\},$$

$$E\left(-\frac{\partial^2 l}{\partial\mu\partial\sigma}\right) = -\frac{n}{\sigma^2\xi}\{p - \Gamma(2+\xi)\},$$

$$E\left(-\frac{\partial^2 l}{\partial\mu\partial\xi}\right) = -\frac{n}{\sigma\xi}\left(q - \frac{p}{\xi}\right),$$

$$E\left(-\frac{\partial^2 l}{\partial\sigma\partial\xi}\right) = -\frac{n}{\sigma\xi^2}\left[1 - \gamma + \frac{\{1 - \Gamma(2+\xi)\}}{\xi} - q + \frac{p}{\xi}\right],$$

where

$$p = (1+\xi)^2 \Gamma(1+2\xi),$$

$$q = \Gamma(2+\xi)\left\{\Omega(1+\xi) + \frac{1}{\xi} + 1\right\},$$

$$\Omega(r) = \frac{\partial \log \Gamma(r)}{\partial r}$$

and $\gamma = 0.5772157\ldots$ is Euler's constant. Escobar and Meeker (1994) provide an algorithm to evaluate these elements.

5. Generalized Pareto (GP) distribution

Balkema and de Haan (1974) and Pickands (1975) show that $F \in D(G)$, where G is GEV(μ, σ, ξ) for some μ, σ and ξ, if and only if

$$\lim_{t \to w(F)} \sup_{0 < x < w(F) - t} \left| \frac{1 - F(t+x)}{1 - F(t)} - \left\{1 + \xi \frac{x}{\sigma_*(t)}\right\}^{-1/\xi} \right| = 0. \tag{12}$$

The second limiting term within the suprema is the survivor function of the Generalized Pareto (GP) distribution and is defined when either $0 < x < \infty$ ($\xi \geq 0$) or $0 < x < -\sigma_*(t)/\xi$ ($\xi < 0$). The case $\xi = 0$ is again interpreted as the limit $\xi \to 0$ and thus we have the exponential distribution with mean $\sigma_*(t)$ as a special case. We again refer to ξ and σ_* as the shape and scale parameters, respectively.

The Generalized Pareto distribution has been studied extensively and many characterizations exist in the literature. Nagaraja (1977, 1988) provides characterizations based on record values. For a random sample $X_1, X_2, \ldots, X_n$ with common d.f. F let $\{X_{L_n}, n \geq 0\}$ denote the sequence of record values, where $L_0 \equiv 1$ and $L_n = \min(i \mid i > L_{n-1}, X_i > X_{L_{n-1}})$, $n \geq 1$. Then Nagaraja (1977) shows that if for some constants p and q

$$E(X_{L_1} | X_{L_0} = x) = px + q \quad \text{a.s.}$$

then except for change in location and scale

$$F(x) = \begin{cases} 1-(-x)^\theta, & -1 < x < 0, & \text{if } 0 < p < 1, \\ 1-\exp(-x), & x > 0, & \text{if } p = 1, \\ 1-x^\theta, & x > 1, & \text{if } p > 1, \end{cases}$$

where $\theta = p/(1-p)$. Nagaraja (1988) extends this result for adjacent record values: namely,

$$E(X_{L_m}|X_{L_{m+1}} = x) = px + q \text{ a.s.} \quad \text{and} \quad E(X_{L_{m+1}}|X_{L_m} = x) = rx + s \text{ a.s.}$$

hold for some real p, q, r and s if and only if F is an exponential d.f. Kotz and Shanbhag (1980) provide interesting characterizations in terms of the expected remaining life function, $\phi(x) = E(X - x \mid X \geqslant x)$. They implicitly show that ϕ is a nonvanishing polynomial on (α, β), where $F(\alpha) = 0$ and $F(\beta_-) = 1$, if and only if one of the following is true:

- $\alpha > -\infty$, $\beta = \infty$ and F is a shifted exponential with left extremity α;
- $\alpha > -\infty$, $\beta = \infty$ and F is a Pareto with left extremity α;
- $\alpha > -\infty$, $\beta < \infty$ and for all $x \in (\alpha, \beta)$

$$F(x) = 1 - \frac{\phi(\alpha_+)}{\phi(x)} \exp\left\{-\int_\alpha^x \frac{dy}{\phi(y)}\right\},$$

with ϕ as a polynomial of the form referred to.

When ϕ is linear the distribution in the last case reduces to

$$F(x) = 1 - \left(1 - \frac{x-\alpha}{\beta-\alpha}\right)^\gamma$$

for some $\gamma > 0$.

One of the recent characterizations due to Dembińska and Wesolowski (1998) uses the order statistics of X_i, say $Y_1 < Y_2 < \cdots < Y_n$. They show that if for some $k \leqslant n - r$ and real p, q

$$E(|Y_{k+r}|) < \infty \quad \text{and} \quad E(Y_{k+r}|Y_k) = pY_k + q$$

hold then only the following three cases are possible: (1) $p = 1$ and F is an exponential d.f.; (2) $p > 1$ and F is a Pareto d.f.; (3) $p < 1$ and F is a power d.f. Most recently, Asadi et al. (2001) have proposed some unified characterizations of the generalized Pareto using a new concept of 'extended neighboring order statistics'. For other results we refer the reader to the two monographs Galambos and Kotz (1978) and Rao and Shanbhag (1994).

The moments of the Generalized Pareto distribution are readily obtained by noting that

$$E\left(1 + \xi \frac{X}{\sigma_*(t)}\right)^r = \frac{1}{1 - r\xi},$$

if $1 - r\xi > 0$. The rth moment exists if $\xi < 1/r$ and thus all moments exist for $-1/2 < \xi < 0$. The mean and the variance are

$$\frac{\sigma_*(t)}{1 - \xi} \quad \text{and} \quad \frac{\sigma_*^2(t)}{(1 - \xi)^2(1 - 2\xi)},$$

respectively.

When t is large enough (12) provides the model

$$F(x) \approx 1 - \omega(t)\left\{1 + \xi \frac{x - t}{\sigma_*(t)}\right\}^{-1/\xi}, \tag{13}$$

where $\omega(t) = 1 - F(t)$ is the probability of exceeding t and either $t < x < \infty$ ($\xi \geq 0$) or $t < x < t - \sigma_*(t)/\xi$ ($\xi < 0$). We denote (13) by GP($\sigma_*(t)$, ξ, t). This model forms the basis for the approach of modeling i.i.d. exceedances over a high threshold and has the advantage of using more of the available data than just the annual maxima. The analogue of the T-year return level for this model is defined as the value x_T such that $F(x_T) = 1 - 1/(n\omega T)$, where n is the number of i.i.d. observations in a year. Inverting (13), we get

$$x_T = t - \frac{\sigma_*(t)}{\xi}\left\{1 - (n\omega T)^\xi\right\}.$$

The maximum likelihood estimate of $\omega(t)$ is the empirical proportion of exceedances over the threshold t. The maximum likelihood estimates of ξ and $\sigma_*(t)$ exist in large samples provided that $\xi > -1$ and they are asymptotically normal and efficient if $\xi > -1/2$. For an account on how the non-regular cases could be treated see Walshaw (1993) and for an algorithm for computing the maximum likelihood estimates see Grimshaw (1993).

Davison and Smith (1990) give details of statistical inference for (13). For instance, the Fisher information matrix for n i.i.d. observations is the inverse of

$$\frac{1 + \xi}{n}\begin{pmatrix} 2\sigma_*^2 & \sigma_* \\ \sigma_* & 1 + \xi \end{pmatrix}.$$

Davison and Smith also provide a graphical diagnostic, now known as the *residual life plot*, for selecting a sufficiently large value for the threshold t. For other diagnostics for threshold selection see Nadarajah (1994) and Dupuis (1999a).

Some practical applications of (13) include the estimation of the finite limit of human lifespan (Zelterman, 1992; Aarssen and De Haan, 1994), the modeling of high concentrations in short-range atmospheric dispersion (Mole et al., 1995), and the estimation of flood return levels for homogeneous regions in New Brunswick, Canada (ElJabi, Ashkar and Hebabi, 1998).

6. Joint distribution of the r-largest order statistics

Suppose $X_1, X_2, \ldots$ are i.i.d. random variables with common d.f. $F \in D(G)$, where G is GEV(μ, σ ξ) for some μ, σ and ξ. Let $M_n^{(i)}$ denote the ith largest of the first n random variables, $i = 1, \ldots, r$. Then, by arguments in Weissman (1978) or by Leadbetter et al. (1983, Chapter 2), the limiting joint d.f. of the r-largest order statistics for $x_1 \geqslant x_2 \geqslant \cdots \geqslant x_r$ is:

$$\lim_{n\to\infty} \Pr\left\{\frac{M_n^{(1)} - b_n}{a_n} < x_1, \frac{M_n^{(2)} - b_n}{a_n} < x_2, \ldots, \frac{M_n^{(r)} - b_n}{a_n} < x_r\right\}$$

$$= \sum_{s_1=0}^{1} \sum_{s_2=0}^{2-s_1} \cdots \sum_{s_{r-1}=0}^{r-1-s_1-\cdots-s_{r-2}} \frac{(\gamma_2 - \gamma_1)^{s_1}}{s_1!} \cdots \frac{(\gamma_r - \gamma_{r-1})^{s_{r-1}}}{s_{r-1}!}$$

$$\times \exp(-\gamma_r). \tag{14}$$

Here, $\gamma_i = -\log \text{GEV}(x_i; 0, 1, \xi)$, and a_n, b_n are the same norming constants as in (1). Dziubdziela (1978), Deheuvels (1986, 1989) and Falk (1989) provide results on the rate of convergence of (14). Serfozo (1982) identifies necessary and sufficient conditions for the convergence in (14) to hold. Goldie and Maller (1996) provide characterizations of the asymptotic properties,

$$\limsup_{n\to\infty} (M_n^{(r)} - M_n^{(r+s)}) \leqslant c \quad \text{a.s.},$$

for some finite constant c, which tell us, in various ways, how quickly the sequence of maxima increase. These characterizations take the form of integral conditions on the tail of F. The related problem of the joint distribution of the $(n-r)$ smallest-order statistics is considered in Finner and Roters (1994).

Hall (1978) gives canonical representations for a sequence of random variables $\{Z_i\}$ having (14) as finite-dimensional distributions. Let γ denote Euler's constant and let Z_k be independent exponential random variables with mean 1. Then, the representations are that:

$$Z_i =^d \sum_{k=i}^{\infty} \frac{Z_k - 1}{k} + \gamma - \sum_{k=1}^{i-1} \frac{1}{k}, \quad i \geqslant 1,$$

if X_1 is in the domain of attraction of the Type I distribution;

$$Z_i =^d \exp\left[\alpha^{-1}\left\{\sum_{k=i}^{\infty} \frac{Z_k - 1}{k} + \gamma - \sum_{k=1}^{i-1} \frac{1}{k}\right\}\right], \quad i \geqslant 1,$$

if X_1 is in the domain of attraction of the Type II distribution; and,

$$Z_i =^d \exp\left[-\alpha^{-1}\left\{\sum_{k=i}^{\infty}\frac{Z_k - 1}{k} + \gamma - \sum_{k=1}^{i-1}\frac{1}{k}\right\}\right], \quad i \geq 1,$$

if X_1 is in the domain of attraction of the Type III distribution. Hall also gives limit theorems about the behavior of Z_i as $i \to \infty$.

Practical applications of (14) proceed by assuming that n is sufficiently large for the limit law to hold. Following the argument that led to Eq. (9), we can express the joint p.d.f. of $(M_n^{(1)}, M_n^{(2)}, \ldots, M_n^{(r)})$ in the generalized form:

$$f(x_1, x_2, \ldots, x_r) = \sigma^{-r} \exp\left\{-\left(1 + \xi\frac{x_r - \mu}{\sigma}\right)^{-1/\xi}\right. \\ \left. - \left(\frac{1}{\xi} + 1\right)\sum_{i=1}^{r}\log\left(1 + \xi\frac{x_i - \mu}{\sigma}\right)\right\} \quad (15)$$

valid for $x_1 \geq x_2 \geq \cdots \geq x_r$ such that $1 + \xi(x_i - \mu)/\sigma > 0$, $i = 1, 2, \ldots, r$.

Smith (1986), Singh (1987) and Tawn (1988a) perform maximum likelihood estimation of (15) and provide algebraic expressions for the associated Fisher information matrix. Dupuis (1997) considers robust estimation. This allows for identification of observations which are not consistent with (15) and an assessment of the validity of the model.

For the asymptotic approximation of (14) to be valid, r has to be small by comparison with n. Actually, as r increases the rate of convergence to the limiting joint distribution decreases sharply. The choice of r is therefore crucial. Wang (1995) proposes a method for selecting r based on a suitable goodness-of-fit statistic. See also Zelterman (1993), Umbach and Ali (1996) and Drees and Kaufmann (1998).

Some practical applications of (15) include the prediction of the extreme pit depths into the future and over large space of exposed metal (Scarf et al., 1992), forest inventory problems (Pierrat et al., 1995), estimation of the extreme sea-level distribution in the Aegean and Ionian Seas (Tsimplis and Blackman, 1997), the prediction of the extreme hurricane wind speeds at locations on the Gulf and Atlantic coasts of the United States (Casson and Coles, 1998) and the prediction of the maximum size of random spheres in Wicksell's corpuscle problem (Takahashi and Sibuya, 1998).

7. A point process characterization

Smith (1989) points out that the three approaches based, respectively, on the Generalized Extreme Value distribution, Generalized Pareto distribution and the joint distribution of the r-largest order statistics can be treated as special cases of the following point process characterization due to Pickands (1971).

Suppose $X_1, X_2, \ldots$ are i.i.d. random variables with common d.f. $F \in D(G)$, where G is GEV$(\mu, \sigma \, \xi)$ for some μ, σ and ξ. Let $Y_{n,i} = (X_i - b_n)/a_n$, $i = 1, 2, \ldots, n$, where a_n, b_n are the same norming constants as in (1). Then, if $\mathcal{P}_n$ is a point process in the plane with points at $(i/(n+1), Y_{n,i})$, we have

$$\mathcal{P}_n \to^d \mathcal{P}, \tag{16}$$

where $\mathcal{P}$ is a nonhomogeneous Poisson process on $[0, 1] \times (0, \infty)$ with intensity measure Σ satisfying

$$\Sigma\{(s_1, s_2) \times (y, \infty)\} = (s_2 - s_1)\{1 + \xi y\}^{-1/\xi}, \quad 0 \leq s_1 \leq s_2 \leq 1.$$

This result has the interpretation that for large finite n the second ordinates of most of the points of $\mathcal{P}_n$ tend to cluster near zero, the lower end point, while away from the boundary the process looks like a nonhomogeneous Poisson process. Thus, in practical applications of the result, we assume, for a large enough threshold t, that, if $\widetilde{\mathcal{P}}_n$ is a point process in the plane with points at $(i/(n+1), X_i)$, we have

$$\widetilde{\mathcal{P}}_n = \widetilde{\mathcal{P}} \tag{17}$$

on $[0, 1] \times (t, \infty)$, where $\widetilde{\mathcal{P}}$ is a nonhomogeneous Poisson process with intensity measure

$$\Sigma\{(s_1, s_2) \times (x, \infty)\} = (s_2 - s_1)\left(1 + \xi \frac{x - \mu}{\sigma}\right)^{-1/\xi}, \quad 0 \leq s_1 \leq s_2 \leq 1.$$

Note that the norming constants a_n and b_n have been absorbed by the generalized form of this intensity measure (cf. Eq. (10)). The statistical inference for this model reduces to that of a Poisson process, and the latter is well established.

Some practical applications of (17) include the detection of trends in ground-level ozone (Smith, 1989), estimation of the return period wind speeds in Italy (Gusella, 1991), estimation of the return period wave heights and wind speeds in the northern North Sea (Morton, Bowers and Mould, 1997), and the estimation of various characteristics of the extreme sea currents, such as speeds and their directions, that are required when designing offshore structures (Robinson and Tawn, 1997).

8. Extremes of stochastic processes

There have also been developments of extreme value theory for stochastic processes that are not i.i.d. sequences of random variables. In this section, we consider the extreme values of moving average processes, Gaussian processes, non-stationary time series, random walks, Markov and related processes, queuing processes, continuous-time processes and a variety of others. Because of the limitation in space, we provide some details of the results only for the first three. For the remaining, we simply provide a literature survey.

8.1. Moving averages

Studies of the extreme values of moving average processes have been carried out by Rootzén (1978, 1986, 1987), Chernick (1981) and Davis and Resnick (1985, 1988, 1991) Here, we state the main results of Rootzén (1986). Consider the extreme values of the infinite moving average process

$$X_t = \sum_i c_{i-t} Z_i, \quad t = 0, \pm 1, \pm 2, \ldots,$$

where the Z_i are i.i.d. random variables and for convenience of notation let Z be another random variable with the same distribution as the Z_i. This process includes the ARMA-process often used in time series analysis. The results require conditions on the *noise variables*, $\{Z_t\}$, conditions on the *weights*, $\{c_t\}$, and conditions involving $\{Z_t\}$ and $\{c_t\}$ simultaneously.

The conditions on the noise variables are the following. Assume that

$$\Pr(Z > z) \sim K z^\alpha \exp(-z^p), \quad \text{as } z \to \infty, \tag{18}$$

where p, K are positive parameters and α is a real parameter, and that the first moment exists, $E|Z| < \infty$, and for $p \geq 1$ in addition that $EZ^2 < \infty$. For $p > 1$, suppose Z has a continuously differentiable p.d.f. f, which satisfies

$$f(z) \sim K' z^{\alpha'} \exp(-z^p) \quad \text{as } z \to \infty, \tag{19}$$

for $\alpha' = \alpha + p - 1$, $K' = Kp$, and that

$$\exp(cz) f'(z) \text{ is bounded for } z \in (-\infty, 0], \tag{20}$$

for some constant $c \geq 0$. Moreover, define

$$D(z) = \begin{cases} f(z) \exp(z^p), & \text{for } z \geq 0, \\ f(z), & \text{otherwise} \end{cases}$$

and assume that

$$\limsup_{z \to \infty} \left| \frac{z D'(z)}{D(z)} \right| < \infty. \tag{21}$$

The conditions on the weights are that at least one c_t is strictly positive, and that

$$|c_t| = O(|t|^{-\theta}) \quad \text{as } t \to \pm \infty, \text{ for some } \theta > 1. \tag{22}$$

For $p > 1$, assume in addition that

$$|c_t| = O(|t|^{-\theta}) \quad \text{as } t \to \pm \infty, \text{ for some } \theta > \max(1, 2/q), \tag{23}$$

where q is the conjugate exponent of p, defined by $1/p + 1/q = 1$. For later use, let $c_+ = \max\{\max(0, c_t); t = 0, \pm 1, \ldots\}$, $c_- = \max\{\max(0, -c_t), t = 0, \pm 1, \ldots\}$ and let $\Pi_+ = \{t_1, \ldots, t_{k_+}\}$ be the set of t's for which $c_t = c_+$, and let $\Pi_- = \{t_1, \ldots, t_{k_-}\}$ be defined similarly with $\Pi_- = \emptyset$ if $c_- = 0$. Further, write $\|c\|_q = \{\sum_t |c_t|^q\}^{1/q}$ and $\|c^+\|_q = \{\sum_t |\max(0, c_t)|^q\}^{1/q}$.

The conditions involving weights and noise variables simultaneously are the following: if $0 < p \leqslant 1$ they are

A.1 (18) and (22) hold, and all c_t are nonnegative;
A.2 (18) and (22) hold, and $\Pr(Z < z) = O(\exp(-|z|^p/\gamma))$ as $z \to -\infty$, where γ satisfies $c_- \gamma^{1/p} < c_+$;
A.3 (18) and (22) hold, and $\Pr(Z < z) \sim K_- |z|^\alpha \exp(-|z|^p/\gamma)$, for some constant $K_- > 0$, where $c_- \gamma^{1/p} = c_+$, and α is the same as in (18),

and, if $p > 1$ then

B.1 (19), (20), (21) and (23) hold, and all c_t are non-negative;
B.2 (19), (21) and (23) hold, and in addition $f(-z)$ satisfies (19) and (21), with p in (19) replaced by some $p' > p$, and possibly with different D, α', K';
B.3 (19), (21) and (23) hold, and in addition $f(-z)$ satisfies (19) and (21), with the same p as in (19), but possibly with different D, α', K'.

With the conditions set as above, Rootzén (1986) gives the following characterizations of the extreme values of X_t. The characterizations differ depending on whether $p > 1$, $p = 1$ or $0 < p < 1$.

If $p > 1$ and one of B.1–B.3 holds then

$$\Pr\{a_n(M_n - b_n) \leqslant x\} \to \exp\{-\exp(-x)\} \quad \text{as } n \to \infty,$$

where

$$a_n = \begin{cases} p\|c\|_q^{-1}(\log n)^{1/q}, & \text{if B.1 or B.3 holds,} \\ p\|c^+\|_q^{-1}(\log n)^{1/q}, & \text{if B.2 holds} \end{cases}$$

and

$$b_n = \begin{cases} \|c\|_q(\log n)^{1/p} + O\big((\log n)^{1/(\theta q) - 1/q}\big), & \text{if B.1 or B.3 holds,} \\ \|c^+\|_q(\log n)^{1/p} + O\big((\log n)^{\max\{1/(\theta q), q'/q\} - 1/q}\big), & \text{if B.2 holds.} \end{cases}$$

For the $p = 1$ case some further notation is needed. With c_+, c_-, Π_+, Π_-, k_+ and k_- as defined above, let

$$k = \begin{cases} k_+, & \text{if A.1 or A.2 holds,} \\ k_+ + k_-, & \text{if A.3 holds} \end{cases}$$

and let

$$\Pi = \begin{cases} \Pi_+, & \text{if A.1 or A.2 holds,} \\ \Pi_+ \cup \Pi_-, & \text{if A.3 holds.} \end{cases}$$

With this notation, define

$$\widehat{\alpha} = \begin{cases} k\alpha + k - 1, & \text{if } \alpha > -1, \\ \alpha, & \text{if } \alpha < -1 \end{cases}$$

and

$$\widehat{K} = \begin{cases} K^k \Gamma^k(\alpha+1) \Gamma^{-1}\bigl(k(\alpha+1)\bigr) E \exp\{\sum_{t \notin \Pi} c_t Z_t / c_+\}, \\ \quad \text{if A.1 or A.2 holds and } \alpha > -1, \\ K^{k_+} \bigl(K_- \gamma^{\alpha/p}\bigr)^{k_-} \Gamma^k(\alpha+1) \Gamma^{-1}\bigl(k(\alpha+1)\bigr) \\ \quad \times E \exp\{\sum_{t \notin \Pi} c_t Z_t / c_+\}, \quad \text{if A.3 holds and } \alpha > -1, \\ kK \{E \exp(Z)\}^{k-1} E \exp\{\sum_{t \notin \Pi} c_t Z_t / c_+\}, \\ \quad \text{if A.1 or A.2 holds and } \alpha < -1, \\ \bigl[k_+ K \{E \exp(Z)\}^{k_+ - 1} \{E \exp(-Z/c_-)\}^{k_-} \\ \quad + k_- K_- \gamma^{\alpha/p} \{E \exp(Z)\}^{k_+} \{E \exp(-Z/c_-)\}^{K_- - 1}\bigr] \\ \quad \times E \exp\{\sum_{t \notin \Pi} c_t Z_t / c_+\}, \quad \text{if A.3 holds and } \alpha < -1. \end{cases}$$

Then, if one of A.1–A.3 holds, and in addition either $\alpha > -1$ or $\alpha < -1$ and $k = 1$,

$$\Pr\{a_n(M_n - b_n) \leqslant x\} \to \exp\{-\exp(-x)\} \quad \text{as } n \to \infty,$$

where

$$a_n = 1/c_+$$

and

$$b_n = c_+ \log n + c_+ \bigl(\widehat{\alpha} \log \log n + \log \widehat{K}\bigr).$$

Finally, if $0 < p < 1$ and one of A.1–A.3 holds then

$$\Pr\{a_n(M_n - b_n) \leqslant x\} \to \exp\{-\exp(-x)\} \quad \text{as } n \to \infty,$$

where

$$a_n = (1/c_+) p (\log n)^{1-1/p}$$

and

$$b_n = \begin{cases} c_+ (\log n)^{1/p} + (c_+/p)\{(\alpha/p) \log \log n + \log K\}/(\log n)^{1-1/p}, \\ \quad \text{if A.1 or A.2 holds,} \\ c_+ (\log n)^{1/p} + (c_+/p)\{(\alpha/p) \log \log n + \log(K + K_- \gamma^{\alpha/p})\} \\ \quad /(\log n)^{1-1/p}, \quad \text{if A.3 holds.} \end{cases}$$

8.2. Gaussian processes

Let $\{X(t)\}$ be a stationary Gaussian process with mean zero, variance one and covariance $r(l)$ satisfying

$$r(l) = 1 - C|l|^\alpha + o(|l|^\alpha)$$

as $l \to 0$, where $0 < \alpha \leqslant 2$ and $C > 0$ are constants. Leadbetter et al. (1983) show that if $r(l) \log l \to 0$ as $l \to \infty$ then

$$\Pr\left\{\sup_{0 \leqslant t \leqslant h} X(t) \leqslant \frac{x}{a_h} + b_h\right\} \to \exp\{-\exp(-x)\}$$

as $h \to \infty$, where

$$a_h = \sqrt{2 \log h}$$

and

$$b_h = \sqrt{2 \log h} + \frac{1}{2\sqrt{\log h}} \left[\frac{2-\alpha}{2\alpha} \log \log h + \log\left\{\frac{(2C)^{1/\alpha} H_\alpha}{2\sqrt{\pi}}\right\}\right].$$

Also see Cramér (1965, 1966), Pickands (1967, 1969a, 1969b), Welsch (1973), Hebbar (1979), McCormick (1980), Nair (1981), Rootzén (1983), Xie (1984), Adler and Samorodnitsky (1987), Hüsler (1990a, 1995, 1999), Konstant and Piterbarg (1993), Choi et al. (1995), Seleznjev (1996) and Kratz and Rootzén (1997) for related results. Here, H_α is an important parameter and is given by

$$H_\alpha = \lim_{h \to \infty} \frac{1}{h} \int_0^\infty \exp(x) \Pr\left\{\sup_{0 \leqslant t \leqslant h} Z(t) > x\right\} dx,$$

where $\{Z(t)\}$ is a non-stationary Gaussian process with mean $-|t|^\alpha$ and covariance $|s|^\alpha + |t|^\alpha - |t-s|^\alpha$. This parameter also appears in the Erdös–Révész type law of the iterated logarithm. Unfortunately, the exact value of H_α is unknown except for the two special cases $H_1 = 1$ and $H_2 = 1/\sqrt{\pi}$. But Shao (1996) gives the following upper and lower bounds of H_α:

$$5.2^{-1/\alpha} 0.625 \leqslant H_\alpha \leqslant \left(\alpha \, \mathrm{e}/\sqrt{\pi}\right)^{2/\alpha}$$

if $1 \leqslant \alpha \leqslant 2$ and

$$(\alpha/4)^{1/\alpha}\{1 - \exp(-1/\alpha)(1 + 1/\alpha)\}$$
$$\leqslant H_\alpha \leqslant \left[\sqrt{\alpha}\{0.77\sqrt{\alpha} + 2.41\sqrt{8.8 - \alpha \log(0.4 + 2.5/\alpha)}\}\right]^{2/\alpha}$$

if $0 < \alpha < 1$. In particular,

$$0.12 \leqslant H_\alpha \leqslant 3.1$$

if $1 \leqslant \alpha \leqslant 2$ and

$$\lim_{\alpha \to 0} \frac{\alpha \log H_\alpha}{\log \alpha} = 1.$$

Shao also proposes two estimators of H_α based on i.i.d. fractional Brownian motion of order α.

8.3. Non-stationary time series

Niu (1997) considers the extreme values of a class of non-stationary time series with the form

$$X_k = \mu_k + \varepsilon_k, \quad \varepsilon_k = \sum_{i=0}^{\infty} c_i Z_{k-i},$$

where $\{Z_i = \sigma_i \eta_i; -\infty < i < \infty\}$, $\{\eta_i; -\infty < i < \infty\}$ is a sequence of i.i.d. random variables satisfying the regular variation conditions:

$$\Pr(|\eta_1| > x) \in RV_{-\alpha}, \quad \alpha > 0,$$

and

$$\lim_{x \to \infty} \frac{\Pr(\eta_1 > x)}{\Pr(|\eta_1| > x)} = \pi_0, \quad \lim_{x \to \infty} \frac{\Pr(\eta_1 < -x)}{\Pr(|\eta_1| > x)} = 1 - \pi_0,$$

where $0 \leqslant \pi_0 \leqslant 1$, and $\{c_i\}$ is a sequence of real constants with $c_0 = 1$ and

$$\sum_{i=0}^{\infty} |c_i|^\gamma < \infty \quad \text{for some } 0 < \gamma < \min(1, \alpha).$$

Furthermore, assume that

$$\frac{1}{n} \sum_{i=1}^{n} \sigma_i^\alpha \to \sigma^\alpha$$

as $n \to \infty$, where $\sigma > 0$, and let a_n be the $(1 - n^{-1})$ quantile of $|\eta_1|$, that is, let

$$a_n = \inf\{x : \Pr(|\eta_1| \leqslant x) \geqslant 1 - n^{-1}\}.$$

Then, provided that $E(X_k)$ is a bounded function on $[0, \infty)$, we have the following characterization of the extreme values of X_k: the point process

$$\{(k/n, X_k/a_n), k = 1, \ldots\} \to \{(t_k, c_i U_k), k = 1, \ldots, i = 0, \ldots\}$$

as $n \to \infty$. Here, $\{(t_k, U_k)\}$ is a non-homogeneous Poisson process with intensity measure $\nu_0 \times \mu$, ν_0 is a Lebesgue measure on $[0, \infty)$ and

$$\mu(\mathrm{d}x) = \sigma^\alpha \big(\pi_0 \alpha x^{-(\alpha+1)} \, \mathrm{d}x \, I_{(0,\infty]}(x)$$
$$+ (1 - \pi_0)\alpha(-x)^{-(\alpha+1)} \, \mathrm{d}x \, I_{[-\infty,0)}(x)\big).$$

8.4. Random walks

Various limit theorems (including necessary and sufficient conditions) on the distribution tail of the maximum of a random walk are given in Cramér (1955), Darling and Erdös (1956), Pakes (1975), Borovkov (1976), Veraverbeke (1977), Embrechts and Goldie (1982), Embrechts and Veraverbeke (1982), Bertoin and Doney (1996) and Korshunov (1997) among others. For instance, Darling and Erdös investigate the asymptotic behavior of the maximum of normalized sums, $\max_{i=1,\ldots,n} i^{-1/\alpha} S_i$, where $S_i = X_1 + \cdots + X_i$, deriving the famous Darling–Erdös limit theorem: if $E(X_1) = 0$, $E(X_1^2) = 1$ and $E(|X_1|^3) < \infty$, then, for every $x \in \Re$,

$$\lim_{n \to \infty} \Pr\left(\max_{i=1,\ldots,n} \frac{S_i}{\sqrt{i}} \leqslant a_n x + b_n\right) = \exp\{-\exp(-x)\},$$

with

$$a_n = 1/\sqrt{2 \log \log n}$$

and

$$b_n = a_n \left\{2 \log \log n + \frac{1}{2} \log \log \log n - \frac{1}{2} \log(4\pi)\right\}.$$

Several extensions of this theorem have been obtained; see Bertoin (1998) for the most recent result. Deheuvels and Devroye (1987) describe the limiting behavior of the following characteristics associated with S_i:

$$T_n = \max_{i=1,\ldots,n} (S_{i+k(i)} - S_i),$$
$$U_n = \max_{i=0,\ldots,n-k} (S_{i+k} - S_i),$$
$$W_n = \max_{i=0,\ldots,n-k} \max_{l=1,\ldots,k} (S_{i+l} - S_i),$$
$$V_n = \max_{i=0,\ldots,n-k} \min_{l=1,\ldots,k} (k/l)(S_{i+l} - S_i),$$

for $k = k(n) = [c \log n]$, where $c > 0$ is a given constant.

8.5. Markov and related processes

Brockwell et al. (1982) study the extreme values of the integral process, $\int_0^t S(x)\,dx$, $t \geqslant 0$, when S is a finite state-space Markov chain. The results involve weak convergence of a sequence of such processes to a Wiener process. Turkman and Walker (1983), Turkman and Oliveira (1992) and Dorea and GonCalves (1999) study the extremal properties of chain-dependent sequences, i.e., sequences $X_1, X_2, \ldots, X_n, \ldots$ with marginal d.f.s $F_{s_1}, F_{s_2}, \ldots, F_{s_n}, \ldots$, respectively, where $s_1, s_2, \ldots, s_n, \ldots$ is a sequence of discrete random variables. Berman (1987) derives limit theorems for the extreme values of a Markov random field on a discrete lattice. O'Brien (1987) investigates the asymptotic behavior of $M_{i,k} = \max(X_{i+1}, \ldots, X_k)$ when X_i is a strictly stationary sequence. Letting $M_n = M_{0,n}$, it is shown under general circumstances that

$$\Pr(M_n \leqslant c_n) - \{\Pr(X_1 \leqslant c_n)\}^{n\Pr(M_{1,p_n} \leqslant c_n | X_1 > c_n)} \to 0,$$

for any sequence (p_n) satisfying certain growth-rate conditions and that there exists a d.f. G such that

$$\Pr(M_n \leqslant c_n) - \{G(c_n)\}^n \to 0$$

for all sequences (c_n). Rootzén (1988) uses close connection with regeneration theory to study the extremal behavior of stationary Markov chains. Perfekt (1994) obtains similar results for a more general class of stationary Markov chains. Alpium (1989), Adke and Chandran (1994) and Alpium et al. (1995) provide extremal limit laws for a specialized Markovian sequence $\{X_i\}$ defined by: $X_i = X_0$ if $i = 0$ and $X_i = k\max(X_{i-1}, Z_i)$ if $i \geqslant 1$, $0 < k < 1$, where Z_i are i.i.d. and X_0 is a random variable independent of the Z_i's. In related developments, Gomes (1996) studies the extreme value theory for $X_i = \max(\beta X_{i-1}, \alpha\beta X_{i-1} + Z_i)$, $0 \leqslant \alpha < 1$, $0 < \beta < 1$, where X_0 and Z_i are independent random variables, and Chrapek (1998) derives the extreme value behavior of $X_i = \max(A_i X_{i-1}, B_i)$, where X_0, A_i and B_i are independent random variables. Haiman et al. (1995) establish bounds for $\Pr\{\max(X_1, \ldots, X_n) \leqslant x\}$ when X_i is a stationary Markov sequence with the transition density satisfying some regularity conditions. Poskitt and Chung (1996) derive a limit law for the extreme values of a hidden Markov chain process.

8.6. Queuing processes

Extreme value theory has also seen applications in queuing processes. In the seminal paper, Iglehart (1972) derives the limiting behaviors of the maximum waiting time, the maximum virtual waiting time and the maximum queuing length for the $GI/GI/1$ queue. For instance, if the mean interarrival time is less than the mean service time and W_n denotes the nth customer's waiting time, then it is shown that

$$\lim_{n\to\infty} \Pr\{\gamma W_n - \log(bn) < x\} = \exp\{-\exp(-x)/m\}, \quad -\infty < x < \infty,$$

for some constants b, γ and m. Cohen (1972) provides analogues of these results for the $M/G/1$ queue. Serfozo (1988a, 1988b) study the asymptotic behavior of the maximum values of birth and death processes over large time intervals: usually these do not converge or converge to a degenerate distribution. However, by allowing the birth and death rates to vary in a certain manner as the time interval increases, he shows that the maxima can have three possible limit distributions; namely, those with d.f.

$$G(x) = \Phi_1(x), \quad x \geqslant 0;$$
$$G(x) = \Lambda(x), \quad -\infty < x < \infty;$$
$$G(x) = \exp[-c/\{\exp(x) - 1\}], \quad x \geqslant 0,$$

for $0 < c < \infty$. In a related development, letting $X(t)$ denote the number of customers in an $M/M/s$ service system at time t, McCormick and Park (1992) study the limiting behavior of $M(t) = \max_{s \leqslant t} X(s)$ as $t \to \infty$ under a 'heavy traffic assumption'. They exhibit normalizing constants $a(t)$ and $b(t)$ such that the d.f. of $(M(t) - b(t))/a(t)$ converges to the classical extreme value d.f. $\exp(-\exp(-x))$ as $t \to \infty$. Sparaggis et al. (1993) consider the problem of routing jobs to parallel queues with identical exponential servers, unequal finite buffer capacities and state-dependent, nondecreasing service rates with respect to queue lengths. They establish the extremal properties of the shortest nonfull queue and the longest nonfull queue policies, in systems with concave/convex service rates. Park et al. (1994) derive a law of large numbers for the maximum queue length for an $M/M/\infty$ queue and a Poisson AR(1)-process. Asmussen et al. (1999) establish tail asymptotics for the steady-state queue length in $GI/GI/1$ queues with subexponential service times.

8.7. Continuous time stochastic processes

Leadbetter (1978) and Leadbetter and Rootzén (1982) develop the central distributional results of classical extreme value theory for the maxima of continuous-time stochastic processes: particularly, proving that the basic result concerning the existence of just three types distributions in such cases, and giving necessary and sufficient conditions for each to apply. Horowitz (1980) shows that the maximum of the process $X(t) = \exp(f(t) + e(t))$, where f is deterministic and $e(t)$ is a sequence of correlated normal random variables, behaves according to the classical Gumbel distribution, after suitable normalization. He applies this to an air quality estimation problem. Albin (1990) develops a method to determine an asymptotic expression for $\Pr\{\sup_{0 \leqslant t \leqslant h} X(t) > u\}$, where $X(t)$ is a continuous-time stochastic process. Albin (1998) shows that this theory can be developed also for self-similar nonstationary processes. Doney and O'Brien (1991) study the extreme values of the shot noise process

$$X(t) = \sum_{k:\tau_k \leqslant t} h(t - \tau_k), \quad t \geqslant 0,$$

where $h:[0,\infty) \to [0,\infty)$ is nonincreasing and $(\tau_k, k \geq 0)$ is the sequence of renewal times of a renewal process. Considering the stationary sequence (X_i) given by $X_i = \sum_{k \leq i} h(\tau_i - \tau_k)$, they show that

$$\Pr\{\max(X_1, \ldots, X_n) \leq u_n\} - \{\Pr(X_0 \leq u_n)\}^n \to 0$$

for real sequences (u_n) for which $\limsup n \Pr(X_0 > u_n) < \infty$. Borkovec and Klüppelberg (1998) investigate the extremal behavior of a diffusion process, $X(t)$, in finance given by the stochastic difference equation:

$$dX(t) = \mu(X(t)) dt + \sigma(X(t)) dW(t), \quad t > 0, \quad X(0) = x,$$

where W is standard Brownian motion, μ is the drift term and σ is the diffusion coefficient. Hooghiemstra and Lopuhaä (1998) study the extremal behavior of the continuous-time stationary processes $X(t) - t$ and $|X(t) - t|$, on increasing intervals $[0, T]$, as $T \to \infty$, where $X(t)$ is the location of the maximum of standard two-sided Brownian motion minus a parabolic drift.

8.8. Miscellanea

Here, we summarize developments, which are not mentioned above.

Leadbetter (1975) surveys the theory for strictly stationary sequences satisfying certain distributional mixing conditions that ensure that their extremes behave as if the sequences were i.i.d. Daley and Hall (1984) study the suprema of the linear-cost-adjusted sequences,

$$Z(\delta) = \sup_{i \geq 1} \{X_i - i\delta\} \quad \text{for } \delta > 0.$$

They show that if there exist norming constants $\{a(\delta)\}$ with $a(\delta) > 0$ and $\{b(\delta)\}$ such that $a(\delta)\{Z(\delta) - b(\delta)\}$ converges weakly to a non-degenerate d.f. J as $\delta \downarrow 0$, then J is of the same type as the three extreme value distributions. Furthermore, if the d.f. F of X_i is in the domain of attraction of Λ, Φ_α with $\alpha > 1$, or Ψ_α with $\alpha > 0$, then, respectively, $a_*(\delta)\{Z(\delta) - b_*(\delta)\}$ converges weakly to Λ, $\Phi_{\alpha-1}$ or $\Psi_{\alpha+1}$ as $\delta \downarrow 0$ for suitably chosen norming constants $\{a_*(\delta)\}$ and $\{b_*(\delta)\}$. Hsing (1986) considers extreme value theory for the weighted suprema $X_i = \sup c_k Z_{i-k}$, where c_k is a sequence of nonnegative constants and Z_k is a sequence of i.i.d. positive random variables whose tail probabilities are regularly varying with a negative index. Horváth (1987) provides an analogous investigation when the Z_k are quantile processes. De Haan and Verkade (1987) consider extreme value theory in the context of independent but not identically distributed random variables: the departure from the case of identical distributions comes from a trend added to i.i.d. observations. Ballerini and McCormick (1989) discuss the asymptotic distribution of the maximum of $X_i = g(i) + h(i)Z_i, i = 1, \ldots, n$, where Z_i is a stationary random sequence, satisfying some mixing conditions, $g(i)$ is a trend function and $h(i)$ a positive, periodic function with integer period as the variance function. Motivated by the autoregressive conditional heteroscedastic (ARCH) processes that are frequently used in econometrics for modeling financial data, De Haan et al. (1989) explore the asymptotic behavior of the extreme values of the process $X_i = A_i + X_{i-1} + B_i$, where $X_0 \geq 0$ and A_i, B_i are i.i.d. random variables in $\mathfrak{R}_+^2$. Révész (1995) proves limit theorems for $\max(X_1, \ldots, X_n)$ when X_i is a sequence of identically distributed 'nearly independent, nearly Poisson' random variables. Borkovec (2000) investigates the extremal behavior of a special class of autoregressive processes with ARCH(1) errors given by the stochastic difference equation $X_i = \alpha X_{i-1} + \sqrt{\beta + \eta X_{i-1}^2} \varepsilon_n$, where ε_n are i.i.d. random variables.

In the remaining sections we deal with multivariate extreme value distributions: Sections 9–12 provide fundamental theoretical results, Sections 13 and 14 detail relevant statistical models (including simulation schemes) and Section 15 is on multivariate extreme value theory for nontrivial stochastic processes.

9. Limit laws for multivariate extremes

By analogy with the univariate case the traditional approach to define multivariate extremes is to base it on componentwise maxima. If $\{(X_{i,1}, \ldots, X_{i,p}), i = 1, \ldots, n\}$ are i.i.d. p-variate random vectors with joint d.f. F and

$$\mathbf{M}_n = (M_{n,1}, \ldots, M_{n,p}) = \left(\max_{1 \leq i \leq n} X_{i,1}, \ldots, \max_{1 \leq i \leq n} X_{i,p}\right)$$

is the vector of maxima of each component, then we seek normalizing constants $a_{n,j} > 0$, $b_{n,j}$, $j = 1, \ldots, p$, such that as $n \to \infty$

$$\Pr\left(\frac{M_{n,1} - b_{n,1}}{a_{n,1}} \leq x_1, \ldots, \frac{M_{n,p} - b_{n,p}}{a_{n,p}} \leq x_p\right)$$
$$= F^n(a_{n,1}x_1 + b_{n,1}, \ldots, a_{n,p}x_p + b_{n,p}) \qquad (24)$$
$$\to G(x_1, \ldots, x_p)$$

for a p-variate d.f. G with nondegenerate marginals. If this holds for suitable choices of a_n and b_n then we say G is a multivariate extreme value d.f. and F is in the domain of attraction of G, written as $F \in D(G)$. By setting all x_j but one to ∞ in (24) we see that $F_j \in D(G_j)$, $j = 1, \ldots, p$, i.e.,

$$F_j^n(a_{n,j}x_j + b_{n,j}) \to G_j(x_j), \quad j = 1, \ldots, p, \qquad (25)$$

where F_j and G_j are the jth marginal d.f.s of F and G, respectively. It follows by the Extremal Types Theorem (see Section 2) that G_j is a Type I, II or III distribution and hence the norming constants $a_{n,j}$, $b_{n,j}$ are precisely those in (4).

The two extreme forms of the limiting multivariate distribution correspond to the case of asymptotic total independence between the componentwise maxima, for which

$$G(x_1, \ldots, x_p) = G_1(x_1) \cdots G_p(x_p), \qquad (26)$$

and the case of asymptotic total dependence between the componentwise maxima for which

$$G(x_1,\ldots,x_p) = \min\{G_1(x_1),\ldots,G_p(x_p)\}.$$

Asymptotic total independence arises if and only if (25) holds and there exists an $\mathbf{x} = (x_1,\ldots,x_p) \in \Re^p$ such that $0 < G_j(x_j) < 1$, $j = 1,\ldots,p$, and

$$F^n(a_{n,1}x_1 + b_{n,1},\ldots,a_{n,p}x_p + b_{n,p}) \to G_1(x_1)\cdots G_p(x_p)$$

as $n \to \infty$ (Takahashi, 1994a, Theorem 2.2) – see Mardia (1964), Srivastava (1967) and Tiago de Oliveira (1980) for other equivalent conditions. Moreover, (26) holds for any $(x_1,\ldots,x_p) \in \Re^p$ if and only if

$$G(0,\ldots,0) = \{\Lambda(0)\}^p = \exp(-p),$$

provided that $G_j = \Lambda$, $j = 1,\ldots,p$; or, equivalently,

$$G(1,\ldots,1) = \Phi_{\alpha_1}(1)\cdots\Phi_{\alpha_p}(1) = \exp(-p),$$

provided that $G_j = \Phi_{\alpha_j}$, $\alpha_j > 0$, $j = 1,\ldots,p$; or, equivalently,

$$G(-1,\ldots,-1) = \Psi_{\alpha_1}(-1)\cdots\Psi_{\alpha_p}(-1) = \exp(-p),$$

provided that $G_j = \Psi_{\alpha_j}$, $\alpha_j > 0$, $j = 1,\ldots,p$ (Takahashi, 1987, Theorems 2.2–2.4). Asymptotic total dependence arises if and only if (25) holds and there exists an $\mathbf{x} = (x_1,\ldots,x_p) \in \Re^p$ such that $0 < G_1(x_1) = \cdots = G_p(x_p) < 1$ and

$$F^n(a_{n,1}x_1 + b_{n,1},\ldots,a_{n,p}x_p + b_{n,p}) \to G_1(x_1)$$

as $n \to \infty$ (Takahashi, 1994a, Theorem 3.1).

To isolate dependence aspects from marginal distributional features it is convenient to transform components so that they have a standard marginal distribution. For technical convenience we standardize G so that its margins are described by the unit Fréchet d.f. $\Phi_1(y) = \exp\{-y^{-1}\}$, $y > 0$, although there are, of course, other possibilities. This standardization does not pose difficulties, as shown by the two following propositions (Resnick, 1987, Proposition 5.10). Throughout the rest of the paper we shall use the notation Y or y to denote Fréchet random variables.

Suppose G is a multivariate d.f. with continuous marginals. Consider the transform

$$G_*(y_1,\ldots,y_p) = G\big((1/(-\log G_1))^{\leftarrow}(y_1),\ldots,(1/(-\log G_p))^{\leftarrow}(y_p)\big),$$

$$y_1 \geqslant 0,\ldots,y_p \geqslant 0.$$

Then, G_* has marginal d.f.s $G_{*j}(y) = \Phi_1(y)$ and G is a multivariate extreme value df if and only if G_* is also a multivariate extreme value d.f. This proposition standardizes the marginal distributions of a multivariate extreme value d.f. to unit Fréchet margins but yet preserves the extreme value property.

The following proposition justifies the standardization by showing that $F \in D(G)$ if and only if $F_* \in D(G_*)$. Define

$$Y_j = U_j(X_j) = -1/\log F_j(X_j), \quad j = 1, \ldots, p. \tag{27}$$

Let F_* be the d.f. of $(Y_1, \ldots, Y_p)$ so that

$$F_*(y_1, \ldots, y_p) = F\big(U_1^{\leftarrow}(y_1), \ldots, U_p^{\leftarrow}(y_p)\big).$$

If $F \in D(G)$ then $F_* \in D(G_*)$ and

$$\Pr\Big\{\max_{1 \leq i \leq n} U_j(X_{i,j})/n \leq y_j, j = 1, \ldots, p\Big\} = F_*^n(ny_1, \ldots, ny_p)$$

$$\to G_*(y_1, \ldots, y_p)$$

as $n \to \infty$. Conversely, if $F_* \in D(G_*)$, (25) holds and G_* has nondegenerate marginals then $F \in D(G)$.

In the next two sections we provide several fundamental results, which characterize the domain of attraction condition, $F \in D(G)$, or equivalently $F_* \in D(G_*)$, and the form of the multivariate extreme value d.f. G_*. These results have been crucial as theoretical underpinnings to recent developments of statistical models for multivariate extremes (see Sections 13 and 14) and their practical applications.

10. Characterizations of the domain of attraction

The concept of domain of attraction received attention in Section 2 for the classical univariate extreme value distributions. The concept is less straight forward for the multivariate case as we see from the several characterizations of the domain of attraction that follow. The characterizations are classified into those that are just necessary (Section 10.1), those that are just sufficient (Section 10.2) and those that are both necessary and sufficient (Section 10.3).

10.1. Necessary characterizations

These characterizations are especially useful for statistical modeling and simulation of multivariate extreme values. We begin with the point process characterization due to de Haan (1985).

10.1.1. De Haan's point process characterization
Suppose $(X_{i,1}, \ldots, X_{i,p})$, $i = 1, 2, \ldots$, are i.i.d. p-variate random vectors with common joint d.f. $F \in D(G)$. Define

$$T(y_1, \ldots, y_p) = \left(\sum_{j=1}^{p} y_j, y_1 / \sum_{j=1}^{p} y_j, \ldots, y_{p-1} / \sum_{j=1}^{p} y_j \right)$$

and let

$$S_p = \left\{ (w_1, \ldots, w_{p-1}) : \sum_{j=1}^{p-1} w_j \leqslant 1, w_j \geqslant 0, j = 1, \ldots, p-1 \right\}$$

be the $(p-1)$-dimensional unit simplex. Then,

$$\mathcal{P}_n = \{(U_1(X_{i,1})/n, \ldots, U_p(X_{i,p})/n), \ i = 1, \ldots, n\} \to \mathcal{P} \tag{28}$$

as $n \to \infty$, where $\mathcal{P}$ is a nonhomogeneous Poisson process on $\mathfrak{R}_+^p \setminus \{(0, \ldots, 0)\}$ with intensity measure μ_* satisfying

$$\mu_* \circ T^{\leftarrow}(\mathrm{d}r, \mathrm{d}\mathbf{w}) = r^{-2} \, \mathrm{d}r \, H_*(\mathrm{d}\mathbf{w}), \quad r > 0, \ \mathbf{w} \in S_p, \tag{29}$$

and H_* is a non-negative measure on S_p with

$$H_*(S_p) = p \quad \text{and} \quad \int_{S_p} w_j H_*(\mathrm{d}\mathbf{w}) = 1, \quad j = 1, \ldots, p-1. \tag{30}$$

In Eq. (29), $\circ$ denotes the composition operation.

An immediate consequence (28) is that we can write

$$G_*(y_1, \ldots, y_p) = \exp\{-V(y_1, \ldots, y_p)\}, \tag{31}$$

where

$$V(y_1, \ldots, y_p) = \mu_*\left(([0, y_1] \times \cdots \times [0, y_p])^C\right)$$
$$= \int_{S_p} \max\left(\frac{w_1}{y_1}, \ldots, \frac{1 - w_1 - \cdots - w_{p-1}}{y_p} \right) H_*(\mathrm{d}\mathbf{w}).$$

We refer to V as the exponent measure function.

The intuitive content of (28) for $p = 2$ can be described as follows. As $n \to \infty$ the scaling by $1/n$ drags down to the origin all points except those with unusually large values of either $X_{i,1}$ or $X_{i,2}$ or both. Points with unusually large $X_{i,1}$ but not large $X_{i,2}$, will move under the scaling to the horizontal boundary of $\mathfrak{R}_+^2$, and those with unusually large $X_{i,2}$ but not large $X_{i,1}$ will move to the vertical boundary of $\mathfrak{R}_+^2$: only points with both components unusually large will survive in $\mathfrak{R}_+^2$ away from the boundaries.

The limiting intensity measure μ_* describes the dependence structure between unusually large values of $X_{i,j}$, $j = 1, \ldots, p$, after standardization by U_j to have the unit Fréchet distribution. However, under the transformation T, which maps the standardized vector $(U_1(X_{i,1}), \ldots, U_p(X_{i,p})) \in \mathfrak{R}_+^p$ into pseudo-polar coordinates in $(0, \infty) \times S_p$, the measure μ_* factorizes into a known function of the radial component, r, and a measure H_* of the angular component, $\mathbf{w}$. Thus, essentially, the measure H_* on S_p embodies the dependence structure of the extremes. If it concentrates its mass in the interior of S_p, then we have strong dependence structures, e.g., total dependence between the extremes of $X_{i,j}$, $j = 1, \ldots, p$, corresponds to H_* having all its mass at $\{(1/p, \ldots, 1/p)\}$, i.e., $H_*(\{(1/p, \ldots, 1/p)\}) = p$. If it concentrates its mass near the boundary of S_p, then we have weak dependence structures, e.g., total independence between the extremes corresponds to H_* having all its mass at the vertices, i.e., $H_*(\{(1/p, \ldots, 0)\}) = \cdots = H_*(\{(0, \ldots, 1/p)\}) = 1$. Although it is an arbitrary finite non-negative measure, the standardization of $X_{i,j}$ constrains H_* to have unit means with respect to each dimension of S_p. Since these are the only constraints on H_*, no finite parameterization exists for the measure.

Nadarajah (1999c) has developed a comprehensive method for simulating bivariate extremes by approximating the limiting point process result (28) – as described below.

When $p = 2$, (28) reduces to the version that as $n \to \infty$, $\mathcal{P}_n$ converges in distribution to a nonhomogeneous Poisson process on $\{[0, \infty) \times [0, \infty)\} \setminus \{(0, 0)\}$ with intensity measure μ_* satisfying

$$\mu_*(T^{\leftarrow}(\mathrm{d}r, \mathrm{d}w)) = r^{-2}\,\mathrm{d}r\,H_*(\mathrm{d}w), \tag{32}$$

where H_* is a non-negative measure on $[0, 1]$ constrained to have unit means, i.e., $\int_{[0,1]} w H_*(\mathrm{d}w) = \int_{[0,1]} (1-w) H_*(\mathrm{d}w) = 1$, and a total mass of 2. Now take a region $A \subset \{[0, \infty) \times [0, \infty)\} \setminus \{(0, 0)\}$ sufficiently away from $\{(0, 0)\}$ and in it assume that $\mathcal{P}_n$ coincides with the Poisson process; this model was originally proposed by Coles and Tawn (1991) and Joe et al. (1992). Thus, simulation of (X_1, X_2) in A, under the model, reduces to simulation of a Poisson process restricted to A. There are standard procedures to simulate from a Poisson process, we consider a simple one. Take $A = A_0 = \{(x_1, x_2): x_1 + x_2 > r_0, x_1 \geq 0, x_2 \geq 0\}$ with r_0 sufficiently large so that the model is valid; to simulate (X_1, X_2) over a region A not having the form A_0, first simulate (X_1, X_2) over an A_0 for which $A \subset A_0$ and then delete those points falling outside of A. The conditional p.d.f. of $(R, W) = (X_1 + X_2, X_1/(X_1 + X_2))$ over $T(A_0)$ is:

$$f_{R,W}(r, w) = r^{-2} H_*(\mathrm{d}w) \Big/ \int_{r_0}^{\infty} \int_0^1 s^{-2}\,\mathrm{d}s\,H_*(\mathrm{d}v) = r_0 r^{-2} 2^{-1} H_*(\mathrm{d}w),$$

using relationship (32). Since this conditional p.d.f. factorizes, i.e., R and W are independent, we can simulate R and W independently of each other. Simulation of the r component is straight-forward: set $\int_{r_0}^{r} r_0 s^{-2}\,\mathrm{d}s = u$ for $u \sim U(0, 1)$ and use the inversion principle to obtain $r = r_0/(1-u)$. For simulation of the w component, assume differentiability of H_* in the interior, $(0, 1)$, and let

$$h^*(w) = \frac{1}{2 - \theta_0 - \theta_1} \frac{\partial H_*([0, w])}{\partial w},$$

where $\theta_0 = H_*(\{0\})$ and $\theta_1 = H_*(\{1\})$ are atoms of mass H_* may have at the end points of $[0, 1]$. Then simulation of w can be performed by the method of composition: set w as 0, belonging to $(0, 1)$ or 1 with probabilities $2^{-1}\theta_0$, $1 - 2^{-1}\theta_0 - 2^{-1}\theta_1$ and $2^{-1}\theta_1$, respectively; if w is set to be in $(0, 1)$, then Nadarajah (1999c) shows that one can simulate it from the p.d.f. h^* using the rejection method with the Beta envelope

$$\frac{w^{s_0}(1-w)^{s_1}}{Be(s_0+1, s_1+1)}, \quad w \in (0, 1),$$

where $s_0 > -1$, $s_1 > -1$ are constants such that $h^*(w) = O(w^{s_0})$ and $h^*(1-w) = O(w^{s_1})$ hold as $w \to 0$, and

$$Be(a, b) = \frac{\Gamma(a)\Gamma(b)}{\Gamma(a+b)}, \quad a > 0, \, b > 0.$$

Nadarajah (1999c) also shows that one can find valid s_0, s_1 for every known bivariate extreme value distribution. Hence, simulation of bivariate extreme values reduces – in essence – to simulation from a Beta$(s_0 + 1, s_1 + 1)$ distribution and routines are widely available for this.

Finally, in this section, we discuss two technical tools for generating parametric models for H_* (Coles and Tawn, 1991) that will be useful later. We need some terminology. For a given $(w_1, \ldots, w_{p-1}) \in S_p$, define

$$H(w_1, \ldots, w_{p-1}) = H_*\big([0, w_1] \times \cdots \times [0, w_{p-1}]\big),$$

the measure function associated with H_*, and let

$$w_p = 1 - w_1 - \cdots - w_{p-1}.$$

Decompose the measure function H into a hierarchy of densities $h_{m,c}$ defined on subspaces $S_{m,c} = \{\mathbf{w} \in S_p : w_k = 0, k \notin c\}$, where $c = \{j_1, \ldots, j_m\}$ is an index variable over the subsets of size m of the set $c_p = \{1, \ldots, p\}$. The subspace $S_{m,c}$ is isomorphic to the $(m-1)$-dimensional unit simplex S_m and $h_{m,c}$ is the $(m-1)$-dimensional density of H on the subspace $S_{m,c}$. The density $h_{m,c}$ describes the dependence structure between the extremes of $X_{i,k}$ for $k = j_1, j_2, \ldots, j_m$. When $m = p$ and $c = c_p$ we simplify the notation by $h \equiv h_{p,c_p}$.

The first tool relates the exponent measure function, V, to H, by expressing the density $h_{m,c}$ for $c = \{j_1, \ldots, j_m\}$ in terms of derivatives of V (Coles and Tawn, 1991, Theorem 1). Namely,

$$\frac{\partial V}{\partial y_{j_1} \cdots \partial y_{j_m}} = -\left(\sum_{l=1}^{m} y_{j_l}\right)^{-(m+1)} h_{m,c}\left(\frac{y_{j_1}}{\sum_l y_{j_l}}, \ldots, \frac{y_{j_{m-1}}}{\sum_l y_{j_l}}\right) \quad (33)$$

on $\{\mathbf{y} \in \Re_+^p : y_k = 0, k \notin c\}$, where we assume differentiability of V. The importance of this result is that densities of all orders for the measure function H may be obtained for any closed form multivariate extreme value d.f. For $p = 2$ the result shows the following: H, a function on the unit interval $[0, 1] = S_2$, decomposes into the density $h_{2,\{1,2\}}$ defined in the interior, $(0, 1)$, and the 'densities' $h_{1,\{1\}}$ and $h_{1,\{2\}}$ defined, respectively, at the end points, $\{1\}$ and $\{0\}$. The two latter 'densities' (these are actually atoms of mass, $H_*(\{1\})$ and $H_*(\{0\})$) are independent components of H in that they are associated with those $(X_{i,1}, X_{i,2})$, which are extreme in only one component. The density $h_{2,\{1,2\}}$ is the dependence component in that it describes the dependence between the extremes of both components.

The second tool generates a form for H_* by transforming an arbitrary density $h_\dagger$ in the interior of S_p into h_{p,c_p} (Coles and Tawn, 1991, Theorem 2). Specifically, if $h_\dagger$ is any positive density in the interior of S_p with finite first moments:

$$m_j = \int_{S_p} w_j h_\dagger(w_1, \ldots, w_{p-1}) \, dw_1 \cdots dw_{p-1}, \quad j = 1, \ldots, p, \quad (34)$$

then a measure H_* on S_p defined by

$$h_{m,c} \equiv 0, \quad \forall c \neq c_p,$$

$$h_{p,c_p}(w_1, \ldots, w_{p-1}) = \frac{1}{m_0} \prod_{j=1}^{p} \frac{m_j}{m_0} h_\dagger\left(\frac{m_1 w_1}{m_0}, \ldots, \frac{m_{p-1} w_{p-1}}{m_0}\right), \quad (35)$$

where $m_0 = \sum_{j=1}^{p} m_j w_j$, is a valid measure satisfying the constraints (30). Hence, this result is useful in generating a rich class of parametric models for H_* in the interior of S_p.

10.1.2. Ledford and Tawn's point process characterization

The characterization expressed in (28) assumes max-stable dependence between the extremes of $(X_{i,1}, \ldots, X_{i,p})$. A generalization of this characterization to cover weaker forms of dependence structures including total independence and negative association is described below (Ledford and Tawn, 1997, Theorem 1). We provide the result for $p = 2$ to the best of our knowledge. As yet it is not known how it generalizes to the multivariate case.

Let $(Y_{i,1}, Y_{i,2})$, $i = 1, 2, \ldots$, be independent random vectors with both Y_1 and Y_2 having the unit Fréchet distribution. Suppose that for y_1 and y_2 simultaneously large

$$\Pr(Y_1 > y_1, Y_2 > y_2) = \mathcal{L}_1(y_1, y_2) y_1^{-c_1} y_2^{-c_2}$$
$$+ \mathcal{L}_2(y_1, y_2) y_1^{-(c_1+d_1)} y_2^{-(c_2+d_2)} + \cdots, \qquad (36)$$

where $c_1 + c_2 = 1/\eta$, $0 < \eta \leq 1$, $d_k \geq 0$ and $\mathcal{L}_k(y_1, y_2) \neq 0$ denotes a bivariate slowly varying function. Suppose too that $\mathcal{L}_2(ty_1, ty_2) = o\{\mathcal{L}_1(ty_1, ty_2)\}$ as $t \to \infty$ if $d_1 = d_2 = 0$ and

$$t_*(w) = \lim_{t \to \infty} \left\{ \frac{\mathcal{L}_1(tw, t(1-w))}{\mathcal{L}_1(t, t)} \right\}$$

is differentiable for all $w \in (0, 1)$. Choose b_n to satisfy

$$\Pr\{b_n^{-1} \max(\min(Y_{1,1}, Y_{1,2}), \ldots, \min(Y_{n,1}, Y_{n,2})) \leq y\} \to \exp(-y^{-1/\eta})$$

as $n \to \infty$ and define $T(y_1, y_2) = ((y_1 + y_2)/b_n, y_1/(y_1 + y_2))$. Then

$$\mathcal{P}_n = \{(Y_{i,1}/b_n, Y_{i,2}/b_n), i = 1, \ldots, n\} \to \mathcal{P} \qquad (37)$$

as $n \to \infty$, where $\mathcal{P}$ is a nonhomogeneous Poisson process on $(0, \infty) \times (0, \infty)$ with intensity measure μ_* satisfying

$$\mu_* \circ T^{\leftarrow}(dr, dw) = r^{-(1+\eta)/\eta} \, dr \, \upsilon_0(w) \, dw, \quad r > 0, \ w \in (0, 1),$$

where $\upsilon_0(w)$ is a rather formidable function:

$$\upsilon_0(w) = \frac{c_1 c_2 t_*(w) + w(1-w) t'_*(w)(2w - 1 + c_1 - c_2) - t''_*(w) w^2 (1-w)^2}{w^{1+c_1}(1-w)^{1+c_2}}.$$

An immediate consequence is that we can generalize the form of (31) by

$$\lim_{n \to \infty} \Pr\left(\max_{1 \leq i \leq n} Y_{i,1} < b_n y_1, \max_{1 \leq i \leq n} Y_{i,2} < b_n y_2 \right)$$

$$= \exp\left\{ -\int_0^1 \int_{\min(y_1/w, y_2/(1-w))} \upsilon_0(w) r^{-(1+\eta)/\eta} \, dr \, dw \right\}$$

$$= \exp\left[-\eta \int_0^1 \upsilon_0(w) \left\{ \max\left(\frac{w}{y_1}, \frac{1-w}{y_2} \right) \right\}^{1/\eta} dw \right],$$

where the integration is over the open interval $0 < w < 1$.

As in De Haan's point process characterization the intensity measure μ_* factorizes into radial and angular components. But here both terms influence the dependence structure with η playing a fundamental role: the $r^{-(1+\eta)/\eta}$ term describes the main decay of probability due to dependence while the $v_0(w)$ term embodies less important features of the dependence. If the common d.f. of $(Y_{i,1}, Y_{i,2})$ belongs to the domain of attraction of G_*, it is then easily verified that $c_1 = c_2 = 1/2$, $d_1 = d_2 = 1/2$, $\eta = 1$ and

$$t_*(w) = \frac{1 - V\{(1-w)^{-1}, w^{-1}\}}{\{2 - V(1,1)\}\sqrt{w(1-w)}}.$$

Thus, $b_n = n$ and (37) reduces to (28).

Ledford and Tawn (1996) refer to η as the *coefficient of tail dependence* as it provides a measure of the dependence between the marginal tails of Y_1 and Y_2. For example, if $1/2 < \eta \leq 1$ the marginal variables are positively associated; when the marginal variables are independent then $\eta = 1/2$; if $0 < \eta < 1/2$ the marginal variables are negatively associated. Also if the marginal variables are asymptotically dependent then $\eta = 1$, and if $\eta < 1$ then there is asymptotic independence.

Peng (1999) proposes the following consistent estimator for η:

$$\hat{\eta}_n = \log 2 / \log \frac{\sum_{i=1}^n I\{Y_{i,1} > Y_{n,n-2k,1} \text{ and } Y_{i,2} > Y_{n,n-2k,2}\}}{\sum_{i=1}^n I\{Y_{i,1} > Y_{n,n-k,1} \text{ and } Y_{i,2} > Y_{n,n-k,2}\}},$$

where $Y_{n,1,j} \leq \cdots \leq Y_{n,n,j}$ denote the order statistics of $Y_{1,j}, \ldots, Y_{n,j}$ for $j = 1, 2$. Peng also establishes asymptotic normality of this estimate by considering the cases $\eta < 1$ and $\eta = 1$ separately. Assume the following variant of (36):

$$\Pr\{Y_1 > -1/\log(1 - ty_1), Y_2 > -1/\log(1 - ty_2)\}$$
$$= c(y_1, y_2) t^{1/\eta} [1 + O(t^\beta)]$$

uniformly on $\{(y_1, y_2): y_1^2 + y_2^2 = 1, y_1 \geq 0, y_2 \geq 0\}$ as $t \to 0$, where $\eta \in (0, 1]$, $\beta > 0$, and $c(y_1, y_2) \neq 0$ for some $y_1, y_2 > 0$. Assume also that $c(y_1, y_2)$ has continuous first-order partial derivatives denoted by

$$c_j(y_1, y_2) = \frac{\partial c(y_1, y_2)}{\partial y_j}, \quad j = 1, 2.$$

Then, for $\eta < 1$

$$\sqrt{\sum_{i=1}^n I\{Y_{i,1} > Y_{n,n-k,1}, Y_{i,2} > Y_{n,n-k,2}\}}$$
$$\times \sqrt{(2^{1/\hat{\eta}_n})/(2^{1/\hat{\eta}_n} - 1)} (\log 2) \hat{\eta}_n^{-2} (\hat{\eta}_n - \eta) \to^d N(0, 1),$$

where $k = k(n)$ is chosen to satisfy

$$k \to \infty, \quad k/n \to 0, \quad k(n/k)^{1-1/\eta} \to \infty, \quad k(n/k)^{1-2\beta-1/\eta} \to 0$$

as $n \to \infty$. For $\eta = 1$

$$2(\log 2)c(1,1)\sqrt{k}(\hat{\eta}_n - 1) \to^d N(0, \sigma^2), \qquad (38)$$

where

$$\sigma^2 = 2c(1,1)\big[1 - 4c_1(1,1) - 4c_2(1,1) + 6c_1(1,1)c_2(1,1)\big]$$
$$+ 4c(1,2)c_1(1,1)\big[1 - c_2(1,1)\big]$$
$$+ 4c(2,1)c_2(1,1)\big[1 - c_1(1,1)\big] + 2c_1^2(1,1) + 2c_2^2(1,1)$$

and $k = k(n)$ is chosen to satisfy

$$k \to \infty, \qquad k = o\big(n^{2\beta/(1+2\beta)}\big)$$

as $n \to \infty$. Obviously, the limit in (38) involves the unknown quantities $c(y_1, y_2)$, $c_1(1, 1)$ and $c_2(1, 1)$. Peng gives the following consistent estimates:

$$\hat{c}(y_1, y_2) = \frac{1}{k}\sum_{i=1}^{n} I\{Y_{i,1} > Y_{n,n-[ky_1],1}, Y_{i,2} > Y_{n,n-[ky_2],2}\},$$

$$\hat{c}_1(1,1) = k^{1/4}\Bigg\{\frac{1}{k}\sum_{i=1}^{n} I\{Y_{i,1} > Y_{n,n-[k(1+k^{-1/4})],1}, Y_{i,2} > Y_{n,n-k,2}\}$$
$$- \frac{1}{k}\sum_{i=1}^{n} I\{Y_{i,1} > Y_{n,n-k,1}, Y_{i,2} > Y_{n,n-k,2}\}\Bigg\},$$

$$\hat{c}_2(1,1) = k^{1/4}\Bigg\{\frac{1}{k}\sum_{i=1}^{n} I\{Y_{i,1} > Y_{n,n-k,1}, Y_{i,2} > Y_{n,n-[k(1+k^{-1/4})],2}\}$$
$$- \frac{1}{k}\sum_{i=1}^{n} I\{Y_{i,1} > Y_{n,n-k,1}, Y_{i,2} > Y_{n,n-k,2}\}\Bigg\}.$$

10.1.3. Nadarajah, Anderson and Tawn's characterization

Here, we state another generalization of (28), by considering the case, where the marginal variables are linearly ordered (Nadarajah, Anderson and Tawn, 1998, Theorem 2), i.e., $X_1 \leqslant X_2 \leqslant mX_1$ for some $m > 1$. Then the limit measure H_* in Eq. (29), defined on $S_2 = [0, 1]$, becomes concentrated in the subinterval

$$\left[\liminf_{y \to \infty}\left\{\frac{l(y)}{y + l(y)}\right\}, \limsup_{y \to \infty}\left\{\frac{y}{y + r(y)}\right\}\right]$$

with

$$\liminf_{y\to\infty}\left\{\frac{r(y)}{y}\right\} \leqslant 1 \leqslant \limsup_{y\to\infty}\left\{\frac{y}{l(y)}\right\},$$

where

$$r(y) = U_2\{U_1^{\leftarrow}(y)\}$$

and

$$l(y) = U_1\left\{\frac{1}{m}U_2^{\leftarrow}(y)\right\}.$$

Consequently, the linear ordering between the marginal variables has the effect of reducing the domain of H_* to $[a, b]$ with $a \leqslant 1/2$ and $b \geqslant 1/2$.

The following construct provides a simple way of generating parametric models for H_* that are concentrated on a given subinterval $[a, b]$ of $[0, 1]$ (Nadarajah, Anderson and Tawn, 1998, Theorem 3). Let $H_*^\dagger$ be an absolutely continuous positive measure on $[0, 1]$ satisfying the constraints (30). Let $h^\dagger$ denote the density of $H_*^\dagger$. Given a subinterval $[a, b]$ of $[0, 1]$ with $a \leqslant 1/2 \leqslant b$, define a measure H_* on $[a, b]$ as follows: let H_* have atoms of mass

$$H_*(\{a\}) = \gamma_1,$$
$$H_*(\{b\}) = \gamma_2$$

at a and b, where

$$0 \leqslant \gamma_1 \leqslant \frac{2b-1}{b-a},$$
$$0 \leqslant \gamma_2 \leqslant \frac{1-2a}{b-a},$$

and let H_* be absolutely continuous in the interior, (a, b), with density

$$h(w) = \frac{(b-a)(\alpha\beta)^2}{\{\alpha(w-a) + \beta(b-w)\}^3} h^\dagger\left\{\frac{\alpha(w-a)}{\alpha(w-a) + \beta(b-w)}\right\}, \quad w \in (a, b),$$

where

$$\alpha = 2b - 1 + \gamma_1(a - b),$$
$$\beta = 1 - 2a + \gamma_2(a - b).$$

Then, H_* satisfies the constraints (30).

10.2. Sufficient characterizations

Sufficient characterizations enable one to examine whether a given d.f. F_* belongs to the domain of attraction of a multivariate extreme value d.f. G_* and to identify the form of G_*. We provide three sufficient characterizations for $F_* \in D(G_*)$. The last two results, in particular, have wide applicability since knowing the limits of some densities enables one to construct the limiting multivariate extreme value distribution.

10.2.1. Campbell and Tsokos's characterization

The first one is based on canonical series expansion of F_* (Campbell and Tsokos, 1973). Suppose $F_*(y_1, y_2)$ satisfies

$$\int_{-\infty}^{\infty}\int_{-\infty}^{\infty} \left[\frac{dF_*(y_1, y_2)}{d\Phi_1(y_1) d\Phi_1(y_2)}\right]^2 d\Phi_1(y_1) d\Phi_1(y_2) < \infty.$$

Suppose too that F_* admits an expansion of the form

$$dF_*(y_1, y_2) = d\Phi_1(y_1) d\Phi_1(y_2) \left\{1 + \sum_{k=1}^{\infty} \rho_k A_k(y_1) B_k(y_2)\right\},$$

where $\{A_k(y_1)\}$ and $\{B_k(y_2)\}$ are the so-called canonical variables defined on $\Phi_1(y_1)$ and $\Phi_1(y_2)$, respectively, and $\{\rho_k\}$ are the canonical correlations defined by

$$\rho_k = \int_{-\infty}^{\infty}\int_{-\infty}^{\infty} A_k(y_1) B_k(y_2) dF_*(y_1, y_2), \quad k = 1, 2, \ldots.$$

(In general, $A_k(y_1)$ and $B_k(y_2)$ are kth order orthonormal polynomials in y_1 and y_2, respectively. By convention, $A_0(y_1) = B_0(y_2) = 1$.) Then, if F_* belongs to the domain of attraction of G_* it must be of the form

$$G_*(y_1, y_2) = \Phi_1(y_1)\Phi_1(y_2)\exp\{V_*(y_1, y_2)\},$$

where

$$V_*(y_1, y_2) = \lim_{t \to \infty} t \sum_{k=1}^{\infty} \rho_k E[A_k(Y_1)|Y_1 \leqslant ty_1] E[B_k(Y_2)|Y_2 \leqslant ty_2].$$

10.2.2. De Haan and Resnick's characterization

The second result uses regular variation of the joint p.d.f. of F_* (De Haan and Resnick, 1987). Suppose F_* has joint p.d.f. f, which is regularly varying with limit function v, i.e., for $\mathbf{u} = (u_1, \ldots, u_p) \in \Re^p_+ \setminus \{(0, \ldots, 0)\}$,

$$\lim_{t \to \infty} t^{p+1} f(tu_1, \ldots, tu_p) = v(u_1, \ldots, u_p).$$

Evidently υ satisfies $\upsilon(t\mathbf{u}) = t^{-(1+p)}\upsilon(\mathbf{u})$ for $\mathbf{u} \in \Re_+^p \setminus \{(0,\ldots,0)\}$. Suppose further that υ is bounded on $B = \{\mathbf{u} \in \Re_+^p : \|\mathbf{u}\| = 1\}$ and that the following uniformity condition holds:

$$\lim_{t\to\infty} \sup_{\mathbf{u}\in B} \left| t^{p+1} f(tu_1,\ldots,tu_p) - \upsilon(u_1,\ldots,u_p) \right| = 0.$$

Then, for any $\varepsilon > 0$,

$$\lim_{t\to\infty} \sup_{\|\mathbf{u}\|>\varepsilon} \left| t^{p+1} f(tu_1,\ldots,tu_p) - \upsilon(u_1,\ldots,u_p) \right| = 0.$$

Also υ is integrable on $[\mathbf{0}, \mathbf{y}]^c$, $\mathbf{y} > \mathbf{0}$ and $F_* \in D(G_*)$, where

$$G_*(y_1,\ldots,y_p) = \exp\left\{ -\int_{[\mathbf{0},\mathbf{y}]^c} \upsilon(\mathbf{u})\,d\mathbf{u} \right\}, \quad \mathbf{y} > \mathbf{0}.$$

10.2.3. Yun's characterization

The final result supposes absolute continuity of F_* (Yun, 1997). For any $c = \{j_1 < \cdots < j_k\} \subset \{1,\ldots,p\}$ with $k \geq 2$, let $f_{j_k|j_1,\ldots,j_{k-1}}(y_{j_k}|y_{j_1},\ldots,y_{j_{k-1}})$ denote the conditional p.d.f. of the j_kth component of F_* given values of the $(j_1,\ldots,j_{k-1})$th components. If, for any $c \subset \{1,\ldots,p\}$ with $k \geq 2$,

$$l_{j_k|j_1,\ldots,j_{k-1}}(u_{j_k}; u_{j_1},\ldots,u_{j_{k-1}})$$
$$= \lim_{t\to\infty} tf_{j_k|j_1,\ldots,j_{k-1}}(tu_{j_k}|tu_{j_1},\ldots,tu_{j_{k-1}}) < \infty$$

and if, in addition, for every fixed $u_{j_1},\ldots,u_{j_{k-1}}$, there exists a $t^*(u_{j_1},\ldots,u_{j_{k-1}}) < \infty$ such that the class

$$\{tf_{j_k|j_1,\ldots,j_{k-1}}(tu_{j_k}|tu_{j_1},\ldots,tu_{j_{k-1}}): t^*(u_{j_1},\ldots,u_{j_{k-1}}) < t < \infty\}$$

of functions of u_{j_k} is locally uniformly integrable over $(0,\infty)$, then $F_* \in D(G_*)$ with

$$G_*(y_1,\ldots,y_p) = \exp\left\{ -\sum_{j=1}^p \frac{1}{y_j} - \sum_{c\subset\{1,\ldots,p\}:\,|c|\geq 2} (-1)^{|c|-1} \right.$$
$$\left. \times \int_{y_{j_1}}^\infty \cdots \int_{y_{j_k}}^\infty \beta_k(u_1,\ldots,u_k)\,du_1 \cdots du_k \right\},$$

where

$$\beta_k(u_1,\ldots,u_k) = \frac{1}{u_1^2}\left(\prod_{j=1}^{k-1} l_{j+1|1,\ldots,j}(u_{j+1}; u_1,\ldots,u_j) \right).$$

10.3. Necessary and sufficient characterizations

Here, we give three results, all necessary and sufficient for $F_* \in D(G_*)$.

10.3.1. Marshall and Olkin's characterization

The first result is due to Marshall and Olkin (1983) and expresses G_* as the limit of the conditional distribution of $\mathbf{Y} = (Y_1, \ldots, Y_p)$ given that at least one component of $\mathbf{Y}$ has exceeded t. Namely, $F_* \in D(G_*)$ if and only if

$$\frac{-\log F_*(ty_1, \ldots, ty_p)}{-\log F_*(t, \ldots, t)} \to \frac{-\log G_*(y_1, \ldots, y_p)}{-\log G_*(1, \ldots, 1)}$$

as $t \to \infty$ for each $y_j > 0$, $j = 1, \ldots, p$.

10.3.2. Resnick's characterization

The second result appears as Proposition 5.17(ii) in Resnick (1987) and involves the limiting intensity measure μ_* in (29). Namely, $F_* \in D(G_*)$ if and only if

$$t \Pr(t^{-1}\mathbf{Y} \in B) \to \mu_*(B) \qquad (39)$$

as $t \to \infty$ for all relatively compact B for which the boundary of B has μ_* measure equals to 0.

10.3.3. Takahashi's characterization

The third result (Takahashi, 1994b, Propositions 2.1 and 2.2) is in terms of

$$D_{F_*}(u_1, \ldots, u_p) = F_*(\Phi_1^{\leftarrow}(u_1), \ldots, \Phi_1^{\leftarrow}(u_p)), \quad (u_1, \ldots, u_p) \in (0, 1)^p,$$

and

$$D_{G_*}(u_1, \ldots, u_p) = G_*(\Phi_1^{\leftarrow}(u_1), \ldots, \Phi_1^{\leftarrow}(u_p)), \quad (u_1, \ldots, u_p) \in (0, 1)^p,$$

which are the copulas of F_* and G_*, respectively. It says that $F_* \in D(G_*)$ if and only if

$$\lim_{t \to \infty} t\{1 - D_{F_*}(\mathbf{u}^{1/t})\} = -\log D_{G_*}(\mathbf{u})$$

for all $\mathbf{u} \in (0, 1)^p$; or, equivalently,

$$\lim_{t \uparrow 1} \frac{1 - D_{F_*}(\mathbf{u}^{1-t})}{1 - t} = -\log D_{G_*}(\mathbf{u})$$

for all $\mathbf{u} \in (0, 1)^p$; or, equivalently,

$$\lim_{t \downarrow 0} \frac{1 - D_{F_*}(\mathbf{u}^t)}{1 - D_{G_*}(\mathbf{u}^t)} = 1$$

for all $\mathbf{u} \in (0, 1)^p$; or, equivalently,

$$d_{j_1,\ldots,j_k}(\mathbf{u}) = \lim_{t \to \infty} t \overline{F}_*\left(0, \ldots, t\Phi_1^{\leftarrow}(u_{j_1}), \ldots, t\Phi_1^{\leftarrow}(u_{j_k}), \ldots, 0\right) < \infty$$

for all $1 \leq j_1 < \cdots < j_k \leq p$ and for all $\mathbf{u} \in (0, 1)^p$, where $\overline{F}_*$ is the joint survivor function of F_*. If any of these statements is satisfied then we can write

$$D_{G_*}(\mathbf{u}) = u_1 \cdots u_p \exp\left\{\sum_{k=2}^{p}(-1)^k \sum_{1 \leq j_1 < \cdots < j_k \leq p} d_{j_1,\ldots,j_k}(\mathbf{u})\right\}.$$

11. Characterizations of multivariate extreme value distributions

Some of the characterizations in the above section also provided characterizations on the form of G. In this section we consider some more characterizations on the form of G.

11.0.4. Gumbel's characterization

The earliest known characterization is that due to Gumbel (1962). Let $G_{B_1}, G_{B_2}, \ldots, G_{B_m}$ be known bivariate extreme value d.f.s with unit Fréchet margins. Then, their geometric mean

$$G_{B_1}^{\beta_1}(y_1, y_2) G_{B_2}^{\beta_2}(y_1, y_2) \cdots G_{B_m}^{1-\beta_1-\beta_2-\cdots-\beta_{m-1}}(y_1, y_2)$$

is also a bivariate extreme value d.f. with unit Fréchet margins.

11.0.5. Max-stable characterization

We can directly generalize the max-stable characterization in (5) to obtain the following. Multivariate extreme value d.f.s, G, in (24) are precisely the class of max-stable d.f.s with nondegenerate marginals, i.e., those d.f.s for which there exist norming constants $\alpha_{n,j} > 0$, $\beta_{n,j}$, $j = 1, \ldots, p$, such that

$$G^n(x_1, \ldots, x_p) = G(\alpha_{n,1} x_1 + \beta_{n,1}, \ldots, \alpha_{n,p} x_p + \beta_{n,p}), \quad n \geq 1. \tag{40}$$

By setting all x_j but one to ∞ we see that

$$G_j^n(x_j) = G_j(\alpha_{n,j} x_j + \beta_{n,j}), \quad j = 1, \ldots, p,$$

and hence the norming constants $\alpha_{n,j}$, $\beta_{n,j}$ are precisely those in (6).

11.0.6. Tiago de Oliveira's characterization

A characterization due to Tiago de Oliveira (1962) is

$$G_*(y_1, y_2) = \{\Phi_1(y_1)\Phi_1(y_2)\}^{\nu(\log y_2 - \log y_1)},$$

where ν is the so-called dependence function. Obretenov (1991) shows that ν is related to H_* through

$$\nu\left(\log \frac{y}{1-y}\right) = \int_{[0,1]} \max\{w(1-y), (1-w)y\} H_*(dw).$$

For more than two variables the characterization generalizes to

$$G_*(y_1, \ldots, y_p) = \{\Phi_1(y_1) \cdots \Phi_1(y_p)\}^{\nu(\log y_2 - \log y_1, \ldots, \log y_p - \log y_1)}$$

with

$$\nu\left(\log \frac{y_1}{y_2}, \ldots, \log \frac{y_1}{y_p}\right) = \int_{S_p} \max\left(\frac{w_1 y_1}{\sum_{j=1}^p y_j}, \ldots, \frac{w_p y_p}{\sum_{j=1}^p y_j}\right) H_*(d\mathbf{w}),$$

where S_p is the $(p-1)$-dimensional unit simplex.

11.0.7. Pickands's characterization

An alternative way of writing the characterization in Eq. (31) is as follows (Pickands, 1981). Considering the case $p = 2$, we can write

$$G_*(y_1, y_2) = \exp\left\{-\left(\frac{1}{y_1} + \frac{1}{y_2}\right) A\left(\frac{y_1}{y_1 + y_2}\right)\right\}, \tag{41}$$

where A is also referred to as a dependence function and is related to H_* through

$$A(w) = \int_{[0,1]} \max\{w(1-q), (1-w)q\} H_*(dq).$$

It can be verified that A has the following properties:

- $A(0) = A(1) = 1, -1 \leqslant A'(0) \leqslant 0, 0 \leqslant A'(1) \leqslant 1$;
- $A''(w) \geqslant 0$ and $\max(w, 1-w) \leqslant A(w) \leqslant 1$ for $0 \leqslant w \leqslant 1$;
- $A(w) = 1$ implies that Y_1 and Y_2 are totally independent;
- $A(w) = \max(w, 1-w)$ implies that Y_1 and Y_2 are totally dependent;
- A is convex, i.e., $A[\varrho y_1 + (1-\varrho)y_2] \leqslant \varrho A(y_1) + (1-\varrho)A(y_2)$;
- and, if A_k are dependence functions, so is $\sum_{k=1}^m \alpha_k A_k$, where $\alpha_k \geqslant 0$ and $\sum_{k=1}^m \alpha_k = 1$.

A useful property of (41) is that the joint d.f. of $Z = Y_2/(Y_1 + Y_2)$ and $V = G_*(Y_1, Y_2)$ is

$$G_{Z,V}(z, v) = v \int_0^z \frac{t(1-t)}{A(t)} \, dA'(t)$$
$$+ (v - v \log v) \left\{ z + z(1-z) \frac{A'(z)}{A(z)} - \int_0^z \frac{t(1-t)}{A(t)} \, dA'(t) \right\}$$

(Ghoudi, Khoudraji and Rivest, 1998). It follows from this that the marginal d.f. of Z is

$$G_Z(z) = z + z(1-z) \frac{A'(z)}{A(z)} \quad (42)$$

and that the conditional d.f. of $V|Z = z$ is

$$\frac{1}{g_Z(z)} \frac{\partial G_{Z,V}(z, v)}{\partial z} = vp(z) + (v - v \log v)\{1 - p(z)\},$$

where

$$p(z) = \frac{z(1-z)A''(z)}{A(z)g_Z(z)}$$

and g_Z is the derivative of G_Z. Thus, given Z, the law of V is uniform on $(0, 1)$ with probability $p(Z)$ and equal to the law of the product of two independent uniforms on $(0, 1)$ with probability $1 - p(Z)$. This fact can be used to simulate (Y_1, Y_2) from (41) as follows:

- simulate Z according to the distribution given by (42);
- having Z, take $V = U_1$ with probability $p(Z)$ and $V = U_1 U_2$ with probability $1 - p(Z)$. Here, U_1 and U_2 are independent uniforms on $(0, 1)$;
- set $Y_1 = V^{Z/A(Z)}$ and $Y_2 = V^{(1-Z)/A(Z)}$.

Capéraà et al. (2000) propose a useful extension of (41) that has joint d.f.

$$\kappa^{\leftarrow} \left[\{\kappa(\exp(y_1)) + \kappa(\exp(y_2))\} A_* \left\{ \frac{\kappa(\exp(y_1))}{\kappa(\exp(y_1)) + \kappa(\exp(y_2))} \right\} \right], \quad (43)$$

where $A_* : [0, 1] \to [1/2, 1]$ is a convex function such that $\max(y, 1 - y) \leq A_*(y) \leq 1$ for all $0 \leq y \leq 1$ and $\kappa : (0, 1] \to [0, \infty)$ is a convex, decreasing function satisfying $\kappa(1) = 0$, with the convention that $\kappa(0) = \lim_{y \to 0^+} \kappa(y)$ and $\kappa^{\leftarrow}(y) = 0$ when $y \geq \kappa(0)$. This extension includes (41) and Archimedian distributions as special cases: when $\kappa(y) = \log(1/y)$, (43) reduces to (41), and when $A_* \equiv 1$, (43) reduces to

$$\kappa^{\leftarrow} \{\kappa(\exp(y_1)) + \kappa(\exp(y_2))\},$$

which is the general form of Archimedian distributions. Thus, (43) is referred to as an Archimax distribution. A worthwhile property of the Archimax distribution is that if $\kappa(1 - 1/y) \in RV_{-m}$ for some $m \geq 1$, then it belongs to the domain of attraction of a bivariate extreme value distribution of the form (41), where

$$A(w) = \{w^m + (1-w)^m\}^{1/m} A_*^{1/m}\left(\frac{w^m}{w^m + (1-w)^m}\right).$$

Furthermore, A and A_* coincide if and only if $m = 1$.

11.0.8. De Haan's characterizations

Our final characterization is a special case of a spectral representation for max-stable processes (De Haan, 1984): there exist non-negative Lebesgue integrable functions $f_j(s)$, $0 \leq s \leq 1$, satisfying

$$\int_{[0,1]} f_j(s)\, ds = 1, \quad j = 1, \ldots, p,$$

such that

$$G_*(y_1, \ldots, y_p) = \exp\left\{-\int_{[0,1]} \max\left(\frac{f_1(s)}{y_1}, \ldots, \frac{f_p(s)}{y_p}\right) ds\right\}. \tag{44}$$

Another characterization due to de Haan (1978) says that $(Y_1, \ldots, Y_p)$ has a multivariate extreme value distribution if and only if every weighted maximum of its components has a univariate extreme value distribution; namely, that

$$Z = \max(\varrho_1 Y_1, \ldots, \varrho_p Y_p)$$

with $\varrho_j > 0$, $j = 1, \ldots, p$, is a multiple of a unit Fréchet random variable. This is the analogue of the well-known result that a random vector has a p-variate normal distribution if and only if every weighted combination of its components has a univariate normal distribution.

12. Rates of convergence

In parallel with Section 3, here we provide two of the recent results on the rate of convergence of the multivariate limit law (24).

12.1. Omey and Rachev's rate of convergence

Suppose $\mathbf{Z}$ is distributed according to G_* and is independent of $\mathbf{Y} = (Y_1, \ldots, Y_p)$ in (27). Omey and Rachev (1991) give the following estimate of the rate of convergence: if

$$\sup\{\min(|y_1|, \ldots, |y_p|)\}^r |F_*(y_1, \ldots, y_p) - G_*(y_1, \ldots, y_p)| < \infty$$

for some $r > 1$, then there exists a constant A such that

$$\sup |F_*^n(ny_1, \ldots, ny_p) - G_*(y_1, \ldots, y_p)| \leqslant An^{1-r}$$

for all $n \geqslant 1$. Furthermore, they determine

$$A = \begin{cases} A^* = C \max\{\rho(\mathbf{Y}, \mathbf{Z}), v_r(\mathbf{Y}, \mathbf{Z}), v_r^{1/(r-1)}(\mathbf{Y}, \mathbf{Z})\}, & \text{if } 1 < r \leqslant 2, \\ C \max\{A^{*r-1}, \rho(\mathbf{Y}, \mathbf{Z})\}, & \text{if } r > 2, \end{cases}$$

where

$$\rho(\mathbf{Y}, \mathbf{Z}) = \sup |F_*(y_1, \ldots, y_p) - G_*(y_1, \ldots, y_p)|,$$
$$v_r(\mathbf{Y}, \mathbf{Z}) = \sup_{t>0} t^r \rho(\max(\mathbf{Y}, t\mathbf{Z}), \max(\mathbf{Z}, t\mathbf{Z})),$$

and C is a constant, which depends only on r and p. In a related development, De Haan and Peng (1997) give rates of convergence of (24) for $p = 2$ in terms of both the total variation metric and the uniform metric.

12.2. Nadarajah's penultimate approximations

Nadarajah (2000) provides penultimate estimates of the limiting measure density h in (29) for the case $p = 2$. Let $R = Y_1 + Y_2$, $W = Y_1/R$ and define

$$h_r(w) = r \partial \Pr(W \leqslant w, R > r)/\partial w, \quad r > 0, \ w \in (0, 1),$$

assuming existence of the derivative. Nadarajah shows that $h_r(w) \to h(w)$ as $r \to \infty$ for every fixed $w \in (0, 1)$. Furthermore, the following estimates for $h_r(w) - h(w)$ hold uniformly on $[0, 1]$ as $r \to \infty$:

$$h_r(w) - h(w) = h(w)\{f_0(rw(1-w)) - 1\} + \alpha_1(w) f_1(rw(1-w)) + o(1)$$

if $H_*(\{0\}) = 0$ and $H_*(\{1\}) = 0$;

$$h_r(w) - h(w) = h(w)\{f_0(rw(1-w)) - 1\} + \alpha_1(w) f_1(rw(1-w)) \\ + \alpha_2(w) f_2(rw(1-w)) + o(1)$$

if $H_*(\{0\}) = 0$ and $H_*(\{1\}) > 0$;

$$h_r(w) - h(w) = h(w)\{f_0(rw(1-w)) - 1\} + \alpha_1(w) f_1(rw(1-w)) \\ + \alpha_3(w) f_2(rw(1-w)) + o(1)$$

if $H_*(\{0\}) > 0$ and $H_*(\{1\}) = 0$; and,

$$h_r(w) - h(w) = h(w)\{f_0(rw(1-w)) - 1\} + \alpha_1(w) f_1(rw(1-w)) \\ + \{\alpha_2(w) + \alpha_3(w)\} f_2(rw(1-w)) + o(1)$$

if $H_*(\{0\}) > 0$ and $H_*(\{1\}) > 0$, where

$$f_i(s) = s\left\{1 - \left(\sum_{j=0}^{i} \frac{s^{-j}}{j!}\right)\exp\left(-\frac{1}{s}\right)\right\}, \quad s > 0, \ i = 0, 1, 2,$$

$$\alpha_1(w) = \frac{1 - \beta(w) - \gamma(w)}{w(1-w)}, \quad w \in [0,1],$$

$$\alpha_2(w) = \frac{2\beta(w)\{1 - \beta(w)\}}{w}, \quad w \in [0,1],$$

$$\alpha_3(w) = \frac{2\gamma(w)\{1 - \gamma(w)\}}{1-w}, \quad w \in [0,1],$$

$$\beta(w) = \int_{[0,w]} v H_*(dv), \quad w \in [0,1],$$

$$\gamma(w) = \int_{(w,1]} (1-v) H_*(dv), \quad w \in [0,1].$$

Evidently the estimates of $h_r - h$ differ according to presence/absence of atoms of mass at the end points. Each estimate is a linear combination of functionals of h with finite coefficients, which, for w in compact subsets of $(0, 1)$, decay to 0 uniformly $(r \to \infty)$. When there are no atoms of mass the linear combination has two terms and with each atom added the number grows by one.

In the next two sections (Sections 13 and 14) we demonstrate the characterizations described in Sections 10 and 11 to develop flexible parametric families for bivariate and multivariate extreme value distributions.

13. Parametric families for bivariate extreme value distributions

The nine families discussed in this section represent the bulk of the distributions for modeling bivariate extremes. No doubt that additional models will be discovered.

13.1. Logistic distributions (Tawn, 1988b)

The d.f. G_* takes the form

$$G_*(y_1, y_2) = \exp\left[-\frac{1-\vartheta_1}{y_1} - \frac{1-\vartheta_2}{y_2} - \left\{\left(\frac{\vartheta_1}{y_1}\right)^q + \left(\frac{\vartheta_2}{y_2}\right)^q\right\}^{1/q}\right], \quad (45)$$

where $0 \leqslant \vartheta_1, \vartheta_2 \leqslant 1$ and $q > 1$. Applying Eq. (33), we have

$$h(w) = (q-1)\vartheta_1^q \vartheta_2^q \{w(1-w)\}^{q-2}\{(\vartheta_2 w)^q + (\vartheta_1(1-w))^q\}^{1/q-2}$$

and $H_*(\{0\}) = 1 - \vartheta_2$, $H_*(\{1\}) = 1 - \vartheta_1$. Thus, this family has mass both in the interior and at the end points. It allows for asymmetry and nonexchangeability through ϑ_1 and

ϑ_2: symmetry and exchangeability arise if and only if $\vartheta_1 = \vartheta_2$. Total independence corresponds to $\vartheta_1 = 0$ or $\vartheta_2 = 0$ or the limit $q \to 1^+$, whereas total dependence corresponds to $\vartheta_1 = \vartheta_2 = 1$ and the limit $q \to \infty$.

A special case for $\vartheta_1 = \vartheta_2 = 1$ is the symmetric logistic distribution having all its mass in the interior:

$$G_*(y_1, y_2) = \exp\{-(y_1^{-q} + y_2^{-q})^{1/q}\}.$$

This distribution appears in the survival analysis literature; see, for example, Hougaard (1986). Alternative parameterizations for this distribution are possibly advisable: for example $s = 1/q$ ($0 \leqslant s \leqslant 1$). The variables of this distribution are exchangeable and have correlation $(q^2 - 1)/q^2$. The Fisher information matrix for this distribution has been derived by Oakes and Manatunga (1992). Shi et al. (1993) describe a scheme for simulating (Y_1, Y_2) from the distribution: namely, defining the transformations $1/Y_1 = Z \cos^{2/q} V$ and $1/Y_2 = Z \sin^{2/q} V$, they note that the joint p.d.f. of (Z, V) factorizes as

$$(q^{-1}z + 1 - q^{-1}) \exp(-z) \sin 2v, \quad 0 < v < \pi/2, \ 0 < z < \infty,$$

which shows that Z and V are independent. It is easily characterized that V may be represented as $(\arcsin \sqrt{U})$, where U is uniform on $(0, 1)$, while Z is the $1 - (1/q) : 1/q$ mixture of a unit exponential random variable and the sum of two independent unit exponential random variables. Hence, this suggests an easy way of simulating.

If $\vartheta_1 = \vartheta_2$ we get a mixture of symmetric logistic and independence. If $q \to \infty$ we have

$$G_*(y_1, y_2) = \exp\left\{-\max\left(\frac{1}{y_1} + \frac{1-\vartheta_2}{y_2}, \frac{1-\vartheta_1}{y_1} + \frac{1}{y_2}\right)\right\}$$

with $\Pr(Y_1 \vartheta_2 = Y_2 \vartheta_1) = \vartheta_1 \vartheta_2/(\vartheta_1 + \vartheta_2 - \vartheta_1 \vartheta_2)$. When $\vartheta_1 = 1$ and $\vartheta_2 = \alpha$, we have the biextremal (α) distribution:

$$G_*(y_1, y_2) = \exp\left[-\frac{1-\alpha}{y_2} - \left\{\left(\frac{1}{y_1}\right)^q + \left(\frac{\alpha}{y_2}\right)^q\right\}^{1/q}\right],$$

whereas when $\vartheta_1 = \alpha$ and $\vartheta_2 = 1$, we have the dual of the biextremal (α) distribution

$$G_*(y_1, y_2) = \exp\left[-\frac{1-\alpha}{y_1} - \left\{\left(\frac{\alpha}{y_1}\right)^q + \left(\frac{1}{y_2}\right)^q\right\}^{1/q}\right],$$

which corresponds to Y_1 and Y_2 being exchangeable. If $\vartheta_1 = \vartheta_2 = \alpha$, we have the Gumbel distribution (Gumbel and Mustafi, 1967):

$$G_*(y_1, y_2) = \exp\left[-\frac{1-\alpha}{y_1} - \frac{1-\alpha}{y_2} - \alpha\left\{\left(\frac{1}{y_1}\right)^q + \left(\frac{1}{y_2}\right)^q\right\}^{1/q}\right].$$

13.2. Negative logistic distributions (Joe, 1990)

The d.f. G_* takes the form

$$G_*(y_1, y_2) = \exp\left[-\frac{1}{y_1} - \frac{1}{y_2} + \left\{\left(\frac{\vartheta_1}{y_1}\right)^q + \left(\frac{\vartheta_2}{y_2}\right)^q\right\}^{1/q}\right],$$

where $0 \leqslant \vartheta_1, \vartheta_2 \leqslant 1$ and $q < 0$. Applying Eq. (33), we have

$$h(w) = (1-q)\vartheta_1^q \vartheta_2^q \{w(1-w)\}^{q-2} \{(\vartheta_2 w)^q + (\vartheta_1(1-w))^q\}^{1/q-2}$$

and $H_*(\{0\}) = 1 - \vartheta_2$, $H_*(\{1\}) = 1 - \vartheta_1$. This family is similar in structure to the logistic family with the special case $\vartheta_1 = \vartheta_2 = 1$ giving a symmetric version of the family and the limiting cases $q \to 0^-$ and $q \to -\infty$ reducing the family to being totally independent and totally dependent, respectively.

13.3. Bilogistic distributions (Joe, Smith and Weissman, 1992)

This family is motivated by the max-stable representation (44). Setting $f_1(s) = (1 - (1/q_1))s^{-1/q_1}$ and $f_2(s) = (1 - (1/q_2))(1-s)^{-1/q_2}$ into (44), we get its d.f. as

$$G_*(y_1, y_2) = \exp\left[-\int_{[0,1]} \max\left\{\frac{(q_1-1)s^{-1/q_1}}{q_1 y_1}, \frac{(q_2-1)(1-s)^{-1/q_2}}{q_2 y_2}\right\} ds\right]$$

for $q_1 > 1$ and $q_2 > 1$. Applying Eq. (33), we have

$$h(w) = \frac{(1-1/q_1)(1-z)z^{1-1/q_1}}{(1-w)w^2\{(1-z)/q_1 + z/q_2\}},$$

where $z = z(w; q_1, q_2)$ is the root of

$$(1-1/q_1)(1-w)(1-z)^{1/q_2} - (1-1/q_2)wz^{1/q_1} = 0, \tag{46}$$

and $H_*(\{0\}) = H_*(\{1\}) = 0$. Thus, this family has all its mass in the interior and is an asymmetric generalization of the logistic family in that setting $q = q_1 = q_2$ gives the symmetric logistic distribution with the two variables being exchangeable. Total independence and total dependence correspond to taking both q_1 and q_2 to 1^+ and ∞, respectively.

It is possible to think of $(q_1 + q_2)/2$ as a dependence parameter, measuring the strength of dependence between the extremes of the two variables, and $(q_1 - q_2)$ as an asymmetry parameter, the case $q_1 - q_2 = 0$ being one in which the two variables are exchangeable. Joe et al. (1992) apply this distribution to estimate likely combinations of sulphate and nitrate levels in acid rain.

13.4. Negative bilogistic distributions (Coles and Tawn, 1994)

This family has the same d.f. as the bilogistic distributions except that $q_1 < 0$, $q_2 < 0$. Applying Eq. (33), we have

$$h(w) = -\frac{(1 - 1/q_1)(1 - z)z^{1-1/q_1}}{(1 - w)w^2\{(1 - z)/q_1 + z/q_2\}}, \quad q_1 < 0, \; q_2 < 0,$$

and $H_*(\{0\}) = H_*(\{1\}) = 0$, where $z = z(w; q_1, q_2)$ is as defined in (46). This family is similar in structure to the bilogistic family and again setting $q = q_1 = q_2$ reduces it to a symmetric and exchangeable version; namely, the symmetric negative logistic family. Now limiting both q_1 and q_2 to 0^- and $-\infty$ correspond to total independence and total dependence, respectively. Coles and Tawn (1994) find this distribution most suitable for estimating the dependence between the extremes of surge and wave height.

13.5. Gaussian distributions (Smith, 1991; Coles, 1993)

The Gaussian distribution is the most prominent distribution in all applications of probabilistic and statistical methodology and it is therefore only natural to find its applications among bivariate extreme value distributions. The joint d.f. has the form:

$$G_*(y_1, y_2) = \exp\left[-\int_{[0,1]} \max\left\{\frac{f_0(s - t_1)}{y_1}, \frac{f_0(s - t_2)}{y_2}\right\} ds\right],$$

where f_0 is the p.d.f. of the Normal $(0, \sigma)$ distribution. This can be rewritten as

$$G_*(y_1, y_2) = \exp\left[-\frac{1}{y_2}\Upsilon\left\{s\left(\frac{y_1}{y_1 + y_2}\right)\right\} - \frac{1}{y_1}\Upsilon\left\{a - s\left(\frac{y_1}{y_1 + y_2}\right)\right\}\right],$$

where

$$s(w) = \frac{1}{2a}\{a^2 + 2\log w - 2\log(1 - w)\},$$

$$a = \frac{1}{\sigma^2}(t_1 - t_2)^2$$

and Υ is the c.d.f. of the standard normal distribution. Smith (1991) and Coles (1993) use this family to model spatial variation of the extreme storms at locations corresponding to t_1 and t_2. This family also appears in Hüsler and Reiss (1989) as the limit distribution of componentwise maxima of independently and identically distributed bivariate normal vectors; namely, if $\{(X_{i,1}, X_{i,2})\}$ are i.i.d. standard normal random vectors and ρ_n is the correlation coefficient between $X_{i,1}$ and $X_{i,2}$ then

$$\Pr\left\{\max_{1 \leq i \leq n} X_{i,1} \leq b_n + b_n^{-1}x_1, \max_{1 \leq i \leq n} X_{i,2} \leq b_n + b_n^{-1}x_2\right\}$$

$$\to G_*(\exp(x_1), \exp(x_2))$$

as $n \to \infty$ provided that $(1 - \rho_n) \log n \to a^2/4 \in [0, \infty]$. The normalizing constant b_n is given by $b_n = n \exp(-b_n^2/2)/\sqrt{2\pi}$. See also Hooghiemstra and Hüsler (1996) for a similar characterization based on the maxima of projections of i.i.d. bivariate normal vectors with respect to two arbitrary directions. An expression for the measure density h can be derived by applying Eq. (33) as usual. The resulting form has all its mass in the interior. The value of a controls the amount of dependence with the limits $a \to \infty$ and $a \to 0$ corresponding to total independence and total dependence, respectively.

13.6. Circular distributions (Coles and Walshaw, 1994)

This family serves as yet another motivation of (44). The joint d.f.

$$G_*(y_1, y_2) = \exp\left[-\int_{[0,2\pi]} \max\left\{\frac{f_0(\theta; \varphi_1, \zeta)}{y_1}, \frac{f_0(\theta; \varphi_2, \zeta)}{y_2}\right\} d\theta\right],$$

where

$$f_0(\theta; \varphi, \zeta) = \frac{1}{2\pi I_0(\zeta)} \exp\{\zeta \cos(\theta - \varphi)\}$$

is the p.d.f. of the well-known von Mises circular distribution with I_0 denoting the modified Bessel function of order 0. Coles and Walshaw (1994) use this distribution to model the dependence between the extremes of wind speeds corresponding to directions φ_1 and φ_2. Suppose without loss of generality that $\varphi_2 \geqslant \varphi_1$, $\varphi_2 - \varphi_1 \leqslant \pi$ and $\tilde{\varphi} = (\varphi_2 - \varphi_1)/2$. Routine calculations then show that we can rewrite

$$G_*(y_1, y_2) = \exp\left\{-\frac{1}{y_1} \int_B f_0(\theta; \tilde{\varphi}, \zeta) d\theta - \frac{1}{y_2} \int_{\tilde{B}} f_0(\theta; -\tilde{\varphi}, \zeta) d\theta\right\},$$

where

$$B = \{\theta \in (0, 2\pi]: \sin\theta > \gamma(w)\}, \qquad \tilde{B} = (0, 2\pi] \setminus B$$

and

$$\gamma(w) = \frac{1}{2\zeta \sin\tilde{\varphi}} \log\left(\frac{w}{1-w}\right).$$

An expression for h can be obtained by straight-forward application of Eq. (33). Like the Gaussian distributions, this family has the mass of h confined to the interior. Here, both ζ and the angular separation, $\tilde{\varphi}$, control the dependence. The strength of dependence decreases with both ζ and $\tilde{\varphi}$. The limits $\zeta = 0$ and $\zeta \to \infty$ give total dependence and total independence, respectively.

13.7. Beta distributions (Coles and Tawn, 1991)

The p.d.f. of the Beta(q_1, q_2) distribution is

$$h_{\dagger}(w) = \frac{\Gamma(q_1 + q_2)}{\Gamma(q_1)\Gamma(q_2)} w^{q_1-1}(1-w)^{q_2-1}, \quad q_1 > 0,\ q_2 > 0,\ w \in (0, 1).$$

This distribution is widely used in modeling of hydrological variables (see, for example, Johnson, Kotz and Balakrishnan, 1995, Vol. 2, p. 236). By Eq. (34) $m_j = q_j/(q_1 + q_2)$, and from Eq. (35) it follows that

$$h(w) = \frac{q_1^{q_1} q_2^{q_2} \Gamma(q_1+q_2+1)}{\Gamma(q_1)\Gamma(q_2)} \frac{w^{q_1-1}(1-w)^{q_2-1}}{\{q_1 w + q_2(1-w)\}^{1+q_1+q_2}}, \quad w \in (0, 1),$$

is the density of a valid measure H_* on $[0, 1]$ that satisfies the constraints (30) with $H_*(\{0\}) = H_*(\{1\}) = 0$. Like the two bilogistic families this is asymmetric, nonexchangeable and has the mass confined to the interior. For the symmetric and exchangeable version, which arises when $q = q_1 = q_2$, both total independence and total dependence are attained as limiting cases by taking $q \to 0^+$ and $q \to \infty$, respectively.

Applying Eq. (31), we have the corresponding d.f.

$$G_*(y_1, y_2) = \exp\left[-\frac{1}{y_1}\left\{1 - Be\left(q_1+1, q_2; \frac{q_1 y_1}{q_1 y_1 + q_2 y_2}\right)\right\} - \frac{1}{y_2} Be\left(q_1, q_2+1; \frac{q_1 y_1}{q_1 y_1 + q_2 y_2}\right)\right],$$

where

$$Be(\alpha_1, \alpha_2; u) = \frac{\Gamma(\alpha_1 + \alpha_2)}{\Gamma(\alpha_1)\Gamma(\alpha_2)} \int_0^u w^{\alpha_1 - 1}(1-w)^{\alpha_2 - 1}\, dw,$$

a normalized incomplete beta function.

13.8. Polynomial distributions (Nadarajah, 1999a)

This distribution encompasses the structure of all known bivariate extreme value distributions. It can be motivated as follows.

One common feature of the distributions in Sections 13.1–13.7 is that their structure is governed by the behavior of h near the end points of $[0, 1]$. For example, for the bilogistic distribution we have $h(w) = O(w^{q_2-2})$ and $h(1-w) = O(w^{q_1-2})$ as $w \to 0$, and knowing these gives an idea of the whole structure of h and hence that of G_*. In general we can write $h(w) = O(w^r)$ and $h(1-w) = O(w^s)$ as $w \to 0$ (Nadarajah, 1994). Thus, a natural choice for h that has all the flexibility of the known distributions is:

$$h(w) = \begin{cases} \alpha w^r, & \text{if } 0 < w < \theta, \\ \beta(1-w)^s, & \text{if } \theta < w < 1 \end{cases}$$

for $\theta \in (0, 1)$ with

$$H_*(\{0\}) = \gamma_0, \qquad H_*(\{1\}) = \gamma_1, \qquad H_*(\{\theta\}) = \gamma_\theta.$$

To ensure non-negativity of h and its continuity at θ we take $\alpha \geq 0$, $\beta \geq 0$ and impose the requirement $\alpha\theta^r = \beta(1-\theta)^s$. To ensure validity of the unit-mean condition (30) we take $r > -1$, $s > -1$ and parameterize the atoms at the end points as:

$$\gamma_0 = 1 - (1-\theta)\gamma_\theta - \frac{\beta}{s+2}(1-\theta)^{s+2} + \alpha\theta^{r+1}\left[\frac{\theta}{r+2} - \frac{1}{r+1}\right],$$

$$\gamma_1 = 1 - \theta\gamma_\theta - \frac{\alpha}{r+2}\theta^{r+2} + \beta(1-\theta)^{s+1}\left[\frac{1-\theta}{s+2} - \frac{1}{s+1}\right] \qquad (47)$$

with $0 \leq \gamma_0, \gamma_1 \leq 1$ and $0 \leq \gamma_\theta \leq \min\{\theta^{-1}, (1-\theta)^{-1}\}$. The resulting distribution has, in total, five free parameters. The parameters α and β represent coefficients of the amount of dependence put by h on either side of θ. Large values of them are associated with strong dependencies. The parameters r and s represent the structure of dependence exhibited by h on either side of θ. Negative values of them are associated with weak dependence structures as in those cases h puts most of its mass near the end points. The parameter θ represents asymmetry of the dependence structure exhibited by h and also enables accommodation of atoms of mass in the interior. The parameter γ_θ is a measure for the mass of H_* to be concentrated at a single point in the interior (for total dependence the mass of H_* is concentrated at the point $w = 1/2$ with probability one). Finally, γ_0 and γ_1 are measures for the mass of H_* to be concentrated at the end points 0 and 1, respectively (for total independence the mass of H_* is concentrated at each end point with probability half).

It is easily checked that the forms of H and V associated with the distribution are:

$$H(w) = \begin{cases} \gamma_0 + \frac{\alpha}{r+1}w^{r+1}, & \text{if } 0 < w < \theta, \\ 2 - \gamma_1 - \frac{\beta}{s+1}(1-w)^{s+1}, & \text{if } \theta \leq w < 1 \end{cases}$$

and for $y_1 > 0$, $y_2 > 0$

$$V(y_1, y_2) = \begin{cases} \frac{1}{y_1} + \frac{\gamma_0}{y_2} + \frac{1}{y_2}\frac{\alpha}{(r+1)(r+2)}\left(\frac{y_1}{y_1+y_2}\right)^{r+1}, & \text{if } y_1/(y_1+y_2) < \theta, \\ \frac{1}{y_2} + \frac{\gamma_1}{y_1} + \frac{1}{y_1}\frac{\beta}{(s+1)(s+2)}\left(\frac{y_2}{y_1+y_2}\right)^{s+1}, & \text{if } y_1/(y_1+y_2) \geq \theta, \end{cases}$$

where γ_0, γ_1 are given by (47).

The distribution has the requirement that h is continuous at θ, i.e., $\alpha\theta^r = \beta(1-\theta)^s$ and this admits valid solutions for θ for all possible signs of r and s (the solution is unique when r and s have the same signs). Thus, continuity of h at θ is a sensible requirement. However, we find that further requirements for the smoothness of h limit applicability of the distribution. For example, h is differentiable at θ only if $0 \leq \theta = r/(r-s) \leq 1$.

We find that symmetry arises if and only if either $\alpha = \beta$ and $r = s$ when $\theta = 1/2$ or $\alpha = \beta$ and $r = s = 0$ when $\theta \neq 1/2$. Thus, asymmetry of dependence structure for the distribution can be attributed to θ not being equal to $1/2$, the polynomial coefficients not being equal or the polynomial powers not being equal. Exchangeability is equivalent to symmetry when $\theta = 1/2$; otherwise, in addition to symmetry, we must have $\gamma_\theta = 0$. Total independence arises as the special case for $\alpha = 0$, $\beta = 0$ and $\gamma_\theta = 0$ while total dependence arises as the special case for $\alpha = 0$, $\beta = 0$, $\theta = 1/2$ and $\gamma_\theta = 2$.

Two further special cases of interest are $\gamma_0 = \gamma_1 = \gamma_\theta = 0$, where H_* has no atoms of mass, and $\alpha = \beta = 0$, where H_* has no mass in the interiors $(0, \theta)$ and $(\theta, 1)$. In the first case, using conditions (47), we can parameterize

$$\alpha = \frac{(r+1)(r+2)\{2(s+1)\theta - s\}}{[(r+2) + (s-r)\theta]\theta^{r+1}}$$

with $\theta \geqslant s/\{2(s+1)\}$ to ensure $\alpha \geqslant 0$ and

$$\beta = \frac{(s+1)(s+2)\{(r+2) - 2(r+1)\theta\}}{[(r+2) + (s-r)\theta](1-\theta)^{s+1}}$$

with $\theta \leqslant (r+2)/\{2(r+1)\}$ to ensure $\beta \geqslant 0$. Then, the continuity requirement on h reduces to the following quadratic equation:

$$\frac{2(r-s)}{(r+2)}\theta^2 + \frac{s^2 - 3rs - 2r}{(r+1)(s+1)}\theta + \frac{s}{s+1} = 0,$$

which admits valid solutions for θ for all possible signs of r and s (the solution is unique when r and s have the same signs). The resulting distribution has, in total, two free parameters. In the second case, the mass of H_* is distributed only at the end points and θ. From (47) we see that $\gamma_0 = 1 - (1-\theta)\gamma_\theta$ and $\gamma_1 = 1 - \theta\gamma_\theta$. Thus, total independence and total dependence arise when the mass at θ takes the values 0 and 2 (with $\theta = 1/2$), respectively. Exchangeability arises when the mass at both the end points are equal, which occurs if and only if either $\theta = 1/2$ or the distribution is totally independent. The end point 0 has no mass if and only if $\gamma_\theta = 1/(1-\theta)$ and $\gamma_1 = (1-2\theta)/(1-\theta)$ with $\theta \leqslant 1/2$. The end point 1 has no mass if and only if $\gamma_\theta = 1/\theta$ and $\gamma_0 = (2\theta-1)/\theta$ with $\theta \geqslant 1/2$.

Since, under weak dependence structures H_* concentrates most of its mass near the end points of $[0, 1]$, natural measures of weakness of dependence are:

$$M_1 = \int_0^\theta \left(1 - \frac{w}{\theta}\right) H_*(dw)$$

$$= 1 - (1-\theta)\gamma_\theta - \frac{\beta}{s+2}(1-\theta)^{s+2} - \frac{\alpha}{r+2}(1-\theta)\theta^{r+1}$$

for the mass in $[0, \theta]$ and

$$M_2 = \int_\theta^1 \left(1 - \frac{1-w}{1-\theta}\right) H_*(dw)$$

$$= 1 - \theta \gamma_\theta - \frac{\alpha}{r+2}\theta^{r+2} - \frac{\beta}{s+2}\theta(1-\theta)^{s+1}$$

for the mass in $(\theta, 1]$. Since, under strong dependence structures H_* concentrates most of its mass in the interior of $[0, 1]$, natural measures of strength of dependence are:

$$M_3 = \int_0^\theta \frac{w}{\theta} H_*(dw) = \gamma_\theta + \frac{\alpha}{r+2}\theta^{r+1}$$

for the mass in $[0, \theta]$ and

$$M_4 = \int_\theta^1 \frac{1-w}{1-\theta} H_*(dw) = \frac{\beta}{s+2}(1-\theta)^{s+1}$$

for the mass in $(\theta, 1]$. It follows that $M_1 + M_2 = 2 - \theta^{-1}(1-\theta)^{-1} + V(\theta, 1-\theta)$ is the measure of overall weakness of dependence with the values of 2 and 0 for total independence and total dependence, respectively. Similarly $M_3 + M_4 = \theta^{-1}(1-\theta)^{-1} - V(\theta, 1-\theta)$ is the measure of overall strength of dependence with the values of 0 and 2 for total independence and total dependence, respectively. If $\theta = 1/2$, then $(M_1 + M_2)/2 + 1 = V(1, 1)$ is the *extremal coefficient* developed by Coles and Tawn (1994) to measure dependence. Clearly larger values of the polynomial coefficients α and β have the effect of strengthening dependence while larger values of the polynomial powers r and s have the reverse effect. Note further that $M_1 + M_3$ and $M_2 + M_4$ are the total mass of H_* in $[0, \theta]$ and $(\theta, 1]$, respectively. Clearly the total mass in each segment becomes inflated and deflated, respectively, with larger values of the polynomial coefficient and power associated with that segment. Note too that $M_1 + M_2 + M_3 + M_4 = 2$, the total mass of H_*.

Some obvious measures of asymmetry are θ, $|r - s|$, $|\alpha - \beta|$, $r/(r+s)$ and $\alpha/(\alpha+\beta)$. Additional measures based on the dependence measures above are $|M_1 - M_2|$, $|M_3 - M_4 - \gamma_\theta|$, $M_1/(M_1 + M_2)$ and $(M_3 - \gamma_\theta)/(M_3 + M_4 - \gamma_\theta)$. We have $M_1 = M_2$ if and only if either $\theta = 1/2$ or the distribution is totally independent while $M_3 - \gamma_\theta = M_4$ if and only if $\theta = (r+2)/(r+s+4)$.

13.9. *Polynomial distributions* (Klüppelberg and May, 1999)

These are analogous to the above distributions, but formulated in terms of the $A(\cdot)$ function in (41). Take

$$A(w) = a_m w^m + a_{m-1} w^{m-1} + \cdots + a_2 w^2 - \left(\sum_{k=2}^m a_k\right) w + 1$$

for $w \in [0, 1]$ with $a_2 \geq 0$, $\sum_{k=2}^{m} a_k \geq 0$, $0 \leq \sum_{k=2}^{m}(k-1)a_k \leq 1$ and $\sum_{k=2}^{m} k(k-1)a_k \geq 0$. Then, the corresponding joint d.f.

$$G_*(y_1, y_2) = \exp\left\{-\frac{1}{y_1} - \frac{1}{y_2} + \sum_{k=2}^{m} a_k \sum_{l=0}^{m-k} \binom{m-k}{l} \frac{y_1^{k+l-1} y_2^{m-k-l-1}}{(y_1+y_2)^{m-1}}\right\}$$

has $(m-2)$ parameters. Applying Eq. (33), we have

$$h(w) = m(m-1)a_m w^{m-2} + (m-1)(m-2)a_{m-1} w^{m-3} + \cdots + 2a_2$$

and $H_*(\{0\}) = H_*(\{1\}) = 0$. Setting $m = 5$, $a_5 = \varphi/20$, $a_4 = \theta/12$, $a_3 = -(\theta + \varphi)/6$ and $a_2 = 1/2$, we have as a special case

$$h(w) = \varphi w^3 + \theta w^2 - (\theta + \varphi)w + 1,$$

the measure density of the asymmetric mixed distribution due to Tawn (1988b).

14. Parametric families for multivariate extreme value distributions

The five specific models of multivariate extreme value distributions discussed here do not of course exhaust all possible configurations. Section 14.6 provides tools for constructing further multivariate extreme value distributions subject to constraints on their marginals.

14.1. Logistic distributions (Tawn, 1990)

The logistic families (Sections 13.1 and 13.2) have direct generalizations to the multivariate case. They have been among the most applied multivariate extreme value distributions in the literature. Some of their applications include the estimation of the dependence between the extreme surge levels at three sites on the English coast (Coles and Tawn, 1991), the estimation of the dependence between the extremes of hydrometeorological variables such as rainfall, wind speed and snowmelt that are relevant to reservoir flood safety (Anderson and Nadarajah, 1993) and flood frequency analysis (EscalanteSandoval, 1998). One possible way of motivating the generalization is as follows.

Let C be an index variable over the set B, the class of all nonempty subsets of $\{1, \ldots, p\}$. Let $Z_{j,C}^{(i)}$ be the size of the ith realization at site j, of the extreme spatial storms of the type, which occur only at the collection C of sites. Here, $Z_{j,C}^{(i)}$ ($i = 1, \ldots, N_c$) are assumed to be conditionally independent given N_C, where the random variable N_C is taken to have a Poisson distribution with mean τ_C. Also α_C denotes the unrecorded covariate information variable, which has a positive stable distribution and characteristic exponent $0 < 1/q_c \leq 1$. The α_C are assumed to be independent.

We say that a storm affects site j only if an observation at site j exceeds a high threshold, t_j, during that storm. Hence, $Z_{j,C}^{(i)} > t_j$ for all $j \in C$ and $Z_{j,C}^{(i)} \leq t_j$ for all

$j \notin C$. As discussed in Section 5, exceedances of a high threshold have the Generalized Pareto distribution, thus for all i we take

$$\Pr\{Z_{j,C}^{(i)} < z | Z_{j,C}^{(i)} > t_j\} = 1 - \left(1 + \xi_j \frac{z - t_j}{\sigma_j}\right)^{-1/\xi_j},$$

where $z > t_j$, $1 + \xi_j(z - t_j)/\sigma_j > 0$, $\sigma_j > 0$ and $\xi_j \in \Re$. If

$$Z_{j,C} = \max(Z_{j,C}^{(1)}, \ldots, Z_{j,C}^{(N_C)})$$

for $N_C > 0$ then it follows that, for $z > t_j$,

$$\Pr\{Z_{j,C} < z\} = \sum_{n=1}^{\infty} \left\{\prod_{i=1}^{n} \Pr(Z_{j,C}^{(i)} < z | Z_{j,C}^{(i)} > t_j)\right\} \Pr(N_C = n)$$
$$+ \Pr(N_C = 0)$$
$$= \sum_{n=0}^{\infty} \left\{1 - \left(1 + \xi_j \frac{z - t_j}{\sigma_j}\right)^{-1/\xi_j}\right\}^n \frac{\tau_C^n \exp(-\tau_C)}{n!}$$
$$= \exp\left\{-\tau_C \left(1 + \xi_j \frac{z - t_j}{\sigma_j}\right)^{-1/\xi_j}\right\}. \qquad (48)$$

As we are interested only in large z, that is $z > t_j$, there is no loss of generality in taking (48) to apply for all z such that $1 + \xi_j(z - t_j)/\sigma_j > 0$. Hence, $Z_{j,C}$ has a Generalized Extreme Value distribution (see Section 4).

Here, interest is in the joint behavior of $X_1, \ldots, X_p$, where for $j = 1, \ldots, p$

$$X_j = \max_{C \in B_{(j)}} (Z_{j,C}), \qquad (49)$$

where $B_{(j)}$ is the subclass of B containing all nonempty subsets, which include j. Thus, here maximization is over all spatial storms of the type that affect site j. For fixed i, the $Z_{j,C}^{(i)}$, $j \in C$, are dependent Generalized Pareto random variables and hence the $Z_{j,C}$, $j \in C$, are dependent Generalized Extreme Value random variables. However, we take $Z_{j,C} | \alpha_C$, $j \in C$, to be independent. From Feller (1971, Chapter 13, Section 6) and (48) this implies that the conditional distribution can be taken as

$$\Pr(Z_{j,C} < z | \alpha_C) = \exp\left[-\alpha_C \left\{\tau_C \left(1 + \xi_j \frac{z - t_j}{\sigma_j}\right)^{-1/\xi_j}\right\}^{q_C}\right]. \qquad (50)$$

Therefore, given the covariate information, the maximum of each type of the extreme spatial storm has a Generalized Extreme Value distribution with parameters different from those of the unconditional distribution, (48). From univariate extreme value theory, (50) is a highly realistic form for the conditional distribution.

The joint d.f., G, for the X's cannot immediately be obtained, but conditionally on the total unrecorded covariate information the X's are independent. From (49) and (50) we have

$$\prod_{j=1}^{p} \Pr(X_j < x_j | \alpha_C, C \in B)$$

$$= \exp\left\{-\sum_{C \in B} \alpha_C \tau_C^{q_C} \sum_{j \in C}\left(1+\xi_j \frac{x_j - t_j}{\sigma_j}\right)^{-q_C/\xi_j}\right\}.$$

Now, integrating over α_C for all $C \in B$ gives

$$G(x_1, \ldots, x_p) = \exp\left[-\sum_{C \in B} \tau_C \left\{\sum_{j \in C}\left(1+\xi_j \frac{x_j - t_j}{\sigma_j}\right)^{-q_C/\xi_j}\right\}^{1/q_C}\right].$$

Letting $Y_j = (\sum \tau_C)^{-1}\{1+\xi_j(X_j - t_j)/\sigma_j\}^{1/\xi_j}$, where the summation is over $C \in B_{(j)}$, the marginal distribution of Y_j is unit Fréchet for $j = 1, \ldots, p$. Also, $Y_1, \ldots, Y_p$ have joint d.f.

$$G_*(y_1, \ldots, y_p) = \exp\left[-\sum_{C \in B}\left\{\sum_{j \in C}\left(\frac{\vartheta_{j,C}}{y_j}\right)^{q_C}\right\}^{1/q_C}\right], \quad (51)$$

where $q_C \geq 1$ and $\vartheta_{j,C} = \tau_C / \sum \tau_C$, the summation being over $C \in B_{(j)}$. With $\vartheta_{j,C} = 0$ if $j \notin C$, then for $j = 1, \ldots, p$, $0 \leq \vartheta_{j,C} \leq 1$ and $\sum_{C \in B} \vartheta_{j,C} = 1$. It can be shown easily that $\vartheta_{j,C}$ is the probability that the maximum value at site j is due to a spatial storm of the type that occurs only at the collection C of sites.

The derivation of (51) shows that it is a valid joint d.f. As (51) satisfies (40) for $\alpha_{n,j} = n$ and $\beta_{n,j} = 0$, it follows that (51) is a multivariate extreme value d.f. with unit Fréchet margins. Applying Eq. (33) to (51), we have the associated measure densities:

$$h_{m,c}(\mathbf{w}) = \left\{\prod_{k=1}^{m-1}(kq_c - 1)\right\}\left(\prod_{k \in c} \vartheta_{k,c}\right)^{q_c}\left(\prod_{k \in c} w_k\right)^{-(q_c+1)}$$

$$\times \left\{\sum_{k \in c}\left(\frac{\vartheta_{k,c}}{w_k}\right)^{q_c}\right\}^{1/q_c - m},$$

which have $2^{p-1}(p+2) - (2p+1)$ parameters. Thus, there is mass in the interior of S_p and on each of its lower-dimensional boundary. For $p = 2$, this distribution reduces to (45), the bivariate logistic distribution; hence, (51) is indeed a multivariate extension of the logistic distribution. Special cases of the distribution include those in Marshall and Olkin (1967), Johnson and Kotz (1972, Chapter 41) and McFadden (1978), which are obtainable as limits of (51) as $q_C \to \infty$ for all $C \in B$. In addition, by letting only certain $q_C \to \infty$, (51) can then handle cases, where only some variables have singular components to their dependence structure.

Setting $\vartheta_{j,c_p} = 1$, $j = 1, \ldots, p$ and $q_{c_p} = q$ into (51), we have the symmetric logistic distribution:

$$G_*(y_1, \ldots, y_p) = \exp\{-(y_1^{-q} + \cdots + y_p^{-q})^{1/q}\}. \tag{52}$$

Because of its simplicity, this distribution has been studied extensively. Its characteristic function (c.f.) has been given by Shi (1995). After transforming margins of G_* to GEV$(\mu_j, \sigma_j, 0)$, it is found that

$$E\{\exp(i\mathbf{t}^T\mathbf{y})\} = \exp(i\mathbf{t}^T\boldsymbol{\mu}) \frac{\Gamma(1 - i\mathbf{t}^T\boldsymbol{\sigma})}{\Gamma(1 - iq^{-1}\mathbf{t}^T\boldsymbol{\sigma})} \prod_{j=1}^{p} \Gamma(1 - iq^{-1}t_j\sigma_j),$$

where $\mathbf{y}$, $\mathbf{t}$, $\boldsymbol{\mu}$ and $\boldsymbol{\sigma}$ denote p-dimensional vectors with jth components y_j, t_j, μ_j and σ_j, respectively. Shi also computes the product moments of the distribution using properties of the c.f. Letting μ_{abcd} denote the $(a+b+c+d)$th order moment,

$$\mu_{abcd} = E(Y_j - EY_j)^a (Y_k - EY_k)^b (Y_l - EY_l)^c (Y_m - EY_m)^d,$$

it is found that

$$E(Y_j) = \mu_j + \gamma \sigma_j,$$

$$\mu_{2000} = \frac{\pi^2 \sigma_j^2}{6},$$

$$\mu_{1100} = \frac{\sigma_j \sigma_k (q^2 - 1)\pi^2}{6q^2},$$

$$\mu_{3000} = 2\sigma_j^3 \eta_3,$$

$$\mu_{2100} = \frac{2\sigma_j^2 \sigma_k (q^3 - 1)\eta_3}{q^3},$$

$$\mu_{1110} = \frac{2\sigma_j \sigma_k \sigma_l (q^3 - 1)\eta_3}{q^3},$$

$$\mu_{4000} = \frac{3\sigma_j^4 \pi^4}{20},$$

$$\mu_{3100} = \frac{\sigma_j^3 \sigma_k (9q^2 + 4)(q^2 - 1)\pi^4}{60q^4},$$

$$\mu_{2200} = \frac{\sigma_j^2 \sigma_k^2 (27q^4 - 20q^2 - 2q^4)\pi^4}{180q^4},$$

$$\mu_{2110} = \frac{\sigma_j^2 \sigma_k \sigma_l (27q^2+2)(q^2-1)\pi^4}{180 q^4},$$

$$\mu_{1111} = \frac{\sigma_j \sigma_k \sigma_l \sigma_m (9q^2-1)(q^2-1)\pi^4}{60 q^4},$$

where $\gamma = 0.5772\ldots$ is Euler's constant, and $\eta_s = \sum_{k=1}^{\infty} 1/k^s$ is the Zeta function (see Abramowitz and Stegun, 1964). Some special values of the Zeta function are $\eta_2 = \pi^2/6$, $\eta_3 = 1.20205690$, $\eta_4 = \pi^2/90$ and $\eta_5 = 1.03692776$. It follows that the correlation coefficient between Y_j and Y_k is $(q^2-1)/q^2$. Here, the parameter q represents the degree of dependence between the two variables and has a simple interpretation as $1/q = 1 - \tau$, where τ is Kendall's coefficient of concordance.

Shi (1995a) derives the Fisher information matrix of the symmetric logistic distribution (52), assuming $\text{GEV}(\mu_j, \sigma_j, \xi_j)$ margins. For a single observation from the distribution, the log-likelihood function is:

$$l = \log \frac{\partial^p G(x_1,\ldots,x_p)}{\partial x_1 \cdots \partial x_p} = -\sum_{j=1}^{p} \log \sigma_j$$

$$+ \sum_{j=1}^{p} (q+\xi_j) \log u_j + (1-pq) \log z - z + \log Q_p(z,q),$$

where

$$u_j = \left(1 + \xi_j \frac{x_j - \mu_j}{\sigma_j}\right)^{-1/\xi_j}, \quad j=1,\ldots,p, \quad z = \left\{\sum_{j=1}^{p} u_j^q\right\}^{1/q}$$

and $Q_p(z,q)$ is a $(p-1)$-order polynomial in z satisfying

$$Q_p(z,q) = \{q(p-1) - 1 + z\} Q_{p-1}(z,q) - z \frac{\partial Q_{p-1}(z,q)}{\partial z},$$

$$Q_1(z,q) = 1.$$

The derivation of the matrix uses the result that Z and $(U_j/Z)^q$, $j = 1, \ldots, p-1$, are independent: Z has the mixed gamma distribution with p.d.f.

$$\frac{q^{1-p}}{(p-1)!} Q_p(z,q) \exp(-z), \quad z > 0,$$

while $((U_1/Z)^q, \ldots, (U_{p-1}/Z)^q)$ has the multivariate beta distribution $\beta_p(1,\ldots,1)$ (Wilks, 1962, p. 177). Define

$$V_1(\xi; p) = E\left\{ Z^{-\xi} \left(\frac{\partial}{\partial Z} \log Q_p(Z,q)\right)^2 \right\},$$

$$V_2(\xi; p) = -q^2 E\left\{Z^{-\xi} \frac{\partial}{\partial Z} \log Q_p(Z,q) \frac{\partial}{\partial q} \log Q_p(Z,q)\right\},$$

$$V_3(\xi; p) = q^4 E\left\{Z^{-\xi} \left(\frac{\partial}{\partial q} \log Q_p(Z,q)\right)^2\right\}$$

and introduce the notation

$$b_k = \frac{\Gamma(p+kq^{-1})}{\Gamma(1+q^{-1})}, \qquad c_k = \Omega(p+kq^{-1}) - \Omega(1+kq^{-1}),$$

$$e_k = \Omega(p+k) - \Omega(1+k), \qquad f_k = \Omega'(p+k) - \Omega'(1+k),$$

where $\Omega(r) = \partial \log \Gamma(r)/\partial r$ denotes the digamma function. The entire Fisher information matrix is too complicated. But the first four elements can be written as

$$E\left(\frac{\partial l}{\partial q}\right)^2 = \frac{1}{q^4} A_0(p),$$

$$E\left(\frac{\partial l}{\partial q} \frac{\partial l}{\partial \mu_j}\right) = -\frac{1}{q^2 \sigma_j} A_1(\xi_j; p),$$

$$E\left(\frac{\partial l}{\partial q} \frac{\partial l}{\partial \sigma_j}\right) = \frac{1}{q^2 \sigma_j \xi_j} \{A_1(\xi_j; p) - A_1(0; p)\},$$

$$E\left(\frac{\partial l}{\partial q} \frac{\partial l}{\partial \xi_j}\right) = \frac{1}{q^2 \xi_j^2} \{A_1(0; p) - A_1(\xi_j; p)\} - \frac{1}{q^2 \xi_j} A_1'(0; p),$$

where

$$A_0(p) = (pq)^2 \{e_0^2 - f_0 - 2(e_1^2 - f_1)\}$$

$$+ \left\{e_2^2 - f_2 + \frac{p-1}{p+1}\left(e_2 + \frac{3}{2} - \frac{\pi^2}{6}\right)\right\} R_1(0; p)$$

$$- p(p-1)q^2 \left(4e_0 + \frac{2}{p} - \frac{\pi^2}{6}\right)$$

$$+ V_3(0; p) + 2e_1 R_2(0; p),$$

$$A_1(\xi; p) = q(q+\xi)\Gamma(1+\xi)\left[(1-p)\left\{\gamma + \Omega\left(2+\frac{\xi}{q}\right)\right\} - p e_{1+q^{-1}\xi}\right]$$

$$- \text{Beta}\left(p, 2+\frac{\xi}{q}\right)\left\{\left[(p-1)\left\{1-\gamma-\Omega\left(2+\frac{\xi}{q}\right)\right\}\right.\right.$$

$$\left.\left. - \left(p+1+\frac{\xi}{q}\right) e_{1+q^{-1}\xi}\right] R_1(\xi; p) - \left(p+1+\frac{\xi}{q}\right) R_2(\xi; p)\right\},$$

$$R_1(\xi; p) = \frac{\Gamma(1+\xi)P_1(\xi; p)}{\Gamma(p)} + V_1(2+\xi; p),$$

$$R_2(\xi; p) = \frac{\Gamma(1+\xi)P_2(\xi; p)}{\Gamma(p)} + V_2(1+\xi; p),$$

$$P_1(\xi; p) = (1+\xi)(2+\xi)(2b_{1+\xi} - b_{2+\xi}) - (1-pq)(1+pq+2\xi)b_\xi,$$

$$P_2(\xi; p) = (1-pq)\{(1-p)q + \xi c_\xi\}b_\xi$$
$$- (1+\xi)\{(1-p)q + (1+\xi)c_{1+\xi}\}b_{1+\xi}.$$

Here, $A_1'(0; p)$ denotes $\partial A_1(\xi; p)/\partial \xi$ evaluated at $\xi = 0$ and so on. Obviously, these results apply only when $q > 1$ and all the $\xi_j < 1/2$.

14.2. Two level logistic distributions (Tawn, 1990)

Here, we discuss a further generalization of the logistic distribution (51). Again we motivate it physically, following the same terminology.

It is possible that for a spatial storm that affects the collection C of sites, the values at a subset of sites D, $D \subset C$, will be more dependent than at the others. Thus, sites in D may be relatively closely grouped. Then, two-stage conditioning is required: the first stage represents coarse information sufficient to account for dependence between relatively widely spaced sites in $C \setminus D$, and the second stage represents finer covariate information, which accounts for dependence within D. Hence, we first condition on α_C, which is taken to give conditional independence within $C \setminus D$ and between $C \setminus D$ and D, but leaves D conditionally dependent. We then condition on $\alpha_{D,C}$ to give conditional independence within D.

For each $j \in C$, $C \in B$ and $D \subset C$ let $Z_{j,D,C}$ be the size of the maximum, of $N_{D,C}$ observations at site j, of storms which affect the collection C of sites, where a stronger dependence exists between sites of the subset D. Here, $N_{D,C}$ is taken to be Poisson with mean $\tau_{D,C}$. Hence, if α is the total covariate information and α_C and $\alpha_{D,C}$ are independent positive stable variables with respective characteristic exponents $0 < 1/q_C \leqslant 1$ and $1 < 1/q_{D,C} \leqslant 1$, we have for $D \subset C$

$$\Pr(Z_{j,D,C} < z | \alpha) = \begin{cases} \exp(-\alpha_C \tau_{D,C}^{q_C} a_j^{q_C}), & \text{if } j \in C \setminus D, \\ \exp\{-\alpha_{D,C}(\alpha_C \tau_{D,C}^{q_C} a_j^{q_C})^{q_{D,C}}\}, & \text{if } j \in D, \end{cases}$$

where $a_j = \{1 + \xi_j(z - t_j)/\sigma_j\}^{-1/\xi_j}$. In each case this distribution is Generalized Extreme Value. In the most general case, we are interested in the joint d.f. of $X_1, \ldots, X_p$, where for $j = 1, \ldots, p$,

$$X_j = \max_{C \in B_{(j)}} \max_{D \in C^*} Z_{j,D,C}$$

and D is an index variable over the set C^*, the class of all nonempty subsets of C. Then, by an analogous approach to the derivation of (51), the joint d.f. of $X_1, \ldots, X_p$ is

$$G(x_1, \ldots, x_p) = \exp\left(-\sum_{C \in B}\left[\sum_{D \in C^*} \tau_{D,C}^{q_C}\left\{\sum_{j \in C \setminus D} a_j^{q_C} + \left(\sum_{j \in D} a_j^{q_C q_{D,C}}\right)^{1/q_{D,C}}\right\}\right]^{1/q_C}\right).$$

Transformation of the margins to unit Fréchet form gives a multivariate extreme value distribution with joint d.f.

$$G_*(y_1, \ldots, y_p) = \exp\left\{-\sum_{C \in B}\left(\sum_{D \in C^*}\left[\sum_{j \in C \setminus D} \left(\frac{\vartheta_{j,D,C}}{y_j}\right)^{q_C}\right.\right.\right.$$
$$\left.\left.\left. + \left\{\sum_{j \in D}\left(\frac{\vartheta_{j,D,C}}{y_j}\right)^{q_C q_{D,C}}\right\}^{1/q_{D,C}}\right]\right)^{1/q_C}\right\}, \tag{53}$$

where $q_C, q_{D,C} \geq 1$ and

$$\vartheta_{j,D,C} = \tau_{D,C}/\left\{\sum_{C \in B_{(j)}}\left(\sum_{D \in C^*} \tau_{D,C}^{q_C}\right)^{1/q_C}\right\}, \tag{54}$$

thus $0 \leq \vartheta_{j,D,C} \leq 1$ and from (54) the ϑ's satisfy a summation restriction. If $q_{D,C} = 1$ for all $D \subset C$, then (53) reduces to (51). Other special cases include distributions in McFadden (1978) and Joe and Hu (1996, Section 5.2). Because of the hierarchical form of the conditioning we call (53) the multivariate two level logistic distribution. Clearly, in theory it is possible to extend this distribution to any hierarchical level.

A special case of (53) that has been studied to a great extent is the nested logistic distribution (Coles and Tawn, 1991) with d.f.

$$G_*(y_1, y_2, y_3) = \exp\left[-\left\{\left(y_1^{-qq_*} + y_2^{-qq_*}\right)^{1/q_*} + y_3^{-q}\right\}^{1/q}\right], \tag{55}$$

where $q, q_* \geq 1$. Note that the symmetric logistic distribution in (52) is a special case of this for $q_* = 1$. Shi and Zhou (1999) find a transformation of (y_1, y_2, y_3) such that the dependence in (55) reduces to mixed independence. Specifically, define the transformation:

$$\frac{1}{y_1} = uv_1^{1/q} v_2^{1/(qq_*)},$$

$$\frac{1}{y_2} = uv_1^{1/q}\left(\frac{v_2-1}{v_2}\right)^{1/(qq_*)},$$

$$\frac{1}{y_3} = u\left(\frac{v_1-1}{v_1}\right)^{1/q}.$$

Then the joint p.d.f. of the transformed (U, V_1, V_2) is

$$\frac{1}{q_*}r_3\left(u, v_1; \frac{1}{q}\right) + \frac{q_* - 1}{q_*}r_2\left(u, v_1; \frac{1}{q}\right), \tag{56}$$

where

$$r_j\left(u, v_1, \frac{1}{q}\right) = (j-1)v_1^{j-2}a_j\left(u; \frac{1}{q}\right), \quad j = 2, 3,$$

$$a_2\left(u; \frac{1}{q}\right) = \frac{q-1}{q}\exp(-u) + \frac{1}{q}u\exp(-u),$$

$$a_3\left(u; \frac{1}{q}\right) = \frac{(q-1)(2q-1)}{2q^2}\exp(-u)$$
$$+ \frac{3(q-1)}{2q^2}u\exp(-u) + \frac{1}{2q^2}u^2\exp(-u).$$

From (56) it can be seen that V_2 is independent of (U, V_1): V_2 has the uniform distribution on the interval $(0, 1)$ while (U, V_1) is the $(1/q_*)$: $1 - (1/q_*)$ mixture of $r_3(U, V_1; 1/q)$ and $r_2(U, V_1; 1/q)$. In each $r_j(U, V_1; 1/q)$, U and V_1 are independent. Hence, the nested logistic distribution under the transformed variables possesses mixed independence and this easily suggests a way to simulate the distribution.

Shi and Zhou (1999) use the mixed independence above to derive algebraic expressions for the c.f. and product moments of the nested logistic distribution. After transforming margins of (55) to GEV$(\mu_j, \sigma_j, 0)$, they derive the c.f. as

$$E\left(\exp(i\mathbf{t}^T\mathbf{y})\right)$$
$$= \exp(i\mathbf{t}^T\boldsymbol{\mu})\frac{\Gamma(1 - i\mathbf{t}^T\boldsymbol{\sigma})}{\Gamma(1 - iq^{-1}\mathbf{t}^T\boldsymbol{\sigma})}\frac{\Gamma[1 - iq^{-1}(t_1\sigma_1 + t_2\sigma_2)]}{\Gamma[1 - iq^{-1}q_*^{-1}(t_1\sigma_1 + t_2\sigma_2)]}$$
$$\times \Gamma\left(1 - iq^{-1}q_*^{-1}t_1\sigma_1\right)\Gamma\left(1 - iq^{-1}q_*^{-1}t_2\sigma_2\right)\Gamma\left(1 - iq^{-1}t_3\sigma_3\right),$$

where $\mathbf{y}$, $\mathbf{t}$, $\boldsymbol{\mu}$ and $\boldsymbol{\sigma}$ denote 3-dimensional vectors with jth components y_j, t_j, μ_j and σ_j, respectively. The product moments of (55),

$$\mu_{jkl} = E(Y_1 - EY_1)^j(Y_2 - EY_2)^k(Y_3 - EY_3)^l,$$

follow from the c.f. The second-order moments are found to be:

$$\mu_{110} = \frac{\sigma_1\sigma_2(q^2q_*^2 - 1)\pi^2}{6q^2q_*^2},$$

$$\mu_{101} = \frac{\sigma_1\sigma_3(q^2 - 1)\pi^2}{6q^2},$$

$$\mu_{011} = \frac{\sigma_2\sigma_3(q^2-1)\pi^2}{6q^2}.$$

Thus, the correlation coefficient between Y_1 and Y_2 is $(q^2 q_*^2 - 1)/(q^2 q_*^2)$ and that between Y_1 and Y_3 (or Y_2 and Y_3) is $(q^2-1)/q^2$. The third-order moments are:

$$\mu_{jkl} = 2\sigma_1^j \sigma_2^k \sigma_3^l (1-\upsilon^3)\eta_3, \quad j+k+l=3, \ 0 \leqslant j,k,l \leqslant 3.$$

Here, $\eta_s = \sum_{k=1}^{\infty} 1/k^s$ is the Zeta function and υ takes the following values: 0 if there are two zeros among j,k,l; $1/q$ if $l \neq 0$ and one or both of j,k are non-zero; and, $1/(qq_*)$ if $l = 0$ and both of j,k are non-zero. Some fourth-order moments are:

$$\mu_{310} = \frac{\sigma_1^3 \sigma_2 (9q^2 q_*^2 + 4)(q^2 q_*^2 - 1)\pi^4}{60 q^4 q_*^4},$$

$$\mu_{130} = \frac{\sigma_1 \sigma_2^3 (9q^2 q_*^2 + 4)(q^2 q_*^2 - 1)\pi^4}{60 q^4 q_*^4}$$

and

$$\mu_{301} = \frac{\sigma_1^3 \sigma_3 (9q^2 + 4)(q^2 - 1)\pi^4}{60 q^4},$$

$$\mu_{031} = \frac{\sigma_2^3 \sigma_2 (9q^2 + 4)(q^2 - 1)\pi^4}{60 q^4},$$

$$\mu_{103} = \frac{\sigma_1 \sigma_3^3 (9q^2 + 4)(q^2 - 1)\pi^4}{60 q^4},$$

$$\mu_{013} = \frac{\sigma_2 \sigma_3^3 (9q^2 + 4)(q^2 - 1)\pi^4}{60 q^4}.$$

Shi and Zhou (1999) note further that the above results on the mixed independence can be derived also for higher-dimensional distributions, e.g.,

$$\exp\left[-\left\{(y_1^{-qq_*} + y_2^{-qq_*} + y_3^{-qq_*})^{1/q_*} + y_4^{-q}\right\}^{1/q}\right],$$

$$\exp\left\{-\left[\left\{(y_1^{-qq_*q_{**}} + y_2^{-qq_*q_{**}})^{1/q_{**}} + y_3^{-qq_*}\right\}^{1/q_*} + y_4^{-q}\right]^{1/q}\right\},$$

$$\exp\left[-\left\{(y_1^{-qq_*} + y_2^{-qq_*})^{1/q_*} + (y_3^{-qq_{**}} + y_4^{-qq_{**}})^{1/q_{**}}\right\}^{1/q}\right],$$

where $q, q_*, q_{**} \geqslant 1$.

14.3. Negative logistic distributions (Joe, 1990)

This has joint d.f.

$$G_*(y_1,\ldots,y_p) = \exp\left[-\sum_{j=1}^{p}\frac{1}{y_j} + \sum_{c\in C:|c|\geqslant 2}(-1)^{|c|}\left\{\sum_{j\in c}\left(\frac{\vartheta_{j,c}}{y_j}\right)^{q_c}\right\}^{1/q_c}\right]$$

with parameter constraints given by $q_c \leqslant 0$ for all $c \in C$, $\vartheta_{j,c} = 0$ if $j \notin c$, $\vartheta_{j,c} \geqslant 0$ for all $c \in C$ and $\sum_{c\in C}(-1)^{|c|}\vartheta_{j,c} \leqslant 1$. Again by Eq. (33),

$$h_{m,c}(\mathbf{w}) = \sum_{d\in C:c\subset d}(-1)^{|d|}\left\{\prod_{k=1}^{m-1}(1-kq_d)\right\}\left(\prod_{k\in c}\vartheta_{k,d}\right)^{q_d}\left(\prod_{k\in c}w_k\right)^{-(q_d+1)}$$

$$\times \left\{\sum_{k\in c}\left(\frac{\vartheta_{k,d}}{w_k}\right)^{q_d}\right\}^{1/q_d - m}.$$

Evidently this has a structure similar to the logistic family with the special case for $\vartheta_{j,c_p} = 1$, $j = 1,\ldots,p$, and $q_{c_p} = q$ giving a symmetric version that has all its mass in the interior of S_p.

The two bilogistic families discussed earlier are asymmetric generalizations of the logistic families, but it is not yet known how they generalize to the multivariate case. However, the family of Beta distributions generalizes to the following Dirichlet distributions.

14.4. Dirichlet distributions (Coles and Tawn, 1991)

The p.d.f. of the Dirichlet $(q_1,\ldots,q_p)$ distribution is

$$h_{\dagger}(\mathbf{w}) = \left\{\prod_{j=1}^{p}\Gamma(q_j)\right\}^{-1}\Gamma(q_1+\cdots+q_p)\prod_{j=1}^{p}w_j^{q_j-1},$$

$q_j > 0$, $j = 1,\ldots,p$, $\mathbf{w} \in S_p$.

By Eq. (34) $m_j = q_j/(q_1+\cdots+q_p)$, and from Eq. (35) it follows that

$$h_{p,c_p}(\mathbf{w}) = \frac{\Gamma(q_1+\cdots+q_p+1)}{(q_1w_1+\cdots+q_pw_p)^{p+1}}\prod_{j=1}^{p}\frac{q_j}{\Gamma(q_j)}$$

$$\times \prod_{j=1}^{p}\left(\frac{q_jw_j}{q_1w_1+\cdots+q_pw_p}\right)^{q_j-1}$$

is the density, in the interior of S_p, of a valid measure H_* that satisfies the constraints (30). This has structure similar to the Beta family, the special case for $p=2$, with symmetry arising when $q_1 = \cdots = q_p$.

The corresponding form for G_* is complicated although numerical computation is feasible. This distribution has been found most suitable for estimating, in continuous space, the spatial dependence within rainfall storms (Coles, 1993) and for estimating the dependence between the extremes of surge, wave height and wave period (Coles and Tawn, 1994).

14.5. Time series logistic distributions (Coles and Tawn, 1991)

Let $Y_1, \ldots, Y_p$ be a first-order Markov process representing a time series such as observations of a propagating sea storm at sites ordered along a coast. Suppose without loss of generality that Y_j have the unit Fréchet distribution. Let $f^{(j)}$ denote the joint p.d.f. of (Y_j, Y_{j+1}). Then, the joint p.d.f. of $Y_1, \ldots, Y_p$ is

$$f(y_1, \ldots, y_p) = \Phi_1'(y_1) \prod_{j=1}^{p-1} \frac{f^{(j)}(y_j, y_{j+1})}{\Phi_1'(y_j)}.$$

Evaluating (39) at $B = [0, y_1] \times \cdots \times [0, y_p]$ and then differentiating it with respect to $y_1, \ldots, y_p$, we have

$$\lim_{t \to \infty} \left\{ t^{p+1} f(ty_1, \ldots, ty_p) \right\} = \left(\sum_{j=1}^{p} y_j \right)^{-(p+1)} h \left\{ \frac{y_1}{\sum_{j=1}^{p} y_j}, \ldots, \frac{y_{p-1}}{\sum_{j=1}^{p} y_j} \right\}.$$

Hence, if we assume that the joint d.f. of (Y_j, Y_{j+1}) belongs to the domain of attraction of a bivariate logistic extreme value d.f. with $\vartheta_1 = \vartheta_2 = 1$ and $q_j = q$ then

$$h(\mathbf{w}) = \frac{1}{w_1^2} \prod_{j=1}^{p-1} \frac{(q_j - 1) w_j^2}{(w_j w_{j+1})^{q_j+1}} \left(w_j^{-q_j} + w_{j+1}^{-q_j} \right)^{1/q_j - 2}, \quad \mathbf{w} \in S_p,$$

for $q_j \geq 1$.

An extension of this family, not examined here, is based on a higher order Markov sequence with the associated joint p.d.f. of consecutive values taken as multivariate extreme value with unit Fréchet margins.

14.6. Distributions closed under margins

Here, we discuss a few technical tools for constructing multivariate extreme value distributions that are closed under margins. We begin with one communicated by Nadarajah (1999b).

Let G_A, G_B and G_C be known bivariate extreme value d.f.s with respective exponent measure functions V_A, V_B and V_C. Assume as usual that all univariate margins are unit Fréchet. The following steps construct a trivariate extreme value d.f. that has G_A, G_B and G_C as its bivariate margins.

Partition the two-dimensional unit simplex, S_3, into three disjoint sets B_j, $j = 1, 2, 3$, chosen as

$$B_1 = \{(w_1, w_2) \in S_3 : w_1 \geq w_2 \text{ and } 2w_1 + w_2 \geq 1\},$$
$$B_2 = \{(w_1, w_2) \in S_3 : w_2 \geq w_1 \text{ and } 2w_2 + w_1 \geq 1\},$$
$$B_3 = \{(w_1, w_2) \in S_3 : 2w_1 + w_2 \leq 1 \text{ and } 2w_2 + w_1 \leq 1\}.$$

Define $H_j : B_j \to \Re_+$ and $\theta_j : \Re_+^3 \to S_3$ to satisfy

$$\int_{B_1} w_1 \, dH_1(w_1, w_2) = \int_{B_2} w_2 \, dH_2(w_1, w_2)$$
$$= \int_{B_3} (1 - w_1 - w_2) \, dH_3(w_1, w_2) = 1,$$
$$\theta_1(y_1, y_2, y_3) + \theta_2(y_1, y_2, y_3) + \theta_3(y_1, y_2, y_3) = 1,$$
$$\theta_j(ky_1, ky_2, ky_3) = \theta_j(y_1, y_2, y_3), \quad j = 1, 2, 3, \ k > 0,$$

and

$$\theta_1(y_1, y_2, \infty) = \theta_2(y_1, \infty, y_3) = \theta_3(\infty, y_2, y_3) = 1.$$

Then,

$$G_*(y_1, y_2, y_3) = \exp\left\{-\sum_{j=1}^{3} \theta_j(y_1, y_2, y_3) V_j(y_1, y_2, y_3)\right\},$$

where

$$V_j(y_1, y_2, y_3) = \int_{B_j} \max\left[w_1 V_A(y_1, y_2), w_2 V_B(y_1, y_3),\right.$$
$$\left.(1 - w_1 - w_2) V_C(y_2, y_3)\right] dH_j(w_1, w_2),$$

is a trivariate extreme value d.f. with

$$G_*(y_1, y_2, \infty) = G_A(y_1, y_2), \quad G_*(y_1, \infty, y_3) = G_B(y_1, y_3),$$
$$G_*(\infty, y_2, y_3) = G_C(y_2, y_3).$$

The second construct that we discuss is due to Marco and Ruiz-Rivas (1992). Suppose that we can express

$$G_j = v_j(F_j), \quad j = 1, 2,$$

where F_j is an m_j-variate d.f. and $v_j : [0, 1] \to [0, 1]$ is continuous to the right and satisfies:

$$v_j(0) = 0, \qquad v_j(1) = 1, \qquad v_j^{(\prime k)} \geq 0, \quad 1 \leq k \leq m_j.$$

Let

$$a_{jk} = \frac{v_j^{(\prime k)}(0)}{k!}, \quad k = 1, \ldots, m_j - 1, \ j = 1, 2,$$

$$c_j = 1 - \sum_{k=1}^{m_j - 1} a_{jk}, \quad j = 1, 2.$$

Define

$$v(s_1, s_2) = \int_0^{s_1} \int_0^{s_2} \left(\frac{s_1 - u_1}{1 - u_1}\right)^{m_1 - 1} \left(\frac{s_2 - u_2}{1 - u_2}\right)^{m_2 - 1} dD(u_1, u_2),$$

where $D(u_1, u_2)$ is any d.f. in $[0, 1]^2$ with marginal densities:

$$\rho_j(u_j) = \frac{1}{c_j} \frac{(1 - u_j)^{m_j - 1}}{(m_j - 1)!} v_j^{(m_j)}(u_j), \quad u_j \in [0, 1], \ j = 1, 2.$$

It is then easily verified that

$$G(\mathbf{x}_1, \mathbf{x}_2) = v\big(F_1(\mathbf{x}_1), F_2(\mathbf{x}_2)\big)$$

is a d.f. with G_1, G_2 as marginals.

We now discuss three more specialized constructs (due to Joe, 1994) that are closed under margins. Let V_m denote the exponent measure function of an m-variate extreme value distribution. Take $V_2(y_1, y_2) = (y_1^{-q} + y_2^{-q})^{1/q}$ and, for $m \geq 3$, define V_m to satisfy the recurrence relation

$$V_m(y_1, \ldots, y_m) = \left[\{V_{m-1}(y_1, \ldots, y_{m-1})\}^{q_{1,m}} + y_m^{-q_{1,m}}\right]^{1/q_{1,m}}, \tag{57}$$

where $q_{1,2} \geq q_{1,3} \geq \cdots \geq q_{1,m} \geq 1$. Then, V_m corresponds to an m-variate logistic extreme value distribution that is closed under margins. With V_k as given by (57), define

$$V_m^*(y_1, \ldots, y_m) = \frac{1}{y_1} + \cdots + \frac{1}{y_m} - \sum_{j_1 < j_2} (y_{j_1}^{-q_{1,j_2}} + y_{j_2}^{-q_{1,j_2}})^{1/q_{1,j_2}}$$

$$+ \sum_{3 \leq k \leq m} (-1)^{k+1} \sum_{j_1 < \cdots < j_k} V_k(y_{j_1}, \ldots, y_{j_k}),$$

where $q_{1,2} \leqslant \cdots \leqslant q_{1,m} \leqslant 0$. Then, V_m^* corresponds to an m-variate negative logistic extreme value distribution that is closed under margins. For the final construct, take

$$V_2(y_1, y_2) = \frac{1}{y_1} \Upsilon\left(\frac{a}{2} + \frac{1}{a}\log\frac{y_2}{y_1}\right) + \frac{1}{y_2} \Upsilon\left(\frac{a}{2} + \frac{1}{a}\log\frac{y_1}{y_2}\right),$$

the exponent measure function corresponding to the bivariate Gaussian extreme value distribution (where Υ denotes the c.d.f. of the standard normal distribution). Let $\rho_{jkl} = (a_{kj}^2 + a_{lj}^2 - a_{kl}^2)/(2a_{kj}a_{lj})$ for j, k, l distinct and, for $m \geqslant 3$, define V_m recursively by

$$V_m(y_1, \ldots, y_m)$$
$$= V_{m-1}(y_1, \ldots, y_{m-1})$$
$$+ \int_0^{1/y_m} \Upsilon\left(\frac{a_{1,m}}{2} + \frac{\log(y_1 s)}{a_{1,m}}, \ldots, \frac{a_{m-1,m}}{2} + \frac{\log(y_{m-1} s)}{a_{m-1,m}}\right) ds.$$

Here, $\Upsilon(\cdot)$ denotes the c.d.f. of an $(m-1)$-variate normal distribution with means 0, correlations ρ_{mkl} ($1 \leqslant k < l \leqslant m-1$), and variances 1. This construct ensures that V_m corresponds to an m-variate Gaussian extreme value distribution that is closed under margins.

15. Extremes of multivariate stochastic processes

In parallel with Section 8, here we discuss some recent developments of extreme value theory for nontrivial multivariate stochastic processes. We provide some details of the results for Gaussian and independent stochastic processes.

15.1. Gaussian processes

Amram (1985) derives the joint limiting distribution of the maxima of a stationary Gaussian multivariate sequence. Namely, if $\{(X_{i,1}, \ldots, X_{i,p}), i = 1, 2, \ldots\}$ is a stationary Gaussian process with mean, variance, and covariance given, respectively, by

$$E(X_{i,j}) = 0, \quad 1 \leqslant j \leqslant p,$$
$$\operatorname{Var}(X_{i,j}) = 1, \quad 1 \leqslant j \leqslant p,$$
$$r_{jk}(l) = \operatorname{Cov}(X_{i,j}, X_{i+l,k}), \quad 1 \leqslant j, k \leqslant p,$$

and if for some $\alpha = 1, 2, \ldots,$

$$\sum_{l=1}^{\infty} |r_{jk}(l)|^\alpha < \infty, \quad 1 \leqslant j \neq k \leqslant p,$$

and

$$\sum_{l=1}^{\infty} |r_{kk}(l)|^{\alpha} < \infty, \quad 1 \leqslant k \leqslant p,$$

then

$$\lim_{n \to \infty} \Pr(M_{n,1} \leqslant a_{n,1} x_1 + b_{n,1}, \ldots, M_{n,p} \leqslant a_{n,p} x_p + b_{n,p})$$
$$= \exp\{-\exp(-x_1)\} \cdots \exp\{-\exp(-x_p)\},$$

where

$$a_{n,j} = 1/\sqrt{2 \log n}$$

and

$$b_{n,j} = \sqrt{2 \log n} - \left(1/(2\sqrt{2 \log n})\right)(\log \log n + \log 4\pi).$$

15.2. Independent stochastic processes

Let $\{X_n(t), t \in \Re\}$ be independent stochastic processes. Characterizations of the maxima of $\{X_n(t), t \in \Re\}$ for Wiener and Ornstein–Uhlenbeck processes are given in Brown and Resnick (1977). Here, we give a result due to Balkema et al. (1993), which expresses the characterization in terms of spectral functions acting on a Poisson point process. Namely, suppose there exist sequences $a_n > 0$ and $b_n \in \Re$ satisfying

$$|\log a_n| + |b_n| \to \infty$$

and

$$\frac{a_{n+1}}{a_n} \to 1, \quad \frac{b_{n+1} - b_n}{a_n} \to 0$$

such that for each $t_1, \ldots, t_k \in \Re, k \geqslant 1$, the sequence of random vectors

$$\frac{\max_{i \leqslant n} X_i(t_1) - b_n}{a_n}, \frac{\max_{i \leqslant n} X_i(t_2) - b_n}{a_n}, \ldots, \frac{\max_{i \leqslant n} X_i(t_k) - b_n}{a_n}$$

converges in distribution ($n \to \infty$) to a non-degenerate distribution. Let $\{X(t)\}$ denote the limiting process and let x^* be the upper end point of its one-dimensional margin. Also let $\{(T_k, Z_k), k = 1, 2, \ldots\}$ be an enumeration of a Poisson point process on $\Re_+ \times \Re_+$ with Lebesgue measure as its intensity measure. Then, if $\{X(t)\}$ is continuous in probability, we can characterize it by

$$X(t) =^d \sup_k f_t(T_k) - Z_k, \quad t \in \Re,$$

for $x^* = \infty$ and by

$$X(t) =^d \sup_k (x^* - \exp[-\{f_t(T_k) - Z_k\}])$$

for $x^* < \infty$, where $f_t : \Re_+ \to [-\infty, \infty)$ is a family of measurable functions satisfying

$$\int_0^\infty (f_t(x) - x_t)_+ \, dx < \infty$$

for some $x_t \in \Re$.

15.3. Miscellanea

Other developments for extremes of nontrivial multivariate stochastic processes include: Feeney and Sen (1985), who consider the extremes of sequences which satisfy the Markov type condition,

$$\Pr(\mathbf{X}_i \in B_i | \mathbf{X}_k \in B_k, k \leq i - 1) = \Pr(\mathbf{X}_i \in B_i | \mathbf{X}_{i-1} \in B_{i-1}),$$

for all B-sets. They establish that the corresponding distribution of $\mathbf{M}_n$ is asymptotically independent; Hsing (1989), Hüsler (1990b) and Wiśniewski (1997), who study the extremes of a strictly stationary sequence of random vectors under distributional mixing conditions. Among other things they provide characterizations of the joint d.f. F and conditions under which the maxima behave as they would if the sequence were i.i.d.; Hüsler (1989), who investigates the properties of all possible limiting distributions for the extreme values in sequences of independent, non-identically distributed random vectors. He finds that the possible limits are more restricted than in the univariate case; Ferreira (1994), who considers the extremes of a T-periodic sequence of random vectors by appropriately extending the mixing conditions in the stationary case; Steinebach and Eastwood (1996), who derive the extreme value asymptotics for a multivariate renewal process via strong invariance together with an extreme value limit theorem for Rayleigh processes; and, Perfekt (1997), who determines the asymptotic distribution of $\mathbf{M}_n$ for a class of stationary Markov chains with values in $\Re^p$. The derivation requires mild dependence restrictions and suitable assumptions on the marginal distribution and the transition probabilities of the chain.

Acknowledgements

The author is grateful to Professor D. N. Shanbhag and to the referee for many useful suggestions, which helped to improve this article. Many thanks also to Professors S. R. Jammalamadaka, R. B. Leipnik and M. Sobel (University of California in Santa Barbara) for their hospitality and support.

References

Aarssen, K. and L. De Haan (1994). On the maximal life span of humans. *Math. Population Studies* **4**, 259–281.

Abramowitz, M. and I. A. Stegun (Eds.) (1964). *Handbook of Mathematical Functions with Formulas, Graphs and Mathematical Tables*. Dover, New York.

Adke, S. R. and C. Chandran (1994). Asymptotic distributions of extremes of extremal Markov sequences. *J. Appl. Probab.* **31**, 256–261.

Adler, R. J. and G. Samorodnitsky (1987). Tail behavior for the suprema of Gaussian processes with applications to empirical processes. *Ann. Probab.* **15**, 1339–1351.

Albin, J. M. P. (1990). On extremal theory for stationary processes. *Ann. Probab.* **18**, 92–128.

Albin, J. M. P. (1998). On extremal theory for self-similar processes. *Ann. Probab.* **26**, 743–793.

Alpium, M. T. (1989). An extremal Markovian sequence. *J. Appl. Probab.* **26**, 219–232.

Alpium, M. T., N. A. Catkan and J. Hüsler (1995). Extremes and clustering of nonstationary max-AR(1) sequences. *Stochastic Process. Appl.* **56**, 171–184.

Aly, E.-E. A. A. and M. A. Shayib (1992). On some goodness-of-fit tests for the normal, logistic and extreme-value distributions. *Comm. Statist. Theory Methods* **21**, 1297–1308.

Amram, F. (1985). Multivariate extreme value distributions for stationary Gaussian sequences. *J. Multivariate Anal.* **16**, 237–240.

Anderson, C. W. (1970). Extreme value theory for a class of discrete distributions with applications to some stochastic processes. *J. Appl. Probab.* **7**, 99–113.

Anderson, C. W. (1980). Local limit theorems for the maxima of discrete random variables. *Math. Proc. Cambridge Philos. Soc.* **88**, 161–165.

Anderson, C. W., S. G. Coles and J. Hüsler (1997). Maxima of Poisson-like variables and related triangular arrays. *Ann. Appl. Probab.* **7**, 953–971.

Anderson, C. W. and S. Nadarajah (1993). Environmental factors affecting reservoir safety. In *Statistics for the Environment I*, pp. 163–182 (Eds. V. Barnett and K. F. Turkman). Wiley, Chichester.

Arnold, B. C. and J. A. Villaseñor (1984). The distribution of the maximal time till departure from a state in a Markov chain. In *Statistical Extremes and Applications*, pp. 413–426. NATO Science Series C: Mathematical and Physical Sciences, Vol. 131. Reidel, Dordrecht.

Asadi, M., C. R. Rao and D. N. Shanbhag (2001). Some unified characterization on generalized Pareto distributions. *J. Statist. Plann. Inference* **93**, 29–50.

Ashour, S. K. and Y. M. El-Adl (1980). Bayesian estimation of the parameters of the extreme value distribution. *Egyptian Statist. J.* **24**, 140–152.

Asmussen, S., C. Klüppelberg and K. Sigman (1999). Sampling at subexponential times, with queuing applications. *Stochastic Process. Appl.* **79**, 265–286.

Auinger, K. (1990). Quasi goodness of fit tests for lifetime distributions. *Metrika* **37**, 97–116.

Balakrishnan, N. and P. S. Chan (1992). Order statistics from extreme value distribution, II: best linear unbiased estimates and some other uses. *Comm. Statist. Simulation and Comput.* **21**, 1219–1246.

Balkema, A. A. and L. de Haan (1974). Residual life time at great age. *Ann. Probab.* **2**, 792–804.

Balkema, A. A. and L. de Haan (1990). A convergence rate in extreme-value theory. *J. Appl. Probab.* **27**, 577–585.

Balkema, A. A., L. de Haan and R. L. Karandikar (1993). Asymptotic distributions of the maximum of n independent stochastic processes. *J. Appl. Probab.* **30**, 66–81.

Ballerini, R. (1987). Another characterization of the type I extreme value distribution. *Statist. Probab. Lett.* **5**, 83–85.

Ballerini, R. and W. P. McCormick (1989). Extreme value theory for processes with periodic variances. *Comm. Statist. Stochastic Models* **5**, 45–61.

Behr, R. A., M. J. Karson and J. E. Minor (1991). Reliability-analysis of window glass failure pressure data. *Structural Safety* **11**, 43–58.

Beirlant, J., J. L. Teugels and P. Vynckier (1998). Some thoughts on extreme values. In *Probability Towards 2000*, pp. 58–73. Lecture Notes in Statist., Vol. 128. Springer-Verlag, New York.

Beirlant, J. and E. Willekens (1990). Rapid variation with remainder and rates of convergence. *Adv. Appl. Probab.* **22**, 787–801.

Berman, S. M. (1987). Poisson and extreme value limit theorems for Markov random fields. *Adv. Appl. Probab.* **19**, 106–122.

Berman, S. M. (1992). *Sojourns and Extremes of Stochastic Processes.* Wadsworth and Brooks/Cole Advanced Books and Software, Pacific Grove, California.

Berred, M. (1995). K-record values and the extreme-value index. *J. Statist. Plann. Inference* **45**, 49–63.

Bertoin, J. (1998). Darling–Erdös theorems for normalized sums of i.i.d. variables close to a stable law. *Ann. Probab.* **26**, 832–852.

Bertoin, J. and R. A. Doney (1996). Some asymptotic results for transient random walks. *Adv. Appl. Probab.* **28**, 207–226.

Borkovec, M. (2000). Extremal behavior of the autoregressive process with ARCH(1) errors. *Stochastic Process. Appl.* **85**, 189–207.

Borkovec, M. and C. Klüppelberg (1998). Extremal behavior of diffusion models in finance. *Extremes* **1**, 47–80.

Borovkov, A. A. (1976). *Stochastic Processes in Queuing Theory.* Springer-Verlag, New York.

Brockwell, P. J., S. I. Resnick and N. Pacheco-Santiago (1982). Extreme values, range and weak convergence of integrals of Markov chains. *J. Appl. Probab.* **19**, 272–288.

Broussard, J. P. and G. G. Booth (1998). The behavior of extreme values in Germany's stock index futures: An application to intradaily margin setting. *Europ. J. Oper. Res.* **104**, 393–402.

Brown, B. G. and R. W. Katz (1995). Regional-analysis of temperature extremes: Spatial analog for climate-change. *J. Climate* **8**, 108–119.

Brown, B. M. and S. I. Resnick (1977). Extreme values of independent stochastic processes. *J. Appl. Probab.* **14**, 732–739.

Burton, P. W. and K. C. Makropoulos (1985). Seismic risk of circum-Pacific earthquakes: II. Extreme values using Gumbel's third distribution and the relationship with strain energy release. *Pure Appl. Geophys.* **123**, 849–866.

Campbell, J. W. and C. P. Tsokos (1973). The asymptotic distribution of maxima in bivariate samples. *J. Amer. Statist. Assoc.* **68**, 734–739.

Capéraà, P., A.-L. Fougères and C. Genest (2000). Bivariate distributions with given extreme value attractor. *J. Multivariate Anal.* **72**, 30–49.

Casson, E. and S. Coles (1998). Extreme hurricane wind speeds: estimation, extrapolation and spatial smoothing. *J. Wind Engineering and Industrial Aerodynamics* **64**, 131–140.

Castillo, E. (1988). *Extreme Value Theory in Engineering.* Statistical Modeling and Decision Science. Academic Press, Boston, MA.

Castillo, E., J. Galambos and J. M. Sarabia (1989). The selection of the domain of attraction of an extreme value distribution from a set of data. In *Extreme Value Theory*, pp. 181–190. Lecture Notes in Statist., Vol. 51. Springer-Verlag, New York.

Castillo, E. and A. S. Hadi (1997). Fitting the generalized Pareto distribution to data. *J. Amer. Statist. Assoc.* **92**, 1609–1620.

Chernick, M. R. (1981). A limit theorem for the maximum of autoregressive processes with uniform marginal distributions. *Ann. Probab.* **9**, 145–149.

Choi, Y. K., K. S. Hwang and S. B. Kang (1995). Extreme values of a Gaussian process. *J. Korean Math. Soc.* **32**, 739–751.

Chowdhury, J. U., J. R. Stedinger and L.-H. Lu (1991). Goodness-of-fit tests for regional generalized extreme value flood distributions. *Water Resources Res.* **27**, 1765–1776.

Chrapek, M. (1998). On the extremal behaviour of some stationary Markov sequence. *Demonstratio Math.* **31**, 289–298.

Christopeit, N. (1994). Estimating parameters of an extreme value distribution by the method of moments. *J. Statist. Plann. Inference* **41**, 173–186.

Cohen, J. P. (1982a). Convergence rates for the ultimate and penultimate approximations in extreme value theory. *Adv. Appl. Probab.* **14**, 833–854.

Cohen, J. P. (1982b). The penultimate form of approximation to normal extremes. *Adv. Appl. Probab.* **14**, 324–339.

Cohen, J. W. (1972). On the tail of the stationary waiting time distribution and limit theorems for the $M/G/1$ queue. *Ann. Inst. H. Poincaré B* **8**, 255–263.

Coles, S. G. (1993). Regional modelling of extreme storms via max-stable processes. *J. Roy. Statist. Soc. B* **55**, 797–816.

Coles, S. G. and J. A. Tawn (1991). Modelling extreme multivariate events. *J. Roy. Statist. Soc. B* **53**, 377–392.

Coles, S. G. and J. A. Tawn (1994). Statistical methods for multivariate extremes: an application to structural design. *Appl. Statist.* **43**, 1–48.

Coles, S. G. and D. Walshaw (1994). Directional modelling of extreme wind speeds. *Appl. Statist.* **43**, 139–157.

Cramér, H. (1955). *Collective Risk Theory*. Esselte, Stockholm.

Cramér, H. (1965). A limit theorem for the maximum values of certain stochastic processes. *Teor. Veroyatnost. i Primenen.* **10**, 137–139.

Cramér, H. (1966). On extreme values of certain stochastic processes. In *Research Papers in Statistics*, pp. 73–78. Wiley, London.

Daley, D. J. and P. Hall (1984). Limit laws for the maximum of weighted and shifted i.i.d. random variables. *Ann. Probab.* **12**, 571–587.

Darling, D. A. and P. Erdös (1956). A limit theorem for the maximum of normalized sums of independent random variables. *Duke Math. J.* **23**, 143–154.

Davis, R. and S. Resnick (1984). Tail estimates motivated by extreme value theory. *Ann. Statist.* **12**, 1467–1487.

Davis, R. and S. Resnick (1985). Limit theory for moving averages of random variables with regularly varying tail probabilities. *Ann. Probab.* **13**, 179–195.

Davis, R. and S. Resnick (1988). Extremes of moving averages of random variables from the domain of attraction of the double exponential distribution. *Stochastic Process. Appl.* **30**, 41–68.

Davis, R. and S. Resnick (1991). Extremes of moving averages of random variables with finite endpoint. *Ann. Probab.* **19**, 312–328.

Davison, A. C. and R. L. Smith (1990). Models for exceedances over high thresholds (with discussion). *J. Roy. Statist. Soc. B* **52**, 393–442.

Deheuvels, P. (1981). Univariate extreme values – theory and applications (with discussion). In *Proceedings of the 43rd session of the International Statistical Institute*, Vol. 2, pp. 837–858, 894–902 (Buenos Aires, 1981). Bulletin of the Institute of the International Statistics, Vol. 49.

Deheuvels, P. (1986). Strong laws for the kth order statistic when $k \leqslant c \log_2 n$. *Probab. Theory Related Fields* **72**, 133–154.

Deheuvels, P. (1989). Strong laws for the kth order statistic when $k \leqslant c \log_2 n$. II. In *Extreme Value Theory*, pp. 21–35. Lecture Notes in Statist., Vol. 51. Springer-Verlag, New York.

Deheuvels, P. and L. Devroye (1987). Limit laws of Erdös–Renyi–Shepp type. *Ann. Probab.* **15**, 1363–1386.

Dekkers, A. L. M., J. H. J. Einmahl and L. De Haan (1989). A moment estimator for the index of an extreme-value distribution. *Ann. Statist.* **17**, 1833–1855.

Dembińska, A. and J. Wesolowski (1998). Linearity of regression for non-adjacent order statistics. *Metrika* **48**, 215–222.

Dietrich, D. and J. Hüsler (1996). Minimum distance estimators in extreme value distributions. *Comm. Statist. Theory Methods* **25**, 695–703.

Doney, R. A. and G. L. O'Brien (1991). Loud shot noise. *Ann. Appl. Probab.* **1**, 88–103.

Dorea, C. C. Y. (1987). Estimation of the extreme value and the extreme points. *Ann. Instit. Statist. Math.* **39**, 37–48.

Dorea, C. C. Y. and C. R. GonCalves (1999). Asymptotic distribution of extremes of randomly indexed random variables. *Extremes* **2**, 95–109.

Drees, H. and E. Kaufmann (1998). Selecting the optimal sample fraction in univariate extreme value estimation. *Stochastic Process. Appl.* **75**, 149–172.

Dubey, S. D. (1966). Characterization theorems for several distributions and their applications. *Industrial Math.* **16**, 1–22.

Dupuis, D. J. (1997). Extreme value theory based on the r largest annual events: a robust approach. *J. Hydrology* **200**, 295–306.

Dupuis, D. J. (1999a). Exceedances over high thresholds: a guide to threshold selection. *Extremes* **1**, 251–261.

Dupuis, D. J. (1999b). Parameter and quantile estimation for the generalized extreme-value distribution: a second look. *Environmetrics* **10**, 119–124.

Dupuis, D. J. and C. A. Field (1998). Robust estimation of extremes. *Canad. J. Statist.* **26**, 199–215.

Dziubdziela, W. (1978). On convergence rates in the limit laws for extreme order statistics. In *Transactions of the Seventh Prague Conference on Information Theory, Statistical Decision Functions and the Eighth European Meeting of Statisticians*, pp. 119–127. Academia, Prague.

ElJabi, N., F. Ashkar and S. Hebabi (1998). Regionalization of floods in New Brunswick (Canada). *Stochastic Hydrology and Hydraulics* **12**, 65–82.

Embrechts, P. and C. M. Goldie (1982). On convolution tails. *Stochastic Process. Appl.* **13**, 263–278.

Embrechts, P., C. Klüppelberg and T. Mikosch (1997). *Modelling Extremal Events. For Insurance and Finance*. Springer-Verlag, Berlin.

Embrechts, P. and N. Veraverbeke (1982). Estimates for the probability of ruin with special emphasis on the possibility of large claims. *Insurance: Mathematical Economics* **1**, 55–72.

Erdös, P. and A. Rényi (1961). On a classical problem of probability theory. *Magyar Tud. Akad. Mat. Kutató Int. Közl.* **6**, 215–220.

EscalanteSandoval, C. (1998). Multivariate extreme value distribution with mixed Gumbel marginals. *J. Amer. Water Resources Assoc.* **34**, 321–333.

Escobar, L. A. and W. Q. Meeker, Jr. (1994). Fisher information matrix for the extreme value, normal and logistic distributions and censored data. *Appl. Statist.* **43**, 533–540.

Falk, M. (1989). Best attainable rate of joint convergence of extremes. In *Extreme Value Theory*, pp. 1–9. Lecture Notes in Statist., Vol. 51. Springer-Verlag, New York.

Falk, M., J. Hüsler and R.-D. Reiss (1994). *Laws of Small Numbers: Extremes and Rare Events*. Birkhäuser, Basel.

Falk, M. and F. Marohn (1993). Von Mises conditions revisited. *Ann. Probab.* **21**, 1310–1328.

Feeney, G. A. and P. K. Sen (1985). Extreme value theory for certain nonstationary sequences. *Austral. J. Statist.* **27**, 236–245.

Feller, W. (1971). *An Introduction to Probability Theory and Its Applications*, Vol. 2, 2nd ed. Wiley, New York.

Ferreira, H. (1994). Multivariate extreme values in T-periodic random sequences under mild oscillation restrictions. *Stochastic Process. Appl.* **49**, 111–125.

Fill, H. D. and J. R. Stedinger (1995). L moment and probability plot correlation coefficient goodness-of-fit tests for the Gumbel distribution and impact of autocorrelation. *Water Resources Res.* **31**, 225–229.

Finner, H. and M. Roters (1994). On the limit behavior of the joint distribution function of order-statistics. *Ann. Instit. Statist. Math.* **46**, 343–349.

Fisher, R. A. and L. H. C. Tippett (1928). Limiting forms of the frequency distribution of the largest or smallest member of a sample. *Proc. Cambridge Philos. Soc.* **24**, 180–190.

Galambos, J. (1982). A statistical test for extreme value distributions. In *Nonparametric Statistical Inference*, pp. 221–230 (Eds. B. V. Gnedenko, M. L. Puri and I. Vincze). North-Holland, Amsterdam.

Galambos, J. (1987). *The Asymptotic Theory of Extreme Order Statistics*, 2nd ed. Robert E. Krieger Publishing Company, Melbourne, Florida.

Galambos, J. (1995a). *Advanced Probability Theory*, 2nd ed. Probability: Pure and Applied, Vol. 10. Marcel Dekker, New York.

Galambos, J. (1995b). The development of the mathematical theory of extremes in the past half century. *Theory Probab. Appl.* **39**, 234–248.

Galambos, J. (1998). Univariate extreme value theory and applications. In *Order Statistics: Theory and Methods*, pp. 315–333. Handbook of Statistics, Vol. 16. North-Holland, Amsterdam.

Galambos, J. and S. Kotz (1978). *Characterizations of Probability Distributions*. Springer, Berlin.

Galambos, J. and A. Obretenov (1987). Restricted domains of attraction of $\exp(-\exp(-x))$. *Stochastic Process. Appl.* **25**, 265–271.

Galambos, J. and Y. Xu (1991). Regularly varying expected residual life and domains of attraction of extreme value distributions. *Annales Universitatis Scientiarum Budapestinensis de Rolando Eötvös Nominatae. Sectio Mathematica* **33**, 105–108.

Ghoudi, K., A. Khoudraji and L. P. Rivest (1998). Statistical properties of couplas of two-dimensional extrema. *Canad. J. Statist.* **26**, 187–197.

Gnedenko, B. (1943). Sur la distribution limite du terme maximum d'une série aléatoire. *Ann. of Math.* **44**, 423–453. Reprinted in *Breakthroughs in Statistics* Vol. 1 (Eds. S. Kotz and N. L. Johnson). Springer-Verlag, 1999.

Goldie, C. M. and R. A. Maller (1996). A point-process approach to almost-sure behaviour of record values and order statistics. *Adv. Appl. Probab.* **28**, 426–462.

Gomes, J. (1996). Extreme value theory for a thermal energy storage model. *Statist. Probab. Lett.* **30**, 25–31.

Gomes, M. I. (1984). Penultimate limiting forms in extreme value theory. *Ann. Instit. Statist. Math.* **36**, 71–85.

Gordon, L., M. F. Schilling and M. S. Waterman (1986). An extreme value theory for long head runs. *Probab. Theory Related Fields* **72**, 279–287.

Grimshaw, S. D. (1993). Computing maximum likelihood estimates for the generalized Pareto distribution. *Technometrics* **35**, 185–191.

Gumbel, E. J. (1962). Multivariate extremal distributions. *Bull. Internat. Statist. Inst.* **39**, 471–475.

Gumbel, E. J. and C. K. Mustafi (1967). Some analytical properties of bivariate extremal distributions. *J. Amer. Statist. Assoc.* **62**, 569–588.

Gusella, V. (1991). Estimation of extreme winds from short-term records. *J. Structural Engineering – American Society of Civil Engineers* **117**, 375–390.

De Haan, L. (1970). *On Regular Variation and Its Application to the Weak Convergence of Sample Extremes.* Mathematical Centre Tract, Vol. 32. Mathematisch Centrum, Amsterdam.

De Haan, L. (1976). Sample extremes: an elementary introduction. *Statistica Neerlandica* **30**, 161–172.

De Haan, L. (1978). A characterization of multidimensional extreme-value distributions. *Sankhyā A* **40**, 85–88.

De Haan, L. (1984). A spectral representation for max-stable processes. *Ann. Probab.* **12**, 1194–1204.

De Haan, L. (1985). Extremes in high dimensions: the model and some statistics (with discussion). In *Proceedings of the 45th Session of the International Statistical Institute*, Vol. 4, pp. 185–192 (Amsterdam, 1985). Bull. Internat. Statist. Inst., Vol. 51.

De Haan, L. and L. Peng (1997). Rates of convergence for bivariate extremes. *J. Multivariate Anal.* **61**, 195–230.

De Haan, L. and S. Resnick (1987). On regular variation of probability densities. *Stochastic Process. Appl.* **25**, 83–93.

De Haan, L. and S. Resnick (1996). Second-order regular variation and rates of convergence in extreme-value theory. *Ann. Probab.* **24**, 97–124.

De Haan, L., S. I. Resnick, H. Rootzén and C. G. de Vries (1989). Extremal behaviour of solutions to a stochastic difference equation with applications to ARCH processes. *Stochastic Process. Appl.* **32**, 213–224.

De Haan, L. and E. Verkade (1987). On extreme-value theory in the presence of a trend. *J. Appl. Probab.* **24**, 62–76.

Haiman, G., M. Kiki and M. L. Puri (1995). Extremes of Markov sequences. Extreme value theory and applications. *J. Statist. Plann. Inference* **45**, 185–201.

Hall, P. (1978). Representations and limit theorems for extreme value distributions. *J. Appl. Probab.* **15**, 639–644.

Hall, P. (1979). On the rate of convergence of normal extremes. *J. Appl. Probab.* **16**, 433–439.

Hall, P. (1980). Estimating probabilities for normal extremes. *Adv. Appl. Probab.* **12**, 491–500.

Hall, W. J. and J. A. Wellner (1979). The rate of convergence in law of the maximum of an exponential sample. *Statistica Neerlandica* **33**, 151–154.

Harlow, D. G., R. I. Smith and H. M. Taylor (1983). Lower tail analysis of the distribution of the strength of load-sharing systems. *J. Appl. Probab.* **20**, 358–367.

Hasofer, A. M. and Z. Wang (1992). A test for extreme value domain of attraction. *J. Amer. Statist. Assoc.* **87**, 171–177.

Hebbar, H. V. (1979). A law of the iterated logarithm for extreme values from Gaussian sequences. *Z. Wahrsch. Verw. Gebiete* **48**, 1–16.

Hill, B. M. (1975). A simple general approach to inference about the tail of a distribution. *Ann. Statist.* **3**, 1163–1174.

Hooghiemstra, G. and J. Hüsler (1996). A note on maxima of bivariate random vectors. *Statist. Probab. Lett.* **31**, 1–6.

Hooghiemstra, G. and H. P. Lopuhaä (1998). An extremal limit theorem for the Argmax process of Brownian motion minus a parabolic drift. *Extremes* **1**, 215–240.

Horowitz, J. (1980). Extreme values from a nonstationary stochastic process: an application to air quality analysis (with discussion). *Technometrics* **22**, 469–482.

Horváth, L. (1987). On the tail behaviour of quantile processes. *Stochastic Process. Appl.* **25**, 57–72.

Hosking, J. R. M. (1984). Testing whether the shape parameter is zero in the generalized extreme-value distribution. *Biometrika* **71**, 367–374.

Hosking, J. R. M., J. R. Wallis and E. F. Wood (1985). Estimation of the generalized extreme-value distribution by the method of probability-weighted moments. *Technometrics* **27**, 251–261.

Hougaard, P. (1986). A class of multivariate failure time distributions. *Biometrika* **73**, 671–678.

Hsing, T. (1986). Extreme value theory for suprema of random variables with regularly varying tail probabilities. *Stochastic Process. Appl.* **22**, 51–57.

Hsing, T. (1989). Extreme value theory for multivariate stationary sequences. *J. Multivariate Anal.* **29**, 274–291.

Hüsler, J. (1989). Limit properties for multivariate extreme values in sequences of independent, non-identically distributed random vectors. *Stochastic Process. Appl.* **31**, 105–116.

Hüsler, J. (1990a). Extreme values and high boundary crossings of locally stationary Gaussian processes. *Ann. Probab.* **18**, 1141–1158.

Hüsler, J. (1990b). Multivariate extreme values in stationary random sequences. *Stochastic Process. Appl.* **35**, 99–108.

Hüsler, J. (1995). A note on extreme values of locally stationary Gaussian processes. Extreme value theory and applications. *J. Statist. Plann. Inference* **45**, 203–213.

Hüsler, J. (1999). Extremes of a Gaussian process and the constant. *Extremes* **2**, 59–70.

Hüsler, J. and R.-D. Reiss (1989). Maxima of normal random vectors: Between independence and complete dependence. *Statist. Probab. Lett.* **7**, 283–286.

Iglehart, D. L. (1972). Extreme values in the $GI/G/1$ queue. *Ann. Math. Statist.* **43**, 627–635.

Joe, H. (1990). Families of min-stable multivariate exponential and multivariate extreme value distributions. *Statist. Probab. Lett.* **9**, 75–81.

Joe, H. (1994). Multivariate extreme value distributions with applications to environmental data. *Canad. J. Statist.* **22**, 47–64.

Joe, H. (1997). *Multivariate Models and Dependence Concepts*. Monographs on Statistics and Applied Probability, Vol. 73. Chapman and Hall, London.

Joe, H. and T. H. Hu (1996). Multivariate distributions from mixtures of max-infinitely divisible distributions. *J. Multivariate Anal.* **57**, 240–265.

Joe, H., R. L. Smith and I. Weissman (1992). Bivariate threshold methods for extremes. *J. Roy. Statist. Soc. B* **54**, 171–183.

Johnson, N. L. and S. Kotz (1972). *Distributions in Statistics: Continuous Multivariate Distributions*. Wiley, New York.

Johnson, N. L., S. Kotz and N. Balakrishnan (1995). *Continuous Univariate Distributions*, Vol. 2, 2nd ed. Wiley, New York.

Kamps, U. (1955). *A Concept of Generalized Order Statistics*. Teubner, Stuttgart.

Khan, A. H. and M. I. Beg (1987). Characterization of the Weibull distribution by conditional variance. *Sankhyā A* **49**, 268–271.

Kinnison, R. (1989). Correlation coefficient goodness-of-fit test for the extreme-value distribution. *Amer. Statist.* **43**, 98–100.

Klüppelberg, C. and A. May (1999). The dependence function for bivariate extreme value distributions – a systematic approach. Submitted.

Konstant, D. G. and V. I. Piterbarg (1993). Extreme values of the cyclostationary Gaussian random process. *J. Appl. Probab.* **30**, 82–97.

Korshunov, D. (1997). On distribution tail of the maximum of a random walk. *Stochastic Process. Appl.* **72**, 97–103.

Kotz, S. and S. Nadarajah (2000). *Extreme Value Distributions: Theory and Applications*. Imperial College Press, London.

Kotz, S. and D. N. Shanbhag (1980). Some new approaches to probability distributions. *Adv. Appl. Probab.* **12**, 903–921.

Kratz, M. F. and H. Rootzén (1997). On the rate of convergence for extremes of mean square differentiable stationary normal processes. *J. Appl. Probab.* **34**, 908–923.

Kuru, G. A., A. G. D. Whyte and R. C. Woollons (1992). Utility of reverse Weibull and extreme value density-functions to refine diameter distribution growth-estimates. *Forest Ecology and Management* **48**, 165–174.

Leadbetter, M. R. (1975). Aspects of extreme value theory for stationary processes – a survey. In *Stochastic Processes and Related Topics*, Vol. 1, pp. 101–132. Academic Press, New York.

Leadbetter, M. R. (1978). Extreme value theory under weak mixing conditions. In *Studies in Probability Theory*, pp. 46–110. MAA Studies in Mathematics, Vol. 18. Mathematical Association of America, Washington, DC.

Leadbetter, M. R., G. Lindgren and H. Rootzén (1983). *Extremes and Related Properties of Random Sequences and Processes*. Springer-Verlag, New York.

Leadbetter, M. R. and H. Rootzén (1982). Extreme value theory for continuous parameter stationary processes. *Z. Wahrsch. Verw. Gebiete* **60**, 1–20.

Leadbetter, M. R. and H. Rootzén (1988). Extremal theory for stochastic processes. *Ann. Probab.* **16**, 431–478.

Ledford, A. W. and J. A. Tawn (1996). Statistics for near independence in multivariate extreme values. *Biometrika* **83**, 169–187.

Ledford, A. W. and J. A. Tawn (1997). Modelling dependence within joint tail regions. *J. Roy. Statist. Soc. B* **59**, 475–499.

Lee, R. U. (1992). Statistical-analysis of corrosion failures of lead-sheathed cables. *Materials Performance* **31**, 20–23.

Lye, L. M., K. P. Hapuarachchi and S. Ryan (1993). Bayes estimation of the extreme-value reliability function. *IEEE Trans. on Reliability* **42**, 641–644.

Madsen, H., P. F. Rasmussen and D. Rosbjerg (1997). Comparison of annual maximum series and partial duration series methods for modeling extreme hydrologic events. 1. At-site modeling. *Water Resources Res.* **33**, 747–757.

Marco, J. M. and C. Ruiz-Rivas (1992). On the construction of multivariate distributions with given nonoverlapping multivariate marginals. *Statist. Probab. Lett.* **15**, 259–265.

Mardia, K. V. (1964). Asymptotic independence of bivariate extremes. *Calcutta Statist. Assoc. Bull.* **13**, 172–178.

Maritz, J. S. and A. H. Munro (1967). On the use of the generalized extreme-value distribution in estimating extreme percentiles. *Biometrics* **23**, 79–103.

Marohn, F. (1998a). An adaptive efficient test for Gumbel domain of attraction. *Scandinavian J. Statist.* **25**, 311–324.

Marohn, F. (1998b). Testing the Gumbel hypothesis via the pot-method. *Extremes* **1**, 191–213.

Marshall, A. W. and I. Olkin (1967). A multivariate exponential distribution. *J. Amer. Statist. Assoc.* **62**, 30–44.

Marshall, A. W. and I. Olkin (1983). Domains of attraction of multivariate extreme value distributions. *Ann. Probab.* **11**, 168–177.

McCormick, W. P. (1980). Weak convergence for the maxima of stationary Gaussian processes using random normalization. *Ann. Probab.* **8**, 483–497.

McCormick, W. P. and Y. S. Park (1992). Approximating the distribution of the maximum queue length for $M/M/s$ queues. In *Queuing and Related Models*, pp. 240–261. Oxford Statistical Science Series, Vol. 9. Oxford University Press, New York.

McFadden, D. (1978). Modelling the choice of residential location. In *Spatial Interaction Theory and Planning Models*, pp. 75–96 (Eds. A. Karlqvist, L. Lundquist, F. Snickers and J. Weibull). North-Holland, Amsterdam.

Mohan, N. R. and S. Ravi (1993). Max domains of attraction of univariate and multivariate p-max stable laws. *Theory Probab. Appl.* **37**, 632–642.

Mohan, N. R. and U. R. Subramanya (1991). Characterization of max domains of attraction of univariate p-max stable laws. In *Proceedings of the Symposium on Distribution Theory*, pp. 11–24. Publication No. 22. Centre for Mathematical Sciences, Thiruvananthapuram, Kerala State, India.

Mole, N., C. W. Anderson, S. Nadarajah and C. Wright (1995). A generalised Pareto distribution model for high concentrations in short-range atmospheric dispersion. *Environmetrics* **6**, 595–606.

Van Montfort, M. A. J. and J. Otten (1978). On testing a shape parameter in the presence of a location and a scale parameter. *Math. Oper. Statist. Ser. Statist.* **9**, 91–104.

Moothathu, T. S. K. (1990). A characterization property of Weibull, exponential and Pareto distributions. *Sankhyā* **28**, 69–74.

Morton, I. D., J. Bowers and G. Mould (1997). Estimating return period wave heights and wind speeds using a seasonal point process model. *Coastal Engineering* **31**, 305–326.

Mukhopadhyay, N. and M. E. Ekwo (1987). A note on minimum risk point estimation of the shape parameter of a Pareto distribution. *Calcutta Statist. Assoc. Bull.* **36**, 69–78.

Nadarajah, S. (1994). Multivariate extreme value methods with applications to reservoir flood safety. Ph.D. thesis, University of Sheffield.

Nadarajah, S. (1999a). A polynomial model for bivariate extreme value distributions. *Statist. Probab. Lett.* **42**, 15–25.

Nadarajah, S. (1999b). Multivariate extreme value distributions based on bivariate structures. *Unpublished Technical Note*.

Nadarajah, S. (1999c). Simulation of multivariate extreme values. *J. Statist. Comput. Simulation* **62**, 395–410.

Nadarajah, S. (2000). Approximations for bivariate extreme values. *Extremes* **3**, 87–98.

Nadarajah, S., C. W. Anderson and J. A. Tawn (1998). Ordered multivariate extremes. *J. Roy. Statist. Soc. B* **60**, 473–496.

Naess, A. (1998). Estimation of long return period design values for wind speeds. *J. Engineering Mechanics – American Society of Civil Engineers* **124**, 252–259.

Nagaraja, H. N. (1977). On a characterization based on record values. *Austral. J. Statist.* **16**, 70–73.

Nagaraja, H. N. (1988). Some characterizations of continuous distributions based on regressions of adjacent order statistics and record values. *Sankhyā A* **50**, 70–73.

Nair, K. A. (1981). Asymptotic distribution and moments of normal extremes. *Ann. Probab.* **9**, 150–153.

Nasri-Roudsari, D. (1996). Extreme value theory of generalized order statistics. *J. Statist. Plann. Inference* **55**, 281–297.

Nasri-Roudsari, D. and E. Cramer (1999). On the convergence rates of extreme generalized order statistics. *Extremes* **2**, 421–447.

Niu, X. F. (1997). Extreme value theory for a class of non-stationary time series with applications. *Ann. Appl. Probab.* **7**, 508–522.

Oakes, D. and A. K. Manatunga (1992). Fisher information for a bivariate extreme value distribution. *Biometrika* **79**, 827–832.

Obretenov, A. (1989). On the rate of convergence for the extreme value in the case of IFR-distributions. In *Stability Problems for Stochastic Models*, pp. 263–269. Lecture Notes in Math., Vol. 1412. Springer-Verlag, Berlin.

Obretenov, A. (1991). On the dependence function of Sibuya in multivariate extreme value theory. *J. Multivariate Anal.* **36**, 35–43.

O'Brien, G. L. (1987). Extreme values for stationary and Markov sequences. *Ann. Probab.* **15**, 281–291.

Omey, E. (1988). Rates of convergence for densities in extreme value theory. *Ann. Probab.* **16**, 479–486.

Omey, E. and S. T. Rachev (1988). On the rate of convergence in extreme value theory. *Theory Probab. Appl.* **33**, 601–607.

Omey, E. and S. T. Rachev (1991). Rates of convergence in multivariate extreme value theory. *J. Multivariate Anal.* **38**, 36–50.

Osella, A. M., N. C. Sabbione and D. C. Cernadas (1992). Statistical-analysis of seismic data from North-Western and Western Argentina. *Pure Appl. Geophys.* **139**, 277–292.

Öztürk, A. (1986). On the W test for the extreme value distribution. *Biometrika* **74**, 347–354.

Öztürk, A. and S. Korukoğlu (1988). A new test for the extreme value distribution. *Comm. Statist. Simulation and Comput.* **17**, 1375–1393.

Pakes, A. G. (1975). On the tails of waiting-time distribution. *J. Appl. Probab.* **12**, 555–564.

Pancheva, E. (1984). Limit theorems for extreme order statistics under nonlinear normalization. In *Stability Problems for Stochastic Models*, pp. 284–309. Lecture Notes in Math., Vol. 1155. Springer-Verlag, New York.

Park, Y., K. Kim and M. Jhun (1994). A law of large numbers for maxima in $M/M/\infty$ queues and INAR(1) processes. *J. Korean Statist. Soc.* **23**, 483–498.
Peng, L. (1999). Estimation of the coefficients of tail dependence in bivariate extremes. *Statist. Probab. Lett.* **43**, 399–409.
Perfekt, R. (1994). Extremal behaviour of stationary Markov chains with applications. *Ann. Appl. Probab.* **4**, 529–548.
Perfekt, R. (1997). Extreme value theory for a class of Markov chains with values in $\Re^d$. *Adv. Appl. Probab.* **29**, 138–164.
Pfeifer, D. (1989). *Einführung in die Extremwertstatistik*. Teubner Skripten zur Mathematischen Stochastik. B. G. Teubner, Stuttgart.
Pickands, J. (1967). Maxima of stationary Gaussian processes. *Z. Wahrsch. Verw. Gebiete* **7**, 190–223.
Pickands, J. (1968). Moment convergence of sample extremes. *Ann. Math. Statist.* **39**, 881–889.
Pickands, J. (1969a). An iterated logarithm law for the maximum in a stationary Gaussian sequence. *Z. Wahrsch. Verw. Gebiete* **12**, 344–353.
Pickands, J. (1969b). Asymptotic properties of the maximum in a stationary Gaussian process. *Trans. Amer. Math. Soc.* **145**, 75–86.
Pickands, J. (1971). The two-dimensional Poisson process and extremal processes. *J. Appl. Probab.* **8**, 745–756.
Pickands, J. (1975). Statistical inference using extreme order statistics. *Ann. Statist.* **3**, 119–131.
Pickands, J. (1981). Multivariate extreme value distributions (with discussion). In *Proceedings of the 43rd session of the International Statistical Institute*, Vol. 2, pp. 859–878, 894–902 (Buenos Aires, 1981). Bull. Internat. Statist. Inst., Vol. 49.
Pickands, J. (1986). The continuous and differentiable domains of attraction of the extreme value distributions. *Ann. Probab.* **14**, 996–1004.
Pierrat, J. C., F. Houllier, J. C. Hervé and R. Salas González (1995). Estimating the mean of the r largest values in a finite population: application to forest inventory. *Biometrics* **51**, 679–686.
Poskitt, D. S. and S.-H. Chung (1996). Markov chain models, time series analysis and extreme value theory. *Adv. Appl. Probab.* **28**, 405–425.
Prescott, P. and A. T. Walden (1980). Maximum likelihood estimation of the parameters of the generalized extreme-value distribution. *Biometrika* **67**, 723–724.
Qi, Y. C. (1998). Estimating extreme-value index from records. *Chinese Ann. of Math. B* **19**, 499–510.
Ramachandran, G. (1982). Properties of extreme order statistics and their application to fire protection and insurance problems. *Fire Safety J.* **5**, 59–76.
Rao, C. R. and D. N. Shanbhag (1994). *Choquet–Deny Type Functional Equations with Applications to Stochastic Models*. Wiley, Chichester.
Rao, N. M., P. P. Rao and K. L. Kaila (1997). The first and third asymptotic distributions of extremes as applied to the seismic source regions of India and adjacent areas. *Geophys. J. Internat.* **128**, 639–646.
Reiss, R.-D. (1993). *A Course on Point Processes*. Springer Series in Statistics. Springer-Verlag, New York.
Reiss, R.-D. and M. Thomas (1997). *Statistical Analysis of Extreme Values. From Insurance, Finance, Hydrology and Other Fields*. Birkhäuser, Basel.
Resnick, S. I. (1987). *Extreme Values, Regular Variation, and Point Processes*. Springer-Verlag, New York.
Révész, P. (1995). Strong theorems on the extreme values of stationary Poisson processes. Extreme value theory and applications. *J. Statist. Plann. Inference* **45**, 291–300.
Robinson, M. E. and J. A. Tawn (1997). Statistics for extreme sea currents. *Appl. Statist.* **46**, 183–205.
Rolski, T., H. Schmidli, V. Schmidt and J. Teugels (1999). *Stochastic Processes for Insurance and Finance*. Wiley, Chichester.
Rootzén, H. (1978). Extremes of moving averages of stable processes. *Ann. Probab.* **6**, 847–869.
Rootzén, H. (1983). The rate of convergence of extremes of stationary normal sequences. *Adv. Appl. Probab.* **15**, 54–80.
Rootzén, H. (1986). Extreme value theory for moving average processes. *Ann. Probab.* **14**, 612–652.
Rootzén, H. (1987). A ratio limit theorem for the tails of weighted sums. *Ann. Probab.* **15**, 728–747.
Rootzén, H. (1988). Maxima and exceedances of stationary Markov chains. *Adv. Appl. Probab.* **20**, 371–390.
Sastry, S. and J. L. Pi (1991). Estimating the minimum of partitioning and floorplanning problems. *IEEE Trans. on Computer-Aided Design of Integrated Circuits and Systems* **10**, 273–282.

Scarf, P. A. (1992). Estimation for a four parameter generalized extreme value distribution. *Comm. Statist. Theory Methods* **21**, 2185–2201.

Scarf, P. A., R. A. Cottis and P. J. Laycock (1992). Extrapolation of extreme pit depths in space and time using the r-deepest pit depths. *J. Electrochemical Society* **139**, 2621–2627.

Seki, T. and S. Yokoyama (1996). Robust parameter-estimation using the bootstrap method for the 2-parameter Weibull distribution. *IEEE Trans. on Reliability* **45**, 34–41.

Seleznjev, O. (1996). Large deviations in the piecewise linear approximation of Gaussian processes with stationary increments. *Adv. Appl. Probab.* **28**, 481–499.

Serfozo, R. (1982). Functional limit theorems for extreme values of arrays of independent random variables. *Ann. Probab.* **10**, 172–177.

Serfozo, R. (1988a). Extreme values of birth and death processes and queues. *Stochastic Process. Appl.* **27**, 291–306.

Serfozo, R. (1988b). Extreme values of queue lengths in $M/G/1$ and $GI/M/1$ systems. *Math. Oper. Res.* **13**, 349–357.

Shao, Q.-M. (1996). Bounds and estimators of a basic constant in extreme value theory of Gaussian processes. *Statistica Sinica* **6**, 245–257.

Shapiro, S. S. and C. W. Brain (1987a). Some new tests for the Weibull and extreme value distributions. In *Goodness-of-fit*, pp. 511–527. Colloquia Mathematica Societatis János Bolyai, Vol. 45. North-Holland, Amsterdam.

Shapiro, S. S. and C. W. Brain (1987b). W-test for the Weibull distribution. *Comm. Statist. Simulation and Comput.* **16**, 209–219.

Shi, D. (1995a). Fisher information for a multivariate extreme value distribution. *Biometrika* **82**, 644–649.

Shi, D. (1995b). Moment estimation for multivariate extreme value distributions. *Appl. Math. J. Chinese Universities B* **10**, 61–68.

Shi, D., R. L. Smith and S. G. Coles (1993). Joint versus marginal estimation for bivariate extremes. *Unpublished Technical Report*.

Shi, D. and S. Zhou (1999). Moment estimation for multivariate extreme value distribution in a nested logistic model. *Ann. Inst. Statist. Math.* **51**, 253–264.

Silbergleit, V. M. (1996). On the occurrence of geomagnetic storms with sudden commencements. *J. Geomagnetism and Geoelectricity* **48**, 1011–1016.

Singh, N. P. (1987). Estimation of Gumbel distribution parameters by joint distribution of m extremes. *Calcutta Statist. Assoc. Bull.* **36**, 101–104.

Smith, R. L. (1982). Uniform rates of convergence in extreme value theory. *Adv. Appl. Probab.* **14**, 600–622.

Smith, R. L. (1985a). Maximum likelihood estimation in a class of non-regular cases. *Biometrika* **72**, 67–90.

Smith, R. L. (1985b). Statistics of extreme values (with discussion). In *Proceedings of the 45th session of the International Statistical Institute*, Vol. 4, pp. 185–192 (Amsterdam, 1985). Bulletin of the Institute of International Statistics, Vol. 51.

Smith, R. L. (1986). Extreme value theory based on the r largest annual events. *J. Hydrology* **86**, 27–43.

Smith, R. L. (1987). Estimating tails of probability distributions. *Ann. Statist.* **15**, 1174–1207.

Smith, R. L. (1989). Extreme value analysis of environmental time series: an application to trend detection in ground-level zone. *Statist. Sci.* **4**, 367–393.

Smith, R. L. (1990). Extreme value theory. In *Handbook of Applicable Mathematics*, Vol. 7, pp. 437–471 (Ed. W. Ledermann). Wiley, Chichester.

Smith, R. L. (1991). Regional estimation from spatially dependent data. *Technical Report*.

Sparaggis, P. D., D. Towsley and C. G. Cassandras (1993). Extremal properties of the shortest/longest nonfull queue policies in finite-capacity systems with state-dependent service rates. *J. Appl. Probab.* **30**, 223–236.

Srivastava, O. P. (1967). Asymptotic independence of certain statistics connected with the extreme order statistics in a bivariate distribution. *Sankhyā A* **29**, 175–182.

Steinebach, J. and V. R. Eastwood (1996). Extreme value asymptotics for multivariate renewal processes. *J. Multivariate Anal.* **56**, 284–302.

Stephens, M. A. (1977). Goodness of fit for the extreme value distribution. *Biometrika* **64**, 583–588.

Suzuki, M. and Y. Ozaka (1994). Seismic risk analysis based on strain-energy accumulation in focal region. *J. Research of the National Institute of Standards and Technology* **99**, 421–434.

Sweeting, T. J. (1985). On domains of uniform local attraction in extreme value theory. *Ann. Probab.* **13**, 196–205.
Takahashi, R. (1987). Some properties of multivariate extreme value distributions and multivariate tail equivalence. *Ann. Inst. Statist. Math. A* **39**, 637–647.
Takahashi, R. (1994a). Asymptotic independence and perfect dependence of vector components of multivariate extreme statistics. *Statist. Probab. Lett.* **19**, 19–26.
Takahashi, R. (1994b). Domains of attraction of multivariate extreme value distributions. *J. Research of the National Institute of Standards and Technology* **99**, 551–554.
Takahashi, R. and M. Sibuya (1998). Prediction of the maximum size in Wicksell's corpuscle problem. *Ann. Inst. Statist. Math.* **50**, 361–377.
Tawn, J. A. (1988a). An extreme value theory model for dependent observations. *J. Hydrology* **101**, 227–250.
Tawn, J. A. (1988b). Bivariate extreme value theory: Models and estimation. *Biometrika* **75**, 397–415.
Tawn, J. A. (1990). Modelling multivariate extreme value distributions. *Biometrika* **77**, 245–253.
Tiago de Oliveira, J. (1962). La représentation des distributions extrémales bivariées. *Bull. Inst. Internat. Statist.* **39**, 477–480.
Tiago de Oliveira, J. (1980). Bivariate extremes: Foundations and statistics. In *Multivariate Analysis*, Vol. V, pp. 349–366. North-Holland, Amsterdam.
Tiago de Oliveira, J. and M. I. Gomes (1984). Two test statistics for choice of univariate extreme models. In *Statistical Extremes and Applications*, pp. 651–668. NATO Science Series C: Mathematical and Physical Sciences, Vol. 131. Reidel, Dordrecht.
Tiku, M. L. and M. Singh (1981). Testing the two parameter Weibull distribution. *Comm. Statist. Theory Methods* **10**, 907–918.
Tsimplis, M. N. and D. Blackman (1997). Extreme sea-level distribution and return periods in the Aegean and Ionian seas. *Estuarine Coastal and Shelf Science* **44**, 79–89.
Tsujitani, M., H. Ohta and S. Kase (1980). Goodness-of-fit test for extreme-value distribution. *IEEE Trans. on Reliability* **29**, 151–153.
Turkman, K. F. and M. F. Oliveira (1992). Limit laws for the maxima of chain-dependent sequences with positive extremal index. *J. Appl. Probab.* **29**, 222–227.
Turkman, K. F. and A. M. Walker (1983). Limit laws for the maxima of a class of quasi stationary sequences. *J. Appl. Probab.* **20**, 814–821.
Umbach, D. and M. Ali (1996). Conservative spacings for the extreme value and Weibull distributions. *Calcutta Statist. Assoc. Bull.* **46**, 169–180.
Veraverbeke, N. (1977). Asymptotic behavior of Weiner Hopf factors of a random walk. *Stochastic Process. Appl.* **5**, 27–37.
De Waal, D. J. (1996). Goodness of fit of the generalized extreme value distribution based on the Kullback–Leibler information. *South African Statist. J.* **30**, 139–153.
Walshaw, D. (1993). An application in extreme value theory of nonregular maximum likelihood estimation. *Theory Probab. Appl.* **37**, 182–184.
Wang, J. Z. (1995). Selection of the k largest order statistics for the domain of attraction of the Gumbel distribution. *J. Amer. Statist. Assoc.* **90**, 1055–1061.
Weissman, I. (1978). Estimation of parameters and large quantiles based on the k largest observations. *J. Amer. Statist. Assoc.* **73**, 812–815.
Welsch, R. E. (1973). A convergence theorem for extreme values from Gaussian sequences. *Ann. Probab.* **1**, 398–404.
Wilks, S. S. (1962). *Mathematical Statistics*. Wiley, New York.
Wiśniewski, M. (1997). Extremes in multivariate mixing sequences. *Demonstratio Math.* **30**, 661–669.
Xapson, M. A., G. P. Summers and E. A. Barke (1998). Extreme value analysis of solar energetic motion peak fluxes. *Solar Phys.* **183**, 157–164.
Xie, S. R. (1984). Extreme values in a stationary Gaussian sequence. *Chinese Ann. of Math. A* **5**, 287–290.
Yasuda, T. and N. Mori (1997). Occurrence properties of giant freak waves in sea area around Japan. *J. Waterway Port Coastal and Ocean Engineering – American Society of Civil Engineers* **123**, 209–213.
Yun, S. (1997). On domains of attraction of multivariate extreme value distributions under absolute continuity. *J. Multivariate Anal.* **63**, 277–295.

Zelterman, D. (1992). A statistical distribution with an unbounded hazard function and its application to a theory from demography. *Biometrics* **48**, 807–818.

Zelterman, D. (1993). A semiparametric bootstrap technique for simulating extreme order statistics. *J. Amer. Statist. Assoc.* **88**, 422, 477–485.

Zempléni, A. (1996). Inference for generalized extreme value distributions. *J. Appl. Statist. Sci.* **4**, 107–122.

Zolotarev, V. M. and S. T. Rachev (1985). Rate of convergence in limit theorems for the max-scheme. In *Stability Problems for Stochastic Models*, pp. 415–442. Lecture Notes in Math., Vol. 1155. Springer-Verlag, New York.

Biological Applications of Branching Processes

Anthony G. Pakes

0. Introduction

A branching process is a stochastic model which describes the growth of a population in terms of reproduction of individual members of the population. The key assumption is that disjoint subsets of the population fluctuate independently of each other. If the population comprises reproducing asexual objects, then these objects live and reproduce independently of each other. This simple idea allows for a wide variety of structural features, such as detailed life histories, objects of different types and hence growth dynamics, growth in discrete or continuous time, and so on. It is assumed that the reader is acquainted with the basic *theory* of branching processes as presented, for example, in the chapter by K. B. Athreya in the companion Handbook volume. Standard monographs on the subject are Asmussen and Hering (1983), Athreya and Ney (1972), Harris (1963) and Sevast'yanov (1974). Comprehensive bibliographies on theory up to about 1992 comprise the union of Harris (1963), Athreya and Ney (1972), Sevast'yanov (1974), Vatutin and Zubkov (1987, 1993), and Sankaranarayanan (1989). Recent additions to the mathematical corpus are Le Gall (1999) and Rahimov (1995).

Branching process models occur in an amazing variety of applications. Here we accept a rather liberal interpretation of the term 'application'. It certainly embraces modelling involving branching process structures designed to further genuine understanding in a scientific situation of interest. It also admits the analysis of mathematical models constructed from a putative 'real' situation, but which may be very much removed from this situation in terms of a faithful portrayal of it. A third class of applications is branching process constructions which are used to gain greater understanding of other mathematical structures such as critical phenomena in percolation.

The questions addressed by the pioneers of the subject arose from demography and genetics. What are the chances that a family line becomes extinguished? What is the chance of survival of a genetic mutation which arises spontaneously in a large population? We will follow this historical precedent by restricting the coverage to biological applications. These are addressed in the form of review essays under the four major headings: (1) History, surnames and sex; (2) Genetics and evolution; (3) Epidemic modelling; and (4) Ecology and conservation modelling. Lack of temporal resourcing has forced the omission of applications in demography, genealogy, and microbiology.

Some of this is treated in the literature. Jagers (1975) and Keiding (1973) describe branching process models for continuous-time demography. Jagers (1991, 1995) are general accounts of branching population dynamics. Jagers (1975) contains a chapter on cell kinetics, and Alexandersson (1999), and Yakovlev and Yanev (1989), are two recent and more complete treatments. Macken and Perelson (1985) is concerned with polymer formation and cellular aggregation, and Macken and Perelson (1989) examines stem cell kinetics. Examples from cell biology, evolution and molecular genetics are treated by Kimmel and Axelrod (2002), and see Durrett (2002) for the last topic.

The non-biological applications which are not considered in this chapter include the following topics:

- Computer Science: Tree structures and tree algorithms.
- Physics and Chemistry: Particle counters, hadron jets, transport processes, polymer structures, chain reactions.
- Engineering: Coding, cracks and breakage mechanisms in materials.
- Queues and storage.
- Economics and finance.
- Statistics: Lagrange distributions, integer-valued time series.
- Mathematics: Branching constructions in random walks, Brownian motion and percolation, fractal constructions, tree indexed stochastic processes, interacting particle systems, superprocesses.

Much of the applications literature is found outside of mathematics and probability journals, although (Harris, 1963) is quite complete up till near the time of publication. There are quite good reviews for some of the above topics. Computer science applications are surveyed by Devroye (1998) and Mahmoud (1992). Pitman (1998) surveys some aspects of Lagrange distributions, random walks and random forests. Pakes (1996) addresses the continuous version of Lagrange distributions. Johnson et al. (1992) survey Lagrange distributions from a more statistical point of view. Abraham and Mazliak (1998) address connections with Brownian motion. For branching process constructions in: Percolation theory, see Grimmett (1989); Interacting particle systems, see Liggett (1999); Super-processes, see Dawson (1993), Dynkin (1994) and Etheridge (2000); Trees, see Chauvin et al. (1996). Finally, current directions of research with some emphasis on applications are indicated in the edited volumes (Athreya and Jagers, 1997), and (Heyde, 1995).

Another substantial area of activity not discussed here is concerned with statistical inference for branching processes. In addition to its intrinsic interest, this work is historically important because it led to extensions of classical likelihood theory to more general classes of stochastic processes, and to the notions of ergodic and non-ergodic processes. See Guttorp (1991) and Dion (1991).

The author is keen to assemble a bibliography of applications which is as complete as possible, and hence he invites readers to send him citations, reports, and reprints.

Some notation is standardized throughout nearly all of this chapter. Deviations from standard usage will be sign-posted. Thus Z_n denotes the size at discrete time n of a simple (Bienaymé–Galton–Watson) branching process, and Z_t denotes the size at time t of any continuous time analogue. On occasion these notations may be used for the corresponding continuous state versions. The offspring distribution, or law, of a one-type process is denoted by $\{p_j;\ j = 0, 1, \ldots\}$, and it has the probability generating function (PGF) $f(s)$. Abbreviations such as PGF will be kept to a minimum. We always use $m = f'(1-)$ for the per-capita mean number of offspring. Vector (or matrix) notation will be used to denote corresponding quantities for multitype branching processes. Thus $\vec{Z}_n$ denotes the vector of population sizes at time n, and $f_i(\vec{s})$ is the PGF of the offspring law of an i-type individual.

1. History, surnames, and sex

1.1. Some history

The simple branching process model has independently arisen several times since 1845 in attempts to determine the influence of chance on the fluctuations in sizes of successive generations of family lines. The most thorough summary of this history is given in a paper by K. Albertsen which has been translated from the Danish and annotated by Guttorp (1995). The main points are as follows.

It had been long noted that family names tend to disappear, and 'reduced fertility' was commonly used to explain this disappearance. During the 19th century some scholars suggested that chance events influence the sizes of generations, and hence that a mathematical approach to the extinction of names ought be feasible. I. J. Bienaymé first did this in 1845, in a brief descriptive paper which was lost to the scholarly world until 1972. See Heyde and Seneta (1977), and Seneta (1998), for a thorough account of Bienaymé's life and work, and Seneta (1979) on the re-discoverey of Bienaymé's paper. The French text of this paper is reproduced at the end of Kendall (1975). In relation to the probability of extinction of the male family line descended from a single ancestor, Bienaymé writes that

> If ... the mean of the number of male children who replace the number of males of the preceding generation were less than unity, it would be easily realized that the families are dying out due to the disappearance of the members of which they are composed. However, the analysis shows further that when this mean is equal to unity families tend to disappear, although less rapidly...
>
> The analysis also shows clearly that if the mean ratio is greater than unity, the probability of extinction of families with the passing of time no longer reduces to certainty. It only approaches a finite limit, which is fairly simple to calculate and which has the singular characteristic of being given by one of the roots of the equation (in which the number of generations is made infinite) which is not relevant to the question when the mean ratio is less than unity.

This is a clear and correct statement of the fundamental criticality theorem. Let $f(s) = p_0 + p_1 s + p_2 s^2 + \cdots$ denote the probability generating function (PGF) of the offspring distribution, and $m = \sum j p_j$ be the per-capita mean number of offspring. For a family tree descended from a single ancestor, the general formulation of the criticality theorem is that the probability of extinction q equals unity iff $m \leqslant 1$. If $m > 1$ then q is the smallest solution in $[0, 1)$ of the critical equation

$$f(s) = s. \tag{1}$$

Bienaymé's paper ends with:

> M. Bienaymé develops various other notions which the elements of the question have suggested to him, and which he proposes to publish in the near future in a special paper.

No such paper has ever been found, and hence Bienaymé's reasoning is not known. This probably is contained in a passage from Cournot (1847), a friend of Bienaymé. See Bru et al. (1992). Cournot describes a game in which a gambler pays one unit to buy a ticket which returns j units with probability p_j, $j = 0, 1, 2$. Winnings are used to buy similar tickets until the gambler is ruined, which occurs with probability q. This solves (1), in this case a quadratic equation, and it always has the solution $s = 1$. The second solution ≥ 1 if $m \leq 1$, and hence $q = 1$. Cournot says that when $m > 1$ the smaller solution, $s = p_0/p_2$, has to be chosen as q because the probability of ruin can be made arbitrarily small by choosing p_0 to be small. This reasoning is not correct – let $p_2 = (1 + \varepsilon)p_0$ – but it gives the correct choice. Cournot goes on to note that when $f(s)$ is a polynomial, then the critical equation still has a unit solution, and another positive solution which is > 1 if $m < 1$, and which is < 1 if $m > 1$. His discussion ends with the footnote

> The problem given in the text is in essence the same as that which has as its aim the determination of the time of survival of males or families, the problem with which M. Bienaymé is concerned.

Independently of this, in 1873 A. de Candolle drew attention to the desirability of a mathematical approach to the problem of family extinction. F. Galton had for many years been interested in quantitative aspects of heredity, and hence was receptive to de Candolle's challenge. Galton rendered the problem in precise mathematical terms and published it in the Educational Times as Problem 4001:

> A large nation, of whom we will only concern ourselves with adult males, N in number, and who each bear separate surnames colonise a district. Their law of population is such that, in each generation, a_0 percent of the adult males have no male children who reach adult life; a_1 have one such child; a_2 have two, and so on up to a_5, who have five.
> Find (1) what proportion of the surnames will have become extinct after r generations; and (2) how many instances there will be of the same surname being held by j persons.

As no correct solution was submitted, Galton asked his friend H. W. Watson (mathematician, rector, and mountaineer (Kendall, 1966)) for help. Watson made considerable progress, publishing his work in the Educational Times later in 1873. Along with preliminary remarks by Galton, this was formally published as (Galton and Watson, 1874). Watson fully answers Galton's specific questions ((1) and (2) in the quotation). It is worthwhile summarizing his contribution. His notation differs from ours.

Watson assumes $f(\cdot)$ is a polynomial, and he clearly realizes that the solution of Galton's problem is determined by the behaviour of its functional iterates, defined by

$$f_0(s) = s, \qquad f_n(s) = f(f_{n-1}(s)) \quad (n \geq 1).$$

He states that the probability $p_{ij}^{(n)}$ that a surname with i representatives in a generation will have j representatives n generations later is the coefficient of s^j in $(f_n(s))^i$. This answers (2).

He addresses (1) using two examples. For an initial large male population of size N and no surnames alike, the expected number of names extinct by time n is $Nf_n(0)$. Watson computes representative numbers for two specific offspring laws. The first is $f(s) = (1 + s + s^2)/3$, for which $m = 1$, and the second is $f(s) = [(3 + s)/4]^5$ for which $m = 1.25$. He notes the rapid disappearance of names in each case, but with a slower rate for the supercritical example.

Watson then proceeds to his famous partial error concerning the value of q. He correctly asserts that whenever $f(s)$ has the binomial form

$$f(s) = \left(\frac{a+bs}{a+b}\right)^v, \tag{2}$$

then

$$f_n(0) = \left(\frac{a+bf_{n-1}(0)}{a+b}\right)^v,$$

and "as n increases indefinitely the value of $f_n(0)$ approaches indefinitely to the value q where

$$q = \left(\frac{a+bq}{a+b}\right)^v$$

that is where $q = 1$". (Emphases added.) He continues (notation altered in parts):

> All the surnames, therefore, tend to extinction in an indefinite time, and this result might have been anticipated generally, for a surname once lost can never be recovered, and there is an additional chance of loss in every successive generation. This result must not be confounded with that of the extinction of the male population; for in every binomial case where $v > 2$, we have $m > 1$, and, therefore an indefinite increase of male population. The true interpretation is that each of the quantities, $p_{11}^{(n)}$, $p_{12}^{(n)}$ etc. tends to become zero, as n is indefinitely increased, but that it does not follow that the product of each by the infinitely large number N is also zero.

Watson proceeds to a long elaboration of the last paragraph. His incorrect conclusion about the value of q when $m > 1$ is thus erroneously reconciled to his intuition that q ought to be less than unity. His work ends with a clear statement as to its restricted generality:

> We are not in a position to assert from *actual calculation* that a corresponding result is true for every form of $f(\cdot)$, but the reasonable inference is that such is the case, seeing that it holds whenever $f(s)$ can be compared with (2) whatever a, b, or v may be.

We may speculate that Watson's failure to perceive the rôle of the second root of the critical equation (1) in the supercritical case emanates from his choice of example – it is perhaps too elaborate for this purpose. Experimentation with other cases of his quadratic example might have lead him to a fully correct solution.

Albertsen's paper (Guttorp, 1995) mentions other scholars who discussed the surname problem after 1874. But the criticality theorem does not re-appear until

Haldane (1927) clearly asserts the supercritical part of it, and with no restriction on the form of $f(\cdot)$. His motivation for considering the problem is an interest in the survival of a rare mutation which has a small selective advantage over the wild type genotype. He refers only to an earlier discussion of Fisher (1922), and he appeals to results about functional iteration.

The first general statement and proof of the criticality theorem arose in circumstances similar to, but independently of, the Galton–Watson collaboration. The Danish pioneer of queueing and teletraffic theory, A. K. Erlang, himself the last scion of one line of the distinguished Krarup family, posed the following problem in a 1929 issue of the Matematisk Tidsskrift:

> When the probability that a person has n children is a_n, where $a_0 + a_1 + a_2 + \cdots = 1$, find the probability that his family dies out.

Erlang obtained a partial and unpublished solution, having died soon after publishing the problem. His manuscript apparently was used by J. F. Steffensen to arrive at a complete treatment of the problem which he published in 1930. See Guttorp (1995) for an English translation. Steffensen highlights assumptions made to reach his solution, such as homogeneity of individual reproduction laws and independence of family lines. These and much else are discussed in a subsequent paper (Steffensen, 1932, English translation available from the author), by which time Steffensen had learned of the Galton–Watson work.

Steffensen's complete formulation and proof of the criticality theorem marks the end of the early development of branching process theory. Consequently, we shall not further pursue its historical development, except to remark that C. M. Christensen submitted a complete solution to the Matematisk Tidsskrift early in 1930. It was not published because it arrived a little after Steffensen's article had been sent to the printer. Christensen's manuscript is reproduced in Guttorp (1995).

1.2. Extinction of surnames

Despite the fact that the extinction of surnames was the prime motivation for inventing branching processes, there is little published application to surname data. One explanation is the restricted circumstances under which the simple branching model is applicable to human data. The simple model may be applicable if interactions induced by varying types of mating are ignored, and if nubile females are numerous enough not to limit marriage of males. It has proven difficult, however, to include sex and mating in branching models – see Section 1.3.

Steffensen (1932) refers to calculations by Lotka (1931a) of the probability of extinction for American males, based on data from the U.S. Census of 1920. See Pollard (1973) for a summary. Lotka found that $\hat{q} = 0.8797$. Later (Lotka, 1931b) he found that the male offspring law is well fitted by a zero-modified geometric law, yielding the revised value $\hat{q} = 0.819$.

A substantial obstacle in applying the theory is that a census counts only the number of offspring at the time it is held. These data are published as the numbers of women who have 0, 1, ... children of each sex. Thus offspring numbers must be reconstructed from this data. Keyfitz (1968) (also Keyfitz and Tyree, 1967) shows how census data for female lines can be used to estimate polynomial forms of $f(\cdot)$. Data from 1960 are thus used to estimate q for several countries.

The use of surnames and population data as biological markers of inbreeding and migration is a highly developed subtopic of population genetics – see Gottlieb (1983) and Lasker (1985). Quantitative studies in this area are based on Wright's theory of inbreeding and models for selectively neutral mutations. See Yasuda et al. (1974) for use of the neutral model. Just one reference to branching processes is to be found in Gottlieb (1983) (p. 263), and then only for historical completeness. However, Yasuda et al. (1974) analyse an extensive set of name data from the upper Parma valley using both neutral mutation and branching process modelling. A monograph summary may be found in Chapter 4 of Cavalli-Sforza and Feldman (1981).

Fox and Lasker (1983) present data obtained from the number of marriages over a twelve month period in 1972/3 for a small region in the UK. Their data gave the observed proportion $\hat{p}(j)$ of surnames having j representatives, to which they were able to fit a discrete Pareto (or zeta function) law. They found that plots of $\log \hat{p}(j)$ against $\log j$ are substantially linear, hence giving support to their somewhat arbitrary decision to use the discrete Pareto law.

Consul (1991) obtained closer fits using the Geeta (two-parameter) family of laws. He tries to justify using these laws by supposing that one individual or family colonizes a district. Growing prosperity of the founder(s) attracts employees and their families, which comprise a first generation of surnames. In turn these family units attract new arrivals, which comprise the second generation of surnames, and so on. Thus the population of surnames comprise the generations of a simple branching process, and the total number of surnames equals the total progeny $Y = \sum_{j \geq 0} Z_j$. Hence the proposed law for the distribution of numbers of surnames is the basic Lagrange law (see Johnson et al., 1992) corresponding to the offspring law. Consul (1991) arbitrarily chooses this to be a negative binomial law, with PGF $[(1-p)/(1-ps)]^{\beta-1}$, where $0 < p < 1$ and $\beta > 1$. This yields

$$P(Y = j) = \frac{(\beta j - 1)_{(j-1)}}{j!} p^{j-1}(1-p)^{\beta j - 1} \quad (j \geq 1).$$

Expected surname frequencies are obtained from maximum likelihood estimates of the parameters. The resulting chi-squared distances are uniformly smaller than those for Fox and Lasker's Pareto fits. Of course, Consul's modelling clearly has arbitrary elements, chosen perhaps to ensure the Lagrange family appears on stage.

Finally, we mention that Islam (1995) independently has used the same model to compare fits of the log series, discrete Pareto, and Geeta laws using surnames from husbands of consanguineous marriages for the years 1930–1959 in 11 Sardinian dioceses.

1.3. Bisexual branching processes

Accounting for sexual reproduction in population models has proven to be a difficult task. Deterministic models were first advanced in the late 1940's. See Pollard (1973) for a review. An important simplifying idea due to D. G. Kendall is that the number of reproductive units is a deterministic function of nubile females and males. This was incorporated into a discrete time branching model much later by Daley (1968). Let F_n and M_n be the number of females and males in the nth generation. These assort themselves in some fashion to produce $Z_n = \zeta(F_n, M_n)$ mating units which independently spawn female and male offspring according to a discrete bivariate law whose PGF is $E(s_1^F s_2^M) = f(s_1, s_2)$. For human populations a realistic simplification is to set

$$f(s_1, s_2) = f(\alpha s_1 + (1 - \alpha)s_2), \tag{1}$$

where the PGF $f(s) = \sum p_j s^j$ governs total number of children of a mating unit and α is the probability of a given offspring being female.

It is evident that $\{Z_n\}$ comprises a Markov chain whose properties will depend crucially on assumptions about the form of the mating function $\zeta(x, y)$. Natural assumptions are that $\zeta(x, 0) = \zeta(0, y) \equiv 0$, and that $\zeta(x, y)$ should be non-decreasing in each argument. The first ensures that the zero state is absorbing for the numbers of mating units. Particular cases are (i) completely promiscuous mating where all females reproduce provided at least one male exists, $\zeta(x, y) = xI(y \geq 1)$, and (ii) polygamy with perfect fidelity where there is a positive integer d such that $\zeta(x, y) = \min(x, yd)$, i.e., each male mates with an exclusive harem of d females. Obviously, $d = 1$ corresponds to monogamous mating, the case most applicable to humans.

Let q_i denote the probability of extinction when $Z_0 = i$. A peculiarity of these models is that the positive states can be reducible in the sense that there is a subset A such that $q_i = 1$ for each $i \in A$, irrespective of the values of growth rates of the sexes. For example, if $d = 1$ in case (ii) and $p_0 + p_1 + p_2 + p_3 = 1$ in (1) then always $q_1 = 1$, but $q_i < 1$ is possible for larger values of i. However, subject to the exclusion of cases like this, $q_i \equiv 1$ iff the mean reproduction of one of the sexes is small, $E(F) \leq 1$ in case (i), and $\min(E(F), E(M)) \leq 1$ in case (ii). See Daley (1968). Karlin and Kaplan (1973) discuss more general extinction criteria for other life-length laws in the case of monogamous mating.

General drift criteria for Markov chains suggest that $q_i \equiv 1$ if $m_i \equiv E(Z_1|Z_0 = i)/i$ is finite for all i and bounded above by unity for all sufficiently large i. Bruss (1984) confirms this. Under a further condition he gives a rather complicated necessary and sufficient condition.

More general results involving only the m_i require some restriction on the nature of the mating function. Hull (1982) introduces a superadditivity condition on the mating function:

$$\zeta(x_1 + x_2, y_1 + y_2) \geq \zeta(x_1, y_1) + \zeta(x_2, y_2).$$

This asserts that the number of allowable mating units is bounded below by the number which can be formed by banning matings between the females and males produced by different parental mating units. Daley et al. (1986) show in this case that $m = \lim_{i \to \infty} m_i$ exists, and that $q_i \equiv 1$ iff $m \leq 1$. Under a further restriction on the mating function Hull (1993) shows when $m > 1$ that $q_i < 1$ iff $P(Z_{n+1} = i | Z_n = i) > 0$.

Questions which remain are how can q_i be computed when it is not equal to unity, and the nature of its asymptotic dependence on i. Daley et al. (1986) derive computable bounds and give some numerical results for the case $\alpha = 1/2$ and Poisson offspring laws, in which case the reproduction of males and females is independent. Their computations compare extinction probabilities for three mating schemes, showing that extinction for monogamous mating is more probable than for promiscuous mating, and this is more probable than for virgin birth (defined by $\zeta(x, y) = x$, giving a simple branching process for female numbers). In addition, dividing the extinction probability for promiscuous mating by that for virgin birthing, gives a ratio which appears to converge as the initial generation size increases. Alsmeyer and Rösler (1996) investigate this in a slightly more general setting. Their results suggest that the behaviour of this ratio depends on the order relation of the probability of zero male offspring and the corresponding probability for females, and the probability of extinction of the female population. Numerical results show a damped oscillatory behaviour of the ratio, and the authors doubt that this vanishes in the limit, and they prove this in Alsmeyer and Rösler (2002a, 2002b).

As for the simple branching process, the components of the bisexual process grow geometrically fast with positive probability when $m > 1$ and additional moment assumptions are satisfied. Suppose the mating function is superadditive, and let $\varepsilon(i) = m - m_i$. Expanding on earlier work, Gonzalez and Molina (1996) show that if $\sum \varepsilon(i)/i < \infty$ and certain secondary moment conditions are satisfied, then $m^{-n} Z_n \to W$ (almost surely and in the mean) where W is not almost surely zero. Consequently $m^{-n} F_n \to E(F)W$. Gonzalez and Molina (1997) consider L_2 versions of these results.

Using far more intricate proofs, Kesten (1970) discusses geometric growth in some very general models with a variety of mating rules. Asmussen and Hering (1983) adapt some of Kesten's work, giving quite complete results for a bisexual discrete-time model using rather different assumptions to those of Daley et al. (1986), and Gonzalez and Molina (1996). Their approach does not introduce a mating function. Instead mating occurs as a random assorting of F females and M males into C couples whose number is determined by a family of conditional distribution functions $H_{F,M}(x) = P(C \leq x | F, M)$. Couples reproduce independently of each, as in the above models. Let ν denote the expected number of offspring produced by a single couple. Let $\mu(F, M) = \int x H_{F,M}(dx) < \infty$, and suppose that

$$\mu(F, M) \to \mu(z) < \infty \quad \text{as } F + M \to \infty \text{ and } M/(F + M) \to z. \tag{2}$$

The relevant criticality parameter is $m(z) = \nu \mu(z)$. Suppose that $m(z) > 1$. Asmussen and Hering (1983) introduce two technical assumptions. Let C_n denote the number of couples produced by the nth generation of females and males. The first assumption is that $P(\sup_{n \geq 0} C_n > x | C_0 = c) > 0$ for all positive c and x. This is an indespensible irreducibility condition. The second assumption is a condition restricting the growth of $\int (x - \mu(F, M))^2 H_{F,M}(dx)$.

Let $N_n = F_n + M_n$ and $A = \{C_n \to \infty\}$, the event of eternal survival. Then almost surely on A, $M_n/N_n \to z$, and both of C_{n+1}/C_n and N_{n+1}/N_n converge to $m(z)$. In addition $N_n/(m(z) + \varepsilon)^n \to 0$ and $N_n/(m(z) - \varepsilon)^n \to \infty$, and similarly for the number of couples, females and males. Adopting a rate of convergence refinement of (2) permits the last set of conclusions to be strengthened to a classical form: $(m(z))^{-n}C_n \to W$ and $0 < W < \infty$ almost surely on A.

Falahati (1999) gives a detailed analysis of some quite intricate generalizations of the Asmussen–Hering model. He allows for additional sources of dependence at the reproduction and mating stages. For example, the reproduction law of couples can depend on the existing number of couples. Again, the mating success of daughters and sons can be influenced by the number of parental couples and the numbers of female and male offspring of each couple.

Asmussen (1980) and Asmussen and Hering (1983) treat a bisexual version of the linear birth process. If there are F_t females and M_t males alive at time t, then independently of previous history, during $(t, t + \mathrm{d}t)$ there is a transition to state $(F_t + 1, M_t)$, or to $(F_t, M_t + 1)$ with probability $fR(F_t, M_t)\mathrm{d}t$, or $mR(F_t, M_t)\mathrm{d}t$, respectively, where $R(x, y) = (x + y)h((x/(x + y)) > 0$ and $h(\cdot)$ satsfies a Lipschitz condition. The form of $R(\cdot, \cdot)$ reflects the notion that the rate of reproduction depends on the sex ratio, and that mates can be chosen only within local sub-populations. See Asmussen and Hering (1983), §XI.1. Proofs are quite similar to the random walk approach for the Markov branching process, for which see Athreya and Ney (1972). Results include exponential growth, and central limit and iterated logarithm rate of convergence results. It should be possible to cope with more general offspring laws.

Hull (1998) constructs a two-type bisexual branching process to account for patrilineal traits such as surnames. Mating units comprise a female–male pair, and they can be one of two types, denoted by **P** and **N**. Offspring inherit the type of their parent mating unit, and newly formed mating units take the type of the male member. So once a type is lost, it can never return. Mating units reproduce independently of type, and according to the binomial scheme (1).

The number of new mating units is determined by a function of four variables, as follows. Suppose reproduction into a generation results in offspring numbers (f_P, f_N, m_P, m_N) where f_P is the number of **P**-type daughters, etc. Formation of mating units for this generation is determined by the function $M(f_P, f_N, m_P, m_N) = (s, t)$ where s (resp. t) is the number of new **P**-type (resp. **N**-type) mating units. Hull gives several examples, such as the **P**-priority function which assigns, as far as is possible, a female mate to each male, giving priority to **P**-type males. Assignment of mates to males ignores the females' type.

Let S_n (resp. T_n) be the number of **P**-type (resp. **N**-type) mating units in the nth generation. Then $\{(S_n, T_n)\}$ is a bivariate Markov chain, and the process determined by $Z_n = S_n + T_n$ is a one-type bisexual branching process, as defined above. Under broad conditions, the criteria for almost sure extinction of **P**-type mating units coincide with the criteria for extinction of $\{Z_n\}$. Let $q(s, t)$ denote the probability of **P**-type extinction when $(S_0, T_0) = (s, t)$. These quantities cannot explicitly be determined, but Hull (1998) shows how to obtain computable bounds, similarly to Daley et al. (1986).

Hull reworks Lotka's (1931b) data using the bisexual model with the above mating function, though it is not clear which population characteristics correspond to his **P**- and **N**-types. He uses Lotka's estimate $\hat{\alpha} = 0.485$ of the female birth proportion to evaluate

$q(1, t)$ for values of t. By arguing that $q(1, t)$ for large t is close to the male extinction probability, he finally arrives at the value $\hat{q} = 0.856$, slightly less than Lotka's estimate.

We end this account of sex by mentioning the discrete-time bisexual exchangeable models of Kämmerle (1991) and Möhle (1994). They may be regarded as bisexual versions of the Moran and Wright–Fisher models of population genetics. The population in each generation comprises n couples which together produce $2N$ children which randomly pair to form the next generation of couples. Let $X_n(N, i)$ count the number of couples in the nth generation in which at least one member is descended from a given set of i ancestor couples. Further structural conditions lead to the conclusion that the generation numbers $(X_0(N, i), \ldots, X_n(N, i))$ converge in law as $N \to \infty$ to the generation sizes of a simple branching process. These results show that simple branching processes can indeed provide good model approximations to the lineages in large populations.

2. Genetics and evolution

Branching theory was first applied in a substantial way by Fisher (1922) to some problems about survival of genetic mutations. Work on this topic still proceeds. The first few sub-sections review some of this. Other sub-sections describe the modelling of situations arising in molecular genetics, and others treat some topics in evolutionary theory.

2.1. Survival of selectively neutral or advantageous genes

Consider a diploid population comprising N individuals and focus attention on the number of A-type genes at a single chromosome locus. The standard Wright–Fisher model for changes in the number of A genes is a Markov chain with states $0, 1, \ldots, 2N$ and one-step transition probabilities

$$p_{ij} = \binom{2N}{j} \left(\frac{i}{2N}\right)^j \left(1 - \frac{i}{2N}\right)^{2N-j}.$$

The probabilities p_{1j} comprise the individual offspring law for this model. Hence if N is very large this law is approximated well by the Poisson law with PGF $\exp(-(1-s))$. If there are small selection pressures such that genotypes AA, Aa, and aa produce offspring in the ratios $1 + \bar{c} : 1 + c : 1$, then under the condition that N is large and the number of A genes is relatively small, the heterozygote selection coefficient c is the only significant one. Consequently the above offspring PGF is replaced by $f(s) = \exp(-(1+c)(1-s))$.

If the A gene appears as a new mutation in a very large population of aa individuals then, being initially rare, carriers of A will mate almost exclusively with aa individuals, and hence they and their offspring will be Aa individuals. Moreover, initially they will reproduce independently of each other. Consequently, early generations of mutant Aa individuals can be described by a simple branching process with the above Poisson offspring law, and the ultimate fate of the mutation can hence be determined.

For the neutral case, $c = 0$, this insight is due to Fisher (1922). He derives approximations for the probability $1 - f_n$ of survival beyond n generations. The first-order term is $1/2n$. These accord with the results of another method of attack which he explores. The genetical consequences of the branching process approximation are explored more fully in his influential monograph (Fisher, 1958). Within the limitations of applicability of the Wright–Fisher model, Fisher's branching process calculations show that for a small initial number of mutants, survival probabilities fall rather quickly. They give a rough idea of the number of generations required for effective loss of the mutant, and the effects of a small selective advantage. His results show that tiny rates of recurrent mutation are capable of ensuring survival of the mutant in the population. A related conclusion is that the fate of a selectively advantageous mutation is decided within the first few generations. It can become established if chance favours its early growth. Finally, calculation from current genotype frequencies permit a rough estimation of how far back a mutation arose.

Fisher's work has been extended in several directions. One is the consideration of similar questions for genetic configurations giving different offspring laws, perhaps with other variations of the simple branching process structure. For example, Fisher himself argued that offspring laws for humans ought to have a variance to mean ratio exceeding unity. Kojima and Kelleher (1962) show that family size data from the U.S. census of 1950 are fitted better by a negative binomial law than the Poisson. Holgate (1966) uses the critical law, $p_0 = p_2 = 1 - p_1 = 1/4$, to investigate the Founder Principle, i.e., to gauge the extent of increased survival of a neutral gene in a population which grows from a small size. Kojima and Schaffer (1964) consider the survival of two linked mutant genes. Here adults reproduce according to a general offspring law with mean M, say. A child acquires the mutant chromosome with probability $(1 - \rho)/2$, where ρ is the recombination fraction. A fraction w of affected children survive to adulthood. Hence the offspring mean for the mutation is $m = wM(1 - \rho)/2$. This shows, among other things, that tight linkage, meaning low recombination, enhances the accumulation of epistatically favourable mutations. Ewens (1968) extends this analysis. Work up to the late sixties on the fate of mutants is reviewed by Schaffer (1970).

Li et al. (1978) give a refined simulation study to assess the mean time for loss of a neutral mutation in a human population. Recent theoretical work is based on diffusion methods, although branching methods have been used for deleterious mutations (see below). All this is based on very simplifying assumptions about the structure and demographic characteristics of the model population. Li et al. (1978) try to inject greater realism. The simulation is designed to faithfully model the characteristics of a Yanomama Amerindian community comprising four interacting villages. The tribulations of a new neutral mutant over a period of 400 years are modelled as a branching process whose offspring law again is a zero-modified geometric. The authors appeal to Kojima and Kellerher (1962) for this choice. A mean time to extinction of 2.82 generations is found, slightly more than for an earlier and cruder simulation, but much less than than the 11.7 generations computed from the diffusion approximation.

According to Gale (1990), branching process models of mutant survival in large populations give reliable results when the selective advantage is not too small, i.e., when evolution proceeds fairly quickly. They are thus a useful complement to diffusion methods which are most accurate when the selection advantage is small. In addition, the above justification for the characteristic independence of lineages requires that the mutation numbers remain relatively small. This may impose limitations on the time horizon for which the approximation is reliable.

These considerations are well illustrated by Haigh and Maynard Smith (1972). They considered the frequency $P_{v,n}$, $v < n$, of electrophoretically recognizable human haemoglobin mutations which arose between v and n generations in the past. Under the neutral selection model with u equal to the mutation rate per locus per generation, they show that $E(P_{v,n}) = (n - v)u$. By making reasonable assumptions, this quantity can be computed and compared with measured frequencies to assess the validity of the neutral selection theory. Good agreement was found for the past 500 generations. The agreement was not good when looking over the 50,000 generations between 10,000 and one million years ago Haigh and Maynard Smith (1972) make the point that estimates of the variance of $P_{v,n}$ are necessary in order to assess the extent of agreement, or otherwise, between theory and measurement. A branching process model is used to compute this variance for the recent generations, and diffusion methods are used for the more remote past. The branching process calculation is for the variance of the total progeny between generations v and n for the zero-modified geometric offspring law. They appeal to Lotka (1931b) for this choice.

Gale and Lawrence (1984) compute median times to extinction for several critical offspring laws. They are concerned with loss of genetic variability under a certain breeding policy for crop plants. The median time to extinction is interpreted as a segregation 'half-life', i.e., the number of generations until 50% of many independent and initially heterozygous loci have become homozygous. These medians are typically quite small, and hence it is important to obtain accurate estimates for deciding whether to embark on a breeding programme for given plant species. Their calculations show that the simple branching models give more reliable results than diffusion approximations.

Suppose the offspring law is Poisson with $m > 1$ and with $m - 1$ small. Fisher (1930, p. 217) obtained an approximation for $1 - q$ in terms of powers of $\log m$. The leading term gives $1 - q \approx 2(m - 1)$, found earlier by Haldane (1927). Subsequently Bartlett indicated (see Bartlett, 1978, p. 43) that for a general offspring law having a finite variance v, Fisher's approximation becomes $q \approx \exp(-2(m - 1)/v)$, giving

$$1 - q \approx 2(m - 1)/v.$$

A considerable effort has been devoted to finding precise conditions under which this holds, deciding when it is misleading, and in finding (as $m \to 1$) asymptotically tight bounds for $1 - q$. Brook (1966) sought bounds using only the first two moments of the offspring law. Later work gradually extended the ambit of his bounds, and tightened them by using more moments. See Narayan (1981) for ultimate results and earlier references. Rather like the extinction problem, independent embellishment of Fisher's approximation arose in the theoretical biology literature, see Eshel (1981). His work is much elaborated by Hoppe (1992a). In particular Hoppe gives several examples showing that $1 - q$ can be proportional to $\sqrt{m - 1}$, or even that $1 - q$ remains bounded away from zero as $m \to 1$. See Athreya (1992) for further developments. Some of the results in this sequence of papers duplicate those in the earlier papers.

Narayan's (1981) analysis is based on showing that $f(s)$ can be bounded by an offspring PGF having the form $a_0 + a_k s^k + a_{k+1} s^{k+1}$ where k is a positive integer. Suppose $0 < \delta \leqslant 1$ and $M < \infty$ are constants, and let $\mathcal{F}(\delta, M)$ denote the set of offspring PGF's such that $p_0 \geqslant 1 - \delta$ and $m \leqslant M$. Heyde and Schuh (1978) show there is a bounding PGF of Narayan's form, with $a_0 = 1 - \delta$ and $k = [M/\delta]$. Consequently the survival probabilities for the class $\mathcal{F}(\delta, M)$ are maximised by the survival probability for this bounding PGF. Lange (1981) considers the case where $\delta = 1$ and m is fixed. The bounding offspring law has least variance among members of $\mathcal{F}(\delta, M)$. These authors discuss evolutionary implications of their results. Thus Heyde and Schuh (1978) suggest reasons for the observed fact that the clutch size of many bird species is either fixed at some number k, or has a two-point distribution on $\{k, k+1\}$.

The rich theory of diffusion approximations for branching processes, valid when m is near unity and the number of ancestors is large, may be viewed as descendants of Fisher's (1958) derivation of the exponential limit law of the population number in a branching process having a Poisson offspring law with m a little more than unity, and conditioned on non-extinction. His object was to determine the law of the number of representatives far into the future of a slightly advantageous mutation, given it was still present in the population. The most direct descendant of Fisher's result are the 'heavy traffic' approximations such as Fahady et al. (1971), and diffusion approximations, as used by Pakes (1998) and (1999) for example.

Fisher (1930) and Haldane (1939) gave an approximate analysis of a genetical problem which leads to the stationary measure of a critical branching process. Moran (1962) and Harris (1963) give more convincing accounts, in contrast to that of Leigh in Haldane (1990), p. 176. We adapt Moran (1962).

Let X_n denote the number of a-gene individuals in a perfectly replicating haploid population of large and constant size N. General properties of Markov chain models used for describing fluctuations of X_n allow us to write its PGF, to a first order of approximation, as

$$G_n(s) = (P_0 - \alpha_0 \lambda^n) + \alpha_1 \lambda^n s + \cdots + \alpha_{N-1} \lambda^n s^{N-1} + (P_1 - \alpha_N \lambda^n) s^N.$$

Here P_0 and P_1 are the probabilities of fixation at frequencies 0 and N, respectively, $\lambda < 1$ is the eigenvalue governing the rate at which heterozygosity is lost, and $\alpha_0, \ldots, \alpha_n$ are constants. In particular $b_j = \alpha_j / \sum_{i=1}^{N-1} \alpha_i$ ($j = 1, \ldots, N-1$) comprise the limiting-conditional probabilities

$$b_j = \lim_{n \to \infty} P(X_n = j | X_n \neq 0, N).$$

Let $B(s) = \sum_{j=1}^{N-1} b_j s^j$.

Assume N is extremely large and that $f(s)$ is the PGF of a-gene offspring numbers. Then assuming that the heterozygous states are in quasi-equilibrium we can conclude that $G_n(s)$ is effectively constant in n, and hence that

$$G_n(s) \approx G_{n+1}(s) = G_n(f(s)).$$

Expanding each side, cancelling common terms, and regarding N as effectively infinite we can conclude that to a very high degree of accuracy

$$B(f(s)) = B(s) + \frac{\alpha_0(1-\lambda)}{\sum_{j=1}^{N-1}\alpha_j}.$$

But $P_0 + P_1 = 1$, whence $\alpha_1 + \alpha_N$ equals the sum in the denominator. Hence writing $\pi(s) = [(\alpha_0 + \alpha_N)/(1-\lambda)\alpha_0]B(s)$ we obtain

$$\pi(f(s)) = 1 + \pi(s) \quad \text{and} \quad \pi(0) = 0,$$

the functional equation satisfied by the generating function of any stationary measure of the branching process whose offspring PGF is $f(\cdot)$. Fisher (1930) derived an equivalent form of this equation for the critical Poisson case, and Haldane (1939) gave a more general treatment. The final conclusion of this calculation is mathematically untenable in that $\pi(\cdot)$ is not a PGF because $\pi(1) = \infty$. However the conclusion is to be interpreted in the sense that if $\{\pi_j\}$ denotes the coefficients of this PGF, then for $0 < j < N$, we have the approximation

$$b_j \approx \pi_j / \sum_{0 < i < N} \pi_i.$$

Fisher (1930) interpreted the b_j as the number among many loci which have j a-genes in the population. This makes more acceptable the idea that $\sum b_j$ is effectively infinite. Fisher computed values of the π_j for small j in the critical Poisson case, finding that they increase from about 0.82 to just less than unity. Subsequently developed theory shows in the general case when $m = 1$ and $v = f''(1) < \infty$ that $\pi_j \to 2/v$; see Pommerenke (1981). This indicates under rather broad conditions that the quasi-equilibrium probabilities b_j are substantially uniform provided j is not too small. This accords with ancient lore of population genetics (specifically with S. Wright's early findings). Finally, we mention that Fisher (1930) showed that allowing for mutation in each generation leads to a functional equation which has received little, if any, general treatment.

What is the probability a gene becomes extinct before a mutant form appears? Mode (1967) numerically explores this question when the normal type has a Poisson offspring law. The question can be put into a more general framework as follows. Suppose individuals in a branching process have a general offspring law and also that independently of each other they may, or may not, exhibit a characteristic χ. In our case χ is mutation of the individual. Let a denote the probability that an individual does not exhibit χ, and define the indicator random variable K_n which equals 1 iff χ does not occur in generation n. Hence $P(K_n = 1|Z_n) = a^{Z_n}$. Let $T_c = \inf\{n: K_n = 0\}$ number the first generation in which χ occurs. Then

$$E(s^{Z_n}; T_c > n) = E(s^{Z_n} a^{B_n}),$$

where $B_n = \sum_{j=1}^n Z_j$ is the total number of births in the first n generations. When $Z_0 = 1$ the right-hand side is denoted by $w_n(s)$ and it is is computed as the nth functional iterate of the defective PGF $f(as)$ with starting value $w_0(s) = s$. Consequently, if $Z_0 = i$, then

$$P(T = n; T_c > n) = (w_n(0))^i - (w_{n-1}(0))^i,$$

and the probability that extinction occurs before χ is ever exhibited is

$$P(T < T_c) = \lim_{n \to \infty} (w_n(0))^i = E(a^B; T < \infty),$$

where B is the total number of births. The right-hand side is computed as $(h(a))^i$ where $h(a)$ is the unique solution of $h(a) = f(ah(a))$. A simple explicit form is not possible when the offspring law is Poisson, although the probabilities $P(B = j)$ comprise a (shifted) Borel–Tanner law and hence do have an explicit form.

Karlin and Tavaré (1982) explored the above model in connection with detecting the appearance of a deleterious genotype in a family line. Pakes (1984) established the connection with the total number of births and he treated the case where $P(K_n = 1|Z_n)$ has a general dependence on Z_n.

More complicated genetical configurations can be handled using multitype branching processes. For example, modelling the numbers of normal and mutant genes typically will be in terms of a two-type process. A significant consideration is whether mutants can mutate back to the normal type. In either case the mutation probabilities are very small and if the models are appropriately parameterized then it is possible to derive Poisson approximations. In the supercritical case time can be used as the parameter and the approximation then holds on the set of non-extinction. Finkelstein and Tucker (1989) and Finkelstein et al. (1990) derive such results for a binary splitting model of mutation of the β-galactosidase gene carried by a bacterial virus. Pakes (1992) explores the case where each type has the same general offspring law. Particular continuous-time mutation models comprise the quantitative basis of fluctuation tests for bacterial mutation. This subject has a quite extensive literature of its own, and which recently has been comprehensively reviewed by Zheng (1999).

Another multitype example is where the viability of a gene is spatially dependent, a situation which can be modelled by supposing the total population evolves within a set of discrete niches (Pollak, 1966a). See Ewens (1968, 1969) and Schaffer (1970) for a variety of situations. We describe in bare detail a couple of examples. Ewens (1968) considers a large population of fixed size comprising N_1 males and N_2 females. A favourable mutation arises for which the offspring of mutant males occur in the ratio

$(1 + \lambda) : 1$ as compared to normal types, and there is a corresponding ratio $(1 + \mu) : 1$ for females. Thus mutant males produce $(1 + \lambda)(N_1 + N_2)/2N_1$ mutant offspring, of which a proportion $N_1/(N_1 + N_2)$ are male. Hence the expected number of mutant male offspring of a mutant father is $m_{11} = (1 + \lambda)/2$, and the other elements of the mean matrix follow in a similar manner. The mutation survives with positive probability iff $\lambda + \mu > 0$.

Assuming the offspring laws are Poisson, e.g.,

$$f_1(s_1, s_2) = \exp\left[-m_{11}(1 - s_1) - m_{12}(1 - s_2)\right],$$

then the survival probabilities for each type, σ_1 and σ_2 say, satisfy

$$-2\log(1 - \sigma_1) = (1 + \lambda)(\sigma_1 + \theta\sigma_2) \quad \text{and}$$
$$-2\log(1 - \sigma_2) = (1 + \mu)(\sigma_1/\theta + \sigma_2),$$

where $\theta = N_2/N_1$. Define the weighted survival probability $\sigma = (\sigma_1 + \theta\sigma_2)(1 + \theta)$. This is maximised when $N_2/N_1 = (1 + \mu)/(1 + \lambda)$, and then $\sigma_1 = \sigma_2$. The optimum sex ratio is close to unity when $\lambda \approx \mu$. Ewens mentions some evolutionary consequences and caveats.

Now let $\mathcal{M}$ denote the mean matrix of a multitype simple branching process, and suppose that $\mathcal{M}$ is positively regular. Then it has a unique positive eigenvalue λ exceeding all others in modulus and associated left and right eigenvectors $\vec{\ell}, \vec{r}$, respectively. They are strictly positive, and normalized so $\sum_i \ell_i = \sum_i \ell_i r_i = 1$. When $\lambda > 1$, the extinction probabilities q_i for a lineage descended from an i-type ancestor are less than unity. Eshel (1984) extends Bartlett's approximation to the multitype case,

$$1 - q_i \sim \frac{2r_i(\lambda - 1)}{\sum_{i,j} \ell_i r_j^2 v_{ij}} \quad (\lambda \to 1+),$$

where v_{ij} is the variance of the number of j-type offspring of an i-type parent, assumed finite. Eshel gives several examples. Hoppe (1992b) tightens Eshel's arguments by giving precise conditions under which the above asymptotic equivalence is valid. Pollak (1992) discusses genetical examples.

Zhivotovsky and Feldman (1993) examine the fate of a selectively neutral mutation growing in a multi-locus background which has attained a stable equilibrium at which there exists linkage disequilibrium. Gametic types are supposed to exist in equilibrium frequencies γ_i. They assume Poisson offspring laws, and they prescribe genotypic viability coefficients of form $w_{ij} - 1 + \varepsilon z_{ij}$ where ε is small. Let $P_{in}(\varepsilon)$ be the probability of extinction of the mutation by n time units after it first appears in the ith gamete. Then $\widehat{P}_n(\varepsilon) = \sum \gamma_i P_{in}(\varepsilon)$ is a measure of the probability of loss of the mutation by time n. Perturbation expansions are used to show that $\widehat{P}_n(\varepsilon) = f_n + \varepsilon^2 \widehat{P}_n''(0)$. Here f_n is the probability of extinction for a certain critical one-type process, and it is interpreted as the probability of loss in the absence of linkage disequilibrium. Since the above second-order derivative term is positive, the authors conclude that the presence of linkage disequilibrium in the background population increases the probability of loss of the mutant, but only to the second order in ε.

Schuster and Sigmund (1984a) ask how to define a notion of gene fixation meaningful within the neutral theory of evolution. They envisage N neutral mutations arising simultaneously, and independently replicating according to a critical branching process, a linear birth and death process, in fact. If the birth and death parameters are the same for each mutant, then the extinction times T_j ($j = 1, \ldots, N$) of the mutant populations are independent and identically distributed, with $P(T_j > t) \sim \text{const}.t^{-1}$ as $t \to \infty$. The corresponding order statistics $T_{(j)}$ are the successive times at which mutant types are lost to the population. For $j = 1, \ldots, N - 1$ the spacings $T_{(j)} - T_{(j-1)}$ have a finite mean, but the duration $T_{(N)} - T_{(N-1)}$ within which only one mutant type is present has infinite mean. This property suggests that this final mutant type be regarded as 'fixed'. Pötscher (1985) extends the scope of the theoretical results, and Watterson (1973) gives related results when $N = 2$.

2.2. Disadvantageous mutations

The investigation of the fate of disadvantageous mutations is of interest when they are deleterious to its carriers, perhaps even lethal. Investigation of relevant questions for this case is of more recent origin. There are two basic situations. The first is where a single mutation arises, just once, and maybe in several individuals. Of interest then is information about the duration for which the mutation persists, and the total number of affected individuals. Let $\bar{t}(i) = \sum_{n \geq 0}(1 - f_n^i)$ denote the mean time to extinction of a subcritical branching process having i ancestors. This quantity is always finite. Nei (1971a) tabulates some numerical values for the Poisson offspring law with differing values of m and i. They are compared with values from a diffusion approximation, though the relation between the two methods of calculation is unclear because the relation between m and the selection coefficient is not specified. Nei finds a good agreement between the two methods, but that it worsens as the selection coefficient increases. This agreement then justifies an illustrative calculation using the diffusion approximation of the expected time till loss of a haemoglobin mutation occurring in the Japanese population which, while not affecting viability, lessens the chances of marriage by 20%. At the time, there were 23 affected individuals, giving an expected time until loss of about 420 years. Nei (1971b) shows that the mean number of affected individuals until loss is 115. Gladstien and Lange (1978a) extend Nei's analysis by finding general bounds for the first two moments of the time T to extinction, and they display the law of the total progeny for the Poisson, binomial and negative-binomial laws. Distributions and moments are computed for Huntington'e chorea, revealing that the mutation will persist on average for 110 years and affect about 15 individuals. The corresponding standard deviations are large.

Gladstien and Lange (1978b) use a two-type branching process for modelling the fate of an X-linked mutation. Affected males are type-1 individuals and carrier females are of type-2. Each type produces offspring numbers having the PGF $Q_i(\cdot)$, $i = 1, 2$, and a proportion a of offspring are males. The daughters of affected males are carriers, and their sons are not viable, so the joint offspring PGF for type-1 is $Q_1(a + (1-a)s_2)$.

Half the offspring of carrier females don't inherit the mutation, and of those that do, the sons are affected and the daughters are carriers. Hence the joint PGF for offspring of type-2 individuals is $Q_2((1+as_1+(1-a)s_2)/2)$. The mutation will ultimately be lost iff $(1-a)Q_2'(1)(1+aQ_1'(1)) \leqslant 2$. Parameter values estimated for Becker's muscular dystrophy ensure eventual loss of this mutation. It is estimated that in a lineage founded by a carrier female, the average time to loss is four generations, and the average numbers of affected males and carrier females is 5.4 and 9.2, respectively.

A second conceivable situation is where normal type individuals occasionally produce mutant offspring which are selectively disadvantageous. This recurring mutation can achieve an equilibrium at a level determined by a balance between mutation and loss of the mutant gene from individual lineages. When the normal type population is very large, generation sizes of the mutant population can be described by a simple branching process with immigration, where normal types produce mutant offspring according to a PGF $h(\cdot)$. If $m < 1$ and $h'(1) < \infty$ then there is a limiting-stationary law whose PGF, $\pi(\cdot)$, uniquely solves the functional equation $\pi(s) = h(s)\pi(f(s))$. See Lange (1982) for numerical computation using the finite Fourier transform. Skellam (1949) was the first to give mathematical expression to these considerations in the case of Poisson immigration and offspring laws, though the general ideas are certainly implicit in Haldane (1935). Gladstein and Lange (1978a) argue that a large number ν of normal females producing mutant autosomal dominant offspring at rates μ_1 and μ_2 for males and females, respectively, is described by a Poisson immigration law whose mean is $2\nu(\mu_1 + \mu_2)$. The fitness of affected individuals is measured by the offspring mean $m < 1$, and hence the equilibrium level of affected individuals is $2\nu(\mu_1 + \mu_2)/(1-m)$. As an example, they find a mean level of 4211 for Huntington's chorea when $\nu = 10^8$. See Lange and Gladstein (1980), and Lange et al. (1978) for some further aspects of the disadvantageous case.

Lange and Fan (1997) propose semi-stochastic models of mutation–selection balance to account for the increasing temporal pace of mutation induced by an increasing background population of normal types. As well as autosomal dominant and X-linked mutations, they model haplotype information on linked markers, as used in the linkage disequilibrium strategy of positional cloning. Their models have elements in common with the Lea–Coulson model of bacterial mutation used for Luria–Delbrück fluctuation tests; see Zheng (1999) for a review of this topic. We consider their simplest one-type model.

Normal population numbers n_t grow exponentially from a very small founder group which existed in the remote past. The precise relation is $n_t = e^{\lambda t}$ where $-\infty < t \leqslant 0$ and $\lambda > 0$ is the nett rate of reproduction. Mutation occurs according to a non-homogeneous Poisson process for which the expected number of mutations in $(-\infty, t]$ equals $(c/\lambda)e^{\lambda t}$, where $c > 0$. Each mutation initiates a clan of mutant descendants modelled as a subcritical Markov branching process. Thus the total number of presently existing mutants (i.e., at time $t = 0$) has the PGF $\exp[-\int_{-\infty}^{0}(1 - F(s, -t))ce^{\lambda t}\,dt]$.

This, and other such formulae, lead to recursive methods for effectively computing distributions and moments of genetically informative quantities such as the number of extant clans and mutants, of the size of the largest clan, and so on. Finnish population data are used to estimate λ and c, and to fit two offspring laws for the mutants. Predictions from the two models are compared.

Direct measurement of a mutation rate μ is a very difficult matter because its typically tiny value requires very extensive epidemiological data to attain an acceptable accuracy of estimation. One method of indirect estimation is based on the relation $\mu = K/2N\hat{t}_0$, where K is the number of segregating codons, N is the population size, and $\hat{t}_0$ is the time to extinction of an allele originally present as a single mutant. Pollak (1986) discusses the rôle of subcritical branching process models in relation to this topic, arguing that use of the estimating equation contains a tacit assumption that the branching process modelling mutant growth is subcritical. This arises from an assumption that the mutant population has achieved equilibrium of the joint laws of allelic numbers and frequencies. Estimates and confidence intervals are given for mutation rates using data related to the study of Li et al. (1978).

2.3. Inhomogeneous mutant offspring laws

A mutant appearing in a population growing toward some (large) limiting value is more likely to survive than if it appeared after the population reaches its maximum. For example, a neutral mutant appearing in a population which doubles in size each generation has an offspring PGF $(1 - (1 - s)/N)^{2N} \approx \exp(2(s - 1))$. Consequently the mutant population dynamics can be described by a supercritical branching process. Conversely, a falling background population size will increase the risk of losing the mutation. Since the eventual fate of a new mutant is determined largely within its first few generations, it is important to assess the effects of fluctuating background population sizes on the probability of extinction.

Kojima and Kellerher (1962) represent the generational changes in total population size as time variation in the first moment of the mutant's offspring law. They use numerical examples to shed light on the extent to which survival of a slightly advantageous mutation is enhanced or reduced by various patterns of initial waxing and waning of the population size. The final population size is the same as the initial value. Not surprisingly, survival is enhanced if expansion precedes contraction, and the largest survival probabilities occur for a continually oscillating population size.

Ewens (1969) explores other aspects of this question. Suppose the normal-type population sizes vary with a basic cycle $N_1, N_2, \ldots, N_k$. Mutants born when the population size is N_j are classified as a j-type mutant, and they have only $(j + 1)$-type offspring (with $N_{k+1} = N_1$). If the heterozygote advantage is $1 + c$, then the mean number of offspring of a j-type is $(1 + c)N_{j+1}/N_j$. This seasonal typology gives a multitype process of mutant generation sizes with a mean matrix which has k dominant eigenvalues with magnitude $1 + c$. Hence the mutation can survive with positive probability iff $c > 0$. Values of the survival probabilities depend on the season in which the mutation first occurs, and numerical examples show that such dependence can be quite strong. Pollak (1966b) examines cyclic variation too.

2.4. Assortative mating

Assortative mating schemes are used to improve the quality of livestock and food crops. These may involve removing all individuals carrying an undesirable recessive gene. Gupta et al. (1992) model this using a simple branching process in which individuals can migrate, or be withdrawn, before being able to reproduce. Individuals in generation n migrate independently of each other with probability μ_n. The process defined by the number in each generation of sedentary individuals is an inhomogeneous branching process. Suppose, for example, that the ancestor generation carries, at a single locus, the alleles A and a with frequencies α and $\beta = 1 - \alpha$, respectively. After random mating the genotypes AA, Aa, and aa have frequencies α^2, $2\alpha\beta$, and β^2. If all the aa's are removed, then the effective frequency of the a allele in the first generation is $\beta/(1+\beta)$. Continuing in this way, the corresponding frequency for the nth generation is $\beta/(1+\beta n)$, and hence the fraction of individuals removed is $\mu_n = [\beta/(1+\beta n)]^2$. One consequence of this is that the AA genotype eventually dominates the population.

Vatutin and Zubkov (1993) review more general migration models.

2.5. Gene inversions

Chromosomes can break and reassemble with gene segments in a different spatial order. In the case where inversion occurrences are very unlikely ever to repeat the original order, Ohta and Kojima (1968) have argued that survival of a new inversion can be modelled as a time varying branching process, where the inhomogeneity derives from recombination occurring within the non-inversion and inversion chromosomes, respectively, but not between the two types. They use a Poisson offspring law with a mean $1+c_n$ at time n, where c_n measures average advantage of inversion heterozygotes. Their approximate analysis shows that the inversion is lost if $\sum c_n < \infty$. Subsequently developed general theory for time varying offspring laws shows that extinction is almost sure iff $\sum 1/M_n = \infty$ where $M_n = \prod_{j=0}^{n-1}(1+c_n)$ is the expected size of the nth generation of inversion heterozygotes. See Jagers (1975). Such criteria, and bounds on extinction probabilities, seem not to have been exploited in genetical contexts. General theory shows, for example, that if survival is possible then the generation sizes grow in proportion to M_n.

2.6. Transposable elements

A transposable element (TE), or transposon, is a segment of so-called selfish DNA which is replicated and inserted back into the genome. This occurs quite commonly; see Barnett (1998) for a popular account.

Moody (1988) models this for a haploid population as follows. An individual genome is classified as i-type if it contains i TE's, $i = 0, 1, \ldots, M$. Each i-type reproduces itself according to an offspring law which may depend on i, and the offspring independently undergo a process of transposition in which they either retain their type, or gain or lose a TE according to probability laws which define a random walk on the type-set. The 0- and M-types are sticky boundaries for this random walk. Hence the generation numbers of the different types comprise a multitype branching process. Moody (1988) focuses on the supercritical case, seeking the long term relative frequencies of the types. For example, he defines neutral selection to mean that the types have the same mean number of offspring, in which case the limiting proportions of types coincides with the limiting-stationary law of the random walk. This model of course is a special case of branching random walk.

2.7. Gene amplification

Daughter cells usually receive exactly one copy of a parental gene. However, some tumour cells pass on replicated copies of genes. Again, such amplification has been observed in cells cultured in the presence of a cytotoxic drug. Increased gene copy number is accompanied by increasing resistance to the drug. This effect is reversible, for if a culture of resistant cells is transferred to a drug-free medium, the gene copies are gradually lost. The proportions of resistant cells decreases with time, but surprisingly, the overall shape of the distribution of copy number appears to remain stable with time.

Kimmel and Axelrod (1990) suggest some simple Markov chain models which describe fluctuations in the total number of gene copies in the nth generation of a randomly chosen lineage. One model is a simple branching process. The stability of count proportions among remaining resistant cells is modelled by means of the limiting-conditional distribution of the Markov chain. Harnevo and Agur (1991) independently proposed models for gene amplification. They classify a cell as type-i if it contains i copies of a given gene. Cells can die, fail to divide, or divide in two. The two daughter cells either preserve the parental type, or one gains an extra copy of the gene. Assuming that cells evolve independently, and that there is a maximum possible copy number, the numbers of different types comprise a multitype branching process. This work is concerned mainly with model formulation and properties such as extinction probabilities and mean growth. Analysis is simplified by counting only the sensitive type-1 cells, since their numbers comprises a one-type branching process.

Kimmel (1997) describes a hierarchical continuous-time model of which gene amplification is a particular case. This model exhibits the slower growth rates of resistant cells observed in cultures. Cells live for Exp(λ) lifetimes, and then divide into two daughters. If a newly born cell has x copies of the gene, then by the time it divides each copy has independently replicated to Y_j copies ($j = 1, \ldots, x$). When the cell divides, the Y_j copies are allocated to the daughter cells according to a joint law having the PGF

$$f_{12}(s_1, s_2) = E\left(s_1^{Y_{1j}} s_2^{Y_{2j}}\right).$$

These allocations are made independently. Thus the two daughters begin their lives with

$$X_1 = \sum_{j=1}^{x} Y_{1j} \quad \text{and} \quad X_2 = \sum_{j=1}^{x} Y_{2j}$$

copies of the gene, respectively.

As above, classify a cell as type-i if it contains i copies at its birth, let X_{it} denote the number of type-i's alive at time t, and define the first-order moments $m_{ij}(t) = E(X_{jt}|X_{r0} = \delta_{ri})$. The mean matrix $M(t) = [m_{ij}(t)]$ satisfies the differential equation

$$M'(t) = \lambda(2A - I)M(t), \qquad M(0) = I,$$

where the matrix $A = [a_j(i)]$ is defined by

$$\sum_{j \geq 0} a_j(i) s^j = (\phi(s))^i \quad \text{and} \quad \phi(s) = f_{12}(s, 1).$$

The total number of living cells $Z_t = \sum_{j \geq 0} X_{jt}$ comprises a linear birth process and hence its mean increases exponentially fast: $E(Z_t|Z_0 = 1) = \exp(\lambda t)$. Kimmel assumes that $\rho = \phi'(1) < 1$, where ρ is the mean number of copies of each parental copy received by a daughter cell. He shows that

$$m_{i0}(t) = e^{\lambda t} + ice^{(2\rho - 1)\lambda t}(1 + o(1)),$$

where $c > 0$ is a certain constant. Hence sensitive cells (type-0) grow in proportion to the total population. But the mean number of resistant cells $m_{ij}(t)$ $(j \geq 1)$ grows in proportion to $\exp((2\rho - 1)\lambda t)$, i.e., more slowly. Moreover, the corresponding proportions

$$m_{ij}(t) / \sum_{k \geq 1} m_{ik}(t) \to b_j \quad (j \geq 1),$$

where $\{b_j\}$ is the limiting-conditional law of the simple branching process whose offspring PGF is $\phi(\cdot)$. These results predict that although the number of cells carrying one or more gene copies diminish as a fraction of the population size, their distribution by copy number tends to a stable shape.

More detailed results are available for a more precisely specified form of $f_{12}(\cdot, \cdot)$. This choice is motivated by the fact that the amplified gene copies are located on DNA fragments residing outside the chromosome. Kimmel and Stivers (1994) suppose that by the time a cell divides, each of its extrachromosomal gene copies independently replicate with probability $\beta < 1$. The resulting two copies are allocated to the same daughter cell with probability α, or one copy goes to each daughter with probability $1 - \alpha$. Parental gene copies which do not replicate are lost. This set-up is a special case of the one above, and with $\rho = \beta$. Closed form expressions in terms of Bessel functions can be found at least for the means $m_{1j}(t)$.

Kimmel et al. (1992) propose a discrete time model for a more detailed description of the mechanism of gene amplification. They suggest there is an entity, which they call an amplicon, which serves as the template for production of additional copies of the gene. Amplicons reside outside the chromosome and carry at least one copy of the target gene and sufficient extra structure to allow their replication. The model employs two basic elements. One of these is an acentric (i.e., lacking a centromere) replicating element (ARE), an extrachromosomal structure containing one or more amplicons. An ARE can reintegrate into a chromosome, and then it is denoted by RE, a reintegrated element. Each element is classified as type-i if it contains i amplicons, $i = 1, 2, \ldots$. Some quite complex rules governing elemental replication and disassociation, and reintegration into a chromosome, lead to a doubly-infinite type branching process. The ancestral generation comprises a single cell containing one acentric element carrying one amplicon copy. As usual, elements evolve independently of each other.

The model is much simplified by counting the total numbers of ARE's and RE's at time n, X_n and Y_n, respectively. Each ARE existing at time n will reintegrate with probability c to produce an RE at time $n+1$, or it reproduces according to a subcritical law whose PGF $\phi(s)$ is a quadratic function. On the other hand, RE's are stable, simply reproducing themselves each generation. The pairs (X_n, Y_n) comprise a two-type process with offspring PGF $E(s_1^{X_1} s_2^{Y_1} | (X_0, Y_0) = (1, 0)) = (1-c)\phi(s_1) + cs_2$ and $E(s_1^{X_1} s_2^{Y_1} | (X_0, Y_0) = (0, 1)) = s_2$. This defines a process where the RE's comprise a final type. Consequently the ARE's eventually vanish, and the RE's accumulate in number with a limiting distribution whose PGF satisfies a functional equation. Parameter values of the model are estimated from experimental data, and the model which thus results gives a good predictive fit to these data. This supports the validity of the biological theory originally adduced to account for gene amplification in the experimental set-up.

See Sun and Waterman (1997) for a very detailed and complex model of gene amplification.

2.8. Replication accuracy

A necessary condition for evolution is that long replicating molecular chains will do so with a high degree of accuracy. Consider a linear polymer comprising ν nucleotides, each of which is correctly copied with probability γ independently of all others. Then the whole chain is replicated without error over a generational time span with probability γ^ν. If the replication is faulty then only the parent molecule survives. The parent molecule has probability w of surviving until such time that it can 'attempt' replication. In essence then, it is ≈ 1 replaced with 0, 1, or 2 identical molecules with probabilities $1-w$, $w(1-\gamma^\nu)$ and $w\gamma^\nu$, respectively. Assuming these chains replicate independently, the population of 'true' chains evolves as a simple branching process. Consequently the chain produces an immortal lineage with positive probability iff it is not too long, i.e., iff

$$\nu < \frac{\log(w^{-1}-1)}{\log \gamma}.$$

If the probability of a one-digit error $1-\gamma$ is very small then the last condition can be rewritten as the approximate inequality

$$\nu \lesssim -\frac{\log(w^{-1}-1)}{1-\gamma}.$$

Either condition can be interpreted as limiting the complexity threshold v by the one-digit error probability γ.

This simple model gives results which are compatible with observation. To take one example, bacterial DNA is copied with $1 - \gamma$ in the range 5×10^{-7} to 5×10^{-6}, and the DNA-molecule in E. coli contains about 4×10^6 nucleotides. See Hofbauer and Sigmund (1988) (§11.3), Schuster and Sigmund (1980, 1984b), and Krawczak et al. (1989) for further discussion.

It is plausible that the replication of polynucleotide chains which are mutable to other species can be modelled using a multitype process. Schuster and Sigmund (1984b) explore this, but ultimately reject it on the following grounds. Mutation rates typically are very small, meaning that the mean matrix of a branching process model is very close to being diagonal. Suppose the type labelling is such that $m_{11} > m_{jj}$ ($j \geqslant 2$). Then the right-hand eigenvector corresponding to the maximal eigenvalue is close to the basis vector $\vec{e}_1$. Hence a surviving population becomes dominated by the type-1 polynucleotide, with the other species subsisting in tiny proportions. Such a configuration is incompatible with the more even representation which is required for viable transmission of genetic information. Hence branching processes appear to be inappropriate descriptions in this context. However, this model deficiency appears also to be a characteristic of Eigen's deterministic hypercycle description of cooperative replication – see (16) in Schuster and Sigmund (1980). The hypercycle model provides interpretations in terms of complexity length similar to those from the branching process description.

There is indeed a close connection between the mean behaviour of a multitype Markov branching process and Eigen's selection equation for autocatalytic replication. Let λ_i^{-1} be the mean lifetime of a type-i individual, M_{ij} be the elements of the mean matrix, and $a_{ij} = \lambda_i(m_{ij} - \delta_{ij})$. The matrix $A = [a_{ij}]$ is the infinitesimal generator of the mean semigroup $\{M(t): t \geqslant 0\}$, where $M(t) = \exp(At)$. Let $\vec{z}$ denote the initial composition of the branching population and $\vec{y}(t) = M(t)\vec{z}$. Then $x_i(t) = y_i(t)/\sum_j y_j(t)$ is a measure of the mean proportion of type-i individuals living at time t. If $\vec{e} = (1, \ldots, 1)^T$, then the vector $\vec{x}(t)$ solves the non-linear differential equation

$$\vec{x}' = A\vec{x} - \vec{x}(\vec{e}^T A\vec{x}),$$

which is Eigen's selection equation. See Hofbauer and Sigmund (1988). This connection echoes the recurring theme that deterministic dynamics is the mean behaviour of a stochastic model. Demetrius et al. (1985) give the most detailed discussion of the relation between the branching process model and Eigen's equation.

Grey et al. (1995) consider the problem of interacting replicators. Survival of a cellular lineage requires that several kinds of replicators inside cells cooperate in the viable functioning of the cell. For this it is necessary that the replicators will coexist in appropriate proportions. However, replicators do compete with each other, and the deterministic dynamics of their interactions predict the extinction of some replicator types, and hence extinction of the cell lineage. Consequently, the mere existence of life implies that there must be some kind of stochastic corrector mechanism which facilitates the survival of cell lineages.

A model for this is suggested which can be viewed as belonging to Kimmel's (1997) division-within-division class of models. At any given time during its life, a cell is regarded as a compartment which contain a number of X and Y (molecular) templates. A compartment is designated as type (n, m) if it contains n (resp. m) X (resp. Y) templates. These numbers satisfy $0 \leqslant n + m \leqslant B$, where B is a constant. A cell can split only if both template types coexist, so it dies when $nm = 0$. Template numbers within a cell fluctuate during the cell lifetime according to a birth and death process where transitions from the state (n, m) are to one of its neighbours $(n \pm 1, m \pm 1)$ (provided all components are non-negative). The transition rates depend on two birth and two death parameters, and they have a complicated nonlinear dependence on n and m which reflects the dynamics of a previously proposed deterministic model. In particular these dynamics incorporate asymetric competition between the X's and Y's which intensifies as their numbers increase.

Thus the life cycle of a cell is viewed as a sequence of events comprising the death of a compartment having the current type and its replacement by a new one having a different type. The times between events have an exponential distribution with a parameter which depends on the type of the dying compartment. This continues until a transition occurs which increases $n + m$ from $B - 1$ to B, at which point the cell divides. Its constituent templates are allocated between the two daughter cells at random and subject to the restriction that each daughter receives some templates.

The number of compartments at any time coincides with the number of living cells, and the numbers of compartments of the various types comprises a Markov branching process with $(B - 1)(B + 2)/2$ types. The key question of interest for Grey et al. (1995) is to determine parameter configurations which ensure the process is supercritical, i.e., that the replication mechanism can produce immortal lineages. Owing to the large number of states there is no simple relation between the maximal eigenvalue λ_M and the defining parameters. An analytic criterion shows that $\lambda_M > 1$ is possible provided the parameters controlling template death rates are sufficiently small. Numerical calculations suggest that λ_M increases quite rapidly as B increases from small values, and then λ_M gradually decreases as B increases further. This accords with the idea that if B is too small, then the splitting of full cells produces too many daughters lacking either X or Y templates, and hence are non-viable. On the other hand, if B is too large then density dependent death rates of templates and the competition between them reduces the efficiency of the reproductive advantage due to cellular splitting.

2.9. Loss of telomere sequences

Shortening of the ends of chromosomes is a putative cause of cellular aging and death. A possible cause of this is incomplete replication in which a repeat of certain DNA fragments is not copied into one daughter of the parental chromosome. This parent can be regarded as comprised of an upper and lower strand (of DNA), each with identifiable left-hand and right-hand ends. Each of the four components typically contain some number of telomere deletion units, the numbers being coded in the form $(a, b; c, d)$ where a and b (resp. c and d) correspond to the left (resp. right) upper and lower ends. Division of the chromosome ends once any of these numbers reaches zero.

The only combinations possible for these numbers are $(n-1, n; m, m)$ or $(n, n; m, m-1)$, which we will denote more compactly by $(\mathcal{D}n, m)$ and $(n, \mathcal{D}m)$, respectively. A parental chromosome with m and n positive can produce daughter types according to

$$(\mathcal{D}n, m) \mapsto (\mathcal{D}n, m) \,\&\, (n-1, \mathcal{D}m)$$

and

$$(n, \mathcal{D}m) \mapsto (n, \mathcal{D}m) \,\&\, (\mathcal{D}n, m-1).$$

In each case the second daughter-type represents the loss of one deletion unit from each end of the lower or the upper strand, respectively. Chromosomes having zero values for n or m are final types:

$$(\mathcal{D}n, 0) \mapsto (\mathcal{D}n, 0) \quad \text{and} \quad (0, \mathcal{D}m) \mapsto (0, \mathcal{D}m).$$

We can characterize a chromosome by its state, defined as the number i of deletion units on its shorter arm, and then the above transition rules abbreviate to

$$i \mapsto i \,\&\, i-1 \quad \text{and} \quad 0 \mapsto 0.$$

Following Arino et al. (1995), assume chromosomes have independent lifetimes with an Exp(λ) distribution. If we identify a cell with a chromosome, and if $X_{ij}(t)$ is the number of cells in state j at time t which are descended from a single i-type cell at time 0 (i.e., in state i), then $\{X_{ij}(t): t > 0, j = 0, 1, \ldots\}$ comprises a pure birth process with infinitely many types. Note that $X_{ij}(t) \equiv 0$ if $j > i$. Arino et al. (1995) show that

$$E(X_{ij}(t)) = \frac{(\lambda t)^{i-j}}{(i-j)!}, \tag{3}$$

a polynomial growth rate for each cell type, and with larger rates for smaller states. In particular, the zero-state cells accumulate at a greater rate than any other type. These authors explore further aspects of this and other models, and they discuss the application of their results to experimental data on the loss of telomeres.

Olofsson and Kimmel (1999) investigate the effect of cell death. Their first model specifies a probability δ of the death of a 0-type cell when it splits. Other types cannot die. This hypothesis preserves the form of (3) when $0 < j \leqslant i$, but now

$$E(X_{i0}(t)) \sim (1-\delta)^{-1} \frac{(\lambda t)^{i-1}}{(i-1)!}.$$

The important change is that 0-types and 1-types now accumulate at the same rate.

An additional hypothesis allows the death of i-type cells ($i \geq 1$) by prescribing a probability η of survival when they split. This causes a dramatic qualitative change, with (3) replaced by

$$E(X_{ij}(t)) = e^{-\lambda(1-\eta)t} \frac{(\eta \lambda t)^{i-j}}{(i-j)!}$$

if $0 < i \leq j$, and differing asymptotic behaviours for $j = 0$, according to the relative sizes of δ and η. If $\delta, \eta < 1$ then all types disappear at an exponentially fast rate.

2.10. Explosive growth of DNA repeats

Some heritable disorders, such as fragile X-syndrome, are associated with abnormally large repeats of DNA-triplets in certain regions of the genome. The number of repeats tends to be much larger for carriers than for normal individuals, and the number is considerably larger still for victims of the disorder.

A similar phenomenon is observed in replication of the bacteriophage T4 DNA. This virus induces production in the host cell of branched networks of concatenated DNA which subsequently resolves into unbranched phage genomes.

Gawel and Kimmel (1996) model this latter situation as a so-called iterated simple branching process $\{X_n\}$. Here X_n is the length of a linear chain of DNA repeats after the nth stage of replication ($n = 1, 2, \ldots$) and $X_0 = i > 1$. A chain with $X_n = \nu$ repeats will replicate as a branched network which is assumed to be a Galton–Watson tree descended from a single ancestor through $\nu - 1$ generations. Thus the replicating chain serves as a template for the height of the daughter tree. This partial tree later resolves into a linear chain of length

$$X_{n+1} = Y_{\nu-1}^{(n)}, \qquad (1)$$

where the processes $\{Y^{(n)}\}_{n \geq 1}$ are independent copies of the total progeny process $\{Y_\nu\}$ of a simple branching process. Hence the sequence $\{X_n\}$ is a Markov chain with state space $\mathbb{N}$, and since $Y_0^{(n)} = 1$, the state 1 is absorbing.

Very little is known about this intriguing model. The following analysis simplifies the proof of the main assertions of Theorem 1 in Gawel and Kimmel (1996). We exclude the trivial case $p_1 = 1$, where $X_n \equiv X_0$. Then $P(\{X_n \to 1\} \cup \{X_n \to \infty\}) = 1$. Let X_∞ denote the almost sure limit of X_n, and let $g(s, \nu)$ denote the PGF of Y_ν. Then $g(s, 0) = s$ and $g(s, \nu + 1) = sf(g(s, \nu))$ (Harris, 1963). It follows from (1) that

$$E(s^{X_{n+1}}) = E(g(s, X_n - 1))$$

and hence in all cases

$$E(s^{X_\infty}) = E(g(s, X_\infty - 1)). \qquad (2)$$

If

$$0 < p_0 < 1 \qquad (3)$$

we may choose $s \in (0, q)$ and then $f(s) > s$. This gives $g(s, 1) > sf(s) > s^2$ and hence, by induction, that $g(s, \nu - 1) > s^\nu$. Since (2) can be written as

$$s + E(s^{X_\infty}, X_\infty > 1) = s + E(g(s, X_\infty - 1), X_\infty > 1),$$

it is clear that this can hold iff $P(X_\infty > 1) = 0$. We conclude that the process is absorbed at unity when (3) holds. Next, if $p_0 = 0$ then $Y_\nu^{(n)} > \nu + 1$ and hence (1) implies $X_{n+1} \geqslant X_n$. So $X_n \uparrow \infty$ if $X_0 \geqslant 2$.

The iterated branching process appears to increase very rapidly when (3) fails. In fact a plausible conjecture is that its growth is governed by the following self-normalization property

$$\frac{\log X_{n+1}}{X_n} \to \log m.$$

Gawel and Kimmel (1996) also formulate a stochastically smaller binomially thinned version of $\{X_n\}$ which is thought to be closer to biological reality. Their simulations show extremely fast growth after a period of relative quiescence.

2.11. Geographical spread of mutants

There is now a large literature devoted to spatial branching processes. The basic idea is that branching particles move around some base space according to a Markov process. The particles are assumed to move independently of each other, but other spatial complications are possible. For example, the branching rate and/or offspring laws can be spatially dependent, and offspring can be born at points different to the parent. The most commonly studied spatial mechanisms are the random walk (Biggins, 1997; Révész, 1994), and diffusion processes, with emphasis on the Brownian motion process (Dawson, 1993; Le Gall, 1999). These models are relevant to questions such as the probability that a new mutation arising at some location will spread to another region, and distributional properties of typical times required to spread beyond a specified distance. We mention only a couple of examples relating to selectively neutral alleles.

For Crump and Gillespie (1976) the base space is discrete, with points representing isolated colonies. At time zero a single mutation event produces a new allele a at some point. Let Z_{jn} denote the number of a genes in the jth colony and the nth generation of its lineage. Mutant individuals have offspring PGF $f(s)$. Offspring born in the ith colony migrate independently to colony j with probability p_{ij}. The a-allele is selectively neutral and capable also of mutating, so the offspring law is effectively subcritical with mean $m = 1 - u < 1$, where u is the probability of mutation.

Let

$$X_j = \sum_{n \geqslant 0} Z_{jn} \quad \text{and} \quad M_{ij} = E(X_j).$$

The M_{ij} solve the linear system

$$M_{ij} = m \sum_k p_{ik} M_{kj} + \delta_{ij}.$$

Suppose colonies form a linear array and that $i = 0$. Let $\pi_j(\nu)$ be the probability that ν mutants reach the jth colony, and that none of their ancestors have reached it. Then $M_{0j} - M_{00} = \sum_{\nu \geq 1} \nu \pi_j(\nu)$ and M_{0j}/M_{00} is the mean number of pioneer settlers in the jth colony. Assume further, that $p_{ii} = 1 - r > 0$ and $p_{i,i\pm 1} = r/2$. Then $M_{ij} = a\lambda^{|i-j|}$ where

$$a = \left(mr\sqrt{\zeta^2 - 1}\right)^{-1}, \quad \lambda = \zeta - \sqrt{\zeta^2 - 1} \quad \text{and} \quad \zeta = [1 - m(1-r)]/mr.$$

Since u typically is very small, $\lambda \approx 1 - \sqrt{2u/r}$, whence the smallest colony number j^* having probability ≤ 0.01 of ever being occupied by the mutation is $j^* \approx 3.23\sqrt{r/u}$.

Using this estimate in conjunction with data on Drosophila populations, Crump and Gillespie (1976) deduce that 99% of neutral mutations will travel no further than 200 miles. They also estimate a probability of under 2.8×10^{-6} that a mutation will traverse continental USA. But since Drosophila gene frequencies are observed to be substantially constant over their specific range, Crump and Gillespie conclude that if these alleles are in fact selectively neutral, then their spatial sample paths are aberrant.

Geographical spread across a continuous domain can be modelled in terms of individuals moving according to independent copies of a diffusion process. Sawyer (1975) discusses in detail the case where individuals move in $\mathbb{R}^d$, and with particular consideration for the driftless Brownian motion process. Attention is directed toward characteristics of the spatial joint distribution of numbers of descendants of a founder population which carries a new mutation. The founder group may be a single individual, or a uniform or random distribution of individuals, or there may be a flux of immigrants or mutations from a background field of normal types. He gives results for arbitrary values of m although, as above, selectively neutral mutation corresponds to $m = 1 - u < 1$. In contrast to other possible modes of geographic spread, the independence of lineages can lead to a heterogeneous spatial distribution of descendants. The most extreme examples occur when $m = 1$ and $d = 1, 2$, and then there is a clumping effect for which there are regions of arbitrary extent containing high local population densities. See Sawyer (1976), §4.2 for more detail, and Sawyer (1991) for a more recent discussion.

Still assuming a Brownian movement process, Sawyer and Fleischman (1979) compute the probability that a single mutant ancestor has a descendant which reaches a remote region. In one dimension the ancestor is located at the origin and $p(x)$ denotes the probability of diffusing as far as x. If $d \geq 2$, the ancestor is located at $\vec{x}$, and $p(\vec{x})$ denotes the probability of a descendant reaching the open ball with a fixed radius $a < \|\vec{x}\|$ centered at the origin. The complementary probability solves a non-linear elliptic partial differential equation.

A typical result appropriate for the neutral case in one dimension is that

$$p(x) \sim 6/vx^2 \quad (x \to \infty),$$

where v is the variance of the offspring distribution, and the limit is taken in such a way that $x\sqrt{u} \to 0$. Provided $u = 0$, this generalizes to higher dimensions,

$$p(\vec{x}) \sim \begin{cases} 2(4-d)/vx^2, & \text{if } d \leqslant 3, \\ [vx^2 \log(x\sqrt{v})]^{-1}, & \text{if } d = 4, \\ C(d)/(x\sqrt{v})^{d-2}, & \text{if } d > 4, \end{cases}$$

where the branching law is binary splitting in the last case, and $C(d)$ is a constant. These estimates are independent of a, the radius of the ball.

The mathematical structure of spatial branching processes is very rich, and the subject has received much attention for over 20 years. See Dawson (1993). A central problem is finding conditions for persistence in the critical case. This means finding compensating conditions on the branching and spatial components which ensure that if the process starts with ancestors distributed in space as a Poisson field, then the branching process has a weak limit, as $t \to \infty$, which has the same intensity as the initial field. This requirement precludes clumping behaviour of the sort mentioned above. The necessary and sufficient condition for this is that the symmetrized spatial process is transient. For Brownian motion, this is equivalent to the dimension restriction $d \geqslant 3$. See Wakolbinger (1995) and López-Mimbella and Wakolbinger (1997) for reviews.

The geographical spread of selectively advantageous mutations was first described by Fisher (1937) using the following one-dimensional deterministic model. Let $p(x, t)$ denote the population frequency of the favourable allele at position x and time t, and let c measure the selection intensity, which may also be spatially dependent. Fisher suggested that $p(\cdot, \cdot)$ satisfies the reaction–diffusion equation

$$\frac{\partial p}{\partial t} = \frac{\partial^2 p}{\partial x^2} + cp(1-p). \tag{1}$$

Ignoring spatial dependence, continuous-time Wright–Fisher dynamics implies that the frequency of favourable alleles increases at the rate $cp(1-p)$. Fisher simply added the diffusion term, apparently in analogy with physical diffusion phenomena subject to Fick's law. A very considerable literature has grown around (1) (Murray, 1989, Chapter 11) and various generalizations. Some of these are intimately related to a functional of branching Brownian motion.

Suppose the spatial process is a standard Brownian motion on the real line, and that the branching component is a Markov branching process with $p_0 = 0$ and splitting rate c. Suppose too that the population starts with a single individual at the origin, and let R_t be the position of the right-most individual alive at time t. Then $p(x, t) = P(R_t > x)$ solves the Kolmogorov–Petrovski–Piscounov generalization of (1),

$$\frac{\partial p}{\partial t} = \frac{\partial^2 p}{\partial x^2} + g(p),$$

where $g(p) = c(1 - p - f(1-p))$. This result is due to McKean (1975). Fisher's Eq. (1) is the case $f(s) = s^2$.

This is yet another example where a probabilistic representation provides a constructive solution for a deterministic equation, and does so under weak assumptions. Chauvin and Rouault (1988) show how analytical properties of $p(x, t)$ are deduced from properties of the branching Brownian process. The constructive solution forms the basis for Monte Carlo solutions of reaction–diffusion equations. See Chauvin and Rouault (1990) and references cited there. Harris (1999) uses purely probabilistic methods to study the asymptotics of travelling wave solutions of (1).

2.12. Evolution of diversity

It is natural to seek specific deterministic causes for observed patterns of speciation in the fossil record. For example, in the Cambrian period trilobite species comprised some 75% of the total, and this fell quite rapidly and vanished in the Permian. Molluscan species rose from 3% of the total to over 60% during the same interval. Must there be a specific cause? Again, is there a common cause for observed simultaneous disappearances of groups?

Raup (1985) asked whether such patterns could arise by randomly occurring speciation and extinction. He explores this by examining simulated sample paths of a binary splitting branching process. Most splits represent species formation within groups, but some splits are classified according to pre-determined rules as forming new groups. The simulation results are represented as ancestral 'pine' trees, similar to those commonly used by paleontologists. Each 'pine cone' is a symmetric histogram representing the history of a single group. The axis of a cone lies in the direction of time, and its width for a given value of time is proportional to the number of species which then comprises the simulated group. Cones are linked by branches showing the ancestral relations of all the groups.

These simulated trees show features which are very similar to many in the paleontological literature. Raup (1985) warns against concluding from this that 'chance' is a sufficient explanation for the observed historical record. For example, one of his trees has a cone similar in form to the observed pattern of trilobite species. This cone quickly reaches a width of 20 (simulated) species and then falls to a much smaller value about which it fluctuates for the remaining two thirds of its lifetime. On the other hand, the number of trilobite species grew very quickly to some 6000, and at this level the estimated half-life of trilobite lineages gives the probability of extinction as 10^{-82}. Non-chance causes of trilobite extinction are clearly on the agenda. Further detail is given in Raup et al. (1973), the first of a series of investigations.

2.13. An infinite alleles model of mutation

A recent and influential conceptual notion in population genetics, the neutral mutation hypotheses, is the idea that mutation results in a completely novel allele, and that all mutant types are selectively equivalent. Thus fixation of novel mutations is due solely to random drift, and not to a fortuitous adaptation to existing conditions. There now exists much theoretical work on infinite alleles versions of traditional genetical models, such as the Wright–Fisher and Moran models. Quantities of interest include the number of living mutant types K_t, the time of the last mutation event, and the mean numbers $\phi(j, t)$ of mutant types having j living representatives at time t. For each t, the $\phi(j, t)$ collectively define the frequency spectrum (Ewens, 1979).

Griffiths and Pakes (1988) integrated the simple branching process with the infinite alleles mechanism. They envisage that with probability u offspring mutate at the time of birth to a novel type, independently of their ancestry and sibship. Selective neutrality is embraced by assuming that individuals have the same offspring law, independent of their allelic type. Thus the joint PGF of the numbers of normal and mutant offspring, ν and μ respectively, is

$$E(s_1^\nu s_2^\mu) = f\big((1-u)s_1 + us_2\big).$$

In particular the numbers by generation of the ancestral type comprise a simple branching process with offspring PGF $H(s) = f(u + (1-u)s)$, whose behaviour is described by standard theory, and classified according to values of the mean number of normal offspring $M = m(1-u)$. In particular, $1 - H_n$ denotes the probability that the lineage of the ancestral type (or a new mutant) survives beyond n generations. Consequently

$$E_i(K_n) = 1 - H_n^i + iu \sum_{r=1}^n m^r (1 - H_{n-r})$$

since the ancestral type contributes the first term, and each of the mean number im^r of individuals alive at time r mutates with probability u into a new allele which then survives to time n with probability $1 - H_{n-r}$. Hence the long-term per capita proportion of mutant alleles is

$$\lim_{n \to \infty} m^{-n} E_i(K_n) = iA,$$

where

$$A = u \sum_{n=0}^{\infty} m^{-n}(1 - H_n).$$

In the supercritical case $m \geq 1$ there is a corresponding law of large numbers:

$$K_n/Z_n \to A \qquad (1)$$

a.s. on the set of non-extinction. Thus the number of mutants grows exponentially fast at the same rate as the generation sizes. The following version holds in the critical case when $v = f''(1) < \infty$:

$$K_n/n | Z_n > 0 \Rightarrow vA\varepsilon/2, \qquad (2)$$

where ε has the standard exponential law. Contrast this with the critical-case Yaglom theorem

$$Z_n/n | Z_n > 0 \Rightarrow v\varepsilon/2.$$

A conditional limit theorem for the subcritical case exists but it is not informative. A much more accessible approximation is valid for a population with many normal-type ancestors and a very small rate of mutation. Specifically, if the offspring variance is finite and $u \to 0$ such that $ui \to \theta < \infty$ then the following Poisson limit law holds:

$$\lim_{i \to \infty} P_i(K_n - 1 = j) = e^{-\mu_n} \frac{\mu_n^j}{j!},$$

where $\mu_n = \theta \sum_{r=1}^{n} m^r(1 - f_{n-r})$.

When $m \leqslant 1$, interest in the last time of a mutation L lies in its size relative to the time of extinction T. The results found by Griffiths and Pakes (1988) suggest that $T - L$ is likely to be large when $m < 1$, and that $L \approx T$ when $m = 1$. The more complete calculations which are possible for the Markov branching version support these conjectures (Pakes, 1989a).

Griffiths and Pakes (1988) show that, no matter what the size of m, there is a limiting frequency spectrum

$$\phi(j) = \lim_{n \to \infty} m^{-n}\phi(j,n) = u \sum_{r \geqslant 0} m^{-r} q_{1j}^{(r)},$$

where $\sum_{j \geqslant 0} q_{ij}^{(n)} s^j = H_n(s)$. Thus $\psi(I) = A^{-1} \sum_{j \in I} \phi(j)$ is a measure of the long-term expected proportion of alleles in the frequency range I. The analogous quantity for the large population limit of the infinite-alleles Wright–Fisher model is $\theta x^{-1}(1-x)^{\theta-1} \, dx$, the number of alleles whose representatives comprise a proportion between x and $x + dx$, where $0 \leqslant x \leqslant 1$ and θ is a constant. No such representation is available for the branching process, although the upper tail behaviour of the frequency spectrum can be determined.

Suppose, for example, that $M > 1$, and let $\rho = (\log m)/(\log M)$ and $\bar{\psi}(x) = \psi([x, \infty))$. For each $x > 0$, the sequence $\{m^{-n}\bar{\psi}(M^n x)\}$ converges to some function of x. Griffiths and Pakes (1988) also conjecture a corresponding result for the individual proportions $\psi(\{j\})$, and confirm it for the fractional linear offspring PGF.

Similarly, when $M = 1$ Griffiths and Pakes (1988) conjecture that $\psi(\{j\}) \sim Cj^{-K} \exp(-\lambda\sqrt{j})$ where C, K and λ are constants which appear to depend in a complicated way on the offspring law. Again this conjecture is valid for the fractional linear case, and then $K = 3/4$.

The analogues of these conjectures for the infinite alleles Markov branching process are fully resolved by Pakes (1989a). In particular, the continuous time setting gives a simpler result for the supercritical case: If $v = (m-1)/(M-1)$ then $j^{1+v}\psi(\{j\}) \to$ const. – the frequency spectrum decays algebraically fast. The tail estimate for the critical case agrees with the above discrete-time conjecture, and it is notable because K depends on the second- and third-order factorial moments of the offspring law.

A substantial generalization of these models is treated in depth by Taib (1992). The basic branching mechanism is the general (or Crump–Mode–Jagers) process in which individuals reproduce independently and their life histories are governed by an arbitrary $\mathbb{N}_0$-valued counting process (Jagers, 1975). No particular mechanism of mutation has to be prescribed, rather it is assumed that individual life-history processes are resolved into a pair of processes counting the epochs of mutant and non-mutant births, respectively. An index is assigned to offspring, 1 if it carries the parental allele, and 0 if not.

A deep insight of Taib (1992) is that all population members carrying the same allele, α say, may be regarded as a generalized individual whose offspring comprise all the new alleles which are ever directly produced by an α-type individual. These macro-individuals comprise a general branching process. Taib's very general formulation provides a framework for answering questions about the allelic composition of the ancestry of a randomly sampled individual, and much else besides. One example is given in the next sub-section.

2.14. The molecular clock of evolution

This term refers to the hypothesis that the number of amino acid or nucleotide substitutions separating a pair of species is roughly proportional to the time since they diverged from a common ancestral species. This clock is a stochastic one, first modelled as a homogeneous Poisson process. Assuming the rate is constant has long been a subject of controversy (Gillespie, 1991). However, Dobzhansky et al. (1977) argue that averaging over the clocks of many proteins provides a reasonably accurate composite evolutionary clock. Much of the theoretical work on molecular clocks is based on structures constructed from classical genetical models.

Jagers (1991) shows that Taib's (1992) macro-individual process provides an approach which admits very general life-history mechanisms and varying population sizes. It is based on a supercritical general branching process, characterized in part by its reproduction measure $\mu(\mathrm{d}x)$, the mean number of offspring born to a mother aged in $(x, x + \mathrm{d}x)$. The Malthusian parameter $\alpha > 0$ is determined by $\hat{\mu}(\alpha) \equiv \int_0^\infty e^{-\alpha x} \mu(\mathrm{d}x) = 1$. If a mother $\mathcal{M}$ is randomly sampled from a long-lived family tree, then her age, the age of her mother $\mathcal{M}_1$ when $\mathcal{M}$ was born, the age of $\mathcal{M}$'s grandmother when $\mathcal{M}_1$ was born, ..., are independent random variables which have the distribution function $\int_0^t e^{-\alpha x} \mu(\mathrm{d}x)$, defining the stable-age law at child bearing. Its expectation β is the mean age at child bearing. Consequently the process of birth times looking backwards from $\mathcal{M}$ is a renewal process.

Now consider the infinite-alleles version of this model where $p(x)$ denotes the probability that the offspring born to a mother aged x is a mutant. The expected number of mutant children born up to age t is

$$\mu_m(t) = \int_0^t p(x) \mu(\mathrm{d}x).$$

The branching process of macro-individuals has a reproduction measure *whose Malthusian parameter is again* α, and whose mean age at child bearing is $\beta_m = \beta/\hat{\mu}_m(\alpha)$.

The backwards process of macro-births is precisely the process of mutation events along $\mathcal{M}$'s lineage, and it is a renewal process with intensity β_m. If this quantity is the same for all species irrespective of their life histories then there is a constant c, the rate of evolution, such that $\beta_m = c$ for all μ. This can occur iff $p'(x) = c$. One consequence of these considerations is as follows (Corollary 1 in Jagers, 1991): assume that mutation results from at least one Poisson stream of genetic events having a small rate λ. Then the resulting mutation probability is $p(x) = 1 - e^{-\lambda x}$ at maternal age x, and it yields an approximately species-independent molecular clock of mutations. The rate of evolution is approximately equal to λ.

Donnelly (1991) gives further commentary on this topic. Jagers et al. (1991) explain how the renewal structure of the backwards mutation process when the mutation probability is constant yields computable normal approximations to the time back to $\mathcal{M}$'s nth mother.

3. Epidemic modelling

3.1. Introduction

A basic model for the spread of a disease through a closed population assumes individuals progress partly or all of the way through the irreversible sequence of states of health, susceptible → infected → removed. Removal can mean physical isolation from the susceptible population, death, or recovery with lasting immunity. A fundamental simplifying assumption is that the population is fixed in size, and that infectious and susceptible individuals mix homogeneously. This interaction between the two sub-populations is mirrored in non-linearities in even the simplest models with the consequence that exact results which can be obtained usually have a very unwieldly form.

Branching process approximations to epidemic models have two principal uses. The first is to evaluate the likelihood of a major outbreak of the disease, by which is meant that a large proportion of the susceptibles have been infected by the time all infectives have been removed. The second use is concerned with the number of original susceptibles ever infected, called the final size of the epidemic. A simple characterization of its law is not possible for even the simplest epidemic models. Under quite broad conditions this law can be approximated by that of the total number of births in an approximating branching process. Up until 1980 branching process approximations were used in an intuitive way. A principal contribution since then has been the clarification of the precise nature of this approximation using modern techniques from the limit theory for stochastic processes.

Bailey (1975) is the classical reference for epidemic modelling, with Daley and Gani (1999) giving a more modern account. Contemporary research is surveyed in the collections Gabriel et al. (1990) and Mollison (1995). Ball (1997) also surveys the rôle of branching processes in epidemic modelling.

3.2. The general epidemic

Denote the numbers of susceptible, infectious and removed individuals at time $t \geq 0$ by S_t, I_t and R_t, respectively, and let $S_0 = N$, $I_0 = a$, and $R_0 = 0$. The classical mathematical expression of homogeneous mixing and removal is to suppose that $(S_t, I_t: t \geq 0)$ is a two-dimensional birth and death process with the following single-jump transitions in the interval $(t, t+h)$:

$$(s,i) \to (s-1, i+1) \quad \text{with probability } \beta s i h + o(h) \tag{1a}$$

and

$$(s,i) \to (s, i-1) \quad \text{with probability } \gamma i h + o(h). \tag{1b}$$

Here β is interpreted as a contact rate and γ is a removal rate. The process ends at the random time $T(N) = \min\{t: I_t = 0\} < \infty$, and the final size of the epidemic is defined to be $\zeta_N = R_{T(N)} - a$, the total number of infected susceptibles. This structure defines the *general* epidemic when $\gamma > 0$ and the *simple* epidemic when $\gamma = 0$. A newly infected individual becomes immediately infectious, remaining so until its removal.

In the simple epidemic eventually all susceptibles are infected, so we consider only the general case. An outbreak is judged to be mild if $R_{T(N)} \ll N$, and severe if $R_{T(N)} = O(N)$. The nonlinear transition rate in (1a) complicates the determination of these matters. For institutions such as schools and cities it usually is the case that a is very small in comparison to the large value of N. In addition, the final severity of the outbreak is determined to a large extent by its early progress during which I_t remains small in comparison to S_t. Thus a useful approximation to the epidemic model is achieved by replacing s in (1a) by N, giving a linear birth and death process for the number of infectives denoted by $\hat{I}_t$. Since the per capita birth and death rates are βN and γ, respectively, the probability of a mild outbreak is unity if the relative removal rate $\rho = \gamma/\beta \geq N$. If however

$$\rho < N \tag{2}$$

the probability of a major outbreak is positive, being given by $1 - (\rho/N)^a$. The reciprocal relative removal rate β/γ is the mean proportion of the susceptible population contacted by a single infective before its removal. Hence N/ρ is the mean number of infections per infective, and the above results are thus rendered plausible.

It is evident from (1a) that the infection rate for the approximating process is always larger than for the true epidemic process. It plausibly follows that $\hat{I}_t$ is stochastically larger than I_t, and hence that the total number of removals (i.e., births) $\hat{B}$ in the approximating process is stochastically larger than ζ_N. Let $0 < p < 1$ and $\pi(p) = P(\zeta_N \leq Np)$ be the probability that the proportion of susceptibles affected by the disease does not exceed p. Consequently, $\pi(p) \geq P(\hat{B} \leq Np) \approx P(\hat{B} < \infty)$, supposing that N is sufficiently large. This gives the lower bound in the inequality

$$\left[\min(1, \rho/N)\right]^a \leq \pi(p) \leq \left[\min\left(1, \rho/N(1-p)\right)\right]^a.$$

This inequality comprises the essence of Whittle's (1955) stochastic threshold theorem. The upper bound is obtained by observing that if $R_t \leq Np$ then $S_t \geq N(1-p) + a - I_t \approx N(1-p)$ when N is large. Consequently an epidemic process for which $R_{T(N)} \leq Np$ can be approximated from below by a linear birth and death process whose birth rate is $\beta N(1-p)$. See p. 108 in Goel and Richter-Dyn (1974) for discussion about the interpretation of this result.

3.3. Life history of infectives

The active life of a typical infective in the general epidemic can be regarded as having an exponential distribution with mean γ^{-1}, and during this life the infective contacts original susceptibles at event times of a Poisson process whose rate is βN. Infective life histories are mutually independent. For the true epidemic model these contacts result in transmission of the infection only if the contacted individual is still susceptible. The large population approximation is realized by supposing all contacts transmit infection, and hence $(\hat{I}_t)$ is a particular and simple instance of a Crump–Mode–Jagers branching process. Offspring of a given infective comprise the susceptibles it directly infects, and the offspring PGF is

$$f(s) = \gamma \int_0^\infty e^{-\beta Nt(1-s)-\gamma t} \, dt = [1+m-ms]^{-1}, \qquad (1)$$

where $m = \beta N/\gamma$ is the offspring mean. The criterion (2.2) for a major outbreak with positive probability is thus immediate. Also, using the fact that $\widehat{B}$ has the Lagrange law corresponding to $f(s)$ and a ancestors (Pakes and Speed, 1977) gives, with almost no calculation,

$$P(\widehat{B}=j) = a\frac{(a+2j-1)!}{(a+j)!j!} \cdot \frac{\theta^{j+a}}{(1+\theta)^{2j+a}} \quad (j \geq 0),$$

where $\theta = 1/m$. This is the large-population approximation of the law of the final size ζ_N of the epidemic. It is defective when $m > 1$.

Members of the population can be classified into generations, with the original infectives comprising the zeroth generation, the susceptibles they directly contact comprise the first generation, and so on. Bartoszyński (1969) is an early proponent of this idea. This classification quite often matches the type of data gathered by public health officials. Infectious contacts can be traced generation by generation to construct the family tree generated by a handful of infectives entering a pristine population. Observed generation numbers can be used in now standard ways (Guttorp, 1991) to estimate m, and to predict the likelihood of a major outbreak. See Becker (1989, Chapter 8) and Heyde (1979) for discussion, examples and further references. Becker (pp. 8, 9) observes that determining the value of m is important because if $m > 1$ then a proportion at least $1 - m^{-1}$ of the susceptible population has to be immunized to prevent a major outbreak.

An important descriptive parameter of epidemics is the *basic reproduction ratio* R_0, informally defined as the mean number of susceptibles directly infected by an infectious individual after being introduced to a pristine and finite population. The parameter m clearly over-estimates R_0. The following alternative model of making contacts ought to give a closer approximation to R_0 (see Daley and Gani, 1999, pp. 87, 88). It is supposed that an infective who is active for time t has probability $1 - e^{-\beta t}$ of contacting a given susceptible independently of other possible contacts. Consequently the PGF of the number of contacts made by an infective before removal has the mixed binomial form

$$f^*(s) = \gamma \int_0^\infty [s + (1-s)e^{-\beta t}]^N e^{-\gamma t}\, dt. \tag{2}$$

If N is large then, as above, we may reasonably suppose that contacts with infected susceptibles occur with negligible probability. Consequently the mean number of infections per infective is

$$m^* = \frac{\beta N}{\beta + \gamma},$$

and $R_0 < m^* < m$. Presumably then, the extinction probability computed from (2) when $m^* > 1$ is a more reliable indicator of a major outbreak than the probability m^{-a} obtained from the Poisson contact model. Daley and Gani (1999) observe that $m^* \approx m$ if $\beta \ll \gamma$. Indeed (1) is obtained from (2) if $N \to \infty$ and $\beta \to 0$ such that $N\beta \to m\gamma > 0$. This indicates a need for more careful investigation of how epidemic models behave as $N \to \infty$.

The binomial contact model has the desirable consistency property that an infective cannot directly infect more than N susceptibles, though of course there is, in the long run, no limit on the possible number of infectives. The correct mathematical representation of how infectives and susceptibles make contact in the Reed–Frost model (defined below) has been discussed by Jacquez (1987), and by Lefèvre and Picard (1989).

An extension of the general epidemic due to Pettigrew and Weiss (1967) allows two types of infected individual. One of these, called infecteds, eventually shows clinical symptoms and is removed by the usual means. The second type is a carrier of the infection who never shows symptoms and who may persist in the population for a considerable time. Both infecteds and carriers are infectious.

Denote the contact rate of infecteds and carriers by β_i and β_c, and their removal rates by γ_i and γ_c, respectively. Let I_t and C_t denote the number of infecteds and carriers at time t. Assuming, as above, that N is so large that all contacts result in an infection, the process $((I_t, C_t): t \geq 0)$ is a two-type Markov branching process. In particular we assume that an infected makes contacts at event times of a Poisson process with rate β_i, and that a contacted susceptible becomes an infected with probability p_i and a carrier with probability $q_i = 1 - p_i$, independently of what happens for all other contacts. Similarly for carriers, but with probabilities p_c and q_c. Thus the offspring PGF for each infectious type is

$$f_i(\vec{s}) = \frac{\rho_i}{\rho_i + N(1 - p_i s_1 - q_i s_2)} \quad \text{and}$$

$$f_c(\vec{s}) = \frac{\rho_c}{\rho_c + N(1 - p_c s_1 - q_c s_2)}, \qquad (3)$$

where, for example, $\rho_i = \gamma_i/\beta_i$.

The mean matrix is

$$\mathcal{M} = N \begin{bmatrix} \rho_i p_i & \rho_i q_i \\ \rho_c p_c & \rho_c q_c \end{bmatrix},$$

and there is positive probability of a major epidemic iff the larger eigenvalue of $\mathcal{M}$ exceeds unity. If the odds ratio for becoming an infected is the same as for becoming a carrier, $p_i/q_i = p_c/q_c$, then $\det \mathcal{M} = 0$ and the condition for a major epidemic with positive probability is that

$$(\rho_i p_i + \rho_c q_c) > 1. \qquad (4)$$

This is in obvious extension of (2.2).

Pettigrew and Weiss (1967) effectively assume that before their removal, infecteds and carriers have similar social behaviours, i.e., $N\beta_i = N\beta_c$. They express this common value as $b_i + b_c$, and $p_i = b_i/(b_i + b_c)$. The interpretation of b_i is that the probability a given susceptible becomes an infected during $(t, t+h)$ is $b_i(I_t + C_t)h + o(h)$. The criterion (4) then takes the form $b_i/\gamma_i + b_c/\gamma_c > 1$, obtained by Pettigrew and Weiss by consideration of mean population sizes. Probabilities of extinction and laws of the final size can in principle be found from (3). Ball and Clancy (1995) extend this model by allowing several types of infective and they obtain far stronger results.

3.4. Some extensions

This embedded branching process technique can be used in similar ways for many other configurations. For example a host-parasite epidemic comprises populations of susceptible host and vector individuals, with initial sizes N_1 and N_2, respectively. An infectious host contacts susceptible vectors at event times of a Poisson process with rate $\beta_2 N_2$ and its time to removal again has an exponential law with mean γ_1. A symmetric situation obtains for infectious vectors, and the corresponding parameters are $\beta_1 N_1$ and γ_2, respectively.

The imbedded branching process is a periodic two-type process with offspring PGF's

$$f_1(\vec{s}) = \frac{\rho_1}{\rho_1 + N_2(1 - s_2)} \quad \text{and} \quad f_2(\vec{s}) = \frac{\rho_2}{\rho_2 + N_1(1 - s_1)},$$

where $\rho_1 = \gamma_2/\beta_1$ and $\rho_2 = \gamma_1/\beta_2$. The mean matrix has elements $m_{11} = m_{22} = 0$, $m_{12} = N_2/\rho_1$ and $m_{21} = N_1/\rho_2$. It follows immediately that a major epidemic can occur with a positive probability iff

$$N_1 N_2 > \rho_1 \rho_2.$$

Under this condition the probabilities of a minor outbreak starting from a single infected host or vector are easily found to be

$$Q_1 = \frac{\rho_1(N_1 + \rho_2)}{N_1(N_2 + \rho_1)} \quad \text{and} \quad Q_2 = \frac{\rho_2(N_2 + \rho_1)}{N_2(N_1 + \rho_2)},$$

respectively. See Bailey (1975) for the conventional analysis.

The assumption of homogeneous mixing across a large population is likely to be satisfied in only a few cases. One way of relaxing it is to suppose that individuals belong to one of K types. Suppose that initially there are N_j j-type susceptibles, that the contact rate for an i-type infective and j-type susceptible is β_{ij}, and that i-type infectives are removed at rate γ_i. This leads to a birth–death process with $2K$ states and bilinear interaction terms.

We obtain a branching process approximation (Griffiths, 1973) by assuming that all the N_j are so large that while infective numbers are relatively small any contact they may have results in an infection. Thus an i-type infective infects susceptibles according to a Poisson process with rate $b_i = \sum_{k=1}^{K} \beta_{ik} N_k$, and the infected susceptible is a j-type with probability $\beta_{ij} N_j / b_i$. Hence the offspring laws of the embedded approximating branching process have the PGF

$$f_i(\vec{s}) = \left[1 + \sum_{j=1}^{K} m_{ij}(1 - s_j)\right]^{-1},$$

where $m_{ij} = \beta_{ij} N_j / \gamma_i$.

The contact rates need not all be positive, but typically it is assumed that the mean matrix $\mathcal{M} = [m_{ij}]$ is positively regular. This amounts to assuming that any type of susceptible can eventually be infected from the introduction of any one type of infective. A major epidemic occurs with a positive probability iff the eigenvalue λ of $\mathcal{M}$ having maximum modulus > 1. Let Q_i be the probability of a major outbreak when a single i-type infective is introduced. These probabilities comprise the smallest positive solution of the system of equations

$$Q_i \left[1 + \sum_{j=1}^{K} m_{ij}(1 - Q_j)\right] = 1.$$

See Daley and Gani (1999) for details.

The final outcome of a multitype epidemic can be characterized as the total numbers of infections of each type of susceptible during the epidemic. The branching process approximation is simply the total number of births of each type. Let $h_i(\vec{s})$ denote their joint PGF, given one original i-type infective. They solve the nonlinear system $h_i = f_i(s_1 h_1, \ldots, s_K h_K)$. The corresponding probabilities can in principle be obtained using a multivariate version of Lagrange's reversion of series. A binomial contact model can be formulated as above (Daley and Gani, 1999, pp. 88, 89) for which $m_{ij} = \beta_{ij} N_j / (\beta_{ij} + \gamma_i)$.

This simple model of heterogeneity is too complicated for obtaining explicit exact results. Various special cases do yield some explicit results. One such is where $\beta_{ii} \equiv \beta$ and $\beta_{ij} \equiv \bar{\beta}$ ($i \neq j$), i.e., within-group and between-group mixings occur at rates independent of the group label (Daley and Gani, 1999, pp. 90–92). Becker and Marschner (1990) study a model with separable contact rates. Here $N_j = N\pi_j$ where $\pi_j > 0$ for all j, parameters α_j measure the susceptibility to infection of a j-type susceptible, and β_i is the relative infectiousness of an i-type infective. These give offspring means of the form $m_{ij} = \theta\beta_i\alpha_j\pi_j$ where θ^{-1} is a relative removal rate. This structure has the pleasing property that $\lambda = \sum_j m_{jj}$. When $\lambda < 1$ the mean total size m_i due to one original i-type infective has the explicit form

$$m_i = 1 + \frac{\theta\beta_i}{1-\lambda} \sum_{j=1}^{K} \alpha_j \pi_j.$$

These and other results are used to compare the spread of an epidemic in the heterogeneous population with that in an 'equivalent' homogeneous population. This is defined by the vanishing of the covariance $\sum_j (\alpha_j - 1)(\beta_j - 1)\pi_j$, and then θ is the threshold parameter.

We mention a selection of the many variants in the literature. Ball (1985) allows $K = N$ but keeps the β_i constant, i.e., variation among susceptibles is allowed, but not among infectives. See also Butler (1994) and Marschner (1992). Ball and Clancy (1993) assume infectives can move among the groups according to a stochastic process, and that contact rates can depend on the infective's group of origin, the group it currently occupies, and the address of the target susceptible. Ball et al. (1997) give a very detailed treatment of the case where susceptibles are classified as local or remote, and the probability that a contact results in infection depends on whether the susceptible is or is not a neighbour of the infected.

An important extension of the embedded generations idea, usually attributed to Ludwig (1975), is that the real-time evolution of the epidemic process is irrelevant to its threshold behaviour and final size law. For these matters the rôle of the removal-time law and the contact model are significant only insofar as they determine the offspring law of an approximating discrete-time branching process. Thus, for the general epidemic with an arbitrary removal-time law whose Laplace–Stieltjes transform is denoted by $\phi(\theta)$, the PGF of the induced offspring law is $\phi(\beta N(1-s))$. More generally still, the social history of an infective could be described by an orderly point process of contact times and a removal time, and these could be dependent. The real-time approximating branching process would then be a Crump–Mode–Jagers process.

A classical case of the binomial contact model, called the Reed–Frost model, assumes removals occur at times $n = 1, 2, \ldots$. Let S_n and I_n be the numbers of susceptibles and infectives at time n, respectively. Assume that each susceptible at time n avoids contact with a given infective over the next unit of time with a probability q, independently of previous history. If the I_n infectives interact with a given susceptible independently of each other, then this susceptible escapes infection with probability q^{I_n}. Next, S_{n+1} is assumed to arise through binomial sampling of the susceptibles at time n, giving $S_{n+1} =_L \text{Bin}(S_n, q^{I_n})$. The I_n infectives are removed just before time $n + 1$, giving $I_{n+1} = S_n - S_{n+1} =_L \text{Bin}(S_n, 1 - q^{I_n})$. These conditions determine the time evolution of the Reed–Frost model.

Suppose for some $\lambda > 0$ that $1 - q \sim \lambda/N$ as $S_0 = N \to \infty$. If $a = I_0 \ll N$ then the conditional law of infective numbers is $I_{n+1} \approx_L \text{Poisson}(\lambda I_n)$, i.e., the process of infectives is approximated by a branching process with the Poisson(λ) offspring law, and once again the final size law is approximated by the Borel–Tanner law.

3.5. Limit theory

Bailey (1975) wrote "So far, very few asymptotic approximations to discrete-time models are available". This began to change from 1980, when weak convergence theory and strong approximations began to be used *inter alia* to clarify the sense in which epidemic processes are approximated by a branching process. Von Bahr and Martin-Löf (1980) initiated this for the Reed–Frost model and certain generalizations. Since then many other generalizations have been formulated. The following so-called collective model embraces many of these, as well as the general epidemic with an arbitrary removal-time law. See Lefèvre and Picard (1995) for a review and references.

The collective model generalizes the binomial contact mechanism of the Reed–Frost model, but it is the same in other respects. The infectives in any generation behave independently of each other. Suppose there are N susceptibles at time n. Each infective fails to contact any given subset of size $k \leqslant N$ with a probability $q(k; N)$ which depends only on k and N. Thus if A_ℓ is the event that susceptible $\ell = 1, \ldots, N$ escapes contact with any of the I_n infectives, then

$$P(A_1 \cap \cdots \cap A_k | (S_n, I_n)) = [q(k; S_n)]^{I_n},$$

and hence the A_ℓ's are exchangeable.

There are many possible contact mechanisms which realize this structure. One supposes that

$$q(k; N) = f(1 - k/N), \tag{1}$$

where $f(\cdot)$ is a PGF. The physical picture here is that any infective makes contact by sampling with replacement a random number ξ of susceptibles, where ξ has the PGF $f(\cdot)$. Thus the Reed–Frost model corresponds to $q(k; N) = q^k$ if $k \leqslant N$, $= 0$ otherwise, whence $f(s) = q^{N(1-s)}$. An extension assumes infective i who is active during $[n, n+1)$ fails to contact any given susceptible with a random probability $Q_{i,n}$ which is an independent copy of a random variable Q. Hence $q(k; N) = E(Q^k)$. The general epidemic with Poisson contact rate β is included by choosing $q(k; N) = E[\exp(-k\beta \mathcal{R})]$, where $\mathcal{R}$ represents a removal time. If $\mathcal{R} \sim \text{Exp}(\gamma)$ then (1) holds with $f(\cdot)$ given by (3.1).

Suppose again that $S_0 = N$, and now let $N \to \infty$ and suppose that (1) holds with the parameters of $f(\cdot)$ chosen so that either it is independent of N, or that it converges to a non-defective PGF, denoted still by $f(\cdot)$. Let $m = f'(1) < \infty$ and

$\tau(N) = \inf\{n: I_n = 0\}$ be the duration of the epidemic. The final size $\zeta_N = N - S_{\tau(N)}$ converges in law to the total number of births $\widehat{B}$ of the simple branching process whose PGF is $f(\cdot)$. When $m > 1$, $P(\widehat{B} = \infty) = 1 - \sigma^a$, where σ denotes the extinction probability of the branching process. This is the probability of a major epidemic, and corresponding to this event,

$$(S_{\tau(N)} - \eta N)/\sqrt{N} \to_L \mathcal{N}(0, V),$$

where η is the least solution in $(0, 1)$ of the equation $z = e^{-m(1-z)}$,

$$V = \frac{\eta}{(1-m\eta)^2}\left[1 - \eta + \eta(\nu^2 - m)(1 - \eta)\right]$$

and $\nu^2 < \infty$ is the variance of the offspring law. The parameter η is the limiting proportional final size $1 - S_{\tau(N)}/N$ in the Reed–Frost model with $q = \exp(-m)$.

The proof of this result for the Reed–Frost model (plus some extensions) was given by von Bahr and Martin-Löf (1980), though they assert that the asymptotic variance is $1 - \eta$. Lefèvre and Picard (1995) give references for various more general configurations, including an unpublished report for the collective model.

The above limit theorem can be extended to multitype versions of Reed–Frost models. See Ball and Clancy (1993), Lefèvre and Picard (1995), and Andersson (1999) for various formulations and proof techniques, as well as references cited in these papers.

The limit theorem for the final size provides insight to a fundamental objection to using a branching process approximation for classifying behaviour of a finite-state epidemic model. Is it really true that unity is the critical value of R_0 at which major epidemics become possible? Is the major/minor epidemic dichotomy even meaningful? Numerical calculations of the law of the final size of the general epidemic suggests that the answer is 'Yes' to the second question. These calculations suggest that there is a critical value of R_0 at which the final-size law changes from a J-shape to a U-shape, i.e., at which a second mode appears at a large value, corresponding to the possibility of a major epidemic. Nåsell (1995) proposes that this bifurcation of shape should form the basis of the threshold criterion for an epidemic model. He further proposes for the general epidemic model that the critical value of the reproduction ratio is $R_0 \approx 1 + KN^{-1/3}$, where K is an unspecified positive constant.

That the critical value of R_0 should exceed unity is plausible because not all contacts with original susceptibles will result in infection, and because the sampled population is finite in size. Ball and Nåsell (1994) explain the $N^{-1/3}$ dependence on N by investigating the limit law of ζ_N when $m > 1$. This is a mixture of two components, and they substitute $m = 1 + KN^{-1/\kappa}$ and determine the 'typical size' of each component for very large N. They find when $\kappa < 1/3$ that the asymptotic size of the minor epidemic component is of order $N^{2\kappa}$, and the asymptotic size of the major epidemic component is $N^{1-\kappa}$. Since $2\kappa < 2/3 < 1 - \kappa$, it follows that the value $\kappa = 1/3$ represents a critical value at which major and minor components bifurcate. This approach also produces an appropriate value of Nåsell's (1995) unspecified constant K.

A third question is whether the reproductive ratio is always the only determinant of threshold behaviour? Barbour (1994) gives examples of population processes where, within certain regions of the parameter space, the analogue of the reproductive ratio plays no rôle in determining whether there is a positive probability of unlimited growth of the population. One model is a host-parasite system which is discussed more fully by Barbour and Kafetzaki (1993). Born (1998) looks at another version where parasites reproduce within hosts according to a Markov branching process.

Ball and Donnelly (1995) make precise the sense in which a continuous time epidemic model is approximated by a branching process. They allow a general life history model for infective contacts, as mentioned above, though here the discussion is restricted to the general epidemic with a general removal-time law. See Ball and Donelly (1990) and Ball (1995) for summaries.

Let (Z_t) denote the Crump–Mode–Jagers branching process with a ancestors and a reproduction process which has, as above, a Poisson(β) process of birth times stopped at time $\mathcal{R}$, a random variable having the removal-time law. Epidemic processes ε_N are constructed in terms of (Z_t) as follows. The epidemic process begins with a infectives and N susceptibles, and the contact rate of infectives is β/N. The process of infectives $(I_t(N))$ begins by following (Z_t). Contacts with original susceptibles occur at the birth times of the branching process. Label the susceptibles $1, \ldots, N$ and let $U_1, U_2, \ldots$ be independent random variables having a standard uniform law. The ith contact is with the susceptible whose label is $1 + [NU_i]$. If this original susceptible is already infected then the newborn individual in the branching process, along with all its progeny, is ignored in ε_N. Ignored individuals are called ghosts.

Thus the epidemic processes are constructed as subtrees of the branching process family tree. Indeed they coincide up to the first time $G(N)$ a ghost appears. Let $\mathcal{E}$ denote the extinction event of the branching process. A consequence of the U_i having a continuous law is that

$$P\big(G(N) = \infty \text{ for all sufficiently large } N | \mathcal{E}\big) = 1.$$

i.e., once N is large enough $(I_t(N))$ coincides with (Z_t). The situation for $\overline{\mathcal{E}}$ is a little weaker:

$$P\Big(\lim_{N \to \infty} G(N) = \infty | \overline{\mathcal{E}}\Big) = 1.$$

These show that if $t < \infty$ and N is sufficiently large then functionals of $(I_t(N))$ are pathwise coincident with the corresponding branching process functional on the time interval [0,1].

Ball and Donnelly (1995) thus prove that $\max_{u \leqslant t} I_u(N) \to_{\text{a.s.}} \max_{u \leqslant t} Z_u$, $\zeta_N \to_{\text{a.s.}} \widehat{B}$, and that $\int_0^t h(I_u(N))\,du \to_{\text{a.s.}} \int_0^t h(Z_u)\,du$ for any measurable function $h(\cdot)$ and $t \leqslant \infty$. They also estimate the total variation distance between the laws of $I_t(N)$ and Z_t. A significant qualitative consequence is that in a major epidemic, $I_t(N)$ grows like Z_t until about $\sqrt{N}$ of the original susceptibles become infected. After this Gaussian process approximations around the deterministic trajectory are more appropriate (Wang, 1977).

Strong and weak approximations have since been developed for certain modifications of the general epidemic. Ball and O'Neill (1994) replace the intensities $\beta si/N$ and γi with general forms $f_N(s,i)$ and $g_N(s,i)$, respectively. A particular case which limits the mixing rate is $f_N(s,i) = \beta si/(s+i)$, called the modified epidemic model. The key technical conditions for general rate functions are that for all i, $f_N(s_N, i) \to \beta i$ and $g_N(s_N, i) \to \gamma i$ as $N \to \infty$ and $|s_N - N| = O(1)$. The epidemic processes are constructed in terms of a strongly approximating linear birth and death process with per capita birth rate β and death rate γ.

O'Neill (1995) studies the modified epidemic where infectives are classified into high- and low-risk groups, distinguished by different rate constants β. High-risk infectives eventually switch to the low-risk group. Ball et al. (1997) is a related reference. O'Neill (1997) allows the contact rate constant in the modified epidemic to depend on the current number of removals. These models are motivated by problems of sexually transmitted diseases.

Ball (1996) considers a population comprising M groups, or households, each with N individuals. Initially one of the MN individuals is infectious. The contact mechanism for infectives is defined by a removal time and two-point processes, accounting for within-group and between-group contacts, respectively. The model is otherwise similar to the general epidemic. Strong convergence to an approximating Crump–Mode–Jagers process is obtained by allowing $M \to \infty$. Ball (1999) modifies this arrangement so that infectives become susceptible when their period of infectiousness ends.

4. Ecology and conservation modelling

4.1. Introduction

An isolated population will eventually become extinct, and ecological field evidence and opinion support the notion that small populations tend toward extinction more quickly than large populations. See, for example, pp. 99–105 in Primack (1993) and pp. 114–117 in Rosenzweig (1995). Indeed Rosenzweig writes that "Ecologists have confidence in the notion that small populations lead to extinction", and "Rarity begets ultimate rarity". These findings from the field appear to be translated into beliefs about the behaviour of models of population growth. For example, Williamson (1981, p. 88) writes "There is a generally accepted proposition, *derived from the theory of stochastic processes*, that small populations are more likely to become extinct by chance". This belief, or derivation, comprises the basic principle of population viability analysis. This is that the probability of surviving beyond any fixed time horizon can be made arbitrarily close to unity by choosing a sufficiently large intitial population. Indeed, Shaffer (1981), in commenting on arbitrary choices to be made in defining a minimum viable population, suggests the possibility that the critical survival probability can be chosen to be unity!!

In the terms that Williamson (1981) states it, some evaluation of the truth of this belief can be gained from a consideration of simple models of population growth. Begin by considering a stochastic model $(X_t: t \geqslant 0)$ of population growth whose state space has the form $\{0, 1, \ldots\}$, and for which 0 is an absorbing state, the model representation of population extinction. This model is assumed to somehow take account at least of demographic uncertainty. Other sources of uncertainty recognized in the ecological literature as important determinants of population growth are environmental uncertainty, uncertainty due to catastrophes, and genetic uncertainty. Genetic uncertainty is held to be less influential in the long term than the others (Allen et al., 1992; Lande, 1988), and hence it will be neglected here. Let $T = \inf\{t: X_t = 0\}$ be the time to extinction. Since real populations eventually vanish, assume for the time being that $q(i) = P_i(T < \infty) \equiv 1$, where $P_i(\cdot) = P(\cdot | X_0 = i)$, with a similar shorthand for expectations. Let $h_i = E_i(T)$.

It is possible to identify three nested properties which population models might be expected to have.

1. For any $t > 0$, $P_i(T > t)$ increases with i, i.e., extinction is less likely for larger initial sizes.
2. $P_i(T > t) \uparrow 1$ as $i \uparrow \infty$, i.e., the probability of survival can be made arbitrarily large.
3. Each of the previous properties persists after the model has been modified to allow for environmental uncertainty, or for uncertainty due to catastrophes, if these factors are not already taken into account.

Much of the stochastic modelling of populations in the ecological literature is in terms of birth and death processes (BDP's), or large-population approximations based on stochastic differential equations. For references and reviews see Brillinger (1981), Foley (1997), and Wissel and Stöcker (1991). These broad approaches allow considerable flexibility and tractability, and they lend themselves to descriptions in terms of experimentally measurable quantities such as state-dependent per capita growth rates and variances. The last 25 years have seen quite a lot of work using such models to investigate the effects of random environments, resource limitation, and large scale decrements arising from mass emigration and/or catastrophes. Most of this work looks at each factor alone.

But populations wax and wane because individuals live and reproduce. Branching process models take direct account of this, at least for ideal conditions of independent life histories. It is therefore of some interest to investigate how these modifications affect extinction phenomena of branching process dynamics. Much of the mathematical theory about branching processes seeks very precise descriptions of population growth from just one, or a few, ancestors. Of much more interest to population biologists are what can be called *extinction phenomena*. This term denotes the properties of extinction time laws when initial sizes and/or the time horizon are moderate to large. In terms of the above general population model, let $q(i) = P_i(T < \infty)$ be the probability of ultimate extinction. Under which conditions is $q(i) \equiv 1$? If this holds then how does $r_i = \lim_{t \to \infty} P_i(T > t)/P_1(T > t)$ vary with i? The quantity r_i is a measure of how dependence-causing influences affect the increase of T with i. How quickly does $h_i = E_i(T)$ increase with i? Indeed, does it increase at all? Are there limit theorems for T? In the supercritical case, $q(i) < 1$, does $q(i)$ decrease to zero as i increases, and if so, how quickly? These are important questions. In connection with the design of nature reserves, Shaffer and Sampson (1985) write

> Thus the key item of information is the relationship of extinction probabilities to population size.

Pimm (1991) offers a stimulating discussion of ecological issues for conservation biology.

4.2. Extinction phenomena for branching processes

Gosselin and Lebreton (1997) argue in favour of discrete-time branching process models as a useful quantitative tool for conservation biology. However the continuous-time Markov branching process (MBP) is mathematically more tractable, and it exhibits essentially the full range of behaviours shown by still more realistic and complicated branching process models provided their individual lifetime laws have an exponentially small right-hand tail. Denote the MBP by $(Z_t: t \geqslant 0)$, its splitting rate by ρ and its offspring law by $\{p_j: j \in \mathbb{N}_0\}$. Adopt the standard conventions that $p_0 > 0$ and $p_1 = 0$. Let $\mathcal{T}_i$ denote the law of T when $Z_0 = i$. If $m = \sum j p_j < 1$, i.e., the MBP is subcritical, then let $\nu = (1 - m)\rho$. The following properties of the MBP provide a benchmark for comparison with other models.

First, the MBP has the property of being stochastically monotone, meaning that for each $j = 0, 1, \ldots$ and $t > 0$, $P_i(Z_t > j)$ increases with i. This entails the property 1 above. In fact, the characteristic independence of family lines implies that $P_i(T \leqslant t) = (P_1(T \leqslant t))^i$, and this entails property 2 above. If $m < 1$ then $r_i = i$, so that deviations from this identity indicate a loss of independence of family lines due to non-demographic sources of uncertainty. If in addition,

$$\sum p_j j \log^+ j < \infty$$

then

$$\lim_{t \to \infty} e^{\nu t} P_i(T > t) = ic, \tag{1}$$

where $c > 0$. Hence $\mathcal{T}_i$ has an exponentially small tail which fattens as i increases. Another manifestation of this fattening is that

$$h_i \equiv E_i(T) = \nu^{-1}\left[\log(ic) + \gamma + o(1)\right] \quad (i \to \infty), \tag{2}$$

where γ is Euler's constant, and that $\{\nu T - \log(ic)\}$ converges in law to the standard Gumbel law. See Pakes (1989b) for a summary of these properties, and for references.

4.3. Random environments

It is accepted in the ecological literature that extinction is more likely when environments are randomly varying. Haccou and Iwasa (1996) review relevant literature. They consider a supercritical discrete-time branching process where the offspring laws are Poisson with random means. Using a combination of analytical approximations and numerical calculations they compare establishment, or survival probabilities with the corresponding values for a constant environment. There is a considerable theoretical literature on branching processes in random environments, but rather little of it bears directly on properties (1) and (2) above, and that which does is concerned with discrete time. Kaplan (1973) considers a continuous-time model.

Some guidance comes from what is known about the discrete-time case with independent and identically distributed environments. In this case there is an environmentally determined per-capita mean number of offspring M, a random variable. Let ℓ be the left extremity of the support of its distribution function. Under mild conditions, $q(i) < 1$ for all i iff $E(\log M) > 0$, and then $q(i) \downarrow 0$ as $i \uparrow \infty$. If $\ell > 1$ then $q(i) = O(\alpha^i)$ for some $\alpha < 1$, giving essentially classical behaviour, though algebraic factors may appear. If $\ell = 1$ then some environments can be critical, and there is a constant $\beta > 0$ such that $q(i)e^{\beta\sqrt{i}}$ grows or decays algebraically fast. A substantial qualitative change occurs when $\ell < 1$. If $\beta > 0$ satisfies $E(M^{-\beta}) = 1$, and other technical conditions hold, then $q(i) \sim \text{const.}i^{-\beta}$ – extinction becomes much more likely when subcritical environments can occur. But note that this represents the worst possible behaviour for the supercritical case. See Grey and Lu (1993) for more precise statements.

Any construction of a subcritical branching process with random environments must have properties (1) and (2) above. This is because the population process, when conditioned on the environmental process, retains the property of independent family lines. If the environmental process is independent then the process of population sizes is still a Markov chain, although the characteristic branching property of independent lines of descent is lost. The subcritical case, $E(\log M) < 0$, is further subdivided according to the sign of $\lambda = E(M \log M)$ which controls the long-term frequency of supercritical environments. The following results hold when $i = 1$. They were stated by Afanas'ev (1979), and much later proofs have been provided by D'Souza and Hambly (1997) and Liu (1996). When $\lambda < 0$, called the strongly subcritical case, the probability ζ_n of survival beyond time n is asymptotically proportional to m^n where $m = E(M)$. This generalizes the case of steady environments. If $\lambda = 0$ then ζ_n is of order $n^{-1/2}m^n$. In the weakly subcritical case, $\lambda > 0$, ζ_n is of order $n^{-3/2}\mu^n$ where $\mu = \inf_{0 \leqslant \theta \leqslant 1} E(M^\theta)$.

Hence in all these cases the extinction time law continues to have an exponentially small right-hand tale. It is likely that similar behaviour occurs in continuous-time models and hence that the qualitative content of (1.1) and the leading term component of (1.2) subsist when environments are temporally random. Putting it another way, if random environments affect only individual fecundity, and not the splitting rate ρ, then gross extinction phenomena will be described as for steady environments with m replaced by $E(M)$. *This conflicts with the conclusions of* Allen et al. (1992), which are not informed by any explicit modelling of environmental effects. On the other hand, it supports property 3 provided catastrophes are excluded. Numerical results of Haccou and Iwasa (1996) suggest that even better approximations could be obtained by using the corresponding constant environment with offspring mean given by $\exp(E(\log M))$.

Mode and Jacobson (1987a, 1987b), and Mode and Root (1988), present an age-class model with random environments constructed with an eye to ecological relevance and feasibility of computer simulation. The model is a discrete-time multitype branching process, with the types corresponding to each of finitely many age classes. There are age dependent survival probabilities from one age class into the next, and age dependent Poisson offspring laws, and all these quantities are linked through logistic transformations to a stationary auto-regressive environmental process. The above references explore how variations of significant parameters (e.g., sign of environmental correlation) affect, respectively, extinction probabilities, minimum viable population sizes, and expected generation sizes conditioned on the environmental process. An important aspect of these studies is the revelation of phenomena not normally revealed by analytic study of extinction probabilities, limit theorems, and so on. For example, Mode and Root (1988) show time series of the conditional generation sizes which rise to a sharp peak and quickly decay. The overall shape is similar for the various computed realizations except for considerable variation in the position of the peak.

4.4. Catastrophes and emigration

These terms are mathematical synonyms for large random decrements. They can diminish populations much more dramatically than random environments, e.g., the knock-out effect of an earthquake, or large fire. The literature recognizes two mechanisms of catastrophe occurence – they occur at rates proportional to instantaneous population sizes, or at a constant rate. Call these Type 1 and 2, respectively. Examples of the Type 1 case are mass emigrations, and other reductions which occur more frequently for larger population sizes. The Type 2 case corresponds to an external and independent Poisson process of catastrophe events such as climatic disasters.

Let d_{ij} denote the probability that the population size is j immediately following a catastrophe when the population size was $i \geqslant j$. In the extreme case $d_{i0} = 1$ for all $i \geqslant 1$ extinction will occur after the first catastrophe. In such a case $q(i) \equiv 1$. In addition T_i is an exponential law for Type 2 occurences, thus violating properties 1 and 2. Moreover this occurs no matter what underlying population model is used. More generally still, these phenomena occur for any catastrophe modification which gives a q-matrix element q_{i0} which is independent of $i \geqslant 1$. The general Markovian situation was explored by Pakes (1995) and independently by Kyriakidis (1993).

The benchmark example of this phenomenon is the MBP with $p_0 = 0$ (no deaths), a Type 1 occurence mechanism, and uniformly distributed decrements $d_{ij} = 1/i$ for $i = 0, 1, \ldots, j - 1$: a catastrophe removes a random proportion of the population. This provides an example of a model population whose rarity or abundance makes no difference to its survival-time law. Brockwell et al. (1982) first observed that T_i can be independent of i in the case where the MBP is the linear birth process. They described this finding as "intriguing" and Anderson (1991) called it a "surprising result". Pakes (1995) analysed the MBP extension (still with $p_0 = 0$) in some detail. Allowing $p_0 > 0$ causes only quantitative changes to these properties. In particular the laws T_i remain tight with respect to i irrespective of the value of m.

The MBP with uniformly distributed decrements and the Type 2 occurence mechanism retains the essential qualitative properties of the MBP. A positive probability of establishment $1 - q(i)$ can occur only if $m > 1$. In this case the criticality parameter is $A = \kappa/(m-1)\rho$, where κ is the occurence rate of catastrophe events. It is known that $q(i) < 1$ iff $A > 1$, and then $q(i+1) \leqslant q(i)$. The exact rate of convergence to zero is not known, but $q(i)$ is approximately of order $i^{-(A-1)}$, and hence this form of catastrophe greatly increases the probability of extinction. If $m \leqslant 1$ or $m > 1$ and $A < 1$, then

$$h_i \sim \frac{\log i}{\kappa - (m-1)\rho}.$$

Hence the logarithmic increase of h_i for the unmodified MBP seen in (2.2) is preserved.

The most thoroughly understood example of model catastrophes is that of *random walk decrements*, meaning that the probability law of the numbers lost by catastrophic mortality is independent of the current population size. More precisely,

$$d_{ij} = \begin{cases} \delta_{i-j} & \text{when } 1 \leqslant j \leqslant i-1 \,\&\, i \geqslant 1, \\ \sum_{k \geqslant i} \delta_k & \text{when } j = 0 \,\&\, i \geqslant 1, \end{cases}$$

where $\{\delta_i\}$ is a discrete law on $\mathbb{N}_0$. See Pakes (1986) and (1989b), and references therein. It is often more convenient to work with a parametrization which extends the representation of the MBP as a Markov process having linear jump rates and a jump chain which is a left-continuous random walk stopped at the origin. Indeed Athreya and Karlin (1967) construct and analyze the supercritical MBP from this point of view. Adding the linear-rate random walk decrement component effectively changes the jump chain to a general random walk on the integers. This approach is developed in Pakes (1986). Denote a typical increment of this random walk by Y, its law by $\{a_j : j \in \mathbb{Z}\}$, and its drift by $D = E(Y)$.

Not surprisingly, $q(i) \equiv 1$ when $D \leqslant 0$ and $q(i) < 1$ when $D > 0$. In the latter, supercritical case, $q(i) \downarrow 0$ as $i \uparrow \infty$ with a rate which is determined by the nature of the left-hand tail of Y. This rate is geometrically fast when $P(Y < -i)$ is exponentially small and the rate is regularly varying (Bingham et al., 1987) when $P(Y < -i)$ is regularly varying (Pakes, 1987).

If $D < 0$ and $V = E[(Y^+)^2] < \infty$ then

$$h_i \sim (-\rho D)^{-1} \log i \quad (i \to \infty). \tag{1}$$

This generalizes the estimate contained in the leading-order term of (2.2) for the MBP, since in that case $a_j = p_{j+1}$ ($j \geqslant -1$), $= 0$ otherwise, and $D = m - 1$. It is likely that (1) holds under the simple condition $-\infty < D < 0$.

Under further restrictions on the increment law, Pakes (1989b) shows that $\{T - h_i\}$ has a non-defective limit law which has, unfortunately, an unmanageably complicated form. The cases for which this behaviour is shown to hold are where: (i) $a_j = \beta p q^{-j-1}$ ($j \leqslant -1$), $0 < \beta, q < 1$ and $p + q = 1$; and (ii) $a_j \equiv 0$ when $j \geqslant 2$. Case (i) corresponds to a geometric decrement law, and (ii) corresponds to a subcritical linear birth-death process with a general random walk decrement law. The point is that these cases preserve the qualitative content of asymptotic results for the subcritical MBP. In particular, the spirit of property 3 subsists in the sense that $T \to_p \infty$ as $i \to \infty$. However this property is lost if $P(Y < -i)$ decreases sufficiently slowly.

Assuming only that $-\infty \leqslant D < 0$, it can be shown that h_i is asymptotically proportional to the product of $\log i$ and the terms of a renewal sequence. When D

is finite this representation preserves the essence of (1). But when $D = -\infty$ (1) still holds, now in the sense that $h_i/\log i \to 0$. This example shows that extinction times can remain relatively small with increasing numbers of founders even though the law of the number of removals does not depend on the current population size.

Similar results hold for random-walk catastrophes with the Type 2 occurence mechanism. Suppose $m > 1$. Then $q(i) < 1$ iff $E(\log^+ Y) < \infty$, and then the $q(i)$ decrease at a rate which again is controlled by the left-hand tail of Y. This rate can be geometrically or algebraically fast, or even slowly varying (Pakes, 1988). Complete results for the subcritical case are available only when the population process is the linear BDP (Pakes, 1989b). In this case the general form of (2.2) is preserved with the catastrophe component affecting only the O(1) term.

The binomial decrement law $d_{ij} = \binom{i}{j} p^j (1-q)^{i-j}$ has received some attention in the literature. It expresses the idea that individuals survive a catastrophe with probability p independently of the fate of other members of the population. Results known for the Type 2 occurence case include criteria for extinction, and exponential growth rates when $q(i) < 1$. Little is known for the Type 1 case. What can be said is that $q(i) \equiv 1$ for any value of m, and that h_i does not converge but rather oscillates around a fixed level as i increases without bound.

Summarising the above, a catastrophe component can affect model population processes more dramatically than random environments. In particular, if extinction probabilities are not identically equal to unity, then they can decline very slowly indeed. However, it seems likely that if the decrement probabilities $d_{i,i-j}$ are substantially independent of i for fixed j, then under reasonable moment conditions, the $q(i)$ will still decay geometrically fast. In other words, the establishment probability of an initially expanding population can be made close to unity with a modest number of founders.

When $m < 1$, it appears that the main qualitative property of the expected time to extinction h_i for the MBP, i.e., that it grows in proportion to $\log i$, subsists for a variety of catastrophe mechanisms. This appears to be a very robust property, and it would be useful if a more precise statement could be found.

4.5. The SLOSS debate

A key question of conservation biology is whether the survival probability of a managed population is greater if it is confined to a single large reserve in contrast to being spread among several smaller reserves of equal total area. This is called the SLOSS question, single large or several small.

Diamond (1975) argued from principles of Island Biogeography in support of single large reserves. The debate began when Simberloff and Abele (1976) warned that these principles are not universally applicable, and hence that costly and damaging mistakes could be made. Many papers have been written supporting one side or the other. Rosenzweig (1995) argues, with reference to species-area curves, that there is no difference between the two sides. These issues are discussed by Primack (1993) and Shafer (1990).

Calculations for simple population models suggest that there is no generally applicable answer, and that such answers as can be obtained depend on small changes in model structure. The results reported in the previous section form the basis for these calculations.

Suppose that a population is well represented, at least while it is not too large, by the stochastic model (X_t) and that its probability of ultimate extinction $q(i) < 1$, i.e., it has a positive probability of establishment from any number of founders. Compare two configurations. The first is for several small reserves: there are N reserves each started with i founder individuals, and these reserves are assumed to wax and wane independently of each other. This independence assumption is a simplification because reserves should be located so that they are weakly coupled by migrations. Typically N is quite small. The second configuration is a single reserve with Ni founders.

The extinction probability for the partitioned arrangement is $Q_p(i) = (q(i))^N$, and that for the single reserve is $Q_s(i) = q(Ni)$. Which is larger? To answer this question assume that the probability of extinction has the general form

$$q(i) = Q^i i^a L(i), \qquad (1)$$

where $0 < Q \leqslant 1$, $a \in \mathfrak{R}$ and $L(\cdot)$ is slowly varying at infinity (Bingham et al., 1987). This choice covers most of the configurations discussed in the previous section. A large value of the quotient

$$R(i) \equiv Q_p(i)/Q_s(i) = \frac{i^{aN}(L(i))^N}{(Ni)^a L(Ni)} = \left(\frac{i^{N-1}}{N}\right)^a \cdot \frac{(L(i))^N}{L(Ni)}$$

favours a single large reserve. Note that it is independent of Q. If i is fairly large, then $R(i)$ is large when $a > 0$, and small when $a < 0$. Thus 'single large' is better when $a > 0$, and 'several small' is better when $a < 0$. The second case always occurs when $Q = 1$ in (1), for example when occasional subcritical environments can occur in an otherwise supercritical random environment model. Catastrophes of sufficient magnitude provide another example. This finding is consistent with a 'bet hedging' strategy to reduce the likelihood of a major collapse of the whole population.

Some emigration structures give rise to the case $a = 0$ and either L is identically constant, or $L(i) \to c$ $(i \to \infty)$ where $0 < c < \infty$. Several small reserves are better when $c < 1$. This case occurs for geometrically distributed random walk decrements with the Type 1 occurence mechanism.

When $a < 0$, or $a = 0$ and $c < 1$, it follows that subdivision should be carried as far as possible. Of course this cannot be a reasonable conclusion. One reason is that the above assumptions about validity of the asymptotic estimates of $q(i)$ are violated if N is unbounded as i becomes large. A similar conclusion is found by Burkey (1989) from numerical calculations using a Leslie matrix model which incorporates density dependence.

A supercritical random environments model in which critical environments occur with a positive probability has

$$q(i) = e^{-\beta\sqrt{i}} i^a L(i).$$

Some algebra yields

$$R(i) = e^{-\beta\sqrt{i}(N-\sqrt{N})} O(i^{a(N-1)}).$$

This is smaller when $N > 1$ and hence a 'several small' configuration is better in this case.

Assume now that $q(i) \equiv 1$. There are several comparisons which can be explored. Let

$$\tau_{N,i}(t) = P(T > t | N \text{ reserves, each with } i \text{ founders}).$$

First compare $\tau_{N,i}(t)$ with $\tau_{1,Ni}(t)$, and determine which is larger when t and/or i are large.

Fix i and assume there are constants r_i such that $\tau_{1,i}(t) \sim r_i \tau_{1,1}(t)$ $(t \to \infty)$. Then

$$\tau_{N,i}(t)/\tau_{1,Ni}(t) = \frac{1 - (1 - \tau_{1,i}(t))^N}{\tau_{1,Ni}(t)} \to Nr_i/r_{Ni}.$$

This is unity for the MBP benchmark, as expected since partitioning does not affect survival of the independently developing family lines. A large value of the limit implies that 'several small' gives a larger value of the probability that T is large.

For the MBP with geometrically distributed random walk decrements it is known that $r_i \sim Ci^a$ as $i \to \infty$, where C is a positive constant and $0 < a < 1$. This suggests that the 'several small' configuration is better when founders are fairly numerous. The same can be true for the linear BDP with random walk decrements, and probably in most cases. But the opposite finding can be true when the decrements have a Poisson distribution.

Another comparison allows i and t to be large. The general form of the extinction-time limit theorem for random walk decrements is that there are constants $\nu, c > 0$ such that

$$\lim_{i \to \infty} P_i\left(T \leqslant \nu^{-1}(x + \log(ic))\right) = A(x), \tag{2}$$

where $A(x)$ is a distribution function which in nearly all cases has a very complicated form (Pakes, 1989b). The result (2) holds under moment conditions which almost certainly are satisfied in real cases. For example, it holds for the MBP when $\sum p_j j \log j < \infty$, and then $\nu = \rho(1 - m)$ and $A(x) = \exp(-e^{-x})$.

If (2) holds then, setting $t = \nu^{-1}(x + \log(ic))$, it follows that

$$\lim_{i \to \infty} \tau_{N,i}(t)/\tau_{1,Ni}(t) = (A(x))^N / A(x - \log N).$$

The limit is identically unity for the MBP case, but in the general case its behaviour as a function of N is not clear.

A more manageable approach uses (2) to make a comparison in terms of expected extinction times

$$h_{N,i} \equiv \int_0^\infty \tau_{N,i}(t)\,dt = \int_0^\infty \left[1 - \left(1 - \tau_{1,N}(t)\right)^N\right] dt$$

$$= v^{-1} \int_{-\log(ic)}^\infty \left[1 - \left(P_i\left(T \leq v^{-1}(x + \log(ic))\right)\right)^N\right] dx$$

$$\approx v^{-1} \int_{-\log(ic)}^\infty \left[1 - \left(A(x)\right)^N\right] dx$$

$$= v^{-1} \left\{\log(ic) - \int_{-\log(ic)}^0 \left(A(x)\right)^N dx + \int_0^\infty \left[1 - \left(A(x)\right)^N\right] dx\right\}.$$

Thus to a first approximation $h_{N,i} \approx v^{-1} \log i$, independent of N. The leading contribution to the discrepancy of this approximation is the last integral, giving

$$h_{N,i} - v^{-1} \log(ic) \approx v^{-1} E(M_N),$$

where M_N is the maximum of N independent copies of a random variable whose distribution function is $A(x)$. Hence

$$h_{N,i} = v^{-1}\left[\log(ic) + E(M_N)\right] + o(1) \quad (i \to \infty),$$

giving the estimate

$$h_{N,i} - h_{1,Ni} = v^{-1}\left[E(M_N) - E(M_1) - \log N\right] + o(1).$$

This could be used in conjunction with known approximations for $E(M_N)$. Roughly speaking, if $1 - A(x)$ decreases faster (respectively, slower) than e^{-x} then the left-hand side tends to have a negative (respectively, positive) value, indicating 'single large' is better (respectively, worse).

4.6. Birth–death processes and environmental uncertainty

Many particular cases of the general birth and death process arise in the ecological literature. Any birth and death process is stochastically monotone and hence possesses property 1 in Section 1. Pakes (1995) observed that property 2 is satisfied iff the upper boundary is at infinity and it is natural (Anderson, 1991, p. 262), i.e., iff the forward Kolmogorov equations have a unique solution. This will be the case except in very unusual modelling situations, though Holgate's (1967a) model of small mammal population explosions and crashes is an exception. The work in previous sections shows plainly that property 3 need not be satisfied. Another example is Pakes' (1995) construction of a pure birth process with a catastrophe component such that $P_i(T > t)$

can either increase with i (property 1) but with a limit less than unity (not property 2); or decrease; or oscillate with no limit.

Shaffer (1981), Allen et al. (1992), and many others, group under the term "environmental uncertainty" factors such as changes in weather or food supplies, competition, predation and parasitism from other species. No separation of these factors is made in the above references, nor in most others. Shafer (1990) is exceptional in separating temporal uncertainty (weather etc.) from factors like competition and so on, but then he explicitly ignores temporal uncertainty. In the branching process literature environmental uncertainty is explicitly represented as a (usually stationary) random process of environmental states which determine the temporal selection of offspring laws (and/or splitting rates) from some collection of such laws.

These remarks serve as a background to a new parametrization and representation of environmental uncertainty for birth and death processes which usually is attributed to Goodman (1987a, 1987b), but which has its roots in Leigh (1975, 1981). Let $(B_t; t \geq 0)$ denote a BDP with birth and death parameters $\lambda(i) = ib(i)$ and $\mu(i) = id(i)$ ($i = 0, 1, \ldots$), respectively. Thus $b(i)$ and $d(i)$ are per-capita rates, and since $\lambda(0) = 0$, the zero state is absorbing. Goodman (1987a), p. 216, recommends working with two rates which he asserts have a more immediate biological interpretation than the birth and death rates. These are the instantaneous growth rate

$$r(i) = b(i) - d(i),$$

and the corresponding variance

$$V(i) = (b(i) + d(i))/i.$$

These quantities are derived directly from (B_t) according to

$$r(i) = \lim_{h \to 0} h^{-1} E\left(\frac{B_{t+h} - B_t}{B_t} \bigg| B_t = i\right)$$

and

$$V(i) = \lim_{h \to 0} h^{-1} E\left(\left(\frac{B_{t+h} - B_t}{B_t}\right)^2 \bigg| B_t = i\right). \tag{1}$$

Thus $V(i)$ is a mean-square error corresponding to the prediction of B_{t+h} by B_t implicit in the definition of $r(i)$.

Goodman (1987a, 1987b) observes that $V(i) = O(1/i)$ for the linear BDP, i.e., that demographic uncertainty fades with increasing i. On the other hand, environmental uncertainty is never negligible and it is not diluted by increasing population size and hence it can be modelled by taking

$$V(i) = V_1/i + V_2, \tag{2}$$

a resolution into an individual and an environmental component respectively, where the V_j are constants. This form appears quite frequently in ecological literature, e.g., Wissel and Stöcker (1991).

Supplementing (2) with $r(i) = c - di$ yields

$$\lambda(i) = i(A + Bi) \quad \text{and} \quad \mu(i) = i(C + Di),$$

where

$$A = (V_1 + c)/2, \quad B = (V_2 - d)/2, \quad C = (V_1 - c)/2 \quad \text{and}$$
$$D = (V_2 + d)/2.$$

The state space is finite if $V_2 < d$, a natural extrapolation of the linear case with finite carrying capacity. If $B \geqslant 0$ the state space can still be made finite, but if it is infinite then the minimal process is unique and hence honest (Anderson, 1991, p. 100). There is a scattered literature on the birth and death process with quadratic rates, some of which can be traced through Ismail et al. (1990), Parthasarathy and Lenin (1997), and Picard (1965). Nearly all of this assumes either that $D \leqslant 0$, or $B = D$, and hence is not directly applicable to the present case.

A question which arises is this: In what sense does V_2 represent environmental uncertainty? One answer might be that it just seems like a good idea – it is algebraically the easiest way of including a term larger than the demographic component. Goodman (1987a) answers obscurely by appealing to a diffusion model which in reality is just the geometric Brownian motion process (GBP) constrained by a reflecting barrier at $K \gg 0$ and an absorbing barrier at unity, representing extinction. He does not establish a connection with the assumption (2).

One could argue as follows. The infinitesimal variance of GBP has the form Vx^2, i.e., V is the relative infinitesimal variance when the process is within its state space. There are many modelling examples in the ecological literature, starting with Lewontin and Cohen (1969), where a quadratic variance term is accepted as a valid representation of the effects of environmental uncertainty. Keiding (1975) sketched an argument for a population growing through discrete non-overlapping generations as follows.

Let M_{n+1} be the per-capita mean number of offspring born into generation $n+1$ in a simple branching process $\{Z_n\}$ evolving under the influence of independent and identically distributed environments. Let v_{n+1} be the corresponding variance, both quantities being random variables. If ζ denotes a realization of the environmental process, then

$$E\big[(Z_{n+1} - Z_n)^2 | Z_n = i, \zeta\big] = i v_{n+1} + i^2 (M_{n+1} - 1)^2. \tag{3}$$

Introduce a sequence $\{Z_n^{(\nu)}\}$ of branching processes indexed by $\nu \to \infty$ such that $Z_0^{(\nu)} \sim \nu k$ $(k > 0)$,

$$E\big(M_n^{(\nu)} - 1\big) \sim r/\nu;$$

$$E(v_n^{(\nu)}) \to v > 0;$$

and

$$E(M_n^{(\nu)} - 1)^2 \sim V/\nu.$$

These choices imply that the fluctuations described by (3) scale in proportion to ν, i.e.,

$$\lim_{\nu \to \infty} \nu^{-1} E\big[(Z_1^{(\nu)} - Z_0^{(\nu)})^2 \big| Z_0^{(\nu)}\big] = vk + Vk^2. \tag{4}$$

If the $M_n^{(\nu)}$ are degenerate then $(M_n^{(\nu)} - 1)^2 \sim (r/\nu)^2$, and hence $V = 0$. The contrary case, $V > 0$, manifests the subsistence of environmental uncertainty under the above limiting regime. On this basis Keiding (1975) conjectured that the sequence of interpolated processes $(Z_{[\nu t]}^{(\nu)}/\nu : t \geq 0)$ converges to a diffusion defined by the stochastic differential equation

$$dK_t = rK_t\, dt + \sqrt{vK_t + VK_t^2} \cdot dW_t, \tag{5}$$

where (W_t) is a standard Wiener (or Brownian motion) process. This conjecture was independently verified by Helland (1981) and Kurtz (1978). Evidently the quadratic component Vx^2 arises from the second term on the right-hand side of (3), an environmental mean-square error induced by the i reproduction events in the nth generation.

The situation in continuous time is less clear. The analogue of (3) for the MBP is

$$E\big[(Z_{t+h} - Z_t)^2 \big| Z_t = i, \zeta\big] = hi\rho_t\big(v_t + (M_t - 1)^2\big) + o(h).$$

Introduce an indexed sequence of processes as above, suppress the ν superscripts, and assume that the splitting rate ρ_t is uncorrelated with (M_t, v_t). This is appropriate if the environment affects fecundity but not the rate of birth–death events. If $E(\rho_t) \to \rho > 0$, $E(v_t) \to v > 0$, and $E((M_t - 1)^2) \to V > 0$ as $\nu \to \infty$, then the analogue of (4) is

$$\nu^{-1} E\big[(Z_{t+h} - Z_t)^2 \big| Z_t = i\big] = \rho hk(v + V) + o(h).$$

Hence there appears to be no room for a quadratic term to occur, which might be expected because in continuous time birth/death events cannot simultaneously occur. In the configuration envisaged here, environmental uncertainty acts only to inflate demographic uncertainty.

However, different assumptions *do* yield the quadratic term in the diffusion limit. Ethier and Kurtz (1986) give a general limit theorem for the MBP with random environments and their limit process is implicitly defined as the solution of a stochastic integral equation. They give an example where the limit process is the diffusion (5). In their example the MBP for index ν is a birth and death process with a death parameter which is the constant 1, and the birth parameter is a stochastic process of the form

$1 + v^{-1/2}(-1)^{\xi(t)}$ where $\xi(t)$ is a standard Poisson process. Here the splitting rate and the offspring laws are strongly dependent. Changing details to make the splitting rate constant does not alter the essence of the limit process. Perhaps more importantly, the inherent scaling of the sequence of processes is different from that used above

This discussion shows there is a sound basis for representing environmental uncertainty by a quadratic term in the infinitesimal variance of a diffusion approximation when discrete generation models are being approximated. Moreover, the source of this component is clear. The situation for discrete-state and continuous-time processes is less clear because the source of environmental variation subsisting in a diffusion approximation is much more subtle. There seems to be no rationale by which appeals to analogies with other classes of processes can be used to accept (2) as describing the effects of environmental uncertainty on BDP models.

On the other hand, the additional term in (2) is compatible with catastrophe effects. In fact this gains support by remarks in §6 of Goodman (1987a) which provide an inkling of how he conceives environmental uncertainty. He speculates that the linear birth and death process "seems to foreclose the possibility of random environmental variation acting on all individuals...". Although such variation acts collectively in the sense of altering the defining probability laws, it cannot alter the temporal isolation of birth/death events. Indeed, no modification of the algebraic form of the birth and death rates can do this.

Now catastrophes *do* act collectively. If the definition (1) is extended to a MBP modified so that catastrophes occur at a rate $\kappa(i)$ when the population size is i and with a general decrement law, then

$$V(i) = \rho v/i + (\kappa(i)/i^2) \sum_{j=1}^{i-1} d_{ij}(j-i)^2,$$

where v is the variance of the offspring law. If $C(i)$ denotes the coefficient of $\kappa(i)$ then

$$C(i) = \begin{cases} 1/3 + 1/2i + O(i^{-2}) & \text{(uniform decrements)}, \\ p(1+i^{-2}) & \text{(binomial decrements)}. \end{cases}$$

For a Type 2 occurence mechanism with $\kappa(i) = \kappa$ when $i \geq 1$, then

$$V(i) = V_2 + V_1/i + O(i^{-2}), \qquad (6)$$

where

$$V_1 = \rho v + \begin{cases} \kappa/2i & \text{(uniform)}, \\ 0 & \text{(binomial)} \end{cases}$$

and

$$V_2 = \begin{cases} \kappa/3 & \text{(uniform)}, \\ p\kappa & \text{(binomial)}. \end{cases}$$

Hence these catastrophe mechanisms in essence satisfy (2), but their effect on the MBP produce mainly quantitative changes in many of its mathematical properties, consistent with the above remarks about the MBP with random environments. On the other hand, when $\kappa(i) = \kappa i$ the term V_2 in (6) is replaced by $V_2 i$ and behaviour then is very different from the MBP.

For random walk decrements with $d_{ij} = \delta_{i-j}$ $(1 \leqslant j \leqslant i)$,

$$C(i) = i^{-2} \sum_{j=1}^{i-1} j^2 \delta_j + \sum_{j \geqslant i} \delta_j.$$

If the variance v_d of the decrement law is finite, then $C(i) \sim \text{const.}/i^2$ and $V(i) \sim V_1/i$ with $V_1 > \rho v$. This is effectively the form for the MBP but with an inflated demographic variance term.

Several modes of behaviour can occur when $v_d = \infty$. For example, if $\delta_i \sim \gamma/i^{2+d}$, where $\gamma > 0$ and $0 < d < 1$, then

$$V(i) = \frac{2\gamma c}{1 - d^2} i^{-d} + (\rho v/i)(1 + o(1)).$$

In this case the imbedded random walk has a finite drift and the variance $V(i)$ still decays with increasing i, and the deviations from MBP behaviour are mainly quantitative ones.

If $\delta_i \sim \gamma/i^{1+d}$ the drift $D = -\infty$ and

$$V(i) = \frac{2\gamma c}{d(2-d)} i^{1-d} (1 + o(1)),$$

giving a variance which increases with i. Finally, the intermediate case $\delta_i = \gamma/i(i+1)$ gives $V(i)$ with leading terms as given by (2). Both of these cases show qualitative deviations from MBP behaviour.

Accepting that eliciting the behaviour of extinction phenomena for branching processes which have been modified to include sources of uncertainty in ways which lead to (6) is very hard in general, a more accessible guide is obtained from the extinction phenomena of the diffusion approximation (5). Its boundary classification (Keiding, 1975) shows that zero is absorbing and attainable, and that ∞ is always unattainable, but it is attracting iff

$$r > V/2, \tag{7}$$

i.e., iff the nett growth rate is large enough to overcome the variability of environmental uncertainty. In other words, the extinction time τ of (K_t) satisfies

$$\widehat{P}_k(\tau < \infty) \equiv 1 \quad \text{iff} \quad r \leqslant V/2,$$

where $\widehat{P}_k(\cdot)$ denotes the law of (K_t) when $K_0 = k$.

The backward equation for the Laplace–Stieltjes transform of τ can be explicitly solved with the following consequences. First, suppose (7) holds and let $A = (2r/V) - 1$. Then

$$\widehat{P}_k(\tau < \infty) = (1 + Vk/v)^{-A}. \tag{8}$$

This can be compared with $\exp(-2rk/v)$ when $V = 0$. The algebraic decay exhibited in (8) is compatible with that which occurs for the simple branching process with random environments in which subcritical environments occur with a positive probability. Conversely, any limiting régime which leads to (5) must admit a proportion of subcritical environments. In particular, the aptness of (5) as an approximation entails a preference for the "several small" configuration.

Now let

$$r < V/2 \tag{9}$$

and $B = V/2 - r$. Then

$$\widehat{E}_k(\tau) = B^{-1}\big[\log k + \log(V/v) + \psi(2+B) + \gamma + o(1)\big],$$

where $\psi(\cdot)$ is the psi-function. This may be compared to

$$\widehat{E}_k(\tau) = (-r)^{-1}\big[\log k + \log(-2r/v) + \gamma + o(1)\big]$$

when $V = 0$. This mimics relationships seen above for the MBP with and without catastrophe components.

Still assuming (9) with $V \geq 0$, if $k \to \infty$ then

$$\tau/\widehat{E}_k(\tau) \to_p 1,$$

and this is refined as follows. If $V = 0$ then

$$\tau - \widehat{E}_k(\tau) \to_L G,$$

the centered Gumbel law whose distribution function is $\exp(-e^{-\gamma - x})$. If $V > 0$ then

$$\frac{\tau - \widehat{E}_k(\tau)}{\sqrt{B^{-3} V \log k}} \to_L \mathcal{N}(0, 1).$$

An approximation for the SLOSS problem is obtained by setting $k = N$ and $v = i$, giving $T \approx i\tau$ for sufficiently large i. This yields

$$h_{N,i} - h_{1,Ni} \approx i\widehat{E}_N(\tau) - i N\widehat{E}_1(\tau) \approx (i/B)\log N + o(1).$$

This implies the same conclusions as for the supercritical case, that "several small" is better.

4.7. Resource limitation

As population size increases, resources become more scarce and population growth should slow down. Standard branching process models can be modified to accommodate this by making the offspring laws depend on the current population size. The simple branching process with this modification can be defined by

$$E(s^{Z_{n+1}}|Z_n = i) = (f(s;i))^i,$$

where $f(s;i)$ denotes the offspring PGF for an individual in a generation of size i. The Markov branching version is defined in a similar way and with a splitting rate which may depend on the population size.

These models can incorporate size or density dependence by allowing the offspring laws to be stochastically decreasing when the population size or density grows beyond some level. Thus Lipow (1977a) considers a Markov branching process which evolves subcritically above a fixed threshold K, and supercritically below it. Her model also allows resurrection from the zero state by independent immigration. The motivation behind Lipow's branching dynamics is similar to that given by Goel and Richter-Dyn (1974), pp. 77–86, for some variants of the basic McArthur and Wilson (1967) model of Island Biogeography (i.e., a linear BDP with a reflecting barrier at $K \gg 0$).

A conceptual attraction of size dependent branching models is that they retain in a very specific way the idea that populations grow and decline because individuals live and die. They are in fact more flexible than the BDP's favoured in the ecological literature, but this comes at a price of greater mathematical difficulties. Nevertheless there is quite a sizeable body of literature which investigates their properties; see Vatutin and Zubkov (1993) for a summary of what was known up to 1990.

Much of the early work concentrates on finding conditions giving behaviour similar to classical branching process models and determining what happens when these conditions are perturbed. Let $m(i) = f'(1;i)$ denote the per capita mean number of offspring when the population size is i and let $v(i)$ denote the associated variance. The classical theory can be extended for demographically stabilizing populations, i.e., for those where $m(i) \to m$ and $v(i)$ is appropriately controlled as $i \to \infty$. For instance, if $m > 1$ it seems likely that the population size Z_n will grow in proportion to m^n. It turns out that this depends strongly on the rate at which $m(i)$ approaches m. Klebaner (1984a) shows that if $m^{-n}Z_n$ has a non-degenerate limit then $\sum |m(i) - m|/i < \infty$. Additional regularity conditions imply the converse. Also see Klebaner (1985).

The situation when $m = 1$ is quite complex. If $m(i) \leqslant 1$ then extinction occurs almost surely. When $m(i) \geqslant 1$ and $\lim_{i \to \infty} i[m(i) - 1] = c \geqslant 0$, Klebaner (1984b) shows that the probability of extinction $q(i) < 1$ if $0 < \lim_{i \to \infty} v(i) < c$. In this case Z_n grows linearly (in law) on the set of immortality, with a limiting gamma law. His conditions for $q(i) \equiv 1$ are more fragmentary. This occurs if, for example, $\limsup i^2[m(i) - 1] < \infty$. This condition is bound up with the proof he uses. Reinhard (1990) gives a more complete analysis for the continuous-time version when $m = 1$. In particular the gap in Klebaner's condition is filled in: If for some $0 \leqslant \beta < 1$, $0 < c = \limsup i^{1+\beta}[m(i) - 1] < \infty$ and $\liminf i^\beta v(i) > 2c$, then $q(i) \equiv 1$. Boiko

(1977, 1980) obtains limit theorems for a Markov branching process which has an arbitrary size-dependent offspring law when the population size $\leqslant K$ but which behaves like the classical model above the threshold K – the extreme case of demographic stabilization. In the critical and super-critical cases the asymptotic behaviour of Boiko's model is qualitatively identical to the classical model.

A case where a fairly complete analysis is possible arises by setting $f(s; i) = (a(s))^{1/i} b(s)$ ($i \geqslant 1$) where $a(\cdot)$ and $b(\cdot)$ are PGF's. In essence this is just the simple branching process with offspring PGF $b(\cdot)$, and with independent numbers of immigrants into each generation which have the PGF $a(\cdot)$, and which is stopped when it first hits the zero state. Höpfner (1985a) obtains the same limit theorems for the case $m = 1$ as Klebaner (1984b) but under more general conditions. See Seneta and Tavaré (1983) for the cases where $m \neq 1$. These can be supplemented by Vatutin (1974), who obtains asymptotic estimates of $q(i)$ when $m > 1$. Höpfner (1985b) looks at the same problem when $m = 1$. Nadkarni (1964, 1971) gives incomplete treatments of the particular case where the offspring laws are Poisson. Note that $m = b'(1)$ and $m(i) = m + a'(1)/i$ for this class of models.

In cases where immortality has a positive probability, it would appear that a more natural approach to determining asymptotic growth behaviour of the population size is to represent the size process as a randomly perturbed discrete dynamical system (cf. (5) below). Kersting (1990) reviews results which have been obtained using this point of view.

The case which most likely is applicable to ecological situations is $m < 1$, about which much less is known. Gosselin (1998a, 1998b) gives conditions which ensure the existence of a limiting-conditional distribution (LCD) for a discrete-time and state Markov chain $\{X_n: n \geqslant 0\}$ whose state space is the positive integers and zero is the single absorbing state, accessible from the positive integers. It is worth pausing to recall this concept because merely knowing that a LCD exists provides valuable information about the law of the extinction time T (i.e., in this context, the hitting time of the zero state). Anderson (1991) and Kijima (1997) are good general references for this topic.

Assume $P_i(T < \infty) \equiv 1$. A law $\mathcal{M}$ on the positive integers is an LCD if for each $i, j > 0$,

$$\mathcal{M}(j) = \lim_{n \to \infty} P_i(X_n = j | T > n). \tag{1}$$

There clearly is at most one LCD (though others may exist for different initial laws). The existence of an LCD implies the existence of

$$R^{-1} = \lim_{n \to \infty} \left(P_i(T > n) \right)^{1/n},$$

that $1 < R < \infty$, and that $\mathcal{M}$ is R-invariant in the sense that

$$R \sum_{i \geqslant 1} \mathcal{M}(i) p_{ij} = \mathcal{M}(j) \quad (j \geqslant 1),$$

where $[p_{ij}]$ denotes the one-step transition matrix of $\{X_n\}$. Iteration of the last relation yields

$$P_{\mathcal{M}}(X_n = j) = R^{-n}\mathcal{M}(j) \quad (j \geqslant 1) \tag{2}$$

and summing on j gives the important identity

$$P_{\mathcal{M}}(T > n) = R^{-n}. \tag{3}$$

The content of (1) and (2) is that if the Markov chain is not extinct after running for a considerable time, then the conditional law of X_n is well-approximated by $\mathcal{M}$, and the remaining time until extinction has a law which is closely approximated by a modified geometric law:

$$P_{\mathcal{M}}(T = n) \approx \left(1 - R^{-1}\right) R^{-(n-1)} \quad (n \geqslant 1).$$

Dividing (3) into (2) shows that $\mathcal{M}$ is stationary-conditional (also called quasi-stationary):

$$P_{\mathcal{M}}(X_n = j | T > n) = \mathcal{M}(j).$$

Parallel results hold for the continuous-time case, but now the law of the remaining time to extinction is approximately exponential with the parameter

$$\mu = \lim_{t \to \infty} \left[-t^{-1} \log P_i(T > t) \right].$$

Computing R or μ in particular cases may be difficult. But these results support a conjecture made by Goodman (1987b). He considers BDP's which satisfy (6.2) when $i < K$, where K represents a finite carrying capacity, defined by $\lambda(K) = 0$. On the basis of numerical calculations Goodman (1987b), p. 25, circumspectly conjectures that if the BDP starts at K, and K is large, then the extinction-time law is close to exponential. Subsequent writers have mistakenly asserted this as a proven result.

The principle results in Gosselin (1998b) provide sufficient conditions for density dependent branching processes to become extinct and to have a LCD. Moreover his results assert that the conditioned law of X_n approaches the LCD geometrically fast. Gosselin's conditions involve a knowledge of the spectral radius of an operator defined by the restriction of $[p_{ij}]$ to the positive states, and hence they are somewhat inaccessible. (Ferrari et al. (1996) give conditions which probably are easier to check.) However all requisite conditions are satisfied by some density-dependent branching processes for cases where the offspring laws are power-series laws derived from the same generator, i.e.,

$$f(s; i) = \frac{g(s\ell(i))}{g(\ell(i))},$$

where $g(\cdot)$ is a PGF and $0 < \ell(i) \leqslant 1$. Let $h(s) = sg'(s)/g(s)$ and $\lambda = \sup\{s \leqslant 1: h(s) = 1\}$. A LCD exists if $\limsup_{i\to\infty} \ell(i) < \lambda$ and $\sup_{i \geqslant 1} \ell(i) < 1$. If $g(s) = e^{-1+s}$ then all offspring laws are Poisson with means $m(i) = \ell(i)$, and $\lambda = 1$. Appropriate choices of $m(i)$ gives density dependent growth functions which often occur in the ecological literature. For example, $m(i) = re^{-\beta i}$ ($r \leqslant 1$ and $\beta > 0$) give Ricker dynamics. Gosselin's conditions are satisfied for these models, showing that extinction occurs and that a LCD exists. However his methods offer no insight as to the form of the LCD. This is not surprising since the LCD of the simple branching process is specified by a functional equation which can be solved in very few cases. Seneta and Tavaré (1983) obtain an expression for the LCD of the Höpfner size-dependent model.

The expected generation sizes of the above density-dependent models satisfy the awkward non-linear relation

$$E(Z_{n+1}) = E(Z_n m(Z_n)).$$

Let $\phi(z) = zm(z)$ and define the deterministic sequence $\{z_n\}$ by $z_0 = Z_0$ and $z_{n+1} = \phi(z_n)$. The ecologically relevant situation is where $\phi(\cdot)$ is a unimodal map with $\phi(0) = \phi(\infty) = 0$, and then it is well known that $\{z_n\}$ can exhibit very complicated behaviour. Is it possible for Z_n to track z_n? This occurs in a limiting sense for a special class of processes introduced by Klebaner (see Klebaner, 1997 for references).

Let $K > 0$ be a threshold parameter, which could represent a carrying capacity. Define branching processes $\{Z_n^K\}$ by

$$Z_{n+1}^K = \sum_{j=1}^{Z_n^K} \xi_n(j, Z_n^K/K), \qquad (4)$$

where for each $x > 0$, the $\xi_n(j, x)$ are independent and identically distributed with mean $m(x)$ (not the same as $m(i)$ above), and they are independent of Z_n^K. Let $\varphi(x) = xm(x)$. Klebaner stops the process if it hits $\{K, K+1, \ldots\}$, and hence $\varphi(\cdot)$ can be taken as defined on $[0, 1]$.

The defining relation (4) can be rewritten as a perturbed dynamical system describing the population density process defined by $X_n^K = Z_n^K/K$,

$$X_{n+1}^K = \varphi(X_n^K) + K^{-1/2}\eta_n^K(X_n^K), \qquad (5)$$

where

$$\eta_n^K(x) = K^{-1/2}\sum_{j=1}^{xK}(\xi_n(j, x) - m(x)).$$

Replacing the z_n's above, let $x_{n+1} = \varphi(x_n)$. If $X_0^K \to_p x$ as $K \to \infty$, then $X_n^K \to_p x_n$ for each n. Thus, when K is large, initial segments of the density process are well approximated by the same segments of the deterministic trajectory. This shadowing is observed to persist for very long times in simulations of particular models.

In addition, deviations from the deterministic trajectory are described by a Gaussian process. Let $v(x) = x\operatorname{Var}(\xi_1(1,x)) < \infty$. Then $\eta_n^K(x) \to_L \mathcal{N}(0, v(x))$. Let $Y_n^K = \sqrt{K}(X_n^K - x_n)$ and suppose that $Y_0^K \to_p y$ as $K \to \infty$. Under a further technical condition the process $\{Y_n^K\}$ converges weakly to a Gaussian process $\{Y_n\}$ satisfying

$$E(Y_n) = y \prod_{j=0}^{n-1} \varphi'(x_j)$$

and

$$\operatorname{Cov}(Y_n, Y_v) = \sum_{k=0}^{n-1} v(x_k) \prod_{j=k}^{n-1} (\varphi'(x_j))^2 \prod_{i=k}^{v-1} \varphi'(x_i) \quad (n \leqslant v).$$

More can be said about the structure of the limit process when $\varphi(\cdot)$ has a periodic attractor.

Suppose this indeed is the case and let ω denote the period of the attractor. Then the ω-fold iterate of φ has ω stable fixed points. Consider one of these, and let B denote its basin of attraction. The exit time from B of the sampled process $\widehat{X}_n = X_{n\omega}^K$ is

$$\tau^{x,K} = \inf\{n > 0 \colon \widehat{X}_n \notin B, \widehat{X}_0 = x \in B\}.$$

Then $\tau^{x,K}$ is exponentially large in K: for any small $\delta > 0$ there is a number $V > 0$ not depending on x such that

$$\lim_{K \to \infty} P(e^{K(V-\delta)} \leqslant \tau^{x,K} \leqslant e^{K(V+\delta)}) = 1, \quad \text{i.e.,} \quad \log \tau^{x,K} \to_p V.$$

This result follows from a large deviation principle, and it provides an estimate of the time that trajectories of the density dependent process track the deterministic trajectory. If $\omega = 1$ this is an assertion about the time to extinction of the density process since then $B = (0, 1)$. Computation of V is via a fairly involved recipe. See Klebaner and Zeitouni (1994) for details, and Pierre-Loti-Viaud (1993) for more on large deviation principles and size-dependent branching processes.

The density-dependent branching process is represented by (5) as a randomly perturbed discrete dynamical system. A "meta-theorem" asserts that weak limits of measures which are invariant for the perturbed system are invariant for the unperturbed system. Högnäs (1997) examines this for the particular case where $P(\xi_1(1, x) = j) = e^{-x} p_j$ ($j \geqslant 1$) and $P(\xi_1(1, x) = 0) = 1 - e^{-x}$, where $\{p_j\}$ is an ordinary offspring law with $p_0 = 0$. Denote its mean by e^r, noting that $r > 0$. Consequently

$$E(X_{n+1}^K | X_n^K = x) = \varphi(x) = xe^{r-x},$$

i.e., the deterministic system $\{x_n\}$ follows Ricker dynamics. The stationary-conditional law $\mathcal{M}_K$ of $\{X_n^K\}$ is supported on $\{1/K, 2/K, \ldots\}$. Högnäs restricts r so that φ has an attracting periodic cycle $a_1, a_2, \ldots, a_\omega$ say, with $\varphi'(a_j) < 1$ ($1 \leq j \leq \omega$). With mild additional assumptions, $\mathcal{M}_K$ converges weakly as $K \to \infty$ to the measure which attributes mass ω^{-1} to the points of the periodic attractor. This result says that if the density-dependent branching process is not extinct after a long time and if K is large, then it takes values close to Ka_j with nearly equal frequency before eventually hitting the zero state. Högnäs speculates that adding random environmental effects of sufficient magnitude could destroy this picture by causing extinction to occur much more quickly.

Klebaner et al. (1998) prove related results with less specific assumptions about the form of the offspring laws. They assume that $\varphi(\cdot)$ has only finitely many fixed points, that at least one of them is stable, and that each deterministic trajectory converges to a stable fixed point. Further technical conditions lead to the conclusion that any weakly convergent sub-family of $\{\mathcal{M}_K\colon K > 0\}$ is an invariant measure for $\varphi(\cdot)$, and that each such limit measure attributes no mass to any unstable fixed point.

The behaviour above, whereby a density-dependent branching process tracks a deterministic path with Gaussian fluctuations around this path, is exhibited by a very wide range of density dependent Markov processes. Ethier and Kurtz (1986) is the standard work (for continuous time) on this subject, though Pollett (1990) provides a more 'user-friendly' account together with more relaxed conditions on the asymptotic form of the generator as the parameter K increases.

A related diffusion approximation may be applicable to moderate sized populations in which the per-capita offspring mean is near unity with deviations from this which may be size dependent. Lipow (1977b) considers a family of Markov branching processes $(Z^K(t)\colon t \geq 0)$ where $Z^K(0) = K$ and the offspring law depends on K and possibly also on population size i. The splitting rate ρ is assumed independent of i and K. The key assumption is that the per-capita offspring mean for population size i is

$$m^K(i) = 1 + \alpha(i/K)/K + \varepsilon(i, K),$$

where $\alpha(\cdot)$ is bounded and continuous on $[0, \infty)$ and $K\varepsilon(i, K) \to 0$ uniformly in i as $K \to \infty$. The corresponding offspring variance $v^K(i) = v + \delta(i, K)$ where $v > 0$ and $K\delta(i, K) \to 0$ uniformly in i.

Lipow (1977b) shows under a further technical condition that the normalized process $(Z^K(Kt)/K)$ converges weakly to the diffusion process whose infinitesimal mean is $\rho x \alpha(x)$ and variance is $\rho v x$. This is just the Feller branching diffusion process when $\alpha(\cdot)$ is constant. Lipow (1980) obtains some elementary properties of the limiting diffusion, such as that it hits the origin if $\alpha(x) \leq 0$ for all sufficiently large x.

A full analysis for ecologically motivated forms of $\alpha(\cdot)$ would be very interesting. Rittgen (1990) has generalized Lipow's work for the multitype Markov branching process, as well as allowing some size-dependent variation in the splitting rates.

4.8. Some ecological examples

Lebreton and Clobert (1991) make some interesting general observations about the utility and limitations of branching process models in assessing extinction probabilities of small populations. The LCD of a branching process model with independent random environments has been estimated by Lebreton (1982) for a decreasing population of White Storks in the Alsace region. He estimates the mean of the LCD as 7 individuals and a yearly probability of extinction of 0.15. Since a branching process with independent random environments is a Markov chain, the existence of an LCD for specific models can be ascertained using the general criteria of Gosselin (1998b) or Ferrari et al. (1996).

The determination of offspring laws which minimise the probability of extinction, mentioned in Section 2.1, arises too in connection with optimal life-histories. The theme of finding ecologically useful offspring laws which minimise the extinction probability for a fixed value of p_0 and m is pursued by Holgate (1967b), Daley (1969) and Goodman (1968).

A more interesting pursuit is to ask why is semilparity (reproduction once in a lifetime) not universally favoured over iteroparity (reproduction on several occasions)? Given that reproduction increases the chance of mortality, some have argued that it is better to forgo the opportunity of several reproductive events in favour of just one when the individual has reached full maturity. Holgate (1967b) investigates this question in a model system of biennial animals which can reproduce in each year of their lives, but where the probability of survival into a second year is ha^j if it produces j offspring in the first year ($0 < h, a \leqslant 1$).

He finds for some geometric and Poisson offspring laws that the probability that the population survives is an increasing function of h, the probability that an individual which is childless in the first year will survive to a second year. Also, setting $a = 1$ and comparing survival probabilities with those for a model of annual animals, where the parameters are chosen so both models have the same rate of geometric increase, it is found that that the annual model has the smaller probability of surviving.

Mountford (1973) discusses two differing explanations of observed clutch-sizes of bird species, viz, whether they have evolved to maximise growth rates or minimise extinction probabilities. Arguments based on bet hedging favour the latter as a way of coping with environmental uncertainty and avoiding direct competition for food. He computes clutch-size probabilities which minimise extinction probabilities for models where environmental variation affects nestling survival through food abundance. Mountford finds that clutches of 2 or 3 eggs are optimal. Heyde (1978) also discusses this problem.

Grey (1980) pursues further aspects of this problem. His general set-up incorporates independent stationary environments and PGF's $f(s; N, \zeta)$ for the number of eggs in a clutch of size N which produce a laying bird when the environmental variable is ζ. The aim is to find a clutch-size law $\{\pi_N\}$ which maximize survival probabilities for differing ancestral numbers. The optimal choice depends on the life-history strategy 'used' by the birds. For example, the clutch-size law could vary from one generation to the next, or it could be chosen and fixed for all generations. Numerical examples illustrate Grey's theoretical development.

Mountford (1988) uses a model of univoltine insects reproducing and dispersing to leaves to discuss the question of whether the detection of density dependence from empirical data is obscured by spatial and stochastic heterogeneity. His model is a Markov chain for the sizes of successive generations of insect. It allows for random dispersal of adult insects among many leaves, a Poisson larval production law for each adult, and size-dependent survival to adulthood of the larva on each leaf. Although not modelled as such, infinite divisibility arguments show this model is a size-dependent branching process. Mountford's discussion is based on the LCD estimated from simulations of the process. He finds that the detection of density dependence is enhanced by the presence of spatial heterogeneity and it is unaffected by stochastic heterogeneity. Mountford (1988), p. 848, makes the following interesting observations about the rôle of the LCD in population studies:

> It is difficult to understand why the concept of the quasi-stationary distribution is not used more often in the theoretical study of regulated populations. It is a complete description of the dynamics of the population; it subsumes the question of stability; the mean population size of the distribution is equivalent to the carrying capacity of the environment.

The interaction of fecundity and sex ratio is investigated by Gabriel and Bürger (1992) using size-dependent branching processes. Individual offspring laws are Poisson with size-dependent means $m(i)$ which take one of three ecologically motivated forms, such as

$$m(i) = e^{r(1-i/K)} \tag{1}$$

(Ricker dynamics again!) where $r > 0$ represents the small-population growth rate and K is the carrying capacity. Numerical values of the expected time to extinction are computed and graphed as functions of K, r and Z_0/K, respectively. Effects of sex are modelled as a two-type process in which the number of females and males born to an individual are independent with the same Poisson law whose mean depends on the number of females j and males k in the parent generation. This mean function has the form

$$\begin{cases} 0, & \text{if } j=0 \text{ or } k=0, \\ m(j+k)\frac{2j}{j+k}, & \text{otherwise.} \end{cases}$$

Allowing for sex in this way typically reduces the mean time to extinction because the above sex ratio factor < 1 unless $j = k$.

Some of the analytical properties of these models are inferred from simpler ones valid for extreme parameter configurations and which reduce the generation-size process to a stationary independent one. A careful analysis seems worthwhile, as does the inclusion of more realistic mating models, such as those discussed in Section 1.3.

Ludwig (1996) conducts a numerical study of a size-dependent model which also allows independent random environments and catastrophic mortality. Let $\{\zeta_n\}$ denote independent environmental random variables and suppose ζ_n/a has a gamma law with shape parameter $a > 0$, whence $E(\zeta_n) = 1$. If the size of generation n is i, then the offspring law is Poisson with mean $\zeta_n i m(i)$ where $m(\cdot)$ is given by (1). Averaging over the environmental law shows that the number B_{n+1} born into generation $n+1$ has PGF

$$E\big(s^{B_{n+1}} | Z_n = i\big) = \sum_{j \geq 0} p_{ij}(R) s^j = \big[1 + (im(i)/a)(1-s)\big]^{-a},$$

a negative-binomial PGF. These newborn individuals are subject to catastrophic mortality with probability c. Let $p_{jk}(C)$ denote the probability of k survivors when $B_{n+1} = j$. This law is taken to be a beta-binomial law. There are corresponding and unspecified probabilities $p_{jk}(S)$ of survival from normal causes of mortality. Thus the one-step transition probabilities of $\{Z_n\}$ are given by

$$p_{ij} = \sum_j p_{ik}(R)[cp_{jk}(S) + (1-c)p_{jk}(C)].$$

Ludwig (1996) plots the probability of surviving n or more generations as functions of n for differing values of K. One conclusion is that survival probabilities for small n ($\leqslant 10$ for his parameter values) are insensitive to changes in K, but quite large differences are seen when n is large. This suggests that survival of small populations is little affected by increasing K (creating larger reserves) unless the population size is simultaneously increased.

By informal reasoning Ludwig (1996) shows his branching process model is approximated by a diffusion under appropriate conditions. These are the usual ones of a large population with small changes over successive generations, and also that K is large in comparison to the typical scale δx of per-generation population change, $K = \kappa/\delta x$ where $\kappa > 0$. The resulting diffusion is given by (6.5) with the factor rK_t multiplied by the logistic correction $1 - K_t/\kappa$.

These examples show that branching process methodology has much to offer theoretical investigation of population regulation and extinction, and that much remains to be done!

References

Abraham, R. and Mazliak, L. (1998). Branching properties of Brownian paths and trees. *Exposition Math.* **16**, 59–73.
Afanas'ev, V. I. (1979). On the nonextinction probability of a subcritical branching process in random environment. Manuscript Dept., *VINITI*, No. 1, 794–799.
Alexandersson, M. (1999). *Branching Processes and Cell Populations*. Matematiskt Centrum, Göteborg.
Allen, E. J., J. M. Harris, and L. J. S. Allen (1992). Persistence-time models for use in viability analysis. *J Theor. Biol.* **155**, 33–53.
Alsmeyer, G. and U. Rösler (1996). The bisexual Galton–Watson process with promiscuous mating: extinction probabilities in the supercritical case. *Ann. Appl. Probab.* **6**, 922–939.
Alsmeyer, G. and U. Rösler (2002a). Asexual versus promiscuous bisexual Galton–Watson processes: the extinction probability ratio. *Ann. Appl. Probab.* **12**, 125–142.
Alsmeyer, G. and U. Rösler (2002b). The Martin entrance boundary of the Galton–Watson process. *Report 13/02-S*, Universität Münster.
Anderson, W. J. (1991). *Continuous-Time Markov Chains*. Springer-Verlag, New York.
Andersson, M. (1999). The asymptotic final size distribution of multitype chain-binomial epidemic processes. *Adv. Appl. Probab.* **31**, 220–234.
Arino, O., M. Kimmel and G. F. Webb (1995). Mathematical modeling of the loss of telomere sequences. *J. Theor. Biol.* **177**, 45–57.
Asmussen, S. (1980). On some two-sex population models. *Ann. Probab.* **8**, 727–744.
Asmussen, S. and H. Hering (1983). *Branching Processes*. Birkhäuser, Boston.

Assaf, D., L. Goldstein and E. Samuel-Cahn (2000). An unexpected connection between branching processes and optimal stopping. *J. Appl. Probab.* **37**, 613–626.

Athreya, K. B. (1992). Rates of decay for the survival probability of a mutant gene. *J. Math. Biol.* **30**, 577–581.

Athreya, K. B. and P. Jagers (Eds.) (1997). *Classical and Modern Branching Processes*. Springer-Verlag, New York.

Athreya, K. B. and S. Karlin (1967). Limit theorems for the split times of branching processes. *J. Math. Mech.* **17**, 257–277.

Athreya, K. and P. E. Ney (1972). *Branching Processes*. Springer-Verlag, Berlin.

Bailey, N. T. J. (1975). *The Mathematical Theory of Infectious Diseases and its Applications*, 2nd ed. Charles Griffin, London.

Ball, F. (1985). Deterministic and stochastic epidemics with several kinds of susceptibles. *Adv. Appl. Probab.* **17**, 1–22.

Ball, F. (1995). Coupling methods in epidemic theory. In *Epidemic Models: Their Structure and Relation to Data*, pp. 34–52 (Ed. D. Mollison). Cambridge University Press, Cambridge.

Ball, F. (1996). The threshold behaviour of stochastic epidemics among a population divided into households. In *Athens Conference on Applied Probability and Time Series. Vol. I: Applied Probability*, pp. 253–266 (Eds. C. C. Heyde, Yu. V. Prohorov, R. Pyke and S. T. Rachev). Springer, New York.

Ball, F. (1997). The threshold behaviour of stochastic epidemics. In *Advances in Mathematical Population Dynamics*, pp. 407–424 (Eds. O. Arino, D. Axelrod and M. Kimmel). World Scientific, Singapore.

Ball, F. (1999). Stochastic and deterministic models for SIS epidemics among a population partitioned into households. *Math. Biosci.* **156**, 41–67.

Ball, F. and D. Clancy (1993). The final size and severity of a generalized stochastic multitype epidemic model. *Adv. Appl. Probab.* **25**, 721–736.

Ball, F. and D. Clancy (1995). The final outcome of an epidemic model with several different types of infective in a large population. *J. Appl. Probab.* **32**, 579–590.

Ball, F. and P. Donnelly (1990). Branching process approximation of epidemic models. *Theory Probab. Appl.* **34**, 119–121.

Ball, F. and P. Donnelly (1995). Strong approximations for epidemic models. *Stochastic Process. Appl.* **55**, 1–21.

Ball, F. and O. D. Lyne (2001). Stochastic multi-type SIR epidemics among a population partitioned into housholds. *Adv. Appl. Probab.* **33**, 99–123.

Ball, F. and I. Nåsell (1994). The shape of the size distribution of an epidemic in a finite population. *Math. Biosci.* **123**, 167–181.

Ball, F. and P. O'Neill (1994). Strong convergence of stochastic epidemics. *Adv. Appl. Probab.* **26**, 629–655.

Ball, F., D. Mollison and G. Scalia-Tomba (1997). Epidemics with two levels of mixing. *Ann. Appl. Probab.* **7**, 46–89.

Barbour, A. D. (1994). Threshold phenomena in epidemic theory. In *Probability, Statistics and Optimization*, pp. 101–116 (Ed. F. P. Kelly). Wiley, Chichester.

Barbour, A. D. and M. Kafetzaki (1993). A host parasite model yielding heterogeneous parasite loads. *J. Math. Biol.* **31**, 157–176.

Barnett, S. A. (1998). *The Science of Life*. Allen & Unwin, St Leonards, Australia.

Bartlett, M. S. (1978). *An Introduction to Stochastic Processes*, 3rd ed. Cambridge University Press, Cambridge.

Bartoszyński, R. (1969). Branching processes and models of epidemics. *Dissertationes Math.* **LXI**, 1–48.

Becker, N. G. (1989). *Analysis of Infectious Disease Data*. Chapman & Hall, London.

Becker, N. G. and I. Marschner (1990). The effect of heterogeneity on the spread of disease. In *Stochastic Processes in Epidemic Theory*, pp. 90–103 (Eds. J.-P. Gabriel, C. Lefrèvre and P. Picard). Springer-Verlag, New York.

Bennies, J. and G. Kersting (2000). A random walk approach to Galton–Watson trees. *J. Theor. Probab.* **13**, 777–803.

Biggins, J. D. (1997). How fast does a general branching random walk spread? In *Classical and Modern Branching Processes*, pp. 19–39 (Eds. K. B. Athreya and P. Jagers). Springer-Verlag, New York.

Bingham, N. H., C. M. Goldie and J. L. Teugels (1987). *Regular Variation*. Cambridge University Press, Cambridge.

Boiko, R. V. (1977). A limit theorem for branching random processes with variable mode (critical case). *Ukrainian Math. J.* **29**, 68–73.

Boiko, R. V. (1980). A supercritical branching process with variable mode. *Ukrainian Math. J.* **32**, 118–122.

Born, E. (1998). A point process model with stochastic intensities for a branching population of two dependent types. *Adv. Appl. Probab.* **30**, 723–739.

Boucher, K., A. Zorin, A. Y. Yakovlev, M. Mayer-Proschel and M. Noble (2001). An alternative stochastic model of generation of oligodendrocytes in cell culture. *J. Math. Biol.* **43**, 22–36.

Brillinger, D. R. (1981). Some aspects of modern population mathematics. *Canad. J. Statist.* **9**, 173–194.

Brockwell, P. J., J. Gani and S. Resnick (1982). Birth, immigration and catastrophe processes. *Adv. Appl. Probab.* **14**, 709–731.

Brook, D. (1966). Bounds for moment generating functions and for extinction probabilities. *J. Appl. Probab.* **3**, 171–178.

Bru, B., Jongmans and E. Seneta (1992). I. J. Bienaymé: Family information and proof of the criticality theorem. *Int. Statist. Rev.* **60**, 177–183.

Bruss, F. T. (1984). A note on the extinction criteria for bisexual Galton–Watson processes. *J. Appl. Probab.* **21**, 915–919.

Burkey, T. V. (1989). Extinction in nature reserves: the effect of fragmentation and the importance of migration between reserve fragments. *Oikos* **55**, 75–81.

Butler, S. M. (1994). The early and final states of an epidemic in a large heterogeneous population with a small initial number of infectives. *Adv. Appl. Probab.* **26**, 671–689.

Cavalli-Sforza, L. L. and M. W. Feldman (1981). *Cultural Transmission and Evolution: A Quantitative Approach*. Princeton University Press, Princeton.

Chauvin, B. and A. Rouault (1988). KPP equation and supercritical branching Brownian motion in the subcritical speed area. Application to spatial trees. *Probab. Theory Related Fields* **80**, 299–314.

Chauvin, B. and A. Rouault (1990). A stochastic simulation for solving scalar reaction-diffusion equations. *Adv. Appl. Probab.* **22**, 88–100.

Chauvin, B., S. Cohen and A. Rouault (Eds.) (1996). *Trees: Workshop in Versailles, June 14–16, 1995*. Birkhäuser, Basel.

Consul, P. C. (1991). Evolution of surnames. *Int. Statist. Rev.* **59**, 271–278.

Cournot, A. A. (1847). *De l'Origine et des Limites de la Correspondance entre l'Algebre et la Géométrie*. Hachette, Paris, §36. Réédition: Vrin, Paris, 1989.

Crump, K. S. and J. H. Gillespie (1976). The dispersion of a neutral allele considered as a branching process. *J. Appl. Probab.* **13**, 208–218.

Daley, D. J. (1968). Extinction conditions for certain bisexual Galton–Watson branching processes. *Z. Wahrsceinlichkeitsth.* **9**, 315–322.

Daley, D. J. (1969). Extinction probabilities in branching processes: a note on Holgate and Lakhani's paper. *Bull. Math. Biophys.* **31**, 35–37.

Daley, D. J. and J. Gani (1999). *Epidemic Modelling*. Cambridge University Press, Cambridge.

Daley, D. J., D. M. Hull and J. M. Taylor (1986). Bisexual Galton–Watson branching processes with superadditive mating functions. *J. Appl. Probab.* **23**, 585–600.

Dawson, D. A. (1993). Measure-valued Markov processes. In *Ecole d'Eté de Probabilités de Saint-Flour XXI – 1991*, pp. 1–260 (Ed. P. L. Hennequin). Lecture Notes in Math., Vol. 1541. Springer-Verlag, Berlin.

Demetrius, L., P. Schuster and K. Sigmund (1985). Polynucleotide evolution and branching processes. *Bull. Math. Biol.* **47**, 239–262.

Devroye, L. (1998). Branching processes and their applications in the analysis of tree structures and tree algorithms. In *Probability Methods for Algorithmic Discrete Mathematics*, pp. 249–314 (Eds. M. Habib et al.). Springer, Berlin.

Diamond, J. (1975). The island dilemma: lessons of modern biogeographic studies for the design of natural reserves. *Biol. Conserv.* **7**, 129–146.

Dion, J.-P. (1991). *Statistical Inference for Discrete Time Branching Processes*. Lecture Notes, 7th International Summer School on Probability and Mathematical Statistics, Varna, Bulgaria.

Dobzhansky, T., F. J. Ayala, G. L. Stebbins and J. W. Valentine (1977). *Evolution*. W. H. Freeman & Co., San Francisco.

Donnelly, P. (1991). Comment. *Statist. Sci.* **6**, 277–279.

D'Souza, J. C. and B. M. Hambly (1997). On the survival probability of a branching process in a random environment. *Adv. Appl. Probab.* **29**, 38–55.

Durrett, R. (2002). *Probability Models for DNA Sequence Evolution*. Springer, New York.

Dynkin, E. B. (1994). *An Introduction to Branching Measure-Valued Processes*. Amer. Math. Soc., Providence, RI.

Eshel, I. (1981). On the survival probability of a slightly advantageous mutant gene with a general distribution of progeny – a branching process model. *J. Math. Biol.* **12**, 355–362.

Eshel, I. (1984). On the survival probability of a slightly advantageous gene in a multitype population: a multidimensional branching process model. *J. Math. Biol.* **19**, 201–209.

Etheridge, A. M. (2000). *An Introduction to Superprocesses*. University Lecture Series, Vol. 20. Amer. Math. Soc., Providence, RI.

Ethier, S. N. and T. G. Kurtz (1986). *Markov Processes*. Wiley, New York.

Ewens, W. J. (1968). Some applications of multi-type branching processes in population genetics. *J. Roy. Statist. Soc. B* **30**, 164–175.

Ewens, W. J. (1969). *Population Genetics*. Methuen & Co., London.

Ewens, W. J. (1979). *Mathematical Population Genetics*. Springer-Verlag, Berlin.

Fahady, K. S., M. P. Quine and D. Vere-Jones (1971). Heavy traffic approximations for the Galton–Watson process. *Adv. Appl. Probab.* **3**, 282–300.

Falahati, A. (1999). Two-sex branching populations. Dissertation, Dept. Math. Statistics, Chalmers Univ. Tech., Göteborg, Sweden.

Farrington, C. and A. D. Grant (1999). The distribution of time to extinction in subcritical branching processes: applications to outbreaks of infectious disease. *J. Appl. Probab.* **36**, 771–779.

Ferrari, P. A., H. Kesten and S. Martínez (1996). R-positivity, quasi-stationary distributions and ratio limit theorems for a class of probabilistic automata. *Ann. Appl. Probab.* **6**, 577–616.

Finkelstein, M. and H. G. Tucker (1989). A law of small numbers for a mutation process. *Math. Biosci.* **95**, 85–98.

Finkelstein, M., H. G. Tucker and J. A. Veeh (1990). The limit distribution of the number of rare mutants. *J. Appl. Probab.* **27**, 239–250.

Fisher, R. A. (1922). On the dominance ratio. *Proc. Roy. Soc. Edinburgh* **42**, 321–341.

Fisher, R. A. (1930). The distribution of gene ratios for rare mutations. *Proc. Roy. Soc. Edinburgh* **50**, 205–220.

Fisher, R. A. (1937). The advance of advantageous genes. *Ann. Eugenics* **7**, 355–369.

Fisher, R. A. (1958). *The Genetical Theory of Natural Selection*, 2nd ed. Dover Publications, New York.

Foley, P. (1997). Extinction models for local populations. In *Metapopulation Biology: Ecology, Genetics and Evolution*, pp. 215–246 (Eds. I. A. Hanski and M. E. Gilpin). Academic Press, San Diego.

Fox, W. R. and G. W. Lasker (1983). The distribution of surname frequencies. *Int. Statist. Rev.* **51**, 81–87.

Gabriel, J.-P., C. Lefèvre and P. Picard (Eds.) (1990). *Stochastic Processes in Epidemic Theory*. Springer-Verlag, New York.

Gabriel, W. and R. Bürger (1992). Survival of small populations under demographic stochasticity. *Theor. Popn. Biol.* **41**, 44–71.

Gale, J. S. (1990). *Theoretical Population Genetics*. Unwin Hyman, London.

Gale, J. S. and M. J. Lawrence (1984). The decay of variability. In *Crop Genetic Resources: Conservation & Evaluation*, pp. 77–101 (Eds. J. H. W. Holden and J. T. Williams). Allen & Unwin, London.

Galton, F. and H. W. Watson (1874). On the probability of the extinction of families. *J. Roy. Anthropol. Inst.* **4**, 138–144.

Gawel, B. and M. Kimmel (1996). The iterated Galton–Watson process. *J. Appl. Probab.* **33**, 949–959.

Geiger, J. (1999). Elementary new proofs of classical limit theorems for Galton–Watson processes. *J. Appl. Probab.* **36**, 301–309.

Geiger, J. (2000). Poisson point process limits in size-biased Galton–Watson trees. *Electronic J. Probab.* **5**, 1–12.

Gillespie, J. H. (1991). *The Causes of Molecular Evolution*. Oxford University Press, New York.

Gladstien, K. and K. Lange (1978a). Number of people and number of generations affected by a single deleterious mutation. *Theor. Pop. Biol.* **14**, 313–321.

Gladstien, K. and K. Lange (1978b). Equilibrium distributions for deleterious genes in large stationary populations. *Theor. Pop. Biol.* **14**, 322–328.

Goel, N. S. and N. Richter-Dyn (1974). *Stochastic Models in Biology*. Academic Press, New York.

González, M. and M. Molina (1996). On the limit behaviour of a superadditive bisexual Galton–Watson branching process. *J. Appl. Probab.* **33**, 960–967.

González, M. and M. Molina (1997). On the L_2-convergence of a superadditive bisexual Galton–Watson branching process. *J. Appl. Probab.* **34**, 575–582.

Goodman, D. (1987a). Consideration of stochastic demography in the design and management of biological reserves. *Nat. Res. Modelling* **1**, 205–234.

Goodman, D. (1987b). The demography of chance extinction. In *Viable Populations for Conservation*, pp. 11–34 (Ed. M. E. Soulé). Cambridge University Press, New York.

Goodman, L. A. (1968). How to minimize or maximize the probabilities of extinction in a Galton–Watson process and in some related multiplicative population processes. *Ann. Math. Statist.* **39**, 1700–1710.

Gosselin, F. (1996). Extinction in a simple source/sink system: application of new mathematical results. *Acta Oecologica* **17**, 563–584.

Gosselin, F. (1998a). Asymptotic behaviour of some discrete-time Markov chains conditional on non-extinction. Part I: Theory. *Rapport #98-04*, Inst. Recherche Agronomique, Montpellier, pp. 1–54.

Gosselin, F. (1998b). Asymptotic behaviour of some discrete-time Markov chains conditional on non-extinction. Part II: Applications. *Rapport #98-05*, Inst. Recherche Agronomique, Montpellier, pp. 1–39.

Gosselin, F. (2001). Asymptotic behaviour of absorbing Markov chains conditional on nonabsorption for applications in conservation biology. *Ann. Appl. Probab.* **11**, 261–284.

Gosselin, F. and J.-D. Lebreton (1997). The potential of branching processes as a modeling tool for conservation biology. In *Quantitative Methods for Conservation Biology* (Ed. S. Ferson). Springer-Verlag, Berlin.

Gottlieb, K. (Ed.) (1983). *Human Biology*, Vol. 55, pp. 209–408.

Grey, D. R. (1980). Minimisation of extinction probabilities in reproducing populations. *Theor. Pop. Biol.* **18**, 430–445.

Grey, D. R. and Z. Lu (1993). The asymptotic behaviour of extinction probability in the Smith–Wilkinson branching process. *Adv. Appl. Probab.* **25**, 263–289.

Grey, D. R., V. Hutson and E. Szathmáry (1995). A re-examination of the stochastic corrector model. *Proc. Roy. Soc. London B* **262**, 29–35.

Griffiths, D. A. (1973). Multivariate birth-and-death processes as approximations to epidemic processes. *J. Appl. Probab.* **10**, 15–26.

Griffiths, R. C. and A. G. Pakes (1988). An infinite-alleles version of the simple branching process. *Adv. Appl. Probab.* **20**, 489–524.

Grimmett, G. (1989). *Percolation*. Springer-Verlag, Berlin.

Gupta S. C., O. P. Srivastava and M. Singh (1992). Branching processes with emigration – a genetic model. *Math. Biosci.* **111**, 159–168.

Guttorp, P. (1991). *Statistical Inference for Branching Processes*. Wiley, New York.

Guttorp, P. (1995). Three papers on the history of branching processes. *Int. Statist. Rev.* **63**, 233–245.

Haccou, P. and Y. Iwasa (1996). Establishment probability in fluctuating environments: a branching process model. *Theor. Pop. Biol.* **50**, 254–280.

Haigh, J. and J. Maynard Smith (1972). Population size and protein variation in man. *Genet. Res. Camb.* **19**, 73–89.

Haldane, J. B. S. (1927). A mathematical theory of natural and artificial selection, Part V: Selection and mutation. *Proc. Cambridge Philos. Soc.* **23**, 838–844.

Haldane, J. B. S. (1935). The rate of spontaneous mutation of a human gene. *J. Genet.* **23**, 317–326.

Haldane, J. B. S. (1939). The equilibrium between mutation and random extinction. *Ann. Eugen.* **9**, 400–405.

Haldane, J. B. S. (1990). *The Causes of Evolution*. (With a new Introduction and Afterword by Egbert G. Leigh, Jr.) Princeton University Press, Princeton.

Hanin, L. G. (2001). Iterated birth and death process as a model of radiation cell survival. *Math. Biosci.* **169**, 89–107.

Harnevo, L. E. and Z. Agur (1991). The dynamics of gene amplification described as a multitype compartmental model and as a branching process. *Math. Biosci.* **103**, 115–138.

Harris, S. C. (1999). Travelling-waves for the FKPP equation via probabilistic arguments. *Proc. Roy. Soc. Edinburgh A* **129**, 503–517.

Harris, T. E. (1951). Some mathematical models for branching processes. In *Second Symposium on Probability and Statistics*, pp. 305–328 (Ed. J. Neyman). University of California Press, Berkeley.

Harris, T. E. (1963). *The Theory of Branching Processes*. Springer-Verlag, Berlin.

Helland, I. S. (1981). Minimal conditions for weak convergence to a diffusion process on the line. *Ann. Probab.* **9**, 429–452.

Heyde, C. C. (1978). On an explanation for the characteristic clutch size of some bird species. *Adv. Appl. Probab.* **10**, 723–725.

Heyde, C. C. (1979). On assessing the potential severity of an outbreak of a rare disease: a Bayesian approach. *Austral. J. Statist.* **21**, 282–292.

Heyde, C. C. (Ed.) (1995). *Branching Processes: Proceedings of the First World Congress*. Springer-Verlag, New York.

Heyde, C. C. and H.-J. Schuh (1978). Uniform bounding of probability generating functions and the evolution of reproduction rates in birds. *J. Appl. Probab.* **15**, 243–250.

Heyde, C. C. and E. Seneta (1977). *I. J. Bienaymé: Statistical Theory Anticipated*. Springer-Verlag, New York.

Hofbauer, J. and K. Sigmund (1988). *The Theory of Evolution and Dynamical Systems*. Cambridge University Press, Cambridge.

Högnäs, G. (1997). On the quasi-stationary distribution of a stochastic Ricker model. *Stochastic Process. Appl.* **70**, 243–263.

Holgate, P. (1966). A mathematical study of the founder principle of evolutionary genetics. *J. Appl. Probab.* **3**, 115–128.

Holgate, P. A. (1967a). Divergent population processes and mammal outbreaks. *J. Appl. Probab.* **4**, 1–8.

Holgate, P. (1967b). Population survival and life history phenomena. *J. Theor. Biol.* **14**, 1–10.

Holgate, P. and K. H. Lakhani (1967). Effect of offspring distribution on population survival. *Bull. Math. Biophys.* **29**, 831–839.

Höpfner, R. (1985a). On some classes of population-size-dependent Galton–Watson processes. *J. Appl. Probab.* **22**, 25–36.

Höpfner, R. (1985b). A note on the probability of extinction in a class of population-size-dependent Galton–Watson processes. *J. Appl. Probab.* **22**, 920–925.

Hoppe, F. M. (1992a). Asymptotic rates of growth of the extinction probability of a mutant gene. *J. Math. Biol.* **30**, 547–566.

Hoppe, F. M. (1992b). The survival probability of a mutant in a multidimensional population. *J. Math. Biol.* **30**, 567–575.

Hull, D. M. (1982). Conditions for extinction in those bisexual Galton–Watson branching processes governed by superadditive mating functions. *J. Appl. Probab.* **19**, 847–850.

Hull, D. M. (1984). A necessary condition for extinction in certain bisexual Galton–Watson branching processes. *J. Appl. Probab.* **21**, 414–418.

Hull, D. M. (1993). How many mating units are needed to have a positive probability of survival? *Math. Magazine* **66**, 28–33.

Hull, D. M. (1998). A reconsideration of Galton's problem (Using a two-sex population). *Theor. Pop. Biol.* **54**, 105–116.

Hull, D. M. (2001). A reconsideration of Lotka's extinction probability using bisexual branching processes. *J. Appl. Probab.* **38**, 776–780.

Islam, M. N. (1995). A stochastic model for surname evolution. *Biom. J.* **37**, 119–126.

Ismail, M. E. H., D. R. Masson, J. Letassier and G. Valent (1990). Birth and death processes and orthogonal polynomials. In *Orthogonal Polynomials*, pp. 229–255 (Ed. P. Nevai). Kluwer Academic Publishers, Dordrecht, The Netherlands.

Jacquez, J. A. (1987). A note on chain binomial models of epidemic spread: what is wrong with the Reed–Frost formulation? *Math. Biosci.* **87**, 73–82.

Jagers, P. (1975). *Branching Processes with Biological Applications*, Wiley, London.

Jagers, P. (1991). The growth and stabilization of populations. *Statist. Sci.* **6**, 269–283.

Jagers, P. (1995). Branching processes as population dynamics. *Bernoulli* **1**, 191–200.

Jagers, P. (1999). Branching processes with dependent but homogeneous growth. *Ann. Appl. Probab.* **9**, 1160–1174.
Jagers, P., O. Nerman and T. Ziab (1991). When did Joe's great ... grandfather live? or: On the time scale of evolution. In *Selected Proceedings of the Sheffield Symposium on Applied Probability*, pp. 118–126 (Eds. I. V. Basawa and R. L. Taylor). IMS Lecture Notes Monograph Ser., Vol. 18. Hayward, CA.
Johnson. N. L., S. Kotz and A. W. Kemp (1992). *Univariate Discrete Distributions*, 2nd ed. Wiley, New York.
Kämmerle, K. (1991). The extinction probability of descendents in bisexual models of fixed population size. *J. Appl. Probab.* **28**, 489–502.
Kaplan, N. (1973). A continuous time branching model with random environments. *Adv. Appl. Probab.* **5**, 37–54.
Karlin, S. and N. Kaplan (1973). Criteria for extinction of certain population growth processes with interacting types. *Adv. Appl. Probab.* **5**, 183–199.
Karlin, S. and S. Tavaré (1982). Detecting particular genotypes in populations under nonrandom mating. *Math. Biosci.* **59**, 57–75.
Keiding, N. (1973). *Lecture Notes on the Stochastic Population Model*. Institute of Mathematical Statistics, Univ. of Copenhagen.
Keiding, N. (1975). Extinction and exponential growth in random environments. *Theor. Pop. Biol.* **8**, 49–63.
Kendall, D. G. (1966). Branching processes since 1873. *J. London Math. Soc.* **41**, 385–406.
Kendall, D. G. (1975). The genealogy of genealogy: branching processes before (and after) 1873. *Bull. London Math. Soc.* **7**, 225–253.
Kersting, G. (1990). Some properties of stochastic difference equations. In *Stochastic Modelling in Biology*, pp. 328–339 (Ed. P. Tautu). World Scientific, Singapore.
Kesten, H. (1970). Quadratic transformations: a model for population growth. *Adv. Appl. Probab.* **2**, 1–82, 179–228.
Keyfitz, N. (1968). *Introduction to the Mathematics of Population*. Addison-Wesley, Reading, MA.
Keyfitz, N. and A. Tyree (1967). Computerization of the branching process. *Behav. Sci.* **12**, 329–336.
Kijima, M. (1997). *Markov Processes for Stochastic Modelling*. Chapman & Hall, London
Kimmel, M. (1997). Quasistationarity in a branching model of division-within-division. In *Classical and Modern Branching Processes*, pp. 157–164 (Eds. K. B. Athreya and P. Jagers). Springer-Verlag, New York.
Kimmel, M. and D. E. Axelrod (1990). Mathematical models of gene amplification with applications to cellular drug resistance and tumorigenecity. *Genetics* **125**, 633–644.
Kimmel, M. and D. E. Axelrod (2002). *Branching Processes in Biology*. Springer, New York.
Kimmel, M., D. E. Axelrod and G. M. Wahl (1992). A branching process model of gene amplification following chromosome breakage. *Mutation Res.* **276**, 225–239.
Kimmel, M. and D. N. Stivers (1994). Time-continuous branching random walk models of unstable gene amplification. *Bull. Math. Biol.* **56**, 337–357.
Klebaner, F. C. (1984a). Geometric rate of growth in population-size-dependent branching processes. *J. Appl. Probab.* **21**, 40–49.
Klebaner, F. C. (1984b). On population-size-dependent branching processes. *Adv. Appl. Probab.* **16**, 30–55.
Klebaner, F. C. (1985). A limit theorem for population-size-dependent branching processes. *J. Appl. Probab.* **22**, 48–57.
Klebaner, F. C. (1997). Population and density dependent branching processes. In *Classical and Modern Branching Processes*, pp. 165–169 (Eds. K. B. Athreya and P. Jagers). Springer-Verlag, New York.
Klebaner, F. C. and O. Zeitouni (1994). The exit problem for a class of density-dependent branching processes. *Ann. Appl. Probab.* **4**, 1188–1205.
Klebaner, F. C., J. Lazar and O. Zeitouni (1998). On the quasi-stationary distribution for some randomly perturbed transformations on an interval. *Ann. Appl. Probab.* **8**, 300–315.
Kojima, K.-I. and T. M. Kellerher (1962). Survival of mutant genes. *Amer. Naturalist* **96**, 329–346.
Kojima, K.-I. and H. E. Schaffer (1964). Accumulation of epistatic gene complexes. *Evolution* **18**, 127–129.
Krawczak, M., J. Reiss, J. Schmidtke and U. Rösler (1989). Polymerase chain reaction: replication errors and reliability of gene diagnosis. *Nucl. Acids Res.* **17**, 2197–2201.
Kurtz, T. G. (1978). Diffusion approximations for branching processes. In *Branching Processes*, pp. 269–292 (Eds. A. Joffe and P. E. Ney). Marcel Dekker, Inc., New York.

Kyriakidis, E. G. (1993). A Markov decision algorithm for optimal pest control through uniform catastrophes. *Europ. J. Oper. Res.* **64**, 38–44.

Lande, R. (1988). Genetics and demography in biological conservation. *Science* **241**, 1455–1460.

Lange, K. (1981). Minimum extinction probability for surnames and favourable mutations. *Math. Biosci.* **54**, 71–78.

Lange, K. (1982). Calculation of the equilibrium distribution for a deleterious gene by the finite Fourier transform. *Biometrics* **38**, 79–86.

Lange, K. and R. Z. Fan (1997). Branching process models for mutant genes in nonstationary populations. *Theor. Pop. Biol.* **51**, 118–133.

Lange, K. and K. Gladstein (1980). Further characterization of the long-run population distribution of a deleterious gene. *Theor. Pop. Biol.* **18**, 31–43.

Lange, K., K. Gladstein and M. Zatz (1978). Effects of reproductive compensation and genetic drift in X-linked lethals. *Am. J. Hum. Genet.* **30**, 180–189.

Lasker, G. W. (1985). *Surnames and Genetic Structure*. Cambridge University Press, New York.

Lebreton, J.-D. (1982). Application of discrete time branching processes to bird population dynamics modelling. In *Anais 10a Conferênca Internacional de Biometria*, pp. 115–133. Empraba, Brasilia.

Lebreton, J.-D. and J. Clobert (1991). Bird population dynamics, management and conservation: the role of mathematical modelling. In *Bird Population Studies*, pp. 105–125 (Eds. C. M. Perrins, J.-D. Lebreton and G. J. M. Hirons). Oxford University Press, Oxford.

Lee, C. (2000). The density of the extinction probability of a time homogeneous linear birth and death process under the influence of randomly occurring disasters. *Math. Biosci.* **164**, 93–102.

Lefèvre, C. and Ph. Picard (1989). On the formulation of discrete-time epidemic models. *Math. Biosci.* **95**, 27–35.

Lefèvre, C. and Ph. Picard (1995). Collective epidemic processes: a general modelling approach to the final outcome of SIR infectious diseases. In *Epidemic Models: Their Structure and Relation to Data*, pp. 53–70 (Ed. D. Mollison). Cambridge University Press, Cambridge.

Le Gall, J.-F. (1999). *Spatial Branching Processes, Random Snakes and Partial Differential Equations*. Birkhäuser, Basel.

Leigh, E. G. (1975). Population fluctuations, community stability, and environmental variability. In *Ecology and Evolution of Communities*, pp. 51–73 (Eds. M. L. Cody and J. M. Diamond). Belknap Press, Cambridge, MA.

Leigh, E. G. (1981). The average lifetime of a population in a varying environment. *J. Theor. Biol.* **90**, 213–239.

Lewontin, R. C. and D. Cohen (1969). On population growth in a randomly varying environment. *Proc. Nat. Acad. Sci.* **62**, 1056–1060.

Li, F. H. F., J. V. Neel and E. D. Rothman (1978). A second study of the survival of a neutral mutant in a simulated amerindian population. *Amer. Naturalist* **112**, 83–96.

Liggett, T. M. (1999). *Stochastic Interacting Systems: Contact, Voter, and Exclusion Processes*. Springer-Verlag, Berlin.

Lipow, C. (1977a). A branching process model with size dependence. *Adv. Appl. Probab.* **5**, 14–24.

Lipow, C. (1977b). Limiting diffusions for population-size dependent branching processes. *J. Appl. Probab.* **14**, 14–24.

Lipow, C. (1980). Behaviour of limiting diffusions for density-dependent branching processes. In *Biological Growth and Spread*, pp. 130–137 (Eds. W. Jäger, H. Rost and P. Tautu). Springer-Verlag, Berlin.

Liu, Q. (1996). On the survival probability of a branching process in a random environment. *Ann. Inst. H. Poincaré* **32**, 1–10.

López-Mimbela, J. A. and A. Wakolbinger (1997). Which critically branching populations persist? In *Classical and Modern Branching Processes*, pp. 203–216 (Eds. K. B. Athreya and P. Jagers). Springer-Verlag, New York.

Lotka, A. (1931a). The extinction of families, I. *J. Washington Acad. Sci.* **21**, 377–380.

Lotka, A. (1931b). The extinction of families, II. *J. Washington Acad. Sci.* **21**, 453–459.

Ludwig, D. (1975). Final size distributions for epidemics. *Math. Biosci.* **23**, 33–46.

Ludwig, D. (1996). The distribution of population survival times. *Amer. Naturalist* **147**, 506–526.

McArthur, R. H. and E. O. Wilson (1967). *The Theory of Island Biogeography.* Princeton University Press, Princeton, NJ.

Mahmoud, H. M. (1992). *Evolution of Random Search Trees.* Wiley, New York.

Macken, C. A. and A. S. Perelson (1985). *Branching Processes Applied to Cell Surface Aggregation Phenomena.* Springer-Verlag, Berlin.

Macken, C. A. and A. S. Perelson (1989). *Stem Cell Proliferation and Differentiation: A Multitype Branching Process Model.* Springer-Verlag, Berlin.

Marschner, I. D. (1992). The effect of preferential mixing on the growth of an epidemic. *Math. Biosci.* **109**, 39–67.

McKean, H. P. (1975). Application of Brownian motion to the equation of Kolmogorov–Petrovski–Piscounov. *Comm. Pure Appl. Math.* **28**, 323–331; **29**, 553–554.

McKenzie, A. and M. Steel (2000). Distributions of cherries for two models of trees. *Math. Biosci.* **164**, 81–92.

Mode, C. J. (1967). On the probability a line becomes extinct before a favourable mutation appears. *Bull. Math. Biophys.* **29**, 343–348.

Mode, C. J. (1971). *Multitype Branching Processes.* Elsevier, New York.

Mode, C. J. and M. E. Jacobson (1987a). A study of the impact of environmental stochasticity on extinction probabilities by Monte Carlo integration. *Math. Biosci.* **83**, 103–125.

Mode, C. J. and M. E. Jacobson (1987b). On estimating critical population size for an endangered species in the presence of environmental stochasticity. *Math. Biosci.* **85**, 185–209.

Mode, C. J. and T. Root (1988). Projecting age-structured populations in a random environment. *Math. Biosi.* **88**, 223–245.

Möhle, M. (1994). Forward and backward processes in bisexual models with fixed population sizes. *J. Appl. Probab.* **31**, 309–332.

Mollison, D. (Ed.) (1995). *Epidemic Models: Their Structure and Relation to Data.* Cambridge University Press, Cambridge.

Moody, M. E. (1988). A branching process model for the evolution of transposable elements. *J. Math. Biol.* **26**, 347–357.

Moran, P. A. P. (1962). *The Statistical Processes of Evolutionary Theory.* Oxford University Press, Oxford.

Mountford, M. D. (1973). The significance of clutch size. In *The Mathematical Theory of the Dynamics of Biological Populations*, pp. 315–323 (Eds. M. S. Bartlett and R. W. Hiorns). Academic Press, London

Mountford, M. D. (1988). Population regulation, density dependence, and heterogeneity. *J. Animal Ecol.* **57**, 845–858.

Müller, J., M. Kretzschmar and K. Dietz (2000). Contact tracing in stochastic and deterministic epidemic models. *Math. Biosci.* **164**, 39–64.

Murray, J. D. (1989). *Mathematical Biology.* Springer-Verlag, Berlin.

Nadkarni, U. G. (1964). On some discrete models in branching processes. *J. Indian Soc. Agric. Statist.* **16**, 72–82.

Nadkarni, U. G. (1971). Generating function of modified branching process. *J. Indian Soc. Agric. Statist.* **23**, 67–76.

Narayan, P. (1981). On bounds for probability generating functions. *Austral. J. Statist.* **23**, 80–90.

Nåsell, I. (1995). The threshold concept in stochastic epidemic and endemic models. In *Epidemic Models: Their Structure and Relation to Data*, pp. 71–83 (Ed. D. Mollison). Cambridge University Press, Cambridge.

Nei, M. (1971a). Extinction time of deleterious mutant genes in large populations. *Theor. Pop. Biol.* **2**, 419–425.

Nei, M. (1971b). Total number of individuals affected by a single deleterious mutation in large populations. *Theor. Pop. Biol.* **2**, 426–430.

Ohta, T. and K.-I. Kojima (1968). Survival probabilities of new inversions in large populations. *Biometrics* **24**, 501–516.

Olofsson, P. (1999). Mathematical modeling of telomere shortening: an overview. *Archives of Control Sci.* **9(XLV)**, 133–141.

Olofsson, P. (2000). A branching process model of telomere shortening. *Comm. Statist. Stochastic Models* **16**, 167–177.

Olofsson, P. and M. Kimmel (1999). Stochastic models of telomere shortening. *Math. Biosci.* **158**, 75–92.

Olofsson, P., O. Schwalb, R. Chakraborty and M. Kimmel (2001). An application of a general branching processes in the study of the genetics of aging. *J. Theor. Biol.* **213**, 547–557.

O'Neill, P. (1995). Epidemic models featuring behaviour change. *Adv. Appl. Probab.* **27**, 960–979.

O'Neill, P. (1997). An epidemic model with removal-dependent infection rate. *Ann. Appl. Probab.* **7**, 90–109.

O'Neill, P. D. (1999). On a branching model of division-within-division. *IMA J. Math. Biol. Medicine* **16**, 395–405.

Pakes, A. G. (1984). The Galton–Watson process with killing. *Math. Biosci.* **69**, 171–188.

Pakes, A. G. (1986). The Markov branching-catastrophe process. *Stochastic Process. Appl.* **23**, 1–33.

Pakes, A. G. (1987). Limit theorems for the population size of a birth and death process allowing catastrophes. *J. Math. Biol.* **25**, 307–325.

Pakes, A. G. (1988). The Markov branching process with density-independent catastrophes I. Behaviour of extinction probabilities. *Math. Proc. Cambridge Philos. Soc.* **103**, 351–366.

Pakes, A. G. (1989a). An infinite alleles version of the Markov branching process *J. Austral. Math. Soc. Ser. A* **46**, 146–170.

Pakes, A. G. (1989b). Asymptotic results for the extinction time of Markov branching processes allowing emigration, I. Random walk decrements. *Adv. Appl. Probab.* **21**, 243–269.

Pakes, A. G. (1992). Limit theorems for the numbers of rare mutants: a branching process model. *Adv. Appl. Probab.* **24**, 778–794.

Pakes, A. G. (1995). Quasi-stationary laws for Markov processes: examples of an always proximate absorbing state. *Adv. Appl. Probab.* **27**, 120–145.

Pakes, A. G. (1996). A hitting time for Lévy processes, with applications to dams and branching processes. *Ann. Faculté des Sciences de Toulouse* **V**, 521–544.

Pakes, A. G. (1998). A limit theorem for the maxima of the para-critical simple branching process. *Adv. Appl. Probab.* **30**, 740–756.

Pakes, A. G. (1999). Revisiting conditional limit theorems for the mortal simple branching process. *Bernoulli* **5**, 969–998.

Pakes, A. G. and T. P. Speed (1977). Lagrange distributions and their limit theorems. *SIAM J. Appl. Math.* **32**, 71–87.

Parthasarathy, P. R. and R. B. Lenin (1997). On the exact transient solution of finite birth and death processes with specific quadratic rates. *Math. Scientist* **22**, 92–105.

Pettigrew, H. M. and G. H. Weiss (1967). Epidemics with carriers: the large population approximation. *J. Appl. Probab.* **4**, 257–263.

Piau, D. (2001). Processus de branchement en champ moyan et réaction PCR. *Adv. Appl. Probab.* **33**, 391–403.

Picard, Ph. (1965). Sur les modèles stochastiques logistiques en démographie. *Ann. Inst. H. Poincaré* **II**, 151–172.

Pierre-Loti-Viaud, D. (1993). Large deviations for random perturbations of discrete time dynamical systems. *Bull. Sci. Math.* **117**, 333–355.

Pimm, S. L. (1991). *The Balance of Nature?* Univ. Chicago Press, Chicago.

Pitman, J. (1998). Enumeration of trees and forests related to branching processes and random walks. In *Microsurveys in Discrete Probability*, pp. 163–180 (Eds. D. Aldous and J. Propp). American Mathematical Society, Providence, RI.

Pollak, E. (1966a). On the survival of a gene in a subdivided population. *J. Appl. Probab.* **3**, 142–155.

Pollak, E. (1966b). Some effects of fluctuating offspring distributions on the survival of a gene. *Biometrika* **53**, 391–396.

Pollak, E. (1986). On three methods for estimating mutation rates indirectly. *Am. J. Hum. Genet.* **38**, 209–227.

Pollak, E. (1992). Survival probabilities for some multitype branching processes in genetics. *J. Math. Biol.* **30**, 583–596.

Pollak, E. (2000). The effective population size of some age-structured populations. *Math. Biosci.* **168**, 39–56.

Pollard, J. H. (1973). *Mathematical Models for the Growth of Human Populations*. Cambridge University Press, Cambridge.

Pollett, P. K. (1990). On a model for interference between searching insect parasites. *J. Austral. Math. Soc. Ser. B* **32**, 133-150.

Pommerenke, Ch. (1981). On the stationary measures of critical branching processes. *Z. Wahrscheinlichkeitsth.* **55**, 305–312.
Pötscher, B. M. (1985). Moments and order statistics of extinction times in multitype branching processes and their relation to random selection models. *Bull. Math. Biol.* **47**, 263–272.
Primack, R. B. (1993). *Essentials of Conservation Biology.* Sinauer Associates Inc., Sunderland, MA.
Rahimov, I. (1995). *Random Sums and Branching Stochastic Processes.* Springer-Verlag, New York.
Raup, D M. (1985). The role of chance in Evolution. In *What Darwin Began: Modern Darwinian and Non-Darwinian Perspectives on Evolution*, pp. 94–112 (Ed. L. R. Godfrey). Allyn and Bacon.
Raup, D. M., S. J. Gould, T. J. M. Schopf and D. S. Simberloff (1973). Stochastic models of phylogeny and the evolution of diversity. *J. Geol.* **81**, 525–542.
Reinhard, I. (1990). The quantitative behaviour of some slowly growing population-dependent Markov branching processes. In *Stochastic Modelling in Biology*, pp. 267–277 (Ed. P. Tautu). World Scientific, Singapore.
Révész, P. (1994). *Random Walks of Infinitely Many Particles.* World Scientific, Singapore.
Rittgen, W. (1990). Diffusion limits of population-dependent Markov branching processes. In *Stochastic Modelling in Biology*, pp. 278–292 (Ed. P. Tautu). World Scientific, Singapore.
Rosenzweig, M. L. (1995). *Species Diversity in Space and Time.* Cambridge University Press, Cambridge.
Sankaranarayanan, G. (1989). *Branching Processes and its Estimation Theory.* Wiley Eastern Ltd., New Delhi.
Sawyer, S. (1976). Branching diffusion processes in population genetics. *Adv. Appl. Probab.* **8**, 659–689.
Sawyer, S. (1991). Comment: the geographical structure of populations. *Statist. Sci.* **6**, 280–281.
Sawyer, S. and J. Fleischman (1979). Maximum geographic range of a mutant allele considered as a subtype of a Brownian branching random field. *Proc. Natl. Acad. Sci. USA* **76**, 872–875.
Schaffer, H. E. (1970). Survival of mutant genes as a branching process. In *Mathematical Topics in Population Genetics*, pp. 317–336 (Ed. K.-I. Kojima). Springer-Verlag, Berlin.
Schuster, P. and K. Sigmund (1980). Self-organization of biological macromolecules and evolutionary stable strategies. In *Dynamics of Synergetic Systems*, pp. 156–169 (Ed. H. Haken). Springer-Verlag, Berlin.
Schuster, P. and K. Sigmund (1984a). Random selection – a simple model based on linear birth and death processes. *Bull. Math. Biol.* **46**, 11–17.
Schuster, P. and K. Sigmund (1984b). Random selection and the neutral theory – sources of stochasticity in replication. In *Stochastic Phenomena and Chaotic Behaviour in Complex Systems*, pp. 186–207 (Ed. P. Schuster). Springer-Verlag, Berlin.
Seneta, E. (1979). Round the historical work on Bienaymé. *Austral. J. Statist.* **21**, 209–220.
Seneta, E. (1998). I. J. Bienaymé [1786–1878]: criticality, inequality, and internationalization. *Int. Statist. Rev.* **66**, 291–301.
Seneta, E. and S. Tavaré (1983). A note on some models using the branching process with immigration stopped at zero. *J. Appl. Probab.* **20**, 11–18.
Sevast'yanov, B. A. (1970). Theory of branching processes. *Progress in Mathematics* **7**, 1–51.
Sevast'yanov, B. A. (1974). *Verzweigungsprozesse.* Akademie-Verlag, Berlin.
Shafer, C. F. (1990). *Nature Reserves.* Smithsonian Institute Press, Washington DC.
Shaffer, M. L. (1981). Minimum population sizes for species conservation. *BioScience* **31**, 131–134.
Shaffer, M. L. and F. B. Sampson (1985). Population size and extinction: a note on determining critical population size. *Amer. Naturalist* **125**, 144–152.
Simberloff, D. and L. Abele (1976). Island biogeographic theory and conservation practice: strategy and limitations. *Science* **191**, 1032.
Skellam, J. G. (1949). The probability distribution of gene-differences in relation to selection, mutation, and random extinction. *Proc. Cambridge Philos. Soc.* **45**, 364–367.
Steffensen, J. F. (1932). Deux problèmes du calcul des probabilités. *Ann. H. Poincaré* **3**, 319–344.
Sun, F. and M. S. Waterman (1997). Whole genome amplification and branching processes. *Adv. Appl. Probab.* **29**, 629–688.
Taib, Z. (1992). *Branching Processes and Neutral Evolution.* Lecture Notes in Biomathematics, Vol. 93. Springer-Verlag, Berlin.
Vatutin, V. A. (1974). The asymptotic probability of the first degeneration for branching processes with immigration. *Theory Probab. Appl.* **XIX**, 25–34.

Vatutin, V. A. and A. M. Zubkov (1987). Branching processes I. *J. Soviet Math.* **39**, 2431–2475.
Vatutin, V. A. and A. M. Zubkov (1993). Branching Processes II. *J. Soviet Math.* **67**, 3407–3485.
von Bahr, B. and Martin-Löf (1980). Threshold limit theorems for some epidemic processes. *Adv. Appl. Probab.* **12**, 319–349.
Wakolbinger, A. (1995). Limits of spatial branching populations (with discussion). *Bernoulli* **1**, 171-189.
Watterson, G. A. (1973). On a recent paper by Cook and Nassar. *Biometrics* **29**, 595–600.
Wang, F. J. S. (1977). Gaussian approximation of some closed stochastic epidemic models. *J. Appl. Probab.* **14**, 221–231.
Wang, H.-X. (1999). Extinction of population-size-dependent branching processes in random environments. *J. Appl. Probab.* **36**, 146–154.
Whittle, P. (1955). The outcome of a stochastic epidemic – a note on Bailey's paper. *Biometrika*, **42**, 571–579.
Wick, D. and S. G. Self (2000). Early HIV infection in vivo: branching-process model for studying timing of immune responses and drug therapy. *Math. Biosci.* **165**, 115–134.
Williamson, M. (1981). *Island Populations*. Oxford University Press, Oxford.
Wissel, C. and Stöcker (1991). Extinction of populations by random influences. *Theor. Pop. Biol.* **39**, 315–328.
Yakovlev, A. Yu. and N. Yanev (1989). *Transient Processes in Cell Proliferation Kinetics*. Springer-Verlag, Berlin.
Yasuda, N., L. L. Cavalli-Sforza, M. Skolnick and A. Moroni (1974). The evolution of surnames: an analysis of their distribution and extinction. *Theor. Pop. Biol.* **5**, 123–142.
Zheng, Q. (1999). Progress of a half century in the study of the Luria-Delbrück distribution. *Math. Biosci.* **162**, 1–32.
Zhivotovsky, L. A. and M. W. Feldman (1993). On the probability of loss of new mutations in the presence of linkage disequilibrium. *J. Math. Biol.* **31**, 177–188.

Markov Chain Approaches to Damage Models

C. R. Rao, M. Albassam, M. B. Rao* and D. N. Shanbhag*

1. Introduction

Damage models were first introduced by Rao (1963), and these have led to numerous interesting and illuminating characterizations of discrete distributions; among various important results in the area are those of Rao and Rubin (1964) and Shanbhag (1977). A damage model can be used to describe a certain type of random phenomenon which may not be observable due to a destructive process as a whole undamaged.

In mathematical terms, a damage model can be described by a random vector (X, Y) of non-negative integer-valued components, with the joint probability law of X and Y having the following structure:

$$P(X = n, Y = r) = S(r|n)g_n, \quad r = 0, 1, 2, \ldots, n; \ n = 0, 1, 2, \ldots,$$

where $\{S(r|n) = P(Y = r|X = n): r = 0, 1, 2, \ldots, n\}$ is a discrete probability law for each $n = 0, 1, 2, \ldots$ and $\{g_n = P(X = n): n = 0, 1, 2, \ldots\}$ is the marginal probability law of X. In the context of damage models, the conditional probability law $\{S(r|n): r = 0, 1, 2, \ldots, n\}$ is called the survival distribution. It is also natural to call Y the undamaged part of X and $X - Y$ the damaged part of X.

Among simple examples to show how one could come across a damage model in practice are the count of insect eggs, some of which may be lost or unrecognizable (see, for example, Rao and Shanbhag, 1982), the number of oil deposits in a given square kilometer of a field, some of which might be missed or undiscovered (given by Bather, 1992) or the number of accidents occurring not all of which may be reported.

One of the central problems on a damage model is the following: knowing the survival distribution $\{S(r|n)\}$ and that Y and $X - Y$ satisfy a certain partial independence condition, what information can we obtain about the form of the distribution of X, i.e., about $\{g_n\}$? Rao (1963) gave the following physical interpretation of the damage model: Let X denote an observation on some natural process (e.g., number of eggs, number of accidents, number of oil deposits in a field, etc.). This observation may be partially destroyed, or may be only partially ascertained. In such circumstances, $\{g_n\}$, the original distribution of X, will be distorted. Rao then pointed

---
*Nato Collaborative Research Grant 960017.

out that if the survival distribution $\{S(r|n)\}$ is known, then we can derive the distribution of the observed values knowing the original distribution. Another type of research in damage model theory is on characterizing the survival distributions under the assumption that the original distribution is known or has a certain form.

Rao and Rubin (1964) initiated research in the area of damage models and their work is followed by, among others, the contributions of Srivastava and Srivastava (1970), Talwalker (1975), Shanbhag (1974, 1977), Alzaid et al. (1986b) and several others. The original result of Rao and Rubin (1964) states that if the survival distribution is binomial with parameter vector (x, p) for almost all x, where $p \in (0, 1)$ and fixed, and $P(X = 0) < 1$, then the Rao–Rubin condition (RR(0))

$$P(Y = r) = P(Y = r|X = Y), \quad r = 0, 1, 2, \ldots,$$

is met if and only if X is Poisson.

The study of the connections between damage models and Markov chains is worth exploring because of the central role played by Markov chains in theoretical as well as practical developments of stochastic processes. Recent advances in damage models reveal that many of the results in the area have links with Markov chains; see, for example, Alzaid, Rao and Shanbhag (1986b, 1987) and Sapatinas (1990). In particular, Sapatinas has devoted a chapter of his dissertation to discuss some results on damage models linked with Markov chains and Rao and Shanbhag (1994, Chapter 7) have reviewed and unified via approaches based on nonnegative matrices, Markov chains, and the integrated Cauchy functional equation among others, a major portion of the literature on damage models. See also, Rao, Rao and Shanbhag (2002) for further links of nonnegative matrices and Markov chains with damage models.

The main aim of the present chapter is to provide an up-to-date review of the existing literature on damage models, where specifically Markov chain approaches are used, and make some further revelations on the links between damage models and Markov chains. In the process of doing this, we give improved or unified versions of many of the existing results in the area of damage models.

2. Modified versions of some basic results on damage models

Most of the results on characterizations of probability distributions based on damage models can be translated into those on homogeneous Markov chains. For a homogeneous Markov chain $\{X_n: n = 0, 1, \ldots\}$ with state space $\{0, 1, \ldots\}$ having stochastic matrix $(S(j|i))$, where $S(j|i) = 0$ if $j > i$, if m is any integer such that $P(X_m = X_{m+1}) > 0$, we can take the condition Rao–Rubin (0) at m, $\text{RR}_m(0)$, as

$$P(X_{m+1} = j) = P(X_{m+1} = j|X_m = X_{m+1}), \quad j = 0, 1, 2, \ldots. \tag{1}$$

With obvious modifications, one can also define the $\text{RR}_m(k)$ conditions in a Markov chain set-up. In view of this, we can usually restate the damage model results in a Markov chain set-up.

2.1. Characterizations of the original distributions

We begin our discussion of characterization results on damage models, in a Markov chain set-up, stating the following version of an important theorem due to Shanbhag (1977):

THEOREM 2.1. *Let $\{(a_i, b_i): i = 0, 1, 2, \ldots\}$ be a sequence of real vectors with $a_i > 0 \; \forall i \geqslant 0$, $b_0, b_1 > 0$ and $b_i \geqslant 0 \; \forall i \geqslant 2$. Let $\{c_i\}$ be the convolution of $\{a_i\}$ and $\{b_i\}$, i.e.,*

$$c_i = \sum_{j=0}^{i} a_j b_{i-j}, \quad i = 0, 1, 2, \ldots.$$

Let $\{X_n\}_{n=0}^{\infty}$ be a discrete homogeneous Markov chain with state space $\{0, 1, \ldots\}$, such that $P(X_{n+1} \leqslant X_n) = 1$ for all n, $P(X_{n_0} = 0) < 1$, where n_0 is fixed, and the corresponding transition probabilities, p_{ij}, satisfy

$$p_{ij} = \frac{a_j b_{i-j}}{c_i}, \quad j = 0, 1, 2, \ldots, i; \; i = 0, 1, \ldots. \tag{2}$$

Then the $\mathrm{RR}_{n_0}(0)$ (i.e., condition (1) with $m = n_0$) is met if and only if the distribution, $\{p_i\}$, of X_{n_0} satisfies

$$\frac{p_i}{c_i} = \frac{p_0}{c_0} \lambda^i, \quad i = 1, 2, \ldots, \text{ for some } \lambda > 0.$$

Also, if $\mathrm{RR}_{n_0}(0)$ holds, then X_{n_0+1} and $X_{n_0} - X_{n_0+1}$ are independent.

The above result follows essentially in view of the theorem referred to, i.e., Theorem 1 of Shanbhag (1977). In the cited reference, the result in question is obtained via a lemma, which is referred to in the literature as Shanbhag's lemma, arrived at essentially using a renewal theoretic argument. Theorem 7.2.3 of Rao and Shanbhag (1994) extends the result of Shanbhag to a multivariate case; this latter result follows via a version of a general result on the integrated Cauchy functional equation or that of de Finetti's theorem. More recently, Rao, Rao and Shanbhag (2002) have unified certain results of relevance to damage models, showing among other things that the aforementioned results including in particular a version of de Finetti's theorem follow easily from Theorem 4.4.1 of Rao and Shanbhag (1994), corresponding to nonnegative matrices. Renewal theory and nonnegative matrices have very strong links with Markov chains and hence these provide us with further illustrations of the importance of the role played by Markov chains in studies of damage models.

It is worth noting here that a multivariate extension of Theorem 2.1 involving a vector-valued Markov chain follows easily in view of Theorem 7.2.3 of Rao and Shanbhag (1994). Chapter 7 of the cited monograph provides us with further scope for linking the results on damage models to Markov chains.

In the light of Theorem 2.1, we can obtain several results on specialized distributions in a Markov chain set-up as corollaries to the theorem. Among these results, we have the first part of the following theorem; we give the theorem here in a slightly extended form because it provides us with some interesting additional information relative to $\{X_n\}$ for all n.

THEOREM 2.2. *Suppose $\{X_n\}_{n=0}^{\infty}$ is a discrete homogeneous Markov chain with state space $\{0, 1, \ldots\}$ and such that $P(X_{n+1} \leq X_n) = 1$ for all n. If $P(X_0 = 0) < 1$ and the corresponding transition probabilities p_{ij} are such that*

$$p_{ij} = \binom{i}{j} \pi^j (1-\pi)^{i-j}, \quad j = 0, 1, 2, \ldots, i; \; i = 0, 1, 2, \ldots, \tag{3}$$

where π is a fixed number in $(0, 1)$, then for a fixed n, say n_0, the $\mathrm{RR}_{n_0}(0)$ holds if and only if X_{n_0} is Poisson, i.e.,

$$P(X_{n_0} = j) = e^{-\lambda} \frac{\lambda^j}{j!}, \quad \text{for some } \lambda > 0.$$

Moreover, we have X_{n_0} to be Poisson(λ) if and only if X_n is Poisson($\lambda \pi^{n-n_0}$) for all n and $\mathrm{RR}_n(0)$ condition is met for each n.

PROOF. We observe that the transition probabilities p_{ij} in (3) are of the form (2) given in Theorem 2.1 with $a_i = \pi^i/i!$ and $b_i = (1-\pi)^i/i!$, $i = 0, 1, 2, \ldots$. Since the corresponding $c_i = 1/i!$ for $i \geq 0$, this means that, according to Theorem 2.1, the $\mathrm{RR}_{n_0}(0)$ condition holds if and only if the distribution of X_{n_0} is Poisson(λ) for some $\lambda > 0$. To prove the second part of the theorem, let, for each n, G_n be the p.g.f. (probability generating function) of X_n. It is now easily seen that for each nonnegative n,

$$G_{n+1}(s) = G_n(\pi s + 1 - \pi), \quad |s| \leq 1.$$

Consequently, we get, for $n \geq 0$,

$$G_n(s) = G_0(\pi^n s + 1 - \pi^n), \quad |s| \leq 1;$$

which implies that if X_{n_0} is a Poisson(λ) random variable, then X_0 is a Poisson($\lambda \pi^{-n_0}$) random variable and hence, for each n, $X_n \sim$ Poisson($\lambda \pi^{n-n_0}$). It is then obvious that the "only if" part of the second assertion of the theorem holds. Since the "if" part of the assertion is trivial, we have then that the assertion referred to holds. □

REMARK 2.1. If the state space of a homogeneous Markov chain $\{X_n\}$ is finite and the corresponding stochastic matrix is nonsingular, then, given these and the marginal distribution of $\{X_n\}$ for any particular n, all finite-dimensional distributions corresponding to the chain are determined. Similarly, if the chain is as in Theorem 2.1, but satisfying additionally that $\{b_n\}$ is geometric, then it easily follows that given the stochastic matrix (with state space as specified), the marginal distribution of $\{X_n\}$ for any particular n determines all finite-dimensional distributions corresponding to the chain.

REMARK 2.2. Essentially, in view of Shanbhag's (1977) result on damage models referred to above, we have that under the assumptions in Theorem 2.1, the following are equivalent:

(i) $RR_{n_0}(0)$ holds and $\{\frac{a_n}{c_n}\}$ is a geometric sequence.
(ii) $RR_n(0)$ holds for $n = n_0, n_{0+1}$.
(iii) $RR_n(0)$ holds for all $n \geq n_0$.
(iv) $X_n - X_{n+1}$ for $n = n_0, n_0 + 1, \ldots$ are all independent.

REMARK 2.3. It is an easy exercise to see that if the Markov chain is as in Theorem 2.1 with $\{b_n\}$ as geometric and the $RR_n(0)$ condition is satisfied for any two successive n, then $RR_n(0)$ is satisfied for all n and, for an appropriate (a, b, q), X_n has the Euler (a, b^n, q) distribution for each n. (Compare this with the second part of Theorem 2.2 where obviously we take the parameter λ to be fixed.)

REMARK 2.4. In general, if the state space, stochastic matrix, and the distribution of X_{n_0} for some $n_0 > 0$ relative to a homogeneous Markov chain $\{X_n\}$ are given, it does not follow that all finite-dimensional distributions for the chain are determined. This remains the case even when the stochastic matrix is of the form in Theorem 2.1 and $RR_n(0)$ condition is met for all $n > 0$; the following interesting example illustrates this, among other things:

EXAMPLE 2.5. Let $\{a_n\}$ and $\{b_n\}$ of Theorem 2.1 be respectively the Heine distribution with parameter vector $(1, q)$ and the binomial distribution with parameter vector $(1, 1/(1 + q))$ respectively, where q is a fixed number lying in $(0, 1)$. Now, if we take the chain $\{X_n\}$ of the theorem such that X_0 is a Heine $(1/q, q)$-distributed random variable, we get that $RR_n(0)$ condition is met for all n, and, for each n, the marginal distribution X_n is Heine with appropriate parameters. On the other hand, if we take the chain to be such that X_0 is the weighted distribution relative to the Heine $(1/q, q)$ distribution with weight function as $w(x) = 1 + (-q)^x$, $x = 0, 1, \ldots$, or as $w(x) = 1 - (-q)^x$, $x = 0, 1, \ldots$, then it follows that $RR_n(0)$ is met for each $n > 0$, and, for each $n > 0$, the marginal distribution of X_n is as in the previous case. However, in this latter case, the distribution of X_0 is not Heine and $RR_0(0)$ is not met.

For ready reference, we may recall here the following information concerning Euler and Heine distributions:

DEFINITION 2.6 (*Euler distribution*). A non-negative integer-valued r.v. X is said to have Euler distribution with parameter vector (λ, q) if its probability function is given by

$$p_x = p_0 \frac{\lambda^x}{\prod_{i=1}^{x}(1-q^i)}, \quad x = 0, 1, 2, \ldots,$$

where $p_0 = \prod_{i=0}^{\infty}(1-\lambda q^i)$, $0 < \lambda < 1$ and $0 < q < 1$.

DEFINITION 2.7 (*Heine distribution*). A non-negative integer-valued r.v. X is said to have Heine distribution with parameter vector (λ, q) if its probability function is given by

$$p_x = p_0 \frac{\lambda^x}{\prod_{i=1}^{x}(1-q^i)} q^{x(x-1)/2}, \quad x = 0, 1, 2, \ldots,$$

where $p_0 = \prod_{i=0}^{\infty}(1+\lambda q^i)^{-1}$, $\lambda > 0$ and $0 < q < 1$.

The probability generating functions of the distributions referred to in the two definitions above are known to be (under appropriate parametric constraints and with appropriate domains of definition) respectively:

$$G(s) = \sum_{x=0}^{\infty} p_x s^x = \prod_{k=0}^{\infty} \left(\frac{1-\lambda q^k}{1-\lambda q^k s} \right),$$

$$H(s) = \sum_{x=0}^{\infty} p_x s^x = \prod_{k=0}^{\infty} \left(\frac{1+\lambda q^k s}{1+\lambda q^k} \right).$$

For a more detailed account of these distributions, see, for example, Benkherouf and Bather (1988).

REMARK 2.8. A multivariate extension of Theorem 2.2 involving a vector-valued Markov chain is now obvious.

REMARK 2.9. Many characterizations of discrete distributions can be derived as corollaries to Theorem 2.1. Specifically, the characterization of negative binomial distribution by Patil and Ratnaparkhi (1977), generalized Poisson distribution by Consul (1975), Generalized Polya–Eggenberger distribution by Janardan and Rao (1982) and lastly Euler and Heine distributions by Albassam (2000) can be modified to arrive at versions in a Markov chain set-up, applying Theorem 2.1. The result of Shanbhag (1977) can be applied, specifying the sequences $\{a_i\}$, $\{b_i\}$ and $\{c_i\}$ appropriately.

2.2. *Characterizations of the survival distributions*

In the previous section, we discussed versions (in a Markov chain set-up) of some well known results in which RR(0) condition is employed to identify the form of the original distribution, assuming that the survival distribution is given. It is now reasonable to touch upon briefly the problem of obtaining analogous versions of the existing results where the condition referred to is used to characterize the form of the survival distribution, given that of the original distribution. Srivastava and Srivastava (1970) provided a partial converse to the Rao and Rubin (1964) result. This states that if $\{(X_\lambda, Y_\lambda): \lambda \in (a, b), 0 \leqslant a < b\}$ is a family of random vectors of non-negative integer-valued components such that for all values of the parameter

$$P_\lambda(X_\lambda = n, Y_\lambda = r) = \exp(-\lambda)\frac{\lambda^n}{n!} S(r|n), \quad r = 0, 1, \ldots, n, \; n = 0, 1, \ldots,$$

where $\{S(r|n): r = 0, 1, \ldots, n\}$ is a discrete probability distribution independent of λ for each $n \geqslant 0$ and $0 < S(n|n) < 1$ for some n, then for all values of the parameter λ

$$P_\lambda(Y_\lambda = r) = P_\lambda(Y_\lambda = r|X_\lambda = Y_\lambda), \quad r = 0, 1, \ldots,$$

if and only if the survival mechanism is binomial with fixed parameter $\pi \in (0, 1)$.

Alzaid (1986) extended the result of Srivastava and Srivastava (1970) to arrive at a characterization for a class of survival distributions. This identified a family of original r.v.'s $\{X_\lambda\}$, with $\{X_\lambda\}$ having a power series distribution, for which the RR(0) condition is met.

The following theorem is a corollary to the result (with appropriate amendments) of Srivastava and Srivastava (1970); incidentally, the result of Srivastava and Srivastava referred to above follows, in turn, from the celebrated Raikov theorem on a characterization of Poisson distributions.

THEOREM 2.3. *Let $\{X_n^{(\lambda)}\}_{n=0}^\infty$ be a family of discrete homogeneous Markov chains with state space $\{0, 1, 2, \ldots\}$ such that*

$$P(X_{n_0}^{(\lambda)} = j) = e^{-\lambda}\frac{\lambda^j}{j!}, \quad j = 0, 1, 2, \ldots,$$

with n_0 fixed and λ varying over a dense subset of an open interval, and the corresponding stochastic matrices are independent of λ, with these as lower triangular (i.e., having their (i, j)th elements equal to zero whenever $i < j$) having at least one diagonal element lying in $(0, 1)$. Then the $\mathrm{RR}_{n_0}(0)$ condition

$$P(X_{n_0+1}^{(\lambda)} = j) = P(X_{n_0+1}^{(\lambda)} = j | X_{n_0}^{(\lambda)} = X_{n_0+1}^{(\lambda)}), \quad j = 0, 1, 2, \ldots, \quad (4)$$

is valid for all λ if and only if, for some fixed π, the common stochastic matrix (p_{ij}) has entries satisfying

$$p_{ij} = \binom{i}{j}\pi^j(1-\pi)^{i-j}, \quad j = 0, 1, 2, \ldots, i; \; i = 0, 1, 2, \ldots.$$

It is worth noting that the original theorem of Srivastava and Srivastava (1970) does not state explicitly that the parameter λ is a variable taking all values in a dense subset of an open interval as mentioned in our version of Theorem 2.3. Shanbhag and Panaretos (1979) proved by giving a counter-example that the above result does not remain valid if λ is taken as a fixed parameter. Moreover, Andrews (1988) has used an extended version of the Shanbhag–Panaretos argument to show that the situation remains unaltered even in the case of a variable parameter if we allow the parameter to take only two values, and, recently, Rao and Shanbhag (1994) have provided a more general counter-example to show that the Srivastava–Srivastava result does not hold if we restrict ourselves to a family with finite index set.

REMARK 2.10. An analogue of Alzaid's (1986) result in a Markov chain set-up is now obvious in the light of our discussion so far. Analogues of the existing results on damage models relative to truncated distributions and those relative to RR(k) also follow as trivial exercises.

3. Characterizations based on modified Rao–Rubin conditions

There are several other results on damage models that have links with certain results on nonnegative matrices and hence implicitly on Markov chains. Among the results on nonnegative matrices implied here are the Perron–Frobenius theorem and the Poisson–Martin representation theorem or some related potential theoretic results based on Choquet's theorem of the form of Theorem 4.4.1 of Rao and Shanbhag (1994) (referred to earlier while making certain revelations on Theorem 2.1 in Section 2); for an interesting account of Choquet's theorem, we refer the reader to Phelps (1966). The results on nonnegative matrices referred to above or certain relevant observations on Martin boundary have appeared or implied in either Seneta (1973, 1981) or Williams (1979); see also Kemeny, Snell and Knapp (1966) for a treatment of potential theoretic results on Markov chains.

Alzaid, Rao and Shanbhag (1984, 1986a) had given arguments based on nonnegative matrices involving the Perron–Frobenius theorem or Shanbhag's (1977) lemma to obtain certain general identifiability results on damage models, relative to RR(k) conditions; some preliminary versions of these have appeared in Alzaid (1983). With a somewhat different formulation for the model, Shanbhag and Taillie (1979) and Alzaid, Lau, Rao and Shanbhag (1988) had given related variations of Shanbhag's (1977) theorem on damage models. Chapter 7 of Rao and Shanbhag (1994) has reviewed and unified most of these results. From the cited chapter or the recent review article of Rao et al. (2002), it is now evident that most of the important results on damage models including, in particular, those of Shanbhag and Taillie (1979) and Alzaid et al. (1988) are indeed corollaries to an identifiability result on damage models, given implicitly as a corollary to the Perron–Frobenius theorem, in Alzaid et al. (1986a), or, Theorem 4.4.1 of Rao and Shanbhag (1994). Incidentally, it is of interest to note here that the Perron–Frobenius theorem referred to above also plays an important role, among others, in certain population studies.

In another direction, a certain other modified version of the Rao–Rubin theorem relative to a damage model with binomial survival distribution, was given by Talwalker (1980) and Rao et al. (1980); this new version allows the binomial distribution parameter on the two sides of the RR(0) condition to be different. Alzaid, Rao and Shanbhag (1987) extended this result via a certain result on stationary measures relative to a discrete branching process. From what appears in Athreya and Ney (1972) and Alzaid, Rao and Shanbhag (1984), it is then clear that this latter result is also a by-product of the Poisson–Martin representation relative to Markov chains or nonnegative matrices.

3.1. Characterizations based on two Rao–Rubin(k) conditions

Inspired by a conjecture of Srivastava and Singh (1975), Patil and Taillie (1979) gave some interesting results on characterization of distributions via RR(k) conditions for $k > 0$; these latter authors showed that if the survival distribution is binomial, then one RR(k) condition does not imply the original distribution to be Poisson but two such conditions (for distinct k) do. The findings of Patil and Taillie, in turn, have led to the contributions cited earlier from Shanbhag and Taillie (1979), Alzaid, Rao and Shanbhag (1984, 1986a), and Alzaid et al. (1988). In the light of the information that we have provided above, the cited literature on damage models is seen to be itself a corollary to the relevant results on Markov chains or nonnegative matrices. However, for the sake of completeness, we record below two results that are respectively versions of the main results in Alzaid, Rao and Shanbhag (1986a) and Shanbhag and Taillie (1979) and are in the spirit of the results discussed in Section 2; these results are indeed obvious corollaries to their predecessors referred to here.

It may be worth pointing out in this place that a major step in the argument to have the general identifiability result of Alzaid, Rao and Shanbhag (1986a) is the one where it is implied that if an eigenvector of a nonnegative primitive matrix (with the matrix as defined in Seneta (1973, 1981)) is nonnull and has all its components to be nonnegative, then it is an eigenvector corresponding to the Perron–Frobenius eigenvalue of the matrix and hence has all its components to be positive. To see this and also to understand as to why the main result of Shanbhag and Taillie (1979) is a corollary to the identifiability result just mentioned, we refer the reader to Alzaid, Rao and Shanbhag (1986a) and Rao and Shanbhag (1994).

THEOREM 3.1. *Let $\{X_n\}_{n=0}^{\infty}$ be a homogeneous Markov chain with lower triangular stochastic matrix $(S(j|i))$ with state space given as $\{0, 1, \ldots\}$, and $k_0 \geqslant 0$ and $k_1 > 0$ be given integers. Define the quantities S_{ij}^* as*

$$S_{ij}^* = \sup\{S(j + lk_1|i + k_0 + mk_1): l, m \geqslant 0, \ k_0 \leqslant j + lk_1 \leqslant i + k_0 + mk_1\}.$$

Let n_0 be a given nonnegative integer and $\{g_i: i = 0, 1, \ldots\}$ be the marginal distribution of X_{n_0}, and let $\{g_i\}$ and $(S(j|i))$ satisfy the following conditions:

(i) $g_i > 0$ *for some* $i \geqslant k_0 + k_1$.
(ii) $S(i + jk_1|i + k_0 + (j + l)k_1) > 0$ *for each* $i = 0, 1, \ldots, k_1 - 1$; $l = 0, 1$ *and* $j = 0, 1, \ldots,$ *and* $S(i|i) > 0$ *for each* $i = 0, 1, \ldots, k_0 - 1$ *(when $k_0 > 0$).*

(iii) *The matrix $S^* = (S^*_{ij})$, $i, j = 0, 1, \ldots, k_1 - 1$, is irreducible.*

Then, under the two $RR_{n_0}(k)$ *conditions, given by*

$$P(X_{n_0+1} = j) = P(X_{n_0+1} = j | X_{n_0} - X_{n_0+1} = k_0)$$
$$= P(X_{n_0+1} = j | X_{n_0} - X_{n_0+1} = k_0 + k_1), \quad j = 0, 1, \ldots,$$

assuming $\lambda = P(X_{n_0} - X_{n_0+1} = k_0 + k_1)/P(X_{n_0} - X_{n_0+1} = k_0)$ *is given, the stochastic matrix* $(S(j|i))$ *determines the* $\{g_i\}$ *uniquely and* $g_i > 0$ *for* $i \geq k_0$.

THEOREM 3.2. *Let* $\{(a_i, b_i): i = 0, 1, 2, \ldots\}$ *be a sequence of vectors with nonnegative real components such that* $a_i > 0 \; \forall i \geq 0$, $b_0 > 0$. *Let* $\{X_n\}_{n=0}^{\infty}$ *be a homogeneous Markov chain with state space given as* $\{0, 1, \ldots\}$ *such that for each* i, *with* $P(X_n = i) > 0$, *we have*

$$P(X_{n+1} = j | X_n = i) = \frac{a_j b_{i-j}}{c_i}, \quad j = 0, 1, 2, \ldots, i,$$

where $\{c_i\}$ *is the convolution of* $\{a_i\}$ *and* $\{b_i\}$. *Assume that for a fixed* n, n_0 *say, we have* $P(X_{n_0} - X_{n_0+1} = k_0) > 0$ *and* $P(X_{n_0} - X_{n_0+1} = k_0 + k_1) > 0$ *with* k_0, k_1 *as considered in Theorem 3.1. Then the following conditions are equivalent*

(i) X_{n_0+1} *and* $X_{n_0} - X_{n_0+1}$ *are independent.*
(ii) *The two* $RR_{n_0}(k)$ *conditions, i.e.,*

$$P(X_{n_0+1} = j) = P(X_{n_0+1} = j | X_{n_0} - X_{n_0+1} = k_0)$$
$$= P(X_{n_0+1} = j | X_{n_0} - X_{n_0+1} = k_0 + k_1), \quad j = 0, 1, \ldots.$$

(iii) *For some* $\lambda > 0$ *and some periodic sequence* $\{q_i: i = 0, 1, \ldots\}$ *with the largest common divisor of the* i *for which* $b_i > 0$ *as one of its periods*

$$P(X_{n_0} = i) = q_i c_i \lambda^i, \quad i = 1, 2, \ldots.$$

REMARK 3.1. Essentially, from the example relative to Heine distributions with a weighted distribution for X_0, in Section 2, it is clear that, given the stochastic matrix (with state space as $\{0, 1, \ldots\}$) and the marginal distribution of X_1, relative to a Markov chain $\{X_n: n = 0, 1, \ldots\}$, it does not follow, even when all $RR_n(k)$ are valid for all nonnegative integer k with $P(X_n - X_{n+1} = k) > 0$ and all $n > 0$ (which is so, if and only if $X_n - X_{n+1}, n > 0$, are all independent), that all finite-dimensional distributions of the chain are determined. Indeed, in the present case, the stochastic matrix is a specialized version of the matrix in Theorem 3.2 with $b_0, b_1 > 0$ and other b_i's equal to zero (which implies that $k_0 = 0$, $k_1 = 1$ is the only possibility); the following example provides us with more substantial information in this connection:

EXAMPLE 3.2. Let $\{X_n: n = 0, 1, \ldots\}$ be as above, and let m be a positive integer. Define $Y_n = X_{mn}$, $n = 0, 1, \ldots$. Note that $\{Y_n: n = 0, 1, \ldots\}$ is a homogeneous Markov chain with state space $\{0, 1, \ldots\}$ with stochastic matrix as in Theorem 3.2, but with $b_n > 0$ for $n = 0, 1, \ldots, m$, and $b_n = 0$ for $n > m$. Clearly here $\{Y_n\}$ possesses all the properties of $\{X_n\}$ referred to above, and, its stochastic matrix is more general than that of $\{X_n\}$ (allowing us more choice for k_0, k_1 in the case when $m > 1$).

REMARK 3.3. If n_0 is a fixed nonnegative integer and $\{X_n: n = 0, 1, \ldots\}$ is a homogeneous Markov chain with state space $\{0, 1, \ldots\}$ and stochastic matrix $(S(j|i))$ with $S(i|i) > 0$ for all i and $S(j|i) = 0$ if $i < j$, and $0 < P(X_{n_0} = X_{n_0+1}) < 1$, then the following are equivalent:

(i) $X_n - X_{n+1}$ and X_{n+1} are independent for $n = n_0, n_0 + 1$.
(ii) $X_n - X_{n+1}$ and X_{n+1} are independent for $n = n_0, n_0 + 1, \ldots$.
(iii) $X_{n_0} - X_{n_0+1}$ and X_{n_0+1} are independent and (iii) of Theorem 3.2 is met with "X_{n_0}", "b_i" and "c_i" replaced respectively by "X_{n_0+1}", "$P(X_{n_0} - X_{n_0+1} = i)$" and "$P(X_{n_0} = i)$".

Obviously, in view of Theorems 2.1 and 3.2 and other relevant results, one can easily see that this observation has some interesting implications.

REMARK 3.4. We have already hinted in Section 2 about the possibility of having extensions of Theorem 2.1, via the relevant results in Chapter 7 of Rao and Shanbhag (1994), to the multivariate case (i.e., the case with the states of the chain as k-vectors, where $k > 1$, having components that are nonnegative integers, instead of with the states themselves as nonnegative integers). Under appropriate assumptions, one could also give extensions of Theorems 3.1 and 3.2 in the same spirit; for example, one could extend the results here, under certain relevant assumptions, via the conditions $RR_{n_0}(c_{ij})$, $i = 1, 2$; $j = 1, 2, \ldots, k$, where c_{ij}'s are distinct with, for each j, the points c_{ij}, $i = 1, 2$, such that they have all co-ordinates, except possibly those in the jth place, to be equal to zero.

REMARK 3.5. Suppose that a and c are positive numbers such that $c < a$, and G is a generating function of a sequence of nonnegative real numbers with at least one of them positive, such that $[0, a)$ is a subset of its domain of definition. Then, for all s lying in $[0, a - c)$, we have $G(s)$ to be proportional to $G(c + s)$ if and only if the members of the sequence are proportional to those of a Poisson distribution. This follows easily from Shanbhag's (1974) argument based on the variance of a distribution concentrated on $\{0, 1, \ldots\}$, or Bernstein's theorem concerning absolutely monotonic functions on $(-\infty, 0)$. (There are also several other alternative ways of presenting essentially the same result; incidentally, Bernstein's theorem referred to above is a corollary to a general result on the integrated Cauchy functional equation, see, for example, Theorem 3.5.1 and its proof in Rao and Shanbhag, 1994, pp. 72–73.) In view of the result referred to here and the second part of Theorem 2.2, we get the following theorem as a simple exercise (of which the first part is a slightly improved version of a result in Sapatinas, 1990 and, essentially, Alzaid, 1979):

THEOREM 3.3. *Let n_0 be a fixed nonnegative integer and $\{X_{n,k}: n = 0, 1, \ldots\}$, $k = 1, 2$, be two discrete branching processes with offspring distributions $\{1 - \pi_k, \pi_k\}$*

($k = 1, 2$, on $\{0, 1\}$) respectively such that $\pi_1 \neq \pi_2$ and such that, for some $n \leqslant n_0$, the random variables $X_{n,k}$, $k = 1, 2$, are identically distributed (obviously with at least one positive support point). Then, for a fixed nonnegative integer x_0 for which $P(X_{n_0,k} = X_{n_0+1,k} + x_0) > 0$,

$$P(X_{n_0+1,k} = j)$$
$$= P(X_{n_0+1,k} = j | X_{n_0,k} = X_{n_0+1,k} + x_0), \quad j = 0, 1, 2, \ldots, \qquad (5)$$

if and only if $X_{0,k}$ are Poisson random variables. Moreover, if we have any one particular $X_{n,k}$ to be a Poisson random variable, then all $X_{n,k}$ are Poisson random variables.

REMARK 3.6. From the property of a generating function mentioned in the previous remark, the Rao–Rubin theorem (1964) for a damage model, of which the first part of Theorem 2.2 is a corollary, follows easily. It may be also noted here that the Markov chain of Theorem 2.2 may be viewed as a branching process where the initial population size X_0 is a random variable (satisfying obviously $P(X_0 = 0) < 1$).

3.2. A generalized Rao–Rubin condition related to stationary measures for a branching process

As implied earlier, the generalized Rao–Rubin(0), GRR(0), condition is a modified version of the identity (RR(0)) in the Rao–Rubin theorem with the parameter π not necessarily the same on its two sides. Talwalker (1975) illustrates that this condition is of relevance to certain toxicological experiments.

Talwalker (1980) and Rao et al. (1980) have studied the problem of identifying the distributions for which GRR(0) holds and have arrived essentially at a partial solution to it. More recently, Alzaid, Rao and Shanbhag (1987) have obtained a more complete and a more satisfying solution to it via an extended version of Spitzer's integral representation theorem given by them, for stationary measures of a discrete branching process; these latter authors have used, amongst other things, Bernstein's theorem cited earlier to get the representation theorem referred to. It is worth noting here that Remark 2 of Alzaid et al. (1987) indicates that the extended representation theorem is indeed a specialized version of the Poisson–Martin integral representation theorem relative to Markov chains. (That both Bernstein's theorem and the Poisson–Martin representation theorem implied above have also links with Choquet's theorem is well known.)

Although it is evident that Alzaid et al. (1987) itself provides us with a Markov chain approach to damage models, it is not pointless to highlight here that the characterization theorem based on GRR(0) condition, given in the cited reference, has the following version of it in a Markov chain set-up, in the spirit of other results given in this paper, as an obvious corollary:

THEOREM 3.4. Let n_0, $\{X_{n,1}\}$ and $\{X_{n,2}\}$ be as in Theorem 3.3 with $X_{n_0,k}$, $k = 1, 2$, be identically distributed. Then,

$$P(X_{n_0+1,1} = j) = P(X_{n_0+1,2} = j | X_{n_0,2} = X_{n_0+1,2}), \quad j = 0, 1, \ldots, \qquad (6)$$

if and only if one of the following is valid:

(i) $\pi_1 = \pi_2$ and $X_{n_0,1}$ is a Poisson random variable (*in which case all $X_{n,k}$ are Poisson random variables*).
(ii) $\pi_1 > \pi_2$ and $X_{n_0,1}$ is a random variable with distribution binomial $(\cdot, \frac{\pi_1-\pi_2}{\pi_1(1-\pi_2)})$ (*in which case all $X_{n,k}$ are binomial random variables*).
(iii) $\pi_1 < \pi_2$ and $X_{n_0,1}$ has its distribution of the following form:

$$g_j \propto \sum_{r=-\infty}^{\infty} \int_{[0,1)} c^{r-t} e^{-(\pi_1/\pi_2)^{(r-t)}} \frac{(\pi_1/\pi_2)^{(r-t)j}}{j!} \left[\frac{\pi_2-\pi_1}{\pi_2(1-\pi_1)}\right]^j d\upsilon(t),$$

$$j = 0, 1, \ldots,$$

where υ is a probability measure on $[0, 1)$ and c is a real number lying in $(0, 1)$ (in which case for each n, k, the distribution of $X_{n,k}$ is of the form of the distribution of $X_{n_0,1}$ but for a modification that the exponent of $e^{-(\pi_1/\pi_2)^{(r-t)}}$ (i.e., the quantity $-(\pi_1/\pi_2)^{(r-t)}$) and the quantity $\frac{\pi_2-\pi_1}{\pi_2(1-\pi_1)}$ appear now with certain constant multipliers that are free of t).

REMARK 3.7. Rao and Shanbhag (1994, 7.3(a), pp. 175–181) have essentially revisited the results in Alzaid et al. (1987), making minor alterations to the arguments in one or two places; incidentally, the monograph simplifies slightly the proof given in the cited paper for the extended Spitzer integral representation theorem by involving in place of (2.11) (of the paper) its modified version (showing obvious alterations) with n varying over all integers instead of just positive ones. It is also interesting, and perhaps useful, to note in this place that a closer scrutiny of the proof referred to implies that the functions U and U^* appearing in the proof are such that, for each s lying in $[0, 1)$, $\{c^n[U((1-f_n(0))s + f_n(0)) - U(f_n(0))]\}$ tends, as n tends to ∞, to $U^*(s)$, where f_n is the nth iterate of the generating function, f, of the probability distribution $\{p_j\}$.

REMARK 3.8. Incidentally, the existing sketch of the argument in Remark 2 of Alzaid et al. (1987) referred to above assumes implicitly that $p_1 > 0$, but, that it holds generally with minor modifications (to line 5 of it) is obvious. For example, it holds if we replace the two denominators involved in line 5 of the remark by the respective numerators with s replaced by p_0, or, by the k^*th derivatives at $s = 0$ of the respective numerators divided by $k^*!$, where k^* is the smallest positive support point of the distribution $\{p_j\}$.

REMARK 3.9. The proof of the extended Spitzer integral representation theorem in Alzaid et al. (1987) uses essentially that the restriction of U^* to $(0, 1)$ is absolutely monotonic and appeals to Bernstein's theorem to conclude about the absolute monotonicity of the relevant extension of U^*. However, that the restriction of U^* to $(0, 1)$ is absolutely monotonic if and only if U^* is the generating function of a nonnegative sequence is well known.

REMARK 3.10. Rao et al. (1980) involves some typos. Also, "(possibly infinite)" appearing in the statement of the theorem in the cited paper is ambiguous, since $\mu((0, \infty)) = \infty$, where the notation is as in the statement.

REMARK 3.11. Using an argument based on, amongst other things, the extended Spitzer integral representation theorem referred to here, Sapatinas (1996) gave a slight variation of the characterization of distributions, based on GRR(0) condition, given by Alzaid et al. (1987); this latter result improves a result established earlier by Talwalker (1986).

REMARK 3.12. For the branching processes $\{X_{n,k}\}$, $k = 1, 2$, in Theorem 3.4, the relations between the generating functions of $X_{n,k}$ and $X_{0,k}$ of the type between the generating functions of X_n and X_0 met in the proof of Theorem 2.2, with appropriate notational changes, hold. Consequently, given k, n_0, π_k, and the distribution of $X_{n_0,k}$, all finite-dimensional distributions relative to $\{X_{n,k}\}$ are determined; also, now, if the distribution of $X_{n_0,1}$ is as in (i) or (ii) or (iii) of Theorem 3.4, one can easily see what form each marginal distribution relative to $X_{n,k}$ has.

4. Characterization via conditional expectations

The following theorem is essentially a version of a result of Sapatinas and Aly (1994), which, in turn, is an extended version of a result of Shanbhag and Clark (1972) and follows via a slightly amended version of the argument used in this prior reference; the argument referred to here involves, among other things, certain power series equations. (Incidentally, in what follows, we denote, for a random variable X, $X(X-1)\ldots(X-r+1)$ by $X^{(r)}$, and, for any real number x, we use the analogous notation. Also, for each Markov chain that appears below the state space is implied to be $\{0, 1, \ldots\}$.)

THEOREM 4.1. *Let n_0 be a nonnegative integer and $\{X_{n,\lambda}: n = 0, 1, \ldots\}$ be a family of homogeneous Markov chains where λ varies over a dense subset of an open interval, $X_{n_0,\lambda}$ follows a power series distribution with parameter λ and the diagonals of the concerned stochastic matrices are independent of λ such that, for each λ and almost all i,*

$$E_\lambda\big(X^{(r)}_{n_0+1,\lambda} | X_{n_0,\lambda} = i\big) = i^{(r)} \pi^r, \quad r = k, k+1, k+2, \tag{7}$$

for some fixed $\pi \in (0, 1)$ independent of λ. (Here we do not require the stochastic matrices to be triangular.) If the factorial moment $E_\lambda(X^{(k+1)}_{n_0+1,\lambda})$ corresponding to $X_{n_0+1,\lambda}$ is non-zero and $0 < P_\lambda(X_{n_0,\lambda} = X_{n_0+1,\lambda})$ for some (and hence all) λ, then for all λ,

$$E_\lambda\big(X^{(r)}_{n_0+1,\lambda}\big) = E_\lambda\big(X^{(r)}_{n_0+1,\lambda} | X_{n_0,\lambda} = X_{n_0+1,\lambda}\big), \quad r = k, k+1, k+2, \tag{8}$$

if and only if for each λ, $X_{n_0,\lambda}$ is a Poisson random variable and the ith diagonal entry of the stochastic matrix is proportional to π^i for all i.

REMARK 4.1. In Patil and Ratnaparkhi's (1977) characterizations of binomial and negative binomial distributions, "$0 < m \leqslant N$" should read "$0 < m < N$", and, in the extended versions of these results in Sapatinas and Aly (1994), "$0 < m + \max\{1, k\} \leqslant N$" should read "$0 < m < N - \max\{0, k - 1\}$". (Note that in the latter case the substitution gives the corrected as well as improved versions of the relevant results in Sapatinas and Aly (1994).)

The Sapatinas–Aly characterizations of binomial and negative binomial distributions enable one to arrive at results analogous to Theorem 4.1 in a Markov chain set-up, characterizing binomial and negative binomial distributions.

Incidentally, it is of interest to note that the versions of the Sapatinas and Aly (1994) results referred to in the remark above do not hold if the condition that "$0 < m < N - \max\{0, k - 1\}$" is dropped. This is illustrated by the following counter examples:

EXAMPLE 4.2. Take, in obvious notation,

$$A(\lambda) = (1 - p\lambda)^{-N} + \alpha$$

and

$$A^*(\lambda) = \sum_{n=0}^{\infty} S(n|n) a_n \lambda^n \propto \{(1 - p\lambda)^{-N+1} + \alpha(1 - p\lambda)\}$$

with α as sufficiently small and positive, $p \in (0, 1)$ and $N > 1$. Note that this example satisfies the assumptions in the result relative to a negative binomial distribution in Sapatinas and Aly but for the stated condition. Also, in this case, in the damage model set-up

$$E_\lambda(Y_\lambda^{(r)}) = E_\lambda(Y_\lambda^{(r)} | X_\lambda = Y_\lambda), \quad r = 2, 3, 4. \tag{9}$$

Obviously in this case X_λ is not negative binomial.

EXAMPLE 4.3. Take, in obvious notation,

$$A(\lambda) = (1 - p + p\lambda)^N + \alpha(1 - p + p\lambda)$$

and

$$A^*(\lambda) = \sum_{n=0}^{\infty} S(n|n) a_n \lambda^n \propto (1 - p + p\lambda)^{N-1} + \alpha$$

with $\alpha > 0$, $p \in (0, 1)$ and N as a positive integer $\geqslant 4$. Note that this example satisfies the assumptions in the result relative to a binomial distribution in Sapatinas and Aly but for the stated condition. Also, in this case, the condition (9) is satisfied. Obviously here X_λ is not binomial.

REMARK 4.4. If $\{X_{n,\lambda}\}$ is as in Theorem 4.1 with a modification that the corresponding stochastic matrices are triangular as in (1) and are independent of the parameter λ, then, provided $P_\lambda(X_{n_0,\lambda} = 0) > 0$ and $1 > P_\lambda(X_{n_0,\lambda} = X_{n_0+1,\lambda}) > 0$ for some (and hence all) λ, we have that $RR_{n_0}(0)$ holds for all λ is equivalent to the assertion that $X_{n_0+1,\lambda}$ has a power series distribution with parameter λ (and given parameter space). This result follows essentially via an Alzaid (1986)-type argument.

REMARK 4.5. In view of Remark 4.4 given above, it follows that, under appropriate assumptions, a family $\{X_{n,\lambda}\}$ of homogeneous Markov chains has $RR_n(0)$ condition met for all n and all parameter values if and only if each $X_{n,\lambda}$ has a power series distribution with parameter λ (and the parameter space as given).

REMARK 4.6. If the situation is as in Remark 4.4 with $RR_{n_0}(0)$ holding for each value of λ (with the relevant assumption met) with (in obvious notation) all $S(i|i) > 0$ and $S(0|1) > 0$, then, provided, for each λ, the support of the distribution of $X_{n_0,\lambda}$ is $\{0, 1, \ldots\}$, the common stochastic matrix relative to the relevant Markov chains is seen to be of the form in Theorem 2.1.

REMARK 4.7. In the light of general results of Sapatinas and Aly (1994) and Albassam et al. (2000), it is possible to have more general results than those appearing in Theorem 4.1. Also, under appropriate modified assumptions, it is possible to give the corresponding results with exponential family of distributions on $\{0, 1, \ldots\}$ in place of power series distributions (in the case especially where reparametrizations transform the problems to those corresponding the power series distributions).

The following theorem is a corollary to Theorem 12.4.2 of Albassam, Rao and Shanbhag (2000), which, in turn, is a version of the celebrated characterization result relative to exponential families given by Morris (1982); see also the remark immediately below the theorem in Albassam et al. (2000), regarding a link between the findings respectively of Morris (1982) and Laha and Lukacs (1960).

THEOREM 4.2. *Let n_0 be a nonnegative integer and $\{X_{n,\lambda}: n = 0, 1, \ldots\}$ be a family of homogeneous Markov chains where λ varies over a dense subset of an open interval, $X_{n_0,\lambda}$ follows a power series distribution with parameter λ and the diagonals of the concerned stochastic matrices are independent of λ such that, for each λ and almost all i,*

$$E_\lambda(X_{n+1,\lambda}^{(r)} | X_{n,\lambda} = i) = i^{(r)} c_r, \quad r = 1, 2,$$

where $0 < c_2 < c_1$ and c_1, c_2 are independent of λ. (Here we do not require the stochastic matrices to be triangular.) Then, for all λ,

$$E_\lambda(X_{n_0+1,\lambda}^{(r)}) = E_\lambda(X_{n_0+1,\lambda}^{(r)} | X_{n_0,\lambda} = X_{n_0+1,\lambda}), \quad r = 1, 2, \tag{10}$$

with $1 > P_\lambda(X_{n_0,\lambda} = X_{n_0+1,\lambda}) > 0$, if and only if (in obvious notation) one of the following holds:

(i) $c_2 = c_1^2$ and then $X_{n_0,\lambda} \sim$ Poisson and $p_{ii}(\lambda) = c_1^i$ for all λ, i.

(ii) $c_2 < c_1^2$, $c_1 < 1$ and then $X_{n_0,\lambda} \sim$ Binomial$(\frac{1}{1-\alpha}, \cdot)$ and $p_{ii}(\lambda) = \frac{[c_1(1-\alpha)^{-1}]^{(i)}}{[(1-\alpha)^{-1}]^{(i)}}$ for all $i \leq \frac{1}{1-\alpha}$ and all λ, where $\alpha = \frac{c_1(c_1-1)}{c_2-c_1}$.

(iii) $c_2 > c_1^2$ and then $X_{n_0,\lambda} \sim$ Negative binomial$(-\frac{1}{1-\alpha}, \cdot)$ and $p_{ii}(\lambda) = \frac{[c_1(1-\alpha)^{-1}]^{(i)}}{[(1-\alpha)^{-1}]^{(i)}}$ for all λ, i, where $\alpha = \frac{c_1(c_1-1)}{c_2-c_1}$.

REMARK 4.8. It is worth noting that the aforementioned Theorem 4.2 is valid with appropriate modifications if the condition (10) is replaced by the condition

$$E_\lambda(X_{n_0+1,\lambda}^{(r)}) = E_\lambda(X_{n_0+1,\lambda}^{(r)} | X_{n_0,\lambda} = X_{n_0+1,\lambda} + k), \quad r = 1, 2,$$

where k is a fixed non-negative integer.

REMARK 4.9. (ii) of Theorem 4.2 with "$c_1 = 1$" in place of "$c_1 < 1$" implies that $X_{n_0,\lambda} = X_{n_0+1,\lambda}$ a.s. for all λ. It is also interesting to note that if we replace (7) and (8) by their modified versions with the right-hand sides as c and c' times respectively, of those of the existing ones, where c and c' are fixed positive constants, Theorem 4.1 does not remain valid unless $c = c'$; we assume here that the left-hand sides of the two identities are unaltered.

Algraian (1992) established a result of moment characterization in the light of Alzaid, Rao and Shanbhag (1987), Rao et al. (1980) and Talwalker (1980) results. We give below a variant of that result linked with Markov chains. Incidentally, the argument given by Algraian to prove the original version of the result is slightly inaccurate. We give below a version of this result (with a slight improvement) in a Markov chain set-up.

THEOREM 4.3. Let $\{X_{n,\lambda}\}$ and $\{Y_{n,\lambda}\}$ be two families of discrete homogeneous Markov chains with λ varying over a dense subset, Λ, of an open interval and for a fixed n, say n_0, $X_{n_0,\lambda}$ and $Y_{n_0,\lambda}$ having the same power series distribution

$$p_i(\lambda) = \frac{a_i \lambda^i}{A(\lambda)}, \quad i = 0, 1, \ldots,$$

with $a_k > 0$, where k is a fixed integer. Also, assume that, for all $\lambda \in \Lambda$ and almost all i,

$$E_\lambda(X_{n_0+1,\lambda}^{(k)} | X_{n_0,\lambda} = i) = i^{(k)} \pi^i, \quad a.s., \lambda \in \Lambda,$$

and the diagonal entries of stochastic matrices relative to $\{Y_{n,\lambda}\}$ equal respectively π'^i for all $i \geq 0$, with $\pi, \pi' \in (0, 1)$ and independent of λ. Then

$$E_\lambda(X_{n_0+1,\lambda}^{(k)}) = E_\lambda(Y_{n_0+1,\lambda}^{(k)} | Y_{n_0,\lambda} = Y_{n_0+1,\lambda}), \quad \lambda \in \Lambda, \tag{11}$$

if and only if $\pi = \pi'$ and $a_i = \frac{c}{i^{(k)}} a_{i-k}$ for some $c > 0$ and $i = k, k+1, \ldots$. (Once again, we do not require here the stochastic matrices corresponding to the relevant Markov chains to be triangular.)

PROOF. We can assume without loss of generality Λ to be an open interval of the type $(0, \beta)$ with $\beta > 0$. To prove the "only if" part of the assertion, we note that Eq. (11) is equivalent then to

$$\pi^k \frac{\sum_i i^{(k)} a_i \lambda^i}{A(\lambda)} = \frac{(\pi'\lambda)^k \sum_i i^{(k)} a_i (\pi'\lambda)^{i-k}}{\sum_i (\pi'\lambda)^i a_i}, \quad \lambda \in (0, \beta),$$

or, to

$$\frac{\pi^k \sum_i i^{(k)} a_i \lambda^{i-k}}{A(\lambda)} = \frac{\pi'^k \sum_i i^{(k)} a_i (\pi'\lambda)^{i-k}}{A(\pi'\lambda)}, \quad \lambda \in (0, \beta).$$

We can rewrite the latter equation as

$$\frac{A^{(k)}(\lambda)}{A(\lambda)} = \left(\frac{\pi'}{\pi}\right)^k \frac{A^{(k)}(\pi'\lambda)}{A(\pi'\lambda)}, \quad \lambda \in (0, \beta), \tag{12}$$

where $A^{(k)}$ is the kth derivative of A. The relation (12) leads to

$$\frac{A^{(k)}(\lambda)}{A(\lambda)} = \left(\frac{\pi'}{\pi}\right)^{ik} \frac{A^{(k)}(\pi'^i \lambda)}{A(\pi'^i \lambda)}, \quad \lambda \in (0, \beta),$$

which, in turn, yields, for $\lambda \in (0, \beta)$,

$$\frac{A^{(k)}(\lambda)}{A(\lambda)} = \lim_{i \to \infty} \left(\frac{\pi'}{\pi}\right)^{ik} \frac{A^{(k)}(\pi'^i \lambda)}{A(\pi'^i \lambda)} = \begin{cases} 0, & \text{if } \pi' < \pi, \\ \infty, & \text{if } \pi' > \pi, \\ c, & \text{if } \pi' = \pi, \end{cases}$$

where c is a constant $\neq 0$.

As, under the assumptions of the theorem, both $A(\lambda)$ and $A^{(k)}(\lambda)$ are in $(0, \infty)$, we cannot have the ratio $A^{(k)}(\lambda)/A(\lambda)$ to be either zero or infinite. Hence it follows that the third case is the only case that is valid; here $\pi' = \pi$ and

$$A^{(k)}(\lambda) = c A(\lambda), \quad \lambda \in (0, \beta). \tag{13}$$

By equating the coefficients of λ^i on both sides, we get

$$a_i i^{(k)} = c a_{i-k}, \quad i = k, k+1, \ldots. \tag{14}$$

This establishes the "only if" part of the theorem. The "if" part of the theorem follows easily and hence we have the theorem. □

References

Albassam, M. (2000). Damage models and their applications. Ph.D. thesis, University of Sheffield.

Albassam, M., C. R. Rao and D. N. Shanbhag (2000). Characterizations of some exponential families based on survival distributions and moments. In *Probability and Statistical Models with Applications*, pp. 209–223 (Eds. Charalambides et al.). Chapman and Hall.

Algraian, M. N. (1992). Characterization results based on mixtures and power-series models. Ph.D. thesis, University of Sheffield.

Alzaid, A. A. (1979). Some characterizations in Rao–Rubin damage models. M.Sc. thesis, University of Sheffield.

Alzaid, A. A. (1983). Some contributions to characterization theory. Ph.D. thesis, University of Sheffield.

Alzaid, A. A. (1986). Some results connected with the Rao–Rubin condition. *Sankhyā, Ser. A* **48**, 104–108.

Alzaid, A. A., C. R. Rao and D. N. Shanbhag (1984). Solutions of certain functional equations and related results on probability distributions. *Unpublished Research Report*, University of Sheffield.

Alzaid, A. A., C. R. Rao and D. N. Shanbhag (1986a). An application of the Perron–Frobenius theorem to a damage model problem. *Sankhyā, Ser. A* **48**, 43–50.

Alzaid, A. A., C. R. Rao and D. N. Shanbhag (1986b). Characterization of discrete probability distributions by partial independence. *Comm. Statist. Theory & Methods, A* **15**, 643–656.

Alzaid, A. A., C. R. Rao and D. N. Shanbhag (1987). An extension of Spitzer's integral representation theorem with an application. *Ann. Probab.* **15**, 1210–1216.

Alzaid, A. A., K. Lau, C. R. Rao and D. N. Shanbhag (1988). Solution of Deny's convolution equation restricted to a halfline via a random walk approach. *J. Multivariate Anal.* **24**, 309–329.

Andrews, C. (1988). Recent advances in damage models. M.Sc. dissertation, University of Sheffield.

Athreya, K. B. and P. E. Ney (1972). *Branching Processes*. Springer, Berlin.

Bather, J. A. (1992). Search models. *J. Appl. Probab.* **29**(3), 605–615.

Benkherouf, L. and J. A. Bather (1988). Oil exploration: sequential decisions in the face of uncertainty. *J. Appl. Probab.* **25**, 529–543.

Consul, P. C. (1975). Some new characterizations of discrete Lagrangian distributions. In *Statistical Distributions in Scientific Work*, Vol. 3, pp. 279–290 (Eds. G. P. Patil, S. Kotz and J. K. Ord). Reidel, Dordrecht.

Janardan, K. G. and B. R. Rao (1982). Characterization of generalized Markov–Polya and generalized Polya–Eggenberger distributions. *Comm. Statist. Theory & Methods, A* **11**, 2113–2124.

Kemeny, J. G., J. L. Snell and A. W. Knapp (1966). *Denumerable Markov Chains*. The University Series in Higher Mathematics, van Nostrand, New York.

Laha, R. and E. Lukacs (1960). On a problem connected with quadratic regression. *Biometrika* **47**, 335–343.

Morris, C. (1982). Natural exponential families with quadratic variance functions. *Ann. Statist.* **10**, 65–80.

Patil, G. P. and M. V. Ratnaparkhi (1977). Characterizations of certain statistical distributions based on additive damage models involving Rao–Rubin condition and some of its variants. *Sankhyā, Ser. B* **39**, 65–75.

Patil, G. P. and C. Taillie (1979). On a variation of the Rao–Rubin condition. *Sankhyā, Ser. A* **41**, 129–132.

Phelps, R. R. (1966). *Lecture Notes on Choquet's Theorem*. Van Nostrand, Princeton.

Rao, C. R. (1963). On discrete distributions arising out of methods of ascertainment. Paper presented at the Montreal Conference on Discrete Distributions. Printed in *Sankhyā, Ser. A* **27** (1965), 311–324.

Rao, C. R. and H. Rubin (1964). On a characterization of the Poisson distribution. *Sankhyā, Ser. A* **26**, 295–298.

Rao, C. R., R. C. Srivastava, S. Talwalker and G. A. Edgar (1980). Characterization of probability distributions based on generalized Rao–Rubin condition. *Sankhyā, Ser. A* **42**, 161–169.

Rao, C. R. and D. N. Shanbhag (1994). *Choquet–Deny Type Functional Equations with Applications to Stochastic Models*. Wiley, Chichester.

Rao, C. R., M. B. Rao and D. N. Shanbhag (2002). Damage models – a Martin boundary connection. Special Issue in Memory of D. Basu. *Sankhyā, Ser. A* **64**, 872–887.

Rao, M. B. and D. N. Shanbhag (1982). Damage models. In *Encyclopedia of Statistical Science*, Vol. 2, pp. 262–265 (Eds. N. L. Johnson and S. J. Kotz). Wiley, New York.

Sapatinas, T. (1990). Damage models and applications in Markov chains. M.Sc. dissertation, University of Sheffield.

Sapatinas, T. (1996). On the generalized Rao–Rubin condition and some variants. *Austral. J. Statist.* **38**, 299–306.

Sapatinas, T. and M. A. H. Aly (1994). Characterizations of some well-known discrete distributions based on variants of the Rao–Rubin condition. *Sankhyā, Ser. A* **56**, 335–346.

Seneta, E. (1973). *Nonnegative Matrices.* Allen and Unwin, London.

Seneta, E. (1981). *Nonnegative Matrices and Markov Chains*, 2nd ed. Springer-Verlag, New York.

Shanbhag, D. N. (1974). An elementary proof for the Rao–Rubin characterization of the Poisson distribution. *J. Appl. Probab.* **11**, 211–215.

Shanbhag, D. N. (1977). An extension of the Rao–Rubin characterization of the Poisson distribution. *J. Appl. Probab.* **14**, 640–646.

Shanbhag, D. N. and R. M. Clark (1972). Some characterizations of the Poisson distribution starting with a power series distribution. *Proc. Cambridge Philos. Soc.* **71**, 517–522.

Shanbhag, D. N. and J. Panaretos (1979). Some results related to the Rao–Rubin characterization of the Poisson distribution. *Austral. J. Statist.* **21**, 78–83.

Shanbhag, D. N. and C. Taillie (1979). An extension of the Patil–Taillie characterization of the Poisson distribution. Unpublished.

Srivastava, R. C. and J. Singh (1975). On some characterizations of the binomial and Poisson distributions based on a damage model. In *Statistical Distributions in Scientific Work*, Vol. 3, pp. 271–277 (Eds. G. P. Patil, S. Kotz and J. K. Ord). Reidel, Dordrecht.

Srivastava, R. C. and A. B. L. Srivastava (1970). On the characterization of the Poisson distribution. *J. Appl. Probab.* **7**, 495–501.

Talwalker, S. (1975). Models in medicine and toxicology. In *Statistical Distributions in Scientific Work*, Vol. 2, pp. 263–274 (Eds. G. P. Patil, S. Kotz and J. K. Ord). Reidel, Dordrecht.

Talwalker, S. (1980). A note on the generalized Rao–Rubin condition and characterization of certain discrete distributions. *J. Appl. Probab.* **17**, 563–569.

Talwalker, S. (1986). Functional equations in characterizations of discrete distributions by Rao–Rubin condition and its variants. *Comm. Statist. Theory & Methods, A* **15**, 961–979.

Williams, D. (1979). *Diffusions, Markov Processes, and Martingales. Vol. 1: Foundations.* Wiley, Chichester.

Point Processes in Astronomy: Exciting Events in the Universe

Jeffrey D. Scargle and Gutti Jogesh Babu

We present an overview of point processes in astronomy, with emphasis on data and data analysis, and on modeling of temporal and spatial astrophysical phenomena using point processes. Following a brief discussion of the naive point process theory sufficient for most real-world applications, we outline the facts behind the mysterious gamma-ray bursts, and then detail two different point-process aspects of this exciting problem. Brief discussions of five other areas of astronomy where point processes play an important role conclude this article.

1. Introduction: what's the point?

Point processes are good mathematical models for many astronomical systems. A simple star map is a set of points scattered about a two-dimensional area. The distribution of stars is a spatial point process, in two dimensions as captured by photographs and maps, but of course three-dimensional in reality. This process is rich with clustering on various scales, and other departures from the simple notion of a purely random – identically and independently distributed, or IID – process. The arrival of photons from one of these stars at the focal plane of a telescope is a fine example of a temporal point process – a simple one, well described as IID events.

Indeed, almost any observational program yields coordinates and other quantitative characterizations for each member of a set, e.g., of material objects such as stars, planets, and galaxies; or observational phenomena such as photons, neutrinos, and spectral lines; or conceptual entities such as point masses, stellar eclipses, or spiral arms of galaxies. Such *events* are most often points in time, space, velocity or possibly some other variable. They may even be points in a higher-dimensional manifold such as space-time.

Two aspects of physical events make them candidates for representation in terms of point processes. The first is that they are point-like in the sense of occupying a small and coextensive volume in the relevant space. Stars appear as points on the sky; photons are registered at the telescope in a very short time interval; and light in a spectral line is confined to a small interval of wavelength. Of course size is relative; stars are really quite huge, and appear point-like only because they are quite distant. Further, an extended object/event may have a well defined centroid or other fiducial that can serve as *the point*.

The second aspect is more fundamental, namely the *discreteness* of the events – that is the degree to which they are distinct entities. Stars in a cluster or a galaxy influence each other only slightly through gravity and radiation, but for most purposes are disconnected – individual bodies with a clear and largely empty region of space separating them. Photons for the most part are also noninteracting, due to the physical nature of light. This leads immediately to the photon arrival being well described by the Poisson process (see Section 4.1.1).

In some cases the astrophysical phenomenon of interest is intrinsically discrete. It is the events themselves that are under study. Astronomers accurately time, e.g., eclipses in a binary star system. (This point process is at least approximately deterministic, a rare bird, seldom discussed in the mathematical literature.) Other examples include studies of the positions of stars, galaxies, clusters of galaxies, or other relatively well defined and localized astronomical objects. In all these cases, the discrete astronomical event or object is of direct interest, although of course so are more global features, such as their distribution in time or space.

On the other hand, in many cases the underlying physical process is more or less continuous, but is only revealed to the observer through discrete events. The discreteness is due to the nature of the observation, and is not intrinsic to the physical process. For example, light curves of variable stars are obtained by counting the photons as they arrive at the focal plane of the telescope. Similar counts of x-ray or gamma-ray photons from low mass x-ray binary stars or gamma-ray bursts provide crucial information about the physical nature of these sources. Such high-energy light is not intrinsically more corpuscular, or *quantum*, in nature than is visible light, infrared radiation, or radio waves. But the nature of physical detectors that are efficient at recording high energy radiation is such that individual photons are more easily identified, recorded, and timed accurately. A detailed example of photon counting observations forms the heart of our discussion in Section 4.1 below.

In fact, the trend in scientific data acquisition in general, for both temporal and spatial observations, is toward discreteness – *digital* electronics in preference to *analog*. Another example: modern imaging technology delivers data automatically divided up into discrete levels of intensity corresponding to discrete *pixels* defined at the hardware level. That is, the specification of both pixel location and intensity level are discrete. Of course it is the essentially continuous physical object that is the subject of the scientific inquiry.

The separation of *intrinsic point processes* from *observation point processes* outlined in the paragraphs above is not always clean. Gamma-ray bursts, for example, are quite confined in both time and space, in the sense that they are very brief, apparently nonrepeating, events coming from very small volumes of extragalactic space (see Section 4). To the extent that these discrete objects are viewed as tracing out a more general, smooth macroscopic distribution of matter in the cosmos, these objects have some element of both kinds of point process.

The word *process* needs a bit of explanation too. Basically it describes any prescription for determining points' locations in time, space, or whatever the variable of interest is. We usually mean a *random process*, but the deterministic case is not excluded. In many contexts the term *process* implies an evolution over time. For temporal point processes this interpretation is usually appropriate, but not so for spatial and other kinds of point processes (see below). The authors of (Stoyan, Kendall and Mecke, 1995) suggest that the term *random point field* be used in spatial cases – to avoid any connotation of unfolding in time. In most spatial point processes temporal order is not relevant, or of secondary importance, but is of paramount importance in others (see Section 5.4). In a similar terminological vein, *stationary* and *homogeneous* refer to the same concept in the temporal and spatial domains, respectively; especial care is needed when using these terms in reference to spatio-temporal point processes, where one has a spatial point field unfolding in time.

The present chapter emphasizes data and data analysis, and modeling of astrophysical phenomena using point processes. Most of our examples are temporal and spatial processes. For a similar review emphasizing general phenomenology, spatial processes, and theoretical considerations, see (Babu and Feigelson, 1996).

2. Unique features of astronomical point processes

Astronomical point processes evince unique and interesting properties that attract the attention of the statistical community. Many such processes in astronomy relate to investigations with profound scientific and philosophical implications. Examples include the detection of planetary systems around other stars, the origin and detection of life in the Universe, the structure of the early Universe, and the origin of the Universe itself.

Furthermore, the measurements are often very expensive, either directly as measured by the money spent on the research program devoted to obtaining and interpreting the data, and/or as indicated by the magnitude of the effort on the part of tens, hundreds, or even thousands of scientists, engineers, and data analysts. The point is that the reduction, analysis and understanding of these data are therefore worthy of the devotion of a large effort, including the development of new, special purpose methods and the expenditure of hugh amounts of computational and human resources. Frequently, computational efficiency is unimportant or of secondary importance, if traded off against optimal extraction of information.

The value of the data, computed per photon, can be incredibly large in situations where there are a small number of photons. A famous example concerns the detection of neutrinos – particles, not photons – from the nearby supernova observed in February of 1987 (Loredo, 1990). Due to the extremely low level of interaction between these elusive particles and ordinary matter, even the combination of a Herculean observational facility and this incredibly rare nearby (and therefore intrinsically bright) supernova explosion yielded only about two dozen neutrinos detected. This hand full of data, due to its great scientific importance, has been subject to an enormous battery of analysis efforts, and has inspired much work by astrophysical theorists.

Undoubtedly the most extreme example of small numbers – other than nondetections, i.e. zero events – is the detection of a single, very high energy (18 GeV) photon some thousands of seconds after the main peak in emission from the gamma ray burst of

February 17, 1994 (Hurley et al., 1994). This one photon can be associated with the burst, even though its location on the sky is not known with high accuracy, because of the very low rate of such photons from the general background or from other known point sources.

Another somewhat special situation is that, in many cases, one knows with great certainty the exact distribution of the observational errors. For example, in the case of astronomical photon counting observations, the only significant source of uncertainty is from the statistical fluctuations in the detection process – precisely described by the Poisson distribution. Some astronomical data are a very pure, idealized example of the mathematician's Poisson process. And the small departures from the ideal are in the form of rather well known deadtime effects in the detector electronics. This makes the process not quite Poisson, but very closely so. If nothing else, this can be a refreshing departure from the boring case where errors are normally distributed and additive. Of course this simplicity is not always achieved. Astronomy frequently suffers the problem shared by most empirical sciences, namely that the total observational error is the result of a complex combination of partially known or completely unknown effects, acting in concert.

In a similar vein, event data represents quite "raw" or "pure" information, uncontaminated by degraded time resolution, etc. Other approaches (binning of data, e.g.) dilute the information and impose erroneous assumptions. In fact, astronomers and other physical scientists have, to one degree or another, been guilty of adopting some false assumptions, or folk lore, about binning, described in (Scargle, 1998) as the *binning fallacies*. It is widely and incorrectly held that point process data must be binned[1] in order to be analyzed at all, and further that the bins must be large enough so that there are enough photons in each to provide a good statistical sample. Of course neither of these is true. The common practice of binning event data throws away a considerable amount of information and introduces dependency of the results on the sizes and locations of the bins. It has been known for some time (Cash, 1979), but not well appreciated by practicing data analysts, that fitting models to unbinned data is quite straightforward. Other errors are committed by applying statistical methods based on the assumption of a normal distribution to data that is highly non-normal, such as Poisson data.

3. Naive point process theory

A comprehensive introduction to point processes, including both theoretical and practical matters, is (Snyder and Miller, 1991), a substantial revision of an earlier book (Snyder, 1975). The review (Brillinger, 1978) takes an interesting and informative tack, comparing and contrasting ordinary time series with point processes. Brillinger further discusses generalized theories, of which both point and ordinary processes are special cases. A highly readable treatment of spatial point processes (under the name *stochastic geometry*) can be found in (Stoyan, Kendall and Mecke, 1995). See also the discussion of Fourier analysis of stationary point processes in Section XIII of the addendum in the expanded edition of (Brillinger, 1981). More technical references are (Murthy, 1984; Andersen et al., 1993).

---

[1] I.e., one must divide the observation into equally spaced intervals and count photons within these *bins*.

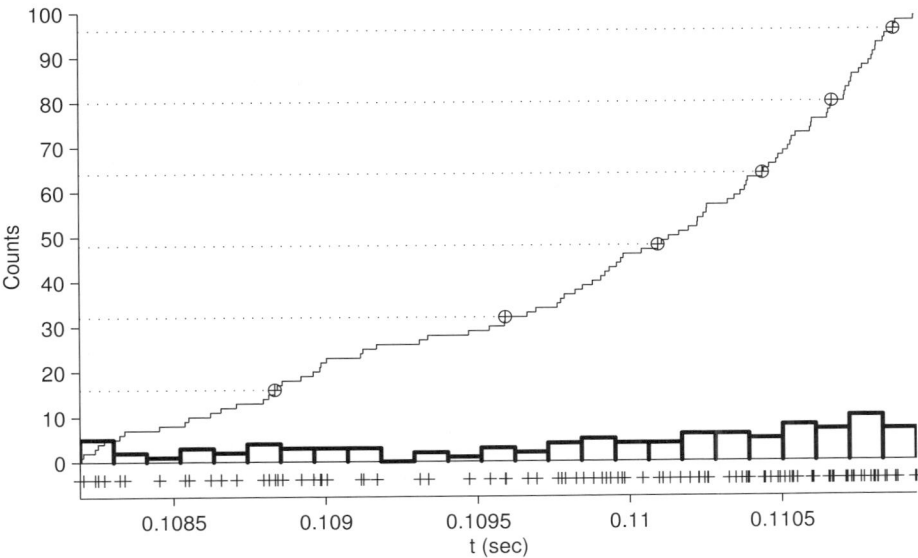

Fig. 1. Three portraits of photon counting data from a Gamma Ray Burst, BATSE Trigger 0551. + = individual event times [$n(t)$]; o = time-to-spill marks (Erlang process); thick line = binned counts; thin line = cumulative counts $N(t)$. These three different ways of graphing the raw data correspond to the three BATSE data modes, as described in the text.

This section gives a brief overview of point process theory at the intuitive level sufficient for most problems in physical and biological science. Real data is always finite. That is, the number of points in any point process actually sampled is finite. And the coordinates, often represented mathematically as real numbers, can always be considered to belong to a discretized, finite space. These facts remove the necessity for many of the delicate measure theoretic, and other issues that make mathematically rigorous random process theory difficult for the practicing scientist. We might call our overview *naive point process theory*, after a similar phrase coined for set theory.

The basic idea behind the notion of a point process is *a set of discrete events occurring at well-defined, but random, points in time*. As in time series analysis in general, *time* really stands for whatever independent variable is relevant. Astronomers deal with cases where the independent variable is time, space, space-time, velocity, redshift, and many others. The only significant difference between temporal processes and the others is the concept of causality – but this is of essential importance in many applications. A key issue here is the question of reversibility. Causality means that the time-reverse of a process can be fundamentally different from the forward version. See (Scargle, 1981) for a discussion of methods for detecting the sense of time, or the *phase information*, in ordinary time series.

As mentioned earlier, in real-world processes one needs only consider the case of a finite number of events – although in some special theoretical contexts infinite sets of events may be relevant. Figure 1 shows a plot of the detection times of 100 photons during a Gamma Ray Burst (GRB), indicated with the + symbols along an undrawn axis at the bottom of the figure. We will return to this figure to point out the graphical representation for other views of the same data. By a *point process*, we mean the mathematical idealization of such a set of events, presumed to be (randomly) distributed in time:

$$E = \{t_n, n = 1, 2, \ldots, N\}. \tag{1}$$

The t_n are just the times at which events occur in a particular realization of the physical process. In real applications the observations extend over a finite interval, say

$$T = \big[\min(t_n), \max(t_n)\big]. \tag{2}$$

It is often convenient to choose the origin of time so that $\min(t_n) = 0$, and sometimes also the unit of time so that $\max(t_n) = 1$. On the other hand, actual data are usually recorded as multiples of a fixed interval, or quantum, of time – never with infinite accuracy. Hence it may be convenient to let the corresponding integer variable represent time. (Of course, precisely the same considerations apply to spatial and other processes.)

It is possible to discuss point processes in terms of probability distributions of these event times, or of the intervals between successive events:

$$I = \{dt_n = t_{n+1} - t_n, n = 1, 2, \ldots, N - 1\}, \tag{3}$$

(the absolute time of one event is also specified). This approach is called specification of the *interval statistics*. Somewhat surprisingly, these rather simple and direct approaches are not very common in the mathematical literature. There are other representations that are even more convenient.

For example, it is often desirable to describe a point process in terms of a random variable defined everywhere on the interval. A common way to do this is in terms of a *counting process*, that is by defining a random variable that counts the number of events that have occurred up to and including a specific moment of time:

$$\begin{aligned} N(t) &= 0 \quad \text{for } t < \min(E), \\ N(t) &\quad \text{increases by 1 at each } t_n. \end{aligned} \tag{4}$$

Obviously with this definition N is a random variable, well defined at any time t and equal to the number of events occurring at or before that time. It is also finite, discontinuous, nondecreasing, and piecewise constant. The thin solid line in Figure 1 is a plot of this *step function*.

A similar procedure is to invoke the increment in the counting process at any time:

$$n(t) = \begin{cases} 1 & \text{for } t \in E, \\ 0 & \text{for } t \notin E. \end{cases} \tag{5}$$

In some contexts this representation is less useful, since $n(t)$ consists of series of delta functions, and is almost everywhere zero. Note that the counting and incremental representations are closely related:

$$N(t) = \int_{-\infty}^{t} n(\tau)\,d\tau, \tag{6}$$

$$n(t) = \frac{dN(t)}{dt}. \tag{7}$$

The + symbols in Figure 1 occur at the locations of the events, and thus form a crude delta function representation of the photon data.

Complete definition of a point process requires a prescription that assigns a probability to all configurations of events that are possible. With the representation in Eq. (1) this can be accomplished by specifying the probability that one or more events will occur in an arbitrary subset of the time axis. In general this is an awkward approach, although it is feasible in special cases.

A big advantage of the continuous representations introduced above is that the usual prescription for ordinary random processes works here too: specify all joint probability distributions of all the relevant random variables. For example, with the counting process in Eq. (4), one needs to specify the probability of the compound events

$$N(t_1) \leqslant N_1, N(t_2) \leqslant N_2, \ldots, N(t_k) \leqslant N_k, \tag{8}$$

for $k = 1, 2, \ldots,$ and for all values of the t_n and the N_k. See (Snyder and Miller, 1991) or any other treatment of random processes for details.

It would be illuminating to continue along these lines, as a comprehensive naive point process theory might aid theoretical statisticians to further appreciate what concepts and relationships are important in astronomical applications. However, the main subject of this article is astronomical applications, to which we now turn.

4. The mystery of Gamma Ray bursts

The remainder of this chapter discusses several examples of point processes in astronomy. We outline the astronomical contexts and mathematical models appropriate for the underlying process. This exposition is not complete, either in exemplifying astronomical point processes, or in reviewing the relevant theory. We do hope to represent the major ideas, with an emphasis on a simplified view concentrating on essential and practical matters. We avoid mathematical refinements, such as those connected with infinite spaces and measure theoretic nuances. Our goal is to pique the interest of statisticians in an exciting array of problems.

Many space observatories – X-ray, Gamma-ray, ultraviolet and visible; past, present and future; launched by the American, Brazilian, Canadian, European, Italian, Japanese, Russian, and other Space Agencies – generate enormous amounts of point process data. From this huge array, we have selected a particular Gamma-ray observatory, BATSE, because of the fascinating problem it addresses, because we have both worked on its data, and because it offers good, clean examples of point process data.

Characterizing the subject of *gamma-ray bursts* (GRBs) as the greatest mystery in modern astrophysics is justified by the fact that it is the only phenomenon known to exist, well-studied observationally, but the nature of which is completely unknown. Very recent progress has suggested that the singular explosive events that generate the bursts are located in very distant galaxies (see Wijers, 1998; other articles in the same issue of *Nature*; and references therein to other work beginning with the breakthrough that occurred in December of 1997). However we are far from knowing the full story on burst locations. The physical mechanism responsible for the enormous release of energy is not understood at all, and the subject of much debate.

The discovery sequence is an interesting point process in its own right, consisting of these events (Bonnell and Klebesadel, 1996):

- A long time ago in a galaxy far, far away: an explosion of unknown nature.
- 1963: the US launch of satellites to monitor the nuclear test ban treaty.
- 1967: a flash of gamma-rays recorded, on July 2, by two of the satellites.
- 1969: the resulting data noticed and analyzed.
- 1973: first public announcement of a Gamma Ray Burst (Klebesadel et al., 1973).

Soon afterward the pace of publication in the subject exploded.

GRBs are now studied using a variety of orbiting high-energy telescopes. The workhorse GRB observatory, launched in 1991 and still operating, is the *Burst and Transient Source Experiment* (BATSE; Kouveliotou, Briggs and Fishman, 1996) on NASA's Compton Gamma-Ray Observatory (CGRO). We frequently refer to data from this instrument in the remainder of this section.

GRBs are seen on average once a day, from random locations on the sky. In fact, the apparently uniform distribution on the sky has been a pivotal observational result – a shocking surprise that ruled out a whole class of models. In fact, the observed homogeneity quickly ruled out what almost all astronomers expected to be the nature of the bursts, namely colliding neutron stars in the Milky Way Galaxy. The distribution of GRBs on the sphere/sky is an interesting two-dimensional spatial point processes.

These explosive events produce a momentary flash of radiation, lasting from a few milliseconds to a few thousand seconds. The detailed time histories of bursts are fascinatingly varied, ranging from the very simple (a single sharp spike, or a period of rising intensity followed by a period of decaying intensity) to the very complex (a seemingly random array of overlapping "pulses," each one of which may be something like the single spikes). This extreme heterogeneity of style of variability is one of the hallmarks of BRGs Joke: *When you've seen one GRB you've seen ... one GRB*. Indeed, this heterogeneity makes it not at all obvious that there is only one type of object here. On the other hand, spectral and other properties are more homogeneous.

Various papers in (Kouveliotou, Briggs and Fishman, 1996) describe the fascinating and varied observational details accumulated over many years and intensively studied by hundreds of astronomers – but all without cracking the basic nut of the problem. Until recently their radiative emission has seemed completely confined to the gamma-ray range of the electromagnetic spectrum, and the objects did not appear associated with any known class of star or galaxy. A major cause of this lack of identification with objects known at other wavelengths are the large positional errors of the BATSE detectors – on the order of several degrees for most bursts, several arcminutes for a few. For a considerable period it was not clear whether gamma-ray bursts occur close to the Sun, in the halo of our Milky Way galaxy, or in distant galaxies far across the universe.

However, in the last few years (Wijers, 1998) several bursts have been seen to be followed by gradually decaying *afterglows* in other wavelengths, from X-rays, to visible light, to radio waves. Indeed, the optical light has been identified with very distant, and faint, galaxies. Astronomers are still sorting out the implications of this observational breakthrough.

In spite of the appearance of over 2,000 published observational and theoretical papers, there is no general agreement on the physical origin of the bursts. Perhaps the favored explanation involves the explosion arising from the collision of two orbiting neutron stars in distant galaxies.

A few specifics about the BATSE instrument and the data it produces are important for what follows. Much of this material is described in detail on the World Wide Web page http://www.batse.msfc.nasa.gov. A catalog of the many bursts is available at http://www.batse.msfc.nasa.gov/data/grb/catalog/ and (http://cossc.hsfc.nasa.gov/cossc/BATSE.html). In addition raw data of the kind discussed below can be obtained at the following location: ftp://cossc.gsfc.nasa.gov/compton/data/batse/.

BATSE consists of 8 detector modules on the corners of the CGRO spacecraft. A *burst trigger* is generated by on-board software if two or more detectors register a significant increase in count rate in any of three time intervals: 64, 256, and 1024 milliseconds. The required increase is specified by command, in terms of standard deviations above the time-dependent background noise, leading to a truncation. The burst triggering is disabled whenever the satellite enters regions of high particle background, or when another burst is active. Sky locations are estimated by comparison of intensities in the eight different detectors, which are pointed in orthogonal directions. (The best positions are determined in those rare instances when the burst is detected by several spacecraft, not just by the BATSE instrument, for then time-of-flight analysis allows accurate triangulation of the location.) The accuracy of estimated locations depends on burst strength and the fortuitous relationship between the location and satellite orientation. Each burst is thus associated with a complicated vector describing truncation and non-homogeneous measurement errors in several variables. The resulting catalogs and time series of individual bursts are available on the World Wide Web sites given above and in the literature (Megan et al., 1996).

The third BATSE catalog (often called the 3B Catalog in the literature) presents data from the Compton Gamma-Ray Observatory for 1122 bursts recorded between April 19, 1991 and September 19, 1994. The database on the Web presents the position, the so-called 50% and 90% burst durations (based on the time over which the indicated fraction of the integrated signal is accumulated), the peak fluxes seen in 64, 256 and 1024 millisecond time bins, and the fluences (number of photons) detected over the entire burst in each of the four energy channels (0–50 keV, 50–100 keV, 100–300 keV and > 300 keV). The errors of measurement are also presented in the catalog. The fourth BATSE catalog (4B) has recently been published (Megan et al., 1998), but most analysis, including that described here, has been carried out with data from the 3B catalog.

As very few gamma-ray burst (GRB) sources have astronomical counterparts at other wavebands, empirical studies of GRBs have been largely restricted to the analysis of their gamma ray properties: bulk properties such as fluence and spectral hardness, and evolution of these properties within a burst event (Fishman and Meegan, 1995). While bursts exhibit a vast range of complex temporal behaviors, their bulk properties appear simpler and amenable to straightforward statistical analyses. Studies fall into two categories: examination of whether GRB bulk properties comprise a homogeneous population or are divided into distinct classes; and search for relationships between bulk properties. Both types of studies may lead to astrophysical insight, just as the distinction between main sequence stars and red giants, and the measurement of a luminosity-mass relation along the main sequence, assisted the development of stellar astrophysics early in the century.

4.1. Counting Gamma Ray photons, one by one

As mentioned in the introduction, measurement of the times of individual photons is quite common, especially in high energy astronomy. Modern photoelectric detectors and data acquisition electronics allow timing individual photons to the nearest microsecond (10^{-6} second). In fact, technology for much greater accuracy is routinely used in high energy physics. However, from general considerations it is expected that astronomical sources cannot vary significantly on much shorter time scales. This physical assumption, coupled with practical implementation details, such as data transmission limitations for space observatories, and signal-to-noise considerations, have limited almost all astronomical timing observations to the microsecond level.

It is thus useful to imagine that there is a underlying clock, constantly ticking away and supervising the timing of the data acquisition. Photon events are thereby tagged with a discrete event time, placing them in very small time intervals. But it would be misleading to think of this as binning of the photon events; we will soon see that the number of events in these intervals is not really counted.

Further, we attempt to carefully distinguish the arrival of the quantum of radiation at the telescope from its detection and recording into the data stream – by purposefully using the terms *photon* for the former and *event* for the latter. This distinction is more important that it at first seems: the realities of the detection system convert the Poisson photon arrival process into a finite Bernoulli lattice process (Stoyan, Kendall and Mecke, 1995). In addition, of course, the instrumental efficiency is not 100% – some photons are simply not detected.

The BATSE Gamma-Ray Observatory (described above, in Section 4) returns data in several modes, the three most important being:

- Time-Tagged Event (TTE).
- Time-to-Spill Event (TTS).
- Binned.

In the TTE mode each photon is tagged with an arrival time, in microseconds. This mode contains the most information, about the shortest time scales, and is in essence the same as the ideal data representation in Eq. (1) above. The TTE mode has the practical limitation that data from only about $32K$ (where K stands for $2^{10} = 1024$) events can be recorded for each burst. That is, long and/or bright bursts often overflow the *data buffer* and some information is lost. The most important advantage of the TTE mode is that it yields more information about variability at short time scales than do the others. We will discuss this mode extensively below.

The time-to-spill (TTS) mode is essentially a thinning of the data, to conserve the precious commodity *downlink bandwidth*. As with any real world experiment, cost factors enter; almost all space observatories have to design clever methods to maximize the data rate to the ground. By recording the time of every Kth event, where say $K = 64$, this mode is very efficient. Statisticians will recognize this as an Erlang process, described in a number of contexts (Brillinger, 1978; Billingsley, 1986; Haight, 1967). Some information is lost – relative to what is intrinsically available at the detector, and what the TTE mode captures. Also the time resolution is variable and dependent on the strength of the signal. However, this mode does contain considerable information. It provides coverage of the entire history of the burst, no matter how long or bright it is, in contrast to the TTE mode. Another price paid for the efficiency of the TTS mode is that the data analysis is rendered somewhat more difficult. Pulse decomposition of TTS data has been accomplished by the methods described in Section 4.1.2 below, as well as others (Lee, Bloom and Scargle, 1998).

There are only a few more practical matters that need to be mentioned. First, detection times *for individual photon events* can be determined with high accuracy – much smaller time intervals than the shortest time scales that are considered astrophysically interesting. However, the times are recorded with finite precision, typically one microsecond as mentioned above. In addition, the data system electronics is usually designed to report only two cases for a given microsecond interval: (1) no photons or (2) one or more photons. Hence the distribution of counts is not Poisson, but a Poisson distribution truncated at one event. This is equivalent to a binomial process, where the event rate is not the probability per unit time of a photon detection, but of one or more such photons. Technically, that is, recorded events form a *finite Bernoulli lattice process* (Stoyan, Kendall and Mecke, 1995). However, the mean counting rates are typically less than 0.01 counts per microsecond "bin," so that the incidence of multiple photons is quite small in any case.

Another complication is that detectors always have a finite *dead time*; that is to say, after a photon has been detected, detection of a subsequent photon is inhibited until the system recovers from the first event. There are two basic types of dead times, depending on the effect of a photon that arrives during the dead interval initiated by a previous event: *paralyzable*, meaning that such a photon extends the dead time further, and *nonparalyzable*, in which case the system more or less ignores such a photon – until, of course, the system has recovered. These are the same as the *Type I* and *Type II* deadtimes discussed in (Brillinger, 1978).

In practice the detector recovery process can be complicated. For example, the dead time may be random – that is the recovery time is not fixed, but varies in accord with physical conditions that are not accessible to the observer, so that the variation is effectively random. The dead time may be a function of the energy of the photon that initiates it. The recovery of the system after an event may be a rather smooth function of time, so that the detection efficiency rises from zero to its maximum value according to a curve that is not precisely known. See (Wen, 1997; Zhang et al., 1995) for discussion of such practical matters, with emphasis on how they affect the estimated power spectrum of a process known or assumed to be Poisson – and therefore possessing a *white*, or flat, power spectrum.

To deal with both the truncation and deadtime limitations of a single detector, many modern high energy observatories contain several detectors, with separate electronic systems that are independent, at least up to the point in the data stream where the photon time stamps have been established. In a situation like this, two photons hitting different detectors can end up tagged with identical arrival times. (For times represented as real numbers this would happen with probability zero, but of course we are dealing with discrete times, represented as integers.) This use of multiple detectors allows information to be derived from the data at time scales less than the detector dead-time, in principle. In practice, this is difficult because the effective signal-to-noise ratio is usually not large enough to allow extraction of very much information.

4.1.1. The Poisson process

Except for the practical complications just mentioned, photon counting (of Gamma Rays, or photons in any energy regime) is closely in the form of the best studied and simplest possible nontrivial point process, *the Poisson process*. This is so because photons traveling through the far reaches of space simply do not interact with each other, to a very high order of accuracy. It is well known that two intense beams of light, such as searchlights, can be pointed so that their beams cross each other at a ninety degree angle, and absolutely no effect of the beams on each other can be detected. In the faint beams of radiation from astronomical objects there is even less in the way of photon–photon interaction. (In fact, photon–photon interactions are not only possible, but they are quite important at the very high densities and energies present in photon beams produced in particle accelerators and some astronomical sources, but the magnitude of the effects for our gamma-ray data can be evaluated and are negligible.) By far the largest departures from independence of the photon detection events, and from a Poisson distribution of same, lies in the detector properties discussed above, mainly dead-time.

The Poisson process is the most important point process, both mathematically and physically, because relatively simple and commonly realized conditions make a process Poisson, and because more complex processes can be easily understood in terms of modifications of the defining characteristics of the Poisson process; the entire book (Snyder and Miller, 1991) is about modifications of the basic Poisson process. Forbidding events prior to the time-origin 0, which is always possible for a real physical process, we have the following

DEFINITION. A counting process $N(t)$ is a Poisson process if

- $N(0) = 0$.

- The counts between any two times, $N(t_1, t_2) = N(t_2) - N(t_1)$ is Poisson distributed, with parameter $\Lambda(t_2) - \Lambda(t_1)$:

$$\Pr[N(t_1, t_2) = k] = \frac{[\Lambda(t_2) - \Lambda(t_1)]^k \, e^{-[\Lambda(t_2) - \Lambda(t_1)]}}{k!}$$

for $0 \leqslant t_1 < t_2$ and $k = 0, 1, 2, \ldots$, and where $0 \leqslant \Lambda(t) < \infty$ is a nondecreasing function of t.
- $N(t)$ has independent increments.

The last property is what really makes the Poisson process; it is something like singling out the important class of *independent, identically distributed* (IID) processes from among all random processes. A fundamental point is that a Poisson process corresponds to points picked randomly in time, perhaps according to some time dependent probability distribution, but *independently of each other*. And in turn, the usefulness of this process in modeling physical systems arises from the fact that this "randomly throwing darts at an axis" is common in the physical world.

An important property of the Poisson process is that a sufficiently small time interval will almost certainly contain no more than one event. This condition, called *orderliness*, nicely represents the detector limitations discussed previously. From a different viewpoint it is essentially the condition that the intensity is nowhere infinite. Another property, called *evolution without aftereffects*, is that if you pick *any* instant of time, events after that time are independent of those before it. In other words, given any time point τ, the distribution of events with $t > \tau$ does not depend in any way on those with $t \leqslant \tau$. The same fact sometimes seems paradoxical when the time selected is that at which an event has happened: an event makes it neither less nor more likely for another one to happen soon after. All of this is easily understood as resulting from the independence of the events in the first place, a condition that is by no means guaranteed for physical processes. Section 5.5, on solar flares, describes a case where there are effects working in both directions – that is, a flare might either trigger subsequent events, or so exhaust the local energy supply as to temporarily prevent the occurrence of same.

The function $\Lambda(t)$ gives the mean rate at which events accumulate over time between 0 and t; specifically, if $\lambda(t)$ is the instantaneous rate of events (events per unit time), then

$$\Lambda(t) = \int_0^t \lambda(\tau) \, d\tau. \tag{9}$$

The important quantity λ, called the instantaneous *Poisson rate* or the *intensity function*, gives the *event rate per unit time*.

The *homogeneous Poisson process*, in which the intensity not a function of time, is a prominent special case. Take a finite time interval T, and assume that a finite number of time points t_n are uniformly and independently distributed within T. This process is said to be *homogeneous* because satisfies the above definition with a constant intensity $\lambda = N/T$. A better term might be *constant*, since homogeneous connotes stationarity rather than constancy. It is well known and interesting that the intervals between events in this process are exponentially distributed.

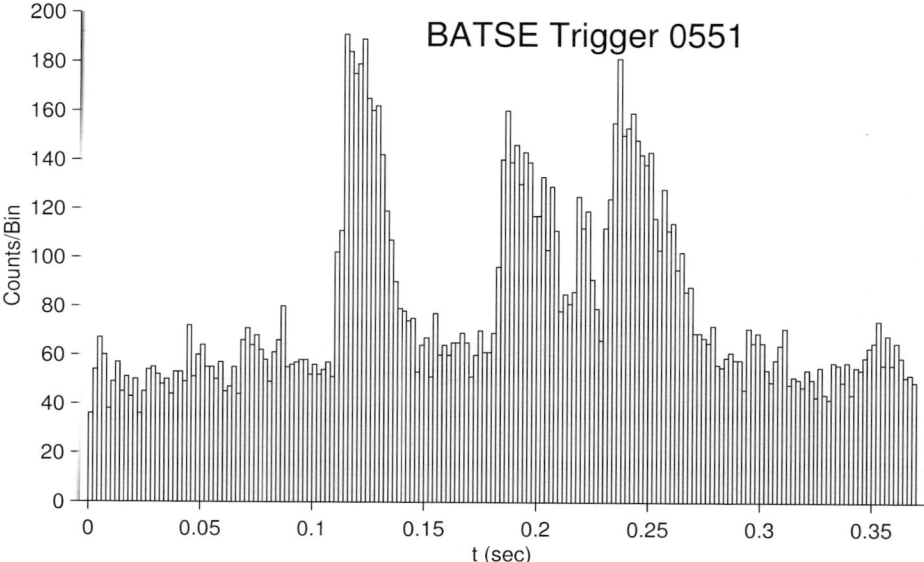

Fig. 2. BATSE data from Trigger 0551. The event times have been counted in somewhat arbitrary time bins, in order to display the character of the variation better that the display methods of Figure 1 can achieve.

4.1.2. Signal estimation in Poisson data

Let us now turn to real data, and to one of the most fundamental data analysis problems in astronomy, namely representation of the time variability of a source, based on data comprising a series of brightness observations spaced in time. We are here interested in this problem in the context of Gamma Ray photons, but the methodology is of general applicability. An exact probabilistic description of the relevant elementary events – photon detection or nondetection – is very simple and leads to explicit solutions of many aspects of the problem.

Figure 2 shows a plot of the gamma-ray intensity of a particular burst as a function of time. The name of the event, *Trigger 0551*, is simply a sequential indicator of the event as detected by the BATSE experiment. The plot shows counts, summed over four different *energy channels vs.* time. (BATSE TTE data includes, for each photon, a tag indicating which of four ranges of energy the photon occupied, as well as one indicating which of the eight detectors it hit. These additional bits of information are useful in actual data analysis, but will not be further discussed here.)

Our goal is an algorithm that takes in the raw data and generates a picture (i.e., a plot, or a model, or some other representation) of the time variation of the brightness of the source, called the object's *light curve*. Figure 2 is an example of the use of simple binning of the events to produce a light curve; we want to do better. Rather than explicit parametric models, we adopt a generic approach of general applicability, namely finding *the most likely piece-wise constant Poisson rate model of the data*. Details, and treatment of other data modes is described in (Scargle, 1998). Since the analysis is based on Bayesian statistics, the structures are called *Bayesian blocks*.

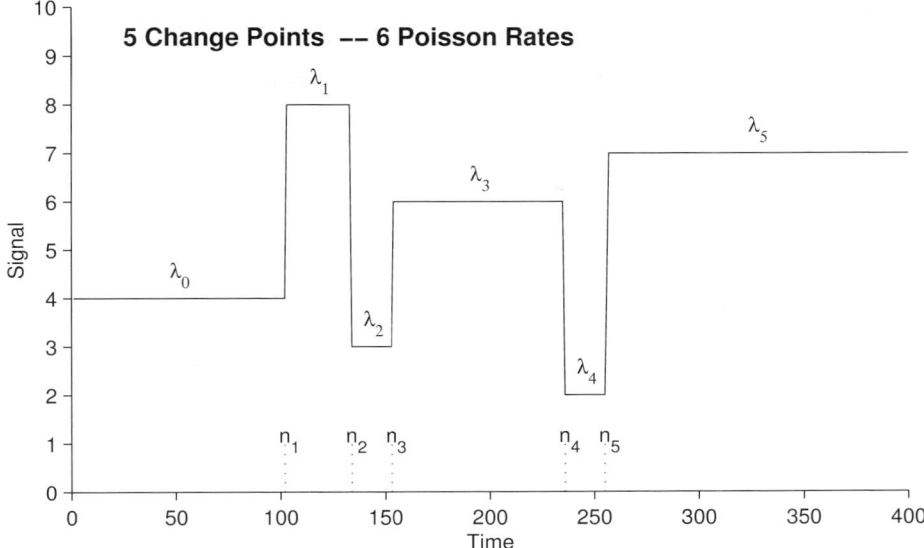

Fig. 3. Piecewise Constant Poisson Model. Cartoon of the model, depicting the segmentation of the interval by change points and intervals of constant event rate between them.

The input data consists of a list of photon detection times, as in Eq. (1). The desired output is the most probable partitioning of the observation interval into segments during which the photon arrival rate was discernibly constant – i.e., had no statistically significant variations. This discontinuous, nonphysical form is adopted because it provides useful, rough information, not as a precise model of the true variability. For example, if the light curve consists of a superposition of pulses which don't overlap too much, the blocks yield reasonable approximations of locations, strengths, and widths of the pulses.

Unlike most, this method does not stipulate, or predetermine, time bins – instead the data themselves determine an effective, non-uniform binning in time. Therefore this *data adaptive* analysis procedure does not of itself impose a lower limit to the time scale on which variability can be detected. The plan of attack is to construct an explicit, *piecewise constant* model for the observed data, marginalize all of the model parameters except for K, the parameter specifying the number of pieces, or blocks; this gives the likelihood as a function of K. Simply finding the maximum of this function gives the maximum likelihood value of K. It is then relatively straightforward to determine the maximum likelihood values of all the other model parameters: the time points where the blocks fit together, and the value of the event rate over these individual pieces.

The precise model we adopt, therefore, is a set of (*Bayesian*) *blocks*, partitioning T and parameterized as follows:

$$M = \{K, n, \lambda\}, \tag{10}$$

where n stands for the K-dimensional array of *change point* locations,

$$n = (n_1, n_2, n_3, \ldots, n_K), \tag{11}$$

where $\min(t_n) \leqslant n_i \leqslant \max(t_n)$. Also λ is the $(K+1)$-dimensional array of Poisson rates over the corresponding partition segments:

$$\lambda = (\lambda_0, \lambda_1, \lambda_2, \ldots, \lambda_K). \tag{12}$$

By the term *rate* we mean either the rate of events, or of photons, per unit time, according to the context. The integer K is the order of the model. Figure 3 depicts the block structure of a sample model (data not shown) of order 5 – with 5 change points and, correspondingly, 6 rates.

By M_K we mean the set of all such piecewise continuous models with K change points, without specification of the change point locations or rates between change points. As usual in Bayesian model comparison, the quantity expressing the evidence provided by the data, D, in favor of such a model class is:

$$p(M_K|D) = p(M_K) \frac{p(D|M_K)}{p(D)}, \tag{13}$$

that is, *Bayes Theorem*. As usual, $P(\bullet)$ is the probability of $\bullet$; and $P(\bullet|\dagger)$ is the probability of $\bullet$, given $\dagger$. In particular, the relative probability of two models of the same data, but of different orders, K and J, is

$$\frac{p(M_K|D)}{p(M_J|D)} = \frac{p(M_K)}{p(M_J)} \frac{p(D|M_K)}{p(D|M_J)}. \tag{14}$$

The prior probability of the data, $p(D)$ in Eq. (13), does not depend on anything having to do with the model, and naturally cancels out. The remaining probabilities must nevertheless be correctly normalized (Jaynes, 1997). The quantity $p(D|M_K)$ in this equation is, to quote the same author, "the fundamental quantity determining the status of model M_K relative to any other" model; it is often called the *Bayes factor*. Eq. (14) translates to "posterior odds ratio = ratio of Bayes factors × prior odds ratio" in English.

The Bayes factor can be found in two steps, the first of which is to marginalize all the model parameters except K, as follows:

$$p(D|M_K) = \int p(D, n, \lambda|M_K) \, dn \, d\lambda. \tag{15}$$

The point is that n and λ are nuisance parameters for the current problem of determining the most likely model order; this integration just takes into account that we don't care about their values. [As described in detail in Chapter 4 of (Sivia, 1996), it is this integration over, or marginalization of, the model parameters that automatically takes into account model complexity. This procedure is usually called the Ockham's Razor, and some authors separate out a special *Ockham factor* in this expression. See (Jaynes, 1997), Chapter 24, for an interesting critique of thinking of this factor as representing simplicity. In any case, without this effect, the likelihood would increase indefinitely with the order of the model. With it, the likelihood achieves a global maximum at the maximum likelihood value of the model order.]

The second step in evaluating the Bayes factor is to apply the product rule for probabilities to transform the integrand in Eq. (15) to that in the following equation:

$$p(D|M_K) = \int p(D|n, \lambda, M_K) p(n, \lambda | M_K) \, dn \, d\lambda. \tag{16}$$

In this expression $p(n, \lambda | M_K)$ is the prior for the model parameters (given the model and its order) – and should not be confused with the prior for the model itself, $p(M_K)$, in Eqs. (13) and (14).

Following the usual Bayesian procedure, lack of any particular prior knowledge about the location of the change points is expressed by adopting a probability distribution that represents the components of n as independent and uniformly distributed over the whole time interval T. The prior for the rates is conveniently expressed in terms of a change of variable, from λ_i to

$$p_i = 1 - e^{-\lambda_i \delta t}. \tag{17}$$

Here δt is the width of the time interval corresponding to the discretization of the recorded photon arrival times, typically on the order of 1μ second, as discussed above. Lack of prior knowledge of the value of p_i, which is just the probability of one event (i.e., one or more photons) in the interval δt, is expressed with the uniform probability distribution function

$$P(p_1) = \begin{cases} 1, & 0 \leqslant p_1 \leqslant 1, \\ 0, & \text{otherwise.} \end{cases} \tag{18}$$

In practice a more stringent prior might be used (e.g., $0 \leqslant p_1 \leqslant 10^{-2}$), but the results do not change significantly. Given these or any other priors, Eq. (16) can be integrated to determine the quantity

$$L(K) \equiv p(D|M_K). \tag{19}$$

For large values of K, this $(2K + 1)$-dimensional integral is not trivial to compute numerically, but can be attacked with the Markov Chain Monte Carlo, or MCMC, method (Gilks, Richardson and Spiegelhalter, 1996). As outlined earlier, finding the peak of this distribution as a function of K cracks open the problem; finding the maximum likelihood value of the number of change points, $K_{\max}$, rather immediately leads to the optimum values of all the other model parameters, using standard nonlinear optimization techniques. This work is still in progress, so we do not show the block decomposition of the data here; the details will appear elsewhere (Scargle et al., 2002).

4.2. Classifying Gamma Ray bursts

Classification has played a key role in the early phases of most empirical sciences, and astronomy is no exception. What may be exceptional is that, even after the long historical development of astronomy, classification still plays an important role. This importance of taxonomy may be the fact that astronomers can almost never experiment with their systems. We must be content to passively observe – and classify – the data that Nature chooses to send our way.

This section starts with an overview of classification in astronomy, and then turns to the point process problem comprised of the search for meaningful classifications of gamma ray bursts.

4.2.1. Classification in astronomy

Astronomers have devoted considerable energies to classifying luminous objects in the sky into various categories, to understand their physical properties and origins. The most successful classification system for stars is the Harvard Spectral Classification, developed around the turn of the century by Annie Jump Cannon and her colleagues. Colors measured in the visible band of the spectrum and spectral lines of stars, are used to place stars in a one-dimensional sequence, crudely OBAFGKM. In this sequence, O stars are the hottest and bluest, while M stars are the coolest and reddest. Later, Morgan and Keenan added a second dimension, the *luminosity class*, indicating whether the star is a *giant, supergiant, main sequence* (like the Sun), *subdwarf*, or *white dwarf*. (These names are somewhat misleading, as they designate luminosity, not size.) These classifications, often represented on the two-dimensional Hertzsprung–Russell, or H–R diagram, led to great insights into stellar structure and evolution. (Joke: Professor: "What is the H–R diagram?" Student: "A plot of H vs. R.")

Stellar spectral classification is traditionally performed by the astronomer's unaided eye-brain, through recognition of spectral line patterns. This method continues today, though some researchers are exploring sophisticated statistical classification procedures. Other types of classifications, however, have been more problematic than visible band spectroscopic classification. X-ray and gamma-ray astronomical sources are frequently variable in intensity, and may thus appear statistically significant in one observation but not in another. Furthermore, new imaging detectors – CCD's in the visible, and other detectors for X-rays and other bands – with good spectral, as well as spatial, resolution, allow the detection and study of sources in specific spectral regions or even in emission lines. Thus, a quite general problem encountered in astronomy is really the study of point processes in multidimensional spaces – that is, source detection and characterization in terms of position on the sky, brightness, and spectral characteristics.

Perhaps the most famous example of use of multidimensional classification was in the discovery by J. Oort in the 1920s of a halo population of stars in the solar neighborhood, in addition to the dominant galactic disk stellar population. The two populations form different asymmetrical distributions in three-dimensional plots of stellar velocity vectors [called *velocity ellipsoid* diagrams; see (Mihalas and Binney, 1981)]. Today, while the existence of galactic halo stars is established, controversy continues over whether a third stellar component, the 'thick disk', is present in the galaxy (Gilmore, Wyse and Kuijken, 1989). A recent discussion of this problem from a Bayesian point of view is (Raftery, 1996). The statistical problem is to determine whether distinct clusters are present in asymmetrical distributions of points in multidimensional spaces that include stellar position, velocity, chemical composition, and other variables derived from the stellar spectra. New, very large astronomical surveys, with 10^6–10^8 objects or measurements, will generate many problems in the analysis of multivariate databases.

Such complex but highly organized distribution in multidimensional space is amenable to modern statistical analysis and classification methods, based on both hierarchical and parametric model-based clustering procedures. Agglomerative hierarchical clustering is a procedure based on the successive merging of proximate pairs of clusters of objects. It produces a clustering tree or dendrogram starting with N clusters of 1 member (or a coarse partition based on prior knowledge) and ending with one cluster of N members. Unfortunately, there are many possible ways to proceed; mathematics provides little guidance among the choices, and no probabilistic evaluation of the results without the imposition of additional assumptions. Distributional assumptions are not required for classification based on a hierarchical clustering, but they are needed in order to perform hypothesis testing and to estimate the statistical significance of the resulting clusters. See (Eisenstein and Hut, 1998) for a discussion of this methodology applied to analysis of data produced by computer simulations of N-body dynamics.

The entire analysis, both clustering and validation, can be conducted within a model-based framework. The Bayesian information criterion can be used to select the 'best' partition among those associated with different numbers of clusters (Dasgupta and Raftery, 1998). Analyses based on the hierarchical clustering procedure, and a maximum likelihood model based clustering procedure, have been conducted by Mukherjee et al. (1998), on gamma ray burst data as described below.

4.2.2. Multivariate classification of Gamma-ray bursts

Let us now turn to categorizing gamma-ray bursts, a difficult task because it is hampered by several things:

- complicated truncation selection biases;
- flux-dependent biases (e.g., durations for faint bursts are shorter because of noise);
- flux-dependent measurement errors in every variable;
- lack of a priori knowledge of the number of classes to use in the classification scheme.

Past efforts, usually based on subjective examination of bivariate plots, have led to two or three class divisions. For example, Kouvelioutou et al. (1993) find two classes in a plot of hardness ratio (i.e., ratio of intensities in different wavebands, equivalent to color) and burst duration.

One of us (G. J. Babu), along with other colleagues has conducted a simple agglomerative hierarchical cluster analysis, followed by multivariate analysis of variance (MANOVA) to confirm the statistical significance of the results (Mukherjee et al., 1998). Of the 1122 bursts in the 3rd BATSE catalog, we retain 797 having complete data in 13 variables concerning location, duration, fluence and spectral hardness ratios.

Ignoring the measurement errors and selection biases, the presence of structure in the data set was examined using average linkage clustering with a Euclidean metric, after taking the logarithm of all variables. The results indicate the presence of three statistically significant clusters of the gamma ray bursters. The larger class reported by Kouvelioutou et al. (1993) appears to have internal structure in higher dimensions than they examined. A principal components analysis of each cluster shows fluence correlated with hardness in two clusters, but anticorrelated in the third.

The visualization tool **XGobi** is useful in exploring the data set. (See the collection of information about this and other statistical software at the Pennsylvania State University WWW site http://www.astro.psu.edu/statcodes/.) A 2-dimensional "grand tour" of the database is provided by displaying various projections of the data, with color brushing of the clusters. These results should assist in the astrophysical interpretation of the three classes. We have also constructed a maximum likelihood model, based clustering procedure and validated with the Bayesian Information Criterion. The two methods yield very similar results. The details are presented in Mukherjee et al. (1998).

The other of us (J. D. Scargle), along with Jay Norris (private communication), has looked for these clustering effects in two data samples, modified from those used by our colleagues and discussed above. In the first data set we replaced their duration measures (those published by the BATSE project team in the 3rd BATSE Catalog) with our own measurements, carried out in a different way. In the second case, we retained the BATSE durations, but used the 4th BATSE Catalog, which extended the number of bursts to 1637, in the period from April 19, 1991 to August 29, 1996. In the first data set we see no evidence for the presence of three clusters, while in the second the triple clustering effect was visually less obvious than that reported by our colleagues and discussed above. This result underscores the difficulties of the duration measurements, as well as the vicissitudes of complex models for relatively small data samples. One of the main goals of the methodology discussed in Section 4.1.2 above is to improve on, and render completely automatic and therefore objective, the determination of GRB durations, in order to improve the data available for this statistical study, and others.

The principal difficulties posed by astronomical databases of this type are due to measurement errors and censoring. The sources are identified by the presence of a significantly high S/N ratio at a given sky location. It is quite common that a star is detected in one or two infrared bands, but not at other bands. Only an upper limit, or censored value, is available at the other bands and the color ratios are consequently also censored. That is, the observed vectors $Y_i = (y_{i1}, \ldots, y_{ik})$ and the actual characteristics of a star $X_i = (x_{i1}, \ldots, x_{ik})$ are related by $y_{ij} = \max(x_{ij}, c_{ij})$ in the censored case, where c_{ij} are censoring variables. In the case of truncation due to limited sensitivity of the instruments, the observations may be recorded only when x_{ij} less than some known t_{ij}. Even sources detected at all bands are subject to measurement errors, $Y_i = X_i + \eta_i$, where the variance of the error η_i is known from the original measurements of signal, background, and signal-to-noise. Thus, associated with each point is a known standard deviation in each coordinate. These errors differ from point to point. Furthermore, some points are censored in one or both axes, and should be added to the analysis.

The statistical challenge is to generalize existing spatial point process clustering and multivariate classification methods to include effects of known heteroscedastic measurement errors, selection biases such as truncation, and possible censoring in each variable. The situation is helped by the astronomers' prior knowledge. Prototypes of each prospective class are already known and well-studied, so classification can be 'supervised'.

The reader interested in more details about truncation and censoring in astronomical contexts should consult the literature of this subject. Of considerable interest was the session on this topic at the first conference on Statistical Challenges in Modern Astronomy (Feigelson and Babu, 1992), and various articles in the second such conference (Babu and Feigelson, 1997). A Bayesian approach can be found in (Loredo and Wasserman, 1995).

5. Other examples of astronomical point processes

The following examples are discussed in less detail than were gamma-ray bursts in the above section, partly because here we are reviewing the work of others, rather than our own.

5.1. Twinkle, twinkle, little star ...

Stars are probably the most obvious astronomical candidates to be relevant to point processes. Although large compared to the human observers and telescopes, they are so distant that they appear to be nearly indistinguishable from points on the sky, with the obvious exception of the Sun. Babu and Feigelson (1996) discussed constellations and stellar statistics, but here we will deal with a different point process that has turned out to be instrumental in measuring the actual apparent diameters of stars.

Even the nearest stars cannot be distinguished from points because of the spreading of their light at the focal plane of the telescope due to *optical diffraction*, and also the random blurring effects of turbulence in the Earth's atmosphere. For most purposes a third source of spreading, that caused by intervening material in space on the way to the Earth, can be neglected. Diffraction results in inherent image sizes on the order of a tenth of an arcsecond, for a large, modern telescope. The atmospheric spreading, called *astronomical seeing*, means that the diameters of all ground-based stellar images are at least an arcsecond – or slightly less under the best conditions at an excellent site. Thus there is a factor of at least ten improvement if the atmospheric effects can be removed.

In fact, there is an imaging technique that largely removes much of the blurring due to the transit through the Earth's atmosphere. If one obtains "snap shots" with a moderately large telescope, of duration on the order of $\frac{1}{100}$ th of a second, the stellar image consists of a somewhat random mosaic of small blobs (Labeyrie, 1978). Each of these *speckles* is essentially an image of the star with resolution limited only by diffraction, but not by seeing. In a sense, the dominant effect of the atmosphere at a given moment is to provide the light a number of different optical paths to the observer; roughly speaking, each path yields a discrete, diffraction limited, star image. These paths change randomly over times longer than a few hundredths of a second, so the average effect is a blurred image.

By combining properly the speckles from many such instantaneous images, one can obtain in effect the image which the same observing time would yield with much less atmospheric turbulence. See (Patience et al., 1998) for a recent application of speckle imaging techniques to an astronomical problem. It may even be possible to deal with speckle images on a photon-by-photon basis (Mertz, 1979).

While other choices are perhaps possible, the most natural model for the process of starlight passing through the Earth's atmosphere into a telescope is the *translated Poisson process* (Snyder and Miller, 1991, Chapter 3). Translation means that the observed value of a quantity is shifted, by some kind of random process, from the true value. This concept is relevant to spatial imaging when there is a random spreading of point sources. Mathematically this is expressed in terms of a *point spread function*. The reference above discusses various estimation problems for such translated Poisson processes.

5.2. Accretion systems: neutron stars, black holes, and white dwarfs

There is a fascinating class of astronomical object, consisting of two stars in a mutual orbit, called *binary stars*. In the cases of interest one of the stars is what is a *compact object* – either a *neutron star*, a *white dwarf*, or a *black hole candidate*. (The word candidate signifies that not all astronomers believe that the existence of black holes has been definitively established.) Material from the other star flows onto the compact star. This process, called *accretion*, is believed to occur in many kinds of binary stars. Because the latter is so small, there is a large release of gravitational potential energy in this process; the result is large amounts of radiation. This light can be in the form of ordinary visible radiation, but is mostly of higher energy, i.e., X-rays or gamma rays.

The flow of matter and radiation in these systems is turbulent. This means that it can be treated as random for the purposes of analysis. That is to say, not only is the random photon arrival process discussed in Section 4.1 operating, but in addition the luminosity of the source varies randomly. See (Shapiro and Teukolsky, 1983) for a complete discussion of the physical processes believed to operate in these systems, and relevant observational matters. Both of the above processes contribute to the fluctuations seen when the brightness of the system is measured as a function of time.

One of the goals of the data analysis is to separate these two sources of variation – one of which is observational noise and the other of which is signal. The latter, or intrinsic luminosity variations, are a fascinating mixture of apparently random changes with power-law shaped power spectra (often called $\frac{1}{f}$-*noise*, although from the astronomer's perspective it is not noise), and quasi-periodic variations (Scargle et al., 1998). See (Abry and Flandrin, 1996) for an interesting discussion of wavelet methods for treating point process data.

When the rate at which independent events occur is itself a random variable, as is the case here, the process is called *doubly stochastic Poisson-process*, sometimes called a *Cox process* after the originator of the idea (Cox, 1955); the terms *conditional, compound*, or *mixed* Poisson process are also used. The reference (Snyder and Miller,

1991) has an extensive discussion of these processes, including many results that are of use to scientists analyzing data of this kind.

5.3. Are we alone?

For some time nearly all astronomers have agreed that planetary systems may be a common accompaniment of star formation. That is to say, perhaps nearly every star in the sky is the center of a system of planets, just as the Sun is the center of our Solar System. However, until quite recently we could not point to any star on the sky and say for sure that it has planetary companions.

The discovery of planets around nearby stars is a very exciting chapter in the history of modern astronomy – one that is still being written. Indeed, we don't yet know if the chapter ends with the discovery of planetary systems inhabited by living organisms – possibly intelligent ones. But many scientists in a variety of research programs, including the Search for Extraterrestrial Intelligence and the NASA Astrobiology Program, are working on related issues.

The discoveries of planets to date are mainly the result of very precise measurements of the radial (i.e., line-of-sight) velocities of stars. Small periodic changes in such velocities can be shown to be caused by the common motion of the star and planet about the center of gravity of the system. These studies have unexpectedly turned up large planets in orbits of small size. On the other hand, the radial velocity method is inherently more sensitive to such planets. Therefore, the current preponderance of large planets in small orbits may be largely an observational selection effect and may go away as more planets are discovered.

Partly to counteract this bias, other planetary detection methods are being pursued, some of which are sensitive to smaller planets or larger orbits. One such is the attempt to observe the passages of planets in front of the parent stars. Detection of this *occultation* event requires that the plane of the putative solar system be aligned with the direction from the star to the Earth. Project Kepler (previously know as FRESIP) is a proposal to make such observations from a dedicated space telescope constantly imaging a single chosen area rich in nearby stars (Borucki et al., 1996). The hope is that by continuously monitoring the brightness of many stars, occultation events from that fraction of systems with the proper alignment can be observed.

Figure 4 shows a field of stars that may be the first one surveyed for planets by the Kepler Project. The first part of the figure is a picture of the actual star field as it might be seen through a telescope. The second panel plots the position of the stars that have been selected for the study. The stars have been selected to lie within a narrow range of apparent brightness – too bright and the star image is saturated, and therefore unmeasurable; too faint, and there is not enough light to measure the star's brightness to the requisite accuracy. In addition, pairs or triplets of stars that are too close together on the sky must be rejected from the program, because the composite image of such star systems also cannot be measured well.

The apparent uniformity of both plots is striking. Especially the second panel looks much too uniformly and smoothly distributed to be a random distribution. Indeed, it is not random at all, but has been chosen based on selection criteria that result in

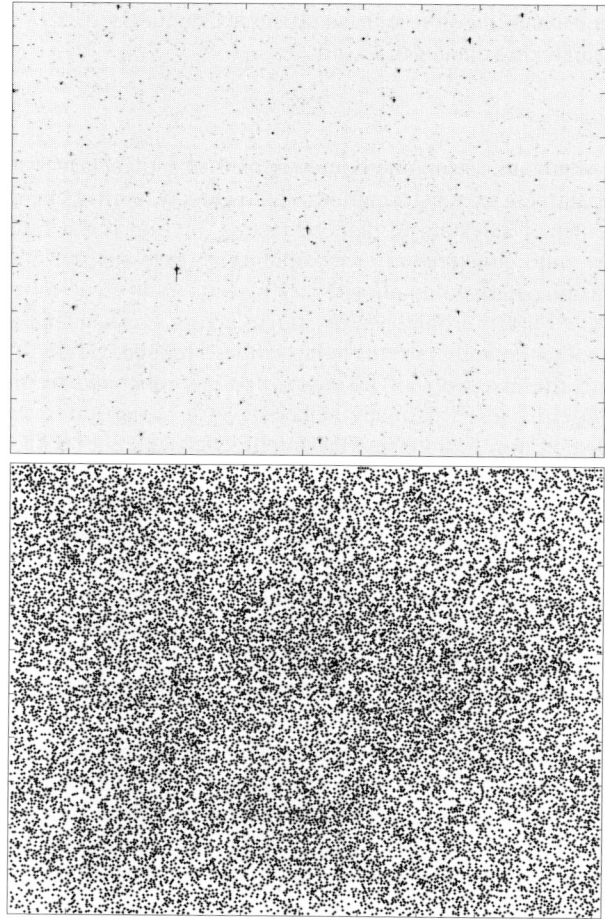

Fig. 4. Star map from the Kepler Project. The first panel shows a region of the sky in the constellation Perseus, selected for a proposed search for earth-like, and therefore presumably habitable, planets around the stars seen here. The second panel shows the positions of the stars from this region that are far enough away (on the sky) from their neighbors that the photometric observations, that may lead to the detection of a number of planets, and be made.

a smooth distribution. Truly random star fields show much more obvious structure – clumps, clusters, chains of stars, holes in the distribution of stars, etc.

In fact, the star field was chosen in the first place partly for its uniform presentation of a large number of stars of the right brightness and in a relatively small region of the sky. The condition applied in selecting the stars for the second part of the figure make this an example of a *hard-core point process* in which points are forbidden to lie closer together to each other than a fixed distance. The reference (Stoyan, Kendall and Mecke, 1995) contains a description of this and other related point processes. One of the complications of real-world examples is that the avoidance is often not an all-or-nothing effect; that is the probability of two points is a function that decreases smoothly as the separation between them gets small.

5.4. Strange attractors

Theoretical astronomy is coming to rely more and more on *numerical simulations*. This term includes numerical solution of differential equations, which has been a staple of astrophysics for many decades. But in the last decade or so, other ways to simulate dynamical systems have become more and more important in the astronomer's bag of analysis tools. Some simulations differ from direct integration of equations of motion in that dynamical evolution is effected by an iterative application of an operator that advances the state by relatively large time intervals. And other simulations are based on novel ways to capture system dynamics or on other principles altogether.

One of the most popular dynamical point processes in astrophysics is generated by locating the intersections of the system's state space trajectory with a sheet of reduced dimensionality. Each time the trajectory of the system in the full, multidimensional phase space, crosses through an imaginary plane of lower dimension, the point of intersection is marked. The resulting plot is called under various circumstances a *Poincare map*, *surface-of-section*, *strange attractor*, or *chaotic attractor* for the system. For example, (Dobrovolskis, 1995) shows several surfaces-of-section for the rotational dynamics of a possibly tumbling satellite of Neptune. Much more secure is the tumbling of Saturn's satellite Hyperion (Klavetter, 1989a, 1989b; Peale, 1984; Wisdom, 1987; Wisdom, Peale and Mignard, 1984) although it is possible that the rotation is not quite so chaotic (Scargle, 1991) as these authors claim.

These point processes pose a number of extremely intesting mathematical problems. In fact, since these operators can be thought of as being applied indefinitely to the initial state of a dynamical system, all of the subtleties associated with infinite, continuous spaces arise in the analysis of these problems. See (Wiggins, 1990) for an introduction to this subject.

Perhaps one of the most overlooked aspect of this subject is that the time-order of the points in the surface of section is important. This is a fundamental difference from the usual view of spatial point processes, where it is presumed that there is no ordering of the points – in time or any other sense. There are several ways to treat this mathematically, including considering the points to be in the higher dimensional space, *space-time*, or as a *marked process* in which *time* is the auxiliary variable, or *mark*, associated with each point. See (Stoyan, Kendall and Mecke, 1995) for a discussion of *marked point processes*.

As a practical matter it is difficult to display the time order of points. Computer screens allow this to be done, in a real-time sense, but presentation of samples of this kind of process in traditional modes of scientific publication is not easy. One of the exciting aspects of JAVA for Internet-based publication is that it makes animation of the time dependence of such dynamical processes a relatively simple matter.

And finally, we mention that the time history of the points on a surface-of-section for a dynamical process is not a trivial or merely cosmetic issue. Crudely stated, if the points in one of these diagrams appear to jump around the attractor randomly as time unfolds, it is a pretty good bet that the system is chaotic. On the other hand, no matter what the strange shape of the attractor, if the points flow gradually from one region to another, the system is almost certainly deterministic, if complex. This remark applies equally well to plots made from empirical as well as theoretical data. [How measurements of a time series can be turned into plots of the dynamical system in a high-dimensional phase space is a fascinating story, well beyond the scope of this article (Ruelle, 1989).]

5.5. Solar flare X-ray bursts

Solar flares are energetic outbursts that occur sporadically on the surface of the sun. The time between the onset of the flare and its maximum intensity (the *rise time*) can be measured, and is quite short. The same goes for the *decay time* – measured from the moment of maximum to the time when the flare has decayed back to invisibility. But because these measured flare durations are quite short (from tens of minutes to hours) compared to the time scales of other physical phenomena on the sun, flares can be treated as points in time for many purposes. Furthermore, flares are confined to small volumes (10^{26}–10^{29} cm^3; the volume of the Sun is about 1.4×10^{33} cm^3) located on the Sun's surface, and therefore are effectively points in space as well. Solar flares can thus be considered point processes in *space-time*, the spatial domain being the surface of a sphere. Indeed, the relation between times and locations of flares is a subject of considerable study.

For example, (Wheatland, Sturrock and McTiernan, 1998) is a recent study of the statistics of the times of solar flares. Their data was obtained with X-ray detectors aboard an International Sun-Earth Explorer spacecraft – a collaboration involving both the United States Space Agency, NASA, and the European Space Agency, ESA. This spacecraft was launched in 1978, primarily to study the interaction of the Solar wind with the Earth's magnetosphere, but its orbit was later changed to fly through the tail of Comet Giacobini-Zinner. The first panel of Figure 5 depicts the times of a small subset of all the flares that are included in this study.

Wheatland et al. addressed the question of whether or not solar flares are independent of each other in a study of the distribution of the time intervals between successive flares, which they call the *waiting-time distribution*. (This experiment did not include the spatial location of the flares on the Sun.) As a stepping stone in their analysis, these authors needed a mathematical model of the variable rate of the flare occurrence process (panel 2 of the figure), which they took to be doubly stochastic Poisson. They constructed it using a modification of the Bayesian Blocks method (Scargle, 1998), which took into account gaps in the data, as well as a known time variability of the experimental efficiency of the detection of flares (plotted in panel 3).

There are possibly at least two competing flare correlations. It is possible that one flare *triggers* others. That is to say, the occurrence of a flare might temporarily increase the probability of others in nearby regions of the Sun's surface. On the other hand, there might also be a *deadtime* or exhaustion effect: a given flare, by depleting the local supply of energy, temporarily diminishes the probability of another one. These effects work in opposite directions – to increase or diminish the population of short intervals between subsequent flares. It is far from obvious what the result would be if both were

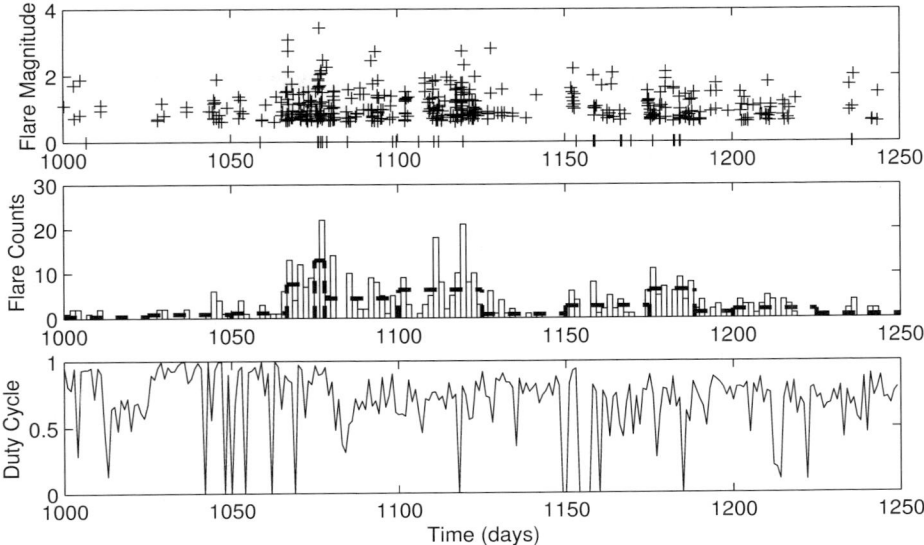

Fig. 5. Solar Flare Data from (Wheatland, Sturrock and McTiernan, 1998). First panel: a plot of a logarithmic measure of flare strength *vs.* the time, in days from an arbitrary origin (each + symbol is one flare). The vertical makes along the x-axis are cases in which the strength could not be measured, but the time was accurate. Second panel: Wheatland et al.'s Bayesian Block analysis of these flare times, on the same time axis; the vertical axis represents rate of flare events (not flare strength). Third panel: A plot of the total *duty cycle* of the experiment, on the same time axis, including intervals of missing data (gaps) as well as reduced effective detection efficiency. This efficiency curve was included in the analysis that led to the representation shown in the second panel of this figure.

present. They may cancel each other out. It is likely that spatial effects might be involved with this issue: a flare might inhibit others nearby but trigger distant ones. *Self-exciting point processes* allow such correlations, since they relax the condition of evolution without aftereffects, mentioned in Section 4.1.1 as part of the defining characteristic of the Poisson process (see Snyder and Miller, 1991).

This flare study illustrates a number of common practical issues that confound analysis of real-world data. In the first place, identification of event times suffers from several ambiguities. Since flares have finite duration, there are several possible definitions of the time interval between events. One could define the waiting time as the interval from the end of one flare to the onset of the next. Three other choices are the intervals between successive times at which the flares begin, reach their peak, or end. Even with some knowledge of the physical processes driving flares, this choice is not obvious.

In addition, the measurement of any of these times from the raw data is not without difficulties and ambiguities. For example, short waiting times may be systematically missed if the emission from the second of a pair of flares has a chance of being lost in the tail of the previous one. This might corrupt correlation studies, foiling attempts to detect triggering or dead-time effects. Event times may well suffer other biases, such as systematic dependence on flare brightness.

Furthermore, orbiting spacecraft cannot observe a given object continuously – there are gaps in the data, spaced roughly at the orbital period of the satellite. Wheaton et al. believe that such gaps were responsible for several previous studies obtaining results that were in poor agreement. In fact, the ISEE satellite was chosen because its interplanetary orbit was far enough away from the Earth that data gaps were minimized. Even so, these authors feel that the presence of some data gaps affected the distribution of the longer waiting times. (These authors' way of treating data gaps was mentioned above.)

The paper discusses some rather sophisticated modeling procedures. The authors developed a 2-dimensional cellular automaton representing a *sandpile* model of the dynamics of the flare phenomenon, in which the *avalanches* are identified with the flares. This kind of model has been developed in a number of contexts (Scargle et al., 1998; Young and Scargle, 1996; Mineshige, Ouchi and Nishimori, 1994).

An important factor is that the overall rate of flaring is quite variable over time ... definitely an inhomogeneous Poisson process. Indeed, the flaring rate has both periodic components, corresponding to the 11 year Solar cycle, as well as a random component. The underlying physical process is probably a good example of a *doubly stochastic process*.

6. Conclusion

We conclude this sketch of astronomical point processes with some general remarks on the selection of the mathematical model appropriate to a given observational or theoretical situation.

We stress the point that there is considerable freedom allowed and judgment required in making this decision, in many cases. A given astrophysical system can be modeled in various ways, depending in part on the point of view of the analysis. For example, we discussed the photon arrival in Gamma-ray Bursts as a Poisson process (Section 4.1), with an variable, but not random, intensity (photon arrival rate). This is because we were thinking of a given burst as being a deterministic unfolding of some (unknown) physical process. If we knew more about the process, we might find that there is a random or possibly chaotic component, suggesting that a doubly stochastic Poisson model is more appropriate. We might even adopt this model in a study of the ensemble of Gamma-ray Bursts, taking the sampling of this ensemble to be a random process. The treatment of accretion systems (Section 5.2) as double stochastic is quite reasonable, because the underlying turbulent flow is presumed random, but not necessarily the only or best model.

Often the use to which the analysis will be put is as important for model choice, or even more so, than is the presumed nature of the underlying physical process. Thus, rather than saying that such and such an astronomical system is a specific kind of point process, it is better to say that we choose to treat it as such. Indeed, *the appropriateness of point process models at all is subject to similar issues.*

And finally, we summarize some peculiar features of astronomical data. We have discussed data gaps. More generally, sampling of astronomical data often occupies a nether region between regular and random spacing. Weather, telescope availability, equipment malfunction, and other accidents often determine the nature of the sampling of time series data. These and other, often uncontrollable, distortions of true signals haunt our observations. We are plagued by runs of data that are too short, or with too much noise, to reveal effects we seek. Finally, *one of the greatest obstacles to astronomical statistics is the fact that there is often only one sample possible, even in principle*. There is only one distribution of galaxies on the sky – at least, as seen from the platform to which we are confined for the foreseeable future. In this case there is no chance to collect either independent or control data. Gamma-ray Bursts, for whatever interesting reason, do not choose to repeat themselves to allow us to get better statistics.

We hope that these problems and challenges inspire statisticians to join astronomers in our study of exciting events in the Universe!

Acknowledgement

J.D.S. wishes to thank Jon Jenkins, Bill Borucki, and Mike Wheatland for helpful discussions and for providing data shown here, and David Brillinger for useful comments.

The research of G. J. Babu is supported in part by NSA grant MDA904-97-1-0023 and NSF grant DMS-9626189.

References

Abry, P. and P. Flandrin (1996). Point processes, long-range dependence and wavelets. In *Wavelets in Medicine and Biology*, pp. 413–437 (Eds. A. Aldroubi and M. Unser). CRC Press, Boca Raton.

Andersen, P., O. Borgan, R. Gill and N. Keiding (1993). *Statistical Models Based on Counting Processes*. Springer Series in Statistics. Springer-Verlag, New York.

Babu, G. J. and E. Feigelson (1996). Spatial point processes in astronomy. *J. Statist. Plann. Inference* **50**, 311–326.

Babu, G. J. and E. Feigelson (Eds.) (1997). *Statistical Challenges in Modern Astronomy II*. Springer-Verlag, New York.

Billingsley, P. (1986). *Probability and Measure*. Wiley, New York.

Bonnell, J. and R. Klebesadel (1996). A brief history of the discovery of cosmic gamma-ray bursts. In *Gamma-Ray Bursts: 3rd Huntsville Symposium*, AIP Conference Proceedings No. 384 (Eds. C. Kouveliotou, M. Briggs and G. Fishman). American Institute of Physics, Woodbury, New York.

Borucki, W., E. Dunham, D. Koch, W. Cochran, J. Rose, K. Cullers, A. Granados and J. Jenkins (1996). FRESIP: a mission to determine the character and frequency of extra-solar planets around solar-like stars. *Astrophys. Space Sci.* **241**, 111–134. See http://www.kepler.arc.nasa.gov/papers.html.

Brillinger, D. (1978). Comparative aspects of the study of ordinary time series and point processes. In *Advances in Statistics*, Vol. 1, pp. 33–133. Academic Press, New York.

Brillinger, R. (1981). *Time Series: Data Analysis and Theory*. McGraw-Hill, New York.

Brillinger, D. (1996). Remarks concerning graphical models for time series and point processes. *Revista de Econometria* **16**, 1–23. http://stat-www.berkeley.edu/users/brill/papers.html.

Brillinger, D. (1997). Some wavelet analyses of point process data. In *Proc. 31st Asilomar Conference on Signals, Systems and Computers.* http://stat-www.berkeley.edu/users/brill/papers.html.

Cash, W (1979). Parameter estimation in astronomy through the application of the likelihood ratio. *Astrophys. J.* **228**, 939–947.

Cox, D. (1955). Some statistical methods connected with series of events. *J. Roy. Statist. Soc. B* **17**, 129–164.

Dasgupta, A. and A. Raftery (1998). Detecting features in spatial point processes with clutter via model-based clustering, *J. Amer. Statist. Assoc.* **93**, 294–302.

Dobrovolskis, A. (1995). Chaotic rotation of nereid?. *Icarus* **118**, 181–198.

Eisenstein, D. and P. Hut (1998). HOP: a new group-finding algorithm for N-body simulations. *Astrophys. J.* **498**, 137.

Feigelson, E. and G. J. Babu (Eds.) (1992). *Statistical Challenges in Modern Astronomy,* pp. 411–428. Springer-Verlag, New York.

Fichtel, C., N. Gehrels and J. Norris (1994). *The Second Compton Symposium.* American Institute of Physics, New York.

Fishman, G. and C. Meegan (1995). Gamma-ray bursts. *Annu. Rev. Astron. Astrophys.* **33**, 415–458.

Gilks, W. R., S. Richardson and D. Spiegelhalter (1996). *Markov Chain Monte Carlo in Practice.* Chapman-Hall, New York.

Gilmore, G., R. Wyse and K. Kuijken (1989). Kinematics, chemistry and structure of the galaxy. *Annu. Rev. Astron. Astrophys.* **27**, 555–627.

Haight, F. A. (1967). *Handbook of the Poisson Distribution.* Wiley, New York.

Hurley, K. et al. (1994). Detection of a gamma-ray burst of very long duration and very high energy. *Nature* **372**, 652–654.

Jaynes, E. (1997). *Probability Theory: The Logic of Science.* Available electronically at http://bayes.wustl.edu.

Jeffrey, A. (1995). *Handbook of Mathematical Formulas and Integrals.* Academic Press, New York.

Klavetter, J. (1989). Rotation of Hyperion. I. Observations, *Astron. J.* **97**, 570–579.

Klavetter, J. (1989). Rotation of Hyperion. II. Dynamics. *Astron. J.* **98**, 1855–1874.

Klebesadel, R. et al. (1973). *Astrophys. J. Lett.* **182**, L85.

Kouveliotou, C., C. Meegan, G. Fishman, N. Bhat, M. Briggs, T. Koshut, W. Paciesas and G. Pendleton (1993). Identification of two classes of gamma-ray bursts. *Astrophys. J. Lett.* **413**, L101.

Kouveliotou, C., M. Briggs and G. Fishman (1996). *Gamma-Ray Bursts. The Third Huntsville Symposium.* American Institute of Physics, New York.

Labeyrie, A. (1978). Stellar interferometry methods. *Annu. Rev. Astron. Astrophys.* **16**, 77–102.

Lee, A., E. Bloom and J. Scargle (1998). Pulse decomposition analysis of BATSE GRB time profiles. In *Gamma-Ray Bursts: 4th Huntsville Symposium,* pp. 261–265 (Eds. C. Meegan, R. Preece and T. Koshut). American Institute of Physics, Woodbury, New York.

Loredo, T. (1990). From Laplace to supernova SN 1987A: Bayesian inference in astrophysics. In *Maximum Entropy and Bayesian Methods,* pp. 81–142 (Ed. P. Fougere). Kluwer Academic Publishers, The Netherlands.

Loredo, T. and I. Wasserman (1995). Inferring the spatial and energy distribution of gamma-ray burst sources. I. Methodology. *Astrophys. J. Supplement Ser.* **96**, 261–301.

Meegan, C. A. A. and 14 others (1996). The third BATSE gamma-ray burst catalog. *Astro. Phys. J. Suppl.* **106**, 65.

Meegan, C. and 14 others (1998). The 4B BATSE Gamma-ray Burst Catalog. In *Gamma-Ray Bursts: 4th Huntsville Symposium,* pp. 3–9 (Eds. C. Meegan, R. Preece and T. Koshut). American Institute of Physics, Woodbury, New York. http://cossc.gsfc.nasa.gov/cossc/BATSE.html.

Mertz, L. (1979). Speckle imaging, photon by photon. *Appl. Optics* **18**, 611–614.

Mihalas. D. and J. Binney (1981). *Galactic Astronomy, Structure and Kinematics.* W. H. Freeman, San Francisco.

Mineshige, S., N. Ouchi and H. Nishimori (1994). On the generation of $1/f$ fluctuations in X-rays from black-hole objects. *Publ. Astron. Soc. Japan* **46**, 97–105.

Mukherjee, S., E. Feigelson, G. J. Babu, F. Murtagh, C. Fraley and A. Raftery (1998). Three types of gamma ray bursts. Preprint.

Murthy, V. (1974). *The General Point Process*. Addison-Wesley.
Patience, J., A. Ghez, I. Reid, A. Weinberger and K. Matthews (1998). The multiplicity of the Hyades and its implications for binary star formation and evolution. Preprint, to appear in the *Astronom. J.*, May 1998.
Peale, S. (1984). The rotation of hyperion. *Philos. Trans. Roy. Soc. London* **313**, 147–156.
Raftery, A. (1996). Hypothesis testing and model selection. In *Markov Chain Monte Carlo in Practice*, Chapter 10, pp. 163–187 (Eds. W. Gilks, S. Richardson and D. Spiegelhalter). Chapman-Hall, London.
Ruelle, D. (1989). *Chaotic Evolution and Strange Attractors: The Statistical Analysis of Time Series for Deterministic Nonlinear Systems*. Cambridge University Press.
Scargle, J. (1981). Studies in astronomical time series analysis. I: Modeling random processes in the time domain. *Astrophys. J. Suppl.* **45**, 1–71.
Scargle, J. (1991). Is hyperion's rotation only mildly chaotic?: evidence for a dense spherical core. *Bull. Amer. Astr. Soc.* **23**, 1171–1172.
Scargle, J. (1998). Studies in astronomical time series analysis. V. Bayesian blocks, a new method to analyze structure in photon counting data. *Astrophys. J.* **504**.
Scargle, J., J. Norris, B. Jackson, T. Hsu, A. Alt, S. Arabhi, D. Barnes, P. Gioumousis, E. Gwin, P. Sangtrakulcharoen, L. Tam and T. Tsai (2002). Studies in astronomical time series analysis. VI. Optimum partition of the interval: Bayesian blocks, histograms and triggers. Submitted to *Astrophys. J.*
Scargle, J., T. Steiman-Cameron, Y. Young, D. Donoho, J. Crutchfield and J. Imamura (1993). The quasi-periodic oscillations and very low-frequency noise of scorpius X-1 as transient chaos: a dripping handrail?. *Astrophys. J. Lett.* **411**, L91–L94.
Shapiro, S. and S. Teukolsky (1983). *Black Holes, White Dwarfs, and Neutron Stars: The Physics of Compact Objects*. Wiley, New York.
Sivia, D. (1996). *Data Analysis: A Bayesian Tutorial*. Clarendon Press, Oxford.
Snyder, D. (1975). *Random Point Processes*. Wiley, New York.
Snyder, D. and I. Miller (1991). *Random Point Processes in Time and Space*. Springer-Verlag, New York.
Stoyan, D., W. Kendall and J. Mecke (1995). *Stochastic Geometry and its Applications*, 2nd ed. Wiley, New York.
Wen, H. C. (1997). 10 microsecond time resolution studies of Cygnus X-1. Ph.D. thesis, Stanford University, SLAC Report 514.
Wheatland, M., P. Sturrock and J. McTiernan (1998). The waiting-time distribution of solar flare hard X-ray bursts. Preprint.
Wiggins, S. (1990). *Introduction to Applied Nonlinear Dynamical Systems and Chaos*. Springer-Verlag, New York.
Wijers, W. (1998). The burst, the burster and its lair. *Nature* **393**, 13.
Wisdom, J. (1987). Rotational dynamics of irregularly shaped natural satellites. *Astronom. J.* **94**, 1350–1360.
Wisdom, J., S. Peale and F. Mignard (1984). The chaotic rotation of hyperion. *Icarus* **58**, 137–152.
Young, K. and J. Scargle (1996). The dripping handrail model: transient chaos in accretion systems. *Ap. J.* **468**, 617–632.
Zhang, W., K. Johoda, J. Swank, J., E. Morgan and A. Giles (1995). Dead-time modifications to fast Fourier transform power spectra. *Astrophys. J.* **449**, 930–935.

D. N. Shanbhag and C. R. Rao, eds., *Handbook of Statistics, Vol. 21*
© 2003 Elsevier Science B.V. All rights reserved

On the Theory of Discrete and Continuous Bilinear Time Series Models

T. Subba Rao and Gy. Terdik

Large number of time series we come across are neither linear nor Gaussian. During the last two decades, several nonlinear time series models have been developed, their statistical properties have been investigated. Nonlinear models often produce better forecasts than linear models. One of the models which received considerable attention is the bilinear model. In this chapter we review the recent developments associated with this model. We discuss both discrete time and continuous time versions of the model. Higher-order moments and cumulants play an important role in the identification of non-linear models, and we discuss various relationships associated with higher-order cumulants for both univariate and multivariate time series. We consider the estimation of the bispectral density function and its use in detection of linearity of the time series. The estimation of univariate and multivariate bilinear models is discussed. We introduce a long memory bilinear model. We also introduce nonstationary models. We study the effect of nonlinearity on spurious regression and cointegration.

1. Introduction

It is often assumed that the time series under consideration for the statistical analysis is linear and perhaps even Gaussian. Recent studies have shown that such an assumption is very unrealistic. If the time series is Gaussian, it is well known that the second-order properties such as second-order correlations and second-order spectra will characterise the entire structure of the series. Otherwise, one has to look to higher-order moments and higher-order cumulant spectra for further information on the time series. Here we review the recent results on higher-order spectra and bilinear time series models and the emphasis is mainly on the contributions made by the authors.

Let $\{X_t\}$ be a discrete parameter time series, and let us assume that the series is second-order stationary in the sense that

(1) $E(X_t) = \mu$ for all t,
(2) $\operatorname{Var} X_t = E(X_t - \mu)^2 = \sigma_x^2 < \infty$,
(3) $\operatorname{Cov}(X_t, X_{t+s}) = R(s) = $ a function of s only $(s = 0, \pm 1, \pm 2, \ldots)$.

We assume further $\sum |R(s)| < \infty$ and define the second-order spectral density function, $f(\omega)$

$$f(\omega) = \frac{1}{2\pi} \sum_{-\infty}^{\infty} R(s) e^{is\omega}, \quad |\omega| \leq \pi. \tag{1.1}$$

If $\{X_t\}$ is real valued, we know $R(s) = R(-s)$, $f(\omega) = f(-\omega)$ and we say the time series $\{X_t\}$ is short memory if the covariances $\{R(s)\}$ satisfy the condition $\sum |R(s)| < \infty$. and is long memory if $\sum |R(s)| = \infty$ (see Beran, 1994). As pointed out above, we need to use higher-order moments when there is a departure from Gaussianity. Let us assume that the time series has finite kth-order moment, i.e., $E|X_t|^k < \infty$, and let

$$m(t_1, t_2, \ldots, t_k) = E(X_{t_1} X_{t_2} \ldots X_{t_k}).$$

If $m(t_1, t_2, \ldots, t_k) = m(0, t_2 - t_1, t_3 - t_1, \ldots, t_k - t_1)$ we say the time series $\{X_t\}$ is kth-order stationary. From now onwards we write $m(t_2 - t_1, t_3 - t_1, \ldots, t_k - t_1)$ for $m(0, t_2 - t_1, t_3 - t_1, \ldots, t_k - t_1)$. In order to define the higher-order cumulants, we proceed as follows (for details see Terdik, 1999). Let

$$M(\theta_1, \theta_2, \ldots, \theta_k) = E\big[\exp(\theta_1 X_{t_1} + \theta_2 X_{t_2} + \cdots + \theta_k X_{t_k})\big]$$

$$= \sum_{m_1, m_2, \ldots, m_k = 0}^{\infty} \frac{\theta_1^{m_1} \theta_2^{m_2} \ldots \theta_k^{m_k}}{m_1! \ldots m_k!} E(X_{t_1}^{m_1} X_{t_2}^{m_2} \ldots X_{t_k}^{m_k}). \tag{1.2}$$

Let

$$\psi(\theta_1, \theta_2, \ldots, \theta_k) = \ln M(\theta_1, \theta_2, \ldots, \theta_k)$$

$$= \sum_{m_1, m_2, \ldots, m_k = 1}^{\infty} \frac{\theta_1^{m_1} \theta_2^{m_2} \ldots \theta_k^{m_k}}{m_1! \ldots m_k!} \mathrm{cum}\big(X_{t_1}(m_1), X_{t_2}(m_2), \ldots, X_{t_k}(m_k)\big),$$

where

$$\mathrm{cum}\big(X_{t_1}(m_1), X_{t_2}(m_2), \ldots, X_{t_k}(m_k)\big)$$
$$= \mathrm{cum}\big(\underbrace{X_{t_1}, X_{t_1}, \ldots, X_{t_1}}_{m_1 \text{ terms}}, \underbrace{X_{t_2}, \ldots, X_{t_2}, \ldots, X_{t_2}}_{m_2 \text{ terms}}, \underbrace{X_{t_k}, X_{t_k}, \ldots, X_{t_k}}_{m_k \text{ terms}}\big)$$

(see Brillinger, 1975; Terdik, 1999). When $m_1 = m_2 = \cdots = m_k = 1$ we have the kth cumulant of the time series $(X_{t_1}, X_{t_2}, \ldots, X_{t_k})$ which we denote by $\mathrm{cum}(X_{t_1}, X_{t_2}, \ldots, X_{t_k})$ and if X_t is kth-order stationary it is a function of $t_2 - t_1, t_3 - t_1, \ldots, t_k - t_1$ and we write it as

$$\mathrm{cum}(X_{t_1}, X_{t_2}, \ldots, X_{t_k}) = C(0, t_2 - t_1, \ldots, t_k - t_1) = C(t_2 - t_1, \ldots, t_k - t_1).$$

From the above we observe that $\mathrm{cum}(X_{t_i}) = E(X_{t_i})$, $\mathrm{cum}(X_{t_i}, X_{t_i}) = \mathrm{var}(X_{t_i})$, $\mathrm{cum}(X_{t_i}, X_{t_j}) = \mathrm{cov}(X_{t_i}, X_{t_j})$, etc. We say the time series $\{X_t\}$ is Gaussian if the p-dimensional random vector $\mathbf{X}' = (X_{t_1}, X_{t_2}, \ldots, X_{t_p})$ has a multivariate normal distribution for all integer values of p. Let $E(X_{t_i}) = \mu_i$, $\sigma_{ij} = \mathrm{cov}(X_{t_i}, X_{t_j})$, $\Sigma = (\sigma_{ij}; i, j = 1, 2, \ldots, p)$. The random vector $\mathbf{X}$ is multivariate normal with mean vector $\boldsymbol{\mu}$ and covariance matrix Σ if the moment generating function of $\mathbf{X}$ is given by

$$M(\theta_1, \theta_2, \ldots, \theta_p) = \exp\left[\sum \theta_i \mu_i + \frac{1}{2}\sum_{i=1}^{p}\sum_{j=1}^{p} \theta_i \theta_j \sigma_{ij}\right]$$

and the cumulant generating function is given by

$$\psi(\theta_1, \theta_2, \ldots, \theta_p) = \sum \theta_i \mu_i + \frac{1}{2}\sum\sum \theta_i \theta_j \sigma_{ij}.$$

From this it follows that the cumulants of all orders greater than two are zero if X_t is Gaussian, i.e.,

$$\mathrm{cum}(X_{t_1}, X_{t_2}, \ldots, X_{t_k}) = 0 \quad \text{for all } k > 2.$$

It is well known that cumulants satisfy many interesting properties. We also have (Brillinger, 1975; Leonov and Shiryaev, 1959)

$$\mathrm{cum}(X_{t_1}, X_{t_2}, \ldots, X_{t_k}) = \sum_{\nu} (-1)^{p-1}(p-1)! \mu_{\nu_1} \mu_{\nu_2} \cdots \mu_{\nu_p},$$

where $\mu_{\nu_i} = E(X_{t_1} X_{t_2} \ldots X_{t_i})$ where $(t_1, t_2, \ldots, t_i)$ is any subset of $(t_1, t_2, \ldots, t_k)$ and $(\nu_1, \nu_2, \ldots, \nu_p)$ are partitions of $(1, 2, \ldots, k)$.

One important application of the higher-order cumulants is in signal processing, where we are interested in the estimation of signals (non-Gaussian) when the signals are corrupted by Gaussian white noise. Suppose we observe a time series $\{Z_t\}$, where for each t, $Z_t = X_t + N_t$. Here the signal $\{X_t\}$ is non-Gaussian and the noise $\{N_t\}$ is Gaussian, and we assume that X_t and N_t are independent. From the definition of cumulants, we have

$$C_Z(t_1, t_2, \ldots, t_k) = C_X(t_1, t_2, \ldots, t_k) + C_N(t_1, t_2, \ldots, t_k),$$

where $C_Z(t_1, t_2, \ldots, t_k) = \mathrm{Cum}(Z_{t_1}, Z_{t_2}, \ldots, Z_{t_k})$. If $\{N_t\}$ is Gaussian, we have

$$C_Z(t_1, t_2, \ldots, t_k) = C_X(t_1, t_2, \ldots, t_k) \quad \text{for } k > 2.$$

Higher-order cumulants play an important role in identifying the noisy systems and also non-Gaussian linear time series models (see Porat, 1994).

Now we assume the kth order cumulants of the time series $\{X_t\}$ satisfy $\sum |C(t_1, t_2, \ldots, t_k)| < \infty$, then the kth order cumulant spectral density function can be defined as

$$f(\omega_1, \omega_2, \ldots, \omega_{k-1}) = \frac{1}{(2\pi)^{k-1}} \sum_{\tau_1 \ldots \tau_{k-1} = -\infty}^{\infty} C(\tau_1, \tau_2, \ldots, \tau_{k-1})$$
$$\times \exp(-i(\omega_1 \tau_1 + \omega_2 \tau_2 + \cdots + \omega_{k-1} \tau_{k-1})),$$

where $C(\tau_1, \tau_2, \ldots, \tau_{k-1}) = \text{cum}(X_{t_1}, X_{t_1+\tau_1}, \ldots, X_{t_1+\tau_{k-1}})$.

One of the important problems considered in the statistical analysis is to test the hypothesis that the residuals (the estimated errors after a model is fitted) are independent. If the data is Gaussian, this can be accomplished by testing for the zero second-order correlation. If the series is non-Gaussian we can test for the serial independence by nonparametric methods, either based on joint cumulative distribution function (see Delgado, 1996; Tjostheim, 1999) or on joint probability density functions which are estimated by the kernel density functions (see Prakasa Rao, 1999; Skaug and Tjostheim, 1993). We can also use higher-order spectra as follows. For example, if $\{X_t\}$ is a set of zero mean independent random variables, we have

$$\text{cum}(X_{t_1}, X_{t_2}, \ldots, X_{t_k}) = 0 \quad \text{for } t_1 \neq t_2 \neq \cdots \neq t_k.$$

This implies that the cumulant spectral density function of $\{X_t\}$

$$f(\omega_1, \omega_2, \ldots, \omega_{k-1}) = \frac{C_X(k)}{(2\pi)^{k-1}} \quad \text{independent of } (\omega_1, \omega_2, \ldots, \omega_{k-1}),$$

where $C_X(k) = \text{cum}(X_{t_1}, X_{t_1}, \ldots, X_{t_1})$ is the kth order (zero lag) cumulant of $\{X_t\}$. In other words the constancy of the higher-order spectral density function of the observations is an indication of independence. In recent years, chaos theory has become an important field of study in the theory of nonlinear dynamical systems and it has become necessary to see whether a series (nonlinear) is chaotic or not. As a part of this study, Grassberger and Procaccia (1983) have constructed a correlation integral as a means of measuring the fractal dimension of deterministic data. In 1987, Brock, Dechart and Scheinkman have proposed correlation dimension for testing the independence of the series and this test is now known as the BDS test. We believe the same purpose can be achieved more effectively by using the higher cumulant spectra. Subba Rao (1992) has shown that the bispectral density function (put $k = 3$ in the above) is useful in distinguishing between a chaotic and non-chaotic time series. In a similar way we can use the trispectral density function ($k = 4$) for testing the independence of the series. We believe that the test based on higher-order cumulant spectra will be very useful, and we hope to pursue these alternative approaches later.

Estimation of cumulant spectra has been considered by Brillinger and Rosenblatt (1957), Rosenblatt and Van Ness (1965), Lii, Rosenblatt and Van Atta (1976), Subba Rao and Gabr (1984) and Terdik (1999). We briefly describe the estimation of the bispectral density function of a stationary time series.

Let $(X_1, X_2, \ldots, X_N)$ be a sample from the stationary time series $\{X_t\}$ which is stationary up to the third order.
Let

$$\overline{X} = \frac{1}{N}\sum X_t, \qquad \widehat{R}(s) = \frac{1}{N}\sum_{t=1}^{N-s}(X_t - \overline{X})(X_{t+s} - \overline{X}), \quad s \geq 0,$$

and

$$\widehat{C}(s_1, s_2) = \frac{1}{N}\sum_{t=1}^{N-\gamma}(X_t - \overline{X})(X_{t+s_1} - \overline{X})(X_{t+s_2} - \overline{X}),$$

where $\gamma = \max(0, s_1, s_2)$, $s_1 \geq 0$, $s_2 \geq 0$.
We estimate the spectral density function $f(\omega)$ by

$$\hat{f}(\omega) = \frac{1}{2\pi}\sum_{-(N-1)}^{N-1}\lambda\left(\frac{\tau}{M}\right)\widehat{R}(\tau)\cos\omega\tau,$$

where $M = M(N)$ and $\lambda(\cdot)$ is a lag window generator if $\lambda(s) = 0$ for $|s| \geq 1$ and M corresponds to the truncation point. We assume that $\lambda(s)$ is a bounded, even, and a square integrable function such that $\lambda(0) = 1$. The integer M is chosen such that as $M \to \infty$, $N \to \infty$ and $\frac{M}{N} \to 0$. It is well known that

$$\mathrm{var}(\hat{f}(\omega)) \simeq \frac{M}{N}f^2(\omega)\int \lambda^2(s)\,ds, \quad \omega \neq 0, \pi.$$

In Table 1, we give some standard lag window generators.
Now consider the estimation of the bispectral density function

$$f(\omega_1, \omega_2) = \frac{1}{(2\pi)^2}\sum_{\tau_1}\sum_{\tau_2} C(\tau_1, \tau_2)\exp[-i(\omega_1\tau_1 + \omega_2\tau_2)]. \tag{1.3}$$

Table 1
Lag window generators

Daniell window	$\lambda_D(s) = \frac{\sin s\pi}{s\pi}$	
Tukey–Hamming window	$\lambda_T(s) = \begin{cases} 0.54 + 0.46\cos s\pi, & \|s\| \leq 1, \\ 0, & \text{otherwise}, \end{cases}$	
Parzen window	$\lambda_P(s) = \begin{cases} 1 - 6s^2 + 6\|s\|^3, & \|s\| \leq \frac{1}{2}, \\ 2(1 - \|s\|)^3, & \frac{1}{2} \leq \|s\| \leq 1, \\ 0, & \text{otherwise}, \end{cases}$	
Bartlett–Priestley window	$\lambda_{BP}(s) = \frac{3}{(\pi s)^2}\left\{\frac{\sin \pi s}{\pi s} - \cos \pi s\right\}.$	

Let $K_0(\theta_1, \theta_2)$ be a bounded and a non-negative function satisfying

(i) $\iint K_0(\theta_1, \theta_2)\, d\theta_1\, d\theta_2 = 1$,

(ii) $\iint K_0^2(\theta_1, \theta_2)\, d\theta_1\, d\theta_2 < \infty$,

$$\iint \theta_i^2 K_0(\theta_1, \theta_2)\, d\theta_1\, d\theta_2 < \infty \quad (i = 1, 2) \tag{1.4}$$

(iii) $K_0(\theta_1, \theta_2) = K_0(\theta_2, \theta_1) = K_0(\theta_1, -\theta_1 - \theta_2)$
$$= K_0(-\theta_1 - \theta_2, \theta_2). \tag{1.5}$$

Condition (iii) is imposed to be consistent with the symmetry conditions the bispectrum satisfies, namely,

$$f(\omega_1, \omega_2) = f(\omega_2, \omega_1) = f(\omega_1, -\omega_1 - \omega_2) = f(-\omega_1 - \omega_2, \omega_2)$$
$$= f^*(-\omega_1, -\omega_2). \tag{1.6}$$

Then the bispectrum $f(\omega_1, \omega_2)$ is estimated by

$$\hat{f}(\omega_1, \omega_2) = \frac{1}{(2\pi)^2} \sum\sum \lambda\left(\frac{\tau_1}{M}, \frac{\tau_2}{M}\right) \widehat{C}(\tau_1, \tau_2) e^{-(i\tau_1\omega_1 + i\tau_2\omega_2)}, \tag{1.7}$$

where M, a function of N, is a window parameter chosen such that $\frac{M^2}{N} \to 0$ as $M \to \infty$, $N \to \infty$. Then (Brillinger and Rosenblatt, 1967; Rosenblatt and Van Ness, 1965) have shown that

$$\operatorname{var}(\hat{f}(\omega_1, \omega_2)) = \frac{M^2}{N} \cdot \frac{V_2}{2\pi} f(\omega_1) f(\omega_2) f(\omega_1 + \omega_2), \quad 0 < \omega_2 < \omega_1,$$

where

$$V_2 = \iint \lambda^2(s_1, s_2)\, ds_1\, ds_2 = (2\pi)^2 \iint K_0^2(\theta_1, \theta_2)\, d\theta_1\, d\theta_2.$$

One choice of the two-dimensional window $\lambda(s_1, s_2)$ is to choose it as a product of three one-dimensional windows of the form

$$\lambda(s_1, s_2) = \lambda(s_1)\lambda(s_2)\lambda(s_1 - s_2).$$

An optimum window which has the smallest mean square error has been obtained by Subba Rao and Gabr (1984), and it is given by

$$K_0(\theta_1, \theta_2) = \begin{cases} \frac{\sqrt{3}}{\pi^3}\left[1 - \frac{1}{\pi^2}(\theta_1^2 + \theta_2^2 + \theta_1\theta_2)\right], & \text{if } (\theta_1, \theta_2) \in G, \\ 0, & \text{otherwise,} \end{cases}$$

where $G = \{(\theta_1, \theta_2); \theta_1^2 + \theta_2^2 + \theta_1\theta_2 \leq \pi^2\}$.

The exact form of the corresponding lag window generator $\lambda(s_1, s_2)$ was given by Saito and Tanaka (1984) and is given by

$$\lambda(s_1, s_2) = \int_{-\pi}^{\pi} \int_{-\pi}^{\pi} e^{is_1\theta_1 + is_2\theta_2} K_0(\theta_1, \theta_2) \, d\theta_1 \, d\theta_2$$

$$= \frac{8}{\alpha^2} J_2(\alpha),$$

where

$$\alpha = \frac{2\pi}{\sqrt{3}} [s_1^2 - s_1 s_2 + s_2^2]^{1/2},$$

and $J_2(x)$ is the Bessel function of the first kind.

Bispectral density function is often used to study the nonlinear characteristics such as nuclear reactor noise (see Saito and Tanaka, 1984), chaos (Subba Rao, 1992) and nonlinear phase decoupling (Nikias and Petropulu, 1994), etc.

2. Linear time series models and cumulant spectra

We say the stationary time series is linear if X_t can be represented in the form

$$X_t = \sum_{u=0}^{\infty} g_u e_{t-u}, \tag{2.1}$$

where $\{e_t\}$ is a sequence of independent, identically distributed random variables with mean zero and variance σ_e^2. Let $k_n(e)$ denote the nth order cumulant of $\{e_t\}$ and we have $k_1(e) = 0$, $k_2(e) = \sigma_e^2$, etc. All the linear models such as AR, ARMA models are special cases of (2.1). We can show that

$$\mathrm{cum}(X_t, X_{t+\tau_1}, \ldots, X_{t+\tau_{n-1}}) = k_n(e) \sum_t g_t g_{t+\tau_1} \cdots g_{t+\tau_{n-1}}. \tag{2.2}$$

The above relation was first derived by Bartlett (1946), and later by Brillinger and Rosenblatt (1967). A multivariate extension of the above result was recently given by Subba Rao and Wong (1999).

By taking the $(n-1)$-dimensional Fourier transform of both sides of (2.2), we can show that the nth order cumulant spectral density function of X_t is

$$f_n(\omega_1, \omega_2, \ldots, \omega_{n-1}) = \frac{k_n(e)}{(2\pi)^{n-1}} \prod_{i=1}^{n} h(\omega_i), \tag{2.3}$$

where $h(\omega) = \sum g_n e^{-in\omega}$ and $\omega_1 + \omega_2 + \cdots + \omega_n = 0 \pmod{2\pi}$. When $n = 2$, we have the second-order spectral density function

$$f(\omega) = \frac{k_2(e)}{2\pi}|h(\omega)|^2, \qquad (2.4)$$

and when $n = 3$, we have the bispectral density function

$$f_3(\omega_1, \omega_2) = \frac{k_3(e)}{(2\pi)^2} h(\omega_1)h(\omega_2)h(-\omega_1 - \omega_2). \qquad (2.5)$$

From (2.3) and (2.4) it is obvious that when X_t is linear, the ratio $\frac{|f_n(\omega_1,\omega_2,\ldots,\omega_{n-1})|^2}{f(\omega_1)f(\omega_2)\ldots f(-\sum_{j=1}^{n-1}\omega_j)}$ is a constant and does not depend on the frequencies $(\omega_1, \omega_2, \ldots, \omega_{n-1})$. If X_t is Gaussian, we have $f_n(\omega_1, \omega_2, \ldots, \omega_{n-1}) = 0$ for $n > 2$ and if X_t is linear but non-Gaussian, the ratio is a constant. This is the basis of statistical tests for Gaussianity and linearity proposed by Subba Rao and Gabr (1980) and Hinich (1982). In recent years several time domain tests have been proposed for testing linearity of time series, and these tests are based on various parametric nonlinear models such as Bilinear models. The tests are constructed to test a specific hypothesis concerning the nonlinear parameters of these models. Under the null hypothesis, the time series defined by these models will have a Gaussian distribution. In other words these tests are strictly for testing Gaussianity rather than linearity. There seems to be some confusion between the tests for linearity and Gaussianity. This is rather unfortunate and the misuse of this terminology is most common in economics literature, and we believe that a clear distinction between the tests for linearity and Gaussianity must be drawn. As far as we are aware the first time domain test for Gaussianity of a stationary time series was proposed by Lomnicki (1961), and recently was extended by Subba Rao and Wong (1998) to multivariate situations. An alternative definition of linearity was proposed by Hannan (1986), and he defined the process to be linear if the linear predictor is optimal and using this definition Terdik and Math (1998) have proposed a new linearity test.

The bispectral density function and the trispectral density function ($n = 4$) have been used by Lii and Rosenblatt (1982) to deconvolve linear non-Gaussian signals.

3. Volterra expansion and bilinear models

The most general form of nonlinear, stationary time series representation is the Volterra representation (Wiener, 1958; Brillinger, 1970) and it is given by

$$X_t = \mu + \sum_{u=-\infty}^{\infty} a_u e_{t-u} + \sum_{u,v=-\infty}^{\infty} a_{uv} e_{t-u} e_{t-v}$$

$$+ \sum_{u,v,w=-\infty}^{\infty} a_{uvw} e_{t-u} e_{t-v} e_{t-w} + \cdots, \qquad (3.1)$$

where $\{e_t\}$ are assumed to be independent, identically distributed random variables. We note that the linear process representation (2.1) is a special case of (3.1). If $\{e_t\}$ are

Gaussian, the above representation can be written in the form of Hermite polynomials (Rugh, 1981) and the expansion is then called the Wiener expansion. Because of orthogonality of Hermite functions it is convenient to use the Wiener expansion rather than the Volterra expansion. We can also write the above expansion in the frequency domain (Schetzen, 1980; Rugh, 1981) leading to Wiener–Itô representation (see Major, 1981; Terdik, 1999). Not all finite parameters nonlinear time series models can be written in the form (3.1). Here we consider bilinear models which have the solutions in the form (3.1) and have been extensively studied in the literature.

Let X_t be a discrete parameter time series satisfying the difference equation

$$X_t + \sum_{j=1}^{p} a_j X_{t-j} = e_t + \sum_{j=1}^{r} c_j e_{t-j} + \sum_{l=1}^{m} \sum_{l'=1}^{k} b_{ll'} X_{t-l} e_{t-l'}, \qquad (3.2)$$

where $\{e_t\}$ is a sequence of independent, identically distributed random variables, each distributed as normal with mean zero and variance σ_e^2.

The existence of bilinear models and conditions for the stationarity in terms of the parameters have been proved (Pham Dinh and Tran, 1981; Liu, 1989; Bhaskara Rao et al., 1983). The above models were introduced by Subba Rao (1977) and Granger and Andersen (1978) in time series literature, but already widely used in control theory. Through simulations (Subba Rao, 1977, 1981a) one can show that these models are useful in describing quick bursts which are common in seismic signals.

We define the model (3.2) as a bilinear model $BL(p, r, m, k)$ and the process $\{X_t\}$ the bilinear process. If we set $b_{ll'} = 0$ for all $l' \leq l$, we get the lower triangular bilinear models. We can show that these models admit the Volterra representation (3.1). To illustrate the ideas consider the model $BL(p, 0, p, 1)$, i.e.,

$$X_t + \sum_{j=1}^{p} a_j X_{t-j} = e_t + \sum_{l=1}^{p} b_{l1} X_{t-l} e_{t-1} \qquad (3.3)$$

which can be written in the "state space" form.

Let us define the matrices

$$\mathbf{A} = \begin{bmatrix} -a_1 & -a_2 & \cdots & -a_p \\ 1 & 0 & \cdots & 0 \\ & & \cdot & \\ & & \cdot & \\ 0 & 0 & \cdots & 0 \end{bmatrix}, \quad \mathbf{B} = \begin{bmatrix} b_{11} & b_{21} & \cdots & b_{p1} \\ 0 & 0 & \cdots & 0 \\ & & \cdot & \\ & & \cdot & \\ 0 & 0 & \cdots & 0 \end{bmatrix},$$

$$\mathbf{C}' = (1, 0, \ldots, 0), \qquad \mathbf{H}' = (1, 0, \ldots, 0), \qquad \mathbf{x}_t' = (X_t, X_{t-1}, \ldots, X_{t-p+1}).$$

Then we can write (3.3) as

$$\mathbf{x}_t = \mathbf{A}\mathbf{x}_{t-1} + \mathbf{B}\mathbf{x}_{t-1} e_{t-1} + \mathbf{C} e_t,$$
$$X_t = \mathbf{H}'\mathbf{x}_t. \qquad (3.4)$$

We define the model (3.4) as a vector bilinear model, and denote it by $VBL(p)$. In fact the above representation is very useful in studying multivariate bilinear models as well. Using the "reversion method", Subba Rao (1981b) has shown how (3.4) can be written in the Volterra form. Briefly we discuss this method which is used later as well. Consider the model

$$\mathbf{x}_t = \mathbf{A}\mathbf{x}_{t-1} + \lambda \mathbf{B}\mathbf{x}_{t-1}e_{t-1} + \lambda \mathbf{C}e_t, \tag{3.5}$$

where λ is a perturbation parameter introduced to facilitate the solution, and we allow λ to tend to unity. We seek the solution of (3.5) in the form

$$\mathbf{x}_t = \sum_{j=1}^{\infty} \lambda^j \mathbf{x}_j(t). \tag{3.6}$$

Substitute (3.6) in (3.5), and then equate powers of λ^j ($j = 1, 2, \ldots$). We obtain

$$\mathbf{x}_1(t) = \mathbf{A}\mathbf{x}_1(t-1) + \mathbf{C}e_t,$$
$$\mathbf{x}_i(t) = \mathbf{A}\mathbf{x}_i(t-1) + \mathbf{B}\mathbf{x}_{i-1}(t-1)e_{t-1} \quad (i = 2, 3, \ldots). \tag{3.7}$$

Assuming $\mathbf{x}_i(0) = 0$ ($i = 1, 2, \ldots$) and $e_t = 0$ ($t \leq 0$) we obtain

$$\mathbf{x}_1(t) = \sum_{j_1=0}^{t} \mathbf{A}^{j_1} \mathbf{C} e_{t-j_1},$$

$$\mathbf{x}_i(t) = \sum_{j_1=0}^{t-1} \mathbf{A}^{j_1} \mathbf{B} \mathbf{x}_{i-1}(t-1-j_1) e_{t-1-j_1} \quad (i = 2, 3, \ldots).$$

Substituting the above solutions in (3.6) we can show that the solution of (3.3) can be written in the form (3.1) with appropriate kernels $\{a_u\}$, $\{a_{uv}\}, \ldots$. In the Volterra expansion (3.1) $\{a_u\}$, $\{a_{uv},\}$ are called Volterra kernels, and we see from the above expansion for the bilinear models, these kernels are separable. In fact one can characterise the bilinear models by their "separability". It is well known that a nonlinear system cannot be characterised on the basis of a single transfer function, and one needs an infinite number of transfer functions for its characterisation. In view of the separability property of the Volterra kernels for the bilinear models, we can see that only the first two transfer functions (linear and quadratic transfer functions) are needed, as all the higher-order transfer functions can be written in terms of these two (see Terdik and Subba Rao, 1989).

An alternative frequency domain expansion which is valid for the lower triangular bilinear model with Gaussian input is given below. The idea is that instead of the Volterra expansion one can use the Hermite polynomial expansion and then use the Wiener–Itô stochastic spectral representation, see Major (1981) for details. Let $\{e_t\}$ be Gaussian and let $H_n(e_t)$ be an nth order Hermite polynomial in e_t. For example,

$H_0(e_t) = 1$, $H_1(e_t) = e_t$, $H_2(e_t) = e_t^2 - \sigma_e^2$, and so on. Since the process $\{e_t\}$ is stationary and Gaussian, we have the spectral representation

$$e_t = \int_{-\pi}^{\pi} e^{it\omega} W(d\omega),$$

where $W(d\omega)$ is an orthogonal random measure. By substituting the above representation for e_t in the Hermite expansion, we obtain:

$$H_n(e_t) = \int_{\mathcal{D}^n} e^{it2\pi \Sigma \omega_{(n)}} W(d\omega_{(n)}), \quad \mathcal{D}^n = [0,1]^n,$$

where the integral is an n-fold Wiener–Itô stochastic integral with respect to the Gaussian stochastic measure $W(d\omega)$, $EW(d\omega) = 0$, $E|W(d\omega)|^2 = \sigma^2 d\omega$.

Now if X_t is a lower triangular bilinear model (3.2) then

$$X_t = \sum_{r=0}^{\infty} \int_{\mathcal{D}^r} e^{it2\pi \Sigma \omega_{(r)}} G_r(\omega_{(r)}) W(d\omega_{(r)}).$$

This representation gives an orthogonal expansion which is very useful for deriving the frequency domain properties of X_t. Again, only the first two transfer functions G_1 and G_2 (linear and quadratic transfer functions) are needed, as all the higher-order transfer functions are given as the recursion of these two (see Terdik and Subba Rao, 1989).

4. Higher-order moments and identification

In this section we briefly consider the statistical properties associated with some special cases of the bilinear models described in the previous section. We refer to the books by Subba Rao and Gabr (1984), Terdik (1999) and Guegan (1994) and the references therein for details.

We consider the bilinear model $BL(p, 0, p, 1)$ and its state space representation (3.4).

The sufficient condition for the second-order stationarity of the model was derived by Subba Rao (1981a, 1981b) and conditions for higher-order stationarity were given by Tang (1987).

The time series satisfying the model $BL(p, 0, p, 1)$ is second-order stationary if $\rho(\mathbf{A} \otimes \mathbf{A} + \sigma_e^2 \mathbf{B} \otimes \mathbf{B}) < 1$, where $\rho(\mathbf{A})$ is the spectral radius of the matrix $\mathbf{A}$, $\mathbf{A} \otimes \mathbf{A}$ is the Kronecker product of $\mathbf{A}$. Suppose $S(\mathbf{A}^i \mathbf{B}^j)$ denotes the summation of all possible Kronecker product combinations of $\mathbf{A}^i$ and $\mathbf{B}^j$, then Tang (1987) has shown that the time series $\{X_t\}$ is lth order stationary if $\rho(\alpha_n) < 1$ for $n = 1, 2, \ldots, l$, where

$$\alpha_n = \sum_{i=0}^{[n/2]} \left(S A^{n-2i} B^{2i}\right) \frac{(2i)!}{2^i i!}. \tag{4.1}$$

We refer to Terdik (1999) for further details. Assuming the conditions are satisfied we can obtain difference equations for the second-order and higher-order moments for the time series. These difference equations in higher-order moments are similar to the well known second-order Yule–Walker equations for the linear ARMA models.

Suppose we define

$$c_{m+1}(s_1, s_2, \ldots, s_m) = E\left[(X_t - \mu)\prod_{i=1}^{m}(X_{t+s_i} - \mu)\right],$$

$$\mu_{m+1}(s_1, s_2, \ldots, s_m) = E\left[X_t \prod_{i=1}^{m} X_{t+s_i}\right], \quad m \geqslant 1,$$

then we can show (Subba Rao, 1981a, 1981b; Sesay and Subba Rao, 1988),

$$c_2(s) + \sum_{j=1}^{p} a_j c_2(s-j) = 0 \quad \text{for } s \geqslant 2,$$

$$c_3(0, s) + \sum_{j=1}^{p} a_j c_3(0, s-j) = 0, \quad s \geqslant 2, \tag{4.2}$$

$$\mu_4(0, 0, s) + \sum_{j=1}^{p} a_j \mu_4(0, 0, s-j) = b_{11}\sigma_e^2 \mu_3(0, 0).$$

The first condition in (4.2) is similar to the Yule–Walker equations for a linear ARMA(p, 1) model, and this suggests that second-order covariances for bilinear models are similar to the linear ARMA models. This shows that second-order moments cannot distinguish between linear ARMA models and bilinear models. This makes it necessary for us to look into higher-order moments for studying nonlinear models.

5. Estimation of higher-order cumulants and the bilinear models

We have shown above that one can derive Yule–Walker-type difference equations for the bilinear models. The difference equations are in terms of higher-order moments and cumulants. Wong (1993) has derived (see also Subba Rao and Wong, 1999) these equations for the multivariate bilinear time series models. The equations can be used for the tentative identification of the order of the bilinear model and also for obtaining estimates of the parameters of the model. This involves the estimation of higher-order cumulants and the moments. The properties of these estimates have been investigated by Kim (1989), Wong (1993) and Chanda (1983). Briefly we describe the result obtained by Kim (1989) as they are relevant to our approach here. We use his notation for convenience.

Let $X_1, X_2, \ldots, X_n$ be a sample from the stationary time series $\{X_t\}$. Let

$$m_k(t_1, t_2, \ldots, t_{k-1}) = E(X_t X_{t+t_1} X_{t+t_2}, \ldots, X_{t+t_{k-1}}),$$

$$c_k(t_1, t_2, \ldots, t_{k-1}) = \text{cum}(X_t, X_{t+t_1}, \ldots, X_{t+t_{k-1}}). \tag{5.1}$$

Let $V = \{v_1, v_2, \ldots, v_p\}$ be a partition of the set $\{0, 1, 2, \ldots, k-1\}$ into p subsets, where $1 \leq p \leq k$. Let $|v_j|$ denote the number of elements in the set $v_j \subset \{0, 1, 2, \ldots, k-1\}$ ($j = 1, 2, \ldots, p$) and we have $\sum_{j=1}^{p} |v_j| = k$. Let v_{jm} denote the mth element of the set v_j ($m = 1, 2, \ldots, |v_j|$). We partition the set $\{0, t_1, \ldots, t_{k-1}\}$ into subsets t_{v_j} so that $\{0, t_1, \ldots, t_{k-1}\} = \{t_{v_1}, \ldots, t_{v_p}\}$ and $t_{v_i} \cap t_{v_j} = \phi$, $i \neq j$. We define the moments and the cumulants

$$m_{v_j} = m_{|v_j|}(t_{v_1}, \ldots, t_{v_{j|v_j|}}),$$

$$c_{v_j} = c_{|v_j|}(t_{v_1}, \ldots, t_{v_{j|v_j|}}) \tag{5.2}$$

for $j = 1, 2, \ldots, p$. We note $m_k(t_1, t_2, \ldots, t_{k-1}) = \mu_k(t_1, t_2, \ldots, t_{k-1})$, and in the definitions of $m_{|v_j|}(\cdot)$ and $c_{|v_j|}(\cdot)$, $|v_j|$ corresponds to the order of the moments and the cumulants. In view of the relations

$$m_k(t_1, t_2, \ldots, t_{k-1}) = \sum c_{v_1}(t_{v_1}) \ldots c_{v_p}(t_{v_p}),$$

$$c_k(t_1, t_2, \ldots, t_{k-1}) = \sum_v (-1)^{p-1}(p-1)! m_{v_1}(t_{v_1}) \ldots m_{v_p}(t_{v_p}), \tag{5.3}$$

we can estimate the cumulants (moments) from the moments (cumulants) using the relations (5.3).

Let $(X_1, X_2, \ldots, X_n)$ be a sample from the stationary time series $\{X_t\}$. We define

$$\hat{m}_{v_j}(t_{v_j}) = \frac{1}{n} \sum_{h \in D(t_{v_j})} X_{h+t_{v_{j1}}} X_{h+t_{v_{j2}}} \ldots X_{h+t_{v_{j|v_j|}}}, \tag{5.4}$$

where

$$D(t_{v_j}) = \begin{cases} \{1 - \min(t_{v_j}), \ldots, n - \max(t_{v_j})\} \\ \quad \text{if } 0 \leq \max(t_{v_j}) - \min(t_{v_j}) \leq n - 1 \text{ and } |t_{v_j}| \leq n - 1, \\ 0, \quad \text{otherwise} \end{cases}$$

as the sample estimate of the t_{v_j}th lagged v_jth moment (about zero). For example, if $v_1 = 1$, $t_{v_j} = 0$, we have the first moment

$$\hat{m}_1(0) = \overline{X} = \frac{1}{n} \sum_{h=1}^{n} X_h,$$

and if $v_j = 2$, $t_{v_1 1} = 0$, $t_{v_2} = t_1$, we have the second moment,

$$\hat{m}_2(t_1) = \frac{1}{n} \sum_{h=1}^{n-t_1} X_h X_{h+t_1}, \tag{5.5}$$

and if $v_j = 3$, $t_{v_1 1} = 0$, $t_{v_1 2} = t_1$, $t_{v_1 3} = t_2$, the third moment

$$\hat{m}_3(t_1, t_2) = \frac{1}{n} \sum_{h=1}^{n-\max(t_1,t_2)} X_h X_{h+t_1} X_{h+t_2}$$

and similarly other sample moments can be obtained.

The estimation of the cumulants can now be obtained by replacing the moments $m_{v_1}(t_{v_1})$ etc. in (5.3) by their estimates given by (5.4). The asymptotic sampling properties of the moment estimates and the sample cumulants have been derived by Kim (1989) by imposing conditions on the maximal correlation coefficient.

Let $\sigma\{X_t; a \leqslant t \leqslant b\}$ denote the σ-field generated by $\{X_a, \ldots, X_b\}$ for some integers a and b, and let $f \in L^2(\sigma(X_t; t \leqslant 0))$, $g \in L^2(\sigma(X_t, t \geqslant n))$,

$$\rho(n) = \sup \frac{|E(f - Ef)(g - Eg)|}{[E(f - Ef)^2 . E(g - Eg)^2]^{1/2}}, \tag{5.6}$$

where sup is the supremum taken over all random variables f and g. The quantity $\rho(n)$ is called the maximal correlation coefficient.

Let $\{X_t\}$ be a stationary ergodic time series and let $E(X_t) = 0$, $E|X_t|^{2k+\eta} < \infty$, for some $\eta > 0$. Suppose $\rho(n) \to 0$ as $n \to \infty$, then as $n \to \infty$. Kim (1989) has shown that

$$\sqrt{n}\bigl(\hat{m}_{v_j}(t_{v_j}) - E(\hat{m}_v(t_{v_j}))\bigr) \to N\bigl(0, v(t_{v_j})\bigr)$$

(the convergence is in distribution). From the above one can deduce that

$$E\bigl(\widehat{C}_k(t_1, \ldots, t_{k-1})\bigr) \to C_k(t_1, \ldots, t_{k-1})$$

as $n \to \infty$ for each $(t_1, t_2, \ldots, t_{k-1})$ and also one can show that $\widehat{C}_k(t_1, t_2, \ldots, t_{k-1})$ is a consistent estimate of $C_k(t_1, \ldots, t_{k-1})$. Using the asymptotic properties one can establish the asymptotic properties of the cumulant spectral density function (see Kim, 1989). We note that in the above derivations, the only assumption to be made is that the time series is stationary, ergodic and satisfies the mixing condition. Chanda (1983) has obtained similar results under the assumption that the process is a bilinear process.

Nonlinear canonical correlation analysis (NCCA)

Principal component analysis (PCA) and canonical correlation analysis are widely used techniques in multivariate analysis. The two techniques are closely related and they are now mainly used in the reduction of dimensions or to find the best linear combinations between two random vectors. Using the canonical factor analysis technique introduced by Rao (1965), Subba Rao (1976) developed a method for the determination of the order and the estimation of the parameters of AR models when the observations are corrupted by noise. This technique is based on the principal component analysis and is widely used in signal processing literature for the estimation of number of sinusoidal signals and the frequencies of multiple signals. The algorithm nowadays known as multiple signal classification (MUSIC) algorithm introduced by Schmidt (1981) in the signal processing literature is similar to the method of Subba Rao (1976) which depends on the PCA performed on the lagged vector of the time series $\{X_t\}$. Recently Subba Rao and Maria da Silva (1992) have shown that one can identify the orders of the bilinear models by performing canonical correlation analysis not only on the lagged vectors of $\{X_t\}$, but also on higher-order powers of $\{X_t\}$, and this technique we called nonlinear canonical correlation analysis (NCCA).

We observe from Eqs. (4.2), that the second-order covariances, third-order covariances, etc., satisfy the Yule–Walker equations. Tsay and Tiao (1985) have also used canonical correlation analysis for identifying linear time series models. We briefly summarise the results of Subba Rao and Maria da Silva (1992) (for details refer to Maria da Silva, 1994).

Consider first the difference equation in (4.2) which is a difference equation in the second-order covariances. This equation suggests that one can perform canonical correlation analysis between the vector $Y_{m,t}^{(1)} = (X_t - \mu, X_{t-1} - \mu, \ldots, X_{t-m} - \mu)'$ and its lagged vector $Y_{m,t-j}^{(1)}$, $j \geq 1$.

Let $\Gamma_2(m)$ be the $(m+1) \times (m+1)$ matrix with

$$\Gamma_2(m) = (\Gamma_2(u,v); \ u,v = 1, 2, \ldots, m+1),$$

where $\Gamma_2(u,v) = c_2(2 + u - v)$. We note $\Gamma_2(m) = E(Y_{m,t-2}^{(1)} Y_{m,t}^{(1)'})$. When $m \geq p$, rank $(\Gamma_2(m)) = m$, and this suggests that there is one zero canonical correlation between $Y_{m,t}^{(1)}$ and $Y_{m,t-2}^{(1)}$.

Let $A_2(m) = \Gamma_{(1)(1)}^{-1}(m)\Gamma_2'(m)\Gamma_{(1)(1)}^{-1}\Gamma_2(m)$ where $\Gamma_{(1)(1)}(m) = E(Y_{m,t}^{(1)} Y_{m,t}^{(1)'})$. Let $N_2(m)$ be the number of zero eigenvalues of $A_2(m)$. For $m \geq p$, $N_2(m) = 1$.

Similar to the above linear canonical correlation analysis one can perform canonical correlation analysis for nonlinear observations. For example, consider the vector $Y_{m,t}^{(1)}$ defined above, and a new vector $Y_{m,t}^{(2)} = ((X_t - \mu)^2, (X_{t-1} - \mu)^2, \ldots, (X_{t-m} - \mu)^2)'$, and let

$$\Gamma_3(m) = (\Gamma_2(u,v); \ u,v = 1, 2, \ldots, m+1)$$

where $\Gamma_3(u,v) = c_3(0, 2 + u - v)$.

Let $A_3(m) = \Gamma_{(1)(1)}^{-1}(m)\Gamma_3'(m)\Gamma_{(2)(2)}^{-1}(m)\Gamma_3(m)$, where $\Gamma_{(2)(2)}(m) = E(Y_{m,t}^{(2)} Y_{m,t}^{(2)'})$. Let $N_3(m)$ be the number of zero eigenvalues of $A_3(m)$. When $m \geq p$, $N_3(m) = 1$.

It is interesting to observe that the difference equations obtained (4.2) suggest a way of performing nonlinear canonical correlation analysis to obtain the order of the bilinear model. In practice one has to estimate these eigenvalues which in turn depend on the estimates of higher-order moments and cumulants. The distribution of the eigenvalues and the resulting test statistic for the significance of the eigenvalues heavily depend on the assumption that the vectors $\{Y_{m,t}\}$ are multivariate normal, and the observations are independent. These two assumptions are not valid in the case of higher-order canonical correlation analysis described above. In order to check the validity of the classical tests in time series situations, Maria da Silva (1994) has performed several simulation studies and these were reported in her Ph.D. thesis. Her conclusion was that it is reasonable to use the test statistic in situations which involve second-order and fourth-order moments. It is well known that the test statistic has approximately a chi square distribution with one degree of freedom. Through simulations, Maria da Silva has shown that the mean and variance of the test statistic are approximately close to one and two respectively which is consistent with the suggested approximation. In the case of third-order moments, that though the mean is close to one, the variance is much smaller than two. For further information we refer to her Ph.D. thesis.

The above equations can also be used for the estimation of the parameters (see Maria da Silva, 1994, for details). Consider the third-order equation given by Sesay and Subba Rao (1991)

$$\mu_3(s, s+1) = -\sum_{s=1}^{p} a_j \mu_3(s, s+1-j) + 2b_{11}\sigma_e^2 \mu_2(s)$$

$$+ \sum_{j=2}^{p} b_{j1}\sigma_e^2 \mu_2(s+1-j).$$

The above equation is nonlinear in the parameters, and is linear if σ_e^2 is known. One can consider the minimisation of

$$Q(\alpha) = \sum_{s=1}^{m} \left[\hat{\mu}_3(s, s+1) + \sum_{j=1}^{p} a_j \hat{\mu}_3(s, s+1-j) - 2b_{11}\sigma_e^2 \hat{\mu}_2(s) \right.$$

$$\left. - \sum_{j=2}^{p} b_{j1}\sigma_e^2 \hat{\mu}_2(s+1-j) \right]^2$$

with respect to $\{a_j\}$, $\{b_{j1}\}$ and σ_e^2. We can linearise the equation and obtain iterative equations to estimate the parameters, and this approach was considered by Maria da Silva (1994). The efficiency of the estimates need to be investigated.

The above minimisation makes use of the total higher-order moments, and an alternative method depending on the conditional second-order moments has been suggested by Grahn (1995). The usual conditional least squares depend on the minimisation of L^2 distances between X_t and its conditional expectation $E(X_t|X_{t-1},\ldots)$, and Grahn (1995) suggested using the L^2 distance between the products $X_t X_{t-s}$ and their conditional expectations $E(X_t X_{t-s}|X_{t-k}, k > s)$ for different $s \geqslant 0$. The asymptotic properties of the estimates have been investigated by Grahn (1995).

6. Multivariate nonlinear time series and higher-order cumulants of random vectors

So far we have considered the analysis of univariate nonlinear time series. In the following we define higher-order moments and cumulants for random vectors and use them to obtain these for multivariate bilinear time series models. For details we refer to the Ph.D. thesis of Wong (1993) and to the paper of Subba Rao and Wong (1999). We give some definitions.

DEFINITION 1. Let $A = (a_{ij})$ be a $m \times n$ matrix, and let $B = (b_{ij})$ be a $p \times q$ matrix. Then the Kronecker product of A and B is defined as

$$A \otimes B = [a_{i_1 j_1} b_{i_2 j_2}], \tag{6.1}$$

where $i_1 j_1$ and $i_2 j_2$ are in lexicographic order with i_2 or j_2 increasing by one when i_1 or j_1 completes a cycle. The matrix $A \otimes B$ is of order $mp \times nq$. The position of the element $a_{i_1 j_1} b_{i_2 j_2}$ in the matrix $A \otimes B$ given by (6.1) corresponds to the coordinate $(i_1 i_2, j_1 j_2)$ where $i_1 i_2$ refers to the $((i_2 - 1)m + i_1)$th row and $j_1 j_2$ refers to $((j_2 - 1)n + j_1)$th column.

DEFINITION 2. Let $X^\ell = [x_{ij}^{(\ell)}]$ be a random matrix of order $m_\ell \times n_\ell$ ($\ell = 1, 2, \ldots, k$). If the kth order scalar joint cumulants $\mathrm{cum}(x_{i_1 j_1}^{(\ell_1)}, x_{i_2 j_2}^{(\ell_2)}, \ldots, x_{i_k j_k}^{(\ell_k)})$ is denoted by $K_{i_1 i_2 \ldots i_k, j_1 j_2 \ldots j_k}^{\ell_1, \ell_2, \ldots, \ell_k}$, then we define the kth order cumulant matrix of order $\prod_{j=1}^{k} m_{1j} \times \prod_{j=1}^{k} n_{\ell j}$

$$\mathrm{cum}\{X^{(\ell_1)}, X^{(\ell_2)}, \ldots, X^{(\ell_k)}\} = [K_{i_1 i_2 \ldots i_k, j_1 j_2 \ldots j_k}^{(\ell_1, \ell_2, \ldots, \ell_k)}], \tag{6.2}$$

where the subscripts $i_1 i_2 \ldots i_k$, and $j_1 j_2 \ldots j_k$ corresponds to the element of the $(i_1 + \sum_{\ell=2}^{k}(i_\ell - 1)m_\ell)$th row, and $(j_1 + \sum_{\ell=2}^{k}(j_\ell - 1)n_\ell)$th column n.

We state the following theorem without proof.

THEOREM 1. *Let $X^{1\ell_1}, X^{2\ell_2}, \ldots, X^{k\ell_k}$ be a sequence of independent random matrices, where each $X^{j\ell_k}$ is of dimension $p_j \ell_j \times q_j \ell_j$. Let $A^{j\ell_j}$ and $B^{j\ell_j}$ be matrices of constants and of dimensions $r_j \ell_j \times p_j \ell_j$ and $q_j \ell_j \times s_j \ell_j$ respectively. Then*

$$\mathrm{cum}\left\{\sum_{\ell_1=1}^{r_1} A^{1\ell_1} X^{1\ell_1} B^{1\ell_1}, \ldots, \sum_{\ell_k=1}^{r_k} A^{k\ell_k} X^{k\ell_k} B^{k\ell_k}\right\}$$

$$= \sum_{\ell_1=1}^{r_1} \cdots \sum_{\ell_k=1}^{r_k} \{A^{1\ell_1} \otimes \cdots \otimes A^{k\ell_k}\}$$

$$\times \mathrm{cum}\{X^{1\ell_k}, \ldots, X^{k\ell_k}\}\{B^{1\ell_1} \otimes \cdots \otimes B^{k\ell_k}\}, \tag{6.3}$$

where $X^{j\ell_j}$ is a $p_j\ell_j \times q_j\ell_j$ random matrix, and A and B's are deterministic matrices. The middle cumulant term of (6.3) is obtained using (6.2).

Eq. (6.3) is a generalisation of the result given by Leonov and Shiryaev (1959) for the univariate time series, and for details see Wong (1993) and Subba Rao and Wong (1998).

THEOREM 2. *Let* $Y^{(\ell)} = X^{(\ell_1)} \otimes \cdots \otimes X^{(\ell_k)}$ ($\ell = 1, 2, \ldots, L$) *where* $\{X^{(\ell)}\}$ *are random matrices. Then*

$$\text{cum}\{Y^{(1)}, Y^{(2)}, \ldots, Y^{(L)}\}$$
$$= \sum_{r=1}^{m} P_r [\text{cum}\{X_{m_n}, m_n \in \nu^{(r)}\} \otimes \cdots \otimes \text{cum}\{X_{m_n}, m_n \in \nu_p^{(r)}\}] Q_r,$$

where P_r and Q_r are permutation matrices, and the ordered set $\nu^{(r)} = \{\nu_1^{(r)}, \ldots, \nu_{p_n}^{(r)}\}$ are the indecomposable partitions with respect to the original partition

$$\nu^* = \{\ell_1, \ell_2, \ldots, \ell_{k_1} | \ldots | \ell_1, \ell_2, \ldots, \ell_{k_L}\}.$$

As a consequence of Definition 2, we have the following which is used to obtain a relation for the cumulant spectra of the linear process $\{\mathbf{X}_t\}$.

Let $\{\mathbf{X}_t\}$ be a $d \times 1$ stationary (up to kth order) vector time series. Let us denote the kth order scalar cumulant cum$\{X_{i_1 t_2}, X_{i_2 t_2}, \ldots, X_{i_k t_k}\}$ by $k_{i_1 i_2 \ldots i_k}(t_1 - t_k, t_2 - t_k, \ldots, t_{k-1} - t_k)$. Then we have the kth order cumulant vector (of order $d^k \times 1$) as

$$\text{cum}(\mathbf{X}_{t_1}, \mathbf{X}_{t_2}, \ldots, \mathbf{X}_{t_k}) = K(t_1 - t_k, \ldots, t_2 - t_k, \ldots, t_{k-1} - t_k)$$
$$= [k_{i_1 i_2 \ldots i_k}(t_1 - t_k, \ldots, t_{k-1} - t_k)],$$

where the subscripts $i_1 i_2 \ldots i_k$ indicate the position in the column vector.

Let $\{\mathbf{X}_t\}$ be a d-dimensional stationary time series. We say $\mathbf{X}_t$ is linear if it can be represented as

$$\mathbf{X}_t = \sum_{j=-\infty}^{\infty} \mathbf{A}_j \mathbf{e}_{t-j}, \qquad (6.4)$$

where $\{\mathbf{e}_t\}$ is a sequence of independent, identically distributed random vectors such that

$$\text{cum}(\mathbf{e}_t) = 0,$$

$$\text{cum}(\mathbf{e}_{t_1}, \mathbf{e}_{t_2}, \ldots, \mathbf{e}_{t_n}) = \begin{cases} \mathbf{c}_n, & \text{if } t_1 = t_2 = \cdots = t_n, \\ 0, & \text{otherwise,} \end{cases}$$

where $\mathbf{c}_n$ is a column vector of order $d^n \times 1$. We can show the following result (see Wong, 1993).

LEMMA. *If* $\mathbf{X}_t$ *satisfies* (6.4) *then*,

(1) $K(s_1, s_2, \ldots, s_{k-1}) = \sum_j \{A_{j+s_1} \otimes A_{j+s_2} \otimes \cdots \otimes A_j\} \mathbf{c}_k$,
(2) *the* kth *order spectral vector is given by*

$$f_k(\omega_1, \omega_2, \ldots, \omega_{k-1}) = \frac{1}{(2\pi)^{k-1}} \{H(\omega_1) \otimes H(\omega_2) \otimes \cdots \otimes H(\omega_k)\} \mathbf{c}_k, \quad (6.5)$$

where

$$\omega_k = -\sum_{j=1}^{k-1} \omega_j, \quad H(\omega) = \sum A_j e^{-ij\omega}.$$

The above result (6.5) *is a multivariate generalisation of* (2.3).

The multivariate bilinear models were introduced by Subba Rao (1985), Stensholt and Tjostheim (1987), Stensholt (1989), Liu (1989) and the detailed statistical analysis was given by Wong (1993) and Subba Rao and Wong (1998). We briefly discuss these results here.

Let $\mathbf{X}_t$ be a d-dimensional random vector with elements $(X_{1t}, X_{2t}, \ldots, X_{dt})$. Let $\mathbf{e}'_t = (e_{1t}, e_{2t}, \ldots, e_{dt})$ where $\{\mathbf{e}_t\}$ is a sequence of independent random vectors with $E(\mathbf{e}_t) = 0$ and $E(\mathbf{e}_t \mathbf{e}'_t) = \Sigma$. Let the ith component X_{it} satisfy the difference equation

$$X_{it} = \sum_{\ell=1}^{p_1} \sum_{j=1}^{d} a_{ij\ell} X_{j,t-\ell} + \sum_{\ell=1}^{q_1} \sum_{j=1}^{d} c_{ij\ell} e_{j,t-\ell}$$

$$+ \sum_{m=1}^{p_2} \sum_{n=1}^{q_2} \sum_{j,v=1}^{d} b_{ijmnv} X_{j,t-m} e_{v,t-n} + e_{it} \quad (i=1,2,\ldots,d), \quad (6.6)$$

or, equivalently, in matrix notation, we have

$$\mathbf{X}_t = \sum_{\ell=1}^{p_1} A_\ell \mathbf{X}_{t-\ell} + \sum_{\ell=1}^{q_1} c_{\ell v} e^v_{t-\ell} + \sum_{m=1}^{p_2} \sum_{n=1}^{q_2} B_{mnv} \mathbf{X}_{t-m} e_{t-n} + D_v e^v_t, \quad (6.7)$$

where $A_\ell = [a_{ij\ell}]$, $B_{mnv} = [b_{ijmnv}]$ are $d \times d$ matrices, $c_{\ell v} = [c_{i\ell v}]$ is a $d \times 1$ vector and D_v is the vth column of $d \times d$ identity matrix. Here $c_{\ell v} e^v_{t-\ell}$ stands for $\sum c_{\ell v} e_{v,t-\ell}$, $D_v e^v_t$ for $\sum D_v e_{v,t}$, etc. The above model is defined as $MDL(d; p_1, q_1, p_2, q_2)$. Some properties of the model $MDL(d; p, 0, p, 1)$ were investigated by Subba Rao and Wong (1998).

As in the univariate case, one can obtain higher-order cumulants of the vector time series $\{\mathbf{X}_t\}$ and also we can show that they satisfy the Yule–Walker type of difference equations.

We now briefly consider the estimation of the parameters of the model (6.6). As in the univariate time series case, we can consider both time domain and frequency domain approaches for estimation of the parameters. The time domain approach of estimation was considered by Stensholt (1989) and also discussed in an unpublished technical report by Stensholt and Subba Rao (1987).

Let $\{\mathbf{X}_1, \mathbf{X}_2, \ldots, \mathbf{X}_n\}$ be a sample for the d-dimensional time series $\{\mathbf{X}_t\}$ generated from the following model.

$$\mathbf{X}_t = \boldsymbol{\mu} + \sum_{\ell=1}^{p_1} A_\ell \mathbf{X}_{t-\ell} + \sum_{n=1}^{q_2} \sum_{m \geqslant n}^{p_2} B_{mnv} \mathbf{X}_{t-m} e^v_{t-n} + \mathbf{e}_t. \tag{6.8}$$

The above model (6.8) is called the lower triangular bilinear model. Here we assume that $\{\mathbf{e}_t\}$ are i.i.d. random vectors and each $\mathbf{e}_t$ is multivariate normal with mean zero and variance covariance matrix $\boldsymbol{\Sigma}$. Let

$$J(\omega) = \frac{1}{\sqrt{2\pi n}} \sum_{t=1}^n \mathbf{X}_t e^{it\omega},$$

and $I(\omega) = J(\omega) J^*(\omega)$ be the periodogram matrix. Let $\mathbf{f}(\omega)$ be the second-order spectral matrix.

The parameters can be estimated by the minimisation of the following criteria

(1) $\sum_{t=1}^n \mathbf{e}'_t \boldsymbol{\Sigma}^{-1} \mathbf{e}_t,$

(2) $\sum \mathbf{e}'_t \mathbf{e}_t,$

(3) $\sum_{j=1}^{[n/2]} \left[\ell n |\mathbf{f}(\omega_j)| + \mathrm{Trace}\big(\mathbf{f}^{-1}(\omega_j) I(\omega_j)\big) \right]; \quad \omega_j = \frac{2\pi j}{n}.$

Stensholt (1989) has shown that the weighted and unweighted sum of squares, given by (1) and (2), lead to the same first-order normal equations. We refer to her Ph.D. thesis for details. Subba Rao and Wong (1998) have used frequency domain criterion (3) for estimating the parameters of several real time series and also considered prediction using the estimated models.

The order of the model can be determined by using either Akaike's Information Criterion (AIC) or Bayesian Information Criterion (BIC) (see Subba Rao and Wong, 1998). Some comments on the estimation and prediction in the case of nonlinear models are in order.

(1) The estimation is very sensitive to the choice of the initial values of the parameters.

(2) In the case of estimation of parameters of time series models, it is usual to omit the first few observations (to correct it for transient effect). This procedure is reasonable for estimating the parameters in the case of linear models whereas for nonlinear time series the estimates can depend on the choice of the starting point and therefore care should be taken. The estimates are very sensitive to the choice of initial starting point.

(3) It is well known that the mean square prediction errors increase as the number of steps of prediction increases. This result can easily be derived for linear predictors (see Priestley, 1981). The same cannot be said about nonlinear predictors. We refer to an empirical observation made by Subba Rao and Gabr (1984, p. 206) in the case of bilinear models fitted to Canadian Lynx data. The estimated mean square error for six steps is smaller than four and five steps prediction errors. We have no theoretical results, but we believe this is true, in general, for all nonlinear models. It is interesting to investigate this phenomenon.

7. Spurious regression and cointegration, nonlinearity

In an interesting study conducted by Granger and Newbold (1974), they have shown that we can observe a spurious correlation between two independent time series y_t and x_t, where y_t and x_t are both nonstationary (and by differencing they can be made stationary). Briefly their analysis is as follows (see Mills, 1999; Maddala and I.-M. Kim, 1998). Suppose both y_t and x_t satisfy the following difference equations.

$$y_t = y_{t-1} + v_t,$$
$$x_t = x_{t-1} + u_t, \qquad (7.1)$$

where $\{v_t\}$ and $\{u_t\}$ are independent, identically distributed random errors, $v_t \sim N(0, \sigma_v^2)$, and $u_t \sim N(0, \sigma_u^2)$. Suppose we consider the linear regression

$$y_t = \alpha + \beta x_t + \eta_t, \qquad (7.2)$$

where $\{\eta_t\}$ are independent, identically distributed random errors. They have shown that about 75% of the times we would reject the null hypothesis that $\beta = 0$ implying that there is a linear relationship between y_t and x_t even though they are strictly independent.

In the following we are interested in conducting a similar study when y_t and x_t are nonlinear (nearly nonlinear, nearly nonstationary). These simulations have been performed by Miss Gillian Hughes, UMIST, and we are very thankful to her for allowing us to report these results.

Here we have generated two bilinear time series $\{x_t\}$ and $\{y_t\}$ where,

$$x_t = a_1 x_{t-1} + b_1 x_{t-1} e_{t-1} + e_t,$$
$$y_t = a_2 y_{t-1} + b_2 y_{t-1} \eta_{t-1} + \eta_t, \qquad (7.3)$$

Table 2
Table of acceptance of H_0

a_1	b_1	a_2	b_2	Acceptance of H_0 (%)
1	0	1	0	26
0.4	0.4	0.4	0.4	85
1	0.4	1	0.4	48
1	0.6	1	0.6	72
1	0.8	1	0.8	74
1	1	1	1	87
0	0.4	0	0.4	90
0	0.6	0	0.6	94
0	0.8	0	0.8	95
0	1	0	1	97

where $\{e_t\}$ and $\{\eta_t\}$ are independent, random errors each distributed normally with mean zero and variance one. The regression equation of the type (7.2) is fitted and the test statistic is constructed based on the t-distribution (sample size chosen is 100). The percentage of times the null hypothesis is accepted is given in the last column of the table. These values are calculated for various values of (a_1, a_2, b_1, b_2) and are given in the following table. We note when $a_1 = a_2 = 1$, $b_1 = b_2 = 0$, the results are similar to the results of Granger and Newbold (1974). About 74% times the null hypothesis is rejected.

Let us look at the results when $a_1 = a_2 = 1$, and other values of b_1 and b_2. In this case as the nonlinearity is increasing remarkably the percentage of acceptance is also increasing, unlike the linear case. This is an interesting result indicating that nonlinearity has a stabilising effect. This phenomenon is similar to one observed in the nonlinear control theory literature, where nonlinear feedback makes an unstable system stable. For this choice of parameters, the series x_t and y_t are both nonlinear and nonstationary. When $a_1 = a_2 = 0$ and b_1 and b_2 are increasing, the percentage of acceptance is also increasing. In other words, nonlinearity in the series seems to lead to the correct conclusion.

In another experiment we generated x_t and y_t using the relations

$$x_t = x_{t-1} + e_t + ae_{t-1}e_{t-2},$$
$$y_t = y_{t-1} + \eta_t + b\eta_{t-1}\eta_{t-2}, \qquad (7.4)$$

where $\{e_t\}$ and $\{\eta_t\}$ are independent, identically distributed as in (7.3). The results are summarized in Table 3.

The conclusions are similar to Granger and Newbold (1974). The interesting aspect of the model (7.4) is that both series $\{x_t\}$, $\{y_t\}$ in (7.4) are nonstationary, but their respective first differences are stationary, but nonlinear. The nonlinear parts ($e_t + ae_{t-1}e_{t-2}$), ($\eta_t + b\eta_{t-1}\eta_{t-2}$) behave like white noise (uncorrelated) sequences, though some of their higher-order moments are non-zero. The important observation in the above analysis is that before we try to relate two or more time series, it is useful to

Table 3

a	b	Acceptance H_0
0.4	0.4	18
0.8	0.8	22
1	1	30
1	0	22
0	1	24

check whether they are linear or not, and higher-order moments need to be calculated to see whether the series under consideration are uncorrelated or independent. This is important as the above calculations show and it is rather unfortunate this preliminary analysis seems to have been neglected in the economics literature.

Recently, Subba Rao (1997) has introduced unit root bilinear models. Here the time series $\{x_t\}$ is nonstationary, but by differencing it is stationary but nonlinear. For illustration purposes, let us assume the series $\{x_t\}$ is nonstationary, but the first differences are stationary (i.e.) let $x_t - x_{t-1} = Z_t$, and let Z_t satisfy the bilinear difference equation

$$Z_t = aZ_{t-1} + bZ_{t-1}e_{t-1} + e_t, \tag{7.5}$$

where a and b satisfy the usual stationarity conditions. The series $\{Z_t\}$ is second-order stationary. By substituting $Z_t = x_t - x_{t-1}$, we can rewrite the above model as

$$x_t - (1+a)x_{t-1} + ax_{t-2} = bx_{t-1}e_{t-1} - bx_{t-2}e_{t-1} + e_t. \tag{7.6}$$

Consider the characteristic equation associated with the AR part of the model (7.6), namely $Q(Z) = Z^2 - (1+a)Z + a$, and the roots of the equation are 1 and a. In other words whatever may be the value of a, there is one root on the unit circle and hence Subba Rao (1997) called (7.6) the unit root bilinear model. These nonstationary, nonlinear models are like ARIMA models for linear time series. So far there is no estimation methodology, nor any detailed statistical analysis of the above class of models. We hope these will be considered in the future.

We generated two series $\{x_t\}$ and $\{w_t\}$, where

$$x_t - (1+a_1)x_{t-1} + a_1 x_{t-2} = b_1 x_{t-1}e_{t-1} - b_1 x_{t-2}e_{t-1} + e_t \tag{7.7}$$

and

$$w_t - (1+a_2)w_{t-1} + a_2 w_{t-2} = b_2 w_{t-1}\eta_{t-1} - b_2 w_{t-2}\eta_{t-1} + \eta_t, \tag{7.8}$$

where $\{e_t\}$ and $\{\eta_t\}$ are generated as before.

The series $\{x_t\}$ and $\{w_t\}$ are generated for various values of a_1, b_1, a_2, b_2 and for each pair of series (η_t, w_t), a model of the form

$$x_t = \alpha + \beta w_t + \zeta_t \tag{7.9}$$

Table 4

a_1	b_1	a_2	b_2	Accept H_0 (5% sig. level)
0.4	0	0.4	0	26
1	0	1	0	8
1	1	1	1	9
1	0.4	1	0.4	0
1	0.8	1	0.8	4
0	0.4	0	0.4	1
0	1	0	1	1
0.4	0.4	0.4	0.4	1
0.8	0.8	0.8	0.8	2
1	1	1	1	15

is fitted by ordinary least squares. The results are summarized in Table 4. In (7.9) we assume $\{\zeta_t\}$ are independent.

The results in the first row are exactly similar to Granger and Newbold. As the nonlinearity increases (by the increase of the parameters b_1 and b_2) we see that there is an increase in percentage of times we observe spurious relations between x_t and w_t. This again confirms earlier results that one must check whether there is any nonlinearity within the series before we ascertain any possible linear relationship (or possibly nonlinear relationship) between the series.

8. Time dependent nonlinear models

In the previous section we considered modelling of nonstationary bilinear models, but the nonstationarity we considered is of a special kind which is very common in economics/financial literature. Here the series can be made stationary by appropriate differencing. In the following section we consider series whose statistical characteristics such as variances, covariances and higher-order moments depend on time but they are finite for all t. This type of nonstationarity was considered by Priestley (1965), Priestley and Gabr (1993), Subba Rao (1970). We follow Priestley's definition of nonstationarity and consider his evolutionary process representation. Let $\{X_t\}$ be a zero mean, third-order nonstationary process and admitting the representation

$$X_t = \int_{-\pi}^{\pi} e^{it\omega} A_t(\omega) \, dZ(\omega) \tag{8.1}$$

where $\{dZ(\omega)\}$ is an orthogonal random process, and for each ω, $A_t(\omega)$ is a deterministic function of ω. We assume further that $A_t(\omega)$ does not oscillate too fast (i.e.) $A_t(\omega) = \int e^{it\theta} \, dk_\omega(\theta)$ where $|dk_\omega(\theta)|$ has an absolute maximum at $\theta = 0$. Priestley (1965) defined the evolutionary spectrum,

$$f_t(\omega) \, d\omega = |A_t(\omega)|^2 \, d\mu(\omega) \tag{8.2}$$

and bispectrum (Priestley and Gabr, 1993) as

$$f_t(\omega_1, \omega_2)\,d\omega_1\,d\omega_2 = A_t(\omega_1)A_t(\omega_2)A_t(-\omega_1-\omega_2)\,d\mu(\omega_1,\omega_2),$$

where $d\mu(\omega) = E|dZ(\omega)|^2$, $d\mu(\omega_1,\omega_2) = E(dZ(\omega_1)\,dZ(\omega_2)\,dZ(\omega_3))$ where $\omega_1 + \omega_2 + \omega_3 = 0 \pmod{2\pi}$.

The estimation of $f_t(\omega)$ and $f_t(\omega_1, \omega_2)$ have been considered by Priestley (1965) and Priestley and Gabr (1993). If the series is Gaussian, we can see $f_t(\omega_1, \omega_2) = 0$ for all ω_1, ω_2 and all values of t. We can also show that if the series is linear, then

$$\frac{|f_t(\omega_1,\omega_2)|^2}{f_t(\omega_1)f_t(\omega_2)f_t(\omega_1+\omega_2)}$$

is a constant, i.e., does not depended on w_1 and w_2.

This can be used for testing for linearity of the nonstationary time series (see for details Ph.D. thesis of Eleni Tsolaki, 2001).

The estimation of time dependent parameters of the models of the form

$$X_t + a_1(t)X_{t-1} + \cdots + a_p(t)X_{t-p} = e_t,$$

$$X_t + a_1(t)X_{t-1} + \cdots + a_p(t)X_{t-p}$$
$$= e_t + b_1(t)e_{t-1} + \cdots + b_q(t)e_{t-q} \tag{8.3}$$

have been considered by Subba Rao (1970), Hussain and Subba Rao (1976) and later Grenier (1983), Dahlhaus (1995) and many others. The estimation technique is based on weighted least squares approach, and Subba Rao (1970) has defined a weighted likelihood function which, in the Gaussian errors case, results in estimates similar to the weighted least squares. In recent years "local log-likelihood" function is used for statistical inference when the observations are not necessarily identically distributed, and the definition of local log-likelihood function is exactly the same as the weighted likelihood function defined much earlier by Subba Rao (1970). Our interest here is to define nonlinear, nonstationary models similar to stationary bilinear models defined earlier. The first-order time dependent bilinear model (TBL(1, 0, 1, 1)) is defined as (Subba Rao, 1997)

$$X_t = a(t)X_{t-1} + b(t)X_{t-1}e_{t-1} + e_t, \tag{8.4}$$

where $\{e_t\}$ is defined as earlier. We can write the solution of the above equation in the form of Volterra expansion with time dependent coefficients. The estimation of the parameters of these models and possible applications to real time series are some of the problems for future consideration. One can also estimate the parameters using wavelet methods (Subba Rao, 1997).

So far we have considered nonlinear time series which are defined in the literature as short memory series. In the following we discuss nonlinear processes whose higher-order cumulants do not satisfy the absolute summability condition and, we define such processes as nonlinear, long range memory series.

9. Long range dependence

It is a well known fact that for an i.i.d. series $X_1, X_2, \ldots, X_n$ with mean *zero* and variance $\text{var}(X_k) = \sigma_X^2$ the variance of the sum of X_k's changes linearly with n, i.e.,

$$\text{var}\left(\sum_{k=1}^{n} X_k\right) = n\sigma_X^2.$$

Asymptotically the above result holds even when $\{X_t\}$ is a stationary time series. Let $\{X_t\}$ be a zero mean stationary time series with spectral density function $f(\omega)$. Then we have

$$\text{var}\left(\sum_{t=1}^{n} X_t\right) = n \sum_{k=-(n-1)}^{n-1} \left(1 - \frac{|k|}{n}\right) \text{cov}(X_t, X_{t+k})$$
$$\simeq nf(0),$$

because by Cesaro summation

$$\lim_{n \to \infty} \sum_{k=-(n-1)}^{n-1} \left(1 - \frac{|k|}{n}\right) \text{cov}(X_t, X_{t+k}) = \sum_{k=-\infty}^{\infty} \text{cov}(X_t, X_{t+k})$$
$$= f(0).$$

Now if one estimates the variance of sums of a time series for different n (usually called as aggregated series) then it is expected that the logarithm of these variances plotted against $\log(n)$ be linear with slope 1. There are several phenomena where it has been found that the slope is different from one, usually it is greater than 1, i.e.,

$$\log \text{var}\left(\sum_{k=1}^{n} X_k\right) \simeq 2H \log(n) + \text{const.},$$

where $H \in (1/2, 1)$. The coefficient H is called the Hurst coefficient, the name goes back to Hurst (1951). This property implies that the classical central limit theorems do not hold any more for these types of processes. This property follows from the behavior of the spectrum at zero, i.e., the behavior of the series $\sum_{k=-\infty}^{\infty} \text{cov}(X_t, X_{t+k})$. A stationary time series $X_t, t = 0, \pm 1, \pm 2, \ldots, \pm n$, will be called *long range dependent* if its spectrum $f(\omega)$ behaves like $|\omega|^{-2h}$ at zero, more precisely

$$\lim_{\omega \to 0} \frac{f(\omega)}{|\omega|^{-2h}} = \text{const.}, \quad h \in (0, 1/2), \tag{9.1}$$

where $h = H - 1/2$. This definition of long range dependence can be stated in terms of autocorrelation function, because (9.1) is equivalent to:

$$\text{cov}(X_t, X_{t+k}) = \sigma_X^2 \, \text{corr}(X_t, X_{t+k}) \simeq k^{2h-1}. \tag{9.2}$$

In other words, the autocorrelation function decays hyperbolically. There are several ways of modelling long range dependence. One way is the fractional Gaussian noise process and the other one is through the discretized fractional Brownian motion. These approaches will be discussed below.

9.1. Fractional (Gaussian) processes

9.1.1. Fractional Gaussian noise process
Define the fractional Gaussian noise process as

$$e_t = (1 - B)^{-h} w_t,$$

where B is the backwardshift operator, i.e., $Bw_t = w_{t-1}$ and w_t is i.i.d. Gaussian white noise process. Now, if $|h| < 1/2$, then e_t is a Gaussian stationary process having both moving average and infinite order AR representations, see Hosking (1981) and Granger and Joyeux (1980). If $0 < h < 1/2$ then e_t is a long range dependent process in the sense that its autocorrelation function decreases hyperbolically (instead of the usual exponential rate), i.e., for large k

$$\text{cov}(e_t, e_{t+k}) = \sigma_e^2 \, \text{corr}(e_t, e_{t+k}) \simeq k^{2h-1}. \tag{9.3}$$

The property (9.3) implies and is implied by the fact that the spectral density of e_t is hyperbolic for low frequencies, i.e.,

$$f_e(z) \simeq \omega^{-2h}, \quad \omega \to 0, \; z = e^{i2\pi\omega}.$$

The exact expressions for the correlation and spectral density are

$$\text{corr}(e_t, e_{t+k}) = \frac{(-1)^k \Gamma^2(1-h)}{\Gamma(1-h+k)\Gamma(1-h-k)},$$

$$\sigma_e^2 = \sigma^2 \frac{\Gamma(1-2h)}{\Gamma^2(1-h)},$$

$$f_e(z) = \sigma^2 |1 - z^{-1}|^{-2h}.$$

The other candidate for modelling long range dependence is the discretized fractional Brownian motion, see Taqqu (1979).

9.1.2. Discretized fractional Brownian motion
The definition of the fractional Brownian motion with parameter $h \in (-1/2, 1/2)$ (Mandelbrot and Van Ness, 1968) is the following:

$$w_t^{(h)} \doteq \frac{1}{\Gamma(1+h)}\left\{\int_{-\infty}^{0}\left[(t-s)^h - (-s)^h\right]\mathrm{d}w_s + \int_0^t (t-s)^h\,\mathrm{d}w_s\right\},$$

$t \in \mathbb{R}$. Formally $w_t^{(h)}$ can be expressed with the help of the fractional integral operator $I_x^{(h)}(f)$ defined as

$$I_x^{(h)}(f) = \frac{1}{\Gamma(h)}\int_{-\infty}^{x}(x-y)^{h-1}f(y)\,\mathrm{d}y,$$

namely, the hth fractional integral process of the Brownian motion, adjusted to zero at zero, i.e.,

$$\begin{aligned}
w_t^{(h)} &= \frac{1}{\Gamma(1+h)}\int_{-\infty}^{t}(t-s)^{(1+h)-1}w_s'\,\mathrm{d}s \\
&\quad - \frac{1}{\Gamma(1+h)}\int_{-\infty}^{0}(-s)^{(1+h)-1}w_s'\,\mathrm{d}s \\
&= I_t^{(1+h)}(w') - I_0^{(1+h)}(w') \\
&= I_t^{(h)}(w) - I_0^{(h)}(w).
\end{aligned}$$

Clearly, $w_t^{(0)} = w_t$.

We shall consider only the case $0 < h < 1/2$ because of the following reasons. First, the long memory or long-range dependence property that we are interested in appears only for positive values of h. It is widely known from experience that fractional Brownian motion with a negative h hardly occurs in practice. Furthermore, the latter case would need a different methodology. Therefore, we shall assume in the sequel that $0 < h < 1/2$.

The most important properties of $w_t^{(h)}$ (see Mandelbrot and Van Ness, 1968) are the following.

- $w_0^{(h)} = 0$,
- $w_t^{(h)}$ is mean square continuous and continuous with probability 1,
- it has stationary increment processes,
- for any $t \in \mathbb{R}$, $w_t^{(h)}$ is not differentiable with probability 1,
- it is self similar with self similarity parameter $h + 1/2$, i.e., the vectors $(w_{ct_1}^{(h)}, \ldots, w_{ct_k}^{(h)})$ and $(|c|^{h+\frac{1}{2}}w_{t_1}^{(h)}, \ldots, |c|^{h+\frac{1}{2}}w_{t_k}^{(h)})$ have the same distribution,
- its first- and second-order moments are

$$\mathsf{E}w_t^{(h)} = 0,$$

$$\text{Cov}(w_t^{(h)}, w_s^{(h)}) = \frac{\kappa(h)}{2}(|t|^{2h+1} + |s|^{2h+1} - |t-s|^{2h+1}), \tag{9.4}$$

$$\text{Var}\, w_t^{(h)} = \kappa(h)|t|^{2h+1}, \tag{9.5}$$

where

$$\kappa(h) \doteq \frac{1}{2\pi} \int_{\mathbb{R}} \left|\frac{e^{i\omega}-1}{i\omega}\right|^2 |\omega|^{-2h}\, d\omega. \tag{9.6}$$

The discretized increment of fractional Brownian motion process $e_k = w_k^{(h)} - w_{k-1}^{(h)}$, $k = 0, \pm 1, \pm 2, \ldots$, is clearly a long range dependent series with Hurst exponent $H = h + 1/2$.

9.1.3. Fractional Gaussian ARMA series

The fractional differenced Gaussian ARMA process can be defined in two ways either through the fractional operator $(1-B)^{-h}$ on the Gaussian ARMA process or through the fractional differenced Gaussian noise process as the noise/input process of an ARMA model (see Beran, 1994 and more references therein). It is easy to see that both approaches leads to the same result. Let

$$e_t = (1-B)^{-h} w_t,$$

$$\alpha(B) X_t = \beta(B) e_t, \quad |h| < 1/2,$$

where α and β are polynomials with roots inside the unit circle. An equivalent long range dependent series X_t can be obtained from the discretized increment of fractional Brownian motion process e_t. The spectral representation of X_t is

$$X_t = \int_0^1 z^t \frac{\beta(z)}{\alpha(z)} (1-z)^{-h} W(d\omega).$$

The spectral density function of X_t is

$$f_X(z) = \sigma^2 \left|\frac{\beta(z)}{\alpha(z)}\right|^2 |1-z|^{-2h}.$$

Using the fact that $|\frac{\beta(z)}{\alpha(z)}|^2$ is bounded, we have the same hyperbolic property for the spectral density $f_X(z)$ at zero as the fractional differenced noise, i.e.,

$$f_X(z) \simeq \omega^{-2h}, \quad \omega \to 0,\ 0 < h < 1/2,$$

and therefore

$$\text{corr}(X_t, X_{t+k}) \simeq k^{2h-1}, \quad k \to \infty,\ 0 < h < 1/2.$$

We now turn our attention to the bilinear model with long range dependence property. This process will account for long range dependency in the nonlinear models.

9.2. Bilinear model with fractional differenced Gaussian noise

Let

$$X_t = aX_{t-1} + bX_{t-1}w_{t-1} + w_t + f_0. \tag{9.7}$$

One can generate the above type of series by either of the following approaches (see Igloi and Terdik, 1997).

(1) Applying the fractional operator $(1 - B)^{-h}$ on the solution of a bilinear equation (9.7) and
(2) using the bilinear model (9.7) with fractional differenced Gaussian noise.

In approach (1) the $\{X_t\}$ is generated from Eqs. (9.7) (we assume that the series is stationary up to third order). The process $\{y_t\}$ given by

$$y_t = (1 - B)^{-h} X_t, \quad |h| < 1/2,$$

is well defined and it will have the long range property if $0 < h < 1/2$.

The spectrum and the bispectrum of y_t is given in terms of the spectrum $f_X(z)$ and the bispectrum $f_{XX}(z_1, z_2)$ of X_t, i.e.,

$$f_y(z) = |1 - z|^{-2h} f_X(z),$$
$$f_{yy}(z_1, z_2) = \left(1 - z_1^{-1}\right)^{-h} \left(1 - z_2^{-1}\right)^{-h} (1 - z_1 z_2)^{-h} f_{XX}(z_1, z_2),$$

where $z = e^{i2\pi\omega}$, $z_j = e^{i2\pi\omega_j}$. Note that the spectrum of X_t is bounded, i.e.,

$$0 < c_1 < f_X(z) < c_2 < \infty$$

therefore $f_y(z)$ behaves like the spectral density of a fractional differenced series and is not Gaussian.

The bispectrum $f_{yy}(z_1, z_2)$ (where $z_1 = e^{2\pi i w_1}$, $z_2 = e^{2\pi i w_2}$) has singularity on the lines $\omega_1 = 0$ and $\omega_1 + \omega_2 = 0$. The higher degree of singularity is at $\omega_1 = \omega_2 = 0$ which is the dominant one. So if $\omega_1, \omega_2 \to 0$ we have

$$f_{yy}(z_1, z_2) \simeq \omega_1^{-h} \omega_2^{-h} (\omega_1 + \omega_2)^{-h}. \tag{9.8}$$

To study the behavior of the third-order cumulants consider

$$\text{cum}(y_t, y_{t+k}, y_{t+l})$$

$$\simeq \int\int_0^1 \left(1 - z_1^{-1}\right)^{-h} \left(1 - z_2^{-1}\right)^{-h} (1 - z_1 z_2)^{-h} z_1^k z_2^l \, \mathrm{d}(w_1) \, \mathrm{d}(\omega_2)$$

$$\simeq \sum_{n=0}^{\infty} \frac{(h)_n}{n!} (-h) \frac{(h)_{k+n}}{(-h)_{k+n+1}} (-h) \frac{(h)_{l+n}}{(-h)_{l+n+1}}$$

$$\cong k^{2h-1}l^{2h-1} \sum_{n=0}^{\infty} \frac{(h)_n}{n!} \frac{(h+k)_n}{(k+1)_n} \frac{(h+l)_n}{(l+1)_n}$$

$$\cong k^{2h-1}l^{2h-1}.$$

This property of the third-order cumulants shows the hyperbolic decay in both lags k and l and we see that $\Sigma\Sigma|\text{cum}(y_t, y_{t+k}, y_{t+l})| = \infty$ for $0 < h < 1/2$, $0 < k < 1/2$.

The situation (2) when the input/noise process of the bilinear model is fractional differenced Gaussian noise is more complicated. Let us consider the following example to illustrate the problems involved.

9.2.1. Symmetric bilinear model with fractional white noise input
To illustrate the ideas we consider a special case of (9.7)

$$X_t = dX_{t-1}e_{t-1} + e_t, \tag{9.9}$$

where e_t is fractional Gaussian white noise with parameter $h \in (0, 1/2)$, i.e., $e_t = (1-B)^{-h}w_t$, $0 < h < 1/2$.

The solution of the above Eq. (9.9) is given by

$$X_t = e_t + \sum_{k=1}^{\infty} d^k e_{t-k}^2 \prod_{j=0}^{k-1} e_{t-j}.$$

For every $d \neq 0$, the expected value of X_t is infinite, more precisely

$$\lim_{n\to\infty} E\left(e_t + \sum_{k=1}^{n} d^k e_{t-k}^2 \prod_{j=0}^{k-1} e_{t-j}\right) = \infty.$$

The proof and details are in Terdik (1999). This example shows that if we consider a scalar bilinear process of the form given above with fractional Gaussian noise, we see that no stationary solution exists. This problem is caused by the fact that the Hermite degree of the scalar bilinear process is always infinite. It has been shown that bilinear realizable process, i.e., the solutions of the multiple bilinear equation might have finite Hermite degree.

9.2.2. Linear model with Hermite degree n
A particular example for long range dependent nonlinear model is the homogeneous Hermite degree n linear model with transfer function

$$G_n(z_{(n)}) = \frac{\beta(z_{(n)}^{t1})}{\alpha(z_{(n)}^{t1})},$$

where $z_{(n)}^{t1} = (e^{it2\pi\omega_1}, e^{it2\pi\omega_2}, \ldots, e^{it2\pi\omega_n})$, i.e., the output series $\{y_t\}$ can be written in the form

$$y_t = \int_{\mathcal{D}^n} z_{(n)}^{t1} \frac{\beta(z_{(n)}^{t1})}{\alpha(z_{(n)}^{t1})} \prod_{j=1}^{n} (1-z_j)^{-h} W(d\omega_{(n)}),$$

or alternatively we can write y_t as

$$\alpha(B) y_t = \beta(B) H_n(e_t), \tag{9.10}$$

where $H_n(e_t)$ is the nth order Hermite polynomial in e_t with leading coefficient one and with spectral representation

$$H_n(e_t) = \int_{\mathcal{D}^n} z_{(n)}^{t1} \prod_{j=1}^{n} (1-z_j)^{-h} W(d\omega_{(n)}).$$

If $\frac{1-1}{2n} < h < \frac{1}{2}$ then y_t is obviously long range dependent with autocorrelation function

$$c_y(k) \simeq k^{n(2h-1)} \quad \text{as } k \to \infty,$$

and spectral density

$$f_y(z) = \sigma^{2n} n! \left|\frac{\beta(z)}{\alpha(z)}\right|^2 \int_{\mathcal{D}^{n-1}} \left|1 - z_{(n-1)}^1 z^{-1}\right|^{-2h} \prod_{j=1}^{n-1} |1-z_j|^{-2h} d\omega_{(n-1)}.$$

It is easy to see that

$$f_y(z) \simeq \omega^{-2nh+n-1}.$$

The assumption above, i.e.,

$$\frac{1}{2} - \frac{1}{2n} < h < \frac{1}{2},$$

is necessary and sufficient for a general homogeneous Hermite-degree n process given by (9.10) to be long range dependent unless its transfer function is bounded.

9.2.3. Bilinear realizable Hermite-degree 2 model with fractional differenced Gaussian noise

One can show (see Terdik, 1991), that under suitable assumptions on the second-order transfer function $G_2(z_1, z_1 z_2)$ a homogeneous Hermite degree 2 process

$$X_t^{(2)} = \int_0^1 \int_0^1 z_{(2)}^{t1} G_2(z_1, z_1 z_2) (1-z_1^{-1})^{-h} (1-z_2^{-1})^{-h} W(d\omega_{(2)})$$

is bilinear realisable. Suppose that G_2 is symmetric and bounded, i.e.,

$$0 < c_1 < |G_2(z_1, z_1 z_2)| < c_2 < \infty.$$

The second-order moments of $X_t^{(2)}$ are given by

(1) $\operatorname{corr}(X_t^{(2)}, X_{t+k}^{(2)}) \simeq k^{2(2h-1)}$ as $k \to \infty$.
(2) The spectral density function of $X_t^{(2)}$ is

$$f_X^{(2)}(z) = 2\sigma_w^4 \int_0^1 |G_2(z_1, z)|^2 |1 - z_1|^{-2h} |1 - zz_1^{-1}|^{-2h} d\omega_1$$

$$\simeq \omega^{1-4h} \quad \text{as } \omega \to 0.$$

Therefore $X_t^{(2)}$ is long range dependent iff

$$\frac{1}{4} < h < \frac{1}{2}.$$

More precisely

$$f^{(2)}(z) \simeq \omega^{1-4h} [e^{2(2h-1)} \Gamma(4h-1) \sin(\pi(1-2h))] \quad \text{as } \omega \to \infty.$$

The third-order moments of $\{X_t^{(2)}\}$ are given by

(1) The third order cumulant of $X_t^{(2)}$ is

$$c_3(k_1, k_2) = \operatorname{cum}(X_t^{(2)}, X_{t+k_1}^{(2)}, X_{t+k_2}^{(2)})$$

$$\simeq k_1^{(2h-1)} k_2^{(2h-1)} c_e(|k_1 - k_2|) \quad \text{as } k_1, k_2 \to \infty.$$

(2) The bispectrum of $X_t^{(2)}$ is given by

$$f^{(2)}(z_1, z_2)$$

$$= 8\sigma_W^6 \int_0^1 G_2(v, v^{-1} z_1) G_2(v z_1^{-1}, v^{-1} z_1 z_3) G_2(v z_1^{-1} z_3^{-1}, v^{-1}) d\lambda \bigg|_{\tilde{z}_{(3)}},$$

where $v = e^{i2\pi\lambda}$, $z_1 z_2 z_3 = 1$ and $|\tilde{z}_{(3)}$ denotes the symmetrization by the variables z_1, z_2, z_3.

$$f^{(2)}(z_1, z_2) \simeq \omega_1^{-2h} \omega_2^{-2h} \quad \text{as } \omega_1, \omega_2 \to 0.$$

For the general H-degree 2 bilinear realizable process

$$y_t = \int_0^1 z_1^t \frac{\beta_{11}(z_1)}{\alpha_{11}(z_1)} (1 - z_1^{-1})^{-h} W(d\omega_1)$$

$$+ \iint_0^1 (z_1 z_2)^t \frac{\gamma(z_1, z_1 z_2)}{\alpha_{21}(z_1) \alpha_{22}(z_1 z_2)} \left(1 - z_1^{-1}\right)^{-h} \left(1 - z_2^{-1}\right)^{-h} W(d\omega_{(2)}).$$

We define the equations

$$\alpha_{11}(B) X_t^{(1)} = \beta_{11}(z_1) e_t,$$

$$\alpha_{21}(B) X_t = \alpha_{22}(B) e_t,$$

$$X_t^{(2)} = \sum_{j=0}^{R} d_{k,j} X_{t-k-j} e_{t-j},$$

then

$$y_t = X_t^{(1)} + X_t^{(2)}.$$

The spectrum of y_t is the sum of the spectrums of $X_t^{(1)}$ and $X_t^{(2)}$ (note $X_t^{(1)}$ and $X_t^{(2)}$ are orthogonal)

$$f_y(z) = \sigma_e^2 \left| \frac{\beta_{11}(z)}{\alpha_{11}(z)} \right|^2 |1 - z|^{-2h}$$

$$+ 2\sigma_e^4 \int_0^1 \left| \frac{\gamma(z_1, z)}{\alpha_{11}(z_1) \alpha_{22}(z)} \right|^2 |1 - z_1|^{-2h} \left|1 - z_1^{-1} z\right|^{-2h} d\omega_1.$$

The autocorrelation function of y_t (as the lag k tends to infinity) is given by

$$\text{corr}(y_t, y_{t+k}) \simeq k^{2h-1} + k^{2(2h-1)}$$

and as expected the linear term is dominant when it is present (Taqqu, 1979).
The consequence of the above result is that the spectral density function is

$$f_y(z) \simeq \omega^{-2h} + \omega^{-4h+1} \quad \text{as } \omega \to 0,$$

i.e., there is singularity at the origin.

The third-order properties can also be obtained in a similar way. For example, the bispectrum is given by

$$f_{yy}(z_1, z_2)$$
$$= 6\sigma_e^4 \, \text{sym}_{z_{(3)}} \left(G_1^*(z_1) G_1^*(z_2) G_2^*(z_1^{-1}, z_2^{-1}) \right)$$
$$+ 8\sigma_e^6 \, \text{sym}_{z_{(3)}} \int_0^1 G_2^*(z_1^{-1} z, z_2^{-1} z^{-1}) G_2^*(z_1 z^{-1}, z) G_2^*(z^{-1}, z_2 z) \, d\omega,$$

where

$$G_1^*(z) = \frac{\beta_{11}(z)}{\alpha_{11}(z)}(1-z^{-1})^{-h},$$

$$G_2^*(z_1, z_2) = \frac{\gamma(z_1, z_1 z_2)}{\alpha_{21}(z_1)\alpha_{22}(z_1 z_2)}(1-z_1^{-1})^{-h}(1-z_2^{-1})^{-h}.$$

The f_{yy} has the same hyperbolic property at zero as (9.8), actually it is $\omega_1^{-2h}\omega_2^{-2h}$ as $\omega_1, \omega_2 \to 0$. Note here that the long range dependence is defined by the second-order structure of the time series under consideration and the third- or higher-order spectra could behave in principle differently. In our examples the Hurst coefficient of the bispectrum is the same as the spectrum.

10. Stationary bilinear process in continuous time

There is a growing interest of stochastic modelling in finance (Bjork, 1998). Among these models the continuous version of bilinear time series plays an important role. We consider below the continuous analog of the bilinear model and give two interpretations, one in the time domain and the other one in the frequency domain. In both cases we concentrate on statistical problems of estimating coefficients.

10.1. Time domain analysis

Throughout this section a single scalar valued input process will be considered. Let us start with the stochastic differential equation

$$dy_t = (\mu + \alpha y_t) dt + (\beta + \gamma y_t) dw_t, \qquad (10.1)$$

where w_t is a Brownian motion with variance σ^2. The σ^2 is set equal to 1 otherwise one can use the transformation w_t/σ, $\sigma\beta$, $\sigma\gamma$. Eq. (10.1) is a linear differential equation, nevertheless it is called bilinear in the system theory to distinguish between the situations when γ is zero and nonzero, i.e., when the solution is Gaussian and non-Gaussian, in other words when the model is linear and nonlinear. The Itô solution of Eq. (10.1) is well known.

$$y_t = e^{(\alpha-\frac{\gamma^2}{2})t+\gamma w_t}\left(y_0 + (\mu-\beta\gamma)\int_0^t e^{-(\alpha-\frac{\gamma^2}{2})s-\gamma w_s} ds\right.$$
$$\left. + \beta \int_0^t e^{-(\alpha-\frac{\gamma^2}{2})s-\gamma w_s} dw_s\right).$$

When $\gamma = 0$, $\alpha < 0$ and $\beta \neq 0$, this provides the stationary Gaussian Ornstein–Uhlenbeck process. We are interested in the stationary non-Gaussian solution of (10.1) therefore it is necessary to assume that $\mu^2 + \beta^2 > 0$, $\gamma \neq 0$ and not only $\alpha < 0$ but

$2\alpha + \gamma^2 < 0$ as well. The starting value y_0 is also well defined and then the stationary physically realizable solution of (10.1) is

$$y_t = (\mu - \beta\gamma) \int_{-\infty}^{t} e^{(\alpha - \frac{\gamma^2}{2})(t-s) + \gamma(w_t - w_s)} \, ds$$

$$+ \beta \int_{-\infty}^{t} e^{(\alpha - \frac{\gamma^2}{2})(t-s) + \gamma(w_t - w_s)} \, dw_s. \tag{10.2}$$

Note that $\mu\gamma \neq \alpha\beta$ must hold, otherwise (10.1) has only the degenerated solution $y_t = -\frac{\beta}{\gamma} = -\frac{\mu}{\alpha}$. From now on we shall assume that $\beta = 0$ and $\mu \neq 0$, i.e., there is no second term in (10.2). This assumption can be fulfilled by the following transformation

$$\tilde{y}_t = \frac{\mu}{\gamma\mu - \alpha\beta} (\beta + \gamma y_t). \tag{10.3}$$

The stochastic differential equation (10.1) without any loss of generality reduces to the form

$$dy_t = (\mu + \alpha y_t) \, dt + \gamma y_t \, dw_t, \tag{10.4}$$

and this equation will be the subject of our investigation.

Unfortunately the solution of (10.4)

$$y_t = \mu \int_{-\infty}^{t} e^{(\alpha - \frac{\gamma^2}{2})(t-s) + \gamma(w_t - w_s)} \, ds,$$

is neither a standardized diffusion process nor a martingale. Our object is to get results with possible statistical applications. We assume γ^2 is known

Suppose we set

$$\gamma^2 = \frac{\langle y \rangle_t}{\int_0^t y_t^2 \, dt},$$

where $\langle y \rangle_t$ is the quadratic variation of y_t, and make the transformation

$$z_t = \frac{1}{\gamma} \log(y_t),$$

where y_t satisfies the differential equation (10.4).

The Itô formula gives the following differential equation for z_t

$$dz_t = \frac{1}{\gamma} \left(\mu \exp(-\gamma z_t) + \alpha - \frac{\gamma^2}{2} \right) dt + dw_t.$$

The Kolmogorov forward equation for the stationary marginal density $p(z)$ of z_t is

$$\frac{2}{\gamma}\frac{\partial}{\partial z}\left[\left(\mu\exp(-\gamma z)+\alpha-\frac{\gamma^2}{2}\right)p(z)\right]=\frac{\partial^2}{\partial z^2}p(z). \tag{10.5}$$

The unique solution of (10.5) among the density functions concentrated on the real line is

$$p(z)=\exp\left\{\frac{1}{\gamma}[2\mu\exp(-\gamma z)-(2\alpha-\gamma^2)z+\text{const}]\right\}.$$

If we use the transformation $y=\exp(-\gamma z)$, we obtain the density of y_t^{-1} and it is Gamma with parameters $1-\frac{2\alpha}{\gamma^2}$ and $\frac{2\mu}{\gamma^2}$. As the density of z_0 is known we can apply the results of absolute continuity of the standardized diffusion processes according to the Brownian motion for the process $z_t - z_0$. The conditional Radon–Nikodym derivative of z_t (when z_0 is given) is

$$p(z_s,\ s\in[0,t])=\exp\left\{-\frac{1}{\gamma^2}\Big[\mu\big(\exp\{-\gamma[z_t-z_0]\}-1\big)\right.$$
$$-\gamma\left(\alpha-\frac{\gamma^2}{2}\right)[z_s-z_0]+\frac{1}{2}\int_0^t\gamma^2\mu\exp(-\gamma[z_s-z_0])$$
$$\left.+\left(\mu\exp(-\gamma[z_s-z_0])+\alpha-\frac{\gamma^2}{2}\right)^2\mathrm{d}s\Big]\right\}.$$

The maximum likelihood estimates of the parameters (μ,α) can be evaluated as follows. Let

$$I_1(t)=\int_0^t\exp(-\gamma[z_s-z_0])\,\mathrm{d}s,$$

$$I_2(t)=\int_0^t\exp(-2\gamma[z_s-z_0])\,\mathrm{d}s.$$

Then the maximum likelihood estimates of the parameters (μ,α) are given by

$$\hat\mu=\frac{1-\exp(-\gamma[z_t-z_0])+\left(\frac{\gamma^2}{2}-\frac{\gamma[z_t-z_0]}{t}\right)I_1(t)}{I_2(t)-I_1(t)^2},$$

$$\hat\alpha=\frac{\gamma^2}{2}-\frac{\hat\mu}{t}I_1(t)+\frac{\gamma[z_t-z_0]}{t}$$

$$=\frac{\gamma^2}{2}+\frac{\gamma[z_t-z_0]}{t}-\frac{1-\exp(-\gamma[z_t-z_0])+\left(\frac{\gamma^2}{2}-\frac{\gamma[z_t-z_0]}{t}\right)I_1(t)}{I_2(t)-I_1(t)^2}\frac{I_1(t)}{t}.$$

10.2. Frequency domain analysis

In this subsection we give explicit formulae for the spectrum and the bispectrum as these are necessary for the identification of the bilinear process in frequency domain. Suppose that there exists a stationary physically realizable solution Eq. (10.4) which is subordinated to the integrated process w_t, see Terdik (1999) for details. The frequency domain representation theorem says (see Major, 1981) that all such solutions can be changed into the so called chaotic spectral representation form

$$y_t = \sum_{k=0}^{\infty} \int_{\mathbb{R}^k} \exp(it \Sigma \omega_{(k)}) F_k(\omega_{(k)}) W(d\omega_{(k)}), \tag{10.6}$$

where $\omega_{(k)} = (\omega_1, \omega_2, \ldots, \omega_k)$, $\Sigma \omega_{(k)} = \sum_{j=1}^{k} \omega_j$ and $W(d\omega_{(k)})$ is the k-dimensional multiple Wiener–Itô spectral measure according to the Wiener process w_t. The representation (10.6) is unique up to the permutation of the variables of the transfer functions F_k. Now under the assumption that we are looking for the stationary Itô solution (10.2) of Eq. (10.4), i.e., putting by definition the principal value for the integral

$$\frac{1}{2\pi} \int_{\mathbb{R}} \frac{1 - e^{-it\omega}}{i\omega} d\omega \doteq 0,$$

it follows from the diagram formula that the following recursion is valid for the transfer functions F_k, see Terdik (1990).

$$F_0 = -\frac{\mu}{\alpha}, \qquad F_k(\omega_{(k)}) = \frac{\gamma F_{k-1}(\omega_{(k-1)})}{i \Sigma \omega_{(k)} - \alpha}, \quad k \geq 1. \tag{10.7}$$

The expectation of y_t can be calculated from (10.4) as

$$\mathsf{E} y_t = -\frac{\mu}{\alpha}.$$

One can easily get the autocovariance function $R(t) = \mathsf{E}(y_t - \mathsf{E} y_t)(y_0 - \mathsf{E} y_0)$ of y_t directly from (10.4) if $t \neq 0$ and the variance $R(0)$ from (10.2),

$$R(t) = R(0) e^{\alpha t} = \frac{-\mu^2 \gamma^2}{\alpha^2 (2\alpha + \gamma^2)} e^{\alpha t}, \quad t > 0,$$

$$R(t) = R(-t), \quad t \leq 0. \tag{10.8}$$

The spectrum of y_t can be calculated from the autocovariance function (10.8), i.e.,

$$f_2(\omega) = \frac{1}{2\pi} \int_{\mathbb{R}} e^{-it\omega} R(t) \, dt = -\frac{\alpha R(0)}{\pi |i\omega - \alpha|^2}, \quad \omega \in \mathbb{R}.$$

It should be noted that there is no difference between the spectrum of an Ornstein–Uhlenbeck process and $f_2(\omega)$ of the above process, therefore it is necessary to consider higher-order spectra for bilinear processes. We will obtain an expression for the bispectrum $f_{yy}(e_1, z_2)$ of the process y_t. The techniques used are the same as in the discrete time case. First consider the quadratic process $q_t = y_t^2$. It satisfies the differential equation

$$dq_t = 2\mu y_t + (2\alpha + \gamma^2) q_t \, dt + 2\gamma q_t \, dw_t,$$

with expectation

$$Eq_t = \frac{2\mu^2}{\alpha(2\alpha + \gamma^2)}.$$

The transfer functions of the chaotic spectral representation form, see (10.6), are given by the following recursion

$$G_0 = Eq_t, \qquad G_k(\omega_{(k)}) = \frac{2\mu F_k(\omega_{(k)}) + 2\gamma G_{k-1}(\omega_{(k-1)})}{i \Sigma \omega_{(k)} - (2\alpha + \gamma^2)}, \qquad k \geq 1.$$

Now we have to calculate the cross-spectrum $f_{qy}(w)$ between q_t and y_t. For this purpose we consider

$$\mathrm{Cum}(q_t, y_0) = \sum_{k=1}^{\infty} \int_{\mathbb{R}^k} \exp(it \Sigma \omega_{(k)}) \, \mathrm{sym}_{\omega_{(k)}} G_k(\omega_{(k)}) \, \mathrm{sym}_{\omega_{(k)}} \overline{F_k(\omega_{(k)})} \, d\omega_{(k)},$$

where $\mathrm{sym}_{\omega_{(k)}} F$ is the symmetrized version of F by the vector $\omega_{(k)}$ and it is the average of those values of F which are taken over all possible permutations of elements of $\omega_{(k)}$. We find that

$$f_{q,y}(\omega) = \frac{2\mu\gamma^2 Ey_t^2}{[i\omega - (2\alpha + \gamma^2)]|i\omega - \alpha|^2} + \frac{c_2}{[i\omega - (2\alpha + \gamma^2)]\overline{(i\omega - \alpha)}},$$

with some appropriate constants c_1 and c_2. The constants are determined by the equation

$$Ey_t^3 = \int_{-\infty}^{\infty} f_{q,y}(\omega) \, d\omega,$$

i.e.,

$$\frac{-2\mu^3}{\alpha(2\alpha + \gamma^2)(\alpha + \gamma^2)} = \frac{2\mu^3\gamma^2}{\alpha^2(2\alpha + \gamma^2)(3\alpha + \gamma^2)} - \frac{c_2}{3\alpha + \gamma^2}.$$

If t and s are nonnegative then the cumulants $\mathrm{Cum}(y_{t+s}, y_t, y_0)$ and $\mathrm{Cum}(y_t, y_0, y_0)$ are determined directly from Eq. (10.4) they are

$$\mathrm{Cum}(y_{t+s}, y_t, y_0) = \exp(\alpha s) \, \mathrm{Cum}(y_t, y_t, y_0),$$

$$\mathrm{Cum}(y_t, y_0, y_0) = \exp(\alpha s) \, \mathrm{Cum}(y_0, y_0, y_0).$$

The bispectrum $f_3(\omega_1, \omega_2)$ is given as

$$f_3(\omega_1, \omega_2) = \frac{6}{(2\pi)^2} \operatorname{sym}_{\omega(3)} \iint_0^\infty \exp(-i(t+s)\omega_1 - it\omega_2)$$
$$\times \operatorname{Cum}(y_{t+s}, y_t, y_0) \, ds \, dt.$$

The function to be symmetrized is

$$\iint_0^\infty \exp(-i(t+s)\omega_1 - it\omega_2) \operatorname{Cum}(y_{t+s}, y_t, y_0) \, ds \, dt$$
$$= \frac{1}{i\omega_1 - \alpha} \int_0^\infty \exp(-it(\omega_1 + \omega_2)) \operatorname{Cum}(y_t, y_t, y_0) \, dt.$$

Let

$$\frac{1}{2\pi} \int_0^\infty \exp(-it\omega) \operatorname{Cum}(y_t, y_t, y_0) \, dt = f_{q,y}^+(\omega),$$

$$\frac{1}{2\pi} \int_0^\infty \exp(-it\omega) \operatorname{Cum}(y_t, y_0, y_0) \, dt = f_{y,q}^+(\omega).$$

It is easy to show that

$$f_{q,y}(\omega) = f_{q,y}^+(\omega) + \overline{f_{y,q}^+(\omega)},$$

therefore

$$f_3(\omega_1, \omega_2) = \frac{6}{2\pi} \operatorname{sym}_{\omega(3)} \left\{ \frac{1}{i\omega_1 - \alpha} \left[f_{q,y}(\omega_1 + \omega_2) - \overline{f_{y,q}^+(\omega_1 + \omega_2)} \right] \right\}$$
$$= \frac{6}{2\pi} \operatorname{sym}_{\omega(3)} \left\{ \frac{1}{i\omega_1 - \alpha} \left[f_{q,y}(\omega_3) - f_{y,q}^+(\omega_3) \right] \right\},$$

where $\Sigma \omega(3) = 0$. The cross-spectrum $f_{y,q}^+(\omega)$ is

$$f_{y,q}^+(\omega) = \frac{1}{2\pi} \frac{\operatorname{Cum}(y_0, y_0, y_0)}{i\omega - \alpha}.$$

The estimation of the parameters of the bilinear process can be based (see Brillinger, 1985) on the spectrum and the bispectrum of the process.

Acknowledgement

One of the authors (Subba Rao) wishes to thank the Hungarian national Science Foundation for a grant (OTKA, no T032658) to visit Debrecen, during which period a part of this paper was written. We are thankful to the referee for the very meticulous reading and for pointing out several errors in this paper which considerably improved the presentation.

References

Bartlett, M. S. (1946). On the theoretical specification of sampling properties of autocorrelated time series. *J. Roy. Statist. Soc. (suppl)* **8**, 27–41.
Beran, J. (1994). *Statistics for Long Memory Process*. Monographs on Statistics and Applied Probability. Chapman & Hall, London.
Bhaskara Rao, M., T. Subba Rao and A. M. Walker (1983). On the existence of some bilinear time series models. *J. Time Series Anal.* **4**, 95–116.
Bjork, T. (1998). *Arbitrage Theory in Continuous Time*. Oxford University Press.
Brillinger, D. R. (1975). *Time Series Data Analysis and Theory*. Holt, Reinehart and Winston Inc., New York.
Brillinger, D. R. (1985). Fourier inference: some methods for the analysis of array and nonGaussian series data. *Water Resources Bull.* **21**, 743–756.
Brillinger, D. R. and M. Rosenblatt (1967). Asymptotic theory of estimates of the kth order spectra. In *Spectral Analysis of Time Series*, pp. 153–188 (Ed. B. Harris). Wiley, New York.
Brillinger, D. R. (1970). The identification of polynomial systems by means of higher-order spectra. *J. Sound and Vibration* **12**, 301–314.
Brock, W. W., Dechert and J. Scheinkman (1987). A test for independence based on the correlation dimension. *Working paper*, Department of Economics, University of Wisconsin.
Chanda, K. C. (1983). Asymptotic properties of serial covariances for nonlinear stationary processes. *J. Multivariate Anal.* **47**, 163–171.
Dahlhaus, R. (1995). Fitting of time series models to nonstationary time series. *Technical Report*, University of Heidelberg.
Delgado, M. (1996). Testing serial independence using the sample distribution function. *J. Time Series Anal.* **17**, 271–286.
Dobrushin, R. L. (1979). Gaussian and their subordinated generalized fields. *Ann. Probab.* **7**(1), 1–28.
Grahn, T. (1995). A conditional least squares approach to bilinear time series estimation. *J. Time Series Anal.* **16**, 509–529.
Granger, C. W. J. and A. P. Andersen (1978). *An Introduction to Bilinear Time Series Models*. Vandenhoek and Ruprecht, Gottingen, Germany.
Granger, C. W. J. and R. Joyeux (1980). An introduction to long memory time series models and fractional differencing. *J. Time Series Anal.* **1**, 15–30.
Granger, C. W. J. and P. Newbold (1974). Spurious regression in econometrics. *J. Econometrics* **2**, 111–120.
Grassberger, P. and J. Procaccia (1983). Characterisation of strange attractors. *Phys. Rev. Lett.* **50**, 346–349.
Grenier, Y. (1983). Time dependent ARMA modeling of nonstationary signals. *IEEE, ASSP* **31**, 899–911.
Guegan, D. (1994). *Series Chronologies Nonlinears a Temps Discreet*. Statistique Mathematique et Probabilitie Economics, Paris.
Haggan, V. and T. Ozaki (1981). Modelling random vibrations using amplitude dependent autoregressive time series models. *Biometrika* **68**, 189–196.
Hannan, E. J. (1986). Remembrance of things past. In *The Craft of Probabilistic Modelling*, pp. 190–212 (Ed. J. Gani). Springer-Verlag.
Hida, T. (1980). *Brownian Motion*. Springer-Verlag, New York.
Hinich, M. (1982). Testing for Gaussianity and linearity of a stationary time series. *J. Time Series Anal.* **3**, 169–176.
Hosking, J. R. M. (1981). Fractional differencing, *Biometrika* **68**, 165–167.
Hurst, H. E. (1951). Long term storage capacity of reservoirs. *Trans. Amer. Soc. Civil Engrs.* **116**, 770–808.
Hussain, M. Y. and T. Subba Rao (1976). The estimation of autoregressive, moving average and mixed autoregressive moving average systems with time dependent parameters of nonstationary time series. *Int. J. Control* **23**, 647–656.

Igloi, E. and Gy. Terdik (1997). Bilinear stochastic systems with long range dependence in continuous time in stochastic differential and difference equations. In *Progress in Systems and Control Theory*, pp. 299–309 (Eds. E. Csiszar and Gy. Michaletzky). Birkhäuser, Boston.

Kim, P. T. (1989). Estimation of product moments of a stationary stochastic process with application to estimation of cumulants and cumulant spectral densities. *Canad. J. Statist.* **17**, 285–299.

Leonov, V. P. and A. N. Shiryaev (1959). On a method of calculation of semi invariants. *Theory Probab. Appl.* **4**, 319–329.

Lii, K. S. and M. Rosenblatt (1982). Deconvolution and estimation of transfer function phase and coefficients for non-Gaussian linear process. *Ann. Statist.* **10**, 1193–1208.

Lii, K. S., M. Rosenblatt and C. Van Atta (1976). Bispectral measurements in turbulence. *J. Fluid Mechanics* **77**, 46–52.

Liu, J. (1989). On the existence of a general multiple bilinear time series. *J. Time Series Anal.* **10**(4), 341–355.

Lomnicki, Z. A. (1961). Tests for departure from normality in the case of linear stochastic processes. *Metrika* **4**, 37–62.

Maddala, G. S. and I.-M. Kim (1998). *Unit Roots, Cointegration and Structural Changes*. Cambridge University Press.

Major, P (1981). *Multiple Wiener–Itô Integrals*. Lecture Notes in Math., Vol. 849. Springer-Verlag, New York.

Mandelbrot, B. and J. W. Van Ness (1968). Fractional Brownian motions, practical noises and applications. *SIAM Rev.* **10**(4), 422–437.

Maria da Silva, E. A. (1994). Some contributions to the analysis of bilinear time series models. Unpublished Ph.D. thesis, UMIST.

Mills, T. C. (1999). *The Econometric Modelling of Financial Time Series*. Cambridge University Press.

Nikias, C. L. and A. P. Petropulu (1994). *Higher Order Spectral Analysis, a Nonlinear Signal Processing Framework*. Prentice-Hall, NJ, USA.

Pham D nh, T. and L. T. Tran (1981). On the first-order bilinear time series models. *J. Appl. Probab.* **18**, 617–627.

Prakasa Rao, B. L. S. (1999). Nonparametric functional estimation. An overview. In *Asymptotics, Nonparametrics and Time Series*, Vol. 158, pp. 461–510 (Ed. S. Ghosh). Marcel Dekker, Inc.

Priestley M. B. (1965). Evolutionary spectra and nonstationary processes. *J. Roy. Statist. Soc. B* **27**, 204–237.

Priestley M. B. (1981). *Spectral Analysis and Time Series*. Academic Press, New York.

Priestley M. B. and M. M. Gabr (1993). Bispectral analysis of nonstationary process. In *Multivariate Analysis: Future Directions* (Ed. C. R. Rao). North-Holland, Amsterdam.

Porat, B (1994). *Digital Processing of Random Signals*. Prentice-Hall Information and Systems Sciences Series.

Rao, C. R. (1965). *Linear Statistical Inference and Its Applications*. Wiley, New York.

Rosenblatt, M. and J. W. Van Ness (1965). Estimation of the bispectrum. *Ann. Math. Statist.* **36**, 1120–1135.

Rugh, W. (1981). *Nonlinear Systems Theory*. John Hopkins Univ. Press, Baltimore, USA.

Saito, K. and T. Tanaka (1984). Exact analytic expression for Gabr–Rao's optimal bispectral two dimensional lag window. *J. Nuclear Science and Technology* **22**(12), 1033–1035. (A publication of the Atomic Energy Society of Japan.)

Schmidt, R. O. (1981). A signal subspace approach to multiple emitter location and spectral estimation. Ph.D. thesis, Stanford University.

Schetzen, M. (1980). *The Volterra and Wiener Theories of Nonlinear Systems*. Wiley, New York.

Sesay, S. A. O. and T. Subba Rao (1988). Yule–Walker difference equations for higher-order moments and cumulants for bilinear time series models. *J. Time Series Anal.* **9**, 385–401.

Sesay, S. A. O. and T. Subba Rao (1991). Difference equations for high order moments and cumulants for bilinear time series model $BL(p, 0, p, 1)$. *J. Time Series Anal.* **11**, 385–401.

Skaug, H. J. and D. Tjostheim (1993). Nonparametric tests for serial independence. In *Developments in Time Series Analysis*, pp. 207–230 (Ed. T. Subba Rao). Chapman & Hall, London.

Stensholt, B. K. and D. Tjostheim (1987). Multiple bilinear time series models. *J. Time Series Anal.* **8**(2), 221–233.

Stensholt, B. K. (1989). Statistical analysis of multivariate bilinear models. Unpublished Ph.D. thesis, submitted to UMIST.

Stensholt, B. K. and T. Subba Rao (1987). On the theory of multivariate linear time series models. *Technical Report no 183*, Dept. of Maths., UMIST.
Subba Rao, T. (1970). The fitting of nonstationary time series models wtih time dependent parameters. *J. Roy. Statist. Soc. Ser. B* **32**, 312–322.
Subba Rao, T. (1976). Canonical correlation analysis and stationary time series models. *Sankhyā* **38**, 256–271.
Subba Rao, T. (1977). On the estimation of linear time series models, *Bull. Int. Statist. Inst.* **41**. Paper presented at the 91st session of ISI meeting, New Delhi.
Subba Rao, T. and M. Gabr (1980). A test for linearity of stationary time series. *J. Time Series Anal.* **1**(2), 145–158.
Subba Rao, T. (1981a). On the theory of bilinear time series models. *J. Roy. Statist. Soc. B* **43**(2), 244–255.
Subba Rao, T. (1981b). The bispectral analysis of nonlinear stationary time series with reference to bilinear time series models. In *Handbook of Statistics*, Vol. 3 (Eds. P. R. Krishnaiah and D. R. Brillinger). North-Holland, Amsterdam.
Subba Rao, T. (1985). Statistical analysis of bivariate bilinear time series models. Paper presented at the International symposium on Advances in multivariate statistical analysis. Indian Statistical Institute, Calcutta, 16–20 Dec.
Subba Rao, T. (1992). Analysis of nonlinear time series (and chaos) by bispectral methods. In *Nonlinear Modelling and Forecasting, SFI Studies in the Sciences and Complexity* (Eds. M. Casdagli and S. Eubank). Addison-Wesley.
Subba Rao, T. (1997). Statistical analysis of nonlinear and nonGaussian time series. In *Stochastic Differential and Difference Equations*, pp. 285–298 (Eds. I. Csiszar and Gy. Michaletzky). Birkhäuser, Boston.
Subba Rao, T. and M. E. Maria Augusto da Silva (1992). Identification of bilinear time series models $BL(p, 0, p, 1)$. *Statistica Sinica* **2**, 465–478.
Subba Rao, T. and M. M. Gabr (1984). *An Introduction to Bispectral Analysis and Bilinear Time Series Models*. Lecture Notes in Statist., Vol. 24. Springer-Verlag.
Subba Rao, T. and W. K. Wong (1998). Tests for Gaussianity and linearity of multivariate stationary time series. *J. Statist. Plann. Inference* **68**, 373–386.
Subba Rao, T. and W. K. Wong (1999). Some contributions to multivariate nonlinear time series and to bilinear models. In *Asymptotics, Nonparametrics and Time Series*, pp. 259–294 (Ed. S. Ghosh). Marcell Dekker Inc., New York.
Tang, Z. (1987). Bilinear stochastic processes and time series. Unpublished Ph.D. thesis, Oregon State University, USA.
Taqqu, M. S. (1979). Convergence of integrated processes of arbitrary Hermite rank. *Z. Wahrsch. Verw. Gebiete* **50**, 53–83.
Terdik, Gy. (1990). Stationary solutions for bilinear systems with constant coefficients. In *Seminar on Stochastic Processes 1989*, pp. 196–206 (Eds. E. Cinlar, K. L. Chung and R. K. Getoor). Progress in Probability. Birkhäuser, Boston.
Terdik, Gy. (1991). Bilinear state space realization for polynomial systems. *Comput. Math. Appl.* **26**(7), 69–83.
Terdik, Gy. (1999). *Bilinear Stochastic Models and Related Problems of Nonlinear Time Series Analysis – A Frequency Domain Approach*. Lecture Notes in Statist., Vol. 142. Springer-Verlag, Heidelberg.
Terdik, Gy. and T. Subba Rao (1989). On Wiener–Itô representation and the best linear predictions for bilinear time series. *J. Appl. Probab.* **26**, 274–286.
Terdik, Gy. (1991). On realization and identification of stochastic bilinear systems. In *Topics in Stochastic Systems: Modelling Estimation and Adaptive Control*, pp. 103–115 (Eds. L. Gerencser and P. E. Canies). Springer-Verlag, New York.
Terdik, Gy. and J. Math (1998). A new test of linearity for time series based on its bispectrum. *J. Time Series Anal.* **19**(6), 737–749.
Tjostheim, D. (1999). Nonparametric specification procedures for time series. In *Asymptotics, Nonparametrics and Time Series*, Vol. 158, pp. 149–199 (Ed. S. Ghosh). Marcel Dekker Inc., New York.
Tong, H. (1990). *Nonlinear Time Series*. Oxford University Press, Oxford.
Tsay, R. S. and G. Tiao (1985). Use of canonical analysis in time series model identification. *Biometrika* **72**, 299–315.

Tsolaki, E. P. (2001). Nonstationary time series. Analysis of monthly global temperature data. Unpublished Ph.D. thesis, submitted to UMIST, Manchester.

Wiener, N. (1958). *Nonlinear Problems in Random Theory*. Wiley, New York.

Wong, W. K. (1993). Some contributions to multivariate stationary nonlinear time series. Unpublished Ph.D. thesis, submitted to UMIST.

Nonlinear and Non-Gaussian State-Space Modeling with Monte Carlo Techniques: A Survey and Comparative Study

Hisashi Tanizaki

Since Kitagawa (1987) and Kramer and Sorenson (1988) proposed the filter and smoother using numerical integration, nonlinear and/or non-Gaussian state estimation problems have been developed. Numerical integration becomes extremely computer-intensive in the higher-dimensional cases of the state vector. Therefore, to improve the above problem, the sampling techniques such as Monte Carlo integration with importance sampling, resampling, rejection sampling, Markov chain Monte Carlo and so on are utilized, which can be easily applied to multi-dimensional cases. Thus, in the last decade, several kinds of nonlinear and non-Gaussian filters and smoothers have been proposed using various computational techniques. The objective of this chapter is to introduce the nonlinear and non-Gaussian filters and smoothers which can be applied to any nonlinear and/or non-Gaussian cases. Moreover, by Monte Carlo studies, each procedure is compared by the root mean square error criterion.

1. Introduction

The Kalman filter proposed by Kalman (1960) and Kalman and Bucy (1961) has been extended to nonlinear and nonnormal cases. The most heuristic and easiest nonlinear filters utilize the Taylor series expansions and the expanded nonlinear measurement and transition functions are directly applied to the standard linear recursive Kalman filter algorithm (see Appendix A for the standard linear recursive prediction, filtering and smoothing algorithms). One of the traditional nonlinear filters is known as the extended Kalman filter, which is discussed in Wishner, Tabaczynski and Athans (1969), Jazwinski (1970), Gelb (1974), Anderson and Moore (1979), Tanizaki (1993a, 1996) and Tanizaki and Mariano (1996). Since we have the inequality $E(g(x)) \neq g(E(x))$ for a nonlinear function $g(\cdot)$ and a random variable x (see Brown and Mariano, 1984, 1989 and Mariano and Brown, 1983, 1989), the state vector estimated by the extended Kalman filter is clearly biased.

In order to avoid the biased estimates, we need to consider the filtering and smoothing algorithms based on the underlying distribution functions. The standard Kalman filter algorithm is represented by the first- and the second-moments (i.e., mean and variance) of the underlying density functions, provided that the measurement and transition equations are linear and normality is assumed for the error terms (see, for example, Harvey, 1989 and Tanizaki, 1996). Unless the distributions of the error terms in the state-space model are normal and/or the measurement and transition equations are linear, we cannot derive an explicit linear recursive expression for the filtering and smoothing algorithms. If we approximate the nonlinear measurement and transition equations by the Taylor series expansions and apply the linearized nonlinear equations directly to the linear recursive filtering and smoothing algorithm, the obtained filtering and smoothing estimates are not appropriate and plausible.

It is known that when the system is linear and normal the Kalman filter estimate is optimal in the sense that it minimizes the mean square error. When the normality assumption is dropped, there is no longer any guarantee that the Kalman filter gives the conditional mean of the state vector. However, it is still an optimal estimator in the sense that it minimizes the mean square error within the class of all linear estimators (see Harvey, 1989). In addition, as Meinhold and Singpurwalla (1989) pointed out, Kalman filter models based on normality assumption are known to be non-robust, which implies that when there is a large difference between the prior density and the observed data, the posterior density becomes unrealistic. Therefore, approximation of the underlying densities, rather than that of the nonlinear functions, is essential to the nonlinear and/or nonnormal filtering problem.

The recursive algorithms on prediction, filtering and smoothing can be obtained from the conditional density functions, which are derived from Bayes' formula. The nonlinear filters and smoothers based on the underlying density functions have been developed to obtain an optimal estimator. Alspach and Sorenson (1972) and Sorenson and Alspach (1971) approximated the densities by a sum of Gaussian distributions, called the Gaussian sum filter. The obtained algorithm is a weighted sum of the extended Kalman filters. Therefore, the sum of the biased estimators also leads to the biased estimator.

To improve the biased filtering and smoothing estimates, Kitagawa (1987) and Kramer and Sorenson (1988) proposed approximating the densities numerically by a piecewise linear function, where each density is represented as number of segments, location of nodes and the value at each node, and it is evaluated through numerical integration. According to the numerical integration approach, however, computational burden increases more than proportionally as the dimension of the state vector increases. Programming is also extremely tedious in multi-dimensional cases.

In order to resolve the problems of the numerical integration procedure, Tanizaki (1993a, 1996), Tanizaki and Mariano (1994) and Mariano and Tanizaki (1995) suggested using Monte Carlo integration with importance sampling to density evaluation, in which a recursive algorithm of the weight functions represented by the ratio of two densities is derived. Geweke (1988, 1989a, 1989b) and Shao (1989) developed an approximation of prior density in Bayesian framework, so-called importance sampling (see Appendix B for several sampling methods). From the point of programming and computational time, the nonlinear and nonnormal filter and smoother based on Monte Carlo integration with importance sampling can be easily extended to the higher-dimensional cases in practice, comparing with the numerical integration procedure, although the importance sampling procedure is inferior to the numerical integration approach for precision of the filtering and smoothing estimates. However, one of the problems in the importance sampling procedure is as follows. For approximation of the target density functions, a researcher has to choose another appropriate density function, called the importance density, which is quite ad hoc. Unless the importance density is plausible, the obtained filtering and smoothing estimates might be biased.

Because both the numerical integration procedure and the importance sampling approach are based on density approximation for each time, where the density approximation at present time depends on that at the past time, accumulation of computational errors possibly become large as time goes by. Therefore, recently, some attempts are made to generate random draws directly from prediction, filtering and smoothing distribution functions. Gordon, Salmond and Smith (1993), Kitagawa (1996, 1998) and Kitagawa and Gersch (1996) utilized the resampling method to generate random draws from prediction, filtering and smoothing densities. For generation of filtering and smoothing random draws, one-step ahead prediction random draws are chosen with the corresponding probabilities. Programming is very easy compared with the above two approaches.

Tanizaki (1996, 1999a), Tanizaki and Mariano (1998), Hürzeler and Künsch (1998), Mariano and Tanizaki (2000) proposed an nonlinear and nonnormal filter and smoother using rejection sampling. For a solution to nonlinear and nonnormal state-space model, we use the random draws to obtain filtering and smoothing estimates. By rejection sampling, a recursive algorithm of the random draws are obtained. Thus, the random draws of the state vector are directly generated from the filtering and smoothing densities. The rejection sampling procedure gives us more precise estimates than the resampling procedure. However, when the acceptance probability in rejection sampling is small, computational time increases.

Carlin, Polson and Stoffer (1992), Carter and Kohn (1994, 1996) and Chib and Greenberg (1996) suggested applying an adaptive Monte Carlo integration technique known as the Gibbs sampler to the density approximation. Geweke and Tanizaki (1999, 2001) extended the approach to more general formulation. That is, the filter and smoother proposed by Geweke and Tanizaki (1999, 2001) can be applied to any nonlinear and non-Gaussian cases, using the Metropolis–Hastings algorithm within Gibbs sampling for random number generation. However, when we apply the Gibbs sampler to the filtering and smoothing framework, convergence of random draws is sometimes very slow, depending on the underlying state-space model, which implies that extremely large number of random draws are required for precision of the filtering and smoothing estimates.

Furthermore, improving computation time on rejection sampling and convergence on the Gibbs sampler, we introduce quasi-filter and quasi-smoother using the Metropolis–Hastings algorithm (see Tanizaki, 2000). As mentioned above, the rejection sampling procedure does not work when the acceptance probability is small and the Markov chain Monte Carlo procedure needs numerous numbers of random draws because of slow convergence of the Gibbs sampler. The quasi-filter and quasi-smoother improve both the problems. However, because the quasi-filter and quasi-smoother utilize some approximations, the obtained estimates are not exactly equal to the true state variables although the estimates are very close to the true values.

Thus, in the past, several kinds of nonlinear and non-Gaussian filters and smoothers have been proposed from various aspects. The objective of this chapter is to introduce the nonlinear and non-Gaussian filters and smoothers which can be applied to any nonlinear and/or non-Gaussian cases. The outline of this chapter is as follows. In the first half of Section 2, some economic applications of the state-space model are discussed and in the last half, two sorts of density-based prediction, filtering and smoothing algorithms are described, i.e., recursive and non-recursive algorithms. Section 3 summarizes recent work on the nonlinear and non-Gaussian filters and smoothers. In Section 4, all the procedures introduced in Section 3 are compared by Monte Carlo studies. Finally, Section 5 makes summary and concluding remarks.

2. State-space model

We consider the nonlinear and non-Gaussian state-space model, which is represented in the following general form:

$$\text{(Measurement equation)} \quad y_t = h_t(\alpha_t, \varepsilon_t), \tag{1}$$

$$\text{(Transition equation)} \quad \alpha_t = f_t(\alpha_{t-1}, \eta_t), \tag{2}$$

for $t = 1, 2, \ldots, T$, where T denotes the sample size. A vector y_t is observable while a vector α_t, called the state variable, is unobserved. The error terms ε_t and η_t are mutually independently distributed, which are typically assumed to be normal but not necessarily. $h_t(\cdot, \cdot)$ and $f_t(\cdot, \cdot)$ are vector functions, which are assumed to be known and may depend on other exogenous variables (however, we omit them for simplicity). Eq. (1) is called the measurement equation, which represents the relationship between the observed data y_t and the unobservable state variable α_t. Eq. (2) indicates the movement of the state variable, which is called the transition equation. Let Y_s be the information set up to time s, i.e., $Y_s = \{y_1, y_2, \ldots, y_s\}$.

The purpose of the state-space model is to estimate the unobserved state variable α_t utilizing the measurement and transition equations. That is, we consider estimating the conditional expectation of α_t using information Y_s, i.e., $a_{t|s} \equiv \mathrm{E}(\alpha_t|Y_s)$. Depending on t and s, the conditional expectation $a_{t|s}$ is called prediction if $t > s$, filtering if $t = s$ and smoothing if $t < s$. Moreover, there are three kinds of smoothing by the relationship between t and s with $t < s$. Let L be the fixed nonnegative integer. Then, $a_{L|t}$ for $t = L+1, L+2, \ldots, T$ is called fixed-point smoothing, which is useful to estimate the initial condition of the system. $a_{t|t+L}$ for $t = 1, 2, \ldots, T-L$ is known as fixed-lag smoothing. $a_{t|T}$ for $t = 1, 2, \ldots, T$ is called fixed-interval smoothing, which is helpful to investigate the past condition of the system.

In this chapter, we focus on L-step ahead prediction (i.e., $a_{t+L|t}$), filtering (i.e., $a_{t|t}$) and fixed-interval smoothing (i.e., $a_{t|T}$) in nonlinear non-Gaussian cases.

2.1. Some applications of state-space model in economics

Some economic applications of the state-space model are discussed in this section, where we consider the following examples: Time Varying Parameter Model (Section 2.1.1), Autoregressive-Moving Average Process (Section 2.1.2), Seasonal Adjustment Models (Section 2.1.3), Prediction of Final Data Using Preliminary Data (Section 2.1.4), Estimation of Permanent Consumption (Section 2.1.5), Markov Switching Model (Section 2.1.6) and Stochastic Variance Models (Section 2.1.7).

2.1.1. Time varying parameter model
In the case where we deal with time series data, the nonlinear regression model can be written as follows:

$$y_t = h_t(x_t, \alpha, \varepsilon_t),$$

for $t = 1, 2, \ldots, T$, where y_t is a dependent variable, x_t denotes a $1 \times k$ vector of the explanatory variables, a $k \times 1$ vector of unknown parameters to be estimated is given by α, and ε_t is the error term. $h_t(\cdot, \cdot, \cdot)$ is assumed to be a known vector function, which is given by $h_t(x_t, \alpha, \varepsilon_t) = x_t\alpha + \varepsilon_t$ in a classical linear regression model. There are some methods to estimate the equation above, for example, the least squares method, the method of moments and so on. In any case, if the unknown parameters to be estimated are constant over time, the model is known as the fixed-parameter model. However, structural changes (for example, the first- and second oil crises), specification errors, proxy variables and aggregation are all the sources of parameter variation; see Sarris (1973), Belsley (1973), Belsley and Kuh (1973) and Cooley and Prescott (1976). Therefore, we need to consider the model such that the parameter is a function of time, which is called the time varying parameter model. Using the state-space form, the model is represented as the following two equations:

(Measurement equation) $\quad y_t = h_t(x_t, \alpha_t, \varepsilon_t),$ $\qquad(3)$

(Transition equation) $\quad \alpha_t = \Psi \alpha_{t-1} + \eta_t,$ $\qquad(4)$

where the movement of the parameter α_t is assumed to be the first-order autoregressive (AR(1)) process, which can be extended to the AR(p) process. The error term η_t, independent of ε_t, is a white noise. Here, Eqs. (3) and (4) are referred to as the measurement equation and the transition equation, respectively. The time varying parameter α_t is unobservable, which is estimated using the observed data y_t and x_t. There are numerous other papers which deal with the time varying parameter model (see, for example, Cooley, 1977; Cooley, Rosenberg and Wall, 1977; Cooper, 1973; Nicholls and Pagan, 1985; Tanizaki, 1989, 1993b; Sant, 1977).

2.1.2. Autoregressive moving average process
It is well known that any autoregressive-moving average (ARMA) process can be written in a state-space form. See, for example, Aoki (1987, 1990), Brockwell and Davis

(1987), Burridge and Wallis (1988), Gardner, Harvey and Phillips (1980), Hannan and Deistler (1988), Harvey (1981, 1989) and Kirchen (1988).

First, consider the following ARMA(p, q) process.

$$y_t = a_1 y_{t-1} + a_2 y_{t-2} + \cdots + a_p y_{t-p} + \eta_t + b_1 \eta_{t-1} + \cdots + b_q \eta_{t-q},$$

where η_t is a white noise. The ARMA(p, q) model above is rewritten as:

$$y_t = a_1 y_{t-1} + a_2 y_{t-2} + \cdots + a_m y_{t-m} + \eta_t + b_1 \eta_{t-1} + \cdots + b_{m-1} \eta_{t-m+1},$$

where $m = \max(p, q+1)$ and some of the coefficients $a_1, a_2, \ldots, a_m, b_1, b_2, \ldots, b_{m-1}$ can be zeros. As it is well known, the ARMA process above is represented as:

(Measurement equation) $y_t = z \alpha_t,$

(Transition equation) $\alpha_t = A \alpha_{t-1} + B \eta_t,$

where z, A and B are defined as:

$$z = \underset{1 \times m}{(1, 0, \ldots, 0)}, \quad A = \underset{m \times m}{\begin{pmatrix} a_1 & & \\ \vdots & I_{m-1} & \\ a_{m-1} & & \\ a_m & & 0 \end{pmatrix}}, \quad B = \underset{m \times 1}{\begin{pmatrix} 1 \\ b_1 \\ \vdots \\ b_{m-1} \end{pmatrix}}.$$

Thus, the state-space model is constructed from the ARMA model, where the first element of α_t represents the time series data to be estimated.

2.1.3. Seasonal adjustment models

A time series consists of seasonal, cyclical, and irregular components. Each component is unobservable and therefore the Kalman filter is applied to estimate each component separately. In this section, two seasonal adjustment models are introduced; one is developed by Pagan (1975) and another is Kitagawa (1996) and Kitagawa and Gersch (1996).

The suggestion by Pagan (1975) is essentially a combination of an econometric model for the cyclical components with the filtering and estimation of the seasonal components formulated in a state-space form (see Chow, 1983). Assume, first, that an endogenous variable y_t is the sum of cyclical, seasonal, and irregular components, as given by:

$$y_t = y_t^c + y_t^s + \varepsilon_t, \tag{5}$$

where y_t^c, y_t^s and ε_t denote the cyclical, seasonal, and irregular components, respectively. Second, the cyclical component y_t^c is represented as the following model:

$$y_t^c = A y_{t-1}^c + C x_t + u_t, \tag{6}$$

where x_t is a $k \times 1$ vector of exogenous variables and u_t denotes a random disturbance. In Eq. (6), the AR(1) model is assumed for simplicity but the AR(p) model is also possible. Finally, an autoregressive seasonal model is assumed for the seasonal component, i.e.,

$$y_t^s = B y_{t-m}^s + w_t, \tag{7}$$

where w_t represents a random residual and m can be 4 for a quarterly model and 12 for a monthly model. Combining Eqs. (5)–(7), we can construct the following state-space form:

(Measurement equation) $y_t = z\alpha_t + \varepsilon_t$,

(Transition equation) $\alpha_t = M\alpha_{t-1} + Nx_t + \eta_t$, $\qquad$ (8)

where z, α_t, M, N and η_t are given by:

$$z = \underbrace{(1, 1, 0, \ldots, 0)}_{1 \times (m+1)}, \qquad M = \underbrace{\begin{pmatrix} A & 0 \\ \hline 0 & I_{m-1} \\ 0 & B & 0 \end{pmatrix}}_{(m+1) \times (m+1)},$$

$$N = \underbrace{\begin{pmatrix} C \\ 0 \end{pmatrix}}_{(m+1) \times k}, \qquad \eta_t = \underbrace{\begin{pmatrix} u_t \\ w_t \\ 0 \end{pmatrix}}_{(m+1) \times 1}.$$

The first and second elements of α_t represent y_t^c and y_t^s, respectively.

Kitagawa (1996) and Kitagawa and Gersch (1996) suggested an alternative seasonal component model, which is represented by Eq. (5) and the following two equations:

$$y_t^c = a_1 y_{t-1}^c + a_2 y_{t-2}^c + \cdots + a_p y_{t-p}^c + u_t, \tag{9}$$

$$y_t^s = -y_{t-1}^s - y_{t-2}^s - \cdots - y_{t-m+1}^s + w_t, \tag{10}$$

where Eq. (9) may depend on the other exogenous variables x_t as in Eq. (6). Eqs. (5), (9) and (10) yield the following state-space model:

(Measurement equation) $y_t = z\alpha_t + \varepsilon_t$,

(Transition equation) $\alpha_t = M\alpha_{t-1} + \eta_t$, $\qquad$ (11)

where z, α_t, M and η_t are given by:

$$z = \underbrace{(1, 0, \ldots, 0, 1, 0, \ldots, 0)}_{1 \times (p+m-1)}, \qquad A = \underbrace{\begin{pmatrix} a_1 \\ \vdots & I_{p-1} \\ a_{p-1} \\ \hline a_p & 0 \end{pmatrix}}_{p \times p},$$

$$B = \begin{pmatrix} -1 & & \\ \vdots & I_{m-2} & \\ -1 & & \\ \hline -1 & 0 \end{pmatrix}, \qquad M = \begin{pmatrix} A & 0 \\ 0 & B \end{pmatrix},$$
$${\scriptstyle (m-1)\times(m-1)} \qquad\qquad {\scriptstyle (p+m-1)\times(p+m-1)}$$

$$\eta_t = (u_t, 0, \ldots, 0, w_t, 0, \ldots, 0)'.$$
$$ {\scriptstyle (p+m-1)\times 1}$$

All the elements of z and η_t are zeros except for the first and $(p+1)$th elements. The cyclical component y_t^c and the seasonal component y_t^s are given by the first element of α_t and the $(p+1)$th element of α_t, respectively. Note that difference between the two systems (8) and (11) is a formulation in the seasonal component, which is described in Eqs. (7) and (10).

2.1.4. Prediction of final data using preliminary data

It is well known that economic indicators are usually reported according to the following two steps: (i) the preliminary data are reported and (ii) thereafter we can obtain the final or revised data (see Table 1). The problem is how to estimate the final data (or the revised data) while only the preliminary data are available.

In the case of annual data on the U.S. national accounts, the preliminary data at the present time are reported at the beginning of the next year. The revision process is performed over a few years and every decade, as shown in Table 1, where an example of the nominal gross domestic product data (GDP, billion dollars) is taken.

In Table 1, the preliminary data of 1988, 1992, 1995 and 1996 are taken from *Survey of Current Business* (January, 1989, January, 1993, July, 1996 and February, 1997), while the rest of the preliminary data and all the revised data are from *Economic Report of the President* (ERP), published from 1984 to 1997. Each column indicates the year when ERP is published, while each row represents the reported data of the corresponding year. The superscripts p and r denote the preliminary data and the data revised in the year corresponding to each column. NA indicates that the data are not available, which implies that the data have not been published yet. For instance, take the GDP data of 1984 (see the corresponding row in Table 1). The preliminary GDP data of 1984 was reported in 1985 (i.e., 3616.3), and it was revised in 1986 for the first time (i.e., 3726.7). In 1987 and 1988, the second and third revised data were published, respectively (i.e., 3717.5 and 3724.8). Since it was not revised in 1989, the GDP data of 1984 published in 1989 is given by 3724.8. Moreover, the GDP data of 1984 was revised as 3777.2 in 1992 and 3902.4 in 1996.

Thus, each data series is revised every year for the first few years and thereafter less frequently. This implies that we cannot really know the true final data, because the data are revised forever while the preliminary data are reported only once. Therefore, it might be possible to consider that the final data are unobservable, which leads to estimation of the final data given the preliminary data.

Table 1
Revision process of U.S. National Accounts (Nominal GDP)

	1984	1985	1986	1987	1988	1989	1990
1979	2375.2	2375.2	2464.4^r	2464.4	2464.4	2464.4	2464.4
1980	2586.4^r	2586.4	2684.4^r	2684.4	2684.4	2684.4	2684.4
1981	2904.5^r	2907.5^r	3000.5^r	3000.5	3000.5	3000.5	3000.5
1982	3025.7^r	3021.3^r	3114.8^r	3114.8	3114.8	3114.8	3114.8
1983	3263.4^p	3256.5^r	3350.9^r	3355.9^r	3355.9	3355.9	3355.9
1984	NA	3616.3^p	3726.7^r	3717.5^r	3724.8^r	3724.8	3724.8
1985	NA	NA	3951.8^p	3957.0^r	3970.5^r	3974.1^r	3974.1
1986	NA	NA	NA	4171.2^p	4201.3^r	4205.4^r	4197.2^r
1987	NA	NA	NA	NA	4460.2^p	4497.2^r	4493.8^r
1988	NA	NA	NA	NA	NA	4837.8^p	4847.3^r
1989	NA	NA	NA	NA	NA	NA	5199.6^p

	1991	1992	1993	1994	1995	1996	1997
1979	2464.4	2488.6^r	2488.6	2488.6	2488.6	2557.5^r	2557.5
1980	2684.4	2708.0^r	2708.0	2708.0	2708.0	2784.2^r	2784.2
1981	3000.5	3030.6^r	3030.6	3030.6	3030.6	3115.9^r	3115.9
1982	3114.8	3149.6^r	3149.6	3149.6	3149.6	3242.1^r	3242.1
1983	3355.9	3405.0^r	3405.0	3405.0	3405.0	3514.5^r	3514.5
1984	3724.8	3777.2^r	3777.2	3777.2	3777.2	3902.4^r	3902.4
1985	3974.1	4038.7^r	4038.7	4038.7	4038.7	4180.7^r	4180.7
1986	4197.2	4268.6^r	4268.6	4268.6	4268.6	4422.2^r	4422.2
1987	4486.7^r	4539.9^r	4539.9	4539.9	4539.9	4692.3^r	4692.3
1988	4840.2^r	4900.4^r	4900.4	4900.4	4900.4	5049.6^r	5049.6
1989	5163.2^r	5244.0^r	5250.8^r	5250.8	5250.8	5438.7^r	5438.7
1990	5424.4^p	5513.8^r	5522.2^r	5546.1^r	5546.1	5743.8^r	5743.8
1991	NA	5671.8^p	5677.5^r	5722.8^r	5724.8^r	5916.7^r	5916.7
1992	NA	NA	5945.7^p	6038.5^r	6020.2^r	6244.4^r	6244.4
1993	NA	NA	NA	6374.0^p	6343.3^r	6550.2^r	6553.0^r
1994	NA	NA	NA	NA	6736.9^p	6931.4^r	6935.7^r
1995	NA	NA	NA	NA	NA	7245.8^p	7253.8^r
1996	NA	NA	NA	NA	NA	NA	7580.0^p

There is a wide literature dealing with the data revision process. Conrad and Corrado (1979) applied the Kalman filter to improve upon published preliminary estimates of monthly retail sales, using an ARIMA model. Howrey (1978, 1984) used the preliminary data in econometric forecasting and obtained the substantial improvements in forecast accuracy if the preliminary and revised data are used optimally.

In the context of the revision process, the filtering and smoothing techniques are used as follows. There is some relationship between the final and preliminary data, because they are originally same data (see, for example, Conrad and Corrado, 1979). This relationship is referred to as the measurement equation, where the final data is unobservable but the preliminary data is observed. The equation obtained by the underlying economic theory is related to the final data, rather than the preliminary data. This equation is taken as the transition equation. Therefore, we can represent the revision problem with the following state-space form:

(Measurement equation) $\quad y_t^p = h_t(y_t^f, \varepsilon_t)$,

(Transition equation) $\quad y_t^f = f_t(y_{t-1}^f, x_t, \eta_t)$,

where y_t^p and y_t^f denote the preliminary data and the final data, respectively. The unobserved state variable is given by y_t^f, while y_t^p is observable. Thus, the state-space model is utilized to estimate y_t^f (see Mariano and Tanizaki, 1995 and Tanizaki and Mariano, 1994).

2.1.5. Estimation of permanent consumption

The next application is concerned with estimation of permanent consumption. Total consumption consists of permanent and transitory consumption. This relationship is represented by an identity equation, which corresponds to the measurement equation. Permanent consumption depends on life-time income expected in the future, i.e., permanent income. The following expected utility function of the representative agent is maximized with respect to permanent consumption (see Hall, 1978, 1990):

$$\max_{\{c_t^p\}} E_0\left(\sum_t \beta^t u(c_t^p)\right), \quad \text{subject to } A_{t+1} = R_t(A_t + y_t - c_t),$$

where $0 < \beta < 1$ and $c_t = c_t^p + c_t^T$. c_t, c_t^p, c_t^f, R_t, A_t, y_t, β, $u(\cdot)$ and $E_t(\cdot)$ denote per capita total consumption, per capita permanent consumption, per capita transitory consumption, the real gross rate of return on savings between periods t and $t+1$, the stock of assets at the beginning of period t, per capita labor income, the discount rate, the representative utility function and the mathematical expectation given information up to t, respectively.

Solving the above maximization problem, we can obtain the transition equation which represents the relationship between c_t^p and c_{t-1}^p. Transitory consumption is assumed to be a random shock with mean zero and variance σ_ε.

Under the above setup, the model to this problem is given by:

(Measurement equation) $\quad c_t = c_t^p + \varepsilon_t$,

(Transition equation) $\quad \dfrac{\beta R_{t-1} u'(c_t^p)}{u'(c_{t-1}^p)} = 1 + \eta_t$.

Note that $c_t^T = \varepsilon_t$, which is assumed to be independent of η_t. c_t is observable while both c_t^p and c_t^T are unobservable, where c_t^p is regarded as the state variable to be estimated by the nonlinear filtering and smoothing technique. Thus, we can estimate permanent and transitory consumption separately. Tanizaki (1993a, 1996) and Mariano and Tanizaki (2000) consider the above example, where the utility function of the representative agent is assumed to be a constant relative risk aversion type of utility function. Also see Diebold and Nerlove (1989) for a concise survey of testing the permanent income hypothesis.

2.1.6. Markov switching model

The Markov switching model was developed by Hamilton (1989, 1990, 1991, 1993, 1994), where the discrete random variable is taken for the state variable. Consider the k-dimensional discrete state variable, i.e., $\alpha_t = (\alpha_{1t}, \alpha_{2t}, \ldots, \alpha_{kt})'$, where we assume that one of the k elements of α_t is one and the others are zeros.

First, consider the following model:

$$y_t = h_t(\mu_t^*, \varepsilon_t), \qquad (12)$$

where μ_t^* is a discrete random variable and $h_t(\cdot, \cdot)$ may depend on the other exogenous variable x_t. Assume that μ_t^* depends on the unobserved random variable s_t^*, which is called the state or regime. Suppose that we have k states (or regimes). If $s_t^* = j$, then the process is in regime j and $\mu_t^* = \mu_j$ is taken. We assume that one of the k states at time t occurs depending on time $t - 1$.

Define $p = (p_1', p_2', \ldots, p_k')'$ as the transition probability matrix, where $p_i = (p_{i1}, p_{i2}, \ldots, p_{ik})$ for $i = 1, 2, \ldots, k$. Note that $\sum_{i=1}^{k} p_{ij} = 1$ should be satisfied for all $j = 1, 2, \ldots, k$. p_{ij} implies the conditional probability of $s_t^* = j$ given $s_{t-1}^* = i$, i.e., $p_{ij} \equiv \mathrm{Prob}(s_t^* = j | s_{t-1}^* = i)$. Such a process is described as an k-state Markov chain with transition probabilities $\{p_{ij}\}_{i,j=1,2,\ldots,k}$. The transition probability p_{ij} gives the probability that state i is followed by state j. Under the above setup, each element of the $k \times 1$ multivariate discrete state variable α_t takes a binary number, i.e.,

$$\alpha_t = \begin{cases} (1, 0, 0, \ldots, 0)', & \text{when } s_t^* = 1, \\ (0, 1, 0, \ldots, 0)', & \text{when } s_t^* = 2, \\ \vdots & \vdots \\ (0, 0, 0, \ldots, 1)', & \text{when } s_t^* = k. \end{cases}$$

Let us define $\mu = (\mu_1, \mu_2, \ldots, \mu_k)$, where each element depends on the regime. Then, μ_t^* is rewritten as: $\mu_t^* = \mu \alpha_t$. Accordingly, the model described in Eq. (12) is represented by the following state-space model:

(Measurement equation) $\quad y_t = h_t(\mu \alpha_t, \varepsilon_t),$

(Transition equation) $\quad \alpha_t = p \alpha_{t-1} + \eta_t,$

for $t = 1, 2, \ldots, T$. μ is a $k \times 1$ vector of unknown parameter to be estimated. η_t is distributed as a k-dimensional discrete random variable. The conditional density of α_t given α_{t-1} is represented by $P_\alpha(\alpha_t | \alpha_{t-1}) = \prod_{i=1}^{k}(p_i \alpha_{t-1})^{\alpha_{it}}$, which implies that the probability which event i occurs at time t is $p_i \alpha_{t-1}$.

Thus, it is assumed in the Markov switching model that the economic situation is stochastically switched from one to another for each time. The Markov switching model is similar to the time varying parameter model in Section 2.1.1 in the sense that the parameter changes over time. From specification of the transition equation, however, the time varying parameter model takes into account a gradual shift in the economic structural change but the Markov switching model deals with a sudden shift because μ_t^* is a discrete random variable which depends on state or regime.

2.1.7. Stochastic variance models

In this section, we introduce two stochastic variance models (see Taylor, 1994 for the stochastic variance models). One is called the autoregressive conditional heteroskedasticity (ARCH) model proposed by Engle (1982) and another is the stochastic volatility model (see Ghysels, Harvey and Renault, 1996).

Let β be a $k \times 1$ vector of unknown parameters to be estimated. y_t and x_t are assumed to be observable variables. The first-order ARCH model is given by the following two equations:

$$\text{(Measurement equation)} \quad y_t = x_t \beta + \alpha_t,$$

$$\text{(Transition equation)} \quad \alpha_t = \left(\delta_0 + \delta_1 \alpha_{t-1}^2\right)^{1/2} \eta_t,$$

for $t = 1, 2, \ldots, T$, where $\eta_t \sim N(0, 1)$, $0 < \delta_0$ and $0 \leqslant \delta_1 < 1$ have to be satisfied. The conditional variance of α_t is represented by $\delta_0 + \delta_1 \alpha_{t-1}^2$ while the unconditional variance is given by $\delta_0/(1 - \delta_1)$. It might be possible to put the error term (say, ε_t) in the measurement equation, i.e., $y_t = x_t \beta + \alpha_t + \varepsilon_t$.

As an alternative stochastic variance model, we can consider the stochastic volatility model, which is defined as follows:

$$\text{(Measurement equation)} \quad y_t = x_t \beta + \exp\left(\frac{1}{2}\alpha_t\right)\varepsilon_t,$$

$$\text{(Transition equation)} \quad \alpha_t = \delta \alpha_{t-1} + \eta_t,$$

for $t = 1, 2, \ldots, T$, where $0 \leqslant \delta < 1$ has to be satisfied. The error terms ε_t and η_t are mutually independently distributed.

For the other applications of the state-space model in economics, we can find estimation of the rational expectation models (for example, see Burmeister and Wall, 1982; Engle and Watson, 1987; McNelis and Neftci, 1983). See Harvey (1987) for a survey of applications of the Kalman filter model.

2.2. Prediction, filtering and smoothing algorithms

In the state-space model shown in Eqs. (1) and (2), the state variable α_t is the unobserved variable to be estimated. As mentioned above, there are three estimation problems, viz., prediction $a_{t+L|t}$, filtering $a_{t|t}$ and smoothing $a_{t|T}$. In this section, we introduce derivation of $a_{t+L|t}$, $a_{t|t}$ and $a_{t|T}$.

Let $P_y(y_t|\alpha_t)$ be the conditional density function derived from Eq. (1) and $P_\alpha(\alpha_t|\alpha_{t-1})$ be the conditional density function obtained from Eq. (2). For prediction, filtering and smoothing, two kinds of algorithms are introduced in this section, i.e., recursive algorithm and non-recursive algorithm, which are described in Sections 2.2.1 and 2.2.2.

2.2.1. Recursive algorithm
The relationships among density-based recursive algorithms on prediction, filtering and smoothing are as follows: (i) the initial value of prediction is given by filtering, (ii) one-step ahead prediction is utilized to obtain filtering, and (iii) smoothing is derived based on one-step ahead prediction and filtering.

Prediction. The L-step ahead prediction algorithm based on the density function is given by the following recursion:

$$P(\alpha_{t+L}|Y_t) = \int P_\alpha(\alpha_{t+L}|\alpha_{t+L-1})P(\alpha_{t+L-1}|Y_t)\,d\alpha_{t+L-1}, \tag{13}$$

for $L = 1, 2, \ldots$. See Kitagawa (1987) and Harvey (1989) for the density-based prediction algorithm. The filtering density $P(\alpha_t|Y_t)$ is assumed to be known, which is the initial density of the density-based L-step ahead prediction algorithm (13). Given $P(\alpha_t|Y_t)$, $P(\alpha_{t+L}|Y_t)$ is recursively obtained given $P(\alpha_{t+L-1}|Y_t)$ for $L = 1, 2, \ldots$.

Filtering. The density-based filtering algorithm is represented by the following two equations:

$$P(\alpha_t|Y_{t-1}) = \int P_\alpha(\alpha_t|\alpha_{t-1})P(\alpha_{t-1}|Y_{t-1})\,d\alpha_{t-1}, \tag{14}$$

$$P(\alpha_t|Y_t) = \frac{P_y(y_t|\alpha_t)P(\alpha_t|Y_{t-1})}{\int P_y(y_t|\alpha_t)P(\alpha_t|Y_{t-1})\,d\alpha_t}, \tag{15}$$

for $t = 1, 2, \ldots, T$, where the initial condition is given by:

$$P(\alpha_1|Y_0) = \begin{cases} \int P_\alpha(\alpha_1|\alpha_0)P_\alpha(\alpha_0)\,d\alpha_0, & \text{if } \alpha_0 \text{ is stochastic,} \\ P_\alpha(\alpha_1|\alpha_0), & \text{otherwise,} \end{cases}$$

where $P_\alpha(\alpha_0)$ denotes the unconditional density of α_0.

Eq. (14) corresponds to one-step ahead prediction, which plays a role of predicting α_t using the past information Y_{t-1}. Eq. (15) combines the present sample y_t with the past information Y_{t-1}. Eq. (14) is called the prediction equation while Eq. (15) is known as the update equation. Based on the two densities $P_\alpha(\alpha_t|\alpha_{t-1})$ and $P_y(y_t|\alpha_t)$, the density-based filtering algorithm is represented as the following two steps: (i) Eq. (14) yields $P(\alpha_t|Y_{t-1})$ given $P(\alpha_{t-1}|Y_{t-1})$ and (ii) $P(\alpha_t|Y_t)$ is obtained given $P(\alpha_t|Y_{t-1})$ from Eq. (15). Thus, repeating predicting and updating for all t, the filtering densities $P(\alpha_t|Y_t)$, $t = 1, 2, \ldots, T$, can be recursively computed.

Smoothing. The density-based fixed-interval smoothing algorithm is represented as (see, for example, Kitagawa, 1987, 1996 and Harvey, 1989):[1]

$$P(\alpha_t|Y_T) = P(\alpha_t|Y_t) \int \frac{P(\alpha_{t+1}|Y_T) P_\alpha(\alpha_{t+1}|\alpha_t)}{P(\alpha_{t+1}|Y_t)} \, d\alpha_{t+1}, \qquad (16)$$

for $t = T - 1, T - 2, \ldots, 1$, which is a backward recursion. Using Eq. (16), $P(\alpha_t|Y_T)$ is obtained from $P(\alpha_{t+1}|Y_T)$, given $P_\alpha(\alpha_{t+1}|\alpha_t)$, $P(\alpha_t|Y_t)$ and $P(\alpha_{t+1}|Y_t)$. Note that the smoothing density at time T (i.e., the endpoint case in the smoothing algorithm) is equivalent to the filtering density at time T. Thus, the fixed-interval smoothing algorithm (16) is derived together with the filtering algorithm given by Eqs. (14) and (15).

Mean, variance and likelihood function. Once the density $P(\alpha_r|Y_s)$ is obtained, the conditional expectation of a function $g(\alpha_r)$ is given by:

$$E\big(g(\alpha_r)|Y_s\big) = \int g(\alpha_r) P(\alpha_r|Y_s) \, d\alpha_r, \qquad (17)$$

for $(r, s) = (t + L, t), (t, t), (t, T)$. The function $g(\cdot)$ is typically specified as: $g(\alpha_r) = \alpha_r$ for mean or $g(\alpha_r) = (\alpha_r - a_{r|s})(\alpha_r - a_{r|s})'$ for variance, where $a_{r|s} = E(\alpha_r|Y_s)$.

When an unknown parameter is included in Eqs. (1) and (2), the following likelihood function is maximized with respect to the parameter:

$$P(Y_T) = \prod_{t=1}^{T} P(y_t|Y_{t-1}) = \prod_{t=1}^{T} \left(\int P(y_t|\alpha_t) P(\alpha_t|Y_{t-1}) \, d\alpha_t \right), \qquad (18)$$

which is called the innovation form of the likelihood function. Recall that $P(y_t|Y_{t-1})$ corresponds to the denominator of Eq. (15) in the filtering algorithm. Therefore, we do not need any extra computation for evaluation of the likelihood function (18).

In the case where the functions $h_t(\alpha_t, \varepsilon_t)$ and $g_t(\alpha_{t-1}, \eta_t)$ are linear and the error terms ε_t and η_t are normally distributed, Eqs. (13)–(16) and (18) reduce to the standard linear recursive algorithms on prediction, filtering and smoothing and the likelihood function. See Appendix A for the standard linear recursive algorithms and the likelihood function.

2.2.2. Non-recursive algorithm

Usually, the prediction, filtering and smoothing formulas are represented as the recursive algorithms based on the density functions. The non-recursive formulas on prediction, filtering and smoothing are described in Tanizaki (1996, 1997) and Tanizaki and Mariano (1998). However, we can easily show equivalence between both the

---

[1] As mentioned above, there are three kinds of smoothing algorithms, i.e., fixed-point smoothing, fixed-lag smoothing and fixed-interval smoothing (see, for example, Anderson and Moore, 1979 and Harvey, 1989). In a field of economics, the economic situations in the past are often analyzed using the data available at the present time. Accordingly, the fixed-interval smoothing algorithm is taken in this section. That is, we consider evaluating the conditional mean $E(\alpha_t|Y_T) = a_{t|T}$ for $t = 1, 2, \ldots, T$.

algorithms (see Appendix C for the proof). In this section, we introduce an alternative solution to a nonlinear and nonnormal prediction, filter and smoothing, which are not conventional recursive algorithms.

Let us define $A_t = \{\alpha_0, \alpha_1, \ldots, \alpha_t\}$, which is a set consisting of the state variables up to time t. Suppose that $P_\alpha(A_t)$ and $P_y(Y_t|A_t)$ are represented as:

$$P_\alpha(A_t) = \begin{cases} P_\alpha(\alpha_0) \prod_{s=1}^{t} P_\alpha(\alpha_s|\alpha_{s-1}), & \text{if } \alpha_0 \text{ is stochastic,} \\ \prod_{s=1}^{t} P_\alpha(\alpha_s|\alpha_{s-1}), & \text{otherwise,} \end{cases} \qquad (19)$$

$$P_y(Y_t|A_t) = \prod_{s=1}^{t} P_y(y_s|\alpha_s). \qquad (20)$$

Based on the two densities $P(A_t)$ and $P(Y_t|A_t)$, prediction, filtering and smoothing formulas can be derived.[2]

Prediction. To derive L-step ahead prediction mean $a_{t+L|t}$, we have to obtain the conditional density of A_{t+L} given Y_t, i.e., $P(A_{t+L}|Y_t)$. First, using Eqs. (19) and (20), note that the joint density of A_{t+L} and Y_t is written as: $P_\alpha(A_{t+L}, Y_t) = P_\alpha(A_{t+L}) P_y(Y_t|A_t)$. Accordingly, $P(A_{t+L}|Y_t)$ is given by:

$$\begin{aligned} P(A_{t+L}|Y_t) &= \frac{P(A_{t+L}, Y_t)}{\int P(A_{t+L}, Y_t) \, dA_{t+L}} \\ &= \frac{P_\alpha(A_{t+L}) P_y(Y_t|A_t)}{\int P_\alpha(A_{t+L}) P_y(Y_t|A_t) \, dA_{t+L}}. \end{aligned} \qquad (21)$$

Thus, from Eq. (21), the L-step ahead prediction density function is represented by:

$$P(\alpha_{t+L}|Y_t) = \int P(A_{t+L}|Y_t) \, dA_{t+L-1}, \qquad (22)$$

which is not a recursive algorithm. However, it is easily shown that Eq. (13) is equivalent to Eq. (22). See Appendix C for equivalence between Eqs. (13) and (22).

Filtering. The conditional density of A_t given Y_t is represented as:

$$\begin{aligned} P(A_t|Y_t) &= \frac{P(A_t, Y_t)}{\int P(A_t, Y_t) \, dA_t} \\ &= \frac{P_\alpha(A_t) P_y(Y_t|A_t)}{\int P_\alpha(A_t) P_y(Y_t|A_t) \, dA_t}, \end{aligned} \qquad (23)$$

where $P(A_t, Y_t) = P_\alpha(A_t) P_y(Y_t|A_t)$ is utilized in the second line of Eq. (23).

---

[2] Tanizaki (1996, 1997) and Tanizaki and Mariano (1998) made an attempt to evaluate the prediction, filtering and smoothing estimates, generating random draws of A_T from $P_\alpha(A_T)$, where $P_y(Y_T|A_T)$ is not utilized to generate the random draws.

Therefore, integrating $P(A_t|Y_t)$ with respect to A_{t-1}, the filtering density is written as:

$$P(\alpha_t|Y_t) = \int P(A_t|Y_t)\,dA_{t-1}. \tag{24}$$

Eq. (24) can be derived from Eqs. (14) and (15). See Appendix C.

Smoothing. The conditional density of A_T given Y_T is obtained as follows:

$$\begin{aligned} P(A_T|Y_T) &= \frac{P(A_T, Y_T)}{\int P(A_T, Y_T)\,dA_T} \\ &= \frac{P_\alpha(A_T)P_y(Y_T|A_T)}{\int P_\alpha(A_T)P_y(Y_T|A_T)\,dA_T}. \end{aligned} \tag{25}$$

Let us define $A_t^* = \{\alpha_t, \alpha_{t+1}, \ldots, \alpha_T\}$, where A_t^* satisfies the following properties: (i) $A_T = A_0^*$ and (ii) $A_T = \{A_t, A_{t+1}^*\}$ for $t = 0, 1, \ldots, T-1$. From Eq. (25), using A_{t+1}^* the smoothing density at time t, i.e., $P(\alpha_t|Y_T)$, is given by:

$$P(\alpha_t|Y_T) = \frac{\iint P_\alpha(A_T)P_y(Y_T|A_T)\,dA_{t-1}\,dA_{t+1}^*}{\int P_\alpha(A_T)P_y(Y_T|A_T)\,dA_T}, \tag{26}$$

for $t = 1, 2, \ldots, T$. Again, note that it is easy to derive the standard density-based smoothing algorithm (16) from Eq. (26). See Appendix C for the proof.

Mean, variance and likelihood function. Using Eq. (21) for prediction, Eq. (23) for filtering and Eq. (25) for smoothing, evaluation of the conditional expectation of a function $g(\alpha_t)$ is given by:

$$\begin{aligned} \mathrm{E}\big(g(\alpha_t)|Y_s\big) &= \int g(\alpha_t)P(A_r|Y_s)\,dA_r \\ &= \frac{\int g(\alpha_t)P(A_r, Y_s)\,dA_r}{\int P(A_r, Y_s)\,dA_r} \\ &= \frac{\int g(\alpha_t)P_\alpha(A_r)P_y(Y_s|A_s)\,dA_r}{\int P_\alpha(A_r)P_y(Y_s|A_s)\,dA_r}, \end{aligned}$$

for $(r, s) = (t+L, t), (t, t), (T, T)$.

In the case where Eqs. (1) and (2) depends on an unknown parameter, the likelihood function to be maximized is written as:

$$P(Y_T) = \int P(A_T, Y_T)\,dA_T = \int P_y(Y_T|A_T)P_\alpha(A_T)\,dA_T,$$

which corresponds to the denominator of Eq. (23) in the filtering formula and moreover it is equivalent to the innovation form of the likelihood function given by Eq. (18).

An alternative estimation method of an unknown parameter is known as the EM algorithm (Expectation–Maximization algorithm), where the expected log-likelihood function is maximized with respect to the parameter, given all the observed data Y_T (see Dempster, Laird and Rubin, 1977; Rund, 1991; Laird, 1993 for the EM algorithm). That is, for the EM algorithm, the following expected log-likelihood function is maximized:

$$E(\log(P(A_T, Y_T))|Y_T) = E(\log(P_y(Y_T|A_T)P_\alpha(A_T))|Y_T)$$
$$= \int \log(P_y(Y_T|A_T)P_\alpha(A_T))P(A_T|Y_T)\,dA_T. \quad (27)$$

As for the features of the EM algorithm, it is known that the convergence speed is very slow but it quickly searches the neighborhood of the true parameter value. Shumway and Stoffer (1982) and Tanizaki (1989) applied the EM algorithm to the state-space model in linear and normal case.

3. Nonlinear and non-Gaussian state-space modeling

Kitagawa (1987) and Kramer and Sorenson (1988) proposed the nonlinear and non-Gaussian filter and smoother using numerical integration. It is well known that the numerical integration procedure is not suitable to the multi-dimensional cases from computational point of view. In order to improve this disadvantage of the numerical integration approach, recently, simulation-based nonlinear and nonnormal filters and smoothers have been investigated. Mariano and Tanizaki (1995), Tanizaki (1993a, 1996) and Tanizaki and Mariano (1994) applied Monte Carlo integration with importance sampling to evaluate each integration. Carlin, Polson and Stoffer (1992), Carter and Kohn (1994, 1996) and Chib and Greenberg (1996) utilized the Gibbs sampler. Gordon, Salmond and Smith (1993), Kitagawa (1996, 1998) and Kitagawa and Gersch (1996) proposed a nonlinear filter using a resampling procedure. Hürzeler and Künsch (1998), Mariano and Tanizaki (2000), Tanizaki (1996, 1999a) and Tanizaki and Mariano (1998) introduced a nonlinear and nonnormal filter with rejection sampling. Furthermore, using the Metropolis–Hastings algorithm within Gibbs sampling, a nonlinear and non-Gaussian smoother was proposed by Geweke and Tanizaki (1999, 2001). See Appendix B for the sampling techniques such as Monte Carlo integration with importance sampling, rejection sampling, Gibbs sampling and Metropolis–Hastings algorithm. In any case, their nonlinear and non-Gaussian filters and smoothers can be applied to any nonlinear and non-Gaussian cases. Generating random draws from the filtering and smoothing densities at each time, the filtering and smoothing means are evaluated. The obtained estimates clearly go to the true means as number of random draws increases.

3.1. Numerical integration

In this section, we introduce the nonlinear and non-Gaussian prediction, filtering and smoothing proposed by Kitagawa (1987) and Kramer and Sorenson (1988), which utilizes numerical integration.

Let $\alpha_{i,t}$, $i = 0, 1, \ldots, N$, be the nodes at time t, which are assumed to be sorted in order of size, i.e., $\alpha_{0,t}$ is the smallest value and $\alpha_{N,t}$ is the largest one. For numerical integration, there are some integration methods such as sum of rectangles, sum of trapezoids, Simpson's formula and so on. In this section, for simplicity of discussion, we introduce the numerical integration prediction, filtering and smoothing with the rectangle rule.[3] However, it can be easily extended to the other integration techniques.

Prediction. Eq. (13) is approximated as[4]:

$$P(\alpha_{i,t+L}|Y_t)$$
$$\approx \sum_{j=1}^{N} P_\alpha(\alpha_{i,t+L}|\alpha_{j,t+L-1}) P(\alpha_{j,t+L-1}|Y_t)(\alpha_{j,t+L-1} - \alpha_{j-1,t+L-1}).$$

Thus, a recursive algorithm of the L-step ahead prediction density is obtained.

Filtering. Eqs. (14) and (15) are numerically integrated as follows:

$$P(\alpha_{i,t}|Y_{t-1}) \approx \sum_{j=1}^{N} P_\alpha(\alpha_{i,t}|\alpha_{j,t-1}) P(\alpha_{j,t-1}|Y_{t-1})(\alpha_{j,t-1} - \alpha_{j-1,t-1}), \quad (28)$$

$$P(\alpha_{i,t}|Y_t) \approx \frac{P_y(y_t|\alpha_{i,t}) P(\alpha_{i,t}|Y_{t-1})}{\sum_{j=1}^{N} P_y(y_t|\alpha_{j,t}) P(\alpha_{j,t}|Y_{t-1})(\alpha_{j,t} - \alpha_{j-1,t})}, \quad (29)$$

where the initial condition is given by:

$$P(\alpha_{i,1}|Y_0) \approx \sum_{j=1}^{N} P_\alpha(\alpha_{i,1}|\alpha_{j,0}) P_\alpha(\alpha_{j,0})(\alpha_{j,0} - \alpha_{j-1,0}),$$

if α_0 is stochastic and

$$P(\alpha_{i,1}|Y_0) = P_\alpha(\alpha_{i,1}|\alpha_0),$$

---

[3] In Monte Carlo experiments of Section 4, numerical integration is performed by the trapezoid rule.

[4] Note on the expression of numerical integration as follows. Let x be a k-dimensional vector, i.e., $x = (x_1, x_2, \ldots, x_k)$, and x_i be the ith node, i.e., $x_i = (x_{i,1}, x_{i,2}, \ldots, x_{i,k})$. Suppose that we have N nodes, which are sorted by size. Numerical integration of $g(x)$ with respect to x is expressed as:

$$\int g(x)\,dx \approx \sum_{i_1=1}^{N} \sum_{i_2=1}^{N} \cdots \sum_{i_k=1}^{N} g(x_{i_1,1}, x_{i_2,2}, \ldots, x_{i_k,k})$$
$$\times (x_{i_1,1} - x_{i_1-1,1})(x_{i_2,2} - x_{i_2-1,2}) \cdots (x_{i_k,k} - x_{i_k-1,k})$$
$$\equiv \sum_{i=1}^{N} g(x_i)(x_i - x_{i-1}).$$

That is, in this paper we write the multivariate numerical integration as the second line of the above equation.

otherwise. $P(\alpha_{i,t}|Y_{t-1})$ for $i = 1, 2, \ldots, N$ is obtained given $P(\alpha_{j,t-1}|Y_{t-1})$ for $j = 1, 2, \ldots, N$ from Eq. (28), while $P(\alpha_{i,t}|Y_t)$ for $i = 1, 2, \ldots, N$ is computed given $P(\alpha_{j,t}|Y_{t-1})$ for $j = 1, 2, \ldots, N$ using Eq. (29). Thus, Eqs. (28) and (29) give us the recursive algorithm.

Smoothing. Eq. (16) is approximated as:

$$P(\alpha_{i,t}|Y_T) \approx P(\alpha_{i,t}|Y_t) \sum_{j=1}^{N} \frac{P(\alpha_{j,t+1}|Y_T) P_\alpha(\alpha_{j,t+1}|\alpha_{i,t})}{P(\alpha_{j,t+1}|Y_t)} (\alpha_{j,t+1} - \alpha_{j-1,t+1}),$$

which is a backward recursive algorithm. That is, given the two densities $P_\alpha(\alpha_{j,t+1}|\alpha_{i,t})$ and $P(\alpha_{j,t+1}|Y_t)$, the smoothing density at time t (i.e., $P(\alpha_{i,t}|Y_T)$) is computed from the smoothing density at time $t+1$ (i.e., $P(\alpha_{j,t+1}|Y_T)$).

Mean, variance and likelihood function. Eq. (17) is evaluated as:

$$\mathrm{E}\big(g(\alpha_r)|Y_s\big) \approx \sum_{j=1}^{N} g(\alpha_{j,r}) P(\alpha_{j,r}|Y_s)(\alpha_{j,r} - \alpha_{j-1,r}),$$

for $(r, s) = (t+L, t), (t, t), (t, T)$.

The likelihood function (18) is computed as:

$$P(Y_T) \approx \prod_{t=1}^{T} \left(\sum_{j=1}^{N} P(y_t|\alpha_{j,t}) P(\alpha_{j,t}|Y_{t-1})(\alpha_{j,t} - \alpha_{j-1,t}) \right),$$

which is obtained from the denominator of the filtering recursion (29).

Some comments. According to the numerical integration approach, clearly we can obtain asymptotically unbiased estimate of $\mathrm{E}(g(\alpha_r)|Y_s)$. Note that an asymptotically unbiased estimate in this case implies that the estimate goes to the true state variable as N increases for all t. However, some problems are encountered. First, when we compute the densities numerically, computation errors are accumulated and therefore integration of $P(\alpha_r|Y_s)$ is not equal to one in practice. In such situations, as L or t increases, we have poor density approximation. To improve this problem, we need ad hoc modification which satisfies the following condition: $\sum_{i=1}^{N} P(\alpha_{i,r}|Y_s)(\alpha_{i,r} - \alpha_{i-1,r}) = 1$, which comes from $\int P(\alpha_r|Y_s)\,\mathrm{d}\alpha_r = 1$. Thus, $P(\alpha_r|Y_s)$ should be re-computed for each time r.

Another problem is choice of the nodes. Density approximation is imprecise when number of nodes is small and/or location of the nodes is not appropriate. If the nodes $\alpha_{i,r}$ for $i = 1, 2, \ldots, N$ are distributed away from the true distribution of α_r given Y_s, the estimate of $\mathrm{E}(g(\alpha_r)|Y_s)$ becomes unrealistic. The nodes $\alpha_{i,r}$ for $i = 1, 2, \ldots, N$ have to be overlapped with $P(\alpha_r|Y_s)$. However, the true distribution of α_r given Y_s is unknown in practice. Let $a^*_{t|s}$ be the extended Kalman filtering mean and $D^*_{t|s}$ be the vector which consists of the diagonal elements of the extended Kalman filtering variance $\Sigma^*_{t|s}$. In the case of filtering, Tanizaki (1993a, 1996), Tanizaki and Mariano (1994) suggested taking the nodes $\alpha_{i,t}$ from the interval: $[a^*_{t|s} - (cD^*_{t|s})^{1/2}, a^*_{t|s} + (cD^*_{t|s})^{1/2}]$, where half of the nodes are taken from $s = t$ and the rest of the nodes are from $s = t - 1$. c is a fixed value and $c \geqslant 1$ is preferable (remember that the probability is 99% when $c = 9$ in the case of normal distribution). Thus, the numerical integration procedure is obtained based on the extended Kalman filter. It might be plausible to consider that mean and variance from the extended Kalman filter are not far from those from the true distribution.

Moreover, when the numerical integration approach is applied to the higher-dimensional cases, it takes an extraordinarily long time and also results in tedious programming. Computational time for $P(\alpha_{i,r}|Y_s)$, $i = 1, 2, \ldots, N$, is proportional to $L \times kN \times N^k$ for L-step ahead prediction, $T \times (kN \times N^k + kN \times N^k)$ for filtering, and $T \times kN \times N^k$ for smoothing.[5] The procedures shown in the proceeding sections can be easily extended to the higher-dimensional cases in the sense of both computational time and programming.

Thus, the numerical integration approach gives us the asymptotically unbiased estimator as number of the nodes increases, but we should keep in mind the following two problems: first, location of nodes has to be set by a researcher, and second, computational time increases more than proportionally as the dimension of the state variable is high.

3.2. Monte Carlo integration with importance sampling

Mariano and Tanizaki (1995), Tanizaki (1993a, 1996, 1999b) and Tanizaki and Mariano (1994) proposed the nonlinear filtering and smoothing algorithms with importance sampling, where recursive algorithms of density functions are converted by those of weight functions.

Define the weight function as:

$$\omega(\alpha_r|Y_s) \equiv \frac{P(\alpha_r|Y_s)}{P_*(\alpha_r)},$$

for $(r, s) = (t + L, t), (t, t), (t, T)$. The density function $P_*(\alpha_r)$ has to be appropriately specified by a researcher, which is called the importance density. Let $\alpha_{i,r}$, $i = 1, 2, \ldots, N$, be the random numbers generated from the importance density $P_*(\alpha_r)$ for $r = 1, 2, \ldots, T$.

Prediction. Using the weight function, the prediction equation (13) is transformed into:

$$\omega(\alpha_{i,t+L}|Y_t) = \int \frac{P_\alpha(\alpha_{i,t+L}|\alpha_{t+L-1})}{P_*(\alpha_{i,t+L})} \omega(\alpha_{t+L-1}|Y_t) P_*(\alpha_{t+L-1}) \, d\alpha_{t+L-1}$$

$$\approx \frac{1}{N} \sum_{j=1}^{N} \frac{P_\alpha(\alpha_{i,t+L}|\alpha_{j,t+L-1})}{P_*(\alpha_{i,t+L})} \omega(\alpha_{j,t+L-1}|Y_t),$$

---

[5] Note that we need $kN \times N^k$ computation to obtain each integration in Eqs. (13)–(16), where k denotes number of elements of the state variable α_t.

for $L = 1, 2, \ldots$. In the second line, integration is approximated using $\alpha_{j,t+L-1}$, $j = 1, 2, \ldots, N$, which are generated from $P_*(\alpha_{t+L-1})$. Therefore, given $\omega(\alpha_{j,t+L-1}|Y_t)$, we can compute $\omega(\alpha_{j,t+L}|Y_t)$. Thus, the weight functions $\omega(\alpha_{j,t+L}|Y_t)$, $L = 1, 2, \ldots$, are recursively obtained.

Filtering. Using the weight functions $\omega(\alpha_t|Y_s)$ for $s = t-1, t$, the density-based filtering algorithm given by Eqs. (14) and (15) is rewritten as follows:

$$\omega(\alpha_{i,t}|Y_{t-1}) = \int \frac{P_\alpha(\alpha_{i,t}|\alpha_{t-1})}{P_*(\alpha_{i,t})} \omega(\alpha_{t-1}|Y_{t-1}) P_*(\alpha_{t-1}) \, d\alpha_{t-1}$$

$$\approx \frac{1}{N} \sum_{j=1}^{N} \frac{P_\alpha(\alpha_{i,t}|\alpha_{j,t-1})}{P_*(\alpha_{i,t})} \omega(\alpha_{j,t-1}|Y_{t-1}), \tag{30}$$

$$\omega(\alpha_{i,t}|Y_t) = \frac{P_y(y_t|\alpha_{i,t}) \omega(\alpha_{i,t}|Y_{t-1})}{\int P_y(y_t|\alpha_t) \omega(\alpha_t|Y_{t-1}) P_*(\alpha_t) \, d\alpha_t}$$

$$\approx \frac{P_y(y_t|\alpha_{i,t}) \omega(\alpha_{i,t}|Y_{t-1})}{(1/N) \sum_{j=1}^{N} P_y(y_t|\alpha_{j,t}) \omega(\alpha_{j,t}|Y_{t-1})}, \tag{31}$$

where the initial condition of the weight function is given by:

$$\omega(\alpha_{i,1}|Y_0) \approx \frac{1}{N} \sum_{j=1}^{N} \frac{P_\alpha(\alpha_{i,1}|\alpha_{j,0})}{P_*(\alpha_{i,1})} \omega(\alpha_{j,0}|Y_0),$$

when α_0 is stochastic and

$$\omega(\alpha_{i,1}|Y_0) = \frac{P_\alpha(\alpha_{i,1}|\alpha_0)}{P_*(\alpha_{i,1})},$$

when α_0 is nonstochastic. Note that $\omega(\alpha_{j,0}|Y_0)$ is given by: $\omega(\alpha_{j,0}|Y_0) \equiv P_\alpha(\alpha_{j,0})/P_*(\alpha_{j,0})$. The recursive algorithm shown in Eqs. (30) and (31) is implemented by the following two steps: (i) given $\omega(\alpha_{j,t-1}|Y_{t-1})$ for $j = 1, 2, \ldots, N$, $\omega(\alpha_{i,t}|Y_{t-1})$ for $i = 1, 2, \ldots, N$ are obtained from Eq. (30), and (ii) given $\omega(\alpha_{j,t}|Y_{t-1})$ for $j = 1, 2, \ldots, N$, $\omega(\alpha_{i,t}|Y_t)$ for $i = 1, 2, \ldots, N$ are given by Eq. (31). Thus, the weight functions $\omega(\alpha_{j,t}|Y_t)$ are recursively computed for $t = 1, 2, \ldots, T$ and $j = 1, 2, \ldots, N$.

Smoothing. Using Monte Carlo integration with importance sampling, Eq. (16) is rewritten as:

$$\omega(\alpha_{i,t}|Y_T) = \omega(\alpha_{i,t}|Y_t) \int \frac{\omega(\alpha_{t+1}|Y_T) P_\alpha(\alpha_{t+1}|\alpha_{i,t})}{\omega(\alpha_{t+1}|Y_t) P_*(\alpha_{t+1})} P_*(\alpha_{t+1}) \, d\alpha_{t+1}$$

$$\approx \omega(\alpha_{i,t}|Y_t) \frac{1}{N} \sum_{j=1}^{N} \frac{\omega(\alpha_{j,t+1}|Y_T) P_\alpha(\alpha_{j,t+1}|\alpha_{i,t})}{\omega(\alpha_{j,t+1}|Y_t) P_*(\alpha_{j,t+1})}, \tag{32}$$

which is a backward recursive algorithm of the weight functions. Thus, based on the filtering formula (30) and (31), $\omega(\alpha_{i,t}|Y_T)$ is computed from $\omega(\alpha_{j,t+1}|Y_T)$ from Eq. (32).

Mean, variance and likelihood function. Using the weight functions $\omega(\alpha_{i,r}|Y_s)$ for $t = 1, 2, \ldots, T$, Eq. (17) is approximated as follows:

$$E(g(\alpha_r)|Y_s) = \int g(\alpha_r)\omega(\alpha_r|Y_s)P_*(\alpha_r)\,d\alpha_r$$

$$\approx \frac{1}{N}\sum_{i=1}^{N} g(\alpha_{i,r})\omega(\alpha_{i,r}|Y_s).$$

From definition of the weight function, we can evaluate the conditional density $P(\alpha_r|Y_s)$ at $\alpha_r = \alpha_{i,r}$, which is represented by:

$$P(\alpha_{i,r}|Y_s) = \omega(\alpha_{i,r}|Y_s)P_*(\alpha_{i,r}),$$

for $(r, s) = (t+L, t), (t, t), (t, T)$.

The likelihood function (18) can be transformed into:

$$P(Y_T) = \prod_{t=1}^{T}\left(\int P_y(y_t|\alpha_t)\omega(\alpha_t|Y_{t-1})P_*(\alpha_t)\,d\alpha_t\right)$$

$$\approx \prod_{t=1}^{T}\left(\frac{1}{N}\sum_{i=1}^{N} P_y(y_t|\alpha_{i,t})\omega(\alpha_{i,t}|Y_{t-1})\right),$$

which corresponds to the denominator of Eq. (31).

Some comments. Compared with the numerical integration procedure, the most attractive features of the importance sampling procedure is less computational burden, especially, in the higher-dimensional cases of the state vector. That is, for numerical integration, the nodes have to be sorted for all elements of α_t and sum of the rectangles has to be performed for all elements of α_t. For Monte Carlo integration, however, computational time does not depend on the dimension of the state vector, which is proportional to $L \times N \times N$ for L-step ahead prediction, $T \times (N \times N + N)$ for filtering, and $T \times N \times N$ for smoothing.[6] Thus, from computational point of view, the importance sampling procedure is more applicable in practice.

The disadvantage of the importance sampling approach is choice of the importance density. It is important that the importance density should not be too different from the target density (see Geweke, 1988, 1989a, 1989b). If the importance density $P_*(\alpha_r)$ is chosen away from $P(\alpha_r|Y_s)$, the weight function $\omega(\alpha_r|Y_s)$ becomes unrealistic. Accordingly, the obtained estimates are biased in such a case. In the case of filtering, the importance density $P_*(\alpha_t)$ needs to cover both $P(\alpha_t|Y_{t-1})$ and $P(\alpha_t|Y_t)$ over the range of α_t. Usually, peak and range of $P(\alpha_t|Y_{t-1})$ is different from those of $P(\alpha_t|Y_t)$. In addition, range of $P(\alpha_t|Y_{t-1})$ is larger than that of $P(\alpha_t|Y_t)$ in general. For the

---

[6] Note that we need $N \times N$ order of computation for each integration in Eqs. (13), (14) and (16), and N order of computation for Eq. (15).

importance sampling filter, it is important that the two densities $P(\alpha_t|Y_s)$, $s = t-1, t$, have to be approximated by one importance density $P_*(\alpha_t)$. From some simulation studies, Tanizaki (1996) pointed out that we should choose the importance density with the following conditions: (i) the importance density should have a wide range of distribution, compared with the original distribution and (ii) center of the importance density should be close to that of the original density but we do not have to pay too much attention to center of the importance density. Tanizaki and Mariano (1994) and Mariano and Tanizaki (1995) suggested using the bimodal distribution for $P_*(\alpha_t)$, i.e.,

$$P_*(\alpha_t) = \frac{1}{2} N(a^*_{t|t-1}, c\Sigma^*_{t|t-1}) + \frac{1}{2} N(a^*_{t|t}, c\Sigma^*_{t|t}),$$

which denotes the average of two normal densities, where $a^*_{t|s}$ and $\Sigma^*_{t|s}$ for $s = t-1, t$ are the extended Kalman filter estimates and c is constant. Since $P_*(\alpha_t)$ should be broader than $P(\alpha_t|Y_{t-1})$ and $P(\alpha_t|Y_t)$, $c \geq 1$ must be chosen. The peak and range of $P(\alpha_t|Y_s)$ are not known in practice, but mean and variance of the state variable can be estimated by the extended Kalman filter even if it is the biased estimator. It is appropriate to consider that the extended Kalman filter estimates are not too far from the true values. Therefore, it might be plausible to take the bimodal distribution for the importance density. Thus, the importance sampling estimator would be improved by utilizing the importance density based on the extended Kalman filter.

Clearly, the Monte Carlo integration yields the simulation errors. Therefore, the numerical integration gives us a better approximation than the Monte Carlo integration. In order to avoid the problem, Geweke (1988) proposed the antithetic Monte Carlo integration and Tanizaki (1999b) suggested using the antithetic Monte Carlo integration in the nonlinear filtering framework.

Moreover, another problem, which is also found in the numerical integration procedure, is that we need to pay attention to accumulation of the computation errors. In order to reduce unreliability of the approximation, we must have the restriction of $(1/N) \sum_{t=1}^{T} \omega(\alpha_{i,r}|Y_s) = 1$ for all r and s, which comes from one of the properties of the density function.

Thus, each integration in the density-based algorithms is evaluated by numerical integration or Monte Carlo integration with importance sampling, where a researcher has to assume the nodes or the importance density. In the next two sections, we introduce the nonlinear filters and smoothers which do not require ad hoc assumption such as choice of the nodes or choice of the importance density.

3.3. Resampling

Using numerical integration or Monte Carlo integration, an attempt has been made to evaluate each integration in Eqs. (13)–(16). Recently, as computer progresses day by day, expectation (17) is evaluated generating random draws directly from the prediction, filtering and smoothing densities, where each integration is not computed explicitly. Gordon, Salmond and Smith (1993), Kitagawa (1996, 1998) and Kitagawa and Gersch (1996) proposed nonlinear filter and smoother using a resampling procedure (see, for example, Smith and Gelfand, 1992 for the resampling procedure). Let us define $\alpha_{i,r|s}$

as the ith random draw of α_r generated from the conditional density $P(\alpha_r|Y_s)$. We consider how to generate random draws $\alpha_{i,r|s}$, $i = 1, 2, \ldots, N$.

Prediction. The prediction estimate is very simple and easy (see, for example, Tanizaki and Mariano, 1998). Suppose that $\alpha_{i,t+L-1|t}$, $i = 1, 2, \ldots, N$, are available, which are the random draws generated from $(L-1)$-step ahead prediction density. Consider generating the random draws from L-step ahead prediction density, given $\alpha_{i,t+L-1|t}$, $i = 1, 2, \ldots, N$.

Given $\alpha_{j,t+L-1|t}$, $j = 1, 2, \ldots, N$, the L-step ahead prediction algorithm (13) is approximately represented as:

$$P(\alpha_{t+L}|Y_t) \approx \frac{1}{N} \sum_{j=1}^{N} P(\alpha_{t+L}|\alpha_{j,t+L-1|t}). \tag{33}$$

Therefore, $\alpha_{i,t+L|t}$ is generated as follows. Pick up $\alpha_{j,t+L-1|t}$ randomly (i.e., pick j with equal probability) and generate a random number of η_{t+L} (i.e., $\eta_{i,t+L}$), and we have a random draw of α_{t+L} (i.e., $\alpha_{i,t+L|t}$) from the transition equation: $\alpha_{i,t+L|t} = f_{t+L}(\alpha_{j,t+L-1|t}, \eta_{i,t+L})$.

Thus, $\alpha_{i,t+L|t}$ is recursively obtained based on $\alpha_{j,t+L-1|t}$ for $L = 1, 2, \ldots$. The initial random draws of the prediction algorithm (i.e., $\alpha_{i,t|t}$, $i = 1, 2, \ldots, N$) are generated in the filtering algorithm shown below.

Finally, note that the resampling procedure is not used to generate random draws from the prediction density (i.e., $\alpha_{i,t+L|t}$, $i = 1, 2, \ldots, N$), which is utilized in the following filtering and smoothing procedures.

Filtering. As discussed above, when $\alpha_{j,t-1|t-1}$ for $j = 1, 2, \ldots, N$ are available, $\alpha_{i,t|t-1}$ for $i = 1, 2, \ldots, N$ can be obtained from the prediction equation: $\alpha_{i,t|t-1} = f_t(\alpha_{j,t-1|t-1}, \eta_{i,t})$, where $\alpha_{j,t-1|t-1}$ is chosen with probability $1/N$ and $\eta_{i,t}$ denotes the ith random draw of η_t.

For filtering, we consider generating random draws α_t from the filtering density $P(\alpha_t|Y_t)$. Based on the random draws $\alpha_{i,t|t-1}$ for $i = 1, 2, \ldots, N$, Eq. (15) is approximately rewritten as follows:

$$P(\alpha_{i,t|t-1}|Y_t) \approx \frac{P_y(y_t|\alpha_{i,t|t-1})}{\sum_{j=1}^{N} P_y(y_t|\alpha_{j,t|t-1})}. \tag{34}$$

Note that $P(\alpha_{i,t|t-1}|Y_{t-1}) \approx 1/N$ is used to derive Eq. (34).[7] Eq. (34) is interpreted as follows. The probability which α_t takes $\alpha_{j,t|t-1}$ is approximately given by $P(\alpha_{j,t|t-1}|Y_t)$. Accordingly, $\alpha_{i,t|t}$ is chosen from $\alpha_{j,t|t-1}$, $j = 1, 2, \ldots, N$, with probability $P(\alpha_{j,t|t-1}|Y_t)$. That is, the ith random draw of α_t from $P(\alpha_t|Y_t)$, i.e., $\alpha_{i,t|t}$, is resampled as:

---

[7] Let x_i, $i = 1, 2, \ldots, N$, be the random draws from the density $P_x(x)$. Then, for all $i \neq j$, the probability which we have x_i should be equal to the probability which x_j occurs. In the case where we have N random draws of x, therefore, $P_x(x_i) \approx 1/N$ is obtained.

$$\alpha_{i,t|t} = \begin{cases} \alpha_{1,t|t-1}, & \text{with probability } P(\alpha_{1,t|t-1}|Y_t), \\ \alpha_{2,t|t-1}, & \text{with probability } P(\alpha_{2,t|t-1}|Y_t), \\ \vdots & \vdots \\ \alpha_{N,t|t-1}, & \text{with probability } P(\alpha_{N,t|t-1}|Y_t). \end{cases} \quad (35)$$

Thus, in order to obtain random draws from the filtering density, first we have to compute $P(\alpha_{j,t|t-1}|Y_t)$ for all $j = 1, 2, \ldots, N$ and next obtain $\alpha_{i,t|t}$ for $i = 1, 2, \ldots, N$ by resampling. In practice, a uniform random draw between zero and one (say, u) is generated and $\alpha_{j,t|t-1}$ is taken as $\alpha_{i,t|t}$ when $\omega_{j-1} \leqslant u < \omega_j$, where $\omega_j \equiv \sum_{m=1}^{j} P(\alpha_{m,t|t-1}|Y_t)$ and $\omega_0 \equiv 0$.

Smoothing. Assume that $\alpha_{j,t+1|T}$, $j = 1, 2, \ldots, N$, are available, which denote the random numbers of α_{t+1} generated from $P(\alpha_{t+1}|Y_T)$. The problem is to generate random draws of α_t from $P(\alpha_t|Y_T)$, given $\alpha_{j,t+1|T}$, $j = 1, 2, \ldots, T$. Eq. (16) is approximated as follows:

$$P(\alpha_t|Y_T) \approx P(\alpha_t|Y_t) \frac{1}{N} \sum_{j=1}^{N} \frac{P_\alpha(\alpha_{j,t+1|T}|\alpha_t)}{P(\alpha_{j,t+1|T}|Y_t)}$$

$$\approx P(\alpha_t|Y_t) \sum_{j=1}^{N} \frac{P_\alpha(\alpha_{j,t+1|T}|\alpha_t)}{\sum_{m=1}^{N} P_\alpha(\alpha_{j,t+1|T}|\alpha_{m,t|t})}. \quad (36)$$

Note that in the first line the integration is approximated by random numbers $\alpha_{j,t+1|T}$ and that in the second approximation we utilize

$$P(\alpha_{j,t+1|T}|Y_t) \approx (1/N) \sum_{m=1}^{N} P_\alpha(\alpha_{j,t+1|T}|\alpha_{m,t|t}).$$

Moreover, $P(\alpha_{m,t|t}|Y_t) \approx 1/N$ is obtained because $\alpha_{m,t|t}$ is a random number of α_t from $P(\alpha_t|Y_t)$. Therefore, Eq. (36) is rewritten as:

$$P(\alpha_{i,t|t}|Y_T) \approx \frac{1}{N} \sum_{j=1}^{N} \frac{P_\alpha(\alpha_{j,t+1|T}|\alpha_{i,t|t})}{\sum_{m=1}^{N} P_\alpha(\alpha_{j,t+1|T}|\alpha_{m,t|t})}. \quad (37)$$

Thus, it is shown from Eq. (37) that $\alpha_{i,t|T}$ is chosen from $\alpha_{j,t|t}$, $j = 1, 2, \ldots, N$, with probability $P(\alpha_{j,t|t}|Y_T)$. That is, the ith random draw of α_t from $P(\alpha_t|Y_T)$, i.e., $\alpha_{i,t|T}$, is resampled as:

$$\alpha_{i,t|T} = \begin{cases} \alpha_{1,t|t}, & \text{with probability } P(\alpha_{1,t|t}|Y_T), \\ \alpha_{2,t|t}, & \text{with probability } P(\alpha_{2,t|t}|Y_T), \\ \vdots & \vdots \\ \alpha_{N,t|t}, & \text{with probability } P(\alpha_{N,t|t}|Y_T). \end{cases} \quad (38)$$

To obtain random draws from the smoothing density, first we compute $P(\alpha_{j,t|t}|Y_T)$ for all $j = 1, 2, \ldots, N$ and next obtain $\alpha_{i,t|T}$ for $i = 1, 2, \ldots, N$ by resampling. Note that $\alpha_{i,T|T}$ is a random draw from the filtering density at time T. As discussed above, a uniform random number has to be compared with $\sum_{m=1}^{i} P(\alpha_{i,t|t}|Y_T)$ when we obtain $\alpha_{i,t|T}$ in practice.

From resampling procedures (35) and (38), both $\alpha_{i,t|t}$ and $\alpha_{i,t|T}$ are chosen from $\alpha_{i,t|t-1}$, $i = 1, 2, \ldots, N$, with different probabilities.

Mean, variance and likelihood function. When the random draws (i.e., $\alpha_{i,r|s}$) are available, the conditional mean (17) is simply computed as:

$$\mathrm{E}\big(g(\alpha_r)|Y_s\big) \approx \frac{1}{N} \sum_{i=1}^{N} g(\alpha_{i,r|s}), \tag{39}$$

for $(r, s) = (t+L, t), (t, t), (t, T)$.

Using one-step ahead prediction random draws, the likelihood function (18) is approximately rewritten as:

$$P(Y_T) \approx \prod_{t=1}^{T}\left(\frac{1}{N} \sum_{i=1}^{N} P_y(y_t|\alpha_{i,t|t-1})\right).$$

Some comments. In both numerical integration and importance sampling procedures, each integration in Eqs. (13)–(16) is evaluated to compute prediction, filtering and smoothing means. However, in the resampling procedure proposed by Gordon, Salmond and Smith (1993), Kitagawa (1996, 1998) and Kitagawa and Gersch (1996), prediction, filtering and smoothing means are computed using the random draws which are directly generated from each appropriate density, where we do not need to evaluate any integration. The obtained algorithm is based on the recursive algorithm of random draws. The resampling procedure improves over the nonlinear filters based on the numerical integration and importance sampling procedures from simplicity of computer programming and no ad hoc assumptions such as choice of the nodes for numerical integration (Kitagawa, 1987; Kramer and Sorenson, 1988) and choice of the importance density for Monte Carlo integration (Tanizaki, 1993a; Tanizaki and Mariano, 1994; Mariano and Tanizaki, 1995). It might be expected that the estimates obtained from the resampling procedure go to the true state vector values as number of the random draws increases.

The disadvantage of the resampling procedure is that computational time extremely increases, especially for smoothing, as number of random draws (i.e., N) increases. Computational burden is given by the order of $L \times N$ for prediction, $T \times (N + N^2)$ for filtering, and $T \times (N^3 + N^2)$ for smoothing.[8]

---

[8] In the case of filtering, we need the following computation: N for (34) and N^2 for comparison between the uniform random draw u and $\sum_{m=1}^{j} P(\alpha_{m,t|t-1}|Y_t)$. In smoothing case, the orders of computational burden are N^3 for Eq. (37) and N^2 for comparison between the uniform random draw u and $\sum_{m=1}^{i} P(\alpha_{m,t|t}|Y_T)$.

3.4. Rejection sampling

An alternative approach to nonlinear and nonnormal prediction, filtering and smoothing algorithms is proposed by Tanizaki (1996, 1999a), Tanizaki and Mariano (1998) and Mariano and Tanizaki (2000). Given random draws of the state vector which are directly generated from the filtering or smoothing density, the filtering or smoothing mean is recursively obtained, where we do not evaluate any integration included in the density-based algorithms (13)–(16). The procedure introduced in this section is similar to the resampling procedure in Section 3.4. However, for random number generation, rejection sampling is adopted in this section.

L-step ahead prediction is derived in the exactly same fashion as in Section 3.3. Therefore, we start with filtering in this section.

Filtering. Suppose that the random draws $\alpha_{i,t-1|t-1}$, $i = 1, 2, \ldots, N$, are available. Then, we consider generating $\alpha_{i,t|t}$, $i = 1, 2, \ldots, N$. By substituting Eq. (14) into Eq. (15), the filtering density at time t, $P(\alpha_t|Y_t)$, is approximated as:

$$P(\alpha_t|Y_t) = \frac{1}{\gamma_t} \int P_y(y_t|\alpha_t) P_\alpha(\alpha_t|\alpha_{t-1}) P(\alpha_{t-1}|Y_{t-1}) \, d\alpha_{t-1}$$

$$\approx \sum_{i=1}^{N} \frac{\gamma_{i,t}}{\gamma_t} \frac{1}{N} \left(\frac{P_y(y_t|\alpha_t) P_\alpha(\alpha_t|\alpha_{i,t-1|t-1})}{\gamma_{i,t}} \right)$$

$$\approx \sum_{i=1}^{N} \frac{\hat{\gamma}_{i,t}}{\hat{\gamma}_t} \frac{1}{N} \left(\frac{P_y(y_t|\alpha_t) P_\alpha(\alpha_t|\alpha_{i,t-1|t-1})}{\gamma_{i,t}} \right)$$

$$\equiv \sum_{i=1}^{N} q_{i,t} \left(\frac{P_y(y_t|\alpha_t) P_\alpha(\alpha_t|\alpha_{i,t-1|t-1})}{\gamma_{i,t}} \right), \tag{40}$$

where γ_t denotes the denominator of Eq. (15). In the second line, approximation is used by generating $\alpha_{i,t-1|t-1}$ from $P(\alpha_{t-1}|Y_{t-1})$. In the third line, γ_t is approximately equal to $\hat{\gamma}_t$, which is rewritten as follows:

$$\gamma_t \equiv \iint P_y(y_t|\alpha_t) P_\alpha(\alpha_t|\alpha_{t-1}) P(\alpha_{t-1}|Y_{t-1}) \, d\alpha_{t-1} \, d\alpha_t$$

$$\approx \frac{1}{N^2} \sum_{j=1}^{N} \sum_{i=1}^{N} P_y(y_t|\alpha_{ji,t|t-1}) \equiv \hat{\gamma}_t,$$

where $\alpha_{ji,t|t-1}$ denotes the jth random draw of α_t generated from $P(\alpha_t|\alpha_{i,t-1|t-1})$. Moreover, $\gamma_{i,t}$ and $\hat{\gamma}_{i,t}$ are represented as:

$$\gamma_{i,t} \equiv \int P_y(y_t|\alpha_t) P_\alpha(\alpha_t|\alpha_{i,t-1|t-1}) \, d\alpha_t \approx \frac{1}{N} \sum_{j=1}^{N} P_y(y_t|\alpha_{ji,t|t-1}) \equiv \hat{\gamma}_{i,t}.$$

Furthermore, in the fourth line, $q_{i,t}$ is defined as $q_{i,t} \equiv \hat{\gamma}_{i,t}/N\hat{\gamma}_t$.

Thus, from Eq. (40), $P(\alpha_t|Y_t)$ is approximated as a mixture of N distributions with probability $q_{i,t}$, $i = 1, 2, \ldots, N$. That is, $\alpha_{i,t-1|t-1}$ is chosen with probability $q_{i,t}$. Therefore, given $\alpha_{i,t-1|t-1}$, the next problem is how to generate a random draw of α_t from the density:

$$P_y(y_t|\alpha_t) P_\alpha(\alpha_t|\alpha_{i,t-1|t-1})/\gamma_{i,t}. \tag{41}$$

Rejection sampling is applied to obtain the random draw.

Let $P_*(z)$ be the proposal density. The acceptance probability $\omega(z)$ is defined as:

$$\omega(z) = \frac{P_y(y_t|z) P_\alpha(z|\alpha_{i,t-1|t-1})/P_*(z)}{\sup_z P_y(y_t|z) P_\alpha(z|\alpha_{i,t-1|t-1})/P_*(z)}, \tag{42}$$

where we require the assumption which the denominator in Eq. (42) exists.

The estimation procedure for the rejection sampling filter is as follows: (i) pick $\alpha_{i,t-1|t-1}$ for i with probability $q_{i,t}$, (ii) generate a random draw z from $P_*(\cdot)$ and a uniform random draw u from the interval between zero and one, (iii) take z as $\alpha_{j,t|t}$ if $u \leqslant \omega(z)$ and go back to (ii) otherwise, (iv) repeat (i)–(iii) N times for $j = 1, 2, \ldots, N$, and (v) repeat (i)–(iv) T times for $t = 1, 2, \ldots, T$.

Note that rejection sampling is utilized in procedures (ii) and (iii). Even though the denominator in Eq. (42) exists (i.e., $\sup_z P_y(y_t|z) P_\alpha(z|\alpha_{i,t-1|t-1})/P_*(z) < \infty$), rejection sampling is very inefficient in the sense of computational time (i.e., number of rejection increases) if the acceptance probability $\omega(z)$ is close to zero. In such a case, (ii) and (iii) are repeated forever.

For choice of the proposal density, we might consider the following candidates, i.e., (i) one is $P_*(\alpha_t) = P_\alpha(\alpha_t|\alpha_{i,t-1|t-1})$ and (ii) another is $P_*(\alpha_t) = N(a^*_{t|t}, c\Sigma^*_{t|t})$, where c is a constant and $a^*_{t|t}$ and $\Sigma^*_{t|t}$ denote the first- and second-moments obtained from the extended Kalman filter. Note that c should be greater than one, because the proposal density should have larger variance than the target density (see Appendix B). However, as mentioned above, because the denominator in Eq. (42) has to exist, it might be easier and better to use the candidate (i) rather than the candidate (ii).[9]

Smoothing. Hürzeler and Künsch (1998), Tanizaki (1996) and Tanizaki and Mariano (1998) proposed the fixed-interval smoothing algorithm using rejection sampling, where a backward recursive algorithm of the random draws is derived. Suppose that $\alpha_{i,t+1|T}$ for $i = 1, 2, \ldots, N$ are available. Then we consider how to generate $\alpha_{i,t|T}$ for $i = 1, 2, \ldots, N$.

In order to obtain the rejection sampling smoothing, first, note that each component in the smoothing algorithm (16) is transformed as follows:

$$P(\alpha_t|Y_t) \approx \frac{1}{\gamma_t}\left(\frac{1}{N}\sum_{i=1}^{N} P_y(y_t|\alpha_t) P_\alpha(\alpha_t|\alpha_{i,t-1|t-1})\right), \tag{43}$$

---

[9] Even if we take the candidate (i) as the proposal density, we possibly have the case where the supremum does not exist, depending on the functional form of $P_y(\alpha_t|y_t)$.

$$\int \frac{P(\alpha_{t+1}|Y_T)P(\alpha_{t+1}|\alpha_t)}{P(\alpha_{t+1}|Y_t)} \, d\alpha_{t+1} \approx \frac{1}{N} \sum_{j=1}^{N} \frac{P_\alpha(\alpha_{j,t+1|T}|\alpha_t)}{P(\alpha_{j,t+1|T}|Y_t)}. \tag{44}$$

In Eq. (43), the integration in the first equality of Eq. (40) is approximated by using $\alpha_{i,t-1|t-1}$, $i = 1, 2, \ldots, N$, which are generated from $P(\alpha_{t-1}|Y_{t-1})$. Moreover, in Eq. (44), the integration is approximated using $\alpha_{j,t+1|T}$, $j = 1, 2, \ldots, N$, which are generated from $P(\alpha_{t+1}|Y_T)$.

Using Eqs. (43) and (44), the smoothing density (16) is approximated as:

$$P(\alpha_t|Y_T)$$

$$\approx \frac{1}{\gamma_t} \left(\frac{1}{N} \sum_{i=1}^{N} P_y(y_t|\alpha_t) P_\alpha(\alpha_t|\alpha_{i,t-1|t-1}) \right) \left(\frac{1}{N} \sum_{j=1}^{N} \frac{P_\alpha(\alpha_{j,t+1|T}|\alpha_t)}{P(\alpha_{j,t+1|T}|Y_t)} \right)$$

$$= \sum_{i=1}^{N} \sum_{j=1}^{N} \frac{\gamma_{ij,t}^*}{N^2 \gamma_t P(\alpha_{j,t+1|T}|Y_t)} \left(\frac{P_y(y_t|\alpha_t) P_\alpha(\alpha_{j,t+1|T}|\alpha_t) P_\alpha(\alpha_t|\alpha_{i,t-1|t-1})}{\gamma_{ij,t}^*} \right)$$

$$\approx \sum_{i=1}^{N} \sum_{j=1}^{N} \frac{\hat{\gamma}_{ij,t}^*}{N \sum_{i=1}^{N} \hat{\gamma}_{ij,t}^*} \left(\frac{P_y(y_t|\alpha_t) P_\alpha(\alpha_{j,t+1|T}|\alpha_t) P_\alpha(\alpha_t|\alpha_{i,t-1|t-1})}{\gamma_{ij,t}^*} \right)$$

$$= \sum_{i=1}^{N} \sum_{j=1}^{N} \frac{q_{ij,t}^*}{N} \left(\frac{P_y(y_t|\alpha_t) P_\alpha(\alpha_{j,t+1|T}|\alpha_t) P_\alpha(\alpha_t|\alpha_{i,t-1|t-1})}{\gamma_{ij,t}^*} \right), \tag{45}$$

for $t = T - 1, T - 2, \ldots, 1$, which is a backward recursion. $\gamma_{ij,t}^*$ is approximated as $\hat{\gamma}_{ij,t}^*$, which is represented by:

$$\gamma_{ij,t}^* \equiv \int P_y(y_t|\alpha_t) P_\alpha(\alpha_{j,t+1|T}|\alpha_t) P_\alpha(\alpha_t|\alpha_{i,t-1|t-1}) \, d\alpha_t$$

$$\approx \frac{1}{N} \sum_{m=1}^{N} P_y(y_t|\alpha_{mi,t|t-1}) P_\alpha(\alpha_{j,t+1|T}|\alpha_{mi,t|t-1}) \equiv \hat{\gamma}_{ij,t}^*.$$

$P(\alpha_{j,t+1|T}|Y_t)$ is approximated as follows:

$$P(\alpha_{j,t+1|T}|Y_t)$$

$$= \int P_\alpha(\alpha_{j,t+1|T}|\alpha_t) P(\alpha_t|Y_t) \, d\alpha_t$$

$$= \frac{1}{\gamma_t} \iint P_y(y_t|\alpha_t) P_\alpha(\alpha_{j,t+1|T}|\alpha_t) P_\alpha(\alpha_t|\alpha_{t-1}) P(\alpha_{t-1}|Y_{t-1}) \, d\alpha_{t-1} \, d\alpha_t$$

$$\approx \frac{1}{\gamma_t} \frac{1}{N} \sum_{i=1}^{N} \int P_y(y_t|\alpha_t) P_\alpha(\alpha_{j,t+1|T}|\alpha_t) P_\alpha(\alpha_t|\alpha_{i,t-1|t-1}) \, d\alpha_t$$

$$\approx \frac{1}{\gamma_t} \frac{1}{N^2} \sum_{i=1}^{N} \sum_{m=1}^{N} P_y(y_t|\alpha_{mi,t|t-1}) P_\alpha(\alpha_{j,t+1|T}|\alpha_{mi,t|t-1})$$

$$= \frac{1}{\gamma_t} \frac{1}{N} \sum_{i=1}^{N} \hat{\gamma}_{ij,t}^*.$$

Moreover, in the fourth line of Eq. (45), $q_{ij,t}^*$ is defined as: $q_{ij,t}^* \equiv \hat{\gamma}_{ij,t}^* / \sum_{i=1}^{N} \hat{\gamma}_{ij,t}^*$. Accordingly, we have the equality: $\sum_{i=1}^{N} q_{ij,t}^* = 1$ for all j.

Thus, in Eq. (45), given $\alpha_{i,t-1|t-1}$ and $\alpha_{j,t+1|T}$, the smoothing density $P(\alpha_t|Y_T)$ is approximated as a mixture of N^2 distributions. $\alpha_{i,t-1|t-1}$ and $\alpha_{j,t+1|T}$ are chosen with probabilities $q_{ij,t}^*$ and $1/N$. The next problem is how to generate a random draw of α_t from the density:

$$P_y(y_t|\alpha_t) P_\alpha(\alpha_{j,t+1|T}|\alpha_t) P_\alpha(\alpha_t|\alpha_{i,t-1|t-1})/\gamma_{ij,t}^*. \tag{46}$$

The random draws from the smoothing density (45) are generated by rejection sampling.
The acceptance probability $\omega(z)$ is defined as:

$$\omega(z) = \frac{P_y(y_t|z) P_\alpha(\alpha_{j,t+1|T}|z) P_\alpha(z|\alpha_{i,t-1|t-1})/P_*(z)}{\sup_z P_y(y_t|z) P_\alpha(\alpha_{j,t+1|T}|z) P_\alpha(z|\alpha_{i,t-1|t-1})/P_*(z)}. \tag{47}$$

As discussed above, we need the assumption that the denominator in Eq. (47) is bounded.

The following procedure is taken for rejection sampling smoother: (i) pick one of $\alpha_{j,t+1|T}$, $j = 1, 2, \ldots, N$, with probability $1/N$ and one of $\alpha_{i,t-1|t-1}$, $i = 1, 2, \ldots, N$, with probability $q_{ij,t}^*$, (ii) generate a random draw z from the proposal density $P_*(\cdot)$ and a uniform random draw u from the interval between zero and one, (iii) take z as $\alpha_{m,t|T}$ if $u \leq \omega(z)$ and go back to (ii) otherwise, (iv) repeat (i)–(iii) N times for $m = 1, 2, \ldots, N$, and (v) repeat (i)–(iv) T times for $t = T-1, T-2, \ldots, 1$.

Mean, variance and likelihood function. As discussed in Section 3.3, utilizing the random draws $\alpha_{i,r|s}$ for $i = 1, 2, \ldots, N$ and $(r, s) = (t+L, t), (t, t), (t, T)$, the expectation of a function $g(\cdot)$ is given by Eq. (39).

From the definition of γ_t, the likelihood function is evaluated as:

$$P(Y_T) = \prod_{t=1}^{T} \gamma_t \approx \prod_{t=1}^{T} \hat{\gamma}_t = \prod_{t=1}^{T} \left(\frac{1}{N^2} \sum_{j=1}^{N} \sum_{i=1}^{N} P_y(y_t|\alpha_{ji,t|t-1}) \right).$$

Some comments. In the rejection sampling procedure, as in Section 3.3, we do not evaluate any integration included in the density-based filtering algorithm, where we utilize the random draws only and the recursive algorithm of random draws are derived. Compared with the numerical integration procedure and the importance sampling approach, the rejection sampling procedure has the advantages from simplicity of computer programming and no ad hoc assumptions. As number of random draws increases, the obtained estimates approach the true state vector values. The rejection sampling procedure does not need as many random draws as the resampling procedure to obtain the same precision of the filtering and smoothing estimates, from difference between the random number generation methods.

Computational burden is proportional to $L \times N$ for prediction, $T \times (N^2 + N \times (N + A))$ for filtering, and $T \times N \times (N^2 + N + A)$ for smoothing, where A denotes the average number of rejection to obtain one random draw.[10] Rejection sampling depends on A, which is related to the acceptance probability. Small acceptance probability implies large number of A. Thus, the random number generator by rejection sampling is inefficient when the acceptance probability $\omega(\cdot)$ is close to zero. That is, for rejection sampling, it sometimes takes a long time, especially when the acceptance probability $\omega(\cdot)$ is small. See, for example, Carlin and Polson (1991) and Carlin, Polson and Stoffer (1992). Thus, the rejection sampling has the disadvantage that we cannot exactly predict computation time.

To improve the rejection sampling procedure in the sense of computation time, we have the following strategies. One is that we may pick another j and/or i in procedure (i) and repeat procedures (ii) and (iii) again when the acceptance probability $\omega(\cdot)$ is too small. Alternatively, we may switch random number generation from rejection sampling to the Metropolis–Hastings algorithm when $\omega(\cdot)$ is too small (see Appendix B for the Metropolis–Hastings algorithm). That is, when repeating procedures (ii) and (iii), we perform the Metropolis–Hastings algorithm in parallel and take the random draw as $\alpha_{j,t|t}$ or $\alpha_{m,t|T}$ for enough large number of iteration if any generated random draw is not accepted by rejection sampling. Furthermore, another strategy is that we may approximately use $\alpha_{i,t|t-1} = f_t(\alpha_{i,t-1|t-1}, \eta_{i,t})$ to evaluate $\hat{\gamma}_t$, $\hat{\gamma}_{i,t}$ and $\hat{\gamma}^*_{ij,t}$, i.e.,

$$\begin{cases} \hat{\gamma}_t \equiv \frac{1}{N} \sum_{i=1}^{N} P_y(y_t|\alpha_{i,t|t-1}), \\ \hat{\gamma}_{i,t} \equiv P_y(y_t|\alpha_{i,t|t-1}), \\ \hat{\gamma}^*_{ij,t} \equiv P_y(y_t|\alpha_{i,t|t-1}) P_\alpha(\alpha_{j,t+1|T}|\alpha_{i,t|t-1}). \end{cases} \quad (48)$$

Under the above re-definitions, computational burden reduces to the order of $T \times (N + N \times (N + A))$ for filtering, and $T \times N \times (N + N + A)$ for smoothing.

3.5. Markov chain Monte Carlo (Metropolis–Hastings algorithm within Gibbs sampling)

Carlin, Polson and Stoffer (1992), Carter and Kohn (1994, 1996) and Chib and Greenberg (1996) introduced the nonlinear and/or non-Gaussian state-space models with Gibbs sampling. They investigated the nonlinear state-space models in the Bayesian framework. Moreover, the state-space models which they used are quite restricted to some functional forms, because they studied the special state-space models such that it is easy to generate random draws from the underlying assumptions. To

---

[10] Filtering computes N^2 to obtain $\hat{y}_t$ and $N \times (N + A)$ to generate N filtering random draws, while smoothing uses $(N^2 + N + A)$ to obtain one smoothing random draw at each time.

improve these problems, Geweke and Tanizaki (1999, 2001) proposed the nonlinear and non-Gaussian smoother using both Gibbs sampling and the Metropolis–Hastings algorithm, which would be suitable to any nonlinear and non-Gaussian state-space model. In this section, the nonlinear and/or non-Gaussian smoother proposed by Geweke and Tanizaki (1999, 2001) is introduced, where the measurement and transition equations are specified in any general formulation and the error terms are not necessarily normal.

Smoothing. We generate random draws of A_T directly from $P(A_T|Y_T)$, shown in Eq. (25). According to the Gibbs sampling theory, random draws of A_T from $P(A_T|Y_T)$ are based on those of α_t from $P(\alpha_t|A_{t-1}, A^*_{t+1}, Y_T)$ for $t = 1, 2, \ldots, T$, which is derived from Eqs. (19) and (20) and represented as the following equation:

$$P(\alpha_t|A_{t-1}, A^*_{t+1}, Y_T)$$
$$= \frac{P(A_T|Y_T)}{P(A_{t-1}, A^*_{t+1}|Y_T)}$$
$$= \frac{P_y(Y_T|A_T)P_\alpha(A_T)}{\int P_y(Y_T|A_T)P_\alpha(A_T)\,d\alpha_t}$$
$$\propto \begin{cases} P_y(y_t|\alpha_t)P_\alpha(\alpha_t|\alpha_{t-1})P_\alpha(\alpha_{t+1}|\alpha_t), & \text{if } t = 1, 2, \ldots, T-1, \\ P_y(y_t|\alpha_t)P_\alpha(\alpha_t|\alpha_{t-1}), & \text{if } t = T \text{ (i.e., endpoint)}, \end{cases} \quad (49)$$

where the third line of Eq. (49) utilizes Eqs. (19) and (20). Thus, Eq. (49) implies that a kernel of $P(\alpha_t|A_{t-1}, A^*_{t+1}, Y_T)$ is given by $P_y(y_t|\alpha_t)P_\alpha(\alpha_t|\alpha_{t-1})P_\alpha(\alpha_{t+1}|\alpha_t)$ when $t = 1, 2, \ldots, T-1$ and $P_y(y_t|\alpha_t)P_\alpha(\alpha_t|\alpha_{t-1})$ when $t = T$ (i.e., endpoint).

Using a kernel of $P(\alpha_t|A_{t-1}, A^*_{t+1}, Y_T)$, we consider generating random draws of A_T directly from $P(A_T|Y_T)$. Here, the Gibbs sampler is applied to random number generation. Let $\alpha_{i,t}$ be the ith random draw of the state vector at time t. Define $A_{i,t}$ and $A^*_{i,t}$ as $A_{i,t} = \{\alpha_{i,0}, \alpha_{i,1}, \ldots, \alpha_{i,t}\}$ and $A^*_{i,t} = \{\alpha_{i,t}, \alpha_{i,t+1}, \ldots, \alpha_{i,T}\}$, respectively, which are the ith random draws of A_t and A^*_t.

Let $P_*(z|x)$ be the proposal density, which is the conditional distribution of z given x. We should choose the proposal density $P_*(z|x)$ such that random draws can be easily and quickly generated. Define the acceptance probability $\omega(x, z)$ as follows:

$$\omega(x, z) = \begin{cases} \min\left(\frac{P(z|A_{i,t-1}, A^*_{i-1,t+1}, Y_T)P_*(x|z)}{P(x|A_{i,t-1}, A^*_{i-1,t+1}, Y_T)P_*(z|x)}, 1\right), \\ \quad \text{if } P(x|A_{i,t-1}, A^*_{i-1,t+1}, Y_T)P_*(z|x) > 0, \\ 1, \quad \text{otherwise.} \end{cases}$$

To generate random draws from $P(A_T|Y_T)$, the following procedure is taken: (i) pick up appropriate values for $\alpha_{1,0}$ and $\alpha_{0,t}$, $t = 1, 2, \ldots, T$, (ii) generate a random draw z from $P_*(\cdot|\alpha_{i-1,t})$ and a uniform random draw u from the uniform distribution between zero and one, (iii) set $\alpha_{i,t} = z$ if $u \leq \omega(\alpha_{i-1,t}, z)$ and set $\alpha_{i,t} = \alpha_{i-1,t}$ otherwise, (iv) repeat (ii) and (iii) for $t = 1, 2, \ldots, T$, and (v) repeat (ii)–(iv) for $i = 1, 2, \ldots, N$.

Note that the Metropolis–Hastings algorithm is used in procedures (ii) and (iii). In procedure (i), typically, the smoothing estimates based on the extended Kalman filter are taken for $\alpha_{0,t}$, $t = 1, 2, \ldots, T$. $\alpha_{i,0}$ for $i = 1, 2, \ldots, N$ depend on the underlying assumption of α_0. That is, $\alpha_{i,0}$ for $i = 1, 2, \ldots, N$ are generated from $P_\alpha(\alpha_0)$ if α_0 is stochastic and they are fixed as α_0 for all i if α_0 is nonstochastic.

Mean, variance and likelihood function. Based on the random draws $\alpha_{i,t}$ for $i = 1, 2, \ldots, N$, evaluation of $E(g(\alpha_t)|Y_T)$ is simply obtained as the arithmetic average of $g(\alpha_{i,t})$, $i = 1, 2, \ldots, N$, which is represented by:

$$E\big(g(\alpha_t)|Y_T\big) \approx \frac{1}{N-M} \sum_{i=M+1}^{N} g(\alpha_{i,t}).$$

Usually, 10–20% of N is taken for M, which implies that the first M random draws are discarded.

For estimation of unknown parameters, the conditional expectation of the log-likelihood function given by equation (27) is maximized (i.e., EM algorithm). Using the random draws generated from $P(A_T|Y_T)$, Eq. (27) is evaluated as follows:

$$E\big(\log(P(A_T, Y_T))|Y_T\big) \approx \frac{1}{N-M} \sum_{i=M+1}^{N} \log\big(P_y(Y_T|A_{i,T}) P_\alpha(A_{i,T})\big).$$

Some comments. For the Markov chain Monte Carlo method, numerous number of random draws have to be generated to obtain the same precision of the smoothing estimates as both the resampling procedure and the rejection sampling procedure. Generally, it is intractable to generate $\alpha_{i,t}$ from $P(\alpha_t|A_{i,t-1}, A^*_{i-1,t+1}, Y_T)$. In such a case, there two ways to generate random draws, i.e., one is rejection sampling and another is the Metropolis–Hastings algorithm. It is known that rejection sampling sometimes takes a long time computationally or it is not feasible in the case where the acceptance probability does not exists. Therefore, in order to generate numerous random draws very quickly, we apply the Metropolis–Hastings algorithm in procedures (ii) and (iii).

The Metropolis–Hastings algorithm has the problem of specifying the proposal density, which is the crucial criticism. Several generic choices of the proposal density are discussed by Tierney (1994) and Chib and Greenberg (1995). We may take the following several candidates for the proposal density function $P_*(z|x)$. First, it might be natural to take the density function obtained from the transition equation (2), i.e., $P_*(z|x) = P_\alpha(z|\alpha_{i,t-1})$. In this case, $P_*(z|x)$ does not depend on x, i.e., $P_*(z|x) = P_*(z)$, which is called the independence chain. Second, it is also possible to utilize the extended Kalman smoothed estimates, i.e., $P_*(z|x) = N(a^*_{t|T}, c\Sigma^*_{t|T})$ (this is also the independence chain), where $a^*_{t|T}$ and $\Sigma^*_{t|T}$ denote the first- and the second-moments (i.e., mean and variance) based on the extended Kalman smoothed estimates at time t and c is an appropriate constant value. Third, we may take the proposal density called the random walk chain, i.e., $P_*(z|x) = P_*(z-x)$, which is written as $P_*(z|x) =$

$N(x, c\Sigma^*_{t|T})$. Fourth, in the case where the state variable α_t lies on an interval, a uniform distribution between the interval might be taken as the proposal density. In Monte Carlo experiments of Section 4, $P_\alpha(z|\alpha_{i,t-1})$ and $N(a^*_{t|T}, c\Sigma^*_{t|T})$ are examined for the proposal density $P_*(z|x)$.

In this section, the filtering problem has not been discussed until now. The filtering procedure might be implemented as follows. Simply, replacing T by t in procedures (i)–(v), the random draws from the filtering density $P(\alpha_t|Y_t)$ are given by $\alpha_{i,t}$, $i = 1, 2, \ldots, N$, where t corresponds to the endpoint in the procedure (i)–(v). Recall that the random draws obtained at the endpoint represent the filtering random draws. In addition to procedures (i)–(v), we should put the following procedure: (vi) repeat (i)–(v) for $t = 1, 2, \ldots, T$. Accordingly, filtering is more computer-intensive than smoothing.[11] For prediction, Eq. (33) is utilized given the filtering random draws. Computational burden is as follows. Number of iteration is given by $T \times N$ for smoothing and $\sum_{t=1}^{T} Nt = NT(T-1)/2$ for filtering. It seems that the Markov chain Monte Carlo procedure is less computational than any other estimators. However, the Markov chain Monte Carlo methods need a lot of random draws, compared with the independence Monte Carlo methods such as importance sampling and rejection sampling, because in the Markov chain Monte Carlo methods we usually discard the first 10%–20% random draws and a random draw is positively correlated with the next random draw in general. Moreover, it is known that convergence of the Gibbs sampler is very slow especially in the case where there is high correlation between α_t and α_{t-1} (see Chib and Greenberg, 1995).

3.6. Quasi-filter and quasi-smoother

The resampling procedure and the rejection sampling approach takes a lot of time and the Markov chain Monte Carlo procedure sometimes has the feature of slow convergence. Improving the problems, Tanizaki (2000) proposed a quasi approach to nonlinear and/or non-Gaussian state estimation.

In this section, let $\alpha_{i,r|s}$ be the ith random draw of α_r from $P(\alpha_r|Y_s)$, which is the same notation as in Sections 3.3 and 3.4. The procedure in this section has the same computational burden as that in Section 3.5. Convergence speed is faster because the Gibbs sampler is not applied in this section.

For L-step ahead prediction, as discussed in Section 3.3, Eq. (33) is utilized based on filtering random draws $\alpha_{j,t|t}$, $j = 1, 2, \ldots, N$.

Filtering. We generate $\alpha_{i,t|t}$ based on $\alpha_{i,t-1|t-1}$, as in Section 3.4. In the filtering algorithm (i)–(v) of Section 3.4, when $q_{i,t} \approx 1/N$ is approximately taken, (iv) can be exchanged with (v). In (ii) and (iii), we can apply the Metropolis–Hastings algorithm to the random number generation.

---

[11] In the standard density-based smoothing algorithm, $P(\alpha_t|Y_{t-1})$ and $P(\alpha_t|Y_t)$ are required. After $P(\alpha_t|Y_{t-1})$ and $P(\alpha_t|Y_t)$ are computed for $t = 1, 2, \ldots, T$, $P(\alpha_t|Y_T)$ are obtained by the backward recursive algorithm. See Section 2.2.1 for the standard algorithms. Thus, clearly smoothing is more computer-intensive than filtering in the conventional density-based recursive algorithm. However, according to the Markov chain Monte Carlo procedure, it is easier to compute smoothing, rather than filtering.

Therefore, given $\alpha_{i,t-1|t-1}$ and $q_{i,t} \approx 1/N$, we generate a random draw of α_t from the density (41) by the Metropolis–Hastings algorithm. Define the acceptance probability as:

$$\omega(x,z) = \begin{cases} \min\left(\frac{P_y(y_t|z)P_\alpha(z|\alpha_{i,t-1|t-1})P_*(x|z)}{P_y(y_t|x)P_\alpha(x|\alpha_{i,t-1|t-1})P_*(z|x)}, 1\right), \\ \quad \text{if } P_y(y_t|x)P_\alpha(x|\alpha_{i,t-1|t-1})P_*(z|x) > 0, \\ 1, \quad \text{otherwise.} \end{cases}$$

The estimation procedure is as follows: (i) given $\alpha_{i,t-1|t-1}$, generate a random draw z from $P_*(\cdot|\alpha_{i-1,t|t})$ and a uniform random draw u from the uniform distribution between zero and one, (ii) set $\alpha_{i,t|t} = z$ if $u \leq \omega(\alpha_{i-1,t|t}, z)$ and set $\alpha_{i,t|t} = \alpha_{i-1,t|t}$ otherwise, (iii) repeat (i) and (ii) for $t = 1, 2, \ldots, T$, and (iv) repeat (i)–(iii) for $i = 1, 2, \ldots, N$. The extended Kalman filter estimates might be chosen for the initial random draws, i.e., $\alpha_{0,t|t} = a_{t|t}^*$ for $t = 1, 2, \ldots, T$. Thus, $\alpha_{i,t|t}$, $i = 1, 2, \ldots, N$, are obtained recursively.

Smoothing. Similarly, consider generating $\alpha_{i,t|T}$ based on $\alpha_{i,t-1|t-1}$ and $\alpha_{i,t+1|T}$. In the smoothing algorithm (i)–(v) of Section 3.4, when $q_{ij,t}^* \approx 1/N$ is approximately taken, (iv) is exchanged with (v). Moreover, in (ii) and (iii), the Metropolis–Hastings algorithm is applied.

Thus, given $\alpha_{i,t-1|t-1}$, $\alpha_{i,t+1|T}$ and $q_{ij,t}^* \approx 1/N$, we generate a random draw of α_t from the density (46) by the Metropolis–Hastings algorithm. Define the acceptance probability as:

$$\omega(x,z) = \begin{cases} \min\left(\frac{P_y(y_t|z)P_\alpha(z|\alpha_{i,t-1|t-1})P_\alpha(\alpha_{i,t+1|T}|z)P_*(x|z)}{P_y(y_t|x)P_\alpha(x|\alpha_{i,t-1|t-1})P_\alpha(\alpha_{i,t+1|T}|x)P_*(z|x)}, 1\right), \\ \quad \text{if } P_y(y_t|x)P_\alpha(x|\alpha_{i,t-1|t-1})P_\alpha(\alpha_{i,t+1|T}|x)P_*(z|x) > 0, \\ 1, \quad \text{otherwise.} \end{cases}$$

The smoothing procedure is implemented as the following backward recursive algorithm: (i) given $\alpha_{i,t+1|T}$ and $\alpha_{i,t-1|t-1}$, generate a random draw z from $P_*(\cdot|\alpha_{i-1,t|T})$ and a uniform random draw u from the uniform distribution between zero and one, (ii) set $\alpha_{i,t|T} = z$ if $u \leq \omega(\alpha_{i-1,t|T}, z)$ and set $\alpha_{i,t|T} = \alpha_{i-1,t|T}$ otherwise, (iii) repeat (i) and (ii) for $t = T - 1, T - 2, \ldots, 1$ (i.e., backward recursion), and (iv) repeat (i)–(iii) for $i = 1, 2, \ldots, N$. The extended Kalman smoothed estimates might be chosen for the initial random draws, i.e., $\alpha_{0,t|T} = a_{t|T}^*$ for $t = 1, 2, \ldots, T$.

Mean, variance and likelihood function. Evaluation of $E(g(\alpha_r)|Y_s)$ is given by:

$$E(g(\alpha_r)|Y_s) \approx \frac{1}{N-M} \sum_{i=M+1}^{N} g(\alpha_{i,r|s})$$

for all $(r, s) = (t + L, t), (t, t), (t, T)$. As discussed in Section 3.5, the first M random draws are excluded because of stability of the random draws.

The likelihood function (18) is represented as:

$$P(Y_T) = \prod_{t=1}^{T} \left(\frac{1}{N-M} \sum_{i=M+1}^{N} P_y(y_t|\alpha_{i,t|t-1}) \right),$$

where $\alpha_{i,t|t-1}$ is obtained from the transition equation $\alpha_{i,t|t-1} = f_t(\alpha_{i,t-1|t-1}, \eta_{i,t})$ given a random draw of η_t (i.e., $\eta_{i,t}$).

Some comments. The rejection sampling procedure discussed in Section 3.4 is slightly modified from the following two points: we impose the approximations of $q_{i,t} = 1/N$ and $q^*_{ij,t} = 1/N$ for all i and j, and the Metropolis–Hastings algorithm is utilized instead of rejection sampling in order to reduce computational burden. Computational time is the order of $T \times N$ for filtering, which implies the least computational burden of all the filters introduced in this paper, and $T \times N$ for smoothing.[12]

For precision the estimates, the quasi-filter and quasi-smoother might be inferior to the resampling and rejection sampling procedures, because the approximations of $q_{i,t} = 1/N$ and $q^*_{ij,t} = 1/N$ are taken. If the approximations are appropriate, the obtained estimates become plausible. However, it is important to note as follows. For smoothing, the numerical integration, the importance sampling, the resampling and the rejection sampling procedures require a great amount of data storage (i.e., the order of $T \times N$) but the Markov chain Monte Carlo procedure and the quasi-filter and quasi-smoother do not need too much storage (i.e., the order of T). From capacity of computer memory, the estimators discussed in Sections 3.5 and 3.6 are more useful than the other estimators.[13]

As pointed out in Section 3.5, we might consider several candidates of the proposal density. In Monte Carlo studies of Section 4, the following two types of the proposal density are examined: $P_*(z|x) = P_\alpha(z|\alpha_{i,t-1|t-1})$ and $P_*(z|x) = N(a^*_{t|s}, c\Sigma^*_{t|s})$ ($s = t$ for filtering and $s = T$ for smoothing).

4. Monte Carlo studies

In this section, by Monte Carlo studies, we compare numerical accuracy for all the estimators introduced in Section 3. The following state-space models are examined.

Simulation I (Linear and normal model). Consider the scalar system: $y_t = \alpha_t + \varepsilon_t$ and $\alpha_t = \delta \alpha_{t-1} + \eta_t$. The initial value α_0 and the error terms ε_t and η_t, $t = 1, 2, \ldots, T$, are assumed to be distributed as: $\alpha_0 \sim N(0, 1)$ and $(\varepsilon_t, \eta_t)' \sim N(0, I_2)$, where I_2

---

[12] For smoothing, the Markov chain Monte Carlo procedure in Section 3.5 is less computational then the quasi-smoother in this section. To obtain the smoothing random draws, the quasi-smoother requires the filtering random draws while the Markov chain Monte Carlo procedure does not utilize them.

[13] However, we should keep in mind that the Markov chain Monte Carlo approach uses a large number of random draws from the convergence property of the Gibbs sampler and that the quasi-filter and the quasi-smoother do not give us the exact solution.

Table 2
Extended Kalman filter and smoother (δ known)

T \ δ	I			II		III		IV
	0.5	0.9	1.0	0.5	0.9	0.5	0.9	
Filtering								
20	0.7292	0.7760	0.7897	0.7016	0.6748	1.1487	2.0909	23.351
40	0.7334	0.7793	0.7928	0.7039	0.6503	1.1577	2.2135	21.275
100	0.7307	0.7747	0.7878	0.7061	0.6439	1.1609	2.2656	22.336
Smoothing								
20	0.7054	0.6855	0.6746	0.7016	0.6748	1.1487	2.0909	19.597
40	0.7096	0.6876	0.6761	0.7039	0.6503	1.1577	2.2135	18.685
100	0.7057	0.6822	0.6705	0.7061	0.6439	1.1609	2.2656	19.079

denotes a 2×2 identity matrix. The exactly same assumptions on the initial value and the error terms are taken in Simulations II and III.

Simulation II (Stochastic volatility model). The system is represented as: $y_t = \exp(\frac{1}{2}\alpha_t)\varepsilon_t$ and $\alpha_t = \delta\alpha_{t-1} + \eta_t$ for $0 \leq \delta < 1$.

Simulation III (ARCH model). Consider the state-space model: $y_t = \alpha_t + \varepsilon_t$ and $\alpha_t = (\delta_0 + \delta\alpha_{t-1}^2)^{1/2}\eta_t$ for $\delta_0 > 0$ and $0 \leq \delta < 1$. In this simulation study, $\delta_0 = 1 - \delta$ is taken.[14]

Simulation IV (Nonstationary growth model). Take the univariate system[15]: $y_t = \alpha_t^2/20 + \varepsilon_t$ and $\alpha_t = \alpha_{t-1}/2 + 25\alpha_{t-1}/(1+\alpha_{t-1}^2) + 8\cos(1.2(t-1)) + \eta_t$, where $\alpha_0 \sim N(0, 10)$, $\varepsilon_t \sim N(0, 1)$, and $\eta_t \sim N(0, 10)$. ε_t and η_t are assumed to be mutually independent.

We compare the extended Kalman filter and smoother[16] and the nonlinear and non-Gaussian filters and smoothers introduced in Section 3. The simulation procedure is as follows: (i) generating random numbers of ε_t and η_t for $t = 1, 2, \ldots, T$, we obtain a set of data y_t and α_t, $t = 1, 2, \ldots, T$, from Eqs. (1) and (2), where $T = 20, 40, 100$ is taken, (ii) given Y_T, perform each estimator, (iii) repeat (i) and (ii) G times and compare the root mean square error (RMSE) for each estimator. RMSE is defined as: $\text{RMSE} = (1/T)\sum_{t=1}^{T} \text{MSE}_{t|s}^{1/2}$, where the mean square error (MSE) is given by: $\text{MSE}_{t|s} \equiv (1/G)\sum_{g=1}^{G}(\hat{\alpha}_{t|s}^{(g)} - \alpha_t^{(g)})^2$ and $\hat{\alpha}_{t|t}$ takes the state variable estimated by each estimator while α_t denotes the artificially simulated state variable.[17] Note that the superscript (g) denotes the gth simulation run, where $G = 1000$ is taken.

---

[14] $\delta_0 = 1 - \delta$ in the transition equation implies that the unconditional variance of α_t is normalized to be one.

[15] This system is examined in Kitagawa (1987, 1996) and Carlin, Polson and Stoffer (1992), which is called the nonstationary growth model in Carlin, Polson and Stoffer (1992).

[16] In the case where the system is linear and normal (i.e., Simulation I), the extended Kalman filter and smoother reduce to the conventional Kalman filter and smoother.

[17] Note that $\text{MSE}_{t|s}$ goes to $\Sigma_{t|s} \equiv \text{Var}(\alpha_t|Y_s)$, as number of random draws (i.e., N) is large for $s = t, T$.

Table 3
Numerical integration (δ known)

T \ δ		N	c	I			II		III		IV
				0.5	0.9	1.0	0.5	0.9	0.5	0.9	
Filtering											
		50	9	0.7292	0.7760	0.7897	0.6855	0.5821	0.9212	1.0931	11.868
			16	0.7292	0.7760	0.7897	0.6854	0.5687	0.9212	1.0930	9.452
	20		25	0.7291	0.7761	0.7898	0.6855	0.5626	0.9212	1.0930	8.336
		100	9	0.7292	0.7760	0.7897	0.6855	0.5810	0.9212	1.0931	11.745
			16	0.7292	0.7760	0.7897	0.6854	0.5681	0.9212	1.0931	8.910
			25	0.7292	0.7760	0.7897	0.6854	0.5621	0.9212	1.0931	7.796
		50	9	0.7334	0.7793	0.7928	0.6889	0.5644	0.9329	1.1077	11.122
			16	0.7334	0.7793	0.7928	0.6889	0.5523	0.9329	1.1076	8.241
	40		25	0.7334	0.7794	0.7928	0.6890	0.5467	0.9329	1.1077	7.591
		100	9	0.7334	0.7793	0.7928	0.6889	0.5635	0.9329	1.1077	10.811
			16	0.7334	0.7793	0.7928	0.6889	0.5518	0.9329	1.1077	7.835
			25	0.7334	0.7793	0.7928	0.6889	0.5463	0.9329	1.1077	6.881
		50	9	0.7307	0.7748	0.7878	0.6903	0.5579	0.9353	1.1138	11.589
			16	0.7307	0.7748	0.7879	0.6903	0.5463	0.9353	1.1138	9.157
	100		25	0.7307	0.7748	0.7879	0.6903	0.5414	0.9353	1.1138	8.141
		100	9	0.7307	0.7747	0.7878	0.6903	0.5570	0.9353	1.1138	11.271
			16	0.7307	0.7747	0.7878	0.6903	0.5458	0.9353	1.1138	8.711
			25	0.7307	0.7748	0.7878	0.6903	0.5411	0.9353	1.1138	7.566
Smoothing											
		50	9	0.7054	0.6855	0.6746	0.6784	0.5663	0.8928	0.9303	12.932
			16	0.7054	0.6855	0.6746	0.6784	0.5506	0.8928	0.9303	9.865
	20		25	0.7053	0.6856	0.6747	0.6784	0.5434	0.8928	0.9303	7.869
		100	9	0.7054	0.6855	0.6746	0.6784	0.5651	0.8928	0.9303	12.642
			16	0.7054	0.6855	0.6746	0.6784	0.5498	0.8928	0.9303	9.555
			25	0.7054	0.6855	0.6746	0.6784	0.5429	0.8928	0.9303	7.437
		50	9	0.7096	0.6877	0.6761	0.6804	0.5494	0.9039	0.9308	12.297
			16	0.7096	0.6877	0.6761	0.6804	0.5353	0.9039	0.9308	8.999
	40		25	0.7096	0.6878	0.6762	0.6805	0.5288	0.9039	0.9308	7.622
		100	9	0.7096	0.6876	0.6761	0.6804	0.5484	0.9039	0.9308	11.988
			16	0.7096	0.6876	0.6761	0.6804	0.5347	0.9039	0.9308	8.514
			25	0.7096	0.6876	0.6761	0.6804	0.5285	0.9039	0.9308	7.039
		50	9	0.7057	0.6823	0.6705	0.6805	0.5416	0.9062	0.9322	12.582
			16	0.7057	0.6823	0.6705	0.6806	0.5279	0.9062	0.9322	9.345
	100		25	0.7057	0.6823	0.6705	0.6806	0.5221	0.9062	0.9323	7.743
		100	9	0.7057	0.6822	0.6705	0.6805	0.5405	0.9062	0.9322	12.227
			16	0.7057	0.6822	0.6705	0.6805	0.5273	0.9062	0.9322	8.764
			25	0.7057	0.6823	0.6705	0.6805	0.5218	0.9062	0.9322	7.160

All the values in Tables 2–8 indicate the RMSE's, defined above, The RMSE's for both filtering and smoothing estimates are reported in all the tables except for Table 7.[18]

---
[18] Table 7 represents the RMSE's for smoothing and not for filtering.

Table 4
Importance sampling (δ known)

T	N	c	I 0.5	I 0.9	I 1.0	II 0.5	II 0.9	III 0.5	III 0.9	IV
Filtering										
20	50	4	0.7365	0.7848	0.7990	0.6920	0.5995	0.9326	1.1216	14.270
		9	0.7398	0.7890	0.8035	0.6939	0.5801	0.9382	1.1366	12.420
		16	0.7437	0.7940	0.8089	0.6965	0.5712	0.9439	1.1493	10.923
	100	4	0.7333	0.7813	0.7954	0.6890	0.5921	0.9280	1.1091	13.517
		9	0.7354	0.7837	0.7978	0.6900	0.5727	0.9308	1.1186	10.831
		16	0.7373	0.7860	0.8003	0.6914	0.5654	0.9334	1.1251	8.977
40	50	4	0.7418	0.7894	0.8035	0.6966	0.5810	0.9446	1.1388	13.808
		9	0.7447	0.7930	0.8074	0.6989	0.5629	0.9495	1.1541	12.053
		16	0.7478	0.7970	0.8117	0.7013	0.5554	0.9544	1.1697	10.602
	100	4	0.7373	0.7839	0.7975	0.6926	0.5731	0.9397	1.1245	13.063
		9	0.7384	0.7852	0.7989	0.6933	0.5562	0.9414	1.1298	10.154
		16	0.7400	0.7872	0.8012	0.6942	0.5494	0.9439	1.1363	8.288
100	50	4	0.7388	0.7847	0.7984	0.6976	0.5745	0.9478	1.1495	14.117
		9	0.7414	0.7880	0.8020	0.6991	0.5573	0.9520	1.1635	12.485
		16	0.7449	0.7924	0.8067	0.7017	0.5506	0.9567	1.1829	11.158
	100	4	0.7348	0.7801	0.7936	0.6942	0.5665	0.9413	1.1289	13.395
		9	0.7359	0.7812	0.7948	0.6949	0.5506	0.9434	1.1345	10.604
		16	0.7373	0.7829	0.7965	0.6958	0.5450	0.9456	1.1422	8.888
Smoothing										
20	50	4	0.7130	0.6962	0.6864	0.6855	0.5855	0.9048	0.9635	14.851
		9	0.7163	0.7009	0.6913	0.6877	0.5633	0.9108	0.9832	13.584
		16	0.7203	0.7066	0.6974	0.6903	0.5527	0.9171	0.9978	11.811
	100	4	0.7096	0.6910	0.6806	0.6821	0.5758	0.9000	0.9477	14.613
		9	0.7119	0.6940	0.6838	0.6833	0.5545	0.9031	0.9571	12.001
		16	0.7140	0.6967	0.6866	0.6846	0.5464	0.9060	0.9659	9.983
40	50	4	0.7187	0.6986	0.6876	0.6887	0.5669	0.9165	0.9691	14.638
		9	0.7217	0.7025	0.6915	0.6911	0.5466	0.9216	0.9842	13.335
		16	0.7249	0.7071	0.6967	0.6937	0.5384	0.9267	1.0008	11.824
	100	4	0.7137	0.6928	0.6814	0.6843	0.5580	0.9122	0.9489	14.192
		9	0.7147	0.6942	0.6831	0.6850	0.5392	0.9138	0.9555	11.596
		16	0.7164	0.6964	0.6855	0.6860	0.5317	0.9166	0.9655	9.458
100	50	4	0.7141	0.6929	0.6818	0.6881	0.5594	0.9200	0.9716	14.872
		9	0.7168	0.6967	0.6860	0.6895	0.5396	0.9250	0.9888	13.592
		16	0.7203	0.7012	0.6908	0.6923	0.5321	0.9303	1.0097	11.947
	100	4	0.7101	0.6880	0.6766	0.6847	0.5500	0.9127	0.9509	14.523
		9	0.7112	0.6897	0.6784	0.6854	0.5321	0.9151	0.9568	11.969
		16	0.7129	0.6920	0.6810	0.6862	0.5258	0.9178	0.9648	9.739

In all the tables, I, II, III and IV denote the corresponding simulation study. T, N and δ denote the sample size, number of random draws and the unknown parameter which is included in Simulations I–III, respectively. c represents range of the nodes in Table 3,

Table 5
Resampling (δ known)

$T \backslash \delta$	N	I 0.5	0.9	1.0	II 0.5	0.9	III 0.5	0.9	IV
Filtering									
20	50	0.7508	0.8067	0.8209	0.7145	0.5934	0.9485	1.1378	5.645
	100	0.7387	0.7907	0.8057	0.7003	0.5655	0.9320	1.1191	5.038
	500	0.7313	0.7794	0.7924	0.6880	0.5491	0.9230	1.0986	4.629
	1000	0.7303	0.7774	0.7912	0.6859	0.5451	0.9220	1.0949	4.547
40	50	0.7557	0.8104	0.8249	0.7182	0.6094	0.9571	1.1537	5.859
	100	0.7440	0.7933	0.8080	0.7043	0.5725	0.9455	1.1289	5.167
	500	0.7355	0.7819	0.7967	0.6924	0.5420	0.9355	1.1129	4.723
	1000	0.7342	0.7806	0.7941	0.6903	0.5384	0.9343	1.1098	4.643
100	50	0.7523	0.8044	0.8206	0.7190	0.5951	0.9588	1.1606	5.882
	100	0.7413	0.7888	0.8042	0.7050	0.5668	0.9477	1.1379	5.227
	500	0.7329	0.7778	0.7911	0.6938	0.5423	0.9382	1.1186	4.740
	1000	0.7320	0.7761	0.7987	0.6919	0.5389	0.9369	1.1166	4.653
Smoothing									
20	50	0.7350	0.7278	0.7195	0.7147	0.5838	0.9310	0.9975	5.186
	100	0.7196	0.7066	0.6964	0.6973	0.5535	0.9107	0.9653	4.514
40	50	0.7404	0.7307	0.7224	0.7183	0.6029	0.9411	0.9990	5.322
	100	0.7236	0.7069	0.6987	0.7010	0.5616	0.9233	0.9694	4.554
100	50	0.7353	0.7230	0.7152	0.7184	0.5865	0.9413	1.0034	5.416
	100	0.7208	0.7020	0.6926	0.7000	0.5539	0.9260	0.9681	4.681

variance of the importance density in Table 4, and variance of the proposal density in Tables 7 and 8. In Tables 7 and 8, M is taken as 20% of N and moreover two types of proposal densities are examined: one is based on the transition equation (2) and another is use of the extended Kalman filter and smoother estimates. In Table 6, the transition equation is utilized for the proposal density.

Since Simulation I represents the linear and normal case, the RMSE's in Simulation I of Table 2 give us the minimum values, compared with those in Tables 3–8. However, for Simulations II–IV, it is easily expected that Table 2 shows the worst RMSE's. Note in Simulations II and III of Table 2 that the filtering estimates are exactly equivalent to the smoothing estimates (i.e., $a^*_{t|s} = 0$, $s = t, T$) because of the functional form of the underlying state-space model (in addition, the initial value of $a_{0|0} = 0$ causes this situation for Simulation III).

The results obtained from the numerical integration procedure are in Table 3, where the trapezoid rule is taken for evaluation of numerical integration (in Section 3.1 each integration is evaluated by the rectangle rule for simplicity of discussion). For Simulation I, the case $\delta = 0.5$ of Simulation II and Simulation III, the RMSE's are unchanged for $N = 50, 100$ and $c = 9, 16, 25$, which implies that the obtained RMSE's are very close to the true RMSE's. However, for the case $\delta = 0.9$ of Simulation II and Simulation IV, the RMSE's are small when N and c increase, i.e., we should take more nodes and larger range of the nodes to obtain the true RMSE's.

Table 6
Rejection sampling (δ known)

T	N	I			II		III		IV
		0.5	0.9	1.0	0.5	0.9	0.5	0.9	
Filtering									
20	50	0.7366	0.7870	0.8014	0.6941	0.5534	0.9319	1.1165	5.233
	100	0.7331	0.7820	0.7964	0.6904	0.5491	0.9281	1.1042	4.864
	500	0.7304	0.7777	0.7910	0.6862	0.5438	0.9224	1.0952	4.590
	1000	0.7296	0.7767	0.7900	0.6854	0.5425	0.9221	1.0940	4.542
40	50	0.7410	0.7891	0.8022	0.6968	0.5476	0.9432	1.1288	5.412
	100	0.7375	0.7851	0.7977	0.6937	0.5429	0.9381	1.1175	4.999
	500	0.7344	0.7804	0.7938	0.6901	0.5384	0.9342	1.1097	4.680
	1000	0.7337	0.7798	0.7936	0.6894	0.5372	0.9333	1.1085	4.640
100	50	0.7381	0.7853	0.7990	0.6988	0.5476	0.9467	1.1337	5.397
	100	0.7351	0.7801	0.7939	0.6948	0.5418	0.9412	1.1243	5.001
	500	0.7315	0.7756	0.7892	0.6912	0.5371	0.9367	1.1158	4.657
	1000	0.7311	0.7753	0.7882	0.6909	0.5363	0.9358	1.1151	4.618
Smoothing									
20	50	0.7137	0.6956	0.6826	0.6887	0.5374	0.9061	0.9512	4.292
	100	0.7098	0.6902	0.6798	0.6846	0.5344	0.8991	0.9407	3.927
40	50	0.7166	0.6975	0.6853	0.6895	0.5328	0.9151	0.9525	4.457
	100	0.7132	0.6927	0.6811	0.6866	0.5274	0.9090	0.9422	3.971
100	50	0.7134	0.6928	0.6806	0.6916	0.5300	0.9172	0.9544	4.414
	100	0.7095	0.6869	0.6753	0.6874	0.5257	0.9126	0.9422	3.989

Table 4 shows the RMSE's obtained from the importance sampling procedure in Section 3.2. In the nonlinear cases of Simulations II–IV, the importance sampling procedure performs much better than the extended Kalman filter and smoother but worse than the numerical integration procedure. To obtain the same precision as the numerical integration approach, more random draws are necessary.

The results of the resampling procedure are in Table 5. Smoothing requires an extremely large computational burden.[19] Therefore, the resampling procedure does not have as small RMSE's as the numerical integration approach. However, for filtering in Simulation I, the case $\delta = 0.5$ of Simulation II and Simulation III, the resampling procedure is very close to the numerical integration approach as N is large. Especially, for the case $\delta = 0.9$ of Simulation II and Simulation IV, the filtering results in Table 5 are much better than those in Tables 2–4.

In Table 6, similarly, smoothing takes an extremely long time computationally, compared with filtering. The approximations based on (48) are taken for less computational burden and the Metropolis–Hastings algorithm are utilized in parallel to avoid repeating procedures (ii) and (iii) forever. Accordingly, when the rejection sampling does not work, a random draw (i.e., $\alpha_{j,t|t}$ or $\alpha_{m,t|T}$) is generated by the Metropolis–Hastings algorithm. We sometimes have the case where it is not clear whether the supremum exists. Since $P_*(z) = P_\alpha(z|\alpha_{i,t-1|t-1})$ is taken in this Monte

---

[19] The case $N = 1000$ is feasible for filtering but even the case $N = 100$ takes a lot of time for smoothing.

Table 7
Markov chain Monte Carlo (δ known)

T	δ N	c	I 0.5	0.9	1.0	II 0.5	0.9	III 0.5	0.9	IV
Smoothing										
$P_*(z\|x) = P_\alpha(z\|\alpha_{i,t-1})$										
20	1000		0.7081	0.6883	0.6778	0.6802	0.5296	0.8945	0.9351	12.944
	5000		0.7061	0.6859	0.6751	0.6786	0.5267	0.8931	0.9318	13.069
40	1000		0.7112	0.6900	0.6787	0.6826	0.5225	0.9062	0.9360	13.009
	5000		0.7099	0.6880	0.6765	0.6810	0.5199	0.9044	0.9321	13.179
100	1000		0.7078	0.6845	0.6734	0.6831	0.5191	0.9082	0.9368	13.038
	5000		0.7060	0.6825	0.6708	0.6808	0.5166	0.9067	0.9334	13.189
$P_*(z\|x) = N(a^*_{t\|t}, c\Sigma^*_{t\|t})$										
20	1000	4	0.7072	0.6896	0.6797	0.6794	0.5635	0.8949	0.9512	15.235
		9	0.7074	0.6903	0.6799	0.6801	0.5436	0.8975	0.9548	14.852
		16	0.7080	0.6910	0.6810	0.6802	0.5392	0.8993	0.9587	14.675
	5000	4	0.7059	0.6878	0.6774	0.6791	0.5538	0.8938	0.9371	15.090
		9	0.7060	0.6877	0.6775	0.6789	0.5405	0.8945	0.9378	14.771
		16	0.7060	0.6880	0.6777	0.6790	0.5364	0.8955	0.9406	14.614
40	1000	4	0.7108	0.6905	0.6793	0.6818	0.5434	0.9069	0.9489	15.008
		9	0.7115	0.6916	0.6807	0.6821	0.5297	0.9079	0.9509	14.649
		16	0.7122	0.6921	0.6816	0.6825	0.5259	0.9092	0.9553	14.488
	5000	4	0.7101	0.6889	0.6776	0.6807	0.5380	0.9053	0.9363	14.846
		9	0.7100	0.6891	0.6780	0.6807	0.5268	0.9056	0.9377	14.563
		16	0.7105	0.6895	0.6785	0.6809	0.5234	0.9057	0.9394	14.448
100	1000	4	0.7068	0.6845	0.6732	0.6815	0.5390	0.9087	0.9503	15.137
		9	0.7074	0.6855	0.6740	0.6818	0.5243	0.9103	0.9544	14.797
		16	0.7079	0.6863	0.6751	0.6823	0.5206	0.9120	0.9652	14.615
	5000	4	0.7059	0.6830	0.6714	0.6808	0.5307	0.9066	0.9354	14.992
		9	0.7060	0.6831	0.6716	0.6808	0.5208	0.9069	0.9365	14.718
		16	0.7062	0.6834	0.6721	0.6809	0.5188	0.9071	0.9388	14.575

Carlo study, the denominators of the acceptance probabilities (42) and (47) are given by $\sup_z P_y(y_t|z)$ and $\sup_z P_y(y_t|z) P_\alpha(\alpha_{j,t+1|T}|z)$, respectively. In the case where we cannot obtain the explicit solution, the supremum is computed by the Newton–Raphson optimization procedure, which implies that the solution is possibly the local supremum. In Simulations II and IV, the acceptance probability is numerically evaluated to obtain smoothing. For all simulation studies, however, the rejection sampling procedure shows a good performance even when N is small. Especially, the resampling procedure requires about two times more random draws to obtain the same RMSE as the rejection sampling procedure.

The Markov chain Monte Carlo approach is taken in Table 7. All the values indicate the RMSE's obtained from smoothing. Two types of the proposal densities are examined. Both proposal densities yield the similar results, but the proposal density of the transition equation is slightly better than that of the extended Kalman smoother.

Table 8
Quasi-filter and quasi-smoother (δ known)

T	N	c	I 0.5	I 0.9	I 1.0	II 0.5	II 0.9	III 0.5	III 0.9	IV
Filtering										
$P_*(z\|x) = P_\alpha(z\|\alpha_{i,t-1\|t-1})$										
20	1000		0.7310	0.7805	0.7954	0.6879	0.5463	0.9227	1.1033	4.708
	5000		0.7300	0.7786	0.7930	0.6855	0.5437	0.9217	1.1014	4.604
40	1000		0.7364	0.7842	0.7981	0.6924	0.5438	0.9346	1.1192	4.774
	5000		0.7341	0.7813	0.7954	0.6899	0.5391	0.9336	1.1180	4.668
100	1000		0.7330	0.7796	0.7942	0.6935	0.5412	0.9374	1.1271	4.785
	5000		0.7313	0.7770	0.7910	0.6911	0.5376	0.9359	1.1250	4.666
$P_*(z\|x) = N(a^*_{t\|t}, c\Sigma^*_{t\|t})$										
20	1000	4	0.7316	0.7956	0.8193	0.6923	0.6337	0.9281	1.2819	17.087
		9	0.7320	0.7974	0.8225	0.6916	0.6170	0.9276	1.2826	15.669
		16	0.7329	0.7988	0.8244	0.6918	0.6071	0.9286	1.2922	14.838
	5000	4	0.7310	0.7945	0.8178	0.6917	0.6336	0.9265	1.2763	16.455
		9	0.7311	0.7963	0.8213	0.6910	0.6170	0.9255	1.2729	15.151
		16	0.7315	0.7971	0.8227	0.6911	0.6064	0.9256	1.2728	14.468
40	1000	4	0.7364	0.7985	0.8215	0.6961	0.6158	0.9410	1.3221	15.954
		9	0.7368	0.8005	0.8253	0.6960	0.6021	0.9409	1.3249	14.959
		16	0.7374	0.8019	0.8270	0.6961	0.5938	0.9416	1.3308	14.383
	5000	4	0.7353	0.7971	0.8198	0.6953	0.6158	0.9395	1.3167	15.606
		9	0.7354	0.7989	0.8230	0.6948	0.6019	0.9384	1.3173	14.716
		16	0.7354	0.7994	0.8240	0.6948	0.5932	0.9385	1.3220	14.232
100	1000	4	0.7332	0.7934	0.8157	0.6975	0.6106	0.9435	1.3579	16.503
		9	0.7339	0.7956	0.8195	0.6973	0.5971	0.9433	1.3563	15.210
		16	0.7345	0.7971	0.8216	0.6976	0.5891	0.9434	1.3640	14.513
	5000	4	0.7326	0.7927	0.8150	0.6970	0.6103	0.9423	1.3512	16.032
		9	0.7329	0.7944	0.8182	0.6965	0.5965	0.9411	1.3487	14.885
		16	0.7331	0.7952	0.8195	0.6964	0.5884	0.9407	1.3521	14.317
Smoothing										
$P_*(z\|x) = P_\alpha(z\|\alpha_{i,t-1\|t-1})$										
20	1000		0.7083	0.6934	0.6838	0.6817	0.5346	0.8978	0.9592	4.329
	5000		0.7067	0.6912	0.6814	0.6791	0.5299	0.8964	0.9567	4.114
40	1000		0.7125	0.6959	0.6848	0.6853	0.5281	0.9106	0.9657	4.340
	5000		0.7107	0.6927	0.6823	0.6817	0.5241	0.9088	0.9633	4.128
100	1000		0.7083	0.6900	0.6796	0.6844	0.5258	0.9120	0.9678	4.356
	5000		0.7070	0.6873	0.6767	0.6822	0.5213	0.9110	0.9656	4.065
$P_*(z\|x) = N(a^*_{t\|T}, c\Sigma^*_{t\|T})$										
20	1000	4	0.7083	0.7072	0.7068	0.6863	0.6275	0.9034	1.1922	15.229
		9	0.7092	0.7103	0.7109	0.6858	0.6085	0.9019	1.1854	14.816
		16	0.7098	0.7114	0.7131	0.6857	0.5973	0.9035	1.1933	14.577
	5000	4	0.7076	0.7065	0.7060	0.6857	0.6273	0.9026	1.1876	15.063
		9	0.7077	0.7086	0.7093	0.6845	0.6083	0.9003	1.1775	14.715
		16	0.7079	0.7095	0.7108	0.6845	0.5963	0.9000	1.1761	14.522

Table 8
(Continued.)

T	N	c	I			II		III		IV
			0.5	0.9	1.0	0.5	0.9	0.5	0.9	
40	1000	4	0.7122	0.7084	0.7066	0.6889	0.6087	0.9168	1.2274	14.950
		9	0.7131	0.7104	0.7100	0.6882	0.5925	0.9151	1.2219	14.574
		16	0.7132	0.7122	0.7125	0.6883	0.5826	0.9155	1.2262	14.363
	5000	4	0.7115	0.7072	0.7053	0.6882	0.6087	0.9154	1.2214	14.807
		9	0.7117	0.7091	0.7085	0.6871	0.5923	0.9129	1.2144	14.498
		16	0.7119	0.7098	0.7097	0.6870	0.5821	0.9123	1.2178	14.320
100	1000	4	0.7086	0.7025	0.7001	0.6899	0.6033	0.9193	1.2604	15.079
		9	0.7094	0.7050	0.7041	0.6892	0.5876	0.9176	1.2512	14.707
		16	0.7101	0.7068	0.7064	0.6893	0.5783	0.9173	1.2558	14.484
	5000	4	0.7077	0.7015	0.6991	0.6892	0.6029	0.9178	1.2540	14.941
		9	0.7079	0.7035	0.7023	0.6881	0.5868	0.9153	1.2434	14.623
		16	0.7082	0.7044	0.7037	0.6878	0.5771	0.9146	1.2443	14.429

The estimator performs well in Simulations I–III but not in Simulation IV. Convergence of the Gibbs sampler is very slow in the case of Simulation IV.

In Table 8, both quasi-filter and quasi-smoother perform better when the proposal density is based on the transition equation. However, the proposal density based on the extended Kalman filter and smoother does not work in spite of c. Accordingly, from the results in Tables 7 and 8, the transition equation should be utilized for the proposal density.

The resampling procedure is the easiest estimator in the sense of programming but it takes an extraordinarily long time for smoothing. The Markov chain Monte Carlo procedure has the least computational burden although it sometimes shows a poor performance. For all the simulation studies, the most accurate estimator is the rejection sampling approach but we should keep in mind that existence of the supremum is required for rejection sampling. The quasi-filter and quasi-smoother gives us the relatively good estimator, which can be applied to any case in the sense of no restriction on the supremum such as rejection sampling.

Next, for each estimator, the order of computation and the computation times are compared in Table 9. The first line in each cell represents the order of computation, which is proportional to computation time. The values in the parentheses indicate the CPU times (the averages from 100 simulation runs, i.e., $G = 100$), which are given in seconds. The case $\delta = 1.0$ in Simulation I are used for comparison of computation times, where $T = 100$ is taken. Computations were performed by 300 MHz Pentium II Processor and WATCOM Fortran 77/32 Compiler (Version 10.6), using double precision. (*) under Rejection Sampling denotes the reduction method of computational burden using (48). Computation times in Rejection Sampling were implemented by (*). Except for Markov Chain Monte Carlo, computations of filtering are necessary for those of smoothing. A denotes number of rejection in Rejection Sampling. The proposal density is based on the transition equation (1) for Rejection Sampling, Markov Chain

Table 9
Comparison of computation times ($T = 100$)

	N	Prediction	Filtering	Smoothing
Numerical integration		LkN^{k+1}	$2TkN^{k+1}$	TkN^{k+1}
	50		(0.37)	(0.65)
	100		(1.43)	(2.53)
Importance sampling		LN^2	$TN(N+1)$	TN^2
	50		(0.28)	(1.03)
	100		(1.03)	(4.02)
Resampling		LN	$TN(N+1)$	$TN^2(N+1)$
	50		(0.05)	(12.2)
	100		(0.10)	(96.9)
Rejection sampling		LN	$TN(2N+A)$	$TN(N^2+N+A)$
(*)		LN	$TN(N+1+A)$	$TN(2N+A)$
	50		(0.24)	(1.95)
	100		(0.52)	(6.58)
Markov chain		LN	$NT(T-1)/2$	TN
Monte Carlo	1000		—	(0.98)
	5000		—	(4.88)
Quasi-filter and		LN	TN	TN
Quasi-smoother	1000		(0.91)	(1.91)
	5000		(4.54)	(9.51)

Monte Carlo and Quasi-Filter and Quasi-Smoother. We choose $c = 16$ for Numerical Integration and $c = 4$ for Importance Sampling.

In Resampling, as expected, smoothing takes an extremely long time computationally, while filtering indicates much less computation. Computation times of Rejection Sampling depend on the functional form of the state-space model and the distribution of the error terms, because the acceptance probability is based on the density function $P_y(y_t|\alpha_t)$ or $P_y(y_t|\alpha_t)P_\alpha(\alpha_{t+1}|\alpha_t)$, while those of the other estimators are not influenced by nonlinear equations and non-Gaussian errors.[20]

In Table 10, comparison between the true parameter and the estimate of δ is shown for each procedure. Note that δ in the table indicates the true value. Given observed data Y_T, the parameter δ in Simulations I–III is estimated using the appropriate likelihood function. AVE, RMSE, 25%, 50% and 75% represent the arithmetic average, the root mean square error, the 0.25th, 0.50th and 0.75th quantiles from 1000 estimates of δ (i.e., $G = 1000$). We take $N = 100$ for Numerical Integration, Importance Sampling, Resampling and Rejection Sampling and $N = 1000$ for Markov Chain Monte Carlo and Quasi-Filter and Quasi-Smoother. Furthermore, we assume $c = 16$ for Numerical Integration and $c = 4$ for Importance Sampling. As in Table 9, the proposal density is based on the transition equation (1) for Rejection Sampling, Markov Chain Monte Carlo and Quasi-Filter and Quasi-Smoother. The maximization of the likelihood function is

---

[20] As for Rejection Sampling, Simulation IV computationally takes much more time than Simulation I. For Simulation IV in the case of $N = 50$ and $T = 100$, filtering and smoothing take 1.19 and 4.15 seconds, respectively.

Table 10
Estimation of unknown parameter ($T = 100$)

		I			II		III	
	δ	0.5	0.9	1.0	0.5	0.9	0.5	0.9
Extended Kalman filter	AVE	0.472	0.878	0.981	0.301	0.652	—	—
	RMSE	0.144	0.065	0.040	0.285	0.313	—	—
	25%	0.400	0.850	0.970	0.140	0.530	—	—
	50%	0.490	0.890	0.990	0.310	0.660	—	—
	75%	0.570	0.920	1.000	0.450	0.790	—	—
Numerical integration	AVE	0.472	0.878	0.981	0.444	0.850	0.440	0.878
	RMSE	0.144	0.065	0.040	0.299	0.168	0.218	0.071
	25%	0.400	0.850	0.970	0.170	0.810	0.330	0.850
	50%	0.490	0.890	0.990	0.490	0.890	0.490	0.890
	75%	0.570	0.920	1.000	0.680	0.950	0.600	0.920
Importance sampling	AVE	0.470	0.877	0.980	0.447	0.845	0.434	0.887
	RMSE	0.146	0.065	0.040	0.304	0.188	0.221	0.078
	25%	0.400	0.850	0.970	0.160	0.810	0.320	0.850
	50%	0.490	0.890	0.990	0.490	0.900	0.490	0.900
	75%	0.570	0.920	1.000	0.695	0.950	0.590	0.940
Resampling	AVE	0.505	0.905	1.002	0.505	0.907	0.505	0.905
	RMSE	0.013	0.013	0.011	0.013	0.018	0.013	0.013
	25%	0.500	0.900	1.000	0.500	0.900	0.500	0.900
	50%	0.500	0.910	1.000	0.510	0.910	0.510	0.910
	75%	0.510	0.910	1.010	0.510	0.910	0.510	0.910
Rejection sampling	AVE	0.505	0.905	1.003	0.506	0.906	0.505	0.904
	RMSE	0.013	0.014	0.011	0.013	0.013	0.013	0.013
	25%	0.500	0.900	1.000	0.500	0.900	0.500	0.900
	50%	0.510	0.910	1.000	0.510	0.910	0.510	0.900
	75%	0.510	0.910	1.010	0.510	0.910	0.510	0.910
Markov chain Monte Carlo	AVE	0.539	0.915	1.003	0.915	0.983	0.515	0.909
	RMSE	0.059	0.024	0.012	0.445	0.086	0.027	0.016
	25%	0.510	0.900	1.000	0.990	0.990	0.500	0.900
	50%	0.530	0.910	1.000	0.990	0.990	0.510	0.910
	75%	0.560	0.920	1.010	0.990	0.990	0.520	0.920
Quasi-filter and quasi-smoother	AVE	0.482	0.903	0.997	0.494	0.907	0.497	0.904
	RMSE	0.106	0.021	0.015	0.138	0.057	0.073	0.016
	25%	0.470	0.890	0.990	0.480	0.890	0.490	0.900
	50%	0.500	0.900	1.000	0.500	0.910	0.500	0.900
	75%	0.530	0.910	1.010	0.530	0.930	0.520	0.910

performed by a simple grid search, in which the likelihood function is maximized by changing the parameter value of δ by 0.01 – in Simulation III of Extended Kalman Filter indicates that the maximum likelihood estimation cannot be performed, because the innovation form of the likelihood function (see Eq. (50) for the likelihood function) does not depend on the unknown parameter δ under the assumption of $E(\alpha_0) = a_{0|0} = 0$.

Resampling, Rejection Sampling and Quasi-Filter and Quasi-Smoother perform better for all the simulation studies I–III. That is, AVE is very close to the true parameter value and RMSE is very small. In Simulation II, Extended Kalman Filter shows the worst estimator and moreover Markov Chain Monte Carlo does not perform good, compared with the other estimators. For all the simulation studies, Numerical Integration is similar to Importance Sampling, while Resampling and Rejection Sampling are close to Quasi-Filter and Quasi-Smoother.

Finally, we can conclude from the simulation studies that Quasi-Filter and Quasi-Smoother which utilize the transition equation for the proposal density might be recommended at the present time (i.e., see Tanizaki, 2000), because precision of the state estimates is quite good (Table 8), computational burden is very small (Table 9) and maximum likelihood estimation works well (Table 10). However, we should keep in mind as follows. The estimator which shows the best performance is Rejection Sampling. Resampling is inferior to Rejection Sampling from computational time and precision of the state estimates. On the other hand, Resampling can be applied to any nonlinear and non-Gaussian cases while Rejection Sampling does not work under some conditions. In the future, therefore, it might be expected that Resampling would be the best estimator because CPU speed of computer becomes faster and faster.

5. Summary and concluding remarks

In this chapter, several nonlinear and non-Gaussian filters and smoothers have been introduced and compared through Monte Carlo studies. Each nonlinear and non-Gaussian estimator has both advantages and disadvantages, which are summarized as follows.

The numerical integration procedure proposed in Kitagawa (1987) and Kramer and Sorenson (1988) has the problems: (i) location of nodes has to be set by a researcher (unless the range of the nodes cover the density $P(\alpha_r|Y_s)$, the density approximation becomes poor), (ii) it is possible that computational errors accumulate because density functions are evaluated at each time, and (iii) computational burden increases more than proportionally as the dimension of the state variable is high. However, the advantage of the estimator is that we can obtain precise estimates of the state variable when the nodes are correctly chosen.

The problems of the Monte Carlo integration procedure with importance sampling developed by Tanizaki (1993a, 1996, 1999b), Tanizaki and Mariano (1994) and Mariano and Tanizaki (1995) are: (i) the importance density has to be appropriately chosen by a researcher (if the random draws generated from the importance density are away from the density $P(\alpha_r|Y_s)$, the approximation of the weight function becomes poor), and (ii) it might be possible that computational errors accumulate because weight functions are evaluated at each time. However, the merit of the importance sampling procedure is that computational burden does not increases too much even in the high-dimensional cases of the state vector.

The resampling procedure by Gordon, Salmond and Smith (1993), Kitagawa (1996, 1998) and Kitagawa and Gersch (1996) has the disadvantage that it requires heavy computation, especially for smoothing. However, this problem will be improved in the future as computer progresses.

The disadvantages of the rejection sampling procedure (Mariano and Tanizaki, 2000; Tanizaki and Mariano, 1998; Tanizaki, 1999a) are: (i) the proposal density has to be appropriately chosen by a researcher (use of the transition equation might be recommended, but not necessarily), (ii) it takes a long time computationally when the acceptance probability is small (i.e., we cannot predict how long the computer program will run), and (iii) sometimes the supremum of the ratio of the target density and the proposal density does not exists (we cannot apply rejection sampling in such a case). If rejection sampling works, i.e., if the supremum exists, the rejection sampling procedure shows the best performance of all the procedures introduced in this paper, because rejection sampling is the random number generation method which generate an exact random draw from any distribution function.

The Markov chain Monte Carlo procedure proposed by Geweke and Tanizaki (1999, 2001) has the following problems: (i) the proposal density has to be appropriately chosen by a researcher (it might be plausible to take the transition equation for the proposal density), and (ii) convergence is very slow because the Gibbs sampler and the Metropolis–Hastings are simultaneously used (remember that the random draw generated by the Markov chain Monte Carlo method is correlated with the next one). To obtain the smoothing estimates from the numerical integration approach, importance sampling procedure, the resampling procedure and the rejection sampling approach, we need at least $N \times T$ data storage, because smoothing is implemented after prediction and filtering. That is, we need to store both prediction and filtering random draws before smoothing. However, the Markov chain Monte Carlo approach uses only T data storage to obtain the smoothing estimates.

The quasi-filter and quasi-smoother using the Metropolis–Hastings (Tanizaki, 2000) have the disadvantages: (i) the proposal density has to be appropriately chosen by a researcher (it is recommended to use the transition equation for the proposal density), and (ii) the obtained estimates do not give us the exact true values, because the filtering and smoothing densities are approximated. This procedure also uses only T data storage. Although the filtering and smoothing estimates are different from the true values, it is too small to ignore the difference between the estimates and the true values, which results come from the simulation studies in Section 4.

Thus, the rejection sampling approach might be the best estimator under the two conditions: (i) the supremum exists and (ii) the acceptance probability is not too small. In the sense of no ad hoc assumptions such as choice of the nodes, choice of the importance density and choice of the proposal density, the resampling procedure in Section 3.3 might be taken as the estimator which gives us an optimal solution for sufficiently large N. However, the resampling procedure takes an extremely long time for smoothing, although it does not take too much time for filtering. For the Markov chain Monte Carlo method, we have the case where the random draws are not correctly generated from the distribution function we want to sample, depending on the underlying nonlinear equations and non-Gaussian error terms. Accordingly, at the present time, the second best estimator might be the quasi-filter and quasi-smoother from computational point of view and getting precise estimates of the state variable.

Recently, the nonlinear non-Gaussian filters and smoothers which are much less computational than the existing ones are proposed by Tanizaki (2001), where the sampling techniques such as rejection sampling, resampling and the Metropolis–Hastings algorithm are utilized. The conventional density-based nonlinear algorithms require the random draws generated from the marginal densities, i.e., $P(\alpha_t|Y_t)$ for filtering and $P(\alpha_t|Y_T)$ for smoothing, but the algorithms proposed in Tanizaki (2001) are based on the joint densities, i.e., $P(\alpha_t, \alpha_{t-1}|Y_t)$ for filtering and $P(\alpha_{t+1}, \alpha_t|Y_T)$ or $P(\alpha_{t+1}, \alpha_t, \alpha_{t-1}|Y_T)$ for smoothing. That is, the random draws of α_t are generated from $P(\alpha_t, \alpha_{t-1}|Y_t)$ for filtering and $P(\alpha_{t+1}, \alpha_t|Y_T)$ or $P(\alpha_{t+1}, \alpha_t, \alpha_{t-1}|Y_T)$ for smoothing. By generating the random draws from the joint densities, much less computer-intensive algorithms on filtering and smoothing can be obtained. Furthermore, taking into account possibility of structural changes and outliers during the estimation period, the appropriately chosen sampling density is possibly introduced into the suggested nonlinear non-Gaussian filtering and smoothing procedures.

Appendix A. Linear and normal system

State-space model. Consider the case where the system is linear and normal, i.e.,

(Measurement equation) $\quad y_t = Z_t \alpha_t + d_t + S_t \varepsilon_t,$

(Transition equation) $\quad \alpha_t = T_t \alpha_{t-1} + c_t + R_t \eta_t,$

$$\begin{pmatrix} \varepsilon_t \\ \eta_t \end{pmatrix} \sim N\left(\begin{pmatrix} 0 \\ 0 \end{pmatrix}, \begin{pmatrix} H_t & 0 \\ 0 & Q_t \end{pmatrix}\right),$$

where Z_t, d_t, S_t, T_t, c_t, R_t, H_t and Q_t are assumed to be known for all time $t = 1, 2, \ldots, T$. Define conditional mean and variance as $a_{r|s} \equiv \mathrm{E}(\alpha_r|Y_s)$ and $\Sigma_{r|s} \equiv \mathrm{Var}(\alpha_r|Y_s)$ for $(r, s) = (t+L, t), (t, t), (t, T)$. Under the above setup, optimal prediction, filtering and smoothing are represented as the standard linear recursive algorithms, which are easily derived from the first- and second-moments of density functions (13)–(18). See, for example, Tanizaki (1996).

Prediction. From the density-based L-step ahead prediction algorithm (13), the following prediction algorithm can be obtained:

$$a_{t+L|t} = T_{t+L} a_{t+L-1|t} + c_{t+L},$$

$$\Sigma_{t+L|t} = T_{t+L} \Sigma_{t+L-1|t} T'_{t+L} + R_{t+L} Q_{t+L} R'_{t+L},$$

for $L = 1, 2, \ldots$. Given filtering mean and variance (i.e., $a_{t|t}$ and $\Sigma_{t|t}$), prediction mean and variance (i.e., $a_{t+L|t}$ and $\Sigma_{t+L|t}$) are obtained recursively.

Filtering. The density-based filtering algorithm given by Eqs. (14) and (15) reduces to the following standard linear recursive algorithm:

$$a_{t|t-1} = T_t a_{t-1|t-1} + c_t,$$

$$\Sigma_{t|t-1} = T_t \Sigma_{t-1|t-1} T'_t + R_t Q_t R'_t,$$

$$y_{t|t-1} = Z_t a_{t|t-1} + d_t,$$
$$F_{t|t-1} = Z_t \Sigma_{t|t-1} Z_t' + S_t H_t S_t',$$
$$K_t = \Sigma_{t|t-1} Z_t' F_{t|t-1}^{-1},$$
$$a_{t|t} = a_{t|t-1} + K_t (y_t - y_{t|t-1}),$$
$$\Sigma_{t|t} = \Sigma_{t|t-1} - K_t F_{t|t-1} K_t',$$

for $t = 1, 2, \ldots, T$. Given the initial values $a_{0|0}$ and $\Sigma_{0|0}$, the filtering mean and variance at time t (i.e., $a_{t|t}$ and $\Sigma_{t|t}$) are recursively computed. See Anderson and Moore (1979), Gelb (1974), Jazwinski (1970) and Tanizaki (1996) for the Kalman filter algorithm.

Smoothing. The first- and the second-moments of the smoothing density (16) give us the following backward recursive algorithm:

$$C_t = \Sigma_{t|t} T_{t+1}' \Sigma_{t+1|t}^{-1},$$
$$a_{t|T} = a_{t|t} + C_t (a_{t+1|T} - a_{t+1|t}),$$
$$\Sigma_{t|T} = \Sigma_{t|t} + C_t (\Sigma_{t+1|T} - \Sigma_{t+1|t}) C_t',$$

for $t = T-1, T-2, \ldots, 1$. Given $a_{t|t}$, $\Sigma_{t|t}$, $a_{t+1|t}$ and $\Sigma_{t+1|t}$, smoothing mean and variance at time t (i.e., $a_{t|T}$ and $\Sigma_{t|T}$) is obtained recursively.

Likelihood function. When Z_t, d_t, S_t, T_t, c_t, R_t, H_t and Q_t depends on an unknown parameter, the following log of the likelihood function is maximized with respect to the parameter:

$$\log P(Y_T) = -\frac{T}{2} \log(2\pi) - \frac{1}{2} \sum_{t=1}^{T} \log |F_{t|t-1}|$$
$$- \frac{1}{2} \sum_{t=1}^{T} (y_t - y_{t|t-1})' F_{t|t-1}^{-1} (y_t - y_{t|t-1}), \tag{50}$$

which is also obtained from Eq. (18). Note that the conditional distribution of y_t given Y_{t-1} is represented as $y_t | Y_{t-1} \sim N(y_{t|t-1}, F_{t|t-1})$, where both $y_{t|t-1}$ and $F_{t|t-1}$ are obtained from the above standard filtering algorithm.

Appendix B. Sampling methods

Monte Carlo integration with importance sampling. When we want to evaluate the expectation of a function $g(\cdot)$, the problem is how to evaluate integration. There are two integration methods: one is numerical integration and another is Monte Carlo integration. To perform Monte Carlo integration, we usually take another appropriate distribution function $P_*(x)$, called the importance density, which is chosen by a researcher. Let x_i, $i = 1, 2, \ldots, N$, be the random draws of x generated from $P_*(x)$. Define $\omega(x)$ as $\omega(x) \equiv P_x(x)/P_*(x)$. Then, in order to approximate integration, importance sampling is performed as follows:

$$E(g(x)) = \int g(x) P_x(x) \, dx = \int g(x) \omega(x) P_*(x) \, dx$$

$$\approx \frac{1}{N} \sum_{i=1}^{N} g(x_i) \omega(x_i) \equiv \bar{g}_N,$$

which is called the Monte Carlo integration method with importance sampling.

Because x_i is a random variable from $P_*(x)$ while x is a random variable from $P_x(x)$, $E(g(x_i)\omega(x_i))$ implies taking the expectation with respect to x_i but $E(g(x))$ is the expectation taken with respect to x. Now, define μ and Σ as:

$$\mu = E(g(x_i)\omega(x_i)) = E(g(x)),$$

$$\Sigma = \text{Var}(g(x_i)\omega(x_i)) = E(g(x_i)\omega(x_i))^2 - \mu^2 = E((g(x))^2 \omega(x)) - \mu^2.$$

$g(x_i)\omega(x_i)$ are mutually independent for $i = 1, 2, \ldots, N$. Therefore, by the central limit theorem, we can easily show:

$$\sqrt{N}(\bar{g}_N - \mu) \to N(0, \Sigma).$$

Moreover, let us define $\overline{\Sigma}_N$ as:

$$\overline{\Sigma}_N = \frac{1}{N} \sum_{i=1}^{N} (g(x_i) - \bar{g}_N)(g(x_i) - \bar{g}_N)' \omega(x_i).$$

Then, we have the following:

$$\overline{\Sigma}_N \to \Sigma.$$

Thus, $\bar{g}_N$ gives us an asymptotically unbiased estimator as N goes to infinity. However, it is shown from the above results that convergence is quite slow as $\sqrt{N}$. See, for example, Geweke (1988, 1989a, 1989b), Koop (1994) and Shao (1989).

Rejection sampling. Rejection sampling is the method which generates random draws from any distribution function under some conditions. In the case where it is not easy to generate a random number from $P_x(x)$, suppose that we want to generate a random draw from $P_x(x)$, called the target density. In such a case, we take another distribution function $P_*(x)$, called the proposal density, which is appropriately chosen by a researcher.

Denote the acceptance probability by $\omega(x) = P_x(x)/aP_*(x)$, where a is defined as $a \equiv \sup_x P_x(x)/P_*(x)$ and the assumption of $a < \infty$ is required for rejection sampling. Under the setup, rejection sampling is implemented as: (i) generate a random draw of x (say, x_0) from $P_*(x)$ and (ii) accept it with probability $\omega(x_0)$. The accepted random draw is taken as a random draw of x generated from $P_x(x)$.

In the case where both $P_x(x)$ and $P_*(x)$ are normally distributed as $N(\mu, \sigma^2)$ and $N(\mu_*, \sigma_*^2)$, it is easily shown that $\sigma_*^2 > \sigma^2$ is required for the condition $a < \infty$, which implies that $P_*(x)$ has to be distributed with larger variance than $P_x(x)$.

Note that $P_x(x)$ is not necessarily a probability density function, i.e., it is possibly a kernel of the target density function, which implies that the target density is proportional to $P_x(x)$, i.e., $cP_x(x)$. Since the supremum a in $\omega(x)$ also includes $P_x(x)$, the constant c is canceled out from the acceptance probability $\omega(x)$.

Using rejection sampling, we can generate a random draw from any distribution function under the condition that $a < \infty$ is satisfied. However, the disadvantages of rejection sampling are: (i) we need to compute a, which sometimes does not exist and (ii) it takes a long time when the acceptance probability $\omega(\cdot)$ is close to zero. See, for example, Knuth (1981), Boswell et al. (1993), O'Hagan (1994) and Geweke (1996) for rejection sampling.

Gibbs sampling. Geman and Geman (1984), Tanner and Wong (1987), Gelfand and Smith (1990), Gelfand et al. (1990) and so on developed the Gibbs sampling theory, which is concisely described as follows (also see Geweke, 1996, 1997).

Consider two random variables x and y. Let $P_{x|y}(x|y)$, $P_{y|x}(y|x)$ and $P_{xy}(x, y)$ be the conditional density of x given y, the conditional density of y given x and the joint density of x and y, which are assumed to be known. Pick up an arbitrary initial value for x (i.e., x_0). Given x_{i-1}, generate a random number of y (i.e., y_i) from the density $P_{y|x}(y|x_{i-1})$. Again, given y_i, generate a random number of x (i.e., x_i) from the density $P_{x|y}(x|y_i)$. Thus, we can generate y_i from $P_{y|x}(y|x_{i-1})$ and x_i from $P_{x|y}(x|y_i)$ for $i = 1, 2, \ldots, N$. From the convergence theory of the Gibbs sampler, as N goes to infinity, we can regard x_N and y_N as random draws from $P_{xy}(x, y)$.

The basic result of the Gibbs sampler is:

$$\frac{1}{N-M} \sum_{i=M+1}^{N} g(x_i, y_i) \to \mathrm{E}\big(g(x, y)\big) = \iint g(x, y) P_{xy}(x, y) \, dx \, dy,$$

as $N \to \infty$, where $g(\cdot, \cdot)$ is a function. We may take $M = 0$ but usually 10–20% of N is taken for M. That is, the first M random draws are discarded from consideration.

Finally, note as follows. It is known that convergence is very slow if correlation between x and y is very high. See Chib and Greenberg (1995).

Metropolis–Hastings algorithm. Smith and Roberts (1993), Tierney (1994), Chib and Greenberg (1995, 1996) and Geweke (1996) discussed the Metropolis–Hastings algorithm, which is also the random number generation method such that we can generate random draws from any density function.

Suppose that we want to generate a random draw of x from $P_x(x)$, which is called the target density function. When it is hard to generate random draws from the target density $P_x(\cdot)$, we can apply the Metropolis–Hastings algorithm to random number generation. The Metropolis–Hastings algorithm utilizes another appropriate distribution function $P_*(z|x)$, which is called the proposal density.

In order to perform the Metropolis–Hastings algorithm, first let us define the acceptance probability $\omega(x, z)$ as:

$$\omega(x,z) = \begin{cases} \min\left(\frac{P_x(z)P_*(x|z)}{P_x(x)P_*(z|x)}, 1\right), & \text{if } P_x(x)P_*(z|x) > 0, \\ 1, & \text{otherwise.} \end{cases}$$

Using the acceptance probability $\omega(x, z)$ defined above, the Metropolis–Hastings algorithm can be implemented as follows: (i) take an initial value of x, which is denoted by x_0, (ii) given x_{i-1}, generate a random draw z from $P_*(\cdot|x_{i-1})$ and a uniform random draw u from the interval between zero and one, (iii) set $x_i = z$ if $u \leqslant \omega(x_{i-1}, z)$ and set $x_i = x_{i-1}$ otherwise, and (iv) repeat (ii) and (iii) for $i = 1, 2, \ldots, N$. Then, x_N is taken as a random draw from $P_x(x)$ for sufficiently large N. The basic result of the Metropolis–Hastings algorithm is as follows:

$$\frac{1}{N-M} \sum_{i=M+1}^{N} g(x_i) \to \mathrm{E}\bigl(g(x)\bigr) = \int g(x) P_x(x) \, dx,$$

as $N \to \infty$, where $g(\cdot)$ is a function.

For choice of the proposal density $P_*(z|x)$, note as follows. The proposal density $P_*(z|x)$ should not have too large variance and too small variance (see, for example, Chib and Greenberg, 1995). That is, the proposal density should be chosen so that the chain travels over the support of the target density. This may fail to occur, with a consequent undersampling of low probability regions, if the chain is near the mode and if candidates are drawn too close to the current value (see Chib and Greenberg, 1996). Moreover, we should take the proposal density such that we can easily and quickly generate random draws. For a functional form of the proposal density, we may take $P_*(z|x) = P_*(z - x)$, called the random walk chain, or $P_*(z|x) = P_*(z)$, called the independence chain.

Note that $P_x(x)$ is not necessarily a probability density function, i.e., it is possibly a kernel of the target density function (remember that we need the ratio of the target and proposal densities to derive $\omega(x, z)$). It is also possible to apply the Metropolis–Hastings algorithm to generate random numbers from $P_{x|y}(x|y)$ and $P_{y|x}(y|x)$ in the Gibbs sampler, when it is difficult to generate a random draw of x from $P_{x|y}(x|y)$ and/or a random draw of y from $P_{y|x}(y|x)$ (see Chib and Greenberg, 1995 for the Metropolis–Hastings algorithm within Gibbs sampling).

Appendix C. Recursive versus non-recursive algorithms

In Sections 2.2.1 and 2.2.2, we introduce two density-based algorithms on prediction, filtering and smoothing The conventional recursive algorithms are represented by

Eqs. (13)–(16) of Section 2.2.1. Eqs. (22), (24) and (26) of Section 2.2.2 indicate the non-recursive algorithms. In this appendix, it is shown that both algorithms are equivalent (i.e., we can derive Eq. (13) from Eq. (22), Eqs. (14) and (15) from Eq. (24) and Eq. (16) from Eq. (26), respectively).

Prediction. Eq. (22) is rewritten as:

$$P(\alpha_{t+L}|Y_t) = \int P(A_{t+L}|Y_t) \, dA_{t+L-1}$$

$$= \int P_\alpha(\alpha_{t+L}|\alpha_{t+L-1}) P(A_{t+L-1}|Y_t) \, dA_{t+L-1}$$

$$= \int\int P_\alpha(\alpha_{t+L}|\alpha_{t+L-1}) P(A_{t+L-1}|Y_t) \, dA_{t+L-2} \, d\alpha_{t+L-1}$$

$$= \int P_\alpha(\alpha_{t+L}|\alpha_{t+L-1}) \left(\int P(A_{t+L-1}|Y_t) \, dA_{t+L-2} \right) d\alpha_{t+L-1}$$

$$= \int P_\alpha(\alpha_{t+L}|\alpha_{t+L-1}) P(\alpha_{t+L-1}|Y_t) \, d\alpha_{t+L-1}, \qquad (51)$$

where the second line in Eq. (51) uses the following two equations: $P(A_{t+L}, Y_t) = F_\alpha(A_{t+L}) P_y(Y_t|A_t) = P_\alpha(\alpha_{t+L}|\alpha_{t+L-1}) P_\alpha(A_{t+L-1}) P_y(Y_t|A_t)$ and $P(A_{t+L}|Y_t) = F_\alpha(A_{t+L}) P_y(Y_t|A_t)/P(Y_t)$ while the fifth line utilizes the first equality. Thus, it can be easily shown that Eq. (13) is equivalent to Eq. (22).

Filtering. Eq. (24) is transformed as:

$$P(\alpha_t|Y_t) = \frac{\int P(A_t, Y_t) \, dA_{t-1}}{\int P(A_t, Y_t) \, dA_t}$$

$$= \frac{\int P_y(y_t|\alpha_t) P_\alpha(\alpha_t|\alpha_{t-1}) P(A_{t-1}, Y_{t-1}) \, dA_{t-1}}{\int\int P_y(y_t|\alpha_t) P_\alpha(\alpha_t|\alpha_{t-1}) P(A_{t-1}, Y_{t-1}) \, dA_{t-1} \, d\alpha_t}$$

$$= \frac{P_y(y_t|\alpha_t) \left(\int P_\alpha(\alpha_t|\alpha_{t-1}) P(A_{t-1}|Y_{t-1}) \, dA_{t-1} \right)}{\int P_y(y_t|\alpha_t) \left(\int P_\alpha(\alpha_t|\alpha_{t-1}) P(A_{t-1}|Y_{t-1}) \, dA_{t-1} \right) d\alpha_t}$$

$$= \frac{P_y(y_t|\alpha_t) P(\alpha_t|Y_{t-1})}{\int P_y(y_t|\alpha_t) P(\alpha_t|Y_{t-1}) \, d\alpha_t}. \qquad (52)$$

Note that $P(A_t, Y_t) = P_y(y_t|\alpha_t) P_\alpha(\alpha_t|\alpha_{t-1}) P(A_{t-1}, Y_{t-1})$ in the second line of Eq. (52). From the second equality of Eq. (51), $P(\alpha_t|Y_{t-1}) = \int P_\alpha(\alpha_t|\alpha_{t-1}) P(A_{t-1}|Y_{t-1}) \, dA_{t-1}$ is substituted into the fourth line. Thus, Eq. (15) is derived from Eq. (24).

Smoothing. Let us define $Y_t^* = \{y_t, y_{t+1}, \ldots, y_T\}$. Suppose that the joint density function of A_{t+1}^* and Y_{t+1}^* is given by:

$$P(A_{t+1}^*, Y_{t+1}^*) = \prod_{s=t+2}^{T} P_\alpha(\alpha_s | \alpha_{s-1}) \prod_{s=t+1}^{T} P_y(y_s | \alpha_s),$$

which implies that the joint density of A_T and Y_T is represented as follows:

$$P(A_T, Y_T) = P(A_t, Y_t) P_\alpha(\alpha_{t+1} | \alpha_t) P(A_{t+1}^*, Y_{t+1}^*),$$

which is utilized in the second and eighth equalities of Eq. (53). Eq. (26) is represented as:

$$\begin{aligned}
P(\alpha_t | Y_T) &= \frac{1}{P(Y_T)} \iint P(A_T, Y_T) \, dA_{t-1} \, dA_{t+1}^* \\
&= \frac{1}{P(Y_T)} \iint P(A_t, Y_t) P_\alpha(\alpha_{t+1} | \alpha_t) P(A_{t+1}^*, Y_{t+1}^*) \, dA_{t-1} \, dA_{t+1}^* \\
&= \frac{1}{P(Y_T)} \int P(A_t, Y_t) \, dA_{t-1} \int P_\alpha(\alpha_{t+1} | \alpha_t) P(A_{t+1}^*, Y_{t+1}^*) \, dA_{t+1}^* \\
&= \frac{P(Y_t)}{P(Y_T)} P(\alpha_t | Y_t) \int P_\alpha(\alpha_{t+1} | \alpha_t) P(A_{t+1}^*, Y_{t+1}^*) \, dA_{t+1}^* \\
&= \frac{P(Y_t)}{P(Y_T)} P(\alpha_t | Y_t) \int \frac{P_\alpha(\alpha_{t+1} | \alpha_t) P(A_{t+1}^*, Y_{t+1}^*)}{\int P_\alpha(\alpha_{t+1} | \alpha_t) P(A_t, Y_t) \, dA_t} \\
&\qquad \times \left(\int P_\alpha(\alpha_{t+1} | \alpha_t) P(A_t, Y_t) \, dA_t \right) dA_{t+1}^* \\
&= P(\alpha_t | Y_t) \iint \frac{P_\alpha(\alpha_{t+1} | \alpha_t) P(A_{t+1}^*, Y_{t+1}^*)}{P(Y_T) P(\alpha_{t+1} | Y_t)} \\
&\qquad \times \left(\int P_\alpha(\alpha_{t+1} | \alpha_t) P(A_t, Y_t) \, dA_t \right) dA_{t+2}^* \, d\alpha_{t+1} \\
&= P(\alpha_t | Y_t) \int \frac{\iint P(A_{t+1}^*, Y_{t+1}^*) P_\alpha(\alpha_{t+1} | \alpha_t) P(A_t, Y_t) \, dA_t \, dA_{t+2}^*}{P(Y_T)} \\
&\qquad \times \frac{P_\alpha(\alpha_{t+1} | \alpha_t)}{P(\alpha_{t+1} | Y_t)} d\alpha_{t+1} \\
&= P(\alpha_t | Y_t) \int \frac{\iint P(A_T, Y_T) \, dA_t \, dA_{t+2}^*}{P(Y_T)} \frac{P_\alpha(\alpha_{t+1} | \alpha_t)}{P(\alpha_{t+1} | Y_t)} d\alpha_{t+1} \\
&= P(\alpha_t | Y_t) \int \frac{P(\alpha_{t+1} | Y_T) P_\alpha(\alpha_{t+1} | \alpha_t)}{P(\alpha_{t+1} | Y_t)} d\alpha_{t+1}. \quad (53)
\end{aligned}$$

In the fourth equality, the first equality of Eq. (52) is utilized (note that $P(Y_t) = \int P(A_t, Y_t)\,dA_t$). The denominator of the sixth equality is obtained from the case $L=1$ in the second equality of Eq. (51). The ninth equality comes from the first equality. Thus, it is shown that Eq. (26) is exactly equivalent to Eq. (16).

Acknowledgements

The author would like to acknowledge an anonymous referee for valuable comments.
   This research was partially supported by Japan Society for the Promotion of Science, Grants-in-Aid for Scientific Research C (#14530033).

References

Alspach, D. L. and H. W. Sorenson (1972). Nonlinear Bayesian estimation using Gaussian sum approximations. *IEEE Trans. Automat. Control* **AC-17**(4), 439–448.
Anderson, B. D. O. and J. B. Moore (1979). *Optimal Filtering*. Prentice-Hall, New York.
Aoki, M. (1987). *State Space Modeling of Time Series*. Springer-Verlag, New York.
Aoki, M. (1990). *State Space Modeling of Time Series*, 2nd, revised and enlarged ed. Springer-Verlag, New York.
Arnold, S. F. (1993). Gibbs sampling. In *Handbook of Statistics*, Vol. 9, pp. 599–625 (Ed. C. R. Rao). North-Holland, Amsterdam.
Belsley, D. A. (1973). On the determination of systematic parameter variation in the linear regression model. *Annals of Economic and Social Measurement* **2**, 487–494.
Belsley, D. A. and E. Kuh (1973). Time-varying parameter structures: an overview. *Annals of Economic and Social Measurement* **2**(4), 375–379.
Boswell, M. T., S. D. Gore, G. P. Patil and C. Taillie (1993). The art of computer generation of random variables. In *Handbook of Statistics*, Vol. 9, pp. 661–721 (Ed. C. R. Rao). North-Holland, Amsterdam.
Brockwell, P. A. and R. A. Davis (1987). *Time Series Theory and Models*. Springer-Verlag, New York.
Brown, B. W. and R. S. Mariano (1984). Residual-based procedures for prediction and estimation in a nonlinear simultaneous system. *Econometrica* **52**(2), 321–343.
Brown, B. W. and R. S. Mariano (1989). Measures of deterministic prediction bias in nonlinear models. *International Economic Review* **30**(3), 667–684.
Burmeister, E. and K. D. Wall (1982). Kalman filtering estimation of unobserved rational expectations with an application to the German hyperinflation. *J. Econometrics* **20**, 255–284.
Burridge, P. and K. F. Wallis (1988). Prediction theory for autoregressive moving average processes. *Econometric Reviews* **7**(1), 65–95.
Carlin, B. P. and N. G. Polson (1991). Inference for nonconjugate Bayesian models using the Gibbs sampler. *Canad. J. Statist.* **19**, 399–405.
Carlin, B. P., N. G. Polson and D. S. Stoffer (1992). A Monte Carlo approach to nonnormal and nonlinear state space modeling. *J. Amer. Statist. Assoc.* **87**, 493–500.
Carter, C. K. and R. Kohn (1994). On Gibbs sampling for state space models. *Biometrika* **81**(3), 541–553.
Carter, C. K. and R. Kohn (1996). Markov chain Monte Carlo in conditionally Gaussian state space models. *Biometrika* **83**(3), 589–601.
Chib, S. and E. Greenberg (1995). Understanding the Metropolis–Hastings algorithm. *Amer. Statist.* **49**(4), 327–335.
Chib, S. and E. Greenberg (1996). Markov chain Monte Carlo simulation methods in econometrics. *Econometric Theory* **12**(4), 409–431.
Chow, G. C. (1983). *Econometrics*. McGraw-Hill.
Conrad, W. and C. Corrado (1979). Application of the Kalman filter to revisions in monthly retail sales estimates. *J. Economic Dynamic and Control* **1**, 177–198.

Cooley, T. F. (1977). Generalized least squares applied to time varying parameter models: a comment. *Annals of Economic and Social Measurement* **6**(3), 313–314.

Cooley, T. F. and E. C. Prescott (1976). Estimation in the presence of stochastic parameter variation. *Econometrica* **44**, 167–183.

Cooley, T. F., B. Rosenberg and K. D. Wall (1977). A note on optimal smoothing for time varying coefficient problems. *Annals of Economic and Social Measurement* **6**(4), 453–456.

Cooper, J. P. (1973). Time-varying regression coefficients: a mixed estimation approach and operational limitations of the general Markov structure. *Annals of Economic and Social Measurement* **2**(4), 525–530.

Dempster, A. P., N. M. Laird and D. B. Rubin (1977). Maximum likelihood from incomplete data via the EM algorithm. *J. Roy. Statist. Soc. Ser. B* **39**, 1–38 (with discussion).

Diebold, F. X. and M. Nerlove (1989). Unit roots in economic time series: a selective survey. In *Advances in Econometrics*, Vol. 8, pp. 3–69. JAI Press.

Engle, R. F. (1982). Autoregressive conditional heteroscedasticity with estimates of variance of U.K. inflation. *Econometrica* **50**, 987–1008.

Engle, R. F. and M. W. Watson (1987). The Kalman filter: applications to forecasting and rational expectations models. In *Advances in Econometrics, Fifth World Congress*, Vol. I. Cambridge University Press.

Gardner, G., A. C. Harvey and G. D. A. Phillips (1980). An algorithm for maximum likelihood estimation autoregressive-moving average models by means of Kalman filtering. *Appl. Statist.* **29**(3), 311–322.

Gelb, A. (1974). *Applied Optimal Estimation*. MIT Press.

Gelfand, A. E., S. E. Hills, H. A. Racine-Poon and A. F. M. Smith (1990). Illustration of Bayesian inference in normal data models using Gibbs sampling. *J. Amer. Statist. Assoc.* **85**(412), 972–985.

Gelfand, A. E. and A. F. M. Smith (1990). Sampling-based approaches to calculating marginal densities. *J. Amer. Statist. Assoc.* **85**(410), 398–409.

Geman, S. and D. Geman (1984). Stochastic relaxation, Gibbs distributions, and the Bayesian restoration of images. *IEEE Trans. Pattern Anal. Mach. Intell.* **Pami-6**(6), 721–741.

Geweke, J. (1988). Antithetic acceleration of Monte Carlo integration in Bayesian inference. *J. Econometrics* **38**, 73–90.

Geweke, J. (1989a). Modeling with normal polynomial expansions. In *Economic Complexity: Chaos, Sunspots, Bubbles and Nonlinearity* (Eds. W. A. Barnett, J. Geweke and K. Shell). Cambridge University Press.

Geweke, J. (1989b). Bayesian inference in econometric models using Monte Carlo integration. *Econometrica* **57**, 1317–1339.

Geweke, J. (1996). Monte Carlo simulation and numerical integration. In *Handbook of Computational Economics*, Vol. 1, pp. 731–800 (Eds. H. M. Amman, D. A. Kendrick and J. Rust). North-Holland, Amsterdam.

Geweke, J. (1997). Posterior simulators in econometrics. In *Advances in Economics and Econometrics: Theory and Applications*, Vol. 3, pp. 128–165 (Eds. D. Kreps and K. F. Wallis). Cambridge University Press.

Geweke, J. and H. Tanizaki (1999). On Markov chain Monte Carlo methods for nonlinear and non-Gaussian state-space models. *Comm. Statist. Simulation Comput.* **28**(4), 867–894.

Geweke, J. and H. Tanizaki (2001). Bayesian estimation of state-space model using the Metropolis–Hastings algorithm within Gibbs sampling. *Comput. Statist. Data Anal.* **37**(2), 151–170.

Ghysels, E., A. C. Harvey and E. Renault (1996). Stochastic volatility. In *Handbook of Statistics*, Vol. 14, pp. 119–191 (Eds. G. S. Maddala and C. R. Rao). North-Holland, Amsterdam.

Gordon, N. J., D. J. Salmond and A. F. M. Smith (1993). Novel approach to nonlinear/non-Gaussian Bayesian state estimation. *IEE Proceedings-F* **140**(2), 107–113.

Hall, R. E. (1978). Stochastic implications of the life cycle-permanent income hypothesis: theory and evidence. *J. Political Economy* **86**(6), 971–987.

Hall, R. E. (1990). *The Rational Consumer*. The MIT Press.

Hamilton, J. D. (1989). A new approach to the economic analysis of nonstationary time series and the business cycle. *Econometrica* **57**, 357–384.

Hamilton, J. D. (1990). Analysis of time series subject to changes in regime. *J. Econometrics* **45**, 39–70.

Hamilton, J. D. (1991). A quasi-Bayesian approach to estimating parameters for mixtures of normal distributions. *J. Business and Economic Statistics* **9**, 27–39.

Hamilton, J. D. (1993). Estimation, inference and forecasting of time series subject to changes in regime. In *Handbook of Statistics*, Vol. 11, pp. 231–260 (Eds. G. S. Maddala, C. R. Rao and H. D. Vinod). North-Holland, Amsterdam.

Hamilton, J. D. (1994). *Time Series Analysis*. Princeton University Press.

Hannan, E. J. and M. Deistler (1988). *The Statistical Theory of Linear System*. Wiley.

Harvey, A. C. (1981). *Time Series Models*. Philip Allen Publishers Limited, Oxford.

Harvey, A. C. (1987). Applications of the Kalman filter in econometrics. In *Advances in Econometrics, Fifth World Congress*, Vol. I. Cambridge University Press.

Harvey, A. C. (1989). *Forecasting, Structural Time Series Models and the Kalman Filter*. Cambridge University Press.

Howrey, E. P. (1978). The use of preliminary data in econometric forecasting. *The Review of Economics and Statistics* **60**, 193–200.

Howrey, E. P. (1984). Data revision, reconstruction, and prediction: an application to inventory investment. *The Review of Economics and Statistics* **66**, 386–393.

Hürzeler, M. and H. R. Künsch (1998). Monte Carlo approximations for general state-space models. *J. Comput. Graphical Statist.* **7**(2), 175–193.

Jazwinski, A. H. (1970). *Stochastic Processes and Filtering Theory*. Academic Press, New York.

Kalman, R. E. (1960). A new approach to linear filtering and prediction problems. *J. Basic Engineering, Trans. ASME, Ser. D* **82**, 35–45.

Kalman, R. E. and R. S. Bucy (1961). New results in linear filtering and prediction theory. *J. Basic Engineering, Trans. ASME, Ser. D* **83**, 95–108.

Kitagawa, G. (1987). Non-Gaussian state-space modeling of nonstationary time series. *J. Amer. Statist. Assoc.* **82**, 1032–1063 (with discussion).

Kitagawa, G. (1996). Monte Carlo filter and smoother for non-Gaussian nonlinear state-space models. *J. Comput. Graphical Statist.* **5**(1), 1–25.

Kitagawa, G. (1998). A self-organizing state-space model. *J. Amer. Statist. Assoc.* **93**(443), 1203–1215.

Kitagawa, G. and W. Gersch (1996). *Smoothness Priors Analysis of Time Series*. Lecture Notes in Statist., Vol. 116. Springer-Verlag.

Kirchen, A. (1988). *Schäzung zeitveränderlicher Strukturparameter in ökono metrischen Prognosemodellen*. Athenäum, Frankfurt/Main.

Kramer, S. C. and H. W. Sorenson (1988). Recursive Bayesian estimation using piece-wise constant approximations. *Automatica* **24**(6), 789–801.

Knuth, D. E. (1981). *The Art of Computer Programming*, Vol. 2 (*Seminumerical Algorithms*, 2nd ed.). Addison-Wesley.

Koop, G. (1994). Recent progress in applied Bayesian econometrics. *J. Economic Surveys* **8**(1), 1–34.

Laird, N. (1993). The EM algorithm. In *Handbook of Statistics*, Vol. 9, pp. 661–721 (Ed. C. R. Rao). North-Holland, Amsterdam.

Mariano, R. S. and B. W. Brown (1983). Asymptotic behavior of predictors in a nonlinear simultaneous system. *Internat. Economic Review* **24**(3), 523–536.

Mariano, R. S. and B. W. Brown (1989). Stochastic simulation, prediction and validation of nonlinear models. In *Economics in Theory and Practice: An Eclectic Approach*, pp. 17–36 (Eds. L. R. Klein and J. Marquez). Kluwer Academic Publishers.

Mariano, R. S. and H. Tanizaki (1995). Prediction of final data with use of preliminary and/or revised data. *J. Forecasting* **14**(4), 351–380.

Mariano, R. S. and H. Tanizaki (2000). Simulation-based inference in nonlinear state-space models: application to testing the permanent income hypothesis. In *Simulation-Based Inference in Econometrics: Methods and Applications*, pp. 218–234 (Eds. R. S. Mariano, M. Weeks and T. Schuermann). Cambridge University Press.

McNelis, P. D. and S. N. Neftci (1983). Policy-dependent parameters in the presence of optimal learning: an application of Kalman filtering to the fair and sargent supply-side equations. *The Review of Economics and Statistics* **65**, 296–306.

Meinhold, R. J. and N. D. Singpurwalla (1989). Robustification of Kalman filter models. *J. Amer. Statist. Assoc.* **84**, 479–486.

Nicholls, D. F. and A. R. Pagan (1985). Varying coefficient regression. In *Handbook of Statistics*, Vol. 5, pp. 413–449 (Eds. E. J. Hannan, P. R. Krishnaiah and M. M. Rao). Elsevier.

O'Hagan, A. (1994). *Kendall's Advanced Theory of Statistics*, Vol. 2B (*Bayesian Inference*). Edward Arnold.

Pagan, A. R. (1975). A note on the extraction of components from time series. *Econometrica* **43**, 163–168.

Rund, P. A. (1991). Extensions of estimation methods using the EM algorithm. *J. Econometrics* **49**, 305–341.

Sant, D. T. (1977). Generalized least squares applied to time varying parameter models. *Annals of Economic and Measurement* **6**(3), 301–311.

Sarris, A. H. (1973). A Bayesian approach to estimation of time varying regression coefficients. *Annals of Economic and Social Measurement* **2**(4), 501–523.

Shao, J. (1989). Monte Carlo approximations in Bayesian decision theory. *J. Amer. Statist. Assoc.* **84**(407), 727–732.

Shumway, R. H. and D. S. Stoffer (1982). An approach to time series smoothing and forecasting using the EM algorithm. *J. Time Series Anal.* **3**, 253–264.

Smith, A. F. M. and A. E. Gelfand (1992). Bayesian statistics without tears: a sampling-resampling perspective. *Amer. Statist.* **46**(2), 84–88.

Smith, A. F. M. and G. O. Roberts (1993). Bayesian computation via Gibbs sampler and related Markov chain Monte Carlo methods. *J. Roy. Statist. Soc. Ser. B* **55**(1), 3–23.

Sorenson, H. W. and D. L. Alspach (1971). Recursive Bayesian estimation using Gaussian sums. *Automatica* **7**, 465–479.

Tanizaki, H. (1989). The Kalman filter model under the assumption of the first-order autoregressive process in the disturbance terms. *Econom. Lett.* **31**(2), 145–149.

Tanizaki, H. (1993a). *Nonlinear Filters: Estimation and Applications*. Lecture Notes in Economics and Mathematical Systems, Vol. 400. Springer-Verlag.

Tanizaki, H. (1993b). Kalman filter model with qualitative dependent variable. *The Review of Economics and Statistics* **75**(4), 747–752.

Tanizaki, H. (1996). *Nonlinear Filters: Estimation and Applications*, 2nd, revised and enlarged ed. Springer-Verlag.

Tanizaki, H. (1997). Nonlinear and nonnormal filters using Monte Carlo methods. *Comput. Statist. Data Anal.* **25**(4), 417–439.

Tanizaki, H. (2000). Nonlinear and non-Gaussian state estimation: a quasi-optimal estimator. *Comm. Statist. Theory Methods* **29**(12), 2805–2834.

Tanizaki, H. (1999a). On nonlinear and nonnormal filter using rejection sampling. *IEEE Trans. Automat. Control* **44**(2), 314–319.

Tanizaki, H. (1999b). Nonlinear and nonnormal filter using importance sampling: antithetic Monte Carlo integration. *Comm. Statist. Simulation Comput.* **28**(2), 463–486.

Tanizaki, H. (2001). Nonlinear and non-Gaussian state-space modeling using sampling techniques. *Ann. Inst. Statist. Math.* **53**(1), 63–81.

Tanizaki, H. and R. S. Mariano (1994). Prediction, filtering and smoothing in nonlinear and nonnormal cases using Monte Carlo integration. *J. Appl. Econometrics* **9**(2), 163–179 (in *Econometric Inference Using Simulation Techniques*, Chapter 12, edited by H. K. van Dijk, A. Manfort and B. W. Brown, pp. 245–261 (1995), Wiley).

Tanizaki, H. and R. S. Mariano (1996). Nonlinear filters based on Taylor series expansions. *Comm. Statist. Theory Methods* **25**(6), 1261–1282.

Tanizaki, H. and R. S. Mariano (1998). Nonlinear and non-Gaussian state-space modeling with Monte Carlo simulations. *J. Econometrics* **83**(1–2), 263–290.

Tanner, M. A. and W. H. Wong (1987). The calculation of posterior distributions by data augmentation. *J. Amer. Statist. Assoc.* **82**(398), 528–550 (with discussion).

Taylor, S. J. (1994). Modeling stochastic volatility: a review and comparative study. *Math. Finance* **4**(2), 183–204.

Tierney, L. (1994). Markov chains for exploring posterior distributions. *Ann. Statist.* **22**(4), 1701–1762.

Wishner, R. P., J. A. Tabaczynski and M. Athans (1969). A comparison of three non-linear filters. *Automatica* **5**, 487–496.

Markov Modelling of Burst Behaviour in Ion Channels

G. F. Yeo, R. K. Milne, B. W. Madsen, Y. Li and R. O. Edeson

1. Introduction

Living organisms are made up of one or more cells, each enclosed by a lipid membrane. Spanning this membrane are specialized protein molecules, called ion channels, which facilitate many cellular processes by selectively controlling the transmembrane movement of small electrically charged ions. Understanding the behaviour of these channels is fundamental to the study of normal and abnormal cell function and regulation.

There are many different types of channel, varying in chemical structure, size, location and function; see, for example, Zimmermann (1993) for an accessible and well-structured review. Ion conduction occurs most commonly through a single aqueous pore with a gate controlled by particular stimuli. Typically, gating appears instantaneous at the usual time scale of observation, so that in the simplest case the gate is effectively closed or fully open. Some well-known channels, such as calcium, chloride, potassium or sodium channels, are classified according to the dominant ion transferred, while others, such as acetylcholine, glycine or glutamate channels, are identified by the neurotransmitter which leads to their activation. Mechanosensitive channels are a more recently discovered class of channels which are activated by physical stresses; Oakley et al. (1999) provide a good review of these. Evidence has been presented also for a class of heat-activated ion channels (Cesare et al., 1999).

Direct observation of channel gating behaviour became possible following development of the patch clamp technique (Neher and Sakmann, 1976; Hamill et al., 1981) which enabled very small currents, of the order of a few pico-amps, to be recorded in continuous time from a single channel; see also the papers in Sakmann and Neher (1995). For their contribution, Sakmann and Neher were awarded the 1991 Nobel prize in Physiology or Medicine; Neher (1992) and Sakmann (1992) give condensed versions of their Nobel lectures.

The evident randomness of single channel activity in these early records and the promise of gaining new insights into the underlying biophysics stimulated development of stochastic models for the gating behaviour of ion channels (e.g., Colquhoun and Hawkes, 1977, 1981, 1982; Fredkin et al., 1985; Labarca et al., 1985; Fredkin and Rice, 1986). These models provide a basis for quantitative inferential studies of ion channel kinetics, though generally development of inference (e.g., Fredkin et al., 1985; Labarca et al., 1985; Fredkin and Rice, 1987, 1992a, 1992b; Yeo et al., 1988; Ball and Sansom, 1989) has lagged behind that of stochastic modelling. Section 3.8 of Guttorp (1995) gives a simple introduction to aspects of channel modelling and inference, prefaced by some background on neurophysiology.

Experimental data indicate that most channels can generally be considered as being, at any particular time, in one of a finite number of kinetically distinct states linked by well defined transition pathways (McManus et al., 1988). Reflecting this, and a lack of memory in the sequence of successive states visited, the gating behaviour of a single ion channel under equilibrium conditions is usually modelled by a continuous-time homogeneous finite-state Markov chain; see, for example, Colquhoun and Hawkes (1982, 1987, 1990, 1995a, 1995b), Fredkin and Rice (1986), Ball and Sansom (1988), Ball and Rice (1992), and other papers in Sakmann and Neher (1995). As discussed by Colquhoun and Sigworth (1995), a system at equilibrium with reversible transitions between states will exhibit the property of (microscopic) reversibility, and this property should be part of any reasonable model.

As the states in a single channel model need not all have distinct conductance levels, it may be that only some of the transitions between states are directly observable. Thus the state space for a Markov model may be *aggregated* by a partitioning into classes, so that at any given time it is possible, on the basis of conductance, to observe only which class the process is in. In the simplest single channel model there are just two classes, conducting (open) and non-conducting (closed), although several intermediate conductance levels (corresponding to so-called sub-conductance states) are needed for modelling behaviour of some types of channel (e.g., potassium channels, Zheng and Sigworth, 1998; or mechanosensitive channels, Sukharev et al., 1999).

Because, in practice, a patch cannot be guaranteed to contain one and only one channel of a given type, a further identification problem occurs. Without additional information, we cannot assign an event to any particular channel; only the superposition process (or total current), is observable (see, for example, Yeo et al., 1989; Fredkin and Rice, 1991). While it is usual to model either a single channel or a multiple channel system as a Markov chain (Ball et al., 1997), it is not generally true that the superposition of Markov chains is also a Markov chain; this problem and the related idea of lumpability have been discussed by Ball and Yeo (1993) and others. Note that records having a similar appearance to those arising from superposition can be generated by some single channels that have more than two conductance levels when these are simple multiples of some unit value.

Markov models for the gating behaviour of ion channels use standard theory of finite state space continuous-time Markov chains (cf. Section 4.8 of Karlin and Taylor, 1975, Chapter II of Asmussen, 1987, or Chapter 3 of Guttorp, 1995), but are non-standard in the particular structure (especially the partitioning) that is imposed on the state space, and often also in the questions and properties of interest. For example, the (joint) properties of class sojourn times when the process is in equilibrium are basic in the study of such ion channel models. These remarks apply both to models for behaviour of a single ion channel and to models for the behaviour of several not necessarily independent channels; see Ball et al. (1997, 2000) for further comments and references.

The power of the theory of aggregated Markov chains has been demonstrated in many papers. Its utility in dealing with the superposition arising from several ion channels, as well as with a single ion channel, was highlighted in Ball et al. (1997). That paper also pointed out related applications of aggregated Markov chains in reliability modelling (although it is not clear whether the phenomenon of bursting has relevance in this area). The semi-Markov approach developed by Ball et al. (1991, 1993) has value as a unified theory which includes the Markov theory as a special case, providing in particular an efficient route to many joint distribution results.

Under the Markovian assumption, sojourn times in a class of states have a distribution which is a linear combination of exponentials (unless two or more eigenvalues of the appropriate submatrix are equal). Fitting of such combinations of exponentials has been used to infer the number of states within a class, and to assess the Markovian assumption (see, for example, Colquhoun and Sigworth, 1995, and Magleby and Weiss, 1990a, 1990b). With an appropriate choice of states, the Markovian assumption appears to be reasonable for a wide variety of channel types.

Recordings of single channels often show periods of repetitive open channel activity, known as *bursts*, which are noticeably separated from other such periods; specifically, a burst is a sequence of times during which the channel is open together with the intervening short closed-times, with individual bursts separated by much longer closed-times. This bursting activity is of special interest because it manifests a certain complexity in the channel kinetics; appropriate study of burst behaviour might therefore be expected to be revealing about the finer structure of the underlying process, for example, the connectivity of open states. Moreover, because charge transfer (which may change transmembrane potential and alter cell function) is greatest during these events, it is likely that burst kinetics are relevant to the intense, non-equilibrium channel activity occurring, say, in synaptic transmission.

Historically, much of the interest in bursts, as opposed to single openings or closures, has followed from recognition that very brief events may be missed in patch clamp data because of finite (although continually improving) recording characteristics; in this situation properties such as burst duration are usually more stable than corresponding properties for individual sojourns. The 'missed brief event' problem, also called 'time interval omission' or 'limited time resolution', has received considerable attention in the literature; see, for example, Roux and Sauvé (1985), Ball and Sansom (1988), Hawkes et al. (1990, 1992), Ball et al. (1993) and Ball and Winch (1999). A further reason for interest in bursts is that channel activity within each burst is usually thought to be due to transitions within a single channel, rather than to possible activation of different channels over longer periods of time.

As we shall discuss more fully later, a burst may be defined formally in two different ways. One of these, which seems natural from a theoretical viewpoint but is of limited practical interest because such bursts are not experimentally observable, assigns closed states to two (mutually exclusive) classes, one that *may* be visited within a burst and one that *must* be visited between bursts. The other approach, which defines a type of burst that can be identified from a channel record, is based on specification of some critical time t_c and use of this to classify closed-times into ones that are short-lived (duration at most t_c) and ones that are long-lived (duration greater than t_c). For a given model, under either definition, various properties of bursts can be derived; these include the number of openings and closures in a burst, the duration of a burst, the total open-time and probability of being open within a burst and, for each nonzero conductance level, the total time spent within a burst at that level.

This chapter reviews and extends aspects of Markov chain based models for burst behaviour of a single ion channel, defining and exploring properties of the two types of burst. The restriction to modelling of bursts was made in order to provide a focus on phenomena relevant to cell physiology and keep the chapter to a manageable size. Section 2 summarizes basic aggregated Markov chain theory in a form that facilitates the study, in Sections 3 and 4 respectively, of these bursts. Although they may look forbidding, the general matrix expressions presented in Sections 2–4 can be employed in conjunction with a matrix-capable package, such as Matlab or S-Plus, for straightforward derivation of properties once a particular model has been specified by assignment of parameter values (usually transition rates). Sections 5–8 consider applications to particular channel types: a five-state model of an acetylcholine receptor channel; a linear sequential model allowing blockade of the channel by a drug; a model showing a concentration-dependent biphasic drug effect; and a model for a double-barrelled chloride channel. For illustrating the methods in each case, values of the transition rates have been chosen to be consistent with some experimental observations. Numerical results are summarized for each of these examples in a way that facilitates comparison between the two types of burst. Most of the time we have employed direct probabilistic arguments, in order to highlight their value in applications, especially as a means of enhancing understanding of results. On occasion these methods lead conveniently to useful transform/generating function results. Other authors, such as Colquhoun and Hawkes (1982), have used transform methods more directly to obtain their results. Random sum methods are used, especially in Sections 6 and 8, to provide illuminating alternative derivations of important properties. A brief overview of approaches to statistical inference based on ion channel data is given in Section 9.

Throughout what follows vectors and matrices are rendered in bold; all vectors are column vectors and $^\top$ denotes transpose, which is often used to express row vectors. Furthermore, **1** denotes a column vector of ones, *I* an identity matrix, and **0** a matrix (vector) of zeros, the dimensions of these being apparent from the context. For a random variable X the mean or expected value of X is denoted by μ_X or $\mathbb{E}(X)$ and its variance by σ_X^2. Although in the past we have made extensive use of Matlab, all computations for numerical examples in the present paper were performed in S-Plus on a personal workstation, with no program taking more than a few seconds to run. Throughout the numerical examples the unit of time is milliseconds (ms).

2. Aggregated Markov chains

2.1. Basic structure

This section summarizes key structure and results from the theory of *aggregated Markov chains*. The theory was developed in the context of ion channel modelling especially by

Colquhoun and Hawkes (1982, 1987) and Fredkin at al. (1985) in the channel literature, and by Fredkin and Rice (1986), Ball and Sansom (1988), and Ball et al. (1997, 2000) in the probability literature. Related applications in reliability are studied by Rubino and Sericola (1989), Sericola (1990) and Csenki (1992, 1994).

Consider an irreducible homogeneous continuous-time Markov chain $\{Y(t)\} = \{Y(t): t \geq 0\}$. (Throughout the paper, we adopt the convention that a continuous-time stochastic process has index set $[0, \infty)$.) Suppose that $\{Y(t)\}$ has state space $\{1, 2, \ldots, m\}$ and transition rate matrix (generator) $Q = [q_{ij}]$. Thus, for $i \neq j$, q_{ij} is the transition rate from state i to state j, and the diagonal elements satisfy $q_{ii} = -d_i$ where $d_i = \sum_{j \neq i} q_{ij}$. Such a process can be regarded as specified by Q or, as in the examples of Sections 5–8, by the associated state space graph in which vertices represent states, directed edges join those states i and j for which q_{ij} is non-zero, and the particular transition rate values are shown associated with each edge; see Figures 1, 3, 5 and 7.

Transitions between states in a continuous-time Markov chain are determined by an associated discrete-time Markov chain, the *jump chain*, having transition (probability) matrix P with diagonal entries all zero and off-diagonal entries $p_{ij} = q_{ij}/d_i$, $i \neq j$. Conditional on the successive states visited, the durations of the sojourns in these states are independent, with each sojourn time in state i having an exponential distribution with parameter d_i (cf. Theorem 1.2 of Asmussen, 1987, or (24) of Colquhoun and Hawkes, 1995a). A well-known fundamental result (cf. p. 152 of Karlin and Taylor, 1975, or Corollary 3.5 of Asmussen, 1987) is that the transition function $P(t) = [P(Y(t) = j|Y(0) = i)]$ is given by

$$P(t) = e^{Qt} \quad (t \geq 0), \tag{2.1}$$

where $e^{Qt} = \sum_{j=0}^{\infty} t^j Q^j/j!$ is the usual matrix exponential.

The state space is assumed to be *aggregated* by a partitioning into classes, so that it is possible to observe only which class the chain is in at any given time, and not the individual state. Such a process is called an *aggregated Markov chain*. Although in principle any finite number of classes could be dealt with (cf. Fredkin and Rice, 1986, or Chapter 4 of Csenki, 1994), for the purposes of the present paper it suffices to consider just three classes, denoted by O, S, L, having respectively n_O, n_S and n_L states, with $n_O + n_S + n_L = m$. Throughout this paper, O denotes the class of all open states and, where relevant, S and L respectively denote classes of short-lived and long-lived closed states. In many examples, it is enough to consider two classes, O and $C = O'$, where $'$ denotes set complementation and C the class of all closed states, with $C = S \cup L$ when subclasses S and L are defined.

The transition rate matrix Q for an aggregated Markov chain whose state space is partitioned into classes O, S, L can be expressed, after appropriate reordering of the states, in a corresponding partitioned form as

$$Q = \begin{bmatrix} Q_{OO} & Q_{OS} & Q_{OL} \\ Q_{SO} & Q_{SS} & Q_{SL} \\ Q_{LO} & Q_{LS} & Q_{LL} \end{bmatrix}. \tag{2.2}$$

In this form, for example, the submatrix $\boldsymbol{Q}_{OO}$ governs transitions between states within O, and $\boldsymbol{Q}_{OS}$ governs transitions from states in O to those in S. Such partitionings are basic for dealing with many problems, particularly derivation of (joint) density functions and moments of class sojourn times and burst properties. For simplicity of exposition we assume that the diagonal submatrices $\boldsymbol{Q}_{OO}, \boldsymbol{Q}_{SS}, \boldsymbol{Q}_{LL}$ have distinct eigenvalues, all with strictly negative real parts.

2.2. Equilibrium properties and reversibility

Since the process $\{Y(t)\}$ has finitely many states and is irreducible, it has an invariant (stationary) distribution $\boldsymbol{\pi} = [\pi_1, \pi_2, \ldots, \pi_m]^\top$ which can be obtained by solving the global balance equations $\boldsymbol{\pi}^\top \boldsymbol{Q} = \boldsymbol{0}$. The process started at time $t = 0$ with $\boldsymbol{\pi}$ as its initial distribution is in equilibrium, that is, it is a (strictly) stationary process. The invariant distribution is also the limiting distribution in the sense that its components satisfy $\pi_k = \lim_{t \to \infty} P(Y(t) = k)$. Just as any partition of the state space induces a partition of $\boldsymbol{Q}$, so it induces a partition of the invariant distribution $\boldsymbol{\pi}^\top = [\boldsymbol{\pi}_O^\top, \boldsymbol{\pi}_S^\top, \boldsymbol{\pi}_L^\top]$ into blocks.

A continuous-time Markov chain $\{Y(t)\}$ is *reversible* essentially if its probabilistic properties remain the same when time is reversed. Any reversible Markov chain is stationary. A stationary Markov chain is reversible if and only if the *detailed balance equations* $\pi_i q_{ij} = \pi_j q_{ji}$, $i, j \in \{1, 2, \ldots, m\}$, are satisfied for some positive numbers π_i summing to one. These and other aspects of reversibility are discussed in Section 1.2 of Kelly (1979). When a process is reversible, as is usually assumed in ion channel models, the invariant distribution is often more conveniently obtained by solving the detailed balance equations together with $\sum \pi_i = 1$ rather than by using the global balance equations. Kolmogorov's criterion (cf. Theorem 1.8 of Kelly, 1979) allows reversibility to be checked by consideration of the transition rates alone: a process is reversible if and only if for any cycle (closed path) of states the product of the transition rates in one direction around the cycle is equal to the corresponding product in the other direction. Guttorp (1995, pp. 168–170) summarizes some useful consequences, due to Fredkin et al. (1985), of the assumption of reversibility for ion channel models.

2.3. Entry processes

For dealing with many aspects of an aggregated Markov chain that is in equilibrium it is helpful to consider properties of certain embedded processes, among which are the *entry processes* for the various classes. For example, for a specified class O, say, consider the point process of times at which the aggregated process $\{Y(t)\}$ commences a sojourn in the class O, and associate with each such time the state which $\{Y(t)\}$ entered at that time. The discrete-time process which records the successive states entered, called the *class O entry process*, or just the O entry process (cf. Ball and Sansom, 1988, and Ball et al., 1991, 1997), is a Markov chain which Fredkin et al. (1985) showed has transition matrix

$$\boldsymbol{P}_O = (-\boldsymbol{Q}_{OO})^{-1} \boldsymbol{Q}_{OO'} (-\boldsymbol{Q}_{O'O'})^{-1} \boldsymbol{Q}_{O'O}. \tag{2.3}$$

Substochastic versions of this result will be discussed in Section 3.1. (Observe that (2.3) is a 'subset' analogue of the result that for fixed states i and j with $i \neq j$, the probability that the chain jumps from i to j and then back to i again is given by $(-q_{ii})^{-1} q_{ij} (-q_{jj})^{-1} q_{ji}$.) The idea of entry processes can be seen also in Colquhoun and Hawkes (1982, 1987); in the latter paper it is the matrix $X_{\mathcal{A}\mathcal{A}}$, given by (1.8), which corresponds to our $\boldsymbol{P}_O$.

Such an entry chain may be aperiodic or, in certain circumstances, periodic. In the aperiodic case the O entry process has an invariant distribution $\boldsymbol{\phi}_O$ given by

$$\boldsymbol{\phi}_O^\top = \boldsymbol{\pi}_{O'}^\top \boldsymbol{Q}_{O'O} / (\boldsymbol{\pi}_{O'}^\top \boldsymbol{Q}_{O'O} \mathbf{1}), \tag{2.4}$$

as in Eq. (3.63) of Colquhoun and Hawkes (1982). The entries of this vector are the stationary probabilities that an O sojourn of the aggregated Markov chain starts in the various states of that class.

In many examples, entry to a given class is possible only through a subset of the states in that class – what have been called the *gateway* states (cf. Ball et al., 1991, 1993). (Note that the term gateway state is used differently by Colquhoun and Hawkes, 1987.) The subset of such gateway states for a given class could then be taken as the state space of the corresponding entry process. With the state space reduced in this manner, entry processes inherit the assumed irreducibility of the continuous-time chain. Furthermore, there is an advantage for numerical computation in that such reduction allows use of matrices (e.g., $\boldsymbol{P}_O$) of smaller size. The disadvantage is some extra notation, matrices denoted by $\boldsymbol{L}$ in Ball et al. (1991, 1993), which is needed to focus the action on the reduced state space. We follow, for example, Colquhoun and Hawkes (1987) in avoiding this additional notation in the present chapter. In situations where there is a single gateway state to the relevant class many simplifications occur. Rather than apply the general theory in such cases, as will be shown in the examples of Sections 6 and 8, it is often easier to employ direct methods based on random sums.

2.4. Class sojourn times

A number of authors, including Fredkin and Rice (1986), Colquhoun and Hawkes (1982, 1995a) and Ball et al. (1991, 1997, 2000) derive various properties of sojourn times in a class of states. This section summarizes properties for a sojourn time in a single class. Extensions to joint properties of a sequence of class sojourn times could be considered, for example as in Ball et al. (1997). However, such extensions will not be dealt with explicitly in the present chapter; rather, the following two sections will discuss joint properties for a sequence of class sojourn times that are part of a burst.

Here it is supposed that the aggregated process is in equilibrium. It is well known (cf. (3.64) of Colquhoun and Hawkes, 1982, or (3.11) of Fredkin et al., 1985) that the density function for an open-time Z, i.e., a sojourn time in the class O of open states, can be expressed as

$$f_Z(t) = \boldsymbol{\phi}_O^\top e^{\boldsymbol{Q}_{OO}t} (-\boldsymbol{Q}_{OO}) \mathbf{1} \quad (t \geqslant 0), \tag{2.5}$$

where ϕ_O is given by (2.4) and $e^{Q_{OO}t}$ is the matrix exponential calculated using the submatrix Q_{OO}. The corresponding distribution is often termed a matrix exponential distribution. When the class O consists of a single open state (2.5) reduces to an ordinary (negative) exponential density. For interpretation of (2.5) it is helpful to use $(-Q_{OO})\mathbf{1} = Q_{OS}\mathbf{1} + Q_{OL}\mathbf{1}$. (Note that, in keeping with our convention stated at the end of Section 1, the vectors written as $\mathbf{1}$ in the latter equation may all have different dimensions, depending on the number of states in the classes O, S and L.)

It is often possible to express (2.5) as a linear combination of exponential densities, as in (1.32) of Colquhoun and Hawkes (1982). This can be done whenever Q_{OO} has a complete set of linearly independent real eigenvectors, as is necessarily the case if either (i) the eigenvalues of Q_{OO} are, as assumed above, all real and distinct, or (ii) $\{Y(t)\}$ is time reversible; see, for example, pp. 168–170 of Guttorp (1995) for a discussion of especially the second case. Moreover, as shown by Kijima and Kijima (1987), in the reversible case the coefficients in the linear combination of exponentials are all positive and the equilibrium distribution of a typical O sojourn time is a mixture of negative exponential distributions.

To derive (2.5), a standard approach (cf. Colquhoun and Hawkes, 1981; Fredkin et al., 1985, and p. 168 of Guttorp, 1995) is to consider a related continuous-time Markov chain in which the states in O' are made absorbing states (by setting to zero all elements of $Q_{O'O}$ and $Q_{O'O'}$), and to determine the density function of the equilibrium time to absorption in this modified chain. Similar techniques can be used to derive the joint density for a sequence of class sojourn times; see (4.3) of Fredkin et al. (1985) or p. 209 of Fredkin and Rice (1986). An alternative approach (cf. Rubino and Sericola, 1989; Csenki, 1992) is to employ the uniformization technique (see, for example, pp. 174–178 of Ross, 1983) to derive the required continuous-time results from their discrete-time analogues.

Moments of the sojourn time Z are given by

$$\mu_Z = \phi_O^\top(-Q_{OO})^{-1}\mathbf{1},$$
$$\mathbb{E}(Z^k) = k!\phi_O^\top(-Q_{OO})^{-k}\mathbf{1} \quad (k=1,2,\ldots), \tag{2.6}$$
$$\sigma_Z^2 = 2\phi_O^\top Q_{OO}^{-2}\mathbf{1} - (\phi_O^\top Q_{OO}^{-1}\mathbf{1})^2 = \phi_O^\top Q_{OO}^{-1}[2I - \mathbf{1}\phi_O^\top]Q_{OO}^{-1}\mathbf{1}. \tag{2.7}$$

The results in (2.6) can be deduced using

$$\int_0^\infty t^k e^{Q_{OO}t}\,dt = k!(-Q_{OO})^{-(k+1)} \quad (k=0,1,2,\ldots), \tag{2.8}$$

which holds whenever, as was assumed above, the eigenvalues of Q_{OO} are distinct and have strictly negative real parts. Ball et al. (1997, p. 64) give further details concerning proof of (2.8). Later we shall need also the results

$$\int_u^\infty e^{Q_{OO}t}\,dt = e^{Q_{OO}u}(-Q_{OO})^{-1} \quad (u \geq 0), \tag{2.9}$$

$$\int_0^\infty e^{-\theta t} e^{\boldsymbol{Q}_{OO} t} \, dt = (\theta \boldsymbol{I} - \boldsymbol{Q}_{OO})^{-1} \quad (\theta \geqslant 0) \tag{2.10}$$

which hold under the same conditions as (2.8).

The distributions described by (2.5) are all of *phase type*, and specifically of the class $\mathcal{PH}_{AT}$ of *phase distributions*, as discussed in Section III.6 of Asmussen (1987); see also Section 2.2 of Neuts (1981). This connection, whilst interesting, does not appear to yield new results in the current context, although, for example, the moment formulae of (2.6) are well-known in the phase distribution literature (cf. (2.2.7) of Neuts, 1981, or Problem 6.7 in Chapter III of Asmussen, 1987).

3. Theoretical bursts

3.1. Definition and preliminary results

Based on an aggregated Markov chain with classes O, S and L of open, short-lived closed and long-lived closed states respectively, as introduced in the previous section, a (*theoretical*) burst is a sojourn in the class $O \cup S$ which starts at the beginning of the first O sojourn following a sojourn in L, with possibly an intervening S sojourn, and ends at the same time as does the last sojourn in O preceding the next sojourn in L. Thus a burst must consist of at least one O sojourn, with any subsequent O sojourn preceded by a sojourn in S. Moreover, the final O sojourn must be followed by an L sojourn, with possibly an intervening S sojourn. The S sojourn times within a burst are often referred to as *gaps*. (Though not ideal, the term 'theoretical' is used because, as explained in Section 1, bursts of this type are not observable experimentally and so have limited practical value.)

Let N_O denote the number of visits to the class O during a burst. Further, let $\boldsymbol{U} = [U_1, U_2, \ldots, U_{N_O}]^\top$ be the random vector of the durations of the successive open-times within the burst, and $\boldsymbol{V} = [V_1, V_2, \ldots, V_{N_O-1}]^\top$ be the random vector consisting of the durations of the intervening short-lived closed-times. The joint density(-probability mass function) of $\boldsymbol{U}$, $\boldsymbol{V}$ and N_O is given, for $\boldsymbol{u} = [u_1, u_1, \ldots, u_n]^\top$ and $\boldsymbol{v} = [v_1, v_2, \ldots, v_{n-1}]^\top$ with non-negative entries and $n = 1, 2, \ldots$ by,

$$f_{\boldsymbol{U}, \boldsymbol{V}, N_O}(\boldsymbol{u}, \boldsymbol{v}, n)$$
$$= \boldsymbol{\psi}_O^\top \left[\prod_{i=1}^{n-1} e^{\boldsymbol{Q}_{OO} u_i} \boldsymbol{Q}_{OS} e^{\boldsymbol{Q}_{SS} v_i} \boldsymbol{Q}_{SO} \right] e^{\boldsymbol{Q}_{OO} u_n} \boldsymbol{K}_{OL} \boldsymbol{1}, \tag{3.1}$$

with the product being unity for $n = 1$. Here, $\boldsymbol{\psi}_O$ is a vector whose entries are the equilibrium probabilities that the burst starts in the various possible states of the class O (see (3.5)), and

$$\boldsymbol{K}_{OL} = \boldsymbol{Q}_{OL} + \boldsymbol{Q}_{OS}(-\boldsymbol{Q}_{SS}^{-1})\boldsymbol{Q}_{SL}$$

is a matrix which accounts for the ways of ending the burst (by moving directly from O to L, or by doing so with an intervening sojourn in S). The vector $K_{OL}\mathbf{1}$ is related to the end of burst vector e_b given in (3.3) of Colquhoun and Hawkes (1982) by $K_{OL}\mathbf{1} = -Q_{OO}e_b$. Let

$$R_O = Q_{OO}^{-1} Q_{OS} Q_{SS}^{-1} Q_{SO} \tag{3.2}$$

be the matrix governing transitions from O to S and back again. Then R_O corresponds to the matrix H_{AA} given by (1.10) of Colquhoun and Hawkes (1987); it is similar to the transition matrix for an entry process, but need not be a stochastic matrix in that some row sums may be less than one. The matrices K_{OL} and R_O are related by

$$(-Q_{OO}^{-1})K_{OL}\mathbf{1} = (I - R_O)\mathbf{1}. \tag{3.3}$$

Analogously to the definition of K_{OL}, the matrix

$$K_{LO} = Q_{LO} + Q_{LS}(-Q_{SS}^{-1})Q_{SO} \tag{3.4}$$

accounts for the ways of moving from L to O, possibly with an intervening sojourn in S.

The vector ψ_O is given by

$$\psi_O^\top = \pi_L^\top K_{LO} / (\pi_L^\top K_{LO}\mathbf{1}) \tag{3.5}$$

(cf. (3.2) of Colquhoun and Hawkes, 1982). This expression, rather than ϕ_O, is needed to take account of additional restrictions on the states of O at the start of a burst, in particular that successive bursts are separated by sojourns in L and that following such a sojourn there may be an intervening sojourn in S before the class O is entered at the start of the burst.

Derivation of (3.1) can be based on the ideas of Fredkin et al. (1985) or Ball et al. (1997), as was mentioned in Section 2.4; see also Section 3 of Colquhoun and Hawkes (1987). If each of the classes consisted of a single state then all vectors and matrices on the right-hand side of (3.1) would be just real numbers and the density itself would reduce to a readily interpretable expression reflecting the basic structure of a continuous-time Markov chain (that transitions between states are determined by the jump chain and that conditional on the successive states visited, the durations of the sojourns in these states are independent, exponentially distributed random variables). An appealing derivation resting on the semi-Markov framework of Ball et al. (1991) has been outlined on p. 63 of Ball et al. (1997).

The result (3.1) is fundamental in that it can be used to obtain many other (joint) densities, probabilities and moments of burst properties, for example by integrating over appropriate variables using (2.8). By taking such a joint density as a starting point it is possible to avoid (Laplace) transform methods in favour of direct and more immediately interpretable calculations. This approach, which is illustrated in several special cases below, has been used previously, for example in Ball et al. (1997, 2000), for convenient derivation of various marginal and joint properties of sojourn times not constrained to be part of a burst.

For example, it follows from (3.1), by summing and integrating over all variables except for U_1, that the (marginal) density function of the first O sojourn time, U_1, within a burst having an arbitrary number of openings, is given by

$$f_{U_1}(t) = \boldsymbol{\psi}_O^\top e^{\boldsymbol{Q}_{OO}t}(-\boldsymbol{Q}_{OO})\mathbf{1} \quad (t \geq 0). \tag{3.6}$$

Furthermore, by analogy with (2.6) the moments of U_1 are

$$\mu_{U_1} = \boldsymbol{\psi}_O^\top(-\boldsymbol{Q}_{OO})^{-1}\mathbf{1},$$
$$\mathbb{E}(U_1^k) = k!\boldsymbol{\psi}_O^\top(-\boldsymbol{Q}_{OO})^{-k}\mathbf{1} \quad (k = 1, 2, \ldots). \tag{3.7}$$

Apart from the initial vector, (3.6) and (3.7) are the same as (2.5) and (2.6) respectively. Conditional on a burst having a single opening, the duration of this opening (which in this case is the burst length, see Section 3.3) has density function

$$f_{U_1|N_O=1}(t) = \boldsymbol{\psi}_O^\top e^{\boldsymbol{Q}_{OO}t} \boldsymbol{K}_{OL}\mathbf{1}/(\boldsymbol{\psi}_O^\top(\boldsymbol{I} - \boldsymbol{R}_O)\mathbf{1}) \quad (t \geq 0), \tag{3.8}$$

where the denominator is the probability that a burst has a single opening, as in (3.10) below. It should be stressed that the marginal density of the first O sojourn time may differ from that of other O sojourn times within a burst, as explained by Colquhoun and Hawkes (1982, pp. 23–26); the general expression is given in their (3.57). Colquhoun and Hawkes (1982, 1987) give also the density for an arbitrary O sojourn time within a burst having a specified number of openings, as well as corresponding results for S sojourn times (gaps) within a burst. For example, for any particular open-time, say U_k, within a burst having n open sojourns, the joint density of U_k and N_O can be determined by integrating (3.1) over all the open-times except the kth and all gaps within the burst to yield, for $n = 1, 2, \ldots$,

$$f_{U_k,N_O}(u, n) = \boldsymbol{\psi}_O^\top \boldsymbol{R}_O^{k-1} e^{\boldsymbol{Q}_{OO}u} \boldsymbol{Q}_{OS}(-\boldsymbol{Q}_{SS})^{-1}\boldsymbol{Q}_{SO}\boldsymbol{R}_O^{n-k}\mathbf{1}$$
$$(u \geq 0, \ k = 1, 2, \ldots, n). \tag{3.9}$$

3.2. Number of openings per theoretical burst

Also easily derived is the distribution of the number of openings N_O per burst. This can be obtained by integrating over the durations of the individual sojourn times in (3.1), and using (3.3), to yield

$$p_n = P(N_O = n) = \boldsymbol{\psi}_O^\top \boldsymbol{R}_O^{n-1}(\boldsymbol{I} - \boldsymbol{R}_O)\mathbf{1} \quad (n = 1, 2, \ldots). \tag{3.10}$$

The distribution (3.10) is of matrix geometric form, as in (3.5) of Colquhoun and Hawkes (1982) or (3.18) of Colquhoun and Hawkes (1987). Such a distribution is also one of phase type, as discussed in Section 2.2 of Neuts (1981). When the eigenvalues of R_O are distinct and positive, (3.10) is a mixture of geometric distributions, as in (3.9) of Colquhoun and Hawkes (1982). In the case where the open class O has a single open state, the distribution (3.10) reduces to an ordinary geometric distribution.

The factorial moments of N_O are given by

$$\mu_{N_O} = \boldsymbol{\psi}_O^\top [I - R_O]^{-1} \mathbf{1},$$

$$\mathbb{E}(N_O^{[k]}) = k! \boldsymbol{\psi}_O^\top R_O^{k-1} (I - R_O)^{-k} \mathbf{1} \quad (k = 1, 2, \ldots), \tag{3.11}$$

where $n^{[k]} = n(n-1) \cdots (n-k+1)$. Hence the variance of N_O is

$$\sigma_{N_O}^2 = 2 \boldsymbol{\psi}_O^\top R_O [I - R_O]^{-2} \mathbf{1} + \mu_{N_O} - \mu_{N_O}^2. \tag{3.12}$$

The factorial moment formulae of (3.11) are known in the phase distribution literature; see, for example, p. 46 of Neuts (1981).

3.3. Burst duration and total open-time per theoretical burst

Let $U_O = U_1 + U_2 + \cdots + U_{N_O}$ denote the total open-time during a burst and similarly let $V_S = V_1 + V_2 + \cdots + V_{N_O-1}$ denote the total closed-time, where V_S is zero for $N_O = 1$. Now consider finding the mean total open-time per burst. For any particular open-time, say the kth within a burst having n openings, the mean $\mathbb{E}(U_k; n)$, $k = 1, 2, \ldots, n$, follows from (3.9), or directly from (3.1), as

$$\mathbb{E}(U_k; n) = \boldsymbol{\psi}_O^\top R_O^{k-1} (-Q_{OO})^{-1} R_O^{n-k} (I - R_O) \mathbf{1}. \tag{3.13}$$

Hence, by summing over all possible values of k and n, the mean total open-time in a burst is

$$\mu_{U_O} = \boldsymbol{\psi}_O^\top K_O^{-1} \mathbf{1}, \tag{3.14}$$

where

$$K_O = -Q_{OO}(I - R_O) = -Q_{OO} - Q_{OS}(-Q_{SS})^{-1} Q_{SO}. \tag{3.15}$$

Note that, by (3.3), $K_O \mathbf{1} = K_{OL} \mathbf{1}$. The mean (3.14) corresponds to (3.26) of Colquhoun and Hawkes (1982) who, in their (3.23), give also the (matrix exponential) density of U_O. Analogous conditional arguments for the second moment yield

$$\sigma_{U_O}^2 = \boldsymbol{\psi}_O^\top K_O^{-1} [2I - \mathbf{1} \boldsymbol{\psi}_O^\top] K_O^{-1} \mathbf{1}. \tag{3.16}$$

This is similar to other variance expressions given in Colquhoun and Hawkes (1987); this, and (2.7), provide the form of the variance for any matrix exponential distribution.

Given n openings during a burst, there are $n - 1$ visits to the short-lived closed states S during the burst. Working similarly to the derivations of (3.13)–(3.14) yields the mean of the total closed-time per burst (cf. (3.41) of Colquhoun and Hawkes, 1982) as

$$\mu_{V_S} = \boldsymbol{\psi}_O^\top K_O^{-1} R_{SS} \mathbf{1} \tag{3.17}$$

and its variance as

$$\sigma_{V_S}^2 = 2\boldsymbol{\psi}_O^\top[(\boldsymbol{K}_O^{-1}\boldsymbol{R}_{SS})^2 + \boldsymbol{K}_O^{-1}\boldsymbol{R}_{SSS}]\mathbf{1} - \mu_{V_S}^2, \quad (3.18)$$

where

$$\boldsymbol{R}_{SS} = \boldsymbol{Q}_{OS}\boldsymbol{Q}_{SS}^{-2}\boldsymbol{Q}_{SO} \quad \text{and} \quad \boldsymbol{R}_{SSS} = \boldsymbol{Q}_{OS}\boldsymbol{Q}_{SS}^{-3}\boldsymbol{Q}_{SO}. \quad (3.19)$$

The mean duration of a burst, μ_B, is clearly the sum of the mean total open-time and mean total closed-time during a burst, i.e., $\mu_B = \mu_{U_O} + \mu_{V_S}$. Similar methods may be used for finding the mean time, here denoted by $\mu_{\bar{B}}$, that the channel is closed between successive bursts. However, a nice probabilistic argument which uses information already available is possible. Let $\mu_{CY} = \mu_B + \mu_{\bar{B}}$ be the mean length of a cycle of a burst and a non-burst. Then $\pi_O = \mu_{U_O}/\mu_{CY}$ is the proportion of time the channel is open, and $p_{OB} = \mu_{U_O}/\mu_B$ is the probability of being open within a burst. Thus $\mu_{\bar{B}} = \mu_{U_O}/\pi_O - \mu_B$, and the mean cycle length can also be obtained.

3.4. Transform methods

From (3.1) it is also possible to obtain many results in terms of transforms. Let $\chi(\theta, \nu, \omega) = \mathbb{E}[e^{-(\theta U_O + \nu V_O)}\omega^{N_O}]$. Then, using (3.1) and (2.10),

$$\chi(\theta, \nu, \omega) = \omega\boldsymbol{\psi}_O^\top[(\theta\boldsymbol{I} - \boldsymbol{Q}_{OO}) - \omega\boldsymbol{Q}_{OS}(\nu\boldsymbol{I} - \boldsymbol{Q}_{SS})^{-1}\boldsymbol{Q}_{SO}]^{-1}\boldsymbol{K}_O\mathbf{1}$$
$$(\theta, \nu \geq 0, \ 0 \leq \omega \leq 1). \quad (3.20)$$

Various special cases can now be deduced, including univariate transform expressions obtained by Colquhoun and Hawkes (1982).

The Laplace transform $\chi_B(\theta) = \mathbb{E}[e^{-\theta B}]$ of the distribution of burst length B is given by $\chi(\theta, \theta, 1)$, corresponding to (3.16) of Colquhoun and Hawkes (1982). The Laplace transform $\chi_{U_O}(\theta) = \mathbb{E}[e^{-\theta U_O}]$ of the distribution of the total open-time per burst U_O is given by $\chi(\theta, 0, 1) = \boldsymbol{\psi}_O^\top[\theta\boldsymbol{I} + \boldsymbol{K}_O]^{-1}\boldsymbol{K}_O\mathbf{1}$, with corresponding density function

$$f_{U_O}(t) = \boldsymbol{\psi}_O^\top e^{-\boldsymbol{K}_O t}\boldsymbol{K}_O\mathbf{1} \quad (t \geq 0). \quad (3.21)$$

These expressions correspond respectively to (3.22) and (3.23) of Colquhoun and Hawkes (1982), though the present derivations seem somewhat simpler. If the matrix $\boldsymbol{K}_O$ (3.15) has real and distinct eigenvalues, this density function will be a linear combination of exponential densities (see the remarks following (2.5)).

The Laplace transform $\chi_{V_S}(\theta) = \mathbb{E}[e^{-\nu V_S}]$ of the distribution of the total closed-time per burst V_S is given by $\chi(0, \nu, 1)$, corresponding to (3.35) of Colquhoun and Hawkes (1982). The corresponding 'density' function

$$f_{V_S}(t) = \boldsymbol{\psi}_O^\top(-\boldsymbol{Q}_{OO})^{-1}[\delta(t)\boldsymbol{K}_O + \boldsymbol{Q}_{OS}e^{-\boldsymbol{K}_S t}\boldsymbol{K}_S(-\boldsymbol{Q}_{SS})^{-1}\boldsymbol{Q}_{SO}]\mathbf{1}$$
$$(t \geq 0), \quad (3.22)$$

as in (3.38) of Colquhoun and Hawkes (1982), where $K_S = -Q_{SS}(I - R_S)$, with $R_S = Q_{SS}^{-1}Q_{SO}Q_{OO}^{-1}Q_{OS}$, is defined by analogy with K_O and $\delta(t)$ is a Dirac delta function. The quantity $\boldsymbol{\psi}_O^\top(-Q_{OO})^{-1}K_O\mathbf{1}$ multiplying the delta function, is the probability that a burst consists of a single opening; it corresponds to an atom at the origin in the distribution of V_S.

The probability generating function, $G_{N_O}(\omega)$, of the number of openings in a burst can be obtained as $\chi(0, 0, \omega)$; after use of (3.3) it follows that

$$G_{N_O}(\omega) = \omega \boldsymbol{\psi}_O^\top [I - \omega R_O]^{-1}(I - R_O)\mathbf{1} \quad (0 \leqslant \omega \leqslant 1). \tag{3.23}$$

This expression could also be obtained directly from (3.10).

3.5. Correlations

Colquhoun and Hawkes (1987) consider correlations between various random variables that can be defined on bursts: for example, the correlation between the first and the kth open (closed) time within a burst with at least n ($\geqslant k$) openings, and the correlation between the durations of successive bursts.

Given the joint density(-probability mass function) (3.1), the form of the expectation in (3.13), or the transform expression (3.20) above, it is possible to determine covariances and correlations between any pair of, for example, total open-time per (theoretical) burst, total closed-time per burst and number of openings per burst, as well as between particular open (closed) times within a burst.

Consider firstly the total open-time and the number of openings per burst. From (3.13)

$$\mathbb{E}(U_O N_O) = \sum_{n=1}^\infty \sum_{k=1}^n n\mathbb{E}(U_k; n)$$
$$= \boldsymbol{\psi}_O^\top [K_O^{-1} R_O K_O^{-1}(-Q_{OO}) + K_O^{-1}(-Q_{OO})K_O^{-1}]\mathbf{1}, \tag{3.24}$$

and hence the correlation is $\rho(U_O, N_O) = [\mathbb{E}(U_O N_O) - \mu_{U_O}\mu_{N_O}]/(\sigma_{U_O}\sigma_{N_O})$.

There is a closure within a burst only if there are at least two openings. Consider the kth opening and the rth closure within a burst having at least $n \geqslant 2$ openings; then

$$\mathbb{E}[U_k V_r] = \boldsymbol{\psi}_O^\top R_O^{k-1}(-Q_{OO})^{-1} R_O^{r-k}(-Q_{OO})^{-1} R_{SS} R_O^{n-k}(I - R_O)\mathbf{1}$$
$$(k \leqslant r \leqslant n), \tag{3.25}$$

$$\mathbb{E}[V_r U_k] = \boldsymbol{\psi}_O^\top R_O^r(-Q_{OO})^{-1} R_{SS} R_O^{k-r-1}(-Q_{OO})^{-1} R_O^{n-k}(I - R_O)\mathbf{1}$$
$$(r < k \leqslant n) \tag{3.26}$$

which yields

$$\mathbb{E}[U_O V_S] = \boldsymbol{\psi}_O^\top [K_O^{-2} R_{SS} + K_O^{-1} R_{SS} K_O^{-1}]\mathbf{1} \tag{3.27}$$

and hence the corresponding correlation $\rho(U_O, V_S)$.

Two successive bursts must be separated by an inter-burst period, including at least one visit to the long-lived closed states L. The joint density function $f_{U_O,U'_O}(t,u)$ of the total amounts of open-time U_O, U'_O in successive bursts can be written down, extending (3.21), as

$$f_{U_O,U'_O}(t,u) = \boldsymbol{\psi}_O^\top e^{-\boldsymbol{K}_O t} \boldsymbol{K}_{OL}(\boldsymbol{I}-\boldsymbol{R}_L)^{-1}(-\boldsymbol{Q}_{LL})^{-1}\boldsymbol{K}_{LO}e^{-\boldsymbol{K}_O u}\boldsymbol{K}_O \mathbf{1}$$
$$(t,u \geq 0), \qquad (3.28)$$

where $\boldsymbol{R}_L = \boldsymbol{Q}_{LL}^{-1}\boldsymbol{Q}_{LS}\boldsymbol{Q}_{SS}^{-1}\boldsymbol{Q}_{SL}$. The covariance and correlation $\rho(U_O, U'_O)$ then follow from

$$\mathbb{E}[U_O U'_O] = \boldsymbol{\psi}_O^\top \boldsymbol{K}_O^{-2}\boldsymbol{K}_{OL}(\boldsymbol{I}-\boldsymbol{R}_L)^{-1}(-\boldsymbol{Q}_{LL})^{-1}\boldsymbol{K}_{LO}\boldsymbol{K}_O^{-1}\mathbf{1}. \qquad (3.29)$$

Milne et al. (1986) considered the interdependence of the burst length and the number of openings per burst by looking at conditional means for two particular three-state models, although explicit expressions for the correlations $\rho(B, N_O)$ were not given.

4. Empirical bursts

4.1. Definition and preliminary results

Although in some circumstances it may seem appropriate to consider separate classes of short-lived and long-lived closed states, the definition of a theoretical burst has limited practical value because it is usually not possible on the basis of conductance to distinguish when a visit to the closed states involves visits to the separate subclasses. An alternative definition of a burst can be given when there is specified a positive constant t_c. Then, although the state space is initially partitioned into just two classes, O consisting of open states and C of closed states, sojourns in C may be classified empirically into those which are at most t_c and those which are greater than t_c. The former sojourns then play the role previously played for a theoretical burst by sojourns in S, and the latter the role of interburst sojourns. Thus, by adapting the definition of a theoretical burst, an *empirical burst* can be defined; gaps within an empirical burst are closed sojourns of duration at most t_c and a burst of this type ends at the same time as does the last open sojourn which precedes the next closed sojourn of duration greater than t_c. With choice of a suitable value of t_c, properties of empirical bursts are close to those of theoretical bursts, as will be shown in examples. Aspects of this choice will be discussed further in the examples of later sections.

Analogously to (3.1) it is possible to write down the joint density of the vectors $U = [U_1, U_2, \ldots, U_{N_O}]^\top$ of open times, $V = [V_1, V_2, \ldots, V_{N_O-1}]^\top$ gaps, and the number N_O of openings during an empirical burst. Note that all closed-times (gaps) within an empirical burst must be of duration at most t_c, and that immediately prior to the start of an empirical burst there must have been a closed-time of more than t_c. The equilibrium probability distribution of the states of O in which an empirical burst starts is

$$\tilde{\boldsymbol{\psi}}_O^\top = \boldsymbol{\pi}_O^\top \boldsymbol{Q}_{OC} e^{\boldsymbol{Q}_{CC} t_c}(-\boldsymbol{Q}_{CC}^{-1})\boldsymbol{Q}_{CO}/(\boldsymbol{\pi}_O^\top \boldsymbol{Q}_{OC} e^{\boldsymbol{Q}_{CC} t_c}(-\boldsymbol{Q}_{CC}^{-1})\boldsymbol{Q}_{CO}\mathbf{1}) \tag{4.1}$$

(rather than $\boldsymbol{\psi}_O$ given by (3.5)) and the joint density(-probability mass function) of U, V and N_O is

$$\tilde{f}_{U,V,N_O}(\boldsymbol{u},\boldsymbol{v},n)$$
$$= \tilde{\boldsymbol{\psi}}_O^\top \left[\prod_{i=1}^{n-1} e^{\boldsymbol{Q}_{OO} u_i} \boldsymbol{Q}_{OC} e^{\boldsymbol{Q}_{CC} v_i} \boldsymbol{Q}_{CO} \right] e^{\boldsymbol{Q}_{OO} u_n} \boldsymbol{Q}_{OC} e^{\boldsymbol{Q}_{CC} t_c}\mathbf{1}, \tag{4.2}$$

where $u_1, u_2, \ldots, u_n$ and $v_1, v_2, \ldots, v_{n-1}$ are non-negative, with the latter variables are each at most t_c, and $n = 1, 2, \ldots$, with the product being unity for $n = 1$. The joint density (4.2) considered as a function of the parameters is just the likelihood function based on observations on the successive class sojourn times within a burst.

Integrating over all possible closed-times within such a burst yields the joint density of U and N_O, for non-negative $u_1, u_2, \ldots, u_n$ and $n = 1, 2, \ldots$, as

$$\tilde{f}_{U,N_O}(\boldsymbol{u};n) = \tilde{\boldsymbol{\psi}}_O^\top \left[\prod_{i=1}^{n-1} e^{\boldsymbol{Q}_{OO} u_i} \boldsymbol{Q}_{OC}(\boldsymbol{I} - e^{\boldsymbol{Q}_{CC} t_c})(-\boldsymbol{Q}_{CC}^{-1})\boldsymbol{Q}_{CO} \right]$$
$$\times e^{\boldsymbol{Q}_{OO} u_n} \boldsymbol{Q}_{OC} e^{\boldsymbol{Q}_{CC} t_c}\mathbf{1}. \tag{4.3}$$

4.2. Number of openings per empirical burst

Integrating (4.2) over all possible open-times and closed-times yields the distribution of the number, N_O, of openings within an empirical burst. This is of matrix geometric form with

$$P(N_O = n) = \tilde{p}_n = \tilde{\boldsymbol{\psi}}_O^\top \boldsymbol{R}_{t_c}^{n-1}(\boldsymbol{I} - \boldsymbol{R}_{t_c})\mathbf{1} \quad (n = 1, 2, \ldots), \tag{4.4}$$

where

$$\boldsymbol{R}_{t_c} = \boldsymbol{Q}_{OO}^{-1}\boldsymbol{Q}_{OC}(\boldsymbol{I} - e^{\boldsymbol{Q}_{CC} t_c})\boldsymbol{Q}_{CC}^{-1}\boldsymbol{Q}_{CO} \tag{4.5}$$

is the analogue of $\boldsymbol{R}_O$, given by (3.2), for a theoretical burst. Similarly to (3.15), let $\tilde{\boldsymbol{K}}_O = (-\boldsymbol{Q}_{OO})(\boldsymbol{I} - \boldsymbol{R}_{t_c})$. Then $\tilde{\boldsymbol{K}}_O \mathbf{1} = \boldsymbol{Q}_{OC} e^{\boldsymbol{Q}_{CC} t_c}\mathbf{1}$, which is the analogue of $\boldsymbol{K}_O \mathbf{1} = \boldsymbol{K}_{OL}\mathbf{1}$ (see Section 3.3).

By analogy with (3.11), the factorial moments of N_O are given by

$$\tilde{\mu}_{N_O} = \tilde{\boldsymbol{\psi}}_O^\top [\boldsymbol{I} - \boldsymbol{R}_{t_c}]^{-1}\mathbf{1},$$
$$\mathbb{E}(N_O^{[k]}) = k!\tilde{\boldsymbol{\psi}}_O^\top \boldsymbol{R}_{t_c}^{k-1}(\boldsymbol{I} - \boldsymbol{R}_{t_c})^{-k}\mathbf{1} \quad (k = 1, 2, \ldots). \tag{4.6}$$

Hence the variance of N_O is

$$\tilde{\sigma}_{N_O}^2 = 2\tilde{\boldsymbol{\psi}}_O^\top \boldsymbol{R}_{t_c}[\boldsymbol{I} - \boldsymbol{R}_{t_c}]^{-2}\mathbf{1} + \tilde{\mu}_{N_O} - \tilde{\mu}_{N_O}^2.$$

4.3. Burst duration and total open-time per empirical burst

Now let $U_O = U_1 + U_2 + \cdots + U_{N_O}$ denote the total open-time during an empirical burst and similarly let $V_S = V_1 + V_2 + \cdots + V_{N_O-1}$ denote the total closed-time. For any particular open-time, say the kth within a burst having n openings, the mean $\mathbb{E}(U_k; n)$, $k = 1, 2, \ldots, n$, follows from (4.2) as

$$\mathbb{E}(U_k; n) = \tilde{\boldsymbol{\psi}}_O^\top \boldsymbol{R}_{t_c}^{k-1}(-\boldsymbol{Q}_{OO}^{-1})\boldsymbol{R}_{t_c}^{n-k}(\boldsymbol{I} - \boldsymbol{R}_{t_c})\mathbf{1}, \qquad (4.7)$$

and hence, by summing over all possible values of k and n, the mean total open-time in a burst is

$$\tilde{\mu}_{U_O} = \tilde{\boldsymbol{\psi}}_O^\top \tilde{\boldsymbol{K}}_O^{-1}\mathbf{1}. \qquad (4.8)$$

Similar conditional arguments for the second moment yield

$$\tilde{\sigma}_{U_O}^2 = 2\tilde{\boldsymbol{\psi}}_O^\top \tilde{\boldsymbol{K}}_O^{-2}\mathbf{1} - (\tilde{\boldsymbol{\psi}}_O^\top \tilde{\boldsymbol{K}}_O^{-1}\mathbf{1})^2. \qquad (4.9)$$

Given n openings during an empirical burst, there are $n - 1$ gaps (visits to the closed states C of at most t_c) during the burst. Working in a similar manner as for (4.7)–(4.8) yields the mean and variance of the total closed-time per empirical burst as

$$\tilde{\mu}_{V_S} = \tilde{\boldsymbol{\psi}}_O^\top \tilde{\boldsymbol{K}}_O^{-1}\boldsymbol{R}_{CC}\mathbf{1}, \qquad (4.10)$$

and

$$\tilde{\sigma}_{V_S}^2 = 2\tilde{\boldsymbol{\psi}}_O^\top [(\tilde{\boldsymbol{K}}_O^{-1}\boldsymbol{R}_{CC})^2 + \tilde{\boldsymbol{K}}_O^{-1}\boldsymbol{R}_{CCC}]\mathbf{1} - \tilde{\mu}_{V_S}^2, \qquad (4.11)$$

where

$$\boldsymbol{R}_{CC} = \boldsymbol{Q}_{OC}[\boldsymbol{I} - e^{\boldsymbol{Q}_{CC}t_c}(\boldsymbol{I} - \boldsymbol{Q}_{CC}t_c)]\boldsymbol{Q}_{CC}^{-2}\boldsymbol{Q}_{CO}$$

and

$$\boldsymbol{R}_{CCC} = \boldsymbol{Q}_{OC}[\boldsymbol{I} - e^{\boldsymbol{Q}_{CC}t_c}(\boldsymbol{I} - \boldsymbol{Q}_{CC}t_c + \boldsymbol{Q}_{CC}^2 t_c^2/2)](-\boldsymbol{Q}_{CC})^{-3}\boldsymbol{Q}_{CO}.$$

The mean duration of an empirical burst $\tilde{\mu}_B$, is the sum of the mean total open-time and mean total closed-time during a burst, i.e., $\tilde{\mu}_B = \tilde{\mu}_{U_O} + \tilde{\mu}_{V_S}$. The same procedure as suggested in Section 3.3 may be used for finding the mean time $\tilde{\mu}_{\bar{B}}$ the system is closed between successive empirical bursts, and the mean cycle length.

4.4. Transform methods for empirical bursts

From (4.2) it is also possible to obtain many results in terms of transforms. Let $\tilde{\chi}(\theta, \nu, \omega) = \mathbb{E}[e^{-(\theta U_O + \nu V_O)}\omega^{N_O}]$. Then, using (4.2) and (2.10),

$$\tilde{\chi}(\theta, \nu, \omega) = \tilde{\psi}_O^T \omega [\theta I - Q_{OO} - \omega Q_{OC}(I - e^{-(\nu I - Q_{CC})t_c})$$
$$\times (\nu I - Q_{CC})^{-1} Q_{CO}]^{-1} \tilde{K}_O 1 \quad (\theta, \nu \geq 0, \ 0 \leq \omega \leq 1).$$
(4.12)

Various special cases can be deduced. The Laplace transform $\tilde{\chi}_B(\theta) = \mathbb{E}[e^{-\theta B}]$ of the distribution of empirical burst length B is given by $\tilde{\chi}(\theta, \theta, 1)$. The Laplace transform $\tilde{\chi}_{U_O}(\theta) = \mathbb{E}[e^{-\theta U_O}]$ of the distribution of the total open-time per empirical burst U_O is given by $\tilde{\chi}(\theta, 0, 1)$, with corresponding density function

$$\tilde{f}_{U_O}(t) = \tilde{\psi}_O^T e^{-\tilde{K}_O t} \tilde{K}_O 1 \quad (t \geq 0).$$
(4.13)

The Laplace transform $\tilde{\chi}_{V_S}(\theta) = \mathbb{E}[e^{-\nu V_S}]$ of the distribution of the total closed-time V_S per empirical burst V_S is given by $\tilde{\chi}(0, \nu, 1)$. Obtaining a corresponding density expression analogous to (3.22) for a theoretical burst does not appear to be straightforward. The probability generating function $\tilde{G}_{N_O}(\omega)$ of the number N_O of openings in an empirical burst is $\tilde{\chi}(0, 0, \omega)$, the corresponding probability function being given by (4.4).

4.5. Correlations for empirical bursts

Given the joint density(-probability mass function) (4.2), the form of the expectation in (4.7), or the transform expression ((4.12) above), it is possible to determine covariances and correlations between any pair of, for example, the total open-time, total closed-time and the number of openings per empirical burst, as well as between particular open-times (closed-times) within an empirical burst.

For the total open-time and the number of openings per empirical burst, it follows from (4.7) that

$$\mathbb{E}(U_O N_O) = \tilde{\psi}_O^T [\tilde{K}_O^{-1} R_{tc} \tilde{K}_O^{-1}(-Q_{OO}) + \tilde{K}_O^{-1}(-Q_{OO})\tilde{K}_O^{-1}]1, \quad (4.14)$$

and hence the correlation is $\tilde{\rho}(U_O, N_O) = [\mathbb{E}(U_O N_O) - \mu_{U_O}\mu_{N_O}]/(\tilde{\sigma}_{U_O}\tilde{\sigma}_{N_O})$.

In a manner similar to that for theoretical bursts, the expectation of the product of the total open-time and total closed-time within an empirical burst is

$$\mathbb{E}[U_O V_S] = \tilde{\psi}_O^T [\tilde{K}_O^{-2} R_{CC} + \tilde{K}_O^{-1} R_{CC} \tilde{K}_O^{-1}]1, \quad (4.15)$$

and hence the corresponding correlation $\tilde{\rho}(U_O, V_S)$ can be written down. Two successive empirical bursts are separated by an inter-burst period, i.e., a sojourn of duration greater than t_c to the closed states. The joint density function $\tilde{f}_{U_O, U_O'}(t, u)$

of the total amounts of open-time U_O, U'_O in successive empirical bursts can be written down, extending (4.13), as

$$\tilde{f}_{U_O, U'_O}(t, u) = \tilde{\boldsymbol{\psi}}_O^\top e^{-\tilde{\boldsymbol{K}}_O t} \boldsymbol{K}_{OL}(-\boldsymbol{Q}_{LL})^{-1}(\boldsymbol{I} - \boldsymbol{R}_L)^{-1} \boldsymbol{K}_{LO} e^{-\tilde{\boldsymbol{K}}_O u} \tilde{\boldsymbol{K}}_O \mathbf{1}$$
$$(t, u \geqslant 0). \tag{4.16}$$

The covariance and correlation $\rho(U_O, U'_O)$ then follow from

$$\mathbb{E}[U_O U'_O] = \boldsymbol{\psi}_O^\top \boldsymbol{K}_O^{-2} \boldsymbol{K}_{OL}(-\boldsymbol{Q}_{LL})^{-1}(\boldsymbol{I} - \boldsymbol{R}_{LS})^{-1} \boldsymbol{K}_{LO} \tilde{\boldsymbol{K}}_O^{-1} \mathbf{1}. \tag{4.17}$$

5. A five-state ligand-activated ion channel model

A five-state scheme that has frequently been considered in the literature as a mechanism for an agonist-gated channel such as the nicotinic acetylcholine receptor is given in Figure 1.

In this scheme, R represents an unbound state of the receptor, A an agonist (e.g., acetylcholine) capable of opening the channel, AR and A_2R denote states in which respectively one and two agonist molecules are bound and the channel remains closed, and AR* and A_2R^* denote corresponding open states. The states are numbered as shown with, for example, 1 and 2 denoting respectively the singly and doubly bound open states. This model is often referred to as the Colquhoun and Hawkes five-state model, in recognition of their pioneering role in describing such systems; see especially Colquhoun and Hawkes (1982, 1995b). The transition rates in Figure 1 (expressed in units of ms^{-1}) are those used by Colquhoun and Hawkes (1982) to ensure model predictions similar to observations described in Colquhoun and Sakmann (1981) and based on the agonist suberyldicholine. (Note, however, that in order to achieve exact reversibility the value given as 0.000667 should be replaced by 2/3000, as was indicated by Ball et al., 1993, p. 368, and Colquhoun and Hawkes, 1995b, p. 593.)

For the given transition rates, R is a long-lived closed state, while AR and A_2R are short-lived closed states and therefore burst behaviour is possible. The classes used in the definition of theoretical bursts are thus $L = \{5\}$, $S = \{3, 4\}$ and $O = \{1, 2\}$. In this model, for each class, every state in that class is a gateway state. For an empirical burst

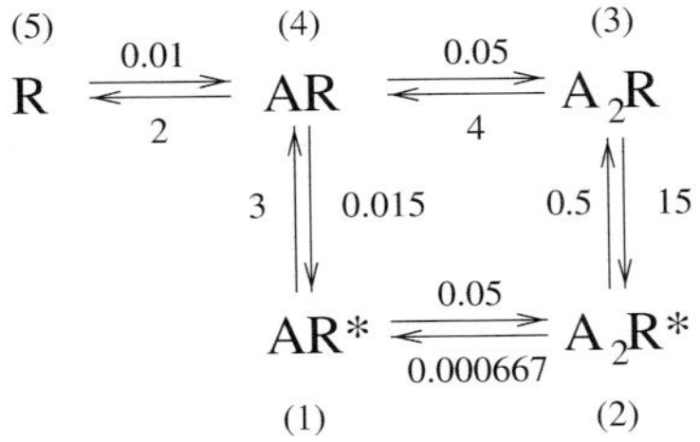

Fig. 1. Five-state model for the acetylcholine receptor channel.

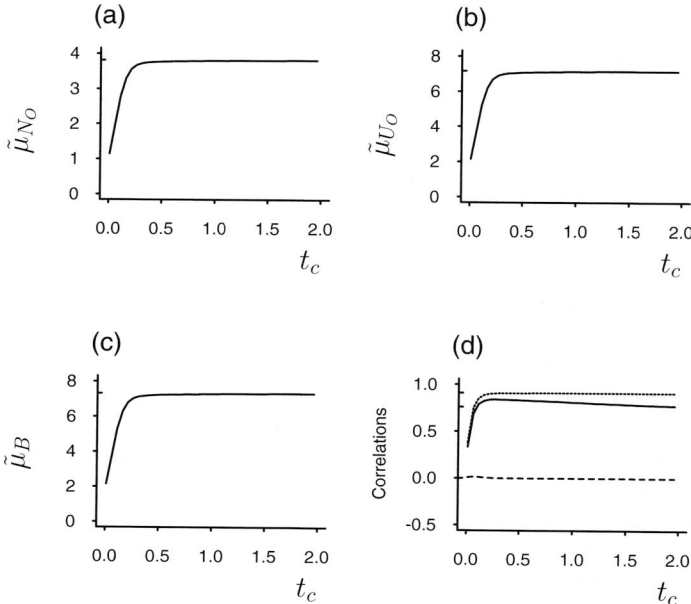

Fig. 2. Predicted empirical burst behaviour for the five state model shown in Figure 1. The mean number of openings ($\tilde{\mu}_{N_O}$), the mean total open-time ($\tilde{\mu}_{U_O}$), and the mean burst duration ($\tilde{\mu}_B$) are shown in (a)–(c) respectively as a function of t_c (in ms). In (d) the solid curve (———) gives the correlation between N_O and U_O, the dotted curve ($\cdots\cdots$) that between U_O and V_S, and the dashed curve (– – –) that between U_O and the total open-time U'_O within the next burst. For comparison, in each of (a)–(d), the corresponding values for a theoretical burst are marked on the right of each vertical axis.

the class of closed states is $C = \{3, 4, 5\}$; in this case state 5 is not a gateway state to the class of open states.

To illustrate their theory, Colquhoun and Hawkes (1982, §4) gave a detailed numerical presentation of many of the features of the five-state model; see also Colquhoun and Hawkes (1995b). We shall give here only a few results, preferring to concentrate on complementary aspects, including a comparison of properties of theoretical and empirical bursts.

The probabilities that a theoretical burst starts in the various open states are given by (3.5). The probability of such a burst starting in state 1 is $\psi_1 = 0.2754$, and so the probability of starting in state 2 is $\psi_2 = 1 - \psi_1 = 0.7246$. The corresponding probabilities for empirical bursts based on $t_c = 1.0$ are $\tilde{\psi}_1 = 0.2753$ and $\tilde{\psi}_2 = 0.7247$.

Results (a)–(c) in Figure 2 show that for an empirical burst the mean number of openings $\tilde{\mu}_{N_O}$, the mean total open-time $\tilde{\mu}_{U_O}$, and the burst duration $\tilde{\mu}_B$ all increase as t_c increases from 0.1 to 2.0 ms. This is because, as t_c increases, more sojourns in the class $\{3, 4, 5\}$ will be classified as gaps within a burst. The slopes are greatest at small values of t_c ($\leqslant 0.25$ ms) and thereafter the means appear to stabilize, and this continues at least until t_c approaches the mean of a sojourn time in the long-lived closed state (100 ms); however, these means still increase as t_c increases further because there is more chance of a visit to the long-lived closed state being classified as a gap. To facilitate comparisons, levels corresponding to the values $\mu_{N_O} = 3.819$, $\mu_{U_O} = 7.166$ and $\mu_B = 7.328$ for a theoretical burst are marked on the right of the relevant vertical axes.

From (d) in Figure 2 it can be seen that the correlations $\tilde{\rho}(U_O, V_S)$ and $\tilde{\rho}(U_O, N_O)$ display similar behaviour as t_c increases, though neither curve is monotonic. The corresponding values $\rho(U_O, V_S) = 0.7605$ and $\rho(U_O, N_O) = 0.9098$ for a theoretical burst are marked on the vertical axis. Furthermore, for all values of t_c, the correlation $\tilde{\rho}(U_O, U'_O)$ for empirical bursts is close to zero, while the value of $\rho(U_O, U'_O)$ for theoretical bursts is zero to at least ten decimal places. Thus the total open-times U_O and U'_O in successive bursts are almost uncorrelated for either type of burst.

Figure 2 shows also that, for the parameter set chosen in Figure 1, values of empirical burst properties will be close to the corresponding properties for theoretical bursts provided a reasonable value is chosen for t_c, although there is as yet no general criterion for this choice. In the simple situation when the two closed classes each have a single state, three methods have been discussed by Colquhoun and Sigworth (1995, pp. 535–536) for selecting a value of t_c; a natural choice then lies between the two inverse eigenvalues of the closed-time distribution. By comparing values of the above discussed theoretical burst properties with those of empirical bursts, we see that an appropriate value of t_c is about 1.0 ms.

6. A linear sequential model with drug blockade

In an agonist-activated channel such as the nicotinic acetylcholine receptor, blockade of ion channel activity by drugs has often been interpreted using linear sequential models (Hille, 1992; Colquhoun and Hawkes, 1995a) with blocked and open states directly connected, as in the following simple scheme:

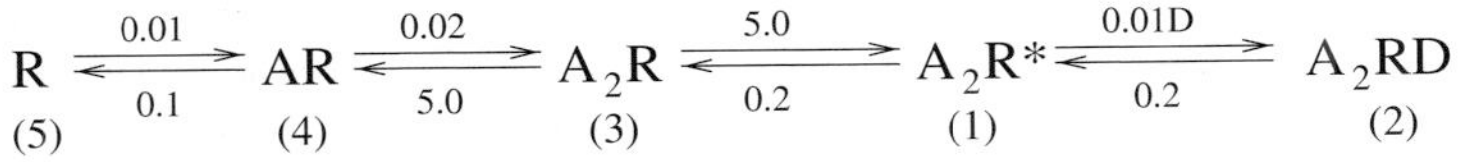

$$R \underset{0.1}{\overset{0.01}{\rightleftarrows}} AR \underset{5.0}{\overset{0.02}{\rightleftarrows}} A_2R \underset{0.2}{\overset{5.0}{\rightleftarrows}} A_2R^* \underset{0.2}{\overset{0.01D}{\rightleftarrows}} A_2RD$$
$$(5) \quad\quad (4) \quad\quad (3) \quad\quad (1) \quad\quad (2)$$

Fig. 3. Sequential ligand-activated ion channel model with drug blockade.

In Figure 3, R represents an unbound state of the receptor, A an agonist capable of opening the channel, AR and A_2R are respectively mono- and bi-liganded states, A_2R^* an open channel form and A_2RD a closed (inhibitory) state formed when drug D binds to the open channel pore. This model is a variant of that considered in Section 5, with several common structural features together with some distinct differences. Reflecting the sequential structure, off-diagonal entries in the transition rate matrix $Q = [q_{ij}]$ are zero unless i and j are neighbouring states. The numerical values of the transition rates given in Figure 3 were suggested by experimental work of Le Dain et al. (1991). In the presence of a classical non-competitive inhibitor (such as histrionicotoxin) at concentration D (in units of micromolar), the only transition rate dependent on D is q_{12}, with this usually assumed proportional to the concentration and given by $q_{12} = k_f D$ where $k_f \, (= 1 \times 10^4 \text{ M}^{-1}\text{ms}^{-1})$ is the binding association rate constant. While this model is interesting in its own right, it is also used in the next section as the basis for an extension where a so-called biphasic effect might occur. Such an effect is not possible for the simple model of Figure 3.

The various classes used in the definition of theoretical bursts are $O = \{1\}$ and $L = \{5\}$ consisting respectively of a single open and a single closed state, and $S = \{2, 3, 4\}$ which consists of the short-lived closed states (3 and 4) and the inhibitory state (2). As for the five-state model in Section 5, for each class, every state in that class is a gateway state.

The durations of sojourns in the individual classes $O = \{1\}$ and $L = \{5\}$ are independent, and have exponential distributions with respective means $\mu_j = 1/d_j$, $j = 1, 5$. Moreover, as indicated in the discussion following (2.5), a sojourn in the class S has a density which is a linear combination of three exponential densities, unless some eigenvalues of the corresponding diagonal submatrix $\boldsymbol{Q}_{SS}$ are equal, or $q_{45} = q_{31}$ (cf. Edeson et al., 1994, p. 27).

A theoretical burst begins with a visit to the open state (1) and ends after the last visit to the open state before the process enters the long-lived closed state (5). Thus $\psi_O = 1$, and for a theoretical burst to end after a particular visit to the open state, the system must reach state 5 before it returns to state 1. Hence, by straightforward probabilistic argument, the probability of the burst ending after a particular visit to state 1 is

$$p = p_{13} p_{34} p_{45}/(1 - p_{43} p_{34}) = q_{13}\xi/d_1, \tag{6.1}$$

where $\xi = p_{34} p_{45}/(1 - p_{43} p_{34})$. This expression could also be obtained by substitution in $\boldsymbol{Q}_{OO}^{-1} \boldsymbol{Q}_{OS} \boldsymbol{Q}_{SS}^{-1} \boldsymbol{Q}_{SL}$, or by appropriate specialization of the general expression $\boldsymbol{Q}_{OO}^{-1} \boldsymbol{K}_{OL} \mathbf{1}$ (see (3.3)).

For a theoretical burst, the number N_O of visits to the open state has the geometric distribution $P(N_O = n) = pq^{n-1}$ $(n = 1, 2, \ldots)$, where $q = 1 - p$, as could be seen also from (3.10). As there is only one open state in this model, the total open-time during a theoretical burst must be a geometric random sum of (independent) exponentially distributed random variables, and so itself has an exponential distribution (see, for example, Appendix 2 of Milne et al., 1988). It follows, either directly from the above discussion or from (3.14), that the mean total open-time within a theoretical burst is $\mu_{U_O} = 1/(pd_1) = 1/(q_{13}\xi)$. Observe that μ_{U_O} does not depend on the concentration D; as D increases the mean number, $1/p$, of visits to the open state increases linearly, while the mean length $1/d_1$ of each visit decreases inversely.

Various other properties of theoretical bursts, in particular the distribution of the burst duration B, can be derived by specialization of the formulae of Section 3. However, although these formulae are valuable in writing general computer programs for numerical computations of model properties based on a specified transition rate matrix, appropriate specialization of them may not offer the most illuminating route to derivation of burst properties for a particular model. For example, for the linear sequential model with blockade a simpler alternative is to use random sums and Laplace transforms. These methods may be employed also for some non-Markov models, such as Markov renewal processes in which sojourns in a state may be more general than exponential but with jump probabilities p_{ij}, $i \neq j$, independent of the durations of sojourns in individual states; see, for example, Edeson et al. (1990).

For any random variable Z, denote by $Z^{(j)}$, $j = 0, 1, \ldots$, a sequence of independent random variables, independent of and with the same distribution as Z. Let X_i be the duration of a sojourn in state i and $\mu_i = \mathbb{E}(X_i)$. Then B can be expressed as X_1 with probability p, $X_1 + X_2 + B^{(1)}$ with probability p_{12}, or as $X_1 + X_3 + \sum_{j=0}^{m}(X_4^{(j)} + X_3^{(j+1)}) + B^{(1)}$ with probability $p_{13}p_{31}(p_{43}p_{34})^m$, $m = 0, 1, \ldots$. Let $\phi_i(\theta) = \mathbb{E}[e^{-\theta X_i}]$, and so $\phi_i(\theta) = d_i/(d_i + \theta)$ ($i = 1, 2, 3, 4, 5$) for the present model. Then the joint generating function (Laplace transform or probability generating function) $\chi(\theta, \nu, \omega) = \mathbb{E}[e^{-\theta U_O - \nu V_S} \omega^{N_O}]$ for the the total open-time, total closed-time and number of openings within a theoretical burst is

$$\chi(\theta, \nu, \omega) = \frac{\omega \xi \phi_1(\theta)[1 - p_{43}p_{34}\phi_4(\nu)\phi_3(\nu)]}{[1 - p_{43}p_{34}\phi_4(\nu)\phi_3(\nu)][1 - \omega p_{12}\phi_1(\theta)\phi_2(\nu)] - \omega p_{31}p_{13}\phi_3(\nu)\phi_1(\theta)}, \quad (6.2)$$

as could also be obtained directly by substitution in (3.20). Moments, such as the mean μ_B and standard deviation σ_B of the duration of a theoretical burst may be evaluated from (6.2) by differentiation, or from formulae such as (3.14) and (3.17). The mean may also be obtained more directly by averaging the above representations for B and solving

$$\mu_B = \mu_1 p + (\mu_1 + \mu_2 + \mu_B)p_{12} + \frac{(\mu_1 + \mu_3 + \mu_B)p_{13}p_{31}}{1 - p_{43}p_{34}}$$
$$+ \frac{(\mu_3 + \mu_4)p_{13}p_{31}p_{43}p_{34}}{(1 - p_{43}p_{34})^2}$$

to give

$$\mu_B = \alpha_B + \beta_B D, \quad (6.3)$$

where $\alpha_B = \mu_{U_O} + p_{31}(\mu_3 + \mu_4 p_{43}p_{34})/[p_{34}p_{45}(1 - p_{43}p_{34})]$ and $\beta_B = k_f \mu_2/(q_{13}\xi) = k_f \mu_2 \mu_{U_O}$ are constants. The mean burst duration increases linearly with D; by contrast, the mean total open-time per burst μ_{U_O} is a constant. Thus, the probability of the channel being open within a burst, $p_{OB} = \mu_{U_O}/\mu_B$, decreases as D increases.

Furthermore, the time $\overline{B}$ between two successive theoretical bursts has a distribution, not dependent on D, which is a geometric random sum of sojourns in state 5 together with possible sojourns in states $\{3, 4\}$. Thus its distribution may be determined in a similar manner to that for burst duration (6.2); this yields

$$\chi_{\overline{B}}(\theta) = \frac{[1 - p_{43}p_{34}]p_{43}p_{31}[\phi_4(\theta)\phi_3(\theta)]^2\phi_5(\theta)}{[1 - p_{43}p_{34}\phi_4(\theta)\phi_3(\theta)][1 - p_{43}p_{34}\phi_4(\theta)\phi_3(\theta) - p_{45}\phi_4(\theta)\phi_5(\theta)]}. \quad (6.4)$$

Its mean $\mu_{\bar{B}}$, may be obtained by differentiation in (6.4) or, as is argued at the end of Section 3.3, by $\mu_{\bar{B}} = \mu_{U_O}/\pi_1 - \mu_B$, where

$$\pi_1^{-1} = 1 + \frac{q_{12}}{d_2} + \frac{q_{13}}{q_{31}} + \frac{q_{13}q_{34}}{q_{31}q_{43}} + \frac{q_{13}q_{34}q_{45}}{q_{31}q_{43}d_5}. \tag{6.5}$$

The duration of a cycle consisting of a burst and an inter-burst period has mean, μ_{CY}, increasing linearly with D, but with the proportion π_1 of open-time decreasing.

For the numerical values in Figure 3, $\xi = 5/11$, $\alpha_B = 12.04$, $\beta_B = 0.55$, $\mu_{N_0} = 2.2 + 0.11D$, $\mu_{U_o} = 11$ (as indicated by the solid line in (b) of Figure 4), $\mu_B = 12.04 + 0.55D$, $\mu_{\bar{B}} = 1209.4$, $\pi_1 = 1/(111.04 + 0.05D)$, $\mu_{V_S} = 1.04 + 0.55D$ and $\mu_{CY} = 1221.44 + 0.55D$.

For an empirical burst there is still only one open state (1), while the class of closed states $C = \{2, 3, 4, 5\}$ has four states. Thus $\tilde{\psi}_O = 1$, and for an empirical burst with critical time t_c it follows from (4.5) that $\widetilde{\boldsymbol{K}}_O = d_1 - \boldsymbol{Q}_{OC}(\boldsymbol{I} - e^{\boldsymbol{Q}_{CC}t_c})(-\boldsymbol{Q}_{CC}^{-1})\boldsymbol{Q}_{CO}$, which is also a scalar. Thus the total open-time per empirical burst is also exponentially distributed, with mean $\widetilde{\boldsymbol{K}}_O^{-1}$. An empirical burst ends at the beginning of a sojourn in the class C that has duration greater than t_c, and so, in particular, may be terminated by a visit of duration greater than t_c to the inhibitory state (2). As D increases, the number of visits to state 2 is likely to be larger, and so more empirical bursts will be ended by such a visit. This makes possible substantial differences between corresponding numerical properties of theoretical and empirical bursts, especially for inappropriately chosen values of t_c, for example, $t_c = 5$ in Figure 4. Numerical results for some means of burst properties, the probability of the channel being open within a burst, and two correlations, are displayed in Figure 4 for empirical bursts for three values of t_c and for theoretical bursts.

As the open class O has a single open state, the total open-times in successive bursts (theoretical or empirical) are independent, so $\rho(U_O, U'_O) = 0$. While the total open-time and the number of openings within a burst have simple distributions, exponential and geometric respectively, they may be (positively) correlated, the degree of correlation in an empirical burst varying with t_c. The mean duration μ_G of a gap within a theoretical burst, that is a visit to the short-lived closed states $\{3, 4\}$ which starts and ends at state 3, is given by $\mu_G = (\mu_3 + \mu_4 p_{43} p_{34})/(1 - p_{43} p_{34})$; for the transition rates in Figure 3, $\mu_G = 0.8667$. A sojourn in the blocked state (2) has mean $\mu_2 = 5$, and one in the long-lived closed state (5) has mean $\mu_5 = 100$. (An inter-burst period has a larger mean, as given above, because it may include several visits to this state, the mean number of visits being $(1 - p_{43} p_{34})/(p_{43} p_{31}) = 11$.) A choice of the critical time t_c which is somewhere between these extremes (1–5 and 100+) could be expected to yield results for empirical burst properties similar to those for theoretical bursts, as there would be a very small probability of misclassification of a gap ending a burst or of a visit to the long-lived closed state being taken as a gap; for example, $t_c = 40$ seems appropriate. For smaller t_c, for example, $t_c = 10$ or $t_c = 5$, there is an increased chance of a gap being regarded as ending a burst, resulting in, as for other examples, a divergence between the corresponding properties of an empirical burst and a theoretical burst.

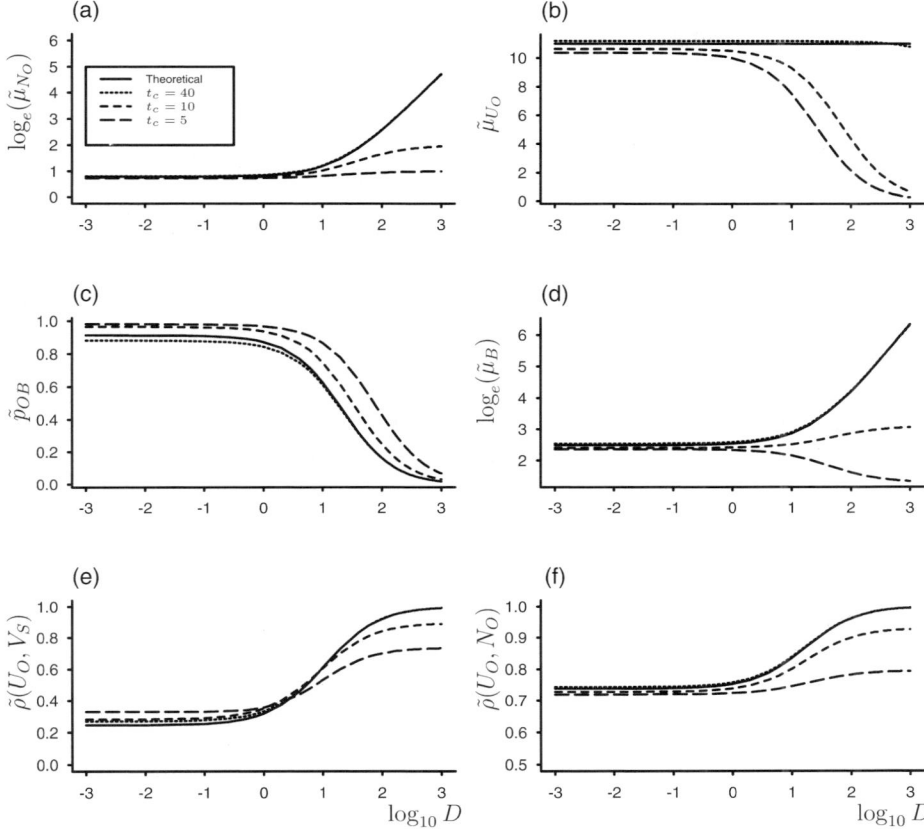

Fig. 4. Predicted empirical burst behaviour for the five state sequential model with drug blockade. The broken curves in (a)–(f) (see inset in (a)) show the mean number of openings ($\tilde{\mu}_{N_O}$), the mean total open-time ($\tilde{\mu}_{U_O}$), the probability of being open within a burst ($\tilde{p}_{OB}$), the mean burst duration ($\tilde{\mu}_B$), the correlations between the total open-time and total closed-time ($\tilde{\rho}(U_O, V_S)$) and between the total open-time and number of openings ($\tilde{\rho}(U_O, N_O)$) per burst for critical times of $t_c = 5$, 10 and 40 (ms), as a function of drug concentration D (units of micromolar), together with their counterparts for theoretical bursts. Log scales (base 10) are used for the drug concentration (horizontal axes), and natural log scales for the mean number of openings per burst and mean burst duration.

7. A model showing biphasic drug effects

Some drugs modulate receptor ion channel activity in a more complex manner than the simple blockade considered in the model of Figure 3; see Arias (1997) and other references cited in Yeo and Madsen (1998). There may be an additional stimulatory effect on channel activity at low concentrations of D, followed thereafter by the more usual inhibitory phenomenon, a so-called 'biphasic' effect. Madsen and Yeo (2000) have considered the following Markov model that is consistent with such behaviour seen in neuronal nicotinic acetylcholine receptors:

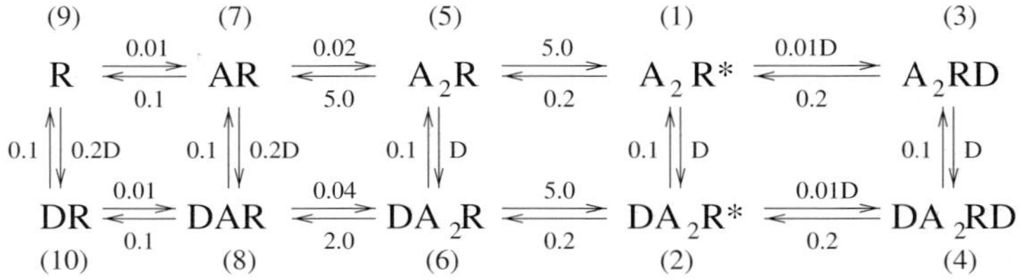

$$
\begin{array}{cccccccccc}
(9) & & (7) & & (5) & & (1) & & (3) \\
& 0.01 & & 0.02 & & 5.0 & & 0.01D & \\
R & \rightleftharpoons & AR & \rightleftharpoons & A_2R & \rightleftharpoons & A_2R^* & \rightleftharpoons & A_2RD \\
& 0.1 & & 5.0 & & 0.2 & & 0.2 & \\
0.1 \updownarrow 0.2D & & 0.1 \updownarrow 0.2D & & 0.1 \updownarrow D & & 0.1 \updownarrow D & & 0.1 \updownarrow D \\
& 0.01 & & 0.04 & & 5.0 & & 0.01D & \\
DR & \rightleftharpoons & DAR & \rightleftharpoons & DA_2R & \rightleftharpoons & DA_2R^* & \rightleftharpoons & DA_2RD \\
& 0.1 & & 2.0 & & 0.2 & & 0.2 & \\
(10) & & (8) & & (6) & & (2) & & (4)
\end{array}
$$

Fig. 5. A model showing biphasic drug effects.

In this model there are two possible types of state of the receptor, a control type (upper level in Figure 5) and a modified type (lower level). The upper row of states depicts agonist receptor activation and channel blockade when an extracellular noncompetitive agonist (NCA) site is not occupied by D (analogous to Figure 3). The lower row depicts the comparable set of states when the NCA site is occupied. Particular values chosen to illustrate the properties of the system were suggested by experimental data for the muscle form of the receptor (Le Dain et al., 1991), with the requirement to maintain detailed balance throughout the model.

There are now two open states in $O = \{1, 2\}$, six short-lived closed states in $S = \{3, 4, 5, 6, 7, 8\}$ and two long-lived closed states in $L = \{9, 10\}$. Although the classes O and L each contain more than one state, the class sojourn times are both single exponential distributions because of the particular assignment of transition rates. (For example, the transition rates between state 1 and its neighbouring states are the same as those between state 2 and its corresponding neighbours.) The mean duration of a sojourn time in L is 100, in the open class O is $1/(0.2 + 0.01D)$ and in the blocked states $\{3, 4\}$ is 5, while that in the short-lived closed class S is rather more complex. For very low values of D there will be very few visits to the blocked states, but these will increase as D increases until for very large values of D most visits to the non-open states will be to state 4.

For empirical bursts the class of closed states is $C = \{3, 4, 5, 6, 7, 8, 9, 10\}$. With a critical time t_c much smaller than 5 there is a significant probability that an empirical burst will appear to end by a visit to the blocked states, but this probability will decrease rapidly as t_c increases substantially above about 5. However, for small values of D there will be very few visits to the blocked states, and hence most empirical bursts will end with a visit to L. On the other hand, as D increases, whether a visit to $\{3, 4\}$ ends an empirical burst is highly dependent on the value of t_c. In the case of theoretical bursts, as D increases there will tend to be fewer visits to L, so the duration of such a burst will increase on average, while the total open-time per burst will not change so rapidly because the mean duration of sojourns in O decreases with increasing D, as is the case in the previous section. For such reasons the total open-time per burst is rather more stable than the burst duration and consequently could be regarded as a more suitable measure of channel activity.

A number of burst properties have been directly evaluated using the results of Sections 3 and 4, in particular for the numerical values given in Figure 5. For $t_c = 5$, 10 and 40, (a)–(d) of Figure 6 show respectively, as a function of $\log_{10} D$, the natural logarithm of the mean number of openings ($\log_e(\tilde{\mu}_{N_O})$), the mean total open-time

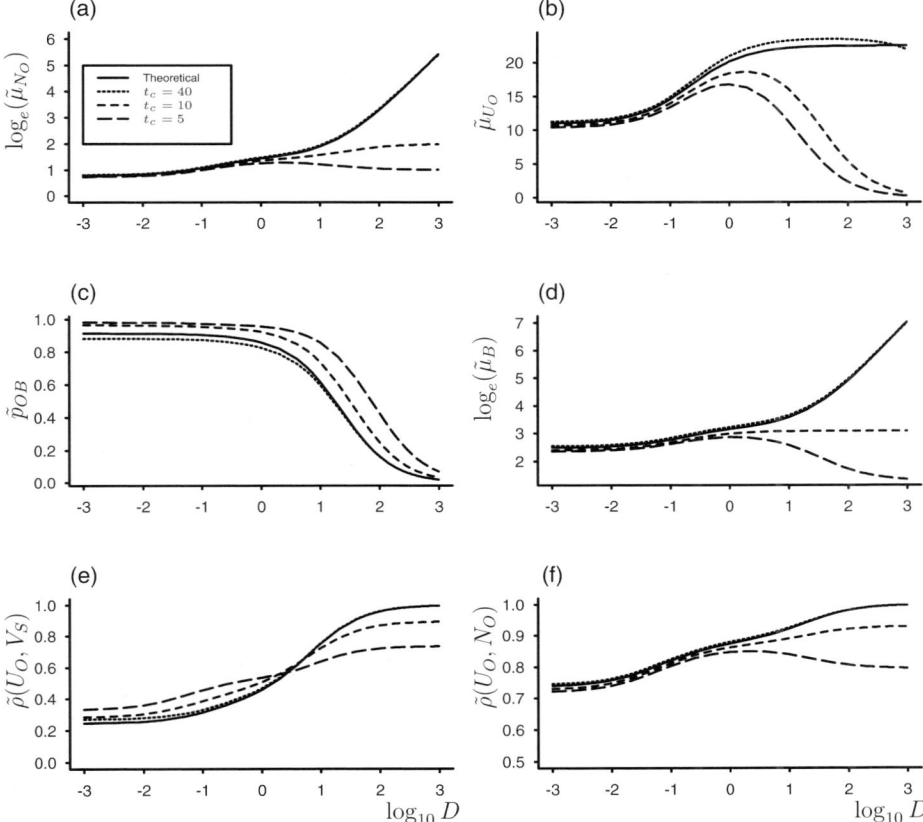

Fig. 6. Predicted empirical burst behaviour for the model showing biphasic drug effects. The broken curves in (a)–(f) (see inset in (a)) show the mean number of openings ($\tilde{\mu}_{N_O}$), the mean total open-time ($\tilde{\mu}_{U_O}$), the probability of being open within a burst ($\tilde{p}_{OB}$), the mean burst duration ($\tilde{\mu}_B$), the correlations between the total open-time and total closed-time ($\tilde{\rho}(U_O, V_S)$) and between the total open-time and number of openings ($\tilde{\rho}(U_O, N_O)$) per burst for critical times of $t_c = 5$, 10 and 40 (ms), as a function of drug concentration D (units of micromolar), together with their counterparts for theoretical bursts. Log scales (base 10) are used for the drug concentration (horizontal axes), and natural log scales for the mean number of openings per burst and mean burst duration.

($\tilde{\mu}_{U_O}$), the probability of the channel being open within a burst ($\tilde{p}_{OB} = \tilde{\mu}_{U_O}/\tilde{\mu}_B$) and the mean burst duration ($\tilde{\mu}_B$) per empirical burst, together with the corresponding quantities (same symbols but without tildes) for a theoretical burst. For some values of t_c, such as 5 and 10, there is a clear biphasic effect with $\tilde{\mu}_{U_O}$ first increasing to a maximum, then decreasing towards zero for large values of D. The mean duration of visits to O decreases with increasing D, while the non-negligible probability that a burst is ended by a sojourn in the blocked states (3, 4) results in shorter mean total open-time per burst. As t_c increases the probability of a sojourn in the blocked states being greater than t_c becomes smaller, and the biphasic effect is much less visible. The effect of D on the correlations $\tilde{\rho}(U_O, V_S)$ and $\tilde{\rho}(U_O, N_O)$ is shown respectively in (e) and (f) of

Figure 6. The correlation $\rho(U_O, U'_O)$ between the total open-time in two successive bursts is small for both theoretical and empirical bursts for all values of D.

Many of the burst properties illustrated in Figure 6 appear similar to those for the simpler five state sequential model shown in Figure 4. This is not entirely unexpected as the transition rates between the classes of states are similar in both cases. An exception is the mean total open-time per burst, where, in particular, a biphasic effect can be seen in the model based on Figure 5 but not that of Figure 3.

8. A model for a supergated double-barrelled chloride channel

The chloride channel from the electric organ of *Torpedo*, ClC-0, is the prototype of a family of voltage-dependent chloride channels found in many species (Jentsch et al., 1999). Unlike most well-studied ion channels that have a single-gated pore, ClC-0 has a 'double-barrelled' structure consisting of two seemingly independent and identically gated pores. Ion flux through each of these is controlled by a 'subgate' together with a 'supergate' that can simultaneously block both pores (Miller, 1982; Bauer et al., 1991). Permeant (chloride) ion concentration and membrane potential affects gating, with the subgates activated by the membrane potential becoming less negative (depolarisation) and the supergate by it becoming more negative (hyperpolarisation); see Foskett (1998). At the usual membrane potentials found in cells, the time scales of gating kinetics are often well separated, the subgates being fast and having mean lifetimes of order several milliseconds, in contrast to the supergate, which has a mean lifetime of order hundreds of milliseconds.

Li et al. (2000) studied a general six-state Markov model incorporating this differential gating; see Figure 7. In this model O, C, o and c indicate that the supergate and subgates are open and closed respectively. There are three observed conductance levels, corresponding to state 2 (two subgates and the supergate open, double conductance level), state 1 (either subgate and the supergate open, single conductance level) and states 3, 4, 5 and 6 (either the supergate closed or both subgates closed, zero conductance level). The transition rates (in units of ms^{-1}) used for this example, which imply means for the subgate and supergate closed-times of the order of 10 ms and 100 ms respectively, were chosen to reflect the kinetic separation often found experimentally for the situation where the supergate and subgates appear to act independently. A possible special case of this is a four-state model where states 4, 5 and 6 are collapsed into one, as was considered by Miller (1982) and Miller and White

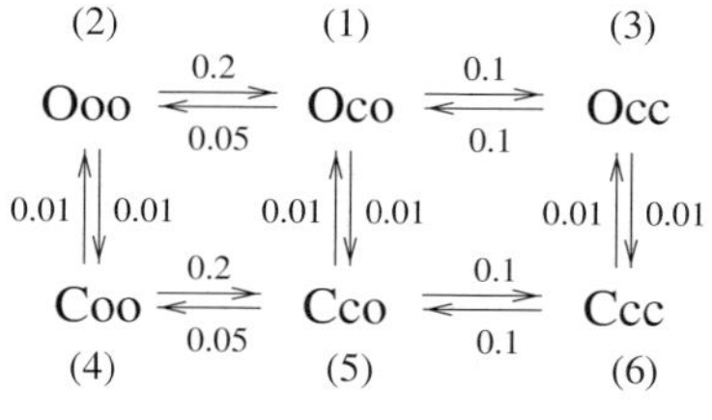

Fig. 7. Six-state model for a double-barrelled chloride channel.

(1984); the latter is likely to be adequate when there is a clear difference in the means of sojourns resulting from closure of the subgates and closure of the supergate, which may happen at some membrane potentials.

The six states in Figure 7 may be partitioned into classes as follows. The open classes, $O_1 = \{1\}$ and $O_2 = \{2\}$ consisting respectively of the open states with conductance levels 1 and 2, together form the open class $O = O_1 \cup O_2$ ($n_O = 2$). As a result of closure of the supergate, sojourns in states 4, 5 and 6 tend to be long, and when the supergate is open, sojourns in state 3 tend to be short. Thus the class of long-lived closed states is defined to be $L = \{4, 5, 6\}$ ($n_L = 3$) and the class of short-lived closed states is $S = \{3\}$ ($n_S = 1$). Again, for this model, every state in a class is a gateway state for that class.

The probabilities that a theoretical burst starts in the various open states are given by (3.5). The probability of the burst starting in state 1 (level 1) is

$$\psi_1 = \frac{\pi_5 q_{51} + \pi_6 q_{63} p_{31}}{\pi_5 q_{51} + \pi_6 q_{63} p_{31} + \pi_4 q_{42}},$$

where, as usual, $p_{ij} = q_{ij}/d_i$ $i, j = 1, \ldots, 6$, is the (conditional) probability that a channel in state i moves to state j. For the numerical values in Figure 7, $\psi_1 = 0.884$, and the probability that a burst starts in state 2 (level 2) is clearly $\psi_2 = 1 - \psi_1 = 0.116$. The corresponding probabilities $\tilde{\psi}_1$ and $\tilde{\psi}_2$ depend on the chosen value of t_c.

For ClC-0 some of the matrix results in Section 3 for theoretical bursts can be written out more simply and explicitly. For example, in (3.14) the matrix $\boldsymbol{K}_O$ reduces to

$$\boldsymbol{K}_O = \begin{bmatrix} d_1(1 - p_{13}p_{31}) & -q_{12} \\ -q_{21} & d_2 \end{bmatrix}. \qquad (8.1)$$

It follows that

$$\boldsymbol{K}_O^{-1} = \frac{1}{d_1 d_2 (1 - p_{12}p_{21} - p_{13}p_{31})} \begin{bmatrix} d_2 & q_{12} \\ q_{21} & d_1(1 - p_{13}p_{31}) \end{bmatrix}, \qquad (8.2)$$

and hence

$$\mu_{U_O} = \boldsymbol{\psi}_O^\top \boldsymbol{K}_O^{-1} \mathbf{1}$$
$$= \frac{\psi_1 + \psi_2 p_{21}}{d_1(1 - p_{12}p_{21} - p_{13}p_{31})} + \frac{\psi_2(1 - p_{13}p_{31})}{d_2(1 - p_{12}p_{21} - p_{13}p_{31})}, \qquad (8.3)$$

which corresponds, as seen below in (8.7), to $\mu_{U_O} = \mu_{T_1} + \mu_{T_2}$, where μ_{T_1} and μ_{T_2} are the mean total open-time per burst at levels 1 and 2 respectively.

For an empirical burst the class of closed states is $C = \{3, 4, 5, 6\}$. For the specified transition rates, empirical bursts consist largely of alternating sojourns among states 1, 2 and 3, as entry to any of (long-lived) states 4, 5 or 6 can usually be expected to terminate such a burst. Empirical bursts are then reasonably well separated, with comparatively long interburst intervals usually caused by supergate closure.

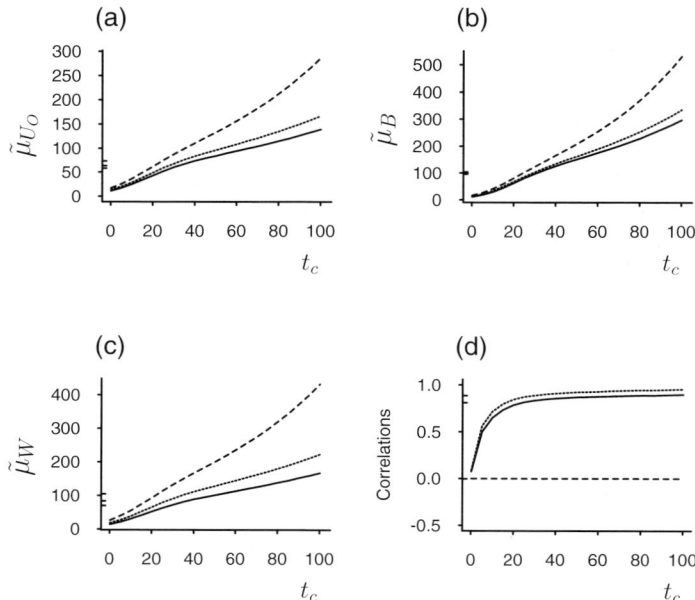

Fig. 8. Predicted empirical burst behaviour for the six-state ClC-0 model. The mean total open-time ($\tilde{\mu}_{U_O}$), the mean total burst duration ($\tilde{\mu}_B$), and the mean total charge transfer ($\tilde{\mu}_W$) in an empirical burst are shown in (a)–(c) respectively as a function of t_c (in ms), solid curve (———) for the independent case, dashed curve (– – –) and dotted curve (· · · · ·) for q_{12} and q_{42} each multiplied by (i) 2 and by (ii) 4 respectively. In (d) the solid curve (———) gives for an empirical burst the correlation between N_O and U_O, the dotted curve (· · · · ·) that between U_O and V_S, and the dashed curve (– – –) that between U_O and the total open-time U'_O within the next burst, all for the independent case. In each of (a)–(d), the corresponding values for a theoretical burst are marked, for comparison, on the vertical axis.

The numerical values in Figure 7 correspond to the supergate and the subgates acting independently. However, it is possible that there could be cooperative interactions between the gates, given that they all operate within a single channel structure. For example, such dependence may be introduced in various ways while maintaining reversibility. If the two types of gate act in a positively cooperative manner, then a shut supergate could be more likely to open if both subgates are open than if one or both subgates are shut (q_{42} increases), and a subgate more likely to open if the supergate and the other subgate are both open (q_{12} increases). If q_{12}/q_{42} remains constant in this situation then reversibility is maintained. As an illustration suppose both q_{12} and q_{42} increase by a multiple of (i) 2 and (ii) 4. The effect is shown in Figure 8 for a selection of burst properties.

The results (a)–(c) in Figure 8 confirm that for an empirical burst the mean total open-time $\tilde{\mu}_{U_O}$, the mean total burst duration $\tilde{\mu}_B$, and the mean total charge transfer $\tilde{\mu}_W$ all increase in all cases as t_c increases from 0.1 to 100 ms. This is because, as t_c increases, more sojourns in {4, 5, 6} will be classified as gaps within a burst. From (d) in Figure 8 we see that both the correlation $\rho(U_O, U'_O) = -0.0020$ for theoretical bursts and $\tilde{\rho}(U_O, U'_O)$, that for empirical bursts, are close to zero. This confirms that the total open-times U_O and U'_O in successive bursts are almost uncorrelated for either type of burst. However, within a burst, the correlations $\tilde{\rho}(U_O, V_S)$ and $\tilde{\rho}(U_O, N_O)$ increase as t_c increases; moreover the slope is greatest at small values of t_c ($\leqslant 20$ ms).

As for the five-state model, Figure 8 shows also that values of empirical burst properties will be close to the corresponding properties for theoretical bursts provided an appropriate value is chosen for t_c; see the last paragraph of the previous section for comments on such a choice. To facilitate comparisons in each of the three situations, levels corresponding to the respective values, $\mu_{U_O} = 57.89, 62.26, 68.98$, $\mu_B = 96.17, 96.57, 97.18$ and $\mu_W = 69.47, 83.02, 103.83$, for a theoretical burst are marked in each case on the right of the respective vertical axes in (a)–(c). The correlation values, $\rho(U_O, V_S) = 0.8113$ and $\rho(U_O, N_O) = 0.8851$, in the independent case for a theoretical burst are marked on the vertical axis in Figure 8(d). For the other two cases of dependence the corresponding values are $\rho(U_O, V_S) = 0.7906, 0.7596$ and $\rho(U_O, N_O) = 0.8683, 0.8443$ respectively. By comparing values of theoretical burst properties with those of empirical bursts, we see that a reasonable value of t_c is about 30 ms, but this changes with the degree of dependence, the appropriate value of t_c being smaller with greater cooperativity.

Although the mean open-time within an empirical burst may be of interest in many situations, it cannot be used in examples like the present one to determine the mean total charge transfer across the membrane during the burst, because the open-times in ClC-0 are of two different types, some at conductance level one and some at level two. Within an empirical burst, denote by T_1 the total open-time at conductance level one and similarly for T_2. Then define a normalised charge transfer $W = T_1 + 2T_2$, having mean

$$\tilde{\mu}_W = \tilde{\mu}_{T_1} + 2\tilde{\mu}_{T_2}. \tag{8.4}$$

Li et al. (2000) obtained exact expressions and some simple approximations for $\tilde{\mu}_{T_1}, \tilde{\mu}_{T_2}$ and $\tilde{\mu}_W$, as well as an approximation to $\tilde{\sigma}_W^2$. In order to derive exact results for T_1, T_2 and W a different partitioning of the transition rate matrix Q is required, using classes O_2, O_1 and C. Li et al. (2000) established multinomial distributions for the number of the various possible types of event within an empirical burst; they showed that the numbers N_1 and N_2 of sojourns at levels 1 and 2 respectively during an empirical burst both have zero-modified geometric distributions (Section 8.2.2 of Johnson et al., 1992), that is, geometric distributions with an altered first term. Thus T_1 and T_2 are each zero-modified geometric random sums of exponential random variables, and it follows that each has a zero-modified exponential distribution, similar to the situation in Section 6. This yields

$$\tilde{\mu}_W = \tau/\left(d_1(1-A)^2\right) + 2\kappa/\left(d_2(1-B)^2\right), \tag{8.5}$$

where

$$\tau = (\tilde{\psi}_1 + \tilde{\psi}_2 v_1)(\tilde{\beta}_1 + \tilde{\beta}_2 v_2), \quad \kappa = (\tilde{\psi}_1 v'_2 + \tilde{\psi}_2)(\tilde{\beta}_1 v'_1 + \tilde{\beta}_2),$$
$$A = \alpha_{101} + \gamma/(1 - \alpha_{202}), \quad B = \alpha_{202} + \gamma/(1 - \alpha_{101}),$$

with

$$\alpha_{i0j} = (-\boldsymbol{Q}_{O_iO_i}^{-1})\boldsymbol{Q}_{O_iC}(\boldsymbol{I} - e^{\boldsymbol{Q}_{CC}t_c})\boldsymbol{Q}_{CC}^{-1}\boldsymbol{Q}_{CO_j} \quad (i,j=1,2),$$
$$\tilde{\beta}_i = (-\boldsymbol{Q}_{O_iO_i}^{-1})\boldsymbol{Q}_{O_iC}e^{\boldsymbol{Q}_{CC}t_c}\boldsymbol{1} \quad (i=1,2),$$
$$v_i = (p_{3-i,i} + \alpha_{3-i,0i})/(1-\alpha_{202}) \quad (i=1,2),$$
$$v'_i = (p_{3-i,i} + \alpha_{3-i,0i})/(1-\alpha_{101}) \quad (i=1,2).$$

In order to study the corresponding quantities T_1, T_2 and W for theoretical bursts the transition rate matrix $\boldsymbol{Q}$ needs to be partitioned into four classes O_2, O_1, S and L. This may appear to require an extension of the results in Sections 2 and 3; however, the particular form of the model for ClC-0 allows many results to be obtained without any difficult extension. During a theoretical burst a visit to the short-lived closed class $S = \{3\}$ can occur only to and from the level 1 open class, $O_1 = \{1\}$; this allows the joint distribution of the number of visits at level 1 and at level 2 during a burst to be evaluated combinatorically in a manner similar to that used to obtain the approximations for the corresponding properties of empirical bursts considered in Li et al. (2000).

Denote the open state in which a burst begins by S^* and that from which it exits by E^*. There are four possibilities, one of which gives, for $0 \leqslant n_2 < n_1 < \infty$,

$$P(N_1 = n_1, N_2 = n_2, E^* = 1 | S^* = 1)$$
$$= \binom{n_1 - 1}{n_2}(p_{12}p_{21})^{n_2}(p_{13}p_{31})^{n_1 - n_2 - 1}\beta_1. \tag{8.6}$$

Here $\beta_1 = p_{15} + p_{13}p_{36}$ is the probability that a channel at level 1 exits the burst; similarly $\beta_2 = p_{24}$. (The quantities β_1 and β_2 are the elements of $\boldsymbol{K}_{OL}\boldsymbol{1}$ mentioned near (3.2).) Expressions similar to (8.6) may be obtained for the other three possibilities of (S^*, E^*), namely $(1, 2), (2, 1)$ and $(2, 2)$. Using the law of total probability and summing over n_2 (or by appropriate use of conditional independence) yields

$$P(N_1 = n_1)$$
$$= \begin{cases} \psi_2 p_{24} & (n_1 = 0), \\ (\psi_1 + \psi_2 p_{21})(1 - p_{12}p_{21} - p_{13}p_{31})(p_{12}p_{21} + p_{13}p_{31})^{n_1 - 1} \\ & (n_1 = 1, 2, \ldots), \end{cases}$$

which is a zero-modified geometric distribution. It follows that

$$\mu_{T_1} = \frac{\mu_{N_1}}{d_1} = \frac{\psi_1 + \psi_2 p_{21}}{d_1(1 - p_{12}p_{21} - p_{13}p_{31})}, \tag{8.7}$$

which is identical to the first part on the right-hand side of (8.3). Similarly it can be shown that μ_{T_2} is given by the second part on the right-hand side of (8.3). Then the mean total charge transfer during a theoretical burst is $\mu_W = \mu_{T_1} + 2\mu_{T_2}$.

For the numerical values of Figure 7 the mean total charge transfers for theoretical and for empirical bursts are the same for t_c about 29. This is similar to the t_c value that yields approximate equivalence for other properties illustrated in Figure 8. As cooperativity increases, by q_{12} and q_{42} increasing, μ_W increases, as does $\tilde{\mu}_W$, with the two being equal for smaller values of t_c. In the cases of two- and four-fold increases of q_{12} and q_{42}, the appropriate values of t_c are about 27 and 23 respectively.

9. Some comments on statistical inference

As remarked in the introduction, statistical inference based on stochastic models for the gating behaviour of ion channels is generally less well developed than is the modelling area itself. Perhaps the simplest approach conceptually is maximum likelihood estimation of the transition parameters of a specified model, using data which is a sequence of alternating open-times and closed-times.

In practice, single channel data does not come in the form of such a sequence, but initially rather as discrete-time digital data obtained by sampling a noisy continuous-time analog record which has been influenced by many experimental variables including thermal and capacitative noise, amplifier characteristics and electronic filtering. To derive sojourn-time information from the sampled noisy record usually requires a 'restoration' in which, in the simplest case, for each sampling time it is decided whether the channel was open or closed. Early restoration methods were based on thresholds; see, for example, Colquhoun and Sigworth (1995). Bayesian methods based on hidden Markov models were introduced to the area by Chung et al. (1990) and Fredkin and Rice (1992a). An inescapable consequence of such approaches is the loss of very brief sojourn times. As was shown by Chung et al. (1990), Bayesian methods based on hidden Markov models can be used also for restoration of multilevel data with known levels, and even for estimation of unknown levels. Ideally, restoration of channel data should also take account of the significant effects of electronic filtering of the primary data, as this attenuates and delays recorded signals and leads to difficulties in discriminating between channel conductance levels and sojourn durations; see Colquhoun and Sigworth (1995).

Subsequent to restoration, estimation of model transition parameters follows a more established route. Studies by Ball and Sansom (1989), Magleby and Weiss (1990a, 1990b), Qin et al. (1997), Sukharev et al. (1999) and Keleshian et al. (2000) illustrate what can be learned about ion channel behaviour using this general two-stage approach in systems of varying complexity. However, it should be remembered that even with perfect restoration the second stage kinetic parameter estimation process may be compromised by model identifiability problems (cf. Kienker, 1989; Edeson et al., 1994).

Hidden Markov models have been used also in an approach to parameter estimation based directly on the sampled noisy record and not requiring an initial restoration. For single channel data this was done by Fredkin and Rice (1992b); multichannel extensions (assuming independent and identically distributed channels) were considered by Albertsen and Hansen (1994) and Klein et al. (1997). Some variations on the traditional hidden Markov approach have been considered by Qin et al. (2000a, 2000b).

A fully Bayesian approach to inference from single channel data is presented in Ball et al. (1999). Their approach, using Markov chain Monte Carlo methods (cf. Gilks et al., 1996; Gamerman, 1997), allows parameter estimation directly from the sampled noisy record, as well as restoration. The computationally intensive technique of Markov chain Monte Carlo is used to estimate the posterior distribution for the model parameters given the sampled noisy record by extensive simulations of a specially constructed Markov chain whose equilibrium distribution is the required posterior distribution. Hodgson (1999) and Hodgson and Green (1999) have considered Markov chain Monte Carlo methods for inference from single channel data where the noise is modelled as an autoregressive process. Some other approaches to statistical inference, in particular methods based on the semi-Markov framework of Ball et al. (1991, 1993), are summarized in the overview given by Ball et al. (1999). A Markov chain Monte Carlo approach to inference for hidden Markov models is presented by Schouten (2000).

Some reasons for focusing on the information contained in ion channel bursts were mentioned in the introduction to this paper. For any specified value t_c, it is possible to determine empirical bursts for this t_c from a restoration of a given channel record. Then the likelihood function based on each empirical burst can be obtained using (4.2). The product of the likelihood functions for each empirical burst yields a type of pseudo-likelihood (cf. Baddeley and Turner, 2000, and references therein). This could be maximized, as if it were a likelihood, to determine estimates of model parameters. Because the closed-times between empirical bursts are often long, the characteristics of separate empirical bursts should be approximately independent and so the pseudo-likelihood should approximate the likelihood based on these (burst) segments of the single channel data. The fact that, in practice, the separate empirical bursts may have come from different channels contributing to the record provides a further reason for using such a pseudo-likelihood.

Methods based on empirical burst analysis require a particular value to be chosen for t_c for a given channel record. Although some suggestions have been described by Colquhoun and Sigworth (1995) in the case of the closed class having just two states, no general method appears to be available. Results from the present paper may partially address this problem. The usual goal in quantitative inferential studies is to obtain some understanding of channel gating properties in terms of transitions between reaction scheme states. Provided initial estimates of transition rates are available, perhaps from analysis of the whole record, comparison of theoretical and empirical burst properties could allow determination of an appropriate value for t_c. Given this value, attention might then be focused on empirical bursts; gating parameter estimation should then be less sensitive to variables such as agonist concentration and the number of channels in the patch. This problem will be considered elsewhere.

Note added in proof

Since this paper was completed significant advances have been made in the study of bursting behaviour in Markov chain models for the behaviour in ion channels. Ball et al. (2002) use a semi-Markov framework to derive, compactly and in a unified manner for theoretical and empirical bursts, fundamental results about some burst properties of multiconductance channels.

Ball et al. (2003) build on that work and derive more detailed results, assuming that the underlying Markov chain is in equilibrium and under the additional assumption this chain is reversible. In particular, the main results of the latter paper show that, when the underlying Markov chain is in equilibrium and reversible, both the autocorrelation function and the distribution of important burst properties, including total charge transfer in a burst and number of openings in a burst, must be of similar structure, that is linear combinations of finitely many geometrically or exponentially decaying components with non-negative coefficients. Ball et al. (2003) also illustrate through examples the power and flexibility of the methodology, and study methods for choosing a suitable critical time for determining empirical bursts in a given data set, which has relevance to practical applications of burst theory.

Acknowledgements

We are grateful to the referee for comments which have led to improvements in the presentation. Support was received from the Australian Research Council (Grant A69801864).

References

Albertsen, A. and U.-P. Hansen (1994). Estimation of kinetic rate constants from multichannel recordings by a direct fit of the time series. *Biophys. J.* **67**, 1391–1403.

Arias, H.R. (1997). Topology of ligand binding sites on the nicotinic acetylcholine receptor. *Brain Res. Rev.* **25**, 133–191.

Asmussen, S. (1987). *Applied Probability and Queues*. Wiley, Chichester, UK.

Baddeley, A. J. and T. R. Turner (2000). Practical maximum pseudolikelihood for spatial point patterns (with discussion). *Aust. N.Z. J. Statist.* **42**, 283–322.

Ball, F. G., Y. Cai, J. B. Kadane and A. O'Hagan (1999). Bayesian inference for ion-channel gating mechanisms directly from single-channel recordings, using Markov chain Monte Carlo. *Proc. Roy. Soc. London A* **455**, 2879–2932.

Ball, F. G., R. K. Milne, I. D. Tame and G. F. Yeo (1997). Superposition of interacting aggregated continuous-time Markov chains. *Adv. Appl. Probab.* **29**, 56–91.

Ball, F. G., R. K. Milne and G. F. Yeo (1991). Aggregated semi-Markov processes incorporating time interval omission. *Adv. Appl. Probab.* **23**, 772–797.

Ball, F. G., R. K. Milne and G. F. Yeo (2000). Stochastic models for systems of interacting ion channels. *IMA J. Math. Appl. Med. Biol.* **17**, 263–293.

Ball, F. G., R. K. Milne and G. F. Yeo (2002). Multivariate semi-Markov analysis of burst behaviour of multiconductance single ion channels. *J. Appl. Probab.* **39**, 179–196.

Ball, F. G., R. K. Milne and G. F. Yeo (2003). A unified approach to burst properties of multiconductance single ion channels. Submitted for publication.

Ball, F. G. and J. A. Rice (1992). Stochastic models for ion channels: introduction and bibliography. *Math. Biosci.* **112**, 189–206.

Ball, F. G. and M. S. P. Sansom (1988). Aggregated Markov processes incorporating time interval omission. *Adv. Appl. Probab.* **20**, 546–572.

Ball, F. G. and M. S. P. Sansom (1989). Ion-channel gating mechanisms: model identification and parameter estimation from single channel recordings. *Proc. Roy. Soc. London B* **236**, 385–416.

Ball, F. G. and T. Winch (1999). A current-based model for time interval omission in Markov single ion channel gating mechanisms. *Comm. Statist. Stochastic Models* **15**, 661–694.

Ball, F. G. and G. F. Yeo (1993). Lumpability and marginalisability for continuous time Markov chains. *J. Appl. Probab.* **30**, 518–528.

Ball, F. G., G. F. Yeo, R. K. Milne, R. O. Edeson, B. W. Madsen and M. S. P. Sansom (1993). Single ion channel models incorporating aggregation and time interval omission. *Biophys. J.* **64**, 357–374.

Bauer, C. K., K. Steinmeger, J. R. Schwarz and T. J. Jentsch (1991). Completely functional double-barreled chloride channel expressed from a single *Torpedo* cDNA. *Proc. Natl. Acad. Sci. USA* **88**, 11052–11056.

Cesare, P., A. Moriondo, V. Vellani and P. A. McNaughton (1999). Ion channels gated by heat. *Proc. Natl. Acad. Sci. USA* **96**, 7658–7663.

Chung, S. H., J. B. Moore, L. Xia, L. S. Premkumar and P. W. Gage (1990). Characterisation of single channel currents using digital signal processing techniques based on Hidden Markov Models. *Proc. Roy. Soc. London B* **329**, 265–285.

Colquhoun, D. and A. G. Hawkes (1977). Relaxation and fluctuations of membrane currents that flow through drug-operated channels. *Proc. Roy. Soc. London B* **199**, 231–262.

Colquhoun, D. and A. G. Hawkes (1981). On the stochastic properties of single ion channels. *Proc. Roy. Soc. London B* **211**, 205–235.

Colquhoun, D. and A. G. Hawkes (1982). On the stochastic properties of bursts of single ion channel openings and of clusters of bursts. *Phil. Trans. Roy. Soc. London B* **300**, 1–59.

Colquhoun, D. and A. G. Hawkes (1987). A note on correlations in single ion channel records. *Proc. Roy. Soc. London B* **230**, 15–52.

Colquhoun, D. and A. G. Hawkes (1990). Stochastic properties of ion channel openings and bursts in a membrane patch that contains two channels: evidence concerning the number of channels present when a record containing only single openings is observed. *Proc. Roy. Soc. London B* **240**, 453–477.

Colquhoun, D. and A. G. Hawkes (1995a). The principles of the stochastic interpretation of ion-channel mechanisms. In *Single-Channel Recording*, 2nd ed. Chapter 18, pp. 397–482 (Eds. B. Sakmann and E. Neher). Plenum Press, New York.

Colquhoun, D. and A. G. Hawkes (1995b). A Q-matrix cookbook: how to write only one program to calculate the single-channel and macroscopic predictions for any kinetic mechanism. In *Single-Channel Recording*, 2nd ed. Chapter 20, pp. 589–633 (Eds. B. Sakmann and E. Neher). Plenum Press, New York.

Colquhoun, D. and B. Sakmann (1981). Fluctuations in the microsecond time range of the current through single acetylcholine receptor channels. *Nature* **294**, 464–466.

Colquhoun, D. and F. J. Sigworth (1995). Fitting and statistical analysis of single-channel records. In *Single-Channel Recording*, 2nd ed. Chapter 19, pp. 483–587 (Eds. B. Sakmann and E. Neher). Plenum Press, New York.

Csenki, A. (1992). The joint distribution of sojourn times in finite Markov processes. *Adv. Appl. Probab.* **24**, 141–160.

Csenki, A. (1994). *Dependability for Systems with a Partitioned State Space. Markov and Semi-Markov Theory and Computational Implementation*. Lecture Notes in Statist., Vol. 90. Springer-Verlag, New York.

Edeson, R. O., F. G. Ball, G. F. Yeo, R. K. Milne and S. S. Davies (1994). Model properties underlying non-identifiability in single channel inference. *Proc. Roy. Soc. London B* **255**, 21–29.

Edeson, R. O., G. F. Yeo, R. K. Milne and B. W. Madsen (1990). Graphs, random sums, and sojourn time distributions, with application to ion-channel modeling. *Math. Biosci.* **102**, 75–104.

Foskett, J. K. (1998). ClC and CFTR chloride channel gating. *Annu. Rev. Physiol.* **60**, 689–717.

Fredkin, D. R., M. Montal and J. A. Rice (1985). Identification of aggregated Markovian models: application to the nicotinic acetylcholine receptor. In *Proceedings of the Berkeley Conference in Honor of Jerzy Neyman and Jack Kiefer*, Vol. 1, pp. 269–289 (Eds. L. M. Le Cam and R. A. Olshen). Wadsworth, Belmont, CA.

Fredkin, D. R. and J. A. Rice (1986). On aggregated Markov processes. *J. Appl. Probab.* **23**, 208–214.

Fredkin, D. R. and J. A. Rice (1987). Correlation functions of a function of a finite-state Markov process with application to channel kinetics. *Math. Biosci.* **87**, 161–172.

Fredkin, D. R. and J. A. Rice (1991). On the superposition of currents from ion channels. *Phil. Trans. Roy. Soc. London B* **334**, 347–356.

Fredkin, D. R. and J. A. Rice (1992a). Bayesian restoration of single-channel patch clamp recordings. *Biometrics* **48**, 427–448.

Fredkin, D. R. and J. A. Rice (1992b). Maximum likelihood estimation and identification directly from single-channel recordings. *Proc. Roy. Soc. London B* **249**, 125–132.

Gamerman, D. (1997). *Markov Chain Monte Carlo: Stochastic Simulation for Bayesian Inference.* Chapman and Hall, London.

Gilks, W. R., S. Richardson and D. J. Spiegelhalter (Eds.) (1996). *Markov Chain Monte Carlo in Practice.* Chapman and Hall, London.

Guttorp, P. (1995). *Stochastic Modeling of Scientific Data.* Chapman and Hall, London.

Hamill, O. P., A. Marty, E. Neher, B. Sakmann and F. J. Sigworth (1981). Improved patch-clamp techniques for high-resolution current recording from cells and cell-free membrane patches. *Pflüg. Arch. Eur. J. Physiol.* **391**, 85–100.

Hawkes, A. G., A. Jalali and D. Colquhoun (1990). The distributions of the apparent open times and shut times in a single channel record when brief events cannot be detected. *Phil. Trans. Roy. Soc. London A* **332**, 511–538.

Hawkes, A. G., A. Jalali and D. Colquhoun (1992). Asymptotic distributions of apparent open times and shut times in a single channel record allowing for the omission of brief events. *Phil. Trans. Roy. Soc. London B* **337**, 383–404.

Hille, B. (1992). *Ionic Currents of Excitable Membranes*, 2nd ed. Sinauer, Sunderland, MA.

Hodgson, M. E. A. (1999). A Bayesian restoration of an ion channel signal. *J. Roy. Statist. Soc. B* **61**, 95–114.

Hodgson, M. E. A. and P. J. Green (1999). Bayesian choice among Markov models of ion channels using Markov chain Monte carlo. *Proc. Roy. Soc. London A* **455**, 3425–3448.

Jentsch, T. J., T. Friedrich, A. Schriever and H. Yamada (1999). The ClC chloride channel family. *Pflügers Arch. Eur. J. Physiol.* **437**, 783–795.

Johnson, N. L, S. Kotz and A. W. Kemp (1992). *Univariate Discrete Distributions*, 2nd ed. Wiley, New York.

Karlin, S. and H. M. Taylor (1975). *A First Course in Stochastic Processes*, 2nd ed. Academic Press, New York.

Keleshian, A. M., R. O. Edeson, G. J. Liu and B. W. Madsen (2000). Evidence for cooperativity between nicotinic acetylcholine receptors in patch clamp records. *Biophys. J.* **78**, 1–120.

Kelly, F. P. (1979). *Reversibility and Stochastic Networks.* Wiley, Chichester, UK.

Kienker, P. (1989). Equivalence of aggregated Markov models of ion-channel gating. *Proc. Roy. Soc. London B* **236**, 269–309.

Kijima, S. and H. Kijima (1987). Statistical analysis of channel current from a membrane patch II. A stochastic theory of a multi-channel system in the steady-state. *J. Theor. Biol.* **128**, 435–455.

Klein, S., J. Timmer and J. Honerkamp (1997). Analysis of multichannel patch clamp recordings by hidden Markov models. *Biometrics* **53**, 870–884.

Labarca, P., J. A. Rice, D. R. Fredkin and M. Montal (1985). Kinetic analysis of channel gating: application to the cholinergic receptor channel and the chloride channel from *Torpedo californica. Biophys. J.* **47**, 469–478.

Le Dain, A. C., B. W. Madsen and R. O. Edeson (1991). Naltrexone modulation of nicotinic acetylcholine receptor activity. *J. Pharmacol. Exp. Ther.* **258**, 551–558.

Li Y., G. F. Yeo, R. K. Milne, B. W. Madsen and R. O. Edeson (2000). Burst properties of a supergated double-barrelled chloride ion channel. *Math. Biosci.* **166**, 23–44.

Madsen, B. W. and G. F. Yeo (2000). Markov modeling of allosteric drug effects on ion channels, with particular reference to neuronal nicotinic acetylcholine receptors. *Archiv. Biochem. Biophys.* **373**, 429–434.

Magleby, K. L. and D. S. Weiss (1990a). Identifying kinetic gating mechanisms for ion channels by using two-dimensional distributions of simulated dwell times. *Proc. Roy. Soc. London B* **241**, 220–228.

Magleby, K. L. and D. S. Weiss (1990b). Estimating kinetic parameters for single channels with simulation. A general method that resolves the missed event problem and accounts for noise. *Biophys. J.* **58**, 1411–1426.

McManus, O. B., D. S. Weiss, C. E. Spivak, A. L. Blatz and K. L. Magleby (1988). Fractal models are inadequate for the kinetics of four different ion channels. *Biophys. J.* **54**, 859–870.

Miller, C. (1982). Open-state substructure of single chloride channels from *Torpedo* electroplax. *Phil. Trans. Roy. Soc. London B* **299**, 401–411.

Miller, C. and M. M. White (1984). Dimeric structure of single chloride channels from *Torpedo* electroplax. *Proc. Natl. Acad. Sci. USA* **81**, 2772–2775.

Milne, R. K., R. O. Edeson and B. W. Madsen (1986). Stochastic modelling of a single ion channel: interdependence of burst length and number of openings per burst. *Proc. Roy. Soc. London B* **227**, 83–102.

Milne, R. K., G. F. Yeo, R. O. Edeson and B. W. Madsen (1988). Stochastic modelling of a single ion channel: an alternating renewal approach with application to limited time resolution. *Proc. Roy. Soc. London B* **233**, 247–292.

Neher, E. (1992). Ion channels for communication between and within cells. *Science* **256**, 498–502.

Neher, E. and B. Sakmann (1976). Single-channel currents recorded from membrane of denervated frog muscle fibres. *Nature* **260**, 799–802.

Neuts, M. F. (1981). *Matrix-Geometric Solutions in Stochastic Models: An Algorithmic Approach*. Johns Hopkins University Press, Baltimore, MD.

Oakley, A. J., B. Martinac and M. C. J. Wilce (1999). Structure and function of the bacterial mechanosensitive channel of large conductance. *Protein Science* **8**, 1915–1921.

Qin, F., A. Auerbach and F. Sachs (1997). Maximum likelihood estimation of aggregated Markov processes. *Proc. Roy. Soc. London B* **264**, 375–383.

Qin, F., A. Auerbach and F. Sachs (2000a). A direct optimization approach to hidden Markov modeling for single channel kinetics. *Biophys. J.* **79**, 1915–1927.

Qin, F., A. Auerbach and F. Sachs (2000b). Hidden Markov modeling for single channel kinetics with filtering and correlated noise. *Biophys. J.* **79**, 1927–1944.

Ross, S. M. (1983). *Stochastic Processes*. Wiley, New York.

Roux B. and R. Sauvé (1985). A general solution to the time interval omission problem applied to single channel analysis. *Biophys. J.* **48**, 149–158.

Rubino, G. and B. Sericola (1989). Sojourn times in finite Markov processes. *J. Appl. Probab.* **26**, 744–756.

Sakmann, B. (1992). Elementary steps in synaptic transmission revealed by currents through single ion channels. *Science* **256**, 503–512.

Sakmann, B. and E. Neher (Eds.) (1995). *Single-Channel Recording*, 2nd ed. Plenum Press, New York. (1st edition, 1983.)

Schouten, J. G. (2000). Stochastic modeling of ion channel kinetics. Doctoral dissertation, Vrije Universiteit, Amsterdam.

Sericola, B. (1990). Closed-form solution for the distribution of total time spent in a subset of states of a homogeneous Markov process during a finite observation period. *J. Appl. Probab.* **27**, 713–719.

Sukharev, S.I., W. J. Sigurdson, C. Kung and F. Sachs (1999). Energetic and spatial parameters for gating of the bacterial large conductance mechanosensitive channel, MscL. *J. Gen. Physiol.* **113**, 525–539.

Yeo, G. F., R. O. Edeson, R. K. Milne and B. W. Madsen (1989). Superposition properties of independent ion channels. *Proc. Roy. Soc. London B* **238**, 155–170.

Yeo, G. F. and B. W. Madsen (1998). Modulatory drug action in an allosteric Markov model of ion channel behaviour: biphasic effects with access-limited binding to either a stimulatory or an inhibitory site. *Biochim. Biophys. Acta* **1372**, 37–44.

Yeo, G. F., R. K. Milne, R. O. Edeson and B. W. Madsen (1988). Statistical inference from single channel records: two-state Markov model with limited time resolution. *Proc. Roy. Soc. London B* **235**, 63–94.

Zheng, J. and F. J. Sigworth (1998). Intermediate conductances during deactivation of heteromultimeric *Shaker* potassium channels. *J. Gen. Physiol.* **112**, 457–474.

Zimmermann, H. (1993). *Synaptic Transmission: Cellular and Molecular Basis*. Georg Thieme Verlag, Stuttgart.

Subject Index

adjacent record values, 621
admissible control, 38
afterglows, 803
agglomerative hierarchical cluster analysis, 813
aggregated Markov chain, 934, 935
aggregation method, 546
Akaike's Information Criterion (AIC), 846
alphabet, 232
alternate probability generating function (a.p.g.f.), 579
annual maximum method, 607, 618
anti-persistent process, 98
antithetic variables, 345
aperiodic state, 537
ARCH model, 882, 907
Archimax distribution, 652
Archimedian distribution, 651, 652
astronomical point processes, 797
asymmetric mixed distribution, 663
asymmetry, 654, 656, 660–662
asymptotic efficiency, 207, 222
asymptotic results, 235
asymptotic variance, 343, 350, 352, 353, 365
asymptotic variance bound, 348, 355, 357
augmented Markov chain, 239
augmented matrix, 239, 240
auto-binomial, 490, 496
auto-logistic, 489
auto-normal, 490, 496
– interaction coefficients, 489
– interaction matrix, 491
autoregressive conditional heteroscedastic (ARCH) processes, 635
autoregressive moving average process, 875
autoregressive processes, 635
auxiliary variable algorithm, 339
availability, 551

batch Markovian arrival process (BMAP), 413, 423
BATSE, 801–803
BATSE catalog, 813
Bayes estimation, 504

Bayes factor, 810
Bayesian blocks, 808, 809, 820
Bayesian information criterion (BIC), 813, 846
Bayesian model, 810
Bernoulli lattice process, 805
Bernoulli trials, 231
Bernstein's theorem, 785, 786
Beta distribution, 659, 673
biextremal (α) distribution, 655
bilinear difference, 849
bilinear equation, 856
bilinear model (s), 827, 834–838, 841, 842, 855–857
bilinear model with fractional differenced Gaussian noise, 856
bilinear process (es), 835, 840, 857, 865
bilinear process in continuous time, 861
bilinear realizable Hermite-degree 2, 858
bilinear realizable process, 857
bilinear time series, 847
bilinear time series models, 827
bilogistic distribution, 656, 657, 659
binomial distribution, 779, 781, 789
binomial random variables, 787
biphasic drug effect, 934, 955, 957
birth and death (B–D) queues, 562
birth and death process, 729, 730, 739, 747
bispectral density function, 830, 831, 834
bispectrum, 851, 856, 859–861, 864, 866
bivariate extreme value distribution (s), 607, 640, 652, 654, 657, 659
bivariate slowly varying function, 642
Blackwell type theorem, 540
BMAP, 423–425
BMAP/G/1 queue, 423, 424
box counting dimension, 375
branching process (es), 693, 786
– bisexual, 699
– catastrophes and emigration, 742
– epidemic models, 728
– extinction of surnames, 698
– general, 727, 730, 734, 737
– infinite alleles, 724

– iterated, 720
– Markov, 711, 717, 740, 742
– multitype, 708, 714, 717, 731
– random environments, 740
– size dependence, 754
– spatial, 721
– varying environments, 712
– with immigration, 711
Brownian bridge, 183
Brownian motion, 389, 394, 397, 854
buffer content (s), 250, 255, 261
buffer content process (es), 243, 250, 255, 261, 263, 264, 276, 278
burst, 933, 934
– empirical, 933, 934, 945, 950, 954–957, 959, 960, 964
– theoretical, 933, 934, 939, 950, 952, 959
burst duration, 942, 947

canonical correlation (s), 646, 841
canonical correlation analysis, 840, 841
canonical factor analysis, 841
canonical gradient, 346
canonical potential, 488
canonical series expansion, 646
canonical variables, 646
central limit theorem, 541
chain binomial model (s), 285, 286, 292, 293, 323
– Greenwood model, 286–288
– Reed–Frost model, 286–288
channel
– acetylcholine, 934, 949, 955
– chloride, 958
– ion, 931
chaos, 833
chaos game, 388
chaotic and non-chaotic time series, 830
chaotic attractor, 819
chaotic spectral representation, 864
characterization of queueing systems, 557, 564
characterization of $M/G/\infty$, 564
characterizations, 777, 782, 783
charge transfer, 961
Chernoff upper bound, 57
chloride channel, 934, 958
chromatic number, 62, 80, 86
circular distribution, 658
class sojourn time, 932, 937
clique, 482
– for irregular sites, 483
– for regular sites, 483
– type of, 483
clique potential, 485, 489
– for auto-normal model, 491

– for MLL, 492
closed-time, 939
– total, 943, 953
cluster analysis, 68, 77
coefficient of tail dependence, 643
cointegration, 827, 847
collective model of epidemics, 735
coloring, 477
componentwise maxima, 635, 636, 657
compound geometric, 570
compound negative binomial, 570
Compton Gamma-Ray Observatory, 802, 803
concentration inequalities, 57
conditional probability, 486
configuration, 477, 483
configuration space, 477
– size of, 477
consolidation method, 546
contextual constraint (s), 473, 479, 489
continuous flow, 1
continuous version/analog of bilinear time series, 861
continuous-time Markov process, 1
control, 288, 308–312
control policies, 8
copulas, 648
corrective maintenance, 36
correctness, 502
correlation, 233
countable state space SMP, 532
counting process, 800
covariance function, 95, 98
covariance ordering, 344
coverage problem, 146
criticality theorem, 695
cumulant (s), 827–830, 838–840, 843, 859
cumulant generating function, 829
cumulant matrix, 843
cumulant spectra, 830
cumulant spectral density, 830, 840
cumulative total time on test statistic, 169, 173

damage models, 775–777, 782
Darling–Erdös limit theorem, 631
data augmentation, 339
de Finetti's theorem, 777
dead time, 805, 820
departure process (es), 557, 563
dependence function, 650
detailed balance, 338
deterministic sweep, 340, 352, 355
differential equation method, 58
diffusion approximation (s), 706, 710, 759
diffusion equation, 394

diffusion model, 749
diffusion process, 634
directed graphs, 79
directly Riemann integrable function, 541
Dirichlet distribution, 673
disadvantageous mutation, 710
discontinuities, 496
discrete branching processes, 785
discrete minification process, 591
discrete self-decomposability, 579
discrete wavelet transformation (DWT), 104
discretized fractional Brownian motion, 854
distribution
– matrix exponential, 938
– matrix geometric, 941
– phase, 939, 941
distributions closed under margins, 674
distributions of order k, 432
DNA replication, 137
DNA sequences, 234, 238
DNA sequencing, 240
DNA strands, 237
domain (s) of attraction, 608, 610, 611, 614–616, 623, 624, 634, 635, 637, 643, 646, 652, 674
domination off the diagonal, 344
dynamic GLM, 599
dynamical system, 757

eclipses, 796
ecology and conservation modelling, 738
efficiency ordering, 344
eigenvalues, 231
EM algorithm, 887
embedded Markov chains, 238
empirical burst, 954–957, 959, 960
empirical estimator, 337, 343, 352, 353, 363, 365
empirical integrated lack-of-memory process, 168, 171
empirical processes, 182
energy, 485
– order of, 489
entry process, 936
environmental uncertainty, 747, 749
epidemic modelling, 285, 728
Erdös–Révész type law of the iterated logarithm, 629
Erdös–Rényi, 52
Erlang process, 805
estimator of the semi-Markov kernel, 553
Ethernet network traffic measurements, 122
Euler distribution, 779, 780
evolution of diversity, 724
evolutionarily stable strategy, 64
evolutionary process, 850
evolutionary spectrum, 850

expected running time of algorithms, 86
explosion time, 526
exponent measure function, 638, 641, 674, 676, 677
exponential, 168, 170
exponential distribution, 620
extended Kalman filter, 871
extended neighboring order statistics, 621
extended Spitzer integral representation theorem, 786, 787
extinction, 738
extinction phenomena, 739, 740
extinction probability, 705, 739
extremal coefficient, 662
extremal types theorem, 607, 608, 635
extreme value theory, 607, 608, 625, 632–635, 664, 677

(**f**, **g**)-process, 542, 543
Fibonacci number (s), 230, 234
filtering, 874, 883, 885, 888, 891, 894, 897, 904, 919, 924
final data, 878
finitely ramified, 399
finitely ramified fractal, 392
first moment method, 57
first recurrence times, 229, 230
first-order logic, 69
first-order property, 53
first-order time dependent bilinear model, 851
Fisher information matrix, 619, 622, 624, 655, 667, 668
fixed-interval smoothing, 874
fixed-lag smoothing, 874
fixed-point smoothing, 874
FKG inequality, 59, 70, 72
flexible manufacturing systems, 1
fluid, 250
fluid model (s), 245, 250, 254, 261, 267, 275, 281–283
fluid queue, 278
fluid queueing models, 243
fluid-flow, 274
fluid-flow model (s), 245, 274
fluid-flow system, 274
fluid-flow traffic models, 245
Fréchet, 609, 614, 618, 636, 637, 639, 642, 649, 652, 665, 670, 674
fractal diffusion, 396, 402
fractal dimension, 101, 102
fractional (Gaussian) processes, 853
fractional Brownian motion, 390, 630, 853–855
fractional Brownian motion process, 855
fractional differenced, 855

fractional differenced Gaussian ARMA, 855
fractional Gaussian ARMA, 855
fractional Gaussian noise, 853
fractional white noise input, 857
FRAME model
– definition, 497
functionals of a semi-Markov process, 543

$G(n, p(n))$, 52
gamma ray burst (s), 796, 797, 802
gamma-ray, 796
gamma-ray observatory, 801, 804
gap, 939
Gate matrix layout problems, 75
gateway state, 937
Gaussian distribution, 657, 658
Gaussian elimination, 416, 417
Gaussian MRF (GMRF), *see* auto-normal
Gaussian Ornstein–Uhlenbeck process, 861
Gaussian process (es), 625, 629, 677
Gaussianity, 834
gene amplification, 714
general epidemic model, 729
generalized extreme value, 608, 618, 619, 624, 664, 669
generalized fARIMA, 97
generalized fractional Brownian motion, 96
generalized fractionally integrated noise, 97
generalized order statistics, 612
generalized Pareto, 608, 620–622, 624, 664
generalized Pareto distribution, 607
generalized Poisson distribution, 780
generalized Polya–Eggenberger distribution, 780
generalized Rao–Rubin condition, 786
generating function (s) (gf), 229, 412, 423
genetics and evolution, 703
geographical spread of mutants, 721
geometric, 168, 170
geometric random sum, 952
$GI/G/s$ queue, 561, 562
$GI/G/1/0$ system, 562
$GI/G/1/L$ system, 563
$GI/G/s/L$ system, 564, 566
$GI/M/1$ paradigm, 419, 422
giant component, 60
Gibbs distribution, 485
Gibbs random field, 485
– hierarchical, 493, 496
– homogeneous, 485
– isotropic, 485
Gibbs sampler, 340, 352, 353, 355, 357, 362, 365
Gibbs sampling, 901, 922
goodness-of-fit, 168
gradient, 346

graph, 51
graph directed constructions, 382
gravity, 796
Greenwood model, 323
growth of DNA repeats, 720
GTH algorithm, 416–418
Gumbel, 609, 610, 615, 616, 618, 633
Gumbel distribution, 655

H–R diagram, 812
Hamilton–Jacobi–Bellman equation, 7
Hamilton–Jacobi–Isac equation, 39
Hammersley–Clifford theorem, 486
Hastings algorithm, 338
Hausdorff dimension, 101, 102, 374
Hausdorff measure, 374
hedging policy, 16
hedging surface, 30
Heine distribution (s), 779, 780, 784
Hellinger differentiable, 346
hepatitis A, 301, 305, 327–328
Hermite degree, 857, 858
Hermite expansion, 837
Hermite functions, 835
Hermite polynomial (s), 835, 836, 858
Hermite polynomial expansion, 836
Hermite-degree n, 858
heterogeneity, 802
hidden Markov model, 963, 964
hierarchical clustering, 813
hierarchical MRF model, 493, 496
hierarchy scheduling planning, 2
higher cumulant spectra, 830
higher-order canonical correlation, 842
higher-order cumulant spectra, 827, 830
higher-order cumulants, 827–829, 838, 842, 846, 851
higher-order moments, 827, 828, 837, 838, 842, 843, 848, 850
higher-order spectra, 827, 830, 861
higher-order spectral density, 830
higher-order transfer, 837
higher-order transfer functions, 836
HIV/AIDS, 290, 294, 312, 314–321
Hölder exponent (s), 386, 389, 391
homogeneous, 486, 490, 797, 802
homogeneous Poisson process, 807
Hurst coefficient, 852, 861

identifiability, 963
image compression, 387
importance sampling, 890, 920
increasing property, 53
increment process, 93, 95

independence, 830
independence Hastings algorithm, 339
independence number, 62, 80
infinite alleles models of mutation, 724
infinite divisibility property, 564
integral statistics, 172, 196
integrated Cauchy functional equation, 777, 785
intensity function, 807
invariant measure, 537
inversion principle, 640
ion channel, 931, 934
irreducible, 540
irreducible MRP, 538, 539
irreducible positive recurrent, 538
Ising model, 490
– generalized, 491
Island Biogeography, 744
iterated function system, 381

Janson inequality, 59
Jensen's algorithm, 414, 415
jump Markov process, 528
jump process, 525

K-dependent Markov chain, 550
K-dependent semi-Markov kernel, 550
K-dependent semi-Markov processes, 549
Kalman filter, 871, 920
Kaluza sequences, 570
Kendall's coefficient of concordance, 667
Key renewal type theorem, 540
Kolmogorov weighted semimetric, 614
Komlós–Major–Tusnády theorem, 182
Kronecker product, 837

Lévy process, 390
label set
– continuity, 476, 483
– continuous, 476
– discrete, 476
– real, 476
labeling of sites, 477
labeling problem, 475, 476, 480
– categories LP1–LP4, 477
– categorization, 477
– under contextual constraint, 479
labelled graphs, 51
Laha–Lukacs property, 570
Laplace operator, 394, 395
Lau–Rao theorem, 168, 562
law of large number, 541
law of the iterated logarithm, 101
layout problem, 67
least squares, 500

life history of infectives, 730
likelihood function, 504, 964
limit distribution, 541
limiting conditional distribution, 755, 761
linear and quadratic transfer functions, 836
linear canonical correlation, 841
linear integral statistics, 196
linear models, 833
linear-cost-adjusted sequence, 634
linearity of the nonstationary time series, 851
linearity of time series, 834
local asymptotic normality, 346
local dimensions, 102
local log-likelihood, 851
local scaling exponent function, 95
locally Lipschitz, 23
locally self-similar process, 95
logarithmic reduction algorithm, 427
logistic distribution (s), 654–656, 663, 665–667, 669, 670
long memory, 828
long memory bilinear model, 827
long range dependence, 93, 96, 852–854, 861,
long range dependence property, 855
long range dependency in the nonlinear models, 855
long range dependent, 852, 853, 858, 859
longitudinal count data, 595, 602
lower triangular bilinear model (s), 835, 836, 846
luminosity-mass, 804
lumping method, 546

$M/E_k/1/L$ systems, 560
$M/G/1$ paradigm, 419, 421, 422
$M/G/1$ queue, 407
$M/G/1/0$ system, 562
$M/G/1/L$ queue, 559, 560, 561
$M/G/\infty$ queue, 558, 559, 561, 564, 566
$M/G/s$ queue, 561
$M/G/s/L$ system, 564
$M/M/s$ system, 558, 559, 566, 569
$MTRGG(n, \rho_n)$, 65
machine setp, 2
maintainability, 552
maintenance control, 30
manufacturing systems, 1
MAP–MRF framework, 474
mapping
– from sites to labels, 477
– with continuous labels, 477
– with discrete labels, 477
marginal distributions
– binary, 588, 593–596, 601
– binomial, 587, 594, 598, 601

– compound Poisson, 583
– geometric, 583
– multinomial, 594
– negative binomial, 585, 598
– Poisson, 580, 581, 594–596, 598, 600
– unspecified, 589
marked point processes, 819
marked process, 819
Markov and related process (es), 625, 632
Markov chain (s), 237, 238, 413, 519, 542, 547, 575, 932, 934, 935
– aggregated, 934, 935
– continuous-time, 932, 935
– reversible, 936, 965
– steady-state probabilities, 415
– transient probabilities, 414
Markov chain embedding, 432, 434, 435, 443
Markov chain method, 239
Markov chain Monte Carlo (MCMC), 499, 901, 964
Markov kernel, 517
Markov process, 484, 516, 518, 533, 674
Markov property, 530
Markov random field, 479, 483
– looks, 475
– coupled, 484
– history, 473
– homogeneous, 484
– in image analysis, 474
– isotropic, 484
– positivity, 483
Markov renewal equation, 528–530, 540
Markov renewal function, 529
Markov renewal matrix, 537, 553
Markov renewal process, 518, 546, 563
Markov renewal theory, 529
Markov switching model, 881
Markov transition function, 517
Markov–Gibbs equivalence, 486
Markov-modulated Poisson process (MMPP), 413
Markovianity, 483
martingale, 58
martingale approximation, 342
martingale methods, 235, 324–326
martingale stopping times, 236
martingale techniques, 238
matrix-geometric distribution, 420
max-stable, 609, 649, 652, 656
maximal correlation coefficient, 840
maximum a posteriori, 474, 505
maximum degree, 66, 72, 80
maximum entropy, 497, 498, 504
maximum likelihood, 503, 505
maximum likelihood estimation, 963
MCMC sampler, 338

mean time to failure, 552
measure function, 640, 641
measurement equation, 874
measures of connectivity, 62
merging algorithm, 547
merging method, 546
method of composition, 640
Metropolis algorithm, 338
Metropolis–Hastings algorithm, 901, 922
microsecond, 804
minimal Markov process, 543
minimal solution, 543, 544
minimum degree, 80
minimum description length, 504
minimum spanning tree, 58, 67
Minkowski dimension, 376
mixed gamma distribution, 667
mixture transition distribution (MTD), 575
ML, *see* maximum likelihood
MLL, *see* multi-level logistic
MMPP, 416, 418, 425
modeling
– geometric, 501
– photometric, 501
models
– based on thinning, 578
– – binomial thinning, 578
– – generalizations of binomial thinning, 590
– – hypergeometric thinning, 587, 591
– – multinomial thinning, 582
– Bayesian, 597, 599
– DARMA, 576
– dynamic GLM, 598, 599
– generalized linear (GLM), 602
– hidden Markov, 597
– hierarchical GLM, 599
– Markov chain, 593–595
– Markov regression, 596
– regression, 594
– state space, 597
modified Rao–Rubin conditions, 782
molecular clock of evolution, 727
Monte Carlo integration, 890, 920
moving average process, 625, 626
MRF–GRF equivalence, *see* Markov–Gibbs equivalence
multi-level logistic, 491
– conditional probability of, 493
– multiple-site clique potential, 492
– pair-site clique potential, 492
– single-site clique potential, 492
multi-resolution computation, 497
multifractal, 377, 399

Subject index

multifractal analysis, 378
multifractal spectrum, 382
multivariate beta distribution, 667
multivariate bilinear models, 827, 836, 845
multivariate bilinear time series, 843
multivariate bilinear time series models, 838
multivariate extension, 833
multivariate extreme value distribution (s), 607, 635, 646, 649, 652, 654, 663, 670, 674
multivariate nonlinear time series, 843
multivariate normal, 829, 842, 846
multivariate time series, 827

N-body dynamics, 813
nature reserves, 744
nearest neighbor interactions, 360, 362
negative bilogistic distribution, 657
negative binomial distribution, 780, 789
negative logistic distribution, 656, 673
neighbor set, 480
neighborhood
– nearest, 479, 481
– shape of, 481
neighborhood system, 476, 480
– 4-neighborhood, 480
– 8-neighborhood, 481
– nearest, 480
– order of, 481
nested fractal, 393, 396
nested logistic distribution, 670, 671
Nile River water levels, 118
non-Gaussian, 829, 830
non-homogeneous $J-X$ process, 545
non-homogeneous Markov chain, 545
non-homogeneous Markov kernel, 545
non-homogeneous Markov kernel Q, 544
non-homogeneous Markov renewal process (es), 544, 545
non-homogeneous semi-Markov process, 544, 545
non-linear models, 827, 834, 838, 846, 847, 849
non-recursive algorithm, 884, 923
non-stationary time series, 625, 630
nonexchangeability, 654
nonexchangeable, 659
nonhomogeneous Poisson process, 559
nonlinear, 830, 847–850
nonlinear canonical correlation, 841
nonlinear canonical correlation analysis, 840, 841
nonlinear dynamical systems, 830
nonlinear predictors, 847
nonlinear processes, 851
nonlinear system, 836
nonlinear time series, 827, 843, 847, 851
nonlinear time series models, 835

nonlinear, long range memory, 851
nonlinear, nonstationary models, 851
nonlinearity, 827, 847, 848, 850
nonnegative primitive matrix, 783
nonstationarity, 850
nonstationary, 847–849
nonstationary bilinear models, 850
nonstationary models, 827
nonstationary process, 850
normal approximation, 59
normal SMP, 529, 537
normalized clique potential, 488
nuclear reactor noise, 833
null recurrent, 538, 539
number of copies of a fixed graph, 61
numerical integration, 887
numerical inversion of generating functions, 412
numerical inversion of Laplace transforms, 410

objective function, 474
Okazaki fragments, 154
open set condition, 381, 383
open-time, 937, 939
– total, 942, 943, 947, 953, 954
optimal control problem, 7
optimal cost function, 10
optimization-based approach, 500
ordering
– of labels, 476
– of sites, 476
original distribution (s), 776, 777
Ornstein–Uhlenbeck process (es), 864, 678
output process (es), 558–561, 563

p-max stable, 610
p-variate normal distribution, 652
packing dimension, 375
parameter estimation, 321, 322
partial differential equation, 391
partition function, 485
patch clamp, 931
pattern of ESSs, 64
pattern (s), 227, 232, 236, 237, 431, 485
– first occurrence, 452
– number of, 440, 448, 452
penultimate approximation (s), 616, 653
percolation model, 385, 392
periodic state, 537
permutation matrices, 844
Perron–Frobenius eigenvalue, 783
Perron–Frobenius theorem, 782
persistent process, 98
perturbation expansion, 346

phase distributions, 939
phase transition, 60, 70, 79, 82
phase-type distributions, 425
piecewise deterministic processes, 4
planets, 795
point process characterization, 624, 637, 638, 641, 642
point processes, 795
point spread function, 816
Poisson approximation (s), 59, 237
Poisson distribution (s), 236, 663, 781, 785
Poisson process, 558, 559, 561, 564, 625, 631, 638, 639, 642, 796, 798, 806, 807, 816
Poisson random variable, 778, 787
policy iteration, 14
polynomial distribution (s), 659, 662
positive recurrent, 538, 539, 540
positive recurrent state, 537
positive stable distribution, 663
positivity, 483
power method, 416
power series distribution, 788, 790, 791
power study, 206
prediction, 874, 883, 885, 888, 890, 894, 919, 924
preliminary data, 878
preventive maintenance actions, 30
preventive maintenance control, 19
principal component analysis, 840, 841
prior
– smoothness, 493
probability of a Hamilton cycle, 61, 80, 84
probability of connectedness, 61
probability of extinction, 695
product of fractals, 379
production control, 19
projection of fractals, 379
proteins, 77
pseudo-likelihood, 964

QBD, 425, 426
QBD process, 419
quadratic integral statistics, 197
Quality-of-Service (QoS), 243, 244, 260, 262, 263, 266–268, 274, 280–282
quantile processes, 634
quasi birth-and-death process, 419, 425
quasi-filter, 904
quasi-renewal equation, 158
quasi-smoother, 904
queuing process, 625, 632

r-largest order statistics, 607, 623, 624
radiation, 796
random events, 227, 235, 237

random field, 477, 483
random geometric graph, 65
random graph process, 52
random graphs, 51
random graphs on the circle, 69
random graphs with prescribed degrees, 83
random hypergraphs, 82
random infinite graphs, 85
random intersection graph, 54
random interval graphs, 80
random lifts of graphs, 85
random numbers, 228
random planar graphs, 83
random randomly coloured graphs, 55, 72
random recursive fractals, 383
random regular graphs, 84
random subgraphs of the cube, 81
random sum
– geometric, 952, 953
random sweep, 340, 353, 357
random tree (s), 84, 383, 384
random triangle model, 71
random walk (s), 625, 631
random-cluster model, 54
Rao–Blackwellized empirical estimator, 348, 349
Rao–Rubin condition, 776
Rao–Rubin theorem, 786
Rao–Rubin(k) conditions, 783
rate (s) of convergence, 607, 612–617, 623, 624, 652, 653
Rayleigh processes, 679
record values, 620
recurrent, 539
recurrent events, 228, 230
recurrent state, 537
recursive algorithm, 883, 923
Reed–Frost, 63
Reed–Frost model, 292, 293, 734, 735
regeneration theory, 632
regression property, 570
regular semi-Markov process, 526, 527
regular SMP, 528, 537
regularly varying functions, 613
rejection method, 640
rejection sampling, 897, 921
reliability, 238, 239, 551, 552, 933, 935
reliability classification of distributions, 176
reliability estimation, 553
reliability estimator, 554
reliability theory, 240
renewal equation, 156
renewal function, 537
renewal output process, 563

Subject index

renewal process, 536, 545, 634
renormalization map, 395
repetitions, 230
replication accuracy of molecular chains, 716
resampling, 893
residual life plot, 622
resistor network, 394
resource limitation, 754
restoration, 963
restricted graph processes, 78
restricted random graph processes, 52
return level (s), 613, 619, 622, 623
return period, 619, 625
reversibility, 932, 936
reversible, 965
RRC graph (s), 54, 55
runs, 227, 228, 238
– overlapping success, 461

scaling exponent, 93
scaling function, 95
seasonal adjustment model, 876
second moment method, 57
second-order covariances, 841
second-order spectral density, 828
second-order stationarity, 837
second-order stationary, 837
seismic signals, 835
selection effect, 817
self-decomposability, 579
self-similar nonstationary process, 633
self-similar process, 93
self-similar set (s), 380, 393
self-similarity parameter, 93
semi-Markov kernel, 516–518, 523–526, 528, 529, 532, 533, 536, 547
semi-Markov matrix, 532, 533
semi-Markov process, 516, 525, 526, 528, 529, 535, 540, 543
semi-Markov random walk, 545
sequences, 235
Shanbhag's lemma, 777, 782
Shanbhag's theorem, 782
shift operator, 533
shift space, 381, 383
short memory, 851
shot noise process, 633
simulation, 285, 388, 391
site
– image lattice, 475
– regular/irregular, 475–483
slice sampler, 339
slices of fractals, 379, 380
SLOSS debate, 744, 753

small world graphs, 82
smoothing, 874, 884, 886, 889, 891, 895, 898, 902, 905, 920, 925
smoothness, 493
smoothness term
– membrane, 494
– plate, 495
– rod, 495
– string, 494
soil, 401
space-time, 795, 799, 819, 820
spacings, 173
spatial models, 301
– diffusion processes, 306–308
– percolation processes, 301–304, 308, 310
spatial point processes, 797
spatio-temporal discrete variate process, 602
spectral density function, 831
spectral dimension, 392, 398
spectral measure, 343
spectral representation, 652
spectrum, 96, 343
spiral arms of galaxies, 795
spurious regression, 827, 847
stable process, 390
standard normal distribution, 657, 677
stars, 795
state-space model, 874, 919
stationary bilinear models, 851
stationary measures, 786
stellar eclipses, 795
step function, 800
stochastic chemistry, 138
stochastic models, 137
stochastic processes, 285
stochastic variance models, 882
stochastic volatility model, 882, 907
strange attractors, 819
Strauss process, *see* multi-level logistic
string overlaps, 232
string (s), 232, 233, 237
strong Markov property, 536
strong unimodality, 568
sub-Markov kernel, 517
subdifferential, 24
subgraph problem in RIC graphs, 74
substitution sampler, 339
success run (s), 433, 458
– multivariate distributions, 457
– non-overlapping, 432, 436, 449, 458
– of fixed length, 432, 449
– of length at least k, 432, 439, 463
– of length exactly k, 432, 440, 451

– overlapping, 432, 438
successive approximation, 14
superdifferential, 24
supernova, 797
survival distribution (s), 775, 776, 781
survival of genes, 703

taboo distributions, 535
taboo transition probabilities, 535
Talagrand's method, 58
telomere sequences, 718
temperature, 485
temporal point process, 795
tests for Gaussianity and linearity, 834
theoretical burst, 952, 959
third-order covariances, 841
third-order moments, 842
threshold for k-GML, 77
threshold theorem, 291–296, 305, 307
thresholds, 52
time dependent coefficients, 851
time dependent parameters, 851
time series logistic distribution, 674
time varying parameter model, 875
Toeplitz matrix, 419
total closed-time, 943, 953
total dependence, 636, 639, 655–662
total independence, 635, 636, 639, 655–662
total open-time, 943, 953, 954
total time on test empirical process, 169, 178
total time on test statistics, 173
total time on test transform, 178
total variation metric, 613, 653
transient, 539
transition density, 398, 400, 401
transition equation, 874
transition kernel, 398
transition probability function, 518, 519
transposable elements, 713
triangular bilinear model, 837
trispectral density function, 830, 834
trivariate extreme value distribution, 608
Tutte polynomial, 70
two level logistic distribution (s), 669, 670

uniform distribution, 671
uniform metric, 613, 653
uniformity test, 174
uniformization procedure, 414
unit root bilinear model (s), 849
unlabelled graphs, 84

variance function, 99
variation in biology, 137
variogram, 95, 99
vector bilinear model, 836
vertical ocean shear measurements, 117
veterinary epidemiology, 313, 326–327
viscosity solutions, 7
Volterra expansion, 835, 836, 851
Volterra expansion and bilinear models, 834
Volterra form, 836
Volterra kernels, 836
Volterra representation, 834, 835
von Mises circular distribution, 658
von Mises condition, 617

waiting times, 239
– distributions, 446, 461
– later, 461
– sooner, 461
walk dimension, 392, 398
wavelength, 795
wavelet methods, 851
wavelet (s), 386, 387
Weibull, 609, 610, 614, 618
weighted likelihood function, 851
when is $\kappa = \delta$?, 62, 66, 69, 81, 84
whp, 53
Wiener expansion, 835
Wiener process, 632, 864
Wiener–Itô representation, 835
Wiener–Itô spectral measure, 864
Wiener–Itô stochastic integral, 837
Wiener–Itô stochastic spectral representation, 836

x-ray binary stars, 796

Yule–Walker equations, 838, 841
Yule–Walker-type difference equations, 838, 846

Handbook of Statistics
Contents of Previous Volumes

Volume 1. Analysis of Variance
Edited by P. R. Krishnaiah
1980 xviii + 1002 pp.

1. Estimation of Variance Components by C. R. Rao and J. Kleffe
2. Multivariate Analysis of Variance of Repeated Measurements by N. H. Timm
3. Growth Curve Analysis by S. Geisser
4. Bayesian Inference in MANOVA by S. J. Press
5. Graphical Methods for Internal Comparisons in ANOVA and MANOVA by R. Gnanadesikan
6. Monotonicity and Unbiasedness Properties of ANOVA and MANOVA Tests by S. Das Gupta
7. Robustness of ANOVA and MANOVA Test Procedures by P. K. Ito
8. Analysis of Variance and Problems under Time Series Models by D. R. Brillinger
9. Tests of Univariate and Multivariate Normality by K. V. Mardia
10. Transformations to Normality by G. Kaskey, B. Kolman, P. R. Krishnaiah and L. Steinberg
11. ANOVA and MANOVA: Models for Categorical Data by V. P. Bhapkar
12. Inference and the Structural Model for ANOVA and MANOVA by D. A. S. Fraser
13. Inference Based on Conditionally Specified ANOVA Models Incorporating Preliminary Testing by T. A. Bancroft and C.-P. Han
14. Quadratic Forms in Normal Variables by C. G. Khatri
15. Generalized Inverse of Matrices and Applications to Linear Models by S. K. Mitra
16. Likelihood Ratio Tests for Mean Vectors and Covariance Matrices by P. R. Krishnaiah and J. C. Lee
17. Assessing Dimensionality in Multivariate Regression by A. J. Izenman
18. Parameter Estimation in Nonlinear Regression Models by H. Bunke
19. Early History of Multiple Comparison Tests by H. L. Harter
20. Representations of Simultaneous Pairwise Comparisons by A. R. Sampson
21. Simultaneous Test Procedures for Mean Vectors and Covariance Matrices by P. R. Krishnaiah, G. S. Mudholkar and P. Subbaiah

22. Nonparametric Simultaneous Inference for Some MANOVA Models by P. K. Sen
23. Comparison of Some Computer Programs for Univariate and Multivariate Analysis of Variance by R. D. Bock and D. Brandt
24. Computations of Some Multivariate Distributions by P. R. Krishnaiah
25. Inference on the Structure of Interaction Two-Way Classification Model by P. R. Krishnaiah and M. Yochmowitz

Volume 2. Classification, Pattern Recognition and Reduction of Dimensionality
Edited by P. R. Krishnaiah and L. N. Kanal
1982 xxii + 903 pp.

1. Discriminant Analysis for Time Series by R. H. Shumway
2. Optimum Rules for Classification into Two Multivariate Normal Populations with the Same Covariance Matrix by S. Das Gupta
3. Large Sample Approximations and Asymptotic Expansions of Classification Statistics by M. Siotani
4. Bayesian Discrimination by S. Geisser
5. Classification of Growth Curves by J. C. Lee
6. Nonparametric Classification by J. D. Broffitt
7. Logistic Discrimination by J. A. Anderson
8. Nearest Neighbor Methods in Discrimination by L. Devroye and T. J. Wagner
9. The Classification and Mixture Maximum Likelihood Approaches to Cluster Analysis by G. J. McLachlan
10. Graphical Techniques for Multivariate Data and for Clustering by J. M. Chambers and B. Kleiner
11. Cluster Analysis Software by R. K. Blashfield, M. S. Aldenderfer and L. C. Morey
12. Single-link Clustering Algorithms by F. J. Rohlf
13. Theory of Multidimensional Scaling by J. de Leeuw and W. Heiser
14. Multidimensional Scaling and its Application by M. Wish and J. D. Carroll
15. Intrinsic Dimensionality Extraction by K. Fukunaga
16. Structural Methods in Image Analysis and Recognition by L. N. Kanal, B. A. Lambird and D. Lavine
17. Image Models by N. Ahuja and A. Rosenfield
18. Image Texture Survey by R. M. Haralick
19. Applications of Stochastic Languages by K. S. Fu
20. A Unifying Viewpoint on Pattern Recognition by J. C. Simon, E. Backer and J. Sallentin
21. Logical Functions in the Problems of Empirical Prediction by G. S. Lbov
22. Inference and Data Tables and Missing Values by N. G. Zagoruiko and V. N. Yolkina
23. Recognition of Electrocardiographic Patterns by J. H. van Bemmel

24. Waveform Parsing Systems by G. C. Stockman
25. Continuous Speech Recognition: Statistical Methods by F. Jelinek, R. L. Mercer and L. R. Bahl
26. Applications of Pattern Recognition in Radar by A. A. Grometstein and W. H. Schoendorf
27. White Blood Cell Recognition by F. S. Gelsema and G. H. Landweerd
28. Pattern Recognition Techniques for Remote Sensing Applications by P. H. Swain
29. Optical Character Recognition – Theory and Practice by G. Nagy
30. Computer and Statistical Considerations for Oil Spill Identification by Y. T. Chien and T. J. Killeen
31. Pattern Recognition in Chemistry by B. R. Kowalski and S. Wold
32. Covariance Matrix Representation and Object-Predicate Symmetry by T. Kaminuma, S. Tomita and S. Watanabe
33. Multivariate Morphometrics by R. A. Reyment
34. Multivariate Analysis with Latent Variables by P. M. Bentler and D. G. Weeks
35. Use of Distance Measures, Information Measures and Error Bounds in Feature Evaluation by M. Ben-Bassat
36. Topics in Measurement Selection by J. M. Van Campenhout
37. Selection of Variables Under Univariate Regression Models by P. R. Krishnaiah
38. On the Selection of Variables Under Regression Models Using Krishnaiah's Finite Intersection Tests by J. L. Schmidhammer
39. Dimensionality and Sample Size Considerations in Pattern Recognition Practice by A. K. Jain and B. Chandrasekaran
40. Selecting Variables in Discriminant Analysis for Improving upon Classical Procedures by W. Schaafsma
41. Selection of Variables in Discriminant Analysis by P. R. Krishnaiah

Volume 3. Time Series in the Frequency Domain
Edited by D. R. Brillinger and P. R. Krishnaiah
1983 xiv + 485 pp.

1. Wiener Filtering (with emphasis on frequency-domain approaches) by R. J. Bhansali and D. Karavellas
2. The Finite Fourier Transform of a Stationary Process by D. R. Brillinger
3. Seasonal and Calendar Adjustment by W. S. Cleveland
4. Optimal Inference in the Frequency Domain by R. B. Davies
5. Applications of Spectral Analysis in Econometrics by C. W. J. Granger and R. Engle
6. Signal Estimation by E. J. Hannan
7. Complex Demodulation: Some Theory and Applications by T. Hasan

8. Estimating the Gain of a Linear Filter from Noisy Data by M. J. Hinich
9. A Spectral Analysis Primer by L. H. Koopmans
10. Robust-Resistant Spectral Analysis by R. D. Martin
11. Autoregressive Spectral Estimation by E. Parzen
12. Threshold Autoregression and Some Frequency-Domain Characteristics by J. Pemberton and H. Tong
13. The Frequency-Domain Approach to the Analysis of Closed-Loop Systems by M. B. Priestley
14. The Bispectral Analysis of Nonlinear Stationary Time Series with Reference to Bilinear Time-Series Models by T. Subba Rao
15. Frequency-Domain Analysis of Multidimensional Time-Series Data by E. A. Robinson
16. Review of Various Approaches to Power Spectrum Estimation by P. M. Robinson
17. Cumulants and Cumulant Spectra by M. Rosenblatt
18. Replicated Time-Series Regression: An Approach to Signal Estimation and Detection by R. H. Shumway
19. Computer Programming of Spectrum Estimation by T. Thrall
20. Likelihood Ratio Tests on Covariance Matrices and Mean Vectors of Complex Multivariate Normal Populations and their Applications in Time Series by P. R. Krishnaiah, J. C. Lee and T. C. Chang

Volume 4. Nonparametric Methods
Edited by P. R. Krishnaiah and P. K. Sen
1984 xx + 968 pp.

1. Randomization Procedures by C. B. Bell and P. K. Sen
2. Univariate and Multivariate Multisample Location and Scale Tests by V. P. Bhapkar
3. Hypothesis of Symmetry by M. Hušková
4. Measures of Dependence by K. Joag-Dev
5. Tests of Randomness against Trend or Serial Correlations by G. K. Bhattacharyya
6. Combination of Independent Tests by J. L. Folks
7. Combinatorics by L. Takács
8. Rank Statistics and Limit Theorems by M. Ghosh
9. Asymptotic Comparison of Tests – A Review by K. Singh
10. Nonparametric Methods in Two-Way Layouts by D. Quade
11. Rank Tests in Linear Models by J. N. Adichie
12. On the Use of Rank Tests and Estimates in the Linear Model by J. C. Aubuchon and T. F. Hettmansperger
13. Nonparametric Preliminary Test Inference by A. K. Md. E. Saleh and P. K. Sen

14. Paired Comparisons: Some Basic Procedures and Examples by R. A. Bradley
15. Restricted Alternatives by S. K. Chatterjee
16. Adaptive Methods by M. Hušková
17. Order Statistics by J. Galambos
18. Induced Order Statistics: Theory and Applications by P. K. Bhattacharya
19. Empirical Distribution Function by F. Csáki
20. Invariance Principles for Empirical Processes by M. Csörgő
21. M-, L- and R-estimators by J. Jurečková
22. Nonparametric Sequential Estimation by P. K. Sen
23. Stochastic Approximation by V. Dupač
24. Density Estimation by P. Révész
25. Censored Data by A. P. Basu
26. Tests for Exponentiality by K. A. Doksum and B. S. Yandell
27. Nonparametric Concepts and Methods in Reliability by M. Hollander and F. Proschan
28. Sequential Nonparametric Tests by U. Müller-Funk
29. Nonparametric Procedures for some Miscellaneous Problems by P. K. Sen
30. Minimum Distance Procedures by R. Beran
31. Nonparametric Methods in Directional Data Analysis by S. R. Jammalamadaka
32. Application of Nonparametric Statistics to Cancer Data by H. S. Wieand
33. Nonparametric Frequentist Proposals for Monitoring Comparative Survival Studies by M. Gail
34. Meteorological Applications of Permutation Techniques Based on Distance Functions by P. W. Mielke, Jr.
35. Categorical Data Problems Using Information Theoretic Approach by S. Kullback and J. C. Keegel
36. Tables for Order Statistics by P. R. Krishnaiah and P. K. Sen
37. Selected Tables for Nonparametric Statistics by P. K. Sen and P. R. Krishnaiah

Volume 5. Time Series in the Time Domain
Edited by E. J. Hannan, P. R. Krishnaiah and M. M. Rao
1985 xiv + 490 pp.

1. Nonstationary Autoregressive Time Series by W. A. Fuller
2. Non-Linear Time Series Models and Dynamical Systems by T. Ozaki
3. Autoregressive Moving Average Models, Intervention Problems and Outlier Detection in Time Series by G. C. Tiao
4. Robustness in Time Series and Estimating ARMA Models by R. D. Martin and V. J. Yohai
5. Time Series Analysis with Unequally Spaced Data by R. H. Jones

6. Various Model Selection Techniques in Time Series Analysis by R. Shibata
7. Estimation of Parameters in Dynamical Systems by L. Ljung
8. Recursive Identification, Estimation and Control by P. Young
9. General Structure and Parametrization of ARMA and State-Space Systems and its Relation to Statistical Problems by M. Deistler
10. Harmonizable, Cramér, and Karhunen Classes of Processes by M. M. Rao
11. On Non-Stationary Time Series by C. S. K. Bhagavan
12. Harmonizable Filtering and Sampling of Time Series by D. K. Chang
13. Sampling Designs for Time Series by S. Cambanis
14. Measuring Attenuation by M. A. Cameron and P. J. Thomson
15. Speech Recognition Using LPC Distance Measures by P. J. Thomson and P. de Souza
16. Varying Coefficient Regression by D. F. Nicholls and A. R. Pagan
17. Small Samples and Large Equations Systems by H. Theil and D. G. Fiebig

Volume 6. Sampling
Edited by P. R. Krishnaiah and C. R. Rao
1988 xvi + 594 pp.

1. A Brief History of Random Sampling Methods by D. R. Bellhouse
2. A First Course in Survey Sampling by T. Dalenius
3. Optimality of Sampling Strategies by A. Chaudhuri
4. Simple Random Sampling by P. K. Pathak
5. On Single Stage Unequal Probability Sampling by V. P. Godambe and M. E. Thompson
6. Systematic Sampling by D. R. Bellhouse
7. Systematic Sampling with Illustrative Examples by M. N. Murthy and T. J. Rao
8. Sampling in Time by D. A. Binder and M. A. Hidiroglou
9. Bayesian Inference in Finite Populations by W. A. Ericson
10. Inference Based on Data from Complex Sample Designs by G. Nathan
11. Inference for Finite Population Quantiles by J. Sedransk and P. J. Smith
12. Asymptotics in Finite Population Sampling by P. K. Sen
13. The Technique of Replicated or Interpenetrating Samples by J. C. Koop
14. On the Use of Models in Sampling from Finite Populations by I. Thomsen and D. Tesfu
15. The Prediction Approach to Sampling Theory by R. M. Royall
16. Sample Survey Analysis: Analysis of Variance and Contingency Tables by D. H. Freeman, Jr.
17. Variance Estimation in Sample Surveys by J. N. K. Rao
18. Ratio and Regression Estimators by P. S. R. S. Rao

19. Role and Use of Composite Sampling and Capture-Recapture Sampling in Ecological Studies by M. T. Boswell, K. P. Burnham and G. P. Patil
20. Data-based Sampling and Model-based Estimation for Environmental Resources by G. P. Patil, G. J. Babu, R. C. Hennemuth, W. L. Meyers, M. B. Rajarshi and C. Taillie
21. On Transect Sampling to Assess Wildlife Populations and Marine Resources by F. L. Ramsey, C. E. Gates, G. P. Patil and C. Taillie
22. A Review of Current Survey Sampling Methods in Marketing Research (Telephone, Mall Intercept and Panel Surveys) by R. Velu and G. M. Naidu
23. Observational Errors in Behavioural Traits of Man and their Implications for Genetics by P. V. Sukhatme
24. Designs in Survey Sampling Avoiding Contiguous Units by A. S. Hedayat, C. R. Rao and J. Stufken

Volume 7. Quality Control and Reliability
Edited by P. R. Krishnaiah and C. R. Rao
1988 xiv + 503 pp.

1. Transformation of Western Style of Management by W. Edwards Deming
2. Software Reliability by F. B. Bastani and C. V. Ramamoorthy
3. Stress–Strength Models for Reliability by R. A. Johnson
4. Approximate Computation of Power Generating System Reliability Indexes by M. Mazumdar
5. Software Reliability Models by T. A. Mazzuchi and N. D. Singpurwalla
6. Dependence Notions in Reliability Theory by N. R. Chaganty and K. Joag-dev
7. Application of Goodness-of-Fit Tests in Reliability by B. W. Woodruff and A. H. Moore
8. Multivariate Nonparametric Classes in Reliability by H. W. Block and T. H. Savits
9. Selection and Ranking Procedures in Reliability Models by S. S. Gupta and S. Panchapakesan
10. The Impact of Reliability Theory on Some Branches of Mathematics and Statistics by P. J. Boland and F. Proschan
11. Reliability Ideas and Applications in Economics and Social Sciences by M. C. Bhattacharjee
12. Mean Residual Life: Theory and Applications by F. Guess and F. Proschan
13. Life Distribution Models and Incomplete Data by R. E. Barlow and F. Proschan
14. Piecewise Geometric Estimation of a Survival Function by G. M. Mimmack and F. Proschan
15. Applications of Pattern Recognition in Failure Diagnosis and Quality Control by L. F. Pau

16. Nonparametric Estimation of Density and Hazard Rate Functions when Samples are Censored by W. J. Padgett
17. Multivariate Process Control by F. B. Alt and N. D. Smith
18. QMP/USP – A Modern Approach to Statistical Quality Auditing by B. Hoadley
19. Review About Estimation of Change Points by P. R. Krishnaiah and B. Q. Miao
20. Nonparametric Methods for Changepoint Problems by M. Csörgő and L. Horváth
21. Optimal Allocation of Multistate Components by E. El-Neweihi, F. Proschan and J. Sethuraman
22. Weibull, Log-Weibull and Gamma Order Statistics by H. L. Herter
23. Multivariate Exponential Distributions and their Applications in Reliability by A. P. Basu
24. Recent Developments in the Inverse Gaussian Distribution by S. Iyengar and G. Patwardhan

Volume 8. Statistical Methods in Biological and Medical Sciences
Edited by C. R. Rao and R. Chakraborty
1991 xvi + 554 pp.

1. Methods for the Inheritance of Qualitative Traits by J. Rice, R. Neuman and S. O. Moldin
2. Ascertainment Biases and their Resolution in Biological Surveys by W. J. Ewens
3. Statistical Considerations in Applications of Path Analytical in Genetic Epidemiology by D. C. Rao
4. Statistical Methods for Linkage Analysis by G. M. Lathrop and J. M. Lalouel
5. Statistical Design and Analysis of Epidemiologic Studies: Some Directions of Current Research by N. Breslow
6. Robust Classification Procedures and their Applications to Anthropometry by N. Balakrishnan and R. S. Ambagaspitiya
7. Analysis of Population Structure: A Comparative Analysis of Different Estimators of Wright's Fixation Indices by R. Chakraborty and H. Danker-Hopfe
8. Estimation of Relationships from Genetic Data by E. A. Thompson
9. Measurement of Genetic Variation for Evolutionary Studies by R. Chakraborty and C. R. Rao
10. Statistical Methods for Phylogenetic Tree Reconstruction by N. Saitou
11. Statistical Models for Sex-Ratio Evolution by S. Lessard
12. Stochastic Models of Carcinogenesis by S. H. Moolgavkar
13. An Application of Score Methodology: Confidence Intervals and Tests of Fit for One-Hit-Curves by J. J. Gart
14. Kidney-Survival Analysis of IgA Nephropathy Patients: A Case Study by O. J. W. F. Kardaun

15. Confidence Bands and the Relation with Decision Analysis: Theory by O. J. W. F. Kardaun
16. Sample Size Determination in Clinical Research by J. Bock and H. Toutenburg

Volume 9. Computational Statistics
Edited by C. R. Rao
1993 xix + 1045 pp.

1. Algorithms by B. Kalyanasundaram
2. Steady State Analysis of Stochastic Systems by K. Kant
3. Parallel Computer Architectures by R. Krishnamurti and B. Narahari
4. Database Systems by S. Lanka and S. Pal
5. Programming Languages and Systems by S. Purushothaman and J. Seaman
6. Algorithms and Complexity for Markov Processes by R. Varadarajan
7. Mathematical Programming: A Computational Perspective by W. W. Hager, R. Horst and P. M. Pardalos
8. Integer Programming by P. M. Pardalos and Y. Li
9. Numerical Aspects of Solving Linear Least Squares Problems by J. L. Barlow
10. The Total Least Squares Problem by S. van Huffel and H. Zha
11. Construction of Reliable Maximum-Likelihood-Algorithms with Applications to Logistic and Cox Regression by D. Böhning
12. Nonparametric Function Estimation by T. Gasser, J. Engel and B. Seifert
13. Computation Using the OR Decomposition by C. R. Goodall
14. The EM Algorithm by N. Laird
15. Analysis of Ordered Categorial Data through Appropriate Scaling by C. R. Rao and P. M. Caligiuri
16. Statistical Applications of Artificial Intelligence by W. A. Gale, D. J. Hand and A. E. Kelly
17. Some Aspects of Natural Language Processes by A. K. Joshi
18. Gibbs Sampling by S. F. Arnold
19. Bootstrap Methodology by G. J. Babu and C. R. Rao
20. The Art of Computer Generation of Random Variables by M. T. Boswell, S. D. Gore, G. P. Patil and C. Taillie
21. Jackknife Variance Estimation and Bias Reduction by S. Das Peddada
22. Designing Effective Statistical Graphs by D. A. Burn
23. Graphical Methods for Linear Models by A. S. Hadi
24. Graphics for Time Series Analysis by H. J. Newton
25. Graphics as Visual Language by T. Selkar and A. Appel
26. Statistical Graphics and Visualization by E. J. Wegman and D. B. Carr

27. Multivariate Statistical Visualization by F. W. Young, R. A. Faldowski and M. M. McFarlane
28. Graphical Methods for Process Control by T. L. Ziemer

Volume 10. Signal Processing and its Applications
Edited by N. K. Bose and C. R. Rao
1993 xvii + 992 pp.

1. Signal Processing for Linear Instrumental Systems with Noise: A General Theory with Illustrations from Optical Imaging and Light Scattering Problems by M. Bertero and E. R. Pike
2. Boundary Implication Results in Parameter Space by N. K. Bose
3. Sampling of Bandlimited Signals: Fundamental Results and Some Extensions by J. L. Brown, Jr.
4. Localization of Sources in a Sector: Algorithms and Statistical Analysis by K. Buckley and X.-L. Xu
5. The Signal Subspace Direction-of-Arrival Algorithm by J. A. Cadzow
6. Digital Differentiators by S. C. Dutta Roy and B. Kumar
7. Orthogonal Decompositions of 2D Random Fields and their Applications for 2D Spectral Estimation by J. M. Francos
8. VLSI in Signal Processing by A. Ghouse
9. Constrained Beamforming and Adaptive Algorithms by L. C. Godara
10. Bispectral Speckle Interferometry to Reconstruct Extended Objects from Turbulence-Degraded Telescope Images by D. M. Goodman, T. W. Lawrence, E. M. Johansson and J. P. Fitch
11. Multi-Dimensional Signal Processing by K. Hirano and T. Nomura
12. On the Assessment of Visual Communication by F. O. Huck, C. L. Fales, R. Alter-Gartenberg and Z. Rahman
13. VLSI Implementations of Number Theoretic Concepts with Applications in Signal Processing by G. A. Jullien, N. M. Wigley and J. Reilly
14. Decision-level Neural Net Sensor Fusion by R. Y. Levine and T. S. Khuon
15. Statistical Algorithms for Noncausal Gauss Markov Fields by J. M. F. Moura and N. Balram
16. Subspace Methods for Directions-of-Arrival Estimation by A. Paulraj, B. Ottersten, R. Roy, A. Swindlehurst, G. Xu and T. Kailath
17. Closed Form Solution to the Estimates of Directions of Arrival Using Data from an Array of Sensors by C. R. Rao and B. Zhou
18. High-Resolution Direction Finding by S. V. Schell and W. A. Gardner
19. Multiscale Signal Processing Techniques: A Review by A. H. Tewfik, M. Kim and M. Deriche

20. Sampling Theorems and Wavelets by G. G. Walter
21. Image and Video Coding Research by J. W. Woods
22. Fast Algorithms for Structured Matrices in Signal Processing by A. E. Yagle

Volume 11. Econometrics
Edited by G. S. Maddala, C. R. Rao and H. D. Vinod
1993 xx + 783 pp.

1. Estimation from Endogenously Stratified Samples by S. R. Cosslett
2. Semiparametric and Nonparametric Estimation of Quantal Response Models by J. L. Horowitz
3. The Selection Problem in Econometrics and Statistics by C. F. Manski
4. General Nonparametric Regression Estimation and Testing in Econometrics by A. Ullah and H. D. Vinod
5. Simultaneous Microeconometric Models with Censored or Qualitative Dependent Variables by R. Blundell and R. J. Smith
6. Multivariate Tobit Models in Econometrics by L.-F. Lee
7. Estimation of Limited Dependent Variable Models under Rational Expectations by G. S. Maddala
8. Nonlinear Time Series and Macroeconometrics by W. A. Brock and S. M. Potter
9. Estimation, Inference and Forecasting of Time Series Subject to Changes in Time by J. D. Hamilton
10. Structural Time Series Models by A. C. Harvey and N. Shephard
11. Bayesian Testing and Testing Bayesians by J.-P. Florens and M. Mouchart
12. Pseudo-Likelihood Methods by C. Gourieroux and A. Monfort
13. Rao's Score Test: Recent Asymptotic Results by R. Mukerjee
14. On the Strong Consistency of M-Estimates in Linear Models under a General Discrepancy Function by Z. D. Bai, Z. J. Liu and C. R. Rao
15. Some Aspects of Generalized Method of Moments Estimation by A. Hall
16. Efficient Estimation of Models with Conditional Moment Restrictions by W. K. Newey
17. Generalized Method of Moments: Econometric Applications by M. Ogaki
18. Testing for Heteroscedasticity by A. R. Pagan and Y. Pak
19. Simulation Estimation Methods for Limited Dependent Variable Models by V. A. Hajivassiliou
20. Simulation Estimation for Panel Data Models with Limited Dependent Variable by M. P. Keane
21. A Perspective Application of Bootstrap Methods in Econometrics by J. Jeong and G. S. Maddala

22. Stochastic Simulations for Inference in Nonlinear Errors-in-Variables Models by R. S. Mariano and B. W. Brown
23. Bootstrap Methods: Applications in Econometrics by H. D. Vinod
24. Identifying Outliers and Influential Observations in Econometric Models by S. G. Donald and G. S. Maddala
25. Statistical Aspects of Calibration in Macroeconomics by A. W. Gregory and G. W. Smith
26. Panel Data Models with Rational Expectations by K. Lahiri
27. Continuous Time Financial Models: Statistical Applications of Stochastic Processes by K. R. Sawyer

Volume 12. Environmental Statistics
Edited by G. P. Patil and C. R. Rao
1994 xix + 927 pp.

1. Environmetrics: An Emerging Science by J. S. Hunter
2. A National Center for Statistical Ecology and Environmental Statistics: A Center Without Walls by G. P. Patil
3. Replicate Measurements for Data Quality and Environmental Modeling by W. Liggett
4. Design and Analysis of Composite Sampling Procedures: A Review by G. Lovison, S. D. Gore and G. P. Patil
5. Ranked Set Sampling by G. P. Patil, A. K. Sinha and C. Taillie
6. Environmental Adaptive Sampling by G. A. F. Seber and S. K. Thompson
7. Statistical Analysis of Censored Environmental Data by M. Akritas, T. Ruscitti and G. P. Patil
8. Biological Monitoring: Statistical Issues and Models by E. P. Smith
9. Environmental Sampling and Monitoring by S. V. Stehman and W. Scott Overton
10. Ecological Statistics by B. F. J. Manly
11. Forest Biometrics by H. E. Burkhart and T. G. Gregoire
12. Ecological Diversity and Forest Management by J. H. Gove, G. P. Patil, B. F. Swindel and C. Taillie
13. Ornithological Statistics by P. M. North
14. Statistical Methods in Developmental Toxicology by P. J. Catalano and L. M. Ryan
15. Environmental Biometry: Assessing Impacts of Environmental Stimuli Via Animal and Microbial Laboratory Studies by W. W. Piegorsch
16. Stochasticity in Deterministic Models by J. J. M. Bedaux and S. A. L. M. Kooijman
17. Compartmental Models of Ecological and Environmental Systems by J. H. Matis and T. E. Wehrly

18. Environmental Remote Sensing and Geographic Information Systems-Based Modeling by W. L. Myers
19. Regression Analysis of Spatially Correlated Data: The Kanawha County Health Study by C. A. Donnelly, J. H. Ware and N. M. Laird
20. Methods for Estimating Heterogeneous Spatial Covariance Functions with Environmental Applications by P. Guttorp and P. D. Sampson
21. Meta-analysis in Environmental Statistics by V. Hasselblad
22. Statistical Methods in Atmospheric Science by A. R. Solow
23. Statistics with Agricultural Pests and Environmental Impacts by L. J. Young and J. H. Young
24. A Crystal Cube for Coastal and Estuarine Degradation: Selection of Endpoints and Development of Indices for Use in Decision Making by M. T. Boswell, J. S. O'Connor and G. P. Patil
25. How Does Scientific Information in General and Statistical Information in Particular Input to the Environmental Regulatory Process? by C. R. Cothern
26. Environmental Regulatory Statistics by C. B. Davis
27. An Overview of Statistical Issues Related to Environmental Cleanup by R. Gilbert
28. Environmental Risk Estimation and Policy Decisions by H. Lacayo Jr.

Volume 13. Design and Analysis of Experiments
Edited by S. Ghosh and C. R. Rao
1996 xviii + 1230 pp.

1. The Design and Analysis of Clinical Trials by P. Armitage
2. Clinical Trials in Drug Development: Some Statistical Issues by H. I. Patel
3. Optimal Crossover Designs by J. Stufken
4. Design and Analysis of Experiments: Nonparametric Methods with Applications to Clinical Trials by P. K. Sen
5. Adaptive Designs for Parametric Models by S. Zacks
6. Observational Studies and Nonrandomized Experiments by P. R. Rosenbaum
7. Robust Design: Experiments for Improving Quality by D. M. Steinberg
8. Analysis of Location and Dispersion Effects from Factorial Experiments with a Circular Response by C. M. Anderson
9. Computer Experiments by J. R. Koehler and A. B. Owen
10. A Critique of Some Aspects of Experimental Design by J. N. Srivastava
11. Response Surface Designs by N. R. Draper and D. K. J. Lin
12. Multiresponse Surface Methodology by A. I. Khuri
13. Sequential Assembly of Fractions in Factorial Experiments by S. Ghosh
14. Designs for Nonlinear and Generalized Linear Models by A. C. Atkinson and L. M. Haines
15. Spatial Experimental Design by R. J. Martin

16. Design of Spatial Experiments: Model Fitting and Prediction by V. V. Fedorov
17. Design of Experiments with Selection and Ranking Goals by S. S. Gupta and S. Panchapakesan
18. Multiple Comparisons by A. C. Tamhane
19. Nonparametric Methods in Design and Analysis of Experiments by E. Brunner and M. L. Puri
20. Nonparametric Analysis of Experiments by A. M. Dean and D. A. Wolfe
21. Block and Other Designs in Agriculture by D. J. Street
22. Block Designs: Their Combinatorial and Statistical Properties by T. Calinski and S. Kageyama
23. Developments in Incomplete Block Designs for Parallel Line Bioassays by S. Gupta and R. Mukerjee
24. Row-Column Designs by K. R. Shah and B. K. Sinha
25. Nested Designs by J. P. Morgan
26. Optimal Design: Exact Theory by C. S. Cheng
27. Optimal and Efficient Treatment – Control Designs by D. Majumdar
28. Model Robust Designs by Y-J. Chang and W. I. Notz
29. Review of Optimal Bayes Designs by A. DasGupta
30. Approximate Designs for Polynomial Regression: Invariance, Admissibility, and Optimality by N. Gaffke and B. Heiligers

Volume 14. Statistical Methods in Finance
Edited by G. S. Maddala and C. R. Rao
1996 xvi + 733 pp.

1. Econometric Evaluation of Asset Pricing Models by W. E. Person and R. Jegannathan
2. Instrumental Variables Estimation of Conditional Beta Pricing Models by C. R. Harvey and C. M. Kirby
3. Semiparametric Methods for Asset Pricing Models by B. N. Lehmann
4. Modeling the Term Structure by A. R. Pagan, A. D. Hall and V. Martin
5. Stochastic Volatility by E. Ghysels, A. C. Harvey and E. Renault
6. Stock Price Volatility by S. F. LeRoy
7. GARCH Models of Volatility by F. C. Palm
8. Forecast Evaluation and Combination by F. X. Diebold and J. A. Lopez
9. Predictable Components in Stock Returns by G. Kaul
10. Interest Rate Spreads as Predictors of Business Cycles by K. Lahiri and J. G. Wang
11. Nonlinear Time Series, Complexity Theory, and Finance by W. A. Brock and P. J. F. deLima
12. Count Data Models for Financial Data by A. C. Cameron and P. K. Trivedi
13. Financial Applications of Stable Distributions by J. H. McCulloch

14. Probability Distributions for Financial Models by J. B. McDonald
15. Bootstrap Based Tests in Financial Models by G. S. Maddala and H. Li
16. Principal Component and Factor Analyses by C. R. Rao
17. Errors in Variables Problems in Finance by G. S. Maddala and M. Nimalendran
18. Financial Applications of Artificial Neural Networks by M. Qi
19. Applications of Limited Dependent Variable Models in Finance by G. S. Maddala
20. Testing Option Pricing Models by D. S. Bates
21. Peso Problems: Their Theoretical and Empirical Implications by M. D. D. Evans
22. Modeling Market Microstructure Time Series by J. Hasbrouck
23. Statistical Methods in Tests of Portfolio Efficiency: A Synthesis by J. Shanken

Volume 15. Robust Inference
Edited by G. S. Maddala and C. R. Rao
1997 xviii + 698 pp.

1. Robust Inference in Multivariate Linear Regression Using Difference of Two Convex Functions as the Discrepancy Measure by Z. D. Bai, C. R. Rao and Y. H. Wu
2. Minimum Distance Estimation: The Approach Using Density-Based Distances by A. Basu, I. R. Harris and S. Basu
3. Robust Inference: The Approach Based on Influence Functions by M. Markatou and E. Ronchetti
4. Practical Applications of Bounded-Influence Tests by S. Heritier and M-P. Victoria-Feser
5. Introduction to Positive-Breakdown Methods by P. J. Rousseeuw
6. Outlier Identification and Robust Methods by U. Gather and C. Becker
7. Rank-Based Analysis of Linear Models by T. P. Hettmansperger, J. W. McKean and S. J. Sheather
8. Rank Tests for Linear Models by R. Koenker
9. Some Extensions in the Robust Estimation of Parameters of Exponential and Double Exponential Distributions in the Presence of Multiple Outliers by A. Childs and N. Balakrishnan
10. Outliers, Unit Roots and Robust Estimation of Nonstationary Time Series by G. S. Maddala and Y. Yin
11. Autocorrelation-Robust Inference by P. M. Robinson and C. Velasco
12. A Practitioner's Guide to Robust Covariance Matrix Estimation by W. J. den Haan and A. Levin
13. Approaches to the Robust Estimation of Mixed Models by A. H. Welsh and A. M. Richardson
14. Nonparametric Maximum Likelihood Methods by S. R. Cosslett
15. A Guide to Censored Quantile Regressions by B. Fitzenberger

16. What Can Be Learned About Population Parameters When the Data Are Contaminated by J. L. Horowitz and C. F. Manski
17. Asymptotic Representations and Interrelations of Robust Estimators and Their Applications by J. Jurečková and P. K. Sen
18. Small Sample Asymptotics: Applications in Robustness by C. A. Field and M. A. Tingley
19. On the Fundamentals of Data Robustness by G. Maguluri and K. Singh
20. Statistical Analysis With Incomplete Data: A Selective Review by M. G. Akritas and M. P. La Valley
21. On Contamination Level and Sensitivity of Robust Tests by J. Á. Visŝek
22. Finite Sample Robustness of Tests: An Overview by T. Kariya and P. Kim
23. Future Directions by G. S. Maddala and C. R. Rao

Volume 16. Order Statistics – Theory and Methods
Edited by N. Balakrishnan and C. R. Rao
1997 xix + 688 pp.

1. Order Statistics: An Introduction by N. Balakrishnan and C. R. Rao
2. Order Statistics: A Historical Perspective by H. Leon Harter and N. Balakrishnan
3. Computer Simulation of Order Statistics by Pandu R. Tadikamalla and N. Balakrishnan
4. Lorenz Ordering of Order Statistics and Record Values by Barry C. Arnold and Jose A. Villasenor
5. Stochastic Ordering of Order Statistics by Philip J. Boland, Moshe Shaked and J. George Shanthikumar
6. Bounds for Expectations of L-Estimates by Tomasz Rychlik
7. Recurrence Relations and Identities for Moments of Order Statistics by N. Balakrishnan and K. S. Sultan
8. Recent Approaches to Characterizations Based on Order Statistics and Record Values by C. R. Rao and D. N. Shanbhag
9. Characterizations of Distributions via Identically Distributed Functions of Order Statistics by Ursula Gather, Udo Kamps and Nicole Schweitzer
10. Characterizations of Distributions by Recurrence Relations and Identities for Moments of Order Statistics by Udo Kamps
11. Univariate Extreme Value Theory and Applications by Janos Galambos
12. Order Statistics: Asymptotics in Applications by Pranab Kumar Sen
13. Zero-One Laws for Large Order Statistics by R. J. Tomkins and Hong Wang
14. Some Exact Properties Of Cook's D_I by D. R. Jensen and D. E. Ramirez
15. Generalized Recurrence Relations for Moments of Order Statistics from Non-Identical Pareto and Truncated Pareto Random Variables with Applications to Robustness by Aaron Childs and N. Balakrishnan

16. A Semiparametric Bootstrap for Simulating Extreme Order Statistics by Robert L. Strawderman and Daniel Zelterman
17. Approximations to Distributions of Sample Quantiles by Chunsheng Ma and John Robinson
18. Concomitants of Order Statistics by H. A. David and H. N. Nagaraja
19. A Record of Records by Valery B. Nevzorov and N. Balakrishnan
20. Weighted Sequential Empirical Type Processes with Applications to Change-Point Problems by Barbara Szyszkowicz
21. Sequential Quantile and Bahadur–Kiefer Processes by Miklós Csörgő and Barbara Szyszkowicz

Volume 17. Order Statistics: Applications
Edited by N. Balakrishnan and C. R. Rao
1998 xviii + 712 pp.

1. Order Statistics in Exponential Distribution by Asit P. Basu and Bahadur Singh
2. Higher Order Moments of Order Statistics from Exponential and Right-truncated Exponential Distributions and Applications to Life-testing Problems by N. Balakrishnan and Shanti S. Gupta
3. Log-gamma Order Statistics and Linear Estimation of Parameters by N. Balakrishnan and P. S. Chan
4. Recurrence Relations for Single and Product Moments of Order Statistics from a Generalized Logistic Distribution with Applications to Inference and Generalizations to Double Truncation by N. Balakrishnan and Rita Aggarwala
5. Order Statistics from the Type III Generalized Logistic Distribution and Applications by N. Balakrishnan and S. K. Lee
6. Estimation of Scale Parameter Based on a Fixed Set of Order Statistics by Sanat K. Sarkar and Wenjin Wang
7. Optimal Linear Inference Using Selected Order Statistics in Location-Scale Models by M. Masoom Ali and Dale Umbach
8. L-Estimation by J. R. M. Hosking
9. On Some L-estimation in Linear Regression Models by Soroush Alimoradi and A. K. Md. Ehsanes Saleh
10. The Role of Order Statistics in Estimating Threshold Parameters by A. Clifford Cohen
11. Parameter Estimation under Multiply Type-II Censoring by Fanhui Kong
12. On Some Aspects of Ranked Set Sampling in Parametric Estimation by Nora Ni Chuiv and Bimal K. Sinha
13. Some Uses of Order Statistics in Bayesian Analysis by Seymour Geisser
14. Inverse Sampling Procedures to Test for Homogeneity in a Multinomial Distribution by S. Panchapakesan, Aaron Childs, B. H. Humphrey and N. Balakrishnan

15. Prediction of Order Statistics by Kenneth S. Kaminsky and Paul I. Nelson
16. The Probability Plot: Tests of Fit Based on the Correlation Coefficient by R. A. Lockhart and M. A. Stephens
17. Distribution Assessment by Samuel Shapiro
18. Application of Order Statistics to Sampling Plans for Inspection by Variables by Helmut Schneider and Frances Barbera
19. Linear Combinations of Ordered Symmetric Observations with Applications to Visual Acuity by Marios Viana
20. Order-Statistic Filtering and Smoothing of Time-Series: Part I by Gonzalo R. Arce, Yeong-Taeg Kim and Kenneth E. Barner
21. Order-Statistic Filtering and Smoothing of Time-Series: Part II by Kenneth E. Barner and Gonzalo R. Arce
22. Order Statistics in Image Processing by Scott T. Acton and Alan C. Bovik
23. Order Statistics Application to CFAR Radar Target Detection by R. Viswanathan

Volume 18. Bioenvironmental and Public Health Statistics
Edited by P. K. Sen and C. R. Rao
2000 xxiv + 1105 pp.

1. Bioenvironment and Public Health: Statistical Perspectives by Pranab K. Sen
2. Some Examples of Random Process Environmental Data Analysis by David R. Brillinger
3. Modeling Infectious Diseases – Aids by L. Billard
4. On Some Multiplicity Problems and Multiple Comparison Procedures in Biostatistics by Yosef Hochberg and Peter H. Westfall
5. Analysis of Longitudinal Data by Julio M. Singer and Dalton F. Andrade
6. Regression Models for Survival Data by Richard A. Johnson and John P. Klein
7. Generalised Linear Models for Independent and Dependent Responses by Bahjat F. Qaqish and John S. Preisser
8. Hierarchial and Empirical Bayes Methods for Environmental Risk Assessment by Gauri Datta, Malay Ghosh and Lance A. Waller
9. Non-parametrics in Bioenvironmental and Public Health Statistics by Pranab Kumar Sen
10. Estimation and Comparison of Growth and Dose-Response Curves in the Presence of Purposeful Censoring by Paul W. Stewart
11. Spatial Statistical Methods for Environmental Epidemiology by Andrew B. Lawson and Noel Cressie
12. Evaluating Diagnostic Tests in Public Health by Margaret Pepe, Wendy Leisenring and Carolyn Rutter
13. Statistical Issues in Inhalation Toxicology by E. Weller, L. Ryan and D. Dockery
14. Quantitative Potency Estimation to Measure Risk with Bioenvironmental Hazards by A. John Bailer and Walter W. Piegorsch

15. The Analysis of Case-Control Data: Epidemiologic Studies of Familial Aggregation by Nan M. Laird, Garrett M. Fitzmaurice and Ann G. Schwartz
16. Cochran–Mantel–Haenszel Techniques: Applications Involving Epidemiologic Survey Data by Daniel B. Hall, Robert F. Woolson, William R. Clarke and Martha F. Jones
17. Measurement Error Models for Environmental and Occupational Health Applications by Robert H. Lyles and Lawrence L. Kupper
18. Statistical Perspectives in Clinical Epidemiology by Shrikant I. Bangdiwala and Sergio R. Muñoz
19. ANOVA and ANOCOVA for Two-Period Crossover Trial Data: New vs. Standard by Subir Ghosh and Lisa D. Fairchild
20. Statistical Methods for Crossover Designs in Bioenvironmental and Public Health Studies by Gail E. Tudor, Gary G. Koch and Diane Catellier
21. Statistical Models for Human Reproduction by C. M. Suchindran and Helen P. Koo
22. Statistical Methods for Reproductive Risk Assessment by Sati Mazumdar, Yikang Xu, Donald R. Mattison, Nancy B. Sussman and Vincent C. Arena
23. Selection Biases of Samples and their Resolutions by Ranajit Chakraborty and C. Radhakrishna Rao
24. Genomic Sequences and Quasi-Multivariate CATANOVA by Hildete Prisco Pinheiro, Françoise Seillier-Moiseiwitsch, Pranab Kumar Sen and Joseph Eron Jr
25. Statistical Methods for Multivariate Failure Time Data and Competing Risks by Ralph A. DeMasi
26. Bounds on Joint Survival Probabilities with Positively Dependent Competing Risks by Sanat K. Sarkar and Kalyan Ghosh
27. Modeling Multivariate Failure Time Data by Limin X. Clegg, Jianwen Cai and Pranab K. Sen
28. The Cost–Effectiveness Ratio in the Analysis of Health Care Programs by Joseph C. Gardiner, Cathy J. Bradley and Marianne Huebner
29. Quality-of-Life: Statistical Validation and Analysis An Example from a Clinical Trial by Balakrishna Hosmane, Clement Maurath and Richard Manski
30. Carcinogenic Potency: Statistical Perspectives by Anup Dewanji
31. Statistical Applications in Cardiovascular Disease by Elizabeth R. DeLong and David M. DeLong
32. Medical Informatics and Health Care Systems: Biostatistical and Epidemiologic Perspectives by J. Zvárová
33. Methods of Establishing In Vitro–In Vivo Relationships for Modified Release Drug Products by David T. Mauger and Vernon M. Chinchilli
34. Statistics in Psychiatric Research by Sati Mazumdar, Patricia R. Houck and Charles F. Reynolds III
35. Bridging the Biostatistics–Epidemiology Gap by Lloyd J. Edwards
36. Biodiversity – Measurement and Analysis by S. P. Mukherjee

Volume 19. Stochastic Processes: Theory and Methods
Edited by D. N. Shanbhag and C. R. Rao
2001 xiv + 967 pp.

1. Pareto Processes by Barry C. Arnold
2. Branching Processes by K. B. Athreya and A. N. Vidyashankar
3. Inference in Stochastic Processes by I. V. Basawa
4. Topics in Poisson Approximation by A. D. Barbour
5. Some Elements on Lévy Processes by Jean Bertoin
6. Iterated Random Maps and Some Classes of Markov Processes by Rabi Bhattacharya and Edward C. Waymire
7. Random Walk and Fluctuation Theory by N. H. Bingham
8. A Semigroup Representation and Asymptotic Behavior of Certain Statistics of the Fisher–Wright–Moran Coalescent by Adam Bobrowski, Marek Kimmel, Ovide Arino and Ranajit Chakraborty
9. Continuous-Time ARMA Processes by P. J. Brockwell
10. Record Sequences and their Applications by John Bunge and Charles M. Goldie
11. Stochastic Networks with Product Form Equilibrium by Hans Daduna
12. Stochastic Processes in Insurance and Finance by Paul Embrechts, Rüdiger Frey and Hansjörg Furrer
13. Renewal Theory by D. R. Grey
14. The Kolmogorov Isomorphism Theorem and Extensions to some Nonstationary Processes by Yûichirô Kakihara
15. Stochastic Processes in Reliability by Masaaki Kijima, Haijun Li and Moshe Shaked
16. On the supports of Stochastic Processes of Multiplicity One by A. Kłopotowski and M. G. Nadkarni
17. Gaussian Processes: Inequalities, Small Ball Probabilities and Applications by W V. Li and Q.-M. Shao
18. Point Processes and Some Related Processes by Robin K. Milne
19. Characterization and Identifiability for Stochastic Processes by B. L. S. Prakasa Rao
20. Associated Sequences and Related Inference Problems by B. L. S. Prakasa Rao and Isha Dewan
21. Exchangeability, Functional Equations, and Characterizations by C. R. Rao and D. N. Shanbhag
22. Martingales and Some Applications by M. M. Rao
23. Markov Chains: Structure and Applications by R. L. Tweedie
24. Diffusion Processes by S. R. S. Varadhan
25. Itô's Stochastic Calculus and Its Applications by S. Watanabe

Volume 20. Advances in Reliability
Edited by N. Balakrishnan and C. R. Rao
2001 xxii + 860 pp.

1. Basic Probabilistic Models in Reliability by N. Balakrishnan, N. Limnios and C. Papadopoulos
2. The Weibull Nonhomogeneous Poisson Process by A. P. Basu and S. E. Rigdon
3. Bathtub-Shaped Failure Rate Life Distributions by C. D. Lai, M. Xie and D. N. P. Murthy
4. Equilibrium Distribution – its Role in Reliability Theory by A. Chatterjee and S. P. Mukherjee
5. Reliability and Hazard Based on Finite Mixture Models by E. K. AL-Hussaini and K. S. Sultan
6. Mixtures and Monotonicity of Failure Rate Functions by M. Shaked and F. Spizzichino
7. Hazard Measure and Mean Residual Life Orderings: A Unified Approach by M. Asadi and D. N. Shanbhag
8. Some Comparison Results of the Reliability Functions of Some Coherent Systems by J. Mi
9. On the Reliability of Hierarchical Structures by L. B. Klebanov and G. J. Szekely
10. Consecutive k-out-of-n Systems by N. A. Mokhlis
11. Exact Reliability and Lifetime of Consecutive Systems by S. Aki
12. Sequential k-out-of-n Systems by E. Cramer and U. Kamps
13. Progressive Censoring: A Review by R. Aggarwala
14. Point and Interval Estimation for Parameters of the Logistic Distribution Based on Progressively Type-II Censored Samples by N. Balakrishnan and N. Kannan
15. Progressively Censored Variables-Sampling Plans for Life Testing by U. Balasooriya
16. Graphical Techniques for Analysis of Data From Repairable Systems by P. A. Akersten, B. Klefsjö and B. Bergman
17. A Bayes Approach to the Problem of Making Repairs by G. C. McDonald
18. Statistical Analysis for Masked Data by B. J. Flehinger[†], B. Reiser and E. Yashchin
19. Analysis of Masked Failure Data under Competing Risks by A. Sen, S. Basu and M. Banerjee
20. Warranty and Reliability by D. N. P. Murthy and W. R. Blischke
21. Statistical Analysis of Reliability Warranty Data by K. Suzuki, Md. Rezaul Karim and L. Wang
22. Prediction of Field Reliability of Units, Each under Differing Dynamic Stresses, from Accelerated Test Data by W. Nelson
23. Step-Stress Accelerated Life Test by E. Gouno and N. Balakrishnan
24. Estimation of Correlation under Destructive Testing by R. Johnson and W. Lu
25. System-Based Component Test Plans for Reliability Demonstration: A Review and Survey of the State-of-the-Art by J. Rajgopal and M. Mazumdar

26. Life-Test Planning for Preliminary Screening of Materials: A Case Study by J. Stein and N. Doganaksoy
27. Analysis of Reliability Data from In-House Audit Laboratory Testing by R. Agrawal and N. Doganaksoy
28. Software Reliability Modeling, Estimation and Analysis by M. Xie and G. Y. Hong
29. Bayesian Analysis for Software Reliability Data by J. A. Achcar
30. Direct Graphical Estimation for the Parameters in a Three-Parameter Weibull Distribution by P. R. Nelson and K. B. Kulasekera
31. Bayesian and Frequentist Methods in Change-Point Problems by N. Ebrahimi and S. K. Ghosh
32. The Operating Characteristics of Sequential Procedures in Reliability by S. Zacks
33. Simultaneous Selection of Extreme Populations from a Set of Two-Parameter Exponential Populations by K. Hussein and S. Panchapakesan